SAUNDERS COLLEGE PUBLISHING
Series of Anatomy & Physiology Titles

ANATOMY & PHYSIOLOGY

ANATOMY & PHYSIOLOGY
   by Arthur C. Guyton

HUMAN ANATOMY & PHYSIOLOGY
   by Eldra P. Solomon and P. William Davis

PHYSIOLOGY

PHYSIOLOGY OF THE HUMAN BODY, 6/e
   by Arthur C. Guyton

ANATOMY

HUMAN ANATOMY
   by E. Marshall Johnson, Richard R. Schmidt, Eldra P. Solomon, P. William Davis

RELATED TEACHING AIDS

HUMAN ANATOMY & PHYSIOLOGY: A LABORATORY MANUAL
   by James Blahnik

CADAVER DISSECTION OVERHEAD TRANSPARENCIES,
   dissections by Edwin Masters of Jefferson Medical College
   and photography by Ken Kasper

ANATOMY
AND PHYSIOLOGY

ANATOMY AND PHYSIOLOGY

Arthur C. Guyton, M.D.

Chairman and Professor
Department of Physiology and Biophysics
University of Mississippi
School of Medicine

 SAUNDERS COLLEGE PUBLISHING

Philadelphia New York Chicago
San Francisco Montreal Toronto
London Sydney Tokyo Mexico City
Rio de Janeiro Madrid

Address orders to:
383 Madison Avenue
New York, NY 10017

Address editorial correspondence to:
West Washington Square
Philadelphia, PA 19105

Text Typeface: Zapf Book Light
Compositor: York Graphic Services, Inc.
Acquisitions Editors: Michael Brown and Edward Murphy
Developmental Editors: Lloyd Black and Leesa Massey
Project Editor: Patrice Smith
Copyeditor: P. L. Smith
Art Directors: Richard L. Moore and Carol C. Bleistine
Art/Design Assistant: Virginia A. Bollard
Text Design: Emily Harste
Cover Design: Lawrence R. Didona
Text Artwork: J & R Technical Services, Iris Nichols, Patricia Johnson
Production Manager: Tim Frelick
Assistant Production Manager: Maureen Iannuzzi
Cover credit: Michelangelo, Nude per la battaglia di Cascina. © Scala/Art Resource , NY

Library of Congress Cataloging in Publication Data

Guyton, Arthur C.
  Anatomy and physiology.

  Includes bibliographies and index.

    1. Human physiology.   2. Anatomy,
Human.   I. Title.
[DNLM: 1. Anatomy.   2. Physiology.
QS 4 G992a]
QP34.5.G867 1984   612   84-10579

ISBN 0-03-063157-2

ANATOMY AND PHYSIOLOGY   ISBN 0-03-063157-2

Library of Congress catalog card number 84-10579

4567   071   987654321

CBS COLLEGE PUBLISHING
Saunders College Publishing
Holt, Rinehart and Winston
The Dryden Press

Preface

The purpose of *Anatomy and Physiology* is to provide an understanding of bodily structure and function for students in the life and health sciences. Studying the structure of the body invariably leads students to a curiosity about how the body functions. And, for those who are especially intrigued with bodily function, an understanding of function is impossible without preliminary knowledge of structure.

PHILOSOPHY AND APPROACH

Throughout my career I have been more identified with physiology than anatomy, yet in practice I have never been able to separate the two. When teaching a physiological subject I usually find it necessary to review anatomy and often to present aspects of functional anatomy that are essential to understanding function. This text acknowledges the interrelation of bodily structure and function.

My interest in anatomy stems partly from my training as a surgeon prior to entering the teaching profession. In surgery, one learns anatomy from a different perspective. The clinical sciences apply biological themes so that anatomy becomes alive, and each structure has a meaning. The textures, tissues, and parts of the body are viewed as active, living, and functioning so that anatomy becomes inseparable from physiology.

There is much to be said for the more rigorous approach to studying the human body through dissection and microscopy. Here again, I had the fortune of serving as a prosector for the Harvard Medical School Anatomy Department. As a prosector I was responsible for making demonstration anatomical dissections used in teaching medical students.

I recount my experiences in anatomy to explain why I came to write a book incorporating both anatomy and physiology. Though I have written several books on human physiology, I must say that the task of writing this present book has been especially satisfying because I have always had the feeling that physiology is not complete without anatomy.

Now, let me speak for a moment of physiology. Physiology studies living mechanisms—how do the cells function to provide the activities of the organs, and how do the organs constitute the totality of the living and active human being? The anatomy of the cells and organs gives us clues to these functions, but physiology goes still farther by attempting to explain as well the chemical and physical principles of activity in both microscopic cellular organelles and macroscopic organ structures. Indeed, the ultimate goal of physiology is to understand life itself, especially to understand the automaticity of the bodily mechanisms that actually make us live.

ORGANIZATION

Because of the great volumes of information that have been amassed in both the anatomical and the physiological fields, it has been necessary to choose carefully those portions of anatomy and physiology to be presented in this book. The anatomical choices have been made to provide the student with a full understanding of human body structure but leaving out those insignificant details that have little meaning for the student's future. And, in the field of physiology, a special attempt has been made to present the basic principles and concepts that allow us to understand how each organ system contributes to total body function. Yet, because some aspects of human physiology are still only partly understood, a strong effort has been made to distinguish fact from theory, but not to burden the student with intricate and insignificant details that more properly belong in a reference textbook.

The text is organized systemically with each system's anatomy covered within that unit. The intention here is to highlight the interplay between structure and function. Additionally, consideration has been given to the interplay of the systems themselves. The nervous, endocrine, and special sensory systems entail a discussion of the bodily systems and functions they regulate. A special chapter on sports physiology is included to illustrate how the many physiological systems function together during the extreme stresses of maximum exercise.

Applications of anatomical and physiological principles to clinical problems are integrated throughout the text to reinforce the importance and usefulness of the basic principles.

ARTWORK

Great efforts have been made to complement the text material pictorially. Students in this course learn by seeing as well as reading. High quality anatomical drawings, many of which are in full color, have been rendered for all parts of the body. Several hundred figures demonstrating physiological functions are provided to help students understand physiological principles rather than simply memorizing.

SUPPLEMENTS

The package of supplementary materials that accompanies this text includes the following: a *Study Guide* for the student that contains an overview of and objectives for each chapter, as well as test questions with answers in the back; an *Instructor's Manual* with a content summary and test items for every chapter; a set of *overhead transparencies* that includes two- and four-color illustrations from the text; and also a set of *cadaver dissection overhead transparencies* including 100 full-color illustrations and a description to accompany each transparency.

ACKNOWLEDGEMENTS

A text such as this requires the cooperative efforts of many different people. For this reason, I wish to express my special appreciation to Mrs. LaVeda Morgan, Ms. Gwendolyn Robbins, and Mrs. Elaine Steed-Davis for their excellent secretarial services. And, for the figures, I am especially indebted to Ms. Tomiko Mita, Ms. Iris Nichols, and Ms. Patricia Johnson.

The individuals who reviewed manuscript for this text in its various stages also deserve special mention. I would like to thank Gail S. Crouse of the Temple University Health Sciences Center; Dale W. Fishbeck of Youngstown State University; Thomas G. Froiland of Northern Michigan University; H. Craig Heller of Stanford University; Frank W. Judd of Pan American University; Wallace Maarsingh of Mesa Community College; Doris Mayner of Middlesex Community College; Charles McCreight of the Bowman Gray School of Medicine at Wake Forest University; Ronald C. Philo of the University of Texas Health Science Center; Judith M. Ryan of New York University; Normand Sasseville of Southeastern Massachusetts University; Donald B. Stratton of Drake University; and Kent Van De Graaff of Brigham Young University.

Finally, I am indebted to the staff of Saunders College Publishing, especially to Mr. Michael Brown, Mr. Lloyd Black, and Ms. Patrice Smith for all the exceptional energy and skill that has gone into the publication of this book.

ARTHUR C. GUYTON

Contents

ANATOMY
AND PHYSIOLOGY

I

INTRODUCTION

1

Introduction to Human Anatomy and Physiology

Overview

Anatomy means the science of structure, and **physiology** means the science of function in living organisms. Study of these two subjects goes a long way toward explaining life itself.

The basic functional unit of the body is the **cell**, of which about 75 trillion function in each human being. Most cells are alive, and most can also reproduce and thereby sustain the continuum of life.

Extracellular fluid fills the spaces between the cells. This fluid is called the *internal environment* of the body—it is in this environment that the cells live. The extracellular fluid contains the nutrients and other constituents necessary for maintenance of cellular life. The functions of most of the organs of the body are geared toward maintaining constant physical conditions and constant concentrations of dissolved substances in this internal environment. This condition of constancy in the internal environment is called *homeostasis*.

The fluid of the internal environment is mixed constantly throughout the body by (1) *pumping of blood* through the circulatory system by the heart and (2) *diffusion of fluid* both outward and inward through the capillary membranes to allow exchanges between that portion of the extracellular fluid in the blood called the **plasma** and that between the cells called the **interstitial fluid.**

Each organ system of the body plays its specific role in homeostasis. For instance, the **respiratory system** controls both the oxygen and the carbon dioxide concentrations in the internal environment. The **kidneys** remove waste products from the body fluids while also controlling the concentrations of the different ions. The **digestive system** provides the appropriate nutrients for the internal environment. The **muscles** and the **skeleton** provide support and locomotion for the body, so that it can seek out its needs, especially to provide the necessary food and drink for the internal environment. The **nervous system** innervates the muscles and also controls the functions of many of the internal organs, and it functions in association with the respiratory system to control the concentrations of carbon dioxide and oxygen. The **endocrine system** controls most of the metabolic functions of the body, including the rates of cellular chemical reactions; the concentrations of glucose, fats, and amino acids in the body fluids; and the synthesis of new substances needed by the cells. Even the **reproductive system** subserves a role in homeostasis since it provides for new human beings and therefore new internal environments as the old ones age and die.

One of the major goals of human scholarship, whether we be biologists, medical scientists, philosophers, writers, or otherwise, is to understand the bases of life itself. **Anatomy** and **physiology** constitute important segments of our knowledge of life. The word "anatomy" means the science of structure. The word "physiology" means the science of function in living organisms. But these two are really inseparable, for it is never possible to understand function without first understanding structure, and anyone who understands the structure of an organism can readily deduce much of its function.

Even the **virus,** one of the simplest forms of life, has its anatomy and physiology, which are far more complex than we once believed. The **bacterium** has still much more complex anatomy and physiology than those of the virus, and the **nucleated cell,** which is the basis of all higher forms of life, has a vastly complex structure and function, the details of which we are only now beginning to understand. In the human body, cells are organized into **tissues,** tissues into **organs,** and organs into **functional systems,** which collectively compose the body.

As small children we begin to wonder what enables persons to move, how it is possible for them to talk, how they can see the expanse of the world and feel the objects about them, what happens to the food they eat, how they derive from food the energy needed for exercise and other types of bodily activity, and by what processes they reproduce other beings like themselves so that life goes on generation after generation. All these activities characterize human life. It is the goal of the sciences of anatomy and physiology to explain these life processes, and indeed to explain life itself.

ROLE OF THE CELL IN THE HUMAN BODY

The basic functional unit of the body is the **cell.** About 75 trillion cells make up the adult human body. Each of these is a living entity in itself, capable of existing, performing chemical reactions, and contributing to the overall function of the body—also capable in many instances of reproducing itself to replenish the cells that die.

The cells are the building blocks of the tissues and organs, and each of the organs performs its own specialized function. One will appreciate the importance of the cell when he realizes that many more millions of years went into the evolutionary development of the cell

than went into evolution from the cell to the human being. Before one can understand how any one of the organs functions or how the organs function together to maintain life, it is necessary that he understand the inner workings of the cell itself. Therefore, a major portion of this book will be devoted to discussion of cellular structure and function as the bases of organ and system operation.

The Internal Environment and Homeostasis

All cells of the body live in a bath of fluid, fluid that weaves its way through the minute spaces between the cells, moves in and out of the blood vessels, and is carried in the blood from one part of the body to another. This mass of fluid is called the **extracellular fluid,** and it is divided into two major portions: (1) the extracellular part of the blood called the **plasma** and (2) the extracellular fluid in the spaces between the tissue cells called the **interstitial fluid** (or **intercellular fluid**).

For the cells of the body to continue living, there is one major requirement: The composition of the extracellular fluid must be controlled exactly from moment to moment and day to day, with no single important constituent ever varying more than a few percent. Indeed, cells can live even after removal from the body if they are placed in a fluid bath that contains the same constituents and has the same physical conditions as those of the extracellular fluid. Claude Bernard, the great nineteenth century medical scientist who originated much of our modern scientific understanding of the body, called the extracellular fluid the **milieu intérieur,** the "**internal environment,**" and the principle of maintenance of constant conditions in these fluids has come to be known as **homeostasis.**

Thus, at the very outset of our discussion of the human body, we are faced with a major question. How does the body maintain the required constancy of the internal environment, that is, the constancy of the extracellular fluid? The answer to this is that almost every organ plays some role in the control of one or more of the fluid constituents. For instance, the **circulatory system,** comprising the heart and blood vessels, transports blood throughout the body. Water and dissolved substances diffuse back and forth between the blood and the fluids that surround the cells. Thus, the circulatory system keeps the extracellular fluid in all parts of the body constantly mixed. This function of the circulatory system is so effective that hardly any portion of fluid in any part of the body remains unmixed with the other fluid more than a few minutes at a time.

The **respiratory system** transfers oxygen from the air to the blood, and the blood in turn transports the oxygen to all the tissue fluids surrounding the cells, thus maintaining the level of oxygen that is required for life by the cells. The carbon dioxide excreted by the cells enters the tissue fluids, then is taken up by the blood, and is finally removed through the lungs.

The **digestive system** performs a similar function for other nutrients besides oxygen; it processes nutrients that are then absorbed into the blood and are rapidly transported throughout the body fluids to where they can be used by the cells. The **liver,** the **endocrine glands,** and some of the other organs participate in what is collectively known as *intermediary metabolism,* which converts many of the nutrients absorbed from the digestive tract into substances that can be used directly by the cells. After the required substances and energy have been removed from the nutrients, the **kidneys** excrete the remaining waste products into the urine.

The **sensory organs** and the **nervous system** provide for hearing, feeling, tasting, smelling, and seeing, all of which aid in the search for and selection of food and also help to protect against dangers so that the almost utopian internal environment in which cells continue their life processes can be perpetuated.

Thus, we can emphasize once again that organ functions of the body depend on individual functions of cells, and sustained life of the cells depends on maintenance of an appropriate environment in the extracellular fluids. In turn, the organs and the cells, in their own ways, play individual roles in maintaining constancy of this internal fluid environment.

ORGANS AND SYSTEMS OF THE HUMAN BODY

Many students will already be familiar with the major organs and systems of the body. But, for those who have only a fragmentary knowledge, it is essential that we present a survey of these here, because understanding them will serve as the basis for most remaining discussions.

The Skeleton

Figure 1–1 illustrates the **skeleton** with some of its attached muscles. The skeleton is composed

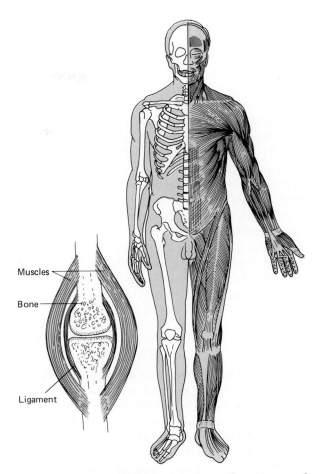

FIGURE 1–1 The skeletal and muscular systems of the human body.

of more than 200 individual bones. For the most part these bones are joined to one another. The skeleton consists of an **axial portion,** comprised of *skull, vertebrae, ribs, sternum,* and *hyoid bone,* and an **appendicular portion,** consisting of the *bones of the upper and lower limbs and the girdles to which they attach, the pectoral and pelvic girdles.* Bones vary greatly in their shapes and sizes. They are spoken of as *long bones, flat bones,* and *bones of irregular shapes.* **Joints** are of three types: *fibrous, fibrocartilaginous,* and *synovial.* A synovial joint, the type most numerous in the skeleton, is enveloped by a **capsule,** and the space within the capsule and between the two respective bones is the **joint cavity,** which is lined with a **synovial membrane.** In the joint cavities is a thick, slippery fluid containing a mucus-like substance, *hyaluronic acid,* which lubricates the joints, promoting ease of movement. On the sides of the capsule there frequently are strong fibrous **ligaments** that keep the joints from pulling apart. Often the ligaments are only on two sides of the joint, which allows the joint to move freely in one direction but not so freely in another. Other joints, particularly those of the spine, hips, and shoulders, having less restrictive ligaments, can move in multiple directions; that is, they can bend forward, backward, and to either side, or they can even be rotated. In these instances, the ligaments merely limit the degree of motion to prevent excessive movement in any one direction.

In summary, the skeleton is a frame of bones that can be contorted into many different configurations. Each bone has its own function, and the limitations of angulation of each joint are decreed by the ligaments and the shapes of the bones. The knee joint bends mainly in one direction, the wrist joint in two, the hip joint in two directions plus an additional rotary motion, and, in general, at least two opposing muscles are available to cause each motion that a joint allows.

The Skeletal Muscles

The **skeletal muscles,** also illustrated in Figure 1–1, are the muscles that are attached to the bones and cause skeletal movements. These muscles are composed of long **muscle fibers.** Usually many thousands of these fibers are oriented side by side like the threads in a skein of wool, and all of them together form the fleshy part of the muscle called the **muscle belly.** At each end of the muscle, the muscle fibers fuse with strong **tendon fibers** that form a bundle called the **muscle tendon.** The muscle tendons in turn penetrate and fuse with the bones on the two sides of the respective joints so that any pull exerted will effect appropriate movement.

Muscles vary greatly in size and appearance. For instance, the smallest skeletal muscle of the body, the *stapedius,* is a minute muscle in the middle ear and is only a few millimeters long; the longest muscle, the *sartorius,* extends more than half a meter down the entire length of the thigh, connecting the bony pelvis with the leg just below the knee. Some muscles, such as those of the abdominal wall, are arranged in sheets. Others are round, cigar-shaped structures: for example, the biceps, which flexes the forearm, and the gastrocnemius, which pulls the foot down when one wishes to stand on tiptoes.

The precise method by which muscle fibers contract is still not completely clear, but we do know that signals arriving in the muscles through nerves cause each fiber to shorten, allowing the entire muscle belly to contract and thereby to perform its function. This will be discussed in detail in Chapter 11.

The skeletal muscles move the limbs and other parts of the body in the directions allowed by the joints and their ligaments. In the case of movement at the knee joint, for instance, one major muscle located anteriorly in the thigh extends the joint while several muscles located posteriorly flex it. The muscles of the vertebral column are especially interesting because, contrary to what is usually believed, the back muscles are not just a few very large muscles but are composed of about 100 different individual muscles that perform specific functions: One rotates a vertebra, another tilts the vertebra forward, a third backward, and so on. This is analogous to the arrangement of the centipede, for each joint can bend independently of all the others. The joint connecting the skull to the vertebral column and the joint between the first two vertebrae are supplied with many additional muscles arranged on all sides so that the head can be rotated or angulated in any direction.

CARDIAC MUSCLE AND SMOOTH MUSCLE

In addition to the skeletal muscles, the body has two other types of muscle, **cardiac muscle** and **smooth muscle.** These types have contractile mechanisms similar to that of skeletal muscle, but their physical characteristics are different.

Cardiac muscle, as its name implies, is the muscle of the heart that pumps blood through the circulatory system. Cardiac muscle fibers are similar to skeletal muscle fibers with the exception of one major characteristic: These fibers interconnect with each other, forming a tight latticework called a *syncytium* that extends throughout all the walls of the heart. Furthermore, the electrochemical signal that causes heart muscle contraction travels over this entire syncytium, so that all walls contract together to pump the blood.

Smooth muscle is vastly different from both skeletal muscle and cardiac muscle, for smooth muscle fibers are extremely small, rarely more than 1 milliliter (ml) in length and usually less than 5 to 10 microns (μ) in diameter. This is in contrast to skeletal muscle fibers that have a diameter more than 10 times as great and a length several hundred times as great. Nevertheless, it is smooth muscle that causes contraction of most of the internal organs of the body, such as contraction of the wall of the gut, the walls of the blood vessels, and even the body of the uterus during birth of the baby. To be able to perform these functions, literally millions of the minute smooth muscle fibers are usually bound tightly together in single large masses or sheets, and their walls are often fused so tightly together that the entire muscle mass contracts together.

All of the muscle masses of the body together—the skeletal muscle, cardiac muscle, and smooth muscle—constitute about 40 percent of the total body weight, which emphasizes the importance of muscle to body function.

The Nervous System

The nervous system, illustrated in Figure 1–2, is composed of the

1. **Brain,** which is the highest level of the nervous system and is located in the cavity of the skull called the **cranial cavity;**
2. **Spinal cord,** which controls many of the local functions of the body and also carries nervous signals both from the periphery to the brain and from the brain back to the periphery; and
3. **Peripheral nerves,** some of which arise from the basal regions of the brain itself, called **cranial nerves,** and others of which arise from the spinal cord, called **spinal nerves.**

The brain and the spinal cord together constitute the **central nervous system,** and the peripheral nerves are called the **peripheral nervous system.**

A major function of the nervous system is to control many of the bodily activities, especially those of the muscles. But to exert this control the brain must be apprised continually of the body's surroundings. Therefore, functionally, the nervous system is composed of two separate portions: the **sensory portion,** which reports and analyzes the nature of conditions around and inside the body, and the **motor portion,** which controls the muscles and glandular secretion.

The sensory portion operates through the senses of sight, hearing, smell, taste, and feeling. The sense of feeling is actually many different senses, for one can feel touch, pin pricks, pressure, pain, vibration, and position of the joints.

Once information has been relayed to the brain from all the senses, the brain then determines what movement, if any, is most suitable, and the muscles are called into action to implement the decision.

One of the most important functions of the nervous system is to control walking. In walking, the body must be supported against gravity, the legs must move rhythmically in a walking motion, equilibrium must be maintained, and the

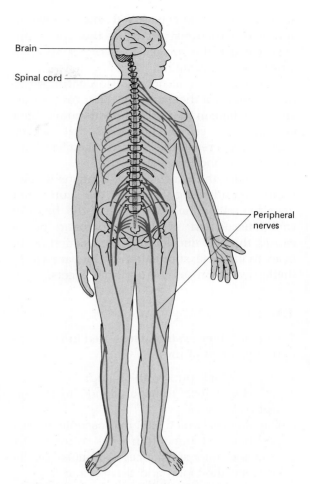

FIGURE 1–2 The nervous system.

direction of movement of the limbs must be guided. Therefore, the initiation and control of locomotion are very complex functions of the nervous system and require the services of major portions of the brain.

The Autonomic Nervous System. The **autonomic nervous system,** which is part of the motor portion of the nervous system, controls many of the subconscious functions of the body. It is divided into two parts: (1) the **parasympathetic system,** which consists of special nerves originating from several cranial nerves and sacral nerves, and (2) the **sympathetic system,** which originates from about half of the spinal nerves. The nerves of both of these systems form their own networks that penetrate into most organs of the body.

Many of the control functions of the autonomic nervous system are achieved by causing contraction or relaxation of smooth muscle in the internal organs—for instance, in the digestive tract, the urinary bladder, the biliary ducts, and the blood vessels, all composed mainly of smooth muscle sheets rolled into tubular or spheroid structures. Some of the autonomic nerves cause the muscle to contract, whereas others cause relaxation.

The autonomic nerves also control secretion by many of the glands in the digestive tract and elsewhere in the body, and autonomic nerve endings even secrete hormones that can increase or decrease the rates of chemical reactions in the body's tissues.

Finally, though the heart contains its own impulse-conducting system, the autonomic nervous system helps to control the heart as well. Stimulation of the sympathetic nerves causes the rate and force of contraction of the heart to increase; stimulation of the parasympathetic nerves causes the opposite effects.

The Circulatory System

The **circulatory system,** illustrated in Figure 1–3, is composed of the

1. **Heart,** which pumps the blood;
2. **Arteries,** which transport blood to the peripheral tissues;
3. **Capillaries,** which are microscopic in size and have extremely thin walls so that nutrients and other substances can diffuse readily between the blood and tissues; and
4. **Veins,** which conduct the blood back to the heart, thus making a complete circuit.

Circulating around and around through the body, the blood acts as a transportation system for conducting various substances from one place to another. It is the circulatory system that carries nutrients to the tissues and then carries excretory products away from the tissues.

The capillaries are highly permeable, allowing fluid and nutrients to diffuse into the tissues and permitting excreta from the cells to reenter the blood.

The Lymphatic System. Large particles that appear for any reason in the tissue spaces, such as old debris of dead tissues, protein molecules, and dead bacteria, cannot be removed from the tissues through the small pores of the blood capillaries. A special accessory circulatory system known as the **lymphatic system** takes care of these materials. Lymph vessels originate in small blind-end **lymph capillaries,** which lie beside the blood capillaries, and lymph, which is fluid derived from the spaces between the cells, flows in these vessels, leading eventually into large veins in the neck. The lymph capillaries are

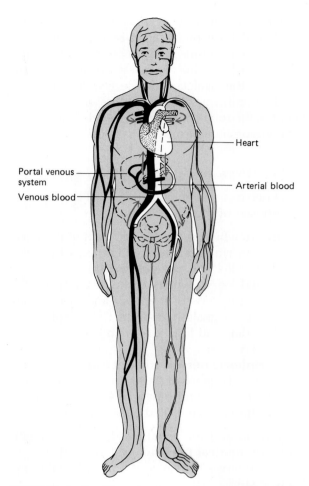

FIGURE 1–3   The circulatory system: heart and major vessels.

extremely porous so that large particles can enter the lymph vessels and be transported by the lymph. At several points along the course of the lymph vessels, the lymph passes through **lymph nodes,** where most large particles are filtered out and where bacteria are engulfed and digested by special cells called **reticuloendothelial cells.**

The Respiratory System

Figure 1–4 illustrates the **respiratory system,** showing the three fundamental portions of this system:

1. **Air passageways**
2. **Lungs**
3. **Blood vessels in the lung**

Air is moved in and out of the lungs by contraction and relaxation of the respiratory muscles, the most important of which is the **diaphragm.** Blood flows continually through the lung blood

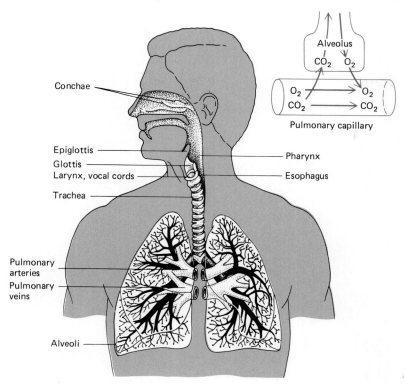

Conchae

Epiglottis
Glottis
Larynx, vocal cords
Trachea

Pharynx
Esophagus

Pulmonary arteries
Pulmonary veins

Alveoli

Alveolus
CO_2 O_2

O_2 O_2
CO_2 CO_2

Pulmonary capillary

FIGURE 1—4   The respiratory system.

vessels, and only a very thin membrane separates the air from the blood. This membrane is very porous to gases, which allows free passage of oxygen into the blood and carbon dioxide from the blood into the air.

Oxygen is one of the nutrients needed by the body's tissues. It is carried by the blood and tissue fluids to the cells, where it combines chemically with foods to release energy. This energy, in turn, is used to promote muscle contraction, secretion of digestive juices, conduction of signals along nerve fibers, and synthesis of many substances needed for growth and function of the cells.

When oxygen combines with foods to liberate energy, carbon dioxide is formed. This diffuses through the tissue fluids into the blood in the lung capillaries and is carried by the blood to the lungs. Then, the carbon dioxide diffuses from the blood into the lung air to be breathed out into the atmosphere.

The Digestive System

The **digestive system** is illustrated in Figure 1–5. Food, after being swallowed, passes successively through the

1. **Stomach,**
2. **Duodenum,**
3. **Jejunum,**
4. **Ileum,**
5. **Colon,**
6. **Sigmoid Colon,**
7. **Rectum,** and
8. **Anus.**

During this passage through the digestive tract, the food is digested, and those portions that are valuable to the body are absorbed into the blood. Along the entire extent of the digestive tract, **digestive glands** secrete *digestive enzymes* and other substances, especially when food is present. In addition to small glands that line the internal surfaces of almost all portions of the digestive tract, several large glands lie beside the digestive tract and deliver their secretions through ducts that empty into the tract. These are the

1. **Salivary glands** (including the *parotid, submandibular,* and *sublingual glands*),
2. **Pancreas,** and
3. **Liver.**

The *digestive enzymes* cause the foods to split into chemicals small enough to pass through the pores of the intestinal membrane and thence into the blood and lymphatic capillaries. Then the digestive products are carried in the blood wherever they are needed in the body.

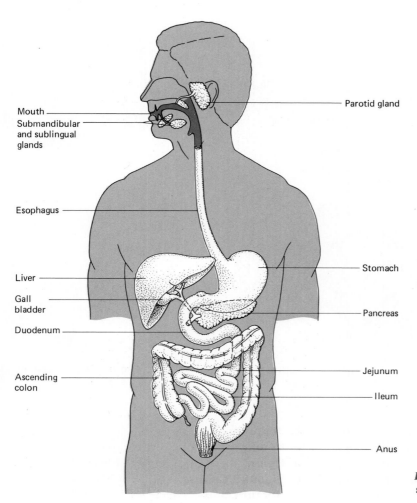

Mouth
Submandibular
and sublingual
glands

Parotid gland

Esophagus

Stomach

Liver

Gall
bladder

Pancreas

Duodenum

Ascending
colon

Jejunum

Ileum

Anus

FIGURE 1–5   The gastrointestinal system.

Metabolic Systems and Associated Organs

Metabolism and Growth.   The term *metabolism* means simply the chemical reactions that occur in the animal organism. These reactions occur inside the individual cells that make up the tissues, and their functions are to provide energy to perform the bodily activities and to build new structures. It is because of the metabolic processes that the cells grow larger and more numerous. The metabolism of special cells allows them to form tissues such as bones and fibrous tissue, enlarging the entire animal. Thus, metabolism is the basis not only for the energy needed by the body but also for growth itself.

Intermediary Metabolism.   Many of the foods entering the blood from the digestive tract can be used by the tissue cells without alterations, but some chemicals not normally found in the food are also needed. To supply these, much of the absorbed food passes to special organs, where it is changed into new substances needed

by the cells. This process is called intermediary metabolism.

The Liver.   The liver, aside from its secretion of one of the juices into the digestive tract, is the most important of all the internal organs for performing functions of intermediary metabolism. It has the special capability to transform carbohydrates, fats, and proteins into various other substances that are required by all the body's cells. The liver also forms products needed for blood coagulation, for transport of fat, for immunity to infection, and for many other purposes. And the liver is capable of storing large quantities of fats, carbohydrates, and proteins and then later releasing these foods into the blood when the tissues need them. An animal can live for only a few hours without a liver.

Control of Metabolism by the Hormones.   Metabolism is an inherent function of every cell of the body. However, the rate of metabolism in each respective cell is very often increased or decreased by the controlling action of *hormones* secreted by **endocrine glands** in different parts of the body. The **thyroid gland,** located in the neck, secretes *thyroxine*, which acts on all cells of the body to increase the rates of most metabolic reactions. *Epinephrine* and *norepinephrine*, two hormones secreted by the **adrenal medullae** located at the upper poles of the two kidneys, also increase the rate of metabolism in all cells. The **ovaries** secrete *estrogens* and *progesterone* and the **testes** secrete *testosterone*, which help to control metabolism in the sex organs of the female and male, respectively. *Insulin*, secreted by the **pancreas,** which is located posterior and inferior to the stomach, increases the utilization of carbohydrates and decreases the utilization of fats in all the tissues. *Adrenocortical hormones*, secreted by the **adrenal cortex,** also located at the upper poles of the kidneys, help to convert proteins to carbohydrates, and they also control the passage of salts and other substances through cell walls. *Parathyroid hormone*, secreted by four minute **parathyroid glands** located behind the thyroid gland in the neck, helps to control the concentration of calcium in the blood and extracellular fluids by removing calcium from the bones when it is needed by the fluids. Finally, at least eight different hormones secreted by the **pituitary gland** located at the base of the brain control a great host of bodily functions such as growth, rates of secretion of many of the other hormones, sexual functions, and excretion of water and electrolytes by the kidneys.

The Excretory System

The excretory system, illustrated in Figure 1–6, is comprised of the

1. Two **kidneys,** one located behind the liver on the right side and the other behind the stomach on the left;
2. **Ureters,** which transport urine from the kidneys to the urinary bladder;
3. **Urinary bladder,** which stores the urine until it is excreted; and
4. **Urethra,** which transports the urine to the exterior.

The principal function of the kidneys is to rid the blood of unwanted substances. Most of these are the end-products of metabolic reactions, mainly urea, uric acid, creatinine, phenols, sulfates, and phosphates. If they were allowed to collect in the blood in large quantities, their mass action would soon stop further metabolic reactions.

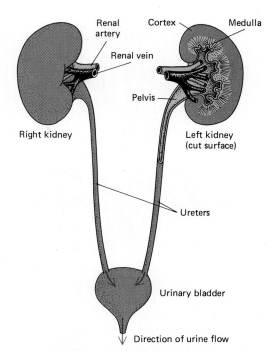

FIGURE 1–6 The kidneys and urinary system.

The kidneys have another very valuable function besides that of excreting the waste products: They regulate the concentrations of most of the ions in the body fluids. A very large proportion of these ions is sodium and chloride, the ions that constitute common table salt. The kidneys continually adjust the concentrations of both sodium and chloride in the blood and tissue fluids; and they also regulate very precisely the concentration of potassium, magnesium, bicarbonate, and many other substances.

The Reproductive Systems

All the functions of the systems of the body that maintain life would be useless were it not for those that provide for life's reproduction. The reproductive systems of the female and male are shown in Figure 1–7.

The principal female organs of reproduction are the

1. **Ovaries,** which form *ova* and release these into the abdominal cavity;
2. **Fallopian tubes (uterine tubes),** which conduct the ovum from the abdominal cavity to the uterus;
3. **Uterus,** in which the fetus grows to babyhood; and
4. **Vagina,** which is the canal from the uterus to the exterior.

FIGURE 1–7 The reproductive systems: (A) female, (B) male.

The principal reproductive organs of the male are the

1. **Testes,** which form the *sperm;*
2. **Vas deferens,** which conducts the sperm away from the testis;
3. **Seminal vesicles** and **prostate gland,** which form special secretions for nourishment or activation of the sperm;
4. **Urethra,** through which the sperm and special secretions, all together called the *ejaculate,* pass to the exterior; and
5. **Penis,** the external male organ.

The female provides the *ovum* (egg) from which a new human being is to develop, but this ovum cannot begin developing until it is fertilized by a sperm from the male. The fertilized ovum derives half of its developmental characteristics from the mother and half from the fertilizing

sperm of the father, so that the offspring owes its characteristics equally to each of the parents.

After the ovum has been fertilized it is at first still a single cell, but soon it divides into two cells, then four cells, and finally into many cells, becoming an *embryo* and then a *fetus.* Gradually, the newly developing cells *differentiate* into the tissues and organs of the body.

The mother provides nutrition for the growing fetus by means of the **placenta,** a structure that attaches to the inner wall of her uterus and through which fetal blood flows from the fetus. Nutrients diffuse from the mother's blood into the baby's blood through the **placental membrane,** which is very much like the membrane of the respiratory system. In turn, excretory products from the baby pass into the mother's blood. Thus, the fetus is nurtured through a period of 9 months in the mother's body until it becomes capable of sustaining life on its own in the outer world. At that time the mother's uterus expels the baby through the vagina.

THE CAVITIES OF THE BODY

Inside the body are a number of cavities, several of which are large and support some of the most important of the body organs. Because we will refer to the larger of these cavities frequently in the various anatomical descriptions of this text, it is important that the student study their locations as illustrated in Figure 1–8:

Cranial Cavity. This is inside the skull and is the location of the brain.

Spinal Cavity. This cavity, also called the *vertebral canal*, lies inside the vertebral column, encased by the arches of the vertebrae. It connects with the cranial cavity above and extends downward all the way to the lower end of the vertebral column. The combination of the cranial and spinal cavities is frequently also called the cerebrospinal cavity.

Pleural Cavities. Each of the lungs is inside a separate cavity called a pleural cavity. The lower border of each pleural cavity is the upper surface of the diaphragm, the principal muscle of respiration, and the two pleural cavities lie on the lateral sides of the heart.

Pericardial Cavity. This is the cavity in which the heart rests. The entire space inside the chest—including the pleural cavities, the pericardial cavity, and the tissues surrounding these cavities—is frequently called the chest cavity or thoracic cavity.

Abdomino-pelvic Cavity. A single large cavity, called the abdomino-pelvic cavity, occupies the large space between the diaphragm and the pelvis (the girdle of bone surrounding the lower trunk where the legs attach). The upper portion of this cavity contains almost all of the digestive system and is called the **abdominal cavity.** The lower portion—that portion inside the pelvic bones—is called the **pelvic cavity.** It contains the rectum, the urinary bladder, the female reproductive organs, and a portion of the male reproductive organs. However, the practicing medical doctor rarely uses the term abdomino-pelvic cavity but instead calls this entire area the "abdominal cavity." Yet he or she does frequently speak of the pelvic cavity separately.

THE LININGS AND FLUIDS OF THE CAVITIES

The wall of each cavity is lined by a thin *membrane.* In addition, the organ or organs within the cavity are lined by a separate membrane. The lining on the outer wall of the cavity is called the **parietal lining,** and that on the surfaces of the organs is called the **visceral lining.** The thin space between these two linings is filled with small amounts of fluid. In the pleural cavities, pericardial cavity, and abdomino-pelvic cavity this fluid is very slick and viscous, providing lubrication for free movement of the organs within the cavity, a condition that is very important for normal function of the lungs, heart, and digestive tract because they enlarge and contract continually. In the cranial and spinal cavities, the fluid has a consistency almost like that of water. The brain and spinal cord literally float in this fluid, thus providing a hydraulic cushion for the central nervous system; this prevents damage to the delicate brain and spinal cord tissues when the body receives a severe blow on its outer surface.

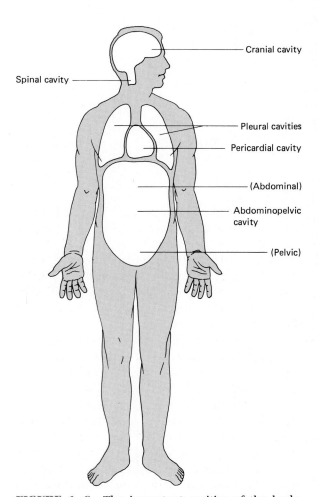

Spinal cavity

Cranial cavity

Pleural cavities

Pericardial cavity

(Abdominal)

Abdominopelvic cavity

(Pelvic)

FIGURE 1–8 The important cavities of the body.

THE PLANES OF THE BODY AND DESCRIPTIVE TERMINOLOGY

In almost all anatomical descriptions it is necessary to describe where in the body an anatomical part is located. For this purpose, the body is described in relation to three separate planes passing through the body as illustrated in Figure 1–9:

The Coronal Plane. A coronal plane passes through the body from side to side and down the entire length of the body as illustrated in the figure.

The Sagittal Plane. A sagittal plane also passes down the entire length of the body but passes through the body from front to back. When a sagittal plane passes down the exact midline of the body, it is called the *medial plane.*

The Horizontal Plane (or Transverse Plane). A horizontal plane transects the body from side to side and from forward to back, as also illustrated in the figure.

Understanding these planes is especially important since drawings of internal body structures usually view the structures in only one plane; therefore, the plane must be known if the drawing is to be understood.

"DIRECTIONAL" TERMS USED IN ANATOMICAL DESCRIPTIONS

Posterior and ***Anterior.*** As also illustrated in Figure 1–9, those parts of the body that are toward the front are said to be anterior; those toward the back are posterior. Sometimes the

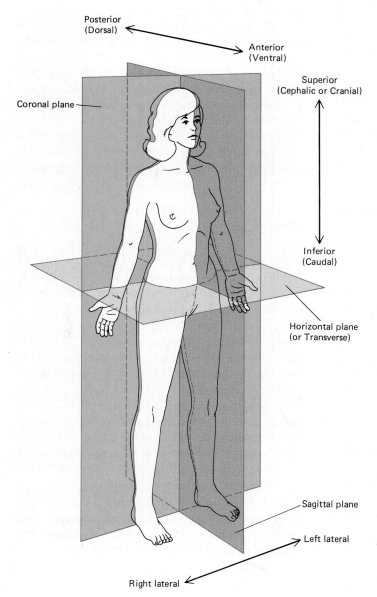

FIGURE 1–9 The three types of planes used in describing sectional views of body anatomy and the directional terms used in connection with the planes.

term *ventral* is used instead of anterior, and the term *dorsal* instead of posterior.

Lateral and Medial. Those parts of the body that are toward one side or the other—away from the midline—are said to be lateral. The two sides of the body are divided into the left lateral and right lateral sides, corresponding to the left and right hands as illustrated in Figure 1–9. Those parts of the body that are near the medial plane—that is, near the plane midway between the two sides—are said to lie *medially*.

Superior and Inferior. The term superior applies to the top portion of the body, and inferior applies to the lower part. The words *cephalic* or *cranial* are also frequently used to describe a superior portion of the body, these two terms meaning simply toward the head. Likewise, the term *caudal* is frequently used instead of inferior, meaning toward the tail.

Superficial and Deep. Superficial means near to or at the surface of the body; deep means at deeper levels.

Proximal and Distal. The term "proximal" means toward the origin of a structure; the term "distal" means farther removed from the origin. For instance, the knee is proximal to the ankle, and the ankle is distal to the knee. Likewise, the small arteries of the hand are distal to the large arteries of the upper arm, and the large arteries of the chest are proximal to those of both the hand and the arm.

Though it is tedious to learn these terms, it is important that they be understood thoroughly because they are necessary in all anatomical descriptions.

QUESTIONS

1. What is the internal environment, and what is the relationship of the extracellular fluid to this?
2. What is homeostasis, and what is its importance to functions of the body?
3. What are the interrelationships between the skeleton, the ligaments, and the muscles? Why are at least two muscles required for each plane of motion at a joint?
4. What are the roles of the sensory and motor portions of the nervous system?
5. What is the function of the autonomic nervous system, and why is the smooth muscle of the body so closely associated with this system?
6. Describe the function of the circulatory system in mixing the body fluids.
7. Describe the transport of oxygen from the air to the peripheral cells and carbon dioxide from these cells back to the air.
8. What types of substances are secreted in the digestive system, and how are these related to digestion and absorption?
9. What roles do the liver and hormones play in the overall function of the body?
10. Besides the excretion of waste products, what other important function do the kidneys subserve?
11. Describe the major cavities of the body.
12. Describe the three planes of the body.

GENERAL ANATOMICAL REFERENCES

Anderson, J.E.: Grant's Atlas of Anatomy, 7th Ed. Baltimore, Williams and Wilkins, 1978.

Bloom, W., and Fawcett, D.W.: A Textbook of Histology, 10th Ed. Philadelphia, W.B. Saunders, 1975.

Copenhaver, W.M., Kelly, D.E., and Wood, R.L.: Bailey's Textbook of Histology, 17th Ed. Baltimore, Williams and Wilkins, 1978.

Figge, F.H.J.: Sobotta/Figge Atlas of Human Anatomy, Vols. I, II, & III. Baltimore, Urban & Schwarzenberg, 1977.

Goss, C.M.: Gray's Anatomy of the Human Body, 29th Ed. Philadelphia, Lea & Febiger, 1973.

Ham, A.W., and Cormack, D.H.: Histology, 8th Ed. Philadelphia, Lippincott, 1979.

Langman, J., and Woerdeman, M.W.: Atlas of Medical Anatomy. Philadelphia, W.B. Saunders, 1978.

Netter, F.H.: The CIBA Collection of Medical Illustrations, Vols. 1–7. Summit, N.J., CIBA Medical Education Division, 1964–1979.

Pernkopf, E.: Atlas of Topographical and Applied Human Anatomy, Vols. I & II. Philadelphia, W.B. Saunders, 1980.

Williams, P.L., and Warwick, R.: Gray's Anatomy, 36th British Edition. Philadelphia, W.B. Saunders, 1980.

GENERAL PHYSIOLOGICAL REFERENCES

Annual Review of Physiology, Palo Alto, Ca. Annual Reviews, Inc. (One book each year.) Each book contains review articles that cover the literature of the preceding year in almost the entire field of animal physiology.

Bernard, C.: Lectures on the Phenomena of Life Common to Animals and Plants. Springfield, Ill., Charles C Thomas, 1974.

Brobeck, J.R.: Best and Taylor's Physiological Basis of Medical Practice, 10th Ed. Baltimore, Williams and Wilkins, 1979.

Cannon, W.B.: The Wisdom of the Body. New York, W.W. Norton, 1932.

Guyton, A.C.: Textbook of Medical Physiology, 6th Ed. Philadelphia, W.B. Saunders, 1981.

International Review of Physiology. (Published every 2 years in eight volumes.) This review covers the entire field of physiology.

Mountcastle, V.B.: Medical Physiology, 14th Ed. St. Louis, C.V. Mosby, 1979.

Physiological Reviews. This journal, published four times a year, contains reviews of most subjects in the field of physiology every few years.

II

CELLS AND TISSUES

Introduction to Cell Structure and Tissues— Connective Tissue and Epithelial Tissue

Overview

The important architectural (and functional) parts of the typical cell are the (1) **cell membrane,** (2) **cytoplasm,** (3) **endoplasmic reticulum,** (4) **ribosomes,** (5) **Golgi apparatus,** (6) **mitochondria,** (7) **lysosomes,** (8) **microtubules,** (9) **centrioles,** (10) **microfilaments,** (11) **nuclear membrane** and **nucleus,** (12) **nucleoplasm,** (13) **chromosomes** and **DNA,** and (14) **nucleolus.** The DNA in the nucleus controls the hereditary characteristics of the cell as well as all its internal functions.

The **blood cells** constitute about one third of all the cells of the body. These are divided into **red blood cells** and **white blood cells,** and the white blood cells in turn are divided into two different groups, the **granulocytes** and the **agranulocytes.** The red blood cells transport oxygen from the lungs to the tissues. The white blood cells defend the body against invading organisms.

A **tissue** is made up of cells and intercellular substance. The relative proportion of the two components varies greatly with different tissues. There are four basic types of tissues in the body: **connective tissue, epithelial tissue, muscular tissue,** and **nervous tissue.**

Connective tissue is abundant and widespread throughout the body. Its principal components are (1) **cells of several types,** but especially including fibroblasts, which are the cells that form most of the intercellular structures in the connective tissue; (2) **collagenous fibers, elastic fibers,** and **reticular fibers,** which give the tissue its strength; and (3) **ground substance,** which is a gel-like material filling the spaces between the fibers and cells. Connective tissue is divided into (1) **loose connective tissue,** found in the subcutaneous tissues and in the spaces between the internal body structures, and (2) **dense connective tissue,** which constitutes the **tendons, ligaments,** and **deep fascia** of the body, all of which provide important elements of structural strength.

Adipose tissue is a special type of loose connective tissue containing a predominance of **adipose cells.** About 95 percent of each adipose cell is pure fat.

Cartilage and **bone** are both special forms of dense connective tissue. Bone contains large quantities of precipitated calcium salts that

19

make them very hard. The **matrix** of each one of these tissues is composed of (1) large numbers of **collagenous fibers** that provide tensile strength and (2) a special type of **ground substance** that cements the fibers together.

Epithelial tissue is represented most typically by **epithelium,** which covers most of the surfaces of the body, both external and internal, and is composed of one or more layers of **epithelial cells.** Different types of epithelia include (1) *simple squamous epithelium,* (2) *simple cuboidal epithelium,* (3) *simple columnar epithelium,* (4) *stratified columnar epithelium,* (5) *transitional epithelium,* and (6) *stratified squamous epithelium.*

Skin is composed of two parts: (1) an outer part called the **epidermis,** which is a type of stratified squamous epithelium with a thick layer of *cornified cells* on its outer surface, and (2) the **dermis,** which is a semidense layer of connective tissue that underlies the epidermis and gives the skin its strength. Some of the specialized structures of the skin include (1) *hairs,* (2) *arrector pili muscles,* (3) *sebaceous glands,* (4) *sweat glands,* (5) *nerve endings,* (6) *fingernails, toenails,* and (7) *teeth.*

All **exocrine glands** and some **endocrine glands** are derivatives of epithelia. The products of the exocrine glands include *saliva, stomach secretions, intestinal secretions, tears, sweat,* and numerous other secretions. Different architectural types of exocrine glands include (1) *single cells,* (2) *tubular glands,* (3) *acinar glands,* and (4) a type of *laminar gland* (the liver). The more complex glands can secrete several different types of substances at the same time, and their **ducts** can reabsorb some substances as well.

Muscle tissue and **nervous tissue** will be described in Chapters 11 and 12.

The **cells** are the basic *living units* of all parts of the body. There are many major types of cells and numerous subtypes. All cells have certain characteristics in common, however, as will become clear not only from the discussions of this chapter but indeed from all the other chapters as well.

Yet, despite the essential role of the cells, they are not the only constituents of the body. Between the cells, in the **extracellular spaces** (also called **intercellular spaces** or **interstitial spaces**) are such components as *fluid, tissue gel, different types of fibers, hard protein matrix in cartilage and bone, calcium salts in bone,* and many other substances or structures. This composite of cells and their surrounding intercellular substances is called a **tissue.** In turn, the **organs** of the body are composed of different types of tissues, each organ having its own characteristic types of cells and intercellular elements. One of the purposes of this chapter is to explain the cellular and noncellular characteristics of several representative types of tissue.

Anatomists classify the different tissues into four basic types: (1) **connective tissue,** which connects the different parts of the body together and also fills the spaces between the parts; (2) **epithelial tissue,** which mainly covers the different internal and external surfaces of the body; (3) **muscular tissue,** which causes movement; and (4) **nervous tissue,** which provides a communicative network among the different organs and tissues. Almost every organ of the body contains varying amounts of each of these four basic tissues. Even so, the different organs have vastly different functions because of a different arrangement of the tissues in each organ and also because the cells in each organ have developed unique anatomical and functional characteristics.

CELL STRUCTURE

It is fitting to begin this discussion of cell structure by viewing the original cell from which the body develops, the **human ovum.** Figure 2–1 illustrates the ovum as seen by the light microscope. It is much larger than most other cells, having a diameter averaging 10 to 15 times as great, but the fertilized ovum is the ancestor of all the other cells. Furthermore, even though the

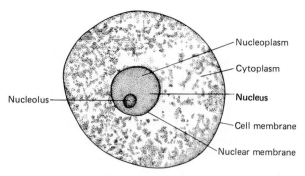

FIGURE 2–1 The basic living cell. This is a picture of an ovum taken from the female ovary and from which the human body develops.

other cells are smaller in size, they still retain the same basic parts illustrated in this figure. In forming the human body, the ovum divides through many successive generations to form the 75 trillion cells of the body. However, to illustrate how closely related the finished body is to the original ovum, if all cells in each generation required the same time to divide, only 47 generations would be required to complete the body.

But as the successive generations of cells divide, many characteristics of the cells also change drastically, a process called *cell differentiation.*

Only a few major features of the cell can be seen readily under the light microscope. These, as labeled in the figure, are (1) the **cell membrane,** (2) the **cytoplasm,** (3) the **nucleus,** (4) the **nuclear membrane,** (5) the **nucleoplasm,** and (6) the **nucleolus** inside the nucleus. We shall say more about these structures momentarily as we discuss the intracellular organization as revealed by the electron microscope.

INTRACELLULAR STRUCTURES AND THEIR ORGANIZATION

Until four decades ago, our knowledge of intracellular structure was limited for the most part to observations with light microscopy. But with the advent of the electron microscope a beautiful world of intracellular organization and architecture was revealed. Figure 2–2 presents a reconstruction of a typical cell, illustrating multiple specialized structures. Some of these deter-

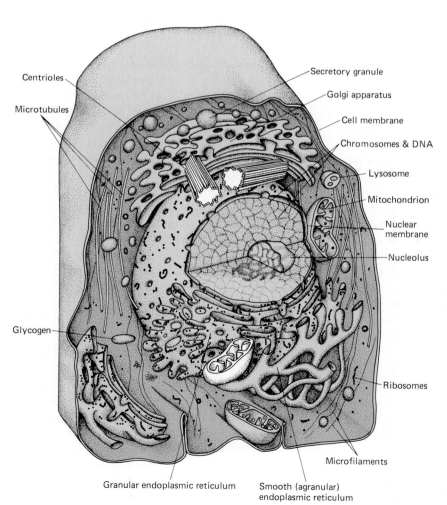

FIGURE 2–2 Reconstruction of a typical cell, showing the internal organelles in the cytoplasm and in the nucleus.

mine the physical features of the cell while others, called **cell organelles,** perform unique cellular functions.

1. **Cell membrane.** This is a very thin envelope that surrounds the entire cell. It is composed mainly of lipid substances (fatty substances) but also contains large numbers of protein molecules floating in the matrix of the membrane. This membrane separates the *intracellular fluid* inside the cell from the *extracellular fluid* that surrounds the cells.

2. **Cytoplasm.** This is the filling substance in the large space enclosed by the cell membrane and surrounding the cell nucleus. It is basically a solution of dissolved nutrients, ions, and many other substances that are important to the life of the cell. It also contains many suspended particles and special functional structures. The more important of these are described in the following paragraphs.

3. **Endoplasmic reticulum.** This is an extensive membranous structure that occurs in widespread areas of the cytoplasm in most cells. The membranes of the endoplasmic reticulum form a closed system of *tubes* and *cisternae* (network of channels). Substances can be transported throughout the cell inside these tubes and cisternae. Also, the membranous walls of the endoplasmic reticulum contain *enzymes* for synthesizing different substances. The **smooth endoplasmic reticulum** synthesizes both carbohydrates and fatty substances. The **granular (rough) endoplasmic reticulum** synthesizes mainly proteins but also some carbohydrates.

4. **Ribosomes.** These are many very small solid protein granular structures. It is on these that protein molecules are synthesized in the cell. Most of the ribosomes are attached to the endoplasmic reticulum, and it is their granular appearance that gives the portion of the endoplasmic reticulum where they attach the name "granular" (or "rough") endoplasmic reticulum.

5. **Golgi apparatus.** This is also a membranous structure similar to the endoplasmic reticulum. Substances synthesized by the endoplasmic reticulum usually pass next into the Golgi apparatus, where they are further processed. The Golgi apparatus in turn releases vesicles called *secretory granules* or *secretory vesicles* into the cytoplasm, containing substances synthesized in the endoplasmic reticulum and in the Golgi appara-

tus. The secretory granules in turn can be utilized for other purposes within the cell or they can be extruded to the exterior through the cell membrane, thus providing special chemical or structural components for the extracellular space, or providing the secretions of the various glands of the body, such as the salivary glands or the stomach glands.

6. **Mitochondria.** Each cell usually contains a hundred or more mitochondria. The mitochondrion is a membranous saclike structure that extracts energy from the foodstuffs as they are metabolized with oxygen, then makes this energy available to the other parts of the cell in the form of the high-energy compound *adenosine triphosphate* (ATP). In turn, it is this substance that energizes the cell's various chemical reactions. Adenosine triphosphate is so important that it will be discussed at length in Chapters 3 and 43.

7. **Lysosomes.** These are many small vesicles that float in the cytoplasm and contain large quantities of *digestive enzymes.* The lysosomes can release these enzymes to digest dead portions of the cell or to destroy abnormal substances such as bacteria that enter the cell.

8. **Microtubules.** These are present in many cells and for several purposes, one of which is to conduct special fluids from one part of the cell to another. They are also quite rigid and therefore function in many cells as an intracellular structural framework.

9. **Centrioles.** Each cell has two centrioles, which are solid structures composed of *microtubules.* Prior to cell division, the centrioles reduplicate themselves and then give rise to the intracellular architectural skeleton called the *mitotic apparatus* that guides the cell through the process of cell division.

10. **Microfilaments.** These are elongated *elastic* or *contractile elements.* In some cells they give strength to the cell membrane, and in special cells, such as muscle cells, they are the basis for *muscle contraction.*

11. **Nuclear membrane and nucleus.** The nucleus is usually a large round or ovoid structure surrounded by its own nuclear membrane. Each cell usually has one nucleus. However, some cells have no nucleus, such as red blood cells, and a few types of cells have multiple nuclei, such as skeletal muscle cells. The nuclear membrane is similar in structure to the cell membrane, except that it is a *double-layered membrane.* The space between the two layers communicates with

the spaces within the endoplasmic reticulum, thus providing a channel for transporting substances between the nucleus and different parts of the cytoplasmic compartment.

12. **Chromosomes and DNA.** The principal structures of the nucleus are the chromosomes. The nucleus of the human cell contains *46 chromosomes* and these in turn are composed principally of **deoxyribose nucleic acid molecules (DNA).** The DNA constitutes the **genes** of the cell, and there are about 100,000 in each cell. The genes determine the hereditary characteristics of the cell as well as the cell's function. This subject will be the topic of the entire Chapter 4.

13. **Nucleolus.** This is a structure within the nucleus that contains a mixture of *protein* and *ribose nucleic acid (RNA).* The material of the nucleolus eventually forms the ribosomes that then leave the nucleus and become distributed throughout the cytoplasm.

In summary, the intracellular architecture of the cell is very complex. It is capable of utilizing foodstuffs for energy and of synthesizing a myriad of special chemical compounds, the most important of which are the different types of proteins. The synthesized substances in turn are used for growth of new intracellular structures or for formation of entirely new cells by the process of *cell division,* or these substances can be extruded to the exterior of the cell to form the structural elements in the spaces between the cells.

The manner in which the intracellular structures function to maintain cellular life will be discussed in more detail in the following chapters.

The Blood Cells

To introduce our study of the diversity of cells, let us begin with the cells that circulate in the blood. Sixty percent of normal blood is composed of a fluid called **plasma,** but the other 40 percent is **blood cells.** These cells, illustrated in Figure 2–3, are of two categories, **red blood cells** and **white blood cells.**

Red Blood Cells (Erythrocytes).
The red blood cells, also called **erythrocytes,** normally constitute 99.8 percent of all the cells in the blood. There are a total of about 25 trillion red blood cells in the adult body. The red blood cell is in the shape of an indented disc, commonly referred to as doughnut-shaped, and it has a diameter of about 7μ. As seen in Figure

2–3, the red blood cells do not contain a nucleus. The nuclei are condensed and then extruded through the cell membrane during the formation of these cells. Without nuclei, the red blood cells can no longer divide to form new cells; therefore, in this sense they no longer possess all the usual characteristics of living cells. They do circulate in the blood for about 120 days, however, and they have the special characteristic of containing very large quantities of the protein *hemoglobin.* It is hemoglobin that gives the blood its red color, but, even more important, hemoglobin combines with oxygen to transport this essential substance from the lungs to the tissues throughout the body.

White Blood Cells (Leukocytes).
The white blood cells, also called **leukocytes,** constitute only two tenths of 1 percent of all the blood cells. Yet, they are exceedingly important because they protect against invading agents such as bacteria and toxins. In contrast to the red blood cells, the white blood cells retain their nuclei. The white cells are divided into two different groups, the **granular leukocytes (granulocytes)** that are all formed in the *bone marrow* and the **agranular leukocytes (agranulocytes)** that are formed in the *bone marrow* or *lymphoid tissue.* Granulocytes exhibit a granular cytoplasm whereas agranulocytes lack cytoplasmic granules.

There are three different types of granulocytes in the circulating blood: (1) **neutrophils,** (2) **eosinophils,** and (3) **basophils.** They are slightly larger than red blood cells and receive their names as a result of the characteristic staining of their cytoplasmic granules. Their nuclei are bilobed or multilobed. The agranulocytes include two types: (1) **monocytes** and (2) **lymphocytes.** Monocytes are large, rounded to ovoid cells with indented nuclei and, like the granulocytes, are formed in the bone marrow. Lymphocytes are usually only slightly larger than red blood cells and contain a relatively large, round nucleus with a thin rim of cytoplasm surrounding it. These cells develop from lymphoid tissue.

Both the *granulocytes* and the *monocytes* are *phagocytic cells,* which means that they can phagocytize (ingest through their membranes) such material as bacteria, toxic proteins, and even dead cells or other types of debris or trash that appear in the blood or tissues.

The *lymphocytes* play an important role in protecting the body against foreign matter in a different way from the phagocytic leukocytes. Lymphocytes function as part of the **immunity system** of the body. In the lymph nodes, they

Red Blood Cells (Erythrocytes)

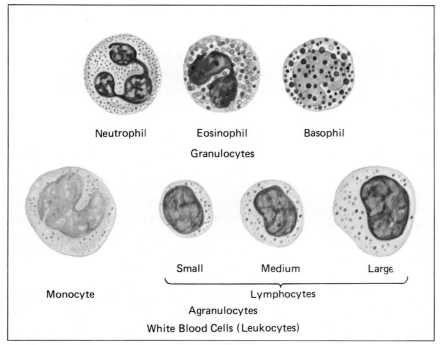

Neutrophil   Eosinophil   Basophil

Granulocytes

Small   Medium   Large

Monocyte

Lymphocytes

Agranulocytes

White Blood Cells (Leukocytes)

FIGURE 2–3 The red and the white blood cells.

can be converted into **plasma cells** that secrete **antibodies,** and the antibodies in turn attach themselves to specific bacteria, toxins, or other harmful agents and either neutralize or help to destroy these agents in other ways. In addition, other lymphocytes are released into the blood and have the ability to attach directly to specific foreign particles. They then secrete enzymes that destroy the incompatible substances.

Both the granulocytes and the agranulocytes can pass readily through the capillary walls into the tissues and therefore perform their protective functions in the tissues as well as in the blood stream.

The great importance of the white blood cells and the entire system of immunity for protection of the body against invading agents will be presented in Chapters 33 and 34.

CONNECTIVE TISSUE

Connective tissue is mainly what its name implies. It connects the different parts of the body together or functions as filler material to fill in the otherwise empty spaces between organs and other body structures. For instance, the tissue that fills the spaces deep to the skin and surrounding the muscles is connective tissue. The *tendons* that connect the muscles to the bones are also connective tissue, and connective tissue is even intermingled with minute smooth muscle cells in the walls of blood vessels to connect the muscle cells together. The *ligaments* that bind the bones together at the joints, the *sheaths* of muscles, the *fibrous capsules* around some of the organs, and the *fibrous globe of the eyeball* are all examples of connective tissues.

Aside from the above types of connective tissue, all of which are called **connective tissue proper,** many other variants of connective tissue also occur, each having special structural and functional properties different from those of connective tissue proper because their cells and intercellular structures are different. Some of these variants are (1) *adipose tissue,* (2) *cartilage,* (3) *bone,* and (4) *bone marrow.* In addition, *blood* is sometimes considered to be a type of connective tissue, even though it does not have a solid structure, or it plus the bone marrow from which most blood cells are derived are considered collectively to be a type of connective tissue.

Connective tissue proper is generally divided into two different types: (1) **loose connective tissue** and (2) **dense connective tissue.**

Lymphocyte Histiocyte Fibroblast Mast cells Eosinophil

Ground substance   Elastic fibers   Collagenous fibers   Capillary

FIGURE 2–4 Typical loose connective tissue.

LOOSE CONNECTIVE TISSUE

Loose connective tissue is soft and easily molded to new shapes. It is the type of tissue found underneath the skin called *subcutaneous tissue.* It is also present as a filler tissue in the spaces between organs and between muscles. Therefore, at many places in this text we will talk about loose connective tissue.

Figure 2–4 illustrates a typical structure of loose connective tissue, composed of a high proportion of intercellular substances supporting multiple cells. Though these cells can be of many different types, those most often present in loose connective tissue are as follows:

Cellular Elements of Loose Connective Tissue

1. **Fibroblasts.** Fibroblasts are the basic cells of connective tissue. One of their functions is to secrete into the intercellular spaces chemical substances that polymerize to form *fibers.* That is, these substances are secreted as individual small molecules still dissolved in the fluid secretions. Soon after secretion, however, the molecules begin to combine with each other to form supermolecules that in turn aggregate to form the fibers. The fibers extend in all directions through the connective tissue and hold it together.

2. **Macrophages (histiocytes).** Macrophages are very large cells found in many tissues, but especially so in loose connective tissue. Most if not all of these are derived from *monocytes* that enter the tissue from the blood. On entering the tissue, the monocytes swell to a large size and take on potent phagocytic proper-

ties, making them capable of engulfing bacteria, dead cells, and any other type of debris in the tissue. Thus, they are a protective type of cell. They are also called "macrophages" because of their large size and intense phagocytic ability. Many of the macrophages move freely through the tissues, especially when infection or inflammation is present and large quantities of bacteria, toxins, or debris are to be phagocytized. On the other hand, in the absence of infection or inflammation, most of the macrophages become attached in the tissues and then are called *histiocytes* (or sometimes simply *nonwandering macrophages*).

3. **Lymphocytes.** Lymphocytes are also present in loose connective tissue. These, too, enter the connective tissue from the blood. They play a special role in providing immunity against bacteria, foreign cells, and some toxins. However, they do this not by phagocytosis but by combining with the foreign agents or by secreting enzymes that destroy them.

4. **Other cells.** In addition to the cells just mentioned that are widely present in loose connective tissue, other types of cells present in less numbers include (a) *neutrophils,* which are rare in normal loose connective tissue but become enormously abundant when the tissue is infected; (b) *eosinophils,* which are also rare normally but appear in very large numbers in tissues experiencing allergic reactions; and (c) *mast cells,* which are similar to or identical with the basophils in the blood and are often collected around the tissue capillaries—these secrete *heparin* into the capillaries, which helps to prevent blood coagulation.

*Intercellular Constituents
of Loose Connective Tissue*

1. **Collagenous fibers.** One type of fiber formed by fibroblasts is *collagenous fibers* (also called simply *collagen fibers*). The fibroblasts actually secrete only simple *collagen molecules*, a type of protein, that are not fibers at the time of secretion. But soon after secretion these molecules polymerize to form long molecular collagen chains, thus becoming the collagenous fibers. It is mainly these collagenous fibers that provide the strength of connective tissue, and the glistening white appearance of the fibers gives most connective tissue a bright white hue. Some types of loose connective tissue have far greater quantities of collagenous fibers than others. For instance, Figure 2–5 illustrates an extensive network of collagenous fibers in the interior of the mesentery, which is a strong fold of tissue that suspends the stomach and intestines within the abdominal cavity. It is this mat of interwoven collagenous fibers that gives the mesentery the strength that it requires.

2. **Elastic fibers.** Another type of fiber in connective tissue is elastic fibers. These are yellow in color, and in those tissues that have large quantities of elastic fibers the tissue itself takes on a yellowish hue. Like the collagenous fibers, the elastic fibers are first secreted in the form of protein molecules, *elastin molecules*, and these also polymerize to form fibers. The elastic fibers are, as their name implies, elastic like rubber bands. Some anatomical structures, such as the walls of most large arteries, have large amounts of elastic fibers and therefore can be stretched like a spring and yet will recoil when the force is removed.

3. **Reticular fibers.** Not illustrated in Figure 2–4 is another type of fiber, reticular fibers, which occurs mainly in the loose connective tissue filling the minute spaces between closely packed cells. These fibers can be demonstrated only with special stains. They form a delicate network both between the adjacent cells and around them, providing the tensile strength that holds the cells together. They are believed to be formed from a mixture of both collagen and elastin molecules.

4. **Ground substance.** The ground substance of connective tissue is the *amorphous material* that fills in the spaces between the cells and fibers. It is composed of two principal constituents: (a) **interstitial fluid** (also called *extracellular* or *intercellular fluid*), which is simply the fluid in which the cells are bathed, and (b) **proteoglycan molecules.** The proteoglycan molecules are composed mainly of *hyaluronic acid* combined with a small amount of *protein*. These molecules are extremely long and are coiled like springs, but they are so thin that they can be seen only with the very best electron microscopes. They intertwine with each other to form an ultramicroscopic meshwork throughout the ground substance that entraps the interstitial fluid to form a *gel;* this has physical characteristics almost identical to those of gelatin. The gel impedes the *flow* of interstitial fluid through the tissues. However, the spaces within the gel are large enough that the various ions of the interstitial fluid, as well as the nutrients for the cells, can *diffuse* readily throughout the tissue—that is, these substances can move by individual molecular diffusive motion (also called kinetic motion) even though the fluid itself flows only very slowly in the tissue. The presence of large amounts of ground substance in loose connective tissue also allows this type of tissue to be very pliable so that it can conform readily to the spaces between the structures of the body.

DENSE CONNECTIVE TISSUE

As its name implies, dense connective tissue has relatively little ground substance but instead is filled almost completely with fibers. Some examples of dense connective tissue follow.

Tendons. Figure 2–6 illustrates a small section of a *muscle tendon*, a typical example of dense connective tissue. The wavy fibers seen in

Collagenous fibers   Elastic fibers

FIGURE 2–5 The collagenous fibers of the mesentery, the fold of tissue that supports the intestines.

FIGURE 2—6 Dense connective tissue removed from a muscle tendon.

this figure are almost entirely *collagenous fibers* lined up in the direction of the pull of the tendon. The waviness of the fibers allows the tendon a small amount of elasticity until the waves are stretched out by the pull of the muscle. But once the waves are stretched out, the collagenous fibers then exhibit very little additional elastic give despite tremendous force applied to the tendon.

Interspersed among the collagenous fibers of the tendon are a few *fibroblasts*. These continue to form new fibers and new ground substance as needed to maintain structural integrity of the tendon.

Ligaments. These hold the bones together at the joints and are also composed of dense connective tissue quite similar to that of tendons.

Deep Fascia. The term **fascia** means a sheet of connective tissue. For instance, the continuous sheet of loose connective tissue underneath the skin in all areas of the body is called the **superficial fascia.** At deeper levels, many sheets of *dense connective tissue* occur called **deep fascia.** These deep fasciae provide support for most of the organs and other structures, as well as provide separation planes between the different body areas. For instance, a deep fascial sheet underlies the superficial fascia over the whole body, providing strong support for deeper parts of the body. In addition, all skeletal muscles are surrounded by deep fascia, and a single muscle or group of muscles may be enclosed entirely in a separate compartment of dense connective tissue fascia called a *muscle sheath.* Also, an inner layer of the abdominal wall is a sheet of deep fascia termed the *transversalis fascia,* which helps to support the contents of the abdominal cavity. Finally, the ends of some of the flat muscles of the abdominal wall are continuous with sheetlike tendinous fasciae called *aponeuroses.* Contraction of these abdominal

wall muscles tightens the aponeuroses around the abdominal contents and helps to make the abdominal wall rigid.

The different planes of deep fascia are especially important to the surgeon because it is mainly the fasciae that he sews together to repair the deep portions of wounds. Most other deep tissues are too weak to provide adequate support for the surgeon's sutures.

Adipose Tissue (Fat Tissue)

Adipose tissue is one of the special variants of loose connective tissue. In addition to containing the usual connective tissue elements, it also contains tremendous numbers of **adipose cells,** also called **fat cells,** illustrated in Figure 2–7. In

Loose connective tissue

Blood vessels

Adipose cells (fat cells)

A

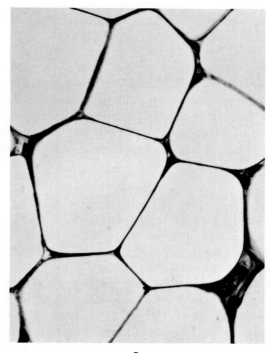

B

FIGURE 2—7 Adipose tissue, showing that the fat cells are large and globular and are filled with almost pure fat. (*B* from Bloom, W., and Fawcett, D.W.: *A Textbook of Histology,* 10th Ed. Philadelphia, W.B. Saunders, 1975.)

most adipose tissue, about 90 percent of the tissue is adipose cells and the remaining 10 percent consists of the other components of loose connective tissue. The connective tissue holds the fat cells together and carries the blood vessels that supply the tissue. In turn, each fat cell contains within its cytoplasmic cavity a single large globule of almost pure *triglyceride*, known commonly as *fat*, that constitutes about 95 percent of the cell mass. Therefore, as illustrated in Figure 2–7, the fat cells have the appearance of large numbers of bubbles stacked on one another.

The fat in adipose tissue serves several functions, the most important of which is to provide a *storehouse of nutrients* that the body can call upon for energy between meals and during long periods of starvation. In the absence of this stored fat, a person could not live more than a few days without food. Another important function of the fat tissue under the skin is to serve as a *heat insulator* to help maintain the warmth of the body in cold weather. And fat tissue in many of the spaces between internal bodily structures serves as packing material to cushion the movements of structures.

Cartilage and Bone

Cartilage and **bone** are both special forms of dense connective tissue. They both contain *ground substance* and large quantities of *collagenous fibers*, but the ground substance in each is quite different from that in the usual forms of connective tissue, which helps to give them their special characteristics. In addition, bone contains massive amounts of deposited calcium salt crystals that give them their rocklike hardness. Occasionally cartilage is also calcified.

CARTILAGE

Cartilage has the toughness of dense connective tissue, but it also has some degree of rigidity similar to that of bone. Yet, unlike bone, it can be bent without breaking.

To understand the nature of cartilage, place a finger inside one nostril and press against the anterior central wall called the nasal septum that separates this nostril from the other one. You will note a rigidity to this septum. But with pressure the lower and anterior part of the septum can be bent slightly from side to side. It is *nasal cartilage* in this portion of the septum that provides this rigidity, and yet this portion of

the nose usually will not break with trauma. The part that breaks is the posterior portion of the septum or the outer structure of the superior nose, both of which are supported by bone.

Other structures of the body where cartilage provides a resilient type of rigidity are the *anterior portions of the ribs* where they attach to the sternum (the breast bone), the *walls of the trachea and bronchi* (the windpipes to the lungs), and the *articulating surfaces* of the bones at joints that provide smooth movement and resilience between the bones.

The Structure of Cartilage. Figure 2–8 illustrates typical cartilage. The upper portion of the cartilage in this figure is *stable cartilage* that remains without change for years on end. The cartilage in the lower portion of the figure illustrates an area of growth and an area of calcification.

The stable cartilage in the upper part of the figure is composed of two major constituents:

1. **Matrix.** This constitutes the great mass of the cartilage, and it in turn is composed of (a) *ground substance* and (b) *collagenous fibers.* The ground substance is quite different from that of usual connective tissue. It contains large amounts of *chondroitin sulphate molecules* (instead of hyaluronic acid molecules) bound with *protein molecules* into a very tight mat. And within this mat are large numbers of collagenous fibers similar to those in dense connective tissue. There are three types of cartilage, classified according to their fibrous content: In **hyaline cartilage** the collagenous fibers usually are not seen in microscopic sections because the ground substance fills the spaces between the fibers so completely that

Matrix
(a) Ground substance
(b) Collagen fibers

Chondrocytes

Area of growth

Area of calcification

Chondroblasts

FIGURE 2–8 Typical cartilage from a growing bone.

the fibers are obscured. Another type, **fibrous cartilage,** contains a much higher ratio of collagenous fibers to ground substance, and the fibers are microscopically visible. *Elastic cartilage,* the third type, as the name implies contains also a component of elastic fibers. It is the matrix of cartilage, composed of the firmly bonded fibers, that gives it its toughness and its resilience.

2. **Chondrocytes.** These are the sparse cells that are found in stable cartilage. During the development of the cartilage, these were responsible for secreting the matrix and the fibers. In the stable state, however, the chondrocytes are mainly dormant, perhaps playing a slight role in maintaining the metabolism of the cartilage. Cartilage is avascular, and nourishment is by diffusion of nutritive elements from surrounding tissue.

In the area of cartilage growth near the bottom of Figure 2–8, the cells are dividing rapidly and secreting large amounts of new matrix. These cells are now called **chondroblasts,** the suffix "-blast" indicating active division and reproduction. A large portion of the skeleton of the developing fetus is formed initially of cartilage that is replaced eventually with bone. Cartilage proliferates near the ends of the long bones of the limbs until adolescence, and this newly forming cartilage is rapidly replaced by bone, which is the mechanism for growth in length of these bones.

At the very bottom of Figure 2–8, an area of calcification is shown, illustrating that under some conditions calcium salts deposit in cartilage and make it considerably more rigid than normally.

BONE

Bone, the structure of which is illustrated in Figure 2–9, is another type of adapted connective tissue. Its three principal components are

1. **Bone matrix.** Like cartilage, bone contains a strong *ground substance* that enmeshes a dense mat of *collagenous fibers*. These together are called the *bone matrix*. Bone matrix is different from that of cartilage in that the collagenous fibers have some unidentified component that causes deposition of calcium salt crystals.
2. **Bone salts.** Deposited within the ground substance of bone matrix adjacent to the collagenous fibers is an abundance of *calcium salt crystals*, composed mainly of the calcium phosphate salt known as *hydroxyapatite* that has a hardness similar to that of marble. It is these hydroxyapatite crystals that give bone both its rigidity and compressive strength. The collagenous fibers give the bone its tremendous tensile strength, and the ground substance holds the calcium crystals and collagenous fibers together.
3. **Bone cells.** Bone has three different types of cells: (a) *osteoblasts*, (b) *osteocytes*, and (c) *osteoclasts.*

The **osteoblasts** are responsible for secreting the substances that make up the bone matrix. They occur in two places in bone, lining (1) the outer surfaces of the bone and (2) many of the surfaces inside the internal cavities of the bone.

The **osteocytes** originate from osteoblasts. When the osteoblasts form the bone matrix, some of these cells become entrapped within small cavities in this matrix called **lacunae,** and these cells become the osteocytes that no longer form new bone but do play a necessary role in maintaining normal bone metabolism.

Osteoclasts are large multinucleated cells that normally line about 3 percent of the surface of the internal bone cavities. They are responsible for removing old bone when it needs repair, and they also are stimulated by one of the endocrine hormones, parathyroid hormone, to cause bone absorption when extra calcium ions are needed in the extracellular fluid. As will be discussed in Chapter 47, bone is continually being resorbed and rebuilt, a process that maintains its structural integrity.

Bone is mainly laid down in concentric layers, or **lamellae,** both on the outer surfaces of the bone and around the internal blood vessels that supply nutrition. Figure 2–9 illustrates this lamellar structure of bone, showing that the lacunae containing the osteocytes lie between the successive lamellae and are interconnected from one layer to another as well as from one lacuna to another by small **canaliculi.** The large canal in the center of each system of concentric lamellae, called the **Haversian canal,** carries the blood vessels, and it too connects with the lacunae via canaliculi. This entire system, composed of the Haversian canal, the lacunae, and the canaliculi, is called the *Haversian system;* it is important in maintaining the nutrition and the calcification of the bone.

To the right in Figure 2–9, the bone is shown to be very dense. This is called **compact**

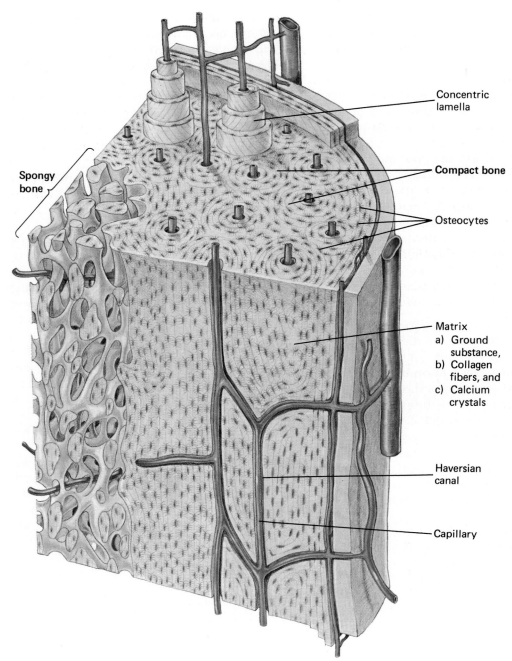

Concentric lamella

Compact bone

Osteocytes

Matrix
a) Ground substance,
b) Collagen fibers, and
c) Calcium crystals

Haversian canal

Capillary

Spongy bone

FIGURE 2–9 The internal structure of bone.

bone. It is this type of bone that provides the structural strength, especially the strength required for bearing the weight of the body.

To the left in Figure 2–9 is illustrated a bony latticework with large spaces in between. This is called **spongy,** or **cancellous, bone.** It is in the spaces within spongy bone that the bone marrow is found for producing both the red blood cells and most of the white blood cells.

EPITHELIAL TISSUE

EPITHELIUM—THE SURFACE LININGS

Every surface of the body is covered with a lining, including the surfaces of the internal body cavities as well as the external surface of the skin. In most instances these linings are composed of **epithelium,** which itself is one or more layers of **epithelial cells** that are attached to each other to form a continuous sheet. However, every type of surface has specialized functional needs, and therefore the epithelia in various areas are often vastly different from one another. Figure 2–10 illustrates some of the different types of epithelia, which are

1. **Simple squamous epithelium.** The word "squamous" means flat, and the term "squamous epithelium" derives from the fact that the cells of this type of epithelium are flat. The term "simple" means that this type of epithelium is only one layer thick. Simple squamous epithelium lines the internal cavities of the body, including the *abdominal cavity*, the *pleural cavity*, and the *pericardial cavity*.
2. **Simple cuboidal epithelium.** The cells of this type of epithelium are cuboidal in shape. This

type of epithelium lines many of the *tubules of the kidney* and also some of the *ducts* leading from various glands.
3. **Simple columnar epithelium.** The cells of this type of epithelium are shaped like elongated columns and are arranged side by side to form a single-layered epithelium. This type of epithelium lines most of the *digestive tract*. The airways of the *respiratory system* also have a modified columnar epithelium with two special features: (a) about 200 *cilia* on the surface of each cell, as illustrated to the left in the figure, and (b) some cells shorter than the others between the bases of the columnar cells; therefore, this epithelium is called **pseudostratified ciliated epithelium.**
4. **Stratified columnar epithelium.** This epithelium also contains mainly columnar cells,

Simple squamous

Basement membrane

Simple cuboidal

Simple columnar

Connective tissue

Stratified columnar

Stratified squamous

Transitional

FIGURE 2–10 Different types of epithelia.

but multiple layers of them. It is rare but may be found in the *urethra*.

5. **Transitional epithelium.** This type of epithelium contains cells of many different shapes that are also multilayered. It lines the *ureter* and the *urinary bladder.* As the bladder stretches, the shapes of the cells change.

6. **Stratified squamous epithelium.** This epithelium contains many layers. The deeper layers are mainly low columnar to cuboidal in shape, but the cells become progressively flattened toward the surface. This type of epithelium occurs in those areas of the body where there is much abrasion. It is found in the

adult *vagina* and in the *esophagus.* This type of epithelium also forms the outer layer of the skin, as we shall discuss later.

Figure 2–11 illustrates some examples of actual epithelia from different areas of the body.

THE BASEMENT MEMBRANE OF THE EPITHELIUM

Note in Figure 2–10 that underneath each type of epithelium is a thin layer called the **basement membrane.** This is composed of a dense layer

Simple squamous
(peritoneum)

Simple cuboidal
(kidney tubules)

Ciliated columnar
(fallopian tube)

Stratified columnar
(urethra)

Stratified squamous
(vagina)

Cornified stratified squamous
(skin)

FIGURE 2–11 Some representative specimens of epithelia from different tissues.

of *collagenous fibers* and *ground substance* containing especially large amounts of *glycoprotein.* It is on the surface of the basement membrane that each type of epithelium rests.

In the stratified types of epithelium, the epithelial cells that rest on the basement membrane are called the **germinal epithelium.** It is from these cells that the outer layers of the cells are derived. The **germinal cells** divide continually, and the successive generations of cells move steadily toward the surface. In the squamous types of stratified epithelium, the successive layers of cells also gradually lose their cytoplasm as well as most of their nuclear substance, so that the outer cells become little more than flattened cell membranes.

When an area of epithelial cells is destroyed but the basement membrane still remains intact, the germinal epithelial cells that surround the denuded area begin to divide rapidly and grow over the denuded surface until the epithelium is repaired.

The Skin

The **skin,** also called the **integument,** is illustrated in Figure 2–12. It is composed mainly of *epithelial tissue* and *connective tissue* but also has small amounts of both *muscular* and *nervous tissues.* The skin is actually considered to be an **organ,** partly because of its complex structure and partly because it performs specific functions for the body. Also, skin illustrates in a simple way the manner in which separate types of tissue can be organized to form an organ. Figure 2–12 shows that the skin consists of two major layers:

1. **Epidermis,** which is the epithelial layer of the skin, and
2. **Dermis,** which is the connective tissue base upon which the epidermis rests.

THE EPIDERMIS

The epidermis is divided into five layers:

1. **Stratum germinativum.** This is the layer of epithelial cells that lies on the basement membrane. It is from these cells that all other cells of the epidermis are derived. Most of the cell division in the epidermis occurs within this stratum, and as new cells are formed they simply push the older cells farther toward the surface.
2. **Stratum spinosum.** This is a multilayer group of spindle-shaped cells that has been pushed outward toward the surface by newly forming cells in the stratum germinativum. The cells of this stratum are living and metabolically active.
3. **Stratum granulosum.** This is a thin layer where the cells of the outer surface of the stratum spinosum acquire granules of *keratohyalin,* which are seen in microscopic sections as dense masses in the cytoplasm. The cells are still living.
4. **Stratum lucidum.** This is a thin layer in thick epidermis such as that of the palms and soles. As the name implies, it stains lightly. The internal cellular structures have largely disappeared. This layer is not present in thin skin; therefore, it is not seen in Figure 2–12.
5. **Stratum corneum.** This is composed of many layers of dried, dead, and flattened epithelial cells called *cornified cells.* The outer layers of the stratum corneum gradually scale away from the skin, and new layers are continually formed from beneath.

THE DERMIS

The **dermis** is a semidense layer of connective tissue that lies underneath the epidermis. One can see in Figure 2–12 that hillocks from the outer portion of the dermis protrude upward

Stratum corneum

Stratum granulosum

Stratum spinosum

Stratum germinativum

Papillary layer of dermis

Vessel

Reticular layer of dermis

Epidermis

Dermis

FIGURE 2–12 A sectional view of typical skin.

into the epidermis in the form of *papillae.* This portion of the dermis is called the *papillary layer,* which is fairly cellular and contains finer fibers. The deeper layer of the dermis contains coarser masses of collagen fibers and is called the *reticular layer* of the dermis.

The importance of the dermis is that it is the locus of all the blood vessels and nerves that serve the skin. None of the blood vessels pene-

trate into the epidermis, and only the very tips of *free nerve endings* enter the deeper layers of the epidermis, none reaching the outer layers.

SPECIALIZED STRUCTURES OF THE SKIN

Figure 2–13 illustrates a three-dimensional reconstruction of the skin. Note in this reconstruction the *blood capillary loops* that extend up-

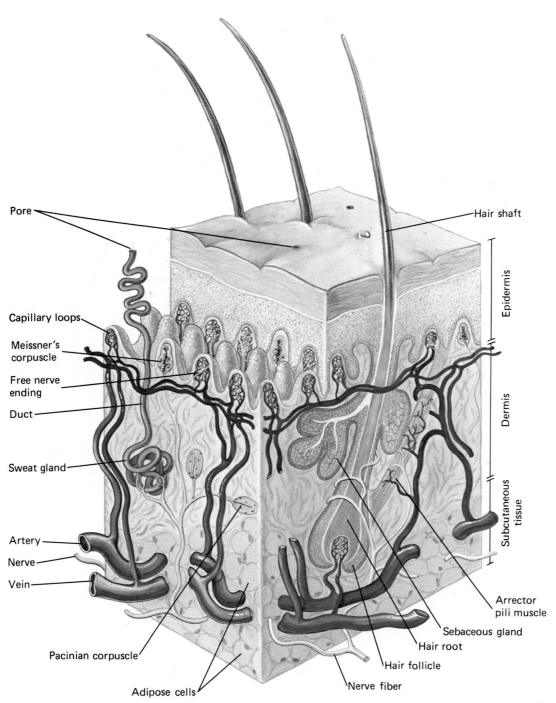

Pore

Capillary loops

Meissner's corpuscle

Free nerve ending

Duct

Sweat gland

Artery

Nerve

Vein

Pacinian corpuscle

Adipose cells

Hair shaft

Epidermis

Dermis

Subcutaneous tissue

Arrector pili muscle

Sebaceous gland

Hair root

Hair follicle

Nerve fiber

FIGURE 2–13 A reconstruction of a block of skin, illustrating especially the skin vasculature, hairs, and other special skin structures.

ward into the papillae; also note the *larger blood vessels* in the reticular layer of the dermis and in the subcutaneous tissue beneath the skin. In addition, note the following specialized structures of the skin:

Hairs. Hairs grow from *hair follicles.* The hair follicles in turn derive from pockets of epidermis that penetrate deeply into the dermis. The epithelial cells at the base of the follicles, called the *matrix,* reproduce rapidly, while moving toward the center of the follicle; there they fuse together to form the hair. As the hair develops deep in the follicle, its terminal portions are pushed progressively outward from the surface of the skin. Hairs vary greatly in their length and diameter. Therefore, we speak of *fine hairs* and *coarse hairs.*

Arrector Pili Muscles. Attached to the side of each hair follicle is a small bundle of smooth muscle fibers that constitutes the arrector pili muscle. The opposite ends of these fibers are anchored in the papillary layer of the dermis. When this muscle contracts, the follicle is pulled to a vertical position, thus causing the hair to "stand on end." In lower mammals, the arrector pili muscles play an important role in fluffing the hair in cold weather, thus providing increased insulation against the cold.

Sebaceous Glands. Also extending from the side of each hair follicle is a short duct that conveys the secretions of the sebaceous gland. The gland is made of large, lightly staining *epithelial cells* that secrete an oily secretion, the *sebum,* into the hair follicle and thence up along the shaft of the hair to the surface of the skin. Sebaceous glands vary greatly in size, and their size is not related directly to the size of the hair.

Sweat Glands. Over the surface of the body are 10 to 20 thousand sweat glands. As illustrated in Figure 2–13, each of these is a coiled *epithelial gland* that penetrates deeply into the dermis. It secretes the watery, salty solution that we all know as *sweat.* This secretion is very important in relation to control of the body temperature, as we shall see in Chapter 44. The sweat gland is composed of two principal parts, a *coiled portion* deep in the dermis where the sweat is initially secreted, and a *straight portion,* the *sweat gland duct,* in which the sweat is modified by absorption of much of the salt before it pours onto the surface of the skin.

Nerve Endings. The dermis of the skin contains many different types of nerve endings. One important type is the nerve ending that twines around the hair root. This ending sends a sensory signal to the brain in response to even extremely light bending of the hair. Therefore, it is one of the important tactile receptors of the body, called the *hair receptor.*

The most common type of nerve ending in the skin is the *free nerve ending.* Many of these penetrate to the peaks of the dermal papillae, and some actually penetrate into the deeper layers of the epidermis. The different free nerve endings detect *pain, temperature,* or *crude touch sensations.*

Finally, many very important **encapsulated nerve endings** occur in the skin. One of these is the *Meissner's corpuscle,* which protrudes outwardly into a dermal papilla. It is the Meissner's corpuscles and the other encapsulated nerve endings that detect the precise area of the skin that is stimulated, a sense called *discriminative touch.*

The different tactile nerve endings and their sensory functions will be discussed in Chapter 16.

FINGERNAILS AND TOENAILS

The nails of the fingers and toes are specialized products of the epidermis. Figure 2–14 illus-

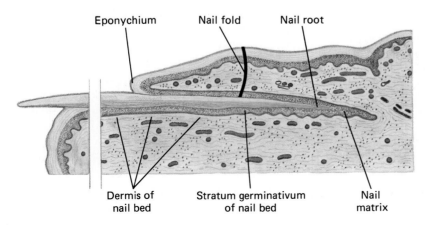

FIGURE 2–14 The base of a fingernail, illustrating the basic structures for nail growth.

trates a typical fingernail and its relationship to the skin. The essential structures are

1. The nail itself, or **body.**
2. The **nail fold,** which is the skin lying on top of the base of the nail. The proximal edge of the nail fold is called the **eponychium.**
3. The **nail matrix.** This is the epithelium in the deep recess underneath the nail fold where the nail is formed. The epithelium in this matrix grows continually, and the older epithelial cells gradually fuse to form the **nail root.** As this root elongates it pushes the body of the nail forward.
4. The **nail bed.** This is the connective tissue that lies underneath the nail. The junction of the distal end of the nail bed with the epidermis is called the **hyponychium.**

If the nail is avulsed from a finger or toe, a new nail will begin immediately to be formed by the nail matrix. However, at times the nail matrix becomes damaged so that the nail grows in the wrong direction, often growing into the tissues of the finger or toe rather than growing forward. When this happens it is important to stop the growth of the nail. To do so requires complete destruction of the nail matrix. If even a portion of the nail matrix is left, small spicules of nail will continue to grow in abnormal directions and cause severe pain. This is the phenomenon of the *ingrown nail.*

Teeth

The **teeth,** like the skin, are simple organs derived mainly from epithelial cells inside the mouth of the fetus months before birth. The tooth-forming epithelial cells invaginate into the *oral margins* of the upper and lower jaw bones, forming **tooth buds** deep inside bony cavities; these cavities later evolve into the **tooth sockets.** The invaginated cells, even though of epithelial origin, take on characteristics very similar to those of osteoblasts, becoming *ameloblasts* and *odontoblasts,* which generate bonelike tissue that becomes the main body of the tooth.

Figure 2–15 illustrates a typical tooth. Its important parts are

1. **Enamel.** This is a thin layer, usually less than 2 mm thick, that covers the exposed portion of the tooth. It is generated by a layer of *ameloblasts* (that function like osteoblasts) on the surface of the tooth before the tooth erupts. However, after the enamel is formed, the ameloblasts disappear, so that the enamel can

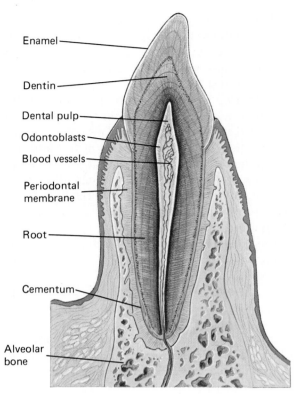

FIGURE 2–15   A longitudinal sectional view of a tooth and its surrounding tooth cavity.

never be replaced once it is broken. Enamel contains very large *hydroxyapatite crystals,* the same as those of bone but much larger and much harder and extremely resistant to destruction. The crystals are held in place by a matrix containing *protein fibers* that are the strongest of almost any known type; additionally, these fibers are also resistant to chemical destruction.

2. **Dentin.** Most of the body of the tooth is composed of dentin. It contains a matrix as well as calcium crystals similar to those found in bone. It contains no cells, however. Instead, the cells that form the dentin are the **odontoblasts** (also similar to osteoblasts) that line the surface of the **pulp cavity** in the middle of the tooth.
3. **Dental pulp.** This is the filler material inside the pulp cavity that extends all the way from the root upward through the middle of the tooth. It contains *blood vessels* and *nerves* imbedded in *loose connective tissue.* The walls of the pulp cavity are lined by the *odontoblasts,* as noted before.
4. **Cementum.** This is a bony substance around the root of the tooth that cements it into the tooth socket.

5. **Periodontal membrane.** This is a dense connective tissue containing mainly *collagenous fibers*. It is anchored in the cementum of the tooth and also into the bone of the tooth socket surrounding the tooth known as the **alveolar ridges** of the upper and lower jaws, thus providing a strong but slightly resilient bridge between the tooth and the bone of the jaws.

Glands and Their Secretions

Among the most important of the functional structures of the body are the many **glands,** most of which are also derived from epithelial cells. It is the glands that provide the digestive juices of the alimentary tract, beginning with *saliva* in the mouth, then proceeding to the *gastric juices* of the stomach, *pancreatic secretion* in the duodenum, and so forth. It is the lacrimal glands of the eye that provide the *tears*, the sweat glands of the skin that provide the *sweat*, and the nasal glands that provide *nasal secretions*. And still other glands secrete various hormones directly into the blood.

The glands of the body are divided into two separate groups:

1. **Exocrine glands,** which form fluid secretions that flow directly into some cavity of the body or onto the surface of the body, including the digestive secretions, sweat secretion, and so forth. The exocrine glands are composed of *epithelial cells* and are derived from out-pocketing of cells from the different epithelia.
2. **Endocrine glands,** which are composed of cells often very similar to those of the exocrine glands. However, they secrete *hormones* through the glandular capillary walls directly into the circulating blood. These glands are so important in controlling the internal functions of the body that they will be the topic of an entire section of this book on the subject of *endocrinology*.

TYPES OF EXOCRINE GLANDS

Figure 2–16 illustrates examples of different types of exocrine glands, which are described as follows:

Single Cell Glands. These are the simplest type of exocrine gland. In many parts of the digestive tract, single glandular cells are interspersed along the surface of the epithelium among the nonglandular epithelial cells. Most of these cells secrete *mucus*, a very slick and chemically inert material that coats the inner lining of the digestive tract and prevents digestion of its walls. Similar cells are also found in the *urethra, urinary bladder, gallbladder, nasal passageways, bronchi*, and other structures of the body that require lubrication and protection. These cells frequently are globular in shape and are called *goblet cells*.

Tubular Glands. These are of several varieties, as illustrated in Figure 2–16. (1) *Simple tubular glands* are straight tubes found throughout both the small intestine and the large intestine. (2) *Simple branched tubular glands* are especially prevalent in the *pyloric region of the stomach*. (3) *Simple coiled tubular glands* are typical of the deeper portions of the *sweat glands*. The deeper portions of these tubular glands are lined with typical secretory cells, but the neck portions are frequently lined with nonsecretory epithelial cells; this neck portion is called the **glandular duct.**

Acinar Glands. These glands have essentially the same function as the tubular glands except that the secretory portions of the glands are arranged in the form of spherical lobules called **acini,** rather than as tubes. The neck portions of these glands also are usually lined with nonsecretory epithelial cells and are the **glandular ducts.** Acinar glands exist with all degrees of complexity. Two types are shown in Figure 2–16: (1) The *simple branched acinar gland* has only one duct but several acini; a typical gland of this type is the *sebaceous gland*, which secretes oils onto the skin. (2) The *compound acinar gland* has a complex system of branching ducts; typical examples of this type of gland are the *salivary glands* of the mouth and the exocrine portion of the *pancreas*, an organ that secretes its juices into the duodenum.

Laminar Gland. The *liver* is the only laminar gland of the body, which means a gland composed of sheetlike layers of cells that secrete into small channels between the layers. However, the liver constitutes about 2 percent of the total body weight. The liver cells are mainly concerned with other metabolic processes besides glandular secretion so that their cells are not at all typical of the usual exocrine secretory cells. As shown in Figure 2–16, lying between adjacent layers of liver cells are many small **canaliculi** into which the liver cells secrete small amounts of its secretion, called *bile*. This bile then passes into **bile ducts** that become successively larger and lead into the duodenum, as we shall see in Chapter 41.

Single cells
(stomach lining)

Simple tubular
(intestinal
glands)

Simple branched
tubular
(stomach pyloric glands)

Simple coiled tubular
(sweat glands)

Simple branched
acinar
(sebaceous glands)

Compound acinar
(pancreas)

Laminar
(liver)

FIGURE 2–16 Different types of exocrine glands.

Typical Exocrine Secretory Cell. Except for the liver cells, the secretory cells of all of the exocrine glands have very similar characteristics, whether they be single cell glands on the epithelial surface of the gastrointestinal tract or compound acinar glands. Figure 2–17 illustrates one of these typical cells, a single cell from the inner surface of the stomach that secretes mucus. The parts of this cell that are important for the secretory process are as follows:

1. The *mitochondria* extract energy from foodstuffs and oxygen and then provide this energy to the remainder of the secretory apparatus of the cells.
2. The *endoplasmic reticulum* synthesizes many of the products that are to be secreted by the cell. The granular portion of this endoplasmic reticulum forms the proteins and some carbohydrates that are to be secreted, and the portions of the endoplasmic reticulum that are not granular, the smooth endoplasmic reticulum, synthesizes lipid substances and some of the carbohydrate substances that are to be secreted.
3. The *Golgi complex* synthesizes still other substances, especially carbohydrate moieties, and also condenses the products from the endoplasmic reticulum before they are secreted. It is the Golgi complex that forms the final secretory products.
4. *Secretory vesicles* are the final products of secretion. They are extruded by the Golgi complex into the apical portion of the cell.

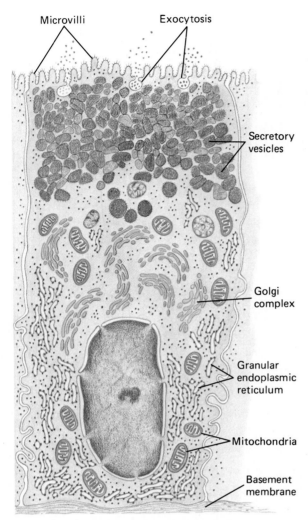

FIGURE 2–17 A typical secretory cell of an exocrine gland.

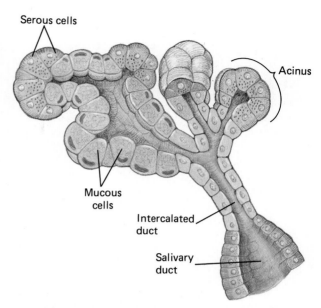

FIGURE 2–18 A small terminal portion of a salivary gland, illustrating different types of epithelial cells in the acini and ducts.

As the secretory vesicles accumulate in the apical portion of the cell, they fuse with the cell membrane and then break open to the outside of the cell, thus spilling their contents into the glandular duct or onto the epithelial surface. This process of fusion and emptying of the vesicles is called *exocytosis*.

Multiple Cell Types in Glands. Almost all of the more complex glands have multiple types of secretory cells and duct cells as we shall discuss at many points in this text. Figure 2–18 is an example, showing the different cell types in the submandibular gland, one of the salivary glands. The principal cells of this gland are

1. *Serous cells.* These line many of the acini and secrete the enzyme *ptyalin*, which begins the digestion of carbohydrates in the mouth.
2. *Mucous cells.* These line some of the acini as well as some of the tubular portions of the gland. These secrete *mucus* that lubricates and protects the mouth, pharynx, and esophagus during the chewing and swallowing process.
3. *Duct cells.* These line the ducts leading from the acini. The duct cells secrete *bicarbonate ions* that make the saliva alkaline, and they also reabsorb sodium and chloride ions so that not many of these ions will be lost from the internal body fluids into the digestive tract.

The other glands have similar specializations.

OTHER TISSUES

The tissues and simple organs discussed in this chapter were chosen mainly to illustrate the manner in which cells and extracellular elements are organized into tissues. We have discussed especially connective tissue and epithelial tissue plus some of their variants and derivative structures. The two other basic types of tissue, *muscular tissue* and *nervous tissue*, have been left for later consideration in relation to their highly specialized function of muscle contraction and nervous conduction of signals. However, Table 2–1 gives a summary of all four basic tissues, their major variants, and some of their derivative structures. If the student will become familiar with these, as he proceeds

TABLE 2-1
Some Representative Types of Tissues

Types	Examples
Connective tissue	
Connective tissue proper	
Loose	Subcutaneous tissue
	Between organs and other body structures
Dense	Tendons
	Ligaments
	Deep fasciae
	Muscle sheaths
	Organ capsules
	Globe of eyeball
Connective tissue variants	
Blood	Bone marrow and circulating blood
Adipose tissue	Fat tissue of body under skin and internally
Bone	Most parts of skeleton
Cartilage	Lower nasal septum
	Anterior ends of ribs
	Surfaces of joints
	In growing areas of long bones
Epithelial tissue	
Epithelium	*Linings of*
Simple squamous	Internal cavities
Simple cuboidal	Ducts, renal tubules
Simple columnar	Digestive tract
Stratified columnar	Urethra
Transitional	Urinary bladder, ureter
Stratified squamous	Vagina, esophagus, outer layer of skin (epidermis)
Epithelial tissue derivatives	
Glands	Exocrine glands, some endocrine glands
Skin derivatives	Hair, nails
Oral epithelial derivative	Bodies of teeth
Muscular tissue	
Skeletal muscle	Muscles mainly attached to skeleton
Cardiac muscle	Muscle of heart
Smooth muscle	Muscle of blood vessels, gut, ducts, uterus, intraocular muscles, arrector pili muscles
Nervous tissue	
Central nervous system tissue	Brain
	Spinal cord
Peripheral nervous tissue	Skeletal nerves
	Sensory nerve receptors
	Autonomic nervous system
	Retina of eye
	Intrinsic nervous system of the gut
Glandular derivatives	Some endocrine glands

through this book he will be able to see how various mixes and various architectural arrangements of the different tissues are employed to perform a vast array of body functions. For instance, which of the different basic systems are found in the eye—the answer is all of them. Also, which of the tissues are found in the liver—here again, the answer is still all of them. But the eye and the liver perform functions that are miles apart. The logic of the body is to employ a few basic elements to build all of its complex structures.

QUESTIONS

1. List the intracellular structures of the typical cell, and tell briefly their functional roles in the cell.
2. What are the different types of blood cells? In what basic ways do the red blood cells differ from the white cells?
3. What are the different elements of connective tissues? How does dense connective tissue differ from loose connective tissue?
4. What are the special characteristics of adipose tissue?
5. How do cartilage and bone differ from other types of connective tissue?
6. What are the characteristics of cartilage, and where in the body is cartilage found?
7. What gives bone its tremendous tensile strength, and what gives it its rigidity and compressive strength?
8. List the parts of the tooth, and give the functional importance of each.
9. List the different types of epithelium.
10. In the skin, what are the specific characteristics of the epidermis and of the dermis? Describe the specialized structures of the skin.
11. How are hairs and fingernails formed?
12. Give the characteristics of the different types of exocrine glands.
13. What parts of the cell are especially important for the secretory process, and what role does each serve?

REFERENCES

Bloom, W., and Fawcett, D.W.: Cells and tissues. *In* A Textbook of Histology, 10th Ed. Philadelphia, W. B. Saunders, 1975, pp. 35–385.

Copenhaver, W.M., Kelly, D.E., and Wood, R.L.: Cells and tissues. *In* Bailey's Textbook of Histology, 17th Ed. Baltimore, Williams and Wilkins, 1978, pp. 1–357.

Fawcett, D.W.: The Cell, 2nd Ed. Philadelphia, W.B. Saunders, 1981.

Fujita, T., Tanaka, K., and Tokunaga, J.: SEM Atlas of Cells and Tissues. New York, Igaku-Shoin, 1981.

Goss, C.M.: The integument. *In* Gray's Anatomy of the Human Body, 29th Ed. Philadelphia, Lea & Febiger, 1973, pp. 1099–1111.

Ham, A.W., and Cormack, D.H.: Cells and tissues. *In* Histology, 8th Ed. Philadelphia, Lippincott, 1979, pp. 35–580.

Hammersen, F.: Cytology and connective tissues. *In* Sobotta/Hammersen Histology. Baltimore, Urban & Schwarzenberg, 1980, pp. 2–67.

Hammersen, F.: Sobotta/Hammersen Histology. Baltimore, Urban & Schwarzenberg, 1980.

Leeson, T.S., and Leeson, C.R.: A Brief Atlas of Histology. Philadelphia, W.B. Saunders, 1979.

Leeson, T.S., and Leeson, C.R.: The cell and tissues. *In* Histology, 4th Ed. Philadelphia, W.B. Saunders, 1981, pp. 19–256.

Netter, F.H.: Endocrine System. The CIBA Collection of Medical Illustrations, Vol. 4, Summit, N.J., CIBA Medical Education Division, 1965.

Reith, E.J., and Ross, M.H.: Atlas of Descriptive Histology. New York, Harper and Row, 1977.

Rhodin, J.G.: Histology. New York, Oxford University Press, 1974.

Weiss, L., and Greep, R.O.: The cell and tissues. *In* Histology, 4th Ed. New York, McGraw-Hill, 1977, pp. 1–372.

Wheater, P.R., Burkitt, H.G., and Daniels, V.G.: The cell and tissues. *In* Functional Histology. New York, Churchill Livingstone, 1979, pp. 2–160.

Williams, P.L., and Warwick, R.: The skin. *In* Gray's Anatomy, 36th British Edition. Philadelphia, W.B. Saunders, 1980, pp. 1216–1227.

(See also General Anatomical References, Chapter 1.)

3

Intracellular Organization and Function

Overview

The cells are composed mainly of five basic substances: (1) *water,* which is present in a concentration between 70 and 85 percent; (2) *proteins,* which normally constitute 10 to 20 percent of the cell mass; (3) *lipids,* which constitute about 2 percent of most cells but as much as 95 percent of "fat cells"; (4) *carbohydrates,* which are about 1 percent of the total cell mass; and (5) various ions, including especially *potassium, magnesium, phosphate, sulfate, bicarbonate,* and small quantities of sodium, chloride, and calcium.

Each cell contains many highly organized physical structures called *organelles.* The characteristics and functions of some of the principal organelles are described in the following.

Cell Membrane. This is a very thin elastic structure, only 7.5 to 10 nanometers (nm) thick. Its basic structure is a thin film of lipids only 2 molecules thick (a *bimolecular layer*) that serves as a barrier to the passage of water and water-soluble substances between the *extracellular fluid* surrounding the cell and the fluid inside the cell called the *intracellular fluid.* Floating in the bimolecular lipid film are large numbers of *protein molecules,* many of which penetrate all the way through the cell membrane and provide passageways called *pores* through which water and water-soluble substances can pass.

Nuclear Membrane. This is similar to the cell membrane except that it separates the *nucleoplasm* from the surrounding *cytoplasm.* It is actually a double membrane, composed of two lipid bilayers each similar to that of the cell membrane. Even so, the nuclear membrane is far more porous than the cell membrane.

Endoplasmic Reticulum. This is a system of interconnected tubular and flat, shelflike chambers that spreads throughout most of the cytoplasm. The membranes of the endoplasmic reticulum are similar to the cell membrane, and it is on the surfaces of this membrane that most of the chemical reactions of the cell take place. Attached to many areas of the endoplasmic reticulum are great numbers of *ribosomes* that synthesize proteins, most of which pass directly from the ribosomes into the internal passageways of the endoplasmic reticulum and then are transported to other parts of the cell.

Golgi Complex. This is similar to the endoplasmic reticulum, and it functions in close association with the endoplasmic reticulum. Generally, proteins and other substances synthesized by the endoplasmic reticulum are passed into the Golgi complex, where they are processed further, forming such additional intracellular components as *secretory vesicles, lysosomes,* and so forth.

Mitochondria. These are elongated enclosed chambers usually about 1 μm long. Many of them are distributed throughout the cytoplasm, sometimes many hundreds in a single cell. These structures are called the "powerhouses" of the cell because they convert food energy into energy stored in the form of *adenosine triphosphate* (ATP). ATP in turn is utilized throughout the cell to energize the different cellular chemical reactions. For instance, ATP energizes (a) the transport of substances through the cell membrane, such as the transport of potassium into cells and sodium out of cells; (b) the synthesis of proteins and other intracellular substances such as phospholipids, cholesterol, and many others; and (c) the muscle contraction that provides all the body's movements.

Lysosomes. These are small round packages of digestive enzymes surrounded by a bilayer lipid membrane. When the membrane is broken, the digestive enzymes are released within the cell, and they digest the local structures. Or, they can also digest foreign substances such as bacteria that enter the cell.

Nucleus. The nucleus is the control center of the cell. It contains the *chromosomes* that in turn are the loci of hundreds of thousands of molecules of *deoxyribonucleic acid* (DNA), which are the *genes.* The genes control the specific chemical functions of the cell and also control reproduction of the cells, which will be discussed in Chapter 4.

Each of the cells in the human being is a living structure. They survive for varying lengths of time, and most of them can reproduce themselves, provided their surrounding fluids contain appropriate nutrients. To understand the function of organs and other structures of the body, it is essential that we first understand the basic organization of the cell and the functions of its component parts.

CHEMICAL CONSTITUENTS OF THE CELL

The different substances that make up the cell are collectively called *protoplasm.* Protoplasm is composed mainly of five basic substances: water, ions, proteins, lipids, and carbohydrates.

Water. The principal fluid medium of the cell is water, which is present in a concentration of between 70 and 85 percent. Many cellular chemicals are dissolved in the water; others are suspended in small particulate form. Chemical reactions take place among the dissolved chemicals or at the surface boundaries between the suspended particles and the water.

Ions. The most important ions in the cell are *potassium, magnesium, phosphate, sulfate, bicarbonate,* and small quantities of *sodium, chloride,* and *calcium.* These will be discussed in much greater detail in Chapter 5, which will consider the interrelationships between the *intracellular* and *extracellular fluids.*

The ions are dissolved in the cell water, and they provide inorganic chemicals for cellular reactions. Also, they are necessary for operation of some of the cellular control mechanisms. For instance, ions acting at the cell membrane allow transmission of electrochemical impulses in nerve and muscle fibers, and the intracellular ions determine the activity of different enzymatically catalyzed reactions that are necessary for cellular metabolism.

Proteins. Next to water, the most abundant substance in most cells is proteins, which normally constitute 10 to 20 percent of the cell mass. These can be divided into two different types, *structural proteins* and *globular proteins,* which are mainly *enzymes.*

To get an idea of what is meant by *structural proteins*, one needs only to note that leather is composed principally of structural proteins, and hair is almost entirely a structural protein. Proteins of this type are present in the cell in the form of long thin filaments that themselves are polymers of many protein molecules. The most prominent use of such intracellular filaments is to provide the contractile mechanism of all muscles, as will be discussed in Chapter 11. However, filaments are also organized into microtubules, which provide the structures of such organelles as cilia and the mitotic spindles of mitosing cells. And, extracellularly, fibrillar proteins are found especially in the collagen and elastic fibers of connective tissue, blood vessels, tendons, ligaments, and so forth.

The *globular proteins*, on the other hand, are an entirely different type of protein, composed usually of individual protein molecules or at most of aggregates of a few molecules in a globular form rather than a fibrillar form. These proteins are mainly the enzymes of the cell and, in contrast to the fibrillar proteins, are often soluble in the fluid of the cell or are absorbed in or adherent to the surfaces of membranous structures inside the cell. The enzymes come into direct contact with other substances inside the cell and catalyze chemical reactions. For instance, the chemical reactions that split glucose into its component parts and then combine these with oxygen to form carbon dioxide and water, in this way releasing energy for cellular function, are catalyzed by a series of protein enzymes. Thus, enzyme proteins control the metabolic functions of the cell.

Special types of proteins are present in different parts of the cell. Of particular importance are the *nucleoproteins* of the nucleus that contain *deoxyribonucleic acid* (*DNA*), which constitutes the *genes;* these control the overall function of the cell as well as transmission of hereditary characteristics from cell to cell. These substances will be considered in detail in Chapter 4.

Lipids. Lipids are several different types of substances that are grouped together because of their common property of being soluble in fat solvents. The most important lipids in most cells are *phospholipids* and *cholesterol*, which constitute about 2 percent of the total cell mass. These are major constituents of the different membranes such as the cell membrane, the nuclear membrane, and the membranes of cytoplasmic organelles, e.g., the endoplasmic reticulum and the mitochondria. The special importance of phospholipids and cholesterol in the cell is that they are either insoluble or only partially soluble in water.

In addition to phospholipids and cholesterol, some cells contain large quantities of *triglyceride*, also called *neutral fat*. In the so-called "fat cells," triglycerides can account for as much as 95 percent of the cell mass. And this fat stored in these cells represents the body's main storehouse of energy-giving nutrient that can be broken down and used for energy whenever the current food intake is insufficient to supply the needed energy.

Carbohydrates. In general, carbohydrates have very little structural function in the cell except as part of glycoprotein molecules, but they play a major role in nutrition of the cell. Carbohydrate, in the form of glucose, is always present in the surrounding extracellular fluid so that it is readily available to the cell. In addition, a small amount of carbohydrate is usually stored in the cells in the form of *glycogen*, about 1 percent of the total cell mass, which is an insoluble polymer of glucose and can be used rapidly to supply the cells' energy needs for up to 12 hours but not for many days of starvation.

PHYSICAL ORGANIZATION OF THE CELL

The Membranous Structures of the Cell

Figure 3–1 illustrates diagrammatically the physical structures of the cell. Essentially all of these are lined by membranes composed primarily of lipids and proteins. The lipids provide a barrier that prevents free movement of water and water-soluble substances from one cell compartment to the other. The protein molecules, on the other hand, interrupt the continuity of the lipid barrier and therefore provide pathways through the protein molecules themselves for passage of various substances through the membrane. The different membranes include the **cell membrane,** the **nuclear membrane,** the **membrane of the endoplasmic reticulum,** and the **membranes of the mitochondria, lysosomes, Golgi complex,** and so forth.

The Cell Membrane. The cell membrane, which completely envelops the cell, is a very thin, elastic structure only 7.5 to 10 nanometers (nm) thick. It is composed almost entirely of proteins and lipids; the approximate composition is proteins, 55 percent; phospholipids, 25 percent; cholesterol, 13 percent; other lipids, 4 percent; and carbohydrates, 3 percent.

The Lipid Barrier of the Cell Membrane. Figure 3–2 illustrates that the basic structure of the cell membrane is a *lipid bilayer*, which is a thin film of lipids only two molecules thick that is continuous over the entire cell surface. Interspersed in this lipid film are large globular protein molecules.

The lipid bilayer is composed almost entirely of phospholipids and cholesterol, which makes this bilayer almost entirely impermeable

Endoplasmic reticulum (granular)

Cellular membrane

Mitochondria

Membranous outpouching

Glycogen

Membranous invagination

Lysosomes

Neutral fat

Agranular endoplasmic reticulum

Nuclear membrane

Nucleus

Endoplasm

Cortex (Ectoplasm)

FIGURE 3—1 Organization of the cytoplasmic compartment of the cell.

to the usual water-soluble substances such as ions, glucose, and others. On the other hand, fat-soluble substances such as oxygen, carbon dioxide, and alcohols can penetrate this portion of the membrane.

A special feature of the lipid bilayer is that it is a *fluid* and not a solid. Therefore, portions of the membrane can literally flow from one point to another in the membrane. Proteins or other substances dissolved in or floating in the lipid

bilayer tend to diffuse to all areas of the cell membrane.

The Cell Membrane Proteins. Figure 3–2 illustrates globular masses floating in the lipid bilayer. These are the cell membrane proteins, most of which are *glycoproteins* (proteins with carbohydrate radicals attached). Two types of proteins occur: *integral proteins*, which protrude all the way through the cell, and *peripheral proteins*, which are attached only to the sur-

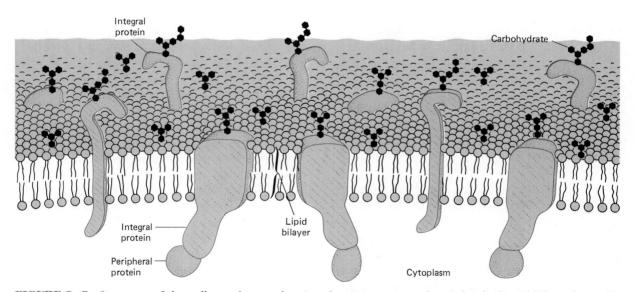

Integral protein

Carbohydrate

Integral protein

Lipid bilayer

Peripheral protein

Cytoplasm

FIGURE 3—2 Structure of the cell membrane, showing that it is composed mainly of a liquid bilayer but with large numbers of protein molecules protruding through the layer. Also, carbohydrate moieties are attached to the protein molecules on the outside of the membrane and additional protein molecules on the inside. (From Lodish and Rothman: *Sci. Am.*, 240:18, 1979. © 1979 by Scientific American, Inc.)

TABLE 3–1
The Cell Structures and Their Basic Functions

Structures	Functions
Major compartments	
Nucleus	Control center of cell
Cytoplasm	Locus of most cellular metabolic activity
Membranous structures	
Cell membrane	Separates intracellular fluid from extracellular fluid
Nuclear membrane	Separates cytoplasmic compartment from nuclear compartment
Endoplasmic reticulum	Transports substances from one part of the cell to another
Granular (rough) portion	Synthesizes proteins and carbohydrates
Smooth portion	Synthesizes lipids and carbohydrates
Golgi apparatus	Compacts endoplasmic reticular secretions and extrudes secretory vesicles
Secretory vesicles (granules)	Contain secretory products
Lysosomes	Vesicles containing digestive enzymes
Mitochondria	Form the high-energy compound ATP, which energizes intracellular functions
Ribosomes	Structure on which the proteins are synthesized; often attached to granular endoplasmic reticulum
Microtubule structures	Transport of substances inside the cell; sometimes serve as intracellular "skeleton"
Centrioles	Microtubular structure that helps in cell division
Cilia	Microtubular protruding structures that move fluid along cell surface
Microfilaments	Tensile elements of cells; add strength to membranes; provide molecular basis for muscle contraction
Chromosomes	Locus of DNA that constitutes the genes
Nucleolus	Origin of ribosomes inside the nucleus before migration into cytoplasm

face of the membrane and do not penetrate. The integral proteins provide structural pathways through which water and water-soluble substances, especially the ions, can diffuse between the extracellular and intracellular fluid, thus providing so-called cell membrane "pores." However, these proteins have selective properties that cause preferential diffusion of some substances more than others. Some of them can also act as enzymes.

The peripheral proteins occur either entirely or almost entirely on the inside of the membrane, and they are normally attached to one of the integral proteins. These peripheral proteins function almost entirely as enzymes that control many of the chemical reactions inside the cell.

The Membrane Carbohydrates. The membrane carbohydrates occur almost invariably on the outside of the membrane; they are the "glyco" portion of protruding glycoprotein molecules. These carbohydrate moieties are the portions of the cell membrane that enter into immune reactions, and they often act as receptor substances for binding hormones, such as insulin, that stimulate specific types of activity in the cells.

The Nuclear Membrane. The nuclear membrane, illustrated in Figure 3–1, is actually two membranes, one surrounding the other with a wide space in between. Each membrane is almost identical to the cell membrane, having a basic lipid bilayer structure with globular proteins floating in the lipid fluid. At many points the two membranes fuse with each other, and at these points the nuclear membrane is so permeable that almost all dissolved or suspended substances, including even very large, newly formed ribosomes, can move with ease between the fluids of the nucleus and the cytoplasm.

The Endoplasmic Reticulum. Figure 3–1 illustrates in the cytoplasm a continuous network of tubular and flat vesicular structures, constructed of lipid bilayer–protein membranes, called the *endoplasmic reticulum*. The total surface area of this structure in some cells—the liver cells, for instance—can be as much as 30 to 40 times as great as the cell membrane area. The detailed structure of this organelle is illustrated in Figure 3–3. The space inside the tubules and vesicles is filled with *endoplasmic matrix*, a fluid medium that is different from the fluid outside the endoplasmic reticulum. Electron micrographs show that the space inside the endoplasmic reticulum is connected with the

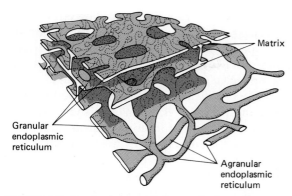

FIGURE 3–3 Structure of the endoplasmic reticulum. (Modified from De Robertis, Saez, and De Robertis: Cell Biology, 6th Ed. Philadelphia, W.B. Saunders, 1975.)

space between the two membranes of the double nuclear membrane.

Substances formed in different parts of the cell enter the spaces of the endoplasmic reticulum and are then conducted to other parts of the cell. Also, the vast surface area of the reticulum, as well as its many enzyme systems, provides the machinery for a major share of the metabolic functions of the cell.

Ribosomes and the Granular Endoplasmic Reticulum. Attached to the outer surfaces of many parts of the endoplasmic reticulum are large numbers of small granular particles called *ribosomes*. Where these are present, the reticulum is called the *granular*, or *rough*, *endoplasmic reticulum*. The ribosomes are composed mainly of ribonucleic acid, which functions in the synthesis of protein in the cells.

The Agranular Endoplasmic Reticulum. Part of the endoplasmic reticulum has no attached ribosomes. This part is called the *agranular*, or *smooth*, *endoplasmic reticulum*. The agranular reticulum functions especially in

FIGURE 3–4 A typical Golgi complex and its relationship to the endoplasmic reticulum and the nucleus.

the synthesis of lipid substances as well as in many other enzymatic processes of the cell.

Golgi Complex. The Golgi complex, illustrated in Figure 3–4, is closely related to the endoplasmic reticulum. It has membranes similar to those of the agranular endoplasmic reticulum. It is usually composed of four or more stacked layers of thin, flat vesicles lying near the nucleus. This complex is very prominent in secretory cells; in these it is located on the side of the cell from which the secretory substances will be extruded.

The Golgi complex functions mainly in association with the endoplasmic reticulum. As illustrated in Figure 3–4, small "transport vesicles" continually pinch off from the endoplasmic reticulum and shortly thereafter fuse with the Golgi complex. In this way substances are transported from the endoplasmic reticulum to the Golgi complex. The transported substances are then processed in the Golgi complex to form secretory vesicles, lysosomes, or other cytoplasmic components.

The Cytoplasm and Its Organelles

The cytoplasm is filled with both minute and large dispersed particles and organelles ranging in size from a few nanometers to 3 μm in size. The clear fluid portion of the cytoplasm in which the particles are dispersed is called *hyaloplasm*; this contains mainly dissolved proteins, ions, glucose, and small quantities of phospholipids, cholesterol, and esterified fatty acids.

Among the large dispersed particles in the cytoplasm are neutral fat globules, glycogen granules, ribosomes, secretory granules, and two especially important organelles—the *mitochondria* and *lysosomes*.

The Mitochondria. The mitochondria are called the "powerhouses" of the cell because they extract energy from nutrients and oxygen and in turn provide the energy in a more usable form to energize essentially all cellular functions. The number of mitochondria per cell varies from less than a hundred to many thousand, depending upon the amount of energy required by each cell. Furthermore, the mitochondria are concentrated in those portions of the cell that are responsible for the major share of its energy metabolism.

The basic structure of the mitochondrion is illustrated in Figure 3–5, which shows it to be composed mainly of two lipid bilayer–protein

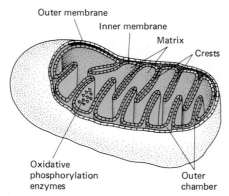

FIGURE 3–5   Structure of a mitochondrion. (Modified from De Robertis, Saez, and De Robertis: Cell Biology, 6th Ed. Philadelphia, W.B. Saunders, 1975.)

membranes: an *outer membrane* and an *inner membrane.* Many infoldings of the inner membrane form *shelves,* onto which the oxidative enzymes of the cell are attached. In addition, the inner cavity of the mitochondrion is filled with a gel *matrix* containing large quantities of dissolved enzymes that are necessary for extracting energy from nutrients. These enzymes operate in association with the oxidative enzymes on the shelves to cause oxidation of the nutrients, thereby forming carbon dioxide and water. The liberated energy is used to synthesize a high-energy substance called *adenosine triphosphate* (*ATP*). ATP is then transported out of the mitochondrion, and it diffuses throughout the cell to release its energy wherever it is needed for performing cellular functions. The function of ATP is so important to the cell that it is discussed in detail later in the chapter.

Mitochondria are self-replicative, which means that one mitochondrion can form a second one, a third one, and so on, whenever there is need in the cell for increased amounts of ATP.

The Lysosomes.   The lysosomes provide an intracellular digestive system that allows the cell to digest and remove unwanted substances and structures, especially damaged or foreign structures such as bacteria. The lysosome, illustrated in Figure 3–1, is 250 to 750 nm in diameter and is surrounded by a typical lipid bilayer membrane. It is filled with large numbers of small granules, which are protein aggregates of hydrolytic (digestive) enzymes. A hydrolytic enzyme is capable of splitting an organic compound into two or more parts by combining hydrogen from a water molecule with part of the compound and combining the hydroxyl portion of the water molecule with the other part of the compound. For instance, protein is hydrolyzed to form amino acids, and glycogen is hydrolyzed

to form glucose. More than 40 different *acid hydrolases* have been found in lysosomes, and the principal substances that they digest are proteins, nucleic acids, mucopolysaccharides, lipids, and glycogen.

Ordinarily, the membrane surrounding the lysosome prevents the enclosed hydrolytic enzymes from coming in contact with other substances in the cell. However, many different conditions of the cell will break the membranes of some of the lysosomes, allowing release of the enzymes. These enzymes then split the organic substances with which they come in contact into small, highly diffusible substances such as amino acids and glucose.

OTHER CYTOPLASMIC STRUCTURES AND ORGANELLES

Secretory Vesicles.   One of the important functions of many cells is secretion of special substances. Almost all such secretory substances are formed by the endoplasmic reticulum–Golgi complex system and are then released from the Golgi complex inside storage vesicles, called *secretory vesicles* or sometimes *secretory granules.* Figure 3–6 illustrates typical secretory vesicles inside pancreatic acinar cells that have formed and stored protein enzymes in them. These enzymes will be secreted later through the outer cell membrane into the pancreatic duct.

Microfilaments   and   Microtubular Structures in the Cell.   The fibrillar proteins of the cell cytoplasm are usually organized into microfilaments or microtubules. These originate as precursor protein molecules synthesized by the ribosomes. At first they are present in dissolved form in the cytoplasm. Then they polymerize to form microfilaments. Large numbers of these microfilaments frequently occur in the

FIGURE 3–6   Secretory granules in acinar cells of the pancreas.

outer zone of the cytoplasm, the zone called the *ectoplasm,* to form an elastic support for the cell membrane. Also, in muscle cells microfilaments are organized into a special contractile machine that is the basis of muscle contraction throughout the body.

Microfilaments are also frequently organized into tubular structures, the *microtubules.* Almost invariably these contain 13 microfilaments lying parallel and in a circle to form a long hollow cylinder 25 nm in diameter and 1 to many microns in length. These are often arranged in bundles, which gives them, *en masse,* considerable structural strength. However, microtubules are stiff structures that break if bent too severely. Figure 3–7 illustrates typical microtubules that were teased from the flagellum of a sperm. Another example of microtubules is the tubular mechanical structure of cilia that gives them structural strength, radiating upward from the cell cytoplasm to the tip of the cilium. Also, the centrioles and the mitotic spindle of the mitosing cell are both composed of stiff microtubules.

Thus, a primary function of microtubules is to act as a *cytoskeleton,* providing rigid physical structures for certain parts of cells. Also, it has been noted that the cytoplasm often *streams* in the vicinity of microtubules, which might result from movement of arms that project outward from the microtubules.

The Nucleus

The nucleus is the control center of the cell. It controls both chemical reactions that occur in the cell and reproduction of the cell. Briefly, the nucleus contains large quantities of *deoxyribonucleic acid,* which we have called *genes* for many years. The genes control the formation of the protein enzymes of the cytoplasm and in this way control cytoplasmic activities. To control reproduction, the genes first reproduce themselves, and after this is accomplished the cell splits by a special process called *mitosis* to form two daughter cells, each of which receives one of the two sets of genes.

The appearance of the nucleus under the microscope does not give much of a clue to the mechanisms by which it performs its control activities. Figure 3–8 illustrates the light microscopic appearance of the interphase nucleus (period between mitoses), showing darkly staining *chromatin material* throughout the *nuclear sap.* During mitosis, the chromatin material becomes readily identifiable as part of the highly structured *chromosomes,* which can be seen easily with the light microscope (illustrated in Chapter 4). Even during the interphase of cellular activity the chromatin material is still organized into fibrillar chromosomal structures, but this is impossible to see except in a few types of cells.

Nucleoli. The nuclei of many cells contain one or more lightly staining structures called *nucleoli.* The nucleolus is a protein structure that contains a large amount of *ribonucleic*

FIGURE 3–7 Microtubules teased from the flagellum of a sperm. (From Porter, K. R.: Ciba Foundation Symposium: Principles of Biomolecular Organization. Boston, Little, Brown, 1966.)

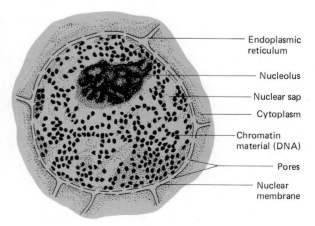

FIGURE 3–8 Structure of the nucleus.

acid that later condenses to form the ribosomes. These then migrate through the nuclear membrane pores into the cytoplasm, where most of them become attached to the endoplasmic reticulum and there play an essential role for the formation of proteins.

COMPARISON OF THE ANIMAL CELL WITH PRECELLULAR FORMS OF LIFE

Many of us think of the cell as the lowest level of animal life. However, the cell is a very complicated organism, which required several billion years to develop after the earliest form of life, some organism similar to the present-day *virus*, first appeared on Earth. Figure 3–9 illustrates the relative sizes of the smallest known virus, a large virus, a *rickettsia*, a *bacterium*, and a *cell*, showing that the cell has a diameter about 1000 times that of the smallest virus and, therefore, a volume about 1 billion times that of the smallest virus. Correspondingly, the functions and anatomic organization of the cell are also far more complex than those of the virus.

The essential life-giving constituent of the very small virus is a *nucleic acid* embedded in a coat of protein. This nucleic acid is similar to that of the cell, and it is capable of reproducing itself if appropriate surrounding conditions are available. Thus, the virus is capable of propagating its lineage from generation to generation and, therefore, is a living structure in the same way that the cell and the human being are living structures.

As life evolved, other chemicals besides nucleic acid and simple proteins became integral parts of the organism, and specialized functions began to develop in different parts of the viral-like organism. A membrane formed around it, and inside the membrane a fluid matrix appeared. Specialized chemicals developed inside the matrix to perform special functions; many protein enzymes appeared that were capable of catalyzing chemical reactions and, therefore, of controlling the organism's activities.

In still later stages, particularly in the rickettsial and bacterial stages, *organelles* developed inside the organism, these representing aggregates of chemical compounds that perform functions in a more efficient manner than can be achieved by dispersed chemicals throughout the fluid matrix. And, finally, in the nucleated cell, still more complex organelles developed, the most important of which is the *nucleus* itself. The nucleus distinguishes this type of cell from all lower forms of life; this structure provides a control center for all cellular activities, and it also provides for very exact reproduction of new cells generation after generation, each new cell having essentially the same structure as its progenitor.

FUNCTIONAL SYSTEMS OF THE CELL

In the remainder of this chapter we will discuss most of the functional systems of the cell that make it a living organism. However, two cellular functions are so important that they will be the subjects of the following two chapters: (1) control of protein synthesis and of other cellular functions by the genes in the nucleus and (2) transport of substances through the cell membrane. But in the present chapter, let us begin by discussing the means by which the cell ingests substances, then consider the energy systems of the cell and synthesis of new substances in the cell.

Ingestion and Digestion by the Cell

Endocytosis—Phagocytosis and Pinocytosis. One of the means by which cells ingest foreign material is by the process of *endocytosis*, in which the cell membrane engulfs particulate matter or extracellular fluid and its contents. Phagocytosis and pinocytosis are both examples of endocytosis.

Phagocytosis means the ingestion of large particulate matter by a cell, such as the ingestion

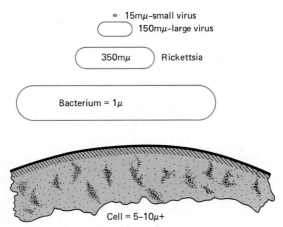

FIGURE 3–9 Comparison of sizes of subcellular organisms with that of the average cell in the human body.

of (1) a bacterium, (2) some other cell, or (3) particles of degenerating tissue. *Pinocytosis*, on the other hand, means ingestion of minute quantities of extracellular fluid and dissolved substances in the form of minute vesicles. The pinocytic vesicles are so small that they were not discovered until the advent of the electron microscope. However, phagocytosis has been known since the time of the earliest studies using the light microscope.

Phagocytosis occurs when certain objects contact the cell membrane. In general, those objects that have an electronegative charge are rejected, whereas those that have an electropositive charge are especially susceptible to phagocytosis. The difference presumably results from the fact that the phagocytic cells themselves normally are electronegatively charged and therefore repel other electronegative objects. Most *normal* particulate objects in the extracellular fluid are also negatively charged; on the other hand, damaged tissues and also foreign invaders that have been especially prepared for phagocytosis by attachment to antibodies (a process called *opsonization*), usually acquire positive charges and are therefore phagocytized.

Pinocytosis also occurs in response to certain types of substances that contact the cell membrane. The two most important are proteins and strong ion solutions. It is especially significant that proteins cause pinocytosis, because pinocytosis is the only means by which significant quantities of proteins can pass through the cell membrane.

Figure 3–10 illustrates the successive steps of pinocytosis, showing first three molecules of protein attaching to the membrane by the simple process of adsorption. The presence of these proteins then causes the surface properties of the membrane to change in such a way that the membrane invaginates and then rapidly closes over the proteins. Immediately thereafter, the invaginated portion of the membrane breaks away from the surface of the cell, forming a *pinocytic vesicle*. *Phagocytic vesicles* are formed in a similar manner.

What causes the cell membrane to go through the necessary contortions for forming the pinocytic and phagocytic vesicles remains a mystery. However, it is known that this process requires energy from within the cell; this is supplied by adenosine triphosphate, the high-energy substance that will be discussed elsewhere in this chapter. Also, endocytosis requires the presence of calcium ions in the extracellular fluid and probably a contractile function by *microfilaments* immediately beneath the cell membrane.

THE DIGESTIVE ORGANELLES OF THE CELL—THE LYSOSOMES

Almost immediately after a pinocytic or phagocytic vesicle appears inside a cell, one or more lysosomes become attached to the vesicle and empty their hydrolases into the vesicle, as illustrated in Figure 3–11. Thus, a *digestive vesicle* is formed in which the hydrolases begin hydrolyzing the proteins, glycogen, lipids, nucleic acids, mucopolysaccharides, and other substances inside the vesicle. The products of digestion are small molecules of amino acids, glucose, fatty acids, phosphates, and so forth that can diffuse through the membrane of the vesicle into the

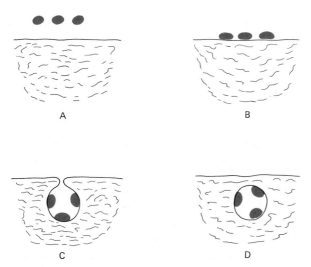

FIGURE 3–10 Mechanism of pinocytosis.

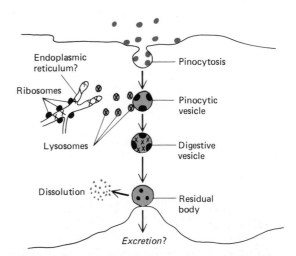

FIGURE 3–11 Digestion of substances in pinocytic vesicles by enzymes derived from lysosomes. (Modified from C. De Duve: Lysosomes, ed. by Reuck and Cameron. Boston, Little, Brown, 1963.)

cytoplasm. What is left of the digestive vesicle, called the *residual body*, represents the undigestible substances. In most instances, this is finally extruded through the cell membrane by a process called *exocytosis*, which is essentially the opposite of endocytosis.

Thus, the lysosomes may be called the *digestive organelles* of the cells.

Regression of Tissues and Autolysis of Cells.   Often, tissues of the body regress to a smaller size. For instance, this occurs in the uterus following pregnancy, in muscles during long periods of inactivity, and in the breasts at the end of a period of lactation. Lysosomes are responsible for most of this regression.

Another special role of the lysosomes is the removal of damaged cells or damaged portions of cells from tissues—cells damaged by heat, cold, trauma, chemicals, or any other factor. Damage to the cell causes lysosomes to rupture, and the released hydrolases begin immediately to digest the surrounding organic substances. If the damage is slight, only a portion of the cell will be removed, followed by repair of the cell. However, if the damage is severe, the entire cell will be digested, a process called *autolysis.*

The lysosomes also contain bactericidal agents that can kill phagocytized bacteria before they can cause cellular damage. And in lysosomes are stored enzymes that, upon release into the cytoplasm, can digest lipid droplets or glycogen granules, making the lipid and glycogen available for use elsewhere in the cell or elsewhere in the body. In the absence of these enzymes, which does occur in occasional genetic disorders, extreme quantities of lipids or glycogen often accumulate in the cells of many organs, especially the liver, and lead to early death.

Extraction of Energy from Nutrients—Function of the Mitochondria

The principal nutrients from which cells extract energy are oxygen and one or more of the foodstuffs. Figure 3–12 shows oxygen and the foodstuffs—glucose, fatty acids, and amino acids—all entering the cell. Inside the cell, the foodstuffs react chemically with the oxygen under the influence of various enzymes that control their rates of reactions and channel the energy that is released in the proper direction.

Formation of Adenosine Triphosphate (ATP).   The energy released from the nutrients

FIGURE 3–12   Formation of adenosine triphosphate in the cell, showing that most of the ATP is formed in the mitochondria.

is used to form adenosine triphosphate, generally called ATP, the formula for which is

ATP is a *nucleotide* composed of the nitrogenous base *adenine*, the pentose sugar *ribose*, and three *phosphate radicals*. The last two phosphate radicals are connected with the remainder of the molecule by so-called *high-energy phosphate bonds*, which are represented by the symbol $\sim$. Each of these bonds contains about 12,000 Calories of energy per mole of ATP under the physical conditions of the body (7300 Calories under standard conditions), which is much greater than the energy stored in the average chemical bond of most other organic compounds, thus giving rise to the term "high-energy" bond. Furthermore, the high-energy phosphate bond is very labile, so that it can be split instantly on demand whenever energy is required to promote other cellular reactions.

When ATP releases its energy, a phosphoric acid radical is split away, and *adenosine diphosphate (ADP)* is formed. Then, energy derived from the cellular nutrients causes the ADP and phosphoric acid to recombine to form new ATP, the entire process continuing over and over again. For these reasons ATP has been called the *energy currency* of the cell, for it can be spent and remade again and again.

Chemical Processes in the Formation of ATP—Role of the Mitochondria. Most of the ATP formed in the cell is synthesized in the mitochondria, a process that is discussed in detail in Chapter 42. However, the mechanism is basically the following: The different foods are first digested in the person's digestive tract to form glucose, fatty acids, and amino acids. Mainly in these forms, they are then delivered to the cells. In the cells they are eventually converted into the compound *acetyl co-A*, which is split inside the mitochondrion into hydrogen atoms and carbon dioxide. The carbon dioxide diffuses out of the mitochondrion and eventually out of the cell. The hydrogen atoms combine with carrier substances and are carried to the surfaces of the shelves illustrated in Figure 3–5 that protrude into the mitochondrion. Attached to these shelves are the so-called *oxidative enzymes* and also protruding globules of *ATPase*, the enzyme that catalyzes the conversion of ADP to ATP. The oxidative enzymes, by a series of sequential reactions, cause the hydrogen atoms to combine with oxygen. During these reactions, the energy released from the combination of hydrogen with oxygen is used to activate the ATPase and drive the reaction to manufacture tremendous quantities of ATP from ADP. The ATP is then transported out of the mitochondrion into all parts of the cytoplasm and nucleoplasm, where its energy is used to energize the functions of the cell.

Uses of ATP for Cellular Function. ATP is used to promote three categories of cellular functions: (1) *membrane transport*, (2) *synthesis of chemical compounds* throughout the cell, and (3) *mechanical work*. These three different uses of ATP are illustrated in Figure 3–13: (a) to supply energy for the transport of sodium through the cell membrane, (b) to promote protein synthesis by the ribosomes, and (c) to supply the energy needed during muscle contraction.

In addition to membrane transport of sodium, energy from ATP is required for membrane transport of potassium ions and, in certain cells, calcium ions, phosphate ions, chloride ions, urate ions, hydrogen ions, and still many other special substances. Membrane transport is

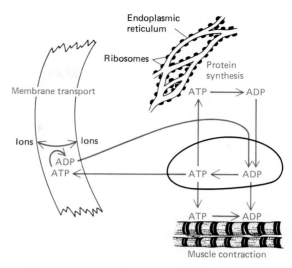

FIGURE 3–13 Use of ATP to provide energy for three of the major cellular functions: (1) membrane transport, (2) protein synthesis, and (3) muscle contraction.

so important to cellular function that some cells, the renal tubular cells for instance, utilize as much as 80 percent of the ATP formed in the cells for this purpose alone.

In addition to synthesizing proteins, cells also synthesize phospholipids, cholesterol, purines, pyrimidines, and a great host of other substances. Synthesis of almost any chemical compound requires energy. For instance, a single protein molecule might be composed of as many as several thousand amino acids attached to each other by peptide linkages; the formation of each of these linkages requires the breakdown of four high-energy bonds; thus many thousand ATP molecules must release their energy as each protein molecule is formed. Indeed, some cells utilize as much as 75 percent of all the ATP formed in the cell simply to synthesize new chemical compounds; this is particularly true during the growth phase of cells.

The final major use of ATP is to supply energy for special cells to perform mechanical work. We shall see in Chapter 11 that each contraction of a muscle fibril requires expenditure of tremendous quantities of ATP. Other cells perform mechanical work in two additional ways, by *ciliary* or *ameboid motion*, both of which will be described later in this chapter. The source of energy for all these types of mechanical work is ATP.

In summary, therefore, ATP is always available to release its energy rapidly and almost explosively wherever in the cell it is needed. To replace the ATP used by the cell, other, much slower chemical reactions break down carbohydrates, fats, and proteins and use the energy derived from these to form new ATP.

Synthesis and Formation of Cellular Structures by the Endoplasmic Reticulum and the Golgi Complex

The synthesis of most intracellular substances begins in the endoplasmic reticulum, but the products formed in the endoplasmic reticulum are then passed on to the Golgi complex, where they are further processed prior to release into the cell. But first, let us note the specific products that are synthesized in the specific portions of the endoplasmic reticulum and the Golgi complex.

Formation of Proteins by the Granular Endoplasmic Reticulum.   The granular endoplasmic reticulum is characterized by the presence of large numbers of ribosomes attached to the outer surfaces of the reticulum membrane. As we shall discuss in the following chapter, protein molecules are synthesized within the structure of the ribosomes. Furthermore, the ribosomes extrude many of the synthesized protein molecules not into the hyaloplasm but instead through the endoplasmic reticular wall into the endoplasmic matrix.

Within the endoplasmic matrix, the protein molecules are further processed during the next few minutes. In the presence of the enzymes in the endoplasmic reticular wall, the simple protein molecules are often folded and are also modified in other ways. In addition, most of them are rapidly conjugated with carbohydrate moieties to form glycoproteins.

Synthesis of Lipids, and Other Functions of the Smooth Endoplasmic Reticulum.   The smooth endoplasmic reticulum synthesizes mainly lipids, including phospholipids and cholesterol, rather than proteins. The phospholipids and cholesterol are rapidly incorporated into the lipid bilayer of the endoplasmic reticulum itself, thus causing the smooth portion of the endoplasmic reticulum to grow continually. However, small vesicles continually break away from the smooth endoplasmic reticulum; we shall see later that these vesicles mainly migrate rapidly to the Golgi apparatus.

Other significant functions of the smooth endoplasmic reticulum are these:

1. It contains the enzymes that control glycogen breakdown when glycogen is to be used for energy.

2. It contains a vast number of enzymes that are capable of detoxifying substances that are damaging to the cell, such as drugs. It achieves this by coagulation, oxidation, hydrolysis, and conjugation with glycuronic acid, and in other ways.

3. It can synthesize a few carbohydrate moieties that are usually conjugated with protein molecules to form glycoproteins.

Synthetic Functions of the Golgi Complex.   Though the major function of the Golgi complex is to process substances already formed in the endoplasmic reticulum, it also has the capability of synthesizing certain carbohydrates that cannot be formed in the endoplasmic reticulum. This is especially true of sialic acid and galactose. In addition, it can cause the formation of saccharide polymers, the most important of which are hyaluronic acid and chondroitin sulfate. A few of the many functions of hyaluronic acid and chondroitin sulfate in the body are these:

1. They are the major components of proteoglycans secreted in mucus and other glandular secretions.

2. They are the major components of the ground substance in the interstitial spaces, acting as a filler between collagen fibers and cells.

3. They are principal components of the organic matrix in both cartilage and bone.

Processing of Endoplasmic Secretions by the Golgi Complex—Formation of Intracellular Vesicles.   Figure 3–14 summarizes the major functions of the endoplasmic reticulum and Golgi complex and also shows the formation of secretory vesicles by the Golgi complex. As substances are formed in the endoplasmic reticulum, especially the proteins, they are transported through the tubules toward the portions of the smooth endoplasmic reticulum that lie nearest the Golgi complex. At this point small "transport" vesicles of smooth endoplasmic reticulum continually break away and diffuse to the *proximal layers* of the Golgi complex, carrying inside the vesicles the synthesized proteins and other products. These vesicles instantly fuse with the Golgi complex, and their contained substances enter the vesicular spaces of the Golgi complex. Here, a few additional carbohydrate moieties are usually added to the secretions, but usually the main function of the Golgi complex is to compact the endoplasmic reticular secretions into highly concentrated packets.

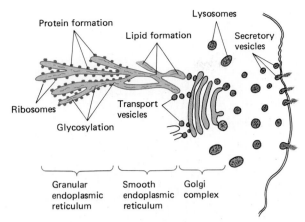

FIGURE 3—14 Formation of proteins, lipids, and cellular vesicles by the endoplasmic reticulum and Golgi complex.

As the secretions pass toward the distal layers of the Golgi complex the compaction and processing proceed, and finally at the distal layer both small and large vesicles continually break away from the Golgi complex, carrying with them the compacted secretory substances, and they then diffuse throughout the cell.

In a highly secretory cell, the vesicles that are formed by the Golgi complex are mainly *secretory vesicles*, containing especially the protein substances that are to be secreted through the surface of the cell. These vesicles diffuse to the surface, fuse with the cell membrane, and empty their substances to the exterior by a mechanism called *exocytosis*, which is essentially the opposite of endocytosis.

On the other hand, some of the vesicles are destined for intracellular use. For instance, specialized portions of the Golgi complex form the *lysosomes* that have already been discussed.

To give one an idea of the timing of these processes, when a glandular cell is bathed in radioactive amino acids, newly formed radioactive protein molecules can be detected in the granular endoplasmic reticulum within 3 to 5 minutes. Within 20 minutes the newly formed proteins are present in the Golgi complex, and within 1 to 2 hours radioactive proteins are secreted from the surface of the cell.

Cell Movement

By far the most important type of cell movement that occurs in the body is that of the specialized muscle cells in skeletal, cardiac, and smooth muscle, which constitute almost 50 percent of the entire body mass. The specialized functions of these cells will be discussed in Chapters 11 and 23. However, two other types of movement occur in other cells, *ameboid movement* and *ciliary movement*.

Ameboid Motion. Ameboid motion means movement of an entire cell in relation to its surroundings, such as the movement of white blood cells through tissues. Typically, ameboid motion begins with protrusion of a *pseudopodium* from one end of the cell. The pseudopodium projects far out away from the cell body, and then the remainder of the cell moves toward the pseudopodium. It is believed that this occurs by the mechanism illustrated in Figure 3–15. The outer portion of the cytoplasm is in a *gel* state and is called the *ectoplasm*, whereas the central portion of the cytoplasm is in a *sol* state and is called *endoplasm*. In the ectoplasm are numerous microfilaments composed of *actomyosin*, which is a highly contractile protein essentially the same as that found in muscle. Therefore, normally there is a continual tendency for the ectoplasm to contract. However, in response to a chemical or physical stimulus, the ectoplasm at one end of the cell becomes thin and weak, thus allowing a pseudopodium to bulge outward in the direction of the chemotactic source. But the ectoplasm at the opposite end of the cell contracts and moves toward the center of the cell. There it changes from a gel state into a solated state, becoming endoplasm, which "streams" into the pseudopodium. On reaching the pseudopodial end of the cell, the endoplasm turns toward the sides of the cell to form new ectoplasm. Therefore, at one end of the cell, ectoplasm is continually being solated while new ectoplasm is being formed at the sides of the pseudopodium. The continuous repetition of this process makes the cell move in the direction in which the pseudopodium projects. One can readily see that this streaming movement inside the cell is analogous to the revolving track of a Caterpillar tractor.

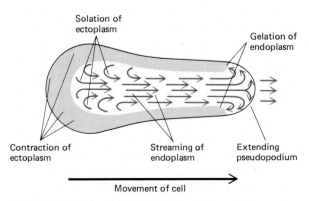

FIGURE 3—15 Ameboid motion by a cell.

Types of Cells That Exhibit Ameboid Motion. The most common cells to exhibit ameboid motion in the human body are the *white blood cells*, which move out of the blood into the tissues in the form of tissue *macrophages* or *microphages*. However, many other types of cells can move by ameboid motion under certain circumstances. For instance, fibroblasts will move into any damaged area to help repair the damage, and even some of the germinal cells of the skin, though ordinarily completely sessile cells, will move by ameboid motion toward a cut area to repair the rent.

Control of Ameboid Motion—"Chemotaxis." The most important factor that usually initiates ameboid motion is the appearance of certain chemical substances in the tissues. This phenomenon is called *chemotaxis*, and the chemical substance causing it to occur is called a *chemotactic substance*.

Movement of Cilia. A second type of cellular motion, *ciliary movement*, is the bending of cilia along the surface of cells in the respiratory tract and in the uterine tubes of the reproductive tract. As illustrated in Figure 3–16, a cilium looks like a minute, sharp-pointed hair that projects 3 to 4 μm (micrometers) from the surface of the cell. As many as 200 cilia can project from a single cell.

The cilium is covered by an outer layer of cell membrane, and it is supported by 11 microtubular filaments, 9 double tubular filaments located around the periphery of the cilium, and 2 single tubular filaments down the center, as shown in the cross-section illustrated in Figure 3–16.

In the inset of Figure 3–16 movement of the cilium is illustrated. The cilium moves forward with a sudden rapid stroke, 10 to 60 times per second, bending sharply where it projects from the surface of the cell. Then it moves backward slowly in a whiplike manner. The rapid forward movement pushes the fluid lying adjacent to the cell in the direction that the cilium moves; then the slow whiplike movement in the other direction has almost no effect on the fluid. As a result, fluid is continually propelled in the direction of the forward stroke. Since most ciliated cells have large numbers of cilia on their surfaces, and since ciliated cells on a surface are oriented in the same direction, this is a very effective means for moving fluids from one part of the surface to another; for instance, for moving mucus out of the lungs or for moving the ovum along the uterine tube.

Mechanism of Ciliary Movement. Though not all aspects of ciliary movement are

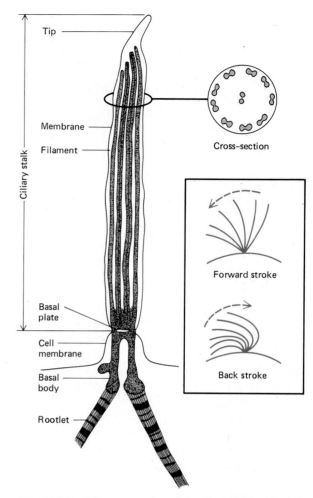

FIGURE 3–16 Structure and function of the cilium. (Modified from Satir: *Sci. Amer.,* 204[2]: 108, 1961.)

as yet clear, the postulated mechanism is the following: First, the nine double tubules and the two single tubules are all linked to each other by a complex of protein cross-linkages. Second, three protein arms with ATPase activity project from each set of peripheral tubules toward the next set. Third, it is believed that release of energy from ATP in contact with the ATPase arms causes these arms to "crawl" along the surfaces of the adjacent tubules. If the front tubules crawl toward the end of the cilium while the back tubules remain stationary, this obviously will cause bending.

Since many cilia on a cell surface contract simultaneously in a wavelike manner, it is presumed that some synchronizing signal—perhaps an electrochemical signal over the cell surface—is transmitted from cilium to cilium.

QUESTIONS

1. List the chemical components of protoplasm.
2. What are some of the important functions of proteins in cells?

3. What are some of the important functions of lipids in cells?
4. Describe the cell membrane.
5. What are the functions of the endoplasmic reticulum?
6. What is the structure of the mitochondrion, and what does it contain?
7. Describe the relationship of the Golgi apparatus to the endoplasmic reticulum.
8. What types of enzymes are found in lysosomes, and what are their functions?
9. What is the most important functional substance found in the nucleus?
10. Describe pinocytosis and phagocytosis.
11. How do lysosomes enter into the formation of the digestive vesicle?
12. Describe the function of adenosine triphosphate in the cell and also its relationship to the mitochondria.
13. What are the roles of the endoplasmic reticulum and Golgi complex in the secretion of proteins?
14. What functions does the endoplasmic reticulum perform in relation to lipid synthesis and glucose release from glycogen?
15. Describe the mechanisms of ameboid and ciliary motion.

REFERENCES

Allen, R.D., and Allen, N.S.: Cytoplasmic streaming in amoeboid movement. *Ann. Rev. Biophys. Bioeng.,* 7:469, 1978.

Chance, B., *et al.*: Hydroperoxide metabolism in mammalian organs. *Physiol. Rev.,* 59:527, 1979.

Cherkin, A., *et al.* (eds.): Physiology and Cell Biology of Aging. New York, Raven Press, 1979.

Dingle, J.T.: Lysosomes in Biology and Pathology. New York, American Elsevier, 1973.

Flickinger, C.J., *et al.*: Medical Cell Biology. Philadelphia, W.B. Saunders, 1979.

Giese, A.C.: Cell Physiology, 5th Ed. Philadelphia, W.B. Saunders, 1979.

Goldman, R.D., *et al.*: Cytoplasmic fibers in mammalian cells: Cytoskeletal and contractile elements. *Ann. Rev. Physiol.,* 41:703, 1979.

Hayflick, L.: The cell biology of human aging. *Sci. Am.,* 242(1):58, 1980.

Hinkle, P.C., and McCarty, R.E.: How cells make ATP. *Sci. Am.,* 238(3):104, 1978.

Jakoby, W.B., and Pastan, I.H. (eds.): Cell Culture. New York, Academic Press, 1979.

Koshland, D.E., Jr.: Bacterial chemotaxis in relation to neurobiology. *Ann. Rev. Neurosci.,* 3:43, 1980.

Lodish, H.F., and Rothman, J.E.: The assembly of cell membranes. *Sci. Am.,* 240(1):48, 1979.

Marchesi, V.T., *et al.* (eds.): Cell Surface Carbohydrates and Biological Recognition. New York, A.R. Liss, 1978.

Reid, E. (ed.): Plant Organelles. New York, Halsted Press, 1979.

Satir, P.: How cilia move. *Sci. Am.,* 231(4):44, 1974.

Singer, S.J.: The molecular organization of membranes. *Ann. Rev. Physiol.,* 43:805, 1974.

Wilkinson, P.C.: Chemotaxis and Inflammation. New York, Churchill Livingstone, 1973.

Williamson, J.R.: Mitochondrial function in the heart. *Ann. Rev. Physiol.,* 41:485, 1979.

4

Genetic Control of Cell Function— Protein Synthesis and Cell Reproduction

Overview

The *nucleus* is the control center of the cell. The basic mechanism of control is the following: Located in the nucleus of each cell are approximately 100,000 different types of *genes*. These are *deoxyribonucleic acid* (DNA) *molecules* that are collected in *23 pairs of chromosomes*. The DNA controls cell function by controlling the formation of proteins in the cell. Some of the proteins, in turn, provide the *enzymes* that promote the specific chemical reactions within the cell, and others are *structural proteins* that provide much of the structure of the cell.

The control of protein synthesis begins with the formation of *ribonucleic acid* (RNA) in the nucleus under the control of the DNA. Actually, three different types of RNA are formed, all of which diffuse out of the nucleus into the cytoplasm and play specific roles in protein formation. *Ribosomal RNA* provides much of the structure of the *ribosomes,* the physical organelles in the cytoplasm where the protein molecules are actually synthesized. *Transfer RNA* attaches to specific amino acids in the cytoplasm and transfers these to the ribosomes, where they are then combined in sequence with each other to form the protein molecules. *Messenger RNA* is a long molecule that literally slides through the ribosome and in so doing determines the sequence of the different amino acids in the synthesized protein molecule.

Each DNA molecule is composed of a long sequence of hundreds to thousands of *nucleotides.* Each three nucleotides in the sequence are a *code word* that leads to the insertion of a single type of amino acid in the final protein that is formed. The RNA molecules are also comprised of sequences of nucleotides, but slightly different types of nucleotides from those of DNA. The code in the DNA for sequencing of amino acids is passed to the messenger RNA as it is formed, again with each three RNA nucleotides, called a *codon,* serving to insert a single amino acid into the protein molecule. This process is called *transcription.* Then when the messenger RNA passes through the ribosome and causes amino acids to combine together to form the protein molecule, the process is called *translation.*

Only a small proportion of the genes are active at any one time in each cell. This *genetic activity* is controlled mainly by internal feedback

58

mechanisms within the cell. For instance, when a particular chemical constituent of the cell becomes scarce, the set of genes that controls its formation, a group of genes called an *operon*, becomes active, and large quantities of the chemical are formed until its appropriate concentration is reestablished. Then the chemical causes *negative feedback repression* of the operon and its formation ceases.

Cell reproduction is also controlled by the genetic mechanism. The first stage in cell reproduction is the formation of a duplicate set of new DNA molecules, a process called *replication of the DNA*. This leads also to *replication of all the chromosomes* in the cell. When a complete set of new chromosomes has been formed, the cell then divides into two new daughter cells, which is the process of *cell mitosis*. To cause the mechanical events of cell division, the *centrioles* of the cell form an extensive system of protein microtubules called the *mitotic apparatus*. This serves as a rigid structure inside the cell that pulls the two sets of chromosomes apart, forms two new nuclei, and eventually causes the cell to split.

Almost everyone knows that the genes control heredity from parents to children, but most persons do not realize that the same genes control the reproduction and the day-by-day function of all cells. The genes control function of the cell by determining what substances will be synthesized within the cell—what structures, what enzymes, what chemicals.

Figure 4–1 illustrates the general schema by which the genes control cellular function. Each gene, which is a nucleic acid called *deoxyribonucleic acid (DNA)*, automatically controls the formation of another nucleic acid, *ribonucleic acid (RNA)*, which spreads throughout the cell and controls the formation of a specific protein. Some proteins are *structural proteins*, which, in association with various lipids, form the structures of the various organelles that were discussed in the preceding chapter. But by far the majority of the proteins are *enzymes*, which catalyze the different chemical reactions in the cells. For instance, enzymes promote all the oxidative reactions that supply energy to the cell, and they promote the synthesis of various chemicals such as lipids, glycogen, adenosine triphosphate, and others.

The Genes

The genes, of which there are about 100,000 different types in human cells, are contained in long, double-stranded, helical molecules of DNA having molecular weights usually measured in the millions. A very short segment of such a molecule is illustrated in Figure 4–2. This molecule is composed of several simple chemical compounds arranged in a regular pattern explained in the following few paragraphs.

FIGURE 4–1   General schema by which the genes control cell function.

FIGURE 4–2   The helical, double-stranded structure of the gene. The outside strands are composed of phosphoric acid and the sugar deoxyribose. The internal molecules connecting the two strands of the helix are purine and pyrimidine bases; these determine the "code" of the gene.

Phosphoric acid:

Deoxyribose:

Bases:

Adenine

Thymine

Guanine

Cytosine

PURINES   PYRIMIDINES

FIGURE 4–3 The basic building blocks of DNA.

The Basic Building Blocks of DNA.

Figure 4–3 illustrates the basic chemical compounds involved in the formation of DNA. These include (1) *phosphoric acid,* (2) a sugar called *deoxyribose,* and (3) four nitrogenous bases: *adenine, guanine, thymine,* and *cytosine.* The phosphoric acid and deoxyribose form the two helical strands of DNA, and the bases lie between the strands and connect them together.

The Nucleotides.

The first stage in the formation of DNA is the combination of one molecule of phosphoric acid, one molecule of deoxyribose, and one of the four bases to form a *nucleotide.* Four separate nucleotides are thus formed, one for each of the four bases: *deoxyadenylic, deoxythymidylic, deoxyguanylic,* and *deoxycytidylic* acids. Figure 4–4 illustrates the chemical structure of deoxyadenylic acid, and Figure 4–5 illustrates simple symbols for all the four basic nucleotides that form DNA.

Note also in Figure 4–5 that the nucleotides are separated into two *complementary*

Adenine

Phosphate

Deoxyribose

FIGURE 4–4 Deoxyadenylic acid, one of the nucleotides that make up DNA.

pairs: (1) Adenylic acid and thymidylic acid form one pair, and (2) guanylic acid and cytidylic acid form the other pair. The *bases* of each pair can attach loosely (by hydrogen bonding) to each other, thus providing the means by which the two strands of the DNA helix are bound together.

Organization of the Nucleotides to Form DNA.

Figure 4–6 illustrates the manner in which multiple numbers of nucleotides are bonded together to form the DNA strands. Note that these are combined in such a way that phosphoric acid and deoxyribose alternate with each other in the two separate strands, and these strands are held together by hydrogen bonding between the respective complementary pairs of bases. Thus, in Figure 4–6 the sequence of complementary pairs of bases is CG, CG, GC, TA, CG, TA, GC, AT, and AT. Because of the looseness of the hydrogen bonds between the bases, the two strands can pull apart with ease, and they do so many times during the course of their function in the cell.

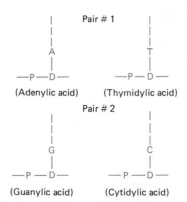

Pair # 1

A   T

—P—D—   —P—D—

(Adenylic acid)   (Thymidylic acid)

Pair # 2

G   C

—P —D—   —P —D—

(Guanylic acid)   (Cytidylic acid)

FIGURE 4–5 Combinations of the basic building blocks of DNA to form deoxyribose nucleotides. (A, adenine; C, cytosine; D, deoxyribose; G, guanine; P, phosphoric acid; T, thymine.) Note that there are four basic nucleotides that make up DNA, and these occur together in two pairs.

—d—d—d—d—d—d—d—d—d—d—d—d—d—d—d—d—d—d—

G   G   C   A   G   A   C   T   T

C   C   G   T   C   T   G   A   A

—P—D—P—D—P—D—P—D—P—D—P—D—P—D—P—D—P—D—

FIGURE 4–6 Combination of deoxyribose nucleotides to form DNA.

Now, to put the DNA of Figure 4–6 into its proper physical perspective, one needs merely to pick up the two ends and twist them into a helix. Ten pairs of nucleotides are present in each full turn of the helix in the DNA molecule, as illustrated in Figure 4–2.

The Genetic Code

The importance of DNA lies in its ability to control the formation of other substances in the cell. It does this by means of the so-called genetic code. When the two strands of a DNA molecule are split apart, this exposes the bases projecting to the side of each strand. It is these projecting bases that form the code.

Research studies in the past few years have demonstrated that the so-called *code words* consist of "triplets" of bases—that is, each three successive bases are a code word. And the successive code words control the sequence of amino acids in a protein molecule during its synthesis in the cell. Note in Figure 4–6 that each of the two strands of the DNA molecule carries its own genetic code. For instance, the top strand, reading from left to right, has the genetic code GGC, AGA, CTT, the code words being separated from each other by the arrows. As we follow this genetic code through Figures 4–7 and 4–8, we shall see that these three code words are responsible for placement of the three amino acids *proline, serine,* and *glutamic acid* in a mol-

ecule of protein. Furthermore, these three amino acids will be lined up in the protein molecule in exactly the same way that the genetic code is lined up in this strand of DNA.

RIBONUCLEIC ACID (RNA)— THE PROCESS OF TRANSCRIPTION

Since almost all DNA is located in the nucleus of the cell and yet most of the functions of the cell are carried out in the cytoplasm, some means must be available for the genes of the nucleus to control the chemical reactions of the cytoplasm. This is achieved through the intermediary of another type of nucleic acid, *ribonucleic acid (RNA)*, the formation of which is controlled by the DNA of the nucleus, the process being called *transcription*. The RNA is then transported into the cytoplasmic cavity, where it controls protein synthesis.

Three separate types of RNA are important to protein synthesis: *messenger RNA, transfer RNA,* and *ribosomal RNA*. Before we describe the function of these different RNAs in the synthesis of proteins, let us see how DNA controls the formation of RNA.

Synthesis of RNA. The code words in DNA cause the formation of *complementary* code words called *codons* in RNA. The stages of RNA synthesis are as follows:

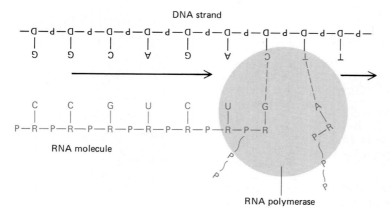

DNA strand

—d—d—d—d—d—d—d—d—d—d—d—d—d—d—d—d—d—

G   G   C   A   G   A   C   T   T

C   C   G   U   C   U   G   A

P—R—P—R—P—R—P—R—P—R—P—R—P—R   P—R

RNA molecule

RNA polymerase

FIGURE 4–7 Combination of ribose nucleotides with a strand of DNA to form a molecule of ribonucleic acid (RNA) that carries the DNA code from the gene to the cytoplasm.

The Basic Building Blocks of RNA. The basic building blocks of RNA are almost the same as those of DNA except for two differences. First, the sugar deoxyribose is not used in the formation of RNA. In its place is another sugar of slightly different composition, *ribose*. Second, thymine is replaced by another base, *uracil*.

Formation of RNA Nucleotides. The basic building blocks of RNA first form nucleotides exactly as described before for the synthesis of DNA. Here again, four separate nucleotides are used in the formation of RNA. These nucleotides contain the bases *adenine, guanine, cytosine,* and *uracil,* respectively, the uracil replacing the thymine found in the four nucleotides that make up DNA.

Activation of the Nucleotides. The next step in the synthesis of RNA is activation of the nucleotides. This occurs by the addition to each nucleotide of two phosphate radicals to form *high-energy phosphate bonds.*

The result of this activation process is that large quantities of energy are made available to each of the nucleotides, and this energy is used in promoting the subsequent chemical reactions that eventuate in the formation of the RNA chain.

Assembly of the RNA Molecule from Activated Nucleotides Using the DNA Strand as a Template—the Process of Transcription. The next stage in the formation of RNA is separation of the two strands of the DNA molecule. Then, one of these strands is used as a template on which the RNA molecule is assembled. It is this strand that contains the genes whereas the other strand remains genetically inactive. Assembly of the RNA molecule is accomplished in the manner illustrated in Figure 4–7 under the influence of the enzyme *RNA polymerase.* The steps of this procedure are (1) temporary bonding of an RNA base with each DNA base, (2) bonding of the successive RNA nucleotides with each other, and (3) splitting of the RNA strand away from the DNA strand.

It should be remembered that there are four different types of DNA bases and also four different types of RNA nucleotide bases. Furthermore, these always combine with each other in the following combinations:

DNA base	RNA base
guanine	cytosine
cytosine	guanine
adenine	uracil
thymine	adenine

Therefore, the code that is present in the DNA

FIGURE 4–8 Portion of a ribonucleic acid molecule, showing three "code" words, CCG, UCU, and GAA, which represent the three amino acids *proline, serine,* and *glutamic acid.*

strand is transmitted in *complementary* form to the RNA molecule.

Once the RNA molecules are formed, they diffuse out of the nucleus and into all parts of the cytoplasm, where they perform further functions.

Messenger RNA

Messenger RNA molecules are long straight strands that are suspended in the cytoplasm. These molecules are usually composed of several hundred to several thousand nucleotides in unpaired strands, and they contain *codons* that are exactly complementary to the code words of the DNA genes. Figure 4–8 illustrates a small segment of a molecule of messenger RNA. Its codons are CCG, UCU, and GAA. These are the codons for proline, serine, and glutamic acid. The transcription of these codons from the DNA molecule was demonstrated in Figure 4–7.

TABLE 4–1
RNA Codons for the Different Amino Acids and for Start and Stop

Amino acid	RNA codons					
Alanine	GCU	GCC	GCA	GCG		
Arginine	CGU	CGC	CGA	CGG	AGA	AGG
Asparagine	AAU	AAC				
Aspartic acid	GAU	GAC				
Cysteine	UGU	UGC				
Glutamic acid	GAA	GAG				
Glutamine	CAA	CAG				
Glycine	GGU	GGC	GGA	GGG		
Histidine	CAU	CAC				
Isoleucine	AUU	AUC	AUA			
Leucine	CUU	CUC	CUA	CUG	UUA	UUG
Lysine	AAA	AAG				
Methionine	AUG					
Phenylalanine	UUU	UUC				
Proline	CCU	CCC	CCA	CCG		
Serine	UCU	UCC	UCA	UCG		
Threonine	ACU	ACC	ACA	ACG		
Tryptophan	UGG					
Tyrosine	UAU	UAC				
Valine	GUU	GUC	GUA	GUG		
Start (CI)	AUG	GUG				
Stop (CT)	UAA	UAG	UGA			

RNA Codons. Table 4–1 gives the RNA codons for the 20 common amino acids found in protein molecules. Note that several of the amino acids are represented by more than one codon; some codons represent such signals as "start manufacturing a protein molecule" or "stop manufacturing a protein molecule." In Table 4–1, these two codons are designated CI for "chain-initiating" and CT for "chain-terminating."

Transfer RNA

Another type of RNA that plays a prominent role in protein synthesis is *transfer RNA*, which transfers amino acid molecules to protein molecules as the protein is synthesized. Each type of transfer RNA can combine specifically with only 1 of the 20 amino acids that are incorporated into proteins. The transfer RNA acts as a *carrier* to transport this specific type of amino acid to the ribosomes where protein molecules are formed.

Transfer RNA, containing only about 80 nucleotides, is a relatively small molecule in comparison with messenger RNA. It is a folded chain of nucleotides with a cloverleaf appearance similar to that illustrated in Figure 4–9. At one end of the molecule is always an adenylic acid that attaches to the amino acid.

The specific prosthetic group in the transfer RNA that allows it to recognize a specific codon in a messenger RNA strand is called an *anticodon*, and this is located approximately in the middle of the transfer RNA molecule (at the bottom of the cloverleaf configuration illustrated in Figure 4–9). During formation of a protein molecule, the anticodon bases combine loosely by hydrogen bonding with the codon bases of the messenger RNA. In this way the respective amino acids are lined up one after another along the messenger RNA chain, thus establishing the appropriate sequence of amino acids in the protein molecule.

Ribosomal RNA

The third type of RNA that is important for protein formation is ribosomal RNA; it constitutes about 60 percent of the ribosome. The remainder of the ribosome is protein, containing as many as 50 different types of protein, both structural proteins and enzymes needed in the manufacture of protein molecules.

The ribosome is the physical and chemical structure in the cytoplasm on which protein molecules are actually synthesized. However, it always functions in association with both of the other types of RNA as well: Transfer RNA transports amino acids to the ribosomes for incorporation into the developing protein molecules, whereas messenger RNA provides the information necessary for sequencing the amino acids in proper order for each specific type of protein to be manufactured.

Formation of Ribosomes in the Nucleolus. The DNA molecules for formation in ribosomal RNA are all located in a single chromosomal pair of the nucleus. However, this chromosomal pair contains many duplicates of these ribosomal genes because of the large amount of ribosomal RNA required for cellular function.

As the ribosomal RNA forms, it collects in the *nucleolus*, a specialized structure lying adjacent to the chromosome. The ribosomal RNA is then specially processed in the nucleolus and combined with "ribosomal proteins" to form granular condensation products that are primordial forms of the ribosomes. These are then released from the nucleolus, and they migrate through the large "pores" of the nuclear membrane to almost all parts of the cytoplasm, most of them eventually attaching to the surfaces of the endoplasmic reticulum.

Formation of Proteins in the Ribosomes— the Process of Translation

When a molecule of messenger RNA comes in contact with a ribosome, it travels through it, beginning at a predetermined end specified by

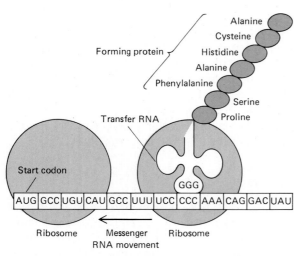

FIGURE 4–9 Postulated mechanism by which a protein molecule is formed in ribosomes in association with messenger RNA and transfer RNA.

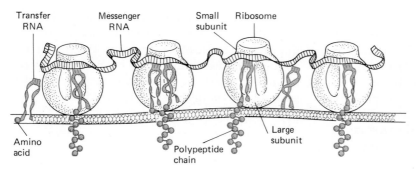

FIGURE 4–10 An artist's concept of the physical structure of the ribosomes as well as their functional relationship to messenger RNA, transfer RNA, and the endoplasmic reticulum during the formation of protein molecules. (From Bloom and Fawcett: A Textbook of Histology, 10th Ed. Philadelphia, W.B. Saunders Company, 1975.)

an appropriate sequence of RNA bases. However, the protein molecule does not begin to form until a "start" (or "chain-initiating") codon enters the ribosome. Then, as illustrated in Figure 4–9, while the messenger RNA travels through the ribosome, a protein molecule is formed—a process called *translation*. Thus, the ribosome reads the code of the messenger RNA in much the same way that a tape is "read" as it passes through the playback head of a tape recorder. Then, when a "stop" (or "chain-terminating") codon slips past the ribosome, the end of the protein molecule is signaled, and the newly formed protein molecule is released from the ribosome.

A messenger RNA can cause the formation of a protein molecule in any ribosome; that is, there is no specificity of ribosomes for given types of protein. The ribosome seems to be simply the physical structure in which or on which the chemical reactions take place.

Figure 4–10 shows the functional relationship of messenger RNA to the ribosomes and also the manner in which the ribosomes attach to the membrane of the endoplasmic reticulum. Note the process of translation occurring in several ribosomes at the same time in response to the same strand of messenger RNA. And note also the newly forming polypeptide chains passing through the endoplasmic reticulum membrane into the endoplasmic matrix, thus generating the protein molecules.

Peptide Linkage. The successive amino acids in the newly forming protein chain combine with each other according to the following typical reaction:

$$\begin{array}{ccccc} NH_2 & O & & H & R \\ | & \| & & | & | \\ R-C & -C-(OH+H)-N-C-COOH \longrightarrow \\ \\ NH_2 & O & H & R \\ | & \| & | & | \\ R-C & -C-N-C & COOH + H_2O \end{array}$$

In this chemical reaction, a hydroxyl radical is removed from the COOH portion of one amino acid while a hydrogen of the NH_2 portion of the other amino acid is removed. These combine to form water, and the two reactive sites left on the two successive amino acids combine with each other, resulting in a single molecule. This process is called *peptide linkage*.

SYNTHESIS OF OTHER SUBSTANCES IN THE CELL

Many thousand protein enzymes formed in the manner just described control essentially all the other chemical reactions that take place in cells. These enzymes promote synthesis of lipids, glycogen, purines, pyrimidines, and hundreds of other substances. We will discuss some of these synthetic processes in relation to carbohydrate, lipid, and protein metabolism in Chapters 42 and 43. It is by means of all these different substances that the many functions of the cells are performed.

CONTROL OF GENETIC FUNCTION AND BIOCHEMICAL ACTIVITY IN CELLS

There are basically two different methods by which the biochemical activities in the cell are controlled. One of these is called *genetic regulation*, in which the activities of the genes themselves are controlled, and the other is *enzyme regulation*, in which the activity rates of the enzymes within the cell are controlled.

Genetic Regulation

Gene function is controlled in several different ways. Some genes are normally dormant but can be activated by *inducer substances*. Other genes are naturally active but can be inhibited by *repressor substances*. As an illustration, let us describe one of the mechanisms for genetic control.

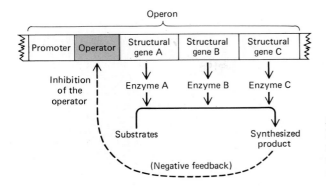

FIGURE 4-11 Function of the operon in controlling biosynthesis. Note that the synthesized product exerts a negative feedback to inhibit function of the operon, in this way automatically controlling the concentration of the product itself.

The Operon and Its Control of Biochemical Synthesis. The synthesis of a cellular biochemical product usually requires a series of reactions, and each reaction is catalyzed by a specific enzyme. Formation of all the enzymes needed for the synthetic process is in turn usually controlled by a sequence of genes all located in series one after the other on the same chromosomal DNA strand. This area of the DNA strand is called an *operon*, and the genes responsible for forming the respective enzymes are called the *structural genes*. In Figure 4-11, three respective structural genes are illustrated in an operon, and it is shown that they control the formation of three respective enzymes utilized in a particular biochemical synthetic process.

The rate at which the operon functions to transcribe RNA, and therefore to set into motion the enzymatic system for the biochemical process, is determined by the presence of two other small segments on the DNA strand called, respectively, the *promoter* and the *operator*, also shown in Figure 4-11. Each of these is a specific sequence of DNA nucleotides, but they do not themselves serve as templates to cause the formation of RNA. Instead, they merely function as control units of the operon.

The promoter first binds with *RNA polymerase*, which is an enzyme that moves along the operon to cause transcription of the appropriate messenger RNAs. However, lying between the promoter and the structural genes is the operator; this is a control gate that can be opened or closed. If the gate is open, the RNA polymerase will travel along the operon and cause the transcription process, but if the gate is closed, the RNA polymerase becomes blocked at the promoter level and the operon remains dormant.

In Figure 4-11, it is shown that the presence of a critical amount of the synthesized product in the cellular cytoplasm will cause negative feedback to the operator to inhibit it, that is, to close the gate. Therefore, whenever there is already enough of the required product, the operon becomes dormant. On the other hand, as the synthesized product becomes degraded in the cell and its concentration falls, the operator gate opens and the operon once again becomes active. In this way, the concentration of the synthesized product is automatically controlled.

Other Mechanisms for Control of Transcription by the Operon. Variations in the basic mechanism for control of the operon have been discovered with rapidity in the last few years. Without going into detail for all these, let us merely list some of the mechanisms of control:

1. An *inducer* substance from outside the cell sometimes activates the operator.
2. A *regulatory gene* elsewhere in the cell nucleus sometimes regulates the operator.
3. Inhibitors or inducers sometimes control many different operators at the same time.
4. Some synthetic processes are controlled not at the DNA level but instead at the RNA level to control the translation process for formation of proteins by messenger RNA.

Control of Enzyme Activity

In the same way that inhibitors and activators can affect the genetic regulatory system, so also can the enzymes themselves be directly controlled by other inhibitors or activators. This, then, represents a second category of mechanisms by which cellular biochemical functions can be controlled.

Enzyme Inhibition. A great many of the chemical substances formed in the cell have a direct feedback effect of inhibiting the respective enzyme systems that synthesize them. Almost always the synthesized product acts on the first enzyme in a sequence, rather than on the subsequent enzymes. One can readily recognize the importance of inhibiting this first enzyme to prevent buildup of intermediary products that will not be utilized.

This process of enzyme inhibition is another example of negative feedback control; it is responsible for controlling the intracellular concentrations of some of the amino acids as well as purines, pyrimidines, vitamins, and other substances.

In summary, there are two principal methods by which the cells control proper proportions and proper quantities of different cellular constituents: (1) the mechanism of genetic regulation and (2) the mechanism of enzyme regulation. The genes can be either activated or inhibited; likewise, the enzymes can be either activated or inhibited. Most often, these regulatory mechanisms function as feedback control systems that continually monitor the cell's biochemical composition and make corrections as needed. But, on occasion, substances from without the cell (especially some of the hormones that will be discussed later in this text) control the intracellular biochemical reactions by activating or inhibiting one or more of the intracellular control systems.

FIGURE 4—12 Stages in the reproduction of the cell. A, B, and C, prophase; D, prometaphase; E, metaphase; F, anaphase; G and H, telophase. (Redrawn from Mazia: *Sci. Am.*, *205*:102, 1961. © by Scientific American, Inc. All rights reserved.)

CELL REPRODUCTION

Cell reproduction is another example of the pervading, ubiquitous role that the DNA-genetic system plays in all life processes. It is the genes and their internal regulatory mechanisms that determine the growth characteristics of the cells and when these cells will divide to form new cells. In this way, this all-important genetic system controls each stage in the development of the human being from the single-cell fertilized ovum to the whole functioning body. Thus, if there is any central theme to life it is the DNA-genetic system.

The Life Cycle of the Cell. The life cycle of a cell is the period of time from reproduction to reproduction. When cells are not inhibited and are reproducing as rapidly as they can, this life cycle lasts for 10 to 30 hours. It is terminated by a series of distinct physical events called *mitosis* that cause division of the cell into two new daughter cells. The events of mitosis are illustrated in Figure 4–12 and will be described later. The actual stage of mitosis, however, lasts for only about 30 minutes, so that more than 95 percent of the life cycle is represented by the interval between mitosis, called *interphase*. In fact, except in those special conditions of rapid cellular reproduction, inhibitory controls almost al-

ways slow or stop the uninhibited life cycle of the cell so that different cells of the body have life cycle periods that vary from as little as 10 hours for stimulated bone marrow cells to an entire lifetime of the human body for nerve and striated muscle cells.

Immediately after new daughter cells are formed, they are still small, but they enter an initial period of rapid growth consisting of rapid formation of RNA, extensive protein synthesis, and accumulation of increased quantities of nucleoplasm and cytoplasm. During rapid cell reproduction, this period lasts for 20 to 40 percent of the total life cycle.

Replication of the DNA

As is true of almost all other important events in the cell, reproduction begins in the nucleus itself. The first step is *replication (duplication) of all DNA in the chromosomes*. It is only after this has occurred that mitosis can take place.

The DNA begins to be duplicated some 5 to 10 hours before mitosis, and this is completed in 4 to 8 hours. The DNA is duplicated only once, so that the net result is two exact *replicates* of all DNA. These replicates, in turn, become the DNA in the two new daughter cells that will be formed at mitosis. Following replication of the DNA there is another period of 1 to 2 hours before mitosis begins abruptly. However, even during this period preliminary changes are already beginning to take place that will lead to the mitotic process.

Chemical and Physical Events of DNA Replication. The DNA is duplicated in almost exactly the same way that RNA is formed from DNA. First, the two strands of the DNA helix of the gene pull apart. Second, each of these strands combines with deoxyribose nucleotides of the four types described early in the chapter as the basic building blocks of DNA. Each base on each strand of DNA attracts a nucleotide containing the appropriate *complementary* base. In this way, appropriate nucleotides are lined up side by side. Third, enzyme mechanisms, employing *DNA polymerases* as well as several other enzymes, cause linkage of the nucleotides to form a new DNA strand. However, a major difference between this formation of the new strand of DNA and the formation of an RNA strand is that the new strand of DNA remains attached to the old strand that has formed it, thus forming a new double-stranded DNA helix. Another major difference is that not only the DNA strand containing the genes but also the complementary DNA strand forms a new double helix. Therefore, the original DNA double helix becomes two double helixes that are exact duplicates of each other, unless an error in replication occurs, which is called a *mutation*.

The Chromosomes and their Replication

The human cell contains 46 chromosomes arranged in 23 pairs. In general, the genes in the two chromosomes of each pair are identical or almost identical with each other, so that it is usually stated that the different genes exist in pairs, though occasionally this is not the case.

Each chromosome consists of two major parts: the DNA and protein. The protein, in turn, is composed mainly of small molecules that serve to fold or otherwise compact the DNA strands into manageable sizes.

Recent experiments indicate that all the DNA of a particular chromosome is arranged in one long double helix and that the genes are attached end-on-end with each other. Such a molecule in the human being has a molecular weight of about 60 billion, and if spread out linearly, would be approximately 7.5 cm long or several thousand times as long as the diameter of the nucleus itself. But, the experiments also indicate that this long double helix is folded or coiled like a spring and is held in this position by its linkages to the protein molecules.

Replication of the chromosomes follows as a natural result of replication of the DNA strand, collecting new protein molecules as needed. However, temporarily, the two chromosomal strands remain attached to each other at a point called the *centromere*, located near the center of each of the strands. At this stage, the two attached chromosomal strands are called chromatids.

Mitosis

The actual process by which the cell splits into two new cells is called mitosis. Once the genes have been duplicated and each chromosome has split to form the two chromatids, mitosis follows automatically, almost without fail, within about an hour.

The Mitotic Apparatus. One of the first events of mitosis takes place in the cytoplasm, occurring during the latter part of interphase and early part of prophase, in or around the small structures call *centrioles.* As illustrated in Figure 4–12, two pairs of centrioles lie close to each other near one pole of the nucleus. Each centriole is a small cylindrical body about $0.4 \mu m$ long and about $0.15 \mu m$ in diameter, consisting mainly of nine parallel, tubule-like structures arranged around the inner wall of the cylinder. The two centrioles of each pair lie at right angles to each other.

Shortly before mitosis is to take place, the two pairs of centrioles begin to move apart from each other. This is caused by protein microtubules growing between the respective pairs and actually pushing them apart. At the same time, other microtubules grow radially away from each of the centriole pairs, forming a spiny star, called the *aster,* in each end of the cell. Some of the spines penetrate the nucleus and will play a role in separating the two chromatids during mitosis. The set of microtubules connecting the two centriole pairs is called the *spindle,* and the entire set of microtubules plus the two pairs of centrioles is called the *mitotic apparatus.*

Prophase. The first stage of mitosis, called *prophase,* is shown in Figure 4–12A, B, and C. While the spindle is forming, the *chromatin material* of the nucleus (the DNA), which in interphase consists of long loose strands, becomes shortened into well-defined chromosomes.

Prometaphase. During this stage (Fig. 4–12D) the nuclear envelope fragments. At the same time, a new set of microtubules begins to extend outward from a small condensed portion of each chromatid called the *kinetochore*, located at or near the centromere where the two chromatids are attached to each other. These new microtubules, in turn, either become attached to or adherent to the microtubules from the two asters of the mitotic apparatus, with one chromatid connecting to one of the asters and the other chromatid connecting to the opposite aster.

Metaphase. During metaphase (Fig. 4–12E) the two asters of the mitotic apparatus are pushed farther apart by additional growth of the mitotic spindle. Simultaneously the chromatids are pulled tightly by the attached microtubules to the very center of the cell, lining up in the *equatorial plate* of the mitotic spindle.

Anaphase. During this phase (Fig. 4–12F) the two chromatids of each chromosome are pulled apart at the centromere. Exactly how the microtubular system causes this is not known, except that it is known that the microtubules contain *actin*, which is one of the contractile proteins of muscle. Therefore, it has been presumed that these microtubules might contract or that they might in some way interact with the microtubules of the aster to cause the pulling force. Regardless of the mechanism, all 46 pairs of chromatids are separated, forming two separate sets of 46 *daughter chromosomes*. One of these sets is pulled toward one mitotic aster and the other toward the other aster at the two respective poles of the dividing cell.

Telophase. In telophase (Fig. 4–12G and H) the two sets of daughter chromosomes are now pulled completely apart. Then the mitotic apparatus dissolves and a new nuclear membrane develops around each set of chromosomes, this membrane being formed from portions of the old nuclear membrane or the endoplasmic reticulum that are already present in the cytoplasm. Shortly thereafter, the cell pinches in two midway between the two nuclei, for reasons mainly unexplained at present, except: (1) It is the two asters that initiate the cell division, each of the two daughter cells being composed of the portion of the cell occupied by a dissoluting aster. And (2) a contractile ring of *microfilaments* composed of *actin* and probably *myosin*, the two contractile proteins of muscle, develops at the juncture of the newly developing cells and pinches them off from each other.

Control of Cell Growth and Reproduction

Cell growth and reproduction usually go together; growth normally leads to replication of the DNA of the nucleus, followed a few hours later by mitosis.

In the normal human body, regulation of cell growth and reproduction is mainly a mystery. We know that certain cells grow and reproduce all the time, such as the blood-forming cells of the bone marrow, the germinal layers of the skin, and the epithelium of the gut. However, many other cells, such as smooth muscle cells, do not reproduce for many years. And a few cells, such as the neurons and striated muscle cells, do not reproduce during the entire life of the person.

If there is an insufficiency of some types of cells in the body, these will grow and reproduce very rapidly until appropriate numbers of them are again available. For instance, seven eighths of the liver can be removed surgically, and the cells of the remaining one eighth will grow and divide until the liver mass returns much of the way back toward normal. The same effect occurs for almost all glandular cells and for cells of the bone marrow, the subcutaneous tissue, the intestinal epithelium, and almost any other tissue except highly differentiated cells such as nerve and muscle cells.

We know very little about the mechanisms that maintain proper numbers of the different types of cells in the body. However, experimental studies have shown that control substances called *chalones* are secreted by the different cells and that these cause feedback effects to stop or slow their growth and reproduction when too many of them have been formed. We know that cells of any type removed from the body and grown in tissue culture can grow and reproduce rapidly and indefinitely if the medium in which they grow is continually replenished. Yet they will stop growing when even small amounts of their own secretions are allowed to collect in the medium, which supports the idea that control substances limit cellular growth.

CANCER

Cancer is a disease that attacks the basic life process of the cell, in almost all instances altering the cell's *genome* (the total genetic complement of the cell) and leading to wild and spreading growth of the cancerous cells. The cause of the altered genome is a *mutation* (alteration) of one or more genes; or mutation of a large segment of a DNA strand containing many genes; or, in some instances, addition or loss of large segments of chromosomes.

But what is it that causes the mutations? When one realizes that many trillions of new cells are formed each year in the human being, this question should probably better be asked in the following form: Why is it that we do not develop literally millions or billions of mutant cancerous cells? The answer is the incredible precision with which DNA chromosomal strands are replicated in each cell before mitosis takes place. Indeed, even after each new strand is formed, the veracity of the replication process is "proofread" several different times. If any mistakes have been made, the new strand is cut and repaired before the mitotic process is allowed to proceed. Yet, despite all these precautions, probably one newly formed cell in every few hundred thousand to every few million still has significant mutant characteristics. We know this because it has been ascertained that each gene in a human offspring has the probability of 1 in 100,000 of being a mutant when compared with the genes of the parents.

Thus, chance alone is all that is required for mutations to take place. However, other factors that increase the probability of mutation include (1) *ionizing radiation*, (2) certain types of chemical substances called *carcinogens*, (3) some *viruses*, (4) *physical irritation*, and (5) *hereditary predisposition*.

Invasive Characteristic of Cancer Cells. Two major differences between the cancer cell and the normal cell are these: (1) The cancer cell does not respect usual cellular growth limits; the reason is that it presumably does not secrete the appropriate *chalones* that are responsible for stopping excess growth. (2) Cancer cells are far less adhesive to each other than are normal cells. Therefore, they have a tendency to wander through the tissues, to enter the bloodstream, and to be transported all through the body, where they form nidi for numerous new cancerous growths.

Why Do Cancer Cells Kill? The answer to this is very simple: Cancer tissue competes with normal tissues for nutrients. Because cancer cells continue to proliferate indefinitely, their number multiplying day by day, one can readily understand that the cancer cells will soon demand essentially all the nutrition available to the body. As a result the normal tissues gradually suffer nutritive death.

QUESTIONS

1. What are the basic building blocks of DNA?
2. How are nucleotides combined to form DNA?
3. How does DNA control the formation of RNA?
4. Describe the role of messenger RNA in the formation of proteins.
5. Describe the function of transfer RNA and the ribosomes in the formation of protein.
6. How are the concentrations of cellular constituents controlled by feedback repression of the genes?
7. Describe the control of enzyme activity in the cell.
8. In cell reproduction what roles do replication of DNA and replication of chromosomes play?
9. Describe the stages of mitosis and the events that take place during each of these stages.
10. What is the cause of cancer?

REFERENCES

Bauer, W.R.: Structure and reactions of closed duplex DNA. *Annu. Rev. Biophys. Bioeng.,* 7:287, 1978.

Butler, J.G., and Klug, A.: The assembly of a virus. *Sci. Am.,* 239(5):62, 1978.

Clark, B.F.C., *et al.* (eds.): Gene Expression: Protein Synthesis and Control, RNA Synthesis and Control, Chromatin Structure and Function. New York, Pergamon Press, 1978.

Dickerson, R.E.: Chemical evolution and the origin of life. *Sci. Am.,* 239(3):70, 1978.

Emmelot, P., and Kriek, E. (eds.): Environmental Carcinogenesis: Occurrence, Risk Evaluation, and Mechanisms. New York, Elsevier/North-Holland, 1979.

Fiddes, J.C.: The nucleotide sequence of a viral DNA. *Sci. Am.,* 237(6):54, 1977.

Foster, R.L.: The Nature of Enzymology, New York, John Wiley & Sons, 1979.

Hiatt, H.H., *et al.* (eds.): Origins of Human Cancer. Cold Spring Harbor, N.Y., Cold Spring Harbor Laboratory, 1977.

Kastrup, K.W., and Neilsen, J.H. (eds.): Growth Factors: Cellular Growth Processes, Growth Factors, Hormonal Control of Growth. New York, Pergamon Press, 1978.

Kornberg, A.: DNA Replication. San Francisco, W.H. Freeman, 1980.

Kouri, R.E. (ed.): Genetic Differences in Chemical Carcinogenesis, Boca Raton, Fla., CRC Press, 1980.

Molineaux, I., and Kohiyama, M. (eds.): DNA Synthesis: Present and Future. New York, Plenum Press, 1978.

Russell, T.R., *et al.* (eds.): From Gene to Protein: Information Transfer in Normal and Abnormal Cells. New York, Academic Press, 1979.

Schopf, J.W.: The evolution of the earliest cells. *Sci. Am.,* 239(3):110, 1978.

Weissman, S.M.: Gene structure and function. *In* Bondy, P.K., and Rosenberg, L.E. (eds.): Metabolic Control and Disease, 8th Ed. Philadelphia, W.B. Saunders, 1980, p. 1.

Wu, R. (ed.): Recombinant DNA. New York, Academic Press, 1979.

5

Fluid Environment of the Cell and Transport Through the Cell Membrane

Overview

The fluid of the body is divided into that outside the cells, called the *extracellular fluid*, and that inside the cells, called *intracellular fluid*. Both of these fluids contain the necessary nutrients for cell metabolism, including such substances as *glucose, amino acids, fatty acids, cholesterol, phospholipids, neutral fat*, and *oxygen*. On the other hand, the ionic compositions of the two types of fluid are very different from each other. The most important difference is that the *extracellular fluid contains large quantities of sodium and chloride ions*. By contrast, the *intracellular fluid contains large quantities of potassium and phosphate ions*.

The cell membrane plays an active role in maintaining the compositional differences between the extracellular and intracellular fluids. It does this by controlling the transport of ions and other substances through the membrane. Transport through the cell membrane occurs by two major processes, (1) *diffusion* and (2) *active transport*.

Diffusion means random motion of molecules or ions. Net diffusion of a substance occurs only from an area of high concentration toward an area of low concentration. For instance, this is the way that oxygen moves from the lungs into the blood and also from the blood to the cells. Diffusion through the cell membrane occurs in two ways: First, lipid-soluble substances such as oxygen, carbon dioxide, alcohol, and so forth diffuse directly through the lipid matrix of the cell membrane. Second, water and water-soluble substances diffuse through *cell membrane pores,* which are channels formed by protein molecules that penetrate through the lipid barrier of the cell membrane.

Some substances can diffuse through the cell membrane only by a special mechanism called *facilitated diffusion*. In this mechanism, the substance first combines with a *carrier protein* in the cell membrane. This makes the substance soluble in the membrane so that it can diffuse to the membrane's inner surface and then be released to the inside of the cell.

Water moves through the cell membrane by *osmosis* when the concentration of solutes in the extracellular fluid is different from the concentration in the intracellular fluid. The solutes dissolved in either of these fluids reduce the concentration of water in the fluid, which in turn slightly reduces the rate of diffusion of water molecules. The rate of diffusion will not be the same in one direction through the cell membrane as in the other direction if the concentration of water is greater on one side of the cell membrane than on the other side. Thus, the net rate of water diffusion will always be from the area of *high water concentration* (the area of *low solute concentration*) toward the area of *low water concentration;* this is the process of osmosis. Osmosis can be prevented if sufficient pressure is applied in the direction opposite to the fluid movement. The amount of pressure required to stop the osmosis completely is called the *osmotic pressure.*

Active transport of a substance through the cell membrane, which is an entirely different transport mechanism from diffusion, means movement of the substance through the cell membrane by a specific membrane chemical mechanism. Active transport is similar to facilitated diffusion except that it can transport a substance even when the concentration of the substance is higher on the side of the membrane toward which it is being transported, which is called "uphill transport." To achieve active transport, the substance first combines with a *carrier protein* in the cell membrane, as also occurs in facilitated diffusion; then, the combination of substance and carrier diffuses to the inner surface of the membrane; and finally, energy derived from ATP is used to split the transported substance away from the carrier even though the concentration of the substance on that side of the membrane may already be very high.

Perhaps the most important active transport mechanism of all cells is the *sodium-potassium pump,* which pumps sodium ions out of the cell and potassium ions into the cell. It is this pump that maintains the low sodium concentration and high potassium concentration in the intracellular fluid. Other ions that are actively transported through the membranes of cells include the ions of *chlorine, calcium, iron,* and *hydrogen. Sugars* and *amino acids* are also actively transported through the membranes of some cells but by a mechanism called *sodium co-transport,* which is slightly different from the carrier-pump mechanism.

The fluid inside the cells of the body, called *intracellular fluid,* is very different from that outside the cells, called *extracellular fluid.* The extracellular fluid circulates in the spaces between the cells and also mixes freely with the fluid of the blood through the capillary walls. Thus, it is the extracellular fluid that supplies the cells with nutrients and other substances needed for cellular function. But before the cell can utilize these substances, they must be transported through the cell membrane.

Differences Between Extracellular Fluid and Intracellular Fluid

Figure 5–1 illustrates the compositions of extracellular fluid and intracellular fluid. Both of these fluids contain reasonable amounts of the usual nutrients required by the cells for metabolism, including such substances as glucose, amino acids, and the lipids cholesterol, phospholipid, and neutral fat. Also, both fluids contain oxygen and carbon dioxide, and the hydrogen ion concentration (which indicates the degree of acidity of the fluids) is only slightly different between the two fluids. On the other hand, some of the other ions are distributed quite differently between the extracellular fluid and the intracellular fluid. Note especially in Figure 5–1 that the *sodium ion concentration* is very high in the extracellular fluid though very low intracellularly. In contrast, the *potassium ion concentration* is very high inside the cell but very low outside. Similarly, *calcium* is high outside and low inside; *magnesium* high inside, low outside; *chloride* high outside, low inside; *bicarbonate* high outside, low inside; *phosphates* high

	Extracellular fluid	Intracellular fluid
Na⁺	142 mEq/L	10 mEq/L
K⁺	4 mEq/L	140 mEq/L
Ca⁺⁺	5 mEq/L	< 1 mEq/L
Mg⁺⁺	3 mEq/L	58 mEq/L
Cl⁻	103 mEq/L	4 mEq/L
HCO₃⁻	28 mEq/L	10 mEq/L
Phosphates	4 mEq/L	75 mEq/L
SO₄⁻⁻	1 mEq/L	2 mEq/L
Glucose	90 mg %	0 to 20 mg %
Amino acids	30 mg %	200 mg % ?
Cholesterol Phospholipids Neutral fat	0.5 g %	2 to 95 g %
Po₂	35 mm Hg	20 mm Hg ?
Pco₂	46 mm Hg	50 mm Hg ?
pH	7.4	7.0
Proteins	2 g % (5 mEq/L)	16 g % (40 mEq/L)

FIGURE 5–1 Chemical compositions of extracellular and intracellular fluids.

inside, low outside; and *sulfates* low in both fluids but higher inside the cell than in the extracellular fluid.

There are two major reasons for the different concentrations between extracellular fluids and intracellular fluids: First, some substances that enter the cells are utilized so rapidly by the cells' metabolic systems that their concentrations become reduced inside the cell in comparison with the outside. For instance, the concentrations of both glucose and oxygen are lower inside than outside because both are continually being used in the metabolic reactions of the cell. These metabolic reactions inside the cell also create new substances, thereby causing the concentrations of these to be greater inside than outside. For instance, large quantities of carbon dioxide are formed in the cells, which means that the carbon dioxide concentration is normally somewhat higher inside than outside. Likewise, other end-products of cellular metabolism, such as urea, creatinine, sulfates, and so forth, are all present in the intracellular fluid in considerably higher concentration than in extracellular fluid.

The second reason for major differences between the concentrations of extracellular and intracellular fluid is selective transport of substances through the cell membrane. The cell membrane is highly permeable to some substances and very poorly permeable to others, which obviously allows some substances to enter or leave the cells more easily than others.

But, in addition to differences in permeability, the cell membrane has the capability of *actively* transporting many substances through the membrane. That is, they can be carried through the membrane by a chemical mechanism and released on the opposite side of the membrane. Furthermore, active transport can occur even when the concentration of the substance is less on the first side of the membrane than on the second. This property of the cell membrane is responsible for many of the body's most important functions, as we shall discuss throughout this text. To give a simple example, it is active transport of sodium and potassium ions through the membrane that causes the concentration differences of these two ions on the two sides of the membrane, as illustrated in Figure 5–1. In turn, these concentration differences are the cause of the electrical potentials that occur in nerve and muscle fibers. And, finally, the electrical potentials are responsible for transmission of nerve impulses and for control of muscle contraction. Therefore, the simple process of selective transport of substances through the cell membrane can lead, first, to differences in the compositions of the intracellular and extracellular fluids and this in turn to many other important functions of the human body.

Movement and Mixing of the Extracellular Fluids Throughout the Body

Fortunately, the extracellular fluids are continually mixed. Were it not for this, the cells would remove all the nutrients from the fluids in their immediate vicinity until none would be left, and cellular excreta would accumulate locally until they would kill the cells. Yet, two different mechanisms provide for continual mixing and movement of the extracellular fluid throughout the body; these are (1) circulation of the blood and (2) the phenomenon of diffusion.

Figure 5–2 illustrates the general plan of circulation, showing continuous flow of blood around the circulatory system, through the lungs, and through the different tissues. The blood picks up oxygen in the lungs and various nutrients in the gut and then carries these to all other areas of the body. On passing through the tissues the blood picks up carbon dioxide, urea, and other excreta from the cells and transports these to the lungs and kidneys, where they are removed from the body. Thus, the circulatory system provides long-distance transport of the extracellular fluids, in this way keeping the

FIGURE 5–2 General schema of fluid flow in the circulation.

fluids in the different parts of the body mixed with each other.

The phenomenon of diffusion will be described in more detail later in the chapter in relation to transport of substances through the cell membrane. However, the basic principle of diffusion is simply that all molecules in the fluid, including both the molecules of water and those of the dissolved substances, are continually moving among each other. This motion allows continual mixing of all substances within the blood and also in the fluids of the tissue spaces, the so-called *interstitial fluid.* Furthermore, since there are large numbers of small openings in the capillaries, the *capillary pores*, the molecules are continually moving both out of the capillaries into the interstitial fluid and then back again through the pores into the blood. And because the blood is continually moving through all parts of the body, one can readily see that all the extracellular fluids of the body become mixed with each other. Indeed, this process is so effective that almost every portion of the extracellular fluid, even that in the minutest tissue space, be-

comes mixed with all the other extracellular fluid of the body at least once every 10 to 30 minutes. Because of this rapidity of mixing, the concentrations of substances in the extracellular fluid of one part of the body are rarely more than a few percent different from those anywhere else. It is in this way that nutrients from the gut, hormones from the endocrine glands, and oxygen from the lungs are transported to the cells, and it is also in this way that the excreta from the cells are carried to either the lungs or the kidneys to be removed from the body.

TRANSPORT OF SUBSTANCES THROUGH THE CELL MEMBRANE

Substances are transported through the cell membrane by two major processes, *diffusion* and *active transport.* Though there are many different variations of these two basic mechanisms, as we shall see later in this chapter, basically, diffusion means movement of substances in a random fashion caused by the normal kinetic motion of matter, whereas active transport means movement of substances as a result of chemical processes that impart energy to cause the movement.

DIFFUSION

All molecules and ions in the body fluids, including both water molecules and dissolved substances, are in constant motion, each particle moving its own separate way. Motion of these particles is what physicists call heat—the greater the motion, the higher is the temperature—and motion never ceases under any conditions except absolute zero temperature. When a moving molecule, A, approaches a stationary molecule, B, the electrostatic and nuclear forces of molecule A repel molecule B, adding some of the energy of motion to molecule B. Consequently, molecule B gains kinetic energy of motion while molecule A slows down, losing some of its kinetic energy. Thus, as shown in Figure 5–3, a single molecule in solution bounces among the other molecules first in one direction, then another, then another, and so forth, bouncing randomly millions of times each second. At times it travels a far distance before striking the next molecule, but at other times only a short distance.

This continual movement of molecules among each other in liquids or in gases is called

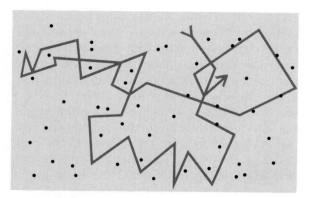

FIGURE 5–3 Diffusion of a fluid molecule during a fraction of a second.

diffusion. Ions diffuse in exactly the same manner as whole molecules, and even suspended colloid particles diffuse in a similar manner, except that because of their very large sizes they diffuse far less rapidly than molecular substances.

Kinetics of Diffusion—Effect of Concentration Difference

If we consider all the different factors that affect the rate of diffusion of a substance from one area to another, they are the following: (1) The greater the concentration differences between the areas, the greater the rate of diffusion. (2) The less the molecular weight, the greater the rate of diffusion. (3) The shorter the distance, the greater the rate. (4) The greater the cross-section of the diffusion pathway, the greater is the rate of diffusion. (5) The greater the temperature, the greater is the molecular motion and also the greater is the rate of diffusion. All these can be placed in the approximate formula for diffusion in solutions shown at the bottom of this page.

Diffusion Through the Cell Membrane

The cell membrane is essentially a sheet of lipid material, called the *lipid matrix*, with interspersed islands of globular protein molecules in the matrix. Some of these protein molecules penetrate all the way through the membrane and thus form membrane "pores," as was discussed in Chapter 3. Therefore, two different methods by which substances can diffuse through the membrane are (1) becoming dis-

solved in the lipid matrix and diffusing through it in the same way that diffusion occurs in water or (2) diffusing through the minute pores that pass directly through the membrane.

DIFFUSION IN THE DISSOLVED STATE THROUGH THE LIPID PORTION OF THE MEMBRANE

A few substances are soluble in the lipid of the cell membrane. These include especially oxygen, carbon dioxide, alcohol, and fatty acids. When one of these comes in contact with the membrane, as illustrated for the oxygen molecule in the top portion of Figure 5–4, it immediately becomes dissolved in the lipid, and the molecule continues its random motion within the substance of the membrane in exactly the same way that it undergoes random motion in the surrounding fluids.

The primary factor that determines how rapidly a substance can diffuse through the lipid matrix of the cell membrane is its solubility in lipids. If it is very soluble, it becomes dissolved in the membrane very easily and therefore passes on through. On the other hand, almost no substances that dissolve very poorly in lipids, such as glucose and the electrolytes, pass through the lipid matrix.

FACILITATED DIFFUSION THROUGH THE LIPID MATRIX

Some substances are very insoluble in lipids and yet can still pass through the lipid matrix by a process called *carrier-mediated* or *facilitated diffusion.*

FIGURE 5–4 Diffusion of substances through the lipid matrix of the membrane. The upper part of the figure shows *free diffusion* of oxygen through the membrane, and the lower part shows *facilitated diffusion* of glucose.

$$\text{Diffusion rate} \propto \frac{\text{Concentration difference} \times \text{Cross-sectional area} \times \text{Temperature}}{\sqrt{\text{Molecular weight}} \times \text{Distance}}$$

fusion. This is the means by which some sugars and amino acids, in particular, cross the membrane. The most important of the sugars is glucose, the membrane transport of which is illustrated in the bottom portion of Figure 5–4. This shows that glucose (G1) combines with a *carrier* substance (C) at point 1 to form the compound CG1. (The carrier substance is a protein, as will be discussed later in the chapter.) This combination is soluble in the lipid, so that it can diffuse to the other side of the membrane where the glucose breaks away from the carrier (point 2) and passes to the inside of the cell. Then the carrier moves back to the outside surface to pick up still more glucose. Thus, the effect of the carrier is to make the glucose soluble in the membrane; without it, glucose cannot pass through the membrane.

The rate at which a substance passes through a membrane by facilitated diffusion depends on the difference in concentration of the substance on the two sides of the membrane, the amount of carrier available, and the rapidity with which the chemical (or physical) reactions can take place.

DIFFUSION THROUGH THE MEMBRANE PORES

Some substances, such as water and many of the dissolved ions, go through holes in the cell membrane called *membrane pores.* Many of these are caused by large protein molecules penetrating all the way through the cell membrane, as discussed earlier, though very small uncharged molecules such as water and urea can also penetrate directly through the interstices between adjacent lipid molecules in the lipid matrix. A large portion of these pores behave as if they were minute round holes approximately 0.8 nanometer (nm) (8 Angstroms) in diameter and as if the total area of the pores equaled approximately $\frac{1}{5000}$ of the total surface area of the cell. Despite this very minute total area of the pores, molecules and ions diffuse so rapidly that the entire volume of fluid in some types of cells—the red blood cell for instance—can easily pass through the pores within one hundredth of a second.

Figure 5–5 illustrates a schematized structure of one type of pore, indicating that its surface may be lined with electrical charges. This figure shows several small particles passing through the pore. The maximum diameter of the molecule or ion that can pass through is approximately equal to the diameter of the pore itself, about 0.8 nm (8 Angstroms).

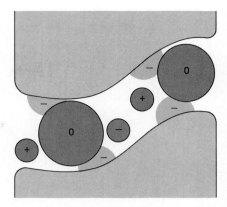

FIGURE 5–5 Postulated structure of the pore in the mammalian red cell membrane, showing the sphere of influence exerted by charges along the surface of the pore. (Modified from Solomon: *Sci. Am. 203:* 146, 1960. © 1960 by Scientific American, Inc.)

Effect of Pore Size on Diffusion Through the Pore—Permeability. Table 5–1 gives the effective diameters of various substances in comparison with the diameter of a pore, and it also gives the relative permeability of the pores for the different substances. *The permeability can be defined as the rate of transport through the membrane for a given concentration difference.* Note that some substances, such as the water molecule, urea molecule, and chloride ion, are considerably smaller than the pore. All these pass through most pores, particularly those of red cells, with great ease. The rates of diffusion of urea and chloride ions through the membrane are somewhat less than that of water, which is in keeping with the fact that their effective diameters are slightly greater than that of water.

Table 5–1 also shows that most of the sugars, including glucose, have effective diameters that are slightly greater than that of the pores. For this reason essentially none of the sugars can pass through the pores; instead, those that do enter the cell pass through the lipid matrix by the process of facilitated diffusion.

Effect of the Electrical Charge of Ions on Their Ability to Diffuse Through Membrane Pores. The electrical charges of ions often markedly affect their ability to diffuse through pores, sometimes impeding and sometimes enhancing their diffusion. For instance, one set of pores, called *sodium channels,* seems especially to allow easy diffusion of sodium ions. It is believed that these pores are lined with strongly *negative* charges that attract sodium ions into the pores and then shuttle the sodium ions from one negative charge to the next, thus allowing rapid movement through the pores. On the other hand, larger-sized positive ions, such

TABLE 5—1
Relationship of Effective Diameters of Different Substances to Pore Diameter and Relative Permeabilities[*]

Substance	Diameter nm	Ratio to Pore Diameter	Approximate Relative Permeability
Water molecule	.3	0.38	50,000,000
Urea molecule	.36	0.45	1,500,000
Hydrated chloride ion			
(red cell)	.386	0.48	500,000
(nerve membrane)	—	—	0.06
Hydrated potassium ion			
(red cell)	.396	0.49	1.1
(nerve membrane)	—	—	0.03
Hydrated sodium ion			
(red cell)	.512	0.64	1
(nerve membrane)	—	—	0.0003
Lactate ion	.52	0.65	?
Glycerol molecule	.62	0.77	?
Ribose molecule	.74	0.93	?
Pore size	.8 (Ave)	1.00	—
Galactose	.84	1.03	?
Glucose	.86	1.04	0.4
Mannitol	.86	1.04	?
Sucrose	.104	1.30	?
Lactose	.108	1.35	?

[*]These data have been gathered from different sources but relate primarily to the red cell membrane. Other cell membranes have different characteristics, as illustrated in the table for the nerve membrane.

as potassium ions, have considerable difficulty passing through these channels, mainly because of their diameters.

Other pores are larger and have less charge and are especially permeable to potassium ions; these pores are therefore called *potassium channels*. And still other pores have specific characteristics that make them function mainly as *water channels*, others as *calcium channels*, and others as other types of channels.

NET DIFFUSION THROUGH THE CELL MEMBRANE AND EFFECT OF A CONCENTRATION DIFFERENCE

From the preceding discussion, it is evident that many different substances can diffuse either through the lipid matrix of the cell membrane or through the pores. It should be noted, however, that substances that diffuse in one direction can also diffuse in the opposite direction. Usually it is not the total quantity of substances diffusing in both directions through the cell membrane that is important to the cell but instead the *net quantity* diffusing in one direction.

In addition to the permeability of the membrane, which has already been discussed, the most important factor that determines the net rate of diffusion of a substance is the *concentration difference of the substance across the membrane.* Figure 5–6 illustrates a membrane with a substance in high concentration on the outside and low concentration on the inside. The rate at which the substance diffuses *inward* is proportional to the concentration of molecules on the outside, for this concentration determines how many of the molecules strike the outside of the pore each second. On the other hand, the rate at which the molecules diffuse *outward* is proportional to their concentration inside the membrane. Obviously, therefore, the rate of net diffusion into the cell is proportional to the concentration on the outside *minus* the concentration on the inside, or

$$\text{Net diffusion} = P(C_o - C_i)$$

FIGURE 5—6 Effect of concentration gradient on diffusion of molecules and ions through a cell membrane pore.

in which C_o is the concentration on the outside, C_i is the concentration on the inside, and P is the permeability of the membrane for the substance.

NET MOVEMENT OF WATER ACROSS CELL MEMBRANES—OSMOSIS ACROSS SEMIPERMEABLE MEMBRANES

By far the most abundant substance to diffuse through the cell membrane is water. It should be recalled again that enough water ordinarily diffuses in each direction through the red cell membrane per second to equal about *100 times the volume of the cell itself.* Yet, *normally*, the amount that diffuses in the two directions is so precisely balanced that not even the slightest *net* movement of water occurs. Therefore, the volume of the cell remains constant. However, under certain conditions, a *concentration difference for water* can develop across a membrane, just as concentration differences for other substances can also occur. When this happens, net movement of water does occur across the cell membrane, causing the cell to either swell or shrink, depending on the direction of the net movement. This process of net movement of water caused by the concentration difference is called *osmosis.*

To give an example of osmosis, let us assume that we have the conditions shown in Figure 5–7, with pure water on one side of the cell membrane and a solution of sodium chloride on the other side. Referring back to Table 5–1, we see that water molecules pass through the cell membrane with extreme ease, whereas sodium and chloride ions pass through only with difficulty. As a result, in the example of Figure 5–7,

more water molecules strike the pores on the left side where there is pure water than on the right side where the water concentration has been reduced. Thus, net movement of water occurs from left to right—that is, osmosis occurs from left to right.

OSMOTIC PRESSURE

If in Figure 5–7 pressure were applied to the sodium chloride solution, osmosis of water into this solution could be slowed or even stopped because the pressure itself can force molecules and ions through the membrane in the opposite direction. The amount of pressure required to stop the osmosis completely is called the *osmotic pressure of the sodium chloride solution.*

The principle of a pressure difference opposing osmosis is illustrated in Figure 5–8, which shows a semipermeable membrane separating two columns of fluid, one containing water and the other containing a solution of water and some solute that will not penetrate the membrane. Osmosis of the water from chamber B into chamber A causes the level of the fluid column in chamber A to rise higher and higher above the level in chamber B, until eventually a pressure difference is developed that is great enough to oppose the osmotic effect. The pressure difference across the membrane at this time is equal to the osmotic pressure of the solution containing the nondiffusible solute.

Osmotic Pressure of the Extracellular and Intracellular Fluids. The ions dissolved

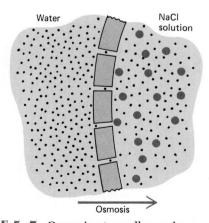

FIGURE 5–7 Osmosis at a cell membrane when a sodium chloride solution is placed on one side and water on the other side of the membrane.

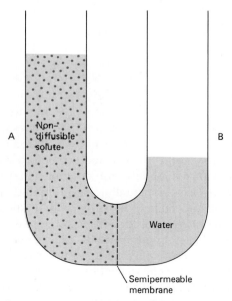

FIGURE 5–8 Demonstration of osmotic pressure on the two sides of a semipermeable membrane.

in the extracellular and intracellular fluids, as well as other substances such as glucose, amino acids, free fatty acids, and so forth, can all cause osmotic pressure at the cell membrane because all of these diffuse through the cell membrane very poorly in comparison with the diffusion of water molecules. The overall concentration of substances in the extracellular and intracellular fluids is enough to create a total osmotic pressure of approximately 5400 mm Hg; that is, if clear water should be placed on one side of the cell membrane while either extracellular or intracellular fluid should be placed on the other side, it would require a pressure of 5400 mm Hg to stop osmosis of water into the fluid. This amount of pressure is approximately equal to the pressure exerted by a column of water 230 feet high. Therefore, one can understand the extreme force that sometimes develops for movement of water through the pores of cell membranes by the phenomenon of osmosis.

Lack of Effect of Molecular and Ionic Mass on Osmotic Pressure; Importance of Numbers of Particles. The osmotic pressure exerted by nonpermeable particles in a solution, whether they be molecules or ions, is determined by the *numbers* of dissolved particles per unit volume of fluid and not the mass of the particles. This holds true whether the particles be ions, small urea molecules, or large protein molecules. The reason for this is that each particle in a solution, regardless of its mass, has almost exactly the same amount of kinetic energy of motion and therefore exerts, on the average, the same amount of pressure against the membrane.

Isotonicity, Hypertonicity, and Hypotonicity of Solutions. A solution that, when placed on the outside of cells, has exactly the same osmotic pressure as the intracellular fluid causes no osmosis through the cell membrane in either direction. Such a solution is said to be *isotonic* with the body fluids. For instance, a 0.9 percent solution of sodium chloride is isotonic. On the other hand, a solution that causes osmosis of fluid out of the cell into the solution is said to be *hypertonic.* Thus, a sodium chloride solution having a concentration greater than 0.9 percent is hypertonic. Finally, a solution that allows osmosis into cells is *hypotonic,* for instance, a sodium chloride solution of less than 0.9 percent concentration. In other words, cells placed in an isotonic solution will maintain their volumes at a constant level, cells placed in a hypertonic solution will shrink, and cells placed in a hypotonic solution will swell.

Osmotic Equilibrium of Extracellular and Intracellular Fluids

Except for very short periods of time, usually measured in seconds, the intracellular and extracellular fluids remain in constant osmotic equilibrium; that is, the osmotic concentrations of nondiffusible substances on the two sides of the cell membranes remain almost exactly equal. The reason for this is that, if ever the concentrations become unequal, osmosis of water through the cell membranes occurs so rapidly that equilibrium will be reestablished within a few seconds. This effect is illustrated in Figure 5–9. Figure 5–9A shows a cell suddenly placed in a very dilute solution *(hypotonic solution)*. Within a few seconds, water passes by osmosis through the cell membrane to the inside of the cell, causing (1) increase in the intracellular fluid volume, and thus swelling of the cell; (2) decrease in the extracellular fluid volume; (3) dilution of the dissolved substances in the intracellular fluid; and (4) increased concentration of the dissolved substances in the extracellular fluid. Once these two fluids have reached the same concentrations of osmotically active substances, osmosis through the membrane ceases. Thus, within a few seconds, a new state of osmotic equilibrium has been reestablished, as shown in Figure 5–9B.

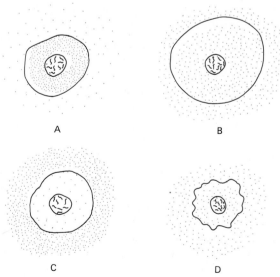

A   B

C   D

FIGURE 5–9 (A) A cell suddenly placed in a hypotonic solution. (B) The same cell after osmotic equilibrium has been established, showing swelling of the cell and equilibration of the concentration of extra- and intracellular fluids. (C) A cell placed in a hypertonic solution. (D) The same cell after osmotic equilibrium has been established, showing shrinkage of the cell and equilibration of fluid concentrations.

Figure 5–9C shows exactly the opposite situation, in which a cell having a dilute intracellular fluid is placed in a concentrated extracellular fluid *(hypertonic solution)*. Water passes by osmosis out of the cell, decreasing the intracellular fluid volume and increasing the extracellular fluid volume. This concentrates the intracellular fluids while diluting the extracellular fluids. Therefore, within a few seconds, the concentrations of the two fluids reach equilibrium, but in the meantime the cell has decreased in size as shown in Figure 5–9D. To illustrate how rapidly osmosis can take place at the cell membrane, a red blood cell placed in pure water gains an amount of water equal to its own volume in less than a second. Therefore, for cells in all parts of the body, one would expect almost complete osmotic equilibrium to take place within 15 to 20 seconds.

ACTIVE TRANSPORT

Often only a minute concentration of a substance is present in the extracellular fluid, and yet a large concentration of the substance is required in the intracellular fluid. For instance, this is true of potassium ions. Conversely, other substances frequently enter cells and must be removed even though their concentrations inside are far less than outside. This is true of sodium ions.

From the discussion thus far it is evident that *no substances can diffuse against a concentration gradient*, or, as is often said, "uphill." To cause movement of substances uphill, energy must be imparted to the substance. This is analogous to the compression of air by a pump. Compression causes the concentration of the air molecules to increase, but to create this greater concentration, energy must be imparted to the air molecules by the piston of the pump as they are compressed. Likewise, as molecules are transported through a cell membrane from a dilute solution to a concentrated solution, energy must be imparted to the molecules. When a cell membrane moves molecules uphill against a concentration gradient (or uphill against an electrical or pressure gradient) the process is called *active transport*.

Among the different substances that are actively transported through cell membranes are sodium ions, potassium ions, calcium ions, iron ions, hydrogen ions, chloride ions, iodide ions, urate ions, several different sugars, and some amino acids.

FIGURE 5–10 Basic mechanism of active transport.

Basic Mechanism of Active Transport

The mechanism of active transport is believed to be similar for most substances and to depend on transport by *carriers*. Figure 5–10 illustrates the basic mechanism, showing a substance S entering the outside surface of the membrane, where it combines with carrier C. At the inside surface of the membrane, S separates from the carrier and is released to the inside of the cell. C then moves back to the outside to pick up more S.

One will immediately recognize the similarity between this mechanism of active transport and that of facilitated diffusion discussed earlier in the chapter and illustrated in Figure 5–4. The difference, however, is that *energy is imparted to the system* in the course of active transport, so that transport can occur *against a concentration gradient* (or against an electrical or pressure gradient).

Though the mechanism by which energy is utilized to cause active transport is not entirely known, we do know some features of this process:

First, the energy is delivered to the inside surface of the membrane from high-energy substances, principally ATP, inside the cytoplasm of the cell.

Second, a specific "carrier" molecule (or combination of molecules) is required to transport each type of substance or each class of similar substances.

Third, a specific enzyme (or enzymes) is required to promote the chemical reactions between the carrier and each transported substance.

A special characteristic of active transport is that the mechanism reaches a maximum rate of transport when the concentration of the substance to be transported is very high. This results from limitation either of quantity of carrier available to transport the substance or of enzymes to promote the chemical reactions. This principle of limitation also applies to carrier-mediated facilitated diffusion, which was discussed earlier in the chapter.

Chemical Nature of Carrier Substances. Carrier substances are proteins, conjugated proteins, or loose physical combinations of more than one protein molecule. Several different carrier systems exist in cell membranes, each of which transports only certain specific substances. One carrier system, for instance, transports sodium to the outside of the membrane and probably transports potassium to the inside at the same time. Another system actively transports sugars through the membranes of intestinal and renal tubular epithelial cells, and still other carrier systems transport different ones of the amino acids.

Energetics of Active Transport. In terms of calories, the amount of energy required to concentrate 1 mol of substance 10-fold is about 1400 calories. Thus, one can see that the energy expenditure for concentrating substances in cells or for removing substances from cells against a concentration gradient can be tremendous. Some cells, such as those lining the renal tubules, expend as much as 80 percent of their energy for this purpose alone.

Active Transport of Sodium and Potassium

Referring back to Figure 5–1, one sees that the sodium concentration outside the cell is very high in comparison with its concentration inside, and the converse is true of potassium. Also, Table 5–1 shows that minute quantities of sodium and potassium can diffuse through the pores of the cell. If such diffusion should take place over a long period of time, the concentrations of the two ions would eventually become equal inside and outside the cell unless there were some means to remove the sodium from the inside and to transport potassium back in.

Fortunately, a mechanism for active transport of sodium and potassium ions is present in all cell membranes of the body. It is called the *sodium-potassium pump.* The basic principles of this pump are illustrated in Figure 5–11. The carrier for this mechanism transports sodium from inside the cell to the outside and potassium from the outside to the inside. This carrier also has the capability of splitting ATP molecules

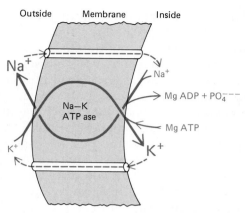

FIGURE 5–11   Postulated mechanism for active transport of sodium and potassium through the cell membrane, showing coupling of the two transport mechanisms and delivery of energy to the system at the inner surface of the membrane.

and utilizing the energy from this source to promote the sodium and potassium transport. Thus, the carrier also acts as an enzyme and is therefore called *sodium-potassium ATPase.* This ATPase is composed of two protein molecules, one a globulin with a molecular weight of 95,000 and the other a glycoprotein with a molecular weight of 55,000. The larger molecule actually binds with both the sodium and potassium ions and also the ATP, but the smaller molecule is also necessary to provide some facilitating function not yet understood.

Note in Figure 5–11 that energy released from ATP at the inner surface of the cell membrane causes potassium ions to split away from the sodium-potassium ATPase carrier molecule and simultaneously causes sodium ions to bind. Then, at the outside surface of the membrane, the sodium ions split away from the carrier while potassium ions bind.

An important feature of the sodium-potassium pump is that it is strongly activated by an increase in sodium ion concentration inside the cell, the activity increasing in proportion to (sodium concentration)^3. This effect is extremely important because it allows even a slight excess buildup of sodium ions inside the cell to activate the pump very strongly and thereby return the intracellular sodium concentration back to its normal low level.

The sodium transport mechanism is so important to many different functioning systems of the body—such as to nerve and muscle fibers for transmission of impulses, various glands for the secretion of different substances, and all cells of the body to prevent cellular swelling—that it is frequently called simply the *sodium pump.* We will discuss the sodium pump at many places in this text.

Active Transport of Other Ions

Calcium and magnesium are probably transported by all cell membranes in much the same manner that sodium and potassium are transported, and certain cells of the body have the ability to transport still other ions. For instance, the glandular cell membranes of the thyroid gland can transport large quantities of iodide ion; the epithelial cells of the intestine can transport sodium, chloride, calcium, iron, hydrogen, and probably many other ions; and the epithelial cells of the renal tubules can transport calcium, magnesium, chloride, sodium, potassium, and a number of other ions.

ACTIVE TRANSPORT THROUGH CELLULAR SHEETS

In many places in the body, substances must be transported through an entire *cellular sheet* instead of simply through the cell membrane itself. Transport of this type occurs through the intestinal epithelium, the epithelium of the renal tubules, the epithelium of all exocrine glands, the membrane of the choroid plexus of the brain, and many other membranes. However, before we can discuss active transport through such cellular sheets, we need first to understand the ways in which cells are joined to each other.

THE ANATOMY OF CELLULAR JUNCTIONS

Cells often are only loosely joined to each other. In these instances, the adjacent cells are generally separated from each other by an accumulation of *glycocalyx* consisting of thin collagen fibers and proteoglycan filaments. The glycocalyx is only loosely attached to each cell membrane, and extracellular fluid can percolate easily through the spaces between the cells, thus allowing the transport of nutrients to all cell surfaces.

On the other hand, in the intestinal epithelium, in the renal tubular epithelium, and elsewhere in the body, it is important for the cells to be strongly joined and also for the junctions to have other properties important to cellular function. Four important types of junctions between cells, illustrated in Figure 5–12, are

1. *Zonula Occludens (Tight Junction).* Figure 5–12 illustrates three adjacent epithelial cells of the intestinal epithelium. Near the luminal surface of the epithelial cells they are joined to each other by a *zonula occludens.* This is also called a *tight junction* because the two cell membranes actually fuse with each other along multiple ridges. These ridges extend the entire width of the respective cells. Therefore, fluids cannot flow past these junctions in the spaces between the cells. Thus, it is the tight junctions that prevent free flow of fluid between the lumen of the intestine and the deeper tissues of the intestinal wall.

Lumen of intestine

Zonula occludens
(Tight junction)

Zonula adherens

Macula adherens
(Desmosome)

Gap junction
(Nexus)

Basement membrane

FIGURE 5–12 Three cells of the intestinal epithelium, illustrating different types of junctions between cells.

2. *Zonula Adherens.* At many points between cells, the cell membranes are moderately adherent to each other even though the membranes do not fuse. Instead, they are held together by a type of cementing substance between the cells. At these points the cell membrane itself is slightly thickened, and ultramicroscopic filaments extend from inside the membrane deep into the cytoplasm of the cell, thus anchoring the cell membrane to the cell's cytoskeleton. These points are called *zonulae adherens.*

3. *Macula Adherens (Desmosome).* A *macula adherens* is similar to a zonula adherens except more adherent. This area of the cell membrane is even more thickened by the presence of minute filaments leading from the membrane into the cytoplasm, and the space between the two cell membranes appears to be filled with a strong cementing substance probably composed of dense glycocalyx. The macula adherens does not extend for long distances along the cell membrane but instead is similar to a spot weld joint where adjacent cells are held tightly together mainly for maintenance of structural strength from one cell to the next.

   A macula adherens is also called a *desmosome,* an older and less descriptive term. Note at the bottom of the cell in Figure 5–12 that there are also two *hemidesmosomes* where the epithelial cell is adherent to the basement membrane.

4. *Gap Junction (Nexus).* *Gap junctions* (also called *nexus*) are special areas where the cell membranes come very close to each other but do not fuse. Instead, small particles, probably protein molecules, project from one membrane to the next, and the membranes are joined to each other by way of these particles. However, much more importantly, the membranes of the two adjacent cells are very highly permeable to ions and molecules that have diameters smaller than 30 nm, thus allowing easy movement of the ions and molecules from cell to cell. It is perhaps through the molecular interstices of the connecting particles that the ions and molecules permeate from one cell to the next. These junctions provide channels of communication from one cell to another, especially easy flow of electrical current as well as movement of hormonal substances from one cell to the next.

   Gap junctions are extremely common between smooth muscle cells and also between cardiac muscle cells. It is by way of these junctions that electrical signals are transmitted through these types of muscles to cause muscle contraction, as will be explained in much more detail in subsequent chapters.

Active Transport of Ions and Water Through an Epithelial Membrane

The general mechanism of active transport of ions and water through an epithelial membrane, such as through the intestinal or renal tubular membrane, is illustrated in Figure 5–13. This figure shows two adjacent cells in a typical membrane. On the luminal surface of the cells is a brush border that is highly permeable to both water and solutes, allowing both of these to diffuse readily from the lumen of the intestine to the interior of the cell. Once inside the cell, some of the solutes are actively transported by the lateral walls of the cell into the spaces between the cells as well as through the base of the cell into the sublying connective tissue. The space between the cells is closed near the brush border of the epithelium by tight junctions, but it is wide open at the base of the cells where the cells rest on the basement membrane. Furthermore, the basement membrane is extremely permeable. Therefore, substances transported into the channel between the cells flow toward the base of the cell.

   Sodium ions, in particular, are actively transported both through the base of the cell and into the channel between the cells. And, since sodium ions are positively charged, this transport of sodium ions out of the cell leaves a deficit of positive charges inside the cell, thus creating strong electronegativity on the inside.

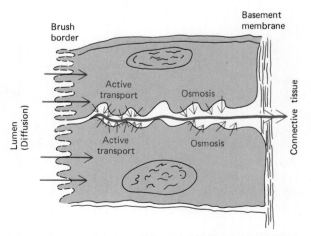

FIGURE 5–13 Basic mechanism of active transport through a layer of cells.

This in turn repels the negatively charged ions, such as chloride ions, from inside the cell so that they too follow the sodium ions through the cellular base and sides of the cell. Finally, the high concentrations of both sodium and chloride ions outside of the membrane now causes osmosis of water through the membrane. Thus, water also follows the ions. Both the water and the ions then diffuse into the blood capillaries of the connective tissue and are carried away from the intestine.

These principles of transport apply generally wherever transport occurs through cellular sheets, whether in the intestine, gallbladder, kidneys, or elsewhere.

Active Transport of Sugars Through Epithelium—the Sodium "Co-Transport" Mechanism

Facilitated diffusion of glucose and certain other sugars occurs in essentially all cells of the body, but active transport of sugars against concentration gradients occurs in only a few places in the body. For instance, in the intestine and renal tubules, glucose and several other monosaccharides are continually transported through the epithelium into the blood even though their concentrations may be minute in the lumens. Thus, essentially none of these sugars is lost in either the intestinal excreta or the urine.

The mechanism of glucose (and other sugar) transport through the epithelial cells of the intestinal mucosa and renal tubules is a curious mixture of diffusion and active transport. This mechanism, called *secondary active transport* or *sodium co-transport*, functions in the following way:

First, let us remember that the epithelial cell has two functionally distinct sides, a brush border that lines the lumen of the gut or renal tubule and a base that lies adjacent to the absorptive capillaries. The basal and lateral walls of the cell transport sodium out of the cell cytoplasm and into the sublying capillaries. This occurs by the usual process of active transport for sodium as described before. The result is marked depletion of sodium ions in the interior of the epithelial cell. This in turn creates a concentration difference for sodium ions across the brush border from the lumen of the intestine toward the interior of the epithelial cell. Consequently, sodium ions attempt to diffuse through the brush border into the cell. However, the brush border is reasonably impermeable to sodium except when the sodium combines with a carrier molecule, one type of which is the so-called *sodium-glucose carrier.* This carrier is peculiar in that it will not transport the sodium by itself but must also transport a glucose molecule at the same time. That is, when bound with both sodium and glucose, the carrier will then diffuse to the inside surface of the cell membrane. One can readily see that it is the sodium concentration difference across the brush border membrane that provides the energy to promote this transport of both the sodium and the glucose. Thus, even when glucose is present in very low concentration, it can still be transported to the interior of the epithelial cell.

Once the glucose has entered the interior of the epithelial cell, it crosses the basal side of the cell by the usual process of facilitated diffusion, in the same manner that glucose crosses essentially all other membranes of the body.

Active Transport of Amino Acids— "Sodium Co-Transport"

Amino acids are the basic building units of proteins, as was discussed in the previous chapter. Most of these, like glucose, are transported to the inside of essentially all cells of the body by facilitated diffusion mechanisms.

Sodium co-transport of amino acids also occurs through a few membranes of the body: the epithelia of the intestines, renal tubules, and some exocrine glands. This involves at least four different carrier systems for transporting different types of amino acids. These transport systems will be discussed further in relation to intestinal absorption in Chapter 42. It should be noted again that in this sodium co-transport mechanism it is the concentration difference of sodium between the intestinal lumen and the interior of the cell that provides the energy for transport of the amino acid molecules, in the same manner that glucose molecules are also transported.

QUESTIONS

1. What are the important ionic differences between the extracellular and intracellular fluids?
2. Give the theory of the diffusion process.
3. What is the difference between facilitated diffusion and simple diffusion?
4. What substances are transported by diffusion through membrane pores, and what by diffusion through the cell membrane matrix?
5. How do electric charges on ions and concentration differences affect the net diffusion rate?
6. Explain the principles of osmosis across cell membranes.

7. Why do extracellular and intracellular fluids remain continually in osmotic equilibrium?
8. What is the difference between active transport and facilitated diffusion?
9. Discuss the active transport of sodium and potassium across the cell membrane.
10. Explain the anatomy and the functions of the different types of intercellular junctions.
11. What is meant by secondary active transport or sodium co-transport of glucose or amino acids?

REFERENCES

Andreoli, T.E., *et al.* (eds.): Membrane Physiology. New York, Plenum Press, 1980.

Ellory, C., and Lew, V.L. (eds.): Membrane Transport in Red Cells. New York, Academic Press, 1977.

Fettiplace, R., and Haydon, D.A.: Water permeability of lipid membranes. *Physiol. Rev., 60:*510, 1980.

Finn, A.L.: Changing concepts of transepithelial sodium transport. *Physiol. Rev., 56:*453, 1976.

Gilles, R. (ed.): Mechanisms of Osmoregulation: Maintenance of Cell Volume. New York, John Wiley & Sons, 1979.

Gregor, H.P., and Gregor, C.D.: Synthetic-membrane technology. *Sci. Am., 239*(1):112, 1978.

Gupta, B.L., *et al.* (eds.): Transport of Ions and Water in Animals. New York, Academic Press, 1977.

Guyton, A.C., *et al.:* Circulatory Physiology II: Dynamics and Control of the Body Fluids. Philadelphia, W.B. Saunders, 1975.

Keynes, R.D.: Ion channels in the nerve-cell membrane. *Sci. Am., 240*(3):126, 1979.

Lodish, H.F., and Rothman, J.E.: The assembly of cell membranes. *Sci. Am., 240*(1):48, 1979.

Macknight, A.D.C., and Leaf, A.: Regulation of cellular volume. *Physiol. Rev., 57:*510, 1977.

Metcalfe, J.C. (ed.): Biochemistry of Cell Walls and Membranes II. Baltimore, University Park Press, 1978.

Schultz, S.G.: Principles of electrophysiology and their application to epithelial tissues. *Int. Rev. Physiol.,* 4:69, 1974.

Ullrich, K.L.: Sugar, amino acid, and Na^+ cotransport in the proximal tubule. *Annu. Rev. Physiol., 41:*181, 1979.

Wallick, E.T., *et al.*: Biochemical mechanism of the sodium pump. *Annu. Rev. Physiol., 41:*397, 1979.

Wright, E.M., and Diamond, J.M.: Anion selectivity in biological systems. *Physiol. Rev., 57:*109, 1977.

THE SKELETAL SYSTEM

6

The Skeletal System and Its Joints: I. General Structure; Bone as Tissue; the Skull

Overview

The **skeleton** is the framework around which the body is built. For their weight, bones are almost as strong as steel. Their compressive strength is achieved by trillions of minute *calcium salt crystals* that are approximately as hard as marble and that are locked together by *bone matrix*. The tensile strength of bone is provided by an extremely strong network of *collagenous fibers*.

In the skeleton the bones are held together at the **joints** by **ligaments,** and the ligaments and joint structure together determine the directions in which the joints can angulate. The joints are classified into three types: (1) *fibrous joints* and (2) *cartilaginous joints,* which usually allow little motion, and (3) *synovial joints,* which usually allow wide range of motion and are the most common. The surfaces of the synovial joints are covered by extremely smooth and slick **articular cartilaginous surfaces.** In addition, the joint space is surrounded by a **joint capsule** lined on the inside with a **synovial membrane** that secretes into the **joint space** an extremely slick fluid that lubricates the joint motions.

Bone tissue can be divided into two different types: *compact bone* and *spongy bone.* Compact bone is extremely dense and strong. It is found mainly near the surfaces of the bones, where bone strength is required. Spongy bone is constructed of many thin bony trabeculae that constitute the interior of most bones. **Red bone marrow** is found in the spongy portions of many of the bones; this marrow forms all of the *red blood cells* and most of the *white bone cells.*

Bone is formed by a type of cell called **osteoblasts.** Another type of cell, the **osteoclasts,** continually removes old and brittle bone. This is followed by regrowth of new bone where the old has been removed, thus continually replacing old and weaker bone with new and stronger bone.

The **skull** is divided into three principal parts: (1) the **cranium,** (2) the **facial bones,** and (3) the **mandible.**

The *cranium* is the vaultlike structure that encases and supports the brain. It, in turn, is formed of eight bones: (1) the **frontal bone,**

(2) the **occipital bone,** (3) and (4) two **parietal bones,** (5) and (6) two **temporal bones,** (7) part of the **sphenoid bone,** and (8) part of the **ethmoid bone.** The cranial cavity is penetrated by multiple openings called **foramina** through which nerves and blood vessels pass. The most important of these is the **foramen magnum,** located at the base of the cranium and through which the spinal cord passes.

The *facial bones* support the structures of the face. They especially form the **nasal cavity** and the **orbit** in which the eye is located. The names of these facial bones are: (1) **maxilla,** (2) **zygomatic bone,** (3) **nasal bone,** (4) part of the **ethmoid bone,** (5) part of the **sphenoid bone,** (6) **lacrimal bone,** (7) **vomer,** (8) **palatine bone,** and (9) **inferior concha.**

The *mandible* provides the framework of the lower jaw. The lower **teeth** grow from an upward projecting ridge of the mandible called the **alveolar border.** The upper teeth grow from a corresponding alveolar border on the inferior surface of the maxilla.

We begin our discussion of the entire human body with the **skeleton**—not because the skeleton is more important than the other organ systems of the body, but because it is the frame around which all else is built. On the other hand, a study of the skeleton is intriguing because we see that evolution has developed a solid bony framework that is incredibly strong, yet also incredibly light for its strength and almost perfectly adapted for the functions that it must perform.

THE SKELETON AS A WHOLE

Let us first study the skeleton as a whole in Figure 6–1. This figure illustrates all the major bones of the body, and if the student will take time to study the names of these different bones, their structures, and their interrelationships, he or she will have gone a long way toward understanding the basic framework of the body and its movements. On the right side of this figure, note the glistening **cartilaginous articulating surfaces** of the **joints,** which, when lubricated with *joint fluid,* allow for a lifetime of smooth motion. To the left, study the **ligaments** that hold the bones together, enveloping the joints and constraining their actions to allow only those movements that are required while preventing motions in undesired directions.

The **shoulder** and **hip joints** can move in all directions, and the ligaments around these joints are of types that allow this wide latitude of movement. The **elbow** and **knee joints** move principally from a straight position to a greatly angulated position, the elbow capable of bending forward and upward in an arc of about 155 degrees and the knee joint backward in an arc of about 135 degrees. The lateral ligaments of these joints prevent sideways movement. Also, ligaments and bony protuberances prevent these joints from extending more than a few degrees beyond the straight position.

The **skull** can rotate on the spinal column, requiring a very special rotational joint that we shall describe later in the chapter. The **vertebral column** can bend gently in the forward or backward direction and slightly to one side or the other; it can rotate slightly as well. But the ligaments of the spinal column are much tighter than those of the other joints of the body. This prevents more than very slight motion between any two adjacent vertebrae, thus preventing injury to the spinal cord in the **vertebral canal** and to the spinal nerves as they leave the vertebral column.

Now, look for a moment at the two bones of the **forearm.** The right arm shows these two bones crossed over each other with the back of the hand facing forward. However, muscles attached to these two bones can pull them either to the parallel position (called *supination*) or the crossed over position (called *pronation*), thus allowing the all important rotation of the forearm and hand.

So, it is no wonder that man with all his ingenuity and with his million-dollar robots has never yet been able to produce a working mechanical model with the efficiency and effectual motions of the human skeleton, much less duplicate the other magical features of the human body.

Cranium
Frontal bone
Parietal bone
Temporal bone
Orbit
Maxilla
Nasal fossa
Mandible
Clavicle
Manubrium of sternum
Scapula
Body of sternum
Ribs
Xiphoid process
Humerus
Vertebrae
Radius
Ilium
Ulna
Sacrum
Pubis
Ischium
Wrist
Carpals
Metacarpals
Phalanges
Femur
Patella
Tibia
Fibula
Tarsals
Metatarsals
Phalanges

FIGURE 6–1 Anterior view of the entire skeleton, showing also to the left the major ligaments of the skeleton.

Finally, before we leave the general concept of the human skeleton, let us consider the *upright stance of man.* This evolutionary result allows a multitude of motions and mechanical functions that otherwise would not be possible. Let us remember especially the multiple and delicate functions of the human hand, a mechanical marvel that can fabricate a watch the size of a dime with literally hundreds of almost microscopic parts, something that would not be possible if the hands were still adapted for running on all fours.

But there is another valuable attribute of the upright stance; it decreases immeasurably the amount of muscle energy required to support the body against gravity. When a person stands totally erect with his legs, hips, and spine extended fully, the forces of gravity are directed almost directly down the linear axis of the supporting bones and joints so that the muscles are needed hardly at all. Therefore, man can remain in the standing position literally all day without significant fatigue of his muscles. Indeed, much of the "fatigue" that a person does feel following such a long stint of standing is often soreness in the ligaments and joints, not the muscles.

THE AXIAL SKELETON AND THE APPENDICULAR SKELETON

For purposes of description, the skeleton is usually divided into two separate parts, (1) the **axial skeleton** and (2) the **appendicular skeleton.** These are summarized in Table 6–1. The axial skeleton consists of most of the bones of the trunk and head, including the

1. **skull;**
2. **hyoid bone,** located in the anterior neck, that helps in the swallowing process and movement of the larynx;
3. **vertebral column,** consisting of 24 movable **vertebrae,** 5 vertebrae fused to form the **sacrum,** and 4 additional vertebrae fused to form the **coccyx;** and
4. **thoracic cage,** consisting of the **ribs** and **sternum.**

The appendicular skeleton consists of the bones of the upper and lower extremities, including the following:

Upper extremity:

1. **Pectoral girdle,** consisting of the **clavicles** and the **scapulae.**
2. **Bones of the arms, forearms, wrists, and hands.**

Lower extremity:

1. **Pelvic girdle,** consisting of the **coxal bones (hip bones).**
2. **Bones of the thighs, legs, ankles, and feet.**

TABLE 6–1
The Axial and Appendicular Skeleton

Axial

Skull

Vertebral Column
  Seven cervical vertebrae
  Twelve thoracic vertebrae
  Five lumbar vertebrae
  Sacrum (consists of five vertebrae fused together)
  Coccyx (consists of four vertebrae fused together)

Thoracic Cage
  Ribs (posterior bony portion and anterior costal cartilage)
  Upper seven called true ribs because they complete the circle from vertebra to sternum
  Lower five called false ribs
  Ribs 8, 9, and 10 attach by cartilage arches to costal cartilage of rib 7
  Ribs 11 and 12 called floating ribs because their ends are unattached

  Sternum
    Manubrium: upper quarter of sternum
    Body: middle and lower three quarters of sternum
    Xiphoid process: very small projection from lower end of the body

Appendicular

Upper Extremity
  Pectoral girdle
    Clavicle
    Scapula
  Arm
    Humerus
  Forearm
    Radius
    Ulna
  Wrist
    Carpals
  Hand
    Metacarpals
    Phalanges
Lower Extremity
  Pelvic girdle
    Coxal bone (Hip bone)
      Ilium, Ischium, and Pubis fused together
  Thigh
    Femur
  Leg
    Tibia
    Fibula
  Ankle
    Tarsals
  Foot
    Metatarsals
    Phalanges

PROPERTIES OF BONE

Before going too far in discussing the skeleton, let us digress to consider the special properties of bone itself that give it extreme strength yet light weight. Then we shall return to the special attributes of the individual bones and joints.

Bone is composed mainly of very dense and hard **calcium salts** that are held together by **bone matrix** containing large numbers of extremely strong **collagenous fibers** and a cement called **ground substance.**

Figure 6–2 illustrates the basic structure of a long bone, section A showing the outside ap-

pearance, section B a cross-section, and sections C, D, and E various details of structure. The main shaft of the bone is called the **diaphysis,** and the two ends are called respectively the **proximal epiphysis** and the **distal epiphysis.** Note especially that the joint surfaces are all covered by a thin layer of very strong, resilient, and smooth **hyaline cartilage.** When lubricated with **synovial fluid,** this cartilage is at least as slick as ice.

FIGURE 6–2 Various views of a long bone. Note especially in Section C the very strong longitudinal collagen fibers of compact bone interlaced with a great abundance of calcium salt crystals arranged like overlapping bricks. Also note in Section D the secretion of bone, a process called osteogenesis, by osteoblasts that line the inner surface of the periosteum.

The Tubular Structure of Long Bones—Compact and Spongy Bone.   Sections A and B of Figure 6–2 illustrate that long bones have a tubular structure. The outer layer of the bone is composed of very strong **compact bone,** but the interior is different. The interior of the epiphyses is filled with **spongy bone;** the interior of the diaphysis is a long cavity called the **medullary cavity** that is lined with spongy bone and is filled with *fatty connective tissue;* it is also sometimes called *yellow bone marrow* because of the yellowness of the fat. The spongy bone is constructed of thin bony trabeculae that add very little to the weight of the bone. Thus, almost the entire weight of the bone is contributed by the compact bone. The spaces among the trabeculae in many bones are filled with **red bone marrow;** it is in this marrow that most of the blood cells are formed.

Compact bone, as its name implies, is extremely dense. Its collagenous fibers have almost as much strength as soft iron wire, and they are oriented almost entirely in the direction of the greatest tensional stress in the bone. The calcium salt crystals are approximately as hard as marble and have the shape of minute plates that are attached to the sides of the collagenous fibers so that they will not slip. These plates are layered in an overlapping fashion, interdigitating in the same manner that bricks are laid in the wall of a building. Thus, compact bone has almost exactly the same structure as reinforced concrete, the collagenous fibers providing almost as much tensional strength as the steel in the concrete and the calcium salt crystals serving the same purpose as the gravel, sand, and cement to provide the compressive strength of the concrete. In fact, weight for weight, compact bone is actually stronger than some reinforced concrete.

The tubular nature of long bones also adds to the strength of these bones, because it is a principle of physics that almost all of the stress in an elongated structure experiencing bending occurs in the outer layers, with almost no stress occurring in the very center. Thus, the tubular structure of bone places the strength where it is needed.

OSTEOGENESIS

Osteogenesis means the formation of bone. This results from the secretory activity of **osteoblasts,** as was explained in Chapter 2. However,

the mechanism of bone formation is somewhat different in the fetus than it is after birth.

Osteogenesis in the Fetus.   Bone formation in the fetus occurs in two ways: (1) by membranous bone formation and (2) by endochondrial bone formation.

Membranous bone formation begins with the development of an appropriately shaped membrane of connective tissue. *Osteoblasts* then invade the surfaces of the membrane, secreting *bone matrix* called **osteoid** that contains both *ground substance* and *collagenous fibers* and then within days thereafter causing *calcium salt deposition.* The membranous bones are mainly the flat and irregular bones of the body, including the bones of the skull, the vertebrae, the hip bones, and many others.

Endochondrial bone formation begins with the development of cartilage replicas of the bones that are to be formed. This is the manner in which the long bones of the extremities originate. After the cartilage replicas have been made, one or more **centers of ossification** occur deep inside the cartilage, usually one in the center of the bone called the *primary center* and one at each end called the *secondary centers.* In each of these centers, some of the *chondroblasts* that had formed the cartilage replica now change into *osteoblasts.* These osteoblasts progressively dissolve small portions of the cartilage while at the same time secreting osteoid and shortly thereafter calcifying this matrix to form bone. That portion of the cartilage that has not yet become ossified continues to grow, however, and this continues until ossification has spread throughout the entire mass of cartilage.

Osteogenesis in the Adult.   In adult bone, osteogenesis also results from the secretory activity of osteoblasts. However, osteogenesis now occurs mainly for the purpose of repairing or remodeling bone, as we shall discuss in the following paragraphs. For this purpose, osteoblasts occur on all surfaces of the bone where new bone is being deposited. The outer surface of the bone is covered by the **periosteum,** which itself is composed of two layers: (1) *an outer strong fibrous tissue layer* about $\frac{1}{2}$ mm thick and (2) *an inner layer of osteoblasts* where the periosteum attaches to the bone. The insides of the bone cavities are also lined with a similar membrane called the **endosteum.**

When forming bone, the osteoblasts secrete osteoid the same as in the fetus, followed within days by calcium salt deposition alongside the collagenous fibers.

Remodeling of Bone. Bone is also being reabsorbed all of the time by collections of large cells called *osteoclasts* that are present within the bone. These actually absorb bone tissue to form cavities in the bone; this process requires about 3 weeks once a collection of osteoclasts appears. At the end of this time, the osteoclasts disappear and *osteoblasts* instead migrate into the cavity and line up around their walls to deposit new bone, layer upon layer in concentric circles. These concentric circles, illustrated at the bottom of Figure 6–2C and also in Figure 6–3, eventually completely fill the *osteoclastic cavity* except for a remaining small central canal called the **Haversian canal.** This system of concentric bony deposits around each Haversian canal is called an **osteon,** or **Haversian system.** The blood vessels that nourish the bone lie in the Haversian canal. Note also in Figure 6–3 the minute cavities called **lacunae** that are dispersed widely throughout the bone. Entrapped in each of these lacunae is a single living bone cell called an **osteocyte** that was derived from an osteoblast that became entrapped in the bone at the time of bone formation. From the many osteocytes, **cellular processes** extend in all directions through multiple small **canaliculi.** These processes link up with each other, forming a network of living cellular tissue that plays an important role in maintaining both a strong bone matrix and a proper degree of calcification of the bone crystals.

This reabsorption and restructuring of bone is called *bone remodeling*. Its purpose is to remove old bone as it becomes weak and brittle and to replace this with new and tougher bone. Remodeling occurs to a tremendous extent in young people but less so in old persons, giving rise to the well-known brittle characteristics of bones in elderly people.

We shall discuss more details of bone function in Chapter 47 when we consider the endocrinological factors that control the calcium in the body, both in the bone and in the body fluids.

Growth of Bone. In early life all bones grow larger in size, and the long bones also grow in length. Most bone growth stops at the time of adolescence or early adult life, though some of the bones, such as those of the lower jaw and nose, grow very slowly long into adult life, which accounts for some of the progressive changes in a person's appearance with age. Especially important, the long bones stop elongating and the person's height ceases to increase at the time of adolescence or shortly thereafter. The reason for this is as follows.

Figure 6–4 illustrates one end of a long bone, showing the diaphysis of the bone separated from the epiphysis by a band of cartilage called the **epiphyseal cartilage.** The epiphyseal cartilage is composed of linear columns of **cartilage cells** surrounded by **cartilage matrix.** The cells at the diaphyseal surface of the cartilage are reproducing rapidly and are also secreting new cartilage. However, almost as soon as the new cartilage is formed, *osteoblasts* at the end of the diaphysis of the bone resorb this newly formed cartilage and simultaneously secrete bone. That is, the diaphyseal surface of the epiphyseal cartilage is resorbed in the same way that old bone is resorbed by osteoclasts, and immediately thereafter the osteoblasts set about the task of forming bone.

Thus, the growth of the epiphyseal cartilage causes progressive elongation of the diaphyseal shaft of the long bone, and the newly formed cartilage shortly thereafter is replaced by true bone. However, in the process of forming new cartilage, the cartilage cells themselves are gradually used up, so that the epiphyseal cartilage becomes progressively thinner. By the time of adolescence, or shortly thereafter, the epiphyseal cartilage has completely disappeared, and the epiphysis at that time joins the diaphysis by bony union, no longer with a cartilage that can grow. Therefore, the epiphysis at each end of the bone becomes sealed with the diaphysis, and the long bone stops growing in length.

Lumen of Haversian canal

Canaliculi

Osteocyte in lacuna

FIGURE 6–3 Concentric circles of bone surrounding a Haversian canal. Note also the network of osteocytes in the lacunae with cellular processes extending throughout the bone inside numerous small canaliculi.

FIGURE 6–4 Growth of a long bone, showing in Section A the epiphyseal cartilage where the shaft of the bone grows in length and in Section B a microscopic picture of growth of the diaphyseal end of the cartilage with subsequent conversion of the newly formed cartilage into bone.

Bone Fracture and Its Repair. As every student already knows, an excessive crushing or bending force applied to a bone can cause a fracture. Bones in young children have a very strong and resilient bone matrix containing healthy and strong collagenous fibers, but they have a smaller proportion of calcium crystals than do the bones of adults. Therefore, fractures of young bones frequently do not cause complete rupture of the bone but instead cause a so-called *greenstick fracture*, as illustrated in Figure 6–5A. Simply bending the bone back in place and holding it there for a few weeks usually results in rapid repair.

By contrast, old bones are much more brittle, and they usually fracture completely, as illustrated in Figure 6–5B, the ends often becoming badly fragmented, giving rise to the so-called *comminuted fracture*. This brittleness of old bone results from a greatly diminished rate of bone remodeling in old age, which allows the bone matrix to become progressively weaker. Also, the diminished remodeling frequently leaves the bones highly porous because of continued osteoclastic resorption without equal rebuilding of new bone. This condition of the bones is called *osteoporosis*. Such osteoporotic bones often fracture simply from the weight of the body; this is especially true of the vertebrae in old age, causing many old persons to become progressively shorter in their later years of life.

Figure 6–5C illustrates the manner in which a broken bone is repaired. The *osteoblasts*

from the broken fragments of bone proliferate rapidly in the fragmented area, while at the same time the *fibroblasts* in the *periosteum* and *endosteum* form *fibrous tissue* that temporarily bridges the gap between the long ends. Then the osteoblasts secrete *osteoid* amongst the fibrous tissue. Within days large numbers of *collagenous fibers* form within this osteoid, and *calcium crystals* begin to grow adjacent to these fibers, thus building new bone. The early overgrowth of fibrous tissue, osteoid, and new bone, illustrated in Figure 6–5C, is called a **callus.** During the succeeding few months in a young person or few years in an old person the callus is resorbed and regenerated again and again until an almost perfectly shaped bone is finally rebuilt. For some reason not yet understood the bone continually reshapes itself to match the stresses applied to it, which is the reason that the bone eventually reattains its original shape. In fact, when a bone heals crookedly, progressive reshaping will sometimes still result in a straightened bone within a few years, especially in children.

FLAT BONES AND IRREGULAR BONES

The skeleton includes other types of bones besides long bones, some of which are *short* such as the vertebrae, some of which are almost *globular* in shape such as the bones of the wrist, and some of which are *flat* such as the bones of the cranium. All of these have an outer shell of *compact bone* and an interior of *spongy bone*. In

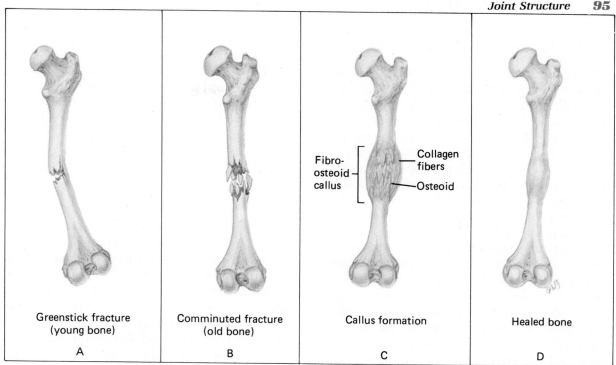

FIGURE 6–5 "Greenstick" and comminuted fractures of young and old bones. Shown also in the right portion of the figure is the healing process of a fracture.

these, as in long bones, it is the compact bone of the outer shells that provides the strength while the interior is mainly a filling that is constructed to be as light as possible. To conserve even more space, *red bone marrow,* where the blood cells are formed, is located in the spaces of the spongy bone of the cranium, the vertebrae, the iliac bones, and some others of the flat and irregular bones. In the fetus and young children, red bone marrow is also present in the long bones.

JOINT STRUCTURE

The joints of the skeleton are classified into three separate types:

1. **Synovial joints,** which are by far the most common and are characterized by a wide range of motion and a **joint space** surrounded by a **synovial membrane.**
2. **Fibrocartilaginous joints,** which have no joint space, but the end of the bones are separated by a *fibrocartilaginous cushion.*
3. **Fibrous joints,** which also have no joint space, and the two bones are bound to each other by *dense fibrous tissue.*

THE SYNOVIAL JOINT

Figure 6–6 illustrates a representative synovial joint. Note that each of the joint surfaces is covered by a **hyaline articular cartilage.** Attached

to the edges of the two articular cartilages is the **joint capsule** with a thicker outer **connective tissue layer** and an inner **synovial membrane.** This completely encapsulates the joint and also folds around the ends of the two articulating bones, forming a closed sac. The total space within the synovial membrane is called the **joint cavity** or the **synovial cavity.** This cavity is filled with **synovial fluid,** which is mucoid in consistency, similar to the very slick mucus of saliva and is secreted into the cavity by the epithelial cells of the synovial membrane itself. The syno-

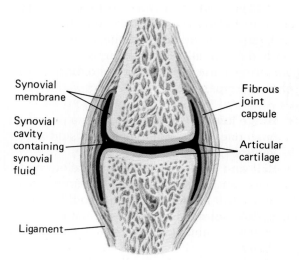

FIGURE 6–6 The principal features of a typical joint.

vial fluid lubricates the joint; without it the joint would quickly become arthritic.

Finally, on the outside of the fibrous joint capsule are the **joint ligaments;** these usually fuse with the fibrous joint capsule. However, the ligaments are often located on some sides of the joints and not on others, thus limiting the motions of the joints in particular directions.

The *fibrous joint capsule*, as well as the *ligaments*, fuses with the periosteum of both of the articulating bones. Also, note especially in Figure 6–6 that the ends of the ligaments actually penetrate through the periosteum and deeply into the substance of the bone itself, which is necessary for them to achieve the strength of attachment that is required for their function. Both the fibrous joint capsule and the ligaments are composed of *dense fibrous connective tissue* consisting mainly of very strong collagenous fibers almost identical with the collagenous fibers of bone and also having the strength of the bone fibers.

Swollen Joints.   Ordinarily, the synovial membranes rest closely against the joint so that the volume of fluid in the synovial cavity remains at a minimum, except for that amount of mucoid fluid required for joint lubrication. This small volume results from a continuous tendency of the surrounding tissues to absorb water and electrolytes from the joint cavity. However, when trauma to the joint occurs or when the tissues surrounding the joint become severely inflamed or infected, adequate absorption of the fluid no longer occurs, and the joint then swells. The amount of fluid in the joint can then increase as much as 10- to 50-fold, giving rise to the well-known swollen joint syndrome. And the swelling goes down only when the joint tissues heal.

Types of Synovial Joints.   Referring again to Figure 6–1, it is clear that the shapes and ranges of motions of the many synovial joints differ greatly. For the purpose of description, these joints are classified into the following principal types:

1. **Hinge joints,** which allow movement around only one axis (example: *elbow* and *finger joints*).
2. **Ball and socket joints,** which allow movement around multiple axes, including rotary movement (example: *hip* and *shoulder joints*).
3. **Saddle joints,** which allow motion in two planes (example: the *proximal joint of the thumb*).
4. **Sliding joints,** which allow slightly curved joint surfaces to slide over each other (exam-

ple: the *joints between the multiple short bones of the wrist and ankle*).
5. **Pivot** (or **rotational**) **joints,** which allow rotation around a central axis (example: rotation of the *first cervical vertebra around a projection of the second vertebra*, thus allowing rotation of the head).

FIBROCARTILAGINOUS AND FIBROUS JOINTS

Fibrocartilaginous and fibrous joints are much less common than synovial joints. However, some important examples follow.

The joints between the vertebral bodies, the **intervertebral joints,** that support the weight of the trunk are *fibrocartilaginous joints.* There is no joint space. Instead, between the two articulating surfaces is a thick mat of *fibrocartilage* called the **vertebral disc.** Strong ligaments encase these intervertebral joints and allow only slight bending and even less rotation. Another fibrocartilaginous joint is the **symphysis pubis,** which is the joint between the two pubic bones at the front of the pelvis. It too allows a slight amount of movement between the two sides of the pelvis.

Fibrous joints occur between the bones forming the *cranial vault and most other parts of the skull.* The edges of these bones fit snugly against each other and are tightly bound at their edges by ligamentous fibers that penetrate deeply into the adjacent bones. Such joints are called **sutures.** In the adult, there is almost no movement at these joints. In fact, the edges sometimes fuse with a bony union. In the newborn child, however, most of the cranial joints are still very loose and broad, and the fibrous tissue of the joints stretches as the brain enlarges inside the cranium. Progressive ossification of the fibrous tissue from the edges of the cranial bones allows growth of the bones.

TYPES OF MOVEMENT AT THE JOINTS

In discussing the different movements, it is important to understand the terms used for specifying direction and type of movement. Some of these, such as flexion and extension, are widely known, but others are not. Therefore, let us list the more important types of movement.

Flexion. This generally means bending an extremity, a finger, or some other part of the body at a joint from the straight position to an angulated position. However, in the case of the shoulder joint it means movement of the arm anterior to the body. A muscle that causes flexion is called a *flexor*.

Extension. This generally means straightening an arm, a leg, a finger, or other part of the body from an angulated position to a straight position. At the shoulder joint it means movement of the arm posterior to the body. A muscle that causes extension is called an *extensor*.

Abduction. This means to pull a thigh, an arm, a finger, or some other part of the body away from the central axis; for example, lateral movement of the thigh is abduction. A muscle that causes abduction is called an *abductor*.

Adduction. This is the opposite of abduction, meaning movement of a part of the body toward a central axis. For instance, pulling the thumb toward the center of the hand is an example of adduction. Adduction is caused by *adductor muscles*.

Rotation. This means revolving some part of the body around an axis. This may be medial or lateral rotation, indicating movement toward or away from the sagittal plane of the body.

Supination. This means rotating the forearm so that the palm is facing anteriorly. The muscle involved is a *supinator*.

Pronation. This means rotating the forearm so that the palm is facing posteriorly. The muscle involved is a *pronator*.

Dorsiflexion. This refers to the ankle joint and means bending the foot upward. The muscle involved is a *dorsiflexor*.

Plantar Flexion. This, too, refers to the ankle joint; it means bending the foot downward. The muscle involved is a *plantar flexor*.

Inversion. This, also, refers to the foot; it means turning the foot so that the sole of the foot faces inward. The muscle involved is an *invertor*.

Eversion. This is the opposite of inversion, that is, turning the foot so that the sole faces outward. The muscle involved is an *evertor*.

Special Motions. Other types of muscles and their motions are: *levator*, which lifts some part of the body; *depressor*, which pulls some part of the body downward; *sphincter*, which constricts some opening of the body; *protractor*, which causes protraction of the jaw forward or of the head forward; and *retractor*, which pulls the jaw or the head backward.

THE SKULL

In the remainder of this chapter and in the next we will discuss the separate bones of the skeleton, beginning with the skull.

The **skull,** illustrated in Figures 6–7 through 6–11, can be divided into three principal parts: (1) the *cranium*, (2) the *facial bones*, and (3) the *mandible*.

The Cranium

The **cranium** is the large, enclosed, vaultlike structure that encases, supports, and protects the brain. Without this support, the brain could hardly survive even the process of walking across the floor, much less jogging, because the tissues of the brain are extremely fragile. For instance, it is very easy to crush the tissues of the exposed brain with only mild pressure of the finger. But, as we shall discuss in more detail in Chapter 13, the brain literally floats inside the cranium, bathed in fluid that fills the space between the brain and the cranium.

As illustrated in Figures 6–7, 6–8, and 6–9 and also summarized in Table 6–2, the cranium is formed of eight bones:

1. **Frontal bone,** located anteriorly.
2. **Occipital bone,** located posteriorly and inferiorly.
3 and 4. **Two parietal bones,** located superiorly and anteriorly to the occipital bone.
5 and 6. **Two temporal bones,** located in the area of the ear on each side.
7. **Sphenoid bone,** located anteriorly in the floor of the cranial cavity, inferior and posterior to the frontal bone but anterior to the two temporal bones.
8. **Extension of the ethmoid bone,** which forms a perforated plate called the **cribriform plate** lying in the midline at the anterior base of the cranial cavity between the sphenoid and frontal bones.

Now, let us consider some of the major features of the cranium itself.

THE THREE MAJOR CRANIAL FOSSAE

On the floor of the cranial cavity are three major depressions called **cranial fossae.** These, illustrated in Figures 6–8 and 6–9, are the

1. **Anterior fossa,** which supports the frontal lobes of the brain;
2. **Middle fossa,** which supports the temporal lobes on each side; and
3. **Posterior fossa,** which supports the cerebellum and brain stem.

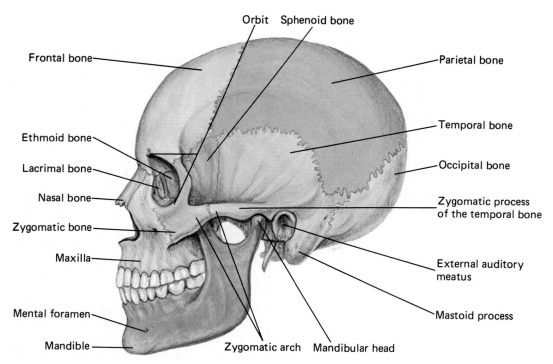

FIGURE 6–7 A lateral view of the skull.

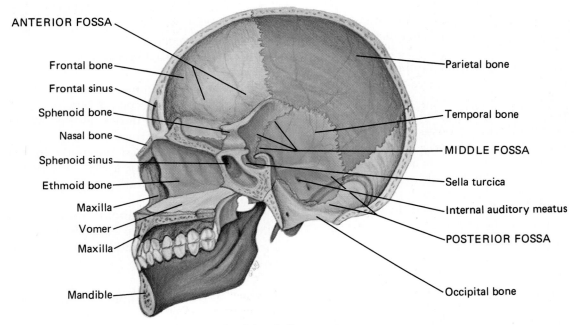

FIGURE 6–8 A lateral view of the inside of the skull.

THE PRINCIPAL OPENINGS OF THE CRANIUM

It is through the openings of the cranium that the spinal cord, cranial nerves, and blood vessels pass between the brain and the peripheral parts of the body. Most of these openings are called **foramina,** though some are called **canals.** They are summarized in Table 6–3. The more important are listed here.

Foramen magnum. This is the very large central opening in the base of the skull illustrated in Figure 6–9. It is through this foramen that the lowest segment of the brain, the medulla, becomes continuous with the spinal cord in the vertebral canal.

Hypoglossal canal (Fig. 6–9), located close to the foramen magnum, transmits the twelfth cranial nerve.

Jugular foramen (Fig. 6–9), also in the floor of

TABLE 6–2
Bones of the Cranium

Bones	Articulations
Frontal	Parietal—posterior Ethmoid—medial, inferior Sphenoid—inferior, posterior Zygomatic—inferior, lateral Nasal—inferior, anterior Maxilla—inferior, anterior Lacrimal—inferior (in orbit)
Parietal (two)	Frontal—anterior Sphenoid—inferior, anterior Temporal—lateral, inferior Occipital—posterior
Temporal (two)	Parietal—superior Sphenoid—anterior, medial Zymogatic—anterior, lateral Occipital—posterior
Occipital	Parietal—lateral, superior Temporal—lateral, inferior Sphenoid—anterior, medial First vertebra (atlas)—inferior
Sphenoid	Frontal—superior, anterior Ethmoid—anterior, medial Temporal—posterior, lateral Parietal—lateral, superior Occipital—posterior, medial Maxilla—anterior Lacrimal—anterior (in orbit) Vomer—anterior, medial (in nose) Palatine—inferior, anterior
Ethmoid	Frontal—superior, lateral Sphenoid—posterior Vomer—inferior (in nose) Maxilla—anterior, lateral Lacrimal—lateral (in orbit)

TABLE 6–3
Principal Openings of Cranium

Opening	Structures that pass through
Foramen magnum	Spinal cord Vertebral arteries and veins
Hypoglossal foramen	Hypoglossal nerve
Jugular foramen	Jugular vein Nerves IX, X, and XI
Internal auditory meatus	Vestibulocochlear nerve (nerve VIII) Facial nerve (nerve VII)
Optic foramen (canal)	Optic nerve Ophthalmic artery
Superior orbital fissure	Ophthalmic division of the trigeminal nerve (nerve V) Nerves III, IV, and VI to extraocular muscles Ophthalmic veins
Foramen ovale	Mandibular division of nerve V
Foramen rotundum	Maxillary division of nerve V
Foramen lacerum	(Covered by cartilage)
Carotid canal	Carotid artery
Perforations of cribriform plate	Olfactory nerve branches

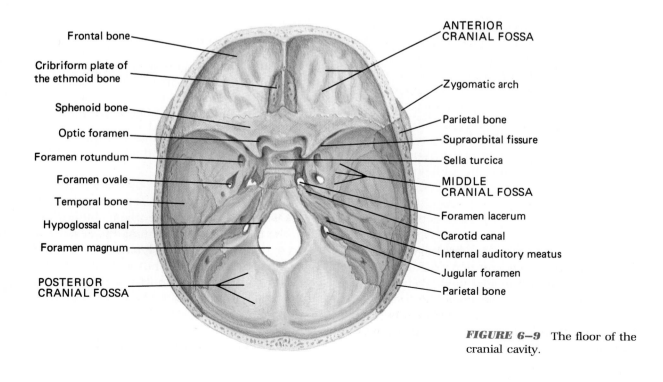

FIGURE 6–9 The floor of the cranial cavity.

the posterior fossa, transmits the internal jugular vein, as well as cranial nerves IX, X, and XI.

External auditory meatus and **internal auditory meatus.** The bony part of the external auditory meatus, illustrated in Figure 6–7, is the canal through which sound enters the ear to strike the *tympanic membrane* (the *ear drum*) located at the end of the meatus. The internal auditory meatus, illustrated in Figures 6–8 and 6–9, is the canal through which the vestibulocochlear, or eighth cranial, nerve from the hearing organ and vestibular apparatus passes into the posterior cranial fossa to join the brain stem. The seventh cranial, or facial, nerve also leaves the cranial cavity via the internal auditory meatus. (Deep to the external auditory meatus, within the body of the *temporal bone,* are located, first, the **middle ear** and, second, the **internal ear.** The latter contains the **hearing organ** and the **vestibular apparatus,** both of which send appropriate nerve impulses into the brain.)

Optic foramen (or **optic canal**) (Fig. 6–9). An optic foramen passes on each side of the midline from the orbit through the sphenoid bone into the middle fossa of the cranial vault. It is through this canal that the optic nerve passes from the eye to the brain. Located immediately behind and between the two optic foramina is a small but deep depression called the **sella turcica,** also often called the **hypophyseal fossa.** The *pituitary gland* (or *hypophysis*) lies in this cavity.

Superior orbital fissure (Fig. 6–9). This communicates between the middle cranial fossa and the orbit and transmits the ophthalmic division of the fifth cranial nerve, as well as the third, fourth, and sixth cranial nerves, all of which innervate extrinsic muscles of the eyeball.

Foramen ovale (Fig. 6–9). One passes on each side through the floor of the middle cranial fossa 2 to 3 cm from the midline. The large mandibular division of the fifth cranial nerve (trigeminal nerve) passes through this foramen. This nerve innervates the jaw muscles to control chewing and also provides sensation for the entire lower jaw region of the facial surface and oral cavity.

Foramen rotundum (Fig. 6–9), slightly more lateral, transmits the maxillary division of the fifth cranial nerve, which innervates structures of the upper jaw and the nasal cavity.

Foramen lacerum (Fig. 6–9). This is a round opening in the midbase of each side of the cranium near the foramen ovale. Nothing of importance passes through this—it is closed in by cartilage.

Carotid canal (Fig. 6–9). This is an elongated opening lateral to the foramen lacerum through which the internal carotid artery enters the cranial cavity to supply approximately 80 percent of all the blood flow to each half of the brain. [It is immediately above this point that many *carotid artery aneurysms* (bulbous enlargements of the artery) occur. They can rupture at any time of life, causing hemorrhagic compression of the brain and usually resulting in death within a few hours if not repaired by emergency surgery.]

Perforations of the cribriform plate of the ethmoid bone (Fig. 6–9). There are about 20 small openings on each side of this plate. It is through these minute openings that the small branches of the olfactory nerves pass into the brain from the olfactory sensory organ (the organ of smell) in the upper part of the nasal cavity. (Occasionally, the membrane covering this ethmoid plate ruptures and allows cerebrospinal fluid to leak continually from the subarachnoid space around the brain into the nose, causing a persistent nasal drip, a drop of fluid forming every few seconds for days on end and rarely stopping until surgical repair is effected.)

The Facial Bones; The Orbit; and The Nasal Cavity

The facial bones are illustrated in Figures 6–7 through 6–11. Some of these are extensions of the same bones that form the anterior portion of the cranial vault. The facial bones are the

1. **Maxillae,** paired bones that form the major mass of the upper jaw and midportion of the anterior face.
2. **Zygomatic bones,** one on each side, that join the lateral side of the maxilla and send a process posteriorly to join the zygomatic process of the temporal bone, forming the zygomatic arch. Underneath this arch passes the tendon of the large temporalis muscle to the mandible to cause upward movement and retraction of the lower jaw, as we shall discuss in Chapter 8.
3. **Small paired nasal bones** that form the upper ridge of the anterior nose.

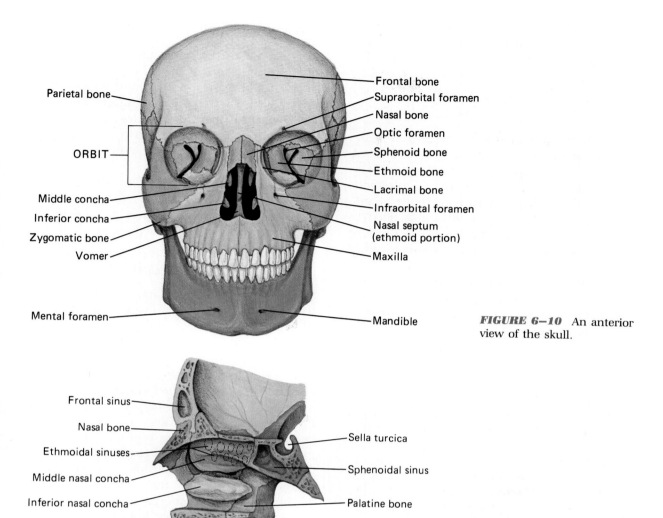

Parietal bone

ORBIT

Middle concha
Inferior concha
Zygomatic bone
Vomer

Mental foramen

Frontal bone
Supraorbital foramen
Nasal bone
Optic foramen
Sphenoid bone
Ethmoid bone
Lacrimal bone
Infraorbital foramen
Nasal septum
(ethmoid portion)
Maxilla

Mandible

FIGURE 6–10 An anterior view of the skull.

Frontal sinus
Nasal bone
Ethmoidal sinuses
Middle nasal concha
Inferior nasal concha
Maxilla

Sella turcica
Sphenoidal sinus
Palatine bone

FIGURE 6–11 An anterior view of the nasal cavity and some of its adjacent sinuses.

4. **Very irregular and complex ethmoid bone** that forms (a) the upper third of the nasal septum; (b) the upper boundary of the nasal cavity, which serves also as the cribriform plate (as discussed earlier); (c) the upper half of the lateral wall of the nasal cavity; and (d) a portion of the medial wall of the orbit.

5. **Sphenoid bone** that forms the posterior surface of the upper nasal cavity and also a portion of both the anterior and the middle cranial fossae, as discussed earlier. In addition, this bone forms a large portion of the posterior orbit.

6. **Paired lacrimal bones** on the sides of the nasal cavity, each of which also forms a small part of the medial wall of the orbit.

7. **Vomer,** a long thin bone that forms the lower third of the bony nasal septum, illustrated in Figures 6–8 and 6–10.

8. **Paired palatine bones,** one on each side, forming the posterior quarter of the bony palate and the posterior third of the lateral wall of the nasal cavity as well as a minute portion of the floor of the orbit.

9. **Paired inferior conchae,** each a thin plate-like bone protruding into the lower nasal cavity from the lateral nasal wall.

FORAMINA IN THE FACIAL BONES

Note on the anterior aspect of each side of the facial bones the following three important foramina:

1. **Supraorbital foramen** (or notch), through which the supraorbital nerve and blood vessels pass to provide sensation and blood supply to the forehead.

2. **Infraorbital foramen,** through which the infraorbital nerve and vessels pass to provide sensation and blood supply to the anterior face.

3. **Mental foramen,** through which a sensory division of the mandibular nerve passes to provide sensation over the anterior lower jaw and lip region.

SPECIAL FEATURES OF THE ORBIT

As illustrated in Figure 6–10 and also in Figure 6–7, the **orbit** (the eye socket) is a deep hemispherical cavity formed by a confluence of several of the different facial bones: (1) *zygomatic bone*, (2) *frontal bone*, (3) *sphenoid bone*, (4) *ethmoid bone*, (5) *lacrimal bone*, (6) *maxilla*, and (7) a small portion of the *palatine bone*. The locations of these bones in the orbit are summarized in Table 6–4.

Leading backward into the cranial vault from the posteromedial aspect of each orbit is the *optic foramen* (or *optic canal*), through which the optic nerve and ophthalmic artery pass.

Lateral to the optic foramen is a slitlike opening, the *superior orbital fissure*, that leads backward into the middle cranial fossa. It is through this slit that the third, fourth, and sixth nerves pass to innervate the extraocular muscles that control the eye movements and the ophthalmic veins pass to join venous channels within the cranial cavity.

The obvious function of the orbit is to provide housing and support for the eye and its muscles. Also, especially important, its deep recess provides protection for the eye.

SPECIAL NASAL STRUCTURES

The bony structures of the **nasal cavity** are illustrated in Figures 6–10 and 6–11, and the locations of the separate nasal bones are summarized in Table 6–4.

The structures that are most evident in the nasal cavity itself are the **nasal septum** and the **conchae.** The septum is formed by the *vomer bone* inferiorly and by a *perpendicular plate of the ethmoid bone* superiorly. In between these bones and anteriorly, the septum is formed of *cartilage*, called the **septal cartilage.**

Three *conchae* protrude into each side of the nasal cavity from the lateral wall. The *superior concha* and *middle concha* are both processes of the ethmoid bone, but the *inferior concha* is a separate bone. The nasal septum and the six conchae all together provide an extensive surface area inside the nasal cavity that is covered with *mucous membrane* and exposed to the air that one breathes. These surface areas humidify and warm the air before it enters the trachea and lungs. Also, the protruding conchae cause the air to become turbulent as it is breathed in, for which reason the conchae are also called *turbinate bones.* This air turbulence causes dust or bacterial particles in the air to be impelled against the nasal surfaces and trapped in the nasal mucus; as a result almost no particles larger than 2 to 3 μ will pass through the nasal cavity into the lower respiratory tract.

The **olfactory epithelium,** which is the nasal organ of smell, lies on the upper surface of each superior concha and upper 1 cm of each side of the septum. An odor can be made to pass more effectively up into these upper reaches of the nose if a person sniffs the air to cause increased air turbulence in the nose.

The Paranasal Air Sinuses. A number of cavities within the facial bones are filled with air.

TABLE 6–4
Bones of the Orbit and Nasal Cavity

Orbit	
Bones	*Location in orbit*
Frontal	Superior
Zygomatic	Lateral
Maxilla	Inferior
Lacrimal	Medial, anterior
Ethmoid	Medial, posterior
Sphenoid	Posterior
Palatine	Inferior, medial (very small area)
Nasal cavity	
Bones	*Location*
Maxilla (two)	Inferior portion of lateral, anterior wall
	Anterior three quarters of inferior wall
Palatine	Posterior quarter of inferior and lateral walls
Inferior concha (two)	Extends from lateral wall into cavity
Ethmoid	Superior portion of posterior lateral wall
	Superior and middle conchae extending from lateral wall
	Posterior, superior plate (cribriform plate)
	Superior two thirds of bony septum
Vomer	Inferior third of bony septum
Sphenoid	Posterior wall
Frontal	Small portion of superior wall
Nasal (two)	Anterior portion of superior wall
(Septal cartilage)	Anterior septum

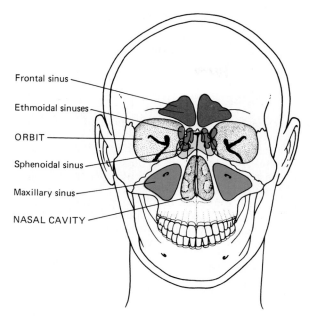

FIGURE 6–12 Positions of the major sinuses in the facial bones.

These, illustrated in Figures 6–11 and 6–12, called the **paranasal sinuses** are

1. **two maxillary sinuses,**
2. **a midline sphenoidal sinus** (usually septated),
3. **multiple small ethmoidal sinuses,** and
4. **two or more frontal sinuses.**

All of these sinuses have small openings through which they communicate with the nasal cavity; it is through these passages that air is supplied to the sinuses. When one or more of the openings becomes occluded, the person usually develops severe headache because absorption of air from the sinus causes a partial vacuum in the cavity. This occurs typically when infection due to a cold occurs in the sinuses.

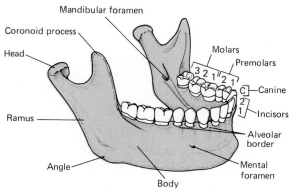

FIGURE 6–13 The mandible and lower teeth.

The Mandible (The Lower Jaw Bone)

Figure 6–13 illustrates the **mandible,** which is the bony portion of the lower jaw. Its special features are the

Body, which contains the lower teeth.
Ramus, which passes superiorly from the body to the head.
Mandibular head (also called the **mandibular condyle**), which provides the articular surface of the temporomandibular joint.
Coronoid process, a large projection extending upward from the ramus anterior to the head, leaving a *notch* between. The temporalis muscle attaches to this process and serves to elevate and retract the lower jaw.
Mandibular foramen, which is a large opening on the inner surface of each side of the posterior portion of the mandible. It is into this foramen that the alveolar nerve passes to supply sensation to the tooth sockets of the lower jaw. The dentist injects a local anesthetic into this nerve where it enters the mandibular foramen to anesthetize the respective half of the lower mandibular region, including the teeth and the tooth sockets.
Alveolar border, which is the superior ridge of the mandible in which the sockets for the teeth are located.

THE TEETH

Note also in Figure 6–13 the relationship of the **teeth** to the mandible. The adult person has 28 to 32 teeth, 32 if all the third molars (wisdom teeth) develop. The names and standardized numbering of the teeth are also shown in the figure: *incisors 1 and 2; canine tooth; premolars 1 and 2;* and *molars 1, 2, and 3.*

A corresponding alveolar border and corresponding sockets and teeth are also present on the bottom of the maxilla. The upper and lower teeth interdigitate with each other so that the projections of the lower teeth fit into the indentations of the upper teeth. This fitting together of the teeth is called *occlusion,* which obviously allows the teeth to perform their cutting and grinding functions.

THE HYOID BONE

The **hyoid bone,** illustrated in Figure 6–14, is the one bone of the body that does not articulate directly with any other bone. Instead, it is sus-

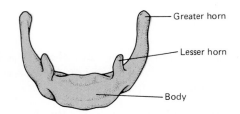

FIGURE 6–14   The hyoid bone.

pended in the midanterior neck between the mandible and larynx by (1) several neck muscles, (2) a ligamentous membrane extending between the hyoid bone and the larynx, and (3) ligaments between the hyoid bone and the two **styloid processes** of the skull, which are two spinelike projections extending downward 2 to 3 cm from the bases of the temporal bones. The function of the hyoid bone is to help lift and direct the movements of the larynx during swallowing and during some types of vocalization.

As illustrated in the figure, the parts of the hyoid bone are the:

1. **Body.**
2. Paired **greater horns,** each of which is 3 to 4 cm long; these horns pass posteriorly from the sides of the body forming a U-shaped arc.
3. Paired **lesser horns,** each about 1 cm long,

projecting superiorly from the junctions between the body and the anterior horns; it is to these that the stylohyoid ligaments attach.

QUESTIONS

1. Describe ball and socket joints and hinge joints.
2. Why is the upright stance in man of value?
3. Describe the structure of bone, and relate this to the different stresses in bone.
4. Discuss osteogenesis and remodeling of bone.
5. How do bones grow?
6. Explain the repair of a bone after a fracture.
7. Discuss the functional importance of the different elements of the joint, especially the articular cartilages, the synovial membrane, and the synovial fluid.
8. Name and describe the separate bones that constitute the cranium.
9. Name and describe the openings of the cranium and their purposes.
10. Name and describe the facial bones.
11. Name and describe the bones of the orbit.
12. Describe the bony structures of the nasal cavity and nasal sinuses.
13. Describe the mandible and the teeth.
14. Describe the hyoid bone.

REFERENCES

See References, Chapter 7.
See also General Anatomical References, Chapter 1.

The Skeletal System
and Its Joints:
II. Vertebral Column;
Thoracic Cage;
and Extremities

Overview

The **vertebral column,** which serves as the axial support of the trunk and head, is comprised of 33 **vertebrae,** 7 in the neck called the *cervical vertebrae,* 12 in the thoracic region called the *thoracic vertebrae,* 5 in the lower back called the *lumbar vertebrae,* and 9 in the pelvic region, 5 of which are fused together to form the **sacrum** and the other 4 fused to form the **coccyx.** This column has two principal functions: (1) to support the weight of the body and (2) to provide a protective canal, the **vertebral canal,** through which the spinal cord passes. The joints between adjacent vertebrae allow slight degrees of forward and backward angulation, called respectively *flexion* and *extension,* as well as small degrees of *rotation,* thus permitting gentle bending of the vertebral column in multiple directions but without any acute bend that could endanger the spinal cord.

Especially interesting is the joint structure between the superior end of the vertebral column and the skull. The first vertebra, which is called the **atlas,** has wide-spreading articular facets that support corresponding facets of the cranium and allows bending of the skull forward or to one side. The joint between the first vertebra and the second vertebra allows rotation of the head. An upward, round projection from the second vertebra, the **dens,** protrudes into a fibrous and bony ring of the first vertebra so that the first vertebra (the atlas) can rotate around this projection. Therefore, the second vertebra is called the **axis,** around which the rotation occurs.

The **thoracic cage** is comprised of 12 **ribs** that articulate posteriorly with the vertebral column, and 10 of them connect anteriorly with the sternum either directly or through the **costal arch.** This cage not only protects the heart and lungs from injury but also provides support for a semivacuum chamber called the **chest cavity.** It is the semivacuum in this cavity that keeps the lungs expanded. Also, when the chest cavity is enlarged by the muscles of respiration, a further increase in the degree of vacuum in this cavity causes inspiration.

The **pectoral girdle** attaches the arm to the trunk. It is comprised of the **scapula** and the **clavicle.** The clavicle is a weak bone that attaches the scapula to the sternum; this bone acts more as a guide for motion of the scapula than as a support structure. The scapula is mainly held in place by many strong muscles that can slide it over the posterolateral rib cage in any direction.

The joint between the scapula and the humerus, the **shoulder joint,** is a *ball and socket type* that allows movement of the **humerus** in almost any direction. By contrast, the **elbow joint** is a *hinge type* that allows the elbow to move only forward or backward. Still another interesting joint structure is that of the joints between the two forearm bones, the **radius** and the **ulna.** Rotational joints are provided in such a way that the distal end of the radius can rotate around the ulna, allowing *pronation* or *supination* (medial or lateral rotation) of the wrist and hand.

There are eight wrist bones, called **carpal bones.** These function as a multistage joint that allows either *flexion* or *extension* of the wrist as well as *abduction* or *adduction.* Extending distally from the wrist bones are the **metacarpals** that constitute the frame for the hand and the **phalanges** that support the fingers.

The **pelvic girdle** is the bony connecting link between the lower limbs and the vertebral column. It is comprised of the two massive **coxal bones** that also serve as the lateral and anterior borders of the bony pelvis.

The **hip joint,** where the upper end of the thigh bone, the **femur,** articulates with the coxal bone, is a *ball and socket joint* similar to that of the shoulder joint. It, like the shoulder joint, allows motion in multiple directions.

The **knee joint** has special characteristics that help it to withstand the severe stresses of knee motions. The articular surfaces of the knee joint are partially separated from each other by knee cartilages, called **menisci,** that help to cushion the heavy shocks that might otherwise damage the knee joint. Also, extremely strong ligaments on the sides of the knees, the **collateral ligaments,** prevent lateral or medial angulation of the knee; special ligaments within the very middle of the knee joint, the **cruciate ligaments,** help to hold the **tibia** tightly against the femur and also to direct the extension and flexion movements of the knee joint.

The ankle bones, the **tarsal bones,** are analogs of the carpal bones of the wrist, and the foot bones, the **metatarsals** and the **phalanges,** correspond to similar bones in the hand. However, these bones, especially the tarsals, are shaped much differently to serve the special movements and stresses of the ankle and foot.

THE VERTEBRAL COLUMN AND INTERVERTEBRAL JOINTS

The **vertebral column,** also called the **spinal column** (or simply the "spine"), is illustrated in Figure 7–1. It serves as the axial support of the trunk and head and also serves another very important function: to protect the spinal cord from injury and to provide a distribution conduit for two-way nerve traffic between the brain and the body.

Note in the figure that the vertebral column is comprised of 33 **vertebrae,** 24 of which are movable, 5 of which are fused to form the **sac-** rum, and 4 of which are often fused to form the **coccyx.** The movable vertebrae in turn are divided into three separate groups: 7 *cervical vertebrae,* 12 *thoracic vertebrae,* and 5 *lumbar vertebrae.* The thoracic vertebrae are in the chest region, and it is to these that the **ribs** attach. The lumbar vertebrae are the principal support structure of the lower back.

Though the **sacrum** in fetal life consists of five separate vertebrae, during development of the fetus these fuse to become a single large and very strong bone. It is to this bone that the pelvic girdle (to be described later) attaches for support of the body.

TABLE 7–1
The Vertebral Column

Numbers of vertebrae in different parts

Cervical   7
Thoracic   12
Lumbar   5
Pelvic   5 sacral (fused as sacrum)
         4 coccygeae (usually fused as coccyx)

Parts of vertebrae

Body, with superior and inferior articular surfaces
Vertebral processes
   1. spinous process
   2. transverse processes
Articular facets
   superior pair
   inferior pair

Special vertebral characteristics

Heavy, strong lumbar vertebrae
Light, less strong, more movable cervical vertebrae
Transverse foramina in cervical transverse processes

Ligaments

Anterior longitudinal: anterior to bodies
Posterior longitudinal: posterior to bodies
Supraspinous: posterior tips and adjacent surfaces of spinous processes
Ligamenta flava: between adjacent posterior arches
Interspinous and intertransverse: between successive spinous and transverse processes

Openings

Vertebral foramina: form vertebral canal
Intervertebral foramina: lateral openings for spinal nerves

Joints

Between adjacent bodies:	fibrocartilaginous joint type
	fibrocartilaginous disc has pulpy center, nucleus pulposus, that frequently ruptures
Between superior and inferior articular facets:	gliding joint type
	allows flexion and extension
Between superior facets of 1st cervical vertebra (the atlas) and facets of occipital bone:	gliding joint type allows flexion, extension, and lateral bending
Between 2nd cervical vertebra (the axis) and 1st cervical vertebra (the atlas):	dens of axis projects into bony fibrous ring of atlas atlas rotates around dens to provide rotation of head

The **coccyx** is four separate vertebrae in fetal life, but these often become fused in late fetal and early infant life to form a single bone. The coccyx is actually a segmented remnant of a tail inherited from our forefathers and has little remaining function in the human being.

Structure of the Individual Vertebrae. Figure 7–2 illustrates a single thoracic vertebra with attachments of a rib on each side. The principal parts of the vertebra are the

1. **body,** which is the massive anterior portion of the vertebra that supports the weight of the body,

2. **arch,** which circles backward from the body to form a large opening, the **vertebral foramen,** through which the spinal cord passes, and

3. three processes of the arch—two **transverse processes** that protrude to the two sides and one **spinous process** that protrudes backward.

Note in Figure 7–1 that the bodies of the lumbar vertebrae in the lower portion of the vertebral column are much larger and stronger than those of the cervical vertebrae near the head. The obvious functional value of this is that

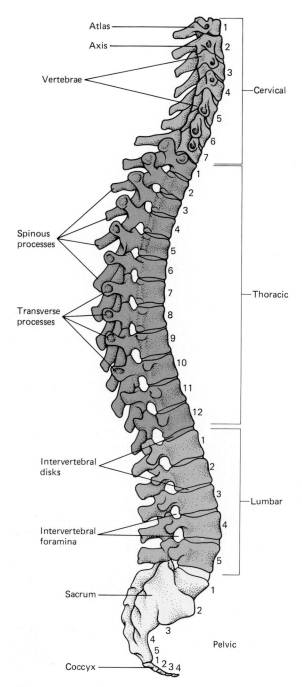

Atlas

Axis

Vertebrae

Cervical

Spinous
processes

Transverse
processes

Thoracic

Intervertebral
disks

Lumbar

Intervertebral
foramina

Sacrum

Pelvic

Coccyx

FIGURE 7-1 The vertebral column.

gether, bending the vertebral column backward, which is called *extension* of the column. Other muscles attach between the laterally extending transverse processes; when these contract, they obviously pull the transverse processes together and cause the vertebral column to bend to that side. Still other muscles pass from the spinous process of one vertebra to a transverse process of the vertebra immediately beneath or above; contraction of these muscles can cause slight rotation of each vertebra on the adjacent one, thus providing the rotary movements of the trunk.

In the transverse processes of the cervical vertebrae is a vertical opening, the **transverse foramen,** through which the vertebral arteries and veins pass to and from the brain, as will be described more fully in Chapter 25.

Observe also in Figure 7–1 four curvatures of the vertebral column: (1) **anterior convexity in the neck,** (2) **posterior convexity in the thoracic region,** (3) **anterior convexity in the lumbar region,** and (4) **posterior convexity in the pelvic region.** The posterior convexities of the thoracic and pelvic regions are said to be primary curvatures of the spine because these are inherited from lower mammals that walk on all four limbs. The anterior convexities of the neck and lumbar regions are called *secondary curvatures* because they developed as the human being assumed the upright stance. An important function of the curvatures is to provide a small degree of extra resilience in the spinal column to soften shock forces that occur when one runs or jumps.

Finally, observe in Figure 7–1 the large lateral openings between the arches of adjacent vertebrae. These openings are called **intervertebral foramina;** it is through these that the spinal nerves leave the vertebral canal to be distributed throughout the body.

THE INTERVERTEBRAL ARTICULATIONS

The successive bodies of the vertebrae lie one on top of the other, and in each space between these bodies is a pad of *fibrocartilage* called the **intervertebral disc.** But note that the vertebrae articulate with each other posteriorly as well. Protruding from the arch on each side of each vertebra is a **superior** and an **inferior articular process,** the surfaces of which, the **articular facets,** form joints with similar processes of the next higher and lower vertebrae. These are *gliding joints,* in which the surfaces slide over each other as the spinal column rotates or bends. Figure 7–2 illustrates three of the articular surfaces

the vertebrae support progressively less weight the higher up the column one proceeds.

Figure 7–1 also shows that the shapes and sizes of both the spinous processes and the transverse processes, as well as their degrees of angulation from the vertebral arches, differ in the separate portions of the vertebral column. These are adapted at each level of the column to function as levers. Many small back muscles attach between the spinous processes, and their contraction pulls the spinous processes to-

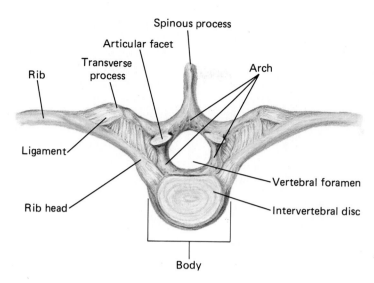

Spinous process

Articular facet

Transverse process

Rib

Arch

Ligament

Rib head

Vertebral foramen

Intervertebral disc

Body

FIGURE 7–2 A typical thoracic vertebra and its attachments to the ends of two ribs.

of a typical thoracic vertebra, showing the intervertebral disc on the body of the vertebra and two superior articular facets protruding from the two sides of the arch.

Articulations with Ribs. Figure 7–2 also illustrates the attachment of two ribs to a thoracic vertebra. The **head** of each rib articulates with the bodies of two adjacent vertebrae, and a small process called the **tubercle** on the side of the rib 3 cm beyond the head articulates with the respective transverse process. Strong ligaments attach each rib to both the vertebral body and the transverse process, forming capsules for the joints and allowing slight upward and downward rotation of the ribs, which is an important motion in respiration.

The Ligaments of the Vertebral Column. The vertebral column is held tightly together by ligaments, as illustrated in Figure 7–3. The principal ones are as follows:

1. The **anterior longitudinal ligament,** a powerful ligament running the entire extent of the anterior surfaces of the vertebral bodies. This ligament prevents overextension (backward bending) of the vertebral column.
2. The **posterior longitudinal ligament,** which runs inside the vertebral canal and attaches to the posterior surfaces of the successive vertebral bodies. This ligament lines the anterior wall of the vertebral canal and helps to prevent excessive flexion (forward bending) of the vertebral column.
3. The **supraspinous ligament,** which runs along the posterior tips and posterolateral surfaces of the spinous processes. This ligament is the most important of all to prevent excessive flexion.

4. The **ligamenta flava,** which extend between the successive posterior arches from one vertebra to the next. These short ligaments form most of the posterior wall of the vertebral canal, and they also help to prevent excessive flexion.
5. Extensive short **interspinous ligaments** between the spinous processes and **intertransverse ligaments** between the transverse processes. These also become taut in flexion.

Note also in Figure 7–3 the additional fibrous coverings of the spinal cord within the vertebral canal; these all together provide additional protection for the spinal cord. These coverings will be discussed more extensively in Chapter 13. Also, note again the intervertebral foramina through which the spinal nerves pass out of the vertebral canal.

The Ruptured Disc. The intervertebral disc is composed of dense fibrocartilage around a pulpy core called the **nucleus pulposus.** This disc serves as a shock pad between the successive vertebrae. Ordinarily, it is held in place by the anterior and posterior longitudinal ligaments of the vertebral bodies. When a person bends forward and lifts a heavy weight at the same time, however, the anterior portions of the vertebrae are compressed tightly together, and a tremendous stress develops to squeeze the intervertebral disc backward against the posterior longitudinal ligament. If the force is great enough, the disc or its pulpy core (the nucleus pulposus) can rupture backward, protruding into the vertebral canal, usually to one side of the ligament and impinging on the nerve leaving through the adjacent intervertebral foramen.

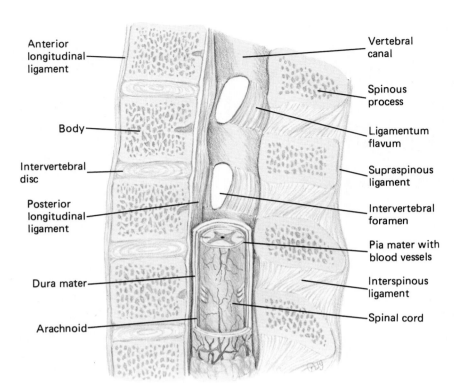

Anterior longitudinal ligament

Body

Intervertebral disc

Posterior longitudinal ligament

Dura mater

Arachnoid

Vertebral canal

Spinous process

Ligamentum flavum

Supraspinous ligament

Intervertebral foramen

Pia mater with blood vessels

Interspinous ligament

Spinal cord

FIGURE 7–3 The intervertebral discs and the principal ligaments of the vertebral column, showing also the vertebral canal and spinal cord in the canal.

This compression of nervous tissue causes severe pain and often loss of nerve transmission. Ruptured discs occur most frequently in the lumbar region, and this may lead to loss of sensation or local paralysis of some muscles in one of the lower extremities.

In old age the intervertebral discs become progressively thinner because of loss of part of their fibrocartilaginous mass and because of dehydration. This is a cause of the decreasing height of older people.

Articulation of the Vertebral Column with the Skull—The "Atlas" and the "Axis." The first two cervical vertebrae of the spinal column, vertebrae C-1 and C-2, are specially adapted for support of the skull and for a wide range of movement of the skull, as illustrated in Figure 7–4. The first cervical vertebra is called the **atlas** because it holds up the skull, and the second vertebra is called the **axis** because a special joint between C-1 and C-2 allows the atlas to rotate around the axis.

In Figure 7–4, study first the principal features of the axis. Protruding upward from the body of the axis is a round projection called the **dens.** It is around the dens that the first vertebra, the atlas, rotates. Now observe the atlas. It has essentially no body (actually the dens is the captured body of the atlas). Instead of a body, there is an open space into which the dens of the axis protrudes enclosed by an **anterior bony arch** and a **posterior fibrous arch** pro-

vided by a very strong transverse ligament passing behind the dens. This arching encirclement allows the atlas to rotate around the dens.

In Figure 7–4, the skull is shown lifted upward from the atlas to show its paired articular facets as well. Now, visualize the articular facets of the skull and those of the atlas approximated to each other, and one can then see that the atlas holds the skull on its "shoulders," allowing forward or backward or lateral bending of the head while the atlas in turn rotates around the dens of the axis.

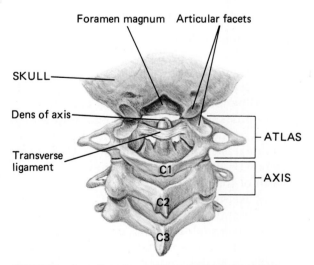

Foramen magnum   Articular facets

SKULL

Dens of axis

Transverse ligament

ATLAS

AXIS

C1

C2

C3

FIGURE 7–4 Support of the skull by the vertebral column, showing also the manner in which the *atlas* rotates on the *axis.*

TABLE 7–2
The Thoracic Cage

Bones	Special features
Ribs (12 in number)	Two joints allow rotation, one with vertebral bodies, one with transverse process
	Ten ribs attach via costal cartilages to sternum—6 directly, 4 by way of costal arch
	Ribs 11 and 12 have no anterior attachments—called floating ribs
Sternum, consisting of	
1. Manubrium	Articulates with clavicle and first two ribs
	Fibrous joint with body of sternum
2. Body	Articulates with ribs 2 through 10 via costal cartilages and costal arch
3. Xiphoid bone	Fibrous joint with body, little function

THE THORACIC CAGE

The **thoracic cage,** also called the **chest cage** or **rib cage,** is illustrated in Figure 7–5. It serves two functions: first, it protects the internal organs of the thorax, including the heart and lungs, and, second, it plays an integral role in respiration. Respiration occurs primarily by suction of air into the lungs. For such action to occur, it is necessary that the lungs be encased in a rigid but expansible structure. In an occasional person who has lost most of his ribs, each time that the diaphragm contracts, which ordinarily would increase the volume of the pleural cavity to cause suction of air into the lungs, the chest cage simply collapses inwardly rather than creating suction, and the person struggles for breath continuously.

Most of the features of the thoracic cage are familiar to the average student, but some explanation is needed.

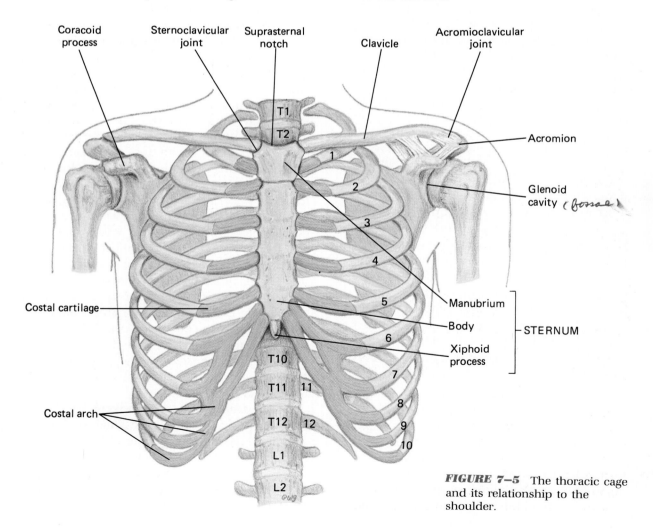

FIGURE 7–5 The thoracic cage and its relationship to the shoulder.

First, we have already discussed the manner in which the ribs are attached to the vertebrae posteriorly. The ribs then curve around the chest wall to attach anteriorly to the sternum. The **sternum** in turn is comprised of

1. the **manubrium,** to which the first rib, part of the second rib, and the medial end of the clavicle all attach;
2. the **body,** to which the second through tenth ribs all attach; and
3. the **xiphoid process,** a small bony extension of the lower end of the body that has little function.

Second, the anterior ends of ribs 1 through 10 all attach to the sternum by way of **costal cartilages,** each of which is 10 or more centimeters long, and bridge the gap between the bony portions of the ribs and the sternum. These cartilages have a moderate degree of flexibility in contrast to the rigidity of the bony portions of the ribs. This flexibility allows upward and downward movement of the sternum without causing significant stresses on the ribs. Note also that the costal cartilages of ribs 7 through 10 form a succession of joined cartilaginous bridges called the **costal arch.** These provide a resilient lower anterior margin to the chest cage.

Third, ribs 11 and 12 do not have any anterior attachments but merely serve as a rigid support for the lowermost portion of the posterior thoracic wall. These ribs are called "floating ribs" because of lack of anterior attachments.

Fourth, the ribs extend downward as well as around the thoracic wall, but they can be rotated upward, which is a motion that is very important for respiration. That is, when the muscles of the chest and neck pull the rib cage upward, this also extends the downward slanting anterior ends of the ribs in the forward direction, thus increasing the anteroposterior diameter of the chest and in doing so causing considerable expansion of the lungs. Indeed, during very heavy exercise, almost one half of the air intake is achieved by increasing this anteroposterior diameter, which acts as a supplement to the action of the downward displacement of the diaphragm.

THE UPPER EXTREMITY

The Pectoral Girdle, Shoulder Joint, and Arm

To the left in Figure 7–5 and in both views of Figure 7–6 are shown the **pectoral girdle** (also called the **shoulder girdle**) and the **shoulder joint.** The pectoral girdle consists of four bones, the paired *scapulae* and *clavicles,* and the shoulder joint is the articulation between the scapula and humerus of the arm.

The Scapula.   The scapula with its intact muscles provides the principal support for the arm. The head of the humerus articulates at the **glenoid fossa** of the scapula to form the shoulder joint. The scapula is triangular in shape, its posterior surface separated into **supraspinous** and **infraspinous fossae** by a posteriorly protruding ridge called the **spine.** The spine projects laterally and superiorly over the shoulder joint to form the **acromion.** The acromion gives attachment to some of the major muscles and ligaments that are closely related to the shoulder joint, and it also articulates anteriorly with the clavicle.

The anterior surface of the scapula is concave to fit the posterior surface of the thorax, so that the scapula can glide smoothly over this surface. Also, a **coracoid process** projects anteriorly and laterally from the scapula, extending over the anterior lip of the glenoid cavity. This also attaches to the clavicle by strong ligaments. Some prominent muscles of the pectoral region and arm attach to the coracoid process, notably the pectoralis minor muscle and the short head of the biceps brachii muscle.

The Clavicle.   The clavicle serves as a strut to hold the scapula in its lateral position. It articulates medially with the manubrium at the **sternoclavicular joint,** which is the only articulation of the upper extremity with the axial skeleton, and laterally with the acromion of the scapula at the **acromioclavicular joint.** But, strangely enough, the shoulder joint can still function even after the clavicle has been separated from the scapula or when a segment of the clavicle has degenerated. The reason for this is that the scapula is still held almost in its normal position by a girdle of powerful muscles that extend in all directions from the scapula to the back, the neck, and the anterior thorax. However, the clavicle is important as an attachment of a large portion of the pectoral muscles (muscles of the anterior chest wall), and it transmits the forces of these muscles to the scapula. It also serves as a guide to help control and limit the motions of the scapula.

The Humerus.   The **humerus** (Fig. 7–6) is a typical long bone with articular surfaces at each end. The articular surface at the upper end is the **head,** and at the elbow joint its lateral articular surface is the **capitulum,** which articulates with the *radius,* and its medial anticular surface the **trochlea,** which articulates with the *ulna.*

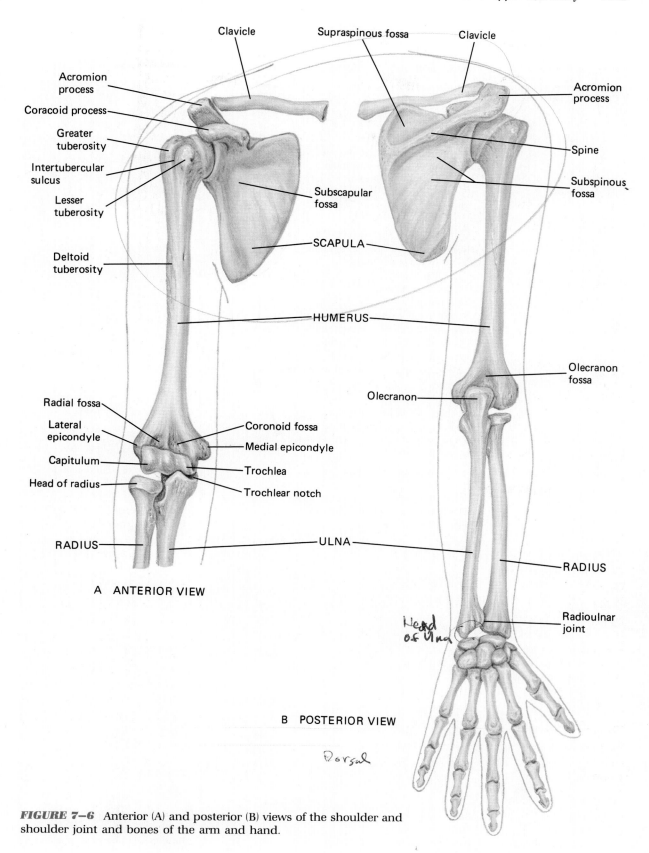

Clavicle

Supraspinous fossa

Clavicle

Acromion process

Coracoid process

Greater tuberosity

Intertubercular sulcus

Lesser tuberosity

Deltoid tuberosity

Acromion process

Spine

Subscapular fossa

Subspinous fossa

SCAPULA

HUMERUS

Radial fossa

Lateral epicondyle

Capitulum

Head of radius

Coronoid fossa

Medial epicondyle

Trochlea

Trochlear notch

Olecranon fossa

Olecranon

RADIUS

ULNA

RADIUS

Radioulnar joint

A ANTERIOR VIEW

B POSTERIOR VIEW

Head of Ulna

Dorsal

FIGURE 7–6 Anterior (A) and posterior (B) views of the shoulder and shoulder joint and bones of the arm and hand.

Note several important protuberances from the shaft of the humerus to which some of the major muscles of the shoulder and elbow joint attach: (1) the **greater tuberosity** at the upper-most point of the humerus lateral to the head, (2) the **lesser tuberosity** located anteriorly and slightly inferior to the head, (3) the **deltoid tuberosity,** on the lateral aspect of the midshaft,

TABLE 7–3
The Upper Extremity

Part and bones	Special features	Articulates with
Pectoral girdle		
Clavicle	Thin bone acts as guide strut	Manubrium
		Acromion of scapula
Scapula	Spine on posterior surface	
	Acromion at lateral end of spine	Clavicle
	Infraspinous fossa below spine	
	Coracoid process, anterior, lateral projection	
	Glenoid fossa, lateral articular cavity for shoulder joint	Head of humerus
Arm		
Humerus	Head, articular surface for shoulder joint	Scapula at glenoid fossa
	Elbow articular surfaces:	
	Trochlea (medial)	Trochlear notch of ulna
	Capitulum (lateral)	Head of radius
	Projections:	
	Greater tuberosity, lateral to head	
	Lesser tuberosity, anterior, inferior to head	
	Deltoid tuberosity, lateral midshaft	
	Lateral epicondyle, lateral to capitulum	
	Medial epicondyle, medial to trochlea	
Forearm		
Ulna	Trochlea notch, proximal end	Trochlea of humerus
	Olecranon, attachment of triceps	
	Radial notch, proximal lateral surface	Head of radius
	Head, distal end	Ulna notch of radius
Radius	Head, round proximal end	Capitulum of humerus and radial notch of ulna
	Radial tuberosity, attachment of biceps	
	Ulna notch, distal medial surface	Head of ulna
	Distal end	Proximal carpals
Wrist		
Carpal bones		
Proximal row:		Distal end of radius
Scaphoid,		
lunate,		
triquetral,		
pisiform		
Distal row:		Bases of metacarpals
Trapezium,		
trapezoid,		
capitate,		
hamate		
Hand		
Metacarpals (five)	Base	Distal carpals
	Shaft	
	Head	Phalanges
Fingers		
Phalanges	Base	Heads of metacarpals
2 for thumb	Shaft	
3 for each finger	Head (except for distal phalanx)	Base to head between phalanges

(4) the **lateral epicondyle** located laterally above the elbow joint, and (5) the **medial epicondyle** located medially above the elbow joint.

Note also the **intertubercular sulcus**, a groove between the greater and the lesser tuberosities adjacent to the head of the humerus. It is

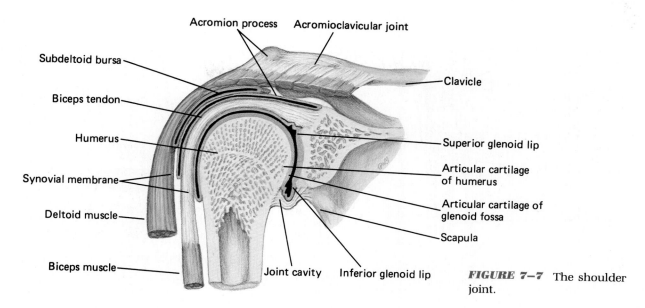

Acromion process Acromioclavicular joint

Subdeltoid bursa

Biceps tendon

Humerus

Synovial membrane

Deltoid muscle

Biceps muscle

Joint cavity Inferior glenoid lip

Clavicle

Superior glenoid lip

Articular cartilage of humerus

Articular cartilage of glenoid fossa

Scapula

FIGURE 7–7 The shoulder joint.

through this sulcus that the tendon of the long head of the biceps brachii muscle passes from its attachment at the superior lip of the glenoid cavity. This tendon and its location in the sulcus play an important role in preventing shoulder joint separation.

THE SHOULDER JOINT

The shoulder joint is illustrated in Figure 7–7. Note especially that the articular surface of the scapula, the **glenoid fossa,** has a very small surface area even though its depth is somewhat increased by the *glenoid lips.* Therefore, it is exceedingly surprising that the shoulder joint is as stable as it is. This is achieved not by the fact that the glenoid fossa and the head of the humerus by themselves form a stable union. Instead, it is partly the strong ligaments and especially the envelopment by strong muscles and tendons around the joint that provides its stability. As illustrated in the figure, the tendon of the long head of the biceps muscle passes within the joint capsule and serves as a superior stabilizer of the shoulder joint. Similarly, other muscles and tendons serve the same function both anteriorly and posteriorly, as will be discussed in detail in Chapter 9.

The Synovial Membrane of the Shoulder Joint, and the Subdeltoid Bursa. Figure 7–7 also shows the extent of the **synovial membrane** of the shoulder joint, illustrating that it surrounds not only the joint itself but also the biceps tendon. In the **synovial space** encased by this synovial membrane is the usual *synovial fluid,* a slick mucoid fluid that lubricates the articular surfaces involved in movement of the

shoulder joints as well as the sliding movement of the biceps tendon over the head of the humerus.

Shown, too, in Figure 7–7 is the **subdeltoid bursa,** which is another synovial space that lies between the deltoid muscle and the shoulder joint. This allows smooth movement of the deltoid muscle over the outer surface of the joint. In addition, there are several other bursae around the shoulder joint, which may or may not communicate with the synovial cavity of the joint. Though most bursae are present from the time of birth, whenever excessive motion occurs between adjacent tissues in any part of the body, a "false bursa" can develop within the connective tissue, and this false bursa then serves the same function as a true bursa. *Bursitis* develops whenever the synovial surfaces of a bursa become inflamed or infected. The inflamed surfaces are exceedingly painful and are made even more so during movement because of failure of the bursa to provide appropriate lubrication under these conditions.

THE FOREARM, AND THE ELBOW JOINT

The forearm bones are the **ulna** and the **radius,** illustrated in Figures 7–6, 7–8, and 7–9. The upper ends of the radius and ulna have articular surfaces for the elbow joint, and the lower ends provide articular surfaces for the wrist joint. Also, the radius, which articulates with the ulna at each end by proximal and distal **radioulnar joints,** can rotate either laterally or medially, so that these two bones can either lie parallel to each other or cross over each other, thus producing *supination* or *pronation* of the hand.

A ANTERIOR VIEW

FIGURE 7–8 The bones of the forearm and hand (A) and of the elbow joint (B).

Figure 7–6 shows that the **head of the radius** articulates with the capitulum whereas the **trochlear notch of the ulna** articulates with the trochlea of the humerus. Figure 7–8 shows especially that it is the articulation of the trochlear notch with the trochlea that provides the principal stability of the elbow joint.

Extension of the forearm is limited by the **olecranon,** which as shown in Figure 7–8B lies behind the posterior lip of the trochlear notch. The olecranon locks in the **olecranon fossa**

(Fig. 7–6) on the back of the humerus when the arm becomes fully extended. It is also to the posterior surface of the olecranon that the tendon of the strong triceps brachii muscle attaches; this muscle extends the forearm by pulling the olecranon in the proximal direction.

In addition to the articulation of the head of the radius with the capitulum on the humerus, this head also has another articular surface called the **annular surface,** a surface that is completely round. This articulates with the **ra-**

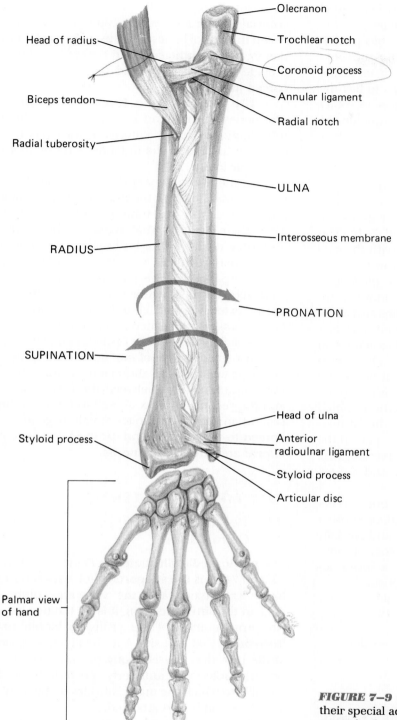

Olecranon

Trochlear notch

Head of radius

Coronoid process

Annular ligament

Biceps tendon

Radial notch

Radial tuberosity

ULNA

Interosseous membrane

RADIUS

PRONATION

SUPINATION

Head of ulna

Anterior radioulnar ligament

Styloid process

Styloid process

Articular disc

Palmar view of hand

FIGURE 7–9 The ligaments of the forearm, showing their special adaptation to allow rotation of the hand, *pronation* in one direction and *supination* in the other direction as illustrated.

dial notch of the ulna, thus allowing rotation of the radius at the same time that the proximal surface of the head articulates with the capitulum. The upper portion of Figure 7–9 shows an **annular ligament** attached to the ulna and surrounding the head of the radius. It is within the encirclement of this ligament that the head of the radius rotates.

At the lower end of the forearm the **head of the ulna** is similar to the head of the radius but located at the opposite end, and this articulates with the **ulnar notch of the radius.** Owing to the C-shape of the ulnar notch, the distal end of the radius rotates in an arc of 180 degrees around the distal end of the ulna. The ulna does not move in this rotation. It is in this way that *pronation* (medial rotation) or *supination* (lateral rotation) of the wrist and hand occurs.

Anterior to the elbow joint, the biceps tendon attaches to the radius at the **radial tuberosity.** This point is located 4 to 5 cm distal to the elbow joint, thus allowing the biceps to flex the forearm. In addition, the biceps acts to rotate the radius laterally, thus serving to supinate the wrist and hand at the same time.

The Wrist and Hand and Their Joints

The wrist and hand bones and their joints are illustrated in Figures 7–8, 7–9, and 7–10. The lower end of the radius articulates with the *carpal bones*, the carpal bones with the *metacarpals*, and the metacarpals with the *phalanges*.

The **carpal bones,** eight in number, are analogous to a bag of greased marbles and allow *flexion* or *extension* of the wrist as well as *abduction* (angulation toward the thumb) or *adduction* (angulation toward the little finger). The names and positions of the separate carpal bones are given in Figure 7–10. The wrist joint is called the **radiocarpal joint.** It is mainly with the *scaphoid* and *lunate* carpal bones that the radius articulates. The ulna takes only an indirect part in the wrist joint, acting through an intervening disc between the distal end of the ulna and the *triquetral* carpal bone of the wrist.

One or more of the carpal bones sometimes becomes fractured or crushed, or at other times one or more may degenerate and partially resorb because of damage to its blood supply. When any of these events occurs, a severe arthritic condition of the wrist can ensue.

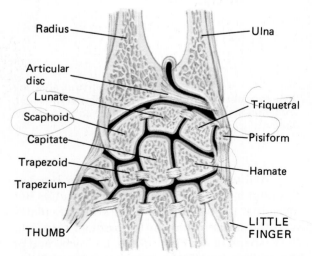

FIGURE 7–10 The bones of the wrist, the *carpals*, and the extensive synovial spaces surrounding these bones.

Note in Figure 7–10 the extensive system of *synovial membranes*, surrounding the joints between the carpal bones. It is the surfaces of these membranes and the enclosed synovial fluid that provide the necessary lubrication for the **intercarpal joints.**

The Bones of the Hand and Fingers. The bones of the hand and fingers need little introduction because almost everyone has had occasion to observe the motions of these bones and their joints.

Of the five **metacarpals,** the first serves as the proximal "joint" of the thumb; the remaining four serve as the structural frame for the hand. Then there are two **phalanges** for the thumb and three phalanges for each finger. The *thumb* metacarpal articulates with the *trapezium*, one of the carpal bones; this joint is a typical *saddle joint*, allowing a wide range of motion, both flexion or extension and abduction or adduction. For the *fingers*, the articulation of the first phalanx of each finger with its respective metacarpal also has a wide range of motion, including flexion or extension and abduction or adduction. But the articulations between the first phalanx and the second phalanx as well as between the second and third phalanges are all *hinge joints*, limited to only flexion and extension by sets of **lateral** and **medial ligaments.**

THE LOWER EXTREMITY

The Pelvic Girdle— The Coxal Bones

The **pelvic girdle,** illustrated in Figures 7–11, 7–12, and 7–13, is the supporting link between the lower extremities and the vertebral column. Also, in the sitting position, it is on the two inferior protuberances of this girdle, the **ischial tuberosities,** that one sits. And, finally, the pelvic girdle provides massive bony prominences superiorly, inferiorly, anteriorly, posteriorly, and laterally to which the major muscles of the trunk and lower limbs are attached.

Figure 7–11 shows that the pelvic girdle is composed of two separate bones, the two **coxal bones** (also called **hip bones**), each of which is formed by fusion during fetal life of three separate bones, the **ilium,** the **pubis,** and the **ischium.**

The ilium is a broad, heavy-winged bone, bounded superiorly by a thickened edge called the **iliac crest.** It is to the two iliac crests that the posterior and lateral trunk muscles attach. On the outer surfaces of the ilia the large gluteal muscles arise and extend to the femur to cause

TABLE 7–4
The Lower Extremity

Part and bones	Special features	Articulates with
Pelvic girdle		
Coxal bone		
1. Ilium	Sacral margin	Sacrum (fibrous joint)
	Iliac crest, thickened superior edge	
2. Pubis	Projects anteriorly and inferiorly from ilium	Pubis of opposite coxal bone (fibrocartilaginous joint)
	Pubic arch joins ischium inferiorly	
3. Ischium	Projects inferiorly from ilium	
	Joins pubic arch anteriorly	
	Ischial tuberosity, inferior prominence of ischium on which we sit	
	Obturator foramen, large opening bounded by ilium, ischium, and pubis	
	Acetabulum, articular socket at confluence of ilium, pubis, and ischium	Head of femur
Thigh		
Femur	Head, articular process	Acetabulum of coxal bone
	Neck, support for head	
	Shaft	
	Greater trochanter, lateral superior surface	
	Lesser trochanter, anterior superior surface	
	Lateral condyle, distal, lateral end	Lateral condyle of tibia
	Medial condyle, distal, medial end	Medial condyle of tibia
Leg		
Tibia	Lateral condyle, proximal, lateral end	Lateral condyle of femur
	Medial condyle, proximal, medial end	Medial condyle of femur
	Tibial tuberosity, anterior, inferior to knee	
	Medial malleolus, medial "ankle bone"	Talus of ankle
	Distal articular surface	Talus of ankle
	Proximal and distal fibular articular surfaces	Fibula (fibrous joints)
Fibula	Proximal and distal tibial articular surfaces	Tibia (fibrous joints)
	Lateral malleolus, lateral "ankle bone"	Talus of ankle
Ankle and foot		
Tarsal bones		
Talus		Tibia and fibula
Calcaneus	"Heel bone"	
Navicular		
Cuboid (four)		Bases of metatarsals 4 and 5
Metatarsals (five)	Base	Bases of metatarsals 1, 2, and 3
	Shaft	Cuneiform and cuboid tarsals
	Head	Bases of phalanges
Toes		
Phalanges	Base	Heads of metatarsals
2 for large toe	Shaft	Base to heads between phalanges
3 for each toe	Head (except distal phalanges)	

extension, rotation, and abduction movements of the thigh. Finally, the inner surface of each ilium is concave, and it is this surface that supports most of the weight of the abdominal contents when the body is in the upright position.

Anteriorly and inferiorly, the ilium fuses with the **pubis.** The pubis then extends medially to the anterior midline, where it articulates with the pubis from the opposite side. This point of articulation is the **symphysis pubis.** The symphysis pubis contains a *fibrocartilaginous disc* held tightly in place by strong ligaments on all sides of the joint. The symphysis gives a moderate degree of resilience to the pelvic girdle. From

FIGURE 7-11 Anterior view of the pelvic girdle.

FIGURE 7-12 The ligaments spanning the anterior and posterior sides of the sacroiliac joint.

the inferior portion of the anterior pubis, an **inferior ramus** passes inferiorly and posteriorly to form the **pubic arch.**

The **ischium** is another strong bony arch that passes inferiorly from the ilium. At its most inferior point is a large bony prominence, the **ischial tuberosity.** It is this that bears our weight when we sit. The ischium then sends a ramus forward to join the pubic arch, forming a large oval opening called the **obturator fora-**

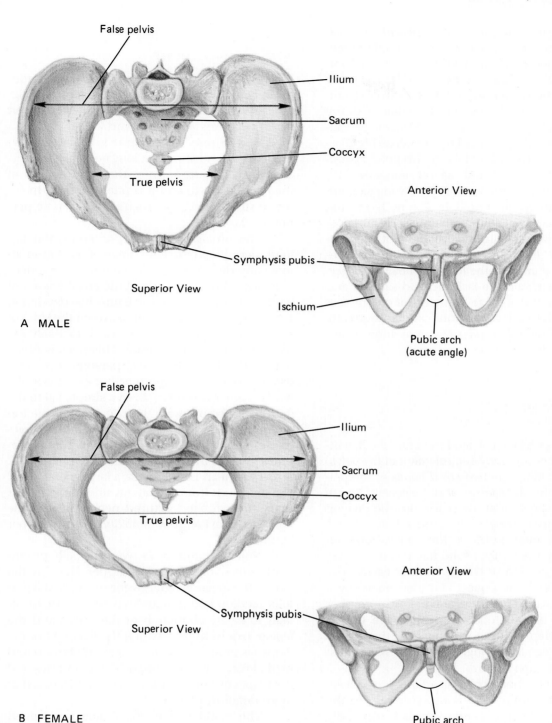

False pelvis

Ilium

Sacrum

Coccyx

True pelvis

Anterior View

Symphysis pubis

Superior View

Ischium

Pubic arch
(acute angle)

A MALE

False pelvis

Ilium

Sacrum

Coccyx

True pelvis

Anterior View

Symphysis pubis

Superior View

B FEMALE

Pubic arch
(obtuse angle)

FIGURE 7–13 Differences between the male and female pelves.

men, bounded superiorly by the ilium, anteriorly by the pubis, and posteriorly and inferiorly by the ischium.

Many important muscles serving hip joint movement or helping to support the floor of the pelvic cavity are attached to the medial, lateral, inferior, and anterior surfaces of the ischium and pubis as well as around the margins of the obturator foramen. And it is to the superior margin of the pubis that the powerful muscles of the anterior abdominal wall attach.

The **socket of the hip joint,** called the **acetabulum,** occurs on the lateral surface of the coxal bone at the point of confluence of the ilium, ischium, and pubis. The acetabulum is a moderately deep fossa into which the ball-shaped head of the femur fits, providing a ball and socket joint and allowing multiple directions of movement, as will be discussed later.

The Sacroiliac Joints.   All the weight of the upper body is supported by the two **sacroiliac joints** where the two ilia articulate with the sacrum. These joints have a small degree of resilience provided by a thick layer of dense fibrocartilaginous material that fills the irregular spaces between the iliac and sacral surfaces. But equally as important are the very strong **sacroiliac ligaments,** illustrated in Figure 7–12, that span the sacroiliac joint both anteriorly and posteriorly. Note that ligaments pass not only between the ilium and the sacrum but also between the ilium and the lower lumbar vertebrae, and also between the ischium and sacrum and coccyx. Tearing of these ligaments or crushing of either of the sacroiliac joints can cause greatly prolonged sacroiliac pain that may limit a person's activities severely.

THE BONY PELVIS AND DIFFERENCES IN THE FEMALE AND MALE

The **bony pelvis,** illustrated in Figure 7–13, consists of the entire bony encirclement of the *pelvic cavity*, including the two *coxal bones* of the pelvic girdle plus the *sacrum* and the *coccyx*. Figure 7–13 also shows that the pelvis may be divided into an upper "false" pelvis and a lower "true" pelvis. The **false pelvis** is the region bounded laterally by the wings of the ilia and is actually the lower portion of the abdominal cavity. The **true pelvis** is the region bounded above by a bony ring, called the **inlet,** and below by another bony and ligamentous ring, called the **outlet.** The importance of the true pelvis is that this is the "birth canal" through which the baby must pass at the time of birth.

The male bony pelvis is adapted especially for powerful support of a strong body. On the other hand, the female bony pelvis is especially adapted for birth of offspring. Note in Figure 7–13 that the inlet of the true pelvis in the female is much larger in diameter than it is in the male. Also, it is ovoid in shape to allow easy passage of the baby, in contradistinction to a more triangular shape of the true pelvis of the male. Also, the distance between the pelvic inlet and pelvic outlet is somewhat shorter in the female than it is in the male. In the male, all the bones of the pelvis are much more massive and stronger than those in the female, though the total width of the male pelvis is about 5 cm less than that of the female pelvis.

The Femur, and the Hip Joint

Figure 7–14A illustrates the bones of an entire lower extremity from the hip to the foot, and Figure 7–14B illustrates specifically the femur and its articulations at the hip and knee.

The Femur.   The **femur** is by far the largest, longest, and strongest long bone of the body. Obviously, it must support the entire weight of the body above it, and it is often exposed to extreme stress when the thigh is hit by some perpendicular force.

The proximal joint of the femur, the hip joint, permits an extreme range of motion in almost any direction, but, since the femur articulates with the lateral side of the coxal bone, not with its inferior extent, the femur has developed a medially projecting **head.** Between the shaft of the femur and this head is also a medially extending **neck of the femur.** This neck is easily fractured, especially in old persons who have osteoporotic bones (porous bones caused by excess bone resorption). Even a simple fall to the floor by a person more than 70 years of age frequently results in a severely comminuted fracture of the femoral neck, and the fracture sometimes extends into the head of the femur as well. But, if the head is still intact, a large metal "nail" can be driven from the lateral side of the femur, then through the fractured neck and into the head to provide new stability for the upper femur.

Note in Figure 7–14 two especially prominent projections from the upper shaft of the femur, the **greater trochanter** to which the gluteus medius and gluteus maximus muscles attach to cause abduction of the thigh, and the **lesser trochanter** to which the ligament of the iliopsoas muscle attaches to provide flexion and medial rotation of the thigh. These muscles and the respective joint motions will be discussed in more detail in Chapter 9.

The distal end of the femur is enlarged both laterally and medially to form the **lateral** and **medial condyles,** which include the articular surfaces of the knee joint. In the space between the two condyles is a **notch** through which important knee ligaments, the *anterior* and *posterior cruciate ligaments*, pass, as we shall discuss later. Anteriorly, the articular surfaces of both of these condyles extend upward to provide an articular surface for the posterior surface of the patella, which in turn is embedded firmly in the quadriceps tendon, thus providing for sliding movement of this tendon over the anterior of the knee joint.

COXAL BONE

Ball and socket joint

Head of femur in acetabulum

Greater trochanter

Neck

Lesser trochanter

FEMUR

Patella

Lateral epicondyle

Medial epicondyle

Lateral condyle

Medial condyle

TIBIA

FIBULA

Metatarsals

Gliding joints

Tarsals

Phalanges

A

B

FIGURE 7–14 Lateral view of the bones of the hip, thigh, leg, and foot (A); anterior view of the hip and thigh bones (B).

THE HIP JOINT

Many of the same features that have already been discussed for the shoulder joint apply also to the **hip joint.** The **acetabulum** (Figs. 7–14 and 7–15), which contains the *articular surface of the coxal bone,* has a much deeper cavity than that of the glenoid cavity of the shoulder joint. Like

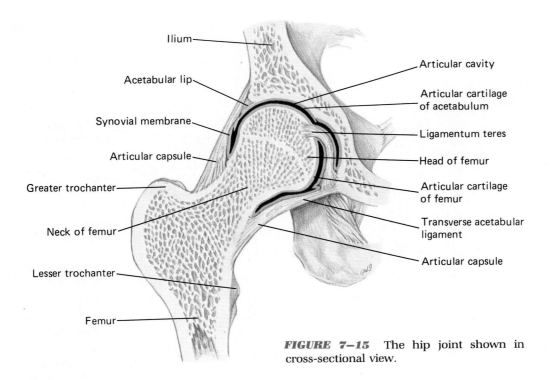

FIGURE 7–15 The hip joint shown in cross-sectional view.

the shoulder joint, the hip joint has a layer of **articular cartilage** on both the acetabular surface and the surface of the *head of the femur.* Finally, the hip joint is surrounded on all sides by a *synovial membrane* covered in turn by a thick fibrous *articular capsule* that envelops the entire joint.

The means for stabilizing the hip joint is also similar to that for stabilization of the shoulder joint but with slight differences. It will be recalled that the shoulder joint is stabilized mainly by powerful muscles and muscle tendons that surround the joint. This is true of the hip joint as well, but, in addition, the hip joint has a much deeper joint cavity. It also has a special ligament called the **ligamentum teres** (Fig. 7–15) that inserts deeply and strongly into the middle of the head of the femur and also extends downward through the articular cavity to attach firmly to the lower lip of the acetabulum. This ligament, in addition to the muscles, helps to prevent upward gliding of the head of the femur and therefore often carries much of the weight of the body itself.

Another special importance of the ligamentum teres is that most of the blood supply to the head of the femur flows through a small artery that accompanies this ligament. Therefore, injury to the ligament and disruption of this blood supply is frequently followed by necrosis (dissolution caused by nutritive death of the tissue) of the femoral head, and this often leads to a joint

that is first severely arthritic and then totally nonfunctional.

The Bones of The Leg, and the Knee Joint

The bones of the leg consist of a very strong and massive **tibia** and a much more slender **fibula,** illustrated in Figure 7–16. These two bones lie parallel and articulate with each other both superiorly and inferiorly but not in the same manner that the radius and ulna of the forearm articulate with each other. Instead, because the leg has become adapted for bearing the entire weight of the body, the **tibiofibular joints** at both ends are very firm *fibrous joints* that allow almost no movement, quite different from the great range of rotary movement at the two ends of the forearm. Instead, the tibia and the fibula, along with their ligaments, function together as if they were a single structural unit.

Note in Figure 7–16 the two separate large prominences at the superior end of the tibia, the **medial** and **lateral tibial condyles.** The articular surfaces of these articulate with the respective surfaces of the *medial* and *lateral condyles of the femur,* thus forming the *knee joint.*

Also, observe at the lower end of the tibia the medial projection called the **medial malleolus** and at the lower end of the fibula the lateral projection called the **lateral malleolus.** These are known in layman's parlance as the two

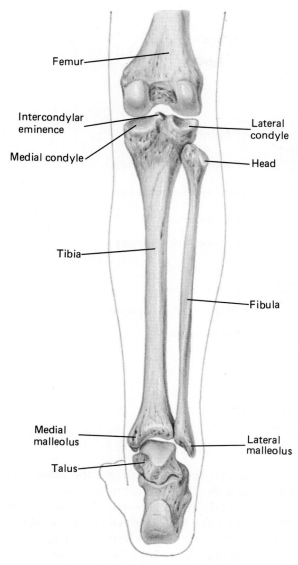

Femur

Intercondylar eminence

Medial condyle

Tibia

Medial malleolus

Talus

Lateral condyle

Head

Fibula

Lateral malleolus

POSTERIOR VIEW

FIGURE 7–16 Bones of the lower leg, and posterior views of the knee joint and ankle joint.

"ankle bones." It is to these that the major medial and lateral ligaments of the ankle are attached.

THE KNEE JOINT

The posterior aspect of the knee is illustrated in Figure 7–16, and an anterior view showing more detail is illustrated in Figure 7–17. This joint has several very special characteristics that deserve our attention, particularly since it is this joint that is most often seriously injured in athletes.

The Lateral and Medial Menisci. In most synovial joints of the body, the articular cartilaginous surfaces of the two opposing bones directly contact each other, thus providing the smooth surfaces for motion. However, as shown

in Figure 7–17, in the knee joint the articular surfaces of the two tibial condyles are for the most part covered by *fibrocartilaginous pads* called the **lateral meniscus** and the **medial meniscus.** These are also commonly called the "knee cartilages." The menisci not only help to provide smooth surfaces for joint motion but also provide extra resilience to the knee joint to help it absorb some of its many shock stresses. Trauma to the knee frequently causes one or both of the cartilages to be torn, which can cause severe pain and limitation of joint motion, usually requiring surgical removal of the torn cartilage. Unfortunately, this may lead in later life to early development of an arthritic knee.

The Anterior and Posterior Cruciate Ligaments. Also shown in Figure 7–17 are two powerful ligaments, the anterior and posterior cruciate ligaments, both of which pass through the notch between the lateral and medial condyles of the femur. These two ligaments cross each other like an "X." The **anterior cruciate ligament** attaches anteriorly to the tibia at the edge of the medial meniscus and then passes backward, upward, and laterally to attach to the femur on the posteromedial surface of the lateral condyle. This ligament prevents anterior displacement of the tibia on the femur and also prevents overextension of the joint.

The **posterior cruciate ligament** attaches anteriorly to the femur in the notch between the two condyles and posteriorly to the tibia at the posterior rim of the lateral meniscus. This ligament helps to prevent the femur from sliding forward on the tibia, as well as preventing overflexion of the joint. (Another important factor that prevents the femur from sliding forward on the tibia is the backward force of the patella against the anterior surface of the femur.) Both cruciate ligaments also provide a strong vertical connection between the femur and the tibia to keep these two bones from pulling apart.

The Collateral Ligaments. On the lateral and medial sides of the knee joint are prominent **collateral ligaments:** the *lateral collateral ligament,* also called the *fibular collateral ligament,* and the *medial collateral ligament,* also called the *tibial collater* ligaments are critical i medial angulation of the

Damage to the Li Joint. Obviously an exce side of the knee joint can ligament on the opposite that forces the femur bac rupture the anterior crucia extreme forces can ruptur

ANTERIOR VIEW OF EXTENDED KNEE JOINT

FIGURE 7–17 Anterior view of the opened knee joint and its ligaments.

ate ligament. Loss of any one of these knee ligaments causes some degree of knee instability, but loss of the cruciate ligaments is especially serious. Fortunately, though, a major part of the stability of the knee joint is provided by muscle tendons that pass both anteriorly and posteriorly to the knee joint. Therefore, the stronger the muscles and the greater the tone of these muscles, the less likely are knee injuries to occur.

The Ankle and Foot

The bones of the ankle and foot, illustrated in Figure 7–18, are similar to those of the wrist and hand. For instance, the **tarsals,** which are the ankle and proximal foot bones, are the analogs to the carpals of the wrist, except that there are only seven tarsals in contrast to the eight carpals. The **metatarsals,** the midfoot bones, are the analogs to the metacarpals. And, finally, the **phalanges** of the toes are structurally similar to the phalanges of the fingers. However, movements of the ankle and foot are considerably dif-[f]erent from those of the wrist and hand, as we [are] all aware, and the bones, especially the tar-[sals a]re correspondingly adapted to different [moveme]nts.

The two largest tarsals are the **talus** and the **calcaneus.** The talus is the bone with which the lower ends of the tibia and fibula articulate. The weight of the body is transmitted downward from the tibia to the talus and thence partly to the calcaneus, the "heel bone," and partly through the other tarsals and metatarsals to the heads of the metatarsals at their distal ends. It is also to the posterior end of the calcaneus that the very powerful **Achilles tendon** attaches, which, when pulled upward by the muscles of the leg, causes the foot to bend downward, an action called *plantar flexion.* The opposite action of lifting the foot is called *dorsiflexion.*

THE ARCHES OF THE FOOT

Note in Figure 7–18 that the foot has two principal arches, a *longitudinal arch* and a *transverse arch.*

The **longitudinal arch** extends from the calcaneus to the heads of the metatarsals, so that the foot of a standing person normally rests on the floor only anteriorly and posteriorly but not in the middle. This provides a high degree of stability for the foot on the plantar surface because the farther apart the two contact points

FIBULA

TIBIA

Calcaneus
(heel bone)

Talus
Navicular
Cuneiforms

TARSALS

TRANSVERSE ARCH

Cuboid

LONGITUDINAL ARCH

METATARSALS   PHALANGES

LATERAL VIEW

FIGURE 7–18 The bones of the ankle and foot, showing also the longitudinal and transverse arches.

are, the more difficult it is to throw the individual off balance. We are all also familiar with the fact that the longitudinal arch is much higher on the medial aspect of the foot than on the lateral aspect.

The **transverse arch** occurs mainly beneath the anterior tarsals and posterior portions of the metatarsals. This arch provides space underneath the foot for appropriate tendons, ligaments, and other structures.

Flat Feet. Some persons do not have high arches, giving rise to the condition called *flat feet* or *fallen arches*. This occurs mainly as a hereditary trait. However, it is often stated that walking barefooted or standing long hours without arch support will also cause flat feet, but these causes are not as common as popularly believed. Pain often accompanies fallen arches because of abnormal compression points among the bones of the feet or because of excessive stresses on the ligaments. Yet, many persons with totally flat feet have no pain at all. It is also remarkable that when flat feet no longer became a cause for rejection by the army, the incidence of pain and disability from flat feet diminished greatly.

QUESTIONS

1. Describe the vertebral column, including its curvatures and the numbers of cervical, thoracic, and lumbar vertebrae.
2. Describe the structure of an individual vertebra, listing its parts and their functions.
3. Describe the articulations of the vertebrae with each other and with the ribs.
4. Discuss the ligaments of the vertebral column and their function, as well as the function of the discs between the vertebrae.
5. Discuss the articulation of the vertebral column with the skull.
6. Describe the thoracic cage and its function in respiration.
7. List the parts of the pectoral girdle and describe its different joints, including the shoulder joint.
8. Describe the bones of the forearm and the forearm joints that allow rotation of the wrist.
9. Describe the bones and joints of the wrist and hand, and discuss the movements that are allowed at these joints.
10. List the bones and joints of the pelvis and discuss their interrelationships; especially, describe the sacroiliac joints.
11. Describe the special features of the hip joint and the role of the ligamentum teres.
12. Discuss in detail the knee joint, the knee cartilages, the ligaments of the knee, and the movements allowed at the knee joint.
13. Discuss the bones and structure of the ankle and foot, paying special attention to the joint motions and the arches of the foot.

REFERENCES

Anderson, J. E.: The back. *In* Grant's Atlas of Anatomy. Baltimore, Williams & Wilkins, 1978, pp. 5–1 to 5–50.

Bloom, W., and Fawcett, D. W.: Cartilage and bone. *In* A Textbook of Histology, 10th Ed. Philadelphia, W. B. Saunders, 1975, pp. 233–287.

Bloom, W., and Fawcett, D. W.: The teeth. *In* A Textbook of Histology, 10th Ed. Philadelphia, W. B. Saunders, 1975, pp. 621–638.

Copenhaver, W. M., Kelly, D. E., and Wood, R. L.: Cartilage and bone. *In* Bailey's Textbook of Histology, 17th Ed. Baltimore, Williams & Wilkins, 1978, pp. 170–205.

Figge, F. H. J.: The skeleton. *In* Sobotta/Figge Atlas of Human Anatomy. Vol. I. Baltimore, Urban & Schwarzenberg, 1977, pp. 7–137.

Goss, C. M.: The skeleton. *In* Gray's Anatomy of the Human Body, 29th Ed. Philadelphia, Lea & Febiger, 1973, pp. 95–367.

Ham, A. W., and Cormack, D. H.: Skeletal tissues. *In* Histology, 8th Ed. Philadelphia, Lippincott, 1979, pp. 367–482.

Langman, J., and Woerdeman, M. W.: Back. *In* Atlas of Human Anatomy. Philadelphia, W. B. Saunders Co., 1978, pp. 475–503.

Leeson, T. S., and Leeson, C. R.: Cartilage and bone. *In* Histology, 4th Ed. Philadelphia, W. B. Saunders, 1981, pp. 137–186.

Liebgott, B.: The Anatomical Basis of Dentistry. Philadelphia, W. B. Saunders, 1982.

Pernkopf, E.: Oral cavity and teeth. *In* Atlas of Topographical and Applied Human Anatomy. Vol. I. Philadelphia, W. B. Saunders, 1980, pp. 156–180.

Pernkopf, E.: The skull. *In* Atlas of Topographical and Applied Human Anatomy. Vol. I. Philadelphia, W. B. Saunders, 1980, pp. 1–28.

Weiss, L., and Greep, R. O.: Teeth. *In* Histology, 4th Ed. New York, McGraw-Hill, 1977, pp. 615–642.

Weiss, L., and Greep, R. O.: The skeletal tissues. *In* Histology, 4th Ed. New York, McGraw-Hill, 1977, pp. 205–250.

Wheater, P. R., Burkitt, H. G., and Daniels, V. G.: The skeletal system. *In* Functional Histology. New York, Churchill Livingstone, 1979, pp. 128–144.

Williams, P. L., and Warwick, R.: The skeletal system. *In* Gray's Anatomy, 36th British Edition, Philadelphia, W. B. Saunders, 1980, pp. 230–505.

(See also General References, Chapter 1.)

IV

THE SKELETAL MUSCLES

8

Muscles and Movement: I. The Head, Neck, and Trunk

Overview

The movements of the body are provided by contraction of skeletal muscles. Each muscle is generally comprised of a muscle belly that is the contractile portion and tendons at the two ends that insert into bones on the two sides of a joint. In the muscle belly are hundreds to many thousands of parallel muscle fibers. Each of these in turn contains billions of minute overlapping filaments of two types, actin and myosin filaments. When signals from a nerve activate a muscle, calcium ions are released inside the muscle fiber, and these cause attractive forces to develop between the actin and myosin filaments, thus causing contraction.

The important classes of muscles in the head, neck, and trunk are as follows:

1. The **superficial facial muscles.** These are responsible for such actions as the facial expressions of smiling, pouting, and so forth. The muscle of the eyelids that causes them to close is the *orbicularis oculi,* a constrictor muscle that encircles the entire eye. The smile is caused by the *risorius* and the *zygomaticus* that pull the angle of the mouth upward and lateralward.
2. The **muscles of mastication.** The most important of these are the *masseter* and *temporalis,* which are responsible for the powerful chewing motions.
3. The **muscles of swallowing.** Swallowing begins with an upward and backward thrust of the *tongue* that pushes the food into the throat. Then it proceeds with progressive constriction of the *pharyngeal muscles* from above, downward toward the esophagus.
4. The **neck muscles.** These control the position of the head and neck with respect to the remainder of the body. Among the more important of these muscles are (a) the *sternocleidomastoid,* which rotates the head laterally and tilts the face sideways; (b) the *scalenus muscles,* which attach between the first rib and the transverse processes of the neck to cause lateral bending and flexion of the neck; (c) the *longus capitis,* which connects the transverse processes of the neck vertebrae with the inferior surface of the skull and causes forward or lateral bending of the head; and (d) the *splenius capitis,* which lies along the back of the neck vertebra and attaches to the back of the skull to cause extension of the neck and head.
5. The **back muscles.** These cause extension of the entire trunk and help to cause extension of the neck and head as well. They pass all the

131

way from the posterior pelvis to the back of the neck and skull. Along the way they attach via muscular slips to the ribs and vertebrae. The most important of the back muscles is the *erector spinae*, which is comprised of multiple small muscles lying side by side. In addition, several large superficial back muscles attach to the shoulder and upper arm, including especially the *deltoid* over the top of the shoulder that lifts the arms, the *trapezius* that attaches the spinous processes of the thoracic and cervical vertebrae to the scapula and clavicle and pulls the entire shoulder upward and backward, and the *latissimus dorsi* that attaches between the spinous processes of the lower vertebrae and the humerus and pulls downward on the humerus.

6. The **abdominal muscles.** These form most of the abdominal wall and support the abdominal organs. They are the (a) *rectus abdominis*, (b) *external oblique muscle*, (c) *internal oblique muscle*, and (d) *transversus abdominis*.

7. The **muscles of the pelvic diaphragm.** These are the *levator ani* and the *coccygeus;* they support the downward thrust of the pelvic and abdominal organs and also play a role in the process of defecation.

8. The **perineal muscles.** These support the anterior part of the pelvic outlet through which the urethra and vagina pass to the exterior.

In this and the following chapter we will discuss the muscles that cause the skeletal movements. These muscles are all **skeletal muscle,** a type of muscle quite different from two other types of muscle that are also of major importance to the body: **cardiac muscle,** which forms the muscular mass of the heart, and **smooth muscle,** which is an integral part of most of the internal organs of the body, such as the intestines, uterus, and blood vessels. These other types of muscle will be discussed in later chapters in relation to their specific organ functions.

STRUCTURAL ORGANIZATION OF SKELETAL MUSCLE

Skeletal muscle usually connects from one bone to another in the manner illustrated in Figure 8–1A. Normally, one of the bones is much more secure than the other so that when the muscle contracts only the second bone does the moving. The attachment of the muscle to the bone that does not move is called the **origin** of the muscle, and the attachment to the bone that does move is called the **insertion** of the muscle.

Muscles have many different sizes and lengths, depending upon the types of movement that they cause. Some are very long, such as the muscles of the thigh, occasionally as long as a half meter. Some are very short, such as the muscles of 1 to 2 cm that attach from one rib to the next, the intercostals. Some are extremely small in diameter, such as the stapedius, a muscle of only 1 mm in diameter inside the middle ear

that attaches to the stapes, one of the ossicles of the ear, and partially immobilizes this ossicle when one is exposed to too intense sound. At the other extreme is the quadriceps femoris muscle on the anterior aspect of the upper thigh, often having a diameter of more than 10 cm.

The shapes of skeletal muscles are also varied. Many of the muscles are *fusiform* in shape (cigar-shaped) similar to the muscle illustrated in Figure 8–1A. It is this type of muscle that causes movement of most of the long bones of the extremities and also some movements of the back, neck, and head. However, many others of the skeletal muscles are in the form of broad or narrow, long or short, thin or thick *sheets*, the most prominent of which are the several layers of sheet muscles that envelop the abdomen, providing support for the internal organs.

Muscle Fibers, Muscle Tendons, and Attachment to Bones. The contractile portion of a muscle is called the **muscle belly.** This is comprised of literally thousands of long parallel **muscle fibers** (Fig. 8–1B) that extend the entire length of the muscle, some as long as a half meter in length. The diameters of these fibers are so small that they can be seen only with the aid of magnification. At their ends, before attaching to the bone, the muscle fibers fuse with an entirely different type of fiber, the extremely strong collagenous **tendon fibers** that have strength equal to that of nylon. It is these fibers that constitute the **muscle tendon.** Some tendons run long distances before they attach to the bone

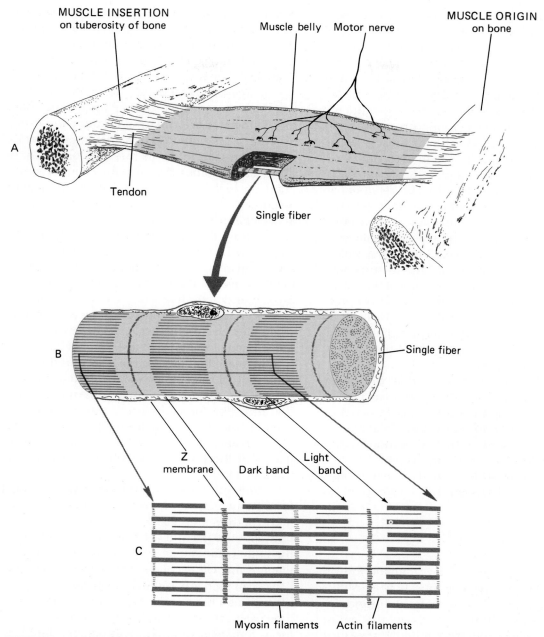

FIGURE 8–1 Basic functional anatomy of muscle. (A) The muscle belly, its attachment to bone, and its nerve supply. (B) A single muscle fiber. (C) Arrangement of actin and myosin filaments within the muscle fiber to cause a sliding motion that leads to contraction.

and others much shorter distances, as illustrated to the left in Figure 8–1A. In other instances a muscle may have broad attachment to a bone and becomes tendinous only a millimeter or so prior to inserting into the bone.

The tendon fibers do not merely attach to the surface of the bone but penetrate deeply into the substance of the bone, as illustrated at both ends of the muscle in Figure 8–1A. These tendon fibers are almost identical to the very powerful collagenous fibers in the matrix of the bone, and they also intermesh with the fibers of the bone, thus providing a bony attachment of the tendon

that is equal in strength to the tendon itself. When injury occurs to a tendon, it is sometimes the tendon itself that ruptures; but at other times the tendon is literally pulled out of the bone, or a small portion of the bone where the tendon attaches chips off the main body of the bone, a condition called *avulsion of the tendon.*

Note also in Figure 8–1A that where the tendon inserts into the long bone, there is often an overgrowth of the bone itself, thus providing an even stronger tendon attachment. This overgrowth is frequently called a **tubercle,** a **tuberosity,** or, if very large, a **trochanter.**

Excitation of Muscle Contraction by the Motor Nerve.   Note in Figure 8–1A the nerve attaching to the muscle belly, called a **motor nerve.** This nerve comes from the spinal cord or from a cranial nerve, and its signals are controlled by both the brain and the cord itself. The motor nerve, like the muscle belly, is comprised of many individual fibers called **nerve fibers.** As these nerve fibers approach the muscle belly, each one of them divides up to 1000 times, so that a single **nerve ending** rests directly on each individual skeletal muscle fiber. The nerve signal is an electrochemical stimulation that spreads along the nerve fibers all the way to their endings. On reaching the ending, the signal causes a small amount of a chemical substance called *acetylcholine* to be released. This in turn acts within $\frac{1}{500}$ second on the membrane of the skeletal muscle fiber to cause another electrochemical signal now to pass over the muscle fibers in the same manner that the nerve signal passed over the nerve fiber. It is this signal that causes the muscle contraction.

Note also in Figure 8–1 that the nerve attaches to the muscle belly approximately in its center portion. This allows the muscle signal to travel toward both ends of the muscle at almost exactly the same time so that all portions of the muscle will contract nearly in unison.

The transmission of signals in nerves and muscles will be discussed fully in Chapter 10.

SKELETAL MUSCLE FIBER AND ITS CONTRACTION

Figure 8–1B illustrates in diagrammatic form a minute section of a skeletal muscle fiber. Multiply this small length of fiber by thousands of times and one can then envisage the entire length of the fiber. Note the banded appearance of the fiber. These bands are called *striations*, and hence skeletal muscle fibers are said to be *striated.* Therefore, skeletal muscle is sometimes called **striated muscle.** However, this designation is somewhat misleading, since cardiac muscle also is striated. These striations are so small that they can be seen only with the microscope.

Actin and Myosin Filaments, and the Contractile Mechanism.   The cause of the banded appearance of striated muscle is its special internal structure (Fig. 8–1C). This structure involves two different types of filaments, **myosin filaments** and **actin filaments.** The darker bands contain the myosin filaments, which are very thick filaments, and these bands also contain the ends of actin filaments; the lighter bands contain only the central portions of actin filaments as illustrated in the figure. Note especially that the actin and myosin filaments interdigitate with each other. During contraction, the actin filaments slide much further in among the myosin filaments, and this is the mechanism by which the muscle fiber shortens, called the **sliding filament mechanism of contraction.** Note also the "Z" membranes that attach the actin filaments to each other in the light bands. It is these Z membranes that keep the actin filaments lined up together and therefore are ultimately responsible for causing the striated appearance of skeletal muscle fibers.

During muscle contraction, the actin filaments are literally pulled inward among the myosin filaments. This is caused by the development of powerful attractive forces between the myosin and actin filaments. The precise nature of these forces is not fully understood. However, when the electrochemical signal spreads over the surface of the muscle fiber, one of its effects is to increase for a fraction of a second the concentration of calcium ions in the fluid inside the muscle fiber. These calcium ions in turn activate the attractive forces between the myosin and actin filaments, thus drawing these filaments together and causing muscle contraction. The details of these processes will be discussed in Chapter 11.

Energy Used for Muscle Contraction.   Even though about 40 percent of the total mass of the body is muscles, under resting conditions only about 20 percent of the total energy used by the body is consumed by the muscles. However, when a muscle becomes very active, its energy usage can increase as much as 300-fold for a few seconds at a time, and it can sustain rates of energy usage at more than 80 times its resting rate for as long as several hours in persons who are well trained, such as marathon runners.

The energy used for muscle contraction during the first few seconds of intense activity comes from so-called "high energy compounds," **adenosine triphosphate** (ATP) and **phosphocreatine,** that are already stored in the muscle cells. However, during maximal muscle activity, these substances will last for only 5 to 10 seconds, and after that time the energy must come from the metabolism of foods. During the next few minutes most of the energy comes from metabolism of *glycogen*, which is a storage form of carbohydrates already present in the muscle fibers themselves. The glycogen is broken down and metabolized with oxygen to cause tremendous new quantities of ATP to be formed, and this then is used by the muscle to energize the contractile process. After exercise has continued for many minutes most of the glycogen will also have been consumed. Thereafter, much of the energy comes from metabolizing fats with oxygen.

These energy processes will be discussed for all cells of the body in Chapters 43 and 44 and specifically for sports activities in Chapter 51. However, let us remark here that strenuous exercise is the most energy-consuming condition that the body ever experiences.

Oxygen Consumption and Cardiac Output in Exercise. Before we leave the subject of energy for muscle contraction, we need to speak about the oxygen that is metabolized with the food during exercise. During a marathon run, the rate of total body oxygen consumption often increases as much as 20-fold, which is approximately also the increase in rate of total energy usage by the body. To provide this vast amount of oxygen, the respiratory system also becomes greatly activated. But, equally important, the blood in the circulatory system must carry all this oxygen to the muscles and carry carbon dioxide away from the muscles after the oxygen has been metabolized. To achieve this, the cardiac output (the amount of blood pumped by the heart each minute) increases as much as sevenfold in the marathon runner. In fact, the major limitation in the speed of running over long periods of time usually is not limitation of muscle function itself but instead limitation of the ability of the heart to pump enough blood with its contained oxygen to the muscles. Thus, the condition of the heart is a major determinant of the time required for a runner to complete a marathon race.

MAJOR SUPERFICIAL MUSCLES OF THE BODY

Many of the most important muscles of the body lie in the superficial layers of the body. These are the muscles that cause the bulges under the skin, and they are the ones that are discussed in everyday parlance as well as in most athletic circles.

The major superficial muscles are illustrated from the front in Figure 8–2 and from the back in Figure 8–3. One should stop at this point and study these two figures, learning the interrelationships among the separate muscles, their relative sizes, their relationships to the joints, and the overall muscular configuration of the entire body. If these two figures alone are learned well, the student will already understand most of the mechanics of bodily motions. Later in this chapter and in the following chapter we will discuss the separate muscles in greater detail, but we present here a survey of the superficial muscles.

Facial Muscles. These are the muscles that provide the different facial movements, including the facial movements of expression. The *frontalis* lifts the eyebrows. The *orbicularis oculi* is a sphincter muscle that constricts around the eyes to close them. The *orbicularis oris* likewise constricts around the lips to cause puckering. And the *zygomaticus* and *risorius* pull the angles of the mouth laterally and upward, thus implementing the act of smiling.

Jaw Muscles. The *temporalis* on the side of the head and the *masseter* on the side of the jaw both attach to the mandible and are the powerful chewing muscles.

Neck Muscles. Obviously, the neck muscles control the position of the head with respect to the body. The two superficial anterior neck muscles shown in Figure 8–2, the *sternocleidomastoid* and the *sternohyoid*, acting together on both sides of the neck, will stiffen the neck and flex it forward slightly, but contraction of the sternocleidomastoid on only one side causes the head to rotate in the opposite direction.

On the back of the neck, it is mainly the deep muscles that cause extension of the neck and head.

Shoulder Muscles. The shoulder muscles are divided into two separate groups: (1) those that move the scapula into position so that it can provide proper support for the arm and (2) those that provide movement of the humerus at the shoulder joint.

Most of the shoulder muscles that provide movement of the scapula are deep muscles, but one of the superficial muscles, the *trapezius* (Fig. 8–3), is a large and powerful muscle that arises from the spinous processes of the entire thoracic and cervical regions and inserts on the spine and acromion of the scapula as well as the lateral third of the clavicle. Contraction of the upper part of the trapezius lifts the scapula, and contraction of all of it together pulls it backward. When the trapezius contracts at the same time as contraction of the *serratus anterior* (Fig. 8–2) that attaches the scapula to the anterior thoracic wall, the scapula becomes fixed in a locked position and provides a firm base for movement of the arm.

FIGURE 8–2   The anterior superficial muscles of the body.

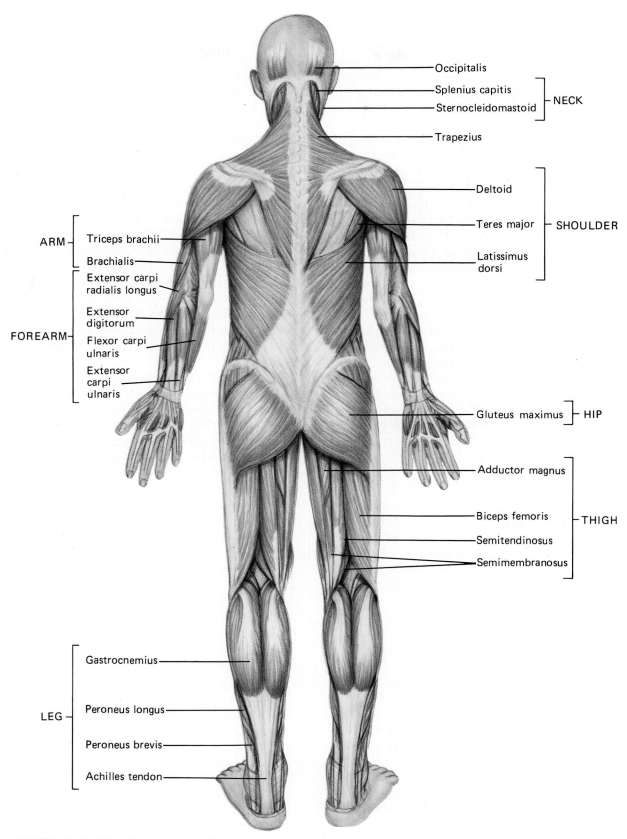

FIGURE 8–3 The posterior superficial muscles of the body.

The superficial muscles that cause movement of the humerus include the following:

1. The *deltoid* envelops the top of the shoulder. The anterior, middle, and posterior portions of the deltoid can be contracted separately from each other, thus (a) raising the arm upward and forward (flexing it), (b) raising it upward and outward (abducting it), or (c) raising it upward and backward (extending it).
2. The *latissimus dorsi* (Fig. 8–3) is one of the most powerful muscles of the body, arising from the lower half of the back and then inserting into the intertubercular groove of the humerus several centimeters below the shoulder joint. When contracted, this muscle causes adduction of the arm toward the body and also pulls the arm posteriorly (extending it) at the same time. This is the "climbing muscle," allowing a person to lift himself by pulling downward on the arm. Another muscle that helps the latissimus dorsi to adduct and extend the arm is the *teres major*, extending from the lateral margin of the scapula to the anterior aspect of the humerus.
3. The *pectoralis major* (Fig. 8–2) arises from the anterior thoracic wall and inserts into the anterior aspect of the humerus. It adducts the arm toward the trunk and rotates it medially as well.

The Arm.   Everyone is familiar with the two major muscles of the upper arm, the *biceps brachii* that mainly causes flexion of the elbow and the *triceps brachii* that causes extension of the elbow. In addition to these two, another important muscle is the *brachialis* that lies mainly beneath the biceps and also causes flexion at the elbow.

Forearm.   The forearm muscles can be divided into three different groups: (1) those that cause motion at the elbow, (2) those that cause flexion of the wrist and fingers, and (3) those that cause extension of the wrist and fingers. Some of these muscles have more than one action. The *brachioradialis* (Fig. 8–2) extends from the lower humerus downward along the radius, and it causes mainly flexion at the elbow, helping the biceps and brachialis in this task.

The *flexor carpi radialis* (Fig. 8–2) and the *flexor carpi ulnaris* (Fig. 8–3) extend from the elbow to the anterior surface of the metacarpal bones and cause flexion at the wrist. The flexors of the fingers are the deep muscles of the anterior compartment of the forearm. These muscles, of course, also flex the wrist.

The superficial muscles on the back of the forearm that extend the wrist, shown in Figure 8–3, are the *extensor carpi radialis longus* and *brevis*, the *extensor carpi ulnaris*, and the *extensor digitorum*. The tendons of the extensor digitorum pass over the wrist and hand to the fingers and also cause extension of the fingers at the same time that they cause extension of the wrist.

Abdominal Muscles.   The abdominal muscles (Fig. 8–2) provide the principal substance of the abdominal wall. These include four sets of paired muscles: (1) *rectus abdominis*, (2) *external oblique*, (3) *internal oblique*, and (4) *transversus abdominis*. These muscles are tightened during exercise, which splints the entire abdominal region of the body and helps to maintain a firm body stance. The two rectus abdominis muscles, the most anterior of the abdominal muscles, extend from the lower margin of the anterior thoracic cage to the pubis. They are especially powerful and when contracted pull the thoracic cage downward, thus flexing the entire trunk.

Hip and Thigh Muscles.   Some of the muscles of the hip and thigh provide movement at the hip joint, some at the knee joint, and some span both joints, causing movement of both at the same time. The *gluteus maximus* (Fig. 8–3) is the major superficial muscle of the buttocks. It originates from the posterior and lateral surfaces of the ilium and sacrum and inserts into the lateral side of the femur below the greater trochanter. Its function is to extend, abduct, and laterally rotate the thigh all at the same time.

One frequently hears of a pulled groin muscle. Several of these muscles are illustrated in Figure 8–2. The more anterior of these, the *iliopsoas* and the *pectineus*, flex the thigh. The medial groin muscles, the *adductor muscles*, and the *gracilis* adduct the leg toward the axis of the body. It is the adductors that allow us to grip the back of a horse with our legs.

The major muscle of the anterior thigh is the *quadriceps femoris*. This muscle has four separate heads, each of which is actually a separate muscle. Three of these heads are superficial and are illustrated in Figure 8–2: the *rectus femoris*, the *vastus lateralis*, and the *vastus medialis*. The fourth head lies under the rectus femoris, the *vastus intermedius*. The quadriceps arises principally from the femur and inserts via the patellar tendon into a tuberosity on the anterior tibia. Therefore, this is the powerful muscle that causes extension of the leg at the knee joint. The rectus femoris portion of the quadriceps, however, arises from the anterior ilium and thus helps to cause flexion of the thigh at the hip joint at the same time that it also helps to extend the knee. Another long muscle, the *sartorius,* extends from the anterior iliac crest over the front of the thigh, passing inferiorly and medially over both the hip and the knee joints to insert on the posterior aspect of the medial condyle of the tibia. When this muscle is contracted, it helps to flex both the hip joint and the knee joint at the same time, and also abducts and rotates the thigh laterally. Thus, this is the muscle that positions the thigh when one sits on the floor with crossed legs.

On the back of the thigh are the *hamstrings,* illustrated in Figure 8–3, the *biceps femoris,* the *semitendinosus,* and the *semimembranosus.* The function of these muscles is both to extend the thigh at the hip joint and at the same time to flex the leg at the knee joint.

The Leg. The most prominent superficial muscle of the leg is the *gastrocnemius,* which forms much of the calf of the leg (Fig. 8–3). This muscle extends all the way from both condyles of the distal end of the femur to the calcaneus, or heel bone, attaching to the calcaneus by the *Achilles tendon.* The gastrocnemius causes mainly plantar flexion (downward movement) of the foot. Lying underneath the gastrocnemius muscle is another large muscle, the *soleus,* which extends from the back of the tibia and fibula also to the Achilles tendon and thence to the heel.

On the anterior side of the leg (Fig. 8–2) are much smaller muscles that function mainly to dorsiflex the foot. The *tibialis anterior* arises from the anterolateral surface of the tibia, and its tendon passes anterior to the ankle joint and inserts into the medial side of the foot, thus dorsiflexing the foot and causing inversion of the foot as well. On the lateral side of the leg the *peroneus longus* and *brevis* arise from the fibula, and their tendons wrap around the lateral side of the foot to provide mainly eversion of the foot. And, finally, the *extensor digitorum longus* and *extensor hallucis longus* send their tendons under ligaments in the anterior ankle region all the way to the toes to cause extension and also, at the same time, to aid in dorsiflexion of the foot.

Recapitulation

Now, let us pause for a few moments and reflect on what we have learned. The superficial muscles that we have discussed represent well over half of the total muscle mass of the body, and they are responsible for major movements of the body. Therefore, review all features of the muscles as illustrated in Figures 8–2 and 8–3. In the remainder of this chapter and in the following chapter we will discuss in much more detail the specific movements of the individual joints and the muscles involved.

SUPERFICIAL FACIAL AND NECK MUSCLES— MUSCLES OF EXPRESSION

The superficial facial muscles, illustrated in Figure 8–4, do not cause movement of bones. Instead, they are imbedded mainly in the subcutaneous tissues of the face immediately beneath the skin and cause movement of the facial skin. These muscles attach to bones only at a very few points, such as along the ridge of the nose, along the infraorbital surface of the maxilla, and on the anterior surface of the mandible. Most other attachments are to the fibrous tissues and fibrous sheaths underneath the skin. Some special features of the superficial facial and neck muscles are as follows:

The Muscles of the Scalp and Forehead— the Furrowed Brow. Note at the top of the scalp a white covering called the **galea aponeurotica** that lies beneath the skin and superficial fascia of the scalp. This is a very strong *aponeurosis* (a broad flat tendon). Attached to the aponeurosis are several superficial muscles, the most important of which is the **frontalis,** which extends over the forehead and downward into the subcutaneous tissue over the orbital ridges. This muscle lifts the eyebrows and also causes the furrowed brow. And at the same time it pulls the galea aponeurotica forward, thereby also shifting the scalp in the forward direction. Extending from the midportion of the lower end of the frontalis is another small muscle lying over the bridge of the nose, the **procerus muscle.** This functions to lift the skin of the nose upward, which usually occurs at the same time that the eyebrows are lifted.

The **occipitalis,** lying over the occipital bone of the skull and attaching to the posterior margin of the galea aponeurotica, can pull the galea backward, thus sliding the whole scalp backward. The **auricularis,** a "vestigial muscle" (a muscle that was used to major advantage in our ancestral animals but not in the human being), has superior, anterior, and posterior divisions all attaching from the galea to the ear. Un-

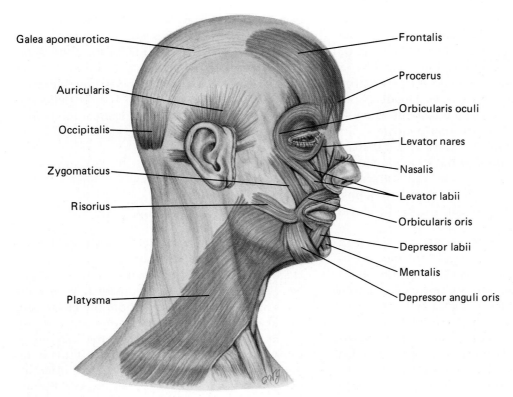

Galea aponeurotica
Auricularis
Occipitalis
Zygomaticus
Risorius
Platysma

Frontalis
Procerus
Orbicularis oculi
Levator nares
Nasalis
Levator labii
Orbicularis oris
Depressor labii
Mentalis
Depressor anguli oris

FIGURE 8–4 The superficial muscles of the face and neck.

doubtedly you will have several friends who can wiggle their ears. These are persons who have inherited voluntary control of the auricularis muscle from their lower animal ancestors.

Muscles of the Eyelid. When the muscles attached to the eyelid are relaxed, the eyelids become almost entirely closed because this is the normal relaxed state of the lids, and a slight amount of constriction by the **orbicularis oculi,** a constrictor muscle that encircles the eye, can then cause complete lid closure. Not shown in the figure is a deeper muscle, the **levator palpebrae,** that lies inside the upper portion of the orbit and inserts into the upper eyelid, elevating it when contracted. In addition, the upper eyelid contains a considerable amount of embedded **smooth muscle** that contracts when stimulated by the sympathetic nervous system (see Chapter 19). When a person is awake, the sympathetic nervous system becomes active, and this smooth muscle in the upper eyelid helps to maintain the wide open eyelids of the awake condition.

The muscles inside the orbit that control movements of the eyeball will be discussed in Chapter 21 in relation to the functions of the eye.

Muscles of the Smile. The principal muscles of the smile are the **risorius,** which pulls the angles of the mouth lateralward, and

the **zygomaticus,** which pulls the angle of the mouth upward toward the zygomatic bone. However, the **levator labii superioris** and **levator nares** also usually contract at the same time, respectively raising the upper lip and lateral border of the nose. On other occasions, the **nasalis muscle** of the lateral border of the nose helps to lift the lateral skin of the nose and to flare the nostrils.

An antismile muscle is the **depressor anguli oris** that pulls the angle of the mouth downward when one is in an unpleasant mood.

Muscles of the Lips. To open the lips, the **levator labii superioris** lifts the upper lip, and the **mentalis** and **depressor labii inferioris** pull the lower lip downward. To keep the mouth closed and also to cause pouting, the **orbicularis oris** constricts around the lips.

Superficial Neck Muscle, the Platysma— the Muscle of Tension. The **platysma muscle** is very thin and lies immediately beneath the skin of the anterior neck, spreading broadly from the clavicle upward over the neck and mandible to insert into the soft tissues of the side of the face. When a person becomes tense, this muscle is apt to become tightened, giving one a taut and urgent appearance. Also, it contracts to take up the slack in the skin when the neck is bent forward.

TABLE 8–1
Superficial Facial and Neck Muscles

Muscle	Function	Origin	Insertion	Nerve
Frontalis	Lift eyebrows	Galea aponeurotica	Skin of eyebrows	Facial (VIIth cranial)
Procerus	Lifts anterior skin of nose	Skin of forehead	Skin of nose	Facial
Levator nares	Lifts lateral skin of nose	Nasal process of maxilla	Lateral cartilage of nose	Facial
Nasalis	Flares the nostril	Maxilla	Lateral cartilage of nose	Facial
Occipitalis	Pulls scalp backward	Occipital bone	Galea aponeurotica	Facial
Auricularis	Moves ears up, forward, backward	Lateral fascia of head	Cartilage of ear	Facial
Orbicularis oculi	Closes the eye	Margins of orbit	Skin around eyes	Facial
Levator palpebrae	Lifts superior eyelid	Upper border of optic canal	Cartilage of upper eyelid	Oculomotor (IIIrd cranial)
Risorius	Lateral pull on angle of mouth	Lateral facial fascia	Angle of mouth	Facial
Zygomaticus	Superior and lateral pull on upper lip	Zygomatic bone	Lateral upper lip	Facial
Levator labii superioris	Lifts upper lip	Nasal process of maxilla	Upper lip	Facial
Depressor labii inferioris	Depresses lateral lower lip	Anterior lower mandible	Lower lip	Facial
Mentalis	Depresses skin below lower lip	Anterior medial mandible	Skin below lower lip	Facial
Orbicularis oris	Constricts mouth	Fascia and skin around lips	Fascia and skin around lips	Facial
Platysma	Tenses skin of anterior neck and lower lateral face	Subcutaneous tissue of upper anterior thorax and shoulder	Lower mandible and lateral facial skin	Facial

NOTE: The nerves of this table are described in Chapter 14.

MUSCLES OF MASTICATION

The principal muscles of mastication are illustrated in Figures 8–5 and 8–6, Figure 8–5 illustrating the superficial muscles and 8–6 the deep muscles. Note especially in Figure 8–5 the two powerful muscles: (1) the *temporalis* and (2) the *masseter*. The masseter muscle is the more superficial.

The **masseter** originates from the outer and lower surfaces of the zygomatic arch (which has been removed in the figure so that the lower end of the temporalis muscle can be seen) and spreads downward over the mandible to insert along the lateral angle and ramus of the mandible.

The **temporalis** is the large muscle arising from the temporal region of the skull. The fibers of this muscle fuse into a tendon that passes deep to the zygomatic arch and inserts into the coronoid process of the mandible.

Contraction of the masseter and the temporalis pulls the lower jaw upward and backward (retraction of the mandible). Thus, it is these two muscles that provide the principal power for closing the jaws in chewing.

The deep muscles of mastication, illustrated in Figure 8–6, consist of the *lateral pterygoid* and the *medial pterygoid*, each of which has two separate heads.

The **lateral pterygoid** arises from the inferior lateral surface of the temporal bone and the posterolateral surface of the sphenoid bone; it inserts into the neck of the mandible immediately inferior to its head. Contraction of this muscle pulls the mandible anteriorly (or pro-

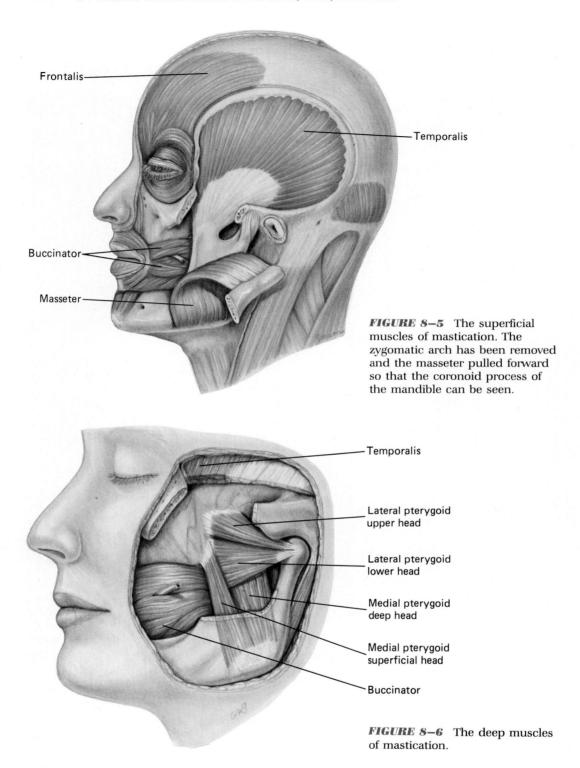

Frontalis

Temporalis

Buccinator

Masseter

FIGURE 8–5 The superficial muscles of mastication. The zygomatic arch has been removed and the masseter pulled forward so that the coronoid process of the mandible can be seen.

Temporalis

Lateral pterygoid upper head

Lateral pterygoid lower head

Medial pterygoid deep head

Medial pterygoid superficial head

Buccinator

FIGURE 8–6 The deep muscles of mastication.

trudes it), thus providing a forward sliding and grinding motion of the lower teeth.

The **medial pterygoid** arises from the posterolateral and inferior surfaces of the sphenoid bone and inserts in the medial surface of the angle of the mandible. When this muscle is contracted on both sides of the jaw at the same time, it helps the temporalis and the masseter muscles close the jaw. However, when it contracts on

only one side at a time, it can slide the jaw from side to side, providing an additional grinding motion.

Finally, the **buccinator muscle,** illustrated in both figures, attaches to both the maxilla and the mandible. This is the principal muscle of the cheek pouch. It is important in the chewing process because it helps to position the food on the teeth.

TABLE 8–2
Muscles of Mastication and Swallowing

Muscle	Function	Origin	Insertion	Nerve
Temporalis	Raises and retracts mandible	Lateral skull over temporal and parietal	Coronoid process of mandible	Mandibular division of trigeminal (Vth cranial)
Masseter	Raises mandible	Zygomatic arch	Angle and ramus of mandible	Mandibular division of trigeminal (Vth cranial)
Lateral pterygoid	Protrudes mandible and side-to-side movement	Surfaces of temporal and sphenoid	Neck of mandible	Mandibular division of trigeminal (Vth cranial)
Medial pterygoid	Raises mandible and side-to-side movement	Lateral sphenoid and posterior maxilla	Medial angle of mandible	Mandibular division of trigeminal (Vth cranial)
Buccinator	Compresses cheek	Superior, medial surfaces of mandible	Alveolar process of mandible	Facial (VIIth cranial)
Genioglossus	Protrudes and depresses tongue	Inside anterior and medial surfaces of mandible	Tongue	Hypoglossal (XIIth cranial)
Mylohyoid	Elevates hyoid bone and larynx, depresses jaw	Inside anterior and medial surfaces of mandible	Body of hyoid	Mandibular division of trigeminal (Vth cranial)
Geniohyoid	Elevates hyoid bone and larynx and pulls these forward, depresses jaw	Inside anterior and medial surfaces of mandible	Body of hyoid	First cervical
Tensor veli palatini	Tenses soft palate	Lateral nasal surface of sphenoid	Soft palate	Mandibular division of trigeminal (Vth cranial)
Levator veli palatini	Raises soft palate	Anterolateral temporal	Soft palate	Vagus (Xth cranial)
Superior pharyngeal constrictor	Constricts superior pharynx	Surface of sphenoid and medial surfaces of mandible	Posterior wall of pharynx	Vagus (Xth cranial)
Middle pharyngeal constrictor	Constricts medial pharynx	Greater horns of hyoid bone	Posterior wall of pharynx	Vagus (Xth cranial)
Inferior pharyngeal constrictor	Constricts inferior pharynx	Posterolateral surfaces of larynx	Posterior wall of pharynx	Vagus (Xth cranial)

NOTE: The nerves of this table are described in Chapter 14.

The anterior neck muscles, to be described later, also play a role in mastication because they pull the jaw downward between bites.

MUSCLES OF SWALLOWING

Figures 8–7 and 8–8 illustrate the principal muscles of swallowing as well as other essential structures involved in the swallowing process. Swallowing begins with the *tongue* (Fig. 8–7), the body of which is a large muscular mass comprised principally of the **genioglossus muscle** that arises from the inside surfaces of the anterior and lateral mandible. Muscle bundles extend into the tongue in several different directions. Contractions of some of these shorten the width of the tongue; others decrease its thickness. When both the width and the thickness are decreased at the same time, the tongue protrudes forward. Contraction of other muscle bundles in the tongue shortens its length while extending its breadth and thickness. And the tongue can be rolled around in the mouth in any manner to help in the chewing process as well as in the swallowing process.

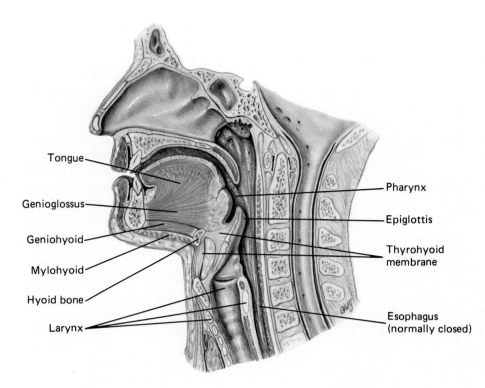

Tongue

Genioglossus

Geniohyoid

Mylohyoid

Hyoid bone

Larynx

Pharynx

Epiglottis

Thyrohyoid membrane

Esophagus (normally closed)

FIGURE 8–7 The tongue and the muscles beneath the tongue; also shown are their relationships to neck structures important in swallowing.

Tensor veli palatini

Levator veli palatini

Superior pharyngeal constrictor

Middle pharyngeal constrictor

Thyrohyoid membrane

Inferior pharyngeal constrictor

Esophagus

Maxilla

Buccinator

Mandible

Mylohyoid

Geniohyoid

Hyoid bone

Larynx

Trachea

FIGURE 8–8 The buccinator and the pharyngeal constrictor muscles.

To cause swallowing, the tongue and the buccinator squeeze the food backwards along the roof of the mouth toward the *pharynx.* Then a rapid sequence of movements in the posterior mouth and pharynx causes the remainder of the swallowing process.

First, before swallowing can occur, the trachea must be closed and the esophagus opened. This is achieved by contraction of muscles that lift the larynx upward and forward. Two of these muscles are the **mylohyoid** and the **geniohyoid** (Fig. 8–7), both of which arise from the inner surface of the mandible and pull the hyoid bone upward and forward toward the mandible. This bone, in turn, is attached by means of a strong ligamentous fascia, the **thyrohyoid membrane,** to the larynx. Therefore, the larynx, too, is pulled upward and forward. Note in Figure 8–7 that the esophagus is normally closed. However, when the larynx moves forward, the attachments of the esophagus to the larynx pull the esophagus open. Note also that the epiglottis is normally in an upward position and allows air to pass into the trachea. However, as the larynx is pulled upward while the tongue is pushing backwards, the epiglottis is drawn inferiorly to the closed position over the opening of the larynx.

At the same time, in the posterior part of the mouth, two muscles, the **tensor veli palatini** and **levator veli palatini** (Fig. 8–8), lift the soft palate upward, which closes the opening of the posterior nares into the pharynx and prevents reflux of food into the nose.

Finally, once food has entered the pharynx, a sequence of nervous signals from the medulla of the brain causes successive contraction of three pharyngeal constrictor muscles (Fig. 8–8) one after the other, first the **superior pharyngeal constrictor,** then the **middle pharyngeal constrictor,** and lastly, the **inferior pharyngeal constrictor.** These muscles are attached anteriorly to the mandible, the hyoid bone, and the larynx, and they wrap around to the back of the pharynx. The entire sequential contraction of these muscles occurs in 1 to 2 seconds, during which time the food is squeezed downward in a peristaltic fashion (a forward moving constriction) into the esophagus.

MOVEMENTS OF THE NECK AND HEAD

The neck muscles and their attachments to the head are illustrated in Figures 8–9 through 8–12. The movements caused by these muscles can be divided into (1) those that cause lateral movement of the head and neck, (2) those that cause forward movement, or flexion, of the head and neck, and (3) those that cause backward movement, or extension, of the head and neck.

Lateral Movement of the Head and Neck. Figure 8–9 illustrates the superficial muscles that cause lateral movement of the head and neck, and Figure 8–10 illustrates the deeper muscles that cause this same movement.

One of the most powerful lateral muscles of the neck is the **sternocleidomastoid** muscle that connects the mastoid process on the lateral side of the skull to the superior end of the sternum and to the medial end of the clavicle. Figure 8–9 shows that this muscle extends inferiorly and anteriorly from the mastoid process of the lateral skull. When it contracts, it rotates the head to the opposite side and at the same time tilts the face slightly upward, thus allowing the person to look to the side. When both sternocleidomastoid muscles are contracted simultaneously along with contraction of muscles in the back of the neck, the entire neck becomes pow-

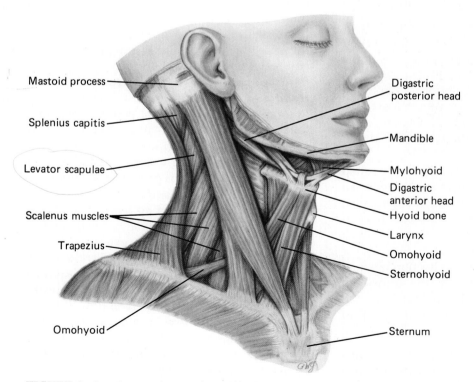

FIGURE 8–9   The more superficial lateral muscles of the neck.

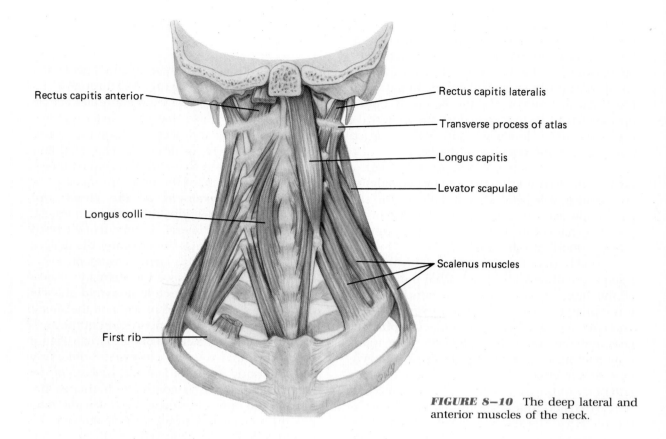

Rectus capitis anterior

Longus colli

First rib

Rectus capitis lateralis

Transverse process of atlas

Longus capitis

Levator scapulae

Scalenus muscles

FIGURE 8–10 The deep lateral and anterior muscles of the neck.

erfully splinted so that it can literally support the weight of the body.

The deeper muscles that move the head and neck to either side (Fig. 8–10) include the

1. **scalenus muscles,** which attach at their lower ends mainly to the first rib and at their upper ends to the transverse processes of the cervical vertebrae;
2. **levator scapulae,** which connects the transverse processes of the cervical vertebrae to the upper medial border of the scapula and pulls the neck to the side at the same time that it lifts the scapula;
3. many small slips of muscle, collectively called the **longus colli,** that lie along the cervical vertebrae themselves, attaching the transverse processes and bodies of the vertebrae to each other, and, when contracted, bend the neck to one side or the other; and
4. **longus capitis,** which is a collection of small parallel muscles connecting the transverse processes of the cervical vertebrae with the lateral inferior surface of the skull medial to the mastoid process.

Note also in Figure 8–10 two prominent muscles that connect each transverse process of the *atlas* with the inferior surface of the skull, the **rectus**

capitis anterior and the **rectus capitis lateralis.** It will be recalled that the atlas is the uppermost vertebra that supports the skull itself. It has especially prominent transverse processes, and these muscles from these transverse processes to the skull control the movement of the skull on the atlas itself, especially causing flexion or lateral bending.

Muscles that Cause Extension of the Head and Neck. Illustrated in Figure 8–9 are two superficial muscles that cause extension of the head and neck. One of these that has a slight extensor action is the **upper portion of the trapezius muscle,** which was illustrated in more detail in Figure 8–3. This portion of the trapezius connects the posterior surface of the occiput to the spinous processes of all the cervical vertebrae as well as to the scapula and clavicle. Secondly, a more important extensor is the **splenius capitis,** which lies deep to the trapezius and extends from the posterior spinous processes of the cervical vertebrae to the posterior base of the skull. When contracted, this muscle provides powerful extension of the head.

The deep muscles of the posterior neck are illustrated in Figure 8–11. These do not need to be named in detail other than to point out that they provide an array of muscles attaching the

TABLE 8–3
Muscles Controlling Neck Movements

Muscle	Function	Origin	Insertion	Nerve
Sternocleidomastoid	Rotates head and flexes neck	Sternum and clavicle	Mastoid process of skull	Accessory (XIth cranial) and cervical plexus
Scalenus muscles	Raises first and second ribs	2nd–6th cervical vertebrae (mainly transverse processes)	First and second ribs	2nd–7th cervical
Levator scapulae	Raises scapula	Transverse processes of 1st–4th cervical vertebrae	Medial border of scapula	3rd and 4th cervical
Longus colli	Flexes cervical vertebrae, and lateral bending	Transverse processes and bodies of 3rd cervical to 3rd thoracic vertebrae	1st–6th cervical vertebrae	All cervical nerves
Longus capitis	Flexes head, and lateral bending	Transverse processes of 3rd–6th cervical vertebrae	Anterior base of occipital	1st–3rd cervical
Rectus capitis anterior	Flexes head, and lateral bending	Lateral body of atlas	Anterior base of occipital	1st and 2nd cervical
Rectus capitis lateralis	Flexes head, and lateral bending	Transverse process of atlas	Anterior base of occipital	1st and 2nd cervical
Splenius capitis	Extends head and neck	Spines of all cervical and 1st–3rd thoracic vertebrae	Posterior base of occipital	3rd–8th cervical
Posterior deep neck muscles	Extends head and neck	Processes and bodies of cervical vertebrae	Processes and bodies of cervical vertebrae	All cervical nerves
Inferior oblique	Rotates head	Spinous process of axis	Transverse process of atlas	2nd cervical
Digastric	Elevates hyoid, depresses mandible	Lower border of mandible and anterolateral temporal	Greater horns of hyoid	Mandibular division of trigeminal (Vth cranial) and facial (VIIth cranial)
Mylohyoid	Elevates hyoid, depresses mandible	Inside anterior and medial surfaces of mandible	Body of hyoid	Mandibular division of trigeminal (Vth cranial)
Geniohyoid	Elevates hyoid, depresses mandible	Inside anterior and medial surfaces of mandible	Body of hyoid	1st cervical
Sternohyoid	Depresses hyoid and mandible	Manubrium of sternum	Body of hyoid	Cervical plexus
Sternohyoid	Depresses hyoid and mandible	Manubrium of sternum	Thyroid cartilage of larynx	Cervical plexus
Omohyoid	Depresses hyoid and mandible	Superior border of scapula	Lateral hyoid bone	Cervical plexus

NOTE: The nerves of this table are described in Chapter 14.

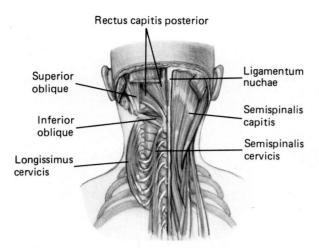

FIGURE 8–11   The deep muscles of the back of the neck.

spinous and transverse processes of the vertebrae of the neck with the posterior base of the skull. However, there are two points about this deep muscular system that do need to be mentioned. First, a very strong ligament called the **ligamentum nuchae** extends from the occipital bone of the skull along the tips of all spinous processes of the cervical vertebrae. This ligament prevents excessive flexion of the head and neck and therefore prevents damage to the spinal cord in the vertebral canal. Second, note the **inferior oblique muscle** that connects the spinous process of the second cervical vertebra (the *axis*) with the transverse process of the *atlas*. It will be recalled from Chapter 6 that the atlas rotates around the upward projecting *dens* of the axis. This inferior oblique muscle and its partner

on the opposite side of the vertebra serve to rotate the head to one side or the other by pulling one or the other transverse process of the atlas toward the spinous process of the axis.

Muscles that Flex the Neck and Head. The muscles that flex the neck and head are mainly those that arise from the sternum and clavicle inferiorly and insert on the mandible superiorly. If these are contracted at the same time that the powerful jaw muscles are contracted, then the whole head and neck will bend forward.

The anterior neck muscles are illustrated in Figure 8–12. Aside from the **sternocleidomastoid muscle,** which has already been discussed, the remainder of the anterior neck muscles are divided into two separate groups, those above the hyoid bone, the **suprahyoid muscles,** and those below, the **infrahyoid muscles.** It will be recalled from the earlier discussion of the muscles of swallowing that the hyoid bone plays an important role in lifting the larynx upward and anteriorly during the swallowing process, thereby helping to open the esophagus at the same time.

The principal **suprahyoid muscles** are as follows:

1. The **digastric muscle** has two bellies, an anterior belly and a posterior belly, with a tendon in between that passes through a fascial stirrup attached to the greater horn of the hyoid bone; the anterior belly arises from the inferior surface of the anterior mandible and the posterior belly arises from anterior to the mastoid process of the temporal bone.

FIGURE 8–12   The anterior muscles of the neck. (The sternocleidomastoid muscle has been removed on the right side.)

2. The **mylohyoid** is a much more powerful muscle that extends from the entire upper surface of the hyoid bone to the medial surface of the mandible on either side.
3. The **geniohyoid muscle** lies deep to the mylohyoid muscle and extends from the inner anterior surface of the mandible to the body of the hyoid bone.

The principal **infrahyoid muscles** are the

1. **sternohyoid muscle,** which extends from the manubrium of the sternum and medial end of the clavicle to the body of the hyoid bone;
2. **sternothyroid muscle,** which extends from the manubrium of the sternum and medial end of the clavicle to the thyroid cartilage of the larynx;
3. **thyrohyoid muscle,** which extends from the larynx to the hyoid bone; and
4. **omohyoid muscle,** which, like the digastric muscle, has an anterior and a posterior muscle belly and a central tendon that passes through a fascial sling attached to the clavicle—the posterior belly arises from the superior border of the scapula and the anterior

belly attaches to the lateral body of the hyoid bone.

Contraction of the anterior muscles of the neck pulls the head and neck forward. These muscles also lower the jaw in the chewing process.

BACK MUSCLES

The superficial muscles of the back—the deltoid, the trapezius, and the latissimus dorsi—were all illustrated in Figure 8–3. However, these muscles have little to do with the movements of the back and trunk of the body. Instead, they are principally concerned with the movements of the shoulder and the arm. They will be discussed further in relation to movements of the extremities in the following chapter.

Beneath the superficial muscles are the group of muscles that are actually called the *back muscles* (illustrated in Fig. 8–13), and these do cause the movements of the back and trunk. The back muscles extend all the way from the sacrum and iliac crest of the coxal bone upward to the back of the neck and even to the back of

Longissimus cervicis

External intercostals

Quadratus lumborum

Sacrum

Semispinalis capitis

Splenius capitis

Spine of the 7th vertebra

Splenius cervicis

Spinalis thoracis

Longissimus thoracis

Ribs

Iliocostalis lumborum

Iliac crest

FIGURE 8–13 The back muscles, showing the more superficial muscles to the right and the deeper muscles to the left.

TABLE 8—4
Muscles Controlling Trunk Movements

Muscle	Function	Origin	Insertion	Nerve
Erector Spinae:				
1. Iliocostalis lumborum	Extends lumbar vertebrae	Crest of ilium	Angles of ribs 6–12	T-6 through L-5
2. Longissimus thoracis	Extends thoracic vertebrae	Lumbar vertebrae and thoracolumbar fascia	Transverse processes of vertebrae T-1 through T-12 and posterior surfaces of ribs 4–12	T-1 through L-5
3. Spinalis thoracis	Extends thoracic vertebrae	Spinous processes of vertebrae T-11 through L-2	Spinous processes of vertebrae T-1 through T-10	T-1 through L-2
Quadratus lumborum	Flexes lumbar vertebrae laterally	Crest of ilium and lumbar vertebrae	Twelfth rib and transverse processes of vertebrae L-1 through L-4	T-12 through L-2
External intercostals	Lifts ribs	Inferior border of rib	Superior border of rib below	T-1 through T-11
Internal intercostals	Depresses ribs	Inferior border of rib	Superior border of rib below	T-1 through T-11
Rectus abdominis	Flexes trunk, depresses ribs	Pubis	Xiphoid process of sternum and costal arch of rib cage	T-7 through L-1
External oblique	Flexes and rotates trunk, compresses abdomen	Inferior costal margin	Crest of ilium and sheath of rectus abdominis	T-7 through L-1
Internal oblique	Flexes and rotates trunk, compresses abdomen	Iliac crest and lumbar aponeurosis	Inferior costal margin and sheath of rectus abdominis	T-7 through L-1
Transversus abdominis	Compresses abdomen	Inferior costal margin, lumbar aponeurosis, iliac crest	Sheath of rectus abdominis	T-7 through L-2

NOTE: The nerves of this table are described in Chapter 14.

the skull. Along the way they attach by small slips of muscle and tendons to (1) the ribs, (2) the transverse processes of the vertebrae, and (3) the spinous processes of the vertebrae. The largest of these muscles, lying on each side of the vertebral column, is a large mass of muscles collectively called the **erector spinae,** consisting of three parallel muscle columns called the

1. **iliocostalis lumborum,** which is a large bundle of muscle slips passing from the iliac crest, sacrum, and lumbar vertebrae to the posterior surfaces of the lower ribs;
2. **longissimus thoracis,** which is another bundle of muscle slips extending for various distances along the posterior thoracic cage, attaching from rib to rib and to the transverse processes of the vertebrae; and
3. **spinalis thoracis,** which connects along the spinous processes and transverse processes

of the thoracic vertebrae as well as to the heads of the ribs.

Several similar deep muscles pass from the ribs, spinous processes, and transverse processes to the posterior neck and inferior surface of the skull; these are the deep neck muscles discussed previously, including the **splenius cervicis, splenius capitis,** and **semispinalis capitis.**

These muscles all acting synergistically can extend the entire back, even causing hyperextension of the spine; or contraction of the muscles on one side only can bend the back, neck, or head to one side or the other.

To the left in Figure 8–13 are shown deeper aspects of some of the back muscles. Important in this figure is the strong muscle passing from the iliac crest to the twelfth rib, the **quadratus lumborum.** The twelfth rib in turn is connected successively to the higher ribs by the **intercostal**

muscles. The **external intercostal muscles** are illustrated in the figure, extending inferiorly and anteriorly from one rib to the next. Deep to the external intercostals is another layer, the **internal intercostals,** which extend inferiorly and posteriorly, thus crisscrossing the external intercostals. These muscles will be discussed in Chapter 36 in relation to respiration. This entire system, comprised of the quadratus lumborum and the intercostals, provides another means for bending the entire trunk to one side or the other.

ABDOMINAL MUSCLES

The general features of the abdominal muscles were discussed earlier in relation to the superficial muscles of the body. Let us now present some of the special characteristics of these muscles.

The abdominal wall is a composite of muscles and fascia. Most of the abdominal fascia is in the form of sheets that lie over or underneath the abdominal muscles and also form fascial muscle sheaths around some of the muscles.

The abdominal wall can be divided into two separate parts, each of which functions somewhat differently: (1) the *anterior abdominal wall* and (2) the *anterolateral walls.*

Anterior Abdominal Wall. The anterior abdominal wall is illustrated in Figure 8–14. Its principal muscles are the two **rectus abdominis muscles** that lie on either side of the midline. Each of these muscles extends from the

lower costal margin (the lower edge of the thoracic cage) and lower end of the sternum to the pubis. Also, each is encased in a strong **fascial sheath,** illustrated to the right in Figure 8–14. The sheaths of the two rectus abdominis muscles fuse with each other in the midline to form the **linea alba,** a strong narrow fascial band that runs from the lower tip of the sternum to the midpoint of the pubis.

The rectus abdominis muscles are the major muscles for flexion of the trunk. They flex the trunk by pulling inferiorly on the costal margin.

During pregnancy, when the abdominal wall becomes greatly stretched, the fascia that forms the linea alba frequently becomes thin and stretched sideways so that the two rectus abdominis muscles are pulled widely apart. Also, the rectus abdominis muscles themselves often become very thin and ineffectual as muscles at the same time. It is for this reason that many women have poor capability for flexing the trunk after pregnancy, making it difficult for them to sit up from the lying position.

Anterolateral Abdominal Wall. The muscles of the anterolateral abdominal wall are illustrated in Figure 8–15. The anterior, posterior, and inferior margins of the muscles insert into broad, flattened tendons called *aponeuroses.* The anterior aponeuroses are continuous with the fascial sheath of the rectus abdominis, and the posterior aponeuroses are continuous with the fascial sheaths of the back muscles. The inferior aponeuroses insert into the iliac crest

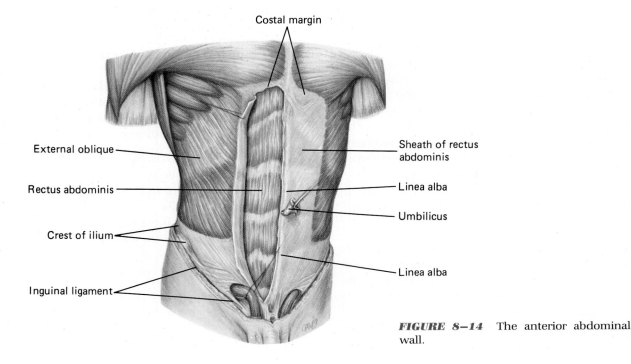

FIGURE 8–14 The anterior abdominal wall.

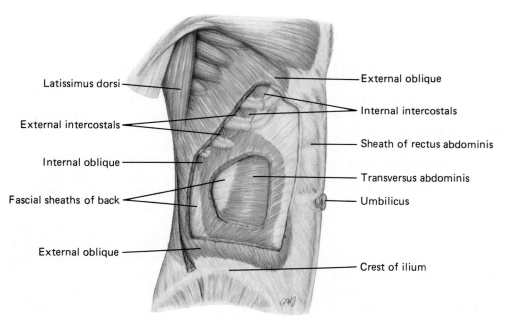

Latissimus dorsi

External intercostals

Internal oblique

Fascial sheaths of back

External oblique

External oblique

Internal intercostals

Sheath of rectus abdominis

Transversus abdominis

Umbilicus

Crest of ilium

FIGURE 8–15   The anterolateral abdominal wall.

and the inguinal ligament that extends between the anterior margin of the iliac crest and the pubis.

The three muscles of the anterolateral wall are the

1. **external oblique,** which passes inferiorly and anteromedially from the lower ribs in an oblique direction, attaching at its other end to the lateral portion of the sheath of the rectus abdominis, to the inguinal ligament, and to the crest of the ilium;
2. **internal oblique,** which arises from the inguinal ligament, the crest of the ilium, and the fascial sheaths of the back muscles and passes superiorly and anteromedially to insert into the lower costal margin and also into the sheath of the rectus abdominis, crossing at right angles to the direction of the external oblique; and
3. **transversus abdominis,** which lies deep to the internal oblique and extends transversely from back fascia, lower ribs, and lateral inguinal ligament around the anterolateral abdominal wall, attaching anteriorly to the lateral edge of the fascial sheath of the rectus abdominis and to the pubis.

When the anterolateral abdominal wall muscles contract in unison on both sides, the abdominal contents are compressed, and this pushes upward on the diaphragm, thus playing an important part in causing expiration. (The diaphragm and other respiratory muscles will be discussed in Chapter 36 in relation to the other respiratory mechanisms.) Contraction of the external and internal obliques on only one side of the body provides a means for lateral body movement in the lumbar region, and contraction of all the abdominal muscles along with contraction of all the back muscles at the same time splints the entire body, which is especially important when the arms are to be used for heavy work.

THE PELVIC DIAPHRAGM AND THE PERINEUM

The pelvic diaphragm, also called the pelvic floor, is composed of the muscles and their fascial coverings that close the inferior outlet of the pelvis; this is the area through which the anus penetrates posteriorly and the urethra and vagina pass anteriorly. The major muscles of the pelvic diaphragm (Fig. 8–16) are as follows:

1. The **levator ani** arises on each side of the pelvis from the inner surfaces of the ilium and pubis. The levator ani muscles from the two sides fuse in the midline, leaving hiatuses for the rectum, vagina, and urethra.
2. The **coccygeus** passes from the coccyx to the ischium and fuses with the posterior border of the levator ani, thus completing the pelvic diaphragm.

These muscles provide the support for both the pelvic and the abdominal contents that

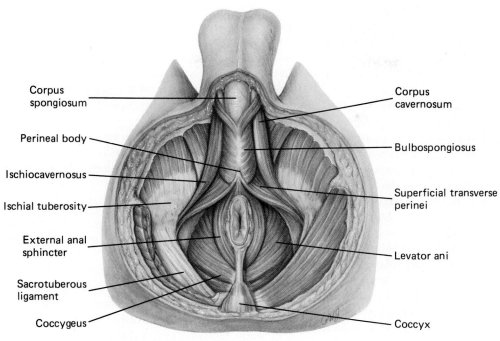

Corpus spongiosum

Corpus cavernosum

Perineal body

Bulbospongiosus

Ischiocavernosus

Ischial tuberosity

Superficial transverse perinei

External anal sphincter

Levator ani

Sacrotuberous ligament

Coccygeus

Coccyx

FIGURE 8–16 The pelvic diaphragm and the perineum.

TABLE 8–5
Muscles of the Pelvic Diaphragm and Urogenital Triangle

Muscle	Function	Origin	Insertion	Nerve
Levator ani	Supports pelvic floor	Inner margins of pubis and ilium	Levator ani from opposite side	S-3 and S-4
Coccygeus	Helps support pelvic floor	Inner margin of ischium	Lateral lower sacrum and coccyx	S-3 and S-4
External anal sphincter	Constricts anus	Tip of coccyx and fascia	Central tendon of perineum	S-3 and S-4
Superficial transverse perinei	Supports central point of perineum	Ischial tuberosity	Central tendon of perineum	Pudendal (S-2, 3, 4)
Ischiocavernosus	Maintains erection of penis or clitoris	Inferior ramus of ischium	Crus penis (or clitoridis)	Pudendal (S-2, 3, 4)
Bulbospongiosus	Constricts bulb of penis or bulbs of vagina	Central tendon of perineum	Around bulb of penis or bulbs of vagina	Pudendal (S-2, 3, 4)
Deep transverse perinei	Supports central point of perineum	Inferior ramus of ischium	Central tendon of perineum	Pudendal (S-2, 3, 4)
Sphincter urethrae	Constricts membranous urethra	Ramus of pubis	Around urethra	Pudendal (S-2, 3, 4)

NOTE: The nerves in this table are described in Chapter 14.

press downward with considerable force when a person is in the upright position.

Note also in Figure 8–16 that, in addition to the levator ani, immediately surrounding the anus is a thick muscle that forms a sphincter around the anus called the **external anal sphincter.** This sphincter normally remains partially contracted all the time and prevents defecation except when it is purposefully relaxed during the defecation process. The levator ani muscles also relax during defecation, thus allowing a downward thrust of the anus and also of the rectum. This straightens the normal angle in the superior portion of the rectum and allows feces from as high up as the sigmoid colon to be expelled.

The Muscles of the Perineal Region. That portion of the pelvic outlet below the pelvic diaphragm, bounded anteriorly and posteriorly by the pubis and coccyx and laterally by the two ischial tuberosities, is called the **perineal region.** The posterior half of this is the **anal triangle** and the anterior half the **urogenital triangle.** The anal triangle is pierced by the anal canal, and the remainder of this region consists almost entirely of skin and subcutaneous tissue.

It is through the **urogenital triangle** that the urethra passes in both the male and the female and the vagina in the female. It is divided by planes into a *superficial space* that lies immediately under the skin and a *deep space* between the superficial space and pelvic diaphragm. The muscles in the superficial space of the male are illustrated in the anterior part of Figure 8–16. These are as follows:

1. The **superficial transverse perinei muscles** arise from the inner border of the two ischii and then insert into a central area of dense fascia called the **perineal body,** located about 2 cm anterior to the anus. These muscles help to support the posterior border of the urogenital triangle.
2. The **ischiocavernosus muscles** arise on each side from the medial border of the ischium and then pass anteriorly to envelop the crura of the penis (or of the clitoris in the female). These crura are part of the root of the penis or clitoris and are composed of erectile tissue that extends anteriorly to become the *corpora cavernosa* forming the major body of these organs.
3. The **bulbospongiosus muscle** wraps around the bulb of the penis or the paired bulbs of the vagina; these bulbs extend anteriorly to become the corpus spongiosum, another portion of the erectile tissue of the penis or clitoris. (The erectile tissue in the male and the female will be discussed more fully in Chapters 48 and 49.)

The bulbospongiosus and ischiocavernosus muscles are said to cause or help to cause erection during intercourse by contracting tonically around the bases of the erectile tissue, thus impeding venous outflow from this tissue and thereby causing the tissue to balloon. It has also been suggested that they contract rhythmically at the termination of intercourse, thus providing the ejaculatory thrusts of the penile erectile tissue, at the same time sweeping the semen forward through the urethra.

The deep space of the urogenital triangle, with its fascial coverings and its muscles, is called the **urogenital diaphragm.** Its muscles are as follows:

1. The **deep transverse perinei** lie deep to the superficial transverse muscles, attaching also laterally to the ischium and medially to the perineal body.
2. The **sphincter urethrae** attaches anteriorly and laterally to the pubis and circles posteriorly around the urethra in both the male and the female and also around the vagina in the female. Contraction of this muscle after voiding expresses the remaining urine from the membranous part of the urethra.

QUESTIONS

1. Describe a typical muscle and its attachment to bone. What is meant by the "origin" of a muscle, and what is meant by its "insertion"?
2. Give a general description of the internal organization of muscle, its excitation by nerves, and its basic mechanism of contraction.
3. What is the source of the energy required for muscle contraction?
4. List the superficial facial muscles and describe the function of each.
5. Describe the muscles of mastication, their origins and insertions, and the jaw movements that they cause.
6. Describe the swallowing mechanism and the sequential roles of the several muscles that cause swallowing.
7. List and describe the muscles that cause movements of the neck and head.
8. Describe the back muscles and the trunk movements that they cause.
9. Describe the abdominal wall, the interrelationships of the muscles in this wall, and the effects of their contraction.
10. Describe the structures and muscles of the pelvic diaphragm and of the perineum.

REFERENCES

See the References, Chapter 9.
See also General Anatomical References, Chapter 1.

9

Muscles and Movement: II. The Extremities

Overview

Though there are more than 100 different muscles in the appendages on each side of the body, a few of these are especially important in causing movement:

Trapezius and *Rhomboid Muscles.* These muscles connect the scapula to the upper vertebral column and help to support the scapula during arm movements.

Pectoralis Major and *Pectoralis Minor.* These pull the shoulder forward and also pull the humerus across the front of the chest.

Latissimus Dorsi. This muscle arises from the vertebrae, fasciae, and ribs of the lower back and extends upward to attach to the humerus. It pulls downward and backward on the humerus and is the most important of all muscles for the climbing motion of the arms.

Muscles Around the Shoulder Joint. Multiple muscles fit tightly around this joint; it is mainly these muscles that prevent separation of the joint.

Biceps Brachii. This muscle extends in the anterior arm all the way from the scapula in the shoulder to the radius in the forearm and provides the principal lifting motion for the forearm.

Triceps Brachii. This muscle attaches from the scapula and humerus to the ulna. It is responsible for extension of the forearm.

Flexor and Extensor Muscles of the Hand and Forearm. Multiple muscles in the anterior portion of the forearm send long tendons through the wrist into the hand and fingers to cause flexion. Similar muscles in the posterior part of the forearm cause extension.

Muscles Within the Hand. More than a dozen muscles within the hand cause such movements as abduction, which means pulling the fingers apart; adduction, which means pulling the fingers together; and opponens movement, which means bringing the thumb into opposition with one or more of the fingers.

Gluteal Muscles. Three separate muscles, the *gluteus maximus,* the *gluteus medius,* and the *gluteus minimus,* attach from the coxal bone to the lateral and posterior shaft of the upper femur and extend the leg, abduct it, or rotate it laterally.

Groin Muscles. These attach mainly from the pubic ramus to the posteromedial side of the femur and cause adduction of the thigh (that is, they pull the two thighs together). It is these muscles that allow a person to grip the back of a horse.

Quadriceps Muscle. This is the very large muscle of the anterior thigh, the largest muscle of the body. It has four separate heads, one of which, the *rectus femoris,* originates from the anterior inferior spine of the coxal bone and is the most important muscle to cause flexion of the thigh. The other three heads originate from the medial, anterior, and lateral surfaces of the femur, and all four heads terminate in the patellar tendon that passes through the patella in front of the knee joint and inserts into the tibia to flex the leg.

Hamstring Muscles. These originate from the ischial tuberosity of the coxal bone; they are the major muscles of the posterior thigh, passing all the way beyond the knee joint to insert into the posterior tibia below the knee. They serve to extend the hip and at the same time to flex the leg.

Gastrocnemius Muscle. This muscle arises by two heads from the posteroinferior end of the femur, and it terminates below in the *Achilles tendon,* which inserts into the heel bone, the calcaneus. This is the powerful muscle that causes downward thrust of the foot during running.

Anterior Tibial Muscle. This lifts the foot upward, called dorsiflexion.

Multiple Muscles Arising Both Anterior and Posterior to the Tibia and Fibula. These send long tendons into the foot to cause flexion and extension of the distal parts of the foot. They are mainly analogous to similar muscles in the forearm that cause flexion and extension of the hand and fingers. In addition, the foot has multiple small muscles that support the foot arches and also help to position the individual toes.

In this chapter we continue our discussion of movements and the muscles that cause these, this time discussing the movements and muscles of the upper and lower extremities. We will begin with the upper extremity (shoulder, arm, forearm, and hand) and then continue to the lower extremity (hip, thigh, leg, and foot).

FIXATION OF AND MOVEMENT OF THE PECTORAL GIRDLE

The **pectoral girdle,** also called the *shoulder girdle,* consists of the scapulae and clavicles, a pair of these bones on either side of the body. As was pointed out in Chapter 7, the clavicle is a weak bone, and the joints connecting the scapula to the clavicle and the clavicle to the sternum have far too little strength to provide great stability of the scapula. Yet, the scapula is the base of operation for the more distal portion of the extremity and does indeed require very strong "fixation" during arm movements. This is provided by a large number of muscles that attach from the trunk to the pectoral girdle, especially to the scapula. Also, contraction of separate ones of these muscles can provide two principal ranges of motion of the scapula as well: (1) *elevating the shoulder or depressing it* and (2) forward thrust of the shoulder, called *protraction,* or backward movement toward the spinal column, called *retraction.* In addition, when the arm is raised, the shoulder muscles can *rotate the scapula* in the upward direction. The principal muscles for causing these movements, as well as for causing fixation of the pectoral girdle, are illustrated in Figures 9–1, 9–2, and 9–3.

Muscles That Elevate the Scapula and Clavicle. The muscles that elevate the scapula and clavicle are illustrated in Figure 9–1; to the left are the superficial muscles and to the right the deeper muscles. These include the following:

1. The **trapezius muscle** originates from the spinous processes of the vertebrae of the entire upper back and attaches to the clavicle and to the spine and acromion of the scapula as illustrated to the left in Figure 9–1. Contraction of the upper portion of the trapezius elevates the scapula and clavicle.
2. The **rhomboideus major** and **rhomboideus minor muscles** pass inferiorly and laterally from the spinous processes of the upper thoracic and lower cervical vertebrae and insert into the medial border of the scapula between the spine and inferior angle.
3. The **levator scapulae** passes from the transverse processes of the cervical vertebrae to the medial border of the scapula superior to the spine.

FIGURE 9–1 The posterior muscles of the pectoral girdle, showing to the left the superficial muscles and to the right the deep muscles.

Muscles That Depress the Scapula. One can see from Figure 9–1 not only that the **trapezius muscle** can elevate the pectoral girdle but also that the lower portion of the trapezius can depress the scapula.

However, by far the most powerful muscle that pulls the pectoral girdle inferiorly is the **latissimus dorsi;** this muscle arises from the lower back, as shown to the left in Figure 9–1. As shown in Figure 9–2, the tendon of the latissimus dorsi does not insert onto the pectoral girdle but instead into a groove on the anterior surface of the upper humerus. When the latissimus dorsi contracts, it pulls the humerus inferiorly, and this in turn pulls the entire girdle inferiorly.

Also important in depressing the pectoral girdle is the **serratus anterior** muscle, illustrated in Figures 9–1 and 9–2, that arises from the anterolateral borders of the first eight ribs and passes posteriorly and superiorly to attach to the medial border of the scapula on its deep surface.

And still another muscle that helps to pull inferiorly on the shoulder girdle is the **pectoralis minor,** a deep muscle of the anterior chest

FIGURE 9–2 The deep muscles of the chest that attach to the anterior surfaces of the pectoral girdle, also showing the attachment of the biceps brachii at the shoulder.

TABLE 9–1
Muscles of Pectoral Girdle and Shoulder Movements

Muscle	Function	Origin	Insertion	Nerve
Trapezius	Lifts scapula, pulls backward and rotates upward	Spines of cervical and thoracic vertebrae and occipital	Lateral clavicle and spine and acromion of scapula	Accessory (cranial XI) and cervical plexus
Rhomboideus major	Lifts and retracts scapula	Spines of vertebrae C-2 through C-5	Medial border of scapula	Dorsal scapular
Rhomboideus minor	Lifts and retracts scapula	Spines of vertebrae C-6 through T-1	Medial border of scapula	Dorsal scapular
Levator scapulae	Lifts scapula	Transverse processes of vertebrae C-1 through C-4	Medial border of scapula superior to spine	C-3 and C-4
Serratus anterior	Draws scapula forward and rotates upward	Anterolateral borders of ribs 1 through 8	Medial border of scapula on deep side	Long thoracic
Pectoralis minor	Draws shoulder forward and downward	Anterior surface of ribs 2 through 5	Coracoid process of scapula	Anterior thoracic
Pectoralis major	Adducts, flexes, and rotates arm medially	Clavicle, sternum, and ribs 1 through 6	Anterior surface of humerus	Anterior thoracic
Latissimus dorsi	Adducts, extends and rotates arm medially	Spines of vertebrae T-1 through L-5, fascia of back, lower ribs	Anterior surface of humerus in bicipital groove	T-1 through T-12
Deltoid	Adducts, flexes, and extends arms	Clavicle, acromion, and spine of scapula	Deltoid tuberosity of humerus	Axillary (cranial XI)
Supraspinatus	Abducts arm	Posterior surface of scapula above spine	Greater tubercle of humerus	Suprascapular
Infraspinatus	Rotates arm laterally	Posterior surface of scapula below spine	Greater tubercle of humerus	Suprascapular
Teres minor	Rotates arm laterally	Inferior, posterior border of scapula	Greater tubercle of humerus	Axillary (cranial XI)
Teres major	Extends, adducts, rotates arm medially	Inferior border of scapula	Anterior surface of humerus	Subscapular
Subscapularis	Rotates arm medially	Deep surface of scapula	Lesser tubercle of humerus	Subscapular
Biceps brachii	Abducts arm, flexes forearm, supinates hand	Long head: scapula above glenoid fossa Short head: coracoid process	Tuberosity of radius	Musculocutaneous
Triceps brachii	Adducts and extends arm, extends forearm	Long head: scapula below glenoid fossa Other heads: shaft of humerus	Olecranon process of ulna	Radial
Coracobrachialis	Flexes, adducts arm	Coracoid process of scapula	Medial surface of humerus	Musculocutaneous

NOTE: The nerves of this table are described in Chapter 14.

GREATER TUBERCLE
Deltoid (cut)
Teres minor
Quadrangular space
HUMERUS
Triangular space
Long head of triceps
Lateral head of triceps
OLECRANON

Supraspinatus
SPINE OF SCAPULA
Infraspinatus
Teres major
Latissimus dorsi

FIGURE 9–3 The deep posterior muscles around the shoulder joint, showing also the relationship of the triceps brachii muscles to the shoulder.

(Fig. 9–2) that arises from the second through the fifth ribs anteriorly and inserts into the coracoid process of the scapula, depressing the point of the shoulder when it contracts.

All these muscles acting together hold the shoulder down when a person supports his weight by downward thrust of his arms (as when walking with crutches).

Muscles That Protract or Retract the Shoulder. Some of the same muscles that elevate or depress the shoulder girdle, when contracted in other combinations, can protract or retract the shoulder. Two muscles that are important for shoulder protraction are the **serratus anterior** and the **pectoralis minor** (Fig. 9–2), both of which also depress the shoulder as discussed before. However, the most powerful protractor of the shoulder is the **pectoralis major** that arises from the entire anterior thoracic wall and inserts into the humerus, as discussed later. When it contracts, it pulls the humerus forward, and the humerus in turn pulls the shoulder forward. It is these protractor muscles that thrust the shoulder forward when a person is practicing pushups.

One can see from Figure 9–1 that the principal muscles for retracting the shoulder are the **trapezius,** the two **rhomboid muscles,** and the **latissimus dorsi.** It is also these muscles that provide backward fixation for the shoulder when one is pulling a heavy load toward the body with the arms.

MUSCLES AND MOVEMENTS AT THE SHOULDER JOINT

The **shoulder joint** is a *ball and socket type* and has the greatest range of motion of any joint of the body, including (1) *flexion* (bringing the arm forward), (2) *extension* (carrying the arm backward), (3) *abduction* (raising the arm outward and upward), (4) *adduction* (pulling the arm toward the body), and (5) *rotation*, both medial and lateral. The muscles that cause these motions are very complex, and their interrelationships are not easy to comprehend. However, the best way to understand them is to begin with the deeper muscles and then return to the more superficial muscles. The deeper muscles are illustrated in Figures 9–2 and 9–3, and the superficial muscles can be seen in Figures 9–1 and 9–4.

Scapular Muscles That Support the Shoulder Joint. Because the socket of the shoulder joint is very shallow, the joint must be supported in some way other than bony joint fixation. This support comes from the muscles and muscle tendons that span the shoulder joint in tight proximity to the joint capsule. Figures 9–2 and 9–3 show that the tendons of some of these muscles actually fuse with and become part of the capsule. Four especially important muscles (and their tendons) support the shoulder joint. Three of these arise from the posterior surface of the scapula: (1) the **supraspinatus,**

FIGURE 9–4 The superficial anterior shoulder and arm muscles.

arising from the surface above the scapular spine; (2) the **infraspinatus,** arising from beneath the spine; and (3) the **teres minor,** arising along the inferior, posterior border of the scapula. The fourth muscle, the **subscapularis** (Fig. 9–2), arises from the anterior surface of the scapula (on its deep side). The tendons of the three posterior scapular muscles all fuse with the superior and posterior portions of the capsule and insert into the greater tubercle of the humerus. The tendon of the anterior muscle, the subscapularis, fuses with the anterior portion of the joint capsule and inserts into the lesser tubercle. In effect, therefore, the tendons of all these muscles serve as a single *musculotendinous cuff* over the joint that provides most of the stability of the shoulder joint.

However, as shown in Figure 9–2, still another muscle plays a role in stabilizing this joint; this is the *long head of the biceps* of the arm. The tendon of this muscle arises from the scapula at the superior border of the glenoid fossa and then passes over the head of the humerus in the intertubercular groove, thus providing an upper sling for the joint.

Anterior Deep Muscles That Connect the Arm with the Pectoral Girdle.
Figure 9–2 illustrates the important anterior muscles that connect the arm to the pectoral girdle. These are as follows:

1. The **biceps brachii muscle** has two separate heads, the *long head* and the *short head.* The long head arises from the supraglenoid tubercle of the scapula and passes over the top of the shoulder joint as previously described; the short head arises from the coracoid process of the scapula. The distal end of the biceps completely bypasses the humerus and inserts into the tuberosity of the radius of the forearm.

2. The **coracobrachialis** lies medial and deep to the short head of the biceps. It also arises from the coracoid process of the scapula, but instead of bypassing the humerus, it inserts into the anteromedial surface of its shaft.

When the biceps and coracobrachialis muscles contract, they lift the arm and bring it forward, the action known as *flexion of the arm.* In addition, the biceps also flexes the forearm at the elbow joint at the same time.

Posterior Muscles Attaching the Arm to the Pectoral Girdle and to the Back.
Figure 9–3 illustrates the posterior muscles attaching the arm to the pectoral girdle and to the back. These are the following:

1. The **latissimus dorsi** is the very powerful muscle arising from the lower back as described earlier, but its tendon passes through the deeper recesses of the axilla (the area underneath the shoulder joint) and eventually inserts into the bicipital groove on the anterior surface of the humerus underneath the tendon of the long head of the biceps. Thus, this muscle pulls the humerus downward and strongly backward, the action known as *extension of the arm.* It also pulls the arm against the body in the *adducted position* while simultaneously causing *medial rotation.* This is the powerful "climbing muscle" that allows the arms to pull the body up a tree or up the side of a mountain, and it is also the "swimming muscle" that provides the motive power for most swimming strokes.

2. The **teres major** arises from the inferior border of the scapula and also passes through the axilla to insert along with the tendon of the latissimus dorsi into the bicipital groove on the anterior surface of the humerus. This muscle supports the *extension, adduction,* and *medial rotation* of the humerus caused by the latissimus dorsi.

3. The **long head of the triceps muscle** arises from the scapula immediately below the inferior lip of the glenoid cavity; the other end of this muscle inserts into the olecranon process of the ulna.

The Superficial Muscles of the Shoulder Joint.
Now let us add the two principal super-

ficial muscles, illustrated from the anterior in Figure 9–2 and from the posterior in Figure 9–1. These are described here:

1. The **deltoid muscle** arises anteriorly from the clavicle, medially from the acromion of the scapula, and posteriorly from the spine of the scapula. This muscle then spreads downward in a wide swath over the entire anterior, lateral, and posterior aspects of the shoulder joint, finally tapering to insert into the lateral border of the humerus at the deltoid tuberosity about one third the distance down the shaft. Contracting the anterior part of the deltoid flexes the arm and rotates it medially; contracting the posterior part extends the arm and rotates it laterally; but most important of all, contracting all parts of the deltoid simultaneously lifts the arm upward and laterally in the action of abduction. (At the same time that this occurs, the serratus anterior, discussed previously, rotates the scapula upward, which provides still additional abduction of the arm.)

2. The **pectoralis major** arises from the anterior surface of much of the chest and also from the medial half of the clavicle. It then inserts into the bicipital groove on the anterolateral border of the upper humerus, slightly above the insertion of the deltoid. This muscle pulls the humerus against the body in adduction and across the chest at the same time. It also rotates the humerus medially, thus aiding in moving the hand toward the opposite side of the body. When it contracts at the same time that the latissimus dorsi contracts, the arm becomes tightly fixed against the lateral wall of the chest.

MUSCLES THAT CAUSE MOVEMENT AT THE ELBOW

The **elbow joint** is of the *"hinge"* type, allowing only forward movement, or *flexion*, and backward movement, or *extension*.

Extension of the Elbow. The only muscle of significance to cause extension of the elbow is the **triceps brachii** (Fig. 9–5A); the long head of this was discussed previously in relation to movement at the shoulder joint. Note in the figure that the triceps has three separate heads: (1) a *long head* that arises from the scapula at a point immediately beneath the glenoid fossa of the shoulder joint, (2) a *lateral head* that arises from the lateral surface of the humerus, and (3) a *medial head* that arises from the medial and posterior surfaces of the humerus. All of these fuse

FIGURE 9–5 The triceps brachii (A) and biceps brachii (B) muscles of the upper arm and their attachments.

TABLE 9–2
Muscles of Arm, Forearm, and Hand

Muscle	Function	Origin	Insertion	Nerve
Triceps brachii	Extends forearm, adducts and extends arm	Scapula beneath glenoid fossa, posterior, lateral, and medial surfaces of humerus	Olecranon process of ulna	Radial
Biceps brachii	Flexes forearm, supinates hand, abducts arm	Long head: scapula above glenoid fossa Short head: coracoid process	Tuberosity of radius	Musculocutaneous
Brachialis	Flexes forearm	Anterior humerus	Anterior, superior ulna	Radial, musculo-cutaneous
Brachioradialis	Flexes forearm	Lateral epicondyle of humerus	Lower end of radius	Radial
Pronator teres	Pronates hand	Medial epicondyle of humerus	Lateral surface of radius	Median
Pronator quadratus	Pronates hand	Anterior surface of distal ulna	Lateral surface of distal radius	Median
Supinator	Supinates hand	Lateral epicondyle of humerus	Medial surface of upper radius	Radial
Palmaris longus	Flexes wrist	Medial epicondyle of humerus	Palmar aponeurosis	Median
Flexor carpi radialis	Flexes, abducts wrist	Medial epicondyle of humerus	Bases of 1st and 2nd metacarpals	Median
Flexor carpi ulnaris	Flexes, adducts wrist	Medial epicondyle of humerus, surface of ulna	Base of 5th metacarpal	Ulnar
Extensor carpi radialis longus and brevis	Extends, abducts wrist	Lateral epicondyle of humerus	Bases of 2nd and 3rd metacarpals	Radial
Extensor carpi ulnaris	Extends, adducts wrist	Lateral epicondyle of humerus and posterior ulna	Base of 5th metacarpal	Radial
Flexor digitorum superficialis	Flexes fingers	Medial epicondyle of humerus, anterior surfaces of ulna and radius	Middle phalanges of fingers	Median
Flexor digitorum profundus	Flexes fingers	Anterior surface of ulna	Distal phalanges of fingers	Ulnar, median

NOTE: The nerves of this table are described in Chapter 14.

at their lower end and form a single tendon that inserts into the olecranon process of the ulna. Thus, the triceps exerts a powerful force on the upper end of the ulna to cause extension of the forearm. This is the muscle that allows one to extend the forearm when pushing a heavy door or performing such functions as boxing.

Muscles That Cause Flexion at the Elbow Joint. One muscle that causes flexion at the elbow joint is so well known that it hardly needs explanation: the **biceps brachii.** This

muscle arises by two tendons from the scapula, the *long head* and the *short head*, as was discussed earlier in relation to the shoulder joint. Then the muscle inserts into the medial surface of the radius at the radial tuberosity about 4 cm in front of the elbow joint. This obviously provides an appropriate mechanical advantage to lift the forearm into a flexed position. The fact that the biceps bypasses the humerus allows the scapula to support much of the lifting force of the forearm without transmitting this force to

Muscle	Function	Origin	Insertion	Nerve
Flexor pollicis longus	Flexes thumb	Anterior surfaces of radius and ulna	Base of thumb distal phalanx	Median
Flexor pollicis brevis	Flexes thumb	Fascia and ligament anterior to carpals	Base of thumb proximal phalanx	Median, ulnar
Flexor digiti minimi brevis	Flexes little finger	Fascia and ligaments anterior to carpals	Medial side of little finger proximal phalanx	Ulnar
Extensor digitorum	Extends fingers	Lateral epicondyle of humerus	Dorsal bases of finger distal phalanges	Radial
Extensor pollicis longus	Extends thumb	Dorsal ulna, interosseous membrane	Dorsal surface of thumb distal phalanx	Radial
Extensor pollicis brevis	Extends thumb	Dorsal radius, interosseous membrane	Dorsal surface of thumb proximal phalanx	Radial
Extensor indicis	Extends index finger	Dorsal ulna, interosseous membrane	Extensor tendon of index finger	Radial
Abductor pollicis longus	Abducts thumb	Dorsal radius and ulna	Lateral side of 1st metacarpal base	Radial
Abductor pollicis brevis	Abducts thumb	Fascia and ligaments anterior to carpals	Lateral side of thumb proximal phalanx base	Median
Abductor digiti minimi	Abducts little finger	Fascia and ligaments anterior to carpals	Medial side of little finger proximal base	Ulnar
Adductor pollicis	Adducts thumb	Anterior carpal fascia and 1st, 2nd, 3rd metacarpals	Medial side of thumb proximal phalanx base	Ulnar
Opponens pollicis	Brings thumb toward fingertips	Anterior carpal fascia	Lateral surface of 1st metacarpal	Median
Opponens digiti minimi	Brings little finger toward thumb	Anterior carpal fascia	Medial surface of 5th metacarpal	Ulnar
Lumbricals (4)	Flex metacarpo-phalangeal joints, extend fingers	Tendons of flexor digitorum profundus	Extensor tendons of 4 fingers	Median, ulnar
Interossei (7)	Abduct, adduct, flex fingers	Sides of metacarpals	Extensor tendons of fingers	Ulnar

the humerus. This helps to prevent downward slippage of the shoulder joint when lifting a heavy object.

In addition to the biceps, two other flexor muscles at the elbow, illustrated in Figure 9–6, are the following:

1. The **brachialis** lies deep to the biceps, arising from the anterior surface of the lower three fifths of the humerus and inserting into the ulna a few centimeters in front of the elbow joint. Thus, the brachialis lifts the ulna at the same time that the biceps lifts the radius.

2. The **brachioradialis** arises from the lateral side of the lower end of the humerus, then extends most of the length of the forearm to insert into the lateral border of the distal shaft of the radius. This muscle is not of much value in initiating the flexion process when the arm is fully extended, but it becomes much more valuable as a flexor once angulation at the elbow joint has begun.

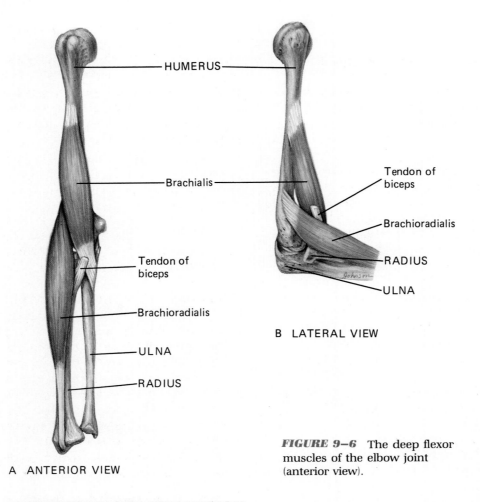

HUMERUS

Brachialis

Tendon of biceps

Tendon of biceps

Brachioradialis

Brachioradialis

RADIUS

ULNA

RADIUS

ULNA

B LATERAL VIEW

A ANTERIOR VIEW

FIGURE 9–6 The deep flexor muscles of the elbow joint (anterior view).

SUPINATION AND PRONATION OF THE FOREARM

One of the most important movements of our body is our ability to rotate the forearm and hand either medially or laterally. Rotation that causes the palm to point posteriorly is called *pronation*. Rotation so that the palm points ante- riorly is called *supination*. Figure 9–7 illustrates the muscles that cause these two motions.

The pronator muscles are as follows:

1. The **pronator teres** originates from the me- dial epicondyle of the humerus, then wraps anteriorly around the radius and inserts into its lateral border about midway along its

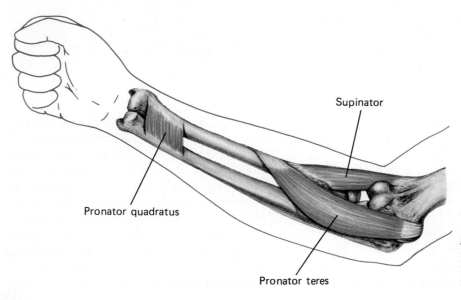

Supinator

Pronator quadratus

Pronator teres

FIGURE 9–7 The pronator and supinator muscles that rotate the forearm (anterior view).

shaft. When it contracts it rotates this lateral border of the radius medially and therefore also rotates the head of the radius within an annular ligament that holds it against the ulna at the superior radioulnar joint. The distal end of the radius at the same time passes in a 90-degree arc around the distal end of the ulna. Therefore, the hand rotates medially as well.

2. The **pronator quadratus,** which is a small muscle arising from the anterior surface of the lower ulna, then wraps around the anterior surface of the radius to insert also into the lateral border of the radius. When it contracts it rotates the distal end of the radius medially, and the radial head pivots in the annular ligament.

The following are the supinator muscles:

1. The **supinator** (Fig. 9–7) arises from the lateral epicondyle of the humerus, then wraps around the radius to insert into the medial border of the radius about one third the distance down the shaft. When this muscle contracts, it brings this medial border of the radius anteriorly and laterally, thus laterally rotating the radius (and therefore the entire forearm as well), which is the process of supination.

2. The **biceps brachii.** It has already been noted that the tendon of the biceps inserts in the radius at the medially located radial tuberosity. Therefore, at the same time that this muscle flexes the elbow, it also pulls the radial tuberosity forward from its medial position, thus rotating the palm of the hand anteriorly in the typical lifting posture.

MUSCLES THAT CAUSE WRIST MOVEMENTS

The muscles of the forearm are about equally divided between those that cause movements at the wrist and those that cause movements of the fingers and thumb. The front and back views of the forearm in Figure 9–8 show that the fleshy portions of almost all these muscles are located in the upper two thirds of the forearm. They transmit their contractile forces by way of long tendons that pass to the lower forearm, wrist, and hand. These tendons are all held in place by strong ligaments, and they are also surrounded by synovial tendon sheaths that provide a mucoid lubricating fluid for slippage of the tendons.

The principal motions of the wrist are *flexion* and *extension*, but the muscles on either side of the forearm can also cause wrist *abduction* or *adduction*.

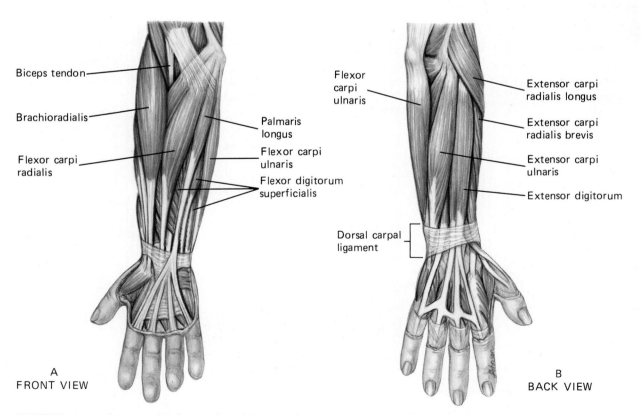

Biceps tendon

Brachioradialis

Flexor carpi radialis

Palmaris longus

Flexor carpi ulnaris

Flexor digitorum superficialis

A
FRONT VIEW

Flexor carpi ulnaris

Extensor carpi radialis longus

Extensor carpi radialis brevis

Extensor carpi ulnaris

Extensor digitorum

Dorsal carpal ligament

B
BACK VIEW

FIGURE 9–8 The superficial muscles of the forearm: (A) anterior view, (B) posterior view.

Flexion of the Wrist. Three of the forearm muscles are especially concerned with flexing the wrist:

1. The **palmaris longus muscle** (Fig. 9–8A) arises mainly from the medial condyle of the humerus, then passes superficially down the front of the forearm to insert into a fascial sheath, the *palmar aponeurosis*, which covers the palm of the hand and in turn is attached mainly to the distal ends of the second through fifth metacarpals. Obviously, contraction of this muscle pulls the whole hand in the direction of the palm, thus flexing the wrist.
2. The **flexor carpi radialis** (Fig. 9–9A) arises from the medial epicondyle of the humerus and inserts into the ligaments at the bases of the first and second metacarpals.
3. The **flexor carpi ulnaris** (Fig. 9–9A) also arises from the medial condyle but inserts into the ligaments at the base of the fifth metacarpal.

Note especially that both of the flexor carpi muscles bypass the bones of the forearm as well as the carpal bones of the wrist. This is another example of muscles providing stability of joints by transmitting the forces of contraction beyond several joints at the same time so that there will not be excessive traction on the joint ligaments themselves.

In addition to the three muscles just mentioned, still other forearm muscles cause flexion of the fingers and thumb, as will be described later. These muscles also help to cause wrist flexion because their tendons pass over the anterior aspects of the wrist on their way to the fingers.

Extension of the Wrist. The muscles principally concerned with extension of the wrist are illustrated in Figure 9–9B. These are the

1. **extensor carpi radialis longus** and **brevis,** which arise from the lateral epicondyle of the humerus and upper posterior shaft of the radius and then insert into the posterior base of the second metacarpal, and
2. **extensor carpi ulnaris,** which also arises from the lateral epicondyle of the humerus but inserts into the posterior base of the fifth metacarpal.

Here again, these muscles bypass the forearm bones as well as the carpal bones.

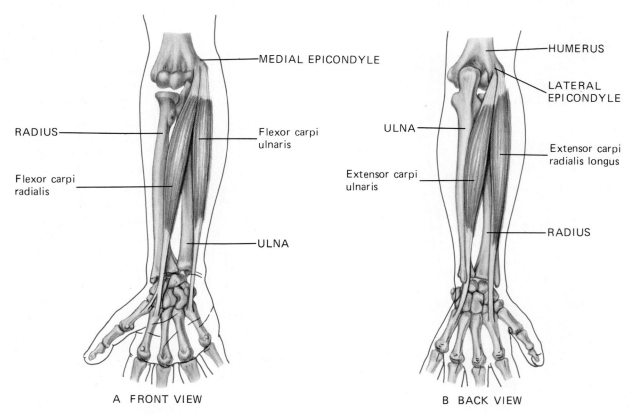

A FRONT VIEW

B BACK VIEW

FIGURE 9–9 The principal flexor and extensor muscles of the wrist: (A) anterior view, (B) posterior view.

In addition to these extensor carpi muscles, the forearm muscles that cause extension of the fingers and thumb, to be described later, also cause wrist extension, because their tendons pass over the posterior aspect of the wrist.

Abduction and Adduction at the Wrist. Obviously, if both the *flexor* and the *extensor carpi radialis* muscles contract at the same time, this will cause the wrist to angulate in the thumb direction, or away from the axis of the body, which is abduction. On the other hand, if both the *flexor* and the *extensor carpi ulnaris* muscles contract at the same time, this will cause the opposite motion of the wrist, away from the thumb and toward the trunk, which is adduction.

MOVEMENTS OF THE FINGERS AND THUMB

The movements of the fingers and thumb are (1) *flexion* and *extension*, (2) *abduction* (pulling the fingers and thumb apart) and *adduction* (pulling the fingers and thumb together), and (3) *opponent contraction* of the thumb and fingers (which means bringing the tip of the thumb together with one or more of the fingers).

The Flexor Muscles. The muscles that cause powerful flexion of the fingers and thumb are all located in the forearm. The three most important are as follows:

1. The **flexor digitorum superficialis,** illustrated in Figure 9–8A, lies underneath the palmaris longus and flexor carpi radialis. This muscle arises from the medial epicondyle of the humerus; at its opposite end it gives rise to four separate tendons that pass through the *carpal tunnel of the wrist* (a fibrous channel that enshrouds all flexor tendons), one tendon going to each finger. As each tendon passes into the finger it divides again into two slips that attach to the two sides of the shaft of the second phalanx. Contraction of this muscle, therefore, flexes the first two phalanges of the fingers as well as flexing the wrist, but it does not flex the distal phalanx.

2. The **flexor digitorum profundus** lies underneath the flexor digitorum superficialis and arises mainly from the anterior surface of the upper ulna. It too gives rise to four separate tendons that extend to the four fingers. The tendon to each finger passes underneath and between the two lateral slips of the tendon of the flexor digitorum superficialis; thus, the two slips of the superficial tendon form a sling for the deep tendon to pass through. The deep tendon then inserts on the anterior surface of the distal phalanx. Contraction of this muscle, therefore, specifically causes flexion of the distal phalanx of each finger. However, because its tendons are constrained by fascial coverings of the fingers and also by the divided tendons of the flexor digitorum superficialis, it also causes flexion of the other phalanges. In addition, its action at the wrist flexes this as well.

3. The **flexor pollicis longus** (Fig. 9–10) arises mainly from the anterior shaft of the radius. Its tendon then passes through the wrist to insert into the distal phalanx of the thumb. This muscle causes flexion of the thumb in a manner similar to the flexion of the fingers caused by the earlier two muscles.

In addition to the powerful flexors of the forearm, several much weaker flexor muscles are in the hand itself. These include the

1. **flexor pollicis brevis** (Figure 9–10A), which arises from the fascia of the radial side of the wrist and inserts into the anterior base of the proximal phalanx of the thumb;

2. **flexor digiti minimi brevis,** which arises from the fascia of the ulnar side of the wrist and inserts into the anterior base of the proximal phalanx of the small finger; and

3. **four separate lumbricals,** which arise from the four tendons of the flexor digitorum profundus muscle and then insert into the extensor expansions (to be described later) of the respective fingers just distal to the metacarpophalangeal joints.

The previous two flexor muscles of the hand provide extra flexion of the thumb and little finger, operating in concert with the more powerful muscles of the forearm. The lumbricals, because they wrap around the fingers, cause mainly flexion of the metacarpophalangeal joints but have an effect to extend the interphalangeal joints because of their attachments to the extensor expansions of the fingers.

Extension of the Fingers and Thumb. As is true for the forearm muscles that cause flexion of the fingers, the extensors also lie in two different planes, a superficial plane and a deep plane, but this time on the back of the forearm. These muscles, illustrated respectively in Figures 9–8B and 9–10B, are described here:

1. The **extensor digitorum** arises mainly from the lateral epicondyle of the humerus, then passes down the back of the forearm in the

Extensor carpi radialis

Flexor carpi ulnaris

Flexor digitorum profundus

Flexor pollicis longus

Abductor pollicis longus

Flexor carpi radialis

Abductor pollicis brevis

Opponens pollicis

Flexor pollicis brevis

Abductor pollicis

Abductor digiti minimi

Opponens digiti minimi

Flexor digiti minimi brevis

Lumbricals

Flexor digitorum superficialis tendon

Brachioradialis

Extensor carpi radialis longus

Biceps

HUMERUS

Supinator

ULNA

RADIUS

Abductor pollicis longus

Extensor pollicis longus

Extensor indicis

Extensor pollicis brevis

Extensor carpi ulnaris

Extensor carpi radialis

Abductor digiti minimi

Interosseous muscles

A   FRONT VIEW   B   BACK VIEW

FIGURE 9–10   The hand muscles and the deep muscles of the forearm: (A) anterior view, (B) posterior view.

superficial plane, giving rise to four separate tendons in the wrist that pass through lubricated synovial tendon sheaths underneath the wrist fascia and thence into the backs of the four separate fingers. There they insert into the bases of the distal phalanges and also into strong fascial sheaths, called *extensor expansions*, that wrap around the more proximal phalanges. Thus, contraction of this muscle provides extension of all the fingers at the same time, including extension at both the metacarpophalangeal joints and the interphalangeal joints. And, in addition, its action at the wrist extends this.

2. The **extensor pollicis longus,** illustrated in Figure 9–10B, lies in the deep muscular plane. This muscle arises from the posterior surface of the upper ulna, and its tendon passes through the wrist, eventually inserting into the dorsal surface of the base of the distal

phalanx of the thumb as well as into a fascial extensor expansion of the thumb. Thus, this muscle performs the same function for the thumb that the extensor digitorum performs for the four fingers.

3. The **extensor pollicis brevis** lies also in the deep plane. This muscle arises from the posterior surface of the radius and inserts into the base of the proximal phalanx of the thumb. It both extends and abducts the thumb.

4. The **extensor indicis,** also one of the deep extensor muscles illustrated in Figure 9–10A, arises alongside the extensor pollicis longus from the posterior surface of the ulna. Its tendon passes through the wrist, but in the back of the hand it joins the extensor tendon to the index finger from the extensor digitorum. Therefore, this muscle provides an extra degree of extensor strength for the index finger.

The Abductors of the Thumb and Fingers. It is often important to spread the thumb and fingers apart. This is achieved by the abductors. And, because the thumb requires extreme abduction power at times, it is provided with the most powerful of the abductors. In addition to the slight abduction of the thumb caused by the extensor pollicis brevis, the other abductor muscles are as follows:

1. The **abductor pollicis longus,** illustrated in Figure 9–10B, arises deep in the forearm mainly from the posterior surface of the radius but also from the fascia that attaches between the radius and ulna, then wraps around the lateral margin of the lower radius and inserts into the lateral side of the base of the first metacarpal. Contraction of this muscle pulls the first metacarpal, and therefore the entire thumb, outward from the remainder of the hand, which is abduction.

2. The **abductor pollicis brevis** arises from the radial side of the carpal bones and their fascial covering and inserts into the lateral side of the base of the first phalanx, thus also causing abduction of the thumb.

3. The **abductor digiti minimi** arises from the ulnar side of the carpal bones and from the fascia covering them and inserts into the medial side of the base of the first phalanx of the little finger, thus causing abduction of the little finger, pulling it away from the remainder of the hand.

In addition to the muscles just listed, seven separate small **interosseous muscles** are located between the metacarpals, as illustrated in Figure 9–10. These insert into the extensor expansions of the phalanges of the four fingers and are so placed that *four of them cause abduction* of the fingers (spreading them apart), while *three others cause adduction.*

The Adduction and Opponent Motions of the Fingers and Thumb. Three of the *interosseous muscles* cause adduction of the fingers, as was just pointed out. In addition, the thumb has an especially powerful adductor muscle, the **adductor pollicis,** that arises from the shaft of the anterior surface of the third metacarpal and passes laterally to insert into the base of the proximal phalanx of the thumb. This muscle is important in providing a powerful handgrip.

The opponent motion applies particularly to movement of the thumb toward one of the fingers; it means rolling the thumb and one of the fingers together over the palm. This motion is especially important for the manipulative skills of the hand, including the pinching action.

Though most of the hand muscles contribute to opponent motions, special opponens muscles for the thumb and little finger are the following:

1. The **opponens pollicis** (Fig. 9–10A) lies deep to the abductor pollicis brevis, so that only a small part of it is illustrated in the figure. It arises from the deep fascial tissue of the wrist and inserts into the entire anteromedial border of the first metacarpal, thus allowing the thumb to be pulled over the palm of the hand toward the small finger.

2. The **opponens digiti minimi** also arises from the deep fascia of the wrist but inserts into the anteromedial margin of the fifth metacarpal. When this muscle contracts, it pulls the metacarpal anteriorly and laterally toward the thumb, thus moving the small finger also toward the thumb.

MOVEMENTS AND MUSCLES AT THE HIP JOINT

Movements at the hip joint are both the most powerful of the body movements and among the most varied in range of motion. The hip joint movements are *flexion, extension, abduction,* and *medial* and *lateral rotation.* The very strong and massive coxal bone (hip bone) is the anchor around which these movements occur, and the thickness and breadth of the muscles of the hip and upper thigh, shown in Figures 9–11 through 9–14, bespeak the extreme forces and mechanical advantages that are at play at this joint. It is no wonder that the hip joint is among the most likely of all the joints of the body to become arthritic or to be severely damaged by trauma, especially in old age.

Flexion at the Hip Joint. All of the thigh muscles that pass over or attach to the anterior surfaces of the coxal bone serve as flexors at the hip joint. The most important of these are listed here:

1. The **sartorius** (Fig. 9–11) arises from the anterior superior iliac spine of the coxal bone, then wraps around the anterior surface of the thigh, and finally inserts into the fascia below the medial condyle of the tibia, thus crossing both the hip joint and the knee joint.

2. The **rectus femoris** (Fig. 9–11) is the longest head of the very large quadriceps femoris muscle of the anterior thigh. It originates underneath the sartorius from the anterior inferior iliac spine of the coxal bone, then passes down the anterior surface of the thigh to fuse with three other heads of the quadri-

CREST OF HIP BONE

Iliacus

ANTERIOR SUPERIOR
ILIAC SPINE

Gluteus medius

Iliopsoas

Tensor fasciae latae

Sartorius

Tractus iliotibialis

TIBIAL TUBEROSITY

Psoas major

Inguinal ligament

PUBIC TUBERCLE

PUBIS

Pectineus

Adductor longus

Gracilis

Quadriceps femoris:

Rectus femoris

Vastus medialis

Vastus lateralis

PATELLA

PATELLAR TENDON

FIGURE 9–11 The superficial anterior hip and thigh muscles.

ceps muscle that in turn form the patellar tendon. This then continues through the patella and inserts into the tibia at the tibial tuberosity a few centimeters below the knee. Though the quadriceps femoris muscle is an especially powerful extensor of the leg, the rectus femoris portion of the muscle also serves as a flexor of the hip.

3. The **psoas major** (Figs. 9–11 and 9–12) arises deep inside the trunk from the lateral surfaces of the lumbar vertebrae, then passes through the bony pelvis along the medial surface of the iliac portion of the coxal bone, and finally emerges into the thigh from underneath the inguinal ligament (a strong ligament that connects the anterior superior iliac spine of the ilium to the pubic tubercle). On entering the thigh (Fig. 9–12) this muscle wraps around the anterior aspect of the neck of the femur and inserts into the lesser tro-

chanter on the anterior surface of the shaft of the femur.

4. The **iliacus** arises from the inner cup of the iliac portion of the coxal bone and then fuses with the lower end of the psoas muscle, the two muscles then forming a single muscle mass often referred to as the iliopsoas. The tendon of the iliacus also is fused with that from the psoas and inserts along with it into the lesser trochanter. Because the iliopsoas tendon lies tightly against the capsule of the hip joint, it helps to provide much of the stability for this joint while also acting as one of the strongest flexors.

In addition to these principal flexors of the hip joint, one can also observe from Figure 9–11 that the *pectineus* and the *tensor fasciae latae*, which are respectively an adductor and an abductor, provide some degree of flexion as well.

TABLE 9—3
Muscles that Cause Movement at the Hip and Knee Joint

Muscle	Function	Origin	Insertion	Nerve
Sartorius	Flexes and laterally rotates thigh	Anterior superior iliac spine	Medial side of proximal tibia	Femoral
Psoas major	Flexes and medially rotates thigh	Lumbar vertebrae and fascia	Lesser trochanter of femur	L-2 and L-3
Iliacus	Flexes and medially rotates thigh	Inner surface of ilium and sacrum	Lesser trochanter of femur	Femoral
Gluteus maximus	Extends, abducts, laterally rotates thigh	Surfaces of ilium, sacrum, coccyx, sacrotuberous ligaments	Gluteal tuberosity on lateral femur	Inferior gluteal
Gluteus medius	Abducts thigh	Lateral surface of ilium	Greater trochanter of femur	Superior gluteal
Gluteus minimus	Abducts thigh	Lateral surface of ilium	Greater trochanter of femur	Superior gluteal
Tensor fasciae latae	Flexes, abducts thigh	Iliac crest	Iliotibial tract of fascia lata	Superior gluteal
Pectineus	Flexes, adducts thigh	Lateral, superior margin of pubis	Upper medial shaft of femur	Femoral and obturator
Adductor longus	Adducts, flexes thigh	Crest of pubis	Posteromedial shaft of femur	Obturator
Adductor magnus	Adducts, extends thigh	Pubic ramus and ischium	Posteromedial shaft of femur	Obturator and sciatic
Adductor brevis	Adducts, flexes thigh	Pubic ramus	Posteromedial shaft of femur	Obturator
Gracilis	Adducts thigh, flexes knee joint	Pubic ramus	Upper medial surface of tibia	Obturator
Lateral rotators: Piriformis Obturator internus Superior gemullus Inferior gemullus Quadratus femoris	Lateral rotation of thigh	Posterior coxal bone	Posterior margin of greater trochanter	Sciatic
Quadriceps femoris: Rectus femoris	Flexes thigh, extends leg	Anterior inferior iliac spine		
Vastus medialis		Medial femur	Patellar tubercle of tibia	Femoral
Vastus lateralis		Lateral femur		
Vastus intermedius		Anterior and lateral femur		
Hamstrings: Biceps femoris	Flexes leg, extends thigh	Ischial tuberosity, upper posterior femur	Lateral condyle of tibia and head of fibula	Tibial, peroneal
Semitendinosus	Flexes leg, extends thigh	Ischial tuberosity	Upper medial tibia	Tibial
Semimembranosus	Flexes leg, extends thigh	Ischial tuberosity	Medial condyle of tibia	Tibial
Gastrocnemius	Flexes knee joint, plantar flexes foot	Posterior surfaces of lateral and medial femoral condyles	Tuberosity of calcaneus	Tibial
Popliteus	Rotates leg medially	Lateral condyle of femur	Posterior surface of tibia	Tibial

NOTE: The nerves of this table are described in Chapter 14.

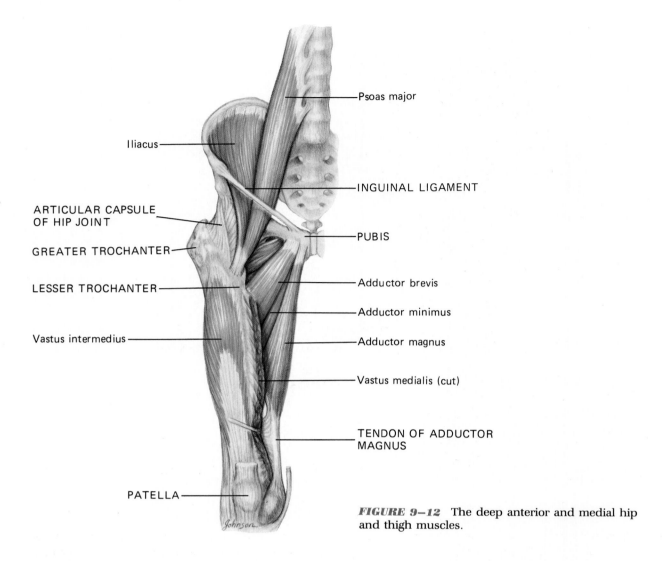

Psoas major

Iliacus

INGUINAL LIGAMENT

ARTICULAR CAPSULE
OF HIP JOINT

GREATER TROCHANTER

PUBIS

LESSER TROCHANTER

Adductor brevis

Adductor minimus

Vastus intermedius

Adductor magnus

Vastus medialis (cut)

TENDON OF ADDUCTOR
MAGNUS

PATELLA

FIGURE 9–12 The deep anterior and medial hip and thigh muscles.

Extension at the Hip Joint. A great mass of muscles in the buttocks and posterior thigh serve as extensors of the thigh. The most important of these are discussed here:

1. The **gluteus maximus** (Fig. 9–13) arises from the entire posterior part of the iliac crest as well as from the upper, outside surface of the posterior ilium, the sacrum, and the ligament from the sacrum to the ischial tuberosity (the sacrotuberous ligament). It then passes superficially over the posterior aspect of the hip joint and inserts into the gluteal tuberosity that extends along the posterolateral shaft of the femur for a distance of about 10 cm. It also inserts into a thickening of the fascial covering of the thigh, the *iliotibial tract*, that itself extends all the way down the lateral side of the thigh to insert into the lateral condyle of the tibia. The gluteus maximus is the very powerful muscle of the buttocks that is especially important for raising the body from a forward bending position to an upright posi-

tion, and it also plays an important role in the backward thrusting motion of the thigh during running as well as in climbing and in walking up stairs.

2. The **long head of the biceps femoris** (Fig. 9–14) lies deep to the gluteus maximus, arising from the ischial tuberosity, then passing through the posterior thigh and wrapping around to the lateral side of the thigh, finally passing beyond the posterior aspect of the knee joint to insert into the head of the fibula. A second head of this muscle arises from the posterior surface of the femur.

3 and 4. The **semitendinosus** and **semimembranous** muscles also arise from the ischial tuberosity but pass down the medial side of the posterior thigh, crossing the posterior medial aspect of the knee joint and inserting into the medial surface of the tibia.

Note that the biceps femoris, the semitendinosus, and the semimembranosus muscles all span both the hip joint and the knee joint.

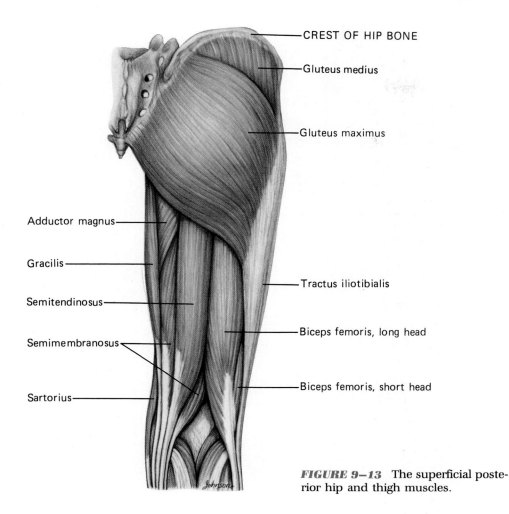

CREST OF HIP BONE

Gluteus medius

Gluteus maximus

Adductor magnus

Gracilis

Semitendinosus

Semimembranosus

Sartorius

Tractus iliotibialis

Biceps femoris, long head

Biceps femoris, short head

FIGURE 9–13 The superficial posterior hip and thigh muscles.

Therefore, they cause hip extension and knee flexion both at the same time, which is the all-important motion for forward propulsion during running. These are the muscles that are known as the hamstrings.

Abduction at the Hip Joint. Abduction at the hip joint, that is, lateral movement of the thigh, is achieved by contracting the lateral hip and thigh muscles. These are as follows:

1. The **tensor fasciae latae** (Fig. 9–11), which also acts as a weak flexor of the thigh as mentioned before, arises from the anterior iliac crest and inserts into the same iliotibial tract along the lateral surface of the thigh that is one of the points of insertion for the gluteus maximus.
2. The **gluteus medius** (Figs. 9–11 and 9–13) arises from the lateral surface of the ilium, then passes over the lateral aspect of the hip joint to insert into the greater trochanter of the femur, thus causing powerful abduction of the femur.
3. The **gluteus minimus** (Fig. 9–14) also arises from the lateral surface of the ilium under-

neath the gluteus medius and also inserts into the greater trochanter, thus serving along with the gluteus medius as the principal abductors of the thigh.

Adduction at the Hip Joint. Adduction of the thigh is one of the most powerful motions of the hip joint, and it is served by a series of fan-shaped muscles that originate mainly from the pubis and insert into the posteromedial surface of the shaft of the femur. These muscles, popularly called the *groin muscles*, are the following:

1. The **pectineus** is a small muscle passing from the superior lateral surface of the pubis to the upper medial shaft of the femur.
2. The **adductor longus** (Fig. 9–11) arises from the anterior crest of the pubis and inserts into the posteromedial shaft of the femur.
3. The **adductor magnus** (Fig. 9–12) arises by a broad origin from the pubis and ischium and inserts into the entire lower three fifths of the medial side of the shaft of the femur and even extends by another tendon to the adductor tubercle on the medial condyle of the femur

Gluteus maximus

Piriformis

Superior gemellus

Inferior gemellus

ISCHIAL TUBEROSITY

Adductor magnus

Gracilis

Semitendinosus

Semimembranosus

MEDIAL CONDYLE
OF FEMUR

MEDIAL CONDYLE
OF TIBIA

Gluteus medius

Gluteus minimus

Gluteus medius

Obturator internus

GREATER
TROCHANTER

Quadratus femoris

FEMUR

Gluteus maximus

Biceps femoris,
long head

Vastus lateralis

Biceps femoris,
short head

LATERAL
CONDYLE
OF FEMUR

HEAD OF FIBULA

FIGURE 9–14 The deep posterior hip and thigh muscles.

at the knee. This is by far the most powerful of the adductor muscles.

4. The **adductor brevis** (Fig. 9–12) extends from the pubis to the medial border of the upper shaft of the femur, lying between the adductor longus and magnus.

5. The **gracilis muscle** (Figs. 9–11 and 9–13) arises from the inferior ramus of the pubis, then passes as a thin muscle down the inner thigh to a tendon that passes medial to the knee joint and inserts into the tibia below its medial condyle.

It is all these adductor muscles that allow a person to grip the back of a horse or to squeeze an opponent in a wrestling match. These muscles also play an important role in the rhythmic, to and fro lateral tilting of the pelvis in walking.

Rotation at the Hip Joint. Many of the same muscles that have already been discussed

function also as important rotators at the hip joint.

Lateral rotation of the thigh is achieved partly by contracting the *sartorius* (Fig. 9–11), which is the muscle that wraps around the anterior surface of the thigh from the anterior superior iliac spine of the coxal bone to the medial side of the tibia. In addition, the lower portions of the *gluteus maximus* act as a powerful lateral rotator at the hip joint (Fig. 9–13). And, several small deep muscles on the posterior side of the hip joint, the **piriformis,** the **obturator internus,** the **superior** and **inferior gemelli,** and the **quadratus femoris** (Fig. 9–14), all inserting into the posterior margin of the greater trochanter, help to cause lateral rotation of the thigh.

Medial rotation of the thigh is provided mainly by the *psoas* and the *iliacus* muscles, the tendons from which wrap around the anterior surface of the hip joint and insert into the lesser trochanter. Likewise, the superficial adductor muscle, the *pectineus*, provides some degree of medial rotation. Finally, the *tensor fasciae latae* not only functions as an abductor but, because of its anterior origin on the iliac crest, also helps to rotate the thigh medially.

In summary, many muscles of the hip and upper thigh cause some of the most powerful movements of the body at the hip joint. The many directions of these movements are the principal basis for the agility of the quarterback of a football team.

MOVEMENT AT THE KNEE JOINT

The knee joint is mainly a "hinge" type joint that can *extend* and *flex* but has the capability of a *slight degree of rotation* as well, as will be discussed later. When the knee joint is fully extended, the extremity is almost straight but with slight displacement of the joint posterior to the straight axis. The ligaments and structure of the knee joint and the tendons of the hamstring muscles prevent further extension than this. However, this slight degree of extension beyond the straight axis, called hyperextension, allows the knee to remain in a "locked" position when a person is standing and to remain so without contraction of any specific muscles. Thus, this characteristic of the knee joint is an important energy saver when a person must stand for a long period of time, and it permits relaxation while standing.

Extension at the Knee Joint—the Quadriceps Muscle and the Patellar Tendon. In contrast to the movements at most joints of the body, the extension movement at the knee joint is achieved entirely by one single multiheaded muscle, the **quadriceps femoris,** that constitutes the great mass of muscle on the anterior thigh and terminates in the **patellar tendon.** The patellar tendon in turn inserts into the patella (commonly known as the knee cap), which rests on the front of the knee joint; then it continues on beyond the patella to insert (via the *patellar ligament*) into the tuberosity of the anterior tibia a few centimeters below the knee. The posterior surfaces of the patella form a typical gliding joint where the patella slides over the anterior surfaces of the femoral condyles. The patella also maintains the patellar tendon in an anteriorly placed position, which gives it an additional degree of mechanical advantage at the knee joint for extending the leg.

The quadriceps femoris muscle is the most powerful single muscle in the body. In an athlete who has a highly developed quadriceps muscle, the amount of force exerted by the patellar tendon on the tibia can be more than 300 kg, or 600 to 700 pounds, and even much greater than this in weight lifters. To provide the extreme power of the quadriceps, this muscle arises from four separate heads, three of which are illustrated in Figure 9–11 and the fourth in Figure 9–12. These heads are the

1. **rectus femoris,** which arises from the anterior inferior iliac spine on the anterior edge of the crest of the ilium as well as the anterior surface of the ilium, as was discussed previously in relation to the hip joint;
2. **vastus medialis,** which arises from the medial side of the shaft of the femur along most of its length;
3. **vastus lateralis,** which arises from the lateral surface of the shaft of the femur, also along most of its length; and
4. **vastus intermedius** (Fig. 9–12), which lies beneath the rectus femoris and arises from the anterior surface of the femoral shaft along almost its entire length.

Flexion at the Knee Joint—the Hamstrings. The knee joint can be flexed approximately 145 degrees, which is an extreme range of motion for any joint. The muscles that cause most of this flexion are the *hamstring muscles*, which have already been described in relation to the hip joint. These muscles are powerful extensor muscles at the hip joint as already noted, but they span the knee joint as well and cause flexion at the knee at the same time. Note specifically in Figure 9–14 the tendons of these muscles as they pass behind the knee joint as follows:

1. The *tendon of the biceps femoris* passes behind the posterolateral aspect of the knee joint and inserts into the posterior surface of the head of the fibula.
2. The *tendon of the semitendinosus* and (3) the *tendon of the semimembranosus* pass behind the posteromedial aspect of the knee joint; the semitendinosus inserts into the medial surface and the semimembranosus into the posteromedial surface of the upper tibia. Thus, because all the tendons of the hamstring muscles pass behind the knee joint, when they contract the knee joint is strongly flexed.

Another thigh muscle that plays a slight role in flexing the knee joint is the *gracilis*, which also functions as an adductor of the hip joint as was discussed earlier. The tendon of this muscle passes slightly behind the axis of the knee joint and also inserts into the posteromedial surface of the upper tibia.

Finally, another powerful muscle not located in the thigh but instead in the leg also causes considerable flexion at the knee joint. This is the **gastrocnemius muscle,** which spans both the knee joint and the ankle joint as will be described later in relation to movement at the ankle. This muscle, illustrated in Figure 9–15, arises by two heads from the posterior surfaces of both the lateral and the medial condyles of the femur and therefore causes flexion of the leg when contracted. It also causes downward movement (plantar flexion) of the foot and therefore serves double function during the running process in the same manner that the hamstrings serve double function at the hip and knee joints.

Rotation at the Knee Joint. When the knee is fully extended, the leg can be rotated at the knee joint only slightly. However, when bent, the leg then has the ability to rotate medially or laterally as much as 20 degrees, making a total range of about 40 degrees. Under these condi-

tions, most of the rotation is caused by differential contraction of the lateral versus medial hamstrings; the *biceps femoris*, the lateral hamstring, causes lateral rotation of the leg, and the two medial hamstrings, the *semimembranosus* and the *semitendinosus*, cause medial rotation. In addition, the **popliteus** (illustrated in Figs. 9–16 and 9–17), a small muscle located in the leg behind and below the knee, can also cause a small degree of medial rotation. This muscle arises from the lateral surface of the lateral condyle of the femur, then wraps around the back of the tibia to insert on the medial side of its upper shaft. Therefore, when it contracts it rotates the tibia medially and thereby rotates the foot medially as well.

MOVEMENTS AT THE ANKLE

The ankle, like the wrist, has a succession of joints involving the tarsal bones. The **ankle joint** proper is the joint between the tibia and the fibula superiorly and the talus inferiorly. This is a *hinge type joint* that allows either upward move-

Sartorius

MEDIAL CONDYLE OF FEMUR

Tendon of gracilis

Tendons of hamstrings

Medial head of gastrocnemius

Soleus

MEDIAL MALLEOLUS

Deep fascia

LATERAL CONDYLE OF FEMUR

Tendon of biceps femoris

Lateral head of gastrocnemius

Soleus

ACHILLES TENDON

LATERAL MALLEOLUS

FIGURE 9–15 The superficial posterior muscles of the leg.

MEDIAL CONDYLE OF FEMUR

Medial head of gastrocnemius

Popliteus

TIBIA

Soleus

Flexor digitorum longus

MEDIAL MALLEOLUS

Biceps femoris

Lateral head of gastrocnemius

HEAD OF FIBULA

Gastrocnemius

ACHILLES TENDON

Peroneus longus

TUBEROSITY OF CALCANEUS

FIGURE 9–16 The posterior muscles of the leg, with the gastrocnemius muscle pulled to one side.

Gastrocnemius

Popliteus

TIBIA

Tibialis posterior

Flexor hallucis longus

Flexor digitorum longus

MEDIAL MALLEOLUS

Tendon of tibialis posterior

Gastrocnemius

Soleus

FIBULA

Peroneus longus

Peroneus brevis

Posterior transverse ligament

TUBEROSITY OF CALCANEUS

FIGURE 9–17 The deep posterior muscles of the leg.

TABLE 9–4
Muscles that Cause Ankle and Foot Movement

Muscle	Function	Origin	Insertion	Nerve
Gastrocnemius	Flexes knee joint Plantar flexes foot	Posterior surfaces of lateral and medial femoral epicondyles	Tuberosity of calcaneus via Achilles tendon	Tibial
Soleus	Plantar flexes foot	Fibula, tibia, and interosseus membrane	Tuberosity of calcaneus via Achilles tendon	Tibial
Tibialis anterior	Dorsiflexes, inverts foot	Anterior tibia, interosseus membrane	Medial cuneiform, 1st metatarsal	Peroneal
Tibialis posterior	Plantar flexes, inverts foot	Posterior surfaces of upper tibia and fibula	Bases of metatarsals and fascia of tarsals	Tibial
Peroneus longus	Everts, plantar flexes foot	Lateral tibial condyle, lateral fibula	Medial cuneiform, 1st metatarsal	Peroneal
Peroneus brevis	Everts, plantar flexes foot	Lateral fibula	5th metatarsal base	Peroneal
Extensor digitorum longus	Dorsiflexes 4 lateral toes and foot	Lateral tibial condyle, anterior fibula, interosseus membrane	Dorsum of distal phalanx of 4 lateral toes	Peroneal
Extensor hallucis longus	Dorsiflexes large toe and foot	Anterior fibula, interosseus membrane	Dorsum of distal phalanx of large toe	Peroneal
Flexor digitorum longus	Plantar flexes 4 lateral toes and foot	Posterior shaft of tibia	Plantar base of distal phalanx of 4 lateral toes	Tibial
Flexor hallucis longus	Plantar flexes large toe and foot	Posterior shaft of fibula	Plantar base of distal phalanx of large toe	Tibial

Intrinsic foot muscles that have origins and insertions similar to those of the hand:

Extensor digitorum brevis	Flexor hallucis brevis	4 lumbricals	Tibial,
Extensor hallucis brevis	Abductor hallucis	7 interossei	peroneal
Flexor digitorum brevis	Abductor digiti minimi		

NOTE: The nerves in this table are described in Chapter 14.

ment of the foot, called *dorsiflexion,* or downward movement, called *plantar flexion.*

Inferior to the talus are the intertarsal joints and the joints between the distal tarsals and the metatarsals. The intertarsal joints, in addition to allowing additional degrees of dorsiflexion and plantar flexion, also allow lateral rotation of the foot so that the sole of the foot faces outward, which is called *eversion,* or medial rotation so that the sole faces inward, which is called *inversion.*

Plantar Flexion at the Ankle—The Power Movement of the Ankle. Plantar flexion at the ankle is the major power movement of the foot that is essential for standing on tiptoes and also for giving the necessary forward thrust during running, as well as the "toe-off" movement in the walking cycle.

Plantar flexion is caused by upward pull of the **Achilles tendon** that inserts into the tuberosity of the calcaneus bone, the most posterior of the tarsal bones. The tension on the Achilles tendon is provided by two muscles, the gastrocnemius and the soleus.

1. The **gastrocnemius muscle,** illustrated in Figure 9–15, is the large superficial muscle of the calf of the leg. As was discussed in relation to the knee joint, it arises via two heads, the medial head and the lateral head, from the posterior surfaces of the medial and lateral condyles of the femur.
2. The **soleus muscle,** illustrated in Figure 9–16, arises mainly from the posterior surface of the upper quarter of the fibula but also from the adjacent regions of the tibia as well. It lies

underneath the two heads of the gastrocnemius muscle, and its tendon fuses with the tendon of the gastrocnemius muscle to form the Achilles tendon.

It is the gastrocnemius muscle that provides the rapid downward, locomotive thrust of the foot during walking and running and also the other rapid powerful thrusting movements of the ankle joint. But, during long periods of standing, the soleus provides most of the continuous force applied to the Achilles tendon. The muscle fibers of the gastrocnemius muscle are of a type called "white" muscle that has the capability of extremely forceful contraction but without much endurance; on the other hand, the soleus muscle fibers are of the "red" type that have less initial contractile power but tremendous endurance.

Dorsiflexion at the Ankle. Dorsiflexion is not one of the power movements at the ankle, but it is important in providing lift of the foot when running or walking to keep the toes from dragging. Persons who have paralysis of the anterior muscles of the leg, as often occurs following poliomyelitis, have "foot drop," which requires that the foot be lifted very high at each step to prevent tripping over the toes.

The principal muscle for causing dorsiflexion is the **tibialis anterior,** illustrated in Figures 9–18 and 9–19. It arises from the lateral condyle of the tibia and from the lateral surface of the upper half to two thirds of the shaft of the tibia. Its tendon passes through the anterior ankle region underneath a strong ligamentous structure called the *extensor retinaculum,* passing anterior to the medial malleolus and then inserting into the medial sides of the distal tarsal bones and first metatarsal. Therefore, when the tibialis anterior contracts it lifts the foot (dorsiflexion) and also causes some inversion.

The other anterior muscles of the leg, lying lateral and deep to the tibialis anterior (Figs. 9–18 and 9–19), cause extension of the toes. These will be discussed in relation to movements of the foot. However, because their tendons pass over

Sartorius

PATELLA

Tractus iliotibialis

HEAD OF FIBULA

TIBIAL TUBEROSITY

Peroneus longus

Medial head of gastrocnemius

Tibialis anterior

Peroneus brevis

Extensor digitorum longus

Extensor hallucis longus

Anterior transverse ligament

LATERAL MALLEOLUS

Extensor hallucis brevis

Tendons of extensor digitorum longus

FIGURE 9–18 The anterior muscles of the leg and foot.

Biceps femoris

Semimembranosus

HEAD OF FIBULA

Gastrocnemius

Soleus

Peroneus brevis

Flexor hallucis longus

LATERAL MALLEOLUS

TUBEROSITY OF CALCANEUS

Iliotibial tract

Patellar tendon

TIBIAL TUBEROSITY

TIBIA

Peroneus longus

Tibialis anterior

Extensor digitorum longus

Anterior transverse ligament

Cruciate ligament

Tendon of extensor hallucis longus

Tendons of extensor digitorum longus

Abductor digiti minimi

FIGURE 9–19 The anterolateral muscles of the leg and foot.

the anterior surface of the ankle and underneath the extensor retinaculum, they, too, help to cause dorsiflexion at the ankle joint.

Inversion of the Foot. It was noted previously that the tendon of the *tibialis anterior* muscle inserts on the medial side of the foot and causes some degree of inversion (rotating the sole of the foot to face inward) at the same time that it causes dorsiflexion. However, the principal inverter of the foot is the **tibialis posterior,** illustrated in Figure 9–17, that arises from the posterior surface of the upper tibia and fibula. The tendon of this muscle passes posterior to the medial malleous and inserts into the medial border of the foot near the insertion of the tibialis anterior, attaching to several of the tarsal and metatarsal bones. The tendon attachments of this muscle have a mechanical advantage that allows considerably more powerful inversion of the foot than is achieved by the tibialis anterior. Also, the tibialis posterior plays a special role in medial stabilization of the ankle during ice skating.

Eversion of the Foot. Eversion of the foot is caused by two muscles:

1. The **peroneus longus,** illustrated best in Figure 9–19, arises from the lateral surface of the upper fibula, and its tendon passes down the posterolateral portion of the leg, then posterior to the lateral malleolus and across the sole of the foot to attach to the medial tarsal and metatarsal bones. Thus, this tendon actually wraps around the lateral side and sole of the foot, so that when the peroneus longus contracts, it rolls the lateral side of the foot upward and the medial side downward, thus everting the foot.

2. The **peroneus brevis** arises from the lateral surface of the midportion of the fibula, and its tendon also passes posterior to the lateral malleolus but inserts into the base of the fifth metatarsal on the lateral side of the foot. Therefore, contraction of this muscle pulls the lateral side of the foot upward, this too everting the foot.

Digital tendon sheaths

Lumbricals

Interossei

Flexor digiti minimi brevis

Abductor minimi brevis

Tendon of flexor hallucis longus

Adductor hallucis

Flexor hallucis brevis

Flexor digitorum brevis

Abductor hallucis

Plantar aponeurosis

MEDIAL MALLEOLUS

TUBEROSITY OF CALCANEUS

FIGURE 9–20 The muscles of the sole of the foot.

It is clear that the peroneus muscles also provide the lateral stabilization of the ankle during ice skating and other similar athletic activities in the same manner that the posterior and anterior tibialis muscles provide this same function for the medial side of the ankle.

MOVEMENTS OF THE DISTAL FOOT AND TOES

The movements of the distal foot and toes are almost analogous to the movements of the hand and fingers, though having much less range of motion. In fact, the different muscles are also almost identical and have similar names except that the term for the thumb, "pollicis," is replaced by the term for the big toe, "hallucis." The principal movements of the distal parts of the foot and toes are *flexion*, which means downward movement, and *extension* (also called dorsiflexion), which means upward movement. The more powerful muscles that cause both flexion and extension of the toes have their muscle bellies located in the leg and not in the foot, and the power is transmitted through the ankle via long tendons, as illustrated in Figures 9–18, 9–19, and 9–20. However, smaller muscles located within the foot itself also help in these movements.

Extension (Dorsiflexion) of the Distal Foot and Toes.   The two principal extensor muscles of the distal foot and toes (illustrated in Figs. 9–18 and 9–19) are the

1. **extensor digitorum longus** and
2. **extensor hallucis longus.** Both of these muscles arise from the anterior surfaces of the fibula and tibia and from the interosseous membrane connecting these bones; multiple tendons from these muscles pass beneath the extensor retinaculum of the ankle and finally insert into the distal phalanges of the toes as well as into fascial sheaths called extensor expansions that wrap around the more proximal phalanges.

Small accessory muscles are also located on the dorsum of the foot, the **extensor digitorum brevis** and the **extensor hallucis brevis,** which send tendons to the dorsal surfaces of the phalanges of the four lateral toes and the big toe, respectively.

Flexion of the Distal Foot and Toes. The principal power for flexion of the distal foot and toes is exerted by two deep muscles of the posterior leg. These are the

1. **flexor digitorum longus,** which arises from the posteromedial surface of the tibia, and
2. **flexor hallucis longus,** which arises from the posterior surfaces of the fibula and tibia in the lower half of the leg.

The tendons of both of these muscles pass posteriorly to the medial malleolus and, as illustrated in Figure 9–20, eventually insert into the plantar surfaces of the distal phalanges, thus causing flexion of the toes when contracted.

On the sole of the foot, the tendons of the long flexor muscles are joined by tendons from two short muscles, the **flexor digitorum brevis** and **flexor hallucis brevis.** In addition, the **abductor hallucis** pulls the big toe to the medial side, and the **abductor digiti minimi** pulls the little toe laterally. Contraction of these two abductors at the same time fans the toes apart. Also, deep in the plantar surface of the foot are **lumbrical** and **interosseus muscles** that perform essentially the same functions in the foot that they perform in the hand.

An important function of the smaller muscles in the sole of the foot is to help maintain the *longitudinal arch* of the foot. But, in addition to these muscles, this arch is also supported by a very strong ligamentous fascia called the **plantar aponeurosis** that covers the intrinsic foot muscles, extending from the tuberosity of the calcaneus bone to the fascial sheaths surrounding the bases of the toes. (The cut edges of this aponeurosis are illustrated in Figure 9–20.)

QUESTIONS

1. Discuss the importance of "fixation" of the scapula during arm movements. Also describe and discuss the muscles that move the scapula to different positions.
2. Name and describe the muscles surrounding the shoulder joint and discuss their importance in providing stability for this joint.
3. Describe especially the motions caused at the shoulder joint by the latissimus dorsi, the deltoid, and the pectoralis major.
4. Discuss the movements at the elbow, and describe the different muscles that cause these movements.
5. What is meant by "pronation" and "supination"? Describe the muscles that cause pronation and supination.
6. Discuss the movements of the wrist and the muscles that cause these.
7. Discuss the movements of the fingers and thumb and give a description of the muscles and their tendons that cause these movements.
8. Describe the muscles around the hip joint and their roles in stabilizing this joint.
9. Describe especially the functions of the following muscles at the hip joint: the rectus femoris, the iliopsoas, the gluteus maximus, the hamstring muscles, and the groin muscles.
10. Discuss the ranges of motion at the hip joint and the different muscles that cause motions in the different directions.
11. How do the hamstrings in the thigh and the gastrocnemius muscle in the lower leg interact at the knee to cause knee flexion?
12. Describe the functions of the quadriceps muscle, the patellar tendon, and the patella in causing extension at the knee joint.
13. Describe the interrelationships between the gastrocnemius and soleus muscles, the Achilles tendon, and the ankle joint in causing plantar flexion at the ankle.
14. Describe the muscles responsible for dorsiflexion at the ankle and also those that cause both inversion and eversion of the foot.
15. Discuss the similarities between the movements of the foot and the hand as well as the muscles involved in the motions.

REFERENCES

Anderson, J. E.: The lower limb. *In* Grant's Atlas of Anatomy. Baltimore, Williams & Wilkins, 1978, pp. 4–1 to 4–129.

Anderson, J. E.: The upper limb. *In* Grant's Atlas of Anatomy. Baltimore, Williams & Wilkins, 1978, pp. 6–1 to 6–124.

Bloom, W., and Fawcett, D. W.: Muscular tissue. *In* A Textbook of Histology, 10th Ed. Philadelphia, W. B. Saunders, 1975, pp. 288–332.

Copenhaver, W. M., Kelly, D. E., and Wood, R. L.: Muscle. *In* Bailey's Textbook of Histology, 17th Ed. Baltimore, Williams & Wilkins, 1978, pp. 240–278.

Figge, F. H. J.: The muscles. *In* Sobotta/Figge Atlas of Human Anatomy. Vol. I. Baltimore, Urban & Schwarzenberg, 1977, pp. 139–259.

Goss, C. M.: The muscles. *In* Gray's Anatomy of the Human Body, 29th Ed. Philadelphia, Lea & Febiger, 1973, pp. 371–520.

Ham, A. W., and Cormack, D. H.: Muscle tissue. *In* Histology, 8th Ed. Philadelphia, Lippincott, 1979, pp. 540–580.

Hammersen, F.: Muscle tissue. *In* Sobotta/Hammersen Histology. Baltimore, Urban & Schwarzenberg, 1980, pp. 68–79.

Langman, J., and Woerdeman, M. W.: Upper and lower limbs. *In* Atlas of Human Anatomy. Philadelphia, W. B. Saunders, 1978, pp. 205–349.

Leeson, T. S., and Leeson, C. R.: Muscle. *In* A Brief Atlas of Histology. Philadelphia, W. B. Saunders, 1979, pp. 77–88.

Leeson, T. S., and Leeson, C. R.: Muscle. *In* Histology, 4th Ed. Philadelphia, W. B. Saunders, 1981, pp. 187–215.

Pernkopf, E.: The lower limb. *In* Atlas of Topographical and Applied Human Anatomy, Vol. II. Philadelphia, W. B. Saunders, 1980, pp. 334–401.

Pernkopf, E.: The upper limb. *In* Atlas of Topographical and Applied Human Anatomy, Vol. II. Philadelphia, W. B. Saunders, 1980, pp. 39–91.

Weiss, L., and Greep, R. O.: Muscular tissue. *In* Histology, 4th Ed. New York, McGraw-Hill, 1977, pp. 251–282.

Wheater, P. R., Burkitt, H. G., and Daniels, V. G.: Muscle. *In* Functional Histology. New York, Churchill Livingstone, 1979, pp. 64–75.

Williams, P. L., and Warwick, R.: The muscles. *In* Gray's Anatomy, 36th British Edition. Philadelphia, W. B. Saunders, 1980, pp. 506–621.

(See also General Anatomical References, Chapter 1.)

V

THE NERVE-MUSCLE UNIT

10

Nerves, Membrane Potentials, and Nerve Transmission

Overview

All nerve signals are transmitted by *nerve fibers*, whether in the brain, the spinal cord, or the peripheral nerves. The long nerve fibers are *axons,* each of which grows outward from a nerve cell body, sometimes extending as long as a meter. The axon is a tubular structure bounded by a typical *cell membrane* and filled on the inside with intracellular fluid called *axoplasm.*

Inside the membrane of all nerve fibers is an *electrical potential of about −90 millivolts,* called the *membrane potential.* It is caused by ionic concentration differences across the cell membrane. Specifically, the *potassium ion concentration inside the membrane is very high* compared with its concentration outside. This concentration difference causes the positively charged potassium ions to leak out of the fiber, leaving on the inside negatively charged protein molecules that cannot leak out, thus creating electronegativity on the inside.

The nerve membrane also has another concentration difference that is essential for transmission of nerve signals. This is a *high concentration of sodium ions on the outside* but a low concentration inside, exactly opposite to the potassium ion concentration difference. When the nerve fiber is appropriately stimulated, *sodium channels* in the membrane become highly permeable, and the positively charged sodium ions now leak in tremendous numbers to the inside of the axon and make the membrane potential suddenly become positive instead of negative. However, this leakage of sodium ions lasts for less than a thousandth of a second, and after it is over potassium ions again leak to the outside and reestablish the negativity inside the membrane. This sequential change of membrane potential from negative to positive and then back to negative is called the *action potential.*

An action potential appearing at any point on the nerve fiber membrane sends *electrical current along the inside of the axon.* This current has an effect to open the sodium channels in the adjacent areas of the axon membrane, thus causing the action potential to spread along the entire extent of the nerve fiber from one end to the other. In this way, nerve signals, called *nerve impulses,* are transmitted from one place in the nervous system to another.

Action potentials are stimulated in nerves in many different ways. For instance, *mechanical sensory nerve endings* throughout the body can open the sodium channels and lead to action potentials. Also, in the

185

central nervous system, signals are transmitted from one nerve fiber to the next by a *hormonal mechanism.* That is, an action potential in the first fiber causes its endings to secrete a small amount of a hormone that acts on the membrane of the next fiber to make it suddenly permeable to sodium ions; this elicits an action potential in this fiber as well. The nerve fibers that control the skeletal muscles secrete a hormone, *acetylcholine,* at the neuromuscular junction, and the acetylcholine opens enough sodium channels in the muscle fiber membrane to elicit action potentials in the muscle fibers as well. The muscle fibers then conduct signals in the same way that nerve fibers conduct them.

The separate nerve fibers are insulated from each other by *Schwann cells.* In some cases these Schwann cells wrap their membranes around the axons, building an insulating sheath around the fibers. This sheath is called the *myelin sheath* because it contains large amounts of the insulating lipid substance *myelin,* and the fibers with such a sheath are called *myelinated fibers.* In other instances, the nerve fibers are simply imbedded in the walls of the Schwann cells. These fibers are called *unmyelinated fibers.* The myelinated fibers, because of their superior insulation, transmit nerve signals very rapidly and with very little use of nerve fiber energy. On the other hand, the unmyelinated fibers are usually much smaller than the myelinated fibers so that far greater numbers of them can be contained in a single small nerve trunk.

The goal of this and the next chapter is to explain the way in which nerves cause muscles to contract. To do this it will be necessary, first, to explain in this chapter the fundamental properties of nerves themselves as well as conduction of signals by nerves. In the following chapter we will then consider the fundamental properties of muscle cells and their contractile mechanism.

The Neuromuscular Unit

Figure 10–1 illustrates the basic neuromuscular unit. All skeletal muscles are controlled by nerve fibers that originate in the spinal cord. In the anterior horns of the spinal cord gray matter (or in comparable areas of the brain stem) are 3 to 10 million large nerve cells called *anterior motoneurons.* Each of these sprouts forth a single nerve fiber that enters a *peripheral nerve trunk* along with hundreds or thousands of other such fibers. These fibers eventually distribute to one or more of the skeletal muscles, as illustrated in Figure 10–1. Also collected in many of these same peripheral nerve trunks are sensory nerve fibers that carry sensory signals from the skin or other parts of the body to the spinal cord or other parts of the central nervous system.

The terminal portion of each muscle controlling nerve fiber branches from 3 to 1000 times, and each branch fiber terminates on a sin-

gle muscle fiber. The junction between the nerve terminal and the muscle fiber is called a *neuromuscular junction.* This has very special anatomic and functional characteristics that will be described in detail later in the chapter.

Physiologic Anatomy of the Nerve Fiber

The motor nerve fiber is composed of two parts, a central portion called the *axon* and an insulating sheath called either the *sheath of Schwann* or the *myelin sheath.* Only a small portion of the myelin sheath is illustrated in Figure 10–1, but this extends the entire distance from the spinal cord to the final terminal nerve endings.

Figure 10–2 illustrates in diagrammatic form the functional parts of the nerve fiber. Note both in this figure and in Figure 10–1 that the axon is a long thin tubular structure bounded by a *membrane* that has exactly the same functions as any other cell membrane, except that it is specifically adapted for transmission of nerve signals, as we shall discuss in subsequent sections of this chapter.

Inside the axon membrane is gelled intracellular fluid called *axoplasm* and outside is interstitial fluid, which is the extracellular fluid that percolates through the spaces between the cells and between the fibers.

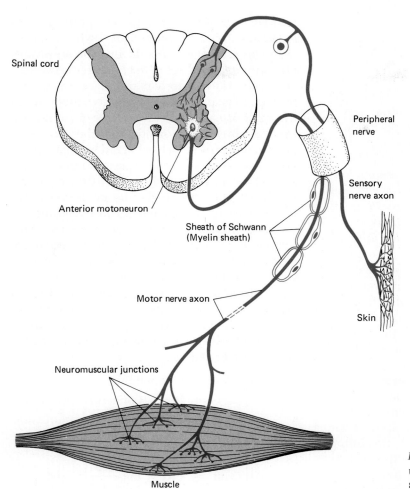

FIGURE 10–1 The neuromuscular unit, illustrating the innervation of skeletal muscle.

Figure 10–2A also shows that the *myelin sheath* is discontinuous at periodic points called *nodes of Ranvier*. Thus, the myelin sheath serves as an excellent electrical insulator of the axon at all points except at these nodes. It is here that the axon membrane comes in direct contact with the interstitial fluid, which is essential for conduction of the nerve impulse.

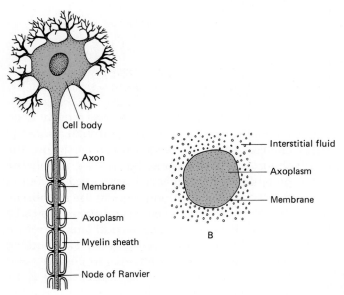

FIGURE 10–2 (A) A nerve cell with its thread-like axon, showing the nerve membrane, the axoplasm, and the myelin sheath. (B) Cross-section of an axon.

Membrane Potentials

All cells of the human body have an electrical potential across the cell membrane called simply the *membrane potential.* Under resting conditions, this potential is negative inside the membrane. The membrane potential is caused by differences between the ion compositions of intracellular and extracellular fluids. Especially important is the fact that *intracellular fluid contains a very high concentration of potassium ions whereas extracellular fluid has only a very low concentration of this ion;* and *the opposite exists for sodium, very high concentration in the extracellular fluid but very low concentration intracellularly.*

Membrane potentials play an essential role in the transmission of nerve signals as well as in the control of muscle contraction, glandular secretion, and undoubtedly still many other cell functions. Therefore, let us explain the mechanism for the development of membrane potentials.

Ionic Concentration Differences Across the Nerve Membrane.

The membrane of the axon has the same *sodium-potassium pump* as that found in all other cell membranes of the body. As was explained in Chapter 5, this pump transports sodium ions from inside the axon to the outside while at the same time transporting potassium ions to the inside. The final effect of these transport processes on the sodium and potassium concentrations outside and inside the axon is illustrated in Figure 10–3, showing the sodium concentration on the outside to be 142 mEq per liter but on the inside only 14 mEq per liter. For potassium, the concentration difference is in exactly the opposite direction with a potassium concentration inside the fiber of 140 mEq per liter but only 4 mEq outside.

Development of the Membrane Potential.

To explain the development of the membrane potential, it is necessary to understand that *the resting axon membrane is very impermeable to sodium ions but very permeable to potassium ions.* Therefore, the highly concentrated potassium ions inside the axon are always attempting to leak out of the fiber to the outside, and many do indeed leak out. Because the potassium ions are positively charged, their loss to the exterior carries positive electricity outward. On the other hand, inside the fiber are large quantities of negatively charged protein molecules that cannot leak to the exterior, and these are left behind inside the fiber. Consequently, the inside of the nerve fiber becomes very negative because of the deficit of positive ions and the excess of negatively charged protein ions. Thus, the membrane potential of the usual large nerve fiber under resting conditions is about −90 millivolts, with the negativity inside the fiber.

The amount of voltage that will develop across a membrane when it is selectively permeable to only one univalent positive ion can be calculated using the following equation, called the *Nernst Equation:*

Millivolts Membrane Potential

$$= -61 \times \log \frac{\text{Concentration inside}}{\text{Concentration outside}}$$

For potassium:

$$\text{Millivolts} = -61 \times \log \frac{140}{4} = -94$$

The true value, about −90 millivolts, is slightly less than the calculated −94 millivolts because the membrane is very slightly permeable to sodium ions, which carry positive charges inward and neutralize a small amount of the intracellular negativity.

THE ACTION POTENTIAL AND THE NERVE IMPULSE

When a signal is transmitted over a nerve fiber, the membrane potential goes through a series of changes called the *action potential.* Before the action potential begins, the resting membrane potential inside the fiber is very negative, but at the outset of the action potential the membrane potential suddenly becomes positive, followed a few ten-thousandths of a second later by return to the very negative resting level. This sudden rise of the membrane potential to positivity and then its return to its normal negative state *is* the action potential; it is also called the *nerve impulse.* The impulse (or action potential) spreads

FIGURE 10–3 Concentration gradients of sodium and potassium at the axon membrane, showing that the membrane in the resting state is permeable only to potassium.

along the nerve fiber, and by means of such impulses the nerve fiber transmits information from one part of the body to another.

Action potentials can be elicited in nerve fibers by any factor that suddenly increases the permeability of the membrane to sodium ions. In the normal resting state the membrane is relatively impermeable to sodium ions even though it is quite permeable to potassium ions. When the fiber is suddenly made permeable to sodium ions, the positively charged sodium ions leak to the inside of the fiber and make this inside positive, thus initiating the action potential. This first stage of the action potential, that is, the initial positive change in the membrane potential, is called *depolarization*. The subsequent return of the potential to its resting negative state is called *repolarization*.

Depolarization of the Membrane, and Transmission of the Nerve Impulse

Figure 10–4A illustrates the normal resting state of a nerve fiber with negative electrical charges on the inside. Now let us see what happens when this fiber is "stimulated" in its middle.

Figure 10–4B illustrates at its centralmost point an area that has suddenly become so permeable that even sodium ions can now diffuse through the membrane with ease. Because the concentration of sodium ions outside the fiber is

10 times as great as inside, the positively charged sodium ions rush to the inside, causing the membrane to become suddenly positive inside and negative outside. This is opposite to the usual resting state of the membrane, and it is called the *overshoot potential*. This sudden loss of the negative membrane potential is also called *depolarization* because the normal polarized state, with positivity on the outside and negativity on the inside, no longer exists.

The Depolarization Wave or Nerve Impulse. In Figures 10–4C and D, the area of depolarization in the middle of the fiber has extended in both directions, and the area of increased permeability has also extended. The cause of this extension is that electrical current flows from the original depolarized area to adjacent areas, and this current, for reasons not yet completely understood, makes the adjacent areas also permeable to sodium. Therefore, sodium ions now diffuse inward through the membrane at these new areas, depolarizing these areas also and causing still more electrical current to spread along the fiber. Thus, the area of increased permeability to sodium ions spreads still farther along the membrane in both directions away from the area of original depolarization. This process repeats itself over and over again. This spread of increased sodium permeability and electrical current along the membrane is called a *depolarization wave* or a *nerve impulse*.

It is obvious from this discussion that once a single point anywhere on the nerve fiber becomes depolarized a nerve impulse travels away from that point in both directions, and each impulse keeps on traveling until it comes to both ends of the fiber. In other words, a nerve fiber can conduct an impulse either toward the nerve cell or away from it.

Repolarization of the Nerve Fiber. Immediately after a depolarization wave has traveled along a nerve fiber, the inside of the fiber has become positively charged as illustrated in Figure 10–4D because of the large number of sodium ions that have diffused to the inside. This positivity stops further flow of sodium ions to the inside of the fiber, and it also causes the membrane to become impermeable to sodium ions again, though the exact mechanism of this also is not understood. Yet, the membrane is still very permeable to potassium. And because of the high concentration of potassium on the inside many potassium ions now diffuse outward carrying positive charges with them. This once again creates electronegativity inside the membrane and positivity outside, a process

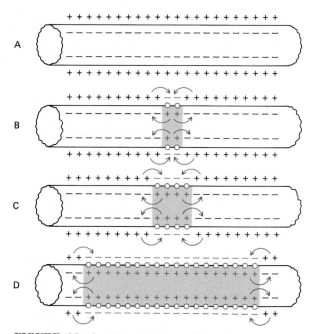

FIGURE 10–4 Transmission of the depolarization wave, the initial event in the nerve impulse.

FIGURE 10-5 The repolarization process, the concluding event in the nerve impulse.

called *repolarization* because it reestablishes the normal polarity of the membrane.

Repolarization usually begins at the same point in the fiber where depolarization had originally begun, and it spreads along the fiber in the manner illustrated in Figure 10-5. Repolarization occurs a few ten-thousandths of a second after depolarization. That is, the whole cycle of depolarization and repolarization takes place in large nerve fibers in less than one-thousandth of a second, and the fiber is then ready to transmit a new impulse.

Refractory Period. When an impulse is traveling along a nerve fiber, the nerve fiber cannot transmit a second impulse until the fiber membrane has been repolarized. For this reason the fiber is said to be in a *refractory* state, and the time that the fiber remains refractory is called the *refractory period*. This varies from about $\frac{1}{2500}$ second for large nerve fibers up to $\frac{1}{250}$ second for small fibers.

Reestablishment of Ionic Concentration Differences After Nerve Impulses are Conducted

After the nerve fiber has become repolarized, the sodium ions that have leaked to the inside of the membrane and the potassium ions that have leaked to the outside must be returned to their original sides of the membrane. This is accomplished by the same sodium-potassium pump discussed previously. That is, this pump pumps the extra sodium ions inside the fiber back to the outside and the potassium ions in the opposite direction. Thus, this process restores the ionic differences to their original levels.

However, it must be emphasized that even when the sodium-potassium pump suddenly fails to act, a hundred thousand or more impulses can still be transmitted over the nerve fiber before transmission will cease. The reason for this is that only a few trillionths of a mole of sodium enter the fiber and approximately the same amount of potassium leaves each time a single impulse is transmitted, so that it takes a hundred thousand or more impulses for the concentration differences across the membrane to be dissipated and therefore to cause cessation of impulse transmission. It is evident, then, that the sodium-potassium pump is not necessary for the initial repolarization of the membrane after each nerve impulse; this is accomplished by the diffusion of potassium outward through the membrane pores. Instead, the sodium-potassium pump simply plods along slowly, reestablishing ionic concentration differences across the membrane whenever a large number of impulses tend to alter these.

Summary of the Steps in the Action Potential

Let us now summarize the essential steps in membrane potential generation and in action potential transmission:

1. The sodium-potassium pump causes a high concentration of sodium on the outside of the membrane but a low concentration on the inside, and high potassium concentration inside but low outside.

2. Because the resting membrane is highly permeable to potassium ions, and because of the high potassium concentration inside the membrane, these ions leak to the outside. This carries positive electrical charges out of the fiber but leaves many negative protein ions on the inside. Thus, the resting membrane develops a negative *membrane potential* of about −90 millivolts inside the fiber.

3. A sudden increase in permeability of the membrane to sodium ions initiates the action potential. Sodium ions rush to the inside of the fiber, carrying positive charges, and this creates positivity inside the membrane at the local point where the membrane has become highly permeable. This is called the *depolarization* process.

4. The positive electricity entering the nerve fiber moves along the inside of the fiber. This has an effect on the adjacent membrane to cause it also to become highly permeable to sodium. Therefore, sodium leaks in here also, and the process is repeated again and again along the nerve fiber. Thus, the nerve impulse travels along the fiber.

5. After the fiber becomes fully depolarized, the membrane now suddenly becomes impermeable to sodium again but remains highly per-

meable to potassium. Because of the high concentration of potassium in the fiber, large quantities of positively charged potassium ions again diffuse to the outside. Loss of these positive charges makes the inside become negative again. This is the *repolarization* process. The nerve fiber is now ready to transmit another impulse.

6. During the period when the fiber had been depolarized, a small number of sodium ions had moved to the inside of the fiber; and during the initial process of repolarization, a small number of potassium ions had moved to the outside. The sodium-potassium pump now begins to work again, continuing to work even between action potentials, and pumps the sodium ions outward and potassium ions inward. This reestablishes the appropriate concentration differences between the inside and the outside.

Recording Action Potentials

A method for recording the potential between the inside and outside of the nerve fiber is illustrated in Figure 10–6. This figure shows a minute glass pipet having a tip less than 1 μ (micrometer) in diameter and filled with a strong solution of potassium chloride that conducts electricity from the tip. The tip pierces the nerve membrane to make contact with the fluid in the center of the fiber. On the outside of the fiber, located anywhere in contact with the extracellular fluid, is another electrode illustrated by the dark rectangle. These two electrodes are connected to a recording meter.

The changing potential across the membrane is illustrated schematically in Figure 10–7. Section A of this figure shows that during the original resting state, a potential of −90 millivolts is recorded inside the fiber with respect to the outside. In Figure 10–7B a depolarization wave

FIGURE 10–7 Principles of recording a monophasic action potential.

has traveled down the fiber until it is directly at the electrodes. At this instant, the rapid influx of positive sodium ions to the inside of the fiber reverses the potential, causing positivity inside the fiber and negativity outside. Then, soon after the impulse has passed the electrodes, repolarization occurs (Fig. 10–7C) and the potential returns again to approximately the original resting potential of −90 millivolts because of diffusion of positive potassium ions to the outside.

Figure 10–8 illustrates a continuous recording of the potential changes in a large nerve fiber as a nerve impulse passes the electrodes. This record is called the *monophasic action potential*. It begins with a normal resting membrane potential of −90 millivolts on the inside of the fiber and during the peak of the action potential overshoots to become about +35 milli-

FIGURE 10–6 Measurement of the membrane potential of the nerve fiber using a microelectrode.

FIGURE 10–8 Graphic record of a monophasic action potential recorded from a large nerve fiber.

volts. Within approximately $\frac{1}{2}$ millisecond, the normal membrane potential recovers, and the record returns almost to its original level. Actually, it undershoots a few millivolts because the sodium pump immediately begins to pump positive sodium ions to the outside, which makes the potential even more negative than the normal value of −90 millivolts. This slight undershoot is called an *after potential*. It lasts for as long as 50 to 100 milliseconds; during this time, much of the sodium that has entered the fiber during the action potential is removed from the fiber.

Use of the Oscilloscope to Record Action Potentials.   The different types of mechanical recording apparatus cannot function quickly enough to record the rapid transient voltages of nerve action potentials. Therefore, a special instrument called the *oscilloscope*, which is similar to a television receiver, is normally used for this purpose. The principal components of the oscilloscope are shown in Figure 10–9. The basic part of this instrument is a cathode ray tube. An *electron gun* at the base of this tube shoots toward the face of the tube a fine beam of electrons having a diameter of about 1 mm. The beam passes between four metal plates, two of which are placed on the two lateral sides of the beam and the other two of which are above and below the beam. On the face of the cathode ray tube is a fluorescent material that glows brightly when the electron beam strikes. Because electrons are themselves charged negatively and therefore are attracted by positive charges or repulsed by negative charges, the electron beam can be moved back and forth or up and down across the face of the tube by applying electrical potentials to the horizontal and vertical plates.

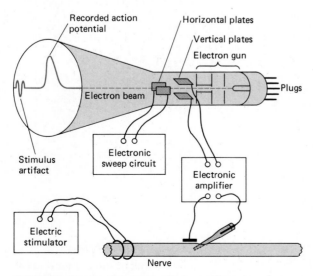

FIGURE 10–9   Diagram of an oscilloscope for recording action potentials from nerves.

An *electronic sweep circuit*, connected to the horizontal plates, causes the electron beam to move from left to right across the face of the tube. The fluorescence along the path of the moving beam gives the appearance of a horizontal line on the face of the tube.

If, while the beam of electrons is moving across the tube, electrical potentials are applied to the vertical plates, the beam can be made to move up or down. In Figure 10–9 the beam of electrons deviates from the line slightly at the point called *stimulus artifact*, and then it deviates again at the point called *recorded action potential*. The electrical voltages across the nerve membrane are responsible for these deviations. When the nerve is stimulated by an electrical stimulator, some of the electrical current from the stimulator spreads through the fluids surrounding the nerve fiber to the pickup electrodes and causes the stimulus artifact. At the same time, an action potential begins traveling down the nerve fiber toward the two pickup electrodes. Then, when the action potential reaches the electrodes, the amplified signal applied to the vertical plates makes the electron beam move first upward and then back down again to record the potential.

Most action potentials of nerves have a total duration of not more than a few ten-thousandths to a few thousandths of a second. The cathode ray oscilloscope fortunately can record an electrical potential that lasts for this short period of time. Therefore, it is quite capable of giving a true record of nerve action potentials.

Types of Stimuli That Can Excite the Nerve Fiber

In the living body, nerve fibers normally are stimulated by both physical and chemical means. For example, pressure on certain of the nerve endings in the skin mechanically stretches these endings, in this way opening the membrane pores to sodium ions and thereby setting off impulses. Heat and cold affect other nerve endings to elicit impulses, and damage to the tissues, such as cutting the skin or stretching the tissues too much, can generate pain impulses.

In the central nervous system, impulses are transmitted from one neuron to another mainly by chemical means. The nerve ending of the first neuron secretes a chemical substance called a "transmitter" substance that in turn excites the second neuron. In this way, impulses are sometimes passed through many hundred neurons before stopping. This will be discussed in detail in Chapter 15.

FIGURE 10–10 A laboratory stimulator.

In the laboratory, nerve fibers are usually stimulated electrically. A typical stimulator is illustrated in Figure 10–10. It emits electrical impulses of any desired voltage either singly or in rapid succession. Usually, a small probe having two wires at its end is placed on either side of a nerve trunk, and the electrical stimulus is applied so that electrical current flows through the fibers. As the current passes through the fiber membranes the permeability is altered, eliciting nerve impulses.

The All or None Law. From the preceding discussion of the nerve impulse it is evident that if a stimulus is strong enough to cause a nerve impulse, this impulse will then travel in both directions along the fiber until the entire fiber fires. In other words, a weak stimulus does not cause only part of the nerve to depolarize; the stimulus is strong enough to depolarize ei-

ther the entire fiber or not any of it. This is called the *all or none law.*

TRANSMISSION OF SIGNALS IN PERIPHERAL NERVES

Types of Nerve Fibers: Myelinated and Unmyelinated

Figure 10–11 illustrates a typical small peripheral nerve. The large white dots in this figure surrounded by black rings are large *myelinated nerve fibers.* The white portions are the axons and the black rings are the myelin sheaths.

However, if one will study the figures more carefully, he will also see a large number of very small white dots lying between the large myelinated fibers. These are *unmyelinated nerve fibers* that do not have a typical myelin sheath; nevertheless, they are insulated from each other in another way that will be discussed later. If one will count the numbers of both myelinated and unmyelinated fibers, he will see that there are about twice as many unmyelinated fibers as myelinated even though this is not evident from first inspection of the figure.

The large myelinated fibers transmit nerve signals extremely rapidly. These signals mainly control rapid muscle activity or transmit very critical sensory signals to the brain. On the other hand, the unmyelinated fibers control such structures as the blood vessels, and they also

FIGURE 10–11 Cross-section of the typical small peripheral nerve, showing (1) large myelinated nerve fibers, (2) small unmyelinated nerve fibers, and (3) the nerve sheath composed of connective tissue.

transmit a large share of the noncritical sensory information to the brain such as crude touch signals from all areas of the skin, pressure signals from the surface of the body, or pain signals of the continuous aching type from anywhere in the body. In our present discussion of muscle control, we are mainly concerned with the myelinated nerve fibers.

The Schwann Cell and Deposition of the Myelin Sheath. Figure 10–12A illustrates a short section of a peripheral nerve fiber. Down the center of this fiber is the axon that transmits the nerve impulse. Surrounding the axon is the *sheath of Schwann* (which is also the *myelin sheath*). This sheath is deposited by *Schwann cells* that are present throughout peripheral nerves and that provide electrical insulation for the axons.

The Schwann cell forms the myelin sheath by first attaching its membrane to the axon and then wrapping the membrane around and around the axon, sometimes forming as many as 20 to 30 layers of concentric cell membranes. As the cell wraps around the axon, the cytoplasm is squeezed out of the concentric rings. Therefore, the sheath that develops around the nerve fiber is, in reality, many layers of almost pure Schwann cell membrane. Because this membrane contains large quantities of the fatty substance *myelin*, the insulating membrane around the axon is called the "myelin sheath." Myelin, because of its fatty nature, is almost totally nonconductive of ions. Therefore, it provides excellent electrical insulation for the axon.

The Node of Ranvier. Each Schwann cell that wraps around an axon spreads lengthwise along the axon for about 1 mm. Therefore, the myelin sheath formed for each Schwann cell is a 1-mm long envelope of myelin insulation. Then, beyond the first Schwann cell, a second one wraps around the axon. The junction between the two Schwann cells is called the *node of Ranvier*. A thin extracellular fluid space exists between the two Schwann cells at this node, and small amounts of ions can flow through this space all the way to the surface of the axon. Therefore, the node of Ranvier is very important, indeed essential, in the transmission of nerve impulses by myelinated nerve fibers.

Role of the Schwann Cell in Insulating Unmyelinated Axons. Even though Schwann cells do not wrap around the unmyelinated axons, they nevertheless do provide insulation for these axons as well. The mechanism of this is

A

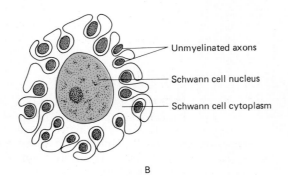

B

FIGURE 10–12 Function of the Schwann cell to insulate nerve fibers: (A) The wrapping of a Schwann cell membrane around a large axon to form the myelin sheath of the myelinated nerve fiber. (Modified from Leeson and Leeson: Atlas of Histology. Philadelphia, W.B. Saunders Company, 1979.) (B) Evagination of the membrane and cytoplasm of a Schwann cell around multiple unmyelinated nerve fibers.

illustrated in Figure 10–12B. The Schwann cells grow into all the spaces between the unmyelinated fibers, and the membranes and cytoplasm of the Schwann cells envelop the unmyelinated axons as illustrated in the figure. Thus, the unmyelinated axons become partially if not totally encircled by at least a single layer of Schwann cell membrane, and the cytoplasm of the Schwann cell forms additional spacing between the unmyelinated axons.

Effect of the Myelin Sheath on Nerve Impulse Transmission—Saltatory Conduction. Figure 10–13 illustrates schematically a myelinated nerve fiber with a myelin sheath broken by a node of Ranvier approximately every millimeter along the length of the fiber. At these nodes, typical membrane depolarization can occur, but beneath the myelin sheath, membrane depolarization does not take place because of the insulator properties of the myelin. Instead, impulses are transmitted along a myelinated nerve by a process called *saltatory conduction*, which may be explained as follows: Referring once again to Figure 10–13, let us assume that the first node of Ranvier becomes depolarized. This causes electrical current to spread, as shown by the arrows, around the outside of the myelin sheath and also down the core of the fiber all the way to the next node of Ranvier, causing it to become depolarized also. Current generated by this node then causes the same effect at the next node; thus, the impulse "jumps" from node to node, which is the process of saltatory conduction.

Saltatory conduction is valuable for two reasons:

1. By causing the depolarization process to jump long intervals along the nerve fiber, it increases the velocity of conduction along the fiber many fold.
2. Perhaps even more important, saltatory conduction prevents the depolarization of large areas of the fiber and thereby prevents leakage of large amounts of sodium to the inside and potassium to the outside of the fiber as each nerve impulse is transmitted. This conserves the energy required by the sodium-potassium pump to expel the sodium and to return potassium to the inside. Therefore, the myelin sheath greatly decreases the amount of energy required by the nerve for impulse transmission.

Velocity of Conduction in Nerve Fibers. The larger the nerve fiber and the thicker the myelin sheath, the more rapidly can the nerve conduct an impulse. The largest myelinated nerve fibers are about $20\,\mu$ (micrometers) in diameter, and the smallest unmyelinated fibers are about $0.5\,\mu$. The very large fibers conduct impulses at a velocity as great as 100 m—about the length of a football field—in 1 second, while the very small fibers conduct impulses at a velocity of only 0.5 m per second, or approximately the distance from the foot to the knee. All sizes of nerve fibers exist between these smallest and largest sizes, so that a wide spectrum of impulse velocities occurs in the different nerves.

Number of Impulses That Can Be Transmitted Per Second. The number of impulses that can be transmitted by any one fiber per second is determined by the "refractory period" of the fiber (which is the duration of time between the beginning of depolarization and the end of repolarization), and this depends also to a great extent on the size of the fiber. Large fibers (15 to $20\,\mu$ in diameter) become repolarized in approximately $\frac{1}{2500}$ second. Therefore, a second nerve impulse can be transmitted $\frac{1}{2500}$ second after the first, or a total of up to 2500 impulses can be conducted each second. At the opposite extreme, the very smallest nerve fibers require as long as $\frac{1}{250}$ second to repolarize, which means that they can transmit no more than 250 impulses per second.

Further descriptions of the different types of fibers and their functions will be presented in connection with the discussions of sensory functions of the nervous system in Chapter 16 and of motor functions in Chapters 17 and 19.

Transmission of Signals of Different Strengths by a Nerve Bundle

A nerve bundle has two means by which it can transmit signals of different strengths—weak, strong, or intermediate. These are to transmit

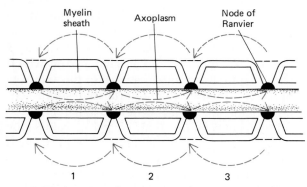

FIGURE 10–13 Saltatory conduction along a myelinated axon.

impulses (1) simultaneously over varying numbers of nerve fibers, which is called *spatial summation*, and (2) at a slow or rapid frequency over the same fiber, which is called *temporal summation*.

As an example of spatial summation, if 100 nerve fibers are connected between the spinal cord and a foot muscle, stimulation of one of these fibers will cause only a weak response in the muscle, but simultaneous stimulation of all 100 fibers will cause a strong contraction. Obviously, any number of fibers between 1 and 100 can be stimulated at a time, giving any one of 100 different strengths of muscle contraction.

Temporal summation means changing the strength of a signal by sending a large or small number of impulses along the same fiber per second. If one impulse is transmitted each second, only a weak effect usually results, but if 5, 15, 25, 75, or more impulses are transmitted per second, the strength of the effect becomes progressively greater.

Ordinarily the nerve trunk transmits signals of different strengths by a combination of both the spatial and temporal methods. That is, when a strong signal is to be transmitted, large numbers of fibers are utilized and large numbers of impulses are also transmitted along each fiber. When a weak signal is to be transmitted, fewer fibers are used and fewer impulses are transmitted.

TRANSMISSION OF IMPULSES BY MUSCLE FIBERS

Transmission by Skeletal Muscle Fibers. Skeletal muscle fibers transmit impulses exactly as nerve fibers do. The normal velocity of transmission in skeletal muscle fibers is about 4 m per second in contrast to 50 to 100 m per second in the very large myelinated nerve fibers and 0.5 m per second in the very small unmyelinated fibers. Because the nerve fibers that control the skeletal muscles are a large type, carrying impulses at about 60 m per second, a signal travels from the brain to the muscle extremely rapidly but then decreases in velocity more than 10-fold as it goes into the muscle itself.

Transmission in Heart Muscle and Smooth Muscle. Transmission of impulses in heart muscle and smooth muscle also occurs like that in nerve and skeletal muscle, but at still lower velocities—about 0.4 m per second in the heart and only a centimeter or so per second in smooth muscle. In heart muscle and in many smooth muscle masses the fibers interconnect with each other to form latticeworks so that stimulation of any one fiber always causes the impulse to travel over the entire muscle mass, resulting in complete contraction of the whole muscle rather than only part of it. This effect as it occurs in the heart is discussed in detail in Chapter 24.

Another difference between these two types of muscle and skeletal muscle is the duration of the action potential. In heart muscle it lasts about 0.3 second and in some smooth muscle as long as a second. So long as the membranes remain depolarized, the muscle fibers remain contracted. Therefore, the duration of contraction of both heart and smooth muscle is unusually long in comparison with that of skeletal muscle.

THE NEUROMUSCULAR JUNCTION

The *neuromuscular junction* is the connection between the end of a large myelinated nerve fiber and a skeletal muscle fiber. In general, each skeletal muscle is supplied with only one neuromuscular junction but rarely more than one.

Physiologic Anatomy of the Neuromuscular Junction. Figure 10–14A and B illustrates a typical neuromuscular junction. The nerve fiber branches at its end to form a complex of branching *axon terminals* called the *end-plate*, which invaginates into the muscle fiber but lies entirely outside the muscle fiber cellular membrane.

Figure 10–14C shows an electron micrographic sketch of the juncture between a single-branch axon terminal and the muscle fiber membrane. The invagination of the membrane is called the *synaptic gutter* or *synaptic trough*, and the space between the terminal and the fiber membrane, about 20 to 30 nanometers (nm) wide, is called the *synaptic cleft*. This cleft is filled with a gelatinous "ground" substance through which extracellular fluid diffuses. At the bottom of the gutter are numerous *folds* of the muscle membrane, which form *subneural clefts* that greatly increase the surface area at which the synaptic transmitter can act. In the axon terminal are many mitochondria that supply energy mainly for synthesis of the excitatory transmitter *acetylcholine* that, in turn, excites the muscle fiber. The acetylcholine is synthesized in the cytoplasm of the terminal but is rapidly absorbed into many small *synaptic vesicles*, approximately 300,000 of which are normally in all

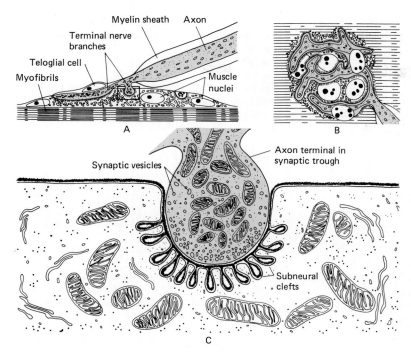

C

FIGURE 10–14 Different views of the motor end-plate. (A) Longitudinal section through the end-plate. (B) Surface view of the end-plate. (C) Electron micrographic appearance of the contact points between one of the axon terminals and the muscle fiber membrane, representing the rectangular area shown in A. (From Bloom and Fawcett, as modified from R. Couteaux: A Textbook of Histology. Philadelphia, W.B. Saunders Company, 1975.)

the terminals of a single end-plate. On the surfaces of the subneural clefts are aggregates of the enzyme *cholinesterase*, which is capable of destroying acetylcholine.

Transmission of the Impulse at the Neuromuscular Junction

Secretion of Acetylcholine. When a nerve impulse reaches the neuromuscular junction, passage of the action potential over the membrane of the nerve terminal causes many of the small vesicles of acetylcholine stored in the terminal to rupture through the terminal membrane into the synaptic cleft between the terminal and the muscle fiber membrane. The acetylcholine then acts on the folded muscle membrane to increase its permeability to sodium ions. This increased permeability in turn allows instantaneous leakage of sodium to the inside of the fiber, which carries positive charges to the inside and immediately depolarizes this local area of the muscle membrane. And this local depolarization sets off an action potential that travels in both directions along the fiber. In turn, the action potential traveling along the fiber causes muscle contraction.

Destruction of Acetylcholine by Cholinesterase. If the acetylcholine secreted by the nerve terminals should remain in contact with the muscle fiber membrane indefinitely, the fiber would transmit a continuous succession of impulses. However, the cholinesterase on the surfaces of the membrane folds in the synaptic gutter enzymatically splits acetylcholine into acetic acid and choline in about $\frac{1}{500}$ second. Therefore, almost immediately after the acetylcholine has stimulated the muscle fiber, the acetylcholine itself is destroyed. This allows the membrane to repolarize and to become ready for stimulation again as soon as a new nerve impulse approaches.

The acetylcholine mechanism at the neuromuscular junction provides an *amplifying system* that allows a weak nerve impulse to stimulate a very large muscle fiber. That is, the amount of electrical current generated by the nerve fiber is not enough by itself to elicit an impulse in the muscle fiber because the nerve fiber has a cross-sectional area only one tenth or less that of the muscle fiber. Instead, the secreted acetylcholine causes the muscle fiber to generate its own impulse. Thus, each nerve impulse actually comes to a halt at the neuromuscular junction, and in its stead an entirely new impulse begins in the muscle.

Paralysis Caused by Myasthenia Gravis

Sometimes a person has very poor transmission of impulses at the neuromuscular junction, an effect that obviously produces paralysis. One cause of this is the condition called *myasthenia gravis*. It is caused by an autoimmune response in which the immune system of the body has developed antibodies against the muscle cell membrane. Reaction of antibodies with the membrane in the synaptic gutter widens the space in the synaptic cleft and also destroys many of the membrane folds. These effects seriously depress the responsiveness of the muscle fiber to acetylcholine. Treatment with *neostigmine,* a drug that prevents the destruction of acetylcholine by cholinesterase, is often dramatic in overcoming the paralysis. This drug allows acetylcholine to accumulate in the neuromuscular junction from one nerve impulse to the next and, therefore, to exert a tremendous effect on the muscle fiber membrane. As a result, persons almost totally paralyzed by myasthenia gravis can sometimes be returned almost to normality within less than 1 minute after a single intravenous injection of neostigmine.

QUESTIONS

1. Describe the anatomy of nervous control of muscle contraction.
2. What is the role of the sodium-potassium pump in establishing ionic concentration differences and membrane potentials across the nerve membrane?
3. Describe the process of depolarization of the nerve membrane.
4. Describe the process of repolarization.
5. Summarize the steps in the action potential.
6. Describe the micropipet method for recording action potentials, as well as use of the oscilloscope for this purpose.
7. How can nerve fibers be excited?
8. What is the role of Schwann cells in insulating both myelinated and unmyelinated nerve fibers?
9. What are the roles of spatial summation and temporal summation in transmission of signals of different strengths?
10. Describe the neuromuscular junction in skeletal muscle.
11. What are the roles of acetylcholine and cholinesterase in transmission of the signal through the neuromuscular junction?

REFERENCES

Carmeliet, E., and Vereecke, J.: Electrogenesis of the action potential and automaticity. *In* Berne, R.M., *et al.* (eds.): Handbook of Physiology. Sec. 2, Vol. 1. Baltimore, Williams & Wilkins, 1979, p. 269.

Ceccarelli, B., and Clementi, F. (eds.): Neurotoxins, Tools in Neurobiology. New York, Raven Press, 1979.

Fozzard, H.A.: Conduction of the action potential. *In* Berne, R.M., *et al.* (eds.): Handbook of Physiology. Sec. 2, Vol. 1. Baltimore, Williams & Wilkins, 1979, p. 335.

Greene, L.A., and Shooter, E.M.: The nerve growth factor: Biochemistry, synthesis, and mechanism of action. *Annu. Rev. Neurosci., 3*:353, 1980.

Hodgkin, A.L.: The Conduction of the Nervous Impulse. Springfield, Ill., Charles C Thomas, 1963.

Hodgkin, A.L., and Huxley, A.F.: Quantitative description of membrane current and its application to conduction and excitation in nerve. *J. Physiol. (Lond.), 117*:500, 1952.

Keynes, R.D.: Ion channels in the nerve-cell membrane. *Sci. Am. 240*(3):126, 1979.

Rall, W.: Core conductor theory and cable properties of neurons. *In* Brookhart, J.M., and Mountcastle, V.B. (eds.): Handbook of Physiology. Sec. 1, Vol. 1. Baltimore, Williams & Wilkins, 1977, p. 39.

Stevens, C.F.: The neuron. *Sci. Am., 241*(3):54, 1979.

Wallick, E.T., *et al.:* Biochemical mechanisms of the sodium pump. *Annu. Rev. Physiol., 41*:397, 1979.

11

Functional Anatomy and Contraction of Muscle

Overview

The body has three different functional types of muscle; **skeletal muscle, cardiac muscle,** and **smooth muscle.** Both skeletal and cardiac muscle are *striated muscle,* and they have similar contractile mechanisms. Smooth muscle, the type of muscle found in most internal organs, has a different internal organization, but the chemical basis of contraction is still the same.

Each skeletal muscle is composed of a few hundred up to many tens of thousands of parallel **skeletal muscle fibers,** each of which runs the entire length of the muscle. In turn, each muscle fiber contains several hundred to several thousand parallel **myofibrils.** And inside the entire length of each myofibril, millions of minute molecular filaments, the **myosin** and **actin filaments,** alternate with each other—first a set of myosin filaments, then a set of actin filaments—this sequence repeating itself the entire length of the myofibril. The ends of the *myosin and actin filaments overlap each other,* and in the presence of *calcium ions* they interact with each other both physically and chemically, causing the actin and myosin filaments to slide together. That is, the ends of the actin filaments are literally pulled in among the ends of the myosin filaments, which is the mechanism of muscle contraction. The midpoints of the actin filaments are all attached to an intracellular membrane, the **Z membrane,** that in turn attaches to the outside membrane of the entire muscle fiber. The portion of the muscle fiber that lies between each two successive Z membranes is called a **sarcomere.** When the actin and myosin filaments slide together, each sarcomere shortens. Thus, the sarcomere is the basic *contractile unit* of skeletal muscle.

Contraction of the muscle fiber is caused by an action potential traveling over the fiber membrane. This action potential also travels deep into the interior of the muscle fiber through many minute **transverse tubules (T tubules),** many of which penetrate all the way through the muscle fiber in each sarcomere. Flow of electrical current to the inside of the fiber during the action potential causes another tubular system inside the muscle fiber, the **sarcoplasmic reticulum,** to release *calcium ions* into the **sarcoplasm,** which is the fluid inside the muscle fiber. It is these calcium ions that initiate the muscle contraction. Within one one-hundredth to one fifth of a second after the calcium ions have been released into the sarcoplasm, a very powerful *calcium pump* in the membrane of the sarcoplasmic reticulum pumps the calcium ions back into the reticulum. Therefore, the duration of contrac-

199

tion of skeletal muscle fibers ranges between one one-hundredth and one fifth of a second.

One theory to explain the attractive forces between the myosin and actin filaments during contraction is the *ratchet mechanism*. Electron micrographs have shown that the myosin filaments project multiple arms called **cross-bridges** toward the actin filaments. At the end of each cross-bridge is an elongated *head* that can bend from side to side. It is believed that this head first bends forward, attaches to the actin filament, pulls the actin filament a short distance, and then releases from the actin. Then the head attaches farther along the actin filament and pulls it still another step. The movements of these heads are energized by the breakdown of *adenosine triphosphate* to adenosine diphosphate, which releases the energy required to promote the muscle contraction.

Smooth muscle is not divided into sarcomeres. Instead, the actin and myosin filaments are intermingled in a much less organized manner throughout the length of the smooth muscle fiber. Also, much if not most of the calcium that causes contraction enters the fiber through the smooth muscle cell membrane at the time of the action potential rather than being released inside the fiber from the sarcoplasmic reticulum. The duration of contraction of smooth muscle is usually 10 to 100 times as long as the duration of skeletal muscle contraction because the calcium ions are pumped outward through the smooth muscle cell membrane very slowly. Many of the actin filaments are attached to the smooth muscle membrane, thus causing shortening of the muscle fiber.

Smooth muscle contraction can also be caused by different hormones and other factors that increase the permeability of the fiber membrane to calcium without the intermediation of action potentials.

All physical functions of the body involve muscle activity. These functions include skeletal movements, contraction of the heart, contraction of the blood vessels, peristalsis in the gut, and many more. Three different types of muscle are responsible for these activities: skeletal muscle, cardiac muscle, and smooth muscle, all of which have some characteristics in common. For instance, the contractile process is the same or nearly the same in each, but, on the other hand, their strengths of contraction, durations of contraction, and other features differ greatly and are especially adapted in each type of muscle for the job to be performed.

The mechanisms of contraction of skeletal muscle and smooth muscle will be discussed in this chapter and cardiac muscle in Chapter 16 in relation to the pumping action of the heart.

PHYSIOLOGIC ANATOMY OF SKELETAL MUSCLE

The Skeletal Muscle Fiber

Figure 11–1 illustrates a single skeletal muscle fiber in both longitudinal and cross-sectional views, and Figure 11–2 gives the different levels

of organization of skeletal muscle, showing that all skeletal muscles are made of numerous fibers ranging between 10 and 80 μ in diameter. Each of these fibers in turn is made up of successively smaller subunits, also illustrated in Figure 11–2, that will be described in subsequent paragraphs.

The Sarcolemma. The **sarcolemma** is the cell membrane of the muscle fiber. However, the sarcolemma consists of a true cell membrane, called the **plasma membrane,** and a

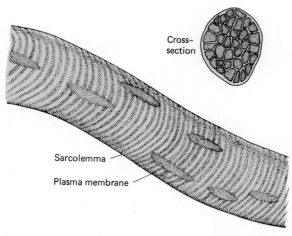

FIGURE 11–1 Longitudinal and cross-sectional views of a skeletal muscle fiber.

Skeletal muscle

Muscle

A

B

Muscle fasciculus

C

Muscle fiber

H Z A I
Band Disc Band Band

Z –Sarcomere– Z

H

Myofilaments

Z   Z

F   G   H   I

G-Actin Molecules
J

F-Actin Filament   K

Myosin Filament   L

Myosin Molecule

M

Light   Heavy
Meromyosin   Meromyosin   N

FIGURE 11–2 Organization of skeletal muscle, from the gross to the molecular level. F, G, H, and I are cross-sections at the points indicated. (Drawing by Sylvia Colard Keene. From Bloom and Fawcett: A Textbook of Histology. Philadelphia, W.B. Saunders Co., 1975.)

layer of polysaccharide material and thin collagen fibrillae that provide strength to the sarcolemma. At the ends of the muscle fibers, this surface layer of the sarcolemma fuses with tendon fibers, which in turn collect into bundles to form the muscle tendons, and thence insert into the bones.

Myofibrils; Actin and Myosin Filaments. Each muscle fiber contains several hundred to several thousand **myofibrils,** which are illustrated by the many small open dots in the cross-sectional view of Figure 11–2C. Each myofibril (Figure 11–2D) in turn has, lying side by side, about 1500 **myosin filaments** and 3000 **actin filaments,** which are large polymerized protein molecules that are responsible for muscle contraction. These can be seen in longitudinal view in the electron micrograph of Figure 11–3 and are represented diagrammatically in Figure 11–2E. The thick filaments are *myosin* and the thin filaments are *actin.* Note that the myosin and actin filaments partially interdigitate and thus

cause the myofibrils to have alternate light and dark bands. The light bands, which contain only actin filaments, are called *I bands.* The dark bands, which contain the myosin filaments as well as the ends of the actin filaments where they overlap the myosin, are called *A bands.* Note also the small projections from the sides of the myosin filaments. These are called **cross-bridges.** They protrude from the surfaces of the myosin filaments along the entire extent of the filament, except in the very center. It is interaction between these cross-bridges and the actin filaments that causes contraction.

Figure 11–2E also shows that the actin filaments are attached to the so-called Z membrane or *Z disc,* and the filaments extend on either side of the Z membrane to interdigitate with the myosin filaments. The Z membrane also passes from myofibril to myofibril, attaching the myofibrils to each other all the way across the muscle fiber.

The portion of a myofibril (or of the whole muscle fiber) that lies between two successive Z

FIGURE 11–3 Electron micrograph of muscle myofibrils, showing the detailed organization of actin and myosin filaments. Note the mitochondria lying between the myofibrils. (From Fawcett: The Cell. Philadelphia, W.B. Saunders Co., 1966.)

membranes is called a **sarcomere.** When the muscle fiber is at its normal fully stretched resting length, the length of the sarcomere is about $2.0\,\mu$. At this length, the actin filaments completely overlap the myosin filaments and are just beginning to overlap each other. It is at this length that the sarcomere also is capable of generating its greatest force of contraction.

The Sarcoplasm. The myofibrils are suspended inside the muscle fiber in a matrix called *sarcoplasm*, which is composed of the usual intracellular constituents. The fluid of the sarcoplasm contains large quantities of potassium, magnesium, phosphate, and protein enzymes. Also present are tremendous numbers of *mitochondria* that lie between and parallel to the myofibrils, a condition that indicates the great need of the contracting myofibrils for large amounts of ATP formed by the mitochondria.

The Sarcoplasmic Reticulum. Also in the sarcoplasm is an extensive *endoplasmic reticulum*, which in the muscle fiber is called the **sarcoplasmic reticulum.** This reticulum has a special organization that is extremely important in the control of muscle contraction, which will be discussed later in the chapter. The electron micrograph of Figure 11–4 illustrates the arrange-

ment of this sarcoplasmic reticulum and shows how extensive it can be. The more rapidly contracting types of muscle have an especially extensive sarcoplasmic reticulum, indicating that this structure is important in causing rapid muscle contraction, as will also be discussed later.

MOLECULAR MECHANISM OF MUSCLE CONTRACTION

Sliding Mechanism of Contraction. Figure 11–5 illustrates the basic mechanism of muscle contraction. It shows the relaxed state of three sarcomeres (above) and the contracted state (below). In the relaxed state, the ends of the actin filaments derived from two successive Z membranes barely overlap each other while at the same time completely overlapping the myosin filaments. On the other hand, in the contracted state these actin filaments have been pulled inward among the myosin filaments so that they now overlap each other to a major extent. Also, the Z membranes have been pulled by the actin filaments up to the ends of the myosin filaments. Thus, muscle contraction occurs by a *sliding filament mechanism.*

FIGURE 11–4 Sarcoplasmic reticulum surrounding the myofibril, showing the longitudinal system paralleling the myofibrils. Also shown in cross-section are the T tubules that lead to the exterior of the fiber membrane and that contain extracellular fluid (arrows). (From Fawcett: The Cell. Philadelphia, W.B. Saunders Co., 1966.)

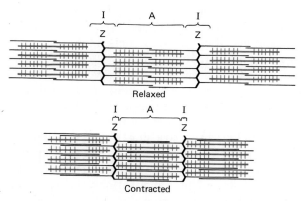

FIGURE 11–5 The relaxed and contracted states of a myofibril, showing sliding of the actin filaments into the channels between the myosin filaments.

But what causes the actin filaments to slide inward among the myosin filaments? This is caused by attractive forces that develop between the actin and myosin filaments. Almost certainly, these attractive forces are the result of mechanical, chemical, and electrostatic forces generated by the interaction of the cross-bridges of the myosin filaments with the actin filaments.

Under resting conditions, the attractive forces between the actin and myosin filaments are neutralized, but when an action potential travels over the muscle fiber membrane, this causes the release of large quantities of calcium ions into the sarcoplasm surrounding the myofibrils. These calcium ions activate the attractive forces and contraction begins. But energy is also needed for the contractile process to proceed. This energy is derived from the high energy bonds of ATP, which is degraded to adenosine diphosphate (ADP) to give the energy required.

In the next few sections we will describe what is known about the details of the molecular processes of contraction. To begin this discussion, however, we must first characterize the myosin and actin filaments themselves.

Molecular Characteristics of the Contractile Filaments

The Myosin Filament. The myosin filament is composed of approximately 200 myosin molecules. Figure 11–6, section A, illustrates an individual molecule; section B illustrates the organization of the molecules to form a myosin filament and also shows its interaction with two

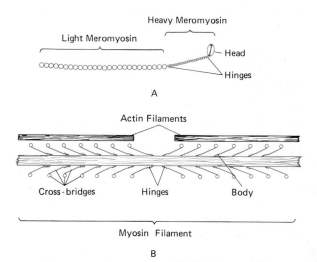

FIGURE 11–6 (A) The myosin molecule. (B) Combination of many myosin molecules to form a myosin filament. Also shown are the cross-bridges and the interaction between the heads of the cross-bridges with adjacent actin filaments.

Active sites (? ADP)   Troponin

F-Actin   Tropomyosin

FIGURE 11–7 The actin filament, composed of two helical strands of F-actin and two tropomyosin strands that lie in the grooves between the actin strands. Attaching the tropomyosin to the actin are several troponin complexes.

actin filaments. Note that the hinged portions of the myosin molecules protrude from all sides of the myosin filament, as illustrated in the figure. These protrusions constitute the *cross-bridges.* The heads of the cross-bridges lie in apposition to the actin filaments; the rod portions of the cross-bridges act as hinged arms that allow the heads to extend far outward from the body of the myosin filament.

The Actin Filament.   The actin filament is also complex. It is composed of three different components: *actin, tropomyosin,* and *troponin.*

The backbone of the actin filament is a double-stranded actin protein molecule, illustrated in color in Figure 11–7. The two strands are wound in a helix, as also illustrated in the figure.

Attached to the actin strand are numerous molecules of ADP. It is believed that these ADP molecules are the active sites on the actin filaments with which the cross-bridges of the myosin filaments interact to cause muscle contraction.

The Tropomyosin Strands.   The actin filament also contains two additional strands comprised of the protein *tropomyosin.* It is believed that each tropomyosin strand is loosely attached to an actin strand and that in the resting state it physically covers the active sites of the actin strands so that interaction cannot occur between the actin and myosin to cause contraction.

Troponin and Its Role in Muscle Contraction.   Attached periodically along each tropomyosin strand is a complex of three globular protein molecules called *troponin.* One of the globular proteins has a strong affinity for actin, another for tropomyosin, and a third for calcium ions. The strong affinity of the troponin for calcium ions is believed to initiate the contraction process, as will be explained in the following section.

Interaction of Myosin and Actin Filaments to Cause Contraction

Inhibition of the Actin Filament by the Troponin-Tropomyosin Complex During Relaxation; Activation by Calcium Ions During Contraction.   A pure actin filament without the presence of the troponin-tropomyosin complex binds strongly with myosin molecules in the presence of magnesium ions and ATP, both of which are normally abundant in the myofibril. But, if the troponin-tropomyosin complex is added to the actin filament, this binding does not take place. Therefore, it is believed that the normal active sites on the actin filament of the relaxed muscle are inhibited (or perhaps physically covered) by the troponin-tropomyosin complex. However, in the presence of calcium ions the inhibitory effect of the troponin-tropomyosin on the actin filaments is itself inhibited. When calcium ions combine with troponin, which has an extremely high affinity for calcium ions even when they are present in minute quantities, the troponin molecule supposedly undergoes a conformational change that in some way "uncovers" the active sites of the actin, thus allowing contraction to proceed. This is believed to occur by physically pulling the tropomyosin strands deeper into the grooves of the actin helix, thus exposing the active binding sites. Therefore, the normal relationship between the tropomyosin-troponin complex and actin is altered by calcium ions—a condition that leads to contraction.

Interaction Between the "Activated" Actin Filament and the Myosin Molecule— the Ratchet Theory of Contraction.   As soon as the actin filament becomes activated by the calcium ions, it is believed that the heads of the cross-bridges from the myosin filaments immediately become attracted to the active sites of the actin filament, and this then causes contraction to occur. Though the precise manner by which this interaction between the cross-bridges and the actin causes contraction is still unknown, a suggested hypothesis is the so-called *ratchet theory of contraction.*

Figure 11–8 illustrates the postulated ratchet mechanism, showing the heads of two cross-bridges attaching to and disengaging from active sites on an actin filament. It is believed that when the head attaches to an active site this attachment simultaneously causes profound changes in the intramolecular forces in the head and arm of the cross-bridge. The new alignment of forces causes the head to tilt toward the arm and to drag the actin filament along with it. This tilt of the head of the cross-bridge is called the

FIGURE 11–8 The ratchet mechanism for contraction of the muscle.

power stroke. Then, immediately after tilting, the head automatically breaks away from the active site and returns to its normal perpendicular direction. In this position it combines with an active site further down along the actin filament; then, a similar tilt takes place again to cause a new power stroke, and the actin filament moves another step. Thus, the heads of the cross-bridges bend back and forth, and step by step pull the actin filament toward the center of the myosin filament. Thus, the movements of the cross-bridges use the active sites of the actin filaments as cogs of a *ratchet*.

Each one of the cross-bridges is believed to operate independently of all others, each attaching and pulling in a continuous, alternating ratchet cycle. Therefore, the greater the number of cross-bridges in contact with the actin filament at any given time, the greater, theoretically, is the force of contraction.

ATP as the Source of Energy for Contraction. When a muscle contracts against a load, work is performed, and energy is required. It is found that large amounts of ATP are cleaved to form ADP during the contraction process. Although we still do not know exactly how ATP is used to provide the energy for contraction, it is believed that once the head of the cross-bridge has completed its power stroke, the tilted position of the head now exposes a site in the head where ATP can bind. Therefore, one molecule of ATP binds with the head, and this binding in turn causes detachment of the head from the active site. In addition, the ATP is itself immediately cleaved by a very potent ATPase activity of the head itself. The energy released supposedly tilts the head back to its normal perpendicular condition and theoretically "cocks" the head in this position. The "cocked" head thus has stored energy derived from the cleaved ATP, and this energizes the next power stroke.

Thus, the process proceeds again and again until the actin filament pulls the Z membrane up against the ends of the myosin filaments or until the load on the muscle becomes too great for further pulling to occur.

INITIATION OF MUSCLE CONTRACTION: EXCITATION-CONTRACTION COUPLING

The Muscle Action Potential

Initiation of contraction in skeletal muscle begins with action potentials in the muscle fibers. These cause release of calcium ions from the sarcoplasmic reticulum. It is the calcium ions that in turn initiate the chemical events of the contractile process.

Almost everything discussed in Chapter 10 regarding initiation and conduction of action potentials in nerve fibers applies equally well to skeletal muscle fibers, except for quantitative differences. Some of the quantitative aspects of muscle potentials are the following:

1. Resting membrane potential: Approximately −90 millivolts in skeletal fibers—the same as in large myelinated nerve fibers.
2. Duration of action potential: 1 to 5 milliseconds in skeletal muscle—about five times as long as in large myelinated nerves.
3. Velocity of conduction: 3 to 5 m per second—about one eighteenth the velocity of conduction in the large myelinated nerve fibers that excite skeletal muscle.

Spread of the Action Potential to the Interior of the Muscle Fiber by Way of the Transverse Tubule System

The skeletal muscle fiber is so large that action potentials spreading along the membrane cause almost no current flow deep within the fiber. Yet, to cause contraction, these electrical currents must penetrate to the vicinity of all the separate myofibrils. This is achieved by transmission of the action potentials along **transverse tubules (T tubules)** that penetrate all the way through the muscle fiber from one side to the other. The T tubule action potentials in turn cause the sarcoplasmic reticulum to release calcium ions in the immediate vicinity of all the myofibrils, and it is these calcium ions that in turn cause contraction. Now, let us describe this system in greater detail.

The Transverse Tubule–Sarcoplasmic Reticulum System. Figure 11–9 illustrates a group of myofibrils surrounded by the trans-

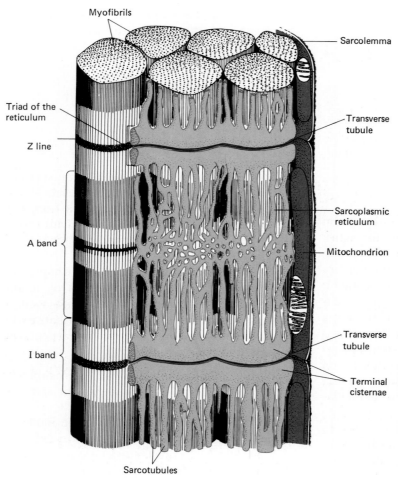

Myofibrils

Sarcolemma

Triad of the reticulum

Transverse tubule

Z line

A band

Sarcoplasmic reticulum

Mitochondrion

I band

Transverse tubule

Terminal cisternae

Sarcotubules

FIGURE 11–9 The transverse tubule–sarcoplasmic reticulum system. Note the *longitudinal tubules* that terminate in large *cisternae*. The cisternae in turn abut the transverse tubules. Note also that the transverse tubules communicate with the outside of the cell membrane. This illustration was drawn from frog muscle, which has one transverse tubule per sarcomere, located at the Z line. A similar arrangement is found in mammalian heart muscle, but mammalian skeletal muscle has two transverse tubules per sarcomere, located at the A–I junctions. (From Bloom and Fawcett: A Textbook of Histology. Philadelphia, W.B. Saunders Co., 1975. Modified after Peachey: *J. Cell Biol.* 25:209, 1965. Drawn by Sylvia Colard Keene.)

verse tubule–sarcoplasmic reticulum system. Note that the transverse tubules, which are themselves extremely small, penetrate all the way from one side of the muscle fiber to the opposite side. Furthermore, where the T tubules originate from the cell membrane they are open to the exterior. Therefore, they communicate with the fluid surrounding the muscle fiber and contain extracellular fluid in their lumens. In other words, the T tubules are internal extensions of the cell membrane. Therefore, when an action potential spreads over a muscle fiber membrane, it spreads along the T tubules to the deep interior of the muscle fiber as well.

Figure 11–9 also shows the extensiveness of the **sarcoplasmic reticulum.** This is composed of two major parts: (1) long **longitudinal tubules** that terminate in (2) large chambers called **terminal cisternae** that abut against the T tubule. This reticulum is also seen in the electron micrograph of Figure 11–4 and is illustrated diagrammatically in Figure 11–10.

In cardiac muscle there is a single T tubule network for each sarcomere, located at the level of the Z membrane as illustrated in Figure 11–9. However, in skeletal muscle there are two T tubule networks for each sarcomere located near the two ends of the myosin filaments, as illustrated in Figure 11–10, which are the points where the actual mechanical forces of muscle contraction are created.

Release of Calcium Ions by the Cisternae of the Sarcoplasmic Reticulum

One of the special features of the sarcoplasmic reticulum is that it contains calcium ions in very high concentration, and many of these ions are released when the adjacent T tubule is excited.

Figure 11–10 shows that the action potential of the T tubule causes current flow through the cisternae where they abut the T tubule. This current flow in turn causes rapid release of calcium ions from the cisternae into the surrounding sarcoplasm. Presumably this release results from the opening of calcium pores similar to the opening of sodium pores at the onset of the

FIGURE 11–10 Excitation-contraction coupling in the muscle, showing an action potential that causes release of calcium ions from the sarcoplasmic reticulum and then re-uptake of the calcium ions by a calcium pump.

action potential, though the actual mechanism is still unknown.

The calcium ions that are thus released from the cisternae diffuse to the adjacent myofibrils, where they bind strongly with troponin, as discussed in an earlier section, and this in turn elicits the muscle contraction.

The Calcium Pump for Removing Calcium Ions at the End of Contraction. Once the calcium ions have been released from the cisternae and have diffused to the myofibrils, muscle contraction will then continue as long as the calcium ions are still present in high concentration in the sarcoplasmic fluid. However, a continually active calcium pump located in the walls of the sarcoplasmic reticulum pumps calcium ions out of the sarcoplasmic fluid back into the vesicular cavities of the reticulum. This pump can concentrate the calcium ions about 2000-fold inside the reticulum, a condition that allows massive buildup of calcium in the sarcoplasmic reticulum and also causes almost total depletion of calcium ions in the fluid of the myofibrils. Therefore, except immediately after an action potential, the calcium ion concentration in the myofibrils is kept at an extremely low level. This level is too little to elicit contraction, but full excitation of the T tubule–sarcoplasmic reticulum system causes enough release of calcium ions to increase the concentration in the myofibrillar fluid high enough to cause maximum muscle contraction. Immediately thereafter, the calcium pump depletes the calcium ions again, and muscle contraction ceases.

The total duration of this calcium "pulse" in the usual skeletal muscle fiber lasts about one thirtieth of a second, though in heart muscle it lasts for as long as 0.3 second. It is during this calcium pulse that muscle contraction occurs. If the contraction is to continue without interruption for longer intervals, a series of such pulses must be initiated by a continuous series of repetitive action potentials, as will be discussed in more detail later in the chapter.

CHARACTERISTICS OF WHOLE MUSCLE CONTRACTION

The Motor Unit

Each motor nerve fiber that leaves the spinal cord usually innervates many different muscle fibers, the number depending on the type of muscle. All the muscle fibers innervated by a single motor nerve fiber are called a *motor unit*. In general, small muscles that react rapidly and whose control is exact have few muscle fibers in each motor unit (as few as 2 to 3 in some of the laryngeal muscles) and have a relatively large number of nerve fibers going to each muscle. On the other hand, the large muscles that do not require very fine degree of control, such as the gastrocnemius muscle, may have several hundred muscle fibers in a motor unit. An average figure for all the muscles of the body can be considered to be about 150 muscle fibers to the motor unit.

The muscle fibers in a motor unit do not necessarily lie side by side but, instead, are divided into many bundles of only a few fibers, each spread throughout the muscle belly. Because of this, stimulation of the motor unit causes a weak contraction in a broad area of the muscle rather than a strong contraction at one specific point.

The Muscle Twitch

One of the laboratory methods for studying muscle contraction is to elicit a *muscle twitch*. To do this, a single instantaneous stimulus is applied to the nerve supplying the muscle. The duration of the resulting contraction, called a muscle "twitch," is between $\frac{1}{5}$ and $\frac{1}{100}$ second, depending on the type of muscle.

Isometric and Isotonic Contraction. Figure 11–11 shows the contraction of a muscle under two different conditions. To the left the muscle lifts weights in a pan, becoming shorter in the process. The total amount of weight applied to the muscle is always the same, for which reason the contraction is called *isotonic*, which means "same force." To the right the muscle is attached to an electronic transducer that will record the tension of muscle contraction on an electrical recorder even though the muscle contracts no more than $\frac{1}{500}$ mm. Stimulation of the muscle under these conditions causes it to tighten but not to shorten significantly, and the contraction is called *isometric*, meaning "same length."

The characteristics of isometric and isotonic muscle contraction are somewhat different. The reasons for this are, first, the isometric system has no inertia whereas the isotonic system does, and, second, during isotonic contraction the shape of the muscle must change so that it can shorten, whereas during isometric contraction the muscle changes its shape very little. Therefore, without inertia and without the necessity for changing shape, the isometric muscle twitch usually has a much shorter duration than the isotonic twitch. In expressing relative abilities of different muscles to contract, the isometric muscle twitch is the usual criterion employed, because its characteristics are dependent only on intrinsic characteristics of the muscle and not at all on extrinsic factors.

FIGURE 11–12 Isometric muscle twitches of ocular, gastrocnemius, and soleus muscles, illustrating the different durations of contraction.

In the human body, muscle contraction is of both the isometric and the isotonic types. When one is simply standing, he tenses his leg muscles to maintain a fixed position of the joints. This is isometric contraction. On the other hand, when he is walking and moving his legs, or when he is lifting his arms, the contraction is more of the isotonic type.

Duration of Contraction of Different Skeletal Muscles. Figure 11–12 shows recordings of isometric contractions by different muscles. The dashed curve of the figure shows the duration of depolarization of the fiber membrane caused by the action potential traveling over the muscle fiber. This is the period when calcium ions are being released into the fluids of the fiber. Immediately thereafter the contraction begins. The isometric contraction of an ocular muscle lasts for about $\frac{1}{100}$ second; that of a gastrocnemius muscle lasts about $\frac{3}{100}$ second; and that of a soleus muscle as long as $\frac{1}{5}$ second. It is evident, then, that different skeletal muscles have widely different durations of contraction. The ocular muscles, which must cause extremely rapid movement of the eyes from one position to another, contract more rapidly than almost any other muscle. The gastrocnemius muscle must contract moderately rapidly because it is used in jumping and in performing other rapid movements of the foot. The soleus usually does not need to contract rapidly at all, because it is used principally for support of the body against gravity.

Effect of Initial Muscle Length on the Force of Contraction

The length to which a muscle is stretched before it contracts makes considerable difference in its

FIGURE 11–11 Methods for recording isotonic and isometric muscle twitches.

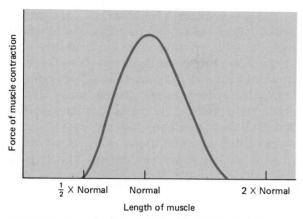

Force of muscle contraction

½ X Normal   Normal   2 X Normal

Length of muscle

FIGURE 11–13 Effect of the initial length of a muscle on the contractile force developed following muscle excitation.

force of contraction. When the length is much less than normal its force of contraction is greatly weakened, and, also, when it is stretched far beyond its normal limits, it fails to contract with as much force as would otherwise be possible. Figure 11–13 illustrates these effects, showing that a muscle in its normal stretched state will usually contract with the greatest possible force.

Fortunately, the normal length of a muscle in its most elongated position is almost exactly optimal for maximal strength of contraction. For instance, when the biceps is at its normal full length, it contracts with its greatest force, whereas, as it progressively shortens, its strength of contraction decreases.

The Lever Systems of the Body. Other factors that determine the force of a movement are (1) the manner in which the contracting muscles are attached to the skeletal system and (2) the structure of the joint at which movement will occur. Figure 11–14, as an example, illustrates movement of the forearm caused by bi-

FIGURE 11–14 The lever system activated by the biceps muscle.

ceps contraction. The fulcrum of the lever system is at the elbow, and the attachment of the biceps is approximately 5 cm in front of the fulcrum. If we assume that the total length of the forearm lever is about 35 cm, one immediately sees that the force of contraction of the biceps must be at least seven times as great as the force of movement of the hand. Thus, if the hand is to lift an object that weighs 50 lb, the total force of contraction of the biceps would have to be about 350 lb. One can readily understand from this tremendous force why muscles sometimes actually pull their tendons out of the bone substance.

Every muscle of the body has its own peculiar shape and length that suits it to its particular function. For instance, the gluteal muscles of the buttocks are extremely broad but do not contract a long distance. They provide tremendous force for movement at the hip joint; even a very slight distance of movement at this joint can cause tremendous movement of the foot. At the other extreme, some of the muscles of the anterior thigh are very long and can shorten as much as 5 cm, pulling the lower leg upward at the knee joint and flexing the upper leg at the hip joint at the same time.

The study of different types of muscle lever systems and their movements is called *kinesiology*; this is a very important phase of human physioanatomy.

Control of Different Degrees of Muscle Contraction— the Mechanism of "Summation"

In performing the different functions of the body, it is quite important that each muscle be able to contract with varying degrees of strength. This is accomplished by "summing" the contractions of varying numbers of muscle fibers at once. When a weak contraction is desired, only a few muscle fibers are contracted simultaneously. When a strong contraction is desired, a great number of fibers are contracted at the same time. In general, the different "gradations" of muscle contraction are achieved by two different methods of summation called *multiple motor unit summation* and *wave summation*.

Multiple Motor Unit Summation. The force of muscle contraction increases progressively as the number of contracting motor units increases. In each muscle, the numbers of muscle fibers and their sizes in the different motor units vary tremendously, so that one motor unit may be as much as 50 times as strong as another. The smaller motor units are far more easily excited than are the larger ones because they are

FIGURE 11-15 Wave summation, showing progressive summation of successive contractions as the rate of stimulation is increased. Tetanization occurs when the rate of stimulation reaches approximately 35 per second, and maximum force of contraction occurs at approximately 50 per second.

innervated by smaller nerve fibers whose cell bodies in the spinal cord have a naturally high level of excitability. This effect causes the gradations of muscle strength during weak muscle contraction to occur in very small steps, whereas the steps become progressively greater as the intensity of contraction increases.

Wave Summation. Figure 11-15 illustrates the principles of wave summation, showing in the lower left-hand corner a single muscle twitch followed by successive muscle twitches at various frequencies. When the frequency of twitches reaches 10 or more per second, the first muscle twitch is not completely over by the time the second one begins. Therefore, since the muscle is already in a partially contracted state when the second twitch begins, the degree of muscle shortening this time is slightly greater than that with the single muscle twitch.

At more rapid rates of contraction, the degree of summation of successive contractions becomes greater and greater, because the successive contractions appear at earlier times following the preceding contraction.

Tetanization. When a muscle is stimulated at progressively greater frequencies, a frequency is finally reached at which the successive contractions fuse together and cannot be distinguished one from the other. This state is called *tetanization*, and the lowest frequency at which it occurs is called the *critical frequency*. Once the critical frequency for tetanization is reached, further increase in rate of stimulation increases the force of contraction only a few more percent, as shown in Figure 11-15.

Asynchronous Summation of Motor Units. Even when tetanization of individual motor units of a muscle is not occurring, the tension exerted by the whole muscle is still continuous and nonjerky because *the different motor units fire asynchronously;* that is, while one is contracting another is relaxing; then another fires, followed by still another, and so forth. Consequently, even when motor units fire as infrequently as five times per second, the muscle contraction, though weak, is nevertheless very smooth.

Muscle Fatigue

Prolonged and strong contraction of a muscle leads to the well-known state of muscle fatigue. This results mainly from inability of the contractile and metabolic processes of the muscle fibers to continue supplying the same work output. The nerve continues to function properly, the nerve impulses usually pass normally through the neuromuscular junction into the muscle fiber, and even normal action potentials spread over the muscle fibers, but the contraction becomes weaker and weaker because of depletion of ATP in the muscle fibers themselves.

Interruption of blood flow through a contracting muscle leads to almost complete muscle fatigue in a minute even when the muscle is not very active because of the obvious loss of nutrient supply.

Effect of Activity in Muscular Development

Exercise and Hypertrophy. The more a muscle is used, the greater becomes its size and strength, though the cause is yet unknown. Physical enlargement of the muscles is called *hypertrophy*. Examples are (1) the intense muscular development of weight lifters, (2) the great hypertrophy of the leg muscles in skaters, (3) the enlargement of the arm and hand muscles of carpenters, and (4) the enlargement of the thigh muscles of runners.

Associated with muscle hypertrophy is usually an increase in the efficiency of muscular contraction, for the hypertrophied muscle stores increased quantities of glycogen, fatty substances, and other nutrients. In addition, the number of contractile myofibrils also increases. All these changes cause the efficiency of the contractile process to increase so that the percentage of energy converted to muscle work becomes considerably greater in the athlete than in the nonathlete.

It should be noted that weak muscular activity, even when sustained over long periods of time, does not result in significant muscle hypertrophy. Instead, hypertrophy is mainly the result of forceful muscle activity even though the activity might occur for only a few minutes each day. For this reason, strength can be developed in muscles more rapidly by using *resistive exercises* or forceful *isometric exercises* than by prolonged mild exercise.

Muscle Denervation and Atrophy. When the nerve supply to a muscle is destroyed, the muscle begins to *atrophy*—that is, the muscle fibers begin to degenerate. In about 6 months to 2 years the muscle will have atrophied to about one-fourth normal size, and the muscle fibers will have been replaced mainly by fibrous tissue.

For some reason, nerve stimulation of a muscle keeps the muscle tissue alive. Even when a person does not use his muscles to a great extent, the weak, intermittent tonic impulses are still sufficient to maintain a relatively normal muscle, but without these impulses the muscle fibers soon atrophy entirely. Perhaps this effect is caused by nutritional changes in the denervated muscle, for action potentials traveling down the fiber membrane alter its permeability markedly, which might be necessary for appropriate transfer of nutrients through the membrane.

The functional integrity of denervated muscle fibers can be maintained quite satisfactorily by daily electrical stimulation. The action potentials produced in this manner take the place of the nerve-induced potentials, and the muscle fibers do not atrophy.

SMOOTH MUSCLE

Most of the internal organs of the body contain *smooth muscle*. The name is derived from the fact that this muscle does not have microscopic striations similar to those in skeletal and cardiac muscle.

Smooth muscle is composed of fibers far smaller than skeletal muscle fibers—usually 2 to 5 μ (micrometers) in diameter and only 50 to 200 μ in length, in contrast to the skeletal muscle fibers that are as much as 20 times as large in diameter and thousands of times as long. Never-

theless, many of the same principles of contraction apply to both smooth muscle and skeletal muscle. Most important, the same chemical substances cause contraction in smooth muscle as in skeletal muscle, but the physical arrangement of smooth muscle fibers is entirely different, as we shall see.

Types of Smooth Muscle

The smooth muscle of each organ is often distinctive from that of other organs in several different ways: physical dimensions, organization into bundles or sheets, response to different types of stimuli, characteristics of its innervation, and function. Yet, for the sake of simplicity, smooth muscle can generally be divided into two major types, which are illustrated in Figure 11–16: *multiunit smooth muscle* and *visceral smooth muscle*.

Multiunit Smooth Muscle. This type of smooth muscle, illustrated in Figure 11–16A, is composed of discrete smooth muscle fibers. Each fiber operates entirely independently of the others and is often innervated by a single nerve ending, as occurs for skeletal muscle fibers. This is in contrast to visceral smooth muscle, which is controlled to a greater extent by non-nervous stimuli. An additional characteristic is that they rarely exhibit spontaneous contractions.

Some examples of multiunit smooth muscle found in the body are the smooth muscle fibers of the ciliary muscle of the eye, the iris of the eye, the nictitating membrane that covers the eyes in some lower animals, and the piloerector muscles that cause erection of the hairs when stimulated by the sympathetic nervous system.

Visceral Smooth Muscle. Visceral smooth muscle fibers, illustrated by the walls of small blood vessels in Figure 11–16B, are usually arranged in sheets, bundles, or tubes, and their cell membranes contact each other at multiple points to form many *gap junctions*, or *nexi*, through which ions can flow with ease from the inside of one smooth muscle fiber to the next one. Therefore, when one portion of visceral muscle is stimulated, the action potential is conducted to the surrounding fibers as well. Thus, the fibers form a *functional syncytium* that usually contracts in unison. Visceral smooth muscle is found in most of the organs of the body, especially the walls of the gut, the bile ducts, the ureters, the uterus, the small blood vessels, and so forth.

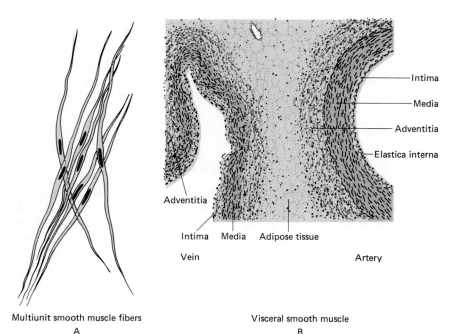

FIGURE 11-16 (A) Discrete multiunit smooth muscle fibers. (B) Organization of visceral smooth muscle fibers in the form of tubular sheets in the walls of a vein and an artery.

The Contractile Process in Smooth Muscle

The Chemical Basis for Contraction. *Actin* and *myosin* filaments derived from smooth muscle interact with each other in the same way as actin and myosin derived from skeletal muscle. Furthermore, the contractile process is activated by calcium ions, and ATP is degraded to ADP to provide the energy for contraction. On the other hand, there are major differences in the physical organization of smooth muscle and skeletal muscle.

The Physical Basis for Smooth Muscle Contraction. The physical organization of the smooth muscle cell is illustrated in Figure 11-17, which shows large numbers of actin filaments attached to so-called *dense bodies.* Some of these bodies in turn are attached to the cell membrane, whereas others are dispersed throughout the sarcoplasm. Interspersed among the actin filaments are a few thick filaments about 2.5 times the diameter of the thin actin filaments. These are assumed to be myosin filaments.

Despite the relative paucity of myosin filaments, it is assumed that they have sufficient cross-bridges to attract the many actin filaments and cause contraction by the sliding filament mechanism in essentially the same way as in skeletal muscle.

Slowness of Contraction and Relaxation of Smooth Muscle. A typical smooth muscle tissue will begin to contract 50 to 100 milliseconds after it is excited and will reach full contraction about half a second later. Then the contraction declines in another 1 to 2 seconds,

FIGURE 11-17 Arrangement of actin and myosin filaments in the smooth muscle cell. Note the attachment of the actin filaments to "dense bodies," some of which are themselves attached to the cell membrane.

giving a total contraction time of 1 to 3 seconds, which is about 30 times as long as the single contraction of skeletal muscle.

Energy Required to Sustain Smooth Muscle Contraction. As little as one five-hundredth as much energy is required to sustain the same tension of contraction in smooth muscle as in skeletal muscle. This presumably results from the very slow activity of smooth muscle myosin ATPase and also from the fact that there are far fewer myosin filaments in smooth muscle than in skeletal muscle.

This economy of energy utilization by smooth muscle is exceedingly important to overall function of the body, because organs such as the intestines, the urinary bladder, the gallbladder, and other viscera must maintain moderate degrees of muscle contractile tone day in and day out.

Membrane Potentials and Action Potentials in Smooth Muscle

Smooth muscle exhibits membrane potentials and action potentials similar to those that occur in skeletal muscle fibers. However, in the normal resting state, the membrane potential is usually about −50 to −60 millivolts, or about 35 millivolts less negative than in skeletal muscle.

Action Potentials in Visceral Smooth Muscle. Typical *spike* action potentials, such as those seen in skeletal muscle, occur in most types of visceral smooth muscle. The duration of this type of action potential is 10 to 50 milliseconds, as illustrated in Figure 11–18A. Such action potentials can be elicited in many ways, such as by electrical stimulation, the action of hormones on the smooth muscle, the action of transmitter substances from nerve fibers, or spontaneous generation in the muscle fiber itself, as discussed later.

Action Potentials with Plateaus. Action potentials with plateaus also occur in some smooth muscle, such as the action potential shown in Figure 11–19. The onset of this action potential is similar to that of the typical spike potential. However, instead of rapid repolarization of the muscle fiber membrane, the repolarization is delayed for several hundred to several thousand milliseconds. Plateaus as long as 30 seconds have been recorded. The importance of the plateau is that it can account for the prolonged periods of contraction that occur in some types of smooth muscle.

Slow Wave Potentials in Visceral Smooth Muscle and Spontaneous Generation of Action Potentials. Some smooth muscle is self-excitatory. That is, action potentials arise within the smooth muscle itself without an extrinsic stimulus. This is usually caused by a basic *slow wave rhythm* of the membrane potential. A typical slow wave of this type is illustrated in Figure 11–18B. The slow wave itself is not an *action* potential, but as illustrated in the figure the peaks of the slow waves are often high enough to elicit one or more true action potentials that spread over the fibers and cause muscle contraction. The slow waves are believed to result from waxing and waning of the pumping of sodium outward through the muscle fiber membrane; the membrane potential becomes more negative when sodium is pumped rapidly and less negative when the sodium pump becomes less active.

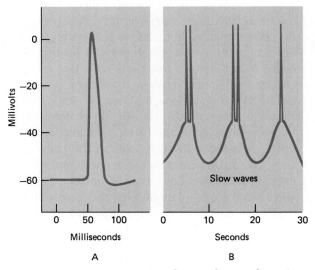

FIGURE 11–18 (A) A typical smooth muscle action potential (spike potential) elicited by an external stimulus. (B) A series of spike action potentials elicited by rhythmical slow electrical waves occurring spontaneously in the smooth muscle wall of the intestine.

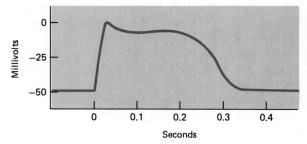

FIGURE 11–19 Monophasic action potential from a smooth muscle fiber.

The importance of the slow waves lies in the fact that they can promote rhythmical contractions of the smooth muscle mass synchronized with the slow waves. Therefore, the slow waves are frequently called *pacemaker waves.* This type of activity is especially prominent in tubular types of smooth muscle masses, such as in the gut, the ureter, and so forth. In Chapter 41 we shall see that this type of activity controls the rhythmical contractions of the gut.

Excitation of Visceral Smooth Muscle by Stretch. When visceral smooth muscle is stretched sufficiently, spontaneous action potentials are usually generated. These result from a combination of the normal slow wave potentials plus a decrease in the membrane potential caused by the stretch itself. This response to stretch is an especially important function of visceral smooth muscle because it allows a hollow organ that is excessively stretched to contract automatically and therefore to resist the stretch. For instance, when the gut is overstretched by intestinal contents, a local automatic contraction sets up a peristaltic wave that moves the contents away from the overstretched intestine.

Excitation-Contraction Coupling—Role of Calcium Ions

In the previous chapter it was pointed out that the actual contractile process in skeletal muscle is activated by calcium ions. This is also true in smooth muscle. However, the source of the calcium ions differs in smooth muscle because the sarcoplasmic reticulum of smooth muscle is poorly developed, in contrast to the extensive sarcoplasmic reticulum of skeletal muscle, which is the source of very nearly 100 percent of the contraction-inducing calcium ions.

In some types of smooth muscle, most of the calcium ions that cause contraction enter the muscle fiber from the extracellular fluid at the time of the action potential. There is a reasonably high concentration of calcium ions in the extracellular fluid, and the action potential itself is caused at least partly by influx of calcium ions along with sodium ions into the muscle fiber.

Yet, in some smooth muscle there is a moderately developed sarcoplasmic reticulum but no T tubules. Instead, the cisternae of the reticulum abut the cell membrane. Therefore, it is believed that the membrane action potentials cause release of calcium ions from these cisternae, thereby providing a greater degree of contraction than would occur on the basis of calcium ions entering through the cell membrane alone.

The Calcium Pump. To cause relaxation of the smooth muscle contractile elements, it is necessary to remove the calcium ions. This removal is achieved by a calcium pump that pumps the calcium ions out of the smooth muscle fiber and back into the extracellular fluid, or pumps the calcium ions into the sarcoplasmic reticulum. However, this pump is very slow-acting in comparison with the fast-acting sarcoplasmic reticulum pump in skeletal muscle. Therefore, the duration of smooth muscle contraction is often in the order of seconds rather than in tens of milliseconds, as occurs for skeletal muscle.

Neuromuscular Junctions of Smooth Muscle

The smooth muscle of the different organs is stimulated by an entirely different division of the nervous system from that which stimulates the skeletal system. This division is called the *autonomic nervous system,* and it normally operates at an entirely subconscious level. This system will be discussed at length in Chapter 19.

Physiologic Anatomy of Smooth Muscle Neuromuscular Junctions. Neuromuscular junctions of the type found on skeletal muscle fibers do not occur in smooth muscle. Instead, the nerve fibers generally branch diffusely on top of a sheet of muscle fibers, as illustrated in Figure 11–20. In a few instances the endings do make direct *contact junctions* with the smooth muscle cells, but usually the fibers form so-called *diffuse junctions* that secrete their transmitter substance into the interstitial fluid a few microns away from the muscle cells; the transmitter substance then diffuses to the cells.

The axons innervating smooth muscle fibers also do not have typical end-feet as observed in the end-plate on skeletal muscle fibers. Instead, the fine terminal axons have multiple varicosities spread along their axes. At these points the Schwann cells are interrupted so that a transmitter substance can be secreted through the walls of the varicosities. In the varicosities are vesicles similar to those present in the skeletal muscle end-plate containing transmitter substance. However, in contrast to the vesicles of

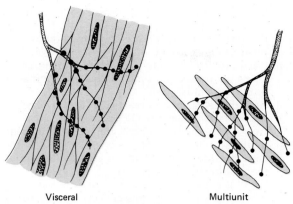

Visceral   Multiunit

FIGURE 11–20 Innervation of smooth muscle.

skeletal muscle junctions that contain only acetylcholine, the vesicles of the autonomic nerve fiber varicosities contain acetylcholine in some fibers and norepinephrine in others.

Excitatory and Inhibitory Transmitter Substances at the Smooth Muscle Neuromuscular Junction. Two different transmitter substances known to be secreted by the autonomic nerves innervating smooth muscle are *acetylcholine* and *norepinephrine*. Acetylcholine is an excitatory substance for smooth muscle fibers in some organs but an inhibitory substance for smooth muscle in other organs. When acetylcholine excites a muscle fiber, norepinephrine ordinarily inhibits it; or when acetylcholine inhibits a fiber, norepinephrine excites it.

It is believed that *receptor molecules* in the membranes of the different smooth muscle fibers determine which will excite them, acetylcholine or norepinephrine. Thus, there are *excitatory receptors* and *inhibitory receptors*.

Excitation of Smooth Muscle. When an action potential reaches the terminal of a fibril and an excitatory transmitter is released, this acts in the same way on the smooth muscle as on skeletal muscle at the end-plate. That is, the muscle membrane becomes highly permeable to sodium ions, and inflowing sodium elicits action potentials that lead to muscle contraction.

Inhibition at the Smooth Muscle Neuromuscular Junction. When a transmitter substance at the nerve ending interacts with an inhibitory receptor instead of an excitatory receptor, the membrane becomes highly permeable to potassium ions. This causes the potential of the muscle fiber to become more negative than ever, for instance, to change from −50 to

−70 millivolts; that is, it becomes *hyperpolarized* and therefore becomes much more difficult to excite than is usually the case. This process of hyperpolarization will be explained in more detail in Chapter 15.

Smooth Muscle Contraction without Action Potentials—Effect of Local Tissue Factors and Hormones

Though we have thus far discussed smooth muscle contraction elicited only by nervous signals and smooth muscle membrane action potentials, we must quickly note that probably half of all smooth muscle control is initiated not by action potentials but by stimulatory factors acting directly on the smooth muscle contractile machinery. Some examples are:

1. *Lack of oxygen* in the local blood vessels causes smooth muscle relaxation in the vascular wall and therefore vasodilatation.
2. *Excess carbon dioxide* causes vasodilatation.
3. Such factors as *lactic acid, high potassium ion* concentration, diminished *calcium ion* concentration, and decreased *body temperature* will also cause local vasodilatation.
4. Some of the more important hormones that either excite or inhibit smooth muscle contraction are norepinephrine, epinephrine, acetylcholine, angiotensin, vasopressin, oxytocin, serotonin, and histamine.

Mechanism of Muscle Control by Local Tissue Factors and Hormones. It is believed that most of the local tissue factors and hormones that cause smooth muscle contraction or relaxation do so by affecting the calcium mechanism for control of the contractile process. Some of these factors change the membrane potential a moderate amount but without necessarily causing an action potential, and this increases or decreases the flow of calcium ions to the interior of the cell. However, most of them can control contraction even when the membrane potential is not altered. In these circumstances calcium ions probably enter through the cell membrane or are released from the sarcoplasmic reticulum to cause contraction, or calcium entry to the cell is inhibited to cause relaxation.

TONE OF SMOOTH MUSCLE

Smooth muscle can maintain a state of long-term, steady contraction that has been called ei-

FIGURE 11–21   Record of rhythmic and tonic smooth muscle contraction.

ther *tonus* contraction of smooth muscle or simply *smooth muscle tone.* This allows prolonged or even indefinite continuance of the smooth muscle function. For instance, the arterioles are maintained in a state of tonic contraction almost throughout the entire life of the person. Likewise, tonic contraction in the gut wall maintains steady pressure on the contents of the gut, and tonic contraction of the urinary bladder wall maintains a moderate amount of pressure on the urine in the bladder.

Figure 11–21 illustrates a record of tonic contraction in smooth muscle. Such contractions can be caused in either of two ways:

1. They are sometimes caused by *summation of individual contractile pulses;* each contractile pulse is initiated by a separate action potential in the same way that tetanic contractions are produced in skeletal muscle.
2. However, most smooth muscle tonic contractions probably result from *prolonged direct smooth muscle excitation* without action potentials, usually caused by local tissue factors or circulating hormones. For instance, prolonged tonic contractions of the blood vessels without the mediation of action potentials are regularly caused by angiotensin, vasopressin, or norepinephrine, and these play an important role in the long-term regulation of arterial pressure, as will be discussed in Chapter 28.

QUESTIONS

1. Describe the functional anatomy of skeletal muscle.
2. Explain the sliding filament mechanism of muscle contraction.
3. Describe the special characteristics of myosin and actin filaments.
4. Discuss the ratchet theory of contraction and explain the mechanism of the "power stroke."
5. What are the separate roles of the transverse tubules and the longitudinal tubules in the initiation of muscle contraction?
6. How do calcium ions react with the myofibrillar filaments to cause muscle contraction?
7. How many muscle fibers are, on the average, found in a motor unit?
8. Describe the effect of initial muscle length on force of muscle contraction.
9. What are the different types of summation by which the strength of muscle contraction can be changed?
10. What is the relationship between exercise and muscle hypertrophy?
11. How does smooth muscle differ anatomically from skeletal muscle?
12. How do "slow waves" cause rhythmic smooth muscle contraction?
13. What are some of the non–action potential factors that control smooth muscle contraction?
14. What is smooth muscle tone, its importance and cause?

REFERENCES

Basmajian, J.V.: Muscles Alive; Their Functions Revealed by Electromyography. Baltimore, Williams & Wilkins, 1978.
Buchthal, F., and Schmalbruch, H.: Motor unit of mammalian muscle. *Physiol. Rev., 60*:90, 1980.
Caputo, C.: Excitation and contraction processes in muscle. *Annu. Rev. Biophys. Bioeng.,* 7:63, 1978.
Costantin, L.L.: Activation in striated muscle. *In* Brookhart, J.M., and Mountcastle, V.B. (eds.): Handbook of Physiology. Sec. 1, Vol 1. Baltimore, Williams & Wilkins, 1977, p. 215.
Curtin, N.A., and Woledge, R.C.: Energy changes in muscular contraction. *Physiol. Rev., 58*:690, 1978.
Daniel, E.E., and Sarna, S.: The generation and conduction of activity in smooth muscle. *Annu. Rev. Pharmacol. Toxicol., 18*:145, 1978.
Endo, M.: Calcium release from the sarcoplasmic reticulum. *Physiol. Rev., 57*:71, 1977.
Hartshorne, D.J., and Gorecka, A.: Biochemistry of the contractile proteins of smooth muscle. *In* Bohr, D.F., *et al.* (eds.): Handbook of Physiology. Sec. 2, Vol. 2. Baltimore, Williams & Wilkins, 1980, p. 93.
Huxley, A.F., and Gordon, A.M.: Striation patterns in active and passive shortening of muscle. *Nature* (*Lond.*), *193*:280, 1962.
Huxley, H.E.: Muscular contraction and cell motility. *Nature,* 243:445, 1973.
Johansson, B., and Somlyo, A.P.: Electrophysiology and excitation-contraction coupling. *In* Bohr, D.F., *et al.* (eds.): Handbook of Physiology. Sec. 2, Vol. 2. Baltimore, Williams & Wilkins, 1980, p. 301.
Murphy, R.A.: Filament organization and contractile function in vertebrate smooth muscle. *Annu. Rev. Physiol., 41*:737, 1979.

Paul, R.J.: Chemical energetics of vascular smooth muscle. *In* Bohr, D.F., *et al.* (eds.): Handbook of Physiology. Sec. 2, Vol. 2. Baltimore, William & Wilkins, 1980, p. 201.

Somlyo, A.P.: Ultrastructure of vascular smooth muscle. *In* Bohr, D.F., *et al.* (eds.): Handbook of Physiology. Sec. 2, Vol. 2. Baltimore, Williams & Wilkins, 1980, p. 33.

Sugi, H., and Pollack, G.H. (eds.): Cross-Bridge Mechanism in Muscle Contraction. Baltimore, University Park Press, 1979.

Tada, M., *et al.*: Molecular mechanism of active calcium transport by sarcoplasmic reticulum. *Physiol. Rev., 58*:1, 1978.

Tregear, R.T., and Marston, S.B.: The crossbridge theory. *Annu. Rev. Physiol., 41*:723, 1979.

THE NERVOUS SYSTEM

12

Anatomy of the Nervous System: I. General Structure, the Cerebrum, and the Diencephalon

Overview

The nervous system has two major divisions: (1) the **central nervous system** and (2) the **peripheral nervous system.** In turn, the central nervous system is divided into the **brain** and the **spinal cord.**

All nervous tissue contains two basic types of cells: (1) **neurons** that conduct the nervous signals and (2) **supporting** and **insulating cells** that hold the neurons in place and prevent unwanted signals from spreading between neurons; these cells and their associated extracellular structures are called the **neuroglia.**

There are about 100 billion neurons in the entire nervous system. Each of these consists of (1) a **cell body** that provides the living functions keeping the neuron alive, (2) multiple **dendrites** that project into the surrounding nervous tissue and receive incoming nerve signals from other neurons, and (3) a single **axon** that extends from a fraction of a millimeter up to as far as a meter outward from the cell body. The axon is the principal conductive structure of the neuron and is called the **nerve fiber.** Nerve signals are transmitted from one neuron to the next through junction points between the neurons called **synapses,** which will be discussed in detail in Chapter 15.

The brain is comprised of six major divisions: (1) the **cerebrum,** (2) the **diencephalon,** (3) the **mesencephalon,** (4) the **pons,** (5) the **medulla oblongata,** and (6) the **cerebellum.** The cerebrum constitutes about three quarters of the entire brain and is the seat of the higher levels of consciousness, sensory perception, and motor control. It is divided into two halves, each half called a **cerebral hemisphere.** These communicate with each other mainly by way of the **corpus callosum,** which is a massive bundle of nerve fibers that interconnect corresponding points in the two hemispheres.

Each cerebral hemisphere is divided into four major lobes and one minor lobe by deep fissures and sulci: (1) the **frontal lobe,** located in the anterior cerebrum, (2) the **occipital lobe,** located most posteriorly in the cerebrum, (3) the parietal lobe, located between the frontal and occipital lobes, (4) the **temporal lobe,** located anterior to the occipital lobe and lateral to the parietal and frontal lobes, and (5) the **insula,** a minor lobe located on a flattened floor deep in the transverse fissure. The cerebrum is also divided into three different types of functional tissue: (1) cerebral cortex, (2) deep nuclear structures, including especially the basal ganglia, and (3) white matter.

The **cerebral cortex** is a grayish appearing layer about 6 mm thick that covers the entire surface of the cerebrum. This layer contains mainly neuronal cell bodies. It is the portion of the brain in which most information, such as one's memories, learned motor patterns for control of muscle activities, spatial information about forms of objects, and so forth, is stored. Different parts of the cerebral cortex serve separate functions: The occipital lobes are primarily related to vision, the parietal lobes are the area for interpreting tactile and proprioceptive sensory information, the temporal lobes are concerned partly with hearing and partly with analyzing all types of incoming information, and the frontal lobes are used partly for motor control and partly for deep thinking processes.

The **basal ganglia** are discrete, large collections of neuronal cell bodies located deep inside each cerebral hemisphere. The most important basal ganglia are the **caudate nucleus,** the **putamen,** and the **globus pallidus.** They function together to control many of the stereotype, subconscious patterns of muscle activity, including such activities as postural movements, turning movements, and even some gross patterns of walking. They also function in association with the cerebral cortex to control the more discrete, learned patterns of muscle activity.

The **white matter** is comprised of great masses of nerve fibers that interconnect the different parts of the cerebrum and also connect the cerebrum with lower brain centers and the spinal cord.

The **diencephalon** is the connecting link between the cerebrum and the lower brain structures. It is located in the midbase of the brain sitting on top of the mesencephalon and enveloped on all other sides by the cerebrum. Its most important structures are the thalamus and hypothalamus.

The **thalamus** serves as the principal sensory "switchboard" for the cerebrum. Most sensory receptors of the body send their signals first to the thalamus, and the signals are then directed to appropriate areas of the brain.

The **hypothalamus** is composed of multiple nuclei that control the so-called vegetative functions of the body such as body temperature, heart rate, blood pressure, and composition of the body fluids.

An area called the **limbic system** is comprised of structures located on the borders of both the cerebrum and the diencephalon. It operates in close association with the hypothalamus, thalamus, and cerebral cortex to control behavior.

The nervous system is the sensing, thinking, and controlling system of our body. To perform these functions, it collects sensory information from all of the body—from a myriad of special sensory nerve endings in the skin, from the deep tissues, from the eyes, the ears, the equilibrium apparatus, and other sensors—and transmits this information through nerves into the spinal cord and brain. The cord and the brain may react immediately to this sensory information and send signals to the muscles or internal organs of the body to cause some response, called a **motor response.** Or, under other conditions, no immediate reaction might occur at all; instead the sensory information is stored in one of the brain's memory banks. There it is compared with other memories already stored; it is combined with other information; and from the various combinations, new thoughts are achieved. Then, perhaps a few minutes later, a month later, or even several years later, this extensive processing of information might at last lead to some motor response, maybe a very simple one or maybe very complex, such as building a house or piloting a space craft. Also, the nervous activation of the internal organs of the body, such as increasing the heart rate or increasing peristalsis in the intestines, may also be part of a motor response.

Thus, the nervous system is said to subserve three principal functions: (1) *sensory function;* (2) *integrative function,* which includes the memory and thinking processes; and (3) *motor function.* However, to understand these requires first a detailed knowledge of the anatomy of the nervous system. This will be presented here and in the following two chapters. Then we will be able to study the detailed functional mechanisms.

THE MAJOR ANATOMICAL DIVISIONS OF THE NERVOUS SYSTEM

Figure 12–1 illustrates the two major divisions of the nervous system:

1. the **central nervous system,** which in turn is comprised of the **brain** and the **spinal cord,** and
2. the **peripheral nervous system.**

The brain is the principal integrative area of the nervous system—the place where memories are stored, thoughts are conceived, emotions are generated, and other functions related to our psyche and to complex control of our body are performed. To perform these complex activities, the brain itself is divided into many

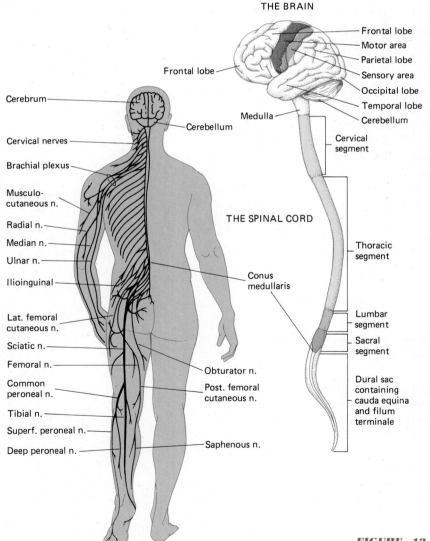

THE BRAIN

Frontal lobe
Frontal lobe
Motor area
Parietal lobe
Sensory area
Occipital lobe
Temporal lobe
Cerebellum

Cerebrum
Cerebellum
Medulla

Cervical segment

THE SPINAL CORD

Cervical nerves
Brachial plexus
Musculo-cutaneous n.
Radial n.
Median n.
Ulnar n.
Ilioinguinal

Thoracic segment

Conus medullaris

Lat. femoral cutaneous n.
Sciatic n.
Femoral n.
Common peroneal n.
Tibial n.
Superf. peroneal n.
Deep peroneal n.

Lumbar segment
Sacral segment

Obturator n.
Post. femoral cutaneous n.
Saphenous n.

Dural sac containing cauda equina and filum terminale

THE MAJOR SPINAL NERVES

FIGURE 12–1 The principal anatomical parts of the nervous system.

separate functional parts, which we shall begin discussing later in this chapter.

The spinal cord serves two functions. First, it serves as a conduit for many nervous pathways to and from the brain. Second, it serves as an integrative area for coordinating many subconscious nervous activities, such as reflex withdrawal of a part of the body from a painful stimulus, reflex stiffening of the legs when a person stands on his feet, and even crude reflex walking movements. Thus, the spinal cord is much more than simply a large peripheral nerve.

The peripheral nervous system is illustrated to the left in Figure 12–1, showing that it is a branching network of nerves that is so extensive that hardly a single cubic millimeter of tissue anywhere in the body is without nerve fibers. The anatomy of a peripheral nerve, containing great numbers of bundled nerve fibers, was discussed in Chapter 10. These fibers are of two functional types: **afferent fibers** for transmission of sensory information into the spinal cord and brain and **efferent fibers** for transmitting motor signals back from the central nervous system to the periphery, especially to the skeletal muscles as we discussed in the last two chapters. Some of the peripheral nerves arise directly from the brain itself and supply mainly the head region of the body. These, called **cranial nerves,** are not illustrated in Figure 12–1, but they will be discussed later. The remainder of the peripheral nerves are **spinal nerves,** one of which leaves each side of the spinal cord through an intervertebral foramen at each vertebral level of the cord.

NERVOUS TISSUE

Nervous tissue, whether it be in the brain, the spinal cord, or the peripheral nerves, contains two basic types of cells:

1. **Neurons** conduct the signals in the nervous system. There are about 100 billion of these in the entire nervous system.
2. **Supporting** and **insulating cells** hold the neurons in place and prevent signals from spreading between the neurons where this is not desired. In the central nervous system these cells and their intercellular structures are collectively called the **neuroglia.** In the peripheral nervous system they are the *Schwann cells.*

We have described the neurons and the Schwann cells in the peripheral nerves in Chapter 10, but let us see how these same principles apply to the central nervous system.

CENTRAL NERVOUS SYSTEM NEURON

Figure 12–2 illustrates a typical neuron of the brain or spinal cord. Its principal parts are the following:

1. **Cell body.** It is from this that other parts of the neuron grow. Also the cell body provides much of the nourishment that is required for maintaining the life of the entire neuron.
2. **Dendrites.** These are multiple branching outgrowths from the cell body. They are the main receptor portions of the neuron. That is, most signals that are to be transmitted by the neuron enter by way of the dendrites, although some enter also through the surface of the cell body. The dendrites of each neuron usually receive signals from literally thousands of contact points with other neurons, points called *synapses,* as we shall discuss later.
3. **Axon.** Each neuron has one axon. This is the portion of the neuron that is called the *nerve*

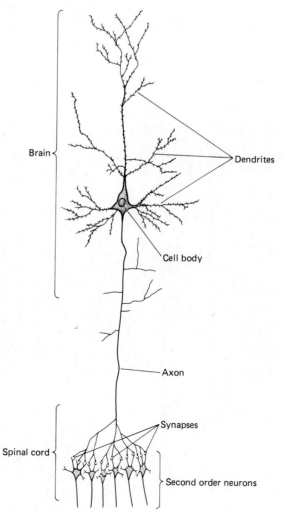

FIGURE 12–2 Structure of a large neuron of the brain, showing its important functional parts.

fiber. It may extend only a few millimeters, as is the case for the axons of many small neurons within the brain, or it may be as long as a meter, in the case of the axons (nerve fibers) that leave the spinal cord to innervate the feet. The axons carry the nerve signals to the next nerve cell in the brain or spinal cord or to muscles and glands in peripheral parts of the body.

4. **Axon terminals and synapses.** All axons branch near their ends many times, often thousands of times. At the end of each of these branches is a specialized axon terminal that, in the central nervous system, is called a **synaptic knob** or **bouton** because of its knoblike appearance. The synaptic knob in turn lies on the membrane surface of a dendrite or cell body of another neuron, thus providing a contact point called a **synapse** through which signals can be transmitted from one neuron to the next. When stimulated, the synaptic knob releases a minute quantity of a hormone called a *transmitter substance* into the space between the knob and the membrane of the neuron, and the transmitter substance then stimulates this neuron as well.

Figure 12–3 illustrates more details of the neuronal cell body. It illustrates a typical *nucleus* with a very prominent *nucleolus.* Also shown are *Nissl bodies*, which are parts of a specialized endoplasmic reticulum that synthesizes sub-

Dendrites

Axon

Glial tissue

FIGURE 12–4 A large neuron of the spinal cord surrounded by its supporting tissue called neuroglia.

stances required to keep the neuron alive. In fact, these substances are even transported into the axon and dendrites through a system of microtubules called *neurofibrils.* Finally, note in Figure 12–3 that the axon arises from a conical pole of the cell body called the *axon hillock.*

THE NEUROGLIA

Figure 12–4 illustrates a large neuron of the spinal cord surrounded by its supporting tissue, the neuroglia. The cells in the neuroglia are called **glial cells.** Many of them function similarly to the fibroblasts of connective tissue; that is, they form fibers that hold the tissue together. But others serve the same function as the Schwann cells of the peripheral nerves to wrap *myelin sheaths* around the larger nerve fibers, thus providing typical *myelinated nerve fibers* that transmit signals at velocities as great as 100 m a second, the same as in peripheral nerves as described in Chapter 10. The very small nerve fibers do not have myelin sheaths and therefore are called *unmyelinated fibers,* but even these are insulated from each other by interposition of glial cells between the fibers, much the same way that Schwann cells insulate the unmyelinated nerve fibers from each other in the peripheral nerves.

Recapitulation of the Functional Structure of the Nervous System

Figure 12–5 gives a composite view of the structural highlights of the complete nervous system. To the left is a typical neuron, a motor neuron located in the spinal cord that sends a large myelinated nerve fiber to a muscle through a peripheral nerve. To the right is shown the course of both a sensory and a motor nerve fiber entering and leaving the spinal cord through a

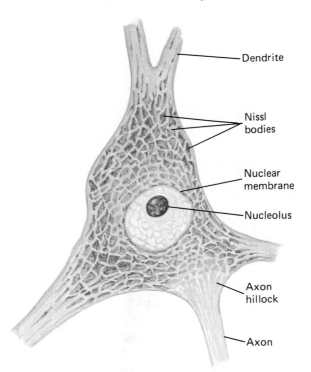

Dendrite

Nissl bodies

Nuclear membrane

Nucleolus

Axon hillock

Axon

FIGURE 12–3 The neuronal cell body.

peripheral nerve. And also shown is an ascending sensory tract in the cord that carries millions of sensory nerve fibers upward from the cord to the brain and a descending motor tract that carries millions of motor fibers downward.

Note especially the junction points, the synapses, between successive neurons. It is at these synapses that the transmitted signals are controlled and their directions of travel in the nervous system are determined, as will be discussed fully in Chapter 15.

THE BRAIN AND ITS DIVISIONS

The brain is that portion of the nervous system located in the cranial cavity. Figure 12–6 illustrates a lateral view of the brain, Figure 12–7 an inferior view (its ventral surface), and Figure 12–8 a sagittal view as seen in the median plane of the brain. Unfortunately, several different terminologies are used to describe the different parts of the brain, three of which are listed in Table 12–1. The most widely used terminology in medical circles is given in the right-hand column of the table in which the brain is divided into six separate parts:

1. the **cerebrum,**
2. the **diencephalon,**
3. the **mesencephalon,**
4. the **cerebellum,**
5. the **pons,** and
6. the **medulla oblongata,** usually called simply the "medulla."

It is important to recognize the relationship of this widely used terminology to the classical terminology given in the left-hand column of the table and also to several anglicized classi-

FIGURE 12–5 The functional components of the nervous system.

Cerebrum

Central sulcus

Frontal lobe

Parietal lobe

Sylvian fissure

Temporal lobe

Occipital lobe

Pons

Medulla

Cerebellum

Spinal cord

FIGURE 12–6 Left lateral view of the brain, showing the principal divisions of the brain and the four major lobes of the cerebrum.

Longitudinal fissure

Frontal lobe

Olfactory bulb and tract

Temporal lobe

Optic nerve

Hypothalamus

Optic chiasma

Mesencephalon

Posterior perforated substance

Cranial nerves

Pons

Medulla

Parietal lobe

Cerebellum

Spinal cord

Occipital lobe

FIGURE 12–7 Basal view of the brain.

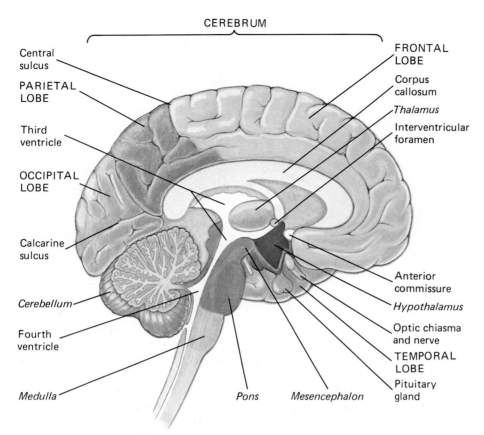

FIGURE 12–8 Medial view of the left half of the brain, showing especially the relationship of the cerebrum to the brain stem and cerebellum.

cal terms given in the middle column. The cerebrum is the same as the *telencephalon,* and the telencephalon and diencephalon together constitute the *prosencephalon,* or *forebrain,* which is the large massive portion of the brain filling the anterior and superior three fourths of the cranial cavity.

The mesencephalon, also called the *midbrain,* is a minute portion of the brain located at the base of the forebrain, as illustrated in Figures

12–7 and 12–8. Yet, despite its small size, it is the only connecting link between the forebrain and all the lower portions of the brain and spinal cord.

The cerebellum, pons, and medulla all lie in the posterior fossa of the cranial cavity, and together they constitute the *rhombencephalon,* or *hindbrain.*

One can see from the figures that the major mass of the brain is the cerebrum, and it is also

TABLE 12–1
Divisions of the Brain

Classical terminology	*Anglicized terminology*	*Most widely used terminology*
Encephalon	*Brain*	*Brain*
Prosencephalon	Forebrain	
Telencephalon		Cerebrum
Diencephalon		Diencephalon (or thalamus, hypothalamus, and surroundings)
Mesencephalon	Midbrain	Mesencephalon
Rhombencephalon	Hindbrain	
Metencephalon		
Cerebellum		Cerebellum
Pons		Pons
Myelencephalon		
Medulla oblongata		Medulla (or medulla oblongata)

clear that the next largest portion of the brain is the cerebellum. This might make one think that the other four parts of the brain—the diencephalon, the mesencephalon, the pons, and the medulla—are of relatively little importance in comparison with the remainder of the brain. But, we shall see later that these parts are absolutely crucial to the maintenance of nervous function, indeed, far more so than any equivalent mass of the cerebrum or cerebellum.

THE CEREBRUM

The Cerebral Hemispheres and the Corpus Callosum. For the next few moments let us study the external views of the cerebrum in Figures 12–6, 12–7, and 12–8 and also a horizontal section of the cerebrum shown in Figure 12–9. The first notable feature that one observes about the cerebrum is that it is composed of two large bilateral masses, the **cerebral hemispheres,** illustrated in Figures 12–6, 12–7, and 12–8. These two hemispheres are connected with each other through several bundles of nerve fibers, the most important of which are discussed here:

1. The **corpus callosum** is illustrated in sagittal section in Figure 12–8 and in horizontal section in Figure 12–9. Note that the corpus callosum is a broad band of fibers with a band width almost half the length of the cerebral hemisphere. Its importance is emphasized by its extremely large number of fibers, about 20 million.

2. The **anterior commissure** is also shown in Figure 12–8. This is a much smaller bundle, probably no more than a million fibers. It is located several centimeters below the anterior third of the corpus callosum, and it interconnects mainly the anterior and medial portions of the two temporal lobes.

Corresponding points in almost all areas of the two hemispheres interconnect with each other in both directions via the fibers in these two bundles, which allows continuous communication between the two hemispheres. When the corpus callosum and anterior commissure are destroyed, each of the two hemispheres functions as a separate brain, even thinking separate thoughts and causing separate reactions in the two sides of the body.

FIGURE 12–9 A horizontal section through the cerebrum at the level of the basal ganglia and thalamus.

TABLE 12–2
The Cerebrum

TABLE 12–2
The Cerebrum

The cerebral lobes	Location and function
Frontal	Anterior superior
Parietal	Superior midportion
Occipital	Posterior
Temporal	Lateral
Insula	Deep in lateral fissure
Principal fissures and sulci	
Longitudinal fissure	Separates the cerebral hemispheres
Central sulcus	Separate frontal and parietal lobes
Lateral fissure	Separates temporal lobe from frontal lobe and part of the parietal lobe
Parieto-occipital sulcus	Separates superior part of parietal lobe from superior part of occipital lobe
Principal structural parts	
Cerebral cortex (gray matter)	Thin layer on surface comprised mainly of billions of neuronal cell bodies
Deep nuclear (also gray matter)	
Basal ganglia	Most important: (1) caudate nucleus, (2) putamen, (3) globus pallidus
Some of the limbic structures	
White matter	Comprised of billions of nerve fibers, mainly myelinated
Functional areas	
Motor areas:	Located in posterior frontal lobe
Motor cortex	Controls discrete muscle activities
Premotor cortex	Controls patterns of coordinate muscle contractions
Broca's area	Controls speech
Somesthetic cortex	Parietal lobe Detects tactile and proprioceptor sensations
Visual area	Occipital lobe Detects visual sensations
Auditory area	Superior temporal lobe Detects auditory sensations
Wernicke's area	Superior posterior temporal lobe Analyzes sensory information from all sources
Short-term memory area	Inferior portions of temporal lobe
Prefrontal area	Anterior half of frontal lobe—"elaboration of thought"

Cerebral Convolutions, Fissures, and Sulci.   The next distinctive feature of the cerebrum is the folds in its surface. These are called **cerebral convolutions,** and each convolution is called a **gyrus.** The grooves between the gyri are called either **fissures** or **sulci,** the larger and deeper ones generally being called fissures whereas the great majority, less deep, are the sulci. Four of the principal fissures or sulci are illustrated in Figures 12–6 and 12–7. They are listed here:

1. The **longitudinal fissure** separates the two cerebral hemispheres from each other in the midsagittal plane of the basin.
2. The **central sulcus** extends in an approximate inferosuperior direction on the lateral side of each hemisphere and divides the cerebrum approximately into an anterior half and a posterior half.
3. The **lateral fissure,** also called the *Sylvian fissure,* extends along the lateral aspect of each cerebral hemisphere for about half its length.
4. The **parieto-occipital sulcus** originates from the side of the longitudinal fissure about one quarter the distance anterior to the posterior pole of the hemisphere and then extends laterally and anteriorly for about 5 cm.

To some extent, these fissures and sulci demarcate separate functional parts of the cerebrum, as we shall discuss.

The Lobes of the Cerebrum.   Figures 12–6 through 12–9 illustrate that the cerebrum is divided into four major **lobes** and a fifth minor one. The major lobes are

1. the **frontal lobe,**
2. the **parietal lobe,**
3. the **occipital lobe,** and
4. the **temporal lobe;** the minor lobe is
5. the **insula.**

The central sulcus separates the frontal lobe from the parietal lobe. The lateral fissure demarcates the frontal lobe and the anterior portion of the parietal lobe from the temporal lobe. And the parieto-occipital sulcus separates the superior part of the parietal lobe from the occipital lobe. The separation between the temporal lobe and the occipital lobe is less distinct. We shall see later that the area where the parietal, temporal, and occipital lobes all meet is the major area of the brain where integration of sensory information occurs, with sensory information from the body feeding into this area through the parietal lobe, visual information through the occipital lobe, and auditory information through the temporal lobe. In contrast, we shall see that the frontal lobe is concerned primarily with control of muscle movement and also with certain types of thinking processes.

The insula cannot be seen from the surface of the cerebrum. Instead, it lies deep in the lateral fissure. The horizontal section of the brain in Figure 12–9 shows that the lateral sulcus has a broad flat bottom covered by overhanging lips from the frontal, parietal, and temporal lobes. This flat bottom is the *insula*, and the lips are called *opercula* of the other lobes. Unfortunately, we know little about the function of the insula except that it probably functions as part of the limbic system (which will be discussed later) to help control behavior.

The Cerebral Cortex—Gray Matter and White Matter

Now let us look inside the cerebrum to see how its internal structure is organized. Figure 12–9 shows a horizontal section through the cerebrum. It is composed of areas that appear gray to the naked eye, called **gray matter,** and other areas that appear white, called **white matter.** The gray matter is collections of great numbers of neuronal cell bodies that all together give it its grayish hue. The white matter is comprised of great bundles of nerve fibers leading to or from the nerve cells in the gray matter; its white appearance is caused by the brilliant white color of the myelin sheaths of nerve fibers.

Figure 12–9 shows especially that a thin shell of gray matter covers the entire surface of the cerebrum, including the fissures and sulci. This is the **cerebral cortex.** One of the principal advantages of having the many fissures and sulci is that they triple the total area of the cerebral cortex; the exposed surface area of the brain is only about two thirds of 1 ft² or 600 cm², but the total area of the cerebral cortex is about 2 ft² or 1800 cm².

The cerebral cortex is the portion of the brain most frequently associated with the thinking process, even though it cannot provide thinking without simultaneous action of most deep structures of the brain. Yet, the cerebral cortex is the portion of the brain in which essentially all of our memories are stored, and it is also the area most responsible for our ability to acquire our many muscle skills. We still do not know the basic physiological mechanisms by which the cerebral cortex stores either memories or knowledge of muscle skills, but what we do know about these will be discussed in Chapter 20.

In most areas, the cerebral cortex is about 6 mm thick, and all together it contains an estimated 50 billion or more nerve cell bodies. Also, perhaps a billion nerve fibers lead away from the cortex as well as comparable numbers leading into it, passing to other areas of the cortex, to and from deeper structures of the brain, and some even all the way to the spinal cord.

FUNCTIONAL AREAS OF THE CEREBRAL CORTEX

Until World War I we knew the function of only a few areas of the cerebral cortex. But, at that time soldiers with bullet wounds in discrete parts of the brain were studied systematically for brain functional changes. Also, in more recent years neurosurgeons and neurologists have carefully documented changes in brain function caused by tumors or other specific lesions. Figure 12–10 illustrates the principal functional areas of the cerebral cortex that have been determined by these studies. These are as follows:

1. ***The Motor Area: Motor Cortex, Premotor Cortex, and Broca's Area.*** The motor area lies in front of the central sulcus and occupies the posterior half of the frontal lobe. It, in turn, is divided into three subdivisions, the motor cortex, the premotor cortex, and Broca's area, all of which are concerned with the control of muscle activity.

The **motor cortex,** located in a strip about 2 cm wide immediately anterior to the central sulcus, controls the specific muscles throughout the body, especially the muscles that cause fine movements, such as the finger and thumb motions and the lip and mouth motions for talking and eating, and to a much lesser extent the fine motions of the feet and toes.

The **premotor cortex,** located anterior to the motor cortex, elicits coordinate movements that involve either sequences of individual muscle movements or combined movements of a number of different muscles at the same time. It is in this area that much of one's knowledge is stored for controlling learned skilled movements such as the special movements required for playing an athletic game.

Broca's area, located anterior to the motor cortex at the lateral margin of the premotor cortex, controls the coordinate movements of the larynx and mouth to produce the words of speech. This area is developed in only one of the two cerebral hemispheres, in the left hemisphere in about 19 out of 20 persons, including all right-handed persons and one half of all left-handed persons.

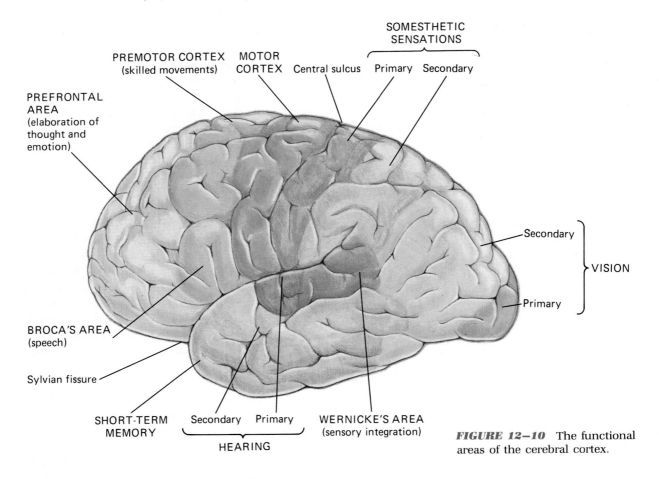

PREFRONTAL AREA (elaboration of thought and emotion)

PREMOTOR CORTEX (skilled movements)

MOTOR CORTEX

Central sulcus

SOMESTHETIC SENSATIONS

Primary   Secondary

Secondary

Primary

VISION

BROCA'S AREA (speech)

Sylvian fissure

SHORT-TERM MEMORY

Secondary   Primary

HEARING

WERNICKE'S AREA (sensory integration)

FIGURE 12–10 The functional areas of the cerebral cortex.

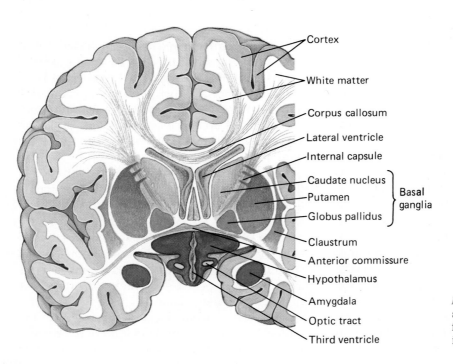

Cortex

White matter

Corpus callosum

Lateral ventricle

Internal capsule

Caudate nucleus

Putamen

Basal ganglia

Globus pallidus

Claustrum

Anterior commissure

Hypothalamus

Amygdala

Optic tract

Third ventricle

FIGURE 12–11 A coronal section of the cerebrum in front of the thalamus, showing especially the basal ganglia.

2. ***The Somesthetic Sensory Area.***
Somesthetic sensations are the sensations from the body such as touch, pressure, temperature, and pain. One can see in Figure 12–11 that the

somesthetic sensory area occupies the entire parietal lobe.

Note that this sensory area is divided into a primary area and a secondary area. This is true also of all the other sensory areas. The primary somesthetic sensory area is the portion of the cortex that receives signals directly from the different sensory receptors located throughout the body. In contrast, signals to the secondary somesthetic sensory area are partly processed in deep brain structures or in the primary somesthetic area before being relayed to the secondary area. The primary area can distinguish the specific types of sensation in discrete regions of the body. The secondary area serves mainly to interpret the sensory signals, not to distinguish them, such as interpreting that a hand is feeling a chair, a table, a ball, or so forth.

3. ***The Visual Area.*** Figure 12–11 shows that the visual area occupies the entire occipital lobe. Most of the primary area for vision is located on the medial surface of the cerebral hemisphere along the course of the calcarine sulcus (Figs. 12–8 and 12–9), but a small portion of the primary visual area projects over the outer pole of the occipital lobe as shown in Figure 12–11. This area detects specific light and dark spots as well as orientations of lines and borders in the visual scene. The secondary visual areas occupy the remainder of the occipital lobe, and their function is to interpret the visual information. For instance, it is in these areas that the meanings of written words are interpreted.

4. ***The Area for Hearing (the Auditory Area).*** The hearing area is located in the upper half of the anterior two thirds of the temporal lobe. The primary auditory area is located in the midportion of the superior temporal gyrus. It is here that specific tones, loudness, and other qualities of sound are detected. The secondary areas occupy the other parts of the hearing area. It is in these areas that the meanings of spoken words are interpreted; and portions of these areas are also important for music recognition.

5. ***Wernicke's Area for Sensory Integration.*** Wernicke's area lies in the posterior part of the superior temporal lobe at the point where the parietal and occipital lobes both come in contact with the temporal lobe. It is here that sensory signals from all of the three sensory lobes—the temporal, occipital, and parietal lobes—all come together. This area is exceedingly important for interpreting the ultimate meanings of almost all the different types of sensory information such as the meanings of sentences and thoughts, whether they be heard, read, felt, or even generated within the brain itself. Therefore, destruction of this area of the brain causes extreme loss of thinking ability. This area is well developed only in one of the cerebral hemispheres, usually in the left hemi-

sphere. This unilateral development of Wernicke's area prevents confusion of thought processes between the two halves of the brain.

6. ***The Short-Term Memory Area of the Temporal Lobe.*** The lower half of the temporal lobe seems to be mainly of importance for storing short-term memories, memories that last from a few minutes to several weeks.

7. ***The Prefrontal Area.*** The prefrontal area occupies the anterior half of the frontal lobe. Its function is less well defined than that of any other part of the cerebrum. It has been removed in many psychotic patients to bring them out of depressive states. These persons function quite well without the prefrontal areas. However, they lose their ability to concentrate for long periods of time and also their abilities to plan for the future or to think through deep problems. Therefore, this area is said to be important for elaboration of thought.

We shall discuss in further detail the intellectual function of the cerebral cortex in Chapter 20.

The Basal Ganglia

The horizontal sectional view of the brain in Figure 12–9 shows several areas of gray matter, called **nuclei,** located deep inside the brain. A nucleus is a mass of nerve cell bodies congregated into a cohesive area. Two separate groups of nuclei are shown in the figure: (1) the *basal ganglia,* which are part of the cerebrum, and (2) the *thalamus,* which is composed of multiple small nuclei and is part of the diencephalon, to be described later in this chapter.

Figures 12–11 and 12–12 show still other views of the brain with the basal ganglia highlighted, Figure 12–11 a coronal view and Figure 12–12 a three-dimensional view to show the locations of the basal ganglia in the cerebral masses and also to show their relationships to the thalamus. The three most important of the basal ganglia are the

1. **caudate nucleus,**
2. **putamen,** and
3. **globus pallidus.**

In addition to these three, anatomists also consider the *claustrum* and the *amygdala* to be basal ganglia. However, the function of the claustrum is unknown, and the amygdala functions as part of the limbic system, which we will discuss later in this chapter; it has very little functional relationship to the remainder of the basal ganglia.

Longitudinal fissure   Caudate nucleus

Tail of caudate

Posterior

Thalamus

Anterior

Lateral

Putamen and globus pallidus   Amygdala   Fibers to and from spinal cord in internal capsule

FIGURE 12–12   Relationship of the basal ganglia to the thalamus, shown in a three-dimensional view.

On the other hand, the basal ganglia of the cerebrum function in very close association with the **subthalamus** of the diencephalon and the **substantia nigra** and **red nucleus** of the mesencephalon. Therefore, physiologists frequently consider these three bodies also to be part of the basal ganglial system.

The Three-Dimensional Location of the Basal Ganglia in the Cerebrum.   Now, let us study in more detail the anatomical locations of the principal basal ganglia. Note in Figure 12–9, in the horizontal section of the brain, that the caudate nucleus, putamen, and globus pallidus lie anterior and lateral to the thalamus. Figure 12–11 shows a coronal section of the brain slightly in front of the anterior end of the thalamus, approximately in the motor region of the cerebrum. In this area the three basal ganglia lie astride one of the major fiber pathways of the cerebrum, called the *internal capsule*. This pathway, which will be described later, is the principal communication link between the cortex and the lower regions of the brain and spinal cord. Many of the fibers in the internal capsule originate in the motor cortex and premotor cortex, and branches of them enter the basal ganglia. Portions of the motor signals are then processed and relayed through these ganglia rather than passing directly from the cerebral cortex to the cord.

Note also in Figure 12–11 the relationship of the basal ganglia to the anterior portions of the two *lateral ventricles*, which are fluid-filled

cavities in the cerebrum. The ventricles lie respectively superior and medial to the caudate nucleus in each of the two cerebral hemispheres.

Finally, the three-dimensional diagram in Figure 12–12 shows especially the relationship of the basal ganglia to the thalamus. One can see the central location of the thalamus in the basal portion of the brain and the location of the basal ganglial system mainly anterior and lateral to the thalamus, but note also the long tail of the caudate nucleus that curls posteriorly through the parietal lobe and thence laterally and inferiorly into the temporal lobe. The amygdala lies at the tip of the caudate nucleus tail in the temporal lobe.

Function of the Basal Ganglia.   If the cerebral cortex is removed in a cat but without removing the basal ganglia, the cat can still perform most of its normal motor activities, including walking, fighting, arching its back, spitting, and almost any other movement. On the other hand, in the human being a similar loss of the cerebral cortex, but with the basal ganglia intact, leaves the person with only crude motor activities such as gross trunk movements and movements of the limbs with a stiff-legged, uncontrolled walking.

Putting this information together, one can deduce that a major function of the basal ganglia in the human being is to control the background gross body movements, whereas the cerebral cortex is necessary for performance of the more precise movements of the arms, hands, fingers, and feet. When the hand is performing some precise activity that requires a background stance of the body, the basal ganglia provide the body movements while the cerebral cortex provides the precise movements.

To achieve the high degree of coordination that is required among the muscles of the body during most motor functions, a very complex circuitry of nerve fibers interconnects (1) the cerebral cortex and basal ganglia in the cerebrum, (2) the thalamus and subthalamus in the diencephalon, (3) the red nucleus and the substantia nigra in the mesencephalon, (4) and the cerebellum in the hindbrain.

The White Matter of the Cerebrum

In almost all areas of the cerebrum besides the cerebral cortex and basal ganglia one finds white matter. The white matter is comprised nearly exclusively of nerve fibers, these often organized into specific bundles of fibers called **fiber tracts.** Three of the principal fiber tracts, each containing millions of fibers, are illustrated in Figure 12–9. These are the following:

1. The **corpus callosum,** which was discussed earlier, connects the respective areas of the cerebral cortex in each cerebral hemisphere with corresponding areas in the opposite hemisphere. The corpus callosum is also seen in sagittal section in Figure 12–8 and in coronal section in Figure 12–11.

2. The **optic radiation** passes from the lateral geniculate body of the thalamus back to the calcarine sulcus area of the occipital lobe. This is the final relay pathway for transmission of visual signals from the eyes to the cerebral cortex.

3. The **internal capsule,** shown in Figure 12–9, is found in the areas between the thalamus, the caudate nucleus, and the putamen. It is through this internal capsule that most signals between the cerebral cortex and the lower brain and spinal cord are transmitted.

Figure 12–13 illustrates even more vividly the gray and white matter of the cerebrum. Shown too is the great mass of nerve fibers that extends upward from the internal capsule to the cerebral cortex through an extensive radiation called the **corona radiata.**

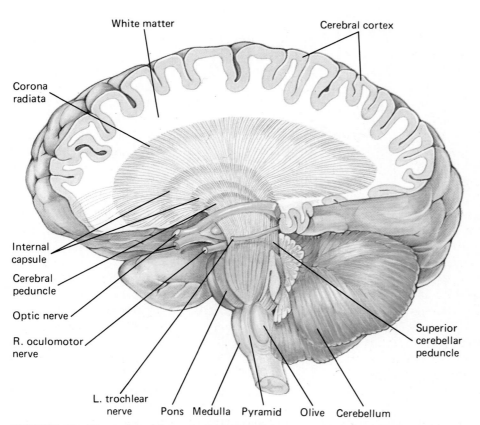

FIGURE 12–13 A deep dissection of the cerebrum showing the radiating nerve fibers, the corona radiata, that conduct signals in both directions between the cerebral cortex and the lower portions of the central nervous system.

THE DIENCEPHALON

The term **diencephalon** means the *between brain*. In primitive animals the diencephalon is a nodular structure, distinct from the remainder of the brain, that links the telencephalon (the cerebrum) with the mesencephalon (the midbrain). In the human being, the diencephalon still provides a similar linkage between the cerebrum and the lower parts of the brain, but anatomically it is so tightly fused with the basal portions of the cerebrum that it is difficult to demarcate its boundaries with the cerebrum. However, the diencephalon is defined as those structures that surround the third ventricle (another fluid-filled cavity in the brain, shown in Figure 12–14 and discussed in the following chapter). The most important of these structures are the

1. **thalamus** and
2. **hypothalamus.**

Both of these are comprised of multiple nuclei that perform many different important nervous functions. In addition to these two, smaller nuclear areas of the diencephalon, located posterior and inferior to the thalamus, are the **epithalamus** and **subthalamus.**

THE THALAMUS

We have already seen in several figures that the thalamus is located in the very center of the brain, enshrouded on all sides except inferiorly by the cerebrum. Figure 12–14 illustrates another view of the thalamus and its location in a coronal section of the brain. Several specific features of the thalamus are described here:

First, the thalamus is comprised of a number of separate discrete nuclei, as illustrated by the multiple gray areas in the figure.

Second, the thalamus rests directly on top of the mesencephalon (also called the "midbrain"); almost all signals from the midbrain and other lower regions of the brain, as well as from the spinal cord, are relayed through synapses in the thalamus before proceeding to the cerebral cortex.

FIGURE 12–14 A coronal view of the cerebrum looking from anteriorly backward. This section was made immediately anterior to the lower brain stem and through the middle of the thalamus.

TABLE 12–3
The Diencephalon and Limbic System

Structure	Location	Function
Diencephalon		
Thalamus	Central base of brain	Relays sensory signals to cortex, sensory analytic functions
Hypothalamus	Inferior to anterior thalamus	Controls internal body functions, stimulates autonomic nervous system
Subthalamus	Inferior to posterior thalamus	Functions with basal ganglia to control subconscious muscle activity
Epithalamus	Posteroinferior to thalamus	Function unknown; includes pineal gland
Limbic system		
Amygdala	Deep inside anterior end of each temporal lobe	Controls behavior for each social occasion
Hippocampus	Medial border of each cerebral hemisphere	Determines which sensory information will be committed to memory
Mammillary body	Posterior to hypothalamus	Perhaps helps to determine mood and degree of wakefulness
Septum pellucidum	Midline of cerebrum anterior and superior to hypothalamus	Perhaps helps to control temper and autonomic nervous system
Limbic cortex: Cingulate gyrus, cingulum, insula, and parahippocampal gyrus	Ring of cerebral cortex in medial part of cerebrum around deeper limbic structures	Conscious components in the control of behavior

Third, the thalamus has numerous two-way connections with all parts of the cerebral cortex, carrying a continual traffic of signals from the thalamus to the cortex and also from the cortex to the thalamus.

Fourth, the thalamus lies in close apposition to the basal ganglia. In fact, the thalamus relays many signals from other lower regions of the brain and spinal cord directly to the basal ganglia. And, in turn, the thalamus also functions as a relay station for signals from some of the basal ganglia to the cortex.

In essence, then, the thalamus is a chief traffic relay station for directing sensory and other signals to appropriate points in both the cerebral cortex and the deeper areas of the cerebrum as well. Some examples of the different types of signals that are relayed through the thalamus include

1. all the *somesthetic sensory signals* from the body (touch, pressure, pain, temperature, and so forth) to the somesthetic cortex of the parietal lobe;
2. *visual signals* to the calcarine sulcus area of the occipital cortex (the part of the thalamus that relays these signals is sometimes classified as the metathalamus, which is the posterior end of the thalamus);
3. *auditory signals* to the superior temporal gyrus (also relayed by the metathalamus); and
4. *muscle control signals* from the cerebellum,

mesencephalon, and other areas of the lower brain stem to the motor cortex and basal ganglia.

Anatomical Relationship of the Thalamus to the Ventricles. Note especially in Figure 12–14 the relationship of the thalamus to several of the ventricles: (1) the two lateral ventricles lie immediately above the two lateral halves of the thalamus and (2) the third ventricle bisects the thalamus into two halves. Each half of the thalamus functions separately with the cerebral hemisphere on the same side, and there is very little direct communication between the two halves of the thalamus.

The Sensory Interpretive Function of the Thalamus. In lower animals, at about the reptile stage, the cerebral cortex is not developed to a great extent, but the thalamus is an established part of the brain. In these animals, the thalamus plays a much greater role in sensory interpretation than it does in man. But, even in man some of its sensory interpretive abilities still persist. This is especially true for pain sensation, for a person can lose most if not all of the somesthetic sensory areas of his cerebral cortex and still retain much if not most of his ability to perceive pain. This is in keeping with the fact that pain is one of the most primitive of our sensations and also that the thalamus is a more primitive portion of the brain than is the cerebrum.

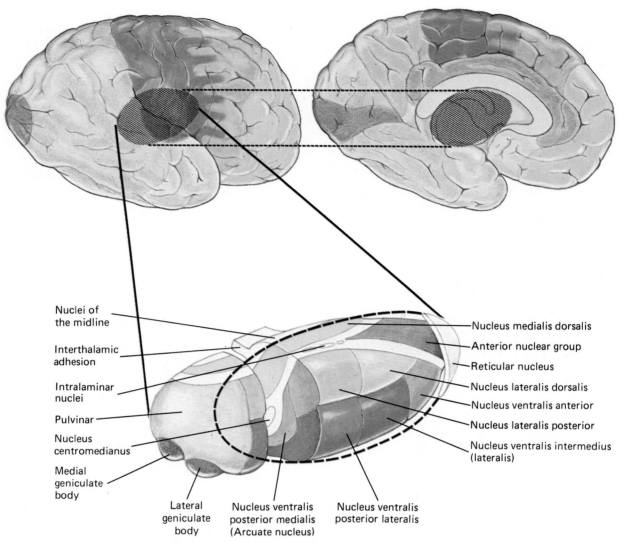

Nuclei of
the midline

Interthalamic
adhesion

Intralaminar
nuclei

Pulvinar

Nucleus
centromedianus

Medial
geniculate
body

Lateral
geniculate
body

Nucleus ventralis
posterior medialis
(Arcuate nucleus)

Nucleus ventralis
posterior lateralis

Nucleus medialis dorsalis

Anterior nuclear group

Reticular nucleus

Nucleus lateralis dorsalis

Nucleus ventralis anterior

Nucleus lateralis posterior

Nucleus ventralis intermedius
(lateralis)

FIGURE 12–15 The relationship between the different nuclei of the thalamus and their connecting areas in the cerebral cortex. (Reprinted from Warwick and Williams: Gray's Anatomy, 35th British Edition. Philadelphia, W. B. Saunders, 1973.)

The Thalamocortical Relationship. In addition to the relay pathways through the thalamus to the cerebral cortex, there are also innumerable two-way connections between the thalamus and all areas of the cortex, with nerve fibers going in both directions. Figure 12–15 illustrates by means of a color code the areas of the thalamus that connect with specific areas of the cerebral cortex. For example, the posterior portion of the thalamus (lateral geniculate body and pulvinar) has two-way connections with the occipital lobe of the cortex. The superior medial part of the thalamus (nucleus medialis dorsalis) connects with the prefrontal area of the frontal lobe. And the posterolateral portion of the ven-

tral nucleus (nucleus ventralis posterior lateralis) connects with the primary somesthetic area of the parietal cortex, and so forth.

But what are the purposes of these two-way connections? First, without the thalamus, the cortex is useless. It is the thalamus that drives the cortex to activity, which is another function of the thalamus in addition to relaying signals to the cortex from other areas of the brain and spinal cord. In fact, one can consider most of the cerebral cortex to be mainly an outgrowth of the thalamus; the cortex provides a great memory storehouse to function at the beck and call of the control centers of the thalamus.

THE HYPOTHALAMUS

Several figures presented earlier in this chapter illustrated a small structure in the middle of the base of the brain called the **hypothalamus.** Study especially Figures 12–7 and 12–8 and locate this structure, but look carefully or otherwise you will miss it. Its small size belies its importance, for it is a major center of the brain for controlling internal body functions.

Figure 12–16 illustrates an enlarged internal view of the hypothalamus, showing in third dimension various nuclei of the hypothalamus. Note that the hypothalamus lies anterior to the red nucleus, which is in the uppermost part of the mesencephalon, and it also lies immediately inferior to the anterior end of the thalamus. There are especially abundant nerve pathways

between the hypothalamus and the anterior thalamus and also between the hypothalamus and the mesencephalon.

Some of the Functions of the Nuclei of the Hypothalamus. At many points in this text we shall discuss the importance of one or more of the hypothalamic nuclei for control of some of the internal body functions, and more details of the function of the hypothalamus as a whole will be presented in Chapter 19. However, let us list here some of the important functions of a few of the hypothalamic nuclei.

The **preoptic nucleus,** located anteriorly, is primarily concerned with body temperature control.

The **supraoptic nucleus,** also located anteriorly but inferiorly as well, lying immediately above the optic nerves, controls the secretion of

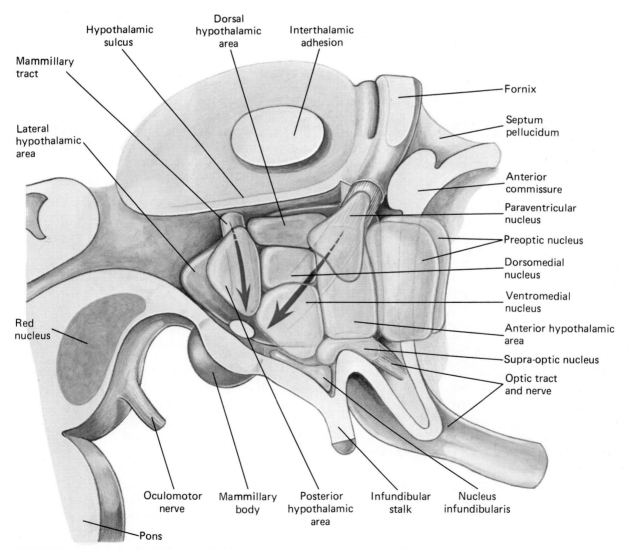

FIGURE 12–16 A three-dimensional view of one side of the hypothalamus, showing its principal nuclei. (From Nauta and Haymaker: The Hypothalamus. Springfield, Ill., Charles C Thomas, 1969.)

antidiuretic hormone; this hormone in turn helps to control the concentration of electrolytes in the body fluids.

The **medial nuclei** of the hypothalamus, when stimulated, gives a person a sense of satiety (that is, he feels satisfied, especially satisfied for food).

Stimulation of the most **lateral regions of the hypothalamus** causes a person to become very hungry, and stimulation anteriorly in the lateral hypothalamus causes a person to become very thirsty.

Stimulation of the **posterior hypothalamus** excites the sympathetic nervous system throughout the body, increasing the overall level of activity of many internal parts of the body, especially increasing heart rate and causing blood vessel constriction.

Finally, stimulation of different areas of the hypothalamus causes its neurons to secrete several hormones called *releasing hormones* that are carried in the venous blood directly to the anterior pituitary gland; here they then cause secretion of the anterior pituitary hormones. The pituitary hormones in turn control such varied activities of the body as the metabolism of carbohydrates, metabolism of proteins, metabolism of fats, functions of the sex glands, and several other functions.

Thus, one cannot help but be impressed by the global importance of this small area of the brain, the hypothalamus, and its many roles in the control of our bodies. Therefore, we will be discussing its functions at many points in this text.

THE LIMBIC SYSTEM

The word "limbic" means border, and the **limbic system,** illustrated in Figure 12–17, is comprised of the border structures of both the cerebrum and the diencephalon that mainly surround the hypothalamus. This limbic system functions especially to control our emotional and behavioral activities. Some of the important parts of the limbic system are as follows:

1. The **amygdala** (also called the *amygdaloid body*) is a small nuclear structure located deep inside each anterior temporal lobe and considered by anatomists to be one of the basal ganglia. However, it functions very closely with the hypothalamus, not with the usual basal ganglia. It is believed that the amygdala helps to control the appropriate behavior of the person for each type of social situation.

2. The **hippocampus,** one on each side, is a primitive portion of the cerebral cortex that lies along the medial-most border of the temporal lobe and folds upward and inward to

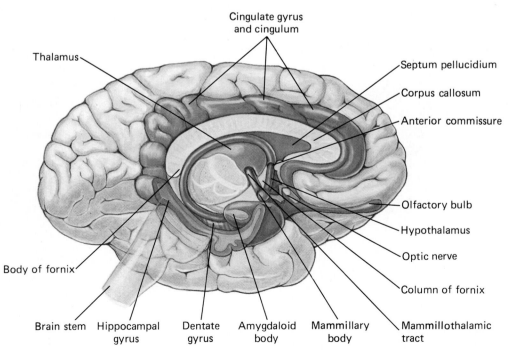

FIGURE 12–17 The limbic system in the midportion of the cerebrum. (From Warwick and Williams: Gray's Anatomy, 35th British Edition, Philadelphia, W. B. Saunders, 1973.)

form the inferior surface of the inferior horn of the lateral ventricle. The hippocampus is believed to interpret for the brain the importance of most of our sensory experiences. If the hippocampus determines an experience to be important enough, then the experience will be stored as a memory in the cerebral cortex. Without the hippocampus, a person's ability to store memories becomes very deficient.

3. The **mammillary bodies** lie immediately behind the hypothalamus and function in close association with the thalamus, hypothalamus, and brain stem to help control many behavioral functions such as the person's degree of wakefulness and perhaps also his feeling of well being.

4. The **septum pellucidum** lies anterior to the thalamus, superior to the hypothalamus, and between the basal ganglia in the median plane of the cerebrum. Stimulation in different parts of this septum can cause many different behavioral effects, including the phenomenon of rage.

5. The **cingulate gyrus,** the **cingulum,** the **insula,** and the **parahippocampal gyrus** all together form a ring of cerebral cortex in each cerebral hemisphere around the deeper structures of the limbic system described in the previous few paragraphs. This ring of cortex is believed to allow association between conscious cerebral behavioral functions and subconscious behavioral functions of the deeper limbic system.

Signals from the limbic system leading into the hypothalamus can modify any one or all of the many internal bodily functions controlled by the hypothalamus. And signals feeding from the limbic system into the mesencephalon can control such behavior as wakefulness, sleep, excitement, attentiveness, and even rage or docility. Yet, the precise manner in which the different parts of the limbic system function together to control all these emotional and behavioral functions of the body is still only slightly understood.

QUESTIONS

1. Give the parts of a neuron and tell the function of each.
2. Discuss the neuroglia and the functions of the glial cells.
3. Name and describe grossly the six separate parts of the brain.
4. Give the anatomy and function of the corpus callosum.
5. Name and describe the major fissures and sulci of the cerebrum.
6. Describe the separate lobes of the cerebrum.
7. Describe the functional areas of the cerebral cortex.
8. What are the anatomical characteristics of the cerebral cortex?
9. Name and describe the basal ganglia as well as their relationships with other structures of the brain.
10. What are the general functions of the basal ganglia?
11. What is the relation of the diencephalon to the cerebrum, and what are its parts?
12. What is the special relationship between the thalamus and the cerebral cortex?
13. Describe the hypothalamus and its relationship to the other brain structures.
14. Describe the limbic system and its relationship to the other brain structures.

REFERENCES

See References, Chapter 14.
See also General Anatomical References, Chapter 1.

13

Anatomy of the Nervous System: II. Brain Stem, Cerebellum, Spinal Cord, and Cerebrospinal Fluid System

Overview

The brain stem is comprised of (1) the **mesencephalon**, (2) the **pons**, and (3) the **medulla oblongata.** Long fiber tracts pass through the brain stem for transmitting sensory signals upward to the thalamus and motor signals downward from the cerebrum to the cord. The most important sensory pathway is the **medial lemniscus**, and the most important motor pathway is the **corticospinal tract.**

The brain stem also has many centers for controlling multiple body functions. For instance, in the mesencephalon and pons are special *centers that control eye movements, turning of the head,* and *rotational movements of the body.* Also, extending the entire length of the brain stem—through the medulla, pons, and mesencephalon—is a system of diffuse interconnected nuclei collectively called the **reticular formation.** The upper portions of this control the level of activity of the entire brain and also control contractions of the *postural muscles* of the body, including the muscle movements required for *equilibrium.* In the lower portions of the reticular formation are located *centers that control respiration, arterial pressure,* and *secretion by the digestive glands* in the digestive tract.

The **cerebellum**, located inferior to the occipital lobe of the cerebrum and posterior to the brain stem, is an important part of the muscle control system. It has extensive nerve fiber connections with the motor cortex, thalamus, basal ganglia, brain stem, and spinal cord. It helps to *coordinate the motor functions* of all these parts of the brain and cord.

The cerebellum has two major functional parts: (1) the **cerebellar cortex** and (2) the **deep nuclei.** One of the functions of the cerebellar cortex is to delay motor signals for short fractions of a second. This allows one set of muscles to contract a short period of time before the second set contracts, thus giving a *controlled sequence of muscle contractions.* The deep nuclei transmit output signals from the cerebellum to the cerebrum, thalamus, brain stem, and spinal cord.

242

The **spinal cord** serves two essential functions. One is its obvious function as a connecting link and distribution system for the sensory and motor nervous pathways between the peripheral structures of the body and the brain. However, equally important are many local control centers within the spinal cord itself. These reside in the gray matter of the cord. In these centers, the spinal cord reflexes are integrated, including reflexes that (1) cause withdrawal of a part of the body when it experiences a painful stimulus, (2) stiffen the legs when one puts pressure on the bottoms of the feet, and (3) even cause crude walking motions.

The **cerebrospinal fluid system** is a protective *flotation system* that surrounds the brain and spinal cord on all sides. The fluid space around the brain and cord is the **subarachnoid space,** and inside the brain are four fluid-filled cavities called **ventricles.** The ventricles all interconnect with each other and also with the subarachnoid space. Most of the cerebrospinal fluid is secreted by **choroid plexuses** in the four ventricles; the fluid then flows into the subarachnoid space around the brain and is absorbed through **arachnoid villi** into **venous sinuses** overlying the cerebrum. Blockage of fluid flow at any point in this pathway can cause *hydrocephalus,* which means excess collection of fluid in the ventricles or in the subarachnoid space surrounding the brain.

THE BRAIN STEM

The brain stem, illustrated in Figure 13–1, is exactly what its name implies: it is the stem of the brain that connects the forebrain with the spinal cord. Its major divisions are

1. the **mesencephalon,**
2. the **pons,** and
3. the **medulla oblongata.**

Some anatomists also consider the **diencephalon,** which was discussed in the previous chapter, to be part of the brain stem because it too is a connecting link.

Several important **fiber tracts** pass both upward and downward through the brain stem, transmitting sensory signals from the spinal cord mainly to the thalamus and motor signals from the cerebral cortex to the cord. In addition, other fiber tracts either originate or terminate in the brain stem, again mainly for the purpose of carrying sensory and motor signals.

However, the brain stem also contains many very important centers that control such physiological variables as respiration, arterial pressure, equilibrium, and others. In fact, centers in the brain stem even determine the level of activity in the cerebrum and also cause the waking-sleeping cycle of the nervous system.

And, finally, the brain stem serves as the connecting link between the cerebellum and the cerebrum superiorly and between the cerebellum and spinal cord inferiorly.

THE MESENCEPHALON

The surface anatomy of the mesencephalon is shown from the left posterolateral side in Figure

FIGURE 13–1 The brain stem.

TABLE 13–1
The Brain Stem and Cerebellum

Structure	Function
Mesencephalon	
Cerebral peduncles	
1. Corticospinal and corticopontine tracts	Motor signals to cord and pons
2. Substantia nigra	Part of basal ganglial motor control system
3. Tegmentum	
a. Red nucleus	Relays signals from cerebellum
b. Reticular formation	Excites whole brain, controls muscular tone
c. Nuclei of cranial nerves III and IV	Control eye movements
d. Medial lemnisci	Sensory signals to thalamus
Tectum	
1. Superior colliculi	Help to control eye movements
2. Inferior colliculi	Cause motor reactions to auditory signals
Pons	
Ventral part	
1. Corticospinal tracts	Pass through ventral pons toward cord
2. Pontine nuclei	Terminis of corticopontine tracts
3. Transverse fibers	Fibers from pontine nuclei to opposite cerebellar hemisphere
Tegmentum	
1. Reticular formation	Same as in mesencephalon, also parts of vasomotor and respiratory centers
2. Nuclei of cranial nerves V, VI, VII, and VIII	Eye and facial movements; facial, auditory, and equilibrium sensations
3. Medial lemnisci	Same as in mesencephalon
Medulla oblongata	
Pyramids and decussation of pyramids	Downward extensions and crossover of corticospinal tracts
Gracile and cuneate nuclei	Origins of fibers in medial lemnisci
Decussation of medial lemnisci	Crossover of the medial lemnisci
Inferior olivary nuclei	Origin of many input fibers to cerebellum
Nuclei of cranial nerves IX, X, XI, and XII	Motor signals to larynx, pharynx, tongue, some neck muscles; sensory signals from viscera; motor signals to parasympathetic nervous system
Reticular formation	
1. Most of vasomotor center	Controls vascular resistance, arterial pressure, heart activity
2. Most of respiratory center	Controls inspiration and expiration
Cerebellum	
Cerebellar peduncles	
1. Inferior peduncle	Extends from medulla; mainly input signals
2. Middle peduncle	Extends from ventral part of pons; entirely input signals
3. Superior peduncle	Extends to mesencephalon; mainly output signals
Vermis	Midline portion of cerebellum; functions with brain stem and spinal cord
Cerebellar hemispheres	Lateral portions of cerebellum; functions mainly with higher motor centers
Cerebellar cortex	Provides delays in motor signals
Deep nuclei	
1. Dentate nuclei	Output nuclei of cerebellar hemispheres
2. Fastigial, globose, and emboliform nuclei	Output nuclei of vermis

13–1, and a horizontal section is illustrated in Figure 13–2. It is divided from anterior to posterior into two major sections: (1) the two **cerebellar peduncles,** which constitute the anterior four fifths of the mesencephalon, and (2) the **tectum,** which comprises the posterior surface structures. Passing inferiorly in the posterior part of the tegmentum is the **cerebral aqueduct,** which is a small tubular canal that connects the third ventricle in the diencephalon with the fourth ventricle in the lower brain stem.

The Cerebral Peduncles. The cerebral peduncles veer anteriorly and laterally at the top of the mesencephalon and then project superi-

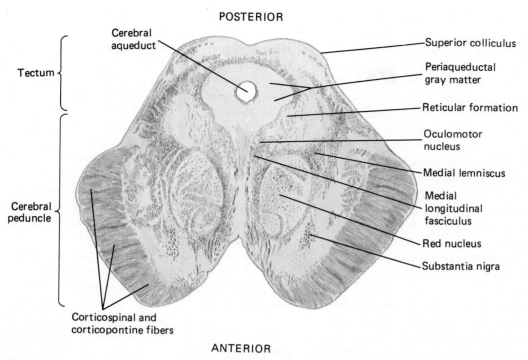

POSTERIOR

Tectum

Cerebral aqueduct

Superior colliculus

Periaqueductal gray matter

Reticular formation

Oculomotor nucleus

Medial lemniscus

Cerebral peduncle

Medial longitudinal fasciculus

Red nucleus

Substantia nigra

Corticospinal and corticopontine fibers

ANTERIOR

FIGURE 13–2 A horizontal section through the mesencephalon at the level of the superior colliculi.

orly into the two lateral halves of the diencephalon. Each cerebellar peduncle is divided into three separate areas:

1. A thick anterolateral surface layer of **corticospinal** and **corticopontine fibers** conducts motor signals from the cortex to the spinal cord and to the pons.
2. A deeper layer of darkly pigmented nerve cell bodies called the **substantia nigra** lies behind the fiber layer. The neurons of the substantia nigra function as part of the basal ganglial system to control subconscious muscle activities of the body. Destruction of these neurons causes Parkinson's disease, in which the person develops continuous muscle spasm and a shaking tremor in part or all of the body, sometimes so severe that muscle functions become useless.
3. The **tegmentum** is the major mass of the cerebral peduncles medial and posterior to the substantia nigra.

The tegmentum contains several important fiber tracts and nuclei that provide specific functions as follows:

1. The **medial lemniscus** is the major fiber tract for transmitting sensory signals from the body to the thalamus.
2. The **medial longitudinal fasciculus** is a tract that connects many of the nuclei of the brain

stem with each other and also with the diencephalon.

3. The **red nucleus,** illustrated in Figure 13–2, occupies a major part of each side of the superior mesencephalic tegmentum. This nucleus functions with the basal ganglia and cerebellum to coordinate muscle movements of the body. It also serves as a relay station for signals transmitted from the cerebellum to the thalamus and cerebrum.
4. The **nuclei of the oculomotor and trochlear nerves** are small collections of nerve cells on each side of the mesencephalon that control most of the muscles for eye movements.
5. The **periaqueductal gray** is a collection of diffuse nuclei around the cerebral aqueduct. This area plays a major role in the analysis of and reaction to pain.
6. The **reticular formation** is composed of many widely dispersed nuclei in large portions of the tegmentum. The reticular formation is not only present in the mesencephalon but actually extends all the way from the superior end of the spinal cord to the diencephalon, passing through the medulla oblongata, the pons, the mesencephalon, and even extending into the middle of the thalamus where it is represented by the thalamic intralaminar nuclei. Various collections of nerve cells within the reticular formation con-

trol many of the stereotyped body movements, such as *turning motions of the trunk, turning and bending motions of the head,* and *postural motions of the limbs.* But, even more important, the reticular formation is the major center of the entire brain for *controlling the brain's overall level of activity.* Generalized stimulation of the mesencephalic and pontine portions of the reticular formation usually causes a high degree of wakefulness in an animal while increasing the tone of the muscles throughout the body. Therefore, the reticular formation, though dispersed rather broadly in the brain stem, is functionally one of the most important of all the brain structures, as we shall discuss much more fully in later chapters.

The Tectum.   The tectum is the posterior fifth of the mesencephalon, and it consists principally of four small nodular bodies, two superior colliculi and two inferior colliculi, that are arranged in a quadrangle on the posterior surface of the mesencephalon, as illustrated in Figure 13–1.

The two **superior colliculi** are located side by side on the superior portion of the posterior mesencephalon, lying immediately beneath the posterior poles of the thalamus. In lower animals, especially in fish, the superior colliculi are the principal brain terminus for vision. In the

human being, the visual functions of these bodies have been lost, but they are still used for *causing eye movements* and even *trunk movements* in response to sudden visual signals such as a flash of light from one side of the field of vision or sudden movement of a person or an animal nearby.

The two **inferior colliculi** are located inferior to the two superior colliculi, also on the posterior surface of the mesencephalon. These serve as way-stations for relaying auditory signals from the ears to the cerebrum. In addition, they play a role in causing a person to turn his or her head or body in response to sounds coming from different directions.

Inferior to the inferior colliculi on the two sides of the mesencephalon are two large bundles of nerve fibers called the **superior cerebellar peduncles** that project inferiorly and posteriorly to connect with the superior portions of the cerebellum. These are one of the major trunk lines between the cerebellum and the remainder of the brain.

THE PONS

The pons, illustrated from the left posterolateral side in Figure 13–1 and in horizontal section in Figure 13–3, has many of the same types of internal structures as the mesencephalon, such as some of the same major fiber pathways that

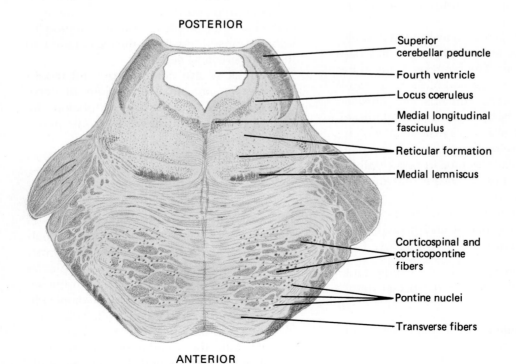

POSTERIOR

- Superior cerebellar peduncle
- Fourth ventricle
- Locus coeruleus
- Medial longitudinal fasciculus
- Reticular formation
- Medial lemniscus
- Corticospinal and corticopontine fibers
- Pontine nuclei
- Transverse fibers

ANTERIOR

FIGURE 13–3   A horizontal section through the pons.

transmit signals both up and down the brain stem and multiple nuclei that perform specific functions. For descriptive purposes, the pons is divided into two parts: (1) the **ventral part** and (2) the dorsal part, also called the **tegmentum of the pons,** which is continuous with the tegmentum of the mesencephalon.

The Ventral Part of the Pons. The ventral part of the pons is the large bulbous anterior protrusion illustrated in Figure 13–1 and seen even better in the basal view of the brain in Figure 12–7 of the previous chapter. Its internal structure is illustrated in Figure 13–3.

The same **corticospinal fibers** and **corticopontine fibers** that pass through the cerebral peduncles of the mesencephalon also descend into the ventral part of the pons. The corticospinal fibers then pass through this part and thence through the medulla into the spinal cord. On the other hand, the corticopontine fibers terminate here, synapsing in multiple **pontine nuclei.** From these, transverse fibers cross immediately to the opposite side of the ventral pons and then circle backwards around the two lateral sides of the pons to form the **middle cerebellar peduncles** that extend posteriorly into the two cerebellar hemispheres. In addition, a few fibers also pass directly backward into these peduncles on the same side. Because of the crossing of the transverse fibers and because most of the output fibers leaving the cerebellum to reenter the brain stem also cross to the opposite side, the right

half of the cerebellum functions mainly with the left half of the cerebrum, and the left half of the cerebellum functions mainly with the right cerebrum.

The Tegmentum of the Pons. Figure 13–3 shows that the pontine tegmentum contains the following three structures that are continuous with those in the mesencephalon: (1) the **medial lemniscus,** (2) the **medial longitudinal fasciculus,** and (3) the **reticular formation.** In addition, it contains the nuclei of several cranial nerves: (1) the **abducens nerve,** which helps to control eye movements; (2) the **facial nerve,** which controls the muscles of expression of the face; (3) the **trigeminal nerve,** which controls the muscles of mastication and also transmits sensory signals from the face, mouth, and scalp; and (4) the **vestibulocochlear nerve,** which transmits sensory signals from the ear and from the vestibular apparatus (the equilibrium apparatus of the inner ear).

THE MEDULLA OBLONGATA

The **medulla oblongata,** usually called simply the **medulla,** is illustrated at the lower end of the brain stem in Figure 13–1, and it is seen in cross-section in Figure 13–4. On the surface are two distinguishing characteristics:

1. On the anterior aspect of the medulla are two protruding longitudinal columns called **pyramids** (seen best in Figure 12–7 of the previous

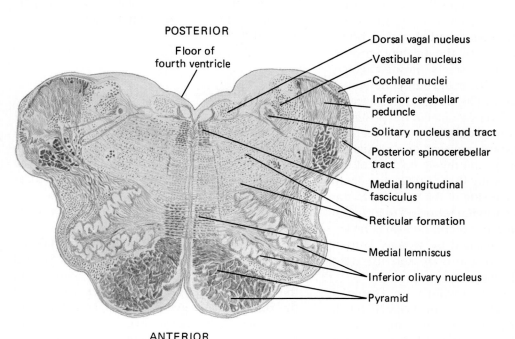

POSTERIOR
Floor of fourth ventricle

Dorsal vagal nucleus
Vestibular nucleus
Cochlear nuclei
Inferior cerebellar peduncle
Solitary nucleus and tract
Posterior spinocerebellar tract
Medial longitudinal fasciculus
Reticular formation
Medial lemniscus
Inferior olivary nucleus
Pyramid

ANTERIOR

FIGURE 13–4 A horizontal section through the medulla oblongata.

chapter but also illustrated in cross-section in Figure 13–4). These carry the same *corticospinal fibers* that pass from the cerebral cortex through the cerebral peduncles of the mesencephalon and through the ventral pons. The fibers in this trunk pass eventually to all levels of the spinal cord, carrying signals that control muscle contraction. In the inferior portion of the medulla, the fibers of the pyramid cross to the opposite side before passing down the spinal cord, which is called the **decussation of the pyramids.** Therefore, the left cerebral cortex controls muscle contraction in the right half of the body while the right cortex controls the left side.

2. An **olive** protrudes from each anterolateral surface of the medulla lateral to the pyramids (Fig. 13–1). Deep to the external projection of the olive is the **inferior olivary nucleus,** which can be seen in cross-section in Figure 13–4. This nucleus functions to relay signals into the cerebellum, functioning similarly to the pontine nuclei. However, it receives its input signals mainly from the basal ganglia and spinal cord and less from the motor cortex. Its outgoing signals go to the contralateral cerebellum through the **inferior cerebellar peduncle** (Fig. 13–1), which is an upward and posteriorly projecting column of nerve fibers extending from the medulla to the cerebellum.

In addition to these specific surface structures, the medulla contains many of the same components as the mesencephalon and pons. Especially prominent are the bilateral *medial lemnisci*, the large fiber tracts through which sensory signals are conducted from the spinal cord to the cerebrum. The fibers in these tracts originate in large bilateral nuclei in the posterior inferior medulla, the **gracile** and **cuneate nuclei,** which themselves receive sensory signals from sensory fibers in the dorsal column of the spinal cord. After leaving these nuclei, the fibers cross to the opposite side of the medulla in the **decussation of the medial lemnisci** to form the medial lemnisci. Because of this crossing, the left side of the brain is excited by sensory stimuli from the right side of the body and the right brain from the left body.

Also present in the medulla are (1) **nuclei for cranial nerves IX, X, XI,** and **XII,** which will be discussed more fully in the following chapter, and (2) the **reticular formation** that makes up a large share of the posterior and lateral medulla.

Some Special Functional Areas of the Reticular Formation in the Medulla and Pons. The reticular formation of the medulla and pons contains two especially important control centers:

1. The **vasomotor center** is comprised of widely dispersed nerve cells in the formation. This center transmits signals to the heart and blood vessels to increase heart pumping activity and to constrict the vessels. These effects acting together can increase the blood pressure greatly.
2. The **respiratory center** is comprised of nerve cells also located widely in the reticular formation. This is an automatic, rhythmically active center that causes the rhythmical respiratory muscle contractions necessary for inspiration and expiration.

In addition to these special centers, the medullary and pontine reticular formation also serves as an important relay station for signals coming from higher brain centers to control many other important internal functions of the body. For instance, signals from the hypothalamus are relayed down the spinal cord to control body temperature, sweating, secretion in the digestive tract, emptying of the urinary bladder, and many other bodily functions. Also, lying in close association with the posterior medullary reticular formation on each side of the midline is the **dorsal motor nucleus of the vagus** (Fig. 13–4). This nucleus relays signals into the vagus nerve to control heart rate, gastric secretion, peristalsis in the gastrointestinal tract, and other internal functions.

THE CEREBELLUM

The **cerebellum** is a large structure of the hindbrain (rhombencephalon) located inferior to the occipital lobe of the cerebrum and posterior to the brain stem. Its location in relation to the remainder of the brain is seen best in Figures 12–6 through 12–8 of the previous chapter, and its specific relationship to the brain stem is illustrated in Figure 13–5.

The cerebellum is an important part of the motor control system. Even though it is located far away from both the motor cortex and the basal ganglia, it interconnects with these through special nerve pathways, and it also interconnects with motor areas in both the reticular formation and the spinal cord. Its primary function is to determine the time sequence of

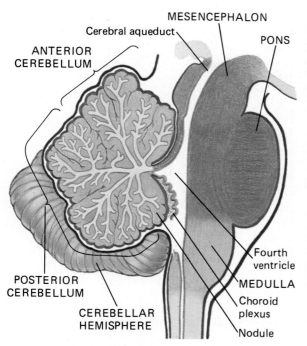

ANTERIOR CEREBELLUM

MESENCEPHALON
Cerebral aqueduct
PONS

POSTERIOR CEREBELLUM

CEREBELLAR HEMISPHERE

Fourth ventricle
MEDULLA
Choroid plexus
Nodule

BLUE: arachnoid mater
RED: pia mater
GREEN: ependyma

FIGURE 13–5 Relationship of the cerebellum to the brain stem.

contraction of different muscles during complex movements of parts of the body, especially when these movements occur extremely rapidly.

Surface Anatomy of the Cerebellum.
Shown in Figure 13–1 are three different cerebellar peduncles that lead from the posterior surface of each side of the brain stem to the cerebellum; these are the

1. **superior cerebellar peduncle,** which connects with the mesencephalon, as discussed earlier;

2. **middle cerebellar peduncle,** which connects with the pons; and
3. **inferior cerebellar peduncle,** which connects with the medulla.

It is through these that signals are transmitted into and away from the cerebellum.

Now, let us study the structure of the cerebellum itself as shown in Figures 13–5 and 13–6. Its major parts are the

1. **vermis,** which is a midline structure 1 to 2 cm wide extending around the entire cerebellum from anterior to posterior and from superior to inferior, and
2. two **cerebellar hemispheres,** located laterally on the two sides of the cerebellum.

Both the vermis and the cerebellum can also be divided into the

1. **anterior lobe of the cerebellum,** which is the superior and anterior third of the cerebellum anterior to the primary fissure; and
2. **middle lobe of the cerebellum,** which is the posterior and inferior two thirds of the cerebellum posterior to the primary fissure.

In the human being, the cerebellar hemispheres make up by far the greater portion of the cerebellar mass. These hemispheres function in concert with the cerebrum to coordinate voluntary movements of the body. The vermis, on the other hand, functions more for coordinating the stereotyped and subconscious body movements, operating mainly in association with the brain stem and spinal cord.

Internal Structure of the Cerebellum.
The internal structure of the cerebellum is illustrated in Figure 13–7. Like the cerebrum, it is comprised of three principal structures: (1) the

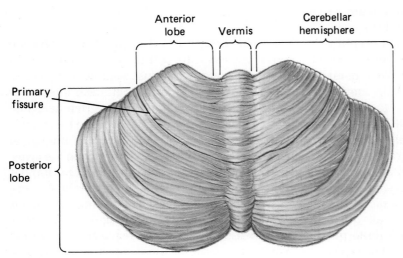

Anterior lobe
Vermis
Cerebellar hemisphere

Primary fissure

Posterior lobe

FIGURE 13–6 Superior view of the cerebellum.

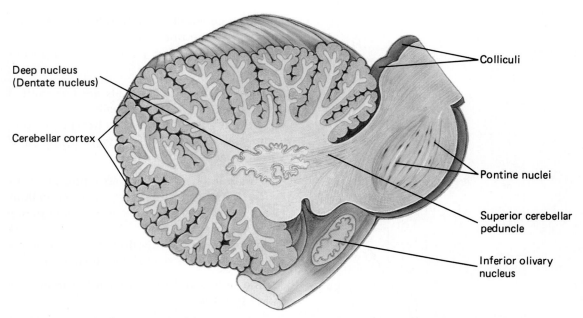

Deep nucleus
(Dentate nucleus)

Cerebellar cortex

Colliculi

Pontine nuclei

Superior cerebellar
peduncle

Inferior olivary
nucleus

FIGURE 13–7   The internal anatomy of the cerebellar hemisphere and related structures.

cerebellar cortex, (2) the **subcortical white matter** comprised almost entirely of nerve fibers, and (3) the **deep nuclei.**

The cerebellar cortex is a 3- to 5-mm thick sheet of nerve cells covering the entire cerebellar surface, containing a total of about 30 billion cells.

The deep nuclei are located in the center of the cerebellar white matter. By far the most prominent of the deep nuclei is the large **dentate nucleus,** which is located in the center of each cerebellar hemisphere. However, three other smaller deep nuclei are located in each side of the vermis: (1) the **fastigial nucleus,** (2) the **emboliform nucleus,** and (3) the **globose nucleus.**

The deep nuclei give rise to the nerve fibers that transmit signals out of the cerebellum to other parts of the nervous system, and the cerebellar cortex is a computing area that receives input information from the cerebral cortex, the basal ganglia, the spinal cord, and the peripheral muscles and integrates all of this to help coordinate the muscle movements.

Note also in Figures 13–5, 13–6, and 13–7 the many folds of the cerebellum called **folia.** If the cerebellar cortex were stretched out to eliminate these folia, it would be represented by a flat sheet about 40 cm in length and 8 cm wide. Each area of the cerebellum has almost exactly the same internal neuronal circuitry as all other areas, suggesting that the cerebellum performs almost exactly the same functions in all its sepa-

rate parts. What it does is to delay signals for short fractions of a second. Thus, when the motor system requires contraction of the biceps to start the arm moving and then contraction of the triceps to stop the movement, the cerebellum determines the appropriate delay before turning off the biceps and simultaneously turning on the triceps. When a person is performing rapid movements, such sequential muscle contractions occur in many separate parts of the body one after the other. Without the appropriate turn on, turn off sequence of these motor signals, the movements become totally uncoordinated, which is what happens when the cerebellum is destroyed.

THE SPINAL CORD

Figure 13–8 illustrates the **spinal cord** and the **spinal nerves** leaving it to be distributed to all parts of the body. Note that a spinal nerve leaves the cord on each side through each *intervertebral foramen* between adjacent vertebrae. Some of these nerves are very large because they innervate large areas of the body, such as the spinal nerves of the lower neck that innervate the arms, forearms, and hands and those of the lumbar and sacral regions that innervate the thighs, legs, and feet. In both of these areas, the cord itself is also enlarged because of the great number of nerve cell bodies that are required to relay the signals, giving rise to the **cervical enlargement**

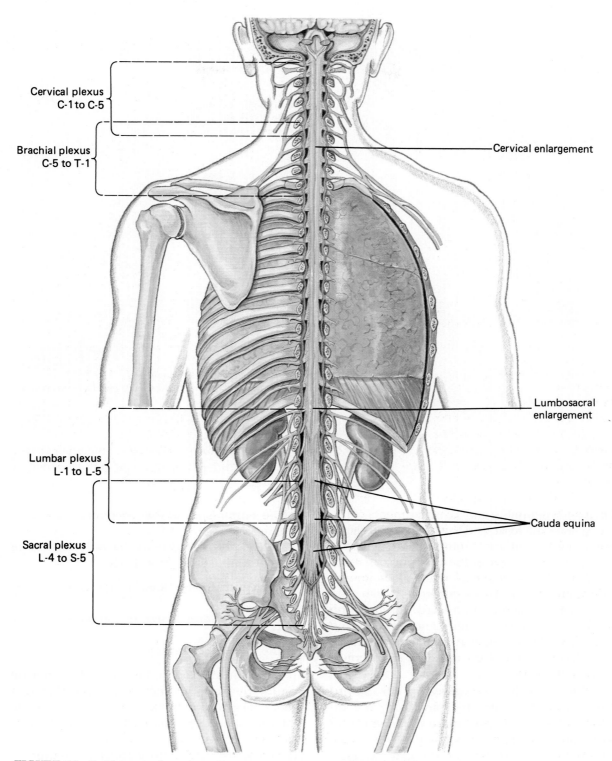

Cervical plexus
C-1 to C-5

Brachial plexus
C-5 to T-1

Cervical enlargement

Lumbosacral
enlargement

Lumbar plexus
L-1 to L-5

Cauda equina

Sacral plexus
L-4 to S-5

FIGURE 13—8 The spinal cord, its relationship to the peripheral nerves, and the spinal nerve plexuses.

of the cord in the lower half of the neck and the **lumbosacral enlargement** at the lower end of the spinal cord.

The spinal cord terminates approximately at the lower end of the second lumbar vertebra.

The reason for this is that during growth of the fetus and young child the spinal cord does not continue to lengthen as the vertebral column lengthens, so that the cord becomes located progressively more superiorly in the vertebral canal.

TABLE 13-2
The Spinal Cord

Structure	Function
Gray matter	
Dorsal horns	Loci of sensory input neurons
Lateral horns	Loci of preganglionic autonomic neurons
Ventral horns	Loci of motor neurons for skeletal muscles
White matter	
Propriospinal tracts	Signals between cord segments
Long motor tracts	
1. Lateral corticospinal	Motor signals from cortex to spinal cord
2. Ventral corticospinal	Same
3. Rubrospinal	Motor signals from brain stem to spinal cord; most are excitatory, a few are inhibitory
4. Reticulospinal	
5. Olivospinal	
6. Vestibulospinal	
7. Tectospinal	
Long sensory tracts	
1. Fasciculus gracilis and fasciculus cuneatus	Discriminatory sensory signals to gracile and cuneate nuclei, thence to thalamus in medial lemnisci
2. Ventral and lateral spinothalamic	Crude touch, pain, and temperature signals to brain stem and thalamus
3. Ventral and dorsal spinocerebellar	Proprioceptor sensory signals to cerebellum
4. Spino-olivary	Cord signals to inferior olivary nuclei, then relayed to cerebellum
Spinal nerve roots	
Dorsal	Sensory input
Ventral	Motor output to muscles and preganglionic output to autonomic nervous system

Yet, the lower lumbar and sacral segments of the spinal cord still exist, and the lumbar and sacral spinal nerves still arise from the cord but they arise from higher up in the vertebral canal because the levels of the cord segments no longer correspond to the levels of the vertebrae. The nerves then course downward through the lower canal as a large bundle of nerves called the **cauda equina,** and each of them finally emerges through its appropriate lumbar or sacral intervertebral foramen.

The Internal Structure of the Spinal Cord. Like the brain, the spinal cord is composed of areas of **gray matter** and areas of **white matter,** though the white matter is on the surface of the cord whereas the gray matter is deep. These are illustrated in the cross-sectional view of the cord in Figure 13–9. The *nerve cell bodies* are in the gray matter along with many short nerve fibers as well. But in the white matter, only *fiber tracts* and *glia* exist. Note in Figure 13–9 that the gray matter has the appearance of multiple horns connected by a crossbridge called the **gray commissure** between the two halves of the cord. Many fiber tracts also pass from one side of the cord to the other through

white commissures that accompany the gray commissure.

The horns of gray matter on each side of the cord are called respectively

1. the **ventral gray horn** (or the *anterior gray horn*),
2. the **dorsal gray horn** (or the *posterior gray horn*), and
3. the **lateral gray horn.**

It is in the ventral horn that the **anterior motor neurons** lie, the nerve cell bodies that send fibers through the spinal nerves to the muscles to cause muscle contraction. In the dorsal gray horn are the nerve cell bodies that receive sensory signals from the spinal nerves. In the lateral gray horn the nerve cells give rise to fibers that lead into the autonomic nervous system, the system that controls many of the internal organs.

Connections of the Spinal Nerves with the Spinal Cord. Note also in Figure 13–9 that each spinal nerve connects with the cord by way of two roots called the **dorsal root** and the **ventral root** (also called *posterior root* and *anterior root*). Each of these roots in turn enters or leaves

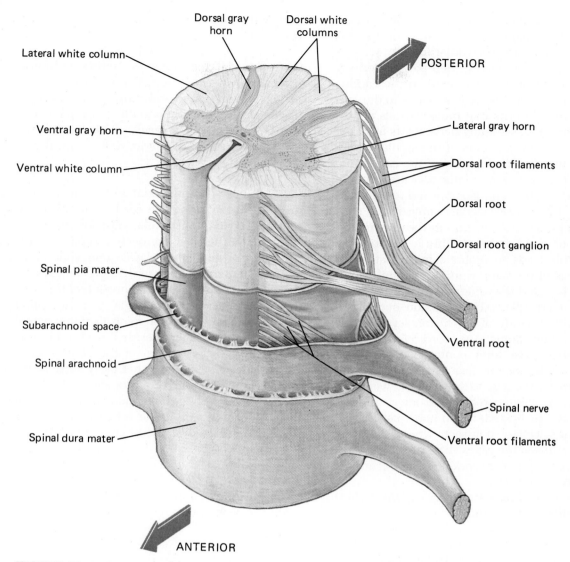

FIGURE 13–9 Structure of the spinal cord, and its connections with the spinal nerves by way of the dorsal and ventral spinal roots. Note also the coverings of the spinal cord, the meninges.

the cord by way of 7 to 10 small **root filaments.** The dorsal root is also called the *sensory root* because it carries only sensory fibers; and the anterior root is called the *motor root* because it carries only motor fibers leading from the cord to the muscles to cause muscle contraction or to the autonomic nervous system to control the activity of the internal organs. The nerve fibers in the ventral root originate from neurons in the ventral and lateral gray horns and then leave the cord along its anterolateral margin in the ventral root filaments. The dorsal root filaments pierce the cord along its posterolateral border, and its nerve fibers then turn either up or down the cord or enter the dorsal gray horn.

On the dorsal root is an enlargement called the **dorsal root ganglion.** This ganglion contains unipolar nerve cells that have no dendrites

but do have the usual single axon. However, immediately after the axon leaves the cell body it divides into two branches, a *peripheral branch* and a *central branch*. The peripheral branch passes through the peripheral portions of the spinal nerve to sensory receptors in the body, and the central branch passes into the spinal cord. About two thirds of the sensory fibers entering the cord terminate in the dorsal gray horn near the point of entry. The other third divides immediately into two branches, one branch terminating in the dorsal horn but the other branch passing all the way up the cord in the white matter and then terminating in the gracile and cuneate nuclei in the lower part of the medulla.

Function of the Spinal Cord Gray Matter. The gray matter of the spinal cord serves two functions. First, its synapses *relay signals* *between the periphery and the brain* in both directions. It is mainly in the dorsal horns that sensory signals are relayed from the sensory roots of the spinal nerves, then pass superiorly in the white matter of the cord to the various sensory areas of the brain. It is mainly in the ventral and lateral horns that motor signals are relayed from the descending nerve tracts from the brain into the motor roots of the spinal nerves.

Second, the gray matter of the cord functions to *integrate some motor activities*. For instance, when the hand is subjected to a painful stimulus, the sensory signals entering the cord cause an immediate reaction in the gray matter of the cord hand region. Within a fraction of a second this leads to motor signals that cause withdrawal of the hand from the painful stimulus. This is called the *withdrawal reflex* (or flexor reflex or pain reflex). This reaction occurs entirely independently of the higher levels of the nervous system. Some other cord reflexes are (1) reflexes that cause tonic contraction of the extensor muscles of the legs when one stands, thus allowing the legs to support the weight of the body; (2) scratch reflexes in lower animals when they are tickled; (3) stretch reflexes that cause muscles to contract when they are stretched (this is the reflex that causes the knee jerk when the patellar tendon is struck); and (4) even walking reflexes.

The Long Fiber Pathways of the Spinal Cord. Figures 13–9 and 13–10 show that the

cord white matter is also divided into columns. These are

1. two **dorsal** (or *posterior*) **white columns** lying between the dorsal gray horns,
2. two **lateral white columns** lying on each side of the cord lateral to the gray matter, and
3. two **ventral** (or *anterior*) **white columns,** lying between and anterior to the ventral gray horns.

All these columns contain fiber tracts that run lengthwise along the cord. Some of these, called **propriospinal tracts,** travel for only a few segments of the cord, connecting separate cord segments of gray matter with one another to help in the performance of the cord reflexes. (A "segment" of the cord is that portion of the cord that corresponds to a single pair of spinal nerves.) The propriospinal tracts lie immediately adjacent to the gray matter, as illustrated in Figure 13–10. The remainder of the white matter contains long fiber tracts that carry sensory information to the brain or motor signals from the brain to the cord. To the left in Figure 13–10 are the motor tracts, and to the right are the sensory tracts. These are described here:

Motor

1. **Lateral corticospinal tract,** from the motor cortex of the brain.

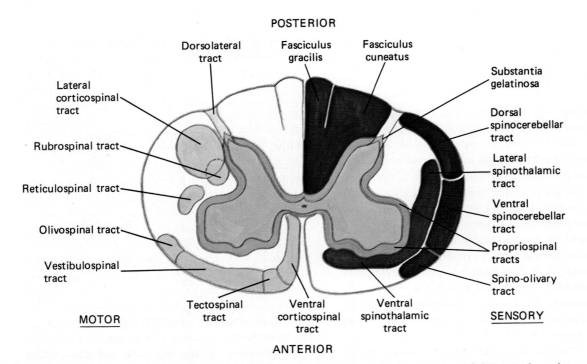

FIGURE 13–10   A cross-sectional view of the major long nerve fiber pathways of the spinal cord.

2. **Ventral corticospinal tract,** also from the motor cortex of the brain.
3. **Rubrospinal tract,** from the red nucleus of the mesencephalon.
4. **Reticulospinal tracts,** from the reticular substance of the mesencephalon, pons, and medulla.
5. **Olivospinal tract,** from the inferior olive of the medulla.
6. **Vestibulospinal tract,** from the vestibular nuclei of the medulla and pons.
7. **Tectospinal tract,** from the tectum of the mesencephalon.

Sensory

1. **Fasciculus gracilis** and **fasciculus cuneatus** (the two of which together make up most of the dorsal white columns), carrying signals directly from the spinal sensory roots all the way to the gracile and cuneate nuclei in the lower end of the medulla.
2. **Ventral** and **lateral spinothalamic tracts,** carrying signals relayed in the posterior gray horn, thence through the anterior white commissure, and finally upward through these tracts on the opposite side of the cord to the brain stem and thalamus.
3. **Ventral** and **dorsal spinocerebellar tracts,** which relay signals from the posterior gray horns upward to the cerebellum.
4. **Spino-olivary tract,** from the posterior gray horns of the cord to the inferior olive of the medulla.

Stimulation of most motor tracts causes either increased muscle tone or actual muscle contraction, and these tracts are said to be *excitatory tracts*; but stimulation of some tracts can decrease muscle tone, and they are called *inhibitory tracts*.

The sensory signals transmitted in the dorsal column pathways (the fasciculus gracilis and fasciculus cuneatus) are mainly those of fine, discriminatory touch that allow one to recognize the surface locations of sensory stimuli on the skin or the positions of the different parts of the body. The sensory signals transmitted in the spinothalamic tracts are those of crude touch, pain, and temperature. The sensory signals transmitted in the spinocerebellar tracts and also in the spino-olivary tract are mainly signals from the muscles and joints that apprise the cerebellum at all times about the movements and positions of different parts of the body so that the cerebellum can help in coordinating the body's movements.

THE CEREBROSPINAL FLUID SYSTEM—A PROTECTIVE FLUID FLOTATION SYSTEM FOR THE BRAIN AND SPINAL CORD

Even though the brain and spinal cord are crucial to the function of our bodies, nevertheless they are extremely delicate structures. For instance, the brain's tissues are so weak that one can push a finger all the way through it with almost no pressure, and the brain can literally be scooped out of the cranial vault with a spoon. Therefore, the brain requires a special protective system. This protection is achieved by the encasement of both the brain and the spinal cord in a rigid, bony vault comprised of the *cranial cavity* in the skull and the *vertebral canal* in the vertebral column. Within this vault, the brain and spinal cord actually "float" in a bath of fluid called the **cerebrospinal fluid.** This flotation system for the brain is illustrated in Figure 13–11. Now, let us describe this system and the mechanisms for maintaining the fluid in the system.

The Ventricular System in the Brain. In several cross-sectional views of the brain presented in the previous chapter and in the first part of this chapter, large fluid cavities called **ventricles** were shown deep inside the cerebrum, diencephalon, and brain stem. These ventricles, four in number, are illustrated in three-dimensional perspective in Figure 13–12. They are as follows:

1. and 2. The **two lateral ventricles.** One of these lies near the median plane in each cerebral hemisphere, extending all the way from the center of the frontal lobe anteriorly to the center of the occipital lobe posteriorly. From the parietal region of each of these ventricles an inferior extension turns laterally and anteriorly into the temporal lobe; this is called the **inferior horn of the lateral ventricle.**
3. The **third ventricle.** This lies between the two lateral halves of the thalamus and also extends anteriorly and inferiorly into the midline plane between the two halves of the hypothalamus.
4. The **fourth ventricle.** This lies in the lower brainstem in the space posterior to the pons and medulla but anterior to the cerebellum.

Now, let us study Figure 13–11 to see how the ventricles interconnect with each other. This figure shows the shadow of one of the lateral ventricles deep within a cerebral hemisphere

FIGURE 13–11   The cerebrospinal fluid system and the meningeal coverings of the brain and cord. Note the directions of flow of cerebrospinal fluid indicated by the arrows.

that connects by way of an **interventricular foramen** (also called *foramen of Monro*) with the anterolateral part of the third ventricle. In turn, the third ventricle connects posteriorly and inferiorly with the **cerebral aqueduct** (*aqueduct of Sylvius*), which is a small tube passing downward through the mesencephalon to enter the fourth ventricle lying behind the pontine and medullary regions of the brain stem. Finally, three openings occur in the outer wall of the fourth ventricle through which fluid can flow onto the surface of the brain. One of these is the **median aperture** (also called the *foramen of Magendie*) in the midline inferior to the cerebellum. The other two, the **lateral apertures** (also called the *foramina of Luschka*), are to the sides of the fourth ventricle.

The Fluid Space Surrounding the Brain and Spinal Cord (Subarachnoid Space), and the Meningeal Coverings of the Brain and Cord. Covering all surfaces of the brain and spinal cord is a thin fluid-filled space several milli-

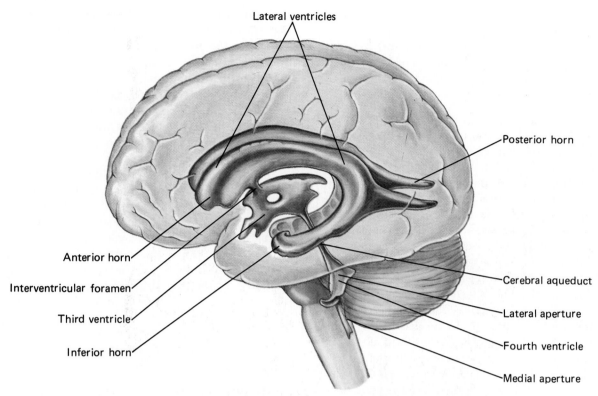

FIGURE 13–12 A three-dimensional representation of the ventricular system of the brain.

meters thick called the **subarachnoid space.** This space is bounded by the coverings of the brain and cord, called the **meninges,** which are illustrated for the spinal cord in Figure 13–9 and shown over a section of the brain in Figure 13–13. There are three layers of meninges:

1. The **dura mater** is a strong fibrous covering that surrounds the entire central nervous sys-

tem; it is bound to the inner surface of the skull but only loosely to the vertebral canal where there is a loose connective tissue space called the **epidural space.**

2. The **arachnoid** is a delicate structure loosely attached to the inner surface of the dura mater. And beneath the arachnoid is the fluid space that surrounds the brain and cord, the *subarachnoid space.* This space is penetrated

FIGURE 13–13 An expanded view of the meninges covering a section of the brain. Note also the venous sinus with arachnoid villi protruding into it.

by large numbers of small **arachnoidal trabeculae** that are part of the arachnoid.

3. The **pia mater** is a thin fibrous and vascular covering of the brain and cord, attached tightly to their surfaces, even dipping into all fissures and sulci.

The blood vessels that serve the brain have special relations to the meninges. First, note in Figure 13–11 the large **venous sinus,** the superior sagittal sinus, that extends along the midline the entire length of the cerebrum from front to back. This large sinus lies within layers of dura mater, and it has a triangular cross-sectional appearance as shown in Figure 13–13. Similar venous sinuses lie over other surfaces of the brain and also in the floor of the cranial cavity. All these sinuses interconnect and eventually give rise to the two internal jugular veins.

Next, note in Figures 13–11 and 13–13 the **arachnoid villi** that protrude into the venous sinuses. These are small penetrations of arachnoidal tissue that have made small openings in the walls of the sinuses. Cerebrospinal fluid can flow through these openings from the subarachnoid space into the venous blood. However, the villi function like valves to prevent blood from flowing backward from the venous sinuses into the subarachnoid space.

Finally, observe also in Figure 13–13 the large artery lying on the surface of the brain. Though this artery protrudes into the subarachnoid space, it is actually covered by the pia mater. Such arteries on the surface of the brain and penetrating branches from them provide nutrition to the brain.

Formation of Cerebrospinal Fluid by the Choroid Plexuses, and Fluid Flow Through the System. Most of cerebrospinal fluid is secreted by special secretory structures called **choroid plexuses** that protrude into each of the four ventricles, as shown in Figure 13–11. The most extensive choroid plexuses lie along the inferior surfaces of the lateral ventricles; therefore, most of the cerebrospinal fluid is formed in the lateral ventricles.

A small section of a choroid plexus is illustrated in Figure 13–14. It has a cauliflower-like growth, with large numbers of small capillary blood vessels embedded in loose connective tissue and covered by a thin layer of cuboidal cells that secrete fluid into the ventricle. The secreted cerebrospinal fluid is a clear, watery fluid that contains nearly the same constituents as the plasma portion of the blood except for the plasma proteins.

Once the fluid has been secreted by the choroid plexuses, it flows through the following pathway:

1. From the two lateral ventricles into the third ventricle by way of the two interventricular foramina.
2. From the third ventricle into the fourth ventricle by way of the cerebral aqueduct.
3. From the fourth ventricle into the subarachnoid space surrounding the brain stem through the median aperture and the two lateral apertures.
4. Through the subarachnoid spaces upward around the surfaces of the brain to the arachnoid villi.

FIGURE 13–14   A microscopic section of a choroid plexus.

5. From the subarachnoid spaces into the venous sinuses through the valvelike structures of the arachnoid villi.

The amount of cerebrospinal fluid formed each day is about 800 ml, and the pressure of this fluid in the cerebrospinal fluid system is about 10 mm Hg, only a very low pressure but enough to support the structures of the brain and spinal cord.

Hydrocephalus. Hydrocephalus is the condition in which there is too much fluid in part or all of the cerebrospinal fluid system of the brain. This can result from any one of several causes: (1) excess formation of fluid by the choroid plexuses, (2) obstruction of its flow at any one of several points in the flow system, or (3) obstruction of its reabsorption through the arachnoid villi into the venous sinuses.

A common form of hydrocephalus occurs in newborn babies caused by congenital obstruction of the cerebral aqueduct ("congenital" refers to a condition that develops during formation of the fetus). Even normally this aqueduct is a very narrow tubular passage through the mesencephalon. When obstruction occurs, all the fluid formed by the choroid plexuses in the two lateral ventricles and in the third ventricle has no escape route from these inner recesses of the brain. As a result, the lateral and third ventricles all become extremely enlarged, distending the brain and thinning the layers of the cerebrum. Obviously, this can cause serious mental disturbances if the condition becomes too severe. Also, in the newborn child with this type of hy-___ ___us, the flat bones of the skull still have ___ ___ach other so that, if the condi-___ ___ed, the skull bones are pushed ___ head becomes tremendous in ___s the size of a watermelon—

with commensurate ballooning of the brain as well. Fortunately, though, a surgical procedure can be performed to make an artificial opening from the ventricular system into the subarachnoid space, into one of the veins, or into one of the cavities of the body to allow escape of the fluid. If this is performed early enough in life, the ravages of the process can often be prevented.

QUESTIONS

1. Describe the surface anatomy of the mesencephalon and the relationship of the mesencephalon to the other parts of the brain.
2. Name the deep structures of the mesencephalon along with their functions.
3. Describe the pons, especially the neuronal pathway synapsing in the pons on its way to the cerebellum.
4. Discuss the relationship of the medulla oblongata to other parts of the central nervous system, and describe especially the vasomotor center, the respiratory center, and the dorsal motor nucleus of the vagus.
5. Give the parts of the cerebellum, and discuss the relationship of the cerebellum to the brain stem.
6. Describe the gross structure of the spinal cord and the connections between the cord and the spinal nerves.
7. Describe and discuss the gray matter of the spinal cord.
8. List the important motor and sensory long tracts in the white matter of the spinal cord.
9. Describe the ventricular system of the brain and the fluid-filled spaces around the brain.
10. Discuss the formation, flow, and absorption of the cerebrospinal fluid.
11. Describe the meninges.

REFERENCES

See References, Chapter 14.
See also General Anatomical References, Chapter 1.

14

Anatomy of the Nervous System: III. The Peripheral Nerves

Overview

The **cranial nerves** arise directly from the ventral surface of the brain and mainly innervate structures of the head. The **spinal nerves** originate from the spinal cord and innervate the somatic areas of the body (those areas below the head).

There are twelve pairs of cranial nerves:

I. The **olfactory nerves** and **olfactory tract** carry smell signals (olfactory signals) into the cerebrum.

II. The **optic nerve** carries visual signals into the diencephalon.

III, IV, and VI. The **oculomotor nerve**, **trochlear nerve**, and **abducens nerve** control the muscle that causes movements of the eyes.

V. The **trigeminal nerve** transmits sensory signals from the face and other anterior parts of the head into the brain stem and also transmits motor signals to control mastication.

VII. The **facial nerve** controls the superficial muscles of the face, mainly the muscles of facial expression but also the buccinator muscle of the cheek.

VIII. The **vestibulocochlear nerve** transmits auditory and equilibrium signals into the brain stem.

IX. The **glossopharyngeal nerve** carries sensory signals from the throat into the brain stem and motor signals from the brain stem to the pharynx mainly to cause swallowing.

X. The **vagus nerve** transmits sensory signals from the pharynx, thorax, and abdominal viscera and motor signals to control such internal bodily functions as heart activity, secretion by the stomach, and peristalsis in most of the digestive tract. The two vagus nerves constitute the major part of the *parasympathetic nervous system.*

XI. The **accessory nerve** controls several muscles in the neck and upper back.

XII. The **hypoglossal nerve** mainly controls the contractions of the tongue.

There are *31 pairs of spinal nerves*—8 cervical (C-1 through C-8), 12 thoracic (T-1 through T-12), 5 lumbar (L-1 through L-5), 5 sacral (S-1 through S-5), and 1 coccygeal (Co). After leaving the vertebral canal through the intervertebral foramina that open between the vertebrae, most of the spinal nerves form interconnecting networks called **plex-**

uses. The four important plexuses on each side of the body are listed here:

1. The **cervical plexus,** spinal nerves C-1 through C-5, supplies the neck, back of the head, and shoulders. In addition, a **phrenic nerve** arises from the cervical plexus on each side of the neck and then passes inferiorly through the thorax all the way to the diaphragm to control diaphragmatic respiratory contractions.
2. The **brachial plexus,** C-5 through T-1, supplies the shoulder area, arm, forearm, and hand mainly through the **musculocutaneous, radial, ulnar,** and **median nerves.**
3. The **lumbar plexus,** L-1 through L-4, supplies mainly the lower abdominal region and anterior and medial thigh. The most important nerves arising from the lumbar plexus are the **obturator** and **femoral nerves.**
4. The **sacral plexus,** L-4 through S-5, supplies the posterior and lateral hip, the entire posterior thigh, and almost the entire leg and foot. The most important nerves arising from this plexus are the **sciatic nerve** and its two major branches, the **tibial** and **common peroneal nerves.**

THE CRANIAL NERVES

In several figures of the last two chapters nerves were shown leaving the basal surfaces of the brain. These, called **cranial nerves,** are illustrated in more detail in Figure 14–1, and their connections both in the brain and peripherally are given in Table 14–1.

There are 12 pairs of cranial nerves, numbered in order of their origin on the basal surface of the brain from anterior to posterior, usually using Roman numerals. Each also has its own

FIGURE 14–1 Origin of the cranial nerves from the ventral surface of the brain.

TABLE 14–1
The Cranial Nerves

Nerve	Connection with brain	Function
I. Olfactory nerves and tract	Anterior ventral cerebrum	Sensory: from olfactory epithelium of s̶ cavity
II. Optic nerve	Lateral geniculate body of the thalamus	Sensory: from retinae of eyes
III. Oculomotor nerve	Mesencephalon	Motor: to four eye-movement muscles an̶ pebrae
		Parasympathetic: smooth muscle in eye
IV. Trochlear nerve	Mesencephalon	Motor: to one eye-movement muscle, oblique
V. Trigeminal nerve		
Ophthalmic branch	Pons	Sensory: from forehead, eye, superior n̶
Maxillary branch	Pons	Sensory: from inferior nasal cavity, face, mucosa of superior mouth
Mandibular branch	Pons	Sensory: from surfaces of jaw, lower teet̶ lower mouth, and anterior tongue
		Motor: to muscles of mastication
VI. Abducens	Pons	Motor: to one eye-movement muscle, th̶ rectus
VII. Facial	Junction pons and medulla	Motor: to facial muscles of expression an̶ cle, the buccinator
VIII. Vestibulocochlear		
Vestibular branch	Junction pons and medulla	Sensory: from equilibrium sensory organ̶ lar apparatus
Cochlear branch	Junction pons and medulla	Sensory: from auditory sensory organ, t̶
IX. Glossopharyngeal	Medulla	Sensory: from pharynx and posterior to̶ ing taste
		Motor: superior pharyngeal muscles
X. Vagus	Medulla	Sensory: much of viscera of thorax and
		Motor: larynx and middle and inferio̶ muscles
		Parasympathetic: heart, lungs, most of ̶ tem
XI. Accessory	Medulla and superior spinal segments	Motor: to several neck muscles, sternoc̶ and trapezius
XII. Hypoglossal	Medulla	Motor: to intrinsic and extrinsic musc̶

individual name, which also appears in the figure. These are the

    I. **olfactory nerves** and **tract,**
   II. **optic nerve,**
  III. **oculomotor nerve,**
  IV. **trochlear nerve,**
   V. **trigeminal nerve,**
  VI. **abducens nerve,**
 VII. **facial nerve,**
VIII. **vestibulocochlear nerve,**
  IX. **glossopharyngeal nerve,**
   X. **vagus nerve,**
  XI. **accessory nerve,** and
 XII. **hypoglossal nerve.**

CONNECTIONS OF THE CRANIAL NERVES INSIDE THE BRAIN

Note in Figure 14–1 that the tract to which the olfactory nerves connect arises from the cere-

brum, the optic nerve from the die̶ and all the remaining 10 from the Because of their special functional for smell and vision, we shall discu̶ nections of the olfactory nerves and the brain in Chapter 22 in relation to ̶ smell and of the optic nerve in Ch̶ relation to vision. The connecting a̶ other cranial nerves in the brain ste̶ trated in Figure 14–2.

    Some of the cranial nerves are e̶ *sory nerves,* some entirely *motor* (t̶ only innervate muscles to cause c̶ and some are combined nerves tha̶ sensory and motor components. To̶ Figure 14–2 are illustrated the **moto̶** which the different motor and combi̶ sensory nerves connect. To the right ̶ **sory nuclei.**

The Motor Nuclei of the B̶

Beginning from the top in Figure 14̶

MOTOR

SENSORY

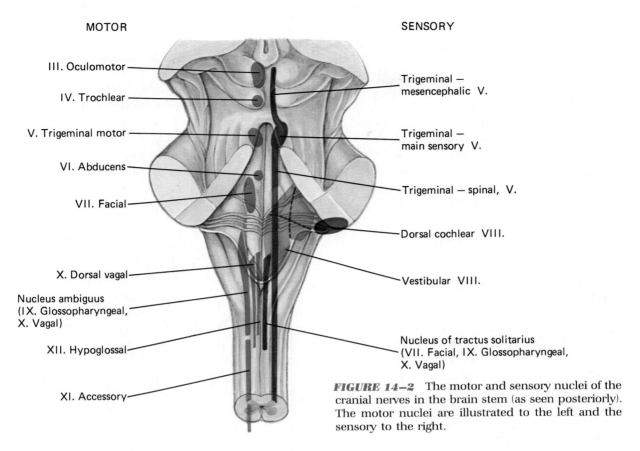

III. Oculomotor

IV. Trochlear

V. Trigeminal motor

VI. Abducens

VII. Facial

X. Dorsal vagal

Nucleus ambiguus
(IX. Glossopharyngeal,
X. Vagal)

XII. Hypoglossal

XI. Accessory

Trigeminal —
mesencephalic V.

Trigeminal —
main sensory V.

Trigeminal — spinal, V.

Dorsal cochlear VIII.

Vestibular VIII.

Nucleus of tractus solitarius
(VII. Facial, IX. Glossopharyngeal,
X. Vagal)

FIGURE 14–2 The motor and sensory nuclei of the cranial nerves in the brain stem (as seen posteriorly). The motor nuclei are illustrated to the left and the sensory to the right.

portant motor nuclei for the cranial nerves are as follows:

The **oculomotor, trochlear,** and **abducens nuclei** (nerves III, IV, and VI) send nerve fibers to the different muscles of the orbit for causing movement of the eye. The upper portion of the oculomotor nucleus is called the **Edinger-Westphal nucleus.** It controls the muscles inside the eye for focusing and for pupillary constriction.

The **trigeminal motor nucleus** (nerve V) controls the muscles of mastication (the muscles for chewing).

The **facial nucleus** (nerve VII) controls the many muscles of expression of the face.

The **dorsal vagal nucleus** (nerve X) is the important nucleus of the parasympathetic nervous system. It controls motor activity in many of the viscera, especially the heart (slowing of the heart), and of the upper digestive tract (increased peristalsis of the stomach and intestines and increased secretion).

The **nucleus ambiguus** sends signals through three different nerves, the glossopharyngeal, the vagus, and the accessory nerves (nerves IX, X, and XI). This nucleus controls such muscles as those for swallowing and the speech muscles of the larynx. The lower

end of the nucleus ambiguus is continuous with the anterior horn of the spinal cord from which signals are transmitted through the spinal roots of the accessory nerve to control portions of the trapezius and sternocleidomastoid muscles.

The **hypoglossal nucleus** (nerve XII) controls primarily the movements of the tongue.

THE SENSORY NUCLEI OF THE BRAIN STEM

To the right in Figure 14–2 are the sensory nuclei of the brain stem. From above downward these are the following:

The **trigeminal nuclei** (nerve V) extends all the way from the mesencephalon downward into the upper part of the spinal cord. It has three major divisions: the *main sensory nucleus*, which subserves principally the function of tactile sensation for the face, mouth, and scalp; the *mesencephalic nucleus*, which receives signals mainly from muscles and other deep structures of the head; and the *spinal nucleus*, which is the principal nucleus for receipt of pain signals from the face, mouth, and scalp.

The **cochlear nucleus** (part of nerve VIII) is the receptor area for sound signals from the ear.

The **vestibular nucleus** (the other part of nerve VIII) receives signals from the vestibular apparatus, the sensory organ for equilibrium.

The **nucleus of the tractus solitarius** is the principal nucleus for receipt of visceral sensory signals from such organs as the heart, stomach, special blood pressure receptors (the baroreceptors), and taste buds of the mouth. This nucleus receives signals through the facial, glossopharyngeal, and vagus nerves (nerves VII, IX, and X).

THE EXTERNAL DISTRIBUTIONS OF THE CRANIAL NERVES

The Olfactory Nerves and Olfactory Tract (I).   The olfactory nerves and olfactory tract are the sensory pathway for smell. The olfactory nerves are about 20 small nerves, each 1 to 2 cm in length, that arise from the olfactory epithelium, the sensory organ for smell located in the superiormost portion of the nasal cavity on the surfaces of the septum and superior concha. These 20 small olfactory nerves pass through an equal number of *foramina* in the *cribriform plate* of the ethmoid bone that forms the superior boundary of the nasal cavity and separates this cavity from the anterior fossa of the cranial cavity. Lying on the superior surface of the cribriform plate and inferior surface of the frontal lobe of the cerebrum is the **olfactory bulb,** and leading posteriorly from this is the **olfactory tract** that terminates in the **olfactory areas of the cerebrum** located in and between the inferior medial portions of the two temporal lobes. After the olfactory nerves pass through the cribriform plate, some of their fibers terminate at synapses in the olfactory bulb, whereas others continue posteriorly in the olfactory tract to terminate in the cerebral olfactory areas, which will be discussed in Chapter 22.

Optic Nerve (II).   Also illustrated in Figure 14–1 is the entire extent of the optic nerve. After leaving the eye, this nerve passes through the posterior recesses of the orbit, then through the *optic foramen* of the sphenoid bone, finally reaching the basal surface of the brain at the posteromedial limit of the frontal lobes. At this point the lateral half of the optic nerve continues posteriorly along the lateral surface of the hypothalamus, while the medial half abruptly bends medially and crosses to the opposite side through the **optic chiasm** lying anterior to the inferior hypothalamus. The crossed medial fibers then combine with the uncrossed lateral fibers on the opposite sides to form the two

optic tracts, which course posteriorl[...] lateral surfaces of the hypothalamus [...] nate in the **lateral geniculate bodies** [...] terior thalamus.

Oculomotor   Nerve   (III),   [...] Nerve (IV), and Abducens Nerve (V[...] are the nerves that control the eye m[...] Note in Figure 14–1 that the oculom[...] leaves the brain stem near the midline[...] terior surface of the mesencephalon. [...] lear nerve arises from the lower pos[...] surface of the mesencephalon and t[...] around its side to its anterior aspect. [...] cens nerve arises from the pons at its [...] junction. All these nerves then cours[...] the *superior orbital fissure* into the [...] innervate the extraocular and intrao[...] cles as illustrated in Figure 14–3. The[...] lar muscles attach to the eyes to cau[...] movements, and the intraocular musc[...] focusing of the eyes and constrict[...] pupil.

Trigeminal Nerve (V).   As illu[...] Figure 14–1, the trigeminal nerve arise[...] anterolateral surface of the midpons. [...] ately enlarges for a distance of 1 cm t[...] ter about two times its original diar[...] portion of the nerve is called the [...] **ganglion;** it contains the cell bod[...] nerve's sensory fibers and therefore is [...] to the dorsal root ganglia of the spi[...] Arising from the trigeminal ganglior[...] major branches of the nerve illustrate[...] 14–4:

1. the **ophthalmic division,**
2. the **maxillary division,** and
3. the **mandibular division.**

The ophthalmic and maxillary d[...] both entirely sensory. The ophtha[...] passes through the upper reaches [...] and branches onto the surface of th[...] superiorly over the forehead, supply[...] nerves to these areas of the face and s[...] the eye itself. It also sends branches [...] cavity and to the air sinuses.

The maxillary nerve leaves the [...] ity through the *foramen rotundu[...]* passes through the inferior orbit and [...] through a bony canal underneath the [...] tribute over the anterior and lateral [...] face to provide sensation. This nerv[...] plies sensation to the upper teeth, [...] portions of the oral mucosa, and the [...] the nasal cavity and nasopharynx. [...]

The mandibular nerve has bot[...] *division* and a *motor division.* This n[...]

FIGURE 14—3 The oculomotor nerve (III), trochlear nerve (IV) and abducens nerve (VI) innervating the eye muscles and also the internal structures of the eyeball through the ciliary ganglion.

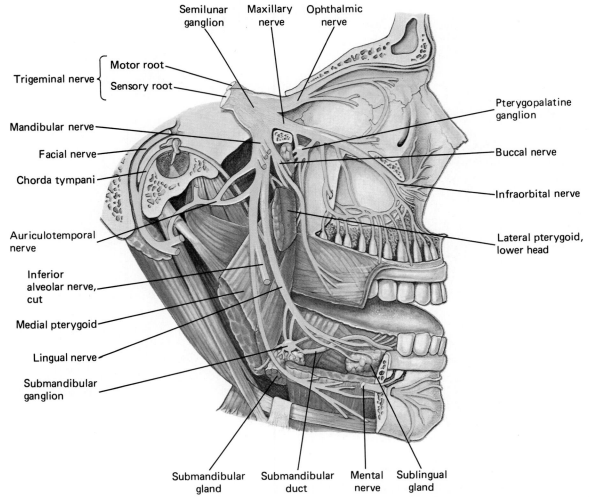

FIGURE 14—4 The trigeminal nerve (V) and its branches. Note also the chorda tympani that connects the lingual nerve (a branch of the trigeminal) to the facial nerve.

through the *foramen ovale* into the space anterior and inferior to the temporal bone and medial to the ramus of the mandible, a space called the **infratemporal fossa.** The sensory division provides sensation to the most lateral portions of the face, the outer surfaces of the lower jaw and chin, the lower teeth, and the lower portions of the oral mucosa including the anterior two thirds of the tongue. The motor division of the mandibular nerve innervates the muscles of mastication: the *temporalis*, the *masseter*, and the *medial* and *lateral pterygoid muscles*.

Facial Nerve (VII). The facial nerve arises from the brain stem at the posterolateral junction of the pons and the medulla (Fig. 14–1). It immediately passes through the *internal auditory meatus* and enters the *facial canal* in the temporal bone (shown in Fig. 14–4) and enters the posterior facial region anterior and inferior to the ear, as illustrated in Figure 14–5. It then

spreads through the superficial layers (tire lateral and anterior facial regions vate all the muscles of facial expressio as the buccinator muscle of the che early course anterior to the ear it passe or adjacent to the parotid gland, or glands for secretion of saliva. This gland sion becomes cancerous, in which ca cial nerve is often destroyed by the can the surgery required to remove the ca person then loses all capability of e expression on that side of the face. He comes unable to close his eye comple cannot keep his lips closed adequately side, and his cheek bulges outward w everytime he eats. This combination i tremely depressing and even debilitatir tion for the patient.

Note also in Figure 14–4 a bran facial nerve called the **chorda tymp**

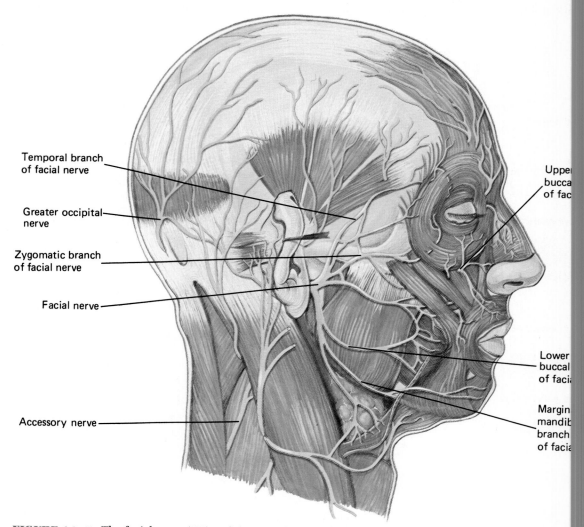

Temporal branch
of facial nerve

Greater occipital
nerve

Zygomatic branch
of facial nerve

Facial nerve

Accessory nerve

Uppe
bucca
of fac

Lower
buccal
of faci

Margin
mandib
branch
of facia

FIGURE 14–5 The facial nerve (VII) and the upper cervical portion of the accessory nerve (XI). Note the many branches of the facial nerve to the muscles of facial expression.

rough the middle ear and eventually s with the *lingual nerve,* one of the of the mandibular nerve. The fibers chorda tympani finally terminate (1) in **mandibular ganglion** from which en extend to the submandibular and l gland to control salivary secretion the anterior two thirds of the tongue to ste sensation. The sensory ganglion of nerve is called the *geniculate ganglion,* ocated in the facial canal.

bulocochlear Nerve (VIII). The ochlear nerve arises from the pons-

medullary junction just lateral to the facial nerve (Fig. 14–1). It is a short nerve that immediately enters the *internal auditory meatus* to innervate both the *vestibular apparatus* (the organ of equilibrium) and the *cochlea* (the organ of hearing). Both of these organs are contained within the petrous portion of the temporal bone itself.

Glossopharyngeal Nerve (IX). The glossopharyngeal nerve arises from the upper lateral border of the medulla (Fig. 14–1) and passes immediately from the cranial vault via the *jugular foramen* into the posterior pharyngeal region. Figure 14–6 illustrates this nerve leaving

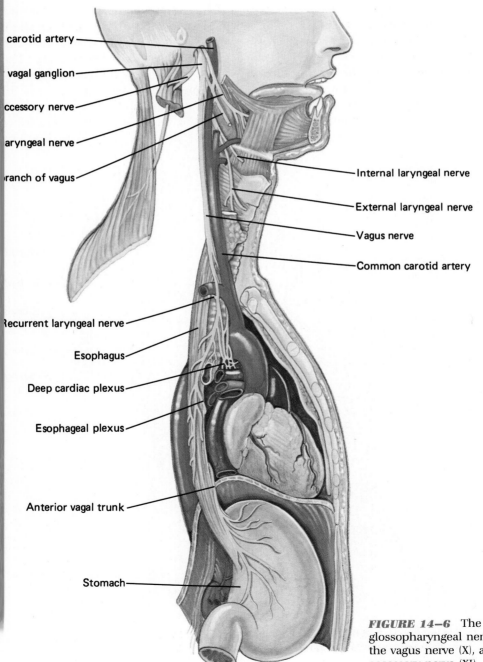

carotid artery

vagal ganglion

ccessory nerve

aryngeal nerve

ranch of vagus

Recurrent laryngeal nerve

Esophagus

Deep cardiac plexus

Esophageal plexus

Anterior vagal trunk

Stomach

Internal laryngeal nerve

External laryngeal nerve

Vagus nerve

Common carotid artery

FIGURE 14–6 The glossopharyngeal nerve (IX), the vagus nerve (X), and the accessory nerve (XI).

the cranial vault along with the vagus and accessory nerves. The glossopharyngeal nerve provides sensory innervation for the mucous membrane of the pharynx as well as for the posterior third of the tongue, including both general sensory and taste sensation from this area. A motor branch of the glossopharyngeal also innervates the superior pharyngeal muscles that are important for swallowing.

Vagus Nerve (X). The vagus nerve arises from the lateral border of the medulla (Fig. 14–1) inferior to the glossopharyngeal nerve. Its entry into the cervical region via the *jugular foramen,* along with the glossopharyngeal and accessory nerves, is illustrated in Figure 14–6. It then courses inferiorly into the thorax alongside the common carotid artery and internal jugular vein. Branches from the vagus nerves in the neck and upper thorax supply the muscles of the larynx for control of speech. At the superior border of the heart, parasympathetic nerve branches from the vagi, together with sympathetic branches from the thoracic sympathetic chains, form the **cardiac plexus** from which nerves then inner-

vate the heart. The distal portions of the vagus nerves continue inferiorly through the thorax alongside the esophagus and pass through the diaphragm to form **anterior** and **posterior gastric nerves.** These give parasympathetic innervation to the stomach, the entire small intestine, the proximal colon, and other viscera of the abdominal cavity. Thus, the vagus nerve carries the majority of the parasympathetic nerve fibers that help to control the internal organs of the body, such as those controlling heart rate, stomach secretion, intestinal peristalsis, and so forth. The distal colon and pelvic organs receive parasympathetic innervation via sacral spinal nerves, as we shall discuss later.

The vagus nerve also conducts sensory nerve fibers to the medulla from all the same visceral areas that receive vagal sympathetic fibers.

The Accessory Nerve (XI). The accessory nerve arises from the lateral border of the inferior medulla as well as from the anterolateral surface of the upper five segments of the spinal cord (Fig. 14–7). It leaves the cranial vault via the *jugular foramen* along with the glossopharyn-

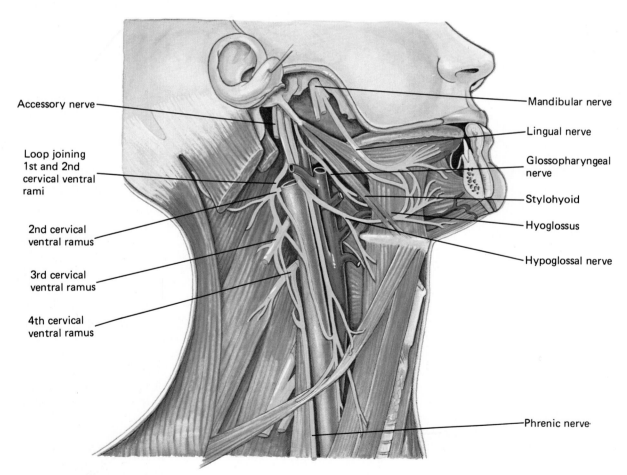

Accessory nerve

Loop joining
1st and 2nd
cervical ventral
rami

2nd cervical
ventral ramus

3rd cervical
ventral ramus

4th cervical
ventral ramus

Mandibular nerve

Lingual nerve

Glossopharyngeal
nerve

Stylohyoid

Hyoglossus

Hypoglossal nerve

Phrenic nerve

FIGURE 14–7 The lingual portion of the trigeminal nerve (V), the glossopharyngeal nerve (IX), the hypoglossal nerve (XII), and branches of the cervical plexus.

geal and vagus nerves, as illustrated in Figure 14–7. Some of its fibers then join the vagus nerve and innervate the muscles of the larynx and the pharynx, but all of the fibers from the spinal roots of the accessory nerve course downward along the posterolateral portion of the neck to innervate portions of the sternocleidomastoid and trapezius muscles as illustrated in Figure 14–6. These muscles also receive fibers from the cervical plexus in the neck region.

The Hypoglossal Nerve. The hypoglossal nerve arises from the lateral border of the lower medulla (Fig. 14–1) anterior to the origins of the vagus and accessory nerves. It leaves the skull via the *hypoglossal foramen.* Figure 14–7 illustrates the entry of the hypoglossal nerve into the inframandibular region of the neck and its distribution to all the muscles of the tongue, including the *hypoglossus, genioglossus, styloglossus,* and *intrinsic tongue muscles.*

THE SPINAL NERVES

The anatomy of the spinal cord was discussed in the previous chapter, as was also the origin of the spinal nerves from the cord. To recapitulate, there is one pair of spinal nerves for each vertebral segment of the cord, and these nerves leave the sides of the vertebral canal through the two **intervertebral foramina** between each two successive vertebrae. The purpose of the remainder of this chapter will be to describe the distributions of the peripheral extensions of the spinal nerves.

Referring again to Figure 13–8 in the previous chapter, showing an overview of the entire spinal cord and its spinal nerves, one can count

Eight pairs of cervical spinal nerves, specified as nerves C-1 through C-8,

Twelve pairs of thoracic spinal nerves, T-1 through T-12,

Five pairs of lumbar spinal nerves, L-1 through L-5,

Five pairs of sacral spinal nerves, S-1 through S-5, and

One pair of extremely small coccygeal spinal nerves, Co.

The thoracic spinal nerves (those of the chest region) are relatively small. However, branches from these control the deep back muscles as well as the very large latissimus dorsi, the "climbing" muscle of the arm. Also, the thoracic spinal nerves give rise to the **intercostal nerves** that course around the body inferior to the ribs to supply the intercostal muscles and also to provide cutaneous innervation for the chest and abdomen. Extensions of the lower intercostal nerves also supply most of the muscles of the anterior abdominal wall.

By contrast, the spinal nerves of the cervical, lumbar, and sacral regions are very large; it is these nerves that provide motor control and sensation for the neck region, the back of the head, the shoulders and upper extremities, the lower trunk, and the lower extremities. Note also that all of these nerves, shortly after leaving the vertebral canal, interconnect among themselves to form four major plexuses:

1. The **cervical plexus,** formed by spinal nerves C-1 through C-5, supplies the neck, the back of the head, portions of the shoulder, and the diaphragm.
2. The **brachial plexus,** formed by C-5 through T-1, supplies most of the shoulder region, the arm, the forearm, and the hand.
3. The **lumbar plexus,** from L-1 through L-4, supplies some muscles of the lower back, the lower abdomen, and the anterior and medial thigh.
4. The **sacral plexus,** from L-4 through S-5, supplies the gluteal region, the posterior and lateral thigh, the leg, and the foot.

Now, let us look at these plexuses individually to see some of the details of their organizations and distributions.

THE CERVICAL PLEXUS

Figure 14–8 illustrates the cervical plexus that originates mainly between C-1 and C-4 but also receives a small nerve bundle from C-5. Two upper branches of the cervical plexus, the **lesser occipital** and the **greater auricular nerves,** supply sensation to the back of the scalp and the region around the ear. Branching from the lower border of the plexus, several **supraclavicular nerves** supply sensation to the lower neck, and branching anteriorly is the **transverse cutaneous nerve** that provides sensation for the anterior neck. The muscles supplied by the cervical plexus are most of the deep neck muscles, the superficial anterior neck muscles, the levator scapulae, and portions of the trapezius and sternocleidomastoid.

The Phrenic Nerve. Note especially the origin of the **phrenic nerve** from the cervical plexus between C-3 and C-5. This is the principal nerve that controls respiration. The phrenic

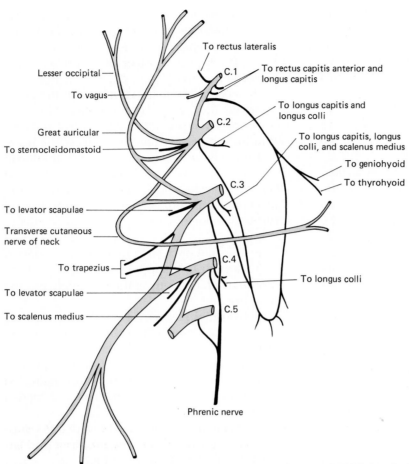

To rectus lateralis

Lesser occipital

To rectus capitis anterior and longus capitis

C.1

To vagus

To longus capitis and longus colli

C.2

Great auricular

To longus capitis, longus colli, and scalenus medius

To sternocleidomastoid

To geniohyoid

To thyrohyoid

C.3

To levator scapulae

Transverse cutaneous nerve of neck

To trapezius

C.4

To longus colli

To levator scapulae

To scalenus medius

C.5

Phrenic nerve

Supraclavicular

FIGURE 14–8 The cervical plexus and its branches.

nerves on the two sides course downward through the neck, then through the thorax on each side of the heart, and finally terminate in the diaphragm to control the breathing movements of this important respiratory muscle. Fractures of the cervical vertebrae, which occur especially in diving accidents, often crush the spinal cord. If the fracture occurs at the fifth through seventh cervical vertebra, as frequently is the case, the connections to the phrenic nerve will remain intact and the person can still breathe. However, all the other spinal nerves below this level will not receive appropriate signals from higher centers, and the entire body except for the neck muscles and this major breathing muscle, the diaphragm, will be paralyzed. This is the condition called *quadriplegia.*

THE BRACHIAL PLEXUS

Figure 14–9 illustrates the brachial plexus, arising from C-5 through T-1. These five spinal nerves are all very large. They unite to form **upper, middle,** and **lower trunks,** each of which splits into an **anterior** and a **posterior**

division. All the divisions pass under the clavicle and over the first rib into the axilla (the armpit) where they again fuse into three large bundles called the (1) **lateral cord,** (2) **posterior cord,** and (3) **medial cord.** All along the course of this plexus multiple nerves are given off to provide both motor and sensory innervation to the shoulder, the superior portion of the anterior and lateral thorax, the arm, the forearm, and the hand. Table 14–2 gives the principal nerve branches of this plexus and the muscles that they innervate. In addition, multiple cutaneous branches conduct sensory signals from skin areas that roughly overlie these muscles.

In the cervical region, several medium-sized nerves are given off from the brachial plexus to cause contraction of the shoulder and anterior thoracic muscles. These nerves are the: (1) **dorsal scapular,** (2) **long thoracic,** (3) **suprascapular,** (4) **subscapular,** and (5) **anterior thoracic.** The muscles that they innervate are given in Table 14–2.

At the termination of the brachial plexus in the axilla, the cords align themselves around the axillary artery and give rise to multiple branch

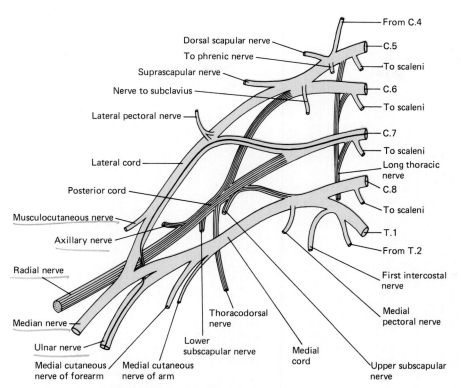

FIGURE 14–9 The brachial plexus and its branches.

nerves that extend into the upper extremity, the most important of which are

1. the **musculocutaneous nerve,**
2. the **radial nerve,**
3. the **median nerve,** and
4. the **ulnar nerve.**

Courses of the Major Nerves in the Upper Extremity. Figure 14–10 illustrates the courses of the four major nerves in the upper extremity. These supply both sensation to the skin and motor signals to the muscles along their courses.

The *musculocutaneous nerve,* upon leaving the plexus, curves laterally through the deep portions of the anterior arm and then continues superficially down the lateral surface of the forearm to provide sensory innervation. As it passes through the arm it innervates the anterior arm muscles listed in Table 14–2, the most important of which is the *biceps brachii,* which causes flexion of the forearm.

The *radial nerve,* after leaving the brachial plexus, curves posteriorly and laterally behind the humerus and enters the forearm over the lateral epicondyle of the humerus. Thereafter it follows mainly the lateral border of the radius and finally continues into the posterior portions of the thumb and first three fingers. Along its course in the arm, forearm, and dorsum of the

hand the radial nerve gives off branches to muscles and to the skin to provide cutaneous sensation. The list in Table 14–2 of muscles innervated by this nerve demonstrates its importance in controlling movements of the upper extremity. Careful study will show that they are the muscles in the posterior arm and dorsal and lateral aspects of the forearm and hand. The principal movements that they cause are (1) extension of the elbow, (2) supination of the forearm and hand, (3) extension of the wrist, fingers, and thumb, and (4) abduction of the thumb.

The *median nerve,* after leaving the brachial plexus, passes down the anteromedial portion of the arm, then distally in the anterolateral portions of the forearm, passing next into the lateral palm of the hand, and into the anterior compartments of the thumb and first two fingers and lateral half of the third finger. Again, along the course of this nerve, cutaneous branches give sensation to the skin. Table 14–2 shows that the median nerve also innervates approximately the lateral two thirds of the muscles in the anterior compartment of the forearm and lateral third of the anterior muscles of the hand. The major movements that these muscles cause are (1) pronation of the forearm and hand, (2) flexion of the wrist, fingers, and thumb, (3) abduction of the wrist, (4) abduction of the thumb, and (5) opponens motion of the thumb.

TABLE 14–2
Major Nerves from the Brachial Plexus, and the Muscles Innervated

Nerves	Spinal cord segments	Muscles
Dorsal scapular	C-5	Rhomboideus major
Long thoracic	C-5,6,7	Serratus anterior
Suprascapular	C-5,6	Supraspinatus Infraspinatus
Subscapular	C-5,6	Teres major Subscapularis
Anterior thoracic	C-5 through T-1	Pectoralis minor Pectoralis major
Musculocutaneous	C-5,6,7	Biceps brachii Coracobrachialis Brachialis
Radial	C-5 through T-1	Triceps brachii Brachialis Brachioradialis Supinator Extensor carpi radialis longus and brevis Extensor carpi ulnaris Extensor digitorum Extensor pollicis longus Extensor pollicis brevis Extensor indicis Abductor pollicis longus
Median	C-6 through T-1	Pronator teres Pronator quadratus Palmaris longus Flexor carpi radialis Flexor digitorum superficialis Flexor digitorum profundus (radial half) Flexor pollicis longus Flexor pollicis brevis (shared with ulnar nerve) Abductor pollicis brevis Opponens pollicis Lumbricals (on radial side of hand)
Ulnar	C-8, T-1	Flexor carpi ulnaris Flexor digitorum profundus (ulnar half) Flexor pollicis brevis (shared with median nerve) Flexor digiti minimi brevis Abductor digiti minimi Adductor pollicis Opponens digiti minimi Lumbricals (on ulnar side of hand) Interossei

NOTE: The above nerves are listed in the approximate order that they leave the brachial plexus.

The *ulnar nerve* passes down the postero-medial portion of the arm, then behind the medial epicondyle of the humerus at the elbow joint, and finally alongside the ulna to enter the medial border of the hand, supplying both the anterior and posterior surfaces of the little finger and the medial half of the third finger. Along the course of this nerve, cutaneous branches provide sensation for the anteromedial surface of the forearm and the surface of the hand medial to the midline of the third finger. Also, Table 14–2 shows that the ulnar nerve innervates approximately the medial third of the muscles in the anterior forearm and the medial two thirds of the muscles in the anterior hand. These muscles cause mainly (1) flexion of the wrist and fingers

Lateral pectoral nerve

Intercostobrachial nerve

Medial pectoral nerve

Long thoracic nerve

Musculocutaneous nerve

Medial cutaneous nerve of arm

Ulnar nerve

Biceps

Medial cutaneous nerve of forearm

Lateral cutaneous nerve of forearm

Median nerve

Radial nerve

Pronator teres { Deep head

Superficial head

Median nerve

Ulnar nerve

FIGURE 14–10 The courses of the musculocutaneous, radial, median, and ulnar nerves down the arm, forearm, and hand.

(shared functions with the median nerve), (2) abduction of the fingers, (3) adduction of the fingers and thumb, and (4) opponens motion of the little finger.

Though it is not important at this point to memorize the exact distributions of all these nerves, surgeons find it essential to know these distributions precisely. When a nerve injury occurs, the surgeon can determine which nerve is damaged as well as where it is damaged by studying the areas of sensory loss on the skin and the specific muscles that are paralyzed. One of the most common points for severe damage is in the brachial plexus itself. For instance, if the arm is pulled upward with tearing force, the medial cord of the brachial plexus is especially likely to be damaged. From Figures 14–9 and 14–10 one will see that this can completely sever the fibers to the ulnar nerve and can destroy many of the fibers to the median nerve as well, mainly paralyzing the muscles along the medial side of the anterior forearm and hand and also causing sensory loss along these surface areas.

THE LUMBAR PLEXUS AND THE FEMORAL NERVE

The lumbar spinal nerves between L-1 and L-4 and a small branch from T-12 form the lumbar plexus, illustrated in Figure 14–11. This plexus lies on the posterior wall of the lumbar region of the abdominal cavity, then sends branches inferiorly along the lateral wall of the pelvis. Near its origin, branches from the plexus innervate a few muscles in the abdominal and back regions, including the low back muscles, the psoas major, the quadratus lumborum, and the most inferior portions of the abdominal muscles. However, this plexus mainly sends nerves into the thigh, the three most important of which are as follows:

1. The **lateral femoral cutaneous nerve** (illustrated in Fig. 14–12A) enters the anterolateral thigh from underneath the inguinal ligament. This then passes downward along the lateral side of the thigh to provide sensory innervation to the skin.
2. The **obturator nerve** (Figs. 14–11 and 14–12A) originates from the lower portion of the lumbar plexus and enters the medial side of the thigh. This is mainly a motor nerve, controlling the large array of adductor muscles of the thigh that are listed in Table 14–3.
3. The **femoral nerve** (Figs. 14–11 and 14–12A) is by far the largest of the nerves from the lumbar plexus. Figure 14–12A shows that this nerve runs for a short distance parallel to the femoral artery in the upper thigh but divides into multiple large branches about 10 cm below the inguinal ligament. Some of these are muscular branches and some cutaneous. As shown in 14–12A, and also in Table 14–3, the **muscular branches** innervate all the muscles of the anterior thigh, the most important of which are all the four heads of the very large quadriceps femoris muscle and also the sartorius muscle. These are both the principal flexors of the thigh and the single massive extensor muscle for the leg. In addition to these muscular branches, there are two principal cutaneous branches: (1) the **anterior femoral cutaneous** that innervates the skin of the anteromedial thigh all the way to the knee and (2) the **saphenous nerve** that innervates the medial surfaces of the leg from the knee all the way to the foot.

THE SACRAL PLEXUS AND THE SCIATIC NERVE

The sacral plexus derives mainly from spinal nerves L-5 through S-3, but also from small branches of L-4 and S-4 to Co, as illustrated in

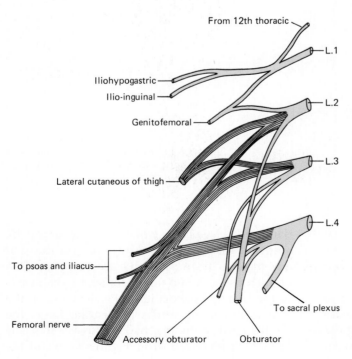

FIGURE 14–11   The lumbar plexus and its branches, especially the femoral nerve.

From 12th thoracic

L.1

Iliohypogastric

Ilio-inguinal

Genitofemoral

L.2

L.3

Lateral cutaneous of thigh

L.4

To psoas and iliacus

To sacral plexus

Femoral nerve

Accessory obturator

Obturator

Lateral cutaneous nerve of thigh

Femoral nerve

Intermediate cutaneous nerve of thigh

Medial cutaneous nerve of thigh

Saphenous nerve

Anterior branch of obturator nerve

Superficial peroneal nerve

Deep peroneal nerve

A

Gluteus maximus

Superior gluteal nerve

Pudendal nerve

Nerve to obturator internus

Posterior cutaneous nerve of thigh

Sciatic nerve

Hamstrings

Tibial nerve

Common peroneal nerve

Nerves to soleus and gastrocnemius

Tendon of biceps femoris

Nerve to popliteus

Sural nerve

Sural communicating branch

Tibial nerve

Medial calcanean nerve

B

FIGURE 14–12 The principal nerves of the lower limbs shown in anterior (A) and posterior (B) views.

TABLE 14–3
Major Nerves from the Lumbar and Sacral Plexuses, and the Muscles Innervated

Nerves	Spinal cord segments	Muscles
Lumbar plexus		
Obturator	L-2,3,4	Pectineus (shares with femoral nerve) Adductor longus Adductor magnus (shares with sciatic) Adductus brevis Gracilis

Femoral	L-2,3,4	Sartorius
		Iliacus
		Pectineus (shares with obturator nerve)
		Quadriceps femoris
		   1. Rectus femoris
		   2. Vastus medialis
		   3. Vastus lateralis
		   4. Vastus intermedius

Other muscular branches	L-2,3	Psoas major Quadratus lumborum
Sacral plexus		
Superior gluteal	L-4,5,S-1	Gluteus medius Gluteus minimus Tensor fasciae latae
Inferior gluteal	L-5,S-1,2	Gluteus maximus
Sciatic	L-4 through S-3	Adductor magnus (shares with obturator nerve) Obturator internus Superior gemellus Inferior gemellus Quadratus femoris
Tibial portion of the sciatic	L-4 through S-3	Biceps femoris (shares with peroneal nerve) Semitendinosus Semimembranosus Gastrocnemius Soleus Popliteus Tibialis posterior Flexor digitorum longus Flexor hallucis longus Plantar and medial foot muscles
Peroneal portion of the sciatic	L-4 through S-2	Biceps femoris (shares with tibial nerve) Tibialis anterior Peroneus longus Peroneus brevis Extensor digitorum longus Extensor hallucis longus Dorsal and lateral foot muscles
Pudendal	S-2,3,4	Muscles of urogenital triangle
Other muscular branches	S-3,4	Levator ani Coccygeus External anal sphincter

Figure 14–13. This plexus lies along the posterior wall of the pelvis. Its principal branches are the following:

1. The **superior** and **inferior gluteal nerves** exit laterally from the pelvis to control mainly the gluteal muscles of the buttock and lateral hip, which cause both extension and abduction at the hip joint.

2. The **posterior femoral cutaneous nerve** passes down the back of the thigh and upper portion of the leg.

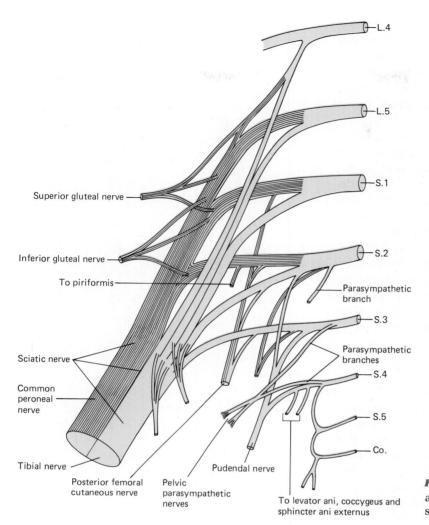

Superior gluteal nerve

Inferior gluteal nerve

To piriformis

Sciatic nerve

Common peroneal nerve

Tibial nerve

Posterior femoral cutaneous nerve

Pelvic parasympathetic nerves

Pudendal nerve

To levator ani, coccygeus and sphincter ani externus

L.4

L.5

S.1

S.2

Parasympathetic branch

S.3

Parasympathetic branches

S.4

S.5

Co.

FIGURE 14–13 The sacral plexus and its branches, especially the sciatic nerve.

3. The **pudendal nerve** passes to the perineum and to the external genital organs, including the penis and scrotum in the male and the vagina in the female, to subserve sexual functions and sensations.

4. **Pelvic parasympathetic nerve branches** derived from sacral spinal nerves S-2 through S-4 pass to the pelvic organs to initiate such functions as defecation (emptying the rectum) and micturition (emptying the bladder), as well as playing roles in the involuntary acts of erection, orgasm, and ejaculation during sexual intercourse.

5. Several small nerves derived from spinal nerves S-3 and S-4 control the voluntary muscle sphincters around the anus and around the external urethra. These allow the person to prevent defecation or micturition when these are inconvenient.

6. The very large **sciatic nerve** is so important that it deserves special consideration as follows.

The Sciatic Nerve. The sciatic nerve, illustrated in Figures 14–12B and 14–13, is by far the largest nerve of the body. It originates in the sacral plexus mainly from spinal segments L-5 to S-2, then leaves the posterior pelvis medial to the ischial tuberosity and courses distally in the posterior compartment of the thigh embedded between the hamstring muscles. Along this course, it supplies muscular branches to all the deep muscles posterior to the hip joint, listed in Table 14–3, and also the hamstring muscles. These muscles all cause extension of the thigh, and the hamstrings are also very strong flexors of the leg.

At the lower end of the thigh, immediately above the knee joint, the sciatic nerve divides into two major branches, the **tibial nerve** and the **common peroneal nerve.** The tibial nerve continues distally in the posterior compartment of the leg, lying in the interval between the tibia and the fibula. It finally enters the medial side of the foot behind the medial malleolus. In this course, it supplies sensory branches to the skin as well as branches to all the muscles of the back of the leg, especially to the soleus, the gastrocnemius, the tibialis posterior, and the flexors of the toes. The principal functions of these muscles are to plantar flex the foot and toes and also to invert the foot.

The common peroneal nerve wraps around the lateral side of the fibula, where it divides into the **superficial** and **deep peroneal nerves.** The superficial peroneal nerve descends in the lateral leg to provide motor innervation to the peroneus muscles and cutaneous innervation to the dorsum of the foot. These muscles are the everters of the foot. The deep peroneal nerve descends in the anterior compartment of the leg in relation to the anterolateral muscles (the tibialis anterior and the extensor muscles of the toes) and controls them. Their principal function is dorsiflexion of the foot.

The Dermatomes

Each spinal nerve provides sensory innervation to a "segmental field" of the skin called a dermatome. This is true even though most of the spinal nerves appear to become mixed up with other spinal nerves as they pass through the plexuses. The different dermatomes for the separate spinal nerves are illustrated in Figure 14–14. However, in this figure, the dermatomes are shown as if there were distinct borders between the adjacent dermatomes. This is only partly true because the distal branches of the nerves invade one another's territory. For this reason, an entire single spinal nerve can often be destroyed without significant loss of sensation in the skin, but when several adjacent spinal nerves are destroyed, one can easily establish the extent of sensory loss and from this determine the segmental level of the nerve injury.

Figure 14–14 shows that the anal region of the body lies in the dermatome of the most distal cord segments, S-4 and S-5. In the embryo, this is the tail region and the most distal portion of the body. The lower limbs develop from the lumbar and upper sacral levels of the embryo rather than from the distal sacral segments, which is also evident from the dermatomal map, for the dermatomes of this limb are L-2 through S-2.

Note also in Figure 14–14 that the face and anterior half of the head are not designated by spinal nerve dermatomes. But remember that sensation in these areas is served by the three branches of the fifth cranial nerve (the trigeminal nerve).

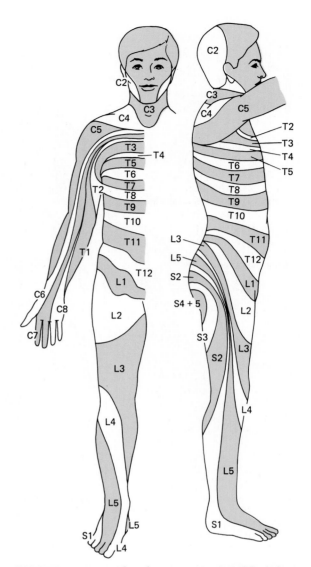

FIGURE 14–14 The dermatomes. (Modified from Grinker and Sahs: Neurology. Springfield, Ill., Charles C Thomas, 1966.)

THE AUTONOMIC NERVOUS SYSTEM

The autonomic nervous system is the collection of nerves that control most of the automatic subconscious functions of the body. It is so important to the function of almost every physiological system of the body that it will be discussed in detail in Chapter 19. However, at this point let us merely note that branches from all of the thoracic spinal nerves, the first three lumbar nerves, sacral nerves S-2 through S-5, and several cranial nerves—especially the vagus nerve—form the autonomic nervous system. It will become evident in Chapter 19 that this system has special properties that are uniquely different from those of all other portions of the peripheral nervous system.

QUESTIONS

1. Name the 12 cranial nerves, and give their respective Roman numerals.
2. Describe the points of exit of the separate cranial nerves from the cerebrum and brain stem.
3. Except for the olfactory and optic nerves, give the locus in the brain stem of the motor and sensory nuclei for each of the cranial nerves.
4. Describe the external distribution and motor and sensory functions of each of the cranial nerves.
5. How many pairs of spinal nerves connect with each of the major divisions of the spinal cord?
6. Describe briefly the origins and locations of the four major spinal nerve plexuses.
7. Describe the distributions of the nerves from the cervical plexus.
8. Describe the distributions of the nerves from the brachial plexus.
9. Describe the distributions of the nerves from the lumbar plexus.
10. Describe the distributions of the nerves from the sacral plexus.

REFERENCES

Anderson, J.E.: The cranial nerves. *In* Grant's Atlas of Anatomy. Baltimore, Williams & Wilkins, 1978, pp. 8-1 to 8-12.

Anderson, J.E.: The head. *In* Grant's Atlas of Anatomy. Baltimore, Williams & Wilkins, 1978, pp. 7-1 to 7-168.

Bloom, W., and Fawcett, D.W.: The nervous tissue. *In* A Textbook of Histology, 10th Ed. Philadelphia, W. B. Saunders, 1975, pp. 333–385.

Carpenter, M.D.: Human Neuroanatomy, 18th Ed. Baltimore, Williams & Wilkins, 1983.

Copenhaver, W.M., Kelly, D.E., and Wood, R.L.: Nervous system. *In* Bailey's Textbook of Histology, 17th Ed. Baltimore, Williams & Wilkins, 1978, pp. 290–357.

Figge, F.H.J.: The central nervous system. *In* Sobotta/Figge Atlas of Human Anatomy. Vol. III. Baltimore, Urban & Schwarzenberg, 1977, pp. 1–131.

Fujita, T., Tanaka, K., and Tokunaga, J.: Muscles, nerves, and brain. *In* SEM Atlas of Cells and Tissues. New York, Igaku-Shoin, 1981, pp. 312–328.

Goss, C.M.: The nervous system. *In* Gray's Anatomy of the Human Body, 29th Ed. Philadelphia, Lea & Febiger, 1973, pp. 777–1089.

Ham, A.W., and Cormack, D.H.: Nervous tissue. *In* Histology, 8th Ed. Philadelphia, Lippincott, 1979, pp. 483–539.

Hammersen, F.: Nervous system. *In* Sobotta/Hammersen Histology. Baltimore, Urban & Schwarzenberg, 1980, pp. 203–216.

Hammersen, F.: Nervous tissue and neuroglia. *In* Sobotta/Hammersen Histology. Baltimore, Urban & Schwarzenberg, 1980, pp. 80–93.

Langman, J., and Woerdeman, M.W.: Head and neck. *In* Atlas of Medical Anatomy. Philadelphia, W.B. Saunders, 1978, pp. 351–472.

Leeson, T.S., and Leeson, C.R.: Nervous tissue. *In* A Brief Atlas of Histology. Philadelphia, W. B. Saunders, 1979, pp. 89–104.

Leeson, T.S., and Leeson, C.R.: Nervous tissue. *In* Histology, 4th Ed. Philadelphia, W. B. Saunders, 1981, pp. 216–256.

Netter, F.H.: Nervous system. *In* The CIBA Collection of Medical Illustrations. Vol. 1, Summit, N.J., CIBA Medical Education Division, 1972.

Pernkopf, E.: Brain and meninges. *In* Atlas of Topographical and Applied Human Anatomy. Vol. I. Philadelphia, W.B. Saunders, 1980, pp. 29–135.

Snell, R.S.: Clinical Neuroanatomy. Boston, Little, Brown, 1980.

Williams, P.L., and Warwick, R.: The nervous system. *In* Gray's Anatomy, 36th British Edition. Philadelphia, W. B. Saunders, 1980, pp. 802–1215.

(See also General Anatomical References, Chapter 1.)

15

Functional Design of the Central Nervous System, the Synapse, and Basic Neuronal Circuits

Overview

The nervous system is comprised of three major functional subsystems: (1) a **sensory axis** that transmits signals from the peripheral sensory nerve endings to almost all parts of the spinal cord, brain stem, cerebellum, and cerebrum; (2) a **motor axis** that carries nerve signals originating in all the central areas of the nervous system to the muscles and glands throughout the body; and (3) an **integrative system** that analyzes the sensory information, stores some of it in memory to be used later, and utilizes both sensory and stored information to determine the appropriate responses.

Many of the *simpler nervous reactions are integrated in the spinal cord,* including such effects as withdrawal of any part of the body from a painful stimulus, reflexes that shorten the muscles wherever they are overstretched, and even signals that elicit walking movements under appropriate conditions. More complex nervous system reactions, such as control of both posture and equilibrium and also control of both respiration and circulatory function, are integrated at the *brain stem level.* The still more complex functions of the nervous system, such as the thought processes, storage of memory, determination of complex motor activities, and so forth, are all *integrated in the cerebrum.* The cerebellum operates in close association with all other parts of the central nervous system to help coordinate sequential motor functions.

The basic control unit of the nervous system is the **synapse**, where signals pass from the terminal nerve fibril of one neuron to the next neuron. The synapse is comprised of a *synaptic knob* at the end of the nerve fibril and the *membrane surface* of the following neuron where the knob rests. The synaptic knob secretes a *transmitter substance* that acts on the membrane and may be either *excitatory* or *inhibitory,* thus allowing incoming signals to cause either excitation or suppression of the next neuron. A neuron can usually be excited only by the simultaneous firing of large numbers of synapses; that is, signals from many separate synapses must *summate* before an action potential will occur in the stimulated neuron. If this neuron is also being inhibited by inhibitory synapses, then still more excitatory signals will be required to achieve a reaction.

The neurons of the central nervous system are organized into many different types of circuits, some of the most important of which are

1. The *diverging circuit,* which allows a signal from an incoming fiber to stimulate not one neuron but many neurons; each of these may then stimulate still many more neurons. For instance, stimulation of a single large neuron in the motor cortex of the cerebrum can sometimes cause as many as 15,000 muscle fibers to contract in one of the peripheral muscles.
2. The *converging circuit,* in which signals from multiple sources in the brain must act at the same time on the same neuron before it will fire. For instance, some of the neurons in the brain that help to analyze incoming signals require not only sensory signals from the periphery but also simultaneous control signals from other portions of the brain before a reaction will occur.
3. The *reverberatory circuit,* which is comprised of a series of neurons transmitting signals in a circular pathway. That is, one neuron stimulates another neuron, then another, and so forth until the signal returns to the first neuron. Thus, the signal passes around and around the circuit, or "reverberates," until the neurons *fatigue.* However, as long as the circuit does reverberate, it sends signals also out of the circuit to various other parts of the brain or to the muscular system to cause prolonged stimulation. For instance, the muscular contractions that cause respiration are the result of a complex pattern of reverberation in the respiratory center of the brain stem.

The brain is actually a computer that receives literally millions of bits of information from the different sensory organs and then integrates all of these to determine the response to be made by the body. Then it controls the muscles, the glands, and other functional systems of the body.

The purpose of this chapter is to present a general outline of the overall mechanisms by which the nervous system performs its functions.

FUNCTIONAL DESIGN OF THE NERVOUS SYSTEM

To perform its multiple functions, the nervous system is composed of three major functional parts: (1) the sensory system, (2) the motor system, and (3) the integrative system, each of which may be described as follows.

The Sensory System. Most activities of the nervous system originate with sensory experience, whether this be visual experience, auditory experience, tactile sensations from the surface of the body, or other sensations. The sensory experience can cause an immediate reaction, or a memory of the experience can be stored in the brain for many weeks or years and can then help to determine bodily actions at some future date.

Figure 15–1 illustrates the general plan of the sensory system, which transmits sensory information from the entire surface and deep structures of the body into the nervous system through the spinal and cranial nerves. This information is conducted into (1) the spinal cord at all levels, (2) the brain stem, including the medulla, pons, and mesencephalon, and (3) also into the higher regions of the brain, including the thalamus and the cerebral cortex. Then signals are relayed secondarily to essentially all other parts of the nervous system to begin analysis and processing of the sensory information.

The Motor System. The most important ultimate role of the nervous system is to control the bodily activities. This is achieved by controlling (1) contraction of skeletal muscles throughout the body, (2) contraction of smooth muscle

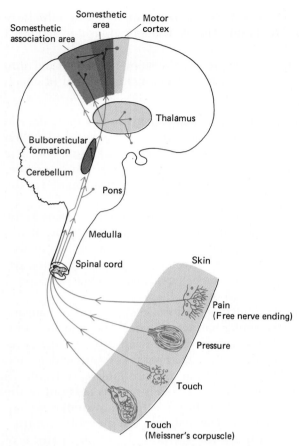

FIGURE 15-1 The sensory division of the nervous system.

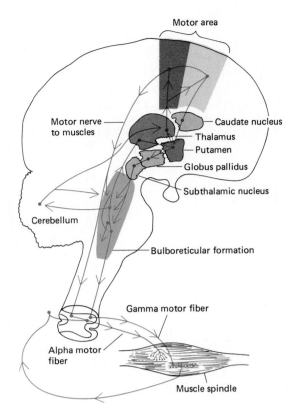

FIGURE 15-2 The motor axis of the nervous system.

in the internal organs, and (3) secretion by both exocrine and endocrine glands in many parts of the body. These activities are collectively called *motor functions* of the nervous system, and the portion of the nervous system that is directly concerned with transmitting signals to the muscles and glands is called the *motor division* of the nervous system.

Figure 15–2 illustrates the general plan of the motor axis for controlling skeletal muscle contraction. Signals originate in (1) the motor area of the cerebral cortex, (2) the basal regions of the brain, or (3) the spinal cord and are transmitted through motor nerves to the muscles. Each specific level of the nervous system plays its own special role in controlling bodily movements, the spinal cord and basal regions of the brain being concerned primarily with automatic responses of the body to sensory stimuli and the higher regions with deliberate movements controlled by the thought processes of the cerebrum.

The Integrative System. The term *integrative* means processing of information to determine the correct and appropriate motor action of the body or to provide abstract think-

ing. Located immediately adjacent to all sensory or motor centers in both the spinal cord and the brain are numerous centers concerned almost entirely with integrative processes. Some of these areas are concerned with the storage of information, which is called *memory*, whereas others assess sensory information to determine whether it is pleasant or unpleasant, painful or soothing, intense or weak, and so forth. It is in these regions that the appropriate motor responses to incoming sensory information are determined; once the determination is made, signals are then transmitted into the motor centers to cause the motor movements.

Reflexes

Many functions of the nervous system are the result of reflexes. A reflex is a motor response that occurs following a sensory stimulus, the response taking place through a *reflex arc* consisting of a *receptor*, a *nerve transmission network*, and an *effector*. A receptor is any type of sensory nerve ending that is capable of detecting one of the usual body sensations such as touch, pressure, smell, sight, or so forth. Once the sensation is detected, a signal is transmitted by the nerve transmission network, which is composed either of a single neuron or of several successive neurons connected in series or parallel with each other or both. Finally, the effector is a skeletal muscle or one of the internal organs such as the heart, gut, or a gland that can be controlled by the nerves.

Let us refer to Figures 15–1 and 15–2 to see how a simple reflex could occur. Assume for instance that one of the free nerve endings in the skin is stimulated by a pain stimulus and that the pain signal is transmitted into the spinal cord. On entering the cord it excites other neurons that eventually send signals back to appropriate muscles to cause withdrawal of that part of the body contacting the painful stimulus. Thus, if a person steps on a nail, his foot is automatically withdrawn as soon as a sensory signal can be transmitted to the spinal cord and then back to the muscles of the leg. This is called, very simply, the *withdrawal reflex*. In other words, many of our automatic muscle functions are controlled primarily by reflexes that are processed in the spinal cord and not by the conscious portions of the brain.

A much more complex reflex would be one in which many different sensory signals pass into the central nervous system from the eyes, ears, skin, and other portions of the sensory nervous system to apprise the person of danger. After a few seconds of integration, an automatic signal is transmitted back to the muscles to make the person run away. This is basically the same type of response as the simple withdrawal reflex except that it involves far more sensory elements, far more integrative elements, and far more motor elements. It also involves memories stored from previous learning, which make one realize the danger.

Thus, if one stretches his imagination far enough, he can explain almost any function of the nervous system on the basis of progressively more complex reflexes. However, most physiologists like to reserve the term reflex for an almost instantaneous automatic motor response to a sensory input signal and to call the more complex activities of the nervous system the *higher functions*.

Three Major Levels of Nervous System Function

In the development of man through the stages of evolution, the nervous system has progressed from a few simple nerve fibers connecting between different parts of the body, as occur in primitive multicellular animals, up to the very complex nervous system containing perhaps 100 billion neurons in the human being. As animalhood reached the multisegmental stage, nerve fibers and neuronal cell bodies aggregated to form a *neural axis* along the body, this neural axis providing the basis for the spinal cord of the human being, which is the *first* level of central nervous system function.

Next there developed in the head end of the axis enlarged aggregates of neurons that transmitted control signals down the neural axis to all parts of the body. These portions of the nervous system became highly developed in fishes, reptiles, and birds, and they are comparable to the brain stem and basal regions of the human cerebrum—the *second* level of nervous system function.

Finally, there bursts forth an overgrowth of nervous tissue surrounding the basal regions of the cerebrum, forming the cerebral cortex, which is the *third* level of function. This occurred primarily in mammals and has reached a tremendous level of development in the human being.

During each of these three stages of development, new functions of the central nervous system were established, and many of these have been inherited all the way from the very lowest animals to the human being. For instance, in the multisegmental worm, the most important nervous achievement was reflex response of the worm to noxious stimuli causing the worm's body to withdraw from any damaging influence. The human being has this same type of withdrawal reflex, as previously described, controlled almost entirely by the spinal cord—an inheritance carried all the way from the worm to the human being.

In the more complex animals, such as fishes and reptiles, it became important for the animal to be oriented properly in space, which initiated the process of equilibrium, and the human being has inherited this same ability. Furthermore, it is in the basal regions of the human brain, comparable to the highest levels of the fish's brain, where equilibrium is still controlled.

Finally, mammals have certain mental abilities over and above those of other animals, including especially the ability to store tremendous quantities of information in the form of memories and to utilize this information throughout life. Associated with this vast storage has also come the thinking process, a faculty that has reached its highest level of expression in the human brain. Along with the thinking process man has developed the monumental art of communication by voice, by sight, and in other ways. All these functions are primarily controlled by the cerebral cortex.

In summary, the three major levels of organization in the central nervous system are (1) the *spinal cord,* which controls many of the basic reflex patterns of the body; (2) the *basal regions of the brain,* which control most of the more complex automatic functions of the body such as equilibrium, eating, gross body movements, walking, breathing, and so forth, and finally (3) the cerebral cortex, which harbors our higher thought processes and controls our voluntary discrete motor activities.

TRANSMISSION OF NERVE SIGNALS FROM NEURON TO NEURON: FUNCTION OF THE SYNAPSE

The *synapse* is the junction between two neurons. It is through the synapse that signals are transmitted from one neuron to another. However, the synapse has the capability of transmitting some signals and refusing other signals, thereby making it a valuable tool of the nervous system for choosing which course of events to follow. It is because of this variable transmission of signals that the synapse is the most important single determinant of central nervous system function.

Physiologic Anatomy of the Synapse

Figure 15–3 illustrates a typical neuron, which is composed of three major parts: the **soma,** or main body of the neuron, and two types of projections, the **dendrites** and the **axon.** Each neuron has only one axon, and signals are transmitted *outward* from the neuron cell body through this axon. The axon may extend only a short distance before it branches into many terminal fibrils that excite neighboring neurons, or it may extend for as long as a meter—for instance, all the way from the spinal cord down to the feet to control the foot muscles. Most neurons have many dendrites, sometimes as many as several hundred, and these transmit signals *into* the neuron cell body. They are usually short projections that extend only a few millimeters in the nervous system where they pick up signals from the axons of other neurons.

Figure 15–3 also shows hundreds of small fibers leading from the axons of other neurons and terminating in small knobs called **synaptic knobs.** These lie on the surfaces of the soma and dendrites. It is these junctions between the synaptic knobs and dendrites or soma that are called the **synapses.** The average neuron in the

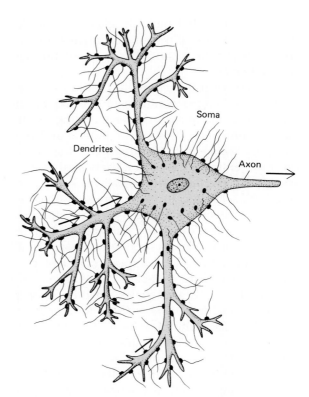

FIGURE 15–3 A typical motoneuron showing synaptic knobs on the neuronal soma and dendrites. Note also the single axon.

central nervous system has several thousand synapses, though this number can be as little as a few hundred on some of the small neurons or greater than 100,000 on some of the very large neurons of the cerebellum.

Transmission of Nerve Signals through the Synapse

Transmission of signals from the synaptic knobs to the dendrites or soma of the neuron occurs in very much the same way that transmission occurs at the neuromuscular junction as explained in Chapter 11. However, there is one major difference: At the neuromuscular junction the nerve endings always secrete acetylcholine, which in turn excites the muscle fiber. At the synapse some synaptic knobs secrete an *excitatory transmitter substance* and others secrete an *inhibitory transmitter substance;* therefore, some of these terminals excite the neuron and some inhibit it.

Excitation of the Neuron—the Excitatory Transmitter and the "Receptor." Figure 15–4 illustrates the structure of a typical synapse, showing a synaptic knob lying adjacent to the membrane of the soma of a neuron. The terminal has many small vesicles containing transmitter substance, and when a nerve impulse

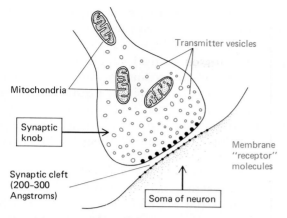

FIGURE 15-4 Physiologic anatomy of the synapse.

reaches the synaptic knob, momentary changes in the membrane structure of the knob allow a few of these vesicles to discharge the transmitter substance into the **synaptic cleft,** a narrow space between the knob and the membrane of the neuron. The transmitter substance then acts on a *receptor* in the membrane to cause excitation of the neuron if the transmitter is excitatory, or inhibition if it is inhibitory. The receptor is a large protein molecule that combines with the transmitter, and this combination in turn alters the membrane permeability, as we shall discuss shortly.

Chemical Nature of Excitatory Transmitters. In recent years it has been discovered that many different excitatory transmitters are secreted in different parts of the nervous system. Furthermore, they have varying characteristics of function, some providing prolonged stimulation of neurons and others providing very rapid and brief stimulation.

It should also be noted that, in general, all the synaptic knobs derived from a single neuron secrete the same type of transmitter substance. Thus, if a neuron located in one part of the brain sends branches of its axon to two separate brain areas, the transmitter substance secreted in both of these areas will still be the same.

One of the excitatory transmitters in the central nervous system is *acetylcholine,* the same excitatory transmitter that transmits signals from motor nerves into muscle fibers. However, a more complete listing of the most common excitatory transmitters includes

1. acetylcholine,
2. norepinephrine,
3. epinephrine,
4. glutamic acid,
5. substance P, and
6. enkephalins and endorphins.

THE MEMBRANE POTENTIAL OF THE NEURON, AND NEURONAL EXCITATION

As illustrated in Figure 15-5, the membrane potential of the central neuronal cell body (also called the *soma*) is much less negative than that of the peripheral nerve fiber, only -70 millivolts (mv) in comparison with -90 mv in the peripheral fiber. The reason for this is that the membrane of the soma is considerably more leaky to sodium ions than the peripheral fiber.

Figure 15-5 also illustrates the distribution of both sodium ions and potassium beween the extracellular fluid and intracellular fluid of the nerve cell body. Note that the concentrations of these ions on both sides of the membrane are approximately the same as those in peripheral nerve fibers, as discussed in Chapter 10. It is important to remember for the discussion in the next few sections that *the sodium concentration outside the membrane is about 10 times as great as inside the membrane*, but that *the potassium ion concentration difference is in exactly the opposite direction—about 35 times as much potassium inside the cell as outside*.

The Excitatory Postsynaptic Potential, and Membrane Excitation. The manner in which an excitatory transmitter excites a neuron can be explained by referring to Figures 15-6A and B. In Figure 15-6A one sees again that the typical intracellular neuronal potential is -70 mv. However, in Figure 15-6B large numbers of excitatory synapses are firing simultaneously and releasing excitatory transmitter into the clefts between the synaptic knobs and neuronal membrane. The transmitter combines with its specific transmitter receptor in the membrane, and in some unknown way this increases the permeability of the membrane. The increased permeability then allows sodium ions, which are in very high concentration in the extracellular fluid, to flow rapidly to the inside of

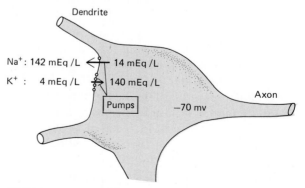

FIGURE 15-5 Distribution of sodium and potassium ions across the neuronal somal membrane; origin of the intrasomal membrane potential.

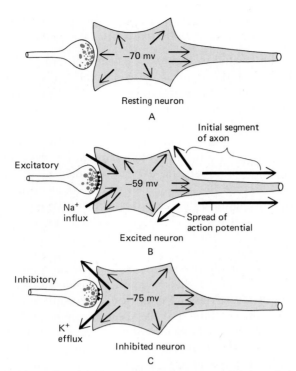

FIGURE 15–6 Three states of a neuron. (A) A resting neuron. (B) A neuron in an excited state, with increased intraneuronal potential caused by sodium influx. (C) A neuron in an inhibited state, with decreased intraneuronal membrane potential caused by potassium ion efflux.

the cell. Since sodium ions carry positive charges, the net result is an increase in positive charges on the inside and consequently an increase in the voltage inside the cell as well, sometimes an increase of as much as $+50$ mv if many excitatory synapses fire at once. This increase in potential is called the *excitatory postsynaptic potential.*

The Threshold for Excitation. When the excitatory postsynaptic potential becomes greater than a certain level (that is, when it *decreases the degree of negativity* inside the cell membrane below a critical level), an action potential is generated in the axon that extends outward from the neuronal soma. The critical membrane potential at which an action potential will occur is called the *threshold for excitation.* For typical neurons, this threshold averages about -59 mv. As illustrated in Figure 15–6B, when the intracellular potential falls below this level, an action potential is automatically generated in the axon. The threshold level of -59 mv is $+11$ mv more positive than the resting membrane potential of -70 mv. Therefore, ordinarily, a minimum excitatory postsynaptic potential of $+11$ mv is required to excite the average neuron, though some neurons are naturally more excitable than this and some less so.

"Summation" of Excitatory Postsynaptic Potentials. Almost invariably, stimulation of a single synaptic knob will not initiate an impulse in the axon. Instead, large numbers of synaptic knobs must become excited at the same time. Referring once again to Figure 15–3, it is readily apparent that hundreds of synaptic knobs could easily be excited simultaneously and that these operating in unison could cause neuronal discharge.

That is, if two knobs release their excitatory transmitter simultaneously, twice as much sodium enters the cell body, and twice as much excitatory postsynaptic potential develops. This is called *summation.* As more and more knobs fire simultaneously, the current flowing through the cell body becomes greater and greater, until finally it is great enough to excite the axon. If the postsynaptic potential increases still more, the rate of action potential firing also becomes faster and faster.

Spatial and Temporal Summation. Two different types of summation occur at the synapse in the same way that two types of summation occur in nerve and muscle fibers. These are spatial and temporal summation. *Spatial summation* means that two or more synaptic knobs fire simultaneously, thereby "summing" their individual effects on the excitatory postsynaptic potential. *Temporal summation* means that the same synaptic knobs fire two or more times in rapid succession, thus adding the effect of the second firing to that of the first before the first effect is over. The effect of the first discharge usually lasts about 15 milliseconds. Therefore, if two successive discharges of the same synaptic knob occur within less than $\frac{1}{70}$ second of each other, temporal summation occurs, and the closer together these discharges, the greater will be the degree of summation.

Repetitive Discharge of the Axon—Significance of the Threshold for Firing. Once sufficient excitatory transmitter has been secreted by the synaptic knobs to raise the excitatory postsynaptic potential above the threshold of the neuron, the axon will fire repetitively and will continue to do so as long as the potential remains above this threshold. Furthermore, as the postsynaptic potential rises higher and higher above the threshold, the more rapidly will the axon fire. For instance, let us assume that an excitatory postsynaptic potential of $+11$ mv is the threshold value. If the potential rises slightly above this, to 12 mv, one would expect the axon to discharge perhaps 5 to 10 times per second. But, if the postsynaptic potential rises to as high as $+33$ mv, 3 times the threshold value, then the axon perhaps would fire as many as 50 to 70 times per second.

FACILITATION AT THE SYNAPSE

When excitatory synaptic knobs discharge but fail to cause an action potential in the axon, the neuron still becomes *facilitated;* that is, even though an action potential does not result, the neuron becomes more excitable to impulses from other synaptic knobs. Let us assume, for instance, that 25 synaptic knobs must fire simultaneously to discharge a neuron. If 20 synaptic knobs fire simultaneously, the neuron will not discharge, but it does become sufficiently "facilitated" so that any 5 additional synaptic knobs anywhere on the surface of the cell could now elicit an impulse. This is the situation that develops in a "nervous" person, for large numbers of his neurons become facilitated though not excited; yet very few extraneous impulses can then cause terrific reactions.

RESPONSE CHARACTERISTICS OF DIFFERENT NEURONS

Some neurons are capable of transmitting as many as 1000 impulses per second over their axons, and some are capable of transmitting no more than 25 to 50 per second. Also, different neurons have different thresholds. All of these variations are fortunate, because neurons in different parts of the brain perform different functions.

Figure 15–7 illustrates typical *response patterns* of three different neurons. Neuron 1 requires a "threshold" postsynaptic potential of 5 mv to discharge it, and it reaches a maximum

discharge rate of 100 per second. The threshold of neuron 2 is 8 mv, and the maximum frequency of discharge is 35. Finally, neuron 3 has a high threshold of 18 mv, but once it becomes excited, its maximum discharge rate is 120 per second.

Inhibition at the Synapse

Some synaptic knobs secrete an *inhibitory transmitter* instead of an excitatory transmitter. The inhibitory transmitter depresses the neuron rather than exciting it. Indeed, sometimes when an excitatory transmitter from other synaptic knobs is exciting the neuron, subsequent stimulation of inhibitory synaptic knobs will actually stop all firing of the neuron. Some of the more common inhibitory transmitters are

1. gamma aminobutyric acid (GABA),
2. glycine,
3. dopamine, and
4. serotonin.

The inhibitory transmitter has an opposite effect on the synapse to that caused by the excitatory transmitter, usually creating a negative potential called the *inhibitory postsynaptic potential*. The manner in which the inhibitory potential is caused is illustrated in Figure 15–6C. The inhibitory transmitter increases the membrane permeability to potassium ions; and, because the concentration of potassium ions inside the membrane is about 35 times as great as outside, an excess of these ions flows outward, leaving a deficit of positive charges inside the neuron. Thus, increased negativity (the inhibitory postsynaptic potential) occurs inside the neuron. In Figure 15–6C the membrane potential is now -75 mv instead of the normal -70 mv; that is, the inhibitory postsynaptic potential is -5 mv.

An example of the way the inhibitory transmitter operates at the synapse is the following: Let us assume that the threshold for stimulation of a neuron is $+10$ mv and that excitatory synaptic knobs are generating a postsynaptic potential of $+20$ mv. The neuron will be firing repetitively at a rate of perhaps 50 to 200 times per second. At this time, stimulation of inhibitory synaptic knobs releases inhibitory transmitter, which creates an inhibitory postsynaptic potential of -15 mv. When this summates with the excitatory potential of $+20$ mv, the net value is only $+5$ mv, a level far below the threshold of the neuron. Therefore, the neuron stops discharging.

Excitatory and Inhibitory Neurons.
The central nervous system is made up of *excitatory neurons*, which secrete excitatory transmit-

FIGURE 15–7 Response characteristics of different types of neurons.

ter at their nerve endings, and of *inhibitory neurons*, which secrete inhibitory transmitter. Certain neuronal centers of the central nervous system are composed entirely of excitatory neurons, others entirely of inhibitory neurons, and still others of both excitatory and inhibitory neurons. Therefore, the central nervous system, unlike the peripheral sensory and motor systems, has two modes of activity, either excitation or inhibition, instead of simply excitation alone.

Some Special Characteristics of Synaptic Transmission

One-way Conduction at the Synapse. Impulses traveling over the soma or dendrites of the neuron cannot be transmitted backward through the synapses into the synaptic knobs. Thus, only one-way conduction occurs at the synapse. This is extremely important to the function of the nervous system, for it allows impulses to be channeled in the desired direction.

Fatigue of the Synapse. Transmission of impulses at the synapse is different from transmission in nerve fibers in a very important respect: The synapse usually *fatigues* very rapidly whereas nerve fibers fatigue hardly at all. Figure 15–8 illustrates this effect, showing that at the onset of an input signal, this particular neuron discharges very rapidly at first but then more and more slowly the longer it is stimulated. Some synapses fatigue very rapidly; others fatigue very slowly.

One might expect this phenomenon of fatigue to be an impediment to the action of the central nervous system, but, on the contrary, it is a necessary feature. Were it not for synaptic fatigue, a person could never stop a thought, a rhythmic muscular activity, or any other prolonged repetitive activity of the nervous system once it had begun. Fatigue of synapses is a means by which the central nervous system allows a nervous reaction to fade away to make way for others. Later in the chapter, specific instances of the importance of fatigue are pointed out.

The "Memory" Function of the Synapse. When large numbers of impulses pass through some synapses (but probably not all), these become "permanently" facilitated so that impulses from the same origin can pass through the synapses with greater ease at a later time. It is believed that this is the means by which memory occurs in the central nervous system. For instance, a given thought initiated by visual, auditory, or any other type of signal causes impulse transmission through given pathways in the brain. If the same thought is repeated over and over again, such as a thought caused by seeing the same view again and again, then the pathways for the thought become permanently facilitated so that subsequent impulses pass through these pathways with the greatest of ease. Therefore, impulses may later enter the same pathways from some source other than the visual apparatus, and the person "remembers" the scene rather than actually seeing it. This process of memory will be discussed in greater detail in Chapter 20.

Other Factors That Affect Synaptic Transmission. Figure 15–9 illustrates the effect of several different factors on the number of impulses transmitted by a synapse in response to a given degree of stimulation of the synaptic knobs. Note that *hypnotics*, *anesthetics*, and *acidosis* all depress the transmission of impulses at the synapse, whereas *alkalosis* and *mental stim-*

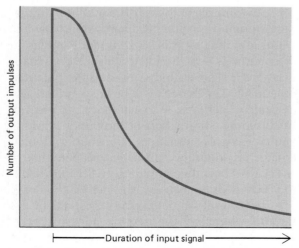

FIGURE 15–8 Effect of fatigue on the number of output impulses from the neuron after it begins to be stimulated.

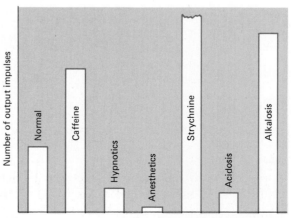

FIGURE 15–9 Effect of different drugs and different physiologic states on the excitability of a neuron.

ulants such as *caffeine, benzedrine,* and *strychnine* all greatly facilitate synaptic transmission. Strychnine facilitates synaptic transmission by inhibiting some of the inhibitory synapses, and when given in sufficiently high dosage, it will cause the neurons to discharge spontaneously even in the absence of a synaptic stimulus. It is in this way that strychnine kills an animal, causing so many neuronal impulses to be transmitted throughout the central nervous system and into the motor system that it causes death by spasm of the respiratory muscles.

The Integrative Function of the Neuron. To summarize the overall function of the neuron, one can call it an "integrator," which means a type of calculator that collects information and sums it all together. Signals reach the neuron by way of excitatory and inhibitory synapses that, in turn, are excited by neurons from other parts of the nervous system. If the resultant sum of all the excitatory and inhibitory effects is above the threshold for excitation, the neuron fires. Some of the neurons fatigue rapidly, others slowly. Some neurons have high thresholds; others have low thresholds. Some fire at rapid rates, others at slow rates. The varying characteristics of the different neurons and their different connections in the nervous system allow the neurons in one portion of the nervous system to control one function of the body and those in another portion to control another function. They allow the sorting of signals to determine their meanings, the performance of special skilled motions, the thinking of specific thoughts, and the modification of these thoughts by signals arriving from other parts of the nervous system.

BASIC NEURONAL CIRCUITS

The Neuronal Pool

The central nervous system is divided into many different anatomic parts, in each of which are located accumulations of neurons called *neuronal pools.* An important feature of the different neuronal pools is that each has a pattern of organization different from all the others. That is, the distribution of nerve fibers within the pool, the number of incoming nerve fibers, the number of outgoing fibers, the types of neurons in the pool, and many other features differ from one pool to another. Each pool is organized to perform a specific function. The purpose of the present section is to discuss, first, the general functions of the neuronal pool, and, second, the characteristics of special types of pools.

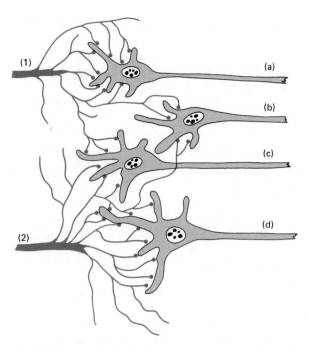

FIGURE 15–10 Schematic organization of four neurons in a neuronal pool.

Synaptic Connections in a Neuronal Pool. A neuronal pool is composed of thousands to millions of neuronal cell bodies. Figure 15–10 shows four typical neurons in a pool stimulated by two incoming nerve fibers. Note that each incoming fiber branches to supply synaptic knobs to several different neurons in the pool. A fiber may deliver one synaptic knob to a given neuron or as many as several hundred. Therefore, when the fiber is stimulated, it will facilitate some neurons and excite others.

Areas of Discharge and Facilitation in the Neuronal Pool. While still observing Figure 15–10, let us expand our imaginations until this pool of four neurons becomes a thousand closely packed neurons with hundreds of thousands of terminal nerve fibrils and synaptic knobs. Near the central point of entry of a fiber to the pool, the fiber usually branches profusely to form many synaptic knobs; this is illustrated in Figure 15–11. But farther away from the center, the number of knobs becomes less and less.

FIGURE 15–11 "Discharge" and "facilitated" zones of a neuronal pool.

Those neurons that lie in the center of the fiber's "field" are usually supplied with enough synaptic knobs that they discharge each time the input fiber is stimulated. Thus, the dark area in Figure 15–11 is called the *discharge zone.*

In the peripheral portion of the fiber field, the number of terminal fibrils ending on any single neuron is usually too few to cause discharge but, nevertheless, still enough to cause facilitation. Therefore, this area is called the *facilitated zone.* When a neuron is facilitated, a stimulus from some other source can excite it more easily.

Simple Circuits in Neuronal Pools

The simplest circuit in a neuronal pool is that in which one incoming nerve fiber stimulates one outgoing fiber. This type of circuit does not exist in a precise form, though occasionally it is approximated. For instance, in transmitting certain types of sensory signals from peripheral nerves into the brain, the incoming impulses into a neuronal pool occasionally cause almost the same number of outgoing impulses.

Such a circuit obviously acts simply as a relay station, relaying on to additional neurons essentially the same information that enters it.

Diverging Circuits. Figure 15–12 illustrates two types of *diverging circuits.* In the first of these, a single incoming fiber progressively stimulates more fibers farther and farther along the pathway. This can also be called an *amplifying circuit* because an input signal from a single nerve fiber causes an output signal in many different fibers. This type of circuit is exemplified by the system for control of skeletal muscles: Under appropriate conditions, stimulation of a single motor cell in the brain sends a signal down to the spinal cord to stimulate perhaps as many as 100 anterior horn cells, and each of these in turn stimulates approximately 150 muscle fibers. Thus, a single motor neuron of the

brain sometimes can stimulate as many as 15,000 muscle fibers.

The second diverging circuit in Figure 15–12 allows signals from an incoming pathway to be transmitted into separate pathways, the same information being relayed in different directions at the same time. This type of circuit is common in the sensory nervous system. For instance, when a limb moves, the sensory information from the joints and muscles caused by the movement is transmitted by such a circuit into (1) neuronal pools of the spinal cord, (2) neuronal pools of the cerebellum, and (3) neuronal pools of the thalamus.

Converging Circuits. Figure 15–13 illustrates two types of *converging circuits*, which are opposites of the diverging circuits. In the figure to the left, different nerve fibers from the same source converge on a single output neuron causing especially strong stimulation.

To the right is another important type of converging circuit in which input fibers from several different sources converge on the same output neuron. This type of circuit allows signals from many different sources to cause the same effect. Thus, the smell of a pigpen might make a person think of a pig; the sight of a pig, the sound of his grunting, touching the pig, or eating pork chops might also lead to the same thought.

Use of Converging Circuits to Perform Complex Functions. Converging circuits can provide functions that are much more complex than can single neurons, for in a converging circuit there may be cells that respond to the incoming signal in several different ways at the same time. For instance, one of the neurons might have a very high threshold but when once stimulated might be so powerful that it causes a very intense reaction. In other parts of the circuit there may be neurons that have low thresholds but when stimulated transmit only weak signals. The output signal from the pool, therefore,

Divergence in same tract

A

Divergence into multiple tracts

B

FIGURE 15–12   Diverging circuits.

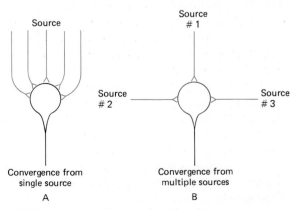

Source

Source
1

Source
2

Source
3

Convergence from
single source

A

Convergence from
multiple sources

B

FIGURE 15–13   Converging circuits.

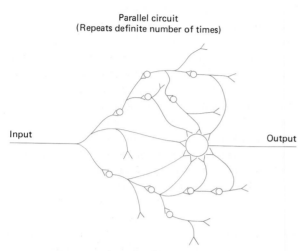

FIGURE 15—14 Reverberating and parallel circuits for repetitive discharge.

might normally be very weak but then suddenly become extremely powerful if a higher threshold of input stimulation should be exceeded. Thus, the converging circuit can be organized to perform almost any type of selective function. The individual neurons of the circuit select which signals shall pass, but the arrangement of the fibers and the combinations of cells determine which incoming signals will have the greater effects, and whether the effects will be excitatory or inhibitory.

Repetitive Circuits. **The Reverberating Circuit.** Figure 15—14 shows two different types of circuits which, when stimulated only once, will cause the output cell to transmit a series of impulses. The upper circuit of this figure is a *reverberating circuit*, which functions as follows: An incoming signal stimulates the first neuron, which then stimulates the second and third. However, branches return to the first neuron and restimulate it. As a result, the signal travels once again through the chain of neurons, this process continuing around and around the circuit indefinitely until one or more of the neurons fail to fire. The usual cause of failure is synaptic fatigue, and until it takes place, the output neuron continues to be stimulated every time the signal goes around the circuit, giving a continuous output signal.

The reverberating circuit is the basis of innumerable central nervous system activities, for it allows a single input signal to elicit a response lasting a few seconds, minutes, or hours. Indeed,

the lifelong respiratory rhythm is probably caused by a reverberating circuit that continues to reverberate without fatiguing as long as the person remains alive. It is probable, also, that a special circuit in the brain causes a person to awaken when it reverberates, and allows him to relax into a state of sleep when it stops reverberating. Almost all rhythmic muscular activities, including the rhythmic movements of walking, are mainly controlled by reverberating circuits.

A few moments of thought about the reverberating type of circuit will emphasize the extreme number of variable functions it can perform. For instance, the number of series neurons in the circuit may be great or small. If the number is great, the length of time required for the signal to go around the circuit will be long; if the number is small, the reverberatory period will be short. As a result, the output signal may repeat itself rapidly or slowly. Furthermore, more than one of the neurons in the circuit can give off output signals. One of the cells in the circuit, for instance, might cause an arm to move upward, then another cell a fraction of a second later in the circuit might cause the arm to move to the right, another cell still later cause it to move downward, and another cell cause it to move to the left. As the reverberatory cycle repeats itself once more, the same motions occur again, causing the arm to move around and around continuously until the cycle stops. When the person wishes to perform this motion, all he has to do is set off the reverberatory cycle.

The Parallel Circuit. The lower part of Figure 15—14 illustrates the *parallel* type of repetitive circuit in which a single input signal stimulates a sequence of neurons that send separate nerve fibers directly to a common output cell. Because a delay of about $\frac{1}{2000}$ second occurs each time an impulse crosses a synapse, the impulse from the first neuron arrives at the output cell $\frac{1}{2000}$ second ahead of the impulse from the second neuron, and impulses continue to arrive at these short intervals until all the neurons have been stimulated. Since there is no feedback mechanism in this circuit, the repetition then ceases entirely.

The parallel type of circuit has certain advantages over the reverberating circuit. For instance, synaptic fatigue is a major factor in determining how long the reverberating circuit will continue to fire, whereas the output duration of the parallel circuit is mainly independent of fatigue. For performing very exact activities such as mathematical calculations, the parallel circuit is perhaps very useful; for rhythmic functions or for greatly prolonged discharges the reverberat-

FIGURE 15–15 Characteristics of signals caused by (1) a single nerve impulse, (2) discharge of a parallel after-discharge circuit, and (3) discharge of a reverberatory after-discharge circuit.

ing circuit is a necessity. Unfortunately, the parallel circuit cannot control functions that last more than a fraction of a second, because one neuron is required in a circuit for each $\frac{1}{2000}$ second. Probably not more than a few dozen neurons are ever organized into parallel circuits, which would limit this type of circuit to activities occurring in less than about $\frac{1}{50}$ second.

Figure 15–15 illustrates the time duration and amplitude characteristics of the output signals from, first, a one-to-one circuit; second, a parallel repetitive circuit; and third, a reverberatory repetitive circuit.

Relationship of the Output Signal to Input Signal of a Neuronal Pool

Now that the characteristics of some of the important neuronal circuits have been described, we can discuss briefly the overall responses of typical neuronal pools. Almost never is a neuronal pool simply a relay station; that is, almost never do the same number of impulses leave a pool as enter it. Usually incoming signals diverge, converge, or are changed into repetitive signals whose durations last long after the input signal is over. Also, many incoming stimuli do not excite the neuronal pool at all but, instead, inhibit it.

An important characteristic of neuronal pools is that the response to input signals from one source can be altered by input signals coming from secondary sources. For instance, sensory nerve impulses from the skin into the spinal cord ordinarily cause no reflex skeletal muscle effects. However, strong facilitatory impulses transmitted from the brain down the spinal cord

to the neurons that control the muscles can make these neurons so excitable that a very light scratch on the skin elicits a strong contraction of the underlying muscle. Thus, a high threshold, low amplification circuit has been changed into a low threshold, high amplification circuit.

It is now up to the imagination of the student to conceive the many possible ways in which individual types of neurons can be organized into different types of neuronal pools and in which different types of neuronal pools can perform an infinite number of reflex and integrative nervous functions. Unfortunately, the precise circuits in most neuronal pools have not yet been worked out in detail. Nevertheless, the types of circuits that are already known can suggest mechanisms by which all the functions of the central nervous system could be performed.

QUESTIONS

1. What is meant by a reflex?
2. What are the three major levels of nervous system function?
3. Describe the physiologic anatomy of the synapse.
4. How does an excitatory transmitter substance cause an excitatory postsynaptic potential?
5. Describe both the spatial and temporal methods for summation of excitatory postsynaptic potentials.
6. Explain the mechanisms by which inhibitory transmitter substances inhibit synaptic transmission.
7. Explain what is meant by diverging circuits and converging circuits.
8. Explain the function of the reverberating circuit and the factors that determine the duration of reverberation.

REFERENCES

Adams, D.J., et al.: Ionic currents in molluscan soma. *Annu. Rev. Neurosci.*, 3:141, 1980.
An der Heiden, U.: Analysis of Neural Networks. New York, Springer-Verlag, 1980.
Anderson, H., et al.: Developmental neurobiology of invertebrates. *Annu. Rev. Neurosci.*, 3:97, 1980.
Ceccarelli, B., and Hurlbut, W.P.: Vesicle hypothesis of the release of quanta of acetylcholine. *Physiol. Rev.*, 60:396, 1980.
Iversen, L.L.: The chemistry of the brain. *Sci. Am.*, 241(3):134, 1979.
Kandel, E.R.: Small systems of neurons. *Sci. Am.*, 241(3):66, 1979.
Pepeu, G., et al. (eds.): Receptors for Neurotransmitters and Peptide Hormones. New York, Raven Press, 1980.
Pinsker, H.M., and Willis, W.D., Jr. (eds.): Information Processing in the Nervous System. New York, Raven Press, 1980.
Stephenson, W.K.: Concepts of Neurophysiology. New York, John Wiley & Sons, 1980.
Stevens, C.F.: The neuron. *Sci. Am.*, 241(3):54, 1979.

16

Somesthetic Sensations and Interpretation of Sensory Signals by the Brain

Overview

The *somesthetic sensations* are those that arise from the body surface or from the deep tissues. These include such sensations as (1) *touch,* (2) *pressure,* (3) *heat,* (4) *cold,* (5) *pain,* and (6) *angulation of the joints.*

The perception of sensation begins with **sensory receptors.** Many of these are **free nerve endings** of the peripheral sensory nerve fibers. The most important example of these are the pain fibers. Other somesthetic sensory receptors are highly specialized, often comprised of a branching nerve ending *encapsulated* in its own tissue covering; these endings respond to specific types of sensation. One such ending, called a **Meissner's corpuscle,** is exceedingly sensitive to very light touch. Large numbers of these endings are found in the fingertips; they endow the fingers with an exceptional ability to detect form, texture, and other characteristics of objects.

After entering the spinal cord from the spinal nerves, sensory signals are transmitted to the brain through two principal pathways, (1) the **dorsal system** and (2) the **spinothalamic system.** In the dorsal system, the signals are carried in very large nerve fibers located mainly in the dorsal columns of the spinal cord, whereas in the spinothalamic system the signals are carried in much smaller fibers located in the anterolateral columns of the cord. Both pathways eventually terminate in the *thalamus,* where their signals then are relayed through still another set of neurons to the **somesthetic area** of the cerebral cortex, called the *somesthetic cortex.* In addition, each of these systems, especially the spinothalamic system, gives off branching fibers in both the spinal cord and the brain stem. Signals from these branches elicit spinal cord reflexes and also brain stem reflexes, especially reflexes that cause postural and equilibrium muscle contractions.

The **thalamus** plays an especially important role in determining the type of sensation, called the *modality of sensation,* that a person feels, that is, whether it be touch, pressure, pain, heat, or cold. The function of the **somesthetic cortex** is mainly to determine where in the body the sensory signals originate.

Pain sensations play an essential role in protecting the body tissues against damage. In fact, it is *tissue damage* itself that stimulates pain

293

nerve endings. When stimulated, the pain sensory system causes multiple responses, beginning with *withdrawal reflexes* integrated in the spinal cord that withdraw individual parts of the body away from the pain stimuli. The pain also elicits an instantaneous high level of excitability in the brain stem and cerebrum followed by such reactions as running, fighting, screaming, and so forth. However, to some extent, a person's degree of stoicism can control his degree of *reactivity to pain.* Part of this control results from a special *pain control mechanism* that transmits signals from the cerebrum and brain stem all the way down to the posterior horns of the spinal cord to inhibit transmission of pain signals at the point where they first enter the cord.

Special types of pain include (1) *visceral pain,* which comes from the internal organs such as the heart, the stomach, the liver, or so forth, and (2) *headache,* which is usually a manifestation of pain originating inside the cranial cavity.

The term *somesthetic sensation* means sensation from the body. Also, physiologists frequently speak of subdivisions of somesthetic sensation, including *exteroceptive* sensation, *proprioceptive* sensation, and *visceral* sensation, though there is much overlap among these different types of sensations.

Exteroceptive sensations are those normally felt from the skin, such as (1) touch, (2) pressure, (3) heat, (4) cold, and (5) pain.

Proprioceptive sensations are those that apprise the brain of the physical state of the body, including such sensations as (1) length of the muscles, (2) tension of the tendons, (3) angulation of the joints, and (4) deep pressure from the bottom of the feet. Note that pressure can be considered to be both an exteroceptive sensation and a proprioceptive sensation.

Visceral sensations are those from the internal organs, including such sensations as (1) pain, (2) fullness, and sometimes (3) the sensation of heat. Thus, the visceral sensations are similar to exteroceptive sensations and are functionally the same except that they originate from inside the body.

General Organization of the Somesthetic Sensory System

Figure 16–1 illustrates the general plan for transmission of somesthetic sensory signals into the brain. The sensations are detected by special nerve endings in the skin, muscles, tendons, or deeper areas of the body. These generate nerve impulses that are transmitted through nerve trunks into the spinal cord. Upon entering the cord, the sensory nerve fibers branch. Some of the branches end in the spinal cord itself to cause *cord reflexes,* which will be discussed in

the following chapter; the others extend to other areas of the cord and brain. The sensory pathways to the brain terminate in several discrete areas as follows: (1) several sensory areas of the brain stem, including the bulboreticular formation and the central gray area; (2) the cerebellum;

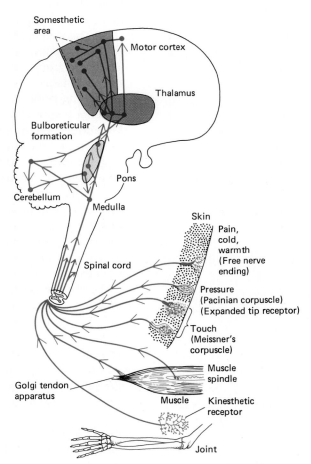

FIGURE 16–1 Transmission of sensory signals to the brain, showing the sensory receptors and the nerve pathways for transmitting these sensations into the brain.

(3) the thalamus and other closely allied structures; and (4) the cerebral cortex. The signals transmitted into each of these different areas subserve specific functions, which will become clear later in the chapter. Signals transmitted into the cerebellum occur entirely at a subconscious level and are concerned with subconscious control of motor function. Therefore, the cerebellar component of the somesthetic sensory system will be discussed in connection with motor activity of the body in Chapter 18 rather than in the present chapter.

Function of Different Nervous System Levels of Somesthetic Sensation. The sensory nerve fibers that terminate in the spinal cord initiate cord reflexes. These cause immediate and direct motor activities such as contraction of muscles to pull a limb away from a painful object, or perhaps rhythmic contraction of the limbs to cause walking movements.

The sensory signals that terminate in the brain stem cause subconscious motor reactions of a much higher and more complex nature than those caused by the cord reflexes. For instance, it is in the brain stem that chewing, control of the body trunk, and control of the muscles that support the body against gravity are all effected.

As the sensory signals travel farther up the brain and approach the thalamus, they begin to enter the level of consciousness. When sensory signals reach the thalamus, their origins are localized crudely in the body, and the types of sensations, which are called the *modalities of sensation,* begin to be appreciated. Thus, one can then determine whether the sensation is touch, heat, or cold, and so forth. Yet for full appreciation of these qualities, and especially for very discrete localization, the signals must pass on into the cortex. The cortex is a large storehouse of information, of memories of past experiences, and it is this stored information that allows the finer details of interpretation to be achieved.

THE SENSORY RECEPTORS

Some sensory nerve endings in the skin and deeper structures of the body are nothing more than small filamentous branches called **free nerve endings,** whereas others are special **end-organs** that are designed to respond only to special types of stimuli. Figures 16–1 and 16–2 illustrate some of the more representative sensory receptors, and their functions may be described as follows:

Free Nerve Endings. Free nerve endings detect the sensations of crude touch, deep pres-

FIGURE 16–2 Several types of somatic sensory nerve endings.

sure, pain, heat, and cold. These sensations occasionally become somewhat confused with each other because the nerve pathways from the free endings interconnect extensively, and the endings themselves also are not always entirely specific for the different types of sensation. For instance, extreme heat or cold is likely to give one a sensation of pain, or very hard pressure might also be confused with pain.

Despite the fact that free nerve endings transmit only crude sensations, they are by far the most common type of nerve ending. They perform most of the general functions of sensation. The specific functions, such as discrimination of very slight differences between degrees of touch, are left to the more specialized receptors.

Specialized Tactile Receptors. Some sensory nerves have specialized endings called **sensory receptors,** several of which are illustrated in Figures 16–1 and 16–2. Those receptors that detect touch, pressure, stretch, or other types of skin or deep tissue deformation are called *tactile receptors.* There are many different types of tactile receptors; some of these, illustrated in Figure 16–2, are

1. the **expanded tip receptor,** which is found in most parts of the body and is capable of prolonged response to very light pressure or light touch on the skin;
2. the **tactile hair receptor,** which consists of nerve fibers entwining the root of each hair of

the body and is stimulated when the hair is bent to one side;

3. the **pacinian corpuscle,** usually found deep in the tissues, responds to rapid compression, rapid stretch, or any other rapid tissue deformation; however, the signal transmitted by the pacinian corpuscle lasts only a small fraction of a second;

4. **Meissner's corpuscle,** a specialized receptor of the fingertips and lips that allows us to discriminate very precise texture and other fine details of objects that we touch;

5. **Krause's corpuscle,** which is found in moderate numbers in the sexual organs and therefore may be concerned with some of the sexual sensations; and

6. **Ruffini's end-organ,** which detects stretch of tissues and joint capsules, thus determining the degree of angulation of joints.

The Muscle Receptors. Also illustrated in Figures 16–1 and 16–2 are two special sensory receptors that transmit proprioceptive information from the muscles. These are the **muscle spindle** and the **Golgi tendon apparatus.** The muscle spindle is composed of nerve filaments wrapped around small muscle fibers. The spindle detects the degree of stretch of the muscle. This information is transmitted to the central nervous system to aid in the control of muscle movements. The Golgi tendon apparatus detects the overall tension applied to the tendon, and, therefore, apprises the central nervous system of the effective strength of contraction of the muscle. The functions of both of these in helping to control muscle contraction will be discussed in the following chapter.

Mechanism of Excitation of the Specialized Sensory Receptors

Each type of sensory receptor has its own means for stimulation. In general, the tactile receptors are stimulated by some type of physical *deformation of the nerve fiber inside the receptor.* For instance, Figure 16–3 illustrates a terminal nerve fiber inside a pacinian corpuscle. When the outside of this corpuscle is compressed, this either bends the tip of the nerve fiber or stretches it, which momentarily increases the permeability of the fiber membrane and allows sodium ions to flow freely to the inside. This in turn sets up a local current flow between the point of high fiber permeability inside the corpuscle and the first node of Ranvier further up the fiber. The current flow creates a voltage across the fiber membrane

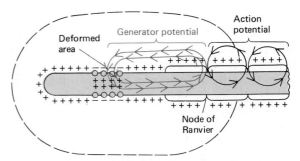

FIGURE 16–3 Excitation of a sensory nerve fiber by a generator potential produced in a pacinian corpuscle. (Modified from Loewenstein: *Ann. N.Y. Acad. Sci.,* 94:510, 1961.)

at this node, called a *generator potential.* The generator potential elicits a series of action potentials one after another as long as the generator potential itself lasts.

Though the other tactile receptors have different physical characteristics, the mechanism of action potential generation is believed to be basically the same as for the pacinian corpuscle. On the other hand, other types of receptors, such as visual receptors, temperature receptors, and some pain receptors, are excited not by mechanical deformation but instead by chemical stimuli acting in or on the receptors, as we shall discuss at later points in this text.

Adaptation of the Sensory Receptors

When a stimulus is suddenly applied to a sensory receptor, the receptor usually responds very vigorously at first, but progressively less so during the next few seconds or minutes. An example of this effect is the sensation felt when one gets into a tub of hot water, which causes an intense burning sensation at first but after a moment's exposure produces a sensation of pleasing warmth. The same is true to varying degrees for all the other sensations. This loss of sensation during prolonged stimulation is called *adaptation* of sensory receptors.

Adaptation occurs to much greater degrees for some sensations than for others. For instance, the sensation of light touch and that for some types of pressure adapt within a few seconds, in some instances in as little as $\frac{1}{100}$ second. This allows one to feel an object when he first touches it but to lose the sensation very soon thereafter. Were it not for rapid adaptation of the light touch receptors, a person would never stop feeling intense touch sensations from all body areas in contact with any object such as the seat of a chair, the shoes, and even the clothing. These sensations would continue to bombard the brain to such an extent that the person would hardly be able to think of anything else. The value of rapid adaptation of most sensations, therefore, is obvious.

The sensations of pain and some types of proprioception usually adapt either very slowly or to only a slight extent. Pain sensations are elicited when tissue damage is occurring. As long as the damage continues to occur, it is important that the person also continues to be apprised of this fact so that he will institute appropriate measures to remove the cause of the damage. The long persistence of some proprioceptive sensations also is desirable because the brain needs to know the physical status of the different parts of the body at all times and not simply immediately after movements have occurred.

Discrimination of Intensity of Sensations — The Weber-Fechner Law

A person can detect the weight of a flea on the tip of his finger or he can detect the weight of a man stepping on the same finger. The difference in weight of these two animals is approximately 20 million times, and yet, from the sensations perceived, the person can estimate the heftiness of the two objects.

The reason why one can discriminate such wide differences in intensities is that the number of impulses transmitted by most sensory receptors is roughly proportional to the *logarithm* of the intensity of sensation rather than to the actual intensity. This logarithmic response of receptors is illustrated in Figure 16–4, which shows the impulse rate from a muscle spindle as different weights are applied to the muscle. Note that the weight scale doubles at each point, and yet the spindle impulses increase linearly. This is a logarithmic type of response; that is, the number of impulses is approximately proportional to the logarithm of the weight rather than to the weight itself.

In the case of the flea and the man, the difference between the logarithms of their weights is only about 7 times even though the difference between the actual weights is 20 million times. This logarithmic detection of sensation applies not only to somesthetic sensations, but to some extent also to the sensations of vision, hearing, taste, and smell; it is called the *Weber-Fechner law.*

FIGURE 16–4 Logarithmic response of the muscle spindle. (Drawn from data in Matthews: *J. Physiol. (Lond.),* 78:1, 1933.)

Another example of the Weber-Fechner law is the following: If one is holding in his hand an object that weighs 1 oz and he exchanges this object for another that weighs 1.1 oz, he barely will be able to discriminate the difference in weight of the two objects. The actual difference in weight is 0.1 oz. Then he picks up an object that weighs 10 lb. To discriminate a difference in weight of another object, the new object must now weigh approximately 11 lb for him to tell the difference. The increase in weight this time must be 1 lb instead of 0.1 oz. In each instance, the increase is 10 percent of the original weight. Therefore, another way of expressing the Weber-Fechner law is: Discrimination of intensity level is on a relative basis rather than on an absolute basis.

Determination of Types of Sensation — The "Modality" of Sensation

Even though different types of nerve receptors are responsible for detecting different types of sensation—pain, touch, pressure, position, and so forth—nevertheless, it is not the receptor itself that determines the type of sensation that a person feels. Instead, it is the point in the brain to which the signal is transmitted. For instance, if a pain nerve fiber is stimulated by crushing it, heating it, bending it, or in any other way, the person will still feel pain regardless of how the pain nerve is stimulated. Likewise, a touch nerve fiber can be stimulated by crushing it, burning it, or bending it, and the person will feel nothing but touch even though the nerve fiber itself is being damaged severely, which one generally considers to cause pain.

Specific Areas of the Brain for Detecting Different Modalities of Sensation. The term "modality" of sensation means the specific quality of sensation felt, that is, whether the feeling be one of pain, touch, pressure, position, vision, hearing, equilibrium, smell, or taste, all of which are different modalities of sensation.

It is frequently stated that the thalamus is the main area of the brain for determining modality of sensation. The reason for stating this is that fiber pathways for almost all of the different modalities terminate at different points in the thalamus, and destruction of the thalamus makes it difficult to distinguish most types of sensory modalities. However, it has now been learned that some pain pathways terminate in even lower areas of the brain, such as in the central gray area of the mesencephalon and in the hypothalamus, and that these areas are principal sites for detection of pain. Furthermore, the cerebral cortex is known to sharpen one's ability to detect the different modalities, even though the lower areas of the brain can provide crude detection.

Thus, it is becoming apparent that widely scattered regions of the brain operate together to determine the different sensory qualities. As the sensory signals enter the brain from the cord, they are picked apart for their different characteristics. One portion of the brain determines whether or not there is a pain element in the sensation, another whether or not there is a touch element, another the area of the body from which the sensation is coming, and so forth.

ANATOMY OF THE SOMESTHETIC SENSORY TRANSMISSION SYSTEM

In previous chapters we have already learned that sensory nerve fibers enter the spinal cord through the posterior roots of the spinal nerves. Some of these fibers terminate almost immediately in the gray matter of the cord near the point of entry and initiate local activity in the neuronal circuits of the cord itself. It is this activity that causes the spinal reflex muscle contractions of the body that will be discussed in detail in the following chapter.

In addition to the nerve fibers that terminate in the spinal cord, others pass up the cord carrying signals to the brain. Figure 16–5 illustrates several pathways for this sensory transmission; these are generally divided into two separate divisions: (1) the **dorsal system** and (2) the **spinothalamic system.**

The Dorsal System. The dorsal system for transmitting sensory signals to the brain is shown in Figure 16–5A. Most of the signals of this system originate in the specialized tactile sensory receptors that excite large myelinated nerve fibers. After these fibers enter the cord through the posterior spinal nerve roots, each fiber divides immediately into two branches. One of these branches terminates in the local cord gray matter and elicits local cord activity and spinal cord reflexes. The other division turns upward in the dorsal white columns (the white matter that lies between the two posterior horns of the cord gray matter) and runs all the way up to the lowermost portion of the medulla oblongata to terminate there in the **dorsal column nuclei.** It is here that the first synapse in this sensory pathway occurs. The neurons of the dorsal column nuclei then give rise to axons that cross immediately to the opposite side of the medulla. These then course upward through the brain stem in a column of fibers called the **medial lemniscus,** finally terminating in the **ventrobasal complex of the thalamus,** an area of the thalamus that lies in its posterolateral ventral region. Here, the fibers form synapses with still new neurons that send their fibers through the **internal capsule** of the cerebrum to the **cerebral cortex,** terminating in the **somesthetic cortex** located in the first 2 cm of the parietal cortex behind the central sulcus.

Thus, the dorsal sensory pathway from the periphery all the way to the cerebral cortex is subserved by three successive neurons. These are called, respectively, the *first order neuron,* the *second order neuron,* and the *third order neuron.* The first order neuron is the peripheral nerve fiber, from the sensory receptor, that enters the cord and goes all the way up the cord to the dorsal column nuclei in the medulla. The second order neuron is the one, located in the dorsal column nuclei, that sends its axon to the ventrobasal complex of the thalamus on the opposite side. And, the third order neuron originates in the thalamus and terminates in the somesthetic cortex.

The Spinocervical Tract. Another portion of the dorsal system for somesthetic signal transmission is the **spinocervical tract,** also illustrated in Figure 16–5A. This tract adds still another order of neuron to the transmission pathway. When the sensory nerve fiber first en-

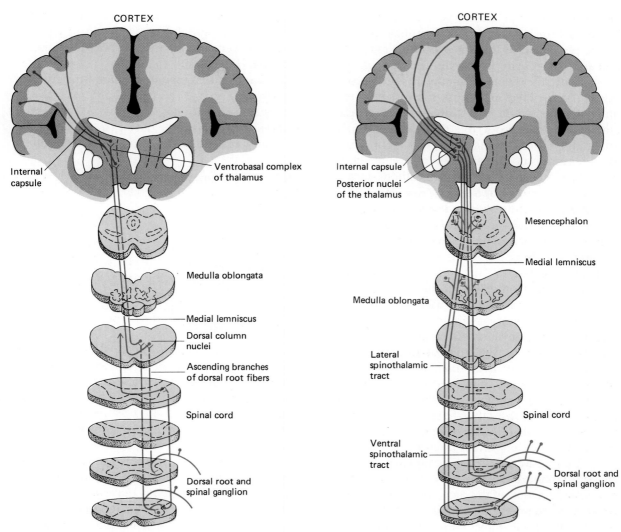

FIGURE 16–5 (A) The dorsal column and spinocervical pathways for transmitting critical types of tactile signals. (B) The spinothalamic pathways for transmitting crude tactile signals, pain signals, and temperature signals. (Modified from Ranson and Clark: Anatomy of the Nervous System. Philadelphia, W.B. Saunders Co., 1959.)

ters the spinal cord, one of its branches almost immediately forms a synapse in the posterior horn of the cord gray matter, and the postsynaptic neuron then sends its fiber into the spinocervical tract in the posterior portion of the lateral white column of the cord. As this tract approaches the cervical region of the cord, its fibers turn back into the gray matter of the cord and form synapses with still a new order of neurons in the posterior horn, and the new fibers cross to the opposite side of the cord to join the medial lemniscus, eventually continuing on to form synapses in the thalamus, the same as in the dorsal column fibers.

The Spinothalamic System. The spinothalamic sensory transmission system is illustrated in Figure 16–5B. In this system, the peripheral sensory nerve fibers are very small,

usually unmyelinated fibers, and most of their endings are free nerve endings or small specialized receptors. After these fibers enter the cord through the posterior roots, they terminate almost immediately in the posterior horns of the cord gray matter. Here they form synapses with several successive neurons, and the last one sends its axon to the opposite side of the cord and thence all the way up to the brain stem and thalamus. Some of these fibers pass upward in the **ventral white columns** of the cord and others in the anterior portion of the **lateral white columns.** Therefore, the spinothalamic pathway has two divisions, the **ventral division** and the **lateral division.**

The spinothalamic fibers branch many times as they enter the brain stem, some of the branches terminating in the reticular substance of the entire brain stem and some joining the medial lemniscus and finally terminating in the thalamus. Some of the fibers that do reach the thalamus terminate in the ventrobasal complex along with the fibers from the dorsal sensory system, but others penetrate more deeply into the thalamus to end in the **intralaminar nuclei,** dispersed throughout the midportions of the thalamus. From the thalamus, connecting neurons continue on to the cerebral cortex, where they, too, terminate in the somesthetic cortex.

Discreteness of Signal Transmission in the Dorsal System Versus the Spinothalamic System.   The dorsal system, as well as the peripheral nerve fibers that connect with it, is composed of very large nerve fibers that transmit signals at velocities of 30 to 110 meters (m) per second. On the other hand, the spinothalamic system and its associated peripheral nerve fibers are composed of very small nerve fibers, some of which are not myelinated at all. These fibers transmit signals at velocities of 10 to 60 m per second. Since the spinothalamic system conducts signals slowly, it is used mainly for information that the brain can afford to receive after a short delay. On the other hand, the dorsal system allows transmission of information to the brain within a very small fraction of a second.

Another major difference between the dorsal and spinothalamic systems is the degree of spatial orientation of the nerve fibers. In the dorsal system, the nerve fibers are oriented very exactly according to their point of origin from the different parts of the body, and there is little crossover of signals anywhere in the tract. Therefore, when a single receptor is stimulated in the skin or other peripheral area of the body, a discrete signal is transmitted to a well-localized point in the thalamus and from there also to a localized point in the cerebral cortex. By contrast, the nerve fibers in the spinothalamic system are less well oriented, and there is far greater sidewise diffusion of nerve signals in the neuronal pools, where the spinothalamic pathways synapse in the spinal cord and thalamus. As a consequence, stimulation of a single nerve receptor exciting the spinothalamic pathway causes excitation of a widely dispersed area in the brain.

Modalities of Sensation Transmitted by the Two Systems.   With the foregoing differences in mind, we can now list the types of sensations transmitted in the two systems.

The Dorsal System

1. Touch sensations requiring a high degree of localization of the stimulus.
2. Touch sensations requiring transmission of fine gradations of intensity.
3. Phasic sensations, such as vibratory sensations.
4. Sensations that signal movement against the skin.
5. Position sensations.
6. Pressure sensations having to do with fine degrees of judgment of pressure intensity.

The Spinothalamic System

1. Pain.
2. Thermal sensations, including both warm and cold sensations.
3. Crude touch and pressure sensations capable of only crude localizing ability on the surface of the body and having little capability for intensity discrimination.
4. Tickle and itch sensations.
5. Sexual sensations.

Within the dorsal system, the dorsal column pathway and the spinocervical pathway each carry different types of signals. The *dorsal columns* carry rapidly changing types of signals, called *phasic signals,* such as vibratory sensations, sensations that signal rapid movement against the skin, sensations that signal the rate of change of positions of the limbs, and so forth. On the other hand, the *spinocervical pathway* carries the so-called *static signals,* that is, steady signals that continue unchanged for long periods such as those used for continuous localization of stimuli on the skin, signals that give long-term information about gradations of signal intensity, and signals from the joint structures of the body that allow us to know the static positions of the different parts of our bodies.

In the spinothalamic system, touch and pressure sensations are transmitted mainly in the *ventral spinothalamic tract,* whereas pain and thermal sensations are transmitted mainly in the *lateral spinothalamic tract,* though there is a tremendous amount of overlap between these two tracts.

Collateral Signals from the Spinothalamic System to the Basal Regions of the Brain.   It is important to note another distinct difference between the spinothalamic and dorsal column systems. That is, a large share of the nerve fibers of the spinothalamic system terminate in the medulla, pons, and mesencephalon even before they reach the thalamus, and many of the fibers that do go all the way to the thalamus give off branches into these lower regions of the brain on the way up. In contrast, the dorsal system passes through these regions with almost no branches. Thus, the spinothalamic system is much more concerned with sensations that elicit subconscious automatic reactions than is the dorsal system. On the other hand, the dorsal system is concerned with very discrete signals that are transmitted primarily into the conscious areas of the brain.

To express this another way, the spinothalamic system is concerned with the older types of sensations, those that occur in even very low forms of animal life; the dorsal system is concerned with types of sensations that have appeared recently in the phylogenetic development of man.

MECHANISM FOR LOCALIZING SENSATIONS IN SPECIFIC AREAS OF THE BODY

Among the most important information that the brain must determine about each somesthetic sensation is its point of origin in the body. Once the location has been determined, it is possible to do something about the condition giving rise to the sensation. To provide this localization ability, the nerve fibers are *spatially oriented* in the nerve trunks, spinal cord, hindbrain, and cerebral cortex. This may be explained as follows:

Spatial Orientation of Nerve Fibers. The sensory nerve fibers arising in the leg are separated in the spinal cord and brain from the fibers arising in the arm. The fibers arising from individual fingers are separated from each other, and even the fibers arising from two areas of skin only 1 cm apart are kept distinct from each other. Also, at each new synaptic relay station along the sensory pathway, the respective sensory signals from adjacent areas of the body are still kept separate from each other all the way to the termination of the signals in the lower brain stem, thalamus, and cerebral cortex.

In general, the thalamus is capable of determining only roughly which part of the body is being stimulated. The thalamus is not organized satisfactorily to localize sensations to very discrete areas of the body; instead, this function is performed primarily in the somesthetic area of the cerebral cortex.

The Somesthetic Cortex

Figure 16–6 illustrates the **somesthetic cortex,** which is located immediately posterior to the central sulcus of the brain and extends from the longitudinal fissure at the top of the brain into

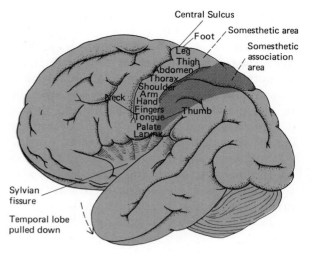

FIGURE 16–6   The somesthetic cortex.

the Sylvian fissure at the side. This figure also shows the areas of the somesthetic cortex that receive sensations from each of the different parts of the body. Sensations from the foot, for instance, excite the somesthetic cortex where it dips over the top of the cortex into the longitudinal fissure that runs forward and backward down the middle of the brain. The leg area is approximately at the point where the somesthetic cortex comes out of the longitudinal fissure; then comes the thigh area, followed by the abdomen, thorax, shoulder, arm, hand, fingers, thumb, neck, tongue, palate, and larynx. Thus, the entire body is spatially represented in the somesthetic cortex, each discrete point in the cortex corresponding to a discrete area only a few millimeters in size on the body surface.

A major function of the somesthetic cortex is to localize very exactly the points in the body from which the sensations originate. Though the thalamus is capable of localizing sensations to very general areas, such as to one arm, to a leg, or to the body, it is not capable of localizing sensations to minute areas of the body. Instead, the thalamus relays the necessary signals into the somesthetic cortex, where a much better spatial representation is available, and there the job of discrete localization is performed.

Not all modalities of sensation are localized equally well. The sensations of aching pain, crude touch, warmth, and cold are localized to general areas of the body rather than to discrete areas. It seems that the thalamus performs most of the localization function for these modalities of sensation. On the other hand, the sensations of light touch, pressure, and position are very discretely localized by the somesthetic cortex. These are the sensations normally detected by the special sensory receptors, for instance

Meissner's corpuscles for light touch and Ruffini's joint receptors for position sensation.

Function of the Somesthetic Cortex in Analyzing Sensory Signals. Even when the somesthetic cortex is completely removed, a person usually still has the ability to detect the type of sensation that is being received, that is, whether it be pain, touch, heat, cold, or so forth, but his appreciation of these sensations is markedly reduced. Therefore, the cerebral cortex is concerned not so much with detection of type of sensation as with *analysis* of the sensory information after it has already been detected.

Interpretation of Somesthetic Sensations—the Somesthetic Association Area

The area of the cortex a centimeter or so behind the somesthetic cortex is called the *somesthetic association area* (see Fig. 16–6), because it is in this region that some of the more complex qualities of sensation are appreciated. Signals are transmitted directly into this area from several sources: (1) from the thalamus, (2) from other basal regions of the brain, and (3) backward from the somesthetic cortex. In ways not understood, the somesthetic association area puts all this information together and determines the following characteristics of sensations: (1) shape of an object; (2) relative positions of the parts of the body with respect to each other, such as the legs, the hands, and so forth; (3) texture of a surface, such as whether it is rough, smooth, or undulating; and (4) orientation of one object with respect to another object—in other words, the spatial orientation of objects that are felt.

Many of the memories of past sensory experiences are also stored in the somesthetic association area, and when new sensations similar to the old ones arrive in the brain, the parallel nature of the two sensations is immediately discerned. It is in this manner that one associates a new sensation with previous ones and thereby recognizes the nature of the sensation. As more and more sensory experiences accumulate, new sensory experiences can be interpreted on the basis of what is remembered from the past.

PAIN

The sensation of pain deserves special comment because it plays an exceedingly important protective role for our body, apprising us of almost any type of damaging process and causing appropriate muscular reactions to remove the body from contact with the damaging stimuli.

The Stimulus That Causes Pain. Pain receptors are stimulated when tissues of the body are *being* damaged. For instance, one feels pain while the skin is *being* cut, but shortly after the cut has been made, the pain generally is gone. Indeed, thousands of soldiers on the battlefield in World War II who had been mortally wounded were asked whether or not they felt pain. In most instances, no pain was actually felt except for a few minutes after the damage had been inflicted or unless the soldier was moved.

Different types of damaging stimuli that can cause pain are *trauma* to the tissues, *ischemia* of the tissues (lack of blood flow), intense *heat* to the tissues, intense *cold* (especially freezing of the tissues), or *chemical irritation* of the tissues. It is believed that as tissues are damaged they release some substance from the cells that stimulates the pain nerve endings. *Bradykinin* has especially been suggested, though not proved, to be this substance.

Perception of Pain. It is frequently said that one person *perceives* pain more intensely than others. However, experiments in which graded intensities of tissue damage were caused in a large number of different persons showed that all normal persons perceived pain at almost precisely the same degree of tissue damage. For instance, when heat is used to cause tissue damage, almost all persons begin to feel pain when the tissue temperature rises to a level between 44°C and 46°C, which is a very narrow range.

Reactivity to Pain. On the other hand, not all people *react* alike to the same pain, for some react violently to only slight pain and others can withstand tremendous pain before reacting at all. This is determined not by differences in sensitivity of the pain receptors themselves but, instead, by differences in the psychic make-up of the individuals. Therefore, when a person is said to be extremely "sensitive" to pain, it is meant that he *reacts* to pain far more than do other persons and not that he perceives far more pain.

Transmission of Pain Signals into the Brain Stem, Thalamus, and Cortex

The Two Types of Pain—Pricking Pain and Burning-Aching Pain. If the student will think for a moment of his own experiences with pain, he will recall that some types of pain are

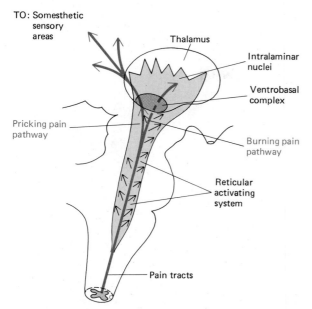

TO: Somesthetic sensory areas

Thalamus

Intralaminar nuclei

Ventrobasal complex

Pricking pain pathway

Burning pain pathway

Reticular activating system

Pain tracts

FIGURE 16—7 Transmission of pain signals into the brain stem, thalamus, and cortex via the "pricking pain" pathway and the "burning pain" pathway.

very sharp and pricking in nature. This type, called *pricking pain*, can be localized very readily to discrete areas of the body. On the other hand, another type of pain, described as a *burning-aching* pain, lingers for long periods of time, as occurs after a scalding burn or in a long-term arthritic joint. Figure 16–7 shows that the pricking pain signals terminate in the **ventrobasal complex of the thalamus** along with the medial lemniscus, and from here successive signals pass on to the somesthetic sensory areas of the cortex. On the other hand, the burning-aching pain fibers terminate differently. They send tremendous numbers of branches into the **reticular substance of the brain stem** at all levels of the medulla, pons, and mesencephalon. Still more of these fibers penetrate upward into the middle of the thalamus to terminate in the **intralaminar nuclei of the thalamus.** These nuclei are actually an upward extension of the reticular substance from the brain stem. The nerve signals terminating in this reticular substance, all of the way from the medulla upward through the thalamus, cause many of our subconscious reactions to pain, such as reactions of foul mood, sometimes increased aggressiveness, or even rage. Thus, the burning-aching pathway is a very primitive system that allows pain to bring the body into a state of subconscious but strongly active response to the pain itself. It can make the person flee or make him fight. And it can make him scream or squirm in agony.

FEEDBACK CONTROL OF PAIN BY THE BRAIN—THE ENKEPHALINS AND ENDORPHINS

Electrical stimulation in several areas of the brain, especially in certain areas of the **hypothalamus** and in the so-called **raphe nuclei** in the midline region of the brain stem, can greatly decrease a person's sensitivity to pain. These same areas of the brain have in them a morphine-like substance, either an *enkephalin* or an *endorphin*. It is believed that specialized nerve endings in these areas of the brain secrete these two morphine-like substances to excite the pain-suppressing brain centers. In fact, it is possible that acupuncture functions in this way. That is, sensory signals from the acupuncture needles could cause the enkephalins and endorphins to be released in the pain-suppressing centers.

The pain-suppressing centers of the hypothalamus and brain stem do not suppress pain by blocking pain transmission in the brain itself. Instead, signals are transmitted down nerve tracts in the spinal cord all the way to the point where the pain enters the cord from the peripheral nerves. The nerve endings of these tracts secrete the synaptic inhibitor substance *serotonin* that then suppresses the pain synapses in the posterior horns, thus greatly decreasing the person's level of pain sensitivity.

The Visceral Sensations and Visceral Pain

The visceral sensations are those that arise from the internal structures of the body, such as the organs of the abdomen and chest, and they also include sensations from inside the head as well as from the muscles, bones, and other deep structures.

The usual modalities of visceral sensation are pain, burning sensations, and pressure (which, in the abdominal organs, is often manifested as a sensation of fullness). Because these modalities are also exhibited by the exteroceptive sensations, the visceral sensations are sometimes considered to be part of the exteroceptive system. Indeed, the pathways of these two types of sensation are closely related, as will be noted later.

Pathways for Visceral Sensations. Ordinarily one is not at all conscious of his internal organs, but an inflamed organ can transmit pain sensations to the central nervous system through two separate pathways, the **parietal pathway** and the **visceral pathway,** both of

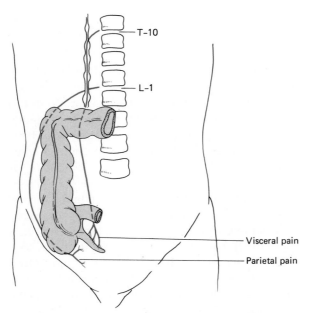

FIGURE 16–8 Parietal and visceral pathways for transmission of pain from the appendix.

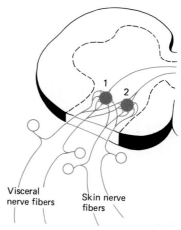

FIGURE 16–9 The neurogenic mechanism responsible for referred pain.

which are illustrated for sensations from the appendix in Figure 16–8. The parietal pathway is the same as the pathway for transmission of exteroceptive sensations. In the case of the appendix, the inflamed appendix irritates the peritoneum overlying the appendix, and pain impulses are transmitted from the peritoneum through the same nerve fibers that carry sensations from the outside of the abdominal wall in this same area, fibers that enter the spinal cord at the L-1 segment.

The visceral pathway utilizes sensory nerve fibers in the autonomic nervous sensory nerve system. Figure 16–8 illustrates these sensory fibers leaving the appendix to enter the sympathetic chain. After traveling upward a few segments, the fibers leave the sympathetic chain to enter the spinal cord at approximately the T-10 segment in the lower thoracic region. Therefore, the visceral sensations from the appendix enter the spinal cord at an entirely different point from the parietal sensations.

Referred Visceral Pain. Pain from an internal organ is often felt on a surface area of the body rather than being localized in the organ itself. This is called *referred pain.* It may be referred to the surface immediately above the organ, or often to an area a considerable distance away. The mechanism of referred pain is probably that illustrated in Figure 16–9. This figure shows two visceral nerve fibers as well as two additional nerve fibers from the skin entering the spinal cord. Both of these sets of fibers synapse with the same two neurons in the spi-

nal cord. Therefore, stimulating either the visceral fibers or the skin fibers will send impulses up the spinal cord along the same pathway. Because the person has never had reason to know the location of his internal organs but is very familiar with the location of the different surface areas, the impulses from the visceral fibers are usually interpreted as coming from the body surface.

When a visceral sensation is transmitted through a parietal pathway, the pain is usually felt on the surface of the body directly over the respective internal organ, but when the sensation is conducted through a visceral pathway, it is usually referred to an area remote from the organ. The reason for this is that fibers of the visceral pathway usually travel a long distance in the sympathetic chain before entering the spinal cord. Figure 16–10 illustrates the surface areas to which pain is referred from many of the different organs. For instance, though the appendix lies far to the right in the lower abdomen, its visceral sensations are referred to an area around the umbilicus. Also, referred pain from the kidney and ureter occurs near the midline on the anterior abdominal wall, though these organs actually lie in the posterior part of the abdomen.

Visceral pain from the heart is among the most important of the referred sensations. Figure 16–10 shows the extensive areas to which cardiac pain is often referred, including the upper thorax, the shoulder, and the medial side of the left arm, particularly along the radial artery.

The surface area to which visceral pain is referred usually corresponds to that portion of the body from which the organ originated during embryonic development. For example, the heart originates in the neck of the embryo and so

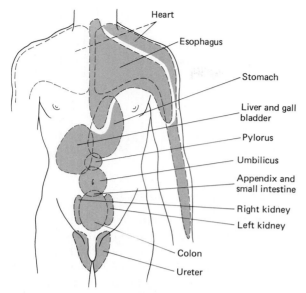

FIGURE 16–10 Surface areas of the body to which pain from different organs is referred.

does the arm; therefore, heart pain is frequently referred to the arm. The appendix, as another example, originates from the portion of the primitive gut that develops near the umbilicus, which explains the reference of appendiceal pain to the umbilical region. The other areas of referred pain illustrated in Figure 16–10 also correspond to the areas of origin of the different organs.

The Stimulus for Visceral Pain. Cutting through the gut, the heart, the liver, the muscle, or other internal organs with a sharp knife causes almost no pain. Instead, pain from these areas is produced much more often by (1) stopping the blood flow to the area, (2) application of an irritant chemical over wide areas, (3) stretching the tissues, or (4) spasm of the muscle in the organ.

One of the most important stimuli for visceral sensation is *ischemia*, which means insufficient blood flow. The pain of a heart attack, for example, is caused by poor flow through the coronaries to the heart muscle. Even the skeletal muscles become extremely painful when their blood supply is diminished for a minute or more during muscle activity. The reason for the pain caused by ischemia has never been determined, though it is believed that the lack of blood flow allows metabolic products such as acids, bradykinin, and so forth to build up in the tissues, and that these in turn produce irritant effects on the pain nerve endings.

Stretching the tissues causes pain either by stretching the nerve endings or by producing ischemia, because overstretching the tissues occludes the blood vessels. Likewise, spasm of

the muscle in an organ can stretch nerve endings as well as compress the blood vessels; spasm also increases the rate of metabolism so that far more than usual quantities of metabolic end-products are dumped into the tissues. These factors together, therefore, could be the reason why spasm of the gut often causes very intense abdominal pain.

Headache

Headache is another type of referred pain, and it is usually caused by irritation or damage occurring in the tissues inside the head. Figure 16–11 illustrates the surface areas of the head to which pain from the deep structures is referred. Pain fibers from inside the eye and from the nasal sinuses are transmitted through the first and second divisions of the fifth cranial nerve, which also supply the skin areas over the lower forehead and around the eye and nose. Therefore, sinus infections or irritation of the eyes caused by their overuse or by intense light can result in dull, aching pain referred diffusely over the frontal and orbital areas of the head.

Irritation occurring inside the skull but above the level of the ears—that is, everywhere around the cerebral cortex—causes headache referred to the frontal and temporal surfaces of the head. The reason for this area of reference is that the third division of the fifth cranial nerve supplies both the upper areas inside the skull and also the surface areas of the skin in the temporal and frontal regions.

Irritative effects in the pocket of the skull where the brain stem and cerebellum reside, that is, beneath the level of the ears, give rise to

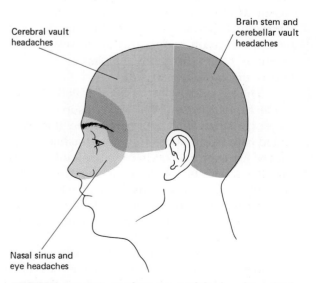

FIGURE 16–11 Surface areas of the head to which different types of headache are referred.

headache localized over the occipital part of the skull, also shown in Figure 16–11. Both the lower areas inside the cranial vault and the occipital skin areas of the skull are supplied by the upper cervical spinal nerves, which explains the reference of this pain to the occipital regions.

Meningeal Headache. Headache originating inside the skull is often caused by irritation of the *meninges*, which are the membranes surrounding the brain. Some causes of very severe headache of this type are (1) meningitis, an infection of the meninges, which causes very intense headache; (2) removal of fluid from the spaces around the brain, allowing the brain to rub freely against the meninges and to irritate them, this too causing very intense headache; and (3) an alcoholic binge (the "hangover"), which probably causes headache by irritation of the meninges, for the meninges almost certainly become reddened and inflamed in the same manner as the whites of the eyes the day after an alcoholic bout.

Many physicians believe that headaches do not often result from pain originating in the substance of the brain itself. In fact, a patient who is having a brain operation under local anesthetic — that is, without being asleep — feels little or no pain when the surgeon cuts through the brain tissue. Yet when he cuts the meninges, or especially when he cuts one of the major blood vessels supplying the meninges, intense headache is experienced.

The Everyday Headache. Despite all that we *do* know about the origin of headaches, the modern physician is still completely befuddled by the common everyday headache. It is generally stated that there are two basic types. One, the *migraine* type, supposedly results from spasm of blood vessels supplying some of the intracranial tissues, followed by intense and painful dilatation of these same vessels lasting for many hours. However, this simple explanation is still quite hypothetical.

The second type is the so-called *tension headache*, which results when a person operates under considerable emotional tension. It is associated with tightening of muscles attached to the base of the skull and perhaps with simultaneous vascular spasm or other effects occurring intracranially. It is possible that the muscles themselves are the source of some of the pain and that the pain is referred to the head, though this theory is extremely hypothetical.

It is also well known that constipation is frequently associated with headache, possibly because of absorption of toxic products from the colon. It is possible that diffuse irritation of the brain by these toxic substances can cause headache, despite the fact that cutting through brain tissue in an awake person does not usually cause much pain.

THE INTERPRETATION OF SENSATIONS BY THE BRAIN

Cortical Localization of Sensations Other Than Somesthetic

The sensations of vision, hearing, taste, and smell, like the somesthetic sensations, are each transmitted from the receptor organs to small circumscribed areas in the cerebral cortex, each of the areas having a different location. These areas are called the *primary areas* for sensations, and from each primary area the impulses go to an *association area*, which actually is an *interpretative area*. The primary area for somesthetic sensation, the somesthetic cortex, and the somesthetic association area were discussed earlier in this chapter, and the peripheral origins of each of the other types of sensation will be discussed in following chapters. However, at this point we will see how all the types of sensations function together to provide conscious interpretation of all sensory experience.

Cortical Areas for Vision. The **primary visual cortex** is located in the posterior part of the brain on the medial side of each hemisphere, as will be discussed in Chapter 21. A small portion of the primary visual cortex extends over the occipital pole, as shown in Figure 16–12, though

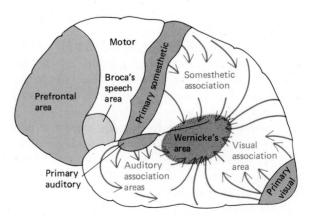

FIGURE 16–12 Organization of the somatic, auditory, and visual association areas into a general mechanism for interpretation of sensory experience. All of these feed into the *general interpretive area* located in the posterosuperior portion of the temporal lobe and the angular gyrus. Note also the prefrontal area, the motor area, and Broca's speech area.

most of it is hidden inside the longitudinal fissure. This is the area where the signals first arrive in the cerebral cortex from the eyes. However, this primary visual area interprets only the more basic meanings of visual sensations such as whether the object is a line, a square, or a star, and the color of the object.

Accessory visual signals pass from the primary visual cortex and also from the thalamus into adjacent areas of the cortex called the **visual association areas,** which are also shown in Figure 16–12. These areas interpret the deeper meaning of the visual signals. That is, they interpret the interrelations of the different objects and identify the objects, and they help to interpret the overall meaning of the scene before the eyes.

Interpretation of written language is one of the most important functions of the visual association areas. To accomplish this feat, the visual cortical system must first discern from the light and dark spots the letters themselves, then from the combination of letters the words, and from the sequence of words the thoughts that they express.

Cortical Areas for Hearing. Auditory sensations are transmitted from the ears to a small area called the **primary auditory cortex** in the upper part of the temporal lobe, which will be discussed in Chapter 22. From this area, signals pass into the surrounding **auditory association areas,** shown in Figure 16–12. The primary auditory cortex interprets the basic characteristics of the sound, such as its pitch and its rhythmicity, whereas the auditory association areas interpret the meaning of the sound. One part of the association areas determines whether the sound is noise, music, or speech; then other parts determine the thoughts conveyed by the sound. To interpret the meaning of speech, the auditory association areas first combine the various syllables into words, then words into phrases, phrases into sentences, and finally sentences into thoughts.

Cortical Areas for Smell and Taste. The *primary cortex for smell* is located on the bottom surface of the brain, in the *pyriform area* and *uncus*, and the **primary cortex for taste** is located at the bottommost end of the primary somesthetic area, deep in the Sylvian fissure. From these primary areas signals pass into surrounding **smell and taste association areas,** and there the sensations are interpreted in the same manner that somesthetic, visual, and auditory sensations are interpreted by their respective association areas.

Wernicke's Area—The Common Integrative Area, or Gnostic Area

Figure 16–12 also illustrates the passage of signals from the somesthetic, visual, and auditory association areas into "Wernicke's" area, located midway between these three respective association areas. Signals are also transmitted into this area from the taste and smell association areas and directly from the thalamus and other basal areas of the brain. It is here that all the different types of sensations are interpreted to determine a common meaning; for this reason, this area is also often called the *common integrative area*. It is also called the *gnostic area*, which means the "knowing area."

If a person were in a jungle and heard a noise in the brush, saw the leaves moving, and smelled the scent of an animal, he might not be able to tell from any one of these sensations exactly what was happening, but from all of them together he could quite readily assess his danger. It is in Wernicke's area that all the thoughts from the different sensory areas are correlated and weighed against each other for deeper conclusions than can be attained by any one of the association areas alone.

Most of the sensory information arriving in the brain finally is funneled through Wernicke's area. For this reason, any damage to this area is likely to leave the person mentally inept; even though the different association areas might still be able to interpret their respective sensations, this information is almost valueless to the brain unless its final meaning can be interpreted. Some signals can be transmitted directly from the sensory association areas to other portions of the brain without going through Wernicke's area, but these are so few that the person who loses his entire Wernicke's area generally becomes seriously confused. This is occasionally the unfortunate result of a *brain tumor* or a *stroke*. (A stroke is caused by sudden loss of blood supply to an area of the brain because of hemorrhage or thrombosis of a blood vessel.)

Dominance of One Side of the Brain. In about 19 out of 20 persons, Wernicke's area is located in the *posterior portion of the superior temporal gyrus of the left cerebral hemisphere*. The corresponding area in the opposite cerebral hemisphere has other functions, such as recognition of faces, interpretation of music, interpretation of form of objects, and so forth. Since these other functions are usually less important than those of Wernicke's area, the side of the brain containing Wernicke's area is said to be dominant over the other side. The corresponding area of the nondominant cerebral hemisphere can function entirely independently of Wernicke's area in the dominant hemisphere, or the information from the nondominant side can be transmitted into the dominant side, where it is combined with all other information.

In most persons whose Wernicke's area is in the left hemisphere, the person is right-handed because information feeds more easily from this area into the motor regions of the left hemisphere, which control the muscles of the right side of the body, than into motor regions of the right hemisphere.

If Wernicke's area is destroyed during the first few years of life, the corresponding area of the opposite hemisphere can develop most of its same functions. However, if the dominant area is destroyed after the person has become an adult, the opposite cerebral hemisphere by that time will have become set in its ways and will never develop significant degrees of general interpretive capability.

Function of Wernicke's Area for Communication and for Symbolism. The most important function of Wernicke's area is to interpret language, whether this language be input to the brain through the auditory system or through the visual system. Therefore, destruction of the dominant hemisphere destroys especially one's capability of understanding either spoken or written language.

Wernicke's area is also very important for other types of symbolism besides language, such as the precepts of philosophy, business, diagnosis, and so forth. Therefore, most complex thinking becomes totally confused when Wernicke's area is destroyed.

Control of Motor Functions by Wernicke's Area. Once Wernicke's area has integrated all incoming sensations into a common thought, signals are then sent into other portions of the brain to cause appropriate responses. If the integrated thought indicates that muscular activity is needed, Wernicke's area sends impulses into the motor portions of the brain to cause muscular contractions.

The basic functions of the cerebral cortex in motor control will be discussed in much more detail in Chapter 18.

Role of the Thalamus and Other Lower Brain Centers in the Interpretation of Sensations

In the preceding sections of the chapter we have discussed cortical functions as if the cortex were operating almost independently of other parts of the brain. This, however, is entirely untrue. Even when major portions of the sensory cortex are destroyed, an animal is still capable of crude degrees of interpretation of sensation. As has already been pointed out, crude localization and interpretation of many somesthetic sensations, especially pain, can be achieved by the thalamus and other related areas. In the case of vision, the cerebral cortex is not needed in some animals to interpret the overall intensity of light, but it is needed to interpret the shapes and colors of objects. In the case of hearing, an animal can detect the existence of sound and to some extent the direction from which sound is coming without the cerebral cortex, though in general he cannot interpret the finer meanings of sound.

Therefore, whenever it is stated that a particular type of sensation is interpreted in a particular region of the cerebral cortex, it is meant simply that this part of the cortex is responsible for the deeper shades of meaning rather than for total interpretation. Indeed, it would be utterly impossible for the cerebral cortex to operate without preliminary processing of information in the lower regions of the brain.

Activation of Specific Portions of the Cerebral Cortex by the Thalamus. During evolutionary development of the brain, the cerebral cortex originated mainly as an outgrowth of the thalamus, for which reason each area of the cerebral cortex is very closely connected with a corresponding discrete area of the thalamus. Thus, the frontal regions of the cerebral cortex have to-and-fro nerve connections directly with the anterior portions of the thalamus. Likewise, occipital portions of the cerebral cortex are connected with the posterior thalamus, and central portions of the cerebral cortex are connected with lateral and mid areas of the thalamus.

Furthermore, the cortex cannot activate itself but must be activated from lower regions of the brain, an effect that will be discussed in Chapter 20. Especially, signals from specific parts of the thalamus activate specific areas of the cerebral cortex. In this way, function in the thalamus presumably calls forth information stored in the memory pool of the cerebral cortex. Therefore, the thalamus is actually a type of *control center* for the cerebral cortex.

Thus, again it is stressed that even though it is conventional to speak of certain types of sensations being interpreted in specific areas of the cerebral cortex, it is really meant that these cortical areas are the loci of the vast memory information associated with these particular types of sensations. Yet, it is still the lower regions of the brain that initiate the signals that control which parts of the cerebral cortex will be activated, that call forth stored information from the cerebral cortex, and that are responsible for channeling sensory signals into appropriate parts of the cerebral cortex. Furthermore, many basic aspects of sensations are detected in the lower regions of the brain even before the signals reach the cerebral cortex.

QUESTIONS

1. Describe the pathways for transmission of sensory signals from the periphery to the cerebral cortex.
2. What are some of the different types of sensory receptors?
3. What is meant by adaptation of sensory receptors?
4. How are different modalities of sensation detected?
5. What types of sensory signals are transmitted through the dorsal system?
6. What types of sensory signals are transmitted through the spinothalamic system?
7. Explain the mechanism for localizing sensations in specific areas of the body.
8. Explain the function of the somesthetic association area in the interpretation of somesthetic sensations.
9. What types of stimuli cause pain?
10. Explain why visceral sensations are frequently referred to areas of the body remote from the viscus.
11. What types of irritation in the cerebral vault will cause headache?
12. Where in the cortex are the primary areas and the association areas for detecting somesthetic sensations, visual sensations, and auditory sensations?
13. What is the role of Wernicke's area of the brain in the interpretation of sensory signals?
14. Discuss the role of the thalamus in the interpretation of sensations.

REFERENCES

Beaumont, A., and Hughes, J.: Biology of opioid peptides. *Annu. Rev. Pharmacol. Toxicol.*, 19:245, 1979.

Brown, E., and Deffenbacher, K.: Perception and the Senses. New York, Oxford University Press, 1979.

Coren, S., *et al.*: Sensation and Perception. New York, Academic Press, 1979.

Fairley, P.: The Conquest of Pain. New York, Charles Scribner's Sons, 1979.

Goldstein, E.B.: Sensation and Perception. Belmont, Cal., Wadsworth Publishing, 1980.

Kenshalo, D.R. (ed.): Sensory Function of the Skin of Humans. New York, Plenum Press, 1979.

Norrsell, U.: Behaviorial studies of the somatosensory system. *Physiol. Rev.*, 60:327, 1980.

Raskin, N.H., and Appenzeller, O.: Headache. Philadelphia, W.B. Saunders, 1980.

Schmidt, R.F. (ed.): Fundamentals of Sensory Physiology. New York, Springer-Verlag, 1978.

Simon, E.J., and Hiller, J.M.: The opiate receptors. *Annu. Rev. Pharmacol. Toxicol.*, 18:371, 1978.

Vallbo, A.B., *et al.*: Somatosensory, proprioceptive, and sympathetic activity in human peripheral nerves. *Physiol. Rev.*, 59:919, 1979.

Zimmerman, M.: Neurophysiology of nociception. *Int. Rev. Physiol.*, 10:179, 1976.

17

Motor Functions of the Spinal Cord and Brain Stem

Overview

The *spinal cord* is the integrating area for multiple motor reflexes that cause local muscle responses. Some of these are listed here:

1. The *muscle stretch reflex* that causes any muscle that becomes over-stretched to contract instantaneously and thereby prevent its length from changing significantly. This reflex is especially important to make the body movements smooth in character, rather than jerky and tremorous.
2. The *muscle tendon reflex* that causes a muscle to relax if the tension on the muscle becomes too great.
3. *Reflexes from the feet* that help to support the body against gravity. For instance, pressure on the bottom of the feet causes the legs to stiffen so that they can support the body.
4. *Reflexes that help to prevent damage to the body.* These are mainly the withdrawal reflexes that cause the appropriate muscle contractions to withdraw a part of the body away from a pain stimulus.
5. *Walking reflexes.* Located in the leg region of the spinal cord are reverberatory circuits that can cause walking movements. Signals from higher centers in the brain control these walking movements so that they may be purposeful in causing locomotion.
6. *Bladder and rectal reflexes.* These cause contraction of the urinary bladder when excess urine volume overstretches the bladder or contraction of the rectum when feces overstretch the rectal wall.

The *brain stem* controls the *subconscious postural muscle contractions* of the body, including contractions that are responsible for *maintenance of body equilibrium.* These control functions take place mainly in the *bulboreticular formation,* which extends the entire distance from the medulla upward through the pons and mesencephalon. When a person is awake and in the standing position, the upper portion of this formation is naturally excitable, and it *excites the extensor muscles of the legs* as well as *the axial muscles of the trunk,* thus allowing the legs and trunk to support the body against gravity.

The degree of contraction of the different muscles in both the legs and trunk is controlled by signals entering the bulboreticular formation from the *vestibular apparatus,* also called the "equilibrium apparatus." This sensory organ contains small *calcified granules,* the *otoliths,* that rest on top of many very excitable sensory nerve endings, the *hair cells,*

310

and the weight of these otoliths causes a specific pattern of nerve signals from the hair cells for each position of the head with respect to the pull of gravity. Thus, the vestibular apparatus signals whether a person is in balance or out of balance, and the bulboreticular formation utilizes this information to contract the appropriate muscles for maintaining equilibrium.

The nervous mechanisms that control the muscles and glands of the body are collectively called the *motor functions* of the nervous system. The purpose of the next few chapters will be to discuss these mechanisms, beginning in the present chapter with a discussion of the motor functions of the spinal cord and lower brain stem.

Though most of us have believed that only the conscious portion of the brain causes muscle movements and other motor activities, this is the farthest from the truth, for the greater proportion of our motor activities is actually controlled by lower regions of the central nervous system, specifically the spinal cord and brain stem, which operate primarily at a subconscious level. Many of these subconscious motor activities occur as simple or complex reflexes, some of the more important of which will be discussed in this chapter.

Neuronal Circuitry of the Spinal Cord

Figure 17-1A illustrates a cross-section of the spinal cord, showing that it is composed of two major portions, the *white* and the *gray matter.* The gray matter, which lies deep inside the cord, has the appearance of double horns protruding anteriorly and posteriorly. The cell bodies of the neurons of the cord are located in the gray matter. The white matter, which comprises all other portions of the cord, is composed of fiber tracts. Several long *descending tracts*, shown on the left side of the cord in Figure 17-1A, originate in the brain and pass down the cord to terminate on the neurons in the gray matter; these are all motor tracts. But several other long tracts, the *ascending tracts*, shown on the right side of the cord in the figure, originate in the cord and then pass upward to the brain; these are the sensory tracts, most of which were discussed in the previous chapter.

In addition to the long descending and ascending tracts, many fibers called *propriospinal fibers* pass from one region of the cord to another, as shown in Figure 17-1B.

Figure 17-1B shows the organization of the neurons in the gray matter of the cord. First, let us study the lowermost segment of the cord in this figure. Note that a sensory nerve fiber enters the posterior horn of the gray matter and that it terminates on an *interneuron*, which is one of the "integrating" neurons of the cord. From here signals are transmitted to other interneurons and then eventually to cells in the anterior horn of the gray matter called *anterior motoneurons.* It is these cells that give rise to the motor axons that leave the spinal cord and pass by way of the motor nerves to the muscles.

Note also in Figure 17-1B the *propriospinal tracts* that interconnect the spinal neurons from one cord segment to the next.

In the gray matter of the spinal cord, one finds almost all the different types of neuronal circuits that were discussed in Chapter 15. Especially important are the *reverberatory circuits*, for these determine many if not most of the spinal cord motor activities. For example, an acute signal entering the spinal cord through a sensory nerve can set a circuit into reverberation in the local cord segment, and this circuit can then transmit a continuous series of impulses to specific muscles to cause a prolonged and intricate motor activity. This is one type of *spinal cord reflex*, as will be described in the following sections.

THE CORD REFLEXES

A nervous reflex is a motor reaction elicited by a sensory signal. For example, a prick of the skin, with subsequent passage of the signal through the spinal cord to cause a muscle jerk, is a reflex. It was pointed out in Chapter 15 that a large number of the functions of the nervous system are mediated through simple or complex reflexes, some of which are integrated entirely in the spinal cord, but some of which utilize signal pathways all the way to the brain and thence back to the muscles or glands to cause motor reactions.

Three essentials must always be present for a reflex to occur: a *receptor* organ, a *nervous transmission system*, and an *effector* organ. All of the sensory nerve receptors are receptor organs for reflexes, and all the muscles of the body,

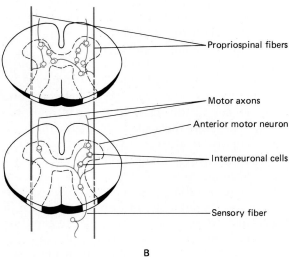

FIGURE 17–1 A, The descending (motor) and ascending (sensory) tracts of the spinal cord. B, Two segments of the cord, illustrating the interneuronal mechanisms that integrate cord reflexes and the propriospinal fibers that interconnect the cord segments.

whether they be smooth muscles or skeletal muscles, are effector organs. Glandular cells that can be stimulated by nerve impulses are also effector organs. The nervous transmission system includes both the sensory nerve fiber from the receptor and the motor fiber to the effector, as well as all the synapses and nerve cell bodies in between.

Spinal Cord Reflexes That Help to Control Muscle Function: the Muscle Spindle Reflex and its Damping Function

Most physical functions of the body are performed by contraction of the skeletal muscles. However, if the muscles are to perform properly, it is essential that the central nervous system know at all times the ongoing state of the muscle, that is, how much it is already contracted, its tension, how rapidly it is contracting, and so forth. Much of this information is provided by two different types of receptors, one called *mus-*

cle spindles, located throughout the belly of the muscle, and the other called *Golgi tendon apparatuses*, located in the tendons. These receptors transmit their information into the spinal cord, and from there information is also relayed to both the cerebellum and the motor area of the cerebrum in the brain. This information is transmitted entirely at a subconscious level, but it helps both the brain and the spinal cord control muscle function.

Figure 17–2 illustrates the basic essentials of the muscle spindle. It consists of several small specially adapted muscle fibers called *intrafusal fibers*. The ends of these intrafusal fibers are connected to the sheaths of the surrounding skeletal muscle fibers. The intrafusal fibers can contract in exactly the same manner as other muscle fibers, and they are excited by a special type of motor nerve fiber called *gamma fibers*.

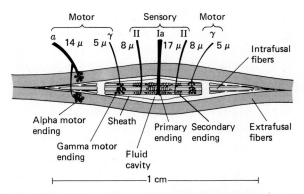

FIGURE 17–2 The muscle spindle, showing its relationship to the large extrafusal skeletal muscle fibers. Note also both the motor and the sensory innervation of the muscle spindle.

However, the middle portion of each intrafusal fiber does not have the ability to contract. Instead, when the end portions of the fiber contract, the middle portion elongates. Or, when the skeletal muscle fibers surrounding the muscle spindle elongate, so also does this stretch the entire spindle and thereby cause the middle of the intrafusal fibers to elongate.

Wrapped around the middle portion of the intrafusal fibers are several nerve endings that are the sensory receptors of the muscle spindle. These transmit signals into the spinal cord to apprise the central nervous system of the degree of elongation of the middle of the spindle.

Function of the Muscle Spindle to "Damp" Muscle Movements—the "Stretch Reflex."

When a muscle is suddenly stretched, the middle of the spindle also is stretched, and this sends an immediate sensory signal into the spinal cord, as illustrated to the right in Figure 17–3. This signal in turn excites the motor nerves that contract the skeletal muscle fibers on all sides of the muscle spindle. Therefore, sudden stretch of the muscle causes an immediate reflex contraction of the same muscle, which automatically opposes further stretch of the muscle. This effect is called the "stretch reflex," and it functions to *damp* changes in muscle length. That is, it prevents the length of the muscle from changing rapidly.

Another feature of the damping process is that it can be turned on or off by stimulation or inhibition of the gamma efferent fibers that supply the intrafusal muscle of the spindle. When gamma efferent signals excite the spindle, the two end portions of the spindle become taut. Under this condition, the spindle reacts rapidly and strongly to any degree of stretch. On the other hand, when the gamma efferent fibers are silent, the muscle spindle becomes flaccid and does not react to stretch. One can understand very readily the importance of this ability to turn on or off the damping mechanism. For instance, if a person is performing a delicate function with his fingers, it is highly important that the muscles of the shoulder and elbow be highly damped so that even the slightest force against the hand will initiate an immediate reflex to prevent the hand from being displaced rapidly from its point of fixation. In contrast, when a person wishes to flail his arm in a wide arc, it is essential that the damping mechanism be suppressed, which can be achieved by simply turning off the signals that pass through the gamma efferent fibers to the muscle spindles.

The Knee Jerk and Other Muscle Reflexes.

One of the best known examples of the stretch reflex is the knee jerk, shown in Figure 17–4. Almost every doctor tests this reflex when he performs a physical examination. A small hammer is used to strike the muscle tendon immediately below the kneecap. The sudden jolt to the tendon stretches the quadriceps muscle in the thigh, which in turn stretches muscle spindles and sends signals to the spinal cord. Then a reflex signal passes back to the same quadriceps muscle, causing a sudden contraction that makes the lower leg jerk forward.

A muscle jerk similar to the knee jerk can be elicited in any muscle of the body by suddenly striking its tendon, or even by striking the muscle itself. The only essential for eliciting such a reflex is to stretch the muscle suddenly.

Muscle reflexes of this type are elicited by physicians for two major purposes: First, if the reflex can be demonstrated, it is certain that both the sensory and motor nerve connections are intact between the muscle and the spinal cord. Second, the degree of reactivity of the mus-

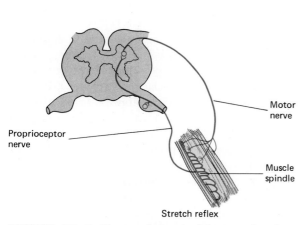

Motor
nerve

Proprioceptor
nerve

Muscle
spindle

Stretch reflex

FIGURE 17–3 Neuronal circuit of the stretch reflex.

FIGURE 17–4 Method for eliciting the knee jerk.

cle reflexes is a good measure of the degree of excitability of the spinal cord. When a large number of facilitatory nerve signals are being transmitted from the brain to the cord, the muscle reflexes will be so active at times that simply tapping the knee tendon with the tip of one's finger might make the leg jump a foot. On the other hand, the cord may be intensely inhibited by inhibitory nerve signals from the brain, in which case almost no degree of pounding on the muscles or tendons can elicit a response.

The "Servo-Assist" Function of the Muscle Spindle. Experiments have shown that when the sensory nerve roots to the spinal cord have been severed, more nervous energy than normal is required to cause a muscle movement. Therefore, it has been suggested that the muscle spindle plays an important role in helping to provide much of the nervous energy needed to cause motor movements. This function of the muscle spindle is believed to operate similarly to the way that "power steering" works in the automobile, utilizing the principle called *servo-assist*. This may be explained as follows:

When a nerve signal is transmitted from the spinal cord to cause contraction of the large skeletal muscle fibers, a simultaneous signal is transmitted through the gamma motor nerve fibers to cause contraction of intrafusal fibers of the muscle spindles as well. This contraction of the intrafusal fibers stretches the sensory elements of the spindles that transmit spindle signals back into the spinal cord. These immediately elicit a muscle spindle reflex that sends large numbers of additional nerve impulses to the large skeletal muscle fibers. Thus, contraction of the skeletal muscles results from two sets of signals: (1) the original signals sent by the spinal cord directly to the large muscle fibers and (2) the additional signals transmitted indirectly by way of the muscle spindle and the muscle spindle reflex. The strength of contraction is, therefore, much greater than it would be in response to the initial signal alone. This is similar to the way that power steering operates; that is, a relatively weak torque on the steering wheel provides a powerful steering effect because of the servo-assistance of the power-steering mechanism.

The Tendon Reflex

Figure 17–5 illustrates the *Golgi tendon receptor*, which is present in all muscle tendons. It detects the amount of tension in the tendon and transmits this information into the spinal cord and from there also into the cerebellum. The information, in turn, is used by the neural mecha-

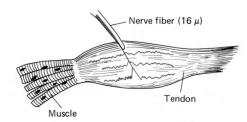

FIGURE 17–5 Anatomy of the Golgi tendon receptor.

nism to help adjust precisely the tension that is needed to perform the required muscle function.

A second function of the tendon receptors is to protect the muscle itself from being overstretched. When the tension detected by these receptors becomes extremely great, great enough that it might cause tearing of the muscle or rupture of the tendon itself, the neuronal centers in the spinal cord automatically and instantly initiate a reflex to *inhibit* the anterior motoneurons that innervate the muscle. The muscle immediately relaxes, and the excessive stretch is removed from the muscle, a very important protective reflex to prevent damage to either muscle or tendon.

The neuronal circuit in the spinal cord that subserves the Golgi tendon reflex is much more complicated than that for the muscle spindle reflex, for it utilizes multiple interneurons in the spinal cord between the input sensory neuron and the anterior motoneuron, in contrast to the muscle spindle reflex in which the sensory neuron makes some direct contacts with the motoneurons without going through any interneurons.

Spinal Cord Reflexes that Help to Support the Body Against Gravity

The Extensor Thrust Reflex. A complex cord reflex that helps support the body against gravity is the *extensor thrust reflex*. Pressure on the pads of the feet causes automatic tightening of the extensor muscles of the legs. This reflex is initiated by pressure receptors in the bottom of the foot. The signal passes first to the interneurons in the cord. Here it is amplified and diverged into an appropriate pattern of impulses to tighten the extensor muscles, causing the animal or person to keep his leg stiffened automatically when standing.

The Magnet Reaction. A reflex closely related to the extensor thrust reflex, but still more complicated, is the so-called *magnet reaction*. In an animal that has had its spinal cord cut so that impulses from the brain will not interfere, one can place the tip of his finger on the pad of the animal's foot, then move his finger in all directions, and the foot will follow the finger. Moving the finger to one side causes appropriate proprioceptor reflexes to make the limb move in the direction of the force. This reaction is an aid to the equilibrium of the animal, for excess pressure on one side of the foot indicates that he is falling in that direction, and automatic stiffening of the leg in that direction helps to prevent the fall.

Reflexes that Help to Prevent Damage to the Body

The Flexor Reflex or Withdrawal Reflex. Pain causes automatic withdrawal of any pained portion of the body from the object causing the pain. This is called a *flexor reflex* or sometimes simply a *withdrawal reflex*.

The neuronal mechanism of the flexor reflex is shown to the left in Figure 17–6. On entering the gray matter of the cord, the pain signal stimulates interneurons that transmit a pattern of impulses through appropriate anterior motoneurons to cause withdrawal of the area of the body that is being pained. In Figure 17–6 the pain stimulus is initiated in the right hand, and the biceps muscle, which is a "flexor" muscle because it flexes the arm, becomes excited and pulls the hand away.

The Crossed Extensor Reflex. When a flexor reflex occurs in one limb, impulses also pass to the opposite side of the cord, where they stimulate the interneurons controlling the extensor muscles of the opposite limb, thus causing it to extend. This is called a *crossed extensor reflex*. For example, a painful stimulus applied to the right hand causes the left arm to extend as shown in Figure 17–6. This reflex pushes the person away from the painful object. That is, at the same time that he withdraws his pained hand on the right, he pushes his entire body away with the left hand.

Withdrawal and cross extensor reflexes are present in all parts of the body, though organized somewhat differently in different areas. For instance, if a needle pricks the small of a person's back, he automatically arches forward, which causes withdrawal. Though this reflex is slightly different from the flexor and crossed extensor reflexes of the limbs, its result is the same.

Reciprocal Inhibition

A very important feature of most reflexes, especially illustrated by the flexor and crossed extensor reflexes, is the phenomenon called *reciprocal inhibition*. That is, when a reflex excites any muscle, it ordinarily inhibits the opposing muscle at the same time. For example, when the flexor reflex of Figure 17–6 excites the biceps muscle, it simultaneously inhibits the opposing triceps muscle. Also, the crossed extensor reflex excites the triceps but inhibits the biceps. In all parts of the body where opposing muscles exist, a corresponding reciprocal inhibition circuit is present in the spinal cord. This obviously allows greater ease in the performance of desired activities.

RHYTHMIC REFLEXES OF THE CORD

The reflexes of the spinal cord are not limited to simple reactions; frequently rhythmic reflexes also occur. That is, "patterns of activity" in the cord can make a portion of the body move back and forth in rhythmic motion. Two especially important types of rhythmic reflexes controlled entirely by the spinal cord are the scratch reflex and the walking reflexes.

The Scratch Reflex

The scratch reflex, though not of importance in the human being, is one of the most important of all protective mechanisms in lower animals. Under appropriate conditions, a dog can actu-

FIGURE 17–6 The flexor reflex, the crossed extensor reflex, and reciprocal inhibition.

ally scratch away a flea or other irritating object even after its spinal cord has been completely transected at the neck level.

The scratch reflex depends on two separate integrative abilities of the spinal cord: First, the cord must provide the rhythmic to-and-fro motion of the leg. This is accomplished by reverberating circuits in the interneurons that send impulses first to one group of muscles, then to the opposing muscles, contracting alternately the two sets of muscles to cause the rhythmic motion.

The second cord mechanism utilized in the scratching act pinpoints the area of the body that needs scratching. For instance, a flea moving across the belly of a sleeping dog is followed by the scratching paw. When the flea crawls across the midpoint of the animal's belly, the paw stops scratching, but the paw on the opposite side of the body immediately finds the flea and begins to scratch. Thus, the scratch reflex is one of the most complicated of all cord reflexes, for it requires, first, localization of the irritation; second, coordinate movement of the paw into position; and third, reverberating motion of the limb.

Walking Reflexes

The spinal cord also is capable of providing the rhythmic to-and-fro movements of the legs that are used in walking. These movements, like the scratching mechanism, are controlled by reverberating circuits in the interneurons that contract alternately the opposing pairs of muscles. Reciprocal inhibition also plays a part, for this keeps the muscles on the two sides of the body operating in opposite phase to each other, causing one leg to move forward while the opposite one moves backward. In lower animals, fiber pathways also pass from the hindlimb region of the cord to the forelimb region to cause appropriate phasing of the hind and forelimb movements. This cord control of walking movements with appropriate phasing of the limbs is illustrated in Figure 17–7, which shows a dog with its spinal cord transected in the neck; it is, nevertheless, still exhibiting continuous, rhythmic, to-and-fro movements of its limbs.

AUTONOMIC REFLEXES OF THE CORD

In addition to the cord reflexes for controlling skeletal muscular action, the spinal cord also harbors reflex circuits that help to control the visceral functions of the body. There reflexes,

FIGURE 17–7 Walking movements in a dog whose spinal cord has been transected at the neck.

called *autonomic reflexes*, are initiated by sensory receptors in the viscera, and the signals are transmitted through sensory nerves to the interneurons of the cord gray matter, where appropriate patterns of reflex responses are determined. Then the signals pass to *autonomic motor neurons* located in the cord gray matter about midway between the anterior and posterior horns. These cells send impulses into the *sympathetic nerves* and back to the viscera to cause autonomic stimulation.

The Peritoneal Reflex

One of the most important of the autonomic reflexes is the *peritoneal reflex*. Whenever any portion of the peritoneum (the surface lining of the gut and abdominal cavity) is damaged, this reflex slows up or stops all motor activity in the nearby viscera. For example, in appendicitis, which almost always irritates the peritoneum, movement of food through the gastrointestinal tract stops almost completely, preventing further irritation of the inflamed appendix and allowing the reparative process of the body to function at optimum efficiency. Without this reflex many persons would die of appendicitis before aid could be given.

Vascular Control by Cord

The resistance to blood flow in most peripheral blood vessels is normally controlled by vasomotor signals transmited from the brain, but after the spinal cord has been transected, these signals can no longer travel from the brain to the peripheral nerves. Yet, cord autonomic reflexes are still capable of modifying local vascular resistance in response to such factors as pain, heat, and cold. For instance, when one of the viscera is pained tremendously by overdistention, such as overfilling of the urinary bladder, the vascular resistance often doubles and thereby raises the arterial pressure to as high as twice normal. Though this effect of cord reflexes on the arterial pressure is not important normally, it does illustrate the capability of the cord to help control many of the involuntary reactions of the autonomic nervous system.

The Bladder and Rectal Reflexes

Among the most important of the autonomic reflexes of the cord are the bladder and rectal reflexes that cause automatic emptying of the urinary bladder and of the rectum when they become filled. When either the bladder or the rectum becomes excessively full, sensory signals are transmitted into the interneurons of the lower end of the cord, as shown in Figure 17–8. Appropriate signals are then transmitted through parasympathetic nerves back to the bladder or colon. Here they excite the main body of the bladder or colon but at the same time inhibit the internal sphincter of the urethra or anus, thus causing emptying.

In persons whose spinal cords have been cut, these automatic reflexes are sometimes effective enough to empty the bladder every half-hour or more and the colon one or more times each day. In the normal person, however, these reflexes are inhibited by signals from the brain until an opportune time arises for emptying. Unfortunately, this inhibition frequently leads to constipation or painful distention of the bladder.

MOTOR FUNCTIONS OF THE BRAIN STEM

The brain stem is composed of the medulla, the pons, and the mesencephalon, the anatomy of which was discussed in Chapter 13. It is in these areas that many of the centers for arterial pressure control, respiration, and gastrointestinal regulation are located, all of which are discussed in detail in other chapters. However, the motor functions of the brain stem that are related to the skeletal muscle system can be divided into two major types: first, its function in helping the person support his body against gravity, and, second, its function in the maintenance of equilibrium. But first we need to describe the *bulboreticular formation*, the brain structure that is mainly responsible for these functions.

The Bulboreticular Formation of the Brain Stem. Figure 17–9 shows the bulboreticular formation, which extends through

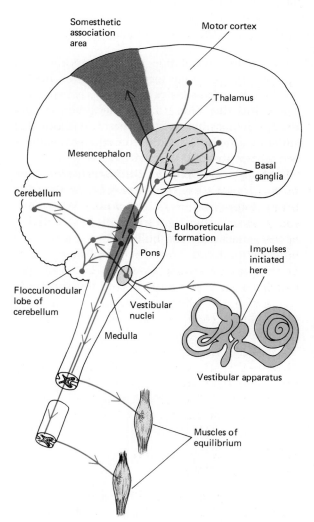

FIGURE 17–9 Nervous mechanisms of body support against gravity and equilibrium.

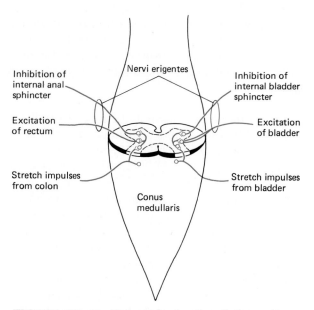

FIGURE 17–8 Neuronal circuits of the urinary bladder- and rectal-emptying reflexes.

the entire brain stem, and also shows some of its connections with other parts of the brain as well as parts of the body. Stimulation of this area transmits signals down several different tracts into the cord to help control the muscles, especially the *reticulospinal* and the *vestibulospinal* tracts.

The bulboreticular area also receives incoming fibers from many sources: fibers directly from all areas of the body through the spinal cord, fibers from the cerebellum, from the equilibrium sensor (the *vestibular apparatus*), from the motor portion of the cerebral cortex, and from the basal ganglia deep in the cerebrum. Thus, the bulboreticular formation is an integrative area for combining and coordinating (1) sensory information from the body, (2) motor information from the motor cortex and basal ganglia, (3) equilibrium information from the vestibular apparatuses, and (4) information about body movements from the cerebellum. With this information available, it controls many of the involuntary muscular activities.

Support of the Body Against Gravity

Widespread stimulation of the middle and upper bulboreticular formation *excites mainly the extensor muscles* of the body, causing the trunk and the limbs to become stiffened, and making it possible to stand erect. Without this stiffening, the body would immediately crumple because of the pull of gravity.

Suppression of the Bulboreticular Area by the Basal Ganglia; Decerebrate Rigidity When Suppression Is Blocked. When one wants to sit rather than to stand, the excitation of the muscles by the bulboreticular formation must be inhibited. This is accomplished mainly by suppressor signals from the basal ganglia, which lie anterior and lateral to the thalamus.

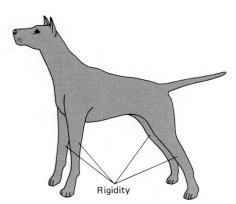

Rigidity

FIGURE 17–10 Rigidity in a dog whose brain has been transected above the bulboreticular formation.

When the basal ganglia are damaged, or when the fiber tracts from them are cut, the bulboreticular formation automatically becomes so active that the animal becomes rigid all over. This phenomenon, called *decerebrate rigidity*, is illustrated in the dog of Figure 17–10. It shows the necessity of the basal ganglia to inhibit the bulboreticular formation when motor functions other than the support of the body are to be performed. (The animal or person also relaxes when resting or when asleep, which also inhibits the bulboreticular area.)

Equilibrium Function of the Brain Stem

In addition to providing the neurogenic mechanisms for support of the body against gravity, the bulboreticular formation is also capable of maintaining equilibrium by varying the degree of tone in the different muscles. Most of the equilibrium reflexes are initiated by the **vestibular apparatuses** located on each side of the head, adjacent to each internal ear.

Anatomy of the Vestibular Apparatus. The vestibular apparatus is closely associated with the ear and is actually considered by many anatomists to be part of the inner ear, which is located in a series of contorted bony canals within the petrous portion of the temporal bone. Inside these bony canals is the so-called **membranous labyrinth** illustrated in Figure 17–11A. The anterior portion of the labyrinth, shown to the right in the figure, is called the **ductus cochlearis** or simply the **cochlea.** This portion functions entirely for the purpose of hearing. (The ear is discussed in Chapter 22.)

The posterior portion of the membranous labyrinth, shown to the left in Figure 17–11A, is the **vestibular apparatus.** It, in turn, is divided into two separate anatomical physiological parts:

1. The **utricle** and **saccule.** These are hollow, fluid-filled chambers that house special sensory structures called *maculae.* The maculae detect the position of the head with respect to the direction of pull of gravity.
2. The **semicircular canals.** These are three separate fluid-filled circular canals that are arranged in the three planes of space. At one end of each of these canals is a bulbous enlargement called the **ampulla,** and inside this is a sensory structure, the **crista ampullaris,** that detects movement of fluid in the semicircular canal, thus detecting rotation of the head in the plane of that canal.

Membranous labyrinth

A

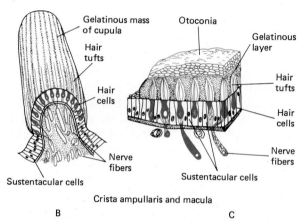

Crista ampullaris and macula

B   C

FIGURE 17-11 Functional structures of the vestibular apparatus. (From Goss: Gray's Anatomy of the Human Body. Philadelphia, Lea & Febiger; modified from Kolmer by Buchanan: Functional Neuroanatomy. Philadelphia, Lea & Febiger.)

The maculae of the utricle and saccule and the cristae ampullaris of all the separate semicircular canals transmit their signals into the central nervous system through the **vestibulocochlear nerve** (cranial nerve VIII) as illustrated in Figure 17-9.

Function of the Maculae. The locations of the maculae on the walls of the utricle and saccule are illustrated in Figure 17-11A, and the microscopic structure is shown in Figure 17-11C. Many nerve cells, called "hair cells," are located at the base of the macula. From these cells, cilia, also called "hairs," project upward into a *gelatinous mass* that overlies the macula. In this mass are many minute bonelike calcified granules, like very small granules of sand, called *otoconia*, which are much heavier than the surrounding fluid or tissues. When the head is bent to one side, the weight of the otoconia pushes

the hairs to that side, thereby stimulating the nerves. The different hair cells are oriented in different directions. Therefore, one pattern of hair cells will be stimulated when the head is bent to the right, another when bent to the left, and so forth. In this way, the maculae of the utricle and saccule supply the equilibrium region of the central nervous system with the information needed to maintain balance.

The maculae also help the person maintain his balance when he suddenly begins to move forward, to one side, or in any other linear direction. When he begins to move forward, the inertia of the otoconia causes them to lag behind the movement of the rest of the body and, therefore, to bend the hairs backward. This gives the sensation of falling off balance in the backward direction. As a result, the person automatically leans forward to correct this imbalance, which explains why an athlete leans forward when he first starts to run.

On the other hand, when one wishes to decelerate, he must lean backward. Again it is the otoconia of the maculae that initiate automatically this backward leaning, for when one brakes himself the momentum of the otoconia keeps them moving forward while the rest of the body slows. This bends the hairs of the maculae in the forward direction, making the person feel as if he were falling head first toward the ground. As a result, the equilibrium mechanism causes the body to lean backward automatically.

Function of the Semicircular Canals. The semicircular canals also are illustrated in Figure 17-11A, and the structure of a crista ampullaris, the sensory structure of each canal, is shown in Figure 17-11B. The three separate semicircular canals are located, respectively, in the three planes of space, one in the horizontal and the other two in the two vertical planes. If one suddenly turns his head in any direction, the fluid in one or more of the semicircular canals lags behind the movement of the head because of the inertia of the fluid. This is the same effect as that observed when one suddenly rotates a glass of water; the glass rotates, but the water remains still. As the fluid moves in the semicircular canal it flows against the *crista ampullaris*, shown in Figures 17-11A and B, which is a valvelike leaflet located at one end of each canal. This structure contains hair tufts (called *cilia*) projecting from hair cells like those in the macula, and bending these cilia to one side or the other gives the person the sensation that his head is *beginning to turn.*

The information transmitted from the semicircular canals apprises the nervous system of sudden *changes in direction of movement.* With this information available, the bulboreticular formation can correct for any imbalance that is likely to occur when running around a corner *even before the imbalance does occur.* This is particularly important when one is changing direction of movement rapidly, as when playing a fast game of almost any type.

Function of the Cerebellum in Equilibrium. In addition to transmitting signals into the bulboreticular formation, the semicircular canals and the maculae also send information to the **flocculonodular lobes** of the cerebellum. Since one of the major functions of the cerebellum is to predict future position of the body in space, which will be discussed in the following chapter, the function of the flocculonodular lobes is probably to predict when a state of imbalance is going to occur. This allows appropriate corrective signals to be given to the bulboreticular formation even before the person falls off balance, and prevents imbalance from occurring rather than necessitating attempts to correct it after it has already occurred. Persons who lose their cerebellum lose the ability to predict, and as a result, must perform all movements slowly or else fall.

The Neck Proprioceptor Receptors and Their Relationship to the Equilibrium Mechanisms. The vestibular apparatuses detect only the position of the *head*, not of the *body*, in relation to the pull of gravity. To translate this information from the head to the whole body, the relationship of the head to the body must be known. This knowledge is provided by proprioceptor receptors (position receptors) in the neck. For instance, when the head bends backward, the vestibular apparatuses send information that the position of the head with respect to gravity is changing, but at the same time the neck proprioceptors send information that the head is angulating backward in relation to the rest of the body. The two sets of information cancel each other, thus allowing the brain to determine that the position of the body with respect to gravity has not changed despite the bending of the head. Therefore, the tautness of the different postural muscles remains exactly the same. In other words, the proprioceptor reflexes from the neck are as necessary for controlling equilibrium as are the complicated reflexes initiated by the vestibular apparatuses.

Peripheral Proprioceptor and Visual Mechanisms of Equilibrium. If the vestibular apparatuses have been destroyed, a person can still maintain his equilibrium provided he moves slowly. This is accomplished mainly by means of proprioceptor information from the limbs and surfaces of the body and visual information from the eyes. If he begins to fall forward, the pressure on the anterior parts of his feet increases, stimulating the pressure receptors. This information transmitted to his brain helps to correct the imbalance. At the same time, his eyes also detect the lack of equilibrium, and this information, too, helps to correct the situation.

Unfortunately, the visual and proprioceptor systems for maintaining equilibrium are not organized for rapid action, which explains why a person without his vestibular apparatuses must move slowly.

OVERALL CONTROL OF LOCOMOTION

From the foregoing discussions, it is now possible to begin constructing the overall pattern of locomotion. First, the person must support himself against gravity. This is accomplished partly by the extensor thrust mechanism of the spinal cord, which allows stiffening of the limbs when pressure is applied to the pads of the feet; in addition to this, the bulboreticular formation also transmits signals to the extensor muscles to keep the body and limbs stiff. The degree of stiffening of the different parts of the body, however, is varied by the equilibrium system so that when one tends to fall over, appropriate muscles are contracted to bring the body back into the upright position.

Once the body is supported against gravity and maintained in a state of equilibrium, locomotion then depends on rhythmic motion of the limbs. Rhythmic circuits in the spinal cord are capable of providing the to-and-fro movements of the limbs, and the movements of the opposing limbs are kept in opposite phase with each other by the reciprocal inhibition mechanism of the cord. Thus, most of the functions of locomotion can be provided by the cord and brain stem, but the cerebral cortex must control these functions in accord with the desires of the individual. When he wishes to move forward, to stop, or to turn to one side, his motor cortex and basal ganglia simply initiate the action, stop it, or change it by sending *command signals.* The cord and brain stem provide the stereotyped actions required to perform the actual movements. In this way, the energy of the conscious portion of the brain is conserved to perform other mental feats.

QUESTIONS

1. Describe the physiologic anatomy of the neuronal circuits in the spinal cord.
2. Describe the function of the muscle spindle and of the muscle spindle reflex.
3. Explain the servo-assist function of the muscle spindle.
4. Explain the function of the tendon reflex.
5. How do cord reflexes help to support the body against gravity?
6. What is the functional importance of the flexor reflex and the crossed extensor reflex?
7. Describe the walking reflexes of the spinal cord.
8. What is the role of the bladder and rectal reflexes?
9. Explain the function of the bulboreticular formation in the support of the body against gravity and in the equilibrium process.
10. Describe the vestibular apparatus and its functions in static equilibrium and in equilibrium during motion.
11. Discuss the overall control of locomotion.

REFERENCES

Burke, R.E., and Rudonmin, P.: Spinal neurons and synapses. *In* Brookhart, J.M., and Mountcastle, V.B. (eds.): Handbook of Physiology. Sec. 1, Vol. 1. Baltimore, Williams & Wilkins, 1977, p. 877.

Creed, R.S., *et al.*: Reflex Activity of the Spinal Cord. New York, Oxford University Press, 1932.

Evarts, E.V.: Brain mechanisms of movement. *Sci. Am.*, *241* (3):164, 1979.

Gallistel, C.R.: The Organization of Action: A New Synthesis. New York, Halsted Press, 1979.

Granit, R., and Pompeiano, O. (eds.): Reflex Control of Posture and Movement. New York, Elsevier Scientific Publishing, 1979.

Hobson, J.A., and Brazier, M.A.B. (eds.): The Reticular Formation Revisited: Specifying Function for a Nonspecific System. New York, Raven Press, 1980.

Precht, W.: Vestibular mechanisms. *Annu. Rev. Neurosci.*, 2:265, 1979.

Sherrington, C.S.: The Integrative Action of the Nervous System. New Haven, Conn., Yale University Press, 1911.

Stein, P.S.G.: Motor systems with specific reference to the control of locomotion. *Annu. Rev. Neurosci.*, *1*:61, 1978.

Talbot, R.E., and Humphrey, D.R. (eds.): Posture and Movement. New York, Raven Press, 1979.

Treischmann, R.B.: Spinal Cord Injuries. New York, Pergamon Press, 1979.

Wilson, V., and Jones, G.M.: Mammalian Vestibular Physiology. New York, Plenum Press, 1979.

18

Control of Muscle Activity by the Cerebral Cortex, the Basal Ganglia, and the Cerebellum

Overview

The more complex motor activities of the body are controlled by the **cerebral cortex, basal ganglia,** and **cerebellum,** with these three areas almost always functioning together rather than separately. However, removal of the *cerebral cortex* in some lower animals has little effect on the animals' ability to *walk, run,* and even *fight.* What they have lost is purposefulness in their motor activities. In the human being, on the other hand, the cerebral cortex is far more highly developed, and correspondingly more of our motor activities are therefore controlled by the cortex. Damage to the cortex in the human being mainly causes *loss of the functional abilities of the hands, fingers,* and *distal parts of the arm,* although the more gross movements of the trunk, legs, and shoulders still remain partially intact. On the other hand, if the *basal ganglia* are damaged severely along with the cortex, even the *gross body movements* of the human being become severely impaired.

The *motor area* of the cerebral cortex is located in the frontal lobe, immediately *in front of the central sulcus.* The posterior part of this motor area, called the *primary motor cortex,* controls either individual muscles or closely associated muscles, especially the discrete muscles of the hands, fingers, and mouth, thus allowing very delicate control. The more anterior portions of the motor area, called the *premotor cortex,* control coordinate contraction of multiple muscle groups to allow performance of many skilled motor activities.

The *cerebellum* functions in association with all other motor areas of the nervous system, including the motor cortex, basal ganglia, and spinal cord, to coordinate mainly *sequential muscle contractions.* The cerebellum has a special type of neuronal circuit that allows signals to be delayed for various fractions of a second. Therefore, when one needs to perform two different movements one in sequence after another, the cerebellum provides the appropriate delay interval between the sequential motor activities. For instance, during walking, each step consists of sequential forward and backward movements of the leg. If the sequencing is not perfect, the person loses his balance, which is exactly what does happen when the cerebellum is severely damaged—a condition called *ataxia.*

The *control of speech* is an especially interesting example of highly complex motor control. The words to be spoken are chosen not by the motor cortex but instead by a part of the sensory cortex called *Wernicke's area,* located in right-handed persons in the *posterior-superior part of the left temporal lobe.* Wernicke's area transmits appropriate signals to *Broca's area,* located in right-handed persons in the left premotor cortex. This area operates in association with the primary motor cortex, the basal ganglia, and the cerebellum to control the sequences of laryngeal, oral, and respiratory muscle contractions necessary for formation of different words.

Although a major share of the motor functions of the body can be performed without involvement of the higher brain centers, one of the distinguishing features of the human being is the ability to carry out extremely complex voluntary muscle activity. He can perform the intricate tasks of talking, writing, using delicate instruments, and achieving specialized patterns of movement required for dance routines, basketball, football, and so forth. All of these activities involve major degrees of control by the higher centers of the brain, especially the cerebral cortex, the basal ganglia, and the cerebellum. It is the goal of this chapter to explain how these higher centers work together to give these special abilities.

Basic Organization of the Higher Centers for Muscle Control

Figure 18–1 illustrates the interplay among the brain centers for muscle control. These centers are composed of four major parts—the bulboreticular formation, which was discussed in the previous chapter, and three additional parts to be discussed here: (1) the motor cortex, (2) the basal ganglia, and (3) the cerebellum.

The **motor cortex** lies slightly anterior in the cerebral cortex, immediately in front of the central sulcus. It is this area that plays the greatest role in control of very fine, discrete muscle movements.

The **basal ganglia** lie deep in the cerebral hemispheres, and they are composed of separate large pools of neurons organized for control of complex semivoluntary movements, such as walking, turning, running, and development of bodily postures for performance of specific functions.

The **cerebellum** is located posteriorly and inferiorly in the brain, lying behind the lower

brain stem. As we shall see later in the chapter, the cerebellum communicates with the other motor areas of the brain through several large neuronal trunks. The cerebellum helps both the motor cortex and the basal ganglia perform their functions. Mainly, it makes the movements smooth rather than jerky. And it also helps to make groups of muscles operate together in a coordinate manner so that very accurate and very fine degrees of muscle control can be achieved. In other words, unlike both the basal ganglia and the motor cortex, the cerebellum is not directly responsible for control of muscle activity but, instead, operates as a helper to the other two areas.

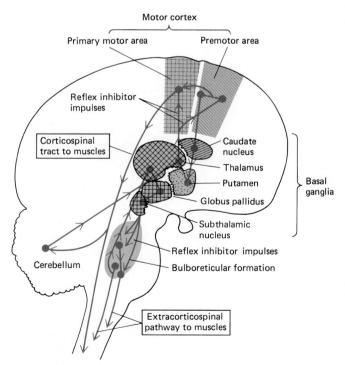

FIGURE 18–1 Control of muscle activity by the motor cortex, basal ganglia, and cerebellum.

Transmission of Motor Signals to the Spinal Cord

The Corticospinal Tract.   Close study of Figure 18–1 will show that motor signals are transmitted from the brain to the spinal cord through two separate pathways, the **corticospinal tract** and the **extracorticospinal pathway.**

The corticospinal tract is a direct pathway that originates in the motor cortex and passes without synapse all the way to the cord. Figure 18–2 illustrates this pathway in more detail, showing that it is composed of a discrete bundle of nerve fibers coming from each side of the brain. In the lower part of the medulla and first few segments of the spinal cord, the fibers from each side of the brain *cross to the opposite side*

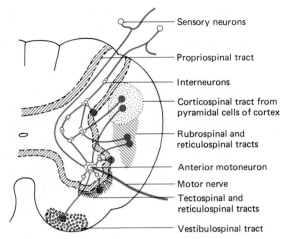

FIGURE 18–3   Convergence of all the different motor pathways on the anterior motoneuron.

and then proceed downward through the two posterolateral portions of the cord. Its location in the cord, lying immediately lateral to the posterior horn of the gray matter, is illustrated in Figure 18–3. Thus, the motor cortex on the left side of the brain controls the muscles of the right side of the body, and the motor cortex of the right side controls those on the left.

The Extracorticospinal Pathway.   The extracorticospinal pathway is composed of all pathways other than the corticospinal tract that transmit motor signals to the cord. These include mainly pathways from the basal ganglia and the bulboreticular formation, as illustrated in Figure 18–1. Figure 18–3 illustrates the location in the spinal cord of the specific tracts from different areas of the brain stem through which the extracorticospinal signals are transmitted. These include the **rubrospinal tract,** two separate **reticulospinal tracts,** the **tectospinal tract,** and the **vestibulospinal tract.** All these tracts are located considerably anterior to the corticospinal tract.

Role of the Spinal Segments in Motor Control

The nerve fibers of both the corticospinal and the extracorticospinal pathways terminate in a special neuronal network of the spinal cord. And it is this network that then sends direct signals to the muscles.

We saw in the previous chapter that local neuronal mechanisms of the spinal cord are responsible for many cord reflexes that elicit muscle activity even without signals from the brain. When signals also come from the brain, these

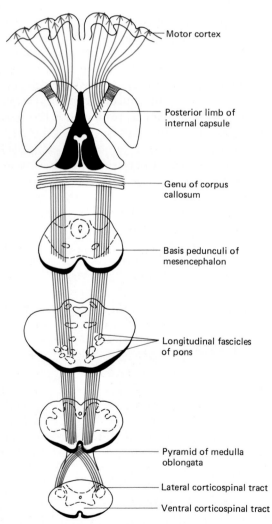

FIGURE 18–2   The corticospinal tract (also called the pyramidal tract), through which motor signals are transmitted from the motor cortex to the spinal cord. (Modified from Ranson and Clark: Anatomy of the Nervous System, 10th Ed. Philadelphia, W.B. Saunders Co., 1959.)

combine with the signals from the sensory nerves entering each spinal segment, and the combined information is utilized by the cord's neuronal network to control muscle activity. Figure 18–3 illustrates this intrinsic network.

The Anterior Motoneuron and the Interneurons. Located in the anterior portion of the spinal cord (illustrated in Figure 18–3) are many very large neurons that send large axons out of the cord into the nerves that supply the muscles. These neurons are called **anterior motoneurons,** and the nerves passing to the muscles are called the **motor nerves.** Note in Figure 18–3 that signals from many different sources impinge on the surface of the anterior motoneuron. Some of the signals arriving at this neuron excite it whereas others inhibit it. Thus, muscle control is much more complex than simply transmitting a signal that always causes muscle contraction.

Figure 18–3 also illustrates many smaller neurons, called **interneurons,** located in the central portions of the spinal cord gray matter. These interneurons receive signals from incoming sensory nerves, from the corticospinal tract, and from the extracorticospinal tracts, and they in turn transmit signals to the anterior motoneuron. Though a few signals bypass the interneurons and go directly to the anterior motoneuron, by far the larger proportion must first be processed by these cells.

It is in the interneurons that many patterns of muscle contraction are determined. For instance, most of the patterns of the spinal reflexes discussed in the previous chapter are determined here, such as the patterns of the flexor reflex, the crossed extensor reflex, the walking reflexes, and different autonomic reflexes.

Control of the Anterior Motoneurons by Brain Signals. When brain signals reach the interneurons of the spinal cord, they do not necessarily cause a muscle contraction. Instead, the brain signal is first combined with both excitatory and inhibitory signals arriving from other sources. If the preponderance of information is in favor of causing muscle contraction, then contraction occurs. On the other hand, if it is not in favor of contraction, then the muscle might remain as is or even be inhibited.

Also important is the ability of the brain signals to "command" the spinal cord neuronal network to perform specific functions. For instance, a command signal from the brain might direct the muscles to perform walking movements. In response, the interneuronal mechanism for achieving walking movements is set into action. It is this mechanism that actually causes the discrete muscle contractions required for walking; the command signal from the brain, on the other hand, has little to do with controlling the individual muscles, but only commands the cord to perform the appropriate function.

If the reader will now allow his imagination to roam, he will understand how motor signals from the brain can command the spinal cord to achieve many complex motor activities without ever directly exciting an individual muscle. Yet, at times it is also important to make single muscles contract, and this too can be achieved. It is signals from the corticospinal tract rather than from the extracorticospinal pathway that have this ability to cause contraction of individual muscles, as will be discussed in more detail later in the chapter.

FUNCTION OF THE BASAL GANGLIA

Now that we have described the general neuronal mechanisms for brain control of muscle function, let us return once again to the higher centers to describe some of the special characteristics of this control. To begin, consider first the functions of the basal ganglia. The basal ganglia are composed of large masses of neurons located deep in the substance of the cerebrum and in the upper part of the mesencephalon. These were shown in Figure 18–1, and their multitude of neuronal connections with other portions of the motor control system are illustrated in Figure 18–4. This multitude of connections allows the basal ganglia to control most of the subconscious stereotyped bodily movements, often involving simultaneous contraction of many muscle groups throughout the body. Though it will not be possible to describe the basal ganglia in detail, the most important of these are the **caudate nucleus,** the **putamen,** the **globus pallidus,** and the **subthalamic nucleus.** In addition, two mesencephalic nuclei, the **substantia nigra** and the **red nucleus,** function in very close association with these basal ganglia of the cerebrum and therefore from a functional point of view are considered to be part of the basal ganglial system.

Control of Subconscious Movements by the Basal Ganglia. Before attempting to discuss the function of the basal ganglia in man, we should speak briefly of their better known functions in lower animals. In birds, for instance, the cerebral cortex is very poorly developed, whereas the basal ganglia are highly developed.

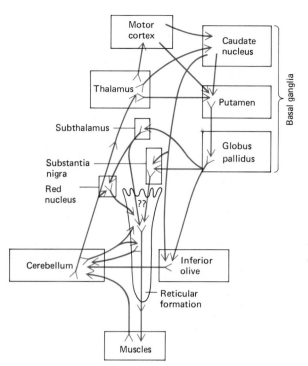

FIGURE 18-4 Interconnections of the basal ganglia with other parts of the motor system.

These ganglia perform essentially all the motor functions, even controlling the voluntary movements in much the same manner that the motor cortex of the human being controls voluntary movements. In the cat, and to a less extent in the dog, removal of the cerebral cortex does not interfere with the ability to walk, to eat perfectly well, to fight, to develop rage, to have periodic sleep and wakefulness, and even to participate naturally in sexual activities. However, if a major portion of the basal ganglia is destroyed, only gross stereotype movements remain, which are controlled by the more primitive areas in the brain stem, especially in the bulboreticular area, which was described in the previous chapter.

In the human being, many of the potential functions of the basal ganglia are suppressed by the cerebral cortex, but if the cerebral cortex becomes destroyed in a very young human being, many voluntary motor functions do develop. The person will never be able to develop very discrete movements, particularly of the hands, but he can learn to walk, to control his equilibrium, to eat, to rotate his head, to perform almost any type of postural movement, and to carry out many subconscious movements. On the other hand, destruction of a major portion of the caudate nucleus almost totally paralyzes the opposite side of the body except for a few stereotyped reflex movements integrated in the cord or brain stem.

Functions of Individual Basal Ganglia

Unfortunately, little is known about the functions of most of the individual basal ganglia other than the fact that they operate together in a closely knit unit to perform many of the subconscious movements. The following are a few of the discrete bits of information that are known about their individual functions:

1. The *caudate nucleus* controls gross intentional movements of the body, these occurring both subconsciously and consciously and aiding in the overall control of body movements.
2. The *putamen* operates in conjunction with the caudate nucleus to help control the gross intentional movements. Both of these nuclei also function in cooperation with the motor cortex to control many other patterns of movement, probably even some of the patterns that are employed in highly complex technical or athletic skills.
3. The *globus pallidus* probably controls the "background" positioning of the gross parts of the body when a person begins to perform a complex movement pattern. That is, if a person wishes to perform a very exact function with one of his hands, he first positions his body appropriately and then tenses the muscles of the upper arm. These functions are believed to be initiated mainly by the globus pallidus.
4. The *subthalamic nucleus* and associated areas possibly control some aspects of the walking process and perhaps other types of gross rhythmic body motions.

Abnormal Muscle Control Associated with Damage to the Basal Ganglia

Even though we do not know all the precise functions of the basal ganglia, we do know many abnormalities that develop when portions of the basal ganglia are destroyed, as follows:

Chorea. Chorea is a condition exhibiting random, uncontrolled sequences of motor movement occurring one after the other. Normal progression of movements cannot occur. Instead, the person may perform one pattern of movement for a few seconds and then suddenly jump to an entirely new pattern, such as changing from straight forward movement of the arm to a rotary movement, and then to a flailing movement, this jumping to a new pattern occurring again and again without stopping. This is caused by widespread damage in the *caudate nucleus* and the *putamen*.

Athetosis.   Athetosis is characterized by slow, writhing movements of peripheral parts of the body. For instance, the hand or arm may undergo wormlike movements, such as twisting of the arm to one side, then to the other side, then backward, then forward, repeating the same pattern over and over again. The damage is almost always in the *globus pallidus*.

Hemiballismus.   Hemiballismus is an uncontrollable succession of violent movements of large areas of the body. For instance, a leg may suddenly kick forward, and this may be repeated once every few seconds or sometimes only once in many minutes. The damage that causes this is in the *subthalamus*.

Parkinson's Disease.   This disease is characterized by *tremor* and *rigidity* of the musculature in either widespread or isolated areas of the body. The typical person with fullblown Parkinson's disease walks in a crouch like an ape, except that his muscles are obviously tense, his face is masklike, and he jerks all over with a violent tremor approximately 6 to 8 times per second. Yet when he attempts to perform voluntary movements, the tremor becomes less or stops temporarily. This disease is caused by destruction of the *substantia nigra*, one of the "functional" basal ganglia that lies anteriorly in the mesencephalon.

FUNCTION OF THE MOTOR CORTEX TO CONTROL MUSCLE MOVEMENTS

As we proceed to a still higher level of motor control, to the cerebral cortex, the types of movements controlled take on several new qualities. First, many more *discrete* movements can be initiated, particularly the very discrete movements by the hands that cannot be controlled by any of the lower centers. Second, the movements are more of the *learned type* rather than of the stereotyped. And, finally, a great majority of the muscle movements controlled by the cerebral cortex are of the *conscious type* rather than of the subconscious type.

The Motor Cortex

The area of the cerebral cortex most directly concerned with muscle control is the area shown in Figure 18–5 called the *motor cortex*. It is located immediately anterior to the central sulcus of the brain, a deep sulcus that separates the front half of the cerebral cortex from the pos-

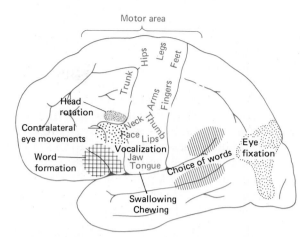

FIGURE 18–5   Representation of the different muscles of the body in the motor cortex, and location of other cortical areas responsible for certain types of motor movements.

terior half. This sulcus also separates the motor cortex from the somesthetic sensory cortex, which lies immediately behind the sulcus.

Stimulation of a point area in the motor cortex causes a specific muscle or a closely allied small group of muscles on the opposite side of the body to contract. Figures 18–5 and 18–6 show the points of the motor cortex for control of muscles in different parts of the body. Note that the muscles of the lower body are controlled by the cortex near the midline, and the muscles of the upper body are controlled far laterally. Thus, there is point-to-point communication between the motor cortex and specific mus-

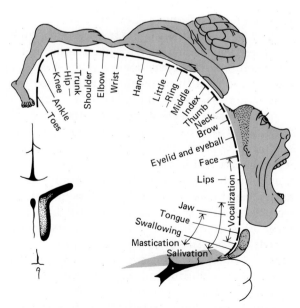

FIGURE 18–6   Degree of representation of the different body muscles in the primary motor cortex. (From Penfield and Rasmussen: The Cerebral Cortex of Man. New York, The Macmillan Co., 1968.)

cles everywhere in the body. It is by means of signals from discrete parts of the motor cortex that discrete muscle movements are achieved, sometimes involving only single muscles, especially in the hands, but more often involving closely allied groups of muscles.

Degree of Representation of Different Muscles in the Motor Cortex. Not all muscles are represented in the motor cortex to the same degree. For instance, stimulation of a single small point in the trunk region of the motor cortex might excite contraction of a large area of back muscles, whereas stimulation of the same amount of cortical tissue in the finger area might cause nothing more than contraction of a single small finger muscle. Figure 18–6 illustrates diagrammatically the different degrees of representation for different types of muscles, showing that the degree of representation of muscles of the thumb and fingers and also of the mouth and throat regions is as much as 100 times that for the trunk muscles. This special representation of the hands allows the cerebral cortex to control with extreme fidelity essentially all of the fine, learned movements of the hands, which is a special capability of the human being and other higher primates. The high degree of representation in the mouth and throat regions accounts for the ability of the human being to talk, an ability that has not been achieved significantly by any lower animals.

Control of Skilled Movements by the Motor Cortex

Essentially all of the skilled movements performed by the human being are learned movements. Furthermore, most skilled movements require function of individual muscles during at least portions of the movement, which involves muscle control by the motor cortex.

Learning Skilled Movements. We all know how awkward a person is when he first begins to learn a highly skilled movement. For instance, a child learning to write letters works meticulously at achieving the appropriate curves and lines. This is a slow and almost painful process that requires a great amount of trial and error. All the while, the sensory nervous system senses the degree of success. The involved sensory signals include those for visual information, somesthetic information, auditory information, and all other types of information that might be useful in determining whether the motor act is successful. After an initial poor degree of success, subsequent attempts to perform the act, combined with sensory correcting signals, fi-

nally lead to perfection. Thus, the sensory system plays a major role in establishing all skilled, learned motor functions in the brain.

Establishment of Rapid Performance of Stereotyped Skilled Movements—the Concept of a "Premotor Area." As one becomes progressively more skilled in the performance of a particular learned act, he also achieves greater and greater speed in its performance. Sometimes it is important that the act be even more rapid than the time it takes for sensory information to be returned to the brain. For instance, in playing basketball a person performs many motor functions several milliseconds before receiving any feedback information from the sensory system to tell whether or not he is performing the acts properly.

To perform these very rapid acts, the cerebral cortex develops a storehouse of "patterns" of motor activities. These patterns can be called forth by the "thinking" part of the brain, and, once initiated, they will cause a discrete set of muscle movements to occur in orderly sequence. For instance, when one is writing rapidly, he simply calls forth the pattern of one letter, then of another, still another, and so forth until each word is written. Most of us at times call for the wrong letter pattern and will write the entire letter before we recognize that we have called the wrong letter. Thus, the process of writing a word or sentence is in reality the calling forth of a sequence of patterns of motor activity.

The precise part of the brain that is utilized for storing the skilled patterns of activity is not known. It probably involves mainly the motor cortex itself, some of the sensory areas, and some of the deeper centers for motor control, such as the basal ganglia, as well. Of special interest is the area 1 to 3 cm wide that lies in the anteriormost part of the motor area of the cerebral cortex. Electrical stimulation of discrete points in this area frequently elicits skilled patterns of movement, especially skilled movements of the hands. Also located in this area are centers for control of eye movements and for formation of words during speech. Areas for control of some of these functions are indicated in Figure 18–5. These areas are frequently called the *premotor cortex*, and they are generally described as a part of the brain that is capable of producing learned patterns of movement that can be performed extremely rapidly.

Role of the Sensory Nervous System for Control of Very Complex Movements. It is already clear from the discussion of learned motor skills that the sensory portion of the brain plays a major role in establishing such skills. But, in addition to helping establish motor skills, the more complex skills can never be performed satisfactorily in the absence of the sensory system, because these skills require a sequence of successive patterns of movement and the memory loci for control of such sequences is located in the memory bank of the sensory system. For instance, the sensory area in Figure 18–5 labeled "choice of words" lies in the region of the brain known as **Wernicke's area,** or the *gnostic area,* which was described in Chapter 16. It is here that a person puts together sensory information from all sources to make complete meaning. Here, also, the person establishes the thoughts and words that he wishes to express either in written or spoken language. Thus, the sequence of words to express a thought is determined strictly by the sensory portion of the cortex.

COORDINATION OF MOTOR MOVEMENTS BY THE CEREBELLUM

Thus far, we have discussed muscle control without considering the role played also by the cerebellum. However, whether the motor movements are caused by signals generated initially in the bulboreticular area, in the basal ganglia, or in the cerebral cortex, the cerebellum always plays an important role in providing coordinate contraction of the individual muscles. The bases of cerebellar function are the following:

Movements of parts of the body are affected greatly by their inertia and momentum.

That is, a limb requires a certain force to start it moving, but once started, it continues moving until an opposing force stops the motion. Neither the cerebral cortex nor the basal ganglia are organized to take these physical factors into consideration. Instead, the *cerebellum* makes the automatic adjustments that keep these factors from distorting the patterns of activity. The cerebellum, whose incoming and outgoing fiber pathways are shown in Figure 18–7, is located posterior to the brain stem. It receives signals from the proprioceptive receptors located in all joints, in all muscles, in the pressure areas of the body, and anywhere else that information about the physical state of the body can be obtained. Signals are transmitted into the cerebellum, too, from the equilibrium apparatus of the ear, and even from the eyes in order to depict the visual relationship of the body to its surroundings. Finally, every time the motor cortex, the basal ganglia, or the reticular formation of the brain stem sends signals to the muscles, a duplicate set of signals is sent into the cerebellum at the same time. In summary, the cerebellum is a collecting house for all possible information on the instantaneous physical status of the body.

Figure 18–7A illustrates the major fiber tracts that transmit the various types of information into the cerebellum. These include (1) the **spinocerebellar tracts,** which carry proprioceptor information from the body; (2) the **corticopontocerebellar tract,** which carries motor information from the motor and premotor cortex to the cerebellum; (3) the **olivocerebellar tract,** which carries signals from the basal ganglia; (4) the **vestibulocerebellar tract,** which

A

B

FIGURE 18–7 (A) Incoming fiber pathways to the cerebellum. (B) Outgoing fiber pathways from the cerebellum.

carries impulses from the equilibrium apparatus; and (5) the **reticulocerebellar tract,** which carries information from the brain stem.

Once the cerebellum has collated its information on the physical status of the body, it transmits its analysis to other areas of the brain through the fiber tracts shown in Figure 18–7B, including especially (1) the **fastigioreticular tract,** which goes from the cerebellum into many of the structures of the brain stem and (2) the **cerebellothalamocortical tract** and related tracts, which pass to the thalamus and then to the motor regions of the cerebral cortex and to the basal ganglia.

Feedback Control of Motor Function. The cerebellum acts as a feedback mechanism to help control the muscle movements initiated by the motor cortex and basal ganglia. This is illustrated for the motor cortex and was illustrated for the basal ganglia in Figure 18–4. Figure 18–8 shows that the cerebellum receives information from the cortex of the muscular movements that it *intends to perform*, while simultaneously receiving proprioceptive information directly from the body apprising it of the movements *actually performed*. After comparing the intended performance with the actual performance, "corrective" signals are sent back to the motor cortex to bring the actual performance in line with the intended one. Similar effects also occur in the basal ganglia system.

Some of the specific functions of the cerebellum are the following:

Damping Function of the Cerebellum— Tremor in the Absence of the Cerebellum. When one moves his hand rapidly to a new position, he can normally stop it exactly at the desired point, but without the cerebellum this cannot be accomplished. If the cerebellum has been removed, the momentum of the arm carries the hand beyond the projected point until some other part of the sensory system besides the cerebellum can detect the fact that the hand has gone too far. The eyes and the proprioceptor mechanisms of the somesthetic cortex finally do detect the overshoot, and they then set into play opposing muscular forces to bring the hand back toward the projected point. Again, momentum is built up during this return motion, and the hand moves too far in the opposite direction. Once more the sensory mechanisms detect the overshoot and bring the hand back to approach the point. Gradually the successive overshoots become less and less until the hand at last reaches the desired point. However, in the meantime the hand goes through a period of oscillation called a "voluntary" tremor because it is initiated by a voluntary muscle movement.

To prevent this effect of momentum, the cerebellum collects information from the moving parts of the body while they are actually moving. Then, even before each part reaches its destination, the cerebellum sends "feedback" signals to the motor cortex and basal ganglia to initiate appropriate "braking" contractions of opposing muscles to slow up and stop the movement at the proper point. In this way, the hand can be brought to rest at a desired position without the annoying overshoot. This overall mechanism to prevent overshoot is called the *damping* function of the cerebellum.

Ataxia. Lack of the damping function of the cerebellum causes the condition called *ataxia*, which means incoordinate contraction of the different muscles. For example, if a person who has lost his cerebellum tries to run, his feet will overshoot the points on the ground necessary for maintenance of equilibrium, and he will fall. Even when walking, his gait will be very severely affected, because placement of the feet can never be precise; he falls first to one side,

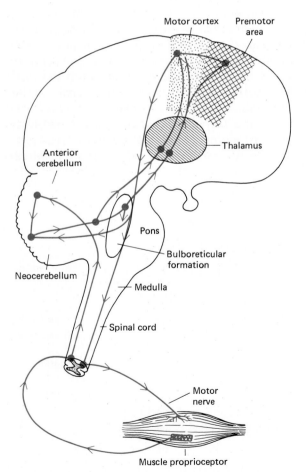

FIGURE 18–8 Feedback circuits of the cerebellum for damping motor movements.

then overcorrects to the other side, and must correct again and again, giving him a broken gait. Ataxia can also occur in the hands and may be so severe that it becomes impossible for the person to write, to hammer a nail, or to perform any other precise movements without overshooting.

Predictive Function of the Cerebellum. Another function of the cerebellum closely allied to the damping function is its ability to predict the position of the different parts of the body. When the leg is moving very rapidly forward while a person is running, the cerebellum, operating in conjunction with the somesthetic cortex, predicts where the leg will be at each instant during the next few hundredths of a second. It is because of this prediction that the person can send appropriate signals to the leg muscles, directing the exact point on the ground where the foot is to be placed to keep him from falling to one side or the other.

The predictive function also applies to the relationship of the body to surrounding objects, for without a cerebellum a person running toward a wall cannot predict accurately when he will reach the wall. Monkeys that have had portions of the cerebellum destroyed have been known to run so rapidly toward walls, being unable to predict how rapidly they are approaching, that they collide with great force.

Equilibrium Function of the Cerebellum. A specific portion of the cerebellum located at its bottommost extent and called the *flocculonodular lobes* is concerned with body equilibrium—that is, with maintenance of the body in the upright position against the pull of gravity. Other aspects of the equilibrium mechanism were considered in detail in the previous chapter. Equilibrium signals are transmitted from the vestibular apparatus into both the vestibular nuclei and the reticular nuclei of the brain stem and also on through the brain stem into the flocculonodular lobes of the cerebellum. These lobes help the person to anticipate that he will lose his equilibrium when he makes a change in direction of movement, and he corrects his movements ahead of time to prevent this from occurring. When the flocculonodular lobes are destroyed, the person loses this ability to predict that he will lose his equilibrium upon changing direction, so he must perform all movements much more slowly than usual. A special type of tumor called the *medulloblastoma* frequently occurs in the flocculonodular region of the cerebellum in young children, and the first indication that a child has such a tumor is often a tendency for the child to lose his equilibrium.

The Basic Neuronal Circuit of the Cerebellum

Up to this point, we have described the gross functions of the cerebellum. But, what is the neuronal circuit in the cerebellum that allows it to perform its integrative functions? The basic essentials of this circuit are illustrated in Figure 18–9.

To the right in the figure are shown the different layers of the **cerebellar cortex,** and deep to the cortex are the **deep cerebellar nuclei.** To the left are illustrated the major cerebellar nervous connections, involving both the deep nuclear cells and the cerebellar cortex. The colored neurons and fibers are all *excitatory*, but one of the neuronal cells of this circuit, the so-called **Purkinje cell,** shown in black, is an *inhibitory cell.*

Note that there are two major inputs to the cerebellum; these are the *climbing fiber input*, which comes from the inferior olive of the medulla oblongata, and the *mossy fiber input*, which comes from all other sources, especially from the pontine nuclei and the reticular nuclei of the brain stem. Now, note two other features of this neuronal circuit. First, both the climbing fibers and the mossy fibers send branch nerve endings directly to the deep nuclear cell to excite it. Thus, just as soon as a muscular movement is initiated by the motor control areas of the cerebrum and brain stem, input signals immediately enter the cerebellum and excite the deep nuclear cells. Output signals from these cells return almost instantaneously into the brain stem and into the thalamo-cortical-basal ganglial pathway that goes to the motor cortex and basal ganglia to reinforce the original excita-

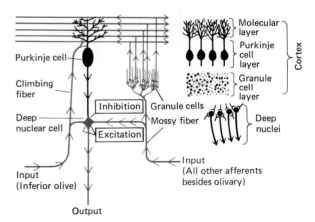

FIGURE 18–9 Basic neuronal circuit of the cerebellum, showing excitatory pathways in color. At right are the three major layers of the cerebellar cortex and also the deep nuclei.

tory signals transmitted from the motor control areas. Thus, at the onset of motor stimulation, the cerebellar activity strongly supports the muscle movements.

The second important point of the neuronal circuit of the cerebellum is the vast numbers of *very minute nerve fibers* that lie in the **molecular layer** on the very surface of the cerebellar cortex. These fibers are so small that they transmit signals extremely slowly. Therefore, they constitute a *delay circuit*. When the signals traveling over these fibers do eventually reach the Purkinje cells, the Purkinje cells in turn send inhibitory signals to the deep nuclear cells, thus turning off the cerebellar output signals. Therefore, after a certain delay time, the initial excitatory signals from the cerebellum are turned off.

Thus, the cerebellum supports motor movements when they first begin, but after a very precise delay time it selectively inhibits further motor activity. This delayed inhibition is believed to provide the damping function of the cerebellum, and it is possibly the time interval between the initial excitatory output and the secondary inhibitory output that helps the cerebellum provide its various predictive functions.

THE CONTROL OF SPEECH

The major characteristic of human beings that sets them apart from other animals is their ability to communicate with one another. This depends on two highly developed functions of the brain that are not shared by any other animals, not even by the apes. These are the ability to interpret speech and the ability to translate thought into speech. These communicative functions require the highest degree of operative perfection in almost all parts of the cerebrum. Therefore, a description of the act of communication provides an overall review of the integrative functions of the cerebral portion of the central nervous system. The principal areas of the cerebral cortex concerned with communication are illustrated in Figure 18–10.

Interpretation of Communicated Ideas. Ideas are generally communicated from one person to another either by sounds or by written words. In the case of sounds, the information enters the *primary auditory cortex* in the superior part of the temporal lobe, as shown in Figure 18–10. The sounds then are interpreted as words and the words as sentences in the *auditory association areas*. The sentences are interpreted as thoughts in **Wernicke's area.** Similarly, combinations of letters seen by the eyes are interpreted as words and the words as sentences by the *visual association areas*, and the sentences again become thoughts in Wernicke's area, which was discussed in more detail in Chapter 16.

Motor Functions of Speech. Wernicke's area also develops thoughts that a person wishes to communicate to someone else. It then operates in association with the lateralmost region of the *somesthetic sensory cortex* to initiate a sequence of signals, each signal representing perhaps a syllable or a whole word, and transmits these into the *motor* and *premotor areas*, which control the larynx and mouth. The premotor area is called *Broca's area*, or it is sometimes called simply the *speech center*. It is actually no more a speech center than is Wernicke's

FIGURE 18–10 Pathways in the brain for communication, showing reception of thoughts by the auditory and visual pathways and control of speech by motor pathways.

area, but it is here that the motor patterns for forming different sounds by the larynx and mouth are controlled. Signals arriving from Wernicke's area set off a sequence of patterns in the speech center, and these in turn form the words. In addition to controlling the larynx and mouth, Broca's area also sends signals into an allied region of the premotor cortex that controls respiration. Therefore, at the same time that the laryngeal and mouth movements occur, the respiratory muscles are contracted to provide appropriate air flow for the speech process.

Thus, the person's thoughts are formulated into speech.

QUESTIONS

1. Describe the basic organization of the higher centers for muscle control.
2. Describe the function of the interneurons and anterior motoneurons in the spinal motor segments.
3. How do the basal ganglia control subconscious movements?
4. Give the functions of the individual basal ganglia: the caudate nucleus, the putamen, the globus pallidus, and the subthalamic nucleus.
5. Describe the function of the motor cortex in the control of muscle movements.
6. How does one learn skilled movements, and what is the function of the premotor area?
7. What role does the cerebellum play in the coordination of motor movements?
8. Why does ataxia occur in patients who have major lesions of the cerebellum?
9. Trace the input of speech to the brain, and give the neuronal motor mechanisms for output of speech.

REFERENCES

Armstrong, D.M.: The mammalian cerebellum and its contribution to movement control. *In* Porter, R. (ed.): International Review of Physiology: Neurophysiology III. Vol. 17. Baltimore, University Park Press, 1978, p. 239.

Evarts, E.V.: Brain mechanisms of movement. *Sci. Am.*, *241*(3):164, 1979.

Gallistel, C.R.: The Organization of Action: A New Synthesis. New York, Halsted Press, 1979.

Granit, R.: The Basis of Motor Control. New York, Academic Press, 1970.

Grillner, S.: Locomotion in vertebrates: Central mechanisms and reflex interaction. *Physiol. Rev.*, *55*:247, 1975.

O'Connell, A.L., and Gardner, E.B.: Understanding the Scientific Bases of Human Movement. Baltimore, Williams & Wilkins, 1972.

Penfield, W., and Rasmussen, T.: The Cerebral Cortex of Man. New York, The Macmillan Co., 1950.

Shik, M.L., and Orlovksy, G.N.: Neurophysiology of locomotor automatism. *Physiol. Rev.*, *56*:465, 1976.

Stein, P.S.G.: Motor systems with specific reference to the control of locomotion. *Annu. Rev. Neurosci.*, *1*:61, 1978.

19

The Autonomic Nervous System and the Hypothalamus

Overview

The **autonomic nervous system** controls the internal functions of the body. It is divided into two separate divisions: (1) the **sympathetic nervous system** and (2) the **parasympathetic nervous system.** Both of these are stimulated by multiple brain centers, located especially in the **hypothalamus** and **brain stem.**

The *peripheral sympathetic nerves,* along with the spinal nerves, originate from the thoracic segments and first two lumbar segments of the spinal cord. These nerves then enter two **sympathetic chains,** one on each side of the vertebral column. From here, terminal sympathetic nerves distribute throughout the body. Most of the sympathetic nerve endings secrete *norepinephrine,* which then exerts the various sympathetic effects on the body.

Especially important among the sympathetic functions are (1) control of the degree of *vasoconstriction* in the skin, thus controlling the rate of heat loss from the body; (2) control of the rate of *sweating* by the sweat glands, which also helps to control heat loss; (3) control of *heart rate;* (4) control of *arterial blood pressure;* (5) inhibition of *gastrointestinal secretion and movements;* and (6) increase in the *metabolism* in most cells of the body.

The *parasympathetic nervous system* originates from several cranial nerves and also from several sacral segments of the spinal cord. The parasympathetic nerve endings all secrete *acetylcholine.* Parasympathetic fibers in the *oculomotor nerve* control *focusing of the eyes* and *dilation of the pupils;* parasympathetic fibers in the vagus and glossopharyngeal nerves control *salivary secretion, heart rate, stomach secretion, pancreatic secretion,* and many of the *contractions of the upper gastrointestinal tract;* and, finally, parasympathetic fibers from the sacral segments control *bladder and rectal emptying.*

The *hypothalamus* is the most important part of the brain for controlling the "vegetative functions" of the body, which is a collective term meaning the subconscious internal bodily functions, and includes control of most of the autonomic nervous system functions. The hypothalamus contains many separate nuclei, as was discussed in Chapter 8.

Some of the regulatory functions of different ones of these nuclei include

1. *regulation of the cardiovascular system,* especially control of heart rate and blood pressure;
2. *regulation of body temperature,* by controlling such phenomena as (a) rate of heat loss from the body by changing the degree of skin vasoconstriction, (b) rate of heat loss from the body by the sweating mechanism, and (c) rate of heat production by the tissues by controlling cell metabolism;
3. *regulation of body water,* by controlling the thirst and drinking mechanisms and also by secreting antidiuretic hormone that causes the kidneys to retain water;
4. *regulation of feeding,* by exciting a hunger center in the hypothalamus when the food stores of the body are depleted;
5. *control of excitement and rage* when the person is threatened in almost any way; and
6. *control of the secretion of almost all the pituitary hormones.* The pituitary hormones in turn control the secretion of about half of all the hormones secreted by the other endocrine glands of the body. Therefore, through these hormonal systems, the hypothalamus controls at least half of the metabolic functions of the body.

In our discussion of the nervous system thus far, we have considered mainly the sensory nervous system and the control of muscle movements. However, the nervous system plays still another entirely different and extremely important role: control of many of the involuntary bodily functions such as arterial blood pressure, heart rate, movements of the gastrointestinal tract, secretion by the digestive glands, and even such functions as dilation of the pupils of the eyes.

The portion of the nervous system that is responsible for these internal control functions is called the **autonomic nervous system.** The purpose of the present chapter is to present the overall function of the autonomic nervous system while also pointing out its special relationships to other nervous system functions.

SYMPATHETIC AND PARASYMPATHETIC DIVISIONS OF THE AUTONOMIC NERVOUS SYSTEM

The autonomic nervous system has two separate divisions; the **sympathetic** and the **parasympathetic.** The differences between these two are: First, the anatomic distributions of the nerve fibers in the two divisions are distinct from each other. Second, the stimulatory effects of the two divisions on the organs are often antagonistic to each other. Third, the types of hormonal trans-

mitter substances secreted at the nerve endings are usually different in the two systems.

Anatomy of the Sympathetic Nervous System

Figure 19–1 shows a diagrammatic representation of the sympathetic nervous system. Note the **sympathetic chain,** which lies on each side of the spinal cord as well as its connections with the cord and with the organs. Periodically along each chain are small bulbous enlargements called **sympathetic ganglia** that contain neuronal cell bodies (to be discussed later). There is one sympathetic ganglion for each thoracic and lumbar spinal cord segment, but there are only three cervical sympathetic ganglia and only two to three very small sacral sympathetic ganglia. Sympathetic fibers spread from the sympathetic chain to all of the viscera of the body, as also illustrated in the figure.

In addition to the ganglia in the sympathetic chain, still other ganglia are found in the sympathetic plexuses in the abdominal cavity. It is from the ganglia in these plexuses that most terminal sympathetic nerve fibers are distributed to the abdominal organs.

Sympathetic nerves enter the sympathetic chain from the spinal cord only from the thoracic and upper three lumbar segments of the cord, and none enter from the neck, lower lumbar, or sacral regions. To supply the head with

Gray rami to C.I
Superior cervical ganglion

Middle cervical ganglion
Anterior ramus, T.I

Inferior cervical
ganglion

Right posterior
pulmonary
plexus

Posterior vagal
trunk

Anterior ramus, L.I

Anterior ramus, S.I

Right inferior
hypogastric plexus

Pelvic
splanchnic
nerves

Coccygeal nerve

Right vagus
nerve

Cardiac branches

Deep cardiac
plexus

Esophageal
plexus

Coronary
plexuses

Anterior vagal
trunk

Gastric plexus

Celiac
plexus (cut)

Superior
mesenteric plexus

Inferior
mesenteric plexus

Vesical and
prostatic
plexuses

FIGURE 19–1 The autonomic nervous system, showing the sympathetic system in black, the communications between the sympathetic chain and the spinal cord in red, and the parasympathetic system in blue. (Reprinted from Warwick and Williams, Gray's Anatomy, 35th British Edition. Philadelphia, W. B. Saunders Co., 1973.)

sympathetic innervation, sympathetic fibers from the thoracic chain extend upward into the cervical portion of the chain and then distribute along the neck and head arteries to the separate structures of the head. Also, sympathetic fibers pass downward from the lumbar portion of the sympathetic chain into the lower abdomen and legs.

The sympathetic chain is linked with the spinal cord in a peculiar manner, as shown in Figure 19–2. Sympathetic fibers leave the cord through the anterior roots of the spinal nerve.

Then, after traveling less than a centimeter, they pass through a small whitish nerve called the *white ramus* into the sympathetic chain. From here, fibers travel in two directions. Some pass into visceral *sympathetic nerves* that innervate the internal organs of the body, while the others return through another small nerve called the *gray ramus* back into the spinal nerve. These latter fibers then travel all through the body along the spinal nerves to supply the blood vessels, the sweat glands, and even the arrector pili muscles, which cause the hairs to stand on end.

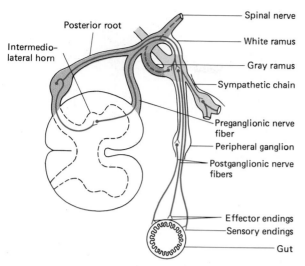

FIGURE 19-2 Connections between the spinal cord, a spinal nerve, and the sympathetic chain.

Visceral Sensory Fibers in the Sympathetic Nerves. Sensory fibers also pass through the sympathetic nerves. These arise in the internal organs, then enter the sympathetic nerves, and finally travel by way of the white rami into the spinal nerves. From here they enter the posterior horns of the cord gray matter and either cause sympathetic cord reflexes or transmit sensations to the brain in the same manner that sensations are transmitted from the surface of the body. These sensations are the visceral sensations that were considered in detail in Chapter 16.

Preganglionic and Postganglionic Neurons of the Sympathetic Nervous System. Sympathetic signals are transmitted from the spinal cord to the periphery through two successive neurons. The cell body of the first neuron is located in the spinal cord in the *lateral gray matter*, which is a slight lateral protrusion of the cord gray matter midway between the anterior and posterior horns. The fiber from this neuron, called the **preganglionic fiber**, passes into the sympathetic system as illustrated in Figure 19-2. It synapses with a second neuron in either a ganglion of the sympathetic chain or one of the more peripheral ganglia. The fiber from the second neuron, called the **postganglionic fiber,** then passes directly to the organ to be controlled. The first neuron, located in the cord, is called the **preganglionic neuron;** the second, located in the ganglion, is called the **postganglionic neuron.** Thus, the sympathetic motor system is different from the skeletal motor system, for skeletal muscles are stimulated through a single neuron rather than through two.

Anatomy of the Parasympathetic Nervous System

Figure 19-3 shows to the right the parasympathetic nervous system. The fibers of this system originate mainly in the tenth cranial nerve, which is the **vagus nerve.** However, a few fibers originate in the third, seventh, and ninth cranial nerves and also in the sacral segments of the spinal cord, mainly S-2 through S-4.

The vagus nerve supplies parasympathetic fibers to the heart, the lungs, and almost all of the organs of the abdomen. The other cranial nerves supply parasympathetic fibers to the head, and the sacral fibers supply the urinary bladder and the lower parts of the colon and the rectum. However, since approximately 70 percent of all the parasympathetic fibers of the body pass through the vagus nerve, most physiologists, in thinking about the parasympathetic system, think almost automatically of the vagus nerve itself.

Preganglionic and Postganglionic Neurons of the Parasympathetic System. The parasympathetic system is like the sympathetic in that signals must travel through both preganglionic and postganglionic neurons before stimulating the various organs. However, there is a different anatomic arrangement of the postganglionic neuron as follows: The cell bodies of the preganglionic neurons are in the brain stem or sacral cord, but their fibers usually pass all the way to the organ to be stimulated instead of to a ganglion as in the sympathetic system. The postganglionic neurons are then located in the wall of the organ, and the postganglionic fibers travel only a few millimeters before they reach their final destination on smooth muscle fibers or glandular cells. However, there are exceptions to this rule, for in the head almost all of the preganglionic parasympathetic fibers terminate in one of four separate small ganglia: (1) the **ciliary ganglion** lying behind the eye and sending postganglionic fibers to the eye structures, (2) the **sphenopalatine ganglion** lying behind the nose and supplying the lacrimal gland and the nasal glands, (3) the **otic ganglion** lying in front of the ear and supplying the parotid gland, and (4) the **submandibular ganglion** lying under the lateral part of the mandible and supplying the submandibular gland.

FIGURE 19–3 Diagrammatic representation of the autonomic nervous system, showing the sympathetic system in red and the parasympathetic system in black. The preganglionic fibers are all represented by solid lines; the postganglionic fibers are represented by broken lines.

The labels within the figure include:

Brain

Ciliary ganglion
Lacrimal gland
Eye
Sphenopalatine ganglion
Parotid gland
Otic ganglion
Sublingual and submandibular glands
Submandibular ganglion

Oculomotor nerve III
Facial nerve VII
Glossopharyngeal nerve IX
Vagus nerve X

Sup. cervical ganglion

T1
2
3
4
5
6
7
8
9
10
11
12
L1
2
3

Greater splanchnic nerve
Lesser splanchnic nerve
Liver
Gallbladder
Least splanchnic nerve
Suprarenal gland
Renal plexus
Kidney
Ovary
Bladder
Spinal cord
Vertebral ganglia (sympathetic trunk)
Uterus
Testis
Urethra

Lung
Heart
Stomach
Celiac plexus
Aortic plexus
Colon
Sup. mesenteric plexus
Small intestine
Inf. mesenteric plexus
Hypogastric plexus
S2
S3
S4
Pelvic nerves
Pelvic plexus

——— Preganglionic sympathetic fibers
- - - - Postganglionic sympathetic fibers
——— Preganglionic parasympathetic fibers
- - - - Postganglionic parasympathetic fibers

Adrenergic Versus Cholinergic Nerve Fibers in the Autonomic Nervous System

One of the major differences between parasympathetic and sympathetic nerves is that the post-ganglionic fibers of the two systems usually secrete different hormonal transmitter substances. The postganglionic neurons of the parasympathetic system secrete *acetylcholine*, for which reason these neurons are said to be *cholinergic*. Those of the sympathetic system secrete mainly *norepinephrine;* therefore, most postganglionic neurons of the sympathetic system are said to be *adrenergic,* a term derived from noradrenalin, which is the British name for norepinephrine. However, a few are cholinergic like the parasympathetic neurons, as is noted later.

Excitatory and Inhibitory Effects of the Autonomic Hormonal Transmitters. Both acetylcholine and norepinephrine have the capability of exciting some internal organs while inhibiting others. Frequently, if one of these hormones excites an organ, the opposite one inhibits it; but this is not always the case. Therefore, in later sections of this chapter we will present the effects of stimulation by both the sympathetic and the parasympathetic nervous systems on most of the important organs and functional systems of the body.

Parasympathetic and Sympathetic Tone

Impulses normally are transmitted at a low rate continuously through all the fibers of both the parasympathetic and sympathetic systems. This allows at least some degree of continuous stimulation of the internal structures, which is called *sympathetic tone* or *parasympathetic tone.* The tone allows nerve stimulation to exert either positive or negative control on a structure. That is, by increasing the number of impulses above the normal value, the stimulation effect can be increased; on the other hand, by decreasing the number of impulses below the normal, the effect can be decreased. As an example, if the sympathetic vasoconstrictor fibers to the blood vessels were normally dormant, it would be possible for sympathetic regulation only to constrict the vessels and never to dilate them; however, because of the normally persistent tone, the blood vessels are always partially constricted so that the sympathetics can either further constrict the vessels by increasing their stimulation or dilate the vessels by decreasing their stimulation. This principle applies throughout the parasympathetic and sympathetic systems and is responsible for a higher degree of effectiveness than would be possible otherwise.

Actions of the Parasympathetic and Sympathetic Nervous Systems on Different Organs

Effect on the Eye. Table 19–1 lists the effects on different organs caused by sympathetic or parasympathetic stimulation. This table

TABLE 19–1
Autonomic Effects on Various Organs of the Body

Organ	Effect of Sympathetic Stimulation	Effect of Parasympathetic Stimulation
Eye:		
Pupil	Dilated	Contracted
Ciliary muscle	Slight relaxation	Excited
Gastrointestinal glands	Vasoconstriction and slight secretion	Stimulation of thin, copious secretion containing many enzymes
Sweat glands	Copious sweating (cholinergic)	None
Heart	Increased rate and strength	Decreased rate and strength
Systemic blood vessels:		
Abdominal	Constricted	None
Muscle	Dilated (cholinergic)	None
Skin	Constricted	None
Lungs:		
Bronchi	Dilated	Constricted
Blood vessels	Mildly constricted	None
Gut:		
Lumen	Decreased peristalsis and tone	Increased peristalsis and tone
Sphincters	Increased tone	Relaxed
Liver	Glucose released	None
Kidney	Decreased output	None
Urinary bladder:		
Body	Inhibited	Excited
Sphincter	Excited	Relaxed
Penis	Ejaculation	Erection
Blood glucose	Increased	None
Basal metabolism	Increased up to 100%	None
Mental activity	Increased	None
Adrenal medullary secretion	Increased	None

shows that the sympathetic system dilates the pupil of the eye, allowing increased quantities of light to enter, and the parasympathetics constrict the pupil, thus decreasing the amount of light. The parasympathetics also control the ciliary muscle that focuses the lens for far or near vision, which will be discussed in more detail in Chapter 21.

Secretion of the Digestive Juices. The secretion of digestive juices by some of the glands of the gastrointestinal tract is controlled mainly by the parasympathetics, whereas the sympathetics have little effect on most of the glands. The salivary glands of the mouth and the gastric glands of the stomach normally are almost entirely controlled by the parasympathetics. On the other hand, the glands in the intestines are controlled only to a slight extent by the parasympathetics, but mainly by local factors in the intestines themselves.

Effects on the Sweat Glands. The sweat glands are stimulated by fibers from the sympathetic nervous system. However, these fibers are different from the usual sympathetic fibers, for they are mainly cholinergic rather than adrenergic. Also, they are stimulated by nervous centers in the brain that normally control the parasympathetics, rather than by the centers that control the sympathetics. Therefore, despite the fact that the fibers supplying the sweat glands are anatomically sympathetic, they can be considered physiologically to function like many parasympathetic fibers.

Effects on the Heart. Stimulation of the sympathetic nervous system increases the heart's activity, sometimes increasing the heart rate as much as threefold and the strength of contraction as much as twofold. On the other hand, parasympathetic stimulation decreases the activity of the heart. In fact, strong stimulation of the vagus nerve to the heart can actually stop the heart for several seconds. But this stimulation will rarely decrease the heart strength each time it does contract more than 30 percent. These effects will be discussed in Chapter 24.

Control of the Blood Vessels. Perhaps the most important function of the sympathetic nervous system is to control the blood vessels in the body. Most vessels are constricted by sympathetic stimulation, though a few, the coronaries for instance, are dilated. By controlling the peripheral blood vessels, the sympathetic nervous system is capable of regulating for short periods of time both the cardiac output and the arterial pressure; constriction of the veins and venous reservoirs increases the cardiac output, and constriction of the arterioles increases the peripheral resistance, which elevates the arterial pressure.

The parasympathetics, when they affect the blood vessels at all, usually dilate them, but this effect is so slight and occurs in so few areas of the body that it can be almost totally ignored.

Effect on the Lungs. The bronchi are dilated by sympathetic stimulation. However, the sympathetic system has almost no effect (very slight vasoconstriction) on the blood vessels of the lungs. This is different from the very strong effect on the blood vessels of the remainder of the body.

Control of Gastrointestinal Movements. About 55 percent of all the parasympathetic nerve fibers are distributed to the gastrointestinal tract, which indicates that by far the most important function of this entire system is regulation of gastrointestinal activities. Parasympathetic stimulation increases peristalsis and at the same time decreases the tone of the gastrointestinal sphincters. Peristalsis propels the food forward while the open sphincters between the different segments of the gastrointestinal tract allow the food to move forward with ease. During extreme parasympathetic stimulation, food can actually pass all the way from the mouth to the anus in approximately 30 minutes, though the normal transmission time is about 24 hours.

Sympathetic stimulation, on the other hand, inhibits peristalsis and tightens the sphincters. This slows the movement of food through the gastrointestinal tract.

Release of Glucose from the Liver. Sympathetic stimulation causes rapid breakdown of glycogen into glucose in the liver and then release of the glucose into the blood. This increased glucose in the blood provides a quick supply of nutrition for the tissue cells, an effect especially valuable during exercise.

Effect on the Kidneys. Sympathetic stimulation causes intense vasoconstriction of the renal blood vessels and greatly decreases the output of urine. This is a very important mechanism for regulation of blood volume and arterial pressure, for when need be, sympathetic stimulation can cause fluid to be retained in the circulatory system, increasing the blood volume and venous return to the heart as well. These effects, over a period of hours or days, also increase the cardiac output and raise the arterial blood pressure.

Emptying of the Urinary Bladder. Emptying of the urinary bladder is caused mainly by parasympathetic stimulation, which excites the muscular wall of the bladder and at the same time inhibits the urethral sphincter that normally holds back the flow of urine. On the other hand, sympathetic stimulation prevents emptying of the bladder. Though the sympathetic effect ordinarily is not important, occasionally when a person has severe peritoneal inflammation in the region of the bladder, a peritoneal reflex excites the sympathetics so greatly that the person becomes unable to urinate.

Control of Sexual Functions. The autonomic nervous system also helps to control the sexual acts of both the male and the female. In the male the parasympathetics cause erection, and the sympathetics cause ejaculation. In the female the parasympathetics cause erection of the erectile tissue around the vaginal opening, which causes tightening, and they also cause the female to secrete large quantities of mucus, which facilitates the sexual act. The effect of the sympathetics on the female sexual act is not well understood, but it is believed that these nerves might initiate reverse uterine peristalsis during the female climax.

Metabolic Effects. Generalized sympathetic stimulation increases the metabolism of all cells of the body. The norepinephrine increases the rates of the chemical reactions in all cells, thereby increasing the overall rate of metabolism of the body. In this way, the sympathetics can keep a person warm when he tends to become too cold, and during exercise or other states of activity they can make the body perform greater quantities of work than would be possible otherwise. These metabolic effects can be brought about in only a few seconds, and they can be stopped in another few seconds when the need for increased metabolism is over.

Stimulation of Mental Activities. Another important effect of sympathetic stimulation is an increase in the rate of mental activity. This probably results partly from the heightened rate of metabolism in the neuronal cells, but norepinephrine also has a direct effect—stimulation of the activating system of the brain.

Sympathetic Stimulation of the Adrenal Medulla—Secretion of Epinephrine and Norepinephrine. The **adrenal medulla** is the central portion of the adrenal gland; one adrenal gland is located on each side of the body immediately above the kidney. Surrounding the medulla is the adrenal cortex, which is an entirely different gland that will be discussed in Chapter 35. The secretory cells of the medulla are modified postganglionic sympathetic neurons, and these cells, when stimulated, secrete *epinephrine* and *norepinephrine* into the blood. These two hormones then circulate in the blood and are distributed to all cells of the body, reaching

many cells that have no direct sympathetic innervation.

The circulating norepinephrine has the same stimulatory or inhibitory effects as the norepinephrine released directly in the tissues by the sympathetic nerves. The epinephrine has almost the same effects, except that epinephrine dilates rather than constricts some blood vessels, and especially it increases cellular metabolism much more than does norepinephrine. Therefore, these two hormones in combination produce almost the same effects on all the organ systems of the body as direct sympathetic nerve stimulation. For instance, they increase the activity of the heart, inhibit peristalsis in the gut, increase the metabolism of all cells, and so forth.

The adrenal medulla mechanism thus provides a second means for causing all or most of the sympathetic activities. During normal sympathetic stimulation, the quantity of epinephrine and norepinephrine secreted by the adrenal medullae is enough to produce between one quarter and one half of all the sympathetic effects in the body. In fact, when all the sympathetic nerves besides those to the adrenal medullae are destroyed, the sympathetic system will still function almost normally because the adrenals compensate by secreting even more hormones. Likewise, loss of adrenal medullary secretion is hardly discernible because of compensation by the direct sympathetic nerves. To stop the action of the sympathetic nervous system entirely, the functions of both the adrenal medullae and all the direct sympathetic nerve fibers to the organs must be blocked at the same time.

Overall Function of the Sympathetic Nervous System

Mass Discharge to Prepare the Body for Activity. In general, when the sympathetic centers of the brain become excited, they stimulate almost all of the sympathetic nerves at once. As a result, the blood pressure rises, the rate of metabolism increases, the degree of mental activity is enhanced, and the glucose in the blood increases. Considering all of these effects together, it can be seen that mass discharge of the sympathetic nervous system prepares the body for activity. When a person's sympathetic nerves have been destroyed, he cannot "generate steam"; in other words, he simply cannot reach the high levels of vigor often necessary to perform rapid, forceful, and excited activities.

Special Functions of the Sympathetic Nervous System. In addition to the mass discharge mechanism of the sympathetic system, certain physiologic conditions can stimulate localized portions of the system independently of the remainder. Two of these are the following:

Heat Reflexes.   Heat applied to the skin causes a reflex through the spinal cord and then back again to the skin to dilate the blood vessels in the heated area. Also, heating the blood flowing through the heat control centers of the hypothalamus in the brain increases the degree of vasodilatation of all the skin blood vessels without significantly affecting the vessels deep in the tissues. The vasodilatation in turn carries increased quantities of heat to the skin and allows it to be lost to the air.

Shift of Blood Flow to the Muscles During Exercise.   During exercise, the increased metabolism in the muscles has a local effect of dilating the muscle blood vessels, but at the same time the sympathetic system becomes strongly activated and constricts the vessels in most other parts of the body. The local vasodilatation in the muscles allows blood to flow through them with great ease, while vasoconstriction elsewhere, except in the heart and brain, decreases almost all other bodily blood flow. Thus, in exercise the sympathetic nervous system helps to cause massive shift of blood flow to the active muscles.

CONTROL OF THE AUTONOMIC NERVOUS SYSTEM BY NERVE CENTERS IN THE BRAIN— ESPECIALLY THE HYPOTHALAMUS

Not much is known about the exact centers in the brain that regulate the discrete autonomic functions. However, stimulation of widespread areas in the reticular substance of the medulla, pons, and mesencephalon can elicit essentially all of the effects of the sympathetic and parasympathetic systems. Also, the hypothalamus can excite most of the functions of the autonomic system. And, finally, stimulation of localized areas of the cerebral cortex can occasionally cause general or sometimes discrete autonomic effects.

Important Autonomic Centers of the Brain Stem

Figure 19–4 shows a tentative outline of major centers in the hindbrain and hypothalamus that regulate different functions of the internal organs. It will be noted from this figure that the

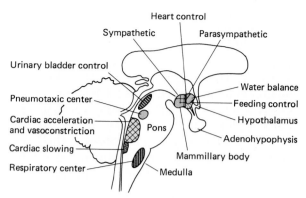

FIGURE 19–4   The major autonomic control centers of the brain stem and hypothalamus.

vascular system and respiration are controlled mainly by areas in the reticular substance of the medulla and pons. The urinary bladder is controlled by centers in the mesencephalon, but these centers are usually suppressed by areas in the cerebral cortex that provide conscious regulation of bladder emptying.

Autonomic Centers of the Hypothalamus

Figure 19–5 gives much more detailed locations of different control centers in the hypothalamus. Some of the specific functions of the hypothalamus are the following:

Cardiovascular Regulation.   In general, stimulation of the **posterior hypothalamus** increases the arterial pressure and heart rate, whereas stimulation of the preoptic area in the anterior portion of the hypothalamus has exactly opposite effects, causing decrease in both heart rate and arterial pressure. These effects are transmitted through the cardiovascular control centers of the brain stem and thence through the autonomic nervous system.

Regulation of Body Temperature. Centers in the **anterior hypothalamus** are directly responsive to blood temperature, becoming more active with increasing blood temperature and less active with decreasing temperature. These areas in turn regulate the body temperature, the details of which will be described in Chapter 44. Basically this mechanism (1) controls the amount of blood flow through the skin, thereby controlling the rate of heat loss from the skin; (2) controls sweating, which cools the body; (3) controls shivering, which greatly increases heat production; (4) controls secretion of norepinephrine and epinephrine by the adrenal medullae, which also stimu-

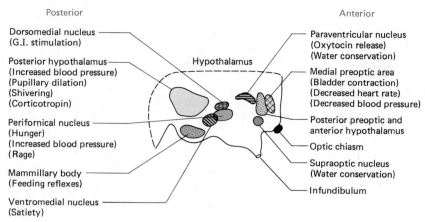

Posterior   Anterior

Dorsomedial nucleus
(G.I. stimulation)

Posterior hypothalamus
(Increased blood pressure)
(Pupillary dilation)
(Shivering)
(Corticotropin)

Perifornical nucleus
(Hunger)
(Increased blood pressure)
(Rage)

Mammillary body
(Feeding reflexes)

Ventromedial nucleus
(Satiety)

Hypothalamus

Paraventricular nucleus
(Oxytocin release)
(Water conservation)

Medial preoptic area
(Bladder contraction)
(Decreased heart rate)
(Decreased blood pressure)

Posterior preoptic and
anterior hypothalamus

Optic chiasm

Supraoptic nucleus
(Water conservation)

Infundibulum

Lateral hypothalamic area (not shown)
(Thirst and hunger)

FIGURE 19–5 Autonomic control centers of the hypothalamus.

late heat production; and (5) controls thyroid hormone production, which has a direct effect on heat production in all cells of the body.

Regulation of Body Water. A center located in the **supraoptic nucleus** controls antidiuretic hormone secretion, which in turn controls rate of water loss through the kidneys. This mechanism will be discussed in detail in Chapter 32 in relation to kidney function.

A closely allied region of the hypothalamus, the **lateral hypothalamus** (not shown in the figure), controls drinking. It does this by transmitting signals into other parts of the brain to create the conscious feeling of thirst, thereby making the person seek water.

Regulation of Feeding. Stimulation of other areas in the *lateral hypothalamus* causes an animal to have a voracious appetite and an intense desire to search for food. This area is frequently called the *hunger center.* Located more medially, in the **ventromedial nuclei,** are areas that when stimulated block all feeling of hunger. Therefore, these areas are called the *satiety center.*

Control of Excitement and Rage. In the very middle of the hypothalamus is a nucleus called the **perifornical nucleus** that, when stimulated, causes an animal to become greatly excited and to develop elevated arterial blood pressure, dilated pupils, and symptoms of rage such as hissing, arching his back, and assuming a stance ready to attack. Thus, the hypothalamus plays a major role in overall behavior of an animal.

Hypothalamic Control of Endocrine Functions. Some areas of the hypothalamus also secrete so-called *neurosecretory substances.* These are then transported through special veins from the hypothalamus down to the anterior pituitary gland where they promote formation of different anterior pituitary hormones. It is in this way that the hypothalamus controls the secretion of most anterior pituitary hormones and thereby controls many of the metabolic functions of the body. These neurosecretory substances and their more specific functions will be discussed in Chapter 45.

Summary. A number of discrete areas of the hypothalamus control many of the internal functions of the body; these are called the "vegetative" functions. However, the hypothalamic areas overlap so much that the separation of different areas for different hypothalamic functions as presented here is partially artificial. Most important of all, the hypothalamus plays a key role in setting the basic tenor of bodily function, the basic degree of excitement, and the basic level of metabolism.

Important Cerebral Centers

Stimulation of the autonomic nervous system by the cerebral cortex occurs principally during emotional states. Many discrete centers, especially in the prefrontal lobes and temporal regions of the cortex, can increase or decrease the degree of excitation of the hypothalamic centers. Also, the thalamus and closely related structures deep in the cerebrum help to regulate the hypothalamus. Therefore, both conscious and subconscious portions of the cerebrum can cause autonomic effects. Sudden psychic shock, which originates in the cerebral cortex, can cause fainting. This usually results from powerful stimulation of cardiac inhibitory fibers (vagus nerves), which causes the arterial pressure to fall precipitously. On the other hand, extreme degrees of psychic excitement, which can result from either conscious or subconscious stimulation in the cerebrum, can stimulate the vasomotor system to increase the arterial pressure and cardiac output.

Unfortunately, our present knowledge of the relation of the cerebrum to the autonomic nervous system is so scanty that it can be only descriptive rather than explanatory of the actual functional relationships.

QUESTIONS

1. Describe the sympathetic and parasympathetic divisions of the autonomic nervous system.
2. Explain the preganglionic and postganglionic neurons of the sympathetic and parasympathetic systems.
3. What is the difference between adrenergic and cholinergic nerve fibers?
4. Give the effects of parasympathetic stimulation and of sympathetic stimulation on the following: (a) the eye, (b) secretion of digestive juices, (c) the sweat glands, (d) the heart, (e) the blood vessels, (f) gastrointestinal movements, (g) emptying of the bladder, and (h) body metabolism.
5. Discuss the role of the adrenal medulla in the function of the sympathetic nervous system.
6. What is the role of the brain stem in the regulation of autonomic activities?
7. Describe the functions of the hypothalamus for control of the autonomic nervous system.

REFERENCES

Aviado, D.M., *et al.* (eds.): Pharmacology of Ganglionic Transmission. New York, Springer-Verlag, 1979.

Bhagat, B.D.: Mode of Action of Autonomic Drugs. Flushing, N. Y., Graceway Publishing Company, 1979.

Carrier, O., Jr.: Pharmacology of the Peripheral Autonomic Nervous System. Chicago, Year Book Medical Publishers, 1972.

Collier, B.: Biochemistry and physiology of cholinergic transmission. *In* Brookhart, J.M., and Mountcastle, V.B. (eds.): Handbook of Physiology. Sec. 1, Vol. 1. Baltimore, Williams & Wilkins, 1977, p. 463.

DeQuattro, V., *et al.*: Anatomy and biochemistry of the sympathetic nervous system. *In* DeGroot, L.J., *et al.* (eds.): Endocrinology. Vol. 2. New York, Grune & Stratton, 1979, p. 1241.

Guyton, A.C., and Reeder, R.C.: Quantitative studies on the autonomic actions of curare. *J. Pharmacol. Exp. Ther.*, *98*:188, 1950.

Hayward, J.N.: Functional and morphological aspects of hypothalamic neurons. *Physiol. Rev.*, *57*:574, 1977.

Kalsner, S. (ed.): Trends in Autonomic Pharmacology. Baltimore, Urban & Schwarzenberg, 1979.

Landsberg, L., and Young, J.B.: Catecholamines and the adrenal medulla. *In* Bondy, P.K., and Rosenberg, L.E. (eds.): Metabolic Control and Disease, 8th Ed. Philadelphia, W.B. Saunders, 1980, p. 1621.

Morgane, P.J., and Panksepp, J. (eds.): Handbook of the Hypothalamus. New York, Marcel Dekker, 1979.

Tucek, S. (ed.): The Cholinergic Synapse. New York, Elsevier/North-Holland, 1979.

20

Intellectual Processes; Sleep and Wakefulness; Behavioral Patterns; and Psychosomatic Effects

Overview

What are *thoughts*? How do we store *memories* in the brain? How are the *intrinsic functions* of the brain itself controlled? These are perhaps the most important functions of the entire brain and yet are functions about which we understand the least.

When a person is thinking a specific *thought* very strongly, many different areas, not merely a single area, of the brain are excited simultaneously. Especially involved are the *thalamus*, widespread portions of the *cerebral cortex*, and even areas of the *brain stem*, especially the *mesencephalon*. It is believed that each of these contributes its own portion to the understanding that makes up the thought. This is called the *holistic theory* of thoughts.

A *memory* is a thought that is stored in a neuronal system of the brain and then recalled at a later time. Some memories last for only a few minutes and are called *short-term memories.* These probably result from continuous activation of neurons, thus keeping the memory of a thought temporarily alive. However, if the thought is a very strong one, especially if it *causes either pain or pleasure*, it will be stored in the memory areas of the brain in the form of a *long-term memory.* This results from some long-term chemical or physical change in the synapses that alters their future ability to transmit signals. That is, those synapses that are excited for a specific thought develop a permanent or semipermanent *facilitation* that allows the same thought to reappear at a later time when elicited by appropriate stimulation.

Another important intellectual process of the brain is its ability to *analyze incoming sensory information and to make it meaningful.* The most important area of the entire cerebral cortex for this function is *Wernicke's area* located in the right-handed person in the posterior, superior part of the left temporal lobe. It is here that information derived from auditory, visual, and somatic experiences all comes together and is interpreted. For instance, the visual areas of the occipital lobe can detect letters and perhaps even words, but without Wernicke's area the meanings of sentences are rarely understood.

345

The **prefrontal regions of the frontal lobes**—those regions anterior to the motor cortex—are an enigma because they can be removed from the brain and yet the person can continue to think and even to perform routine mathematical calculations. However, without the prefrontal areas the person loses his ability to tackle complex thought problems and especially to maintain continuous thought on a single subject for more than a few minutes at a time. Therefore, it is frequently said that the prefrontal areas allow especially deep *elaboration of thoughts.*

But what is it that turns on the thinking processes, or to state this differently, what is it that causes *wakefulness* and *attention* to specific functions in different parts of the brain? We are beginning to learn part of the answers to these questions. The upper portions of the **reticular formation in the brain stem**, particularly those portions located in the mesencephalon and upper pons, can, when excited, stimulate the entire brain to activity, causing the state of *wakefulness.* This excitatory system is called the **reticular activating system.** It transmits most of its signals to the cerebral cortex through the **thalamus**, and stimulation of specific areas in the thalamus can excite specific areas of the cerebral cortex, perhaps in this way directing one's *attention* to specific categories of thought.

The *reticular activating system itself can be excited* by (1) signals from almost all the sensory systems, whether they be somatic, visual, auditory, or other types; (2) signals returning from the cerebral cortex to the mesencephalic portion of the reticular activating system; and (3) signals from the hypothalamus and portions of the limbic system. After many hours of excess activity of the reticular activating system, this system becomes fatigued and its level of activity becomes depressed to the point that *sleep* ensues. After an appropriate period of rest, the fatigue disappears and wakefulness begins again.

Most of the *behavioral patterns* of a person are elicited by excitation in the *hypothalamus* and its surrounding structures, collectively called the *limbic system.* Stimulation in some of these areas will cause severe *pain and feelings of punishment;* of other areas, *pleasure and reward.* And stimulation elsewhere in the limbic system will cause extreme excitability, development of rage, initiation of the fighting posture, and other behavioral responses. Still other areas elicit extreme docility, tameness, and perhaps even love.

We know least and yet want to know the most about these functions of the brain.

Another title for this chapter could be "How do the different parts of the brain function together?" That is, what makes us think? What makes our levels of consciousness increase and decrease during sleep and wakefulness? What influences our inner feelings? How do our intellectual processes express themselves in the form of bodily function?

Unfortunately, it is these integrative aspects of the brain that are least well understood, for which reason much of what we know is based on inference from psychological tests rather than from direct experiments in the brain itself.

ROLE OF THE CEREBRAL CORTEX IN THE THINKING PROCESS

At several points we have discussed various aspects of cerebral function and its relation to some phases of the thinking process. To review briefly, the cerebral cortex is in one sense a large outgrowth of the thalamus and related basal areas of the brain, and there are direct communications back and forth between respective portions of the cerebral cortex and thalamus. Whenever a particular part of the thalamus is stimulated, a corresponding part of the cerebral cortex usually also becomes stimulated. Therefore, we believe that the thalamus has the ability to call forth activity in specific portions of the cerebral cortex. In addition, it is in the cerebral cortex that most of the neurons of the central nervous system are located. Therefore, we must ask ourselves the question: What is the purpose of this great mass of neurons in the cerebral cortex? The answer to this seems to be that the cerebral cortex is a vast storehouse for memories, a place where data can be collected and held for days, months, or years until needed at a later time.

Interpretation of Incoming Signals.

The cerebral cortex is a large sheet of thin folded tissue lying on the surface of the brain and composed of six separate neuronal cell layers, as illustrated in Figure 20–1, each layer performing a different function. Some of these layers send nerve fibers into deeper areas of the brain; others send fibers to adjacent regions of the cortex. Some receive nerve signals from the sensory systems, and some receive signals from the mesencephalon and thalamus, which control the overall level of activity of the cortical area. Finally, it is believed that some of the layers act as *comparators* to compare new incoming signals with memories stored from the past. That is, if the incoming information fits exactly with past memory information that is already stored, one theoretically will immediately identify the new

FIGURE 20–1 Structure of the cerebral cortex, showing to the left the layers of cell bodies and to the right the layers of connecting fibers. (From Ranson and Clark (after Brodmann): Anatomy of the Nervous System. 10th ed. Philadelphia, W. B. Saunders, 1959.)

information as something that has been experienced before. Furthermore, if the previous experience has been associated with some specific quality of sensation such as pain or happiness, displeasure or pleasure, then this too might be remembered.

Nature of Thoughts.

We all know intuitively what a thought is, but, even so, a thought is very difficult to define neurophysiologically. To attempt this, let us first describe what happens in the brain when a person sees with his eyes for the first time a new, exciting visual scene. Upon opening his eyes to the scene, nerve signals are transmitted by way of the optic nerve first into many areas at the base of the brain, including the **lateral geniculate body,** the **superior colliculus,** the **lateral thalamus,** and **midregions of the mesencephalon.** From the lateral geniculate body, secondary signals are transmitted immediately to the visual cortex located in the occipital lobe, as described in Chapter 21. These first signals to the cortex terminate in the **primary visual cortex,** but a few milliseconds later signals reach the visual **association cortex,** and soon thereafter they reach **Wernicke's area of the temporal lobe,** the **hippocampus,** the **prefrontal cortex,** and other parts of the brain. It is the totality of all these signals that gives one the thought expressed by the visual scene. Therefore, a thought can be described as an overall pattern of signal transmission throughout the brain. This is called the *holistic theory* of thoughts.

Every thought has specific qualities. It may be pleasant or it may be unpleasant. It may have an element of luminosity, or it may have an element of sound. It may have a sensation of tactile feeling, or it may have a sensation of taste. It may have a characteristic of repetitiveness or of uninterrupted continuity. There is much reason to believe that specific areas in the brain, many of which are located in the basal regions of the brain, interpret these individual qualities. For instance, the feeling of pleasure or displeasure seems to be interpreted by several closely associated but antagonistic areas located in the midportions of the mesencephalon and hypothalamus and surrounding areas. When a visual scene or other type of sensory input to the brain is pleasant, the pleasure areas of these basal regions receive a signal. If they are unpleasant, the displeasure areas receive a signal.

Finally, every thought has its own specific details, such as contrasts between light and dark areas in the visual scene or contrasts between different frequencies of sound or different intensities of tactile sensations from one second to the next. Memories of these details are mainly stored in the cerebral cortex; therefore, it is presumably the cerebral cortex that is responsible for identification of the fine details of thoughts.

Memories. Memory is the ability to recall thoughts that were originally initiated by incoming sensory signals. Probably most of the memory process occurs in the cerebral cortex, primarily because three quarters of the neurons of the brain are located there. Yet, we know that essentially every area of the central nervous system can participate in the phenomenon of memory. Indeed, experiments have shown that even the spinal cord can hold crude memories for at least as long as a few minutes to perhaps even a few hours.

But what is the nature of a memory? And what causes a memory to persist sometimes for a short time and sometimes for a very long time? There appear to be at least two major types of memory, which can be called *short-term memory* and *long-term memory*.

Short-term Memory. Short-term memory is defined as persistence of an incoming thought for a few seconds or a few minutes without causing any permanent imprint on the brain. Some short-term memories are possibly caused by continued reverberation of signals within the brain for a short time after the initiating sensation is gone. That is, the incoming thought stimulates neuronal cells connected in reverberating circuits. These cells stimulate other cells that then stimulate still other cells, and finally the signal gets back to the original cells. Thus, the signal goes around and around the circuit for seconds or minutes after the incoming sensation is over, and as long as these reverberations persist, the person still retains the thought in his mind. In support of this concept is the fact that such reverberating signals can be demonstrated for as long as half an hour back and forth between the cerebral cortex and the thalamus after a strong sensory signal barrages the brain. Furthermore, these reverberating signals are localized to certain areas of the cerebral cortex depending upon the type of sensation that is experienced.

Long-term Memory. We all know that some memories last for years even though reverberating signals in the brain cannot possibly persist longer than about an hour at most. These long-term memories almost certainly result from the following mechanism: When a signal passes through a particular set of neuronal synapses, these synapses become *facilitated* for passage of similar signals at a later date. Therefore, when a thought enters the brain, it facilitates those synapses that are used for that particular thought, and this makes it easier for one to recall that same thought at some later date.

Yet, passage of a signal through a synapse only one time usually will not cause sufficient facilitation for the thought to be remembered. Therefore, we must ask the question: How is it that a single sensory experience lasting for only a few seconds can sometimes be remembered for years thereafter? One possible answer to this seems to be that the reverberating signals of the short-term memory mechanism send this same thought through the same synapses many thousands of times for up to an hour after the initial sensory experience is over. A reason for believing this is that if an animal is struck on the head immediately after experiencing a very strong sensory experience, the blow to the head can block the reverberating signals and thereby stop the short-term memory. In such an instance, the long-term memory also fails to develop. Therefore, a persisting short-term memory for a period of at least a few minutes seems to be essential to develop the long-term memory "imprint" or "engram." This development of the long-term memory engram is called *consolidation* of the memory. Once this engram has been established, almost any stray signal in the brain can at some later date set off a sequence of signals exactly like those originally initiated by the incoming sensation, whereupon the person experiences the same original thought.

Role of "Rehearsal" in Transferring Short-term Memory into Long-term Memory. Psychological studies have shown that rehearsal of the same information again and again accelerates and potentiates the degree of transfer of short-term memory into long-term memory. This fits with the above reverberatory theory of short-term memory, because each reverberation is actually a form of continued rehearsal of the same information. Indeed, the brain has a natural tendency to rehearse newly found information, especially that which catches the mind's attention. The importance of rehearsal for establishing long-term memories explains why a person can remember information that is studied in depth far better than he can remember large amounts of information studied only superficially. And it also explains why a person who is wide awake will "consolidate" memories far better than will a person who is in a state of mental fatigue.

Codification of Memories During the Storage Process. While memories are being rehearsed and stored, they are also codified.

That is, similar memories that are already stored are called forth and compared with the new memory. And the process involves storing not only the new memory but also its differences from and similarities to the previous memories. Thus, memories are not stored randomly in the mind, but instead are stored in direct association with other memories of the same type and presumably in unique areas of the cortex. This is obviously necessary if one is to be able to scan the memory store at a later date to find the required information.

Determination of Which Memories to Remember—Role of the Hippocampus and of Pain or Pleasure.

Determination of whether or not a memory will be stored for years or almost immediately forgotten seems to be made by some of the basal regions of the brain and not by the cerebral cortex. The **hippocampus,** a very old portion of the cerebral cortex located bilaterally on the floor of each lateral ventricle, is essential for storage of many if not most memories. If one has experienced the same thought many times before and, therefore, has become habituated to it, the hippocampus fails to be stimulated. On the other hand, if the thought evinces either pain or pleasure or some other very strong quality, then the hippocampus does become stimulated. In ways not yet understood, this elicitation of signals in the hippocampus is believed to cooperate with other basal regions of the brain to cause the permanent storage of memory.

Because of the importance of the hippocampus for storage of long-term memories, one can easily understand that large brain lesions involving both hippocampi will cause a person to be unable to, or at least find it difficult to, store long-term memories. Therefore, he forgets everything as rapidly as he learns it. This effect is called *anterograde amnesia.*

Recall of Memories After They Have Been Stored.

Another great mystery about the memory process is: How do we recall memories once they have been stored? It was pointed out above that memories are codified as they are stored, and they seem to be stored in close association with other memories of the same type. It is also known that the brain has mechanisms whereby its attention can be focused on a succession of stored memories, one after another. This process is called searching or *scanning* of the memory storehouse.

There are several reasons for believing that the **thalamus** is strongly involved in the searching process. First, the thalamus has point-to-point connections with essentially all parts of the cerebral cortex, which would provide easy access to specific memory storage loci. Second, it is known that waves of excitation travel through the thalamic nuclei and that these cause similar waves of excitation also to travel through the cerebral cortex; one might imagine that these play some role in the searching process. Third, and most important of all, lesions in the thalamus frequently cause the person to experience difficulty or inability in recalling memories that are known to have been stored at an earlier date. This is called *retrograde amnesia.* It is presumed that these lesions in the thalamus have interfered seriously with the searching process. Sometimes, damage in the thalamic area (or closely associated lower cerebral areas) will cause retrograde amnesia that lasts for days or months and then disappears, presumably because of recovery of the searching process.

Knowledge.

Many persons will be surprised to know that the human being is born already having certain types of knowledge. For instance, a newborn baby knows to suck on the breast and even to search for the breast. He knows to cry when pained and to smile, even without being trained to do so, when pleased. Some of the lower animals are born with still other types of knowledge such as the ability to stand and walk and the ability to search out food. Indeed, a major share of the useful knowledge of some lower animals is inherited.

Yet, the human mind is born with much less knowledge than that of many lower animals. Instead, knowledge accumulated in the mind of the human being is generally of an adaptive type based on previous experience, made up almost entirely of memories rather than based on inherited neuronal connections. It is this difference between the lower animal mind and the human mind that gives the human being his great breadth of abilities, one person becoming a mathematical genius, another a vast storehouse of linguistic data, another a depository of jokes, and so forth.

But we also know that the process of forgetting allows the character of knowledge in the mind to change with time so that a person's mind may be a great depository of book learning in his school days and yet in later years be filled with practical experience that takes the place of forgotten book learning. Psychological tests show that the overall quantity of stored knowledge in a person's mind generally increases during the first 39 years of his life, reaching a peak at this time. Beyond that age the total amount of stored knowledge gradually declines. This does not mean, though, that the amount of stored knowledge of a specific subject might not continue to increase on into very old age.

The Cellular Mechanism of Memory— Long-Term Facilitation of the Synapse. In the above discussion of memory consolidation, it was pointed out that long-term memories can be stored only when the same sequence of signals passes through the involved synapses many times, probably thousands or even millions of times. Because of this, it is believed that the basic neuronal mechanism of memory is a process of long-term facilitation of the synapses. This concept is also supported by animal experiments. For instance, prolonged repetitive electrical stimulation of nerve fibers entering a neuronal pool will eventually cause that neuronal pool to develop a high degree of sensitivity to subsequent stimulation. This phenomenon is called *posttetanic facilitation*, but one can readily see that it is actually a memory process.

On the other hand, the way in which synapses become increasingly sensitive to successive incoming signals is still a big question. There are two basic theories. The first of these is that prolonged stimulation causes the terminal nerve fibrils in a synaptic pool to sprout forth new terminals and these in turn to provide progressively more synaptic knobs on the surface of the postganglionic neuron. A variant of this theory is that the synaptic knobs themselves enlarge or change their chemical or physical characteristics in such a way that they provide increased quantities of excitatory transmitter substance. This theory that the synaptic knobs increase in number or size is supported by the fact that electron microscopic counts of terminal fibrils and synaptic knobs are greater in brain areas that have been purposely stimulated over long periods of time than in brain areas not so stimulated.

The second theory is that the neuronal membrane on the postsynaptic side of the synapse becomes altered, thereby changing the sensitivity of the neuron to incoming signals. One variant of this theory is that synaptic sensitivity is heightened as a result of an increase in ribonucleic acid in the neuronal cell. Here again, experiments have actually demonstrated such an increase in ribonucleic acid, but it is still to be proved that this substance is actually the basis of the memory process rather than some nonspecific effect of greater neuronal activity.

However, more than likely, both of these two theories are much too naive. Recent studies in a large snail called the *Aplysia* have demonstrated a somewhat more complex mechanism of memory, even in this very primitive animal.

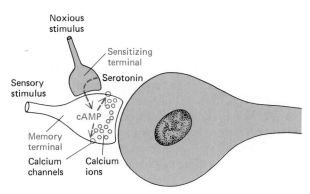

FIGURE 20–2 A memory system that has been discovered in the snail *Aplysia*.

This mechanism, illustrated in Figure 20–2, is comprised of a neuronal cell stimulated by two separate synaptic knobs, one called the *memory terminal* and the other the *sensitizing terminal*. The memory terminal rests directly on the neuronal cell whereas the sensitizing terminal rests instead on the memory terminal itself. Stimulation of the memory terminal will at first excite the neuronal cell, but continued stimulation without simultaneous stimulation of the sensitizing terminal will cause the synapse between the memory terminal and neuronal cell to become progressively less excitable rather than more excitable. This phenomenon is called *habituation*. On the other hand, if the sensitizing terminal is stimulated by a pain signal at the same time that the memory terminal is also stimulated, then exactly the opposite effect occurs: that is, the synapse between the memory terminal and the neuronal cell becomes highly facilitated. Furthermore, once this high degree of facilitation has taken place, it is not necessary to continue stimulating the sensitizing terminal to maintain the high level of excitability. Thus, a memory circuit has been established.

Therefore, as we stand at present, our knowledge of perhaps the single most important function of the brain—memory—is very scanty. When we do understand this process more fully, our total understanding of how the mind works will certainly be greatly enhanced.

Function of Wernicke's Area of the Cerebral Cortex in the Thought Process

It was pointed out in Chapter 16 that most of the sensory input signals to the brain eventually funnel information into *Wernicke's area*, located in the posterior portion of the *temporal lobe* in the brain's dominant hemisphere, usually the left hemisphere. The primary and association areas for somesthetic, visual, auditory, taste, and smell sensations are all located very near to Wernicke's area. Therefore, it is ideally located for integrating the information from all these different sources.

Destruction of Wernicke's area of an adult almost completely destroys his intellect and his ability to perform useful functions. For this reason, this portion of the cortex is perhaps the one most important area of all. In a child, destruction of this area is not nearly so serious because the temporal lobe in the opposite hemisphere of the brain can then develop the same functions, but in the adult this opposite hemisphere has already become developed for other purposes and cannot be significantly reeducated.

The fact that Wernicke's area is very important to intellectual function of the brain does not mean that it is in this area that all important memories are stored. It merely means that this is one of those key points in the brain where, if the neuronal connections are destroyed, the other parts of the brain cannot function satisfactorily. Other aspects of sensory function in Wernicke's area were discussed in Chapter 16. Its function of helping control muscle movements was discussed in Chapter 18.

Function of the Prefrontal Areas of the Cerebral Cortex in the Thought Process

The prefrontal areas of the cerebral cortex are the most anterior portions of the brain, lying in the front 5 to 7 cm of the frontal lobes, anterior to the motor control areas discussed in Chapter 18. These areas are frequently called "silent" areas because no gross overt effects occur when the prefrontal areas are destroyed. Instead, the person simply loses a major share of what we generally call his intellectual ability, especially his ability for abstract thought.

Function of the Prefrontal Areas in Abstract Thinking. An animal that has lost its prefrontal areas loses the ability to keep small bits of information in its mind for longer than a few seconds at a time. For instance, if food is placed on one side of the cage and its attention is drawn to the other side, it forgets the locus of the food when it is no longer looking at it. It is believed that the prefrontal areas in the human being perform this same function, holding many small bits of information for short periods of time. In performing mathematical problems, it is necessary to store information temporarily during each stage of the logical process into a small corner of the mind and then to come back to this information a few seconds or a few minutes later. Thus, after storing and processing several hundred bits of information, the problem can be solved. It is believed that the prefrontal areas are the storehouse of these short-lived memories.

Essentially the same mechanisms are used in abstract thinking related to legal processes, diagnosis of diseases by the physician, or analysis of complex business problems.

The Prefrontal Areas as the Locus of Ambition, Conscience, Planning, and Worrying. Obviously, without the ability to think through the consequences of one's actions a person might act in too great haste and perform activities that he would regret in the future. Thus, it is well demonstrated that persons who have lost their prefrontal areas appear also to have lost much of their conscience.

Likewise, planning for the future, worrying about one's activities, or developing ambition are dependent upon his ability to think through interrelationships of all his actions. Here again, the person who loses his prefrontal areas loses all these qualities. On the other hand, he at least is often exempt from prolonged worry for the remainder of his life.

Effect on Behavior Caused by Loss of the Prefrontal Areas. Occasionally the prefrontal areas are completely destroyed by disease or trauma, or sometimes they have been purposely destroyed surgically because of harmful patterns of thought that have developed and cannot be stopped in other ways. When the prefrontal areas are gone, Wernicke's area is left by itself in charge of the thought patterns of the brain, without the sobering influences of the prefrontal areas. The person is then likely to exhibit extreme reactions, some of which are very happy in nature whereas others have the characteristics of raging temper. Often the responses are rapid and are likely to lead to disastrous results, responding emotionally to many situations but without much thought. When provoked, he is likely to fall into a state of rage that cannot be easily quelled at the moment but that will be totally forgotten a few minutes later.

SLEEP AND WAKEFULNESS

Another of the important mysteries of brain function is its diurnal cycle of sleep and wakefulness. Even when a person remains in total darkness or in total light he still maintains approximately the same sleep-wakefulness cycle, with a periodicity of once every 24 hours. Ordinarily the nervous system shows signs of fatigue shortly before sleep ensues, and it shows signs of having

FIGURE 20–3 The reticular activating system, showing by the arrows passage of impulses from the reticular substance of the mesencephalon upward through the thalamus to all parts of the cortex. (From Lindsley: Reticular Formation of the Brain. Boston, Little, Brown.)

become considerably rested at the termination of sleep. It seems, then, that neuronal fatigue plays an important part in causing sleep, and that sleep, in turn, relieves the fatigue.

Electrical studies of the brain indicate that while a person is awake many nerve impulses pass continuously through the nervous system, never ceasing. However, during most stages of sleep, much fewer impulses are present. Thus, the state of wakefulness seems to be caused by a high degree of activity in the cerebrum, whereas the state of sleep is caused by a low degree of activity. Therefore, any theory to explain sleep and wakefulness must also explain these changing degrees of cerebral activity during the two states.

Stimulation of the Brain by the "Reticular Activating System." Stimulation of portions of the mesencephalon (midbrain) and thalamus greatly increases the activity of the cortex, an effect shown diagrammatically in Figure 20–3. Stimulation anywhere along the arrows will cause impulses to spread upward and eventually to excite the cortex. This system is called the **reticular activating system,** and it is divided into two separate parts, the **mesencephalic part** and the **thalamic part.**

The mesencephalic part of the reticular activating system is composed mainly of the reticular substance in the mesencephalon and upper pons. Stimulation of this region causes very diffuse flow of impulses upward through widespread areas of the thalamus and thence to widespread areas of the cortex, causing generalized increase in cerebral activity.

The thalamic portion of the reticular activating system differs from the mesencephalic part in that stimulation here activates *localized*

regions of the cerebral cortex. Stimulation posteriorly in the thalamus activates posterior parts of the cerebral cortex and stimulation anteriorly activates anterior parts of the cortex. Thus, signals from specific parts of the thalamus can call forth activity in specific parts of the cortex rather than activating the entire cortex.

It is likely that the function of the thalamic portion of the reticular activating system is to direct one's attention to memories or thoughts stored in specific parts of the cerebral cortex and thereby to allow his consciousness to consider individual thoughts one at a time. On the other hand, *activation of the mesencephalic portion of the reticular activating system causes generalized wakefulness.*

The Arousal Reaction. The reticular activating system must itself be stimulated to action by input signals from other sources. When an animal is in very deep sleep, the reticular activating system is in an almost totally dormant state; yet almost any type of sensory signal can immediately activate the system. For instance, pain stimuli from any part of the body or proprioceptor signals from the vestibular apparatuses, from the joints, and so forth can all cause immediate activation of the reticular activating system. This is called the *arousal reaction,* and it is the means by which sensory stimuli awaken us from deep sleep.

Cerebral Stimulation of the Reticular Activating System. Signals from the cerebral cortex can also stimulate the reticular activating system and thereby increase its activity. Fiber pathways to both the mesencephalic and thalamic portions of this system are particularly abundant from the *somesthetic sensory cortex,* the *motor cortex,* and the *frontal cortex.* Whenever any of these cortical regions becomes excited, impulses are transmitted into the reticular activating system, thereby increasing the degree of activity of the reticular activating system.

The Feedback Theory of Wakefulness and Sleep

From the above discussion we can see that activation of the reticular activating system intensifies the degree of activity of the cerebral cortex, and, in turn, greater activity in the cerebral cortex increases the degree of activity of the reticular activating system. Therefore, it is fairly evident that a so-called "feedback loop" could develop whereby the reticular activating system excites the cortex and the cortex in turn reexcites the reticular activating system, thus setting up a cycle that causes continued excitation of both of these regions. Such a cycle is shown in Figure 20–4.

The reticular activating system is also involved in another feedback loop, as follows: It sends impulses down the spinal cord to activate the muscles of the body. In turn, the muscle acti-

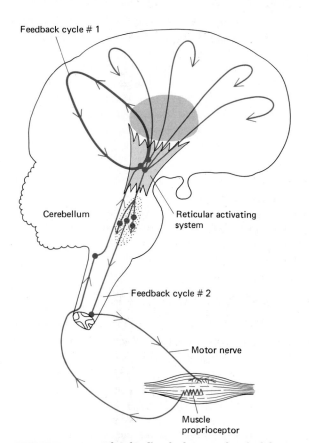

FIGURE 20–4 The feedback theory of wakefulness, showing two feedback cycles passing through the reticular activating system: one to the cerebrum and back (#1) and another to the peripheral muscles and back (#2).

vation excites proprioceptors that send sensory impulses upward to reexcite the reticular activating system. Thus, here again, activity of the reticular activating system sets off another cycle that in turn further increases the activity in the reticular activating system. This cycle is also shown in Figure 20–4.

One can readily see, therefore, that once either one or both of these feedback loops involving the reticular activating system should become excited, the resulting increased activity in the reticular activating system would stimulate all other parts of the brain, thereby creating a state of wakefulness. This is one of the plausible theories to explain the state of wakefulness.

On the other hand, one might wonder how *sleep* could possibly occur once the different feedback loops should become activated. A possible answer to this is that all synapses of the central nervous system eventually fatigue, so that they either stop or nearly stop transmitting impulses. Therefore, after a prolonged period of wakefulness, one would expect the synapses in the feedback loops to become fatigued and, consequently, the feedback loops to stop functioning, thus allowing the reticular activating system to become dormant, which is the state of *sleep*.

Other Theories of Wakefulness and Sleep

Although the feedback factors discussed previously might be very important in maintaining the wakefulness state, many neurophysiologists believe that it is not these feedbacks that are normally responsible for shifting from wakefulness to sleep or from sleep to wakefulness. Especially widespread are various theories that some chemical factor or factors are secreted into the reticular activating system or into other areas of the brain to control the wakefulness-sleep cycle. For instance, it is known that the **median raphe nuclei,** located in the very midline of the brain stem, contain neurons that secrete *serotonin,* an inhibitory transmitter substance. The fibers from these nuclei distribute to many areas of the cerebrum, brain stem, and even spinal cord, and the secreted serotonin inhibits synaptic transmission in all these areas. Therefore, activation of these raphe nuclei could easily cause sleep. In fact, stimulation in them can actually put an animal to sleep, or their destruction can cause insomnia.

On the other hand, excitation of two small bilateral nuclei in the pons, each called the **locus ceruleus,** can cause intense wakefulness and generalized increase in excitability in an animal. The nerve fibers from these nuclei likewise spread all through the cerebrum in addition to their strong activation of the reticular activating system itself. The endings of these fibers secrete *norepinephrine*, a hormone that is known to increase the excitability of most of the central nervous system synapses.

Finally, experiments have also shown that when an animal is kept awake for many hours, still another chemical substance, a sleep-producing *polypeptide*, appears in the cerebrospinal fluid of the animal. Furthermore, this peptide will put an animal to sleep within minutes after it is injected into the ventricular system of the brain. Therefore, it is possible that fatigue of brain neurons causes the formation of a sleep-producing substance that, when it reaches a certain concentration, will lead to sleep.

In summary, cyclic changes in the hormonal or biochemical states of local neurons in the reticular activating system or in other parts of the brain could well be the cause of the increased and decreased activity of the nervous system during the wakefulness-sleep cycle. Or, more likely, it is possible that this cycle is caused by a complex combination of the feedback system described previously and the chemical factors.

REM Sleep

Periodically during sleep a person passes through a state of dreaming associated with mild involuntary muscle jerks and rapid eye movements. The rapid eye movements have given this stage of sleep the name *REM sleep.* Electroencephalograms recorded during these periods show considerable activity in the brain even though it is usually more difficult to arouse the person from REM sleep than from other stages of deep sleep. Muscle tone throughout the body at this time is diminished almost to zero, the heart rate may be as many as 20 beats below normal, and the arterial pressure may be 30 mm Hg below normal. Thus, the person seems physiologically to be in very deep sleep rather than light sleep despite the fact that he is dreaming and despite his active electroencephalogram.

REM sleep usually occurs three to four times during each night at intervals of 80 to 120 minutes, each occurrence usually lasting from 5 to 30 minutes. As much as 50 percent of an infant's sleep cycle is composed of REM sleep, and in the adult approximately 20 percent is REM sleep.

The cause of REM sleep and its cyclic pattern is still unknown. It has been claimed by some psychologists that in the absence of REM sleep a person develops severe psychic instability. However, this theory seems now to have been disproved. Instead, it seems that serious deficiency of the deep, non-REM sleep is, instead, the culprit in many psychic disorders.

A postulated function of REM sleep is that it represents a periodic test of the degree of excitability of the nervous system. That is, approximately every $1\frac{1}{2}$ hours this cycle recurs. And if the neurons of the brain are sufficiently rested, the increased neuronal activity during REM sleep can theoretically awaken the person. This theory is supported by the fact that dreams do frequently awaken persons and also by the fact that REM sleep occurs much more frequently and for more prolonged periods of time as morning approaches.

Effects of Wakefulness

Wakefulness is associated with three major effects: (1) increase in the degree of activity of the cerebrum, (2) increased transmission of signals from the reticular activating system to the muscles, and (3) excitation of the sympathetic nervous system.

The increase in cerebral function is caused by the great number of impulses transmitted from the wakefulness center of the reticular activating system upward through the thalamus to the cerebral cortex. These impulses continually impinge on the cerebral neurons, facilitating them to activity.

Stimulation of the reticular areas of the brain stem also increases the degree of tone in all the muscles of the body. This makes the muscles more excitable than would otherwise be true and prepares them for immediate activity, which explains why a person who is wide awake has the feeling of muscle readiness.

The sympathetic stimulation caused by wakefulness elevates the blood pressure slightly; it increases the rate of metabolism in all the tissues of the body; and, in general, it simply makes the body ready to perform increased amounts of work.

Effects of Sleep

Sometimes it is hard to understand why a person needs to sleep at all. Certain parts of the body such as the heart never rest and still are capable of functioning throughout life. One might reason that sleep is a measure to conserve the energies of most parts of the body when they are not needed. Some animals, as a matter of fact, have carried this principle of conservation so far that they pass into a state of very prolonged and deep sleep called *hibernation*, which lasts throughout the entire winter.

A special psychologic value of sleep seems to be to reestablish an appropriate balance of excitability among the various portions of the nervous system. As a person becomes progressively fatigued, some parts of his central nervous system lose excitability more than others, so that one part may overbalance the others. In fact, extreme nervous fatigue can even precipitate severe psychotic disturbances. Yet after prolonged sleep, all parts of the nervous system usually will have returned once again to appropriate degrees of excitability and to a state of serenity.

Lack of sleep does not *directly* affect the intrinsic functions of the different organs. However, lack of sleep often causes severe autonomic disturbances, and these in turn *indirectly* lead to gastrointestinal upsets, loss of appetite, and other detrimental effects. In this way, loss of sleep can affect the whole body as well as the nervous system itself.

BRAIN WAVES

Electrical impulses called *brain waves* can be recorded from all active parts of the brain and even from the outside surface of the head. The character of these waves is determined to a great extent by the level of sleep and wakefulness at the time that the waves are recorded. When a person is awake but not thinking hard, continuous waves at a rate of approximately 10 to 12 per second can be recorded from almost all parts of the cerebral cortex. These are called *alpha waves*. The brain signals that cause them probably originate in the reticular activating system and then spread into the cerebral cortex. These are believed to be part of the signals that keep the cortex facilitated during wakefulness. The alpha waves are illustrated by the recording at the top of Figure 20-5.

When any part of the brain becomes very active—for example, the motor region initiating muscular activities—additional waves having a frequency sometimes as high as 50 cycles per second, and intensities often greater than those of the alpha waves, take the place of the normal alpha waves. These, called *beta waves*, are shown by the second recording of Figure 20-5.

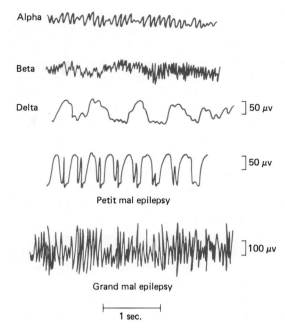

FIGURE 20-5 Brain wave patterns in the normal person and in two persons with different types of epilepsy.

During very deep sleep the alpha and beta waves are replaced by a few straggling waves occurring approximately once every 1 to 2 seconds. These are the "sleep waves" or *delta waves*, as shown by the third recording in Figure 20-5.

Abnormal Brain Wave Patterns in Epilepsy. Various abnormalities of the brain can cause strange brain wave patterns. Two of these, caused by different types of *epilepsy*, are shown in the fourth and fifth records of Figure 20-5. The fourth record shows a "spike and dome" pattern that occurs in *petit mal* epilepsy. In this disease the person suddenly becomes unconscious for 3 to 10 seconds at the same time that the spike and dome pattern occurs in the brain waves. Such episodes may occur every few minutes, every few hours, or only once in many months. When they do occur, the person usually continues, while unconscious, whatever physical activity he is already doing, even though he might be walking across a crowded street. Petit mal epilepsy seems to result from some abnormality of the reticular activating system of the brain. The transmission of the normal alpha waves to the cerebral cortex is temporarily stopped. Instead, the spike and dome pattern is transmitted, and the person falls partially asleep for a few seconds until the alpha wave pattern picks up again.

The bottom record in Figure 20-5 shows the brain wave pattern in *grand mal* epilepsy. In this condition the cerebral cortex becomes extremely excited, and many very strong signals spread over the brain at the same time. When these reach the motor cortex they cause rhythmic movements, called *clonic convulsions*, throughout the body. Grand mal epilepsy possibly results from abnormal reverberating cycles developing in the reticular activating system. That is, one portion of the system stimulates another portion, which stimulates a third portion, and this in turn restimulates the first portion, causing a cycle that continues for 2 to 3 minutes, until the neurons of the system fatigue so greatly that the reverberation ceases. At the beginning of a grand mal attack a person may experience very violent hallucinatory thoughts, but once most of the brain is involved, he no longer has any conscious thoughts at all, because signals are then being transmitted in all directions rather than through discrete thought circuits. Therefore, even though his brain is violently active, he becomes unconscious of his surroundings. Following the attack his brain is so fatigued that he sleeps at least a few minutes and sometimes as long as 24 hours.

BEHAVIORAL FUNCTIONS OF THE BRAIN: THE LIMBIC SYSTEM

Behavior is a function of the entire nervous system, not of any particular portion. However, most involuntary aspects of behavior are controlled by the so-called **limbic system,** which is illustrated in block diagram form in Figure 20–6. This figure shows most of the central structures of the basal cerebrum with a surrounding ring of cerebral cortex called the **limbic cortex.** However, the great majority of the cerebral cortex

lies still beyond this ring and is not part of the limbic system. This ring of limbic cortex consists of (1) the **uncus,** the **pyriform area,** and the **hippocampal gyrus** on the very bottom of the brain; (2) the **cingulate gyrus,** lying deep in the longitudinal fissure of the brain; and (3) the **insular** and **orbital frontal areas** lying in the midanterior and bottom anterior portions of the brain. All of these areas of the cerebral cortex are phylogenetically old—that is, they were among the earliest portions of the cerebral cortex to evolve in primitive animals.

Perhaps the most important part of the limbic system, from the point of view of behavior, is the **hypothalamus,** even though many anatomists do not list the hypothalamus as one of the limbic structures. Many of the surrounding portions of the limbic system, including especially the hippocampus, the amygdala, and the thalamus, transmit major portions of their output signals into the hypothalamus to cause varied effects in the body, such as to stimulate the autonomic nervous system or to participate in causing such feelings as pain, pleasure, or sensations related to feeding, sex, anger, and so forth. Most of these functions of the hypothalamus were discussed in the previous chapter.

Some aspects of limbic control are transmitted through the endocrine system, for it will be remembered that the hypothalamus, in addition to controlling the autonomic nervous system, also controls secretion of many of the pituitary hormones. This will be discussed in detail in Chapter 45, but for the time being it should be noted that the limbic system operating through the hypothalamus and anterior pituitary gland can control (1) the rates of secretion of all sex hormones, which together control the various sexual drives of the person; and (2) the rates of secretion of thyroid hormone, growth hormone, and various adrenocortical hormones, which together control most of the person's day-by-day cellular metabolic functions.

Therefore, the limbic system can be said to control the inner being of the person.

With this background in mind, now let us discuss some of the specific mechanisms for control of behavior by the limbic system.

Pleasure and Pain; Reward and Punishment

One of the most important recent discoveries in the field of behavior is the so-called "pleasure and pain" or "reward and punishment" system of the brain. Certain areas in the mesencephalon, hypothalamus, and other closely associated

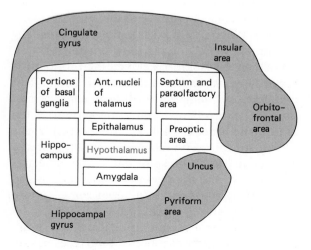

FIGURE 20–6 The limbic system.

FIGURE 20–7 Technique for localizing reward and punishment centers in the brain of a monkey.

areas, when stimulated, make the animal feel intense punishment as if he were being severely pained. Yet stimulation of other nearby areas causes exactly the opposite effect, making the animal appear to be experiencing extreme pleasure.

One of the experimental methods for studying the reward and punishment centers is to implant an electrode in one or the other of these two areas and then to allow the animal itself to press a lever to control the stimulus, as illustrated in Figure 20–7. If the electrode is implanted in a reward area, the animal will press the lever continually. Indeed, the animal would rather stimulate its reward center than eat, even though it might be starving. On the other hand, if the electrode is placed in a punishment area, it will avoid stimulation by all means possible.

Relationship of Reward and Punishment to Learning. Experiments on memory have shown that an animal remembers sensory stimuli that cause either reward or punishment but fails to remember sensory stimuli that do not excite either the reward or punishment area. For instance, a food that is very pleasant to the taste is remembered, or, likewise, a food that is exceedingly unpleasant is remembered. On the other hand, food that causes neither pleasure nor displeasure is rapidly forgotten. Similarly, a very painful stimulus, such as touching a hot iron, is remembered well, whereas simply touching a book or stick of wood is forgotten in a few seconds.

Therefore, in the process of memory, two different components must be present for a sensory experience to be remembered. The first component is the sensory experience itself, and the second component is an experience of either reward or punishment—that is, an experience of pleasure or pain.

Function of the Limbic Cortex

Even though the limbic cortex is the oldest part of the cerebral cortex, its precise function is least understood of almost all portions of the brain. Electrical stimulation in different parts of the limbic cortex can cause such effects as excitement, depression, increased movement, decreased movement, on rare occasions the phenomenon of rage, on other occasions intense degrees of docility, and so forth. However, these effects cannot be elicited repeatedly from specific points in the limbic cortex. Therefore, it is very difficult to state precise functions for the limbic cortex.

The probable function of the limbic cortex is to act as an *association area* for control of most of the behavioral functions of the body. It presumably stores information about past experiences such as pain, pleasure, appetite, various smells, sexual experiences, and so forth. This store of information is then presumably combined with other information channeled into the limbic areas from surrounding regions of the cerebral cortex, such as from the prefrontal areas and from the sensory areas of the posterior part of the brain. This association of information then presumably provides stimuli for initiating appropriate behavioral responses to the animal's or person's surrounding environment, whether this behavior be rage, docility, excitement, lethargy, or so forth. Thus, we believe this to be the part of the cerebral cortex that plays the greatest role in controlling the emotions and other patterns of behavior.

The Defense Pattern—Rage

Stimulation of several different areas in the thalamus, hypothalamus, or mesencephalon, but especially in the **perifornical nuclei of the hypothalamus,** located in the very middle of the hypothalamus, gives an animal an extreme sensation of punishment and simultaneously causes it to (1) develop a defense posture, (2) extend its claws, (3) lift its tail, (4) hiss, (5) spit, (6) growl, and (7) develop piloerection, wide-open eyes, and dilated pupils. Furthermore, even the slightest provocation causes an immediate savage attack. This is the pattern of behavior that has been called the *defense pattern* or simply *rage*. It can occur in decorticated animals, illustrating that the basic behavioral patterns for defense and rage are controlled from the lower regions of the brain, especially from the hypothalamus and the mesencephalon.

Exactly opposite emotional behavioral patterns occur when the reward centers are stimulated, namely, docility and tameness. During such stimulation, the animal becomes completely amenable to almost any type of treatment.

Functions of the Amygdala

The **amygdala** is a complex nucleus located immediately beneath the surface of the cerebral cortex in the anterior pole of each temporal lobe. In lower animals, the amygdala is concerned primarily with smell stimuli, but in human beings it is much larger and operates in very close association with the hypothalamus to control many behavioral patterns. It is believed that the normal function of the various amygdaloid nuclei is to help control the overall pattern of behavior demanded for each social occasion.

Stimulation of various parts of the amygdala can transmit signals through the hypothalamus to cause (1) increase or decrease in arterial pressure; (2) increase or decrease in heart rate; (3) increase or decrease in gastrointestinal activity; (4) defecation or urination; (5) pupillary dilatation or constriction; (6) piloerection, which means hair standing on end; or (7) secretion of various pituitary hormones.

In addition, the amygdala can transmit signals to areas of the lower brain stem to cause (1) changes in the degree of muscle tone throughout the body; (2) postural movements, such as raising the head or bending the body; (3) circling movements; (4) rhythmic movements; or (5) movements associated with eating, such as licking, chewing, and swallowing.

Finally, excitation of other portions of the amygdala can cause sexual excitement, including erection, copulatory movements, ejaculation, ovulation, uterine activity, and premature labor.

From this foregoing list of functions, one can well understand how the amygdala can play a major role in controlling the body's overall pattern of behavior.

PSYCHOSOMATIC EFFECTS

A psychosomatic effect is a bodily (*somatic*) effect produced by psychological stimulation. The brain can produce such effects in three general ways: (1) by transmission of signals through the

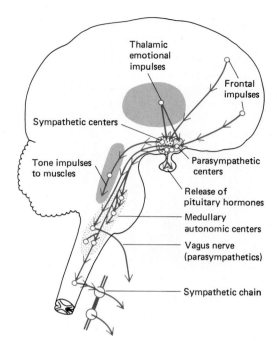

FIGURE 20–8   Neuronal circuits responsible for psychosomatic effects.

autonomic nervous system, (2) by transmission of signals to the muscles through the bulboreticular area, and (3) by control of certain of the endocrine glands. Figure 20–8 illustrates the general neuronal mechanisms believed to be responsible for psychosomatic effects. It shows conscious signals beginning either in the frontal cortex or in the thalamus, then transmitted to the hypothalamus, and finally through the centers of the hindbrain to the cord and thence to the body.

Psychosomatic Effects Transmitted Through the Autonomic Nervous System. Almost any emotion can affect the autonomic nervous system. For example, very intense agitation will increase the excitability of the reticular activating system, and this system in turn excites sympathetic activity throughout the body. Therefore, generalized sympathetic stimulation of the organs is one of the most common of all psychosomatic effects. Many emotions such as excitement, anxiety, or rage often discharge the sympathetics *en masse*, causing marked increase in arterial pressure, palpitation of the heart, and cold chills over the skin.

Psychological effects also often stimulate the parasympathetic centers of the hypothalamus. The emotions of worry, depression, and lethargy, all of which have effects opposite to those that excite the sympathetic system, often stimulate the parasympathetics. On occasion, however, both of the systems may be stimulated simultaneously. Fear, for instance, can cause extreme sympathetic stimulation resulting in elevation of arterial pressure, while at the same time stimulating the parasympathetics to elicit such intense gastrointestinal activity that the person has uncontrolled diarrhea.

Transmission of Psychosomatic Effects Through the Skeletal Muscles. The reticular activating system sends signals downward through the spinal cord and thence directly to the muscles. Therefore, the same emotions that excite the sympathetics usually increase the tone of the muscles throughout the body as well. Sometimes the tone becomes so intense that it causes muscle tremor, which explains why certain emotions can culminate in actual shaking.

On the other hand, the emotions that normally stimulate the parasympathetics usually decrease the activity of the bulboreticular formation. As a result, the muscular tone decreases to a very low level, which explains the muscular *asthenia* (muscular weakness) that is characteristic of some psychic states.

Transmission of Psychosomatic Effects Through Endocrine Glands. The nervous system controls several of the endocrine glands either completely or partially. For instance, the sympathetic nervous system controls the adrenal medulla, and the hypothalamus controls almost all the activities of the pituitary gland. The hypothalamus achieves this control of the anterior pituitary gland by secreting *neurosecretory substances* within the substance of the hypothalamus; these are then absorbed into the local capillaries and carried through minute veins from the hypothalamus to the anterior pituitary gland where they cause secretion of several major hormones: *growth hormone, corticotropin, thyrotropin, prolactin,* and the *gonadotropins.* These hormones, in turn, control growth rate, protein metabolism, overall rate of metabolism, lactation, and most sexual functions. Also, the supraoptic nuclei of the hypothalamus control the secretion of *antidiuretic hormone;* this hormone, in turn, controls the degree of retention of water by the kidneys.

Obviously, therefore, many psychosomatic effects can be mediated through the endocrine glands. For instance, psychic effects that overly stimulate the hypothalamus can produce hyperthyroidism, causing the thyroid gland to secrete excess thyroid hormone and thereby increase the rate of metabolism of all cells of the body. Likewise, psychic signals can affect the output of sex hormones and thus cause failure of ovulation, excess menses, diminished menses, infertility, or other sexual abnormalities.

Psychosomatic Diseases

Perhaps the most common psychosomatic disease is extreme nervous tension associated simultaneously with increased heart rate, elevated arterial pressure, gastrointestinal disturbances, and excess muscle tone. This is often described simply as "nervousness" or more properly as *anxiety.*

Psychosomatic effects can also cause abnormal function of individual organs. For instance, stimulation of the sympathetics can so decrease gastrointestinal activity that constipation results. On the other hand, excessive stimulation of the parasympathetics can increase the degree of gastrointestinal activity so greatly that severe diarrhea results. Another very common psychosomatic disorder is palpitation of the heart caused by excitement, anxiety, or other emotional states.

Occasionally the dysfunction caused by a psychosomatic disorder is so great that tissues are actually destroyed. For instance, stimulation of the parasympathetics to the stomach can cause so much secretion of gastric juices that they eat a hole into the wall of the stomach or upper intestine. This causes the condition called *peptic ulcer.* Such patients can be treated by operative removal of portions of the stomach, by neutralizing the gastric juices or blocking their secretion with special drugs, or in some instances by psychiatric treatment to alleviate the emotional condition that is initiating the excessive secretion of gastric juices.

Psychosomatic Pain. Many psychosomatic disorders can lead to pain that is called simply psychosomatic pain. For instance, a stomach ulcer causes intense burning in the pit of the stomach. Also, spasm of the gut caused by excess parasympathetic stimulation can cause cramps in the abdomen. And, occasionally, overstimulation of the heart can even cause cardiac pain. If the psychic condition that causes the functional abnormality can be corrected, then the pain likewise will be corrected.

Myths About Psychosomatic Disease. Despite the many different ways in which psychosomatic disease can come about, this subject has been greatly overemphasized by newsmen and authors of fiction. It is a common myth that a person can worry so much about the function of one of his organs that he thereby creates disease in that particular organ. Except in a few instances, this is not true. Most psychosomatic diseases exhibit regular patterns such as general states of tension, constipation or diarrhea, ulcer, and others of a similar nature.

QUESTIONS

1. What are the differences between short-term memory and long-term memory?
2. What determines which memories will be remembered? What are the roles of rehearsal and codification in the consolidation of memory?
3. What functions do the thalamus and hippocampus perform in relation to memory, especially in the establishment of memories and recall of memories, and in the thinking process in general?
4. What is the cellular mechanism of memory?
5. What happens to a person when he loses the prefrontal areas of his cerebral cortex?
6. Give the feedback theory of wakefulness and sleep.
7. When recording brain waves, what causes alpha waves, beta waves, and delta waves?
8. What is the function of pleasure and pain in the establishment of memories and in overall brain function?
9. Under what conditions will an animal exhibit rage?
10. Describe the different means by which the brain can cause psychosomatic abnormalities.

REFERENCES

Bekhtereva, N.P.: The Neurophysiological Aspects of Human Mental Activity. New York, Oxford University Press, 1978.

Daniloff, R., *et al.:* The Physiological Bases of Verbal Communication. Englewood Cliffs, N.J., Prentice-Hall, 1980.

Drucker-Colin, R., *et al.* (ed.): The Functions of Sleep. New York, Academic Press, 1979.

Geschwind, N.: Specializations of the human brain. *Sci. Am.,* 241(3):180, 1979.

Gillin, J.C., *et al.:* The neuropharmacology of sleep and wakefulness. *Annu. Rev. Pharmacol. Toxicol.,* 18:563, 1978.

Hobson, J.A., and Brazier, M.A.B. (eds.): The Reticular Formation Revisited: Specifying Function for a Nonspecific System. New York, Raven Press, 1980.

Hubel, D.H.: The brain. *Sci. Am.,* 241(3):44, 1979.

Ito, M., *et al.* (eds.): Integrative Control Functions of the Brain. New York, Elsevier/North-Holland, 1978.

Jones, N.B., and Reynolds, V.: Human Behaviour and Adaptation. New York, Halsted Press, 1978.

Kelly, D.: Anxiety and Emotions: Physiological Basis and Treatment. Springfield, Ill., Charles C Thomas, 1979.

McFadden, D. (ed.): Neural Mechanisms in Behavior. New York, Springer-Verlag, 1980.

Moskowitz, B.A.: The acquisition of language. *Sci. Am.,* 239(5):92, 1978.

Snyder, S.: Biological Aspects of Mental Disorder. New York, Oxford University Press, 1980.

Stern, R.M., *et al.:* Psychophysiological Recording. New York, Oxford University Press, 1980.

Uttal, W.R.: The Psychobiology of Mind. New York, Halsted Press, 1978.

Wassermann, G.D.: Neurobiological Theory of Psychological Phenomena. Baltimore, University Park Press, 1978.

VII

THE SPECIAL SENSORY SYSTEMS

21

The Eye

Overview

The principal components of the visual system are (1) the **eye,** which functions as a camera to focus a visual image on the retina; (2) the **retina,** the part of the eye that converts the visual image into a pattern of nerve impulses transmitted through the optic nerve into the brain; (3) the *brain mechanism for interpreting the visual signals;* and (4) the *brain mechanism for controlling the motor functions of the eyes* such as focusing, control of light entering the eyes, and directing the eyes toward the object of interest.

The *optical system* of the eye, like that of a camera, has a *lens,* a *muscular mechanism for focusing* the lens, and a *diaphragm* (the pupil), which controls the amount of light entering through the lens. The eyes of children can focus on objects as near as 7 cm and as far away as infinite distance, but by the time a person reaches the age of about 45 years, most of his focusing ability has been lost.

The light-sensitive elements of the *retina* that convert the visual image into nerve impulses are the rods and cones. The **rods** detect *black and white* images; the **cones** detect *colors.* There are about 125 million rods and cones in each retina but only about 1 million nerve fibers leading from the eye back to the brain. In the peripheral portions of the retina, large numbers of rods and cones are connected to the same optic nerve fiber so that the acuity of vision in these portions of the retina is very poor. In the very center of the retina is a small area about 0.5 millimeter (mm) in diameter called the **fovea** that is comprised entirely of slender cones that connect with approximately one optic nerve fiber for each cone. Because of these properties, this central area of the retina has excellent acuity of vision and is also the part of the retina most capable of detecting and analyzing color.

Visual signals are transmitted from the retina through the **optic nerve** and **optic tract** to the **lateral geniculate body** in the **thalamus.** From here the signals are further relayed to the **visual cortex** in the **occipital lobe.**

The neuronal mechanisms for vision are especially adapted to respond to *changes* in the visual scene. That is, every time the light intensity increases or decreases, this causes extremely strong signals in the visual system. Likewise, everywhere in the visual scene where there is a change from a dark area to a light area or vice versa, this too causes strong visual signals. Therefore, the visual system is especially geared to respond to such effects as *flashes of light* and *contrast borders* in the visual scene. It is in this way that visual information is dissected apart to determine its meaningful highlights.

The eyes are directed toward the object of attention by a system of three separate pairs of *eye muscles* attached to each eyeball and controlling (1) *upward and downward movement,* (2) *lateral* and *medial movement,* and (3) *rotational movement.* Focusing of the eyes is controlled by a circular constrictor muscle inside the eyeball, the **ciliary muscle,** that surrounds the eye lens. The amount of light that enters the eye is controlled by the degree of constriction of the **pupil.** All these muscle functions of the eye are controlled by *neuronal centers in the brain stem,* but in turn appropriate signals from the visual cortex of the occipital lobe direct the control functions of the brain stem centers.

The eye is a highly specialized receptor organ of the nervous system. Indeed, the retinal portion of the eye is actually formed from neural tissue that grows outward from the brain during development of the fetus. Furthermore, the retina contains neurons, the same as the brain and spinal cord. The purpose of the present chapter will be, first, to describe the anatomy of the eye, second, to describe the specialized sensory function of the eye, and finally, to discuss the neurophysiology of vision.

ANATOMY OF THE EYE AND ITS SURROUNDINGS

Figure 21–1 illustrates the eye and its surrounding structures. The bony cavity in which the eye is located is the **orbit,** which was described in Chapter 6. The other important structures lying within the orbit include the following:

1. The **optic nerve** passes backward from the eye and eventually through the bony **optic canal** into the cranial vault at the base of the brain.
2. The **central retinal artery** penetrates the optic nerve about 1 cm behind the eye and then runs forward in the middle of this nerve

into the interior of the eye. Inside the eye it branches over the entire inner surface of the retina to supply it with nutrition. The retina is the visual receptor portion of the eye.
3. The **extraocular muscles** attach to all sides of the eye and cause its movements in the directions required.
4. The **fat pad of the orbit,** which is a mixture of adipose tissue and connective tissue, fills all the orbital space not occupied by the other structures. This fat pad supports the eye and provides a loose sling in which the eye can rotate as needed. In patients who have excess production of thyroid hormone, the fat pad frequently becomes greatly overdeveloped, and this causes forward protrusion of the eyes, a condition called *exophthalmos.*

The Eyelids and Their Functional Structures. Figure 21–1 also illustrates the eyelids. The two angles where the upper and lower lids come together are called, respectively, the **lateral canthus** and the **medial canthus.** The important structures of the eyelids are as follows:

1. The **tarsal plates.** These are flat cartilaginous plates, one in the upper eyelid and another in the lower eyelid. They are about half a centi-

Superior levator palpebrae

Extraocular muscles

Orbicularis oculi

Conjunctiva

Tarsal plates

Fat pad of orbit

Optic nerve

Sclera

Central retinal artery

FIGURE 21–1 A section through the orbit and eyeball as seen from the side.

meter wide and extend horizontally about 1.5 cm, providing a semisolid skeletal structure for the eyelids.

2. The **orbicularis oculi muscle.** This muscle is shown in cross-section in Figure 21–1, and it was shown in a frontal view of the facial muscles in Figure 8–4. It circles all the way around the eye and functions as a sphincter. Therefore, when it is contracted the eyelids are literally squeezed together. This is the manner in which the eyelids are closed.

3. The **levator palpebrae superioris muscle.** This muscle is illustrated at the very top of the orbit in Figure 21–1. It originates in the posterior recess of the orbit, and its anterior tendon attaches to the connective tissue within the upper eyelid. Therefore, when contracted it pulls this lid upward. In addition to this muscle, there are also moderate amounts of *smooth muscle* in the superior eyelid that contract and help to open the eyelids when the sympathetic nervous system is stimulated. Therefore, when a person is excited or very attentive, which activates the sympathetic nervous system, the eyelids tend to open widely, but when he is sleepy they tend to close.

4. The **conjunctiva.** This is the lining membrane on the inside of the eyelids, and it also folds over the sides of the eyeball. In other words, it is the "skin" of the side surfaces of the eye and of the posterior surfaces of the eyelid. The upper and lower recesses between the eyelids and the eye are called respectively the **superior conjunctival fornix** and **inferior conjunctival fornix.**

The Lacrimal Gland and the Tear Duct System.
The eyes are continually washed and lubricated by tears. The tears are secreted by the **lacrimal glands,** illustrated in Figure 21–2, that lie in the lateral orbit underneath the superior eyelid. It secretes tears into the superior conjunctival fornix through five or six small **excretory ducts.** After the tears wash through the superior and inferior conjunctival fornices, they flow into two small canals, the **lacrimal ducts,** near the medial canthus of the eyelids. The openings of these ducts, each called a **punctum lacrimale,** can be seen as two small red dots on the medial edges of the superior and inferior eyelids. The two lacrimal ducts, each less than 1 cm in length, lead into a large **lacrimal sac** that in turn empties downward through the **nasolacrimal duct** into the lateral side of the nasal cavity underneath the inferior nasal concha. The lacrimal ducts, the lacrimal sac, and the

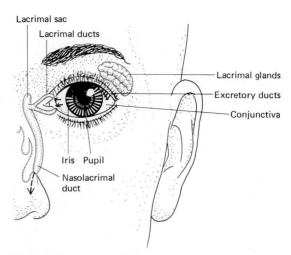

FIGURE 21–2 The lacrimal gland and the tear duct system.

nasolacrimal duct are all lined with mucous membrane similar to that of the nose, and, when a cold develops in the nose, swelling of this membrane often blocks the tear duct system. Yet, the lacrimal glands continue to secrete tears so that the eyes "water" because the tears have no avenue of escape.

The Extraocular Muscles.
Vision is just as much dependent upon proper positioning of the eyes as it is on the visual receptor apparatus itself. Positioning of the eyes requires three separate eye motions and therefore three separate sets of opposing muscles. These motions and their muscles, illustrated in Figure 21–3, are as follows:

1. *Upward and downward deviation.* The **superior rectus muscle** rolls the eye upward; the **inferior rectus muscle** rolls it downward.

2. *Lateral and medial deviation.* The **lateral rectus muscle** rolls the eye lateralward; the **medial rectus muscle** rolls it toward the midline.

3. *Rotation of the eye.* The **superior oblique muscle** rotates the right eyeball in a clockwise direction; the **inferior oblique muscle** rotates the right eyeball in a counterclockwise direction. Rotation of the left eyeball is controlled opposite to the right eyeball by these two muscles.

Note that all of the extraocular muscles besides the inferior oblique originate from a **common tendinous ring** that surrounds the channel through which the optic nerve passes from the posterior orbit into the optic canal. The **levator palpeprae superioris muscle** also originates from this tendinous ring. The remaining muscle, the inferior oblique, originates from the medial border of the orbit.

Inferior oblique
Trochlea for
superior oblique
Superior oblique
Medial rectus
Inferior rectus
Lateral rectus
Superior rectus
Superior levator
palpebrae
Common tendinous ring

FIGURE 21–3 The extraocular muscles as seen from the top.

Note also in Figure 21–3 that all the extraocular muscles wrap around the eyeball for a short distance before inserting into the strong fibrous wall of the eyeball, the sclera. And, finally, note that the superior oblique muscle makes a right angle turn through a fibrous stirrup called the *trochlea for the superior oblique muscle.* This allows the superior oblique to rotate the eyeball even though this muscle actually pulls backward.

THE STRUCTURE OF THE EYEBALL

Much of the internal structure of the eyeball will be discussed in more detail later in the chapter as we consider the process of vision; however, let us study here the gross structures of the eyeball that are illustrated in Figure 21–4. The eyeball is also called the **bulb** of the eye, and in most areas it has three separate layers or "tunics":

1. an **outer fibrous tunic,** comprised of the **sclera** and the **cornea;**
2. an **intermediate vascular tunic,** consisting of the **choroid, ciliary body,** and **iris;** and
3. an **inner nervous tissue tunic,** which is the **retina.**

The Fibrous Tunic—the Sclera and the Cornea. The **sclera** is the strong fibrous encasement of the posterior five sixths of the bulb, and the **cornea** is the anterior clear window of the eyeball that constitutes the anterior one sixth of the bulb. Both the sclera and the cornea are about 1 mm in thickness, and the connective tissue comprising these structures is extremely strong, like that of muscle tendons. Pressure of fluids inside the eyeball stretches the sclera and

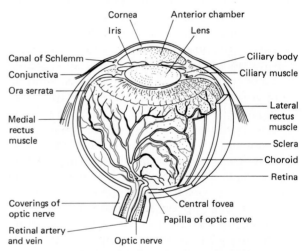

Cornea
Iris
Anterior chamber
Lens
Canal of Schlemm
Conjunctiva
Ora serrata
Ciliary body
Ciliary muscle
Medial
rectus
muscle
Lateral
rectus
muscle
Sclera
Choroid
Retina
Coverings of
optic nerve
Retinal artery
and vein
Central fovea
Papilla of optic nerve
Optic nerve

FIGURE 21–4 The principal structures of the eyeball as seen from above.

cornea and causes the eyeball to be almost spherical.

The cornea has special features that are important for its function as the window to the eye. First, its outer surface is covered by an **epithelium** comprised of multiple layers of polyhedral epithelial cells. This provides a reasonable degree of protection for the outer surface of the cornea against abrasion.

Second, the midportion of the cornea, called the **substantia propria,** is comprised of dense connective tissue. This tissue is different from most dense connective tissue, however, because it contains large amounts of interstitial cement substance containing a high concentration of proteoglycan molecules similar to those found in less concentration in the ground substance of other types of connective tissue. This cement substance helps to make the cornea clear.

Third, the inner surface of the cornea is lined with a single-layered **endothelium,** but an endothelium with special properties. It pumps sodium out of the cornea to the interior of the eyeball, and the removal of the sodium in turn causes osmosis of fluid out of the cornea, thereby compacting the collagenous fibers and the interstitial cement substance so tightly that the cornea is perfectly clear. Blockage of this pumping action of the endothelium causes the cornea to become opaque, which often occurs when the eye is badly traumatized or infected.

The Vascular Tunic—the Choroid, Ciliary Body, and Iris. Immediately inside the sclera is the **choroid,** which is an extensive plexus of small blood vessels embedded in loose connective tissue. These vessels supply nutrition to the entire outer surface of the retina that lies on the inner surface of the choroid.

At the anterior margin of the choroid, for a distance of about 4 mm behind the corneoscleral junction, the vascular tunic becomes greatly thickened and projects medially into the eyeball toward the margins of the lens capsule. This portion of the vascular tunic is the **ciliary body.** From its surface about 70 folds called **ciliary processes** project still farther toward the lens. These processes are highly vascular and are covered with **epithelial cells** that secrete fluid into the eyeball, which is the source of the fluid that maintains the pressure in the eye.

Anterior to the ciliary body, the vascular tunic becomes thin once again and constitutes the **iris** of the eye. This projects medially into the eyeball, lying over the anterior surface of the lens, and the round opening in the middle of the iris is the **pupil.** The iris contains a **circular smooth muscle sphincter** around the pupil that can constrict the pupil to a small size, thus limiting the amount of light that passes through the lens of the eye. The iris also contains **radial smooth muscle fibers** extending from the border of the pupil to the outer border of the iris; when these contract, they open the pupil to a very large size.

The Inner Tunic—the Retina. The **retina** lines the inside surface of the posterior two thirds of the eyeball, lying on the inner surface of the choroid. It is composed of neural tissue derived from the brain itself during development of the embryo. The retina is the sensory organ of the eye, containing light-sensitive cells called **rods** and **cones** that detect light and color and then send appropriate signals to the brain to cause vision. At the posterior pole of the eyeball

is a slightly depressed yellowish area of the retina, about 1.5 mm in diameter, called the **macula lutea,** and in the center of this is a deeper recess about 0.4 mm in diameter called the **fovea centralis.** The fovea has large numbers of highly specialized cones that provide exceptionally clear vision. It is depressed because the inner layers of the retina have been pulled to the side so that light can pass unimpeded to the cones. Thus, the fovea is the most important part of the entire retina for vision, as we shall discuss later.

The **optic nerve** enters the posterior surface of the eyeball about 15 degrees medial to the fovea. At this point the sclera is perforated by a large number of very small holes to allow entry of the nerve fibers into the eyeball, giving this area of the sclera the name **cribriform area.** There are no rods and cones in this cribriform area, only nerve fibers, which makes the area have an appearance of a whitish round spot called the **optic disk.** This disk area, without rods or cones, has no vision, so that it causes a "blind spot" in the visual field. The optic nerve fibers distribute radially in all directions from the optic disk over the inner surface of the retina to innervate the rods and cones.

Entering the eyeball in the middle of the optic nerve are the **retinal artery** and **vein.** These vessels branch also over the inner surface of the retina to supply it with nutrition, supplying approximately the inner three quarters of the thickness of the retina; the blood vessels of the choroid supply approximately the outer one quarter of the retina.

The Lens, the Lens Ligaments, and the Ciliary Muscle. The **lens** of the eye is comparable to the lens of a camera. In the young person, it is a transparent structure encased in a strong **fibrous lens capsule** containing inside a mass of pliable, clear cells composed mainly of high concentrations of protein. Its curved anterior and posterior surfaces allow focusing of the visual image on the retina.

The capsule of the lens is connected around its peripheral border to the ciliary body by about 70 **ligaments of the lens.** These ligaments are normally tense so that the lens capsule is pulled into an ovoid shape. In the deeper layers of the ciliary body is a small muscle, the **ciliary muscle,** that is comprised of smooth muscle fibers. This muscle is so arranged that when it contracts it loosens the ligaments of the lens. When this happens, the elasticity of the lens capsule causes the lens to become rounder than normal and therefore focused for near vision, as will be explained more fully later.

THE FLUID SYSTEM OF THE EYE

The fluid system of the eye maintains a constant pressure in the eyeball of almost exactly 15 mm Hg. This pressure, along with the very strong wall of the eyeball (the sclera and cornea), maintains the eyeball in a relatively rigid shape, keeping the distances between **cornea, lens,** and **retina** always constant within fractions of a millimeter. This is essential to the functioning of the eye's optical system for focusing images on the retina, as discussed later in the chapter.

Aqueous Humor and Vitreous Humor. The fluids of the eye are divided, as shown in Figure 21–5, into two separate types: the *aqueous humor*, which is present in the *anterior chamber* of the eye between the front of the lens and the cornea and also on the sides of the lens, and the *vitreous humor*, which is behind the lens, between it and the retina. The vitreous humor is a gelatinous mass, and fluid in it can *diffuse* through the mass but cannot *flow* through it with ease. On the other hand, the aqueous humor flows freely.

Formation of Aqueous Humor by the Ciliary Body. Fluid is continually formed by the *ciliary processes* of the ciliary body. These processes are small glandlike protrusions of epithelial cells that lie to the sides and slightly behind the lens, as illustrated in Figure 21–5. From here the fluid passes between the ligaments of the lens and into the anterior chamber of the eye. Then it flows, as illustrated by the arrows in the figure, into the angle between the cornea and the iris. There it leaves the eye through minute spaces called the **spaces of Fontana** and enters the **canal of Schlemm,** a circular vein that passes all the way around the eye at the juncture of the cornea and the sclera.

The ciliary body is made up of a large number of small folds of secretory epithelium that have a total surface area in each eye of about 6 cm². This epithelium secretes sodium ions continually into the aqueous humor, which in turn produces the following effects: (1) The sodium ions cause a positive electrical charge to develop in the aqueous humor. This positive charge in turn pulls negative ions, especially chloride ions, through the epithelium, thus increasing the number of these ions in the aqueous humor as well. (2) The increased quantity of sodium and chloride ions in the aqueous humor causes osmosis of water through the epithelium. Thus, indirectly, the secretion of sodium causes a continuous but very slow flow of fluid, about 3 *mm³ per minute*, into the aqueous humor from the ciliary body.

Control of Pressure in the Eyeball. The pressure in the eye is controlled by the minute structures lining the spaces of Fontana at the angle between the cornea and the iris where the fluid leaves the eyeball and enters the canal of Schlemm. If the pressure rises too high, increased amounts of fluid leave the eye. On the other hand, if the pressure falls too low, the outflow of fluid from the eyeball automatically decreases. That is, the spaces of Fontana act like a release valve. When the pressure is above a critical value, the valve structures open up, and when the pressure is below the critical valve, the spaces close. It is in this way that the pressure in the eyeball is regulated to almost exactly 15 mm Hg, rarely rising above or falling below this value more than a few millimeters Hg.

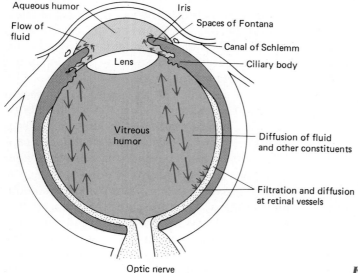

FIGURE 21–5 The fluid system of the eye.

Glaucoma. Glaucoma is a condition in which the pressure in the eye becomes very high, often high enough to cause blindness. The usual cause of the condition is failure of fluid to flow normally through the spaces of Fontana into the canal of Schlemm. When this happens, the continued formation of fluid by the ciliary body causes the eye pressure to rise sometimes to tremendous levels. For instance, *infection* or *inflammation* can cause debris or white blood cells to plug the openings of the spaces of Fontana. Also, many people develop narrowing of these spaces for reasons that are unknown.

In rare instances the pressure in the eye rises to as high as 60 to 70 mm Hg. These extremely high pressures can cause blindness within a few days by damaging the optic nerve where it enters the eyeball and by compressing the blood vessels against the retina so severely that the nutritive blood flow is cut off. When the eyeball pressure is elevated only moderately, on the other hand, blindness can develop gradually over a period of years.

THE EYE AS A CAMERA

Figure 21–6 illustrates the optical system of the eye for focusing an image on the *retina*. This optical system is identical to that of a camera, and the retina corresponds to the photographic film. Light passes first through the clear cornea, through the aqueous humor, then the lens, and finally the vitreous body before impinging on the retina. The retina in turn translates the image into nerve impulses and transmits these into the brain through the optic nerve.

The Lens System of the Eye. The lens system of the eye is composed of the cornea and the crystalline lens. Because the cornea is curved on its outside, light rays passing from the air into the cornea are *refracted* (bent) in the same way that any optical lens refracts rays. After the rays pass through the cornea and aque-

ous humor they strike the curved anterior surface of the crystalline lens, where still more bending occurs, and as they pass through the posterior surface of the lens the rays bend again. Thus, light rays are refracted (bent) at three different major interfaces in the eye. This is analogous to the refraction of light at the different surfaces in a compound lens system of a camera, for in the camera the light rays are also refracted at each interface between the lenses and the air.

Function of a Convex Lens in Forming an Image. Figure 21–7 shows the focusing of light rays from a distant source and also from two points near the lens. Note that the light rays originating from the left-hand side of the figure and striking the outer edges of the convex lens are bent toward the center, whereas those that strike the lens exactly in the center pass on through without being bent. The reason for this is that the outer rays enter the substance of the lens at an angle, which causes refraction, and those striking the center enter the lens perpendicular to its surface, which does not cause refraction. The light rays from the edges of the lens bend inward to meet those passing through the center, and they all focus on a common point called the *focal point.*

Figure 21–8A shows the focusing by a single lens of light rays from two different point sources. Note that the rays from the two lights are each focused to focal points on the opposite side of the lens directly in line with the lens center.

Figure 21-8B illustrates the focusing of light rays from different point sources on a human being's body. Those parts of the body that are

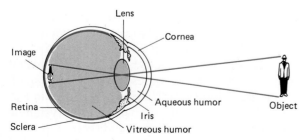

FIGURE 21–6 General structure of the eye, showing the function of the eye as a camera.

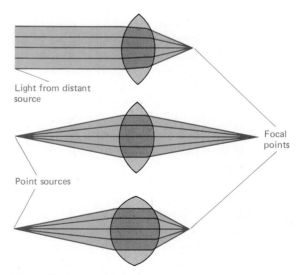

FIGURE 21–7 Focusing of parallel light rays and light rays from point sources by convex lens.

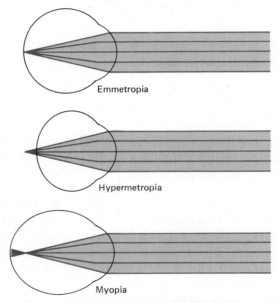

FIGURE 21-8 (A) Focusing of two points of light to two separate focal points. (B) Formation of an image by a convex lens focusing light rays from an object.

very bright are point sources of light, whereas those parts that are dark represent the black spaces between the point sources of light. So far as the eyes are concerned, all objects are mosaics of point sources of light. Light from each source is focused by the lens, and the focal point of each is always directly in line with the center of the lens and the original source. Consequently, an inverted image is formed by the focal points, as shown in the figure.

Abnormalities of the Lens System

The normal eye focuses parallel light rays exactly on the retina. This normal focusing is called *emmetropia,* and it is illustrated at the top of Figure 21-9. However, three different abnormalities frequently occur to prevent focusing of light rays precisely on the retina. These are *hypermetropia, myopia,* and *astigmatism.*

Hypermetropia. Hypermetropia, or *far-sightedness,* is caused by failure of the lens to bend the light rays enough to bring them to a focal point on the retina, which usually occurs because the eyeball is too short, as shown in Figure 21-9. Instead, the light rays are still diffuse when they reach the retina, and thus vision of even distant objects is blurred, and it is even more blurred for near objects. Hypermetropia is called far-sightedness because with this type of vision objects can be seen more clearly at a distance than near at hand.

Myopia. Myopia, which is called *near-sightedness,* is caused by too strong a lens system for the distance of the retina behind the lens, which is usually caused by an eyeball that is too long. That is, the light rays are focused before they reach the retina, and by the time

FIGURE 21-9 Parallel light rays focus on the retina in emmetropia, behind the retina in hypermetropia, and in front of the retina in myopia.

they do reach the retina they have spread apart again as shown in the figure, causing fuzziness of each point in the image.

Myopia is called near-sightedness because the myope can see objects near him with complete clarity, while not being able to focus any objects that are at a far distance.

Astigmatism. Astigmatism occurs when the lens system becomes ovoid (egg-shaped) rather than spherical. This effect is shown in Figure 21-10. Either the cornea or the crystalline lens is more elongated in one direction in comparison with the other direction. Because the radius of curvature is greater in the elongated direction than in the short direction, the light

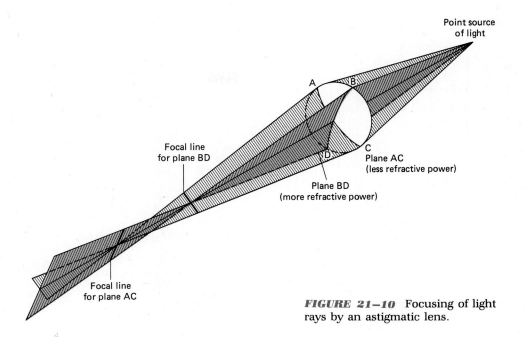

Point source
of light

A B

C
Plane AC
(less refractive power)

Plane BD
(more refractive power)

Focal line
for plane BD

Focal line
for plane AC

FIGURE 21–10 Focusing of light rays by an astigmatic lens.

rays entering the lens along this lengthened curvature are focused behind the retina, whereas those entering along the shortened curvature are focused in front of the retina. In other words, the eye is far-sighted for some of the light rays and near-sighted for the remainder. Therefore, the person with astigmatic eyes is unable to focus any object clearly, regardless of how far the object is away from the eyes; for when the near-sighted light rays are in focus, the far-sighted ones are out of focus and vice versa.

Correction of the Optical Abnormalities of the Eye with Glasses. Glasses with properly prescribed lenses can be used to correct most abnormalities of the lens system of the eye. Glasses bend the light rays before they enter the eye in an appropriate manner to correct for the excess or deficient refractive power of the eye. Figure 21–11 shows the correction of myopia and hypermetropia. In the myopic person the light rays normally focus in front of the retina. To prevent this, a concave lens is placed in front of the eye. This type of lens bends the light rays outward and, therefore, compensates for the excess inward bending of the myopic lens system. By prescribing the appropriate curvature to the concave lens, a myopic person's vision can be made completely normal.

In the hypermetropic eye the lens system normally fails to bend the light rays sufficiently. To correct this abnormality, a convex lens is placed in front of the eye so that the light rays will be partially bent even before they reach the eye. With the aid of this preliminary convergence, the lens system of the eye can then bring the rays to a focal point on the retina.

FIGURE 21–11 Correction of myopia with a concave lens, and correction of hypermetropia with a convex lens.

The lens that must be used to correct astigmatism is somewhat more complicated than those used to correct either myopia or hypermetropia, for it must be fashioned with more curvature in one direction than the other. However, by prescribing a lens with precisely ground curvatures in exactly the right "axes" in front of the eye, the abnormal refraction of light rays by each portion of the astigmatic eye can be corrected appropriately.

FUNCTION OF THE RETINA

Retinal Structure. Figure 21–12 illustrates the anatomy of the retina, showing that it is composed of several different layers of cells. The light rays enter from the *bottom* of the figure

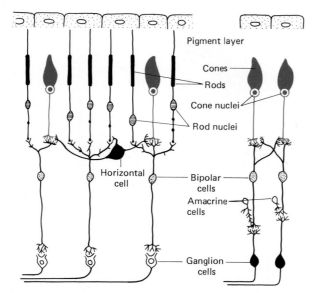

FIGURE 21-12   Functional anatomy of the retina.

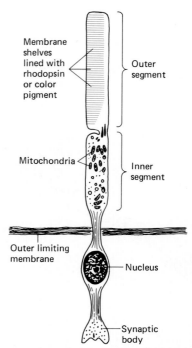

FIGURE 21-13   Schematic drawing of the functional parts of the rods and cones.

and pass all the way through the retina to the top, finally striking the *rods* and *cones* and the pigment layer. The rods and cones are nerve receptors that are excited by light. These cells change the light energy into neuronal signals that are transmitted into the brain.

The pigment layer of the retina contains large quantities of a very black pigment called *melanin.* The function of the melanin is to absorb the light rays after they have passed through the retina and thereby to prevent light reflection throughout the eye. *Albino* persons, who are unable to manufacture melanin in any part of their bodies, have a complete lack of pigment in this layer of the retina. As a result, after passing through the retina the light rays are not absorbed but instead are reflected so intensely in all directions that they cause images to become bleached out with excess light. The albino's vision is usually about three times less acute than that of a normal person and he is so blinded by bright sunlight that he must wear dark glasses to see at all.

By far the greater number of light receptors in the retina are rods. Light of all colors stimulates the rods, whereas the cones are stimulated selectively by different colors. Therefore, the cones are responsible for color vision, in contradistinction to the rods, which provide only black and white vision.

ANATOMY OF THE RODS AND CONES AND THEIR NEURONAL CONNECTIONS

Figure 21–13 illustrates a typical **rod.** The light receptor portion of the rod is at the top, called the **outer segment.** This segment is character-

ized by large numbers of infoldings of the cell membrane that form disklike shelves extending deeply into the cytoplasm. On these shelves is a light-sensitive chemical called *rhodopsin* that comprises about 60 percent of the total mass of the outer segment. When light strikes the rhodopsin, the permeability of the cell membrane changes, and this changes the electrical potential inside the rod as well. The potential is then transmitted downward through the entire rod to its opposite end, to the **synaptic body,** which forms synapses with retinal neurons called **bipolar** and **horizontal cells** (Fig. 21–12). The visual signals are transmitted through these cells to still another set of neurons, the **ganglion cells,** shown at the bottom of Figure 21–12. These give rise to the optic nerve fibers that finally transmit the visual signals into the brain.

The basic structure of cones is very similar to that of rods except that the cones are shorter and the outer segment is conical in shape rather than cylindrical. Also, the light-sensitive substances on the disks of the outer segment are slightly different from rhodopsin, making the different cones sensitive to different colors as will be explained later in the chapter.

The Chemistry of Vision

Chemistry of Rod Excitation.   Figure 21–14 shows the basic chemical changes that occur in the rods, both when light strikes the

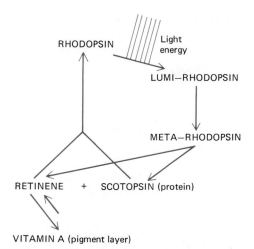

FIGURE 21–14 The retinene-rhodopsin chemical cycle responsible for light sensitivity of the rods.

retina and during periods between light stimulations. Vitamin A is the basic chemical utilized by both the rods and cones for synthesizing substances sensitive to light. On being absorbed into a rod, vitamin A is converted into a substance called *retinene.* This then combines with a protein in the rods called *scotopsin* to form the light-sensitive chemical *rhodopsin.* If the eye is not being exposed to light energy, the concentration of rhodopsin builds up to an extremely high level.

When a rod is exposed to light energy, some of the rhodopsin is changed immediately into *lumirhodopsin.* However, lumirhodopsin is a very unstable compound that can last in the retina only about a tenth of a second. It decays almost immediately into another substance called *metarhodopsin,* and this compound, which is also unstable, decays very rapidly into retinene and scotopsin.

Thus, in effect, light energy breaks rhodopsin down into the substances from which the rhodopsin itself had been formed, retinene and scotopsin. In the process of splitting rhodopsin, the rods become excited, probably by ionic charges that develop momentarily on the splitting surfaces of the rhodopsin. These charges last for only a split second. During this slight interval, nerve signals are generated in the rod and transmitted into the optic nerve and thence into the brain.

After rhodopsin has been decomposed by light energy, its decomposition products, retinene and scotopsin, are recombined again during the next few minutes by the metabolic processes of the cell to form new rhodopsin. The new rhodopsin in turn can be utilized again to provide still more excitation of the rods. Thus, a continuous cycle occurs: Rhodopsin is being formed continually, and it is broken down by light energy to excite the rods.

Chemistry of Cone Vision. Almost exactly the same chemical processes occur in the cones as in the rods except that the protein scotopsin of the rods is replaced by one of three similar proteins called *photopsins.* The chemical differences among the photopsins make the three different types of cones selectively sensitive to different colors.

Persistence of Images and Fusion of Flickering Images. Following a sudden flash of light that lasts only one millionth of a second, the eye sees an image of the light that lasts for approximately one-tenth second. The duration of the image is the length of time that the retina remains stimulated following the flash.

The persistence of images in the retina allows flickering images to *fuse* when one views a moving picture or television screen. The moving picture flashes 16 to 30 pictures per second, and the television screen provides 60 pictures per second. The image on the retina persists from one picture to the next, which gives one the impression of seeing a continuous picture.

Light and Dark Adaptation. It is common experience to be almost totally blinded when first entering a very bright area from a darkened room and when entering a darkened room from a brightly lighted area. The reason for this is that the sensitivity of the retina is temporarily not attuned to the intensity of the light. To discern the shape, the texture, and other qualities of an object, it is necessary to see both the bright and dark areas of the object at the same time. Fortunately, the retina automatically adjusts its sensitivity in proportion to the degree of light energy available. This phenomenon is called light and dark adaptation.

The mechanism of *light adaptation* can be explained by referring once again to Figure 21–14. When large quantities of light energy strike the rods, large amounts of rhodopsin are broken into retinene and scotopsin, and, because rhodopsin formation is a relatively slow process, requiring minutes or even hours at times, the concentration of rhodopsin in the rods falls to a very low value as the person remains in the bright light. Essentially the same effects occur in the cones. Therefore, the sensitivity of the retina soon becomes greatly depressed in bright light.

The mechanism of *dark adaptation* is opposite to that of light adaptation. When the person enters a darkened room from a lighted area, the quantity of rhodopsin in his rods (and color-sensitive chemicals in his cones) is at first very slight. As a result, he cannot see anything. Yet,

FIGURE 21–15 Dark and light adaptation.

the amount of light energy in the darkened room is also very slight, which means that very little of the rhodopsin formed in the rods is broken down. Therefore, the concentration of rhodopsin builds up during the ensuing minutes or hours until it finally becomes high enough for even a very minute amount of light to stimulate the rods.

During dark adaptation, the sensitivity of the retina can increase as much as 1000-fold in only a few minutes, and as much as 100,000 times in an hour or more. This effect is illustrated in Figure 21–15, which shows the retinal sensitivity increasing from an arbitrary light-adapted value of 1 up to a dark-adapted value of 100,000 within 1 hour after the person has left a very bright area and moved into a completely darkened room. Then, on reentering the bright area, light adaptation occurs, and retinal sensitivity decreases from 100,000 back down to 1 within another 10 minutes, which is a more rapid process than dark adaptation.

Function of the Cones— Color Vision

The cones are different from the rods in several respects. First, they respond selectively to certain colors, some to one color and others to other colors. Second, cones are considerably less sensitive to light than are rods, for which reason they cannot provide vision in very dim light. Third, many of the cones are connected, one cone to one optic nerve fiber, which provides greater acuity of vision than the rods provide. Ten to 200 rods usually connect with the same optic nerve fiber, which means that impulses transmitted by the rods to the brain do not necessarily originate from one very discrete point on the retina.

Detection of Different Colors by the Cones. The retina contains three different types of cones, each of which responds to a different spectrum of colors. This is illustrated in Figure 21–16, which shows light wave lengths at which the three different types of cones—the blue cone, the green cone, and the red cone—respond. Note that the blue cone responds maximally at a wave length of 430 millimicrons (millimicrometers or mμ), which is a blue color; the green cone at 535 mμ, a greenish-yellow color; and the red cone at 575 mμ, an orange color. The so-called "red" cone is called the red cone not because its maximal response is in the red range, but because it is the only cone that has any significant response at all above 600 mμ, which is red.

Determination of the Intermediate Colors by the Blue, Green, and Red Cones. It is quite easy to understand how the blue cones determine that an object is blue, how the green cones determine that an object is green, and how the red cones determine that an object is red, but it is more difficult to understand how these cones detect the intermediate colors between the three primary ones. This is accomplished by utilizing a combination of cones. For instance, yellow light stimulates the red and green cones approximately equally. When both of these types of cones are stimulated equally, the brain interprets the color as yellow. Also, when the red cones are stimulated about $1\frac{1}{2}$ times as strongly as the green cones, which occurs when light with a wave length of 580 mμ strikes the retina, the brain interprets the color as orange. If both the red and green cones are

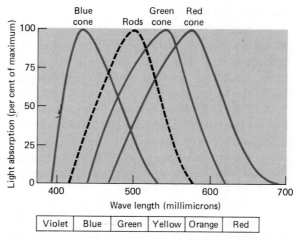

FIGURE 21–16 Spectral sensitivity curves for blue, green, and red cones and also for rods. (Drawn from curves recorded by Marks, Dobelle, and MacNichol, Jr.: *Science*, 143:1181, 1964, and by Brown and Wald: *Science*, 114:45, 1964.)

stimulated, but the green more than the red cones, the color is interpreted as a greenish-yellow. Likewise, when both the green and blue cones are stimulated, the color is interpreted as a bluish-green. Thus, by combining the degrees of stimulation of the different cones, the brain can distinguish not only among the three primary colors but also among the other colors having intermediate wave lengths.

Colorblindness. Colorblindness also can be understood very readily on the basis of Figure 21–16. Occasionally one of the three primary types of cones is lacking because of failure to inherit the appropriate gene for formation of the cone. The color genes are sex-linked and are found in the female sex chromosome, the X chromosome. Since females have two of these chromosomes, they almost never have a deficiency of a color gene, but because males have only one female chromosome, one or more of the color genes is absent in about 8 percent of all males. For this reason, almost all colorblind people are males.

If a person has complete lack of red cones, he is able to see green, yellow, orange, and orange-red colors by use of his green cones. However, he is not able to distinguish satisfactorily between these colors because he has no red cones to contrast with the green ones. Likewise, if a person has a deficit of green cones, he is able to see all the same colors, but he is not able to distinguish between green, yellow, orange, and red colors, because the green cones are not available to contrast with the red. Thus, loss of either the red or the green cones makes it difficult or impossible to distinguish between the colors of the longer wave lengths. This is called *red-green colorblindness.*

In very rare instances a person has a deficiency of blue cones, in which case he has difficulty distinguishing violet, blue, and green from each other. This type of colorblindness is frequently called *blue weakness.*

NEURONAL CONNECTIONS OF THE RETINA WITH THE BRAIN

Figure 21–17 illustrates the connections of the retina with the brain, showing that the right halves of the retinas of the two eyes are connected with the right visual cortex and the left halves with the left visual cortex. The *optic nerve* fibers from the nasal half of each retina cross in the *optic chiasm,* located on the bottom surface of the brain, and join the fibers from the tempo-

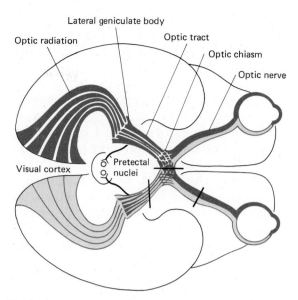

FIGURE 21–17 Optic pathways for transmission of visual signals from the retinas of the two eyes to the optic cortex. (From Polyak: The Retina. The University of Chicago Press, 1957.)

ral half of the opposite retina. Then the combined fibers pass backward through the *optic tract,* synapse in the *lateral geniculate body,* and finally spread through the *optic radiation* into the *visual cortex.*

In addition, fibers pass directly from the optic tract into the *pretectal nuclei;* these fibers carry signals for control of pupillary size in response to light intensity, as will be discussed later in the chapter.

Discrimination of the Visual Image at the Retinal Level. Even at the level of the retina the visual image begins to be analyzed. Therefore, the pattern of stimulation that is transmitted to the visual cortex is considerably different from the actual visual image on the retina. The retina breaks the visual image into two components. The first component is the level of *luminosity.* That is, some of the optic nerve fibers transmit signals to the brain to indicate the general light intensity of the observed scene. The second component is a series of signals to indicate *changes* in light intensity. Indeed, two different types of changes in light intensity are transmitted in the optic fibers. One of these denotes changes in light intensity at contrast borders in the visual image. For instance, the contrast between a white piece of paper and a black line on the paper gives an intense signal that is transmitted in the optic fibers. The second denotes changes in amount of light entering the lens of the eye and striking a spot on the retina. For example, a person is looking toward a dark wall and suddenly a bright insect flies across his field of vision. At each point where the image of the insect appears on the retina there is a sudden flash of light. And this excites a succession of optic nerve fibers very powerfully.

Thus, the messages that are sent from the retina in the optic nerve fibers do not transmit a true mosaic pattern of the visual image but, instead, transmit one type of signal to denote the general level of illumination and others to denote where in the image there are changes in light intensity. And besides these signals, the respective color cones send additional signals to indicate color contrasts.

Function of the Lateral Geniculate Body in the Analysis of the Visual Image. At the lateral geniculate body, several other aspects of the visual scene begin to be interpreted. One of these is probably *depth perception*, for it is here that signals from the two eyes first come together. As we shall see later in the chapter, depth perception depends upon comparing minute differences in shapes of objects as seen by the two separate eyes. The lateral geniculate body is admirably suited for this because it has six layers of neurons, the visual signals from one eye terminating in layers 1, 4, and 6 and those from the other eye in layers 2, 3, and 5; thus, the signals from the two eyes are intimately interconnected.

It is also possible that the lateral geniculate body plays a role in *color vision*. A reason for believing this is that one can look at a red light with the left eye and a green light with the right eye and see a yellow color, thus indicating that combination of colors from the two separate eyes occurs at least to some extent in the brain, perhaps at the lateral geniculate level.

Function of the Visual Cortex in Discriminating the Visual Image. Once the visual image reaches the visual cortex, it has by that time been changed to a pattern of stimulation that is considerably different from the actual image on the retina. This is illustrated in Figure 21–18. To the left is the retinal image of a heavy colored cross; to the right is the pattern of stimulation in the visual cortex. Note that stimulation occurs only at the edges of the cross. This is caused by a succession of mechanisms in the retina, the lateral geniculate body, and the visual cortex that allows a cortical point to be stimulated if there is a contrast border between a light area and a dark area. If there is no contrast, the neuron will not be stimulated. Thus, the visual neuronal processing system brings out borders and thereby determines the shapes of images. This explains why a simple line drawing of

Retinal image   Cortical stimulation

FIGURE 21–18 Pattern of stimulation in the visual cortex caused by observing a heavy, colored cross.

someone's face can be recognized as a picture of that person. In fact, the visual processing system actually converts the image of a person to a type of line drawing anyway.

Another feature of discrimination by the visual cortex is that it also determines directions of orientation of the lines and borders in the image. For instance, if the cross shown in Figure 21–18 is straight up and down, neurons at one depth in the cortex are stimulated. But, if the cross is leaning, neurons at another depth are stimulated.

From the primary visual cortex, secondary signals are transmitted into the visual association areas located to the lateral sides of the primary visual cortex. In these areas, the finer meanings of the visual signals are interpreted. For instance, the picture of a letter is believed to be interpreted here as letter A, B, C, or so forth, and a combination of letters is interpreted as a word. Still further away from the primary visual cortex, the combination of words is interpreted as a thought.

Fields of Vision

A means used to determine the extent of a person's normal vision, and also to detect abnormalities of vision, is to plot his *fields of vision*. The person is asked to close one eye and to look straight forward with the other eye. A small spot of light is then moved first above his central point of vision as far as he can see, then below as far as he can see, then to the right, to the left, and in all directions. In this way, his ability to see in all areas away from his centralmost point of vision is plotted.

Figure 21–19 shows the normal field of vision of the right eye. To the lateral side one can actually see objects at right angles (90 degrees) from the direction in which the eye is looking. To the nasal side, the nose is in the way, and the person can see objects only 50 degrees away from the central point of vision. Likewise, in the

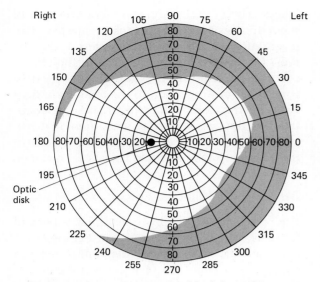

FIGURE 21-19 The visual field of the right eye.

upward direction the orbital ridge is in the way, and in the downward direction the cheekbone is in the way. Were it not for these structures around the eye, the field of vision would be considerably greater.

In the field of vision is a *blind spot* caused by the *optic disk*, which is the point where the optic nerve enters the eyeball. At this point no rods or cones are present. The blind spot of each eye is located approximately 15 degrees to the lateral side of the central point of vision. However, the blind spots of the two eyes are on opposite sides in the respective fields of vision so that when the images from the two eyes fuse there is no part of the visual scene that is not covered. Fortunately, also, the blind spots normally are not noticeable in one's vision.

Diagnosing Lesions in the Visual System from the Fields of Vision. Plotting the fields of vision provides a means for determining and locating damage in the retina or in the neuronal tracts from the retina to the brain. Referring back to Figure 21-17, it will be noted that three dark lines have been placed respectively across _____ he optic chiasm _____ the right optic r _____ eld of vision of _____ in other words, _____ d. If the optic c _____ asal half of each _____ that the *lateral* _____ lind, because t _____ verts the image on the retina. Finally, if the right optic tract is destroyed, the left halves of the visual fields of both eyes will be blind. Obviously, also, damage in any portion of the visual cortex or at

any other point in the visual transmission system will cause loss of vision in respective areas in the fields of vision. Thus, the point in the eye or in the brain at which damage has occurred can often be discerned from the pattern of visual loss.

Visual Acuity

Visual acuity means the degree of detail that the eye can discern in an image. The usual method for expressing visual acuity is to compare the vision of the person being tested with the vision of a normal person. If at a distance of 20 feet he can barely read letters of exactly the size that the normal person can barely read, he is said to have 20/20 vision, which is normal. If the letters must be as large as those that the normal person can read at a distance of 40 feet, his vision is said to be 20/40, or, in other words, his vision is one-half normal. If he can barely read letters at a 20-foot distance that the normal person can read at 100 feet, his vision is 20/100, or, in other words, his vision is one-fifth normal. An occasional person has better than normal vision, so that he can read at 20 feet what the normal person can read only at 15 feet, in which case his vision is said to be 20/15.

Importance of the Fovea for Good Visual Acuity. A person normally has very clear vision only in the central area of his visual field. The reason for this is that the small central area of the retina, only 0.4 mm in diameter, called the *fovea*, is especially adapted for high acuity vision. No rods are present in the fovea, and the cones there are considerably smaller in diameter than those in the peripheral portions of the retina so that very small points in the visual image can be detected. Also, the nerve fibers and blood vessels are all pulled to one side, so that light can pass with ease directly to deep layers of the retina where the cones are located. Finally, and especially important, each of the cones of this region connects through an almost direct pathway with the brain, one cone to one optic nerve fiber, so that the impulses from each cone do not become confused with impulses from other cones.

Depth Perception

The eyes determine the distance of an object by two principal means. These are, first, by the *size of the image* on the retina and, second by the phenomenon called *stereopsis*. To determine the distance of an object from the eyes by the image size, the person must have had previous experience with the object and know its actual

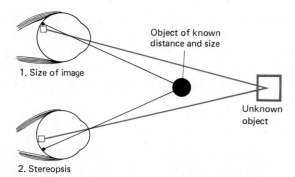

FIGURE 21–20   Mechanisms of depth perception.

size. For example, previous experience with other persons makes one remember that their average height is somewhere between 5 and 6 feet. Therefore, even when using one eye, one can determine relatively accurately the distance of a person from the size of his image on the retina. If he is nearby, his image is very large, but, if he is far away, his image is very small.

The second means for determining the distance of an object, *stereopsis*, depends on slight differences between the shapes and positions of the images on the retinas of the two eyes. This effect, shown in Figure 21–20, occurs because the two eyes are set several inches apart. In this figure the two eyes are observing a ball near the eyes and a block at a distance. In the left eye, the image of the block lies to the right of the image of the ball, but in the right eye, the image of the block lies to the left of the image of the ball. In other words, the images of these two objects are actually reversed on the retinas because of the angles from which the two eyes observe them, and the brain interprets their relative distances by the degree to which they are reversed. Obviously, this is an extreme example, but even when looking at a single object the images in the two eyes are slightly different. It must be noted particularly that this mechanism interprets only the *relative distances* of objects and not the actual distances. However, if the distance of one of the objects is known, that of the second object is also known. For practical considerations, a person always has his hands or other parts of his body to use as objects of known distance, which can be used as reference points for determining the distances of unknown objects.

NEUROGENIC CONTROL OF EYE MOVEMENTS

If the eyes are to function satisfactorily as a camera, their line of sight must be appropriately directed so that the most important portion of the

image will fall exactly on the fovea, the area of the retina designed for the most accurate vision. In addition to this, the lens system of the eye must be focused for the distance of the object, and the pupil of the eye must be enlarged or contracted in proportion to the amount of light available, thus aiding in the light and dark adaptation of the eyes. It can be seen, then, that several different sets of eye muscles must be controlled very exactly to attain visual effectiveness.

Positioning of the Eyes

Each eye is positioned by three different pairs of muscles, which are illustrated in Figure 21–21 and were shown in greater anatomical detail in Figure 21–3. One pair of these muscles is attached superiorly and inferiorly to the eyeball to move it up or down; another pair is attached horizontally to move the eye from side to side; and another pair is attached around the eyeball respectively on the bottom and top so that it can be rotated in either direction.

To control the eye movements, the visual association areas of the optic cortex must determine first whether or not the eyes are pointing toward the object; if not, impulses are transmitted through *oculomotor centers* in the brain stem to move the eyes in the appropriate direction. The visual association areas also determine whether or not the two eyes are receiving the same image on the corresponding portions of the two retinae—that is, whether or not the images are *fused* with each other. If not, one or

FIGURE 21–21   The extraocular muscles.

both of the eyes are adjusted up or down, to one side or the other, or rotated so that the images will fuse. These minute adjustments of the eye positions are so important to vision that almost as much of the visual association areas in the brain is concerned with eye movements as with the interpretation of the meaning of visual signals.

When the eyes do not fuse, crosseyedness occurs as shown at the bottom of Figure 21–21.

Focusing of the Eyes

The eyes are focused for clear vision by changing the curvature of the eye lens as the object's distance changes. This is accomplished by the following neuromuscular mechanism:

Figure 21–22 shows the lens of the eye suspended from the sides of the eyeball by the *suspensory ligaments* of which there are about 70. The lens of a young person is an elastic structure having a clear envelope and clear viscous cells in its center. When the ligaments are not pulling on the lens it assumes a spherical shape, but when the ligaments are tightened it flattens out. Normally, the ligaments are tight and the lens is flattened.

A smooth muscle called the *ciliary muscle* attaches to the suspensory ligaments of the lens where they connect with the eyeball. This muscle is composed of *meridional fibers* and *circular fibers.* The meridional fibers extend from the ends of the suspensory ligaments anteriorly to the sclerocorneal junction; when they contract they pull the ends of the suspensory ligaments forward and loosen them. This allows the flat elastic lens to become thickened and the curvature of the lens to become greater, increas-

ing the focusing power. The circular fibers extend all the way around the eye. When they contract they act as a sphincter, tightening progressively into a smaller and smaller circle, thus in still another way loosening the suspensory ligaments so that the lens becomes still more convex and develops more focusing power.

Focusing, like positioning of the eyes, is accomplished by signals initiated in the visual association areas. When the image on the retina is out of focus, the fuzziness of the image initiates appropriate reactions to cause a change in the tension in the ciliary muscle. As the focus becomes better and better, it finally reaches a point at which the image is seen with greatest acuity. At this point, the focusing mechanism "locks in" and holds until the distance of the observed object changes again. Then the focusing mechanism proceeds to the new focus in about one-half second.

Control of the Pupil

The pupil of the eye is the round opening of the iris through which light passes to the interior. The iris can constrict until the pupillary diameter is no greater than 1.5 mm, or it can open widely until the diameter becomes 8 to 9 mm. Constriction of the pupil is caused by contraction of the *pupillary sphincter*, a circular muscle in the *iris* surrounding the pupillary opening that is controlled by the parasympathetic nerves to the eye. Dilation of the pupil is caused by relaxation of this sphincter and by contraction of *radial muscle fibers* that extend from the edge of the pupil to the outer border of the iris; this muscle is controlled by the sympathetic nerves to the eye.

Because the amount of light that enters the eye is proportional to the area of the pupil and not to its diameter, the amount of light entering the eye changes with the square of the diameter, and can be varied between its two extremes approximately 30-fold. This provides a mechanism for light and dark adaptation in addition to the retinal mechanism for light and dark adaptation. Changes in pupillary size can occur in less than 1 second, in contrast with retinal adaptation, which requires several minutes.

The size of the pupil is regulated by a *pupillary light reflex*. When the retina is stimulated by light, some of the optic nerve impulses, instead of passing all the way to the visual cortex, leave the optic tract as shown in Figure 21–17 and pass to the pretectal nuclei of the mesencephalon. From here, signals are transmitted to the

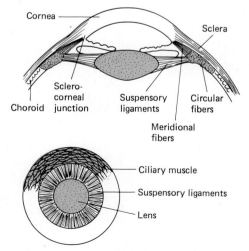

FIGURE 21–22 The focusing (accommodation) mechanism of the eye.

oculomotor centers of the brain stem that control the muscles of the eye and finally back to the iris. When the light intensity in the eyes is great, the size of the pupillary opening diminishes. On the other hand, when the light intensity becomes slight, the opening of the pupil increases.

The pupillary light reflex is frequently absent in a person with syphilis of the central nervous system, for this disease has a special predilection for destroying the pretectal nuclei or their connections.

QUESTIONS

1. Describe the optical system of the eye and the formation of an image on the retina.
2. What causes hypermetropia, myopia, and astigmatism?
3. Describe the structure of the rod.
4. Describe the rhodopsin cycle and its relationship to vision.
5. Explain light and dark adaptation.
6. Give the mechanism for color vision.
7. Describe the neuronal connections of the retina with the brain.
8. In what forms are the visual signals transmitted from the retina back to the cerebral cortex?
9. Explain the importance of the fovea in visual acuity.
10. Explain depth perception.
11. Explain the mechanism for focusing the eyes and control of this by the brain.
12. Give the mechanism of the pupillary light reflex.
13. Describe the fluid system of the eye and the control of intraocular pressure.

REFERENCES

Brindley, G.S.: Physiology of the Retina and Visual Pathway, 2nd ed. Baltimore, Williams & Wilkins, 1970.

Daw, N.W.: Neurophysiology of color vision. *Physiol. Rev.*, *53*:571, 1973.

Fine, B.S., and Yanoff, M.: Ocular Histology: A Text and Atlas. Hagerstown, Md., Harper & Row, 1979.

Fraser, S.E., and Hunt, R.K.: Retinotectal specificity: Models and experiments in search of a mapping function. *Annu. Rev. Neurosci*, *3*:319, 1980.

Gogel, W.C.: The adjacency principle in visual perception. *Sci. Am.*, *238*(5):126, 1978.

Goldchrist, A.L.: The perception of surface blacks and whites. *Sci. Am.*, *240*(3):112, 1979.

Hubbell, W.L., and Bownds, M.D.: Visual transduction in vertebrate photoreceptors. *Annu. Rev. Neurosci.*, *2*:17, 1979.

Hubel, D.H., and Wiesel, T.N.: Brain mechanisms of vision. *Sci. Am.*, *241*(3):150, 1979.

Kaneko, A.: Physiology of the retina. *Annu. Rev. Neurosci.*, *2*:169, 1979.

Michaelson, I.C.: Textbook of the Fundus of the Eye. New York, Churchill Livingstone, 1980.

Miller, D.: Ophthalmology: The Essentials. Boston, Houghton Mifflin, 1979.

Polyak, S.: The Vertebrate Visual System. Chicago, University of Chicago Press, 1957.

Records, R.E.: Physiology of the Human Eye and Visual System. Hagerstown, Md., Harper & Row, 1979.

Rodieck, R.W.: Visual pathways. *Annu. Rev. Neurosci.*, *2*:193, 1979.

Safir, A. (ed.): Refraction and Clinical Optics. Hagerstown, Md., Harper & Row, 1980.

Toates, F.M.: Accommodation function of the human eye. *Physiol. Rev.*, *52*:828, 1972.

Walsh, T.J.: Neuro-Ophthalmology: Clinical Signs and Symptoms. Philadelphia, Lea & Febiger, 1978.

22

The Ear; and the Taste and Smell Systems

Overview

The weakest whisper that we can hear has only about one-millionth the sound energy in the normal spoken voice, illustrating extreme sensitivity of the ear for detecting sound. An important ingredient in providing for this sensitivity is the mechanical system of the external and middle ear, first, for collecting sound and, second, for transmitting this into the inner ear. *Sound* is a series of repetitive *compression waves* that travel in air at a speed of about one mile every five seconds. When the waves impinge on the **tympanic membrane,** the membrane vibrates with the waves. The very center of the tympanic membrane is connected to a series of three bony levers, called the **auditory ossicles,** that conduct the vibrations to the inner ear. The three auditory ossicles are the **malleus,** the **incus,** and the **stapes.**

In the inner ear the sensory organ that converts the sound into nerve signals is the **cochlea.** The basic structure of this organ is a system of *coiled tubes* in which the sound resonates, causing nodes of high intensity vibrations, one node for each tone in the sound. In the coiled tubes is an elastic membrane called the **basilar membrane.** Along the surface of this are located a total of 20 to 30 thousand **hair cells** that convert the vibratory energy of the resonance nodes into nerve impulses. The frequency of each tone is determined by the *place* on the basilar membrane that resonance occurs. The nerve fibers leading from the hair cells into the brain are spatially oriented so that for each place stimulated on the basilar membrane, corresponding places are excited in the different auditory areas of the brain.

The auditory signals enter the brain through the cochlear portion of the eighth cranial nerve, the **vestibulocochlear nerve,** which terminates in the **cochlear nuclei** of the brain stem. From here, the signals are transmitted (1) through the **superior olivary nucleus** and **inferior colliculus** of the brain stem, (2) through the **medial geniculate body** of the thalamus, and finally (3) into the **auditory cortex** located in the midportion of the superior gyrus of the temporal lobe. The *brain stem auditory centers* play especially important roles in helping to determine the *direction from* which the sound comes and also in causing the head and eyes to move toward the direction of the sound. In the *auditory cortex* the tonal *characteristics* and *meanings of the sounds* are analyzed.

The *taste sensations* are detected by **taste buds** located mainly on the tongue but also to a lesser extent on the walls of the posterior mouth and anterior pharynx. Psychologically, we can detect four types of taste

381

called the *primary taste sensations*. These are (1) *salty*, (2) *sweet*, (3) *bitter*, and (4) *sour*. Different taste buds have preferential sensitivities for one or more of these primary taste sensations. The taste signals from the taste buds are transmitted first into the *brain stem* and from there to the **taste area of the cerebral cortex** in the lateralmost part of the parietal lobe. The taste sensations play a special role in helping us to choose the food that we eat. Especially important is the bitter taste that usually causes rejection of food, fortunately so because many bitter foods are poisonous.

The *sense of smell*, though extremely highly developed in some lower animals, is almost rudimentary in man. Unfortunately, we understand relatively little of its function.

The sensory organ for detecting smell is the **olfactory epithelium**, located in the upper reaches of the nasal cavity. In this epithelium are numerous nerve receptors called **olfactory cells.** Specific odors excite different ones of these cells, but, thus far, it has not been possible to determine how many types of olfactory cells there are for different odor specificities. Physiological research studies have suggested as few as 7 different types up to as many as 50 different types.

The nerve signals of smell are transmitted into the base of the brain through the **olfactory tracts.** These terminate in two major areas of the brain: (1) the **medial olfactory area,** located mainly in front of the hypothalamus; and (2) the **lateral olfactory area,** an important part of which is the **amygdala** in the temporal lobe. The medial olfactory area excites *primitive responses to smell* such as salivation, licking the lips, and so forth. The lateral olfactory area excites mainly *learned olfactory responses* such as appetite for special foods.

GENERAL ANATOMY OF THE EAR

The function of the ear is to convert sound into nerve impulses. Figure 22–1 illustrates the general organization of the ear, which is divided into three parts:

1. the **external ear,**
2. the **middle ear,** and
3. the **inner ear.**

The *external ear* also consists of three parts:

1. The **auricle** is the funnel-like external structure that collects the sound waves.
2. The **external auditory meatus** (*ear canal*) conducts the sound waves into the inner recesses of the ear. In the skin lining the meatus are both *sebaceous glands* and *ceruminous glands* that are a modified type of sweat gland. These together secrete *cerumen*, which protects the canal from becoming too dry; cerumen is commonly called "ear wax."
3. The **tympanic membrane** (commonly called the *eardrum*) is the thin membrane at the depth of the external auditory meatus that vibrates with the sound waves and then di-

rects the sound through small bony levers, the ossicles, into the middle and inner ear as shown in Figure 22–2.

The Middle Ear and the Ossicular System. Behind the tympanic membrane is a cavity containing air called the **tympanic cavity.** This cavity and the structures contained in it are the **middle ear.**

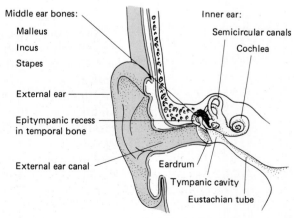

Middle ear bones:
Malleus
Incus
Stapes

External ear

Epitympanic recess in temporal bone

External ear canal

Inner ear:
Semicircular canals
Cochlea

Eardrum
Tympanic cavity
Eustachian tube

FIGURE 22–1 An overview of the ear and vestibular apparatus.

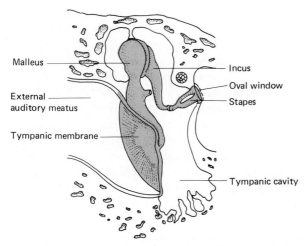

FIGURE 22–2   The tympanic membrane and the ossicular system of the middle ear.

The most important structures in the middle ear are the small bones called the **auditory ossicles,** illustrated in Figures 22–1 and 22–2. The ossicles are three separate bones:

1. The **malleus** is a lever suspended at its middle by ligaments from the walls of the middle ear. One end of this lever attaches to the middle of the tympanic membrane. Sound vibrations on the tympanic membrane cause a rocking motion of the malleus.
2. The **incus** is connected to the opposite end of malleus by strong fibrous attachments. The rocking motion of the malleus causes the incus also to rock.
3. The **stapes** is attached by fibrous attachments to the opposite end of the incus; this too rocks with the sound vibrations. The *footplate of the stapes* protrudes into a small bony opening called the **oval window** that leads into the inner ear. Therefore, the rocking movements of the stapes in the oval window transmit the sound into the inner ear.

Two muscles attach to the ossicles. One of these, illustrated in Figure 22–1, is called the **tensor tympani.** It pulls the malleus in the direction away from the tympanic membrane and thus tenses this membrane. This changes the degree of sensitivity of the ear for sound. The other muscle, the **stapedius,** an extremely small muscle, attaches to the stapes and pulls it in the direction away from the oval window. This partially immobilizes the stapes and in this way also changes the degree of sensitivity of the ear for sound.

Leading from the tympanic cavity into the posterior nasal cavity is a tube called the **audi-**

tory tube (or *eustachian tube*). This tube is normally closed, but when the person swallows, some of the swallowing muscles stretch the tube in such a way that it opens for about 1 second. This allows air to flow freely between the throat and the tympanic cavity, thus bringing the air pressure inside the tympanic cavity into equilibrium with the surrounding atmospheric pressure. It is essential to reequilibrate these pressures when a person rises rapidly in an airplane or elevator, or otherwise a pressure differential will cause the tympanic membrane to bulge either excessively inward or outward and sometimes actually to rupture. When a person has a severe cold, the membranes of the auditory tube are frequently swollen closed, and severe pain will then develop when flying.

The Inner Ear—The Cochlea.   The sensory receptor structure for hearing, the **cochlea,** is located in the inner ear. However, in addition to the cochlea, the inner ear is sometimes considered to include the vestibular apparatus as well. The cochlea and the vestibular apparatus are located in a series of contorted bony tubes and chambers called the **bony labyrinth** that lies within the substance of the *petrous portion of the temporal bone* as illustrated in Figures 22–1 and 22–3. Inside the bony labyrinth is a complex of membranous tubes and sensory receptors, called the **membranous labyrinth,** that is separated from the walls of the bony labyrinth by a thin layer of fluid called *perilymph*. The tubes of the membranous labyrinth are filled with a different fluid, the *endolymph*. It is in the membranous labyrinth that the auditory and equilibrium signals arise and are then transmitted to the brain. These signals are transmit-

FIGURE   22–3   The bony and membranous labyrinths of the inner ear and their cochlear and vestibular apparatus portions.

ted through the eighth cranial nerve, the **vestibulocochlear nerve,** also illustrated in Figure 22–3.

The anterior portion of the inner ear is the **cochlea,** which converts sound vibrations into auditory nerve signals. The posterior part is the **vestibular apparatus** that detects the state of equilibrium of the body and sends appropriate signals to the brain, as was explained in Chapter 17.

The *cochlea* consists of three tubes wound together into a tight spiral. As will be explained later in this chapter, sounds of different frequencies resonate at different points in this spiral and thereby excite different nerve fibers. It is in this way that the different sound frequencies are detected.

Sound is conducted into the cochlea through the *oval window* by vibration of the footplate of the stapes against the fluid in the membranous labyrinth of the cochlea. Another small bony window, the *round window,* also opens between the middle ear and the cochlea anterior to the oval window; it, like the oval window, is covered by a membrane to keep the fluid in the cochlea. This window allows the sound waves entering through the oval window to pass all the way through the cochlea and then back out the round window, which is essential to the resonating process.

The Vestibular Apparatus. Figure 22–3 shows that the tubular structure of the vestibular apparatus is continuous with that of the cochlea and that the **vestibular nerve** from the vestibular apparatus joins the **cochlear nerve** from the cochlea to form the **vestibulocochlear nerve,** the eighth cranial nerve, which enters the lateral border of the medulla, as described in Chapter 14. The function of the vestibular apparatus in helping to maintain body equilibrium is discussed in Chapter 17.

TRANSMISSION OF SOUND TO THE INNER EAR

Characteristics of Sound and Sound Waves. Sound is a series of compression waves traveling through the air. The transmitter of the sound, whether it be another person's voice, a radio speaker, or some noisemaking device, creates the sound by alternately compressing the air and then relaxing the compression. For instance, a vibrating violin string creates sound by moving back and forth. When the string moves forward it compresses the air, and when

it moves backward it decreases the amount of compression below normal. This alternate compression and decompression of the air produces sound.

Sound waves travel through the air in much the same way that waves travel over the surface of water. Thus, compression of the air adjacent to a violin string builds up extra pressure in this region, and this in turn causes the air a little farther away to become compressed. The pressure in this second region then compresses the air still farther away, and this process continues on and on until the wave finally reaches the ear.

Function of the Tympanic Membrane (Tympanum) and Ossicles

When sound waves strike the tympanic membrane, the alternate compression and decompression of the air adjacent to the membrane cause the membrane to move backward and forward. The center of this membrane is connected to the handle of the *malleus;* as described earlier, this in turn is connected to the *incus,* and the incus to the *stapes.* Movement of the handle of the malleus, therefore, causes the stapes also to move back and forth against the *oval window* of the cochlea, thus transmitting the sound into the cochlear fluid.

Transformation of the Pressure of the Sound Waves by the Ossicular Lever System. If sound waves were applied directly to the oval window, they would not have enough pressure to move the fluid in the cochlea backward and forward to produce adequate hearing, because fluid has many times as much inertia as air, and a correspondingly greater amount of pressure is required to cause movement of fluid. The tympanic membrane and the ossicular system transform the pressure of the sound waves into a usable form by the following means:

The sound waves are collected by the large tympanic membrane, the area of which is approximately 55 square millimeters (mm^2), or 17 times that of the oval window, which has an area of only 3.2 mm^2. Therefore, 17 times as much sound energy is collected as could be collected by the oval window alone, and all of this is transmitted through the ossicles to the oval window. Then, in addition, the ossicular lever system multiplies the pressure at the stapes another 1.3-fold. Thus, the pressure of movement of the foot of the stapes is increased to about 22 times that which could be effected by applying sound waves directly to the oval window. This pressure is now sufficient to move the fluid of the cochlea backward and forward.

Transmission of Sound into the Cochlea by the Bones of the Head

The cochlea lies in a bony chamber inside the temporal bone of the skull. Therefore, any vibration of the skull also can cause vibration of the fluid in the cochlea. Ordinarily, sound waves in the air cause almost no vibration in the skull bones, but clicking of the teeth or holding a vibrating device such as a tuning fork or a special sound vibrator against the skull can cause bone vibrations. In this way, instead of the fluid vibrating inside the cochlea, the cochlea vibrates around the fluid, allowing the cochlea to react to the sound as if the sound had entered via the tympanic membrane and ossicles.

FUNCTION OF THE COCHLEA

Physiologic Anatomy of the Cochlea. The cochlea is a membranous device formed of coiled tubes. The outside appearance of the cochlea is illustrated in Figure 22–3, and a cross-section of its structure is shown in Figure 22–4. Finally, a functional diagram of the cochlea is illustrated in Figure 22–5, showing the coiled tubes of the cochlea stretched out linearly. The total length of this uncoiled cochlea is about 3.5 cm.

The cross-sectional diagram of Figure 22–4 shows that the cochlea is actually composed of three separate tubes lying side by side, called the **scala vestibuli,** the **scala tympani,** and the **scala media.** All of these tubes are filled with fluid, and they are separated from each other by membranes. The membrane between the scala vestibuli and scala media is so thin that it never obstructs the passage of sound waves. Its function is simply to separate the fluid of the scala media from that of the scala vestibuli. The two fluids have separate origins, and the chemical differences between them are important for proper operation of the sound receptor cells.

The membrane separating the scala media from the scala tympani, called the **basilar membrane,** is very strong, and it does impede the sound waves. It is supported by about 20 thousand reedlike, thin spines that project into the membrane from one side. These spines, called **basilar fibers,** protrude most of the dis-

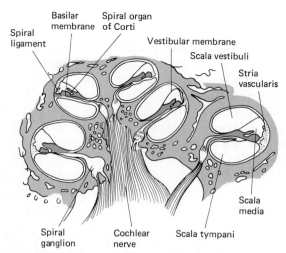

FIGURE 22–4 Cross-section of the cochlea, showing its coiled tubes composed of the scala vestibuli, the scala tympani, and the scala media. The cochlear nerve leading from the basilar membrane is also illustrated. (From Goss: Gray's Anatomy of the Human Body. 29th Am. Ed. Philadelphia, Lea & Febiger, 1973.)

tance across the membrane. Finally, located on the surface of the membrane is the **organ of Corti,** the sound receptor portion of the cochlea. This organ contains multiple sound receptor cells, called the **hair cells,** that convert sound vibrations into nerve signals, as we will discuss later.

In the uncoiled cochlea depicted in Figure 22–5, the membrane between the scala vestibuli and the scala media has been eliminated because it does not affect sound transmission in the cochlea in any way. So far as the transmission of sound is concerned, the cochlea is composed of two separate tubes rather than three, and the two tubes are separated by the basilar membrane. The reedlike basilar fibers near the oval window at the base of the cochlea are short, but they become progressively longer, like the reeds of a harmonica, farther and farther up the cochlea, until at the tip of the cochlea the fibers are about $2\frac{1}{2}$ times as long as those near the base.

FIGURE 22–5 Movement of fluid in the cochlea when the stapes is thrust forward.

Resonance of Sound in the Cochlea

Conduction of Sound in the Cochlea. As each sound vibration pushes the stapes forward against the oval window, this also pushes the fluid of the scala vestibuli deeper into the cochlea, as indicated by the arrows in Figure 22–5. The increased pressure in the scala vestibuli bulges the basilar membrane into the scala tympani; this pushes fluid in this chamber toward the round window and causes it to bulge outward into the middle ear. Then, as the sound vibration causes the stapes to move backward, the procedure is reversed, with the fluid moving in the opposite direction along the same pathway and the basilar membrane bulging now into the scala vestibuli.

Resonance in the Cochlea. A phenomenon called *resonance* occurs in the cochlea and causes each sound frequency to vibrate a different section of the basilar membrane. These vibrations are similar to those that occur in many musical instruments and can be explained as follows: When the string of a violin is pulled to one side it becomes stretched a little more than usual, and this stretch makes the string then move back in the other direction. However, as it moves, it builds up momentum and does not stop moving when it reaches its normal straight position. The momentum causes the string to become stretched once again but this time in the opposite direction; then the string moves back in the first direction. The cycle continues over and over again so that once the string starts vibrating it continues for a while.

Two factors determine the frequency at which a string will vibrate. First, the greater the *tension* of the string, the more rapidly it turns around and moves in the opposite direction, and the higher is the frequency of vibration. The second factor is the *mass* of the string. The greater its mass, the greater is the momentum developed during vibration, and the longer it will take for the string to change direction of movement; therefore, the lower will be the sound frequency. These same two factors, mass and elastic tension, apply to resonance in the cochlea. The vibrating mass in the cochlea that vibrates back and forth is the fluid between the oval and round windows; the elastic tension is mainly the tension developed by bending the basilar fibers.

When *high frequency* sound enters the oval window, it resonates near the base of the cochlea. There are two reasons for this: First, the mass of fluid between the oval and round windows and the first portion of the basilar membrane is very slight. Second, the basilar fibers,

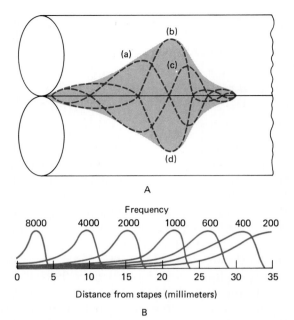

FIGURE 22–6 (A) Amplitude pattern of vibration of the basilar membrane for a medium frequency sound. (B) Amplitude patterns for sounds of frequencies between 200 and 8000 per second, showing resonance at different points for the different frequencies.

which provide the elastic tension for the vibrating system, are much shorter and also have far greater rigidity in the first portion of the membrane than toward its tip. The combined effect of these two factors, the low mass and the greater rigidity, causes the basilar membrane to resonate at very high frequencies near its base.

On the other hand, when a sound wave travels all the way to the tip of the basilar membrane, the mass of fluid that must move is very great. Also, the basilar fibers are longer and less rigid near the tip of the cochlea. The combined effect of these two factors, the great mass and the low rigidity of the elastic component, makes the basilar membrane near this end resonate at a very low frequency. Similarly, intermediate frequency sounds resonate at intermediate points along the membrane. Figure 22–6A illustrates the amplitude of vibration of different parts of the basilar membrane for a medium frequency sound wave, and Figure 22–6B represents the degree of vibration of the membrane for several different sound frequencies from very low to very high.

THE ORGAN OF CORTI AND CONVERSION OF SOUND VIBRATIONS INTO NERVE SIGNALS

The **organ of Corti,** illustrated in Figures 22–4 and 22–7, is the receptor organ that generates nerve impulses in response to vibration of the basilar membrane. Note that the organ of Corti lies on the surface of the basilar fibers and basilar membrane. The actual sensory receptors in the organ of Corti are two types of **hair cells**—a single row of **internal hair cells,** numbering about 3500, and three to four rows of **external hair cells,** numbering about 20,000. The bases and sides of the hair cells are enmeshed by a network of cochlear nerve endings. These lead to the **spiral ganglion of Corti,** which lies in the *modiolus* (the central core) of the cochlea. The spiral ganglion in turn sends axons into the **cochlear nerve** and thence into the central nervous system at the level of the upper medulla. The relationship of the organ of Corti to the spiral ganglion and to the cochlear nerve is illustrated in Figure 22–4.

Excitation of the Hair Cells. Note in Figure 22–7 that minute hairs, or cilia, project upward from the hair cells and either touch or are embedded in the surface gel coating of the **tectorial membrane,** which lies above the cilia in the scala media. These hair cells are similar to the hair cells found in the maculae and cristae ampullaris of the vestibular apparatus, which were discussed in Chapter 17. Bending of the hairs excites the hair cells, and this in turn excites the nerve fibers synapsing with their bases.

Upward movement of the basilar fiber rocks the hair cells upward and *inward*. Then, when the basilar membrane moves downward, the cells rock downward and *outward*. The inward and outward motion causes the hairs to shear back and forth against the tectorial membrane, thus exciting the cochlear nerve fibers whenever the basilar membrane vibrates.

DETERMINATION OF PITCH— THE PLACE PRINCIPLE

From the earlier discussions it is already apparent that low-pitch (or low-frequency) sounds cause maximal activation of the basilar membrane near the apex of the cochlea; sounds of high pitch (or high frequency) activate the basilar membrane near the base of the cochlea; and intermediate frequencies activate the membrane at intermediate distances between these two extremes. Furthermore, there is a spatial organization of the cochlear nerve fibers from the cochlea to the **cochlear nuclei** in the brain stem, the fibers from each respective area of the basilar membrane terminating in a corresponding area in the cochlear nuclei. This spatial organization also continues all the way up the brain stem to the cerebral cortex. The recording of signals from the auditory tracts in the brain stem and from the auditory receptive fields in the cer-

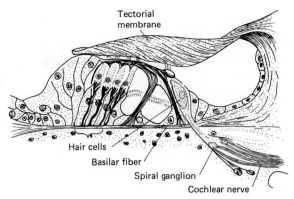

FIGURE 22–7 The organ of Corti, showing the hair cells and the tectorial membrane pressing against the projecting hairs. (Modified from Bloom and Fawcett: A Textbook of Histology, 10th Ed. Philadelphia, W. B. Saunders, 1975.)

ebral cortex shows that specific neurons are activated by specific pitches. Therefore, the method used by the nervous system to detect different pitches is to determine the position along the basilar membrane that is most stimulated. This is called the *place principle* for determination of pitch.

TRANSMISSION OF SOUND INTO THE CENTRAL NERVOUS SYSTEM

Figure 22–8 shows the nervous pathway for transmission of sound impulses from the cochlea into the central nervous system. After passing through the **cochlear nerve** (and thence through the combined vestibulocochlear nerve), the impulses are transmitted through at least

FIGURE 22–8 Pathways for transmission of sound impulses from the cochlea into the central nervous system.

five different levels in the brain: (1) the **dorsal** and **ventral cochlear nuclei,** (2) the **trapezoid body** and **superior olivary nucleus,** (3) the **inferior colliculus,** (4) the **medial geniculate body,** and (5) the **auditory cortex.**

Function of Lower Auditory Centers— Determination of Sound Direction. The function of the lower auditory centers is not well understood, but several of them are especially involved in the localization of the direction from which sound is coming. For instance, if sound comes from the left, it reaches the left ear before it reaches the right. If the sound is coming from directly in front, it reaches both ears simultaneously. That is, as the sound direction changes from directly ahead to one side, the time lag becomes progressively greater. This time lag is interpreted in the trapezoid body and superior olivary nucleus to determine the direction from which the sound is coming.

Another function of the auditory centers in the brain stem is reflex production of rapid motions of the head, of the eyes, or even of the entire body in response to auditory signals. That is, auditory signals pass directly into the medulla, pons, and cerebellum to alter a person's equilibrium, to make his head jerk to one side, or to make his eyes turn in one direction. Most of these rapid responses can occur even though the cerebral portions of the auditory system have been destroyed.

Figure 22–8 shows that the auditory signals from each ear are transmitted approximately equally into the auditory pathways on both sides of the brain stem and cerebral cortex. Therefore, destruction of one of the pathways will not greatly affect one's conscious ability to hear sound in either ear.

Function of the Auditory Cortex. The function of the auditory cortex has already been discussed briefly in Chapter 16. The *primary auditory cortex* is located in the middle of the superior gyrus of the temporal lobe, as illustrated in Figure 22–9. This area receives the sound signals and interprets them as different sounds. However, the signals must also be transmitted into surrounding **auditory association areas** before their meanings become clear. Finally, the signals are transmitted into the *common integrative center* in **Wernicke's area** of the cortex, where the overall meaning of all combined auditory, visual, and other types of sensation is determined. (This was discussed in detail in Chapter 16.)

Destruction of one auditory cortex reduces hardly at all one's ability to hear, but destruction

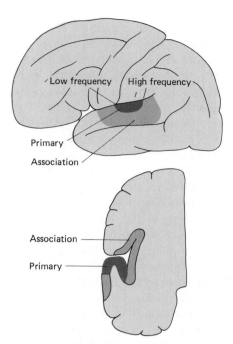

FIGURE 22–9 The auditory cortex.

of both cortices greatly depresses the hearing. It is especially interesting, though, that even with both auditory cortices destroyed, an animal is still capable of hearing sounds of high intensity. It seems that the medial geniculate bodies (which are in reality extended protrusions of the thalamus) are the regions that interpret these sounds, for destruction of these areas then causes total deafness. This illustrates again that many sensory functions of the brain can be performed by the thalamus and its related structures independently of the cerebral cortex.

DEAFNESS AND HEARING TESTS

Deafness is generally separated into two categories: *conduction deafness* and *nerve deafness.* The term conduction deafness means deafness caused by failure of sound waves to be conducted from the tympanic membrane through the ossicular system into the cochlea. Nerve deafness, on the other hand, means failure of auditory nerve signals to reach the auditory cortex because of damage to the cochlea itself or to any portion of the neurogenic transmission system for sound.

One of the most common causes of *conduction deafness* is repeated blockage of the *auditory tube.* This tube connects the middle ear with the nasopharynx, and it normally opens temporarily every time one swallows. Its function is to keep the pressure inside the middle ear equal to the pressure of the surrounding atmosphere, so that no pressure difference will exist between the two sides of the tympanic membrane. If this tube becomes plugged because of a cold, because of allergic swelling of the nasal membranes, or for some other reason, then the air in the middle ear becomes absorbed, and in its place a serous fluid collects. Also, the tympanic membrane is pulled inward because of lowered pressure in the middle ear. Fibroblasts then grow into the serous fluid and cause fibrous tissue between the ossicles and the walls of the middle ear. If this process continues long enough, the ossicles finally become so firmly bound to the walls of the middle ear that sound conduction by way of the ossicles into the cochlea becomes almost nil.

Nerve deafness is characteristic of old age; almost all older people normally develop at least some degree of this type of deafness, especially for very high frequency sounds. This is probably caused by the aging process in the cochlea itself. Exposure to excessively loud sounds—as occurs in rock music, boiler factories, wartime explosions, and so forth—can also cause nerve deafness. This results from actual disruption of the organ of Corti by the excessively strong vibrations of the basilar membrane.

Hearing Tests

Almost any type of sound instrument can be used to check a person's hearing; the most common of these for many years has been the *tuning fork* or the tick of a watch. After striking a tuning fork, the sound can usually be heard for about 30 seconds if the fork is held near the normal ear. If the person has *conduction deafness*, the ear is unable to hear the sound, but placing the butt of the fork against the skull will allow transmission of vibrations to the cochlea via the bones of the skull. If the cochlea and the neural transmission system are both still functioning properly, the sound is then heard even though it could not be heard by air conduction. On the other hand, if the person has *nerve deafness*, he will still be unable to hear the tuning fork even by bone conduction.

The Audiometer.   In audiology clinics, a special sound emitter, called an audiometer, is used to measure degree of deafness. This is an electronic apparatus capable of generating sound of all frequencies in an earphone or in a vibrator placed against the bone of the skull. The apparatus is calibrated so that the zero mark corresponds to the intensity of sound required

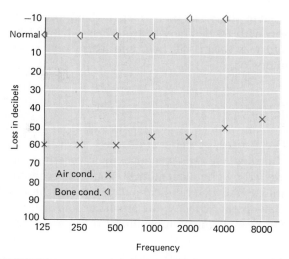

FIGURE 22–10   An audiogram from the ear of a person with conduction deafness.

for a normal person barely to hear it. If the person is deaf or partially deaf for sound of a particular frequency, the amount of extra sound energy that must be added is said to be the *hearing loss* for that particular frequency.

Figure 22–10 shows an *audiogram* from a person who has conduction deafness. Approximately 40 to 60 decibels of extra sound energy had to be transmitted into the earphone at each frequency for the person to hear the sound. However, when the skull vibrator, which causes bone conduction, was used instead of the earphone, no extra energy was required; indeed, at the high frequencies the hearing was even better than normal for bone conduction. This illustrates that bone conduction was at least normal, which means that the cochlea and auditory pathway were also normal. However, conduction of sound through the ossicular system was greatly impaired.

The Decibel System for Expressing Sound Energy.   Note in Figure 22–10 that the degree of hearing loss is expressed in *decibels*. The decibel system for expressing sound energy is a logarithmic rather than a linear scale. Thus, a change of 10 decibels is a 10-fold change in sound energy, 20 decibels is a 100-fold change, 30 decibels is 1000-fold, and 60 decibels is 1,000,000-fold. For example, in the audiogram of Figure 22–10, the hearing loss at most frequencies was about 60 decibels, which means that this deaf person had an actual hearing loss of about 1,000,000-fold.

Sounds that the normal ear hears frequently vary in intensity 100 million–fold or more. For instance, the intensity of sound in a very noisy factory is about 1,000,000 times as great as that of a quiet whisper. Therefore, a person with 60 decibels hearing loss—a 1,000,000-fold loss—can still hear sounds of very strong intensity.

TASTE

The senses of taste and smell are called the **chemical senses** because their receptors are excited by chemical stimuli. The taste receptors are excited by chemical substances in the food that we eat, whereas the smell receptors are excited by chemical substances in the air.

The Taste Bud

The sensory receptor for taste is the **taste bud,** shown in Figure 22–11. It is composed of epithelioid **taste receptor cells** arranged around a central pore in the mucous membrane of the mouth. Several very thin hairlike projections called *microvilli,* each several microns long, protrude from the surface of each taste cell and thence through the pore into the mouth. It is these microvilli that detect the different taste sensations.

Interweaving among the taste cells of each taste bud, and even invaginating into creases in the cell membranes, is a branching network of two or three taste nerve fibers that are stimulated by the taste cells.

Before a substance can be tasted, it must be dissolved in the fluid of the mouth, and it must diffuse into the taste pore around the microvilli. Therefore, highly soluble and highly diffusible substances such as salts or other small molecular compounds generally cause greater degrees of taste than do less soluble and less diffusible substances such as proteins or others of very large molecular size.

The Four Primary Sensations of Taste. Psychologically, we can detect four major types of taste, called the *primary taste sensations.* They are: (1) salty, (2) sweet, (3) bitter, and (4) sour.

Until recent years it was believed that four entirely different types of taste buds existed, each type detecting one particular primary taste sensation. It has now been learned that every taste bud has some degree of sensitivity for all the primary taste sensations. However, each bud usually has a greater degree of sensitivity to one or two of the taste sensations than to the others. The brain detects the type of taste by the ratio of stimulation of the different taste buds. That is, if a bud that detects mainly saltiness is stimulated to a higher intensity than buds that respond more to the other tastes, the brain interprets this as a sensation of saltiness even though the other buds are stimulated to a lesser extent at the same time.

Substances That Stimulate the Different Taste Buds. The *salty buds* in general determine the concentration of *salts* and other *ionic substances* present in food. The most familiar salt that stimulates these buds is sodium chloride, common table salt. Therefore, we normally associate the function of these buds almost entirely with the content of table salt in food.

The *sweet taste buds* detect the amount of *sugars* in food. This is one of the means by which animals determine whether or not fruits are ripe and whether or not unknown foods are nutritious. Essentially all wild foods that are sweet are safe to eat and contain a considerable amount of nutrition.

The *bitter taste buds* provide a protective function, for they detect principally the *poisons* in wild plants. The poisonous alkaloidal compounds of wild herbs, for instance, are among the poisons detected by the bitter taste buds. These same alkaloids, when given in small quantities, are frequently very valuable drugs. For instance, quinine, though of very bitter taste, can have a beneficial effect in the treatment of malaria when used properly, and yet can be lethal when used in too great a quantity.

The *sour taste buds* detect the degree of acidity of food—that is, they detect the *concentration of hydrogen ion* in the mouth. If the degree of acidity is slight, such as that of dilute vinegar, the food is usually very palatable; if the acidity is extremely strong, the taste is very unpleasant, and the food is rejected.

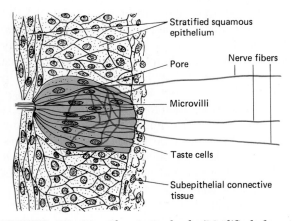

Stratified squamous epithelium

Pore

Nerve fibers

Microvilli

Taste cells

Subepithelial connective tissue

FIGURE 22–11   The taste bud. (Modified from Bloom and Fawcett: A Textbook of Histology, 10th Ed. Philadelphia, W. B. Saunders, 1975.)

Regulation of the Diet by the Taste Sensations

The taste sensations obviously help to regulate the diet. For example, the sweet taste is usually pleasant, causing an animal to choose foods that are sweet. On the other hand, the bitter sensation is almost always unpleasant, causing bitter foods, which are often poisonous, to be rejected. The sour taste is sometimes pleasant and sometimes unpleasant, and, likewise, the salty taste is sometimes unpleasant. The pleasantness of these types of taste is often determined by the momentary state of nutrition of the body. If a person has been long without salt, then, for reasons that are not yet understood, the salty sensation becomes extremely pleasant. If a person has been eating an excess of salt, the salty taste is very unpleasant. The same is true for the sour taste and to a lesser extent for the sweet taste. In this way the quality of the diet is automatically varied in accord with the needs of the body. That is, lack of a particular type of nutrient often intensifies one or more of the taste sensations and causes the person to choose foods having a taste characteristic of the deficient nutrient.

Importance of Smell to the Sensation of Taste. Much of what we call taste is actually smell, because foods entering the mouth give off odors that spread into the nose. Often, a person who has a cold states that he has lost his sense of taste, but on testing for the four primary sensations of taste they are all still present.

The smell sensations, which are discussed in later sections of the chapter, function along with taste sensations to help control the appetite and the intake of food.

Transmission of Taste Signals to the Central Nervous System

Figure 22–12 shows the pathways for transmission of taste signals into the brain stem and then into the cerebral cortex. The signals pass from the taste buds in the mouth to the **tractus solitarius** located in the medulla. Then they are transmitted to the **thalamus,** and from there to the **primary taste cortex** of the opercularinsular region, as well as into surrounding taste association areas and finally into the common integrative area (Wernicke's area), which integrates all sensations.

Taste Reflexes. One of the functions of the taste apparatus is to provide reflexes to the salivary glands of the mouth. To do this, impulses are transmitted from the tractus solitar-

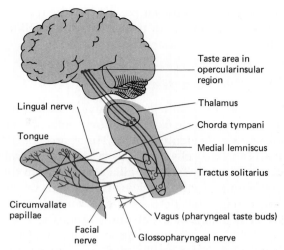

FIGURE 22–12 Transmission of taste impulses into the central nervous system.

ius in the brain stem into nearby *salivatory nuclei* that control secretion by the parotid, submandibular, and other salivary glands. When food is eaten, the quality of the taste sensations, operating through these reflexes, helps to determine whether the output of saliva will be great or slight.

SMELL

The sense of smell is vested in the **olfactory epithelium,** located on each side in the upper reaches of the nasal cavity. The location of these areas is illustrated in Figure 22–13, which shows

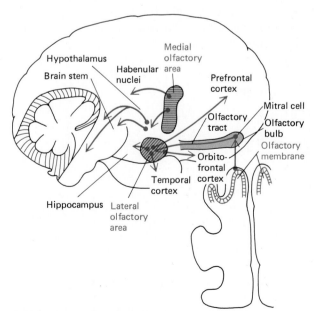

FIGURE 22–13 Organization of the olfactory system, and pathways for transmission of olfactory impulses into the central nervous system.

a cross-section of the air passages of the nose as well as connections of the olfactory epithelium with the nervous system.

The Olfactory Receptors

The olfactory epithelium contains numerous nerve receptors, called **olfactory cells,** which are shown in Figure 22–14. These are special types of nerve cells that project small microvilli called **olfactory hairs** or **cilia** outward from the epithelium into the overlying mucus. It is the olfactory hairs that detect the different odors.

The means by which odors excite the olfactory hairs are not well understood. However, those odors most easily smelled are, first, the very *highly volatile substances*, and, second, *substances that are highly soluble in fats.* The necessity for volatility is easily understood, for the only means by which an odor can reach the high spaces of the nose is by air transport to this region. The reason for the fat solubility factor seems to be that the olfactory hairs themselves are outcroppings of the cell membrane of the olfactory cell, and all cell membranes are composed mainly of fatty substances. Presumably, then, the odoriferous substance becomes dissolved in the membranes of the olfactory hairs, and this changes the membrane potential and produces nerve impulses in the olfactory cells.

The Primary Sensations of Smell. It has been very difficult to study individual olfactory cells, for which reason we are not yet sure what primary chemical stimuli excite the various types of olfactory cells. Yet on the basis of crude experiments, the following primary sensations have been postulated:

1. camphoraceous,
2. musky,
3. floral,
4. pepperminty,
5. ethereal,
6. pungent, and
7. putrid.

Thus, smell, like all the other sensations, is probably subserved by a few discrete types of cells that give rise to specific primary olfactory sensations. However, the preceding listing is mainly conjecture and probably is in error, for other studies have indicated as many as 50 discrete types of smell sensation.

Adaptation of Smell. Smell, like vision, can adapt tremendously. On first exposure to a very strong odor, the smell may be very strong, but after a minute or more the odor will hardly be noticeable. The olfactory receptors apprise the person of the presence of an odor, but do not keep belaboring him with its presence. This is especially valuable when one must work in pungent surroundings.

Masking of Odors. Unlike the eye's ability to see a number of different colors at the same time, the olfactory system detects the sensation of only a single odor at a time. However, the odor may be a combination of many different odors. If both a putrid odor and a floral odor are present, the one that dominates the other is the one that has the greater intensity, or, if both are of about equal intensity, the sensation of smell is between that of floralness and putridness. The ability of a high intensity odor to dominate is called *masking.* This effect is used in hospitals, toilets, and other areas to make pungent surroundings pleasant; incense may be burned or some odoriferous but pleasant-smelling substance may be evaporated into the air to mask the less desirable odors.

Transmission of Smell Signals into the Nervous System

Because smell is a subjective phenomenon that can be studied satisfactorily only in human beings, very little is known about transmission of smell signals into the brain. Smell pathways terminate in two major areas of the brain called the **medial olfactory area** and the **lateral olfactory area** respectively, both illustrated in Figure 22–13. The medial olfactory area lies in the very middle of the brain, anterior and slightly superior to the hypothalamus; the lateral olfactory area is located on the underside of the brain, spreading laterally into the base of the anterior temporal lobe.

Olfactory tract

Olfactory bulb

Mitral cell

Glomerulus

Bowman's gland

Sustentacular cells

Olfactory cell

Olfactory hairs

FIGURE 22–14 The olfactory membrane, showing especially the olfactory cells with their cilia protruding into the overlying mucus of the nose. (Modified from Bloom and Fawcett: A Textbook of Histology, 10th Ed. Philadelphia, W. B. Saunders, 1975.)

The medial olfactory area is responsible primarily for the primitive functions of the olfactory system, such as eliciting salivation in response to smell, licking the lips, and causing an animal to stalk a juicy meal.

On the other hand, the lateral olfactory area includes portions of the amygdala in the temporal lobe, and it is very closely associated with higher functions of the nervous system; direct pathways pass from this area into the temporal cortex, the hippocampus, and the prefrontal cortex, all important regions of cortical function. The lateral olfactory area is concerned with complicated responses to olfactory stimuli. Thus, recognition of a certain type of smell as belonging to a particular animal is a function believed to be performed by this area. And recognition of various delectable or detestable foods based on previous experience presumably is also a function of this area. In human beings, brain tumors located in this region frequently cause the person to perceive very abnormal smells which may be of any type, pleasant or unpleasant, and continuing sometimes for months on end.

QUESTIONS

1. Describe the ossicular system of the ear and its transmission of sound.
2. Describe the physiologic anatomy of the cochlea.
3. How does resonance of sound occur in the cochlea?
4. How do the hair cells convert sound energy into nerve impulses?
5. Trace the nerve signals from the ear to the cortex.
6. Explain how the taste bud converts taste sensations into nerve signals.
7. What are the four primary sensations of taste, and what chemical substances stimulate each?
8. Give the characteristics of function by the olfactory receptors.
9. Trace the transmission of smell signals into the nervous system.

REFERENCES

Aitkin, L.M.: Tonotopic organization at higher levels of the auditory pathway. *Int. Rev. Physiol.,* 10:249, 1976.

Alberts, J.R.: Producing and interpreting experimental olfactory deficits. *Physiol. Behav.,* 12:657, 1974.

Beagley, H.A. (ed.): Auditory Investigation: The Scientific and Technological Basis. New York, Oxford University Press, 1979.

Dastoli, F.R.: Taste receptor proteins. *Life Sci.,* 14:1417, 1974.

Douek, E.: The Sense of Smell and Its Abnormalities. New York, Churchill Livingstone, 1974.

Evans, E.F., and Wilson, J.P. (eds.): Psychophysics and Physiology of Hearing. New York, Academic Press, 1977.

Kare, M.R., and Maller, O.: The Chemical Sense and Nutrition. Baltimore, Johns Hopkins Press, 1967.

Ohloff, G., and Thomas, A.F. (eds.): Gustation and Olfaction. New York, Academic Press, 1971.

Sataloff, J., *et al.*: Hearing Loss, 2nd Ed. Philadelphia, J.B. Lippincott, 1980.

Scheich, O.C.H., and Schreiner, C. (eds.): Hearing Mechanisms and Speech. New York, Springer-Verlag, 1979.

Singh, R.P.: Anatomy of Hearing and Speech. New York, Oxford University Press, 1980.

Van Hattum, R.J.: Communication Disorders. New York, Macmillan, 1980.

THE CIRCULATORY SYSTEM

23

Introduction to the Circulatory System; the Heart; and the Coronary Vessels

Overview

The circulatory system serves as an internal transport mechanism for the body. As the blood passes through the capillaries of the different organs, much of the fluid in the blood, as well as dissolved substances, diffuses rapidly back and forth between the blood and tissue fluids. In this way nutrients are transported to the cells and excreta transported away from the cells. For instance, oxygen is carried to the cells from the lungs, and food from the intestinal tract. Also, carbon dioxide, formed in all the cells of the body, is transported to and excreted by the lungs, whereas most of the other waste products are carried to the kidneys and excreted into the urine.

The **heart,** which provides the propulsive force for moving the blood through the circulation, is actually two separate pumps: (1) the **right heart,** which pumps the blood through the lungs, and (2) the **left heart,** which pumps the blood through the remainder of the body. Each of these two hearts is divided into two separate chambers: (1) the **atrium** and (2) the **ventricle.** The two atria are primer pumps that force extra blood into the two ventricles ahead of ventricular contraction. Then the ventricles contract with great force a fraction of a second later and pump the blood through the lungs or systemic circulation. Therefore, the ventricles are called the **power pumps.**

The heart has four separate valves that allow forward flow of blood but prevent backflow. Two of these valves, the **atrioventricular valves,** function as inflow valves to the two respective ventricles. The other two, the **semilunar valves,** serve as outflow valves for the ventricles. When the ventricle contracts, back pressure on the atrioventricular valves causes these to snap closed so that the blood in the ventricles will not flow back into the atria. At the same time, compression of the blood by the ventricular walls forces the blood forward against the semilunar valves, opening these valves and allowing the blood to flow into the **pulmonary artery** or **aorta.** Then, when the ventricles relax, the high blood pressure in the great arteries forces blood backward against the semilunar valves and closes these, thus keeping the blood that has been pumped into the arteries from flowing back into the ventricles. At the same time, blood returning to the heart from the systemic veins pushes the atrioventricular valves open and fills the ventricles again in preparation for the next pumping cycle.

The heart, like all other muscles of the body, requires a nutrient blood supply; this is provided by the **coronary blood vessels.** Two **coronary arteries,** the **right coronary** and the **left coronary,** arise from the base of the aorta immediately beyond the aortic valve. The right coronary artery supplies mainly the right ventricle but also part of the left ventricle. The left coronary artery is restricted almost entirely to the left ventricle. About 85 percent of the coronary blood flow goes to the left ventricle because the left ventricular muscle is much more massive and requires more nutrition than the right ventricular muscle; it pumps the blood against a very high pressure to force it through the entire body in contrast to a low pressure pumped by the right ventricle to force blood through the lungs. When a person has a serious heart attack, it is almost always caused by severe blockage of one of the coronary artery branches to the left ventricle.

The rate of blood flow through the coronary blood vessels is controlled mainly by the *autoregulation mechanism.* That is, when the heart works extremely hard and utilizes extra amounts of oxygen and other nutrients, the coronary blood vessels automatically dilate and allow increased blood flow, thus supplying the nutrients needed for the extra heart work.

The function of the circulatory system, the gross features of which are illustrated in Figure 23–1, is to transport substances from one part of the body to another. But what are some of these substances? The most abundant is the oxygen that is transported from the lungs to all cells of the body. After the oxygen is used in the cells, forming carbon dioxide, the carbon dioxide is transported back from the cells to the lungs and then exhaled. Other nutrients are transported from the digestive tract and liver to the cells, especially (1) glucose, the principal digestive product of carbohydrates; (2) fatty acids, the principal digestive product of fats; and (3) amino acids, the principal digestive products of proteins. As these are used by the cells, such waste products as urea, uric acid, phosphate compounds, and sulphate compounds are formed, and these are then transported to the kidneys where they are excreted in the urine. Also, water and ions such as sodium, potassium, calcium, magnesium, chloride, bicarbonate, and others are all carried from the gastrointestinal tract to the tissues to maintain an appropriate fluid environment around the cells and to allow the cells to obtain whatever amounts of water and ions are needed for existence. The excesses of both water and the various ions are then removed from the body fluids by the kidneys, thus maintaining almost exactly constant concentrations of the different ionic substances in the fluids of the blood and tissues.

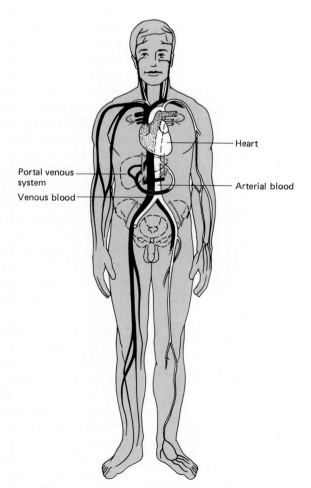

FIGURE 23–1 An anatomical overview of the circulatory system.

Heart

Portal venous system

Venous blood

Arterial blood

Lack of blood flow to any part of the body leads to dire effects. Let us think for a moment about failure of coronary blood flow to a portion of the heart muscle when someone is running. Immediately, the muscle cells begin to suffer for lack of oxygen, as well as other nutrients. However, for a few seconds, the muscle can continue to utilize stored nutrients. In so doing, various waste products are formed in the tissues that are not removed readily by the blood, especially large amounts of carbon dioxide, urea, and various acids; as these accumulate in the tissues they begin to damage the cells. Therefore, for lack of nutrients as well as excess of waste products, the muscle becomes progressively nonfunctional, so that after a few seconds of no coronary blood flow, the heart contraction becomes very weak or ceases altogether. Fortunately, the muscle cells themselves will not die for another 30 minutes to several hours. Therefore, if the circulation is reestablished in that period of time, hopefully no permanent damage will ensue. However, some tissues are much more sensitive to lack of blood flow than others. For instance, lack of blood flow to the brain causes unconsciousness in 4 to 5 seconds, and permanent destruction of brain tissue in as little as 5 to 10 minutes.

THE PRINCIPAL CARDIOVASCULAR STRUCTURES

The principal anatomical structures of the cardiovascular system are illustrated in Figure 23–1, and a functional diagram is pictured in Figure 23–2. This system has four major components:

1. the **heart,** which is the pump;
2. the **arteries,** which transport the blood to the tissues;
3. the **veins,** which transport the blood from the tissues back to the heart; and
4. the **capillaries,** which are minute porous vessels connecting the smallest arteries with the smallest veins in all tissues of the body where nutrient substances and waste products pass between the blood and tissue fluids.

We must also note that the heart is actually two separate pumps:

1. the **right heart,** which pumps blood through the lungs, and
2. the **left heart,** which pumps blood through the remainder of the body.

Likewise, the vascular system is divided into two separate divisions:

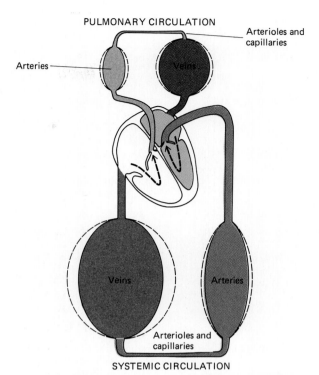

FIGURE 23–2   Schematic representation of the circulation, showing the two sides of the heart and the pulmonary and systemic circulations.

1. the **pulmonary circulation,** which includes the arteries, capillaries, and veins of the lungs; and
2. the **systemic circulation,** which includes the arteries, capillaries, and veins of the remainder of the body.

We will discuss the pulmonary circulation in Chapter 36 in connection with the lungs, and the systemic circulation in Chapter 25 after discussing the pumping action of the heart in this and the following chapter.

AN OVERVIEW OF THE HEART AS A PUMP

THE ATRIA AND THE VENTRICLES— THE PUMPING CHAMBERS OF THE HEART

Figure 23–3 shows the functional details of the heart as a pump. Blood entering the right atrium from the two large veins, the **superior** and **inferior venae cavae,** is forced by atrial contraction through the **tricuspid valve** into the **right ventricle.** The right ventricle then pumps the blood through the **pulmonary valve** into the **pulmonary artery** and thence through the lungs and **pulmonary veins** into the **left atrium.** Similarly, left atrial contraction forces blood through the **mitral (bicuspid) valve** into the **left ventricle,**

TABLE 23–1
The Parts of the Heart

Structure	Function
The chambers	
Right atrium	Primer pump for right ventricle
Left atrium	Primer pump for left ventricle
Right ventricle	Pumps blood through lungs
Left ventricle	Pumps blood through systemic circulation
The septa	
Interatrial septum	Muscular wall between two atria
Interventricular septum	Muscular wall between two ventricles
The valves	
Tricuspid (3 cusps)	Prevents backflow from right ventricle into right atrium
Mitral (2 cusps)	Prevents backflow from left ventricle into left atrium
Pulmonary (3 cusps)	Prevents backflow from pulmonary artery into right ventricle
Aortic (3 cusps)	Prevents backflow from aorta into left ventricle
Papillary muscles	Tense cusps of the tricuspid and mitral valves during heart contraction
Pericardium	
Visceral	Mesothelial covering on surface of heart
Parietal	Outside fibrous sac with inside mesothelium— heart moves freely within this sac

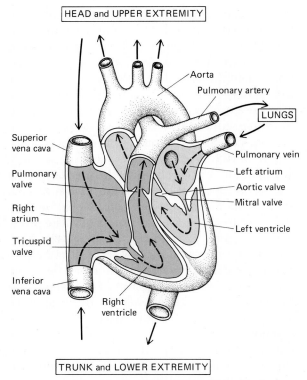

FIGURE 23–3 The functional parts of the heart and directions of blood flow through its chambers.

from whence it is pumped through the **aortic valve** into the **aorta** and through the systemic circulation.

The two atria are primer pumps that force extra blood into the respective ventricles immediately before ventricular contraction. This makes the ventricles about 30 percent more efficient as pumps than they would be if they had no special filling mechanism. However, the ventricles are so powerful that they can still pump large quantities of blood even when the atria fail to function. The ventricles are frequently called the *power pumps.*

Anatomical Differences Between the Atria and the Ventricles. An important difference between the atria and the ventricles is the thickness of the muscular wall, and therefore strength of contraction as well. The thickness of the ventricular wall is generally five or more times that of the atrial wall. Later in the chapter we shall see that the ventricles pump blood at pressures 5 to 12 times as great as do the atria, thus explaining the greatly thickened walls of the ventricles.

A second important difference between the atria and the ventricles is that the ventricles have valves to prevent blood from flowing backward into the atria, but the atria do not have valves to prevent blood from flowing backward from the atria into the veins. Therefore, when the atria contract they not only force blood forward into the ventricles, but a small amount of blood also flows backward. This is partially offset by the fact that the direction of thrust of the contracting atrial walls is toward the valve openings leading into the ventricles so that, even without atrial input valves, most of the blood does go to the ventricles rather than in the reverse direction.

Differences Between the Right Heart and the Left Heart. Though, basically, the right and the left hearts have similar functions and structures, the left ventricle pumps against about five times as much pressure in the systemic arteries as the right heart pumps against

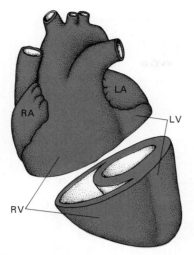

FIGURE 23-4 The anatomical relationship of the right ventricle to the left ventricle, showing the globular shape of the left ventricle and the half-moon shape of the right ventricle as it wraps around the left ventricle.

in the pulmonary arteries. As illustrated in Figure 23–4, this causes two principal differences between the right and the left hearts, especially of the ventricles:

1. The thickness of the right ventricular muscle is only two-fifths that of the left ventricular muscle.

2. Because the left ventricle contracts with extreme force in comparison with the right ventricle, the left ventricle assumes a globular shape, and the ventricular septum, the muscular wall between the two ventricles, protrudes into the right ventricle. And, since each side of the heart must pump the same quantity of blood, the external wall of the right ventricle bulges far outward and drapes around a large portion of the left ventricle, giving the interior of the right ventricle a large quarter-moon shape that allows it to accommodate about the same volume of blood as the left ventricle. The draping, thin-walled nature of the right ventricle makes it considerably easier for the right atrium to prime this ventricle with blood than for the left atrium to prime the left ventricle. Therefore, the left atrium also has a thicker wall than the right atrium.

One of the consequences of the tremendous pumping load under which the left ventricle must work is that extensive damage to the left ventricle, as often occurs after a heart attack, is very likely to cause death, whereas similar damage to the right ventricle is rarely life threatening.

SURFACE ANATOMY OF THE HEART AND THE GREAT VESSELS

Figure 23–5 illustrates the anterior surface of the heart and the heart's relationship to the great vessels. Observe especially that the heart is rotated around the axis running from base to apex about 30 degrees toward the left, bringing the right heart more anterior and the left heart more posterior. Therefore, the *right ventricle* dominates the greater part of the anterior surface of the heart. The superior end of the right ventricle has the appearance of an inverted funnel called the **conus arteriosus.** This leads immediately into the **pulmonary artery** that divides after another 4 to 5 cm into the **right** and **left pulmonary arteries,** which pass to the respective lungs. In its course to the right lung, the right pulmonary artery passes posterior to the ascending aorta and the superior vena cava and ventral to the right main bronchus.

The *right atrium* is seen to the right side of the right ventricle (left side of the figure). Leading into this, on the far right side of the heart, are the two great veins, the superior and inferior venae cavae.

Note also that a wrinkled projection of the right atrium extends over the upper surface of the right ventricle. This projection is called the **right auricle** because it has the appearance of an ear. During each heart beat the end of the auricle contracts toward the center of the atrium, giving it the characteristic of a wiggling ear. The **right coronary artery** and its accompanying vein lie in the groove between the right atrium and the right ventricle.

The *left ventricle* lies mainly along the left side of the heart but also on the posterior aspect of the heart. Therefore, in the anterior view of Figure 23–5, only a border of the left ventricle can be seen. Also, because the muscle mass of the left ventricle is much greater than that of the right ventricle, the left ventricle extends inferiorly toward the diaphragm to form the **apex** of the heart. When one feels his heart beat over the left chest, it is usually this apex that he feels, because the apex protrudes anteriorly to contact the inside of the thoracic wall.

To visualize the outflow tract of the left ventricle, observe once again Figure 23–3, which shows that the left ventricle opens into the *aorta* behind the juncture of the right ventricle and pulmonary artery. The aorta then projects upward for a few centimeters as the *ascending aorta* before curving to the left and posteriorly as the *aorta arch.* The large arteries that supply

Left subclavian artery

Brachiocephalic veins

Left common carotid artery

Brachiocephalic artery

ARCH OF AORTA

Superior vena cava

Ligamentum arteriosum

Right pulmonary artery

Left pulmonary artery

ASCENDING AORTA

Right pulmonary veins

Left pulmonary veins

Right coronary artery

Pulmonary artery

Auricle of right atrium

Auricle of left atrium

Right atrium

Conus arteriosus

Descending branch of left coronary artery

Right ventricle

Left ventricle

Apex

Inferior vena cava

DESCENDING AORTA

FIGURE 23–5 An anterior view of the heart and its relationships to the great vessels.

the head, neck, and upper extremities arise from the superior border of the aortic arch. The aorta then curves inferiorly between the right and left pulmonary arteries and along the posterior surface of the left ventricle, finally leaving the thorax to enter the abdomen.

The *left atrium* lies almost entirely on the superior aspect of the posterior surface of the heart, and only a small portion of it can be seen on the anterior surface. This small portion is the **left auricle** which, like the right auricle, is an earlike extension. Its tip, too, contracts toward the center of the atrium, thus giving it the same wiggling excursion as that of the right auricle. Two *left pulmonary veins* and two *right pulmonary veins* enter the left atrium from the respective lungs, making a total of four veins entering this chamber.

The Thoracic Surroundings of the Heart—The Mediastinum

The space within the thoracic cage that lies between the two lungs, above the diaphragm and

inferior to the front ribs, is called the **mediastinum.** This area, illustrated in Figure 23–6, is further divided into the *inferior mediastinum* where the heart is located and the *superior mediastinum*, much of which is occupied by great vessels leading from the heart. From anterior to posterior the mediastinum is also divided into the *anterior mediastinum* in which the anterior great vessels are found, the *middle mediastinum* in which most of the heart is located, and the *posterior mediastinum* that is posterior to the heart.

In the superior portion of the posterior mediastinum are the trachea and right and left main bronchi that lead to the lungs. And, posterior to the trachea, the esophagus courses all the way from the pharynx to the stomach, passing posterior to the heart and through the diaphragm. Though it is not shown in Figure 23–6, we shall see in later chapters that many lymph vessels and lymph nodes, as well as the vagus and phrenic nerves, are also in the mediastinum; and in the superior mediastinum, anterior to the great vessels, one finds the thymus gland, a

FIGURE 23—6 Relationship of the heart and great vessels to other structures of the thorax, showing also the pericardial sac around the heart.

modified lymphoid gland that is especially important in the development of immunity against disease, as we shall discuss in Chapter 34.

Note especially in Figure 23–6 the relationship of the heart to the lungs and diaphragm. The two lungs lie immediately adjacent to the two sides of the heart. Because the heart projects mainly to the left in the chest, the left ventricle forms a prominent impression on the inferior portion of the left lung called the *cardiac notch.* A portion of the left lung also folds posterior to the heart, and both lungs overlap much of the anterior surface of the heart as well, though the lungs are pulled to the side in the figure so that the entire heart can be seen.

The heart lies directly superior to the diaphragm. Also, the apex of the heart is displaced about 30 degrees from the base so that much of the right border of the heart rests on the diaphragm.

THE PERICARDIUM AND THE PERICARDIAL SAC

During each pumping cycle of the heart, the heart first enlarges and then contracts, which means that its surface area becomes greater, then smaller. This requires that the surfaces of the heart move in relation to their surroundings. If the heart were attached to the surroundings, this would be difficult. However, the heart lies mainly free inside a sac called the **pericardial sac,** also illustrated in Figure 23–6. The heart literally hangs inside this sac by its great vessels, and the great mass of the heart moves without hindrance. The wall of the pericardial sac is composed of a relatively strong *outer fibrous coat* and a thin but very slick *inner mesothelial membrane lining* called the **parietal pericardium.** At the bases of the great vessels the mesothelial layer is reflected back onto the surface of the heart and is called the **visceral pericardium.** The space between the visceral pericardium and the parietal pericardium is the **pericardial space.** This contains a small amount of viscid but very slick lubricating fluid with a consistency similar to that of dilute mucus. Thus, unimpeded motion of the heart can occur within this lubricated pericardial sac.

On occasion, the surfaces of the pericardium become infected or inflamed, a condition called *pericarditis.* This can be very painful. It can also lead to the formation of fibrous attachments between the parietal and visceral pericardial surfaces, which compromises the pumping capability of the heart. In fact, fibrous tissue around the heart sometimes becomes so thick and so contracted that it seriously limits the filling of the heart with blood, a condition called *constrictive pericarditis.* In its most serious form, this can decrease the pumping capability of the heart to the point of lethality.

THE HEART VALVES

Each ventricle has both an inflow valve and an outflow valve. The inflow valves are the **tricuspid** and the **mitral valves,** which are also called the **atrioventricular valves,** so-called because they function between the atria and the ventricles. The outflow valves, the **pulmonary** and **aortic valves,** are called the **semilunar valves** because the cusps (or leaflets) of these valves resemble half moons. Both types of valves are illustrated in different views of Figures 23–7, 23–8, and 23–9.

Each one of the valves has three cusps except the left atrioventricular (mitral) valve that has only two cusps.

The Atrioventricular Valves. The tricuspid and mitral valves must open easily to allow blood flow from the atria into the ventricles because the atria generate little pressure. Also, these valves must prevent backflow of blood from the ventricles even when the ventricles contract with tremendous force. To achieve these seemingly disparate goals, the atrioventricular valves are complex structures. Figures 23–7 and 23–9 illustrate the basic anatomy of the cusps of the tricuspid and mitral valves. Note that each of these is a very thin, filmlike structure that can flap easily in the flowing blood. Each is constructed of a very strong fibrous membrane covered on all surfaces by a slick lining of *endocardium* that prevents clotting of blood on its surface. The proximal ends of the cusps are attached to *fibrous rings* around the atrioventricular openings. The tip of each cusp is

Superior vena cava

Right pulmonary arteries

Pulmonary valve cusp

Pulmonary veins

Right atrium

Tricuspid valve cusp

Right ventricle

Inferior vena cava

Aorta

Left pulmonary arteries

Pulmonary artery

Pulmonary veins

Left atrium

Mitral valve cusp

Aortic semilunar valve cusp

Chordae tendineae

Papillary muscles

Left ventricle

Aorta

FIGURE 23–7   A transected view of the heart, illustrating especially the heart valves.

ANTERIOR

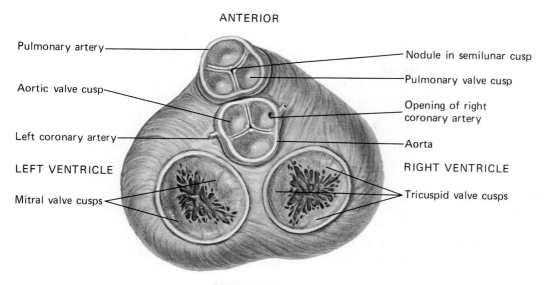

Pulmonary artery ——
Aortic valve cusp ——
Left coronary artery ——
LEFT VENTRICLE
Mitral valve cusps ——

—— Nodule in semilunar cusp
—— Pulmonary valve cusp
—— Opening of right coronary artery
—— Aorta
RIGHT VENTRICLE
—— Tricuspid valve cusps

POSTERIOR

FIGURE 23—8 The valves of the heart as seen from above after removal of the atria and transection of the pulmonary artery and aorta.

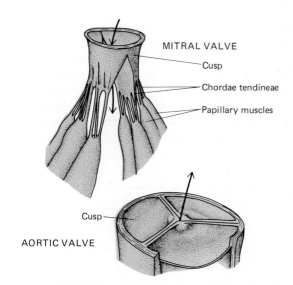

MITRAL VALVE
—Cusp
—Chordae tendineae
—Papillary muscles

Cusp—
AORTIC VALVE

FIGURE 23—9 Structure of the mitral and aortic valves.

attached by 8 to 12 strong fibrous strands called **chordae tendineae** to the ends of several small muscles that project from the inner surfaces of the two ventricles, the **papillary muscles.** When the ventricles contract and blood is forced backward from the ventricles toward the atria, the filmy valve cusps are immediately thrust backward over the valve openings. But, at the same time that the ventricles contract, so do the papillary muscles contract and tense the ends of the cusps, preventing their backward protrusion into the atria. Thus, the valve cusps form hammock-like slings that press against one another

and prevent backward flow. In summary, the filmy nature of the valves allows ease of blood flow from the atria into the ventricles, and the combination of the chordae tendineae and the papillary muscles prevents backflow of blood despite the very strong forces generated by ventricular contraction.

The Semilunar Valves. Now, observe the semilunar valves, the aortic and pulmonary valves, in Figures 23–7, 23–8, and 23–9. The cusps of these valves are much thicker than those of the atrioventricular valves, and they do not have any chordae tendineae to prevent backward protrusion. Instead, each cusp hangs from the inside wall of the aorta or pulmonary artery like a cup. When the ventricles contract, blood is forced through the valves, pushing the cusps upward and outward so that they fold flat against the inside walls of the arteries, thus opening the valves. Then, after ventricular contraction is over, a small amount of the blood that has been expelled under high pressure flows backward toward the ventricles. Immediately, the cusps are caught in the back-flowing blood, causing them to balloon away from the vessel wall; thus their edges all come together centrally in the lumen of the artery to close the valve. Until the next ventricular contraction, these cusps literally hang from the walls of the arteries and form a tight seal against backflow. In the middle of the free edge of each semilunar cusp is a small *nodule*; these nodules fill in the small gap where the three cusps come together in the center of the artery.

Now, let us consider the functional differences between the semilunar valves and the atrioventricular valves. First, the semilunar valves do not have to open with extreme ease because the ventricles generate high pressures and therefore can push open even stiff valves. But, second, the semilunar valves must withstand much more trauma than the atrioventricular valves because the blood flows through these valves with extreme force and high velocity, causing severe shearing forces along the cusp edges. Also, the cusps must withstand a sudden backward-snapping thrust when they close. Therefore, one can understand the functional structure of these valves: they are very strong to withstand the traumatic forces that they experience, but fortunately they do not significantly impede the outflow of blood from the ventricles because the large pressures created by ventricular contraction can overcome the resistance that does exist.

The Heart Sounds

When one listens with a stethoscope to the beating heart he normally hears two sounds that are aptly described "lub, dub; lub, dub; lub, dub." The "lub" is called the **first heart sound** and the "dub" the **second heart sound.** The first heart sound is caused by closure of the A-V valves when the ventricles contract, and the "dub" sound is caused by closure of the aortic and pulmonary valves at the end of contraction. This relationship to the cycle of heart beat is illustrated in Figure 23–10A, which shows a phonocardiogram for the normal heart and is a graphic representation of the heart sounds. In this figure the term *systole* means the period of the heart beat when the ventricles are contracting, and *diastole* means the period when they are relaxed. When the ventricles contract, the increasing pressures in the two ventricles force the cusps of the A-V valves closed. The sudden stoppage of backflow from the ventricles into the atria creates vibration of the blood and of the heart walls; these vibrations are transmitted to the chest wall to be heard as the first sound, the "lub" sound. Immediately after the ventricles have discharged their blood into the arterial systems, the subsequent ventricular relaxation allows blood to begin flowing backward from the arteries toward the ventricles, thereby causing the aortic and pulmonary valves to close suddenly. This also sets up vibrations, this time in the blood and walls of the aorta and pulmonary artery as well as the ventricles. These vibrations also are transmitted to the chest wall, causing the "dub" sound.

Valvular Heart Disease

The most frequent cause of valvular heart disease is *rheumatic fever*, a disease that results from an immune reaction to toxin secreted by streptococcal bacteria as follows: The acute phase of the rheumatic fever usually occurs 2 to 4 weeks after a person has had a streptococcal sore throat, scarlet fever, a streptococcal ear infection, or some other streptococcal infection. Antibodies formed by the body's immunity system in response to the streptococcus toxin attack not only the toxin but also the valves, causing small, cauliflower-like growths on their edges. These growths also erode the valves and cause ingrowth of fibrous tissue. The valve sometimes is eaten away completely, or at other times it becomes so constricted and hardened by fibrous tissue that it cannot close. Obviously, such a valve is then merely a constricted hole rather than a valve.

Valvular Stenosis. Sometimes in rheumatic fever the valve openings are so greatly narrowed by fibrous scar tissue that blood can flow through the opening only with much difficulty. This is called *stenosis.* If the aortic valve becomes stenosed, blood dams up in the left ventricle. If the mitral valve becomes stenosed,

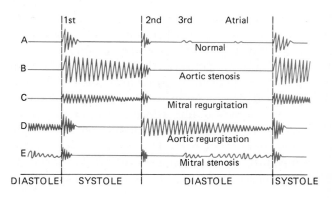

FIGURE 23–10 Phonocardiograms from the normal heart and from hearts with aortic or mitral valvular disease.

blood dams up in the left atrium and lungs. Likewise, stenosis of the pulmonary or tricuspid valves causes blood to dam up in the right ventricle or systemic circulation.

Valvular Regurgitation. Often the valves do not become stenotic but instead are eroded so much that they cannot close. As a result, blood leaks backward through the valve that should be stopping this backward flow; this effect is called *regurgitation.* For instance, in aortic regurgitation much of the blood pumped into the aorta by the left ventricle returns to the ventricle at the end of heart contraction rather than flowing on through the systemic circulation. Likewise, failure of the mitral valve to close when the left ventricle contracts allows blood to flow backward into the left atrium rather than being pumped forward into the aorta. Thus, leaking valves are as disastrous to cardiac function as are narrowed valvular openings. Often a valve is both leaky and narrowed, decreasing the effectiveness of cardiac pumping in two ways.

Pulmonary Edema in Heart Valvular Disease. The valve most frequently affected by rheumatic fever is the mitral valve, though the aortic valve is affected almost as often. On the other hand, the valves of the right heart are rarely damaged severely by rheumatic fever, though they are often abnormal because of malformation prior to birth (called a congenital malformation). Severe damage to the mitral valve leads to excessive damming of blood in the left atrium and lungs. Aortic valvular disease causes blood to dam up in the left ventricle and in severe cases in the left atrium and lungs as well. Therefore, after damage to either of these valves blood often engorges the lungs, causing fluid to leak from the pulmonary capillaries into the pulmonary tissues and alveoli, resulting in severe pulmonary edema and often drowning the patient in his own fluid. This will be explained in detail in Chapter 29 in the discussion of heart failure.

Heart "Murmurs" in Valvular Heart Disease. The usual way to diagnose valvular heart disease is to listen with the stethoscope for abnormal heart sounds. Sections B through E of Figure 23–10 illustrate phonocardiograms from patients with either stenosis or regurgitation of the aortic or mitral valve, showing abnormal sounds resulting from blood flowing with force through a stenotic valve or leaking backward through a regurgitating valve. These abnormal sounds are called *heart murmurs.* For instance in Section B of the figure, blood is forced through a stenotic aortic valve into the aorta during ven-

tricular contraction (that is, during systole). Therefore, the heart "murmur" of aortic stenosis is a "systolic" murmur.

Now we will leave it to the student to think about the other murmurs illustrated in Sections D, C, and E of Figure 23–10 and to determine why

1. the murmur of mitral regurgitation occurs during systole,
2. the murmur of aortic regurgitation occurs during diastole, and
3. the murmur of mitral stenosis occurs during diastole.

THE BLOOD SUPPLY OF THE HEART— THE CORONARY VESSELS

The heart, like all other muscles of the body, requires nutrition and removal of waste products. However, contrary to what might be thought, the heart muscle does not receive its blood supply directly from the heart chambers. Instead, the heart is supplied by its own special blood vessels, the **coronary arteries** and **cardiac veins** that lie on its surface, the branches of which penetrate inward to supply the muscle. Some of these coronary vessels were illustrated in the surface views of the heart in Figures 23–5 and 23–6. The functional arrangements of the coronary arteries are shown in an anterior view of the heart in Figure 23–11, and both the veins and arteries are seen in a posterior view in Figure 23–12.

THE CORONARY ARTERIES

The heart is supplied by two coronary arteries, the **right coronary** and the **left coronary.** Referring back to Figure 23–8, one can see that both arise from the aorta, from cuplike cavities called **sinuses** behind two of the cusps of the aortic valve, the right coronary from a sinus behind the anterior cusp and the left coronary from a sinus behind the left cusp. Now, let us observe the courses of the two coronary arteries over the surface of the heart.

The Right Coronary Artery. Figure 23–11 shows that the right coronary artery, after arising from the anterior sinus of the aorta, courses along the groove between the right atrium and right ventricle and circles around the right side of the heart to the posterior surface. After giving off several small branches to supply the anterior wall of the right atrium and upper

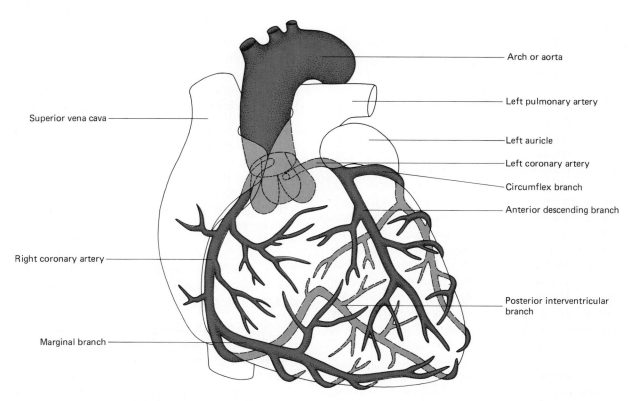

Arch or aorta

Left pulmonary artery

Left auricle

Left coronary artery

Circumflex branch

Anterior descending branch

Posterior interventricular branch

Superior vena cava

Right coronary artery

Marginal branch

FIGURE 23-11   A front view of the coronary arterial system.

Aorta

Superior vena cava

Right pulmonary veins

RIGHT ATRIUM

Inferior vena cava

Coronary sinus

Right coronary artery

Middle cardiac vein

RIGHT VENTRICLE

Pulmonary arteries

Left pulmonary veins

Circumflex artery

Great cardiac vein

LEFT ATRIUM

LEFT VENTRICLE

FIGURE 23-12   The posterior surface of the heart, illustrating especially the coronary sinus and the major veins draining into this.

TABLE 23–2
The Coronary Vessels

Vessel	Portion of heart supplied or drained
Coronary arteries and branches	
Left coronary	
1. Anterior descending branch	Anterior left and right ventricles
2. Circumflex branch	Wall of left ventricle
Right coronary	
1. Marginal branch	Wall of right ventricle
2. Posterior descending branch	Posterior right and left ventricle
Cardiac veins and tributaries	
Coronary sinus	
1. Great cardiac vein	Anterior and left wall of heart; mostly drains left ventricular wall
2. Middle cardiac vein	Posterior wall of heart; drains portions of both ventricles
3. Small cardiac vein	Right side of heart; drains wall of right ventricle
Anterior cardiac veins	Anterior wall of right ventricle
Thebesian veins	Walls of both atria

part of the right ventricle, it gives off two major branches to the right lateral and posterior portions of the heart:

1. A **marginal branch** courses along the lateral margin of the right ventricle toward the apex of the heart.
2. A **posterior descending branch** (or *posterior interventricular branch*) runs in the posterior groove between the right and left ventricles and is also directed toward the apex of the heart.

The Left Coronary Artery. The left coronary artery, after arising from the left sinus of the aorta, passes posterior to the pulmonary trunk and divides almost immediately into two major branches:

1. An **anterior descending branch** (also called the *anterior interventricular branch*) courses directly down the anterior surface of the heart toward the apex, lying in the groove between the two ventricles.
2. A **circumflex branch** circles around the left side of the heart between the left ventricle

and left atrium, supplying blood to the left lateral margin of the heart and to part of its posterior wall.

The circumflex branch anastomoses with the terminal portion of the right coronary artery, and the anterior descending branch anastomoses near the apex of the heart with the posterior descending branch through many very small arteries.

THE CARDIAC VEINS

The cardiac veins differ from the coronary arteries. The coronary arteries arise from the aorta rather than from the heart. The cardiac veins do not empty into one of the great veins but instead directly into the right atrium. The cardiac veins can be divided mainly into two separate groups, the **coronary sinus system** and the **anterior cardiac veins.**

The Coronary Sinus System. The coronary sinus system is seen best from the posterior side of the heart, illustrated in Figure 23–12. This system returns approximately 85 percent of the blood from the heart muscle to the right atrium.

The *coronary sinus* itself is a large bulbous vein, illustrated in Figure 23–12, that lies on the posterior surface of the heart in the groove between the atria and the ventricles. It empties directly into the right atrium about 2 cm medial and slightly superior to the entry point of the inferior vena cava into this atrium. The coronary sinus receives venous blood from the heart through many small tributaries as well as through three major cardiac veins:

1. The **great cardiac vein** begins in the anterior interventricular groove, paralleling the anterior descending branch of the left coronary artery, and then winds around the left side of the heart in the groove between the left atrium and left ventricle. This vein drains blood from numerous small veins arising from both ventricles and the left atrium.
2. The **middle cardiac vein** ascends the posterior surface of the heart in the groove between the left ventricle and right ventricle and parallels the posterior descending branch of the right coronary artery. This vein, along with other small tributaries to the coronary sinus, drains blood from the posterior portions of the right ventricle and right atrium as well as from part of the left ventricle.
3. The **small cardiac vein** lies along the course of the marginal branch of the right coronary artery and aids in draining the wall of the right ventricle.

The Anterior Cardiac Veins. About 15 percent of the cardiac venous blood drains directly into the right atrium through many small veins called *anterior cardiac veins*, most of which lie on the anterior surface of the heart. These drain mainly the anterior portions of the right ventricle and right atrium.

There are also many tiny veins in the walls of the atria called *thebesian veins* that open individually into the atria.

FLOW OF BLOOD THROUGH THE CORONARY VESSELS

The amount of blood flowing through the coronaries each minute is approximately 225 ml, which amounts to about 4 to 5 percent of all the blood pumped by the heart.

In contrast to the flow of blood in other portions of the circulation, flow in most of the coronaries is greater during diastole than during systole. The reason for this is that systolic contraction of the heart muscle compresses the coronary vessels where they penetrate through the muscular wall of the heart. This partially or totally occludes these vessels for a short time during systole, though this occlusion is not present during diastole. Therefore, the rate of coronary flow is determined more by the diastolic level of arterial pressure than by either the mean pressure or the systolic pressure. This has importance in some circulatory diseases in which diastolic pressure is especially low. For instance, in aortic regurgitation the aortic valve is destroyed, which allows blood to empty during diastole from the aorta backward into the left ventricle, making the aortic diastolic pressure very low and the blood flow to the heart greatly curtailed.

Regulation of Coronary Blood Flow. Blood flow through the coronaries is regulated mainly in proportion to the need of the heart for nutrition. For example, when the heart pumps very hard during exercise to supply the needed blood flow to the skeletal muscles, the coronary blood flow also increases as much as three- to fourfold. The most important method of regulation is the *autoregulation mechanism*, which will be discussed in Chapter 26. That is, when the metabolic need for nutrients—especially for oxygen—becomes greater than the supply, the arterioles automatically dilate. As a result, blood flow through the coronaries increases until the level of nutrition matches the degree of cardiac activity.

The autonomic nervous system also helps slightly to control blood flow through the coronaries. Increased sympathetic stimulation increases coronary flow and decreased sympathetic stimulation decreases flow. However, most of this effect results from the fact that sympathetic stimulation tremendously increases the degree of activity of the cardiac muscle, and this in turn increases coronary blood flow by the autoregulation mechanism just discussed.

At any rate, nervous control of coronary blood flow is relatively unimportant in comparison with the autoregulatory regulation of flow.

CORONARY OCCLUSION AND THE HEART ATTACK

About one third of all persons in the affluent societies of Western culture die from coronary occlusion. And in almost all instances, coronary occlusion results from atherosclerosis.

Atherosclerosis is a degenerative disease of the arteries caused by development of fatty, fibrotic deposits in the arterial wall. The coronary arteries are especially susceptible to this disease.

In its early stages, atherosclerosis is believed to begin with mild damage or deterioration of the very thin endothelial cells that line the inside surface of the artery. The endothelial cell damage in turn causes a twofold effect: (1) The smooth muscle cells that lie deep to the endothelium begin to multiply and to protrude against the endothelial cell lining. (2) Fatty substances, especially *cholesterol*, begin to deposit in and around the multiplying smooth muscle cells. The combination of overgrowth of cells and deposit of cholesterol causes the development of a fatty-fibrotic lesion called an *atherosclerotic plaque* that eventually protrudes into the vessel lumen and impedes blood flow. Over a period of years the plaque grows larger, and more and more fibrous tissue grows into the plaque to cause progressive narrowing of the artery. In very late atherosclerosis, many of the plaques become calcified. This development of calcified plaques and the growth of excessive fibrous tissue in the arterial walls gives the disease in its late stages the name *arteriosclerosis*, which is also known as "hardening of the arteries."

Coronary occlusion can result from atherosclerosis in two different ways:

1. The progressive narrowing of the coronary arteries can gradually cause poorer and poorer blood supply to the heart, with eventual occlusion of one or more of the larger coronary vessels. To some extent this happens in almost all older persons, leading to

progressive decrease in what is called the *coronary reserve*, that is, decrease in the coronary blood flow that can be supplied when heavy activity of the heart during exercise demands greatly increased flow.

2. The second way in which atherosclerosis leads to heart debility is to cause *acute coronary occlusion*, the condition called the "heart attack." The occlusion occurs almost invariably at a point in the coronary artery where an arteriosclerotic plaque protrudes into the flowing blood, forming a rough surface on which a blood clot develops. A clot can sometimes grow enough in only a few hours to cause complete occlusion of the vessel. Or, at other times, a clot forms on a plaque in one of the larger coronary vessels, then breaks off and flows in the blood to a smaller branch where it causes complete occlusion, this occurring in a matter of seconds.

When an acute coronary occlusion occurs, the muscle beyond the point of occlusion becomes *ischemic*, which means lack of sufficient blood flow to provide appropriate nutrition. Within 30 seconds to a minute, any totally ischemic muscle becomes nonfunctional. And, if a large enough portion of the heart becomes nonfunctional, the person can die within minutes. But, if the area of affected heart muscle is small, especially if it does not involve the left ventricular wall, the heart attack then causes only partial cardiac debility, and the person usually recovers in a few weeks to a few months because of enlargement of the remaining coronary vessels and even growth of new small vessels into the damaged muscle area. The recovery sometimes is complete but at other times only partial.

Conditions that Predispose to Heart Attacks. In recent years intensive research has been directed toward understanding heart attacks, and we now know a few definitive factors that predispose to atherosclerosis and coronary occlusion. The most important of these are discussed here:

1. *Obesity* and a diet that has *excessive amounts of fats and cholesterol* in it. Research has also shown that fats with a high content of saturated fats (fats that contain excessive amounts of hydrogen atoms) are much more likely to cause cholesterol deposits in the arterial walls than diets containing instead mainly unsaturated fats. This is the reason for the nutritional fad to promote diets containing unsaturated fats.

2. Persons who smoke *cigarettes* have more than two times as great a chance of developing lethal heart attacks as persons who do not smoke. The reason smoking causes this predisposition is not entirely understood. It may result from coronary vessel spasm that occurs during smoking, or it may result from damaging substances in the smoke such as carbon monoxide, or it may result from toxic substances in the smoke. At any rate, the statistical damnation of smoking as a cause of heart attacks is very clear.

3. *Lack of exercise* also predisposes to heart attacks. Most people believe that this predisposition results from excessive fat deposition in the arteries of the nonexercising person because he simply does not burn up enough of the fatty foods that he eats when he does not exercise. But it may also be that some of the hormones secreted during exercise in some way help to reduce fatty deposition in the arterial walls.

4. The *male sex hormone testosterone* has a potent effect to increase the amount of circulating fat in the blood and also to cause excessive deposition of atherosclerotic plaques. This accounts for the especially high incidence of heart attacks in young to middle-aged men, about four times as great as in women. However, in old age, the level of testosterone secretion becomes greatly reduced, and the incidence of heart attack in men then becomes no greater than it is in women.

5. *The tendency to heart attacks is highly hereditary.* That is, heart attacks run in families. This results mainly from different capabilities of different persons to transport fats in the blood. When the transport mechanisms do not function properly, excessive amounts of cholesterol become deposited in the vascular walls.

QUESTIONS

1. Describe the atria and the ventricles, including wall thickness and patterns of blood flow through these chambers.
2. Why is the left ventricle globular in shape whereas the right ventricle has a quarter moon shape?
3. Why is the right ventricle more prominently represented on the anterior surface of the heart than is the left ventricle?
4. What are the structures of the mediastinum?
5. Discuss the manner in which the heart is supported in the pericardial sac as well as the motion of the heart within this sac.
6. Contrast the structures and the functions of the atrioventricular and semilunar valves.
7. Give the relationship between the heart sounds and the contractile cycle of the heart.
8. Describe what is meant by valvular stenosis and valvular regurgitation.

9. Describe the course of the right coronary artery over the surface of the heart.
10. Describe the course of the left coronary artery over the surface of the heart.
11. Describe the coronary sinus system of the heart.
12. Describe the regulation of coronary blood flow.
13. What is the mechanism of the usual heart attack, and what are the important factors that predispose to this?

REFERENCES

Berne, R.M., and Rubio, R.: Coronary circulation. *In* Berne, R.M., *et al.* (eds.): Handbook of Physiology. Sec. 2, Vol. I. Baltimore, Williams & Wilkins, 1979, p. 873.

Figge, F.H.J.: The heart. *In* Sobotta/Figge Atlas of Human Anatomy. Vol. II. Baltimore, Urban & Schwarzenberg, 1977, pp. 1–26.

Goss, C.M.: The heart. *In* Gray's Anatomy of the Human Body. Philadelphia, Lea & Febiger, 1973, pp. 527–560.

Guyton, A.C., *et al.:* Circulatory Physiology: Cardiac Output and Its Regulation, 2nd Ed. Philadelphia, W.B. Saunders, 1973.

Hamby, R.I.: Clinical-Anatomical Correlates in Coronary Artery Disease. Mt. Kisco, N.Y., Futura Publishing Co., 1979.

Klocke, F.J., and Ellis, A.K.: Control of coronary blood flow. *Annu. Rev. Med.,* 31:489, 1980.

Netter, F.H.: Heart. The CIBA Collection of Medical Illustrations. Vol. 5. Summit, N.J., CIBA Medical Education Division, 1969.

Pernkopf, E.: The heart. *In* Atlas of Topographical and Applied Human Anatomy. Vol. II. Philadelphia, W.B. Saunders, 1980, pp. 137–163.

Rapaport, E. (ed.): Current Controversies in Cardiovascular Disease. Philadelphia, W.B. Saunders, 1980.

Roskamm, H., and Schumuziger, M. (eds.): Coronary Heart Surgery: A Rehabilitation Measure. New York, Springer-Verlag, 1979.

Sommer, J.R., and Johnson, E.A.: Ultrastructure of cardiac muscle. *In* Berne, R.M., *et al.* (eds.): Handbook of Physiology. Sec. 2, Vol. I. Baltimore, Williams & Wilkins, 1979, p. 113.

Wenger, N.K. (ed.): Exercise and the Heart. Philadelphia, F.A. Davis, 1978.

Williams, P.L., and Warwick, R.: The heart. *In* Gray's Anatomy, 36th British Edition. Philadelphia, W.B. Saunders, 1980, pp. 635–672.

See also References, Chapter 25.
See also General Anatomical References, Chapter 1.

24

Heart Muscle, Heart Rhythm, and the Cardiac Cycle

Overview

In many ways cardiac muscle is very similar to skeletal muscle, but it has two characteristics that make it especially adapted for the pumping function of the heart: First, the *cardiac muscle fibers are interconnected* with each other so that an action potential originating in any part of the muscle mass travels everywhere and causes the entire muscle mass to contract in unison. This allows the heart muscle in all walls of each heart chamber to contract at the same time and therefore to squeeze the blood in a forward direction. Second, the *action potential of cardiac muscle lasts for about three tenths of a second,* which is ten or more times as long as the action potential in most skeletal muscle. Therefore, the duration of contraction of cardiac muscle also lasts for about three tenths of a second, which is the time required for the blood to flow out of the heart into the arteries.

The heart also has a special rhythmical control system consisting of (1) the **sinoatrial node** *(SA node),* located in the wall of the right atrium near the entry point of the superior vena cava; (2) the **atrioventricular node** *(AV node),* located in the atrial septum near the point where the two atria attach to the ventricles; and (3) a system of rapidly conducting large cardiac muscle fibers called **Purkinje fibers** that conducts the cardiac impulses rapidly from the AV node to all parts of both ventricles.

In a normal heart, the SA node controls the rate of heart beat and therefore is called the *pacemaker of the heart.* When a person is at rest, the SA node contracts rhythmically at about 72 beats per minute, and the action potential, called the *cardiac impulse,* generated in this node spreads throughout the heart—first through the atrium, then into the AV node, next through the Purkinje system into the ventricles, and finally through the ventricular muscle itself.

As the cardiac impulse passes through the AV node, it is *delayed for a little more than a tenth of a second* because the conducting fibers of this node are extremely small and conduct very slowly. This delay is especially important to heart function, because it allows the atria to beat a fraction of a second ahead of ventricular contraction, thus allowing blood to flow into the ventricles ahead of the ventricular pumping cycle.

When the heart becomes damaged, as often occurs when one of the coronary blood vessels is occluded, part of the conduction system of

413

the heart may be blocked so that normal conduction no longer occurs. For instance, blockage of conduction from the AV node into the ventricular Purkinje system, a condition called *heart block,* is a frequent occurrence in old age. When this occurs, the atria continue to beat at their normal rhythmical rate of about 72 beats per minute, but the atrial signals fail to be conducted into the ventricles. Instead, the large Purkinje fibers of the ventricular system now begin to emit impulses at a rhythmical rate of 15 to 40 beats per minute, and this focus now becomes the pacemaker of the ventricles. Thus, the ventricles beat at their own separate rate from that of the atria so that the atria and ventricles are no longer synchronized with each other.

Another conduction abnormality of the heart, a much more disastrous one, is *ventricular fibrillation,* in which the cardiac signal travels around and around in devious circuits through the ventricles in such a way that the signal never stops. Therefore, portions of ventricles remain contracted all the time, and there is never a period of relaxation during which the ventricles can fill with blood. Consequently, all pumping by the heart ceases, and the person dies within a few seconds.

In the previous chapter we discussed mainly the anatomy of the heart along with some of the gross principles of heart pumping. In the present chapter we will present the dynamics of heart function, including especially the contractile properties of heart muscle, the generation and control of heart rhythmicity, and the pumping cycle of the heart.

CARDIAC MUSCLE: ITS EXCITATION AND CONTRACTION

The Syncytial Character of Cardiac Muscle

Figure 24–1 illustrates a microscopic section of cardiac muscle. Note that the fibers have the same striated appearance that is characteristic of skeletal muscle. This is caused by the presence of the same type of *actin and myosin sliding filament contractile mechanism* in both cardiac and skeletal muscle. However, note also that unlike skeletal muscle the cardiac muscle fibers interconnect with each other, forming a latticework called a *syncytium.* This arrangement is similar to that which occurs in visceral smooth muscle where the smooth muscle fibers also fuse to form an interconnected mass of fibers also called a syncytium.

In the heart there are two separate muscle syncytia. One of these is the cardiac muscle that wraps around the two atria, and the other is the cardiac muscle that wraps around the two ventricles. These two muscle masses are separated from each other by fibrous tissue between the atria and the ventricles.

The importance of the two separate syncytial muscle masses is the following: When either one of the two muscle masses is stimulated, the action potential spreads over the entire syncytium and therefore causes the entire muscle mass to contract. Thus, when the atrial muscle mass is stimulated at any single point, the action potential spreads over both the right and the left atria and this causes the whole complex of atrial walls to contract in unison, thereby squeezing the atrial blood into the ventricles through the tricuspid and mitral valves. Then when the action potential spreads into the ventricles, here again it excites the entire ventricular muscle syncytium. Therefore, all of the ventricular walls now contract in unison, and the blood in their

FIGURE 24–1   Cardiac muscle, showing the lattice arrangement of the fibers.

chambers is appropriately pumped through the aortic and pulmonary valves into the arteries. Later in the chapter we will discuss the manner in which the heart controls the contraction of both the atria and the ventricles.

Automatic Rhythmicity of Cardiac Muscle. Some cardiac muscle fibers will contract rhythmically even when not stimulated by nerves. This is especially true of a group of small cardiac fibers located in the superior wall of the right atrium called the *sinoatrial node* or, more simply, the *SA node.* We will discuss the sinoatrial node in great detail later in this chapter because it is the normal controller of the heart beat. However, first let us discuss the mechanism by which rhythmic contraction occurs.

Figure 24–2 illustrates rhythmic action potentials generated by an SA nodal fiber. The cause of this rhythmicity is the following: The membranes of the SA fibers, even in the resting state, are very permeable to sodium. Therefore, large numbers of sodium ions leak to the interior of the fiber, causing the resting membrane potential to drift continually toward a more positive value as illustrated in the figure. Just as soon as the membrane potential reaches a critical level, called the "threshold" level, an action potential suddenly occurs. At the end of the action potential, the membrane is temporarily less permeable to sodium ions but more than normally permeable to potassium ions, and leakage of potassium ions out of the fiber carries positive charges to the outside. Therefore, the inside membrane potential now becomes more negative than ever, a state called *hyperpolarization,* because of loss of the extra positive charges. This condition persists for a fraction of a second and then disappears because the permeability to both sodium and potassium ions returns to its normal state, whereupon the natural leakiness of the membrane to sodium ions then elicits

another action potential. This process continues over and over throughout life, thereby providing rhythmic excitation of the SA nodal fibers at a normal resting rate of about 72 times per minute and for a total of about 2 billion heart beats during the life of the person.

Normally, action potentials originating in the SA node spread throughout the entire heart and thereby elicit the rhythmic contractions of the heart. However, if the SA node fails to generate rhythmic impulses, some other area of the heart will begin to generate impulses and will then take over control of the heart beat, as we shall discuss more fully later in the chapter.

Prolonged Duration of the Cardiac Muscle Action Potential and of Cardiac Contraction. Another important difference between cardiac muscle and skeletal muscle is that cardiac muscle contraction lasts for a prolonged period of time in comparison with skeletal muscle, averaging a duration about 10 to 15 times that of the average skeletal muscle.

The cause of the prolonged cardiac muscle contraction is that the cardiac muscle action potential also lasts for a long period of time. This is illustrated in Figure 24–3. Instead of the action potential being a sharp spike and then returning immediately to the baseline, as occurs in large nerve fibers and in skeletal muscle fibers, this potential remains on a *plateau,* as illustrated in the figure, for almost 0.3 second before returning to the resting level. The cause of this plateau is slowness of the membrane to repolarize once it has become depolarized. Let us explain this more fully. It will be recalled from the discussion in Chapter 6 that the action potential is caused by two separate processes: first, the process of depolarization of the membrane resulting from rapid influx of sodium ions into the fiber, and second, the repolarization process caused by rapid efflux of potassium ions from inside the fiber to the outside. The depolarization process causes the action potential to begin, and the re-

FIGURE 24–2 Rhythmic discharge of an SA nodal fiber.

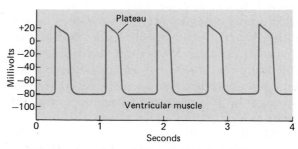

FIGURE 24–3 Rhythmic action potentials from a ventricular muscle fiber recorded by means of microelectrodes.

polarization process causes it to end. In large nerve fibers, skeletal nerve fibers, and some smooth muscle fibers, the repolarization process follows almost instantly after the depolarization process, thereby giving a very short spikelike action potential. In cardiac muscle and in some smooth muscle, the repolarization process is delayed. The cause of the delay is mainly the following: In these types of muscle fibers not only do sodium ions enter the fiber during depolarization but so do large numbers of calcium ions as well. Furthermore, the calcium ions continue to enter the fiber for many hundredths of a second after the sodium ions have stopped entering. Because the calcium ions carry positive charges, this maintains a positive state inside the fiber for the duration of the plateau, thus preventing repolarization. However, after a few tenths of a second, the inflow of calcium ions ceases, the repolarization process then takes place, and the action potential returns to the baseline.

In atrial muscle the total duration of the action potential as well as of the contraction is about 0.15 second, and in ventricular muscle about 0.3 second.

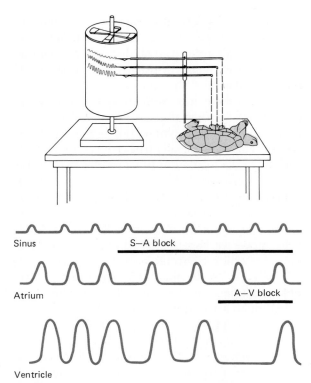

FIGURE 24—4   Method for recording the separate contractions of the sinus, atrium, and ventricle of the turtle heart. Below are illustrated the effects of SA block and AV block on impulse transmission.

REGULATION OF CARDIAC RHYTHMICITY

The inherent rhythmicity of the heart beat is seen beautifully by simply observing an exposed heart. An especially instructive experiment is the one illustrated in Figure 24–4, which shows recordings of muscle contraction in different portions of a turtle heart. The turtle heart has three chambers instead of the two that are present in each side of the human heart. The first chamber is called the *sinus* and the next two are the same as in the human heart, the *atrium* and the *ventricle*. The top record in Figure 24–4 shows contraction of the sinus, the middle record contraction of the atrium, and the bottom record contraction of the ventricle. No nerves are attached to the different portions of the heart in this preparation to make them contract, and no other signals arrive from outside the heart to cause the rhythmicity. In other words, the rhythmicity of the heart is vested in the heart itself, and if portions of the heart are removed from the body, they will continue to contract rhythmically as long as they are provided with sufficient nutrition.

The Sinoatrial Node as the Pacemaker of the Human Heart.   Referring once again to Figure 24–4, it will be observed at the beginning

of the record that the sinus of the turtle heart contracts slightly ahead of the atrium, and the atrium slightly ahead of the ventricle. That is, every time the sinus contracts, an action potential spreads along the muscle fibers from the sinus to the atrium and then to the ventricle, making these parts of the heart contract in succession.

The effects of blocking the action potential between the sinus and atrium (SA block) and then between the atrium and ventricle (AV block) are shown to the right in the figure. Note that the atrium and the ventricle then beat at their own natural rhythmic rates.

The human heart is different from that of the turtle, because no distinct sinus exists. However, as mentioned earlier, located in the superior wall of the right atrium near the point of entry of the superior vena cava is a small area known as the **sinoatrial node** (SA node), which is the embryonic remnant of the sinus from lower animals. The rhythmic rate of contraction of muscle fibers in the human SA node is approximately 72 times per minute, whereas the human atrial muscle has a rhythm of 40 to 60 times per minute and the ventricle about 20 times per minute. Because the SA node has a faster rate of rhythm than any other portion of the heart, impulses originating in the SA node spread into the atria and ventricles, stimulating these areas so rapidly that they can never slow down to their natural rates of rhythm. As a result, the rhythm of the SA node becomes the rhythm of the entire heart, for which reason the SA node is called the *pacemaker* of the heart.

Conduction of the Impulse through the Heart

THE PURKINJE SYSTEM

Even though the cardiac impulse can travel perfectly well along cardiac muscle fibers, the heart has a special conduction system called the Purkinje system, composed of specialized cardiac muscle fibers called **Purkinje fibers** that transmit impulses at a velocity approximately 5 times as rapidly as that in normal heart muscle, approximately 2 meters per second (m/sec) in contrast to 0.4 m/sec in cardiac muscle.

Figure 24–5 illustrates the organization of the Purkinje system. It originates in the **sinoatrial node** (SA node), which has already been discussed. From here, several very small bundles of Purkinje fibers, called **internodal pathways,** pass in the wall of the atrium to a second node, the **atrioventricular node** (AV node), which is also in the right atrial wall but located in the lower part of the posterior atrium, and toward the center of the heart. From this

node a large bundle of Purkinje fibers, called the **AV bundle,** arises and immediately passes out of the atria into the ventricles, entering first the ventricular septum. After traveling a short distance in the septum, the AV bundle divides into two large bundles; a **left bundle branch,** which continues along the inner surface of the left ventricle, and a **right bundle branch,** which continues along the inner surface of the right ventricle. In the ventricles, the bundles divide into many smaller Purkinje fiber branches that eventually make direct contact with the cardiac muscle in all areas. Therefore, an impulse traveling along the Purkinje fibers is conducted rapidly and directly into the cardiac muscle.

Role of the Purkinje System to Cause Coordinate Contraction of Cardiac Muscle. The major function of the Purkinje system is to transmit the cardiac impulse rapidly throughout the atria and, after a short pause at the AV node, then also rapidly throughout the ventricles. Rapid conduction of the impulse will cause all portions of each cardiac muscle syncytium—the atrial syncytium and then the ventricular syncytium—to contract in unison so that it will exert a coordinated pumping effort. Were it not for the Purkinje system, the impulse would travel much more slowly through the muscle, allowing some of the muscle fibers to contract long before others and then also to relax before the others. Obviously, this would result in decreased compression of the blood and, therefore, decreased pumping efficiency.

Sequence of Impulse Transmission through the Heart—Impulse Delay at the AV Node. After the cardiac impulse originates in the SA node, it travels first throughout the atria, causing the atrial muscle to contract. A few hundredths of a second after leaving the SA node the impulse reaches the AV node. However, the AV node delays the impulse another few hundredths of a second before allowing it to pass on into the ventricles. This delay allows time for the atria to force blood into the ventricles prior to ventricular contraction. After this delay, the impulse then spreads very rapidly through the Purkinje system of the ventricles, causing both ventricles to contract with full force within the next few hundredths of a second.

The AV node delays the cardiac impulse by the following mechanism: The fibers in this node are extremely small, which is very different from the remainder of the Purkinje system, but it allows these special fibers to conduct the cardiac impulse very slowly, at a velocity only one tenth that of normal cardiac muscle fibers and only one fiftieth that of the large Purkinje fibers.

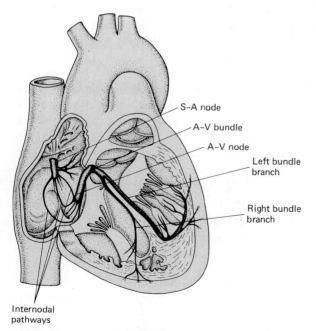

FIGURE 24–5 Transmission of the cardiac impulse from the SA node into the atria via internodal pathways, then into the AV node, and finally through the AV bundles to all parts of the ventricles.

Therefore, the cardiac impulse travels at a snail's pace through this node, causing a delay of more than 0.1 second between contraction of the atria and the ventricles.

BLOCK OF IMPULSE CONDUCTION IN DAMAGED HEARTS

Occasionally the cardiac impulse is blocked at some point in its pathway because of damage to the heart. For instance, a portion of heart muscle or of the Purkinje system may be destroyed and replaced by fibrous tissues that cannot transmit the impulse. Returning to Figure 24–4, note the effect of artificially blocking the impulse at two critical points in the turtle heart. First, a *sino-atrial block* was effected by tying a string tightly around the heart between the sinus and the atrium. This stopped impulse conduction from the sinus into the atrium, and as is evident from the illustration, the sinus continued to beat at its own natural rate whereas the atrium and ventricle assumed a rate equal to that of the natural rate of the atrium. In other words, the atrium became the pacemaker for the ventricle because the ventricle's natural rate of rhythmicity was much slower than that of the atrium. Another ligature was then tied tightly around the heart, this time between the atrium and the ventricle, to cause *atrioventricular block*. After this, the sinus, the atrium, and the ventricle all beat at their own respective natural rates of rhythm. This experiment shows that the portion of the heart that beats most rapidly controls the rate of rhythm of the remainder of the heart only so long as functioning conductive fibers exist between the different areas.

In the human heart, block rarely occurs between the SA node and the atrial muscle, but block does occur very frequently in the AV bundle. *It is only through this bundle that the normal impulse can pass from the atria into the ventricles,* because elsewhere the atria are connected to the ventricles not by conductive fibers, but instead by fibrous tissue that cannot conduct impulses. Therefore, whenever the AV bundle is blocked, the atria will beat at the rhythm of the SA node, and the ventricles at their own natural rate. In other words, the rate of the atria will remain at approximately 72 beats per minute, but that of the ventricles will decrease to 15 to 40 beats per minute. Despite this asynchrony of the atria and ventricles, the heart still operates as a pump, though its pumping ability may be decreased as much as 50 percent. Nevertheless, it is evident that the atria are not absolutely essen-

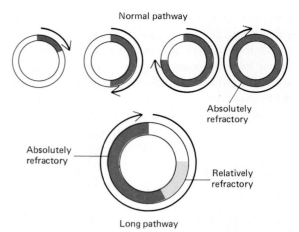

FIGURE 24–6 The principles of the circus movement.

tial for the heart to pump blood through the circulatory system.

THE REFRACTORY PERIOD, AND CESSATION OF THE CARDIAC IMPULSE AT THE END OF EACH HEART BEAT

Normally, when an impulse spreads along the membranes of the heart muscle fibers, a new impulse cannot spread along these same membranes until approximately 0.30 second later because this is the duration of the cardiac muscle action potential as explained earlier in the chapter. During this time the heart is said to be *refractory* to new impulses.

After an impulse enters the ventricles from the atria, it spreads all the way around the ventricles in about 0.06 second. Since the ventricular fibers cannot conduct again for 0.30 second, the impulse completely stops. The upper part of Figure 24–6 illustrates this principle diagrammatically. It shows a circular strip of cardiac muscle in which an impulse starts at the 12 o'clock point, then travels around the heart and finally returns to the 12 o'clock point. When the impulse reaches the starting point, the entire heart is still refractory, which causes the impulse to die.

THE CIRCUS MOVEMENT

Occasionally conditions become sufficiently abnormal that the cardiac impulse does not die at the end of the heart beat but instead continues on and on around the heart, never stopping. For instance, in the lower part of Figure 24–6, the length of the pathway has been increased so greatly by the larger circle (which represents an enlarged heart) that, after traveling all the way around the heart, the impulse now returns to the 12 o'clock position more than 0.30 second after it starts. By this time, the originally stimulated portion of the muscle is no longer refractory, which allows the impulse to travel around again. As the impulse proceeds, the refractory state recedes ahead of it, allowing it to continue indefinitely. This effect is known as a *circus movement*, and it is often lethal, as explained later.

Causes of Circus Movements. Circus movements can result from any of four different abnormalities of the heart. First, as explained in the preceding example, a circus movement is likely to occur when the heart becomes greatly enlarged, thus creating a long pathway. A second cause is slow conduction of the impulse through the heart. For instance, failure of the Purkinje system causes the impulse then to be transmitted by the cardiac muscle itself. This slows impulse transmission about fivefold, often causing the impulse to return to the starting point after the originally excited muscle is no longer refractory. A third cause of a circus movement may be decreased refractory period of the heart muscle. This sometimes results from increased cardiac excitability caused by epinephrine, sympathetic stimulation, or irritation of the heart as a result of disease. Fourth, probably the most common cause is conduction of the impulse through the heart muscle in an odd pathway such as in figure 8's, or in zigzags, the impulses sometimes traveling deep in the muscle and then later traveling at shallow levels, recrossing the same area that had already been stimulated at a deeper level. In this way the pathway becomes extremely lengthened, thus allowing development of an odd-shaped circus movement. Such a pathway, as illustrated in Figure 24–7B, causes ventricular fibrillation, which is discussed below.

Effect on Cardiac Pumping. A circus movement is disastrous to the pumping action of the heart because normal pumping requires that the muscle relax as well as contract. A period of total ventricular relaxation is not possible if the impulse continually travels through the muscular mass. Therefore, the entire muscle mass never relaxes or contracts simultaneously, never allowing an alternating filling and squeezing action.

Atrial Flutter and Atrial Fibrillation. Occasionally a circus movement occurs around and around the two atria, as shown in Figure

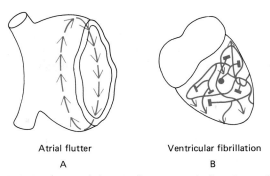

Atrial flutter

A

Ventricular fibrillation

B

FIGURE 24–7 (A) Impulse transmission in *atrial flutter.* (B) Impulse transmission in *ventricular fibrillation.*

24–7A, at a rate of 200 to 400 times per minute, but not involving the ventricles. This causes *atrial flutter*, during which the atria are "fluttering" rapidly but pumping almost no blood. At other times, an odd-shaped, zigzag, or figure 8 pattern of circus movement occurs in the atria at such a rapid rate that one can see only minute fibrillatory movements of the muscle. This is called *atrial fibrillation*, and when it occurs the atria are of no use whatsoever as primer pumps for the ventricles, but the ventricles can still pump blood with about two-thirds normal effectiveness.

Ventricular Fibrillation. A circus movement having the very odd pattern illustrated in Figure 24–7B frequently develops suddenly in the ventricles, in which case impulses travel in all directions and cause *ventricular fibrillation*. The impulses go around refractory areas of muscle, dividing into multiple wave fronts in doing so. Some of the impulses die, but for every impulse that does die, another impulse divides to form two impulses going in separate directions in the heart. This type of circus movement has been called the *chain-reaction movement*, for it is similar to the chain reaction that occurs in nuclear bomb explosions.

Ventricular fibrillation causes the ventricles to contract continuously in very fine, rippling fibrillatory movements. The contracting areas are widespread in the ventricles, interspersed with relaxing areas. As a consequence, the ventricles are incapable of pumping any blood, and the person becomes unconscious within a few seconds after fibrillation begins and dies within the next few minutes.

Ventricular fibrillation is frequently caused by electric shock, particularly by 60 cycle alternating current; this causes impulses to go in many different directions at once in the heart, setting up the odd-shaped pattern of impulse transmission illustrated in Figure 24–7B. Ventricular fibrillation also occurs in diseased ventricles that (a) become overly excitable, (b) develop a damaged Purkinje system, or (c) become greatly enlarged.

Fortunately, if the doctor works fast enough after ventricular fibrillation begins, he can sometimes save the person's life by sending a single very strong electric shock through the chest. This will stop the fibrillatory impulses and, if lucky, allow a normal cardiac impulse to resume.

The Electrocardiogram

The electrocardiogram is a very important tool for assessing the ability of the heart to transmit the cardiac impulse. When the impulse travels through the heart, electrical current generated by the action potential of the heart muscle spreads into the fluids surrounding the heart,

and a minute portion of the current actually flows as far as the surface of the body. By placing electrodes on the skin over the heart or on any two sides of the heart, such as on the two arms, and connecting these to an appropriate recording instrument, the electrical voltages generated during each heart beat can be recorded.

In the normal electrocardiogram illustrated in Figure 24–8A, the small hump in the recording labeled "P" is caused by electrical voltage generated by passage of the impulse through the atria. The spikes marked "Q," "R," and "S" are caused by passage of the impulse through the ventricles, and the hump "T" is caused by return of the membrane potential in the ventricular muscle fibers to its normal resting level at the end of contraction.

When cardiac abnormalities are caused by various diseases, the electrocardiogram often becomes changed from the normal. Figure 24–8B shows the effect when some of the ventricu-

A — Normal electrocardiogram

B — Depressed S-T segment due to damaged muscle of heart

C — Ratio of S to R wave abnormal due to enlargement of one side of heart

D — Prolonged QRS due to failure of impulse conduction in the Purkinje system

E — Dissociated atrial and ventricular beats due to heart block

F — Early beat due to irritable focus in heart

FIGURE 24–8 The normal and several abnormal electrocardiograms.

lar muscle has been damaged. In this record the portion of the electrocardiogram between the S and T waves is depressed. This results from abnormal leakage of electrical current from the heart between heart beats. It indicates damage to the membranes of the ventricular muscle fibers, which often occurs when one has had an acute heart attack.

Figure 24–8C illustrates the effect seen when one side of the heart is enlarged more than the other. The record shows abnormal enlargement of the S wave and diminishment of the R wave, indicating more current flow from the left side of the heart than from the right. High blood pressure very frequently causes this type of electrocardiogram because of excessive pressure load on the left ventricle.

Figure 24–8D shows the electrocardiogram in a person who has a partially blocked Purkinje system. In this instance, the impulse is transmitted through much of the ventricles by way of slowly conducting cardiac muscle fibers rather than rapidly conducting Purkinje fibers, so that the QRS complex lasts for a prolonged period of time and also develops an abnormal shape.

Figure 24–8E illustrates the effect of blocking the impulse at the AV bundle. The P waves occur regularly, and the QRST waves also occur regularly but in no definite relationship to the P waves. The atria are beating at the natural rate of rhythm of the SA node, at 72 beats per minute, whereas the ventricles have assumed their own rate of rhythm at 38 beats per minute.

Finally, Figure 24–8F illustrates a record, indicated by the arrow, of a *premature contraction* of the heart. The only abnormality here is that the impulse occurs too soon after the previous heart beat. This is usually caused by an irritable heart resulting from such factors as too much smoking, too much coffee drinking, or lack of sleep.

A study of Figure 24–8 shows how various abnormalities of heart function can be discovered from electrocardiographic recordings and why this diagnostic procedure is used in all persons with heart disease.

THE CARDIAC CYCLE

Now that the physiological anatomy and rhythmic control of the heart have been described, it is possible to synthesize this information into a sequence of events called the *cardiac cycle*. Figure 24–9 illustrates the major events of this cycle, showing curves from above downward for (1) pressure changes in a ventricle, (2) the electrocardiogram, and (3) the phonocardiogram. The cardiac cycle rightfully begins with initiation of the rhythmic impulse in the SA node. Then transmission of the impulse through the heart causes the muscle fibers to contract. Thus, as shown in the figure, the P wave of the electrocardiogram occurs immediately before onset of the atrial pressure wave caused by atrial contraction. Approximately 0.16 second after the P wave begins, the electrical impulse has com-

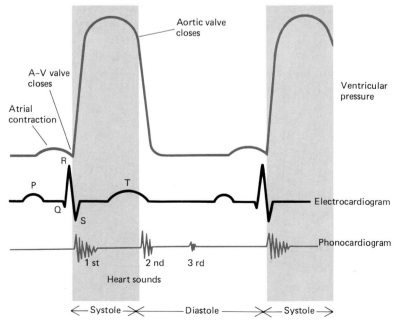

FIGURE 24–9 Relationship of ventricular pressure to the electrocardiogram and phonocardiogram during the cardiac cycle, showing also the periods of systole and diastole.

pleted its passage through the atria, the AV node, and the AV bundle. Then it begins to spread rapidly over the ventricles, causing the QRS wave of the electrocardiogram and stimulating the ventricular muscle to contract. The rising ventricular pressure closes the mitral and tricuspid valves, thereby also generating the first heart sound, and it opens the aortic and pulmonary valves a small fraction of a second later. The ventricles remain contracted approximately 0.30 second and then relax. At the onset of relaxation, ions retransfer through the fiber membranes to reestablish the normal negative electrical charge inside the cardiac muscle fibers. This causes the T wave of the electrocardiogram. Immediately after the ventricular muscle relaxes, a small amount of blood flows backward from the arteries toward the ventricles, closing the aortic and pulmonary valves, which elicits the second heart sound. Following ventricular relaxation, no further contraction occurs until a new electrical impulse is initiated in the SA node.

Systole and Diastole

The period during the cardiac cycle when the ventricles are contracting is called *systole*, and the period of relaxation is called *diastole*. A clinician examining the heart can note the periods of systole and diastole either from the electrocardiogram or from the heart sounds; systole begins with the QRS wave and ends with the T wave, or it begins with the first heart sound and ends with the second sound. Diastole, on the other hand, begins with the T wave and ends with the QRS wave, or it begins with the second heart sound and ends with the first sound.

Sometimes it is quite important to distinguish between systole and diastole. This is particularly true when one is studying valvular disorders or abnormal openings between the two sides of the heart. For instance, leakage of the

aortic or the pulmonary valve causes a "swishing" sound (a *murmur*) during diastole, as was discussed earlier. On the other hand, a murmur caused by leakage of an AV valve occurs during systole, because that is the period when these valves leak if they are abnormal.

Pressure Changes During the Cardiac Cycle

Figure 24–10 shows the changes in pressure in the left atrium, left ventricle, and aorta during a typical cardiac cycle. During diastole, the left atrial pressure is a little higher than that of the left ventricle because blood continually flows from the pulmonary veins into the atrium. This obviously causes blood to flow also from the left atrium into the left ventricle. Toward the end of diastole, contraction of the atrium elevates the atrial pressure to an even higher level and forces an extra quantity of blood into the ventricle. Then, suddenly, the ventricle contracts, the mitral valve closes, and the ventricular pressure rises rapidly. When this pressure rises higher than that in the aorta, the aortic valve opens and blood flows into the aorta during the entire remainder of systole. When the ventricle relaxes, the ventricular pressure falls precipitously, allowing a slight backflow of blood that immediately closes the aortic valve. Throughout diastole the aortic pressure remains high because a large quantity of blood has been stored in the very distensible arteries during systole. This blood runs off slowly through the systemic capillaries back to the right atrium, allowing the aortic pressure to fall from a peak during systole of approximately 120 mm Hg down to a minimum of approximately 80 mm Hg by the end of diastole. Therefore, the normal systemic arterial blood pressure is said to be 120/80, meaning by this 120 mm Hg *systolic pressure* and 80 mm Hg *diastolic pressure*.

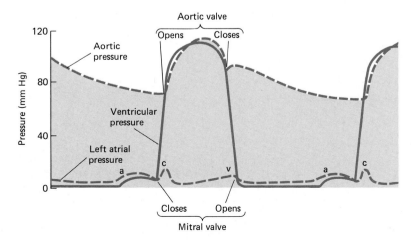

FIGURE 24–10 Pressures in the aorta, left ventricle, and left atrium during the cardiac cycle.

THE LAW OF THE HEART

The amount of blood pumped by the heart is normally determined by the amount of blood flowing from the veins into the right atrium. This principle is called the "law of the heart," or often the "Frank-Starling" law in honor of the physiologists who discovered it. That is, the heart is simply an automaton that continues to pump all of the time, and whenever blood enters the right atrium it is pumped on through the heart. Of course, there is a maximum rate at which the heart can pump, for which reason, to be completely accurate, the law of the heart is more correctly stated as follows: *Within physiologic limits, the heart pumps all of the blood that flows into it, and it does so without significant damming of blood in the veins.* Thus, the heart is analogous to a sump pump, because any time fluid enters the chamber of a sump pump, it is pumped out of the chamber immediately. In the case of the heart, whenever blood enters one of the atria, it is immediately pumped on into a ventricle and then on into the arteries.

Cardiac muscle has a special characteristic that gives the heart this ability to pump varying amounts of blood in response to changing rates of venous inflow. When cardiac muscle is stretched beyond its normal length, it contracts with greater force than when it is not stretched. Therefore, when only a small quantity of blood enters the heart, the muscle fibers are not stretched greatly, and the force of contraction is weak. On the other hand, if large quantities of blood enter, the heart chambers dilate greatly, the muscle fibers stretch, and the force of contraction becomes very great. As a result, the increased quantity of blood returning to the heart is pumped on through.

The law of the heart often fails when the heart has been damaged, for then even normal quantities of blood returning to the heart are more than the heart can cope with. As a consequence, blood begins to dam up in the veins of either the lungs or the systemic circulatory system. In this case the heart is said to be *failing.* The subject of heart failure is so very important that it will be discussed in detail in Chapter 29.

The Heart-Lung Preparation. One method frequently used in the physiology laboratory to demonstrate the automatic ability of the heart to pump blood is the heart-lung preparation illustrated in Figure 24–11. In this preparation, blood flow from the heart is channeled from the aorta, through an external system, and then back into the right atrium, rather than through the animal's body. Artificial respiration

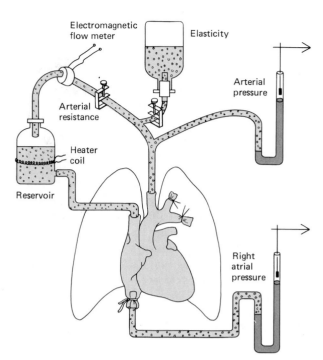

FIGURE 24–11 The heart-lung preparation.

is supplied to the lungs to keep the blood appropriately oxygenated, and nutrient in the form of glucose is added to the blood. The heart will continue to beat for many hours, with the external flow system taking the place of the systemic circulation. One can vary the resistance of the external circuit by tightening or loosening a screw clamp. He can measure the blood flow with a flowmeter, and he can change the amount of blood flowing into the right atrium by raising or lowering the venous reservoir.

Several principles that demonstrate the independent function of the heart can be illustrated beautifully with the heart-lung preparation. First, it can be shown that, within physiologic limits, raising or lowering the venous reservoir is the main factor that determines how much blood will be pumped. In other words, *the greater the input pressure forcing blood from the veins into the heart, the greater the volume of blood pumped,* which is another expression of the "law of the heart."

Second, it can be demonstrated that changing the resistance of the external circuit within normal limits *does not* greatly affect the amount of blood pumped. On the other hand, the pressure in the aorta does increase approximately in proportion to the increase in resistance. That is, *the amount of resistance to blood flow through the circulatory system changes arterial pressure greatly, but hardly affects the amount of blood pumped by the heart* unless the resistance becomes so great that the heart simply cannot pump with enough force to overcome it.

A third interesting effect that can be demonstrated is that caused by changing the distensibility of the arterial system. This can be done by connecting an air bottle to the arterial system. Blood flows into the bottle and compresses the air when the pressure rises during systole, but during diastole the compressed air forces blood back into the arteries, thereby helping to maintain a relatively high level of arterial pressure even between heart beats. Therefore, with the bottle in the system, the arterial pressure pulsates far less than when it is out of the system. This effect illustrates that *distensibility of the arterial walls is very important to smooth out pulsations in the arterial pressure.*

Finally, the heart-lung preparation can be used to demonstrate the effect of temperature, drugs, or abnormal blood constituents on the heart. For instance, *increasing the temperature of the blood 10°F increases the heart rate approximately 100 percent.* Also, increasing the amount of *calcium* in the blood makes the heart contract with increased vigor, but increasing the amount of *potassium* decreases the vigor of heart contraction. Finally, when the nutrients of the blood fall to low values, addition of glucose or other nutrients can greatly enhance the function of the heart.

NERVOUS CONTROL OF THE HEART

Though the heart has its own intrinsic control systems and can continue to operate without any nervous influences, the efficacy of heart action can be changed greatly by regulatory impulses from the central nervous system. The nervous system is connected with the heart through two different sets of nerves, the *parasympathetic nerves* and the *sympathetic nerves.* These were discussed in detail in Chapter 19 in connection with function of the nervous system. The connections of the parasympathetic (vagi) and sympathetic nerves with the heart are shown in Figure 24–12.

Parasympathetic Stimulation. Stimulation of the parasympathetic nerves causes the following three important effects on the heart: (1) decreased rate of heart beat, (2) decreased force of contraction of the arterial muscle, and (3) delayed conduction of impulses through the AV node, which lengthens the delay period between

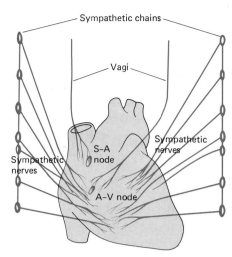

FIGURE 24–12   Innervation of the heart.

atrial and ventricular contraction. All of these effects may be summarized by saying that *parasympathetic stimulation decreases all activities of the heart.* Usually, heart activity is reduced by the parasympathetics during periods of rest; this allows the heart to rest at the same time that the remainder of the body is resting. This preserves the resources of the heart; without such periods of rest the heart undoubtedly would wear out at a much earlier age than it normally does.

Sympathetic Stimulation. Stimulation of the sympathetic nerves has essentially the opposite effects on the heart: (1) increased heart rate, (2) increased vigor of cardiac contraction, and (3) increased rapidity of conduction of the cardiac impulse through the heart. These effects can be summarized by saying that *sympathetic stimulation increases the activity of the heart as a pump,* sometimes increasing the ability to pump blood as much as 100 percent. This stimulation of the heart is necessary when a person is subjected to stressful situations such as exercise, disease, excessive heat, and other conditions that demand rapid blood flow through the circulatory system. Therefore, the sympathetic effects on the heart are a standby mechanism held in readiness to make the heart beat with extreme vigor when necessary.

QUESTIONS

1. What are the differences between cardiac muscle contraction and skeletal muscle contraction?
2. Why is it important that cardiac muscle have a syncytial structure?
3. What causes the automatic rhythmicity of cardiac muscle?
4. Explain why the sinoatrial node is normally the

pacemaker of the heart, but at times some other area of the heart might become the pacemaker.

5. How does the Purkinje system increase the effectiveness of the heart as a pump?

6. Explain the circus movement and how it can cause flutter or fibrillation of the heart.

7. In the electrocardiogram, what does the "P" wave represent? The "QRS" wave? The "T" wave?

8. In the cardiac cycle, what is the relationship between left ventricular pressure and aortic pressure?

9. Explain the "law of the heart" and its significance.

10. What are the effects of sympathetic and parasympathetic stimulation of the heart?

REFERENCES

Alpert, N.R., *et al.:* Heart muscle mechanics. *Annu. Rev. Physiol., 41:*521, 1979.

Braunwald, E., and Ross, J., Jr.: Control of cardiac performance. *In* Berne, R.M., *et al.* (eds.): Handbook of Physiology. Sec. 2, Vol. I. Baltimore, Williams & Wilkins, 1979, p. 533.

Chung, E.K.: Electrocardiography: Practical Applications With Vectorial Principles, 2nd Ed. Hagerstown, Md., Harper & Row, 1980.

Cranefield, P.F., and Wit, A.L.: Cardiac arrhythmias. *Annu. Rev. Physiol., 41:*459, 1979.

Fozzard, H.A.: Heart: Excitation-contraction coupling. *Annu. Rev. Physiol., 39:*201, 1977.

Guyton, A.C.: Determination of cardiac output by equating venous return curves with cardiac response curves. *Physiol. Rev., 35:*123, 1955.

Guyton, A.C., *et al.:* Circulatory Physiology: Cardiac Output and Its Regulation, 2nd Ed. Philadelphia, W.B. Saunders, 1973.

Guyton, A.C., and Crowell, J.W.: A stereovectorcardiograph. *J. Lab. Clin. Med., 40:*726, 1952.

Guyton, A.C., and Satterfield, J.: Factors concerned in electrical defibrillation of the heart, particularly through the unopened chest. *Am. J. Physiol., 167:*81, 1951.

Irisawa, H.: Comparative physiology of the cardiac pacemaker mechanism. *Physiol. Rev., 58:*461, 1978.

Jones, P.: Cardiac Pacing. New York, Appleton-Century-Crofts, 1980.

Nobel, D.: The Initiation of the Heartbeat. New York, Oxford University Press, 1979.

Pick, A., and Langendorf, R.: Interpretation of Complex Arrhythmias. Philadelphia, Lea & Febiger, 1980.

Sarnoff, S.J.: Myocardial contractility as described by ventricular function curves. *Physiol. Rev., 35:*107, 1955.

Vasselle, M.: Electrogenesis of the plateau and pacemaker potential. *Annu. Rev. Physiol., 41:*425, 1979.

25

Anatomy of the Systemic Circulation

Overview

The **systemic circulation** begins with the **aorta** that arises from the outflow tract of the left ventricle, then passes through smaller vessels in all the tissues of the body besides the lungs, and finally terminates with the **superior** and **inferior venae cavae** entering the right atrium.

The major branches of the aorta are the (1) **brachiocephalic artery,** (2) **left common carotid artery,** (3) **left subclavian artery,** (4) **celiac artery,** (5) **superior mesenteric artery,** (6) **right** and **left renal arteries,** (7) **inferior mesenteric artery,** and (8) **right** and **left common iliac arteries.**

The blood supply to the *neck, face,* and *brain* is carried through the *two common carotid arteries* plus several smaller arteries that arise from the *subclavian arteries,* the most important of which are the **vertebral arteries.** In the upper neck, each common carotid divides into the **external carotid** and **internal carotid.** Many arterial branches from the external carotid supply blood to the anterior neck, face, and crown of the head. The internal carotid enters the cranial cavity through an opening at the base of the skull called the *carotid canal* and almost immediately divides into arteries that supply the eye and the cerebrum, the **ophthalmic artery** and **cerebral arteries.** The cerebral arteries lie on the surface of the brain; only small branches penetrate inward to supply the cerebral tissues. The brain stem is supplied by the two *vertebral arteries* that enter the cranial cavity through the foramen magnum alongside the spinal cord.

The *upper extremity* is supplied with blood from the *subclavian artery,* the name of which becomes the **axillary artery** as it passes through the axilla, then the **brachial artery** as it passes through the upper arm. In the forearm, the brachial artery divides into the **radial** and **ulnar arteries.**

In the *abdomen,* the **celiac artery** supplies the superior abdominal organs, including the liver, stomach, spleen, and part of the duodenum and pancreas. The **superior mesenteric** artery supplies most of the small intestine and the right two thirds of the colon; the **inferior mesenteric artery** supplies the left one third of the colon and part of the rectum.

The *pelvic structures* receive their blood supply almost entirely from the **internal iliac artery,** and the *lower extremity* receives its blood supply almost entirely from the **external iliac artery.** As the external iliac enters the thigh its name changes to the **femoral artery,** and as it passes behind the knee it changes to the **popliteal artery.** Shortly below the knee, the popliteal artery divides into the **anterior tibial** and **posterior tibial arteries** that supply the leg and foot.

In general, for each arterial distribution system there is a corresponding venous drainage system carrying blood back toward the heart. However, in three portions of the systemic circulation, venous drainage is quite different from the arterial distribution.

First, in the *brain,* the major venous channels are large triangular sinuses called **dural sinuses** that are encased in the dural membranes where these membranes fold into brain fissures and sulci. These sinuses empty into the two **internal jugular veins** that then flow downward through the neck.

Second, in the *abdomen,* essentially all of the blood flow from the gastrointestinal tract and spleen drains into the **portal vein** rather than into the inferior vena cava. The blood in the portal vein in turn flows into the liver where it passes through an extensive network of **hepatic sinuses,** finally emptying from the liver into the inferior vena cava through two **hepatic veins.**

Third, in both the upper and lower extremities a major share of the venous blood flows into **superficial veins** that lie just underneath the skin rather than into the **deep veins.** These superficial veins empty into the deep venous system respectively in the upper arm or axilla and upper leg.

The structures of the arterial, venous, and capillary walls are all adapted to the specific functions of these respective vessels. Both the *arterial* and *venous walls* are constructed mainly of *smooth muscle,* though the arterial wall is thick and contains large amounts of *elastic fibers* as well, which makes the arteries very distensible to accommodate the arterial blood pressure pulsations. The venous walls are thin because of the low venous pressure, and the volumes of the veins are much greater than the volumes of their corresponding arteries, thus accommodating the massive blood volume of the veins. The *capillaries* are extremely minute, but their vast number—about 10 billion—gives them a total surface area approaching that of a football field. Also, their walls are very thin and porous so that cellular nutrients and excreta can diffuse readily between the tissue fluids and the blood in the capillaries.

In this chapter we will describe the anatomy of the **systemic circulation** and in the following chapter discuss its function. Later, in Chapter 36 in connection with the respiratory function of the lungs, we will present both the anatomy and the function of the pulmonary circulation.

The systemic circulation is divided into **arteries, capillaries,** and **veins,** but for convenience of description, it is also often divided into the following:

1. The **central vessels.** These are the large systemic vessels in the thorax that are closely associated with the heart. On the arterial side they include the *thoracic aorta* and its major branches and on the venous side the *thoracic portion* of the *superior* and *inferior venae cavae* and their major tributaries.
2. The **distribution vessels,** which are the intermediate-sized and small vessels leading from the tissues.

THE CENTRAL VESSELS

THE THORACIC AORTA AND ITS MAJOR BRANCHES

Note in Figure 25–1 that the **aorta** arises from the outflow tract of the left ventricle, as was discussed in Chapter 23. The first 5 to 7 cm of the aorta are directed cranially, but then it arches posteriorly and inferiorly to run posterior to the heart and eventually through the abdominal cavity along its posterior wall. The thoracic aorta can be divided into the following parts:

1. the **ascending aorta,** the short portion that ascends cranially,
2. the **arch of the aorta,** and
3. the **thoracic descending aorta.**

Branches of the Ascending Aorta—the Coronary Arteries. The only branches of the ascending aorta are the **right** and **left coronary**

TABLE 25–1
The Arteries

Artery.	Area supplied
Ascending aorta	
Coronary artery	Heart muscle
Aortic Arch	
Brachiocephalic	
Right common carotid	
External carotid	
Superior thyroid	Anterior neck and thyroid gland
Lingual	Tongue region
Facial	Superficial face
Occipital	Posterior neck and scalp
Maxillary	Deep facial, mouth, and nasal structures
Middle meningeal	Lateral and superior dura
Superficial temporal	Lateral and superior scalp
Internal carotid	
Ophthalmic	Eye and orbital structures
Middle cerebral	Lateral and superior cerebrum
Anterior cerebral	Anteromedial cerebrum
Right subclavian and axillary	
Vertebral	Lower brain stem
Basilar	Middle brain stem
Cerebellar	Cerebellum
Posterior cerebral	Posteromedial cerebrum
Internal thoracic	Inner anterior thoracic wall
Cervical	Neck
Lateral thoracic	Lateral outer thoracic wall
Subscapular	Scapular muscles
Brachial	
Profunda brachii	Posterior and lateral arm
Radial	Lateral forearm
Ulnar	Medial forearm
Palmar arches	
Dorsal arch	
Metatarsal	Hand and fingers
Digital	
Left carotid	
Left subclavian	Same as right side
Thoracic descending aorta	
Posterior intercostals	Posterior thoracic wall
Bronchial	Bronchi and lung septal structures
Esophageal	Esophagus
Superior phrenic	Middle diaphragm

arteries that supply blood to the heart itself. These arise at the very base of the ascending aorta from sinuses behind two of the aortic valve cusps and supply the heart muscle with nutrition as was described in Chapter 23 in connection with the anatomy of the heart.

Branches of the Aortic Arch. Three major branches arise in rapid succession from the summit of the aortic arch, as illustrated in Figure 25–1. These are, from right to left,

1. the **brachiocephalic artery,** which supplies the right side of the head and neck as well as the right upper extremity;
2. the **left common carotid artery,** which supplies most of the left side of the neck and head; and
3. the **left subclavian artery,** which supplies the left shoulder and upper extremity.

Branches of the Thoracic Descending Aorta. The thoracic descending aorta has no large branches, but it does have a series of small branches on each side, including

TABLE 25–1   **(Continued)**

Artery	Area supplied
Abdominal aorta	
Celiac	
Common hepatic	Liver
Right gastric	Right lesser curvature of stomach
Right gastroepiploic	Right greater curvature of stomach
Superior pancreaticoduodenal	Pancreas and duodenum
Splenic	Spleen
Left gastroepiploic	Left greater curvature of stomach
Left gastric	Left lesser curvature of stomach
Superior mesenteric	
Inferior pancreaticoduodenal	Pancreas and duodenum
Jejunal	Jejunum
Ileal	Ileum
Ileocolic	Distal ileum and cecum
Right colic	Ascending colon
Middle colic	Transverse colon
Right and left renal	Kidneys
Right and left gonadal	Ovaries or testicles
Inferior mesenteric	
Left colic	Descending colon
Sigmoidal	Sigmoid colon
Superior rectal	Proximal rectum
Lumbar	Lateral and posterior abdominal wall
Right and left common iliac	
Internal iliac	
Iliolumbar	Lateral inner wall of pelvis
Superior and inferior gluteal	Buttocks and gluteal muscles
Obturator	Posteromedial superior thigh muscles
Uterine	Uterus, fallopian tube, ovary
Superior and inferior vesical	Urinary bladder
Internal pudendal	Perineum, distal rectum, external sexual organs
External iliac	
Inferior epigastric	Anterior abdominal wall
Femoral and popliteal	Anterior and medial thigh
Profunda femoris	Posterior and lateral thigh
Anterior tibial	Anterior medial leg
Dorsalis pedis	Dorsum of foot
Posterior tibial	Posterior medial leg
Peroneal	Lateral leg
Dorsal arch	Dorsum of foot
Plantar arch	Plantar surface of foot
Digital	Toes

1. the **posterior intercostal arteries** that course around the thoracic cage between the successive ribs,
2. usually one **right bronchial artery** and two **left bronchial arteries** that supply the bronchial tissues of the lungs,
3. four or five small **arteries that supply the esophagus** that lies adjacent to the descending aorta, and
4. several small **superior phrenic arteries** that supply the central portion of the diaphragm.

THE THORACIC VENAE CAVAE AND THEIR MAJOR TRIBUTARIES

Though the heart pumps all of its blood into a single systemic artery, the aorta, it receives its returning blood from two separate systemic veins, as illustrated in Figure 25–2:

1. the **superior vena cava** and
2. the **inferior vena cava.**

Both these venae cavae enter the right atrium, the superior vena cava entering from

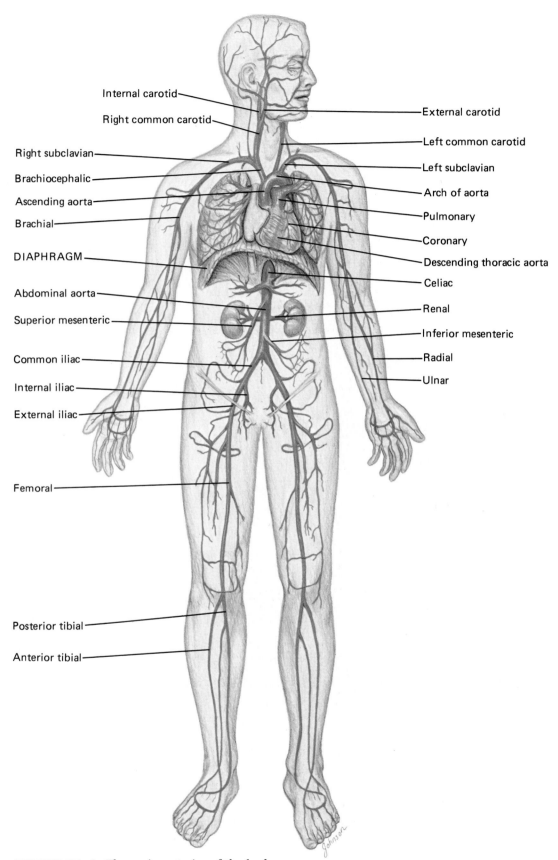

Internal carotid
Right common carotid
Right subclavian
Brachiocephalic
Ascending aorta
Brachial
DIAPHRAGM
Abdominal aorta
Superior mesenteric
Common iliac
Internal iliac
External iliac
Femoral
Posterior tibial
Anterior tibial

External carotid
Left common carotid
Left subclavian
Arch of aorta
Pulmonary
Coronary
Descending thoracic aorta
Celiac
Renal
Inferior mesenteric
Radial
Ulnar

FIGURE 25–1 The major arteries of the body.

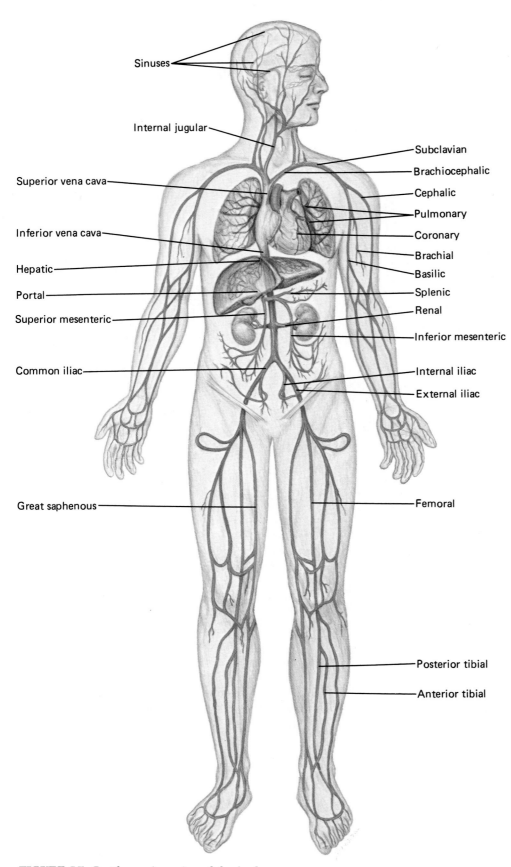

Sinuses

Internal jugular

Subclavian

Brachiocephalic

Superior vena cava

Cephalic

Pulmonary

Coronary

Inferior vena cava

Brachial

Basilic

Hepatic

Splenic

Portal

Renal

Superior mesenteric

Inferior mesenteric

Common iliac

Internal iliac

External iliac

Great saphenous

Femoral

Posterior tibial

Anterior tibial

FIGURE 25–2   The major veins of the body.

TABLE 25-2
Special Venous Channels that Are Different from Arteries

Special veins	Area drained
Venae cavae	
Superior	
Azygos	Posterior thoracic and lumbar walls
Right and left brachiocephalic	
Internal jugular	Brain, upper neck, and head
Subclavian	Low neck, shoulder, upper extremity
External jugular	Lateral and posterior superficial neck and head
Inferior	
Right and left hepatic	Liver
Portal	Drains gastrointestinal veins into liver
Right and left renal	Kidneys
Right and left common iliac	Pelvis and lower extremities
Dural sinuses	
Superior sagittal	Superior cerebrum
Inferior sagittal	Medial inferior cerebrum
Cavernous	Orbital structures, upper nasal, anterior inferior cerebrum
Straight	From inferior sagittal and cavernous sinuses
Transverse, sigmoid, and internal jugular vein	From superior sagittal and straight sinuses
Superficial veins of upper extremity	
Cephalic	Lateral forearm, dorsum of hand, and inferior arm
Basilic	Medial forearm and inferior arm
Median antebrachial	Anterior forearm
Superficial veins of lower extremity	
Great saphenous	Anterior and medial leg and foot; most of thigh
Small saphenous	Posterolateral leg and foot

above, the inferior vena cava from below. Because of this arrangement, there is no arch in either vena cava similar to the arch of the aorta. There are no tributaries to the thoracic portion of the inferior vena cava, which is only a few centimeters long between its penetration of the diaphragm and its entry to the right atrium.

    The Major Tributaries of the Superior Vena Cava. In general, the tributaries of the superior vena cava are approximately the same as those of the arch of the aorta, but the detailed anatomy is somewhat different. Three major branches are as follows:

1 and 2. The **right** and **left brachiocephalic veins.** Each of these is formed by (a) an **internal jugular** vein that courses down the side of the neck and corresponds to the carotid artery and (b) a **subclavian vein** that follows a course from the upper extremity parallel to the subclavian artery.

3. The **azygos vein** enters the superior cava from posteriorly, for which reason it cannot be seen in Figure 25-2. This vein collects blood from many small **intercostal, bronchial, esophageal,** and **phrenic** veins in the thorax and small **segmental** veins in the upper abdominal region that drain mainly the lateral and posterior abdominal wall.

THE BLOOD VESSELS OF THE NECK, HEAD, AND BRAIN

In most of the remainder of this chapter we will describe the distribution arteries and veins in specific body regions. Let us begin with the neck and head.

THE ARTERIES OF THE NECK AND EXTERNAL HEAD

    The Brachiocephalic, Subclavian, and Common Carotid Arteries. Returning to Figure 25-1, note that all of the blood going to the right side of the neck and head leaves the aorta through the **brachiocephalic artery,** which is the first large branch arising from the arch of the aorta. Figure 25-3 shows that as this artery enters the neck it branches into the **right common carotid** and **right subclavian arteries** that in turn supply the right side of the neck and head as well as the upper extremity.

    The **left common carotid** and **left subclavian arteries** arise separately from the arch of the aorta and thereafter run essentially the same courses as the corresponding arteries on the right side.

    The Common Carotid Artery and Its Branches. Figure 25-3 illustrates the common

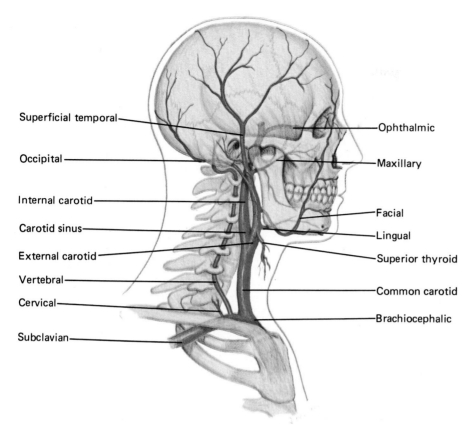

FIGURE 25—3 The arterial supply of the neck and head.

carotid artery and its branches. It has no branches until midneck level where it divides into

1. the **external carotid artery,** which supplies the more superficial parts of the neck and head, and
2. the **internal carotid artery,** which passes through the deeper levels of the neck and enters the cranial vault to supply the brain and the eye.

The *external carotid artery* rapidly divides into numerous branches to supply the anterior neck and entire side of the head, the most important of which are listed here:

1. The **superior thyroid artery** supplies the thyroid gland and adjacent anterior neck structures, including part of the larynx.
2. The **lingual artery** supplies the tongue and the tissues inferior to the chin.
3. The **facial artery** arises from the external carotid artery inferior to the angle of the mandible and then branches over the outer surface of the mandible to supply the anterior chin, the cheeks, and the upper face.
4. The **occipital artery** courses posteriorly beneath the mastoid process of the temporal

bone and thence over the back of the head to supply the posterior scalp.

5. The **maxillary artery** courses deep to the ramus of the mandible and supplies the muscles of mastication, then passes forward to supply both the internal and the external structures of the maxillary region, including the teeth and the nasal sinuses. It also sends an important branch, the **middle meningeal artery,** into the cranial cavity to supply the central portion of the dura mater that covers the brain.
6. The **superficial temporal artery** passes anterior to the ear and divides into multiple branches to supply much of the scalp on the side and crown of the head.

The *internal carotid artery* does not have any branches in the neck region. Instead, it passes deep to the mandible and enters the cranial cavity through an opening in the base of the skull called the *carotid canal.* Soon after entering the cranium it gives rise to the **ophthalmic artery** that runs anteriorly into the orbit to supply blood to the eye and its associated structures. However, by far the greater proportion of the blood flowing through the internal carotid artery goes to the brain, as we shall discuss later.

Note especially in Figure 25–3 that the initial portion of the internal carotid artery is slightly enlarged. This enlarged area is called the **carotid sinus.** In its wall are special nerve receptors, called **baroreceptors,** that detect the degree of stretch of the sinus, which is a measure of the pressure in the artery. The nerve fibers from these baroreceptors signal this pressure information to the brain, which in turn uses the information to help control the blood pressure.

Neck Arteries Arising from the Subclavian Artery—the Vertebral Artery. Figure 25–3 shows several small **cervical arteries** arising from the neck portion of the subclavian artery to supply the lower neck and scapular regions. However, an especially important artery arising from this part of the subclavian is the vertebral artery.

The **vertebral artery** arises from the subclavian artery within a centimeter or two of its origin. It then courses posteriorly in the neck to the vertebral column and passes cranially along this column through small foramina, called the **transverse foramina,** in the transverse processes of the six upper cervical vertebrae. Along this course, it provides branches to the spinal cord. At its superior end, it passes through the

foramen magnum into the posterior fossa of the cranial cavity along with the spinal cord. Here it serves as one of the major arteries supplying the brain stem, as we shall discuss in the following section.

THE ARTERIAL SUPPLY OF THE BRAIN

Note in Figure 25–4 that all the arteries that supply the brain lie on its surface. Only small terminal arteries penetrate into the brain substance. All the blood supply of the brain comes from two sources:

1. the **internal carotid arteries** and
2. the **vertebral arteries.**

Branches of the Internal Carotid Artery. Each internal carotid artery enters the cranial vault through the carotid canal at the base of the brain on each side of the hypothalamus. Its first branch is the ophthalmic artery that was discussed earlier; it then divides immediately into two major cerebral branches:

1. The **middle cerebral artery,** the larger branch, courses through the lateral fissure of the cerebral hemisphere to supply the entire lateral portion of the hemisphere.

Middle cerebral

Circle of Willis

Cerebellar

Basilar

Cerebellar

CEREBELLUM

FRONTAL LOBE

Anterior cerebral

Internal carotid

PITUITARY GLAND

Posterior communicating

Posterior cerebral

TEMPORAL LOBE

Vertebral

OCCIPITAL LOBE

FIGURE 25–4 The arteries of the brain.

2. The **anterior cerebral artery** runs anteriorly into the longitudinal fissure between the two frontal lobes of the cerebral hemispheres. Then it curves superiorly and posteriorly in this fissure to supply the anterosuperior aspects of the medial surfaces of the cerebral hemispheres.

The Vertebral and Basilar Arteries.
The two vertebral arteries, after entering the cranial cavity through the foramen magnum on the two sides of the spinal cord, curve anteriorly and superiorly along the anterior surface of the medulla and join with each other in the anterior midline to form the **basilar artery.** This artery then runs superiorly along the midline of the pons. At the top of the pons, it divides into two **posterior cerebral arteries** that supply the posteroinferior aspects of the medial surfaces of the two cerebral hemispheres.

Along their courses, the vertebral and basilar arteries give rise to many branches that supply the cerebellum, medulla, pons, mesencephalon, and internal ear and vestibular organs. The most important of these are the three moderate-sized **anterior inferior, posterior inferior,** and **superior cerebellar arteries** and the **internal auditory** or **labyrinthine artery.**

The Circle of Willis.
Note especially the arterial structure that encircles the pituitary gland and hypothalamus. A small **anterior communicating artery** connects the two anterior cerebral arteries, and **posterior communicating arteries** connect the middle cerebral and the posterior cerebral arteries that in turn communicate with each other where they leave the basilar artery. Thus, a complete arterial circle surrounds the pituitary gland and hypothalamus. This is called the **circle of Willis.** This structure obviously provides a safety factor for the blood supply to the brain, because one or more of the four major brain arteries, the two carotids and the two vertebrals, sometimes become occluded and the connections of the circle allow for alternative (or collateral) blood flow. This is especially true in old age when the arteries become arteriosclerotic.

THE VEINS OF THE HEAD AND NECK

The Extracranial Venous System.
Figure 25–5 illustrates the veins of the head and neck. Most of the extracranial veins have the same names and courses as the extracranial arteries, especially the following veins:

1. the **facial veins,**
2. the **superficial temporal vein,**
3. the **occipital vein,**
4. the **thyrocervical vein,** and
5. the **external jugular vein.**

However, the courses of these veins through the neck are somewhat different from those of the arteries. The *superficial temporal vein* and other veins from the crown and lateral surfaces of the scalp drain into the **retromandibular vein** that passes in front of the ear and behind the ramus and angle of the mandible. An anterior division of this vein joins the *facial* and *thyrocervical veins* to form a common trunk that empties into the *internal jugular vein.* Then at the lower extent of the neck the internal jugular vein and the subclavian vein combine to form the brachiocephalic vein that leads to the superior vena cava. The superior end of the internal jugular vein drains the blood from almost the entire brain, as will be discussed later.

Immediately inferior to the ear, the posterior division of the retromandibular vein joins the **posterior auricular vein** to form the **external jugular vein,** which passes inferiorly just underneath the skin. In the lower part of the neck it penetrates deep to join the subclavian vein.

A **vertebral vein** follows the same course as the vertebral artery through the foramina of the vertebral transverse processes, and it also empties into the subclavian vein.

Note especially the many communications between the separate neck veins as well as between the external and internal jugular systems.

The Venous Drainage from the Brain—The Dural Sinuses.
Like the arteries of the brain, the small and medium-sized cerebral veins lie on the surfaces of the cerebral hemispheres parallel to the cerebral arteries. However, the drainage of these veins does not parallel the internal carotid arteries. Instead, they drain into venous sinuses, called dural sinuses, in the dura mater, illustrated by the light-colored vessels inside the cranial vault in Figure 25–5. The cranial dura mater has two layers that separate the intervals to form these sinuses. Some of the important dural sinuses are the following:

1. The **superior sagittal sinus** runs from anterior to posterior overlying the longitudinal fissure between the two cerebral hemispheres.
2. The **inferior sagittal sinus** runs from anterior to posterior deep in the longitudinal fissure and empties into the **straight sinus;** this then continues backward to join the posterior end of the superior sagittal sinus, forming a dilated junction called the **confluence of the sinuses.**

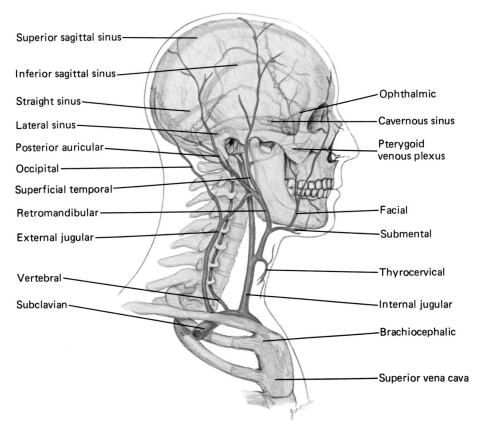

Superior sagittal sinus

Inferior sagittal sinus

Straight sinus

Lateral sinus

Posterior auricular

Occipital

Superficial temporal

Retromandibular

External jugular

Vertebral

Subclavian

Ophthalmic

Cavernous sinus

Pterygoid
venous plexus

Facial

Submental

Thyrocervical

Internal jugular

Brachiocephalic

Superior vena cava

FIGURE 25–5 The venous drainage of the brain, head, and neck.

3. Two **transverse sinuses** originate bilaterally from the confluence of the sinuses. Each of these courses laterally around the lower edge of the occipital lobe of the cerebrum in a groove in the occipital and temporal bones to become the **sigmoid sinus,** which leaves the base of the cranial cavity through the *jugular foramen* in the temporal bone. This is the origin of the large **internal jugular vein** that then courses downward through the neck.

4. Two **cavernous sinuses** are present, one lying on each side of the body of the sphenoid bone behind the eye. This receives venous drainage from structures around the eye in the orbit, from some of the paranasal sinuses, and from the base of the brain. It empties eventually into the dural sinuses in the region of the jugular foramen. A cavernous sinus sometimes becomes the focus of infection feeding backward from the orbit or the paranasal sinuses to the base of the brain; such infections were often lethal prior to the days of antibiotic drugs. Likewise, the transverse sinus sometimes is invaded by infection from the ear that lies in the temporal bone adjacent to the course of this sinus. This, too, was often lethal prior to antibiotics.

THE BLOOD VESSELS OF THE UPPER EXTREMITY

THE ARTERIES

The Subclavian, Axillary, and Brachial Arteries. The upper extremity receives all of its blood supply from the **subclavian artery** as illustrated in Figure 25–6. On leaving the neck, this artery passes underneath the clavicle and after crossing the first rib, enters the axilla where it becomes the **axillary artery.** After passing through the axilla and into the arm, it then is called the **brachial artery.** The brachial artery courses deep within the medial side of the arm and enters the forearm from the anteromedial side of the elbow joint. Immediately below the elbow it divides into a lateral branch, the **radial artery,** and a medial branch, the **ulnar artery,** the two of which continue down the sides of the anterior forearm into the hand.

Smaller Branches of the Subclavian System. In addition to the vertebral and cervical branches that leave the subclavian artery in the neck and that have already been described, the following are additional important branches of the subclavian system:

FIGURE 25–6 The arterial supply of the shoulder and upper extremity.

1. The **internal thoracic artery** originates in the neck from the inferior side of the initial portion of the subclavian artery and then passes inferiorly along the inner surface of the anterior thoracic cage about 2 cm lateral to and parallel to the sternum, giving rise along its course to **anterior intercostal arteries** that course around the chest between the ribs and anastomose with (join) the posterior intercostal arteries from the aorta. Thus, this artery helps to supply the anterior wall of the thoracic cage.
2. The **lateral thoracic artery** arises from the initial portion of the axillary artery and supplies the lateral wall of the thorax, especially the ventral thoracic muscles.
3. The **subscapular artery** arises a little more distally, also from the axillary artery, and then passes underneath the shoulder joint to sup-

ply muscles on the lateral thoracic wall and on both surfaces of the scapula.
4. A number of muscular branches arise from the brachial artery in the upper arm, especially the **profunda brachii** that courses around the back of the humerus.
5. The **common interosseous artery** originates in the proximal part of the forearm from the ulnar artery and its **posterior interosseous branch** then penetrates to the back of the forearm through the interosseous fascia between the radius and ulna.
6. **Superficial** and **deep arterial palmar arches** in the anterior compartment of the wrist and hand (and a similar **dorsal arch** on the dorsum of the hand) form arching networks that give rise to **metatarsal branches.**
7. **Digital arteries** for each finger and thumb arise from the palmar and dorsal metatarsal branches of the hand and pass down the medial and lateral sides of each digit.

Note especially in Figure 25–6 the many collateral arterial branches that encircle the shoulder, elbow, and wrist joints. These provide tremendous safety in preserving arterial blood flow when one of these joints is involved in serious traumatic injury.

THE VEINS OF THE UPPER EXTREMITY

Though there is some similarity between the venous system of the upper extremity and its arterial system, this is only partly true. Instead, the venous system is divided into a *deep system* and a *superficial system.*

The Deep Venous System. The darker veins of Figure 25–7 illustrate the deep venous system of the arm. This is similar to the arterial system, beginning in the **palmar venous arches** of the hand that drain into the **radial** and **ulnar veins** of the forearm, then passing up the extremity to become successively the **brachial, axillary,** and **subclavian veins.** However, note that all of these veins are small in comparison with the veins that constitute the superficial venous system.

The Superficial Venous System. The superficial veins are those that one can see under the skin in the arm, forearm, and back of the hand. Most of the venous blood from the hand and forearm drains into these subcutaneous veins and then flows upward along the superficial system. This system begins in the **dorsal venous arch** of the hand, which is a plexus of superficial veins on the dorsum of the hand.

FIGURE 25–7 The venous drainage of the upper extremity and shoulder.

In the distal forearm, this plexus drains into three major veins:

1. the **cephalic vein,** which lies laterally;
2. the **basilic vein,** which lies medially; and
3. the **medial antebrachial vein,** which courses up the midline of the anterior forearm and then enters the **median cubital vein** anterior to the elbow joint.

In the arm, all the superficial veins, including the median cubital vein, coalesce mainly into the cephalic and basilic veins. The basilic vein then enters the deep system above the elbow, and the cephalic vein enters the axillary vein in the axilla.

Note the extreme numbers of anastomosing channels among the many veins of the forearm. Following injury, blood frequently clots in some of the veins, but the other veins can almost

always serve equally as well, so that this is rarely a problem.

THE ABDOMINAL AORTA, THE INFERIOR VENA CAVA, AND THE ABDOMINAL VASCULATURE

BRANCHES OF THE ABDOMINAL AORTA

The abdominal aorta, which is illustrated in Figure 25–1, gives rise to several large branches that, under resting conditions, carry about one half of all the arterial blood flow. These branches are as follows:

1. The very large **celiac artery** arises anteriorly from the aorta a few centimeters below the dome of the diaphragm. This artery divides immediately into smaller arteries that supply the lower esophagus, the liver, the spleen, the stomach, the pancreas, and the initial part of the small intestine.
2. The **superior mesenteric artery** also arises from the anterior surface of the aorta, about 2 cm below the origin of the celiac artery. This vessel supplies the remainder of the small intestine not supplied by the celiac artery, as well as the first two thirds of the colon.
3 and 4. The **right** and **left renal arteries** arise from the two sides of the aorta about 1 cm below the superior mesenteric artery. These two arteries supply blood to the kidneys.
5 and 6. The **gonadal arteries** arise just inferior to the origins of the renal arteries and supply the ovaries or testes.
7. The **inferior mesenteric artery** arises from the anterior surface of the aorta about 7 cm below the two renal arteries. This artery supplies the last third of the colon, the sigmoid colon, and part of the rectum.
8. A series of four pairs of **lumbar arteries** supply the posterior abdominal wall.
9 and 10. The **right** and **left common iliac arteries.** Just before the abdominal aorta reaches the brim of the pelvis, it divides into the two common iliac arteries. The point of division is called the *bifurcation of the aorta.* Each common iliac artery then divides at the pelvic brim into the **external iliac artery,** which courses inferiorly to supply the thigh, leg, and foot, and the **internal iliac artery,** which passes along the posterior wall of the pelvis to supply the pelvic and adjacent structures, including the lower rectum, urinary bladder, uterus, and other male or female internal and external sexual organs.

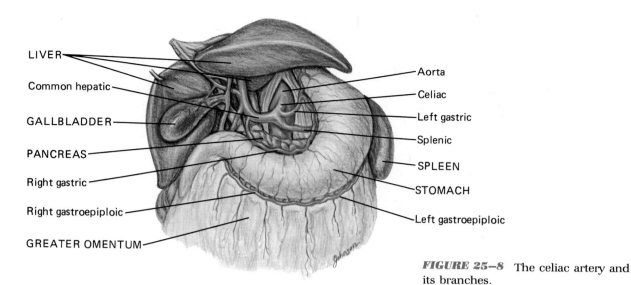

LIVER

Common hepatic

GALLBLADDER

PANCREAS

Right gastric

Right gastroepiploic

GREATER OMENTUM

Aorta

Celiac

Left gastric

Splenic

SPLEEN

STOMACH

Left gastroepiploic

FIGURE 25–8 The celiac artery and its branches.

Branches of the Celiac Artery. The celiac artery, shown in Figure 25–8, divides almost immediately into three major branches:

1. The **common hepatic artery** runs only a short distance to the right and superiorly to the inferior surface of the liver to supply it with arterial blood.
2. The **splenic artery** passes far to the left side of the abdominal cavity deep to the stomach to supply blood to the spleen.
3. The **left gastric artery** supplies blood to the left end of the lesser curvature of the stomach as well as to the most inferior portion of the esophagus.

The stomach also receives arterial blood (in addition to that from the left gastric artery) from the following three arteries:

1. The **right gastric artery** arises from the first part of the common hepatic artery, then courses above the pylorus of the stomach and along the stomach's lesser curvature to join the left gastric artery.
2. The **right gastroepiploic artery** also arises from the common hepatic artery but passes inferior to the pylorus to the greater curvature of the stomach, then courses to the left along the greater curvature to supply blood to this area.
3. The **left gastroepiploic artery** arises near the distal end of the splenic artery, then courses to the right along the greater curvature of the stomach to join the right gastroepiploic artery, thus sharing with this artery the blood supply to the lower half of the stomach.

In addition, the **superior pancreatico-duodenal artery** from the common hepatic artery supplies blood to the pancreas and first half of the duodenum.

The Superior Mesenteric Artery. The superior mesenteric artery supplies blood to almost all of the small intestine and to about two thirds of the colon. This artery, illustrated in Figure 25–9, arises from the anterior surface of the abdominal aorta about 2 cm inferior to the origin of the celiac artery, then passes deep to the pancreas and stomach and courses into the mesentery, giving rise to an array of arches that supply blood to the various intestinal areas as illustrated in the figure. Some of the larger branches are the following:

1. The **inferior pancreaticoduodenal artery** arises from the initial portion of the superior mesenteric artery and supplies parts of the pancreas and duodenum, sharing this with the branch from the common hepatic artery as already explained.
2. The **jejunal arteries** supply the jejunum (which is the second portion of the small intestine after the duodenum).
3. The **ileal arteries** supply the ileum (which constitutes the distal two thirds of the small intestine).
4. The **ileocolic artery** supplies the distal end of the ileum, the cecum, the appendix, and the lower part of the ascending colon.
5. The **right colic artery** supplies the upper part of the ascending colon.
6. The **middle colic artery** supplies the transverse colon.

The Inferior Mesenteric Artery. The inferior mesenteric artery, illustrated in Figure 25–10, is only one-half as large as the superior mesenteric artery. It arises from the midportion

TRANSVERSE COLON

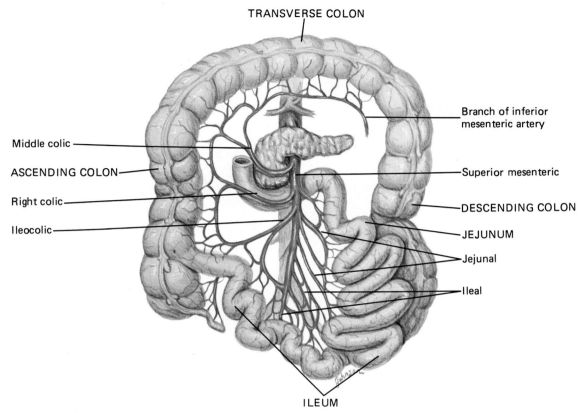

Middle colic

ASCENDING COLON

Right colic

Ileocolic

Branch of inferior
mesenteric artery

Superior mesenteric

DESCENDING COLON

JEJUNUM

Jejunal

Ileal

ILEUM

FIGURE 25–9   The superior mesenteric artery and its branches.

TRANSVERSE COLON

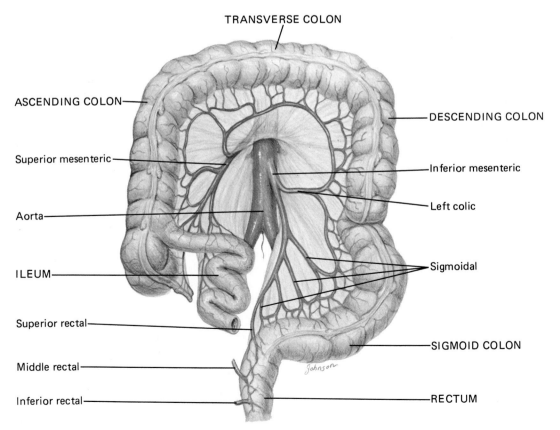

ASCENDING COLON

Superior mesenteric

Aorta

ILEUM

Superior rectal

Middle rectal

Inferior rectal

DESCENDING COLON

Inferior mesenteric

Left colic

Sigmoidal

SIGMOID COLON

RECTUM

FIGURE 25–10   The inferior mesenteric artery and its branches.

of the abdominal aorta. The following are its principal branches:

1. The **left colic artery** supplies the descending colon: an upper branch of this artery arches upward to anastomose with the middle colic artery from the superior mesenteric artery.
2. The **sigmoidal arteries** supply the sigmoid colon.
3. The **superior rectal artery** supplies the upper end of the rectum. A branch of the superior rectal artery anastomoses with the middle and inferior rectal arteries that also supply the rectum and that originate from the internal iliac and internal pudendal arteries in the pelvis and perineum, respectively.

Note especially in Figures 25–9 and 25–10 the tremendous numbers of arching branches that interconnect the arteries supplying all areas of the gastrointestinal tract. These arches provide an extreme safety factor for delivery of blood and nutrition in case of injury, including such injuries as gunshot wounds, stab wounds of the abdomen, crushing wounds that subsequently lead to clotting of major arteries, and so forth.

TRIBUTARIES OF THE INFERIOR VENA CAVA

The inferior vena cava corresponds to the abdominal aorta, but its tributaries vary considerably from the branches of the abdominal aorta as follows:

1 and 2. Two **hepatic veins** drain blood from the two sides of the liver. Most of the blood that flows from the liver into the hepatic veins comes into the liver from the gastrointestinal tract through the portal venous system. This will be described later.

3 and 4. The **right** and **left renal veins** collect blood from the two kidneys. These enter the inferior vena cava from its two sides about midway in the abdomen.

5. The **right gonadal vein. The left gonadal vein** drains into the left renal vein.

6 and 7. The **right** and **left common iliac veins** correspond to the two common iliac arteries. These two veins join just above the brim of the pelvis and just below the aortic bifurcation to form the lower end of the inferior vena cava.

THE VENOUS DRAINAGE OF THE GASTROINTESTINAL TRACT— THE PORTAL SYSTEM

The venous drainage from the gastrointestinal system and spleen is different from the arterial supply in one very important way: The venous drainage does not pass directly from these organs back into the inferior vena cava. Instead, it passes first through a so-called **venous portal system,** then through a great array of venous sinusoids within the liver, and finally from the liver into the inferior vena cava through the *hepatic veins.* This total system is illustrated in Figure 25–11.

The Portal Vein and Its Tributaries. The major venous trunk that carries blood from the portal venous system into the liver is called simply the **portal vein.** Its tributaries mainly parallel the arteries that supply blood to the stomach, the spleen, and the intestines. These veins are as follows:

1. The **superior mesenteric vein** collects blood from most of the small intestine and from two thirds of the colon.
2. The **inferior mesenteric veins** collect blood from the descending colon, the sigmoid colon, and the upper portion of the rectum.
3. The **left** and **right gastric veins** derive from the lesser curvature of the stomach, and the **left** and **right gastroepiploic veins** from the greater curvature of the stomach.
4. The **splenic vein** comes from the spleen.
5. Several small veins come from the pancreas and the duodenum.

Liver Blood Flow and the Hepatic Veins. Approximately three quarters of the blood flowing into the liver comes from the portal vein and about one quarter from the hepatic artery. The hepatic artery supplies mainly the structural portions of the liver such as the fibrous septa and the bile ducts, whereas the portal vein blood flows through large liver **venous sinusoids** that lie between plates of **hepatic cells** (or **liver cells**) that play a special role in processing the portal venous blood arriving from the gastrointestinal tract. On entering the liver this blood contains large amounts of nutrients such as glucose and amino acids that have been absorbed from the gastrointestinal tract following digestion of food. A major share of these is absorbed from the portal blood into the hepatic cells and processed into other substances that are stored and then returned as needed to the blood and carried to the rest of the body. The blood from the gastrointestinal tract contains some debris as well, including bacteria. In the liver sinusoids are also special macrophage cells attached to their walls, called **Kupffer cells,** that phagocytize this debris before allowing the blood to pass back into the remainder of the systemic circulation. Finally, the blood from the

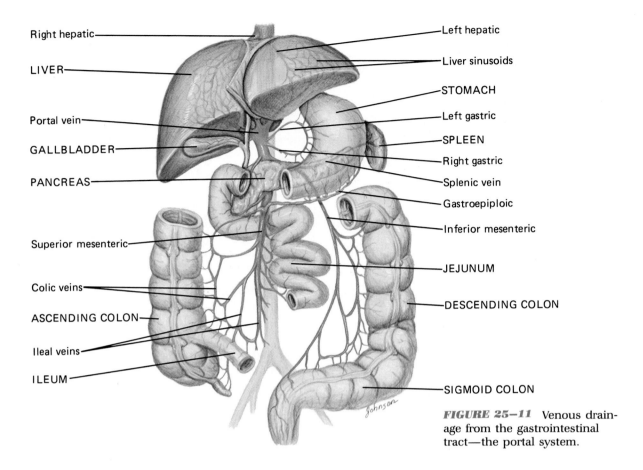

Right hepatic

LIVER

Portal vein

GALLBLADDER

PANCREAS

Superior mesenteric

Colic veins

ASCENDING COLON

Ileal veins

ILEUM

Left hepatic

Liver sinusoids

STOMACH

Left gastric

SPLEEN

Right gastric

Splenic vein

Gastroepiploic

Inferior mesenteric

JEJUNUM

DESCENDING COLON

SIGMOID COLON

FIGURE 25–11 Venous drainage from the gastrointestinal tract—the portal system.

hepatic artery also flows into and through the liver sinusoids, some of it directly and some after it has delivered nutrients to the structural elements of the liver.

After passing through the liver sinusoids, the blood is collected into two major hepatic veins:

1. the **right hepatic vein** from the right side of the liver and
2. the **left hepatic vein** from the left side.

These two veins then enter the inferior vena cava on its two anterolateral sides.

THE VASCULAR SYSTEMS OF THE KIDNEYS, ADRENAL GLANDS, AND GONADS

The Kidney Vessels. The two kidneys lie against the posterior wall of the abdominal cavity at the sides of the vertebral column as illustrated in Figure 25–12. Both the abdominal aorta and the inferior vena cava pass between the two kidneys. This figure also shows that the two renal arteries arise from the sides of the aorta just inferior to the origins of the celiac and superior mesenteric arteries.

The **renal artery** enters the kidney through the area of the kidney called the **hilum.**

It is also through the hilum that the **renal vein** leaves the kidney. The **ureter** that collects urine from the kidney also begins as a continuation of the *renal pelvis* in the hilum and passes down the posterior abdominal and pelvic walls to the urinary bladder.

A principal point to observe in Figure 25–12 is that the renal veins do not empty into the portal system. Instead, they empty directly into the inferior vena cava. Approximately 22 percent of all of the blood pumped by the heart under normal resting conditions passes through the two kidneys. In the kidneys this blood is cleansed of waste products that are eliminated in the urine.

The Vessels Supplying the Adrenal Glands, Testicles, and Ovaries. The two adrenal glands are supplied with arterial blood through small **adrenal branches** from the aorta, renal arteries, and phrenic arteries, and the venous drainage from each gland passes through a single **adrenal vein,** that on the left empties into the left renal vein and on the right into the inferior vena cava.

Most of the blood supply to the gonads— the testicles or the ovaries—originates by way of the **testicular** or **ovarian arteries** from the anterolateral aspects of the aorta. One might won-

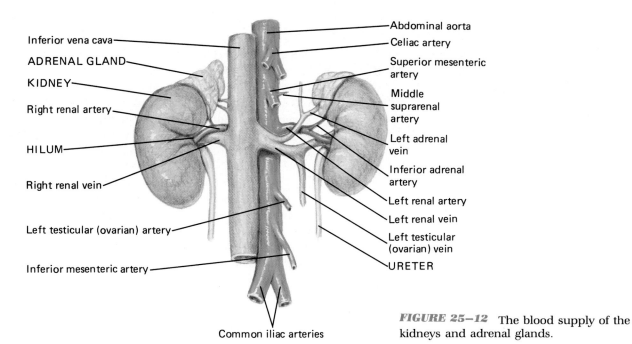

Inferior vena cava

ADRENAL GLAND

KIDNEY

Right renal artery

HILUM

Right renal vein

Left testicular (ovarian) artery

Inferior mesenteric artery

Common iliac arteries

Abdominal aorta

Celiac artery

Superior mesenteric artery

Middle suprarenal artery

Left adrenal vein

Inferior adrenal artery

Left renal artery

Left renal vein

Left testicular (ovarian) vein

URETER

FIGURE 25–12 The blood supply of the kidneys and adrenal glands.

der why these organs receive their blood supply from so high up in the arterial tree, far above the level of the gonads themselves. The reason is that during embryonic development, these organs originate in this part of the abdomen and later migrate into the pelvis in the case of the ovaries or into the scrotum in the case of the testicles. The **left gonadal vein** drains into the left renal vein and the **right gonadal vein** into the inferior vena cava.

THE VASCULATURE OF THE PELVIS AND LOWER EXTREMITIES

THE ARTERIES

The Common Iliac Artery and Its Major Branches. A few centimeters above the brim of the pelvis, the aorta divides into the two **common iliac arteries** that then supply the respective sides of the pelvis and the lower extremities as shown in Figure 25–13. The common iliac artery divides into the internal and external iliac arteries.

1. The **internal iliac artery** supplies the pelvic walls and viscera, the gluteal region, the perineum, and a small portion of the lower extremity.
2. After passing under the inguinal ligament and into the anteromedial thigh, the **external iliac artery** becomes the **femoral artery.** This descends in the anterior thigh and curves around to the back of the femur in the lower deep medial thigh muscles and then

passes behind the knee joint. Where this artery passes behind the knee joint, it is called the **popliteal artery.**

Immediately below the knee joint, the popliteal artery divides into

1. the **anterior tibial artery,** which immediately penetrates the interosseous fascia between the tibia and the fibula to emerge anteriorly and run down the anterior leg, and
2. the **posterior tibial artery,** which runs down the posteromedial border of the leg. About 10 cm below the knee joint, the **peroneal artery** arises from the posterior tibial artery and passes down the posterolateral border of the leg.

Branches of the Internal Iliac Artery. As the internal iliac artery passes posteriorly into the pelvis, within a distance of about 12 cm it gives rise to multiple arteries, some of the more important of which are the following:

1. The **iliolumbar artery** supplies the inside lateral wall of the pelvis.
2. The **superior** and **inferior gluteal** arteries pass posteriorly between the sacrum and the ilium into the buttocks and posterolateral thigh to supply the gluteal and other muscles in these regions.
3. The **obturator artery** passes through the obturator foramen into the superior portion of the posteromedial thigh, mainly supplying muscles in these areas.
4. The **uterine artery** (in the female) supplies

FIGURE 25-13 The arterial supply of the hip and lower limb.

the uterus, fallopian tube, and ovary (shared with the gonadal artery from the aorta).

5. The **superior** and **inferior vesical arteries** supply the urinary bladder.

6. The **internal pudendal artery** supplies mainly the structures of the perineum and the external genitalia. This artery is much larger in the male than in the female because it carries all the blood for the penis and testicles.

Major Branches from the External Iliac, Femoral, Anterior Tibial, Posterior Tibial, and Peroneal Arteries. Immediately superior to the inguinal ligament, the **inferior epigastric artery** leaves the external iliac and passes into the anterior abdominal wall, then ascends cranially behind the rectus abdominis muscle, finally anastomosing with the internal thoracic artery at the lower margin of the thoracic cage.

From the femoral artery, large numbers of muscular branches supply the vast musculature of the thigh. The most important deep branch is the **profunda femoris** that courses posteriorly and laterally to supply the posterior and lateral thigh muscles.

The distal end of the anterior tibial artery passes onto the dorsum of the foot where it becomes the **dorsalis pedis artery.** This artery is important because it lies immediately under the skin, and one can easily feel a pulse in it whenever the blood supply to the leg is good. When the pulse is absent, one fears for the nutritional status of the leg.

Finally, the inferior ends of the anterior tibial, posterior tibial, and peroneal arteries form arterial arches in the foot, a **dorsal arch** on the dorsum of the foot and a **plantar arch** on the bottom of the foot. From these arches, **digital arteries** pass into both the lateral and the medial sides of the toes.

Finally, note the extensive collateral vessels around the hip, the knee, and the ankle. These vessels, as is true for the upper extremities, provide an extreme measure of safety for the blood supply in case of joint injuries that destroy or occlude major arterial vessels.

THE VENOUS SYSTEM OF THE PELVIS AND LOWER EXTREMITY

The lower extremity, like the upper extremity, has both a deep venous system and a superficial system. And, as is also true in the upper extremity, the superficial system, illustrated in Figure 25-14B, is more highly developed than the deep system, illustrated in Figure 25-14A.

The Deep Venous System. Note in Figure 25-14A that the deep veins of the foot and lower leg are all small. It is only in the upper leg and thigh where many muscular veins enter this deep system that the vessels reach large size. The names of the veins in the deep system are similar in most instances to those of the arteries, and for the most part they accompany the arteries. Some of the larger veins are

1. the **anterior** and **posterior tibial veins,**

External iliac

Femoral

Common iliac

Internal iliac

Great saphenous (cut)

Popliteal

Posterior tibial

Anterior tibial

Dorsal venous arch

Femoral

Great saphenous

Great saphenous

Superficial dorsal arch

A

B

FIGURE 25–14 (A) The deep venous system of the lower limb. (B) The superficial venous system of the lower limb.

2. the **popliteal vein,**
3. the **femoral vein,**
4. the **internal** and **external iliac veins,** and
5. the **common iliac vein,** which joins the common iliac vein from the opposite side to form the inferior vena cava immediately above the brim of the pelvis.

The Superficial Venous System. Figure 25–14B illustrates the anteromedial portion of the superficial venous system, which consists of an extensive plexus of veins lying underneath the skin of the entire extremity. These veins drain mainly into the **great saphenous vein,** which begins along the medial border of the foot, then passes superiorly along the antero-medial border of the leg, next along the medial side of the knee and up the anteromedial thigh, finally emptying into the femoral vein about 3 cm below the inguinal ligament.

In addition to the great saphenous venous system there is a smaller drainage system, not shown in Figure 25–14, the **small saphenous vein** and its tributaries from the posterolateral leg. This vein empties into the deep popliteal vein at the back of the knee.

THE VENOUS VALVES

Many of the veins of the body have **valves** inside them to prevent backflow of blood, which is not true of the arteries. Figure 25–15 shows a large vein of the lower extremity and one of its tributaries, both split open and flattened to show typ-

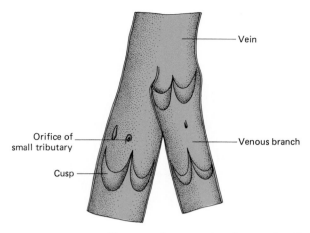

FIGURE 25–15 Venous valves seen in a large vein of the lower extremity and one of its tributaries after opening and flattening the vessels.

ical venous valves. Each valve has two cuplike leaflets that lie opposite each other and are called **cusps,** the same term used for the leaflets of the heart valves as discussed in Chapter 23. When blood attempts to flow backward, the cups of the two cusps fill with blood, causing their edges to balloon together in the middle of the vein and thus block the backflow.

Valves are not present in the venae cavae nor in the veins draining blood from the head. But they are found about every 3 to 4 cm in most veins of the upper and lower extremities where blood must flow uphill or against gravity. They not only prevent backflow of blood but also help to propel the blood toward the heart in the following way: Every time a muscle contracts or a joint is bent so that the tissues compress a vein, blood is squeezed forward along the vein because the valves prevent the blood from ever moving backward. Thus, progressive propulsion of the blood toward the heart occurs. This is called the *venous pump*, the function of which will be discussed in Chapter 29.

Varicose Veins. The superficial veins of the lower extremity are often affected in the clinical condition called *varicose veins*. In this condition, the veins become extremely large, on rare occasions so large that in the aggregate they may hold as much as a liter of blood. The cause of this condition is failure of the venous valves to function properly. As long as the valves prevent backflow of blood down the superficial veins, each movement of the limb will pump blood upward along the leg and thigh and finally into the deep venous system. But when the valves become incompetent (nonfunctional), the blood pumped upward with each leg movement flows back down between movements. As a conse-

quence, the venous pressure becomes many times as great as normally, and the veins often distend to extreme sizes. Also, the high back-pressure in the peripheral veins blocks much of the blood flow from the skin and deep tissues, so that the leg often becomes severely edematous and the blood flow badly compromised. The final result is occasionally serious ulceration and sometimes even gangrene of the skin and deeper tissues.

VENOUS THROMBOSIS

Another clinical condition that frequently occurs in the veins of the lower extremity is *venous thrombosis*, which means clotting of the blood in the veins. The clot sometimes becomes infected as well, which makes the clotted veins extremely painful. And the clotting often leads to painful swelling of the limb because of inadequate drainage. This condition occurs very commonly in older persons, and also in pregnant women because of compression of the veins from the legs where they pass through the pelvic region posterior to the enlarged uterus.

Venous thrombosis occurs in many patients in hospitals whose legs are propped over pillows because this causes stasis and clotting of the venous blood. The clots then frequently break away from the leg veins and flow all the way to the lungs where they block lung blood flow, a condition called *pulmonary embolism* that often results in death.

THE CAPILLARY SYSTEM

The capillaries are the very minute blood vessels that connect the arterial system to the venous system. It is through the walls of these that nutrients pass to the cells and excreta from the cells pass back into the blood. This function of the capillaries will be discussed later in detail. Yet, here we need to fit the capillaries anatomically into the pattern of the circulating blood.

Figure 25–16 illustrates a composite view of the total functional circulatory system, showing both the pulmonary circulation and the systemic circulation. Note that the blood is bright red in the systemic arteries (because oxygenated blood is red) but then becomes a dusky bluish-red color in the systemic veins after giving up its oxygen to the tissues in the capillaries. When the blood passes through the lung capillaries, it becomes oxygenated again and returns to bright red.

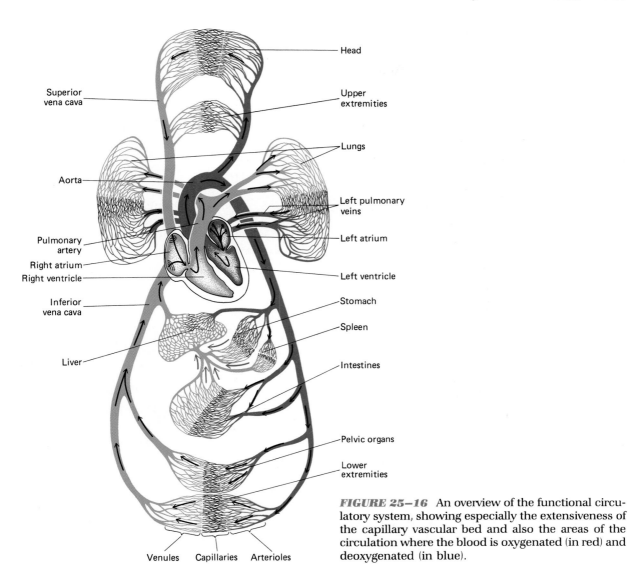

FIGURE 25–16 An overview of the functional circulatory system, showing especially the extensiveness of the capillary vascular bed and also the areas of the circulation where the blood is oxygenated (in red) and deoxygenated (in blue).

Some Physical Attributes of the Capillaries. There are about 10 billion capillaries in the body. The average diameter of the capillaries is 5 to 8 microns (μ), barely large enough for the blood cells to pass through. In general, their lengths are less than 1 mm. Yet, because of their tremendous number, their total surface area in the lungs alone is equal to about half the surface area of a tennis court, and their total surface area in the remainder of the body may approach the surface area of a football field.

Yet, despite the tremendous surface area of the aggregate capillary system, the individual capillaries are so small and so short that no more than 5 to 10 percent of the total blood volume is in the capillaries at any one time. When blood passes through them it normally remains there no longer than a few seconds at most. It is during this extremely short time that all the transfer of oxygen and other substances occurs.

STRUCTURES OF THE VASCULAR WALLS

Vascular walls come in all sizes, from the aorta and venae cavae down to the very small capillaries. The microscopic structures of the walls of different-sized vessels are illustrated in Figure 25–17. Let us begin with the simplest, the capillary.

Structure of the Capillary Wall. The capillary wall, illustrated in Figure 25–17C, is only a single cell layer thick, no more than 1 μm in thickness. It is composed of flat **endothelial cells.** In the middle of each of these cells is a **cell nucleus** that provides metabolic factors essential for maintaining the living processes in the remainder of the cell. Between the edges of the cells, where they attach to each other, small slits pass all the way through the capillary wall. These slits are the **capillary pores,** and it is

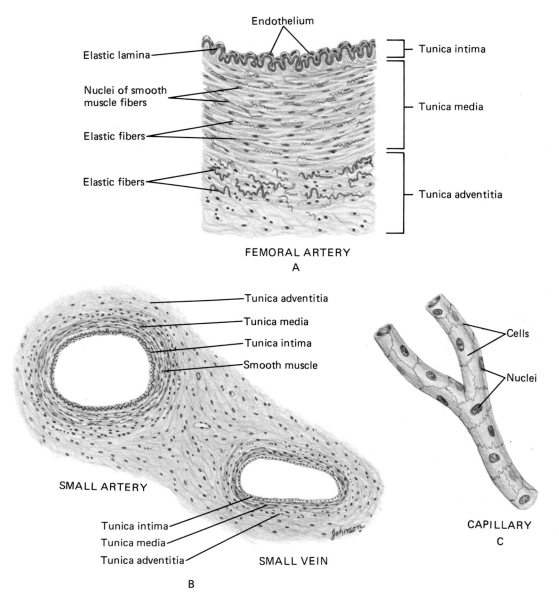

FIGURE 25–17 The wall structures of blood vessels: (A) A large artery (the femoral artery). (B) A small artery and a small vein each approximately 2 mm in diameter. (C) A capillary.

through them that many substances pass between the blood and the tissue fluids.

The inside surface of the capillary, where the endothelial cells are exposed to the blood, is extremely smooth. This is an important characteristic because it prevents blood clotting.

Now, to put the capillary into perspective, place literally billions of these in an interconnected network carrying blood from the arteries to the veins. Then visualize trillions of tissue cells lying among the capillaries. Rarely is a tissue cell more than 30 μm away from the nearest capillary, a distance about equal to four diameters of a red blood cell. Therefore, nutrients and cell excreta can diffuse with ease between the tissue cells and the blood in the capillaries.

The Walls of the Small Arteries and Veins. As the blood vessels become larger, they are no longer simple endothelial cell tubes. Yet, their inner surfaces are still lined with single-layered *endothelial cells* almost exactly like the endothelial cells of the capillary. But, surrounding these endothelial cells are layers of smooth muscle fibers and fibrous tissue that provide structural strength to these larger vessels.

Figure 25–17B illustrates a small artery approximately 2 mm in diameter and a similarly sized vein. Each of these vessels has three layers:

1. The **tunica intima.** This is a thin layer, the inner surface of which is lined with endothelial cells where it is exposed to the blood. Out-

side these endothelial cells is a thin coat of fibrous tissue containing two types of fibers: (a) *collagenous fibers* similar to the collagenous fibers in muscle tendons and in bone and (b) *elastic fibers* that are elastic, like the elasticity of rubber. The layer of elastic fibers in the tunica intima is called the **elastic lamina.**

2. The **tunica media.** This is comprised almost entirely of *smooth muscle fibers*, but it contains small amounts of *collagenous fibers* as well. When the smooth muscle is stimulated by the sympathetic nervous system, the muscle contracts and the diameter of the vessel becomes smaller, a process called vasoconstriction. It is in this way that the nervous system controls the blood flow through the different tissues.

3. The **tunica adventitia.** This layer consists of fibrous tissue containing mainly *collagenous fibers* but some *elastic fibers* as well.

Note in Figure 25–17B that the wall of the small artery is several times as thick as that of the small vein, the artery having an especially large amount of smooth muscle in the tunica media. This allows the artery to withstand the distending force caused by the high arterial pressure, in contrast to the low pressure in the vein.

The Walls of the Large Veins and Arteries. The walls of the large veins, such as the venae cavae and the major veins of the limbs, are not greatly different from the walls of the small veins except that they are thicker and stronger. They have a tunica intima, a tunica media, and a tunica adventitia the same as the small veins.

On the other hand, the large arteries, such as the aorta and its major arterial branches, have very thick walls, illustrated in Figure 25–17A, that are different from the walls of the smaller arteries in important respects: The *tunica media* is prominent and is composed of many concentric layers of *smooth muscle fibers* interspersed with *intervening elastic layers*. Also, the *tunica adventitia* of the large arteries has an abundant supply of *elastic fibers*. The importance of the elastic fibers is the following: The large arteries are subjected to pulsatile injection of blood each time the heart beats. Therefore, these arteries must distend and contract billions of times in a person's lifetime. The elastic fibers allow this pulsatile resilience.

Finally, to put the large arteries into perspective, the aorta is about the same size as a small garden hose, and its wall also has about the same thickness and strength as a cheap plas-

tic hose. But, fortunately, the pressure inside the aorta is usually only about one-twelfth the pressure in the garden hose. Even so, the fibrous tissue in the aorta sometimes degenerates in older age, and the pressure then causes it to distend to several times its normal size. This is the condition called *aortic aneurysm*. When this occurs the aorta can easily rupture and cause death even at the pressure level at which the arterial system normally operates. But, fortunately, the surgeon can replace the diseased aorta with an artificial dacron aorta that usually serves well for the remainder of the person's life.

QUESTIONS

1. Describe the arch of aorta and its branches.
2. Describe the entry of the superior and inferior venae cavae into the right atrium, and also list the major veins that empty into the venae cavae.
3. Give the blood supply of the neck, listing the principal arteries and veins.
4. Describe the arterial supply of the brain.
5. How does the venous system draining blood from the brain differ from the brain's arterial system?
6. Trace the arteries of the upper extremity, giving the branches to the different parts of the extremity.
7. Discuss the differences between the deep and the superficial venous systems of the upper extremity.
8. Describe the arterial supply to the different abdominal structures.
9. Give the distinguishing features of the portal system.
10. Describe the blood supply of the kidneys.
11. Trace the arterial supply, including the branches, to the pelvic organs and lower extremity.
12. Contrast the deep and the superficial venous systems of the lower extremity.
13. Contrast the structure of the capillary with the structures of the intermediate and large blood vessels.
14. Describe the layers of the intermediate and large blood vessel walls.

REFERENCES

Bloom, W., and Fawcett, D.W.: Blood and lymph systems. *In* A Textbook of Histology, 10th Ed. Philadelphia, W.B. Saunders, 1975, pp. 386–499.

Copenhaver, W.M., Kelly, D.E., and Wood, R.L.: The circulatory system. *In* Bailey's Textbook of Histology, 17th Ed. Baltimore, Williams & Wilkins, 1978, pp. 358–421.

Figge, F.H.J.: Peripheral vessels. *In* Sobotta/Figge Atlas of Human Anatomy. Vol. III. Baltimore, Urban & Schwarzenberg, 1977, pp. 199–337.

Figge, F.H.J.: The vascular trunks. *In* Sobotta/Figge Atlas of Human Anatomy. Vol. II. Baltimore, Urban & Schwarzenberg, 1977, pp. 27–59.

Fujita, T., Tanaka, K., and Tokunaga, J.: The circulatory system. *In* SEM Atlas of Cells and Tissues. New York, Igaku-Shoin, 1981, pp. 46–99.

Goss, C.M.: The vascular system. *In* Gray's Anatomy of the Human Body. Philadelphia, Lea & Febiger, 1973, pp. 561–767.

Ham, A.W., and Cormack, D.H.: The circulatory system. *In* Histology, 8th Ed. Philadelphia, Lippincott, 1979, pp. 581–613.

Hammersen, F.: Blood and vascular system. *In* Sobotta/Hammersen Histology. Baltimore, Urban & Schwarzenberg, 1980, pp. 94–113.

Leeson, T.S., and Leeson, C.R.: The circulatory system. *In* A Brief Atlas of Histology. Philadelphia, W.B. Saunders, 1979, pp. 105–132.

Leeson, T.S., and Leeson, C.R.: The circulatory system. *In* Histology, 4th Ed. Philadelphia, W.B. Saunders, 1981, pp. 257–307.

Weiss, L., and Greep, R.O.: The cardiovascular and lymph systems. *In* Histology, 4th Ed. New York, McGraw-Hill, 1977.

Wheater, P.R., Burkitt, H.G., and Daniels, V.G.: Circulatory system. *In* Functional Histology. New York, Churchill Livingstone, 1979, pp. 76–86.

Williams, P.L., and Warwick, R.: The vascular system. *In* Gray's Anatomy, 36th British Edition. Philadelphia, W.B. Saunders, 1980, pp. 622–801.

(See also General Anatomical References, Chapter 1.)

26

Blood Flow Through the Systemic Circulation and Its Regulation

Overview

Blood flow through the systemic circulation is caused by *pressure* in the arteries, and the rate at which the blood flows each minute is determined by the total *resistance*, known as the *total peripheral resistance*, in all the different systemic blood vessels. The following formula gives the relationship between blood flow, pressure, and resistance:

$$\text{Blood Flow} = \frac{\text{Pressure}}{\text{Resistance}}$$

The pressure in the arteries is caused by the *pumping of blood into the aorta from the left ventricle.* The resistance to blood flow through the systemic circulation is caused by *friction of the blood* as it slides over the surfaces of the vascular walls. Most of the resistance is in the very small blood vessels, especially in the *arterioles* and *capillaries,* because *resistance to flow is inversely proportional to the fourth power of the vessel diameter.* Thus, the resistance of a blood vessel of 1 mm in diameter is 16 times as great as the resistance of a vessel 2 mm in diameter.

The *normal mean blood pressure* in the aorta is about *100 mm Hg,* and the total blood flow through the entire systemic circulation, called the *cardiac output,* is about 5 liters per minute (L/min) under resting conditions. However, even in the average person, the cardiac output can increase to as much as 15 to 20 L/min during heavy exercise and in some well-trained athletes to twice these values.

Blood flow through the tissue is controlled mainly by a mechanism called *local autoregulation of blood flow;* this means control principally by the rate at which nutrients are used in the different tissues. For instance, when a tissue uses excess amounts of *oxygen,* the local blood vessels dilate and allow increased blood flow to the tissue. The reason for this dilatation is not completely known; it may result from some *vasodilator substance* released by the tissues, especially the vasodilator substance *adenosine* that is released from hypoxic cells. Or it may result from simple lack of enough oxygen to maintain contraction of the blood vessel walls.

In some conditions, the *sympathetic nervous system* also plays a very important role in controlling blood flow. For instance, during heavy exercise, when the muscles require extreme amounts of blood flow, the

451

sympathetic nervous system constricts the blood vessels in almost all the nonmuscular tissues to divert the blood flow to the muscles.

About three quarters of all the *blood volume* in the circulatory system is in the veins. When a person bleeds severely, the veins can constrict, causing adequate amounts of blood still to fill the other vessels of the body. Therefore, the veins, in addition to serving as a conduit for return of blood from the peripheral tissues, are also called the *blood reservoir* of the body.

The blood flowing through the systemic circulation provides nutrition to the tissues, transport of excreta away from the tissues, cleansing of the blood as it passes through the kidneys, absorption of nutrients from the gastrointestinal tract, and mixing of all the fluids of the body, as explained in Chapter 1. The rate of blood flow to each respective tissue is almost exactly the amount required to provide adequate function, no more, no less. The purpose of the present chapter, therefore, is to describe, first, the basic principles of blood flow through the circulation and, second, the mechanisms that control the blood flow to each respective tissue in proportion to its needs.

HEMODYNAMICS

The study of the physical principles that govern blood flow through the vessels and the heart is known as *hemodynamics*. The heart forces blood into the aorta, distending it and creating pressure within it. This pressure then pushes the blood through the arteries, arterioles, capillaries, venules, veins, and finally back to the heart. As long as the animal remains alive, this flow of blood around the continuous circuit never ceases.

The small arteries, arterioles, capillaries, venules, and small veins have such small diameters that blood flows through them with considerable difficulty. In other words, the vessels are said to offer *resistance* to blood flow. Obviously, the smaller the vessel the greater is the resistance, and the larger the vessel the less the resistance.

In essence, the discussion in the present chapter centers on the effects of pressure and resistance on blood flow.

Blood Flow and Cardiac Output

The amount of blood pumped by the heart when a person is at rest is approximately 5 L/min. This is called the *cardiac output*, and it can increase to as much as 25 to 35 L/min during the most extreme exercise in athletes, or it can decrease following severe hemorrhage to as low as 1.5 L/min without causing immediate death, though this will cause death if it lasts for more than an hour or two.

Blood flow to the different parts of the body during rest is given in Table 26–1. One will note that the brain receives approximately 14 percent of the total blood flow, the kidneys 22 percent, the liver 27 percent, and the muscles, which comprise almost half of the body, only 15 percent. However, during exercise quite a different picture develops, for then essentially all of the increase in blood flow occurs in the muscles, this then representing as much as 75 to 80 percent of the total blood flow.

Methods for Measuring Blood Flow. Many different methods have been used for measuring blood flow, most of which require cutting the blood vessel and then allowing the blood to flow through some physical device that measures the rate of flow. However, there is much advantage in using types of flowmeters that do not require opening the vessels. Two such flowmeters are the *electromagnetic flowmeter* and the *Doppler flowmeter*.

TABLE 26–1

*Blood Flow to Different Organs and Tissues Under Basal Conditions**

	Percent	*ml/min*
Brain	14	700
Heart	4	200
Bronchi	2	100
Kidneys	22	1100
Liver	27	1350
Portal	(21)	(1050)
Arterial	(6)	(300)
Muscle	15	750
Bone	5	250
Skin (cool weather)	6	300
Thyroid gland	1	50
Adrenal glands	0.5	25
Other tissues	3.5	175
Total	100.0	5000

*Based mainly on data compiled by Dr. L. A. Sapirstein.

FIGURE 26–1   The electromagnetic flowmeter.

The Electromagnetic Flowmeter. Figure 26–1 illustrates the electromagnetic flowmeter. The electromagnet creates a very strong magnetic field that passes through the blood vessel. As the blood flows through this field, it "cuts" the magnetic lines of force and develops an electrical potential at right angles to these lines of force, and this potential is proportional to the rate of flow. Two electrodes touching the surface of the vessel at right angles to the magnet, therefore, can pick up this potential and transmit it to an appropriate electronic recording apparatus. Obviously, the blood vessel does not have to be opened when this type of flowmeter is used. Also, the response of the instrument is so rapid that it can measure even very transient changes in flow that take place in as little as 0.01 second. Because of these advantages, this type of flowmeter has become very widely used in physiologic studies of the circulation.

The Ultrasonic Doppler Flowmeter. Another type of flowmeter that can be applied to the outside of the vessel and that has many of the same advantages as the electromagnetic flowmeter is the ultrasonic Doppler flowmeter, illustrated in Figure 26–2. A minute piezoelectric crystal is mounted in the wall of the device. This crystal, when energized with an appropriate electronic apparatus, transmits sound of a frequency of several million cycles per second downstream along the flowing blood. A portion of the sound is reflected by the flowing red blood cells, and the reflected sound waves travel backward from the blood toward the crystal. However, these reflected waves have a lower frequency than the transmitted wave because the

red cells are moving away from the transmitter crystal. This is called the Doppler effect. It is the same effect that one experiences when a train approaches a listener and passes by while blowing the whistle. Once the whistle has passed by, the pitch of the sound from the whistle suddenly becomes much lower than when the train is approaching, and the faster the train is moving, the lower the pitch. An electronic apparatus determines the frequency difference between the transmitted wave and the reflected wave, thus determining the velocity of blood flow.

Velocity of Blood Flow in Different Portions of the Circulation. The term *blood flow* means the actual *quantity* of blood flowing through a vessel or group of vessels in a given period of time. In contradistinction to this, the *velocity* of blood flow means the distance that the blood travels along a vessel in a given period of time.

If the quantity of blood flowing through a vessel remains constant, the velocity of blood flow obviously decreases as the size of the vessel increases. The aorta as it leaves the heart has a cross-sectional area of approximately 2.5 square centimeters (cm²). Then it branches into large arteries, small arteries, and capillaries, with a portion of the aortic blood flowing into each of these vessels. The total cross-sectional area of the branching vessels is considerably greater than that of the aorta; in the capillaries, for instance, it is 1000 times that in the aorta. As a consequence, the velocity of blood flow is greatest in the aorta and least in the capillaries, where it is only 1/1000 that in the aorta. Numerically, the velocities are approximately the following: aorta, 30 cm/sec; arterioles, 1.5 cm/sec; capillaries, 0.3 mm/sec; venules, 3 mm/sec; and venae cavae, 8 cm/sec.

Transit Time for Blood in the Capillaries. The velocity of blood flow in the capillaries is particularly significant because it is here that oxygen, other nutrients, and excreta pass back and forth between the blood and the tissue spaces. The length of the average capillary is about 0.5 to 1 mm. Therefore, at a velocity of blood flow of 0.3 mm/sec, the length of time that blood remains in each capillary averages 1 to 3 seconds. Despite this short time, an extremely large proportion of the substances in the blood can transfer through the capillary membrane into or out of the tissues. This signifies the extreme rapidity of nutrient and excreta exchange through the capillary walls between the blood and the tissue fluids. Without this exchange, all the other hemodynamic functions of the circulation would be useless.

FIGURE 26–2   An ultrasonic Doppler flowmeter.

FIGURE 26–3 The historical method for recording arterial pressure using a mercury manometer.

Blood Pressure

The pressure in a blood vessel is the force that the blood exerts against the walls of the vessel. This force distends the vessel because all blood vessels are distensible, the veins eight times as much as the arteries. Pressure also makes blood attempt to leave a vessel by any available opening, which means that the normally high pressure in the arteries forces blood through the small arteries, then through the capillaries, and finally into the veins. The importance of blood pressure, then, is that it is the force that makes the blood flow through the circulation.

Measurement of Blood Pressure—the Mercury Manometer. Historically, the standard device for measuring blood pressure has been the mercury manometer shown in Figure 26–3. A tube is attached to a blood vessel by means of a cannula inserted into the vessel. The blood pressure in the vessel pushes on the fluid in the tube, and this presses downward on the left ˇcolumn of mercury, which then forces the right column upward. A float resting on the mercury rises with the level of the mercury, and a recording arm connected to the float records the pressure on a moving paper. If the level of mercury in the right column is 100 mm above the level in the left column, the blood pressure is said to be 100 millimeters of mercury (mm Hg). If the level to the right is 200 mm above that on the left, then the pressure is 200 mm Hg.

Because of the convenience of the mercury manometer for measuring pressure, almost all pressures of the circulatory system are ex-

pressed in millimeters of mercury. However, pressure can also be expressed as water pressure because one can equally well connect an artery to a column of water and determine how high the water rises. Mercury weighs 13.6 times as much as water, which means that the level of water in a manometer will rise this many times as high as the level of mercury in a manometer. Therefore, 13.6 mm or 1.36 cm of water pressure equals 1 mm Hg pressure.

High Fidelity Methods for Recording Blood Pressure. Unfortunately, the mercury in the mercury manometer has so much *inertia* that it cannot rise and fall rapidly. For this reason, this type of manometer, though excellent for recording *mean* pressure, cannot respond to pressure *changes* that occur more rapidly than one cycle every 2 to 3 seconds. Also, it is a cumbersome apparatus to use. Therefore, in modern physiology a type of pressure recorder such as one of those illustrated in principle in Figure 26–4 is used. Each of the three "transducers" illustrated in this figure converts pressure into an electrical signal that is then recorded on a high-speed recorder. A syringe needle is connected to each transducer; this needle is inserted directly into the vessel whose pressure is to be measured (or the transducer is con-

FIGURE 26–4 Principles of operation of three different electronic transducers for recording rapidly changing blood pressure.

FIGURE 26-5 Effect of pressure on flow.

nected to a catheter that is inserted into the vessel), and the pressure is transmitted into a chamber bounded on one side by a thin membrane. In transducer A this membrane forms one plate of a variable capacitor. As the pressure rises, the membrane moves closer to the other plate and changes the electrical *capacitance*, which is detected and recorded by an electronic instrument. In transducer B a small iron slug on the membrane moves upward into a coil and changes the coil's *inductance*, which likewise is detected and recorded electronically. In transducer C the membrane is connected to a thin, stretched wire. As the membrane moves, the the length of the wire changes, which changes its *resistance*, once again allowing the pressure to be recorded by an electronic recorder.

Relationship of Pressure to Blood Flow. When the blood pressure is high at one end of a vessel and low at the other end, blood will attempt to flow from the high toward the low pressure area because the two forces are unequal. *The rate of blood flow is directly proportional to the difference between the pressures.* Figure 26-5 illustrates this principle, showing that the flow from the spout becomes greater in proportion to the water level in the chamber.

It should be noted that it is not the actual pressure in a vessel that determines how rapidly blood will flow, but the *difference* between the pressures at the two ends of the vessel. For instance, if the pressure at one end of the vessel is 100 mm Hg and 0 mm Hg at the other end, the pressure difference is 100 mm Hg. If the pressure at the two ends of this same vessel are suddenly changed to 5000 mm Hg at one end and 4950 mm Hg at the other end, the pressure difference now is only 50 mm Hg, and the flow will

be reduced to one half as much (provided the vessel's diameter does not become changed by the pressure).

Resistance to Blood Flow

Resistance is essentially the same as friction, for it is friction between the blood and the vessel walls that creates impediment to flow. The amount of resistance is dependent on the length of the vessel, the diameter of the vessel, and the viscosity of the blood.

Effect of Vessel Length on Resistance to Blood Flow. The longer a vessel, the greater the vascular surface along which the blood must flow, and consequently the greater the friction between the blood and vessel wall. For this reason, the *resistance to blood flow is directly proportional to the length of the vessel.*

Effect of Vessel Diameter on Resistance to Blood Flow. Fluid flowing through a vessel is retarded mainly along the walls. For this reason, the velocity of blood flow in the middle of a vessel is very great whereas the velocity along the surface is very low, and the larger the vessel the more rapidly can the central portion of blood flow. Because of this effect, the blood flow through a vessel—that is, the total quantity of blood that passes through the vessel each minute—increases very markedly as the diameter of the vessel increases. In fact, if all other factors stay constant, the blood flow through the vessel is directly proportional to the *fourth power* of the diameter. Thus, in Figure 26-6 three vessels of the same length but with diameters of 1, 2, and 4 are shown. Note that even these small changes in diameter increase the flow 256-fold. Because of this extreme dependence of flow on diameter, very slight changes in vascular diameter can affect blood flow to different vascular regions tremendously. The diameter of most vessels can change approximately 4-fold, and can therefore change the blood flow as much as 256-fold.

Effect of Viscosity on Resistance to Blood Flow. The more viscous the fluid attempting to flow through a vessel, the greater the

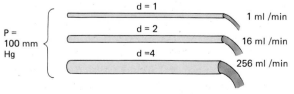

FIGURE 26-6 Effect of vessel diameter on flow.

FIGURE 26-7 Effect of viscosity on flow.

friction with the wall, and, consequently, the greater the resistance. This effect is illustrated in Figure 26-7, which shows water, plasma, and normal blood attempting to go through spouts of equal dimensions. In each instance the amount of fluid flowing through the spout is 100 ml/min, but, because of different viscosities of the three fluids, the pressure required to force each through the spout is also different. The level of fluid in the vertical tube, which is an indicator of the pressure required to cause the flow, is 1 cm when water is flowing, 1.5 cm for plasma, and 3 cm for blood. In other words, the *relative* viscosities of water, plasma, and blood are 1, 1.5, and 3.

The most important factor governing blood viscosity is the *concentration of red blood cells.* The viscosity of normal blood is about 3 times that of water. However, when the red blood cell concentration falls to one-half normal, the viscosity becomes only 2 times that of water, and when the concentration rises to approximately 2 times normal, the viscosity can increase to as high as 15 times that of water. For these reasons, the flow of blood through the blood vessels of anemic persons—that is, persons who have low concentrations of red blood cells—is extremely rapid, whereas blood flow is very sluggish in persons who have excess red blood cells (polycythemia).

Interrelationships Between Pressure, Flow, and Resistance

It is obvious from the preceding discussions that pressure and resistance oppose each other in affecting blood flow, pressure tending to increase flow and resistance tending to decrease flow. One can state this relationship mathematically by the following formula:

$$\text{Blood flow} = \frac{\text{Pressure}}{\text{Resistance}}$$

The same formula can be expressed in two other algebraic forms:

$$\text{Pressure} = \text{Blood flow} \times \text{Resistance}$$

$$\text{Resistance} = \frac{\text{Pressure}}{\text{Blood flow}}$$

These formulas are basic in almost all hemodynamic studies of the circulatory system and therefore must be thoroughly understood prior to any attempt to analyze the operation of the circulation.

Poiseuille's Law. By inserting the various factors that affect resistance into the preceding formulas, one can derive still another formula known as Poiseuille's law:

$$\text{Blood flow} = \frac{\text{Pressure} \times (\text{Diameter})^4}{\text{Length} \times \text{Viscosity}}$$

This formula expresses the ability of blood to flow through any given vessel, showing that the rate of blood flow is directly proportional to the pressure difference between the two ends of the vessel, directly proportional to the fourth power of the vessel diameter, and inversely proportional to the vessel length and blood viscosity.

REGULATION OF BLOOD FLOW THROUGH THE TISSUES

The Arterioles as the Major Regulators of Blood Flow

The rate of flow through each respective tissue is controlled mainly by the arterioles. There are three separate reasons for this:

First, approximately one half of all the resistance to blood flow in the systemic circulation is in the arterioles, which can be seen by carefully studying Figure 26–8. This figure shows the minuteness of the flow channel through the arteriole, which gives this vessel its high resistance. Therefore, a slight change in arteriolar diameter can change the total resistance to flow through any given tissue far more than can a similar change in diameter in any other vessel.

Second, the arterioles have a very strong muscular wall constructed in such a manner that their diameters can change as much as three- to fivefold. Remembering once again that resistance is inversely proportional to the *fourth power* of a vessel's diameter, it immediately becomes obvious that the resistance to blood flow through the arterioles can be changed as much

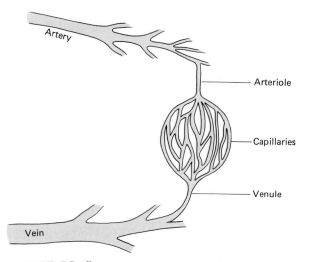

FIGURE 26–8 Demonstration of the minute size of the arteriole relative to the remainder of the vascular system, which explains its tremendous resistance.

as several hundred- or even a thousandfold by simply relaxing or contracting the smooth muscle walls.

Third, the smooth muscle walls of the arterioles respond to two different types of stimuli that regulate blood flow. First, they respond to the local needs of the tissues to increase blood flow when the supply of nutrients to the tissues falls too low and to decrease the flow when the nutrient supply becomes too great. This mechanism is called *autoregulation*. Second, autonomic nerve signals, particularly *sympathetic signals*, have a profound effect on the degree of contraction of the arterioles.

Autoregulation of Blood Flow— Role of Oxygen

The term autoregulation means automatic adjustment of blood flow in each tissue to the need of that tissue. In most instances the need of the tissue is nutrition, but in a few instances other factors that depend on blood flow are needed even more than nutrition. For instance, in the kidneys, the need to excrete end-products of metabolism and electrolytes is prepotent, and the concentrations of these substances in the blood play a major role in controlling renal blood flow. In the brain it is essential that carbon dioxide concentration remain very constant because reactivity of the brain cells increases and decreases with the amount of carbon dioxide; in this tissue it is primarily the need to remove carbon dioxide that determines the rate of blood flow.

However, in most tissues it is the need of the tissue for the nutrient *oxygen* that seems to be the most powerful stimulus for autoregulation. For instance, if a tissue becomes more active than usual, the need for oxygen might increase as much as 5- to 10-fold, and the blood flow automatically increases to help supply this extra oxygen.

Mechanism By Which Oxygen Deficiency Causes Arteriolar Dilatation. Even though we know that blood flow through most tissues is controlled in proportion to their need for oxygen, we still do not know the exact mechanism by which oxygen need can affect the degree of vascular constriction. There are two basic theories for arteriolar vasodilatation in response to oxygen need in the tissues. One of these, the "oxygen demand" theory, is illustrated in Figure 26–9. This shows that the tissue cells lie in close proximity to the arterioles and that they are in competition with the arterioles for the oxygen, which is represented by the small dots. It is as-

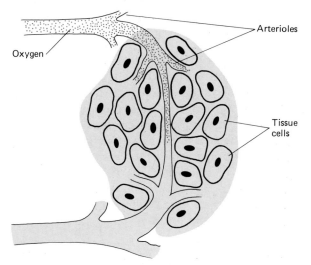

FIGURE 26–9 Postulated mechanism of autoregulation.

sumed that the smooth muscle of the arteriolar wall requires oxygen to contract. Should the oxygen supply diminish, the tissue cells would deplete the oxygen; therefore, the vessels would dilate simply because of failure to receive adequate oxygen to maintain contraction. If the tissue should utilize an increased quantity of oxygen, the amount of oxygen available to the arterioles would again decrease, this too causing increased blood flow. The resulting increase in blood flow, with the transport of more oxygen to the tissues, would then increase the tissue oxygen concentration back toward a normal level.

Another theory is that oxygen lack causes *vasodilator substances* to be formed in the tissue. That is, decreased availability of oxygen to the metabolic systems of the cells supposedly causes them to release a chemical substance or substances that then act directly on the blood vessels to cause active vasodilatation. The most likely vasodilator is the substance *adenosine*, which is always released from very active or hypoxic cells. However, some other suggested vasodilator substances have been *histamine, lactic acid, carbon dioxide*, and *potassium ions*, all of which are known to be released from hyperactive tissues and all of which are known to cause at least some dilatation of the arterioles.

Long-term Autoregulation. If the needs of the tissues for oxygen and other nutrients are more than the acute autoregulatory mechanism can supply, a long-term autoregulatory mechanism, requiring several weeks or months to develop, can cause still more increase in blood flow. This mechanism causes the actual number of blood vessels to increase and also causes en-

largement of the existing vessels. For instance, when a blood vessel such as a coronary vessel of the heart has become partially occluded, new vessels grow into the area of poor blood supply, often returning the blood supply toward normal after a few weeks. Also, if a person ascends to a high altitude and remains there for many months, he gradually develops enlarged vessels as well as an increased number of blood vessels in all his tissues, thereby supplying the tissues with increased amounts of oxygen. Conversely, when he comes back to a lower altitude, some of the vessels actually disappear after many more months, thus illustrating that this long-term vascular phenomenon helps to autoregulate blood flow in response to the tissues' needs.

To state this mechanism another way, in the long run, most tissues of the body try to adjust their vascular supplies—their actual numbers of vessels—to their needs for nutrients.

Nervous Control of Blood Flow

All the arteries, arterioles, and veins of the systemic circulation are supplied by nerves from the sympathetic nervous system, as illustrated in Figure 26–10. Stimulation of the sympathetic nerves causes especially the arterioles and veins, and the large arteries to a lesser extent, to constrict.

Vasoconstrictor Tone of the Systemic Vessels. The sympathetic vasoconstrictor nerves normally transmit a continual stream of impulses to the blood vessels, maintaining the vessels in a moderate state of constriction all of the time. This is called *vasomotor tone*. Then, when the sympathetic nervous system is required to constrict the vessels more than their normal state of constriction, it does so by increasing the number of sympathetic impulses

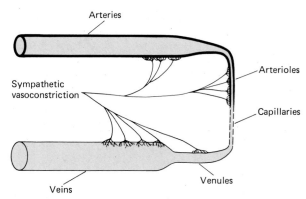

FIGURE 26–10 Innervation of the peripheral vascular system, showing especially the abundant innervation of the arterioles.

above normal. And the mechanism by which the sympathetic nervous system dilates the vessels is to decrease the impulses to less than normal. In this manner the sympathetic system can cause both vasoconstriction and vasodilatation. Therefore, special nerves are not required to cause vasodilatation in almost all parts of the circulation.

Importance of Nervous Control of the Blood Vessels. In contrast to the role of autoregulation, nervous control of the blood vessels is not generally concerned with regulating the delivery of nutrients to the tissues. Instead, nervous control is concerned with distribution of blood flow to major sections of the body. For instance, when one exercises, the muscles require such an extreme increase in blood flow that the heart often cannot pump an adequate quantity of blood both through the dilated muscle vessels and at the same time through such areas as the skin, the kidneys, and the gastrointestinal tract, areas which normally receive three fourths of the total cardiac output. Therefore, during exercise, sympathetic impulses cause vasoconstriction in these areas. But, at the same time, the muscle vessels become dilated because of local muscle vasodilator mechanisms mainly initiated by greatly increased oxygen usage by the exercising muscle. As a result, a major portion of the total blood flow shifts to the muscles, thereby allowing much more muscular activity than could otherwise occur.

The sympathetic nervous system can also shift large amounts of blood flow to the skin to help regulate body temperature. When the body temperature rises too high, a decrease in sympathetic stimulation dilates the arterioles of the skin, increasing the flow of warm blood to the skin; this promotes heat loss until the temperature returns to normal. On the other hand, when the body temperature falls below normal, sympathetic-induced vasoconstriction occurs in the skin vessels, skin blood flow decreases, heat loss becomes less, and body temperature rises back toward normal.

A third instance of important nervous regulation of blood flow occurs when the circulatory system has been damaged so greatly that the heart pumps an insufficient quantity of blood to supply all portions of the body. Under these conditions sympathetic stimulation causes vasoconstriction in the less vital areas such as the skin, muscles, and gut; organs such as the brain and heart continue to receive adequate flow.

And, finally, nervous control of the arterioles plays a very important role in reflex regula-

tion of arterial pressure. This effect is so important that it will be discussed in detail in Chapter 28. Briefly, when the arterial pressure begins to fall as a result of hemorrhage or cardiac damage or for any other cause, pressure-sensitive detectors in the walls of several large arteries, called *baroreceptors*, detect the falling pressure and transmit signals to the brain. The brain in turn reflexly transmits nerve signals back to the heart to increase heart activity and to the arterioles throughout the body to cause vasoconstriction, thereby elevating the arterial pressure back toward normal.

DISTRIBUTION OF BLOOD VOLUME IN THE BODY

Figure 26–11 shows the approximate percentages of the total blood volume in the different portions of the circulatory system. It is evident from this figure that about three fourths of the blood is in the systemic circulation and one fourth in the heart and pulmonary circulation. Also, a far higher proportion is in the veins than in the arteries. For instance, the arteries of the systemic circulation contain only 13 percent of the total blood, whereas the veins, venules, and venous sinuses contain approximately 50 per-

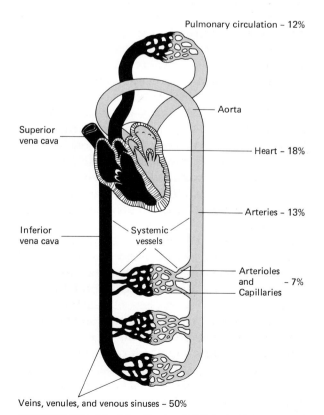

FIGURE 26–11 Distribution of blood volume in the different portions of the circulatory system.

cent. The arterioles and capillaries, despite their extreme importance to the circulation, contain only a few percent of the blood.

The Blood Reservoirs—The Reserve Supply of Blood

When the volume of blood in the heart and peripheral blood vessels falls low enough that the vessels are no longer adequately filled, blood cannot circulate normally through the tissues. For this reason, it is important to have an extra supply of blood. The entire venous system acts as a *blood reservoir,* for the veins exhibit a plastic quality whereby their walls can distend and contract in response to the amount of blood available in the circulation. Also, the veins are supplied with nerves from the sympathetic nervous system so that any time the tissues begin to suffer for lack of blood flow, nervous reflexes cause a large number of sympathetic impulses to pass to the veins, constricting them and translocating blood into the heart and other vessels. It is mainly this contractile and expansile quality of the venous system that protects the circulation against the disastrous effects of blood loss.

Certain portions of the venous system that are especially important for storing blood are called the *blood reservoirs.* They are as follows:

First, the *large veins of the abdominal region* are particularly distensible and, therefore, normally hold a tremendous amount of extra blood. Yet they can contract when the blood is needed elsewhere in the circulation.

Second, the *venous sinuses of the liver* can expand and contract many times over, so that the liver under certain circumstances may hold as much as a liter and a half of blood, but at other times only a few hundred milliliters.

Third, the *spleen* normally contains approximately 200 ml of blood, but can expand to hold as much as 1 L, or can contract to hold as little as 50 ml.

The *venous plexuses of the skin* are a fourth important blood reservoir. Normally the blood in these plexuses is used to regulate the heat of the body—the more rapidly blood flows through them, the greater is the loss of heat. However, when the vital organs need extra blood flow, the sympathetic nervous system can markedly contract the skin's venous plexuses, transferring the stored blood into the main stream of flow.

A fifth blood reservoir is the *pulmonary vessels.* Approximately 12 percent of the blood is normally in the pulmonary circulation, but much of this can be displaced into other portions of the circulation without impairing the function of the lungs. Therefore, the lungs also act as a source of blood in times of need.

QUESTIONS

1. Describe the methods for measuring blood flow.
2. Describe the methods for measuring blood pressure.
3. How is resistance to blood flow calculated?
4. Why does the diameter of the blood vessel play such an important role in determining vascular resistance?
5. If the blood pressure is decreased to one-half normal but the diameter of the blood vessel is increased at the same time to two times normal, what happens to the blood flow through the vessel?
6. Explain why the arterioles are the major regulators of blood flow.
7. What is meant by autoregulation of blood flow? Explain its significance as well as the role of oxygen in its mechanism.
8. Discuss the role of the sympathetic nervous system in the control of local blood flow throughout the body.
9. Name the major blood reservoirs in the body.

REFERENCES

Bevan, J.A., *et al.* (eds.): Vascular Neuroeffector Mechanisms. New York, Raven Press, 1980.

Duling, B.: Oxygen, metabolism, and microcirculatory regulation. *In* Kaley, G., and Altura, B.M. (eds.): Microcirculation. Vol. II. Baltimore, University Park Press, 1977.

Gow, B.S.: Circulatory correlates: Vascular impedance, resistance, and capacity. *In* Bohr, D.F., *et al.* (eds.): Handbook of Physiology. Sec. 2, Vol. 2. Baltimore, Williams & Wilkins, 1980, p. 353.

Gross, J.F., and Popel, A. (eds.): Mathematics of Microcirculation Phenomena. New York, Raven Press, 1980.

Guyton, A.C.: Arterial Pressure and Hypertension. Philadelphia, W.B. Saunders, 1980.

Guyton, A.C., *et al.*: Circulation: Overall regulation. *Annu. Rev. Physiol.,* 34:13, 1972.

Keatinge, W.R., and Harman, M.C.: Local Mechanisms Controlling Blood Vessels. New York, Academic Press, 1979.

Murphy, R.A.: Mechanics of vascular smooth muscle. *In* Bohr, D.F., *et al.* (eds.): Handbook of Physiology. Sec. 2, Vol. 2. Baltimore, Williams & Wilkins, 1980, p. 325.

Pedley, T.J.: The Fluid Mechanics of Large Blood Vessels. New York, Cambridge University Press, 1979.

Schneck, D.J., and Vawter, D.L. (eds.): Biofluid Mechanics. New York, Plenum Press, 1980.

Shepherd, J.T., and Vanhoutte, P.M.: The Human Cardiovascular System: Facts and Concepts. New York, Raven Press, 1979.

Sparks, H.V., Jr.: Effect of local metabolic factors on vascular smooth muscle. *In* Bohr, D.F., *et al.* (eds.): Handbook of Physiology. Sec. 2, Vol. 2. Baltimore, Williams & Wilkins, 1980, p. 475.

Walker, J.R., and Guyton, A.C.: Influence of blood oxygen saturation on pressure-flow curve of dog hindleg. *Am. J. Physiol.,* 212:506, 1967.

Wolf, S., and Werthessen, N.T. (eds.): Dynamics of Arterial Flow. New York, Plenum Press, 1979.

27

Special Areas of the Circulatory System

Overview

The vascular system in each organ of the body is specifically adapted to serve that organ's functions. Some of the more important special areas of circulatory function are as follows:

1. *Muscle Blood Flow.* Though the skeletal muscles comprise almost 40 percent of the body mass, their total blood flow under resting conditions is only about *1 liter per minute* (L/min). Yet, during heavy exercise this flow can increase to 20 L or more per minute. Thus, muscle blood flow is extremely changeable and is related almost entirely to the increased need of the muscles for nutrients during activity, especially for oxygen. In fact, the *blood flow is almost directly related to the usage of oxygen* by the muscles, which is one of the major factors controlling blood flow, as explained in the previous chapter.

   To supply the great increase in blood flow required by the muscles during exercise, all the remainder of the circulatory system also responds. For instance, the brain transmits an extreme barrage of *sympathetic signals* to the circulation during exercise to cause (1) *increased heart rate,* (2) *increased strength of heart contraction,* (3) *constriction of the blood vessels in the nonmuscle tissues* to divert blood flow into the muscles, and (4) *constriction of the large veins* so that extra amounts of blood are made available to the heart for pumping to the muscles.

2. *The Cerebral Circulation.* The total blood flow to the brain averages about 700 milliliters per minute (ml/min). This rate of flow remains relatively constant under most conditions, which helps to maintain very constant concentrations of nutrients and ions in the fluid environment of the brain cells. The three factors that are most important in controlling cerebral blood flow are the brain concentrations of (1) *carbon dioxide,* (2) *oxygen,* and (3) *hydrogen ions.* The blood flow increases whenever oxygen is needed, and it also increases to carry away excess carbon dioxide or hydrogen ions whenever either of these is in excess. This control of brain blood flow is essential for proper function of the brain because the nerve cells are highly responsive to abnormalities of any one of these three substances.

3. *The Portal Circulation and Liver Blood Flow.* Almost all of the venous blood flow from the gastrointestinal tract flows first into the portal vein, then through the liver, passing through an extensive network of

461

minute *hepatic sinuses* before emptying into the inferior vena cava. The blood coming from the gastrointestinal tract contains large quantities of food substances absorbed from the intestines, especially *glucose, other monosaccharides,* and *many amino acids* that are the digestive products of proteins. Lining the hepatic sinuses on all sides are continuous walls of *liver cells* that remove most of these absorbed food substances and store them temporarily. Then, they are meted out to the blood during the inbetween meal times so that a steady flow of nutrients is available to the tissues.

Also, along the walls of the hepatic sinuses are many large *macrophages* called *Kupffer cells* that remove debris from the blood. Especially important, the portal blood usually contains large numbers of *colon bacilli* that have made their way from the large intestine into the blood, and the Kupffer cells remove almost all of these before they can enter the general circulation.

4. *Skin Blood Flow.*   Normally, one tenth or less of the blood flow through the skin is for the purpose of supplying nutrition to the skin tissues. Instead, most of this flow is *for the purpose of controlling body temperature.* To serve this purpose, the skin has an extensive *venous plexus* lying a few millimeters underneath its surface. When the small arteries supplying blood to this venous plexus are dilated, large quantities of warm blood flow into the plexus from the central core of the body, and this blood is thus cooled before it returns to the central core. This obviously also cools the entire inner body. Nervous centers in the brain detect the body temperature and control the skin blood flow accordingly. When the body is cold, almost no blood flows into the skin venous plexus so that little body heat is lost. But, when the body is hot, as much as 2 to 3 L of blood can flow through the skin per minute so that tremendous amounts of heat are then eliminated.

Though most parts of the circulation are basically similar, some organs have special circulatory needs and functions. For instance, blood flow to the heart muscle through the coronary vascular system is related to the beating heart and therefore has its own special peculiarities, as was discussed in Chapter 23. In another example, blood flow through the lungs has the special purpose of exchanging respiratory gases between the air of the lungs and the circulating blood. Therefore, the lungs have a vascular system that is adapted to this purpose, especially a vast array of capillaries that lie in close association with the air sacs of the lungs. This will be presented in detail in Chapter 36 in conjunction with the anatomy and function of the respiratory system.

In the present chapter we will discuss the functions of still other special vascular systems, including (1) muscle blood flow, especially during exercise; (2) the cerebral circulation; (3) blood flow in the gastrointestinal system and liver; and (4) the blood flow of the skin.

BLOOD FLOW IN SKELETAL MUSCLES AND ITS REGULATION IN EXERCISE

Very strenuous exercise is the most stressful condition the normal circulatory system faces. This is true because the blood flow in muscles can increase more than 20-fold (a greater increase than in any other tissue of the body) and also because there is such a very large mass of skeletal muscle in the body. The total muscle blood flow can become great enough to increase the cardiac output in the normal young adult to as much as five times normal and in the well-trained athlete to as much as six to seven times normal.

Rate of Blood Flow in Muscles

During rest, blood flow in skeletal muscle averages 3 to 4 ml/min per 100 g of muscle. However, during extreme exercise this rate can increase as much as 15- to 25-fold, rising to 50 to 80 ml per 100 g of muscle.

FIGURE 27–1 Effects of muscle exercise on blood flow in the calf of a leg during strong rhythmic contraction. The blood flow is much less during contraction than between contractions. (From Barcroft and Dornhorst: *J. Physiol., 109:*402, 1949.)

Intermittent Flow During Muscle Contraction.

Figure 27–1 illustrates a study of blood flow changes in the calf muscles of the human leg during strong rhythmic contraction. Note that the flow increases and decreases with each muscle contraction, decreasing during the contraction phase and increasing between contractions. At the end of the rhythmic contractions, the flow remains very high for 1 to 2 minutes and then gradually fades toward normal.

The cause of the decreased flow during sustained muscle contraction is compression of the blood vessels by the contracted muscle. During strong *tetanic* contraction, blood flow can sometimes be almost totally stopped.

Opening of Muscle Capillaries During Exercise.

During rest, only 20 to 25 percent of the muscle capillaries are open. But during strenuous exercise all the capillaries open up, which can be demonstrated by studying histologic specimens removed from muscles appropriately stained during exercise. It is this opening up of dormant capillaries that allows most of the increased blood flow. It also diminishes the distance that oxygen and other nutrients must diffuse from the capillaries to the muscle fibers and contributes a much increased surface area through which nutrients can diffuse from the blood.

Control of Blood Flow in Skeletal Muscle

Local Regulation.

The tremendous increase in muscle blood flow that occurs during skeletal muscle activity is caused primarily by local effects in the muscles acting directly on the arterioles to cause vasodilatation. One of the most important of these is reduced oxygen in the muscle tissues, because during muscle activity the muscle utilizes oxygen very rapidly, thereby decreasing its concentration in the tissue fluids. This in turn causes vasodilatation either because the vessel walls cannot maintain contraction in the absence of oxygen or because oxygen deficiency causes release of vasodilator substances. The vasodilator substance that has been suggested most widely in recent years has been adenosine.

Other vasodilator substances released during muscle contraction include potassium ions, acetylcholine, adenosine triphosphate, lactic acid, and carbon dioxide. Unfortunately, we still do not know quantitatively how great a role each of these plays in increasing muscle blood flow during muscle activity.

Nervous Control of Muscle Blood Flow.

In addition to the local tissue regulatory mechanism, the skeletal muscles are also provided with sympathetic nerves. However, these have so much less effect on muscle blood flow than do the local control factors that they are usually unimportant. They do help to maintain the arterial pressure by constricting the arterioles, following severe hemorrhage.

Circulatory Readjustments During Exercise

Three major effects occur during exercise that are essential for the circulatory system to supply the tremendous blood flow required by the muscles. These are (1) mass discharge of the sympathetic nervous system, (2) increase in cardiac output, and (3) increase in arterial pressure.

Mass Sympathetic Discharge.

At the onset of exercise, signals are transmitted not only from the brain to the muscle to cause muscle contraction but also from the higher levels of the brain into the vasomotor center in the medulla oblongata to initiate mass sympathetic discharge. Two major circulatory effects result. First, the heart is stimulated to greatly increased heart rate and pumping strength. Second, all the blood vessels of the peripheral circulation are strongly contracted except the vessels in the active muscles, which are strongly vasodilated by the local vasodilator effects in the muscles themselves. Thus, the heart is stimulated to supply the increased blood flow required by the muscles, and blood flow through most nonmuscular areas of the body is temporarily reduced, thereby temporarily "lending" their blood supply to the muscles. However, two of the organ circulatory systems, the coronary and cerebral systems, are spared this vasoconstrictor effect because both of these circulatory areas have very poor vasoconstrictor innervation—fortunately so, because both the heart and the brain are as essential to exercise as are the skeletal muscles themselves.

Increase in Cardiac Output. The increase in cardiac output that occurs during exercise results mainly from the intense local vasodilatation in the active muscle. As will be explained in Chapter 29, in relation to the basic theory of cardiac output regulation, local vasodilatation increases the venous return of blood back to the heart. The heart in turn pumps this extra returning blood immediately back to the muscles through the arteries. Thus, it is the degree of vasodilatation in the muscles themselves that determines the amount of increase in cardiac output—up to the limit of the heart's ability to respond.

Another factor that greatly helps to cause the large increase in venous return is the strong sympathetic stimulation of the veins. This stimulation constricts all the venous blood reservoirs and mobilizes extra quantities of blood out of the veins and toward the heart.

Mechanisms By Which the Heart Increases Its Output. One of the principal mechanisms by which the heart increases its output during exercise is the Frank-Starling mechanism, which was discussed in Chapter 24. Via this mechanism, when increased quantities of blood flow from the veins into the heart and dilate its chambers, the heart muscle contracts with increased force, thus also pumping an increased volume of blood with each heart beat. However, in addition to this basic intrinsic cardiac mechanism, the heart is also strongly stimulated by the sympathetic nervous system. The net effects are greatly increased heart rate (occasionally to as high as 200 beats per minute) and near doubling of the cardiac muscle strength of contraction. These two effects combine to make the heart capable of pumping about 100 percent more blood than would be true based on the Frank-Starling mechanism alone.

Increase in Arterial Pressure. The mass sympathetic discharge throughout the body during exercise and the resultant vasoconstriction of most of the blood vessels besides those in the active muscles almost always increase the arterial pressure during exercise. This increase can be as little as 20 mm Hg or as great as 80 mm Hg, depending on the conditions under which the exercise is performed. For instance, when a person performs exercise under very tense conditions but uses only a few muscles, the sympathetic response still occurs throughout the body, and vasodilatation occurs

in only a few muscles. Therefore, the net effect is mainly one of vasoconstriction, often increasing the mean arterial pressure to as high as 180 mm Hg. Such a condition occurs in a person standing on a ladder and nailing with a hammer on the ceiling above. The tenseness of the situation is obvious, and yet the amount of muscle vasodilatation is relatively slight.

On the other hand, when a person performs whole-body exercise, such as running or swimming, the increase in arterial pressure is usually only 20 to 40 mm Hg. The lack of a tremendous rise in pressure results from the extreme vasodilatation occurring in large masses of muscle.

In rare instances, persons are found in whom the sympathetic nervous system is absent, either because of congenital absence or because of surgical removal. When such a person exercises, instead of the arterial pressure rising, the pressure actually falls, and as a result, the cardiac output rises only about one third as much as it does normally.

Importance of the Arterial Pressure Rise During Exercise. In the well-trained athlete, it has been calculated, the muscle blood flow can increase at least 20-fold. Although most of this increase results from vasodilatation in the active muscles, the increase in arterial pressure also plays an important role. If one remembers that an increase in pressure not only forces extra blood through the muscle because of the increased pressure itself but also dilates the blood vessels, he can see that a rise in pressure of as little as 20 to 40 mm Hg can at times actually double blood flow.

The Cardiovascular System as the Limiting Factor in Heavy Exercise. The capability of an athlete to enhance cardiac output and consequently to deliver increased quantities of oxygen and other nutrients to his muscles is the major factor that determines the degree of prolonged heavy exercise that the athlete can sustain. For instance, the speed of a marathon runner is almost directly proportional to his ability to enhance cardiac output. Therefore, the ability of the circulatory system to adapt to exercise is equally as important as the muscles themselves in setting the limit for the performance of muscle work.

THE CEREBRAL CIRCULATION

Figure 27–2 illustrates an overview of the arterial system on the base of the brain, showing the two *vertebral arteries* passing upward along the

FIGURE 27—2 The cerebral circulation.

medulla and the two *internal carotid arteries* approaching the brain on its ventral surface. Communicating branches between these four major arteries protect the brain against damage should any one or even two of them become occluded. However, the smaller arteries that spread over the surface of the brain do not have many communications from one to another. Therefore, occlusion of any one of them usually causes destruction of the respective brain tissue.

Cerebral Blood Flow and Its Regulation

The total blood flow to the brain averages about 700 ml/min, which is about 14 percent of the blood pumped by the heart.

Autoregulation of Cerebral Blood Flow.
Cerebral blood flow is autoregulated better than that in almost any other area of the body except the kidneys. Even though the arterial pressure might fall to as low as 40 mm Hg or rise to as high as 200 mm Hg, the blood flow through the brain still varies no more than a few percent.

Autoregulation of blood flow in the brain results mainly from a carbon dioxide autoregulation mechanism. The cerebral circulation is different from that of other tissues in that it responds much more to carbon dioxide concentration than do the other tissues. When the carbon dioxide concentration rises above normal, the cerebral vessels dilate and cerebral blood flow automatically increases. This then washes carbon dioxide out of the tissues, bringing the carbon dioxide concentration back toward normal. Conversely, decreased carbon dioxide in the brain decreases the blood flow, which allows the carbon dioxide being formed in the tissues

to accumulate and therefore to increase the carbon dioxide concentration toward normal.

The carbon dioxide mechanism is much more powerful in the cerebral circulation than is the oxygen deficiency mechanism of autoregulation (which is far more powerful in most other tissues), and it is mainly because of this carbon dioxide mechanism that cerebral blood flow is autoregulated to an almost exactly constant level all the time. The only exception to this constancy of blood flow occurs when the brain itself becomes overly active. Under these conditions cerebral blood flow can on rare occasions rise to as high 30 to 50 percent above normal, but even this increase in flow is a manifestation of autoregulation, the flow increasing in proportion to the increase in carbon dioxide formation in the brain.

The unvarying rate of cerebral blood flow is very advantageous to cerebral function, because neurons must never become excessively excitable nor excessively inhibited if the brain is to function properly, and, obviously, changes in the nutritional status of the neurons could quite easily change their degree of excitability. For instance, if the carbon dioxide concentration in the tissues should rise too high, the tissue fluids would become excessively acidic because carbon dioxide reacts with the water of the tissues to form carbonic acid, and this tissue acidity is known to depress neuronal function severely.

Venous Dynamics in the Cerebral Circulation

Function of the veins in the cerebral circulation is somewhat different from that of veins in the remainder of the body because the cerebral veins are noncollapsible. They exist as *sinuses* in the fibrous lining, the *dura mater*, of the cerebral vault, and the sides of the sinuses are so firmly attached to the skull and other tissues that they cannot collapse completely. Also, because the head is in an upright position most of the time, the weight of the venous blood makes it run rapidly downward toward the heart, creating a semivacuum in the cerebral veins. If one of these veins is inadvertently punctured while a person is in the upright position, air is often sucked into the venous system and occasionally may even flow so rapidly to the heart that it blocks the action of the heart valves and kills the person. Because the veins in all other parts of the body are flimsy and collapse very easily whenever suction is applied to them, any sucking action that might occur simply closes the veins rather than sucking air.

Cerebral Vascular Accidents — The "Stroke"

When a blood vessel supplying blood to an important part of the brain suddenly becomes blocked or when a cerebral vessel ruptures, the person is said to have had a *cerebral vascular accident* or a "stroke." This is the cause of death in as many as 10 percent of all people. In about one fourth of all cases the damage is caused by a blood clot developing on an atherosclerotic fatty deposit in a major cerebral artery. In the other three fourths, an artery ruptures because of excessively high arterial pressure or because the vessel itself has been weakened by the atherosclerosis disease process. Such hemorrhage into the tissues of the brain often compresses the neuronal cells enough to destroy them or at least to stop their functioning. Also, a blood clot develops in the hemorrhagic area and extends backward into the artery, blocking further flow inside the vessel. Therefore, still more brain tissue dies for lack of nutrition. Thus, regardless of whether the cerebral vascular accident is initially caused by simple vascular occlusion or by hemorrhage, the results are essentially the same, that is, destruction of a major part of the brain itself.

The effects of a stroke depend on which vessel is blocked. Frequently it is the middle cerebral artery supplying the area of the brain that controls muscular function, in which case the opposite side of the body becomes paralyzed. Blockage of posterior branches of the middle cerebral artery that supply Wernicke's area (the area in the posterior temporal lobe that integrates the highest level of thought) will cause the

person to become demented, unable to understand any incoming communication or to speak a coherent sentence. Blockage of the posterior cerebral artery, on the other hand, causes partial blindness. Blockage of arteries in the hindbrain region is likely to destroy nerve tracts connecting the brain with the spinal cord, causing any number of abnormalities such as paralysis, loss of sensation, loss of equilibrium, and so forth.

BLOOD FLOW IN THE GASTROINTESTINAL TRACT AND LIVER — THE PORTAL CIRCULATORY SYSTEM

Figure 27–3 illustrates diagrammatically the vascular system of the gastrointestinal tract, the portal veins, and the liver. After blood flows through the gastrointestinal vessels, it then flows into a special set of veins called the *portal venous system*, from there through the liver, and then through a vast array of sinusoids in the liver called *hepatic sinuses*. After passing through the liver, the blood empties directly into the inferior vena cava through two large *hepatic veins*.

As the blood flows through the liver sinusoids, it is processed by the liver cells in two ways: First, the blood comes into contact with special reticuloendothelial phagocytic cells that line the sinuses, the *Kupffer cells*, which are capable of removing abnormal debris. These cells,

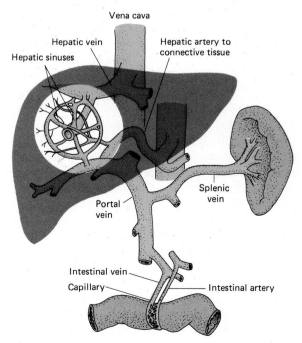

FIGURE 27–3 The gastrointestinal and portal circulation.

therefore, cleanse the intestinal blood before it reenters the general circulation. This is quite important because a few of the literally trillions of bacteria in the fecal material of the gastrointestinal tract make their way into the portal blood each minute. The Kupffer cells are so effective in removing the bacteria that probably not one in a thousand escapes through the liver into the general circulation.

Second, the liver tissue cells, the *hepatic cells*, remove various absorbed nutrients from the circulation. For instance, these cells normally remove two thirds of the glucose absorbed into the portal blood from the intestines and as much as one half of the proteins before the blood ever reaches the general circulation. These nutrients are stored in the liver cells until they are needed later by the remainder of the body and then are released back into the blood for use elsewhere. The storage of glucose and proteins "buffers" the blood concentrations of these nutrients, that is, prevents sudden increases in their concentrations after a meal and also prevents low concentrations during the long intervals between meals.

Vascular Dynamics of Gastrointestinal and Portal Blood Flow

The blood flow through the entire gastrointestinal tract plus the spleen accounts for about 21 percent of the blood pumped by the heart. This is one fifth of all the resting blood flow of the body, slightly greater than 1 L/min. This blood then flows into the portal system and finally through the liver. An additional 6 percent of the blood pumped by the heart flows into the liver from the hepatic artery, which directly supplies the tissues of the liver, making a total of 27 percent of all the cardiac output passing through this route. Thus, the blood flow through the gastrointestinal tract, portal system, and liver accounts for more than one quarter of the normal resting blood flow of the body and therefore is, under resting conditions, perhaps the single most important peripheral vascular bed.

Control of Blood Flow in the Gastrointestinal Blood Vessels. As we shall learn in Chapter 41, two functions of the gastrointestinal tract require major amounts of blood flow. These are (1) the contractions of the smooth muscle wall of the gastrointestinal tract that cause the gastrointestinal movements, such as the peristaltic movements that propel food through the gut, and (2) formation of the gastrointestinal secretions, which requires a large amount of nutri-

tion for the glandular cell secretory process. When the gastrointestinal tract is very active, the rate of blood flow increases in direct proportion to this activity. This results mainly from increased oxygen usage by the tissues during the increased activity. As more oxygen is used, for reasons discussed in the previous chapter, the local blood vessels also dilate more. Thus, blood flow in the gastrointestinal tract is *autoregulated* in proportion to the needs of the tissues in the same way that this occurs in most other parts of the body.

Portal Vascular Pressure and Ascites. The small vessels in the liver offer a small amount of resistance to blood flow, and this causes the portal venous pressure normally to average 8 to 10 mm Hg. Ordinarily, this elevated portal pressure is of no major concern, but when some abnormality causes the systemic venous pressure to rise, the portal pressure increases a corresponding amount, always remaining 8 to 10 mm Hg greater than the systemic venous pressure. Consequently, the portal capillary bed is one of the first to suffer high capillary pressure as a result of excessively high systemic venous pressure, a condition that occurs most frequently when the heart fails, and this in turn causes blood to dam up in the intestinal veins. The high capillary pressure forces fluid to leak through the capillary walls into the intestinal tissues and peritoneal cavity. In addition, high pressure in the small liver sinuses causes fluid also to leak profusely from the surface of the liver. Therefore, instead of the norm of only a few milliliters of fluid in the abdomen, many liters may accumulate. This accumulation of intra-abdominal fluid is called *ascites*.

Portal Obstruction. Occasionally the portal vein leading from the intestines and spleen to the liver becomes totally occluded by a large blood clot. And even more frequently the small vessels in the liver become blocked or severely constricted because of alcoholic liver disease called *liver cirrhosis*. In either event, the portal venous pressure rises, which in turn causes high portal capillary pressure, edema of the intestinal walls, enlargement of the spleen, and ascites (free fluid in the abdomen, sometimes as much as 20 L).

Function of the Spleen

Reservoir Function. As was pointed out in Chapter 26, the spleen, which is part of the portal system, is a blood reservoir. It is capable of enlarging in some pathologic conditions up to a total volume of more than 1000 ml or of contract-

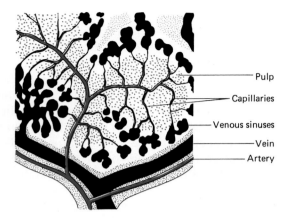

FIGURE 27–4   Internal organization of the spleen. (Modified from Bloom and Fawcett: A Textbook of Histology, 8th Ed.)

ing down to a minimum volume of less than 50 ml. The internal structure of a small segment of the spleen is illustrated in Figure 27–4, which shows several small arteries leading into capillaries in the *pulp* of the spleen and finally connecting with large venous sinuses. The blood flows from the venous sinuses into the large circumferential veins and then back into the general circulation. Much of the blood, including the red blood cells, leaks from the capillaries into the pulp and then squeezes through this area, finally reentering the general circulation through the walls of the venous sinuses. Large numbers of cells are often trapped in the pulp in this manner, whereas the plasma is returned to the circulation. Therefore, the spleen stores a much higher percentage of red blood cells than plasma, but when it contracts, the cells are forced from the pulp back into the circulation. Hence, the spleen is often said to be a *red blood cell reservoir* rather than a general blood reservoir.

In lower animals, sympathetic stimulation causes smooth muscle in the capsule of the spleen to contract, which expels blood from both the splenic pulp and the splenic sinuses into the general circulation. In the human being there is very little smooth muscle in the capsule. Therefore, the blood reservoir function of the spleen in the human being is relatively unimportant except in those pathological conditions in which it becomes greatly enlarged.

Phagocytic Function of the Spleen. Another major function of the spleen is to cleanse the blood. This results from phagocytosis by reticuloendothelial cells, which line the splenic sinuses and are located throughout the splenic pulp.

BLOOD FLOW THROUGH THE SKIN

Blood flow through the skin has two functions: first, to regulate the temperature of the body and, second, to supply nutrition to the skin itself.

Relationship of Skin Blood Flow to Body Temperature Regulation. Figure 27–5 illustrates the vascular architecture of the skin. Note especially the extensive *venous plexus* lying a few millimeters beneath the surface of the skin. The rate of blood flow through this plexus can be altered up or down as much as 50-fold. When the arteries that supply blood to the plexus are constricted, the blood flow to the skin of the entire body may be as little as 50 ml/min, whereas when the arteries are completely dilated, the blood flow to the skin may be as great as 2 to 3 L/min.

When the skin blood flow is very rapid, the amount of heat transmitted by the blood from the internal structures to the surface of the body is very great, and large amounts of heat are then lost from the skin. Conversely, when the quantity of blood flowing to the skin is slight, the amount of heat lost from the body will also be slight. Special temperature regulatory mechanisms, controlled by the hypothalamus in the brain and acting through the sympathetic nerves, can either vasoconstrict or vasodilate the skin vessels, thereby helping to regulate body temperature. This will be discussed in much more detail in Chapter 44.

Certain areas of the body, such as the hands, feet, and ears, have special very muscular vessels called *arteriovenous anastomoses*, which connect the arteries directly with the venous plexus. When these anastomoses are dilated, blood bypasses the capillary system and flushes extremely rapidly into the plexus. In this way the hands, feet, and ears receive tremendous

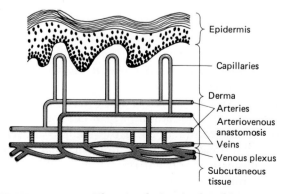

FIGURE 27–5   The circulation in the skin.

amounts of blood flow when they are exposed to excessive cold, and the increased blood flow obviously acts to protect these tissues from freezing.

The Nutrient Blood Flow. Ordinarily the blood flow through the skin is some 20 to 30 times that required to supply the necessary amount of nutrition to the skin tissues. Yet, when the skin becomes extremely cold, the body temperature mechanisms constrict the skin blood flow so greatly that nutrition can then become impaired. And this also occurs occasionally in diseases of the vasculature supplying the skin—for instance, in atherosclerosis of arteries leading to the skin, especially in the hands and feet. It is only under such conditions as these that one need be concerned with the skin nutrient vascular supply.

Figure 27–5 illustrates capillary loops that carry blood into all the tufts beneath the skin. The nutritive blood flow is subject to the same local autoregulatory mechanism for blood flow as that found elsewhere in the body. For instance, when a person compresses an area of the skin so hard that the blood flow is completely blocked out for a few minutes, the skin blood flow becomes far greater than normal immediately after release of the pressure, causing the skin to have a red blush for several minutes thereafter. This automatically makes up for the deficiency in nutrients during the blood flow blockage. Fortunately, the skin can be without blood flow for many hours before serious damage occurs.

QUESTIONS

1. Discuss the relative roles of local regulation and nervous regulation of the blood flow in muscles.
2. How does the heart increase the cardiac output during exercise?
3. How does the arterial pressure change during exercise, and what role does this play in increasing muscle blood flow?
4. What is the major mechanism for regulation of cerebral blood flow, and what is the role of carbon dioxide in this?
5. Discuss the causes of "stroke."
6. Describe the portal circulatory system and explain its relationship to the liver.
7. Discuss the control of gastrointestinal blood flow.
8. Distinguish between blood flow through the skin for the purpose of controlling body temperature and blood flow through the skin to provide nutrition.

REFERENCES

Abramson, D.I. (ed.): Circulatory Diseases of the Limbs: A Primer. New York, Grune & Stratton, 1978.

Apple, D.F., Jr., and Cantwell, J.D.: Medicine for Sport. Chicago, Year Book Medical Publishers, 1979.

Betz, E.: Cerebral blood flow: Its measurement and regulation. *Physiol. Rev., 52*:595, 1972.

Boullin, D.J. (ed.): Cerebral Vasospasm. New York, John Wiley & Sons, 1980.

Helwig, E.B., and Mostofi, F.K. (eds.): The Skin. New York, R.E. Kreiger, 1980.

Hutchins, G.M.: Pathological changes in aortocoronary bypass grafts. *Annu. Rev. Med., 31*:289, 1980.

Juergens, J.L., *et al.* (eds.): Allen, Barker, Hines Peripheral Vascular Disease, 5th Ed. Philadelphia, W.B. Saunders. 1980.

McHenry, L.C., Jr.: Cerebral Circulation and Stroke. St. Louis, W.H. Green, 1978.

Rappaport, A.M.: Hepatic blood flow: Morphologic aspects and physiologic regulation. *In* Javit, N.B. (ed.): International Review of Physiology: Liver and Biliary Tract Physiology I. Vol. 21. Baltimore, University Park Press, 1980, p. 1.

Svanik, J., and Lundgren, O.: Gastrointestinal circulation. *Int. Rev. Physiol., 12*:1, 1977.

Wilkins, R.H. (ed.): Cerebral Vasospasm. Baltimore, Williams & Wilkins, 1980.

28

Systemic Arterial Pressure and Hypertension

Overview

The *arterial pressure* is *pulsatile* because a small amount of blood is pumped by the heart into the aorta with each heart beat. After each heart contraction, the pressure normally rises to about *120 mm Hg*, which is called the *systolic pressure*. Then, between heart beats, because blood continues to flow out of the large arteries and through the systemic circulation, the pressure falls to about *80 mm Hg*, which is called the *diastolic pressure*. The usual method for writing these pressures is *120/80*. The *mean arterial pressure*, which is the average pressure during the complete heart cycle, is the pressure that determines the average rate at which blood will flow through the systemic vessels. The mean pressures at different points in the circulation are approximately the following: aorta, 100 mm Hg; at the junction between the small arteries and arterioles, 85 mm Hg; at the junction between the arterioles and capillaries, 30 mm Hg; at the venous ends of the capillaries, 10 mm Hg; and in the venae cavae, where they empty into the right atrium, 0 mm Hg.

To ensure that blood flow in the systemic circulation does not increase and decrease because of varying pressures, it is exceedingly important that the mean arterial pressure be regulated to a very constant level. This is achieved by a complex array of mechanisms involving (1) the nervous system, (2) the kidneys, and (3) several hormonal mechanisms.

1. *Nervous Control.* Short-term control of arterial pressure, over a period of seconds or minutes, is achieved almost entirely by *nervous reflexes*. One of the most important of these is the *baroreceptor reflex*. When the arterial pressure becomes too high, this stretches and excites special nerve receptors, the *baroreceptors, in the walls of the aorta and internal carotid arteries*. The baroreceptors then send signals to the *medulla oblongata* in the brain stem, which in turn sends signals from the medulla through the *autonomic nervous system* to cause (a) slowing of the heart, (b) decreased strength of contraction of the heart, (c) dilatation of the arterioles, and (d) dilatation of the large veins, all of which work together to decrease the arterial pressure back toward normal. Exactly opposite effects occur when the blood pressure falls too low and the baroreceptors lose their stimulation.

470

2. *Control by the Kidneys.* The kidneys are responsible almost entirely for *long-term control of arterial pressure.* They function through two important mechanisms for controlling pressure, one of which is a *hemodynamic mechanism* and the other a *hormonal mechanism,* which will be described under hormonal control below. The hemodynamic mechanism is a very simple one. When the arterial pressure rises above normal, excess pressure in the renal arteries causes the kidneys to filter extra amounts of fluid and therefore also to excrete increased amounts of water and salt out of the body. Loss of this water and salt decreases the blood volume, thus decreasing the blood pressure back to normal. Conversely, when the pressure falls below normal, the kidneys retain water and salt until the pressure rises again back to normal.

3. *Hormonal Control.* Several different hormones play important roles in pressure control, but by far the most important is the *renin-angiotensin hormonal system* of the kidneys. When the blood pressure falls too low to maintain normal blood flow through the kidneys, the kidneys secrete the substance *renin.* This is an *enzyme* that acts on one of the proteins in the plasma to split away the hormonal substance *angiotensin.* The angiotensin in turn constricts the arterioles throughout the body, which helps to raise the blood pressure back up to a normal level.

The term *hypertension* means *high arterial pressure.* On rare occasions the mean arterial pressure rises to two times normal, as high as 200 mm Hg. In some patients, the cause of the hypertension is *abnormal kidney function,* which prevents normal excretion of water and salt. In other instances, hypertension is caused by (a) *excess sympathetic nervous activity;* (b) *excess secretion of hormones by the adrenal cortex,* these hormones then acting on the kidneys to cause salt and water retention; or (c) *excess secretion of renin by the kidneys.* However, about 95 percent of all persons with hypertension have *essential hypertension,* which means hypertension of unknown origin. Recent research has shown that several aspects of kidney function are abnormal in essential hypertension, which may be the cause of the hypertension.

When the left ventricle contracts, it forces blood into the systemic arteries, thereby creating pressure that drives the blood through the systemic circulation. The *regulation of arterial pressure* is among the most important subjects of circulatory physiology, and the disease *hypertension,* which means high arterial pressure and is commonly called simply "high blood pressure," is one of the most common of all disorders of the body. These two topics receive special emphasis in the present chapter.

PULSATILE ARTERIAL PRESSURE

Instead of pumping a continuous stream, the heart pumps a small quantity of blood with each beat. As a result, the arterial pressure rises during systole but falls during diastole. Figure 28–1 shows graphically this pulsatile pressure under normal conditions and also under two abnormal conditions that will be discussed later.

Systolic and Diastolic Pressures

The pressure at its highest point during the pressure cycle is called the *systolic pressure,* and the pressure at its lowest point is called the *diastolic pressure.* As shown by the top curve of Figure 28–1, systolic pressure in a normal young adult is approximately 120 mm Hg and the diastolic pressure is approximately 80 mm Hg. The usual method for writing these pressures is 120/80. As another example, if a person's blood pressure is stated to be 210/125 this means that the systolic pressure is 210 mm Hg and the diastolic pressure 125 mm Hg.

The systolic and diastolic pressures change with age, as demonstrated graphically in

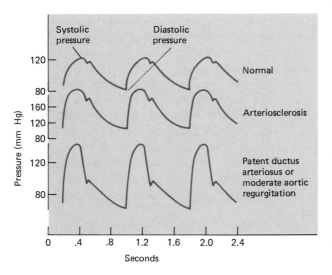

FIGURE 28–1 Pressure pulse contours in the normal circulation, in arteriosclerosis, in patent ductus arteriosus, and in moderate aortic regurgitation.

Figure 28–2. In a newborn baby the systolic pressure is about 90 mm Hg and the diastolic about 55 mm Hg. The pressure usually reaches 120/80 by young adulthood, and in old age the average is about 150/90.

Measurement of Systolic and Diastolic Pressures. Rapid-Response Direct Recording Methods. The arterial pressure varies so rapidly during the cardiac cycle that special high-speed recorders designed to follow the rapid changes in pressure must be used to measure it. Several electronic transducers, used with appropriate electronic recorders, were described in Chapter 26. The basic principle of all these transducers is the following: The vessel from which the pressure is to be measured is connected by way of a nondistensible tube with a small chamber bounded on one side by a thin elastic membrane. When the pressure rises, the membrane bulges outward, and when the pressure falls, the bulge decreases. The bulge in turn activates an electric sensor, the voltage output

from which is amplified and recorded on a moving strip of paper.

Appropriately designed membrane transducers can record pressure changes that occur in less than 0.01 second, which is more than adequate for recording any significant pressure change in the circulatory system. The arterial pressure pulse curves illustrated in Figure 28–1 were recorded using a membrane type of electronic transducer.

Indirect Measurement of Systolic and Diastolic Pressures by the Auscultatory Method. Figure 28–3 illustrates the method usually employed in the doctor's office for measuring systolic and diastolic pressures. An inflatable blood pressure cuff is placed around the upper arm and connected by a tube to a mercury manometer. When the pressure in the cuff is elevated above the pressure in the brachial ar-

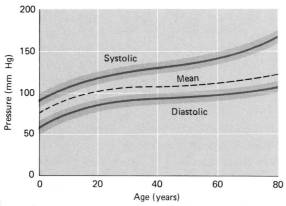

FIGURE 28–2 Changes in systolic, diastolic, and mean pressures with age. The shaded areas show the normal range.

FIGURE 28–3 The auscultatory method for measuring systolic and diastolic pressures.

tery, the wall of the artery collapses because more pressure is then exerted against the outside of the arterial wall than against the inside by the blood. If the pressure in the cuff is gradually decreased until it falls below systolic pressure, small spurts of blood then begin to jet intermittently through the artery each time the arterial pressure rises to its systolic value, a value now great enough to oppose the outside pressure and to open the vessel intermittently at the peaks of the pressure waves. However, during diastole, blood still does not flow through the artery. This intermittent flow of blood causes sound vibrations in the arteries of the lower arm each time the flow suddenly turns on, and this sound can be heard with a stethoscope as shown in the figure. Therefore, to determine systolic pressure, one simply inflates the cuff until the pressure rises to a high value and then deflates it gradually until the intermittent sounds can be heard. At this instant the pressure recorded by the mercury manometer is a reasonably accurate measure of systolic pressure.

To determine diastolic pressure, the cuff pressure is reduced still more. When the pressure falls below the diastolic value, blood then flows through the artery all of the time and no longer in jetlike bursts. As a result, the vibrations caused by the jetting blood disappear, and sounds can no longer be heard with the stethoscope. The pressure recorded at this instant is a close estimate of the diastolic pressure. The intensity of the sounds heard through the stethoscope at different cuff pressure levels is shown graphically at the top of Figure 28–3.

Factors That Affect Pulse Pressure

The arterial *pulse pressure* is the pressure difference between the systolic and diastolic pressures. The higher the systolic pressure and the lower the diastolic pressure, the greater is the pulse pressure, but the more closely these two pressures approach each other, the less becomes the pulse pressure. Thus, when the arterial pressure is normal, at a level of 120/80, the pulse pressure is 40 mm Hg. Two important factors affect the pulse pressure in the arterial system: (1) stroke volume output of the heart and (2) distensibility of the arterial system.

Stroke Volume Output. The stroke volume output is the amount of blood pumped by the heart with each beat. Normally, it is approximately 70 ml, but it can on occasion fall to as low as 10 to 20 ml or can rise to as high as 200 ml. Obviously, if only a small quantity of blood is pumped into the arterial system with each beat, the pressure will not rise and fall greatly during each cardiac cycle. But if the stroke volume output is great, the pressure will rise and fall tremendously, thereby giving an extremely high pulse pressure. In other words, the pulse pressure is roughly proportional to the stroke volume output.

The slower the rate of the heart, the greater must be the stroke volume output to maintain adequate blood flow through the body. Very slow heart beats are quite often found in well-trained athletes, the slowness actually indicating efficient hearts. The large stroke volume output is accompanied by a high pulse pressure. On the other hand, persons with fever, toxicity, or weakened hearts very frequently have fast heart rates with accompanying small stroke volume outputs. The total amount of blood pumped per minute may be the same as that pumped by the athlete's heart, but because of the small stroke volume output the pulse pressure is very slight.

Distensibility of the Arterial System. The more distensible the arteries, the greater the quantity of blood that can be compressed into the arterial system by a given amount of pressure. Therefore, each beat of the heart causes far less pressure rise in very distensible arteries than in nondistensible arteries.

Conditions That Cause Abnormal Pulse Pressure. In old age, the distensibility of the arterial system decreases tremendously because of arteriosclerotic changes in the walls of the vessels. *Arteriosclerosis* (hardening of the arteries) is usually the end result of atherosclerosis, which was described in Chapter 27. In many older persons who have arteriosclerosis, the arterial walls even become calcified, bone-hard tubes. As a result, the arterial system cannot stretch adequately during systole and cannot recoil to a smaller size during diastole. Therefore, pulses of blood entering the arterial tree cause tremendous changes in pressure, as illustrated by the middle curve of Figure 28–1, sometimes causing a pulse pressure of 100 mm Hg or more.

Aortic regurgitation occurs when the aortic valve has been partially destroyed. After blood has been pumped by the heart into the arterial system during systole, the valve fails to close so that much of the blood flows backward into the relaxed left ventricle during diastole, thus causing the arterial diastolic pressure to fall very low—often to zero, as shown by the third curve in Figure 28–1. Consequently, the pulse pressure becomes the greatest possible, sometimes as great as 160 mm Hg.

FIGURE 28–4 Progressive movement of the pressure pulse wave along the aorta.

Transmission of Pressure Pulses to the Smaller Vessels

When the blood pumped by each beat of the heart is suddenly thrust into the aorta, the resulting increase in pressure at first distends only the initial portion of the aorta, that portion adjacent to the heart as shown in Figure 28–4. A short period of time is required for some of this blood to push forward along the arterial tree and build up the pressure further peripherally. This movement of the pressure along the arteries is called *transmission of the pressure pulse.*

Damping of the Pressure Pulses. The pressure pulse becomes less and less intense as the wave spreads toward the smaller vessels. This is called *damping* of the pulsations, and it results mainly from (1) the resistance to blood flow in the vessels and (2) the distensibility of the vessels. Figure 28–5 illustrates this damping of the pulse wave as it spreads peripherally, showing especially the almost complete absence of pulsations in the capillary.

PRESSURES AT DIFFERENT POINTS IN THE SYSTEMIC CIRCULATION

Figure 28–6 shows the pressures in the different vessels from the aorta to the venae cavae. From this chart it is obvious that about one half of the entire fall in pressure is in the arterioles and about one fourth in the capillaries. The reason for this is that about one half of the resistance to blood flow is in the arterioles and about one fourth in the capillaries.

In the normal person, the pressure in the aorta, though highly pulsatile, averages approximately 100 mm Hg, and it falls to about 85 mm Hg at the ends of the small arteries. The pressure then falls another 55 mm Hg in the arterioles, decreasing to 30 mm Hg at the beginning of the capillaries. At the end of the capillaries the pressure normally is about 10 mm Hg, and it gradually decreases in the venules, small veins, and large veins until it reaches 0 mm Hg in the right atrium.

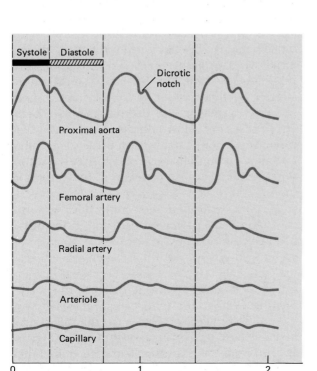

FIGURE 28–5 Changes in the pulse pressure contour as the pulse wave travels toward the smaller vessels.

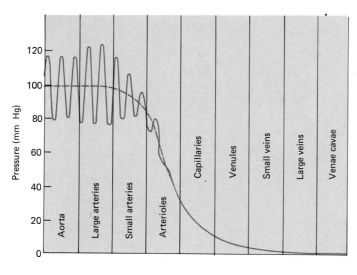

FIGURE 28-6 Pressures in different parts of the systemic circulation.

MEAN ARTERIAL PRESSURE

The *mean arterial pressure* is the arterial pressure averaged during a complete pressure pulse cycle. The mean pressure usually is not exactly equal to the average of systolic and diastolic pressures, for during each pressure cycle the pressure usually remains at systolic levels for a shorter time than at diastolic levels. Therefore, when the pressures at all stages of the pressure cycle are averaged, the resulting mean pressure is slightly nearer diastolic pressure than systolic pressure.

So far as flow of blood in the circulatory system is concerned, *the mean arterial pressure is much more important than is either systolic or diastolic pressure*, because it is the mean pressure that determines the average rate at which blood will flow through the systemic vessels. For this reason, in most physiologic studies it is not necessary to record the pressure changes throughout the pressure cycle but instead simply to record the mean arterial pressure.

REGULATION OF ARTERIAL PRESSURE

Relationship of Mean Arterial Pressure to Total Peripheral Resistance and Cardiac Output

Recalling once again the basic formula given in Chapter 26 for the relationship of arterial pressure to blood flow and resistance, one can derive the following formula, which is basic to understanding arterial pressure regulation:

$$\text{Arterial pressure} = \text{Cardiac output} \times \text{Total peripheral resistance}$$

The *cardiac output* is the rate at which blood is pumped by the heart, and the *total peripheral resistance* is the total resistance of all the vessels in the systemic circulation, from the origin of the aorta back to the veins entering the right atrium. Applying this formula, one can see that either generalized constriction of all the small resistance blood vessels, mainly the arterioles, or an increase in cardiac output will raise the arterial pressure. Conversely, dilatation of the resistance vessels or a decrease in cardiac output will lower the pressure.

Basic Mechanisms of Pressure Regulation

Under resting conditions, the mean arterial pressure is normally regulated at a level of almost exactly 100 mm Hg. The body has four major types of arterial pressure regulatory systems that are responsible for maintaining the normal pressure. These are (1) nervous mechanisms that regulate arterial pressure by controlling the degree of blood vessel constriction as well as heart pumping activity, (2) a capillary fluid shift mechanism that regulates arterial pressure by altering the blood volume, (3) a kidney excretory mechanism that also regulates arterial pressure by altering the blood volume, and (4) hormonal mechanisms that regulate either the blood volume or the degree of arteriolar constriction.

Nervous Regulation of Arterial Pressure

The anatomy of the nervous system for control of the circulation is illustrated in Figure 28-7. The inset to the left shows the medullary portion of the lower brain stem in which is located the so-called *vasomotor center*, which controls (a)

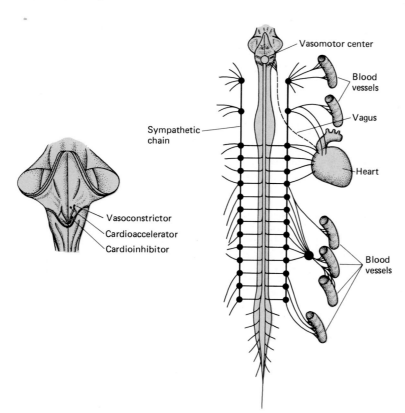

Vasomotor center

Blood vessels

Vagus

Heart

Blood vessels

Sympathetic chain

Vasoconstrictor
Cardioaccelerator
Cardioinhibitor

FIGURE 28–7 The vasomotor center and its connections with the circulatory system through the sympathetic nervous system.

the degree of vasoconstriction of the blood vessels and (b) the heart rate (cardioacceleration and cardioinhibition).

Sympathetic Effects on the Circulation. The vasomotor center controls the circulation mainly through the sympathetic nervous system, which is illustrated to the right in Figure 28–7. Nerve impulses are transmitted down the spinal cord into the sympathetic chains and then to the heart and blood vessels. These impulses increase the *rate of heart beat* and also *strength of heart contraction*, both of which tend to increase the arterial pressure. They also cause *vasoconstriction* of the arterioles and small arteries in most parts of the body, which increases the total peripheral resistance and, therefore, increases the arterial pressure. Finally, they constrict the venous reservoirs, which forces extra blood into the heart and increases cardiac output, this also increasing the arterial pressure.

Parasympathetic Effects on the Heart. The vasomotor system also helps to control the circulation through the vagus nerves which carry *parasympathetic* fibers to the heart. Usually, when the sympathetics are stimulated, the parasympathetic fibers in the vagi are inhibited. Since parasympathetic stimulation has opposite effects on the heart to those of sympathetic stimulation, that is, it decreases the heart's activity, parasympathetic inhibition has the same effect on heart pumping as does sympathetic stimulation.

Except for this effect on the heart, the parasympathetics play very little role in controlling the circulation.

Vasomotor Tone. Even normally, the sympathetic nervous system transmits impulses at a low but continuous rate to the vascular system, thus maintaining a moderate degree of vasoconstriction in the blood vessels. The sympathetic system then decreases the arterial pressure by simply decreasing the number of impulses transmitted to the blood vessels, thus allowing the arterioles to dilate and reduce the pressure. Conversely, it increases the arterial pressure by increasing the number of impulses above those normally transmitted, thus further constricting the arterioles.

Figure 28–8 demonstrates the importance of vasomotor tone in the regulation of arterial pressure. At the beginning of this record, the mean arterial pressure was 100 mm Hg; then total spinal anesthesia was instituted by injecting a local anesthetic all the way up the spinal canal. This blocked all nerves coming from the spinal cord, including the sympathetic nerves. Note that the arterial pressure immediately fell to 50 mm Hg, which is the pressure in the normal circulatory system when there is no vasomotor tone. After a few minutes, a small amount

FIGURE 28—8 Effect on arterial pressure of greatly decreased vasomotor tone caused by spinal anesthesia and of greatly increased vasomotor tone caused by injection of norepinephrine.

of norepinephrine (the substance normally secreted at the sympathetic nerve endings) was injected into the blood. This caused excessive vasoconstriction throughout the body, as well as increased heart activity, which immediately elevated the arterial pressure up to 150 mm Hg. Yet, when the effect of the norepinephrine wore off after another minute, the arterial pressure returned to its basal level. Thus, we see from this experiment that arterial pressure can be reduced far below normal by decreasing the degree of vasomotor tone, and it can be elevated far above normal by increasing the tone.

THE BARORECEPTOR PRESSURE CONTROL SYSTEM

The nervous pressure regulatory system that has been studied more than any other is the *baroreceptor system* illustrated in Figure 28–9. In the arch of the aorta and in the walls of the internal carotid arteries near their origins in the neck (areas called the *carotid sinuses*) are many small nerve receptors that detect the degree of stretch of the arteries caused by the pressure. These receptors are called *baroreceptors* or *pressoreceptors.* The number of impulses transmitted from the baroreceptors increases as the arterial pressure rises. On passing to the brain, the impulses *inhibit* the vasomotor center, thereby dilating the systemic blood vessels and reducing the pumping activity of the heart; both of these effects decrease the arterial pressure back toward normal. On the other hand, when the pressure in the arteries falls too low, the baroreceptors lose their stimulation, so that the vasomotor center now

becomes excessively excited, elevating the arterial pressure again back toward normal.

One can readily see that the baroreceptor system opposes either a rise or a fall in pressure. For this reason, it is sometimes called a *moderator system* or a *buffer system.* As an example of this system's effectiveness, if 500 ml of blood is injected into a person's circulation over a period of about 30 seconds, the arterial pressure will normally rise no more than 15 mm Hg. However, injection of the same quantity of blood into the

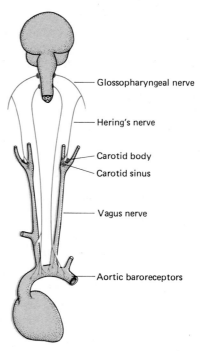

FIGURE 28—9 The baroreceptors (pressoreceptors) and their connections with the vasomotor center.

circulation of someone whose baroreceptors have been inactivated will cause the pressure to rise several times as much, or as much as 40 mm Hg.

Function of the Baroreceptors When a Person Stands. When a person stands, the blood vessels in the lower body stretch because they are then exposed to the weight of the vertical columns of blood in the vessels that run the length of the body. This obviously causes loss of blood from the heart while *pooling* the blood in the lower vessels, and the net result is also a marked decrease in both heart pumping and arterial pressure. But the falling pressure automatically excites the baroreceptor reflex, which in another few seconds returns the pressure almost to normal. One of the values of this constancy of pressure is that it helps to maintain an adequate blood supply to the brain regardless of the position of the body.

EFFECT OF ISCHEMIA OF THE VASOMOTOR CENTER ON ARTERIAL PRESSURE

The term *ischemia* means insufficient blood flow to supply the normal needs of a tissue. The vasomotor center can become ischemic because of too low arterial pressure to cause adequate blood flow to the brain or because the brain is compressed by a brain tumor or by too much pressure in the cerebrospinal fluid. When ischemia does occur, the vasomotor center becomes greatly excited, thereby raising the arterial pressure; this in turn usually relieves the ischemia. This effect is illustrated by the experiment of Figure 28–10, which shows fluid being injected under pressure into the space around the brain. Compression of the brain by the fluid also compresses the arteries that supply the vasomotor

center in the medulla, and the resultant ischemia excites the center, thereby elevating the arterial pressure, this occurring within a few seconds.

The medullary ischemia mechanism normally protects the brain from being injured because of insufficient pressure to supply adequate blood flow to the brain tissue. That is, the resulting rise in pressure increases the brain blood flow back toward normal.

This vasomotor ischemia mechanism for raising the arterial pressure is by far the most powerful of all the nervous controls of pressure. Indeed, severe ischemia of the brain can increase the mean arterial pressure to as high as 260 mm Hg, the highest pressure that the normal heart can achieve.

The Capillary Fluid Shift Mechanism for Pressure Regulation

Because of the rapidity with which the various nervous control systems can return arterial pressure toward normal when it becomes abnormal, it is very tempting to explain arterial pressure regulation entirely on the basis of nervous mechanisms alone. However, the nervous mechanisms have a major failing that prevents their continuing to control arterial pressure over long periods of time. This failing is that the nervous mechanisms *reset* after a while to the new level of pressure and soon stop sending their signals indicating abnormality. Therefore, after a few days, most nervous regulatory mechanisms become ineffective. By that time, nonnervous mechanisms take over the pressure regulation. One of these is the *capillary fluid shift mechanism*, which is particularly important in helping to regulate arterial pressure when the blood vol-

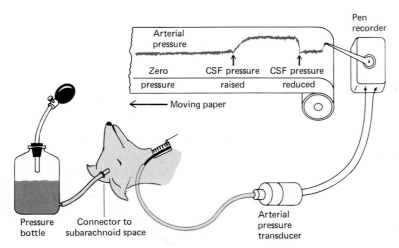

FIGURE 28–10 Elevation of arterial pressure caused by increased cerebrospinal fluid pressure.

ume tends to become either too little or too great.

An increase in blood volume, such as might occur following a transfusion, increases the pressures in all parts of the systemic circulation, including the pressure in the capillaries. However, in the next few minutes to several hours, even in the absence of nervous control of pressure, the pressures in all parts of the circulation return back toward normal. Much of the decline in pressure is caused by shift of fluid out of the circulation through the capillary membranes into the interstitial spaces, thereby decreasing the blood volume back toward normal.

The mechanism of the capillary fluid shift is simply the following: Too much blood volume increases the pressure in the capillaries as well as in the arteries. The increased capillary pressure causes fluid to leak out of the circulation into the interstitial spaces. Conversely, when the blood volume becomes too low, the capillary pressure falls, and now fluid is pulled by osmosis from the interstitial spaces into the circulation because of the osmotic pressure effect of the proteins in the plasma (which will be discussed in Chapter 30).

The capillary fluid shift mechanism is much slower to regulate arterial pressure than are the nervous mechanisms, for it requires from 10 minutes to several hours to readjust the arterial pressure back toward normal.

Regulation of Arterial Pressure by the Kidneys

Kidney Control of Pressure by Controlling Blood Volume.   The kidneys, like the capillaries, can regulate arterial pressure by increasing or decreasing the blood volume. After even a slight decrease in arterial pressure, the kidneys often stop or almost stop forming urine, because the rate of urine formation by the kidneys is determined to a major extent by the pressure in the renal arteries. Therefore, fluid and electrolytes taken in by mouth gradually accumulate in the body—that is, they are not excreted by the kidneys—until the blood volume rises enough to reestablish normal arterial pressure.

Conversely, when the arterial pressure rises too high, the urinary output increases, and over a period of hours, the blood volume decreases, causing the arterial pressure again to return toward normal.

Formation of Renin by the Kidneys, and Function of the Renin-Angiotensin System in Arterial Pressure Regulation.   The kidneys also have a hormonal mechanism for regulating

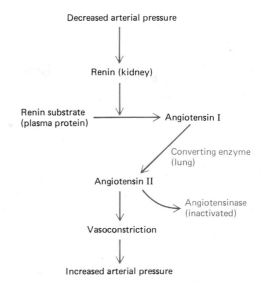

FIGURE 28–11   The renin-angiotensin-vasoconstrictor mechanism for arterial pressure control.

arterial pressure. When the pressure falls to a lower than normal value, diminished flow through the kidneys causes them to secrete the substance *renin* into the blood. Renin in turn acts as an enzyme to convert one of the plasma proteins, *renin substrate*, into the hormone *angiotensin I*. This hormone has relatively little effect on the circulation, but it is rapidly converted into a second hormone, angiotensin II, by another enzyme called *converting enzyme* found mainly in the small blood vessels of the lung. The angiotensin II lasts in the blood for only 1 to 3 minutes because it is inactivated within a few minutes by still other enzymes in both the blood and the tissues, collectively called *angiotensinase*. Nevertheless, during the time that the angiotensin II does circulate in the blood, it causes vasoconstriction of the arterioles, which increases the arterial pressure back toward normal.

This entire sequence, from decreased arterial pressure to increased formation of angiotensin II, vasoconstriction, and return of the arterial pressure back toward normal, is illustrated in Figure 28–11. This system sometimes causes hypertension when it functions abnormally, as we shall discuss later in the chapter.

Importance of Kidney Regulation of Arterial Pressure.   *The kidneys are by far the most important of all organs of the body for long-term regulation of arterial pressure.* This is shown very dramatically by the experiment of Figure 28–12. If clamps are placed on the arteries of both kidneys so that the renal blood supply is diminished, the arterial pressure begins to rise. In several days the mean pressure

FIGURE 28—12 Effect of placing a constricting clamp on the renal artery of one kidney after the other kidney has been removed. Note the changes in systemic arterial pressure, renal artery pressure distal to the clamp, and rate of renin secretion.

may rise to twice its original value. The elevated pressure will then force normal blood flow through the kidneys despite the continued constriction of the arteries, and the kidneys will function normally again. Note that increased renin secretion is one of the factors that increases the arterial pressure during the early part of the experiment. However, the prolonged elevation of the pressure is caused by the fluid volume retention mechanism of the kidneys.

When the constrictions are removed, the arterial pressure falls back to normal within another few days. Therefore, it is obvious that the kidneys are capable of protecting themselves against diminished blood flow; that is, *diminished flow through the kidneys causes a general arterial pressure rise until the kidneys again receive an adequate blood supply.* On the other hand, if the blood flow through the kidneys becomes excessive, the arterial pressure falls until the renal flow again returns to normal.

Hormonal Regulation of Arterial Pressure— the Effect of Aldosterone

In addition to the renin-angiotensin hormonal mechanism of the kidneys, one other major hormonal system is also involved in the regulation of arterial pressure. This is the secretion of aldosterone by the *adrenal cortex.* The adrenal cortex is the outer shell of two small endocrine glands, the adrenals, located immediately above the two kidneys. This cortex secretes adrenocortical hormones, one of which, *aldosterone,* controls the kidney output of salt and water, as we shall discuss in much more detail in Chapter 46. The amount of salt and water in the body in turn helps to determine the volumes of blood and interstitial fluid.

The manner in which the aldosterone mechanism enters into blood pressure regulation is the following: When the arterial pressure falls very low, lack of adequate blood flow through the body's tissues causes the adrenal cortices to secrete aldosterone. One of the causes of this effect is stimulation of the adrenal glands by the angiotensin II that is formed when the arterial pressure falls. The aldosterone then has a direct effect on the kidneys to decrease the excretion of both salt and water in the urine. As a consequence, both of these substances are retained in the blood, which builds up the blood volume and returns the arterial pressure to normal. Conversely, an elevated arterial pressure reverses this mechanism so that fluid volumes, and consequently the arterial pressure, decrease.

Regulation of Mean Arterial Pressure During Exercise

One of the most severe types of stress to which the circulation can be subjected is strenuous exercise, for active muscles need tremendous blood supply. One of the means for increasing blood flow to the muscles during exercise is to elevate the arterial pressure, the pressure usually rising from the normal mean value of 100 mm Hg up to about 130 mm Hg in most instances of moderate to heavy exercise. The mechanisms causing this are the following: First, increased muscular metabolism during exercise increases the concentrations of carbon dioxide, lactic acid, and other metabolites in the blood, and these circulate to the vasomotor center and excite it, elevating the pressure. Second, and probably even more important, the vasomotor center is stimulated by nerve impulses generated in the muscle-controlling "motor" area of the brain. Therefore, at the same time that the motor area causes the muscles to contract, it also excites the vasomotor center and thereby elevates the arterial pressure.

HYPERTENSION

Hypertension, which is the same as "high blood pressure," means high arterial pressure, and it occurs in approximately one out of every five persons before the end of life, usually in middle or old age. The excessive arterial pressure of hypertension can rupture blood vessels in the brain to cause "strokes"; in the kidney to cause "renal failure"; or in other vital organs to cause blindness, deafness, heart attacks, or so forth. Also, it can place excessive strain on the heart and cause it to fail. For these reasons one of the most important problems in physiology is to determine the causes of hypertension. Some of the known causes follow.

Hypertension Caused by Abnormal Function of the Pressure Regulatory Mechanisms

RENAL HYPERTENSION

Many conditions that damage the kidneys can cause *renal hypertension*. For instance, constriction of the renal arteries, as illustrated in Figure 28–12, causes the arterial pressure to rise, and the greater the degree of constriction, the greater the elevation of pressure. Diseases of the kidneys such as kidney infections, sclerosis of the renal arterioles, kidney inflammation, and so forth can all elevate the pressure.

A particularly interesting type of renal hypertension occurs when the aorta is occluded above the kidneys, a condition called *coarctation of the aorta*. This occurs in an occasional person because of abnormal formation of the aorta during fetal life before birth. In this condition blood finds its way through many smaller arteries of the body wall from the upper segment of the aorta into the lower segment, but the resistance to flow through these small arteries is so great that the arterial pressure in the lower aorta is much lower than that in the upper aorta. The insufficient blood flow to the kidneys from the lower aorta causes the pressures throughout the body to rise until the pressure in the lower aorta

returns to normal. Obviously, by this time the pressure in the upper aorta will have risen to a very high level. Thus, the kidneys automatically maintain a normal arterial pressure at the kidney level even at the expense of the arterial pressure in the upper part of the body.

Excess Blood Volume as One of the Major Mechanisms of Renal Hypertension. One mechanism of renal hypertension is development of too much blood volume when renal function becomes abnormal. Obviously, if renal output becomes reduced as a result of renal disease or poor blood supply, the retention of salt and water would be expected to increase both the interstitial fluid volume and the blood volume. Experiments have recently demonstrated that removal of large portions of the kidney plus a high daily intake of water and salt does indeed increase both of these volumes considerably. These greater volumes elevate the arterial pressure in the following ways: (1) The heightened blood volume increases the blood flow into the heart, which obviously raises the cardiac output above normal. (2) The excess cardiac output causes hypertension. And it also causes too much blood to flow through the tissues. (3) The excess tissue blood flow causes the tissue vessels to constrict during the next few days because of the long-term tendency of each tissue to reset its own blood flow back to normal. (4) This constriction of the tissue vessels increases the total peripheral resistance far above normal, which automatically returns both the blood volume and the cardiac output back toward normal even though the arterial pressure remains elevated.

Thus, the initial increase in blood volume leads to hypertension, but a further sequence of events returns both the blood volume and the cardiac output almost back to normal even though the total peripheral resistance remains very high. In fact, after the first few days, the blood volume and the cardiac output remain only about 5 to 8 percent above normal even in pure volume-loading hypertension. These small increases can rarely be measured, so that most persons with volume-loading hypertension are said to have normal blood volume and cardiac output but high total peripheral resistance and arterial pressure.

Role of the Renin-Angiotensin System in Renal Hypertension. Sometimes, damaged kidneys secrete large amounts of renin, which in turn leads to formation of angiotensin and this to vasoconstriction and increased total peripheral resistance. This sequence occurs especially in the condition called *malignant hypertension.* The vasoconstriction obviously can act in concert with the excess blood volume effect that occurs in kidney disease to cause very severe hypertension.

HORMONAL HYPERTENSION

Occasionally the adrenal *cortices* secrete excessive quantities of aldosterone, either because of an aldosterone-secreting tumor in one of the adrenal glands or because of excessive stimulation of the adrenals by the anterior pituitary gland. In any event, the increased production of aldosterone causes the kidneys to retain excessive quantities of water and salt. The fluid volumes throughout the body increase, and the arterial pressure often rises considerably above normal.

A second type of hormonal hypertension is that caused by a *pheochromocytoma.* This is a tumor of the adrenal *medulla,* which is the central portion of the adrenal gland, entirely distinct from the adrenal cortex. Because the medulla is part of the sympathetic nervous system, the pheochromocytoma secretes large quantities of epinephrine and norepinephrine, the hormones normally secreted by sympathetic nerve endings. These hormones circulate in the blood to all the vessels and promote intense vasoconstriction. As a result, the mean arterial pressure may rise to as high as 200 mm Hg or higher. The pheochromocytoma usually secretes its hormones only when stimulated by the sympathetic nervous system. This means that when a person who has this type of tumor becomes excessively excited, the secretion of epinephrine and norepinephrine may be tremendous, causing the blood pressure to become very high. Between periods of excitement the pressure may be essentially normal.

NEUROGENIC HYPERTENSION

Many clinical doctors believe that excessive *nervous tension* can lead to hypertension. Also, animal experiments have demonstrated that lesions in certain parts of the brain stem or in the hypothalamus can sometimes cause hypertension lasting for at least several weeks. But, how do these various nervous factors cause the hypertension? The usual thesis is that the degree of sympathetic activity is increased and that this constricts the peripheral blood vessels and increases heart activity, thus increasing the arterial pressure.

But, if increased sympathetic activity should be the usual cause of hypertension, then removing the sympathetic nervous system should cause the arterial pressure to return to normal. Unfortunately, in thousands of patients who had such operations for treatment of hypertension 30 to 40 years ago, statistical studies showed no significant reduction of pressure. Therefore, if neurogenic mechanisms do cause permanent hypertension, it is likely that they do so by some other means besides chronic sympathetic stimulation. One suggestion is then that *extreme intermittent sympathetic activity* occurring at repeated intervals may cause permanent changes in the kidneys and that it is these changes that lead to the chronic hypertension.

Essential Hypertension

The types of hypertension described above all have known causes. Yet in approximately 95 percent of all hypertensive persons the cause is unknown, and they are said to have *essential hypertension,* the word "essential" meaning simply "of unknown cause." Many efforts have been made to prove that essential hypertension is caused by a kidney abnormality, a glandular abnormality, or excessive activity of the vasomotor center. Authentic proof that essential hypertension is caused by any one of these factors is yet lacking. However, recent studies have demonstrated several different abnormalities of renal function in patients with essential hypertension. The most important of these is that the kidneys of these patients require a very high arterial pressure to make them excrete normal amounts of salt and water. That is, when the mean arterial pressure falls to the normal level of 100 mm Hg, the kidneys form almost no urine, so that salt and water accumulate in the blood and interstitial fluids until the arterial pressure rises to a hypertensive level. Only then will the kidneys begin to form normal amounts of urine.

Yet, it still is not known why the kidneys of essential hypertensive patients require such high arterial pressures to excrete normal amounts of salt and water. Most likely, the cause is an abnormality of the renal blood vessels, but it could also be nervous or hormonal influences on the kidneys that diminish excretion.

Essential hypertension is mainly a hereditary disease. That is, the hypertensive person usually inherits the abnormality that causes hypertension from one or both of his parents, or occasionally from a grandparent or other ancestor.

QUESTIONS

1. What is meant by systolic pressure, diastolic pressure, and pulse pressure?
2. How do the stroke volume output of the heart and the distensibility of the arterial system affect the pulse pressure?
3. In the systemic circulation, what are the approximate normal pressures in the aorta, the arterioles, the capillaries, the venules, and the venae cavae?
4. If the arterial pressure increases 50 percent while the cardiac output increases twofold, how much does the total peripheral resistance change?
5. Explain the nervous mechanism for regulation of arterial pressure, including the baroreceptor control system.
6. How does shift of fluid into or out of the capillaries act as a mechanism for pressure regulation?
7. Explain how the kidneys help to control arterial pressure by increasing or decreasing blood volume.
8. Explain how hypertension can result from abnormal kidney function.
9. What is the role of the renin-angiotensin system in the control of arterial pressure and also in hypertension?
10. What is meant by "essential hypertension"?

REFERENCES

Bianchi, G., and Bazzato, G. (eds.): The Kidney in Arterial Hypertension. Baltimore, University Park Press, 1979.
Coleman, T.G., and Guyton, A.C.: Hypertension caused by salt loading in the dog. III. Onset transients of cardiac output and other circulatory variables. *Circ. Res.*, 25:153, 1969.
Downing, S.E.: Baroreceptor regulation of the heart. *In* Berne, R.M., *et al.* (eds.): Handbook of Physiology. Sec. 2, Vol. 1. Baltimore, Williams & Wilkins, 1979, p. 621.
Guyton, A.C.: Arterial Pressure and Hypertension. Philadelphia, W.B. Saunders, 1980.
Guyton, A.C., *et al.*: Arterial pressure regulation: Overriding dominance of the kidneys in long-term regulation and in hypertension. *Am. J. Med.*, 52:584, 1972.
Guyton, A.C., *et al.*: A systems analysis approach to understanding long-range arterial blood pressure control and hypertension. *Circ. Res.*, 35:159, 1974.
Guyton, A.C., *et al.*: Integration and control of circulatory function. *Int. Rev. Physiol.*, 9:341, 1976.
Guyton, A.C., *et al.*: Salt balance and long-term blood pressure control. *Annu. Rev. Med.*, 31:15, 1980.
Hall, J.W., and Guyton, A.C.: Changes in renal hemodynamics and renin release caused by increased plasma oncotic pressure. *Am. J. Physiol.*, 231:1550, 1976.
Mancia, G., *et al.*: Reflex control of circulation of heart and lungs. *Int. Rev. Physiol.*, 9:111, 1976.
Manning, R.D., Jr., *et al.*: Essential role of mean circulatory filling pressure in salt-induced hypertension. *Am. J. Physiol.*, 236:R40, 1979.
Peach, M.J.: Renin-angiotensin system: Biochemistry and mechanisms of action. *Physiol. Rev.*, 57:313, 1977.
Scriabine, A. (ed.): Pharmacology of Antihypertensive Drugs. New York, Raven Press, 1980.
Ziegler, M.G.: Postural hypotension, *Annu. Rev. Med.*, 31:239, 1980.

29

Cardiac Output, Venous Pressure, Cardiac Failure, and Shock

Overview

The *cardiac output* is the rate at which the heart pumps blood. For a normal adult this is usually about 5 L/min, but it can rise to as high as 20 L/min in a normal young adult during strenuous exercise and sometimes to as high as 35 to 40 L/min in the well-trained marathon runner.

Two basic factors determine the level of cardiac output: One of these is the *ability of the heart to pump blood.* The second is the *ability of the blood to flow through the systemic circulation.* Normally, the ability of the heart to pump blood is far greater than the actual cardiac output. Therefore, the heart's pumping capability usually is not a limiting factor in determining cardiac output. Instead, the cardiac output is normally determined by factors in the peripheral circulation that control blood flow through the peripheral vessels, mainly the following two factors: (1) the resistance to blood flow through the systemic blood vessels and (2) the mean systemic filling pressure.

The *resistance to blood flow through the systemic blood vessels,* in turn, is controlled mainly by the mechanism of *autoregulation,* which means the ability of each tissue to control its own blood flow according to its needs, mainly its need for oxygen as explained in Chapter 26. That is, an insufficiency of oxygen decreases local resistance and therefore increases the local blood flow. The total amount of blood returning by way of the veins to the heart, called the *venous return,* is equal to the sum of all the blood flows through the separate parts of the body. A normal heart will pump all of this blood returning to the heart. Therefore, cardiac output is normally determined indirectly by the usage in the different tissues of nutrients, that is, by the metabolism of the tissues—mainly the usage of oxygen.

The *mean systemic filling pressure* is the average pressure in all parts of the systemic circulation, and it is a measure of the tightness with which the vasculature is filled with blood. It can be increased by increasing the blood volume or by constricting the blood vessels around the blood that is already present. The *rate of venous return is directly proportional to the mean systemic pressure,* so that this too is one of the factors that is important in the control of venous return and cardiac output.

The term *cardiac failure* means depressed pumping effectiveness of the heart. Most frequently, it is caused by *obstructive coronary artery disease* that has decreased the blood supply to the heart muscle. In cardiac failure, one or more of three separate abnormalities of circulatory function occur: (1) *low cardiac output,* leading to insufficient blood flow to all parts of the body; (2) *congestion of blood in the lungs* because of failure of the left heart to pump the blood from the lungs back into the systemic circulation—this frequently leads to severe *pulmonary edema* and to death because of suffocation; and (3) *congestion of blood in the systemic circulation* because of failure of the right heart to pump adequately—this frequently leads to very severe edema throughout the body.

Circulatory shock means a decrease in cardiac output to a level so low that tissues throughout the body suffer for lack of blood supply. Basically, there are two different types of circulatory shock: (1) *cardiac shock* and (2) *low venous return shock.* Cardiac shock usually occurs in persons who have had an *acute myocardial infarction,* which means acute blockage of one of the major arteries to the heart muscle and consequent severe acute depression of the pumping capability of the heart. Low venous return shock usually results from *loss of blood volume,* which *reduces the mean systemic filling pressure* of the circulation to too low a value for adequate venous return. The decreased blood volume can result from *hemorrhage, extreme dehydration* caused by excess sweating, or *loss of plasma* out of the blood vessels into traumatized tissues.

A major problem in circulatory shock, whatever its initiating cause, is that once it becomes severe enough it becomes *progressive.* That is, the circulatory shock itself causes many components of the circulatory system to begin to deteriorate, and this leads to still more circulatory shock. For instance, circulatory shock causes weakness of the heart, and the weakness of the heart causes still more circulatory shock. Once the shock becomes progressive, the person will usually die unless strong measures of treatment are instituted.

CARDIAC OUTPUT

Cardiac output is the rate at which the heart pumps blood. It is usually expressed in liters of flow per minute (L/min). Because the function of the circulatory system is to supply adequate nutrition to the tissues, the subject of cardiac output is one of the most important in the field of physiology.

NORMAL VALUES OF CARDIAC OUTPUT

The average cardiac output in an adult lying down and in a state of complete rest is about 5 L/min. If he walks, it rises to perhaps 7.5 L/min. If he performs strenuous exercise, it might rise to as high as 20 to 25 L/min in the normal person or to as high as 35 L/min in the well-trained marathon runner.

One can see, therefore, that the cardiac output varies in proportion to the degree of activity of the person. The ability to regulate the cardiac output in accord with the needs of the body is one of the most remarkable feats of the circulation. Much of the present chapter is devoted to discussing this regulation.

Regulation of Cardiac Output

The amount of blood that the heart pumps each minute is determined by two major factors: (1) the pumping effectiveness of the heart itself and (2) the ease with which blood can flow through the body and return to the heart from the systemic circulation after it is pumped. To understand cardiac output regulation it is essential that one be familiar with the way in which each of these factors affects the circulation.

The Pumping Effectiveness of the Heart. Normally the heart is an automatic pump that pumps any blood that flows into it from the veins. If the amount flowing from the veins increases, the heart becomes stretched and automatically adjusts to accommodate the extra blood. This ability of the heart to pump any amount of blood, within limits, is called the *law*

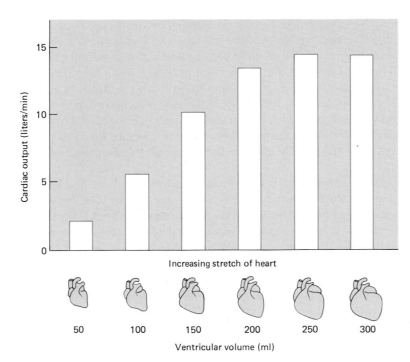

FIGURE 29–1 Effect of increased cardiac filling on the heart's ability to pump blood.

of the heart, which was explained in detail in Chapter 24. Figure 29–1 illustrates this principle pictorially and graphically, showing that as the heart becomes progressively stretched by greater and greater inflow of blood, it automatically pumps increasing amounts of blood until the heart muscle finally becomes stretched beyond its physiologic limits.

Effect of Nervous Stimulation on the Effectiveness of the Heart as a Pump. Stimulation of the sympathetic nerves to the heart greatly enhances the effectiveness of the heart as a pump, whereas stimulation of the parasympathetic nerves (the vagi) greatly decreases the pumping effectiveness. Under normal resting conditions parasympathetic stimulation keeps the heart activity depressed, which probably prolongs the life of the heart. But during exercise and other circulatory states in which the cardiac output needs to be enhanced very greatly, sympathetic stimulation takes over and makes the heart a very effective pump, about two times as effective as the normal unstimulated heart.

Therefore, the pumping effectiveness of the heart can be increased in two separate ways: (1) by local adaptation of the heart caused by increased inflow of blood, the mechanism called the law of the heart, and (2) by a shift from parasympathetic to sympathetic stimulation.

Flow of Blood into the Heart (Venous Return). Because the heart pumps whatever amount of blood enters its chambers (up to its limit to pump), cardiac output normally is regulated mainly by the amount of blood that returns to the heart from the peripheral vessels. This flow of blood into the heart is called *venous return*. When a person is at rest, venous return averages 5 L/min. Therefore, this is also the amount of blood that is pumped, and it is also the amount of cardiac output. However, if 40 L/min, an abnormally large amount, should attempt to return to the heart of a normal person, even a heart strongly stimulated by the sympathetics would be able to pump only 20 to 25 L because this is the heart's limit. Therefore, much of the returning blood would dam up in the veins rather than being pumped into the arteries.

The three important factors that determine the amount of venous return per minute are (1) the resistance of blood flow through the systemic vessels, (2) the average pressure of the blood in all parts of the systemic circulation, which is called the *mean systemic filling pressure*, and (3) the right atrial pressure.

Effect of Vascular Resistance on Venous Return. The factor that most often controls venous return is the vascular resistance. That is, if a large share of the peripheral circulatory vessels becomes widely opened so that the vascular resistance decreases, large amounts of blood will flow from the arteries into the veins. This obviously increases the rate of venous return and also augments the cardiac output. Venous return is actually the sum of all the blood flows through the different peripheral tissues. Therefore, every time the blood flow in any single peripheral tissue increases, the venous return likewise increases.

The phenomenon of autoregulation, which has already been discussed, plays a major role in determining blood flow in each tissue. Therefore, this mechanism also figures significantly in determining the rate of venous return. For instance, if the rate of metabolism in a tissue becomes increased, the autoregulatory mechanism will automatically decrease the resistance, thereby increasing the blood flow through that tissue to supply the needed nutrients. This in turn boosts the rate of venous return by the same amount.

Another instance in which changes in peripheral resistance play a major role in determining venous return is in the condition called *arteriovenous fistula*. This is a condition in which an abnormal direct opening develops between an artery and a vein. It occurs frequently in persons who have a gunshot wound that ruptures simultaneously an artery and a vein that lie side by side. Upon healing, the arterial blood flows directly into the vein. This greatly decreases the peripheral resistance and also allows tremendous flow of blood directly from the arterial tree into the veins. One can well understand that this condition greatly increases the venous return and, therefore, also greatly increases cardiac output.

Figure 29–2 illustrates the effect on venous return and cardiac output of changing the systemic vascular resistance. Note in Figure 29–2A that when all the peripheral resistances are decreased to one-half normal, the cardiac output rises from its normal value of 5 L/min up to double this value—that is, to 10 L/min. Conversely, Figure 29–2B shows that a twofold increase in resistances everywhere in the peripheral circulation will decrease the cardiac output to approximately one-half normal, or to 2.5 L/min. Thus, it is clear that changing the resistance to blood flow in the peripheral circulatory system is a potent means for controlling cardiac output.

Effect of Mean Systemic Filling Pressure and Right Atrial Pressure on Venous Return. The *mean systemic filling pressure* is the average pressure in all the distensible vessels of the systemic circulation. It is this average pressure that continually pushes blood from the peripheral vessels toward the right atrium.

The *right atrial pressure*, on the other hand, is the input pressure to the heart; this acts as a back pressure to impede blood flow into the heart. When the heart becomes weakened so that it cannot pump blood out of the right atrium with ease, or whenever the heart becomes overloaded for any reason, the right atrial pressure rises and tends to keep blood from entering the heart from the systemic circulation.

Thus, the mean systemic filling pressure is the average pressure in the periphery tending to push blood toward the heart, and the right atrial pressure is the input pressure against which the blood must flow. Therefore, venous return is determined to a great extent by the difference between these two pressures, *mean systemic filling pressure* minus *right atrial pressure*. This difference is called the *pressure gradient for venous return.*

Figure 29–3 illustrates the effect of both mean systemic filling pressure and right atrial pressure on the control of venous return and cardiac output. In Figure 29–3A, the mean systemic pressure is 7 mm Hg and the right atrial pressure 0 mm Hg. The pressure gradient for venous return in this case is 7 − 0, or 7 mm Hg, and the venous return and cardiac output are 5 L/min. In Figure 29–3B, the mean systemic pressure is 18 mm Hg, and the right atrial pressure, because the heart is now beginning to be overloaded, has risen to 4 mm Hg. In this case, the pressure gradient for venous return is 18 − 4, or 14 mm Hg. This pressure gradient is twice that in Figure 29–3A. Likewise, the cardiac output and venous return are now twice that in Figure 29–3A, or 10 L/min, showing the importance of the pressure gradient for venous return in determining cardiac output.

Cardiac output = 10 liters/min

Cardiac output = 2.5 liters/min

Resistance = ½ × normal

Resistance = 2 × normal

A   B

FIGURE 29–2 Effect of systemic vascular resistance on venous return and cardiac output.

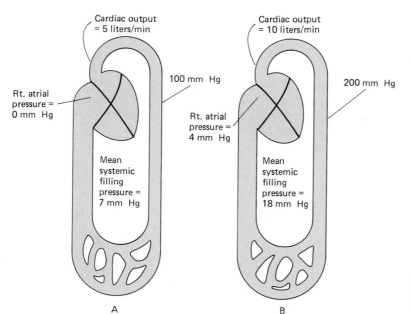

FIGURE 29–3 Effect of mean systemic filling pressure and right atrial pressure on venous return.

Effect of Exercise on Venous Return and Cardiac Output. Exercise greatly increases the venous return and cardiac output for several reasons. First, exercise is one of the most potent stimulators of metabolism that is known; indeed, intense muscle activity occasionally causes as much as a 50-fold increase in total body metabolism for short periods of time and a 20-fold increase for long periods of time. This obviously increases the rate of utilization of nutrients, including greatly enhanced use of oxygen. Almost instantly, the blood vessels of the muscles open widely to supply the tissue with needed nutrients. This greatly increases venous return and cardiac output.

However, in addition to the vasodilatation that occurs in exercise, the mean systemic filling pressure also increases for two reasons. First, sympathetic activity is greatly heightened during exercise, and sympathetic impulses constrict the arteries and veins throughout the body. This squeezes the blood and elevates the mean systemic filling pressure. Second, the tightening of the muscles themselves also squeezes the blood vessels, thus further raising the mean systemic filling pressure. Therefore, these effects also help to increase the cardiac output during exercise.

Thus, by utilizing all these mechanisms for dilating the muscle blood vessels and elevating mean systemic filling pressure during exercise, the venous return and cardiac output can be increased to as high as 20 to 25 L/min in the normal young adult and to as high as 30 to 35 L/min in the well-trained athlete.

Summary of Cardiac Output Regulation—Role of Tissue Need in Controlling Cardiac Output. Under normal conditions the heart is capable of pumping far more blood than the body needs. Therefore, the cardiac output is normally controlled very little by the heart itself. Instead, it is determined almost entirely by the rate at which blood flows into the heart from the peripheral circulation. That is, normally, the cardiac output is determined by the venous return to the heart.

In turn, the venous return is determined by the sum of the blood flows through all the different tissues of the body. And each tissue is, in general, capable of controlling its own blood flow in relation to its need for flow. Most tissues control this flow in proportion to their need for oxygen, but this is not true of all tissues. For instance, the brain controls its blood flow in proportion to the amount of carbon dioxide that it needs to remove from the brain tissue; the kidneys control their blood flow in proportion to the excretory products in the blood that need to be excreted; and the skin blood flow is controlled in proportion to the need of the body to eliminate heat.

Therefore, in effect, the cardiac output is normally controlled in proportion to the needs for blood flow in all the respective tissues of the body. When the needs are great, the local tissue control mechanisms increase the flow of blood from the arteries to the veins and this in turn increases the venous return and cardiac output simultaneously.

Conditions Under Which the Heart Plays a Major Role in Controlling Cardiac Output. At times the heart is unable to pump all of the blood that attempts to return to it. This occurs when too much blood attempts to flow from the arteries to the veins through a very large *arteriovenous fistula* or occasionally when the *metabolism of the tissues is so great* that venous return is greater than the amount the heart can pump. Under these conditions the heart becomes overloaded. Obviously, it is the heart then, instead of the venous return, that determines the cardiac output. This, however, is a rare state of events so long as the heart is normal.

A much more common condition in which the heart becomes the controller of cardiac output is *cardiac failure.* Cardiac failure simply means reduced effectiveness of the heart as a pump, usually occurring because of damaged heart muscle or heart valves. In cardiac failure, even the normal venous return might be too great for the heart to pump. Therefore, the limiting factor becomes the heart and not the venous return. The subject of heart failure is so important that it will be discussed in much more detail later in this chapter.

Measurement of Cardiac Output

A *flowmeter* can be applied to the major arteries or veins of an experimental animal in such a manner that the rate of blood flow through the heart, the cardiac output, will be recorded. Such procedures, however, are not very useful even in animal experimentation because of the extreme disruption of normal function caused by the necessary surgery. And, obviously, in the human being they can hardly be used at all. Therefore, for measuring cardiac output in human beings and intact animals the following indirect procedures have been developed.

The Fick Method. One indirect method for measuring cardiac output is based on the so-called Fick principle, which is explained by the example in Figure 29–4. The volume of oxygen in a sample of venous blood is analyzed chemically in 160 ml of oxygen per liter of blood, and in the arterial blood, 200 ml. Thus as each liter of blood passes through the lungs it picks up 40 ml of oxygen. At the same time, using appropriate respiratory apparatus, the amount of oxygen absorbed into the blood through the

FIGURE 29–4 The Fick principle for indirectly measuring cardiac output.

lungs is measured to be 200 ml per minute. From these values, one can calculate the cardiac output in the following manner: If the amount of oxygen absorbed per minute is 200 ml and the amount of oxygen picked up by each liter of blood flowing through the lungs is 40 ml, by dividing 200 by 40 one can conclude that five 1-liter portions of blood must pass through the lungs each minute. Thus, the total cardiac output is calculated to be 5×1 L or 5 L/min.

Expressing the Fick principle mathematically, the following formula applies.

$$\text{Cardiac output in liters per minute} = \frac{\text{Total oxygen absorbed per minute}}{\text{Oxygen absorbed by each liter of blood}}$$

The Dye Method. Another simple procedure for measuring cardiac output is to inject rapidly into a vein a small quantity of foreign substance such as brightly colored dye. As the substance passes through the heart and lungs into the arterial system, its concentration in the arterial blood is recorded by a photoelectric "densitometer" or by taking periodic blood samples for the next 20 to 30 seconds and analyzing these for their content of dye. The more rapidly the blood flows through the heart, the more rapidly the dye appears in the arteries and the sooner it will disappear. From the area under the recorded time-concentration curve of the dye as it passes through the arteries one can calculate the cardiac output quite accurately. That is, the less the area under this curve, the greater is the cardiac output. If the reader will think about it for a moment, he will be able to understand intuitively the principle of this measurement, though the mathematics of the calculation is more than can be discussed here.

VENOUS PRESSURE

RIGHT ATRIAL PRESSURE

The pressure in the venous system is determined mainly by the pressure in the right atrium. Normally, the right atrial pressure is approximately zero—that is, it is almost exactly equal to the pressure of the air surrounding the body. This does not mean, though, that the force distending the walls of the right atrium is zero, because the pressure in the chest cavity surrounding the heart is about 4 mm Hg less than atmospheric pressure. This partial vacuum in the chest actually pulls the walls of the atrium outward so that blood is normally sucked into the atrium from the veins.

If the heart becomes weakened, blood begins to dam up in the right atrium, thereby causing the right atrial pressure to rise. Or, if excessive quantities of blood attempt to flow into the heart from the veins, and the heart cannot pump the extra blood, again the right atrial pressure rises. Therefore, the right atrial pressure, though normally almost exactly zero, often rises above zero in abnormal conditions.

PERIPHERAL VENOUS PRESSURE

The pressure in a peripheral vein is determined by five major factors: (1) the right atrial pressure; (2) the resistance to blood flow from the vein to the right atrium; (3) the rate of blood flow along the vein; (4) pressure caused by weight of the blood itself, called "hydrostatic pressure"; and (5) the venous pump. The effects of most of these factors are self-evident, but some of them need special explanation.

Effect of Right Atrial Pressure on Peripheral Venous Pressure. Since blood flows in the veins always toward the heart, the pressure in any peripheral vein of a person in the lying position must always be as great as or greater than the pressure in the right atrium. If considerable resistance exists between the peripheral vein and the heart, and the flow of blood is at a reasonable rate, then the peripheral venous pressure will be considerably higher than right atrial pressure. Many of the veins are compressed where they pass abruptly over ribs, between muscles, between organs of the abdomen, or so forth. At these points the venous resistance is usually great, impeding blood flow and increasing the pressure in the distal veins. On the average, the pressure in a vein of the arm or leg of a person lying down is approximately 6 to 8 mm Hg in contrast to zero pressure in the right atrium.

Effect of Hydrostatic Pressure on Peripheral Venous Pressure. Hydrostatic pressure is the pressure that results from the weight of the blood itself. Figure 29-5 illustrates the human being in a standing position, showing that for blood to flow from the lower veins up to the heart, considerable extra pressure must develop in these veins to drive the blood uphill. The weight of the blood from the level of the heart down to the bottom of the foot is great enough that when all other factors affecting venous pressure are nonoperative the venous pressure in the foot will be 90 mm Hg.

The Venous Pump. To prevent the extremely high venous pressures that hydrostatic pressure can cause, the venous system is provided with a special mechanism for propelling blood toward the heart. This is called the *venous pump* or sometimes the *muscle pump*. All peripheral veins contain valves that allow blood to flow only toward the heart. Every time a muscle contracts or every time a limb moves, the moving

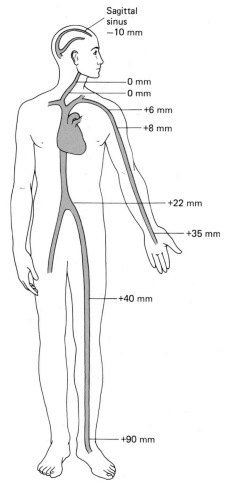

Sagittal
sinus
−10 mm

0 mm
0 mm
+6 mm
+8 mm

+22 mm

+35 mm

+40 mm

+90 mm

FIGURE 29-5 Hydrostatic pressures in various parts of the venous system of a person standing quietly so that the venous pump is inactive.

tissues compress at least some of the veins. Since the valves prevent backward flow, this always pushes the blood toward the heart, in this way emptying the veins and decreasing the peripheral venous pressure.

The hydrostatic pressures shown in Figure 29–5 occur only when a person is standing completely still or when some disease condition has destroyed the valves of his veins. Ordinarily, the venous pump is so effective even when a person walks very slowly that pressures in the leg veins are only 15 to 30 mm Hg. But when the valves are destroyed, such high pressures (80 to 90 mm Hg) develop in the leg veins that these veins become progressively distended to diameters four and five times normal, causing the condition known as *varicose veins.*

Measurement of Venous Pressures

The most important of all venous pressures is the right atrial pressure, for one can usually tell from this how well the heart is pumping. Right atrial pressure is usually measured through a cardiac catheter in the manner illustrated in Figure 29–6A. The catheter is threaded into a vein of the arm, then upward through the veins of the shoulder, into the thorax, and finally into the right atrium. The pressure transmitted through the catheter is recorded by a water manometer located at the patient's side. This same technique can be used for recording pressures in other parts of the heart or pulmonary circulation simply by sliding the catheter through the tricuspid valve into the right ventricle, or on through the pulmonary valve into the pulmonary artery.

For measuring peripheral venous pressures, either a catheter or a needle can be inserted directly into a peripheral vein and the pressure measured as just described. However, one can usually estimate peripheral venous pressures quite satisfactorily using the simple technique illustrated in Figure 29–6B of raising and lowering the arm above or below the level of the heart. When the veins are lower than the heart they become full of blood and stand out beneath the skin. Then, as the arm is raised, the veins normally collapse at a level approximately 9 cm above the level of the heart. This means that the pressure in the arm veins, when the arm is at heart level, is about 9 cm of water (or about 7 mm of mercury, because 1 cm of water equals .74 mm Hg). If the veins do not collapse until they rise to a level 20 cm above the heart, then the peripheral venous pressure is approximately +20 cm of water (or 15 mm Hg).

CARDIAC FAILURE

The term *cardiac failure* means *depressed pumping effectiveness of the heart.* The most frequent cause of cardiac failure is actual damage to the heart itself caused by some disease process. For instance, the coronary arteries might become blocked because of *atherosclerosis* and the strength of the heart correspondingly reduced, which is by far the most common cause of heart failure, or sometimes the valves of the heart are destroyed by rheumatic heart disease. In either case, the effectiveness of the heart as a pump is decreased so that even the normal amount of blood is not pumped as well as usual.

Low Cardiac Output Failure. A failing heart often, but not always, fails to pump adequate quantities of blood to the tissues. This is called *low cardiac output failure.* It can be caused by weakness of any part of the heart or of the whole heart, because failure of any one part can often hinder satisfactory pumping.

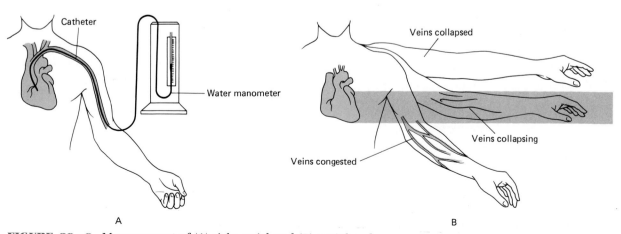

A   B

FIGURE 29–6 Measurement of (A) right atrial and (B) peripheral venous pressures.

Pulmonary Congestion Resulting from Cardiac Failure. When it is primarily the left side of the heart that is failing, the right heart continues to pump blood into the lungs with normal vigor and the left heart is unable to move the blood on into the systemic circulation. The resulting accumulation of blood in the lungs increases the pressures in all the pulmonary vessels and engorges them with blood. If the pulmonary capillary pressure rises above 25 to 30 mm Hg, fluid will then leak out of the capillaries into the lung tissue and even into the air spaces of the lungs (the alveoli), resulting in *pulmonary edema.* Indeed, many patients who die from failure of the left heart do not die because of decreased cardiac output, but because of failure of the water-soaked lungs to aerate the blood.

Peripheral Congestion and Peripheral Edema Resulting from Cardiac Failure. If the right side of the heart fails or if the entire heart fails (which is the usual case), the right atrial pressure rises, causing much of the blood attempting to return to the heart to be dammed in the peripheral veins. As a result, the pressures throughout the entire venous system rise. The veins of the neck become greatly distended, as illustrated in Figure 29–7, and the venous reservoirs such as the liver and spleen become engorged with blood. Also, the capillary pressure throughout the entire systemic circulatory system may become so great that fluid leaks continually into the tissue spaces, resulting in extreme *generalized peripheral edema.* One of the conditions occasionally encountered in severe untreated heart disease is swelling from head to toe as the result of excessive interstitial fluid, a condition called *dropsy.*

The edema in cardiac failure is caused not only by high pressure in the capillaries but also by two other factors: First, decreased cardiac output (with slight decrease in arterial pressure as well) decreases the amount of urine formed by the kidneys, causing excessive quantities of water and electrolytes to remain in the body, thus enhancing the volume of extracellular fluid. Second, diminished cardiac output also causes the adrenal cortex to secrete large quantities of aldosterone, which is still another factor that makes the kidneys retain water and salt, thus further increasing the total extracellular fluid volume.

One of the most essential features in the treatment of cardiac failure is to control the amount of salt that the person eats and the amount of water that he drinks. Also, very powerful drugs called *diuretics,* which increase the output of urine by the kidneys, are frequently administered.

CIRCULATORY SHOCK

Circulatory shock is the condition resulting when the cardiac output becomes so reduced that the tissues everywhere in the body fail to receive adequate blood supply. As a result, the tissues suffer from inadequate nutrition and inadequate removal of cellular excretory products because of the reduced blood flow.

Any condition that decreases the cardiac output to a very low level can cause circulatory shock. Therefore, shock can result from weakness of the heart itself or from diminished venous return, for which reason it can be classified into two main types: (1) *cardiac shock* and (2) *low venous return shock.*

CARDIAC SHOCK

Cardiac shock is the same as very severe low cardiac output failure caused by greatly decreased effectiveness of the heart as a pump. This type of shock occurs most frequently immediately after a severe heart attack because this often causes the heart's ability to pump blood to decrease manyfold in only a few minutes. The person frequently dies because of diminished blood supply to his tissues before the heart can begin to recover.

SHOCK CAUSED BY LOW VENOUS RETURN

Any of the factors that decrease the tendency for blood to return to the heart can cause shock. These are (1) decreased blood volume, which causes *hypovolemic shock;* (2) increased size of

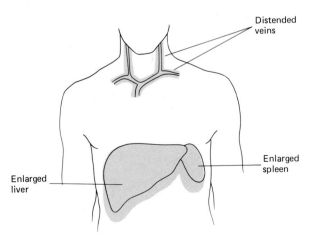

FIGURE 29–7 Engorgement of the peripheral venous system and blood reservoirs in cardiac failure.

the vascular bed so that even normal amounts of blood will not fill the vessels adequately—this is called *venous pooling shock;* or (3) obstruction of blood vessels, particularly veins.

Hypovolemic Shock. Hypovolemic shock most frequently results from blood loss, such as loss from a bleeding wound or a bleeding stomach ulcer. However, the blood volume can be decreased as a result of plasma loss through exuding wounds or burns; loss of plasma into severely crushed tissues; or dehydration caused by extreme sweating, lack of water to drink, or excessive loss of fluids through the gut or kidneys. In any of these conditions, the diminished blood volume decreases the mean systemic filling pressure so greatly that inadequate quantities of blood return to the heart, and the diminished venous return causes shock.

Venous Pooling Shock. If the blood vessels lose their vasomotor tone, their diameters may increase so greatly that the blood collects, or "pools," in the highly distensible veins. As a result, the pressures in the systemic circulation fall so low that venous return becomes slight and circulatory shock ensues.

Neurogenic Shock. A special type of venous pooling shock is *neurogenic circulatory shock* caused by sudden cessation of sympathetic impulses from the central nervous system to the peripheral vascular system. The result is loss of normal vasomotor tone, diminished pressures everywhere in the systemic circulation, and consequently diminished venous return. Emotional fainting is an acute example of this type of shock.

Allergic Shock. Extreme allergic reactions can also cause venous pooling shock. This type of shock is known as *anaphylactic shock,* and its probable mechanism is that illustrated in Figure 29-8, which may be explained as follows: When a person becomes immune to a foreign substance, such as the protein of a type of bacterium or a protein toxin, his body manufactures *antibodies;* these are special proteins that will destroy the bacteria or toxins, as explained in Chapter 34. However, when the person is then exposed to another large dose of the same protein to which he has become immune, the reaction between the antibodies and the protein in the circulating blood can cause the release of several substances that produce shock. One of these substances is *histamine.* The histamine circulates to all the peripheral blood vessels and promotes vasodilatation, especially dilatation of the veins, thus causing venous pooling and diminished venous return, culminating immedi-

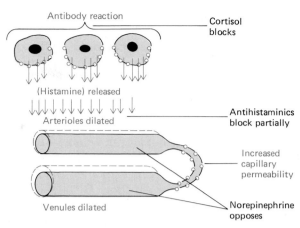

FIGURE 29-8 Mechanism of anaphylactic shock.

ately in a state of anaphylactic shock. Indeed, venous pooling can sometimes result so rapidly and to such an extreme extent during anaphylaxis that the person may die before therapy can be started.

Stages of Shock

The severity of shock varies tremendously, depending upon the degree of the abnormality that causes it. If it is very mild, the different control systems of the body for maintenance of normal arterial pressure and of blood flow through the tissues can overcome the effects of the shock, and the person will recover very quickly. This is called *compensated shock.* However, if the shock is more severe it will continue to progress, giving rise to a stage of shock called *progressive shock.* And, if it is extremely severe, the shock will progress to a point at which, even though the person is still alive, all types of known therapy still cannot save his life. This is called *irreversible shock.*

Compensated Stage. When circulatory shock is caused by hemorrhage, the blood loss is compensated by nervous constriction of the veins and venous reservoirs. Consequently, even though blood is lost, venous return of blood to the heart continues almost as usual. The cardiac output usually remains almost normal until 15 percent of the total blood volume (about 750 ml) is removed. However, beyond this point, the blood loss can no longer be fully compensated, and cardiac output begins to fall. This early stage of shock is called the *compensated stage.* During this stage the person is not in imminent danger of dying, and unless the cause of the shock is intensified, the condition will be corrected automatically by the usual circulatory control mechanisms, and the person will return to normal within a few hours or certainly within a day or two.

Progressive Stage of Shock. If the degree of shock becomes very severe, regardless of the initial cause of the shock, *the shock itself promotes more shock.* In other words, certain vicious cycles develop, and the shock becomes progressively more severe even though the original cause of the shock does not worsen. Some of the reasons for the increased severity are shown in Figure 29–9 and may be explained as follows:

First, if the degree of shock is so great that the heart fails to pump enough blood to supply its own coronary vessels, the heart itself becomes progressively weakened. This further diminishes the cardiac output, which weakens the heart even more and diminishes the cardiac output again, leading to a vicious cycle that eventually kills the person.

Second, very poor blood flow to the brain causes damage to the vasomotor and respiratory centers. Failure of the vasomotor center allows venous pooling, resulting in even more extensive shock. Also, respiratory failure causes diminished oxygenation of the blood and, therefore, malnutrition of the tissues, which also makes the shock worse.

Third, peripheral vascular failure causes a vicious cycle. Ischemia of the blood vessels can make the vascular musculature so weak that the vessels dilate, resulting in still more vascular pooling of blood, diminished venous return, increased shock, increased ischemia, and so forth, the vicious cycle repeating itself again and again until the vessels are fully dilated or the person is dead.

Fourth, recent experiments have shown that greatly diminished blood flow causes minute blood clots to develop in the small vessels. As a consequence of the plugged vessels, venous return decreases, and the progressively more sluggish flow of blood through the peripheral vessels causes more and more clotting, creating still another vicious cycle.

Fifth, very severe shock also damages the oxidative and energy systems of the tissues, particularly the adenosine triphosphate system discussed in Chapter 3, which causes all the circulatory functions to diminish further.

Thus, because several vicious cycles develop, once shock has reached a certain degree of severity, progressive deterioration of the circulatory system causes the shock to become more and more severe until death ensues, unless appropriate therapy is instituted.

Irreversible Shock. In the early stages of progressive shock, the person's life can usually be saved by appropriate and rapid treatment. For instance, if the shock is initiated by hemorrhage, rapid transfusion of blood can restore normal circulatory dynamics unless the damage to the circulation has already become too great. However, if the shock has progressed for a long period of time, such extreme damage may have occurred that no amount of treatment can be successful in restoring the cardiac output enough to sustain life. When this state has been reached, the patient is said to be in *irreversible shock.* The heart, for instance, may have become damaged beyond repair by this time so that it is incapable of continued pumping at an adequate rate of cardiac output regardless of what therapy is instituted. Or so many peripheral vessels may have become plugged with small clots that blood can no longer flow rapidly enough to repair the damage. In these instances, despite any amount of therapy, the vicious cycles of the progressive stage of shock continue on and on until death of the person.

FIGURE 29–9 Factors that produce progression of shock and finally cause it to become irreversible.

Treatment of Shock

The treatment of shock depends on the cause. In the case of cardiac shock, the best therapy is to increase the pumping effectiveness of the heart. Usually this is not easily accomplished, and almost 85 percent of persons with this type of shock die. One can sometimes help a person with cardiac shock by judiciously transfusing blood or by administering norepinephrine to constrict the peripheral arterioles and to increase the arterial pressure. The increased pressure then promotes greater flow of blood in the coronary arteries to the heart muscle.

If the shock results from hypovolemia, the blood volume can be increased by administering blood, plasma, or sometimes salt solution. In fact, even drinking water is often life-saving; and fortunately the person with shock, if he is still conscious, experiences intense thirst to make him drink. Return of the blood volume to normal increases the pressures throughout the systemic vessels, which brings the venous return back to normal and thereby alleviates the shock.

In both neurogenic and anaphylactic shock, the main difficulty is venous pooling. Administration of drugs that act like the sympathetic nervous system in constricting the blood vessels—such as norepinephrine—compresses the blood in the vessels and forces it along the veins toward the heart, usually bringing the person out of shock.

QUESTIONS

1. What is the normal resting cardiac output, and how is this affected by exercise?
2. Explain the mechanisms by which the cardiac output can be increased?
3. Explain the role of "venous return" in controlling cardiac output.
4. What is meant by "mean systemic filling pressure," and how does it affect venous return and cardiac output?
5. Discuss the role of tissue need for blood flow as the basic stimulus for control of cardiac output.
6. Explain the Fick method for measuring cardiac output.
7. What is the effect of hydrostatic pressure on peripheral venous pressure?
8. Why does cardiac failure often cause low cardiac output, pulmonary congestion, or peripheral congestion, or any combination thereof?
9. What is meant by circulatory shock, and what factors can cause low venous return shock?
10. What is meant by the following stages of shock: compensated, progressive, and irreversible?

REFERENCES

Bradley, R.D.: Studies in Acute Heart Failure. London, Edward Arnold, 1977.

Braunwald, E.: Heart failure—Pathophysiological considerations. *In* Dickinson, C.J., and Marks, J. (eds.): Developments in Cardiovascular Medicine. Lancaster, England, MTP Press, 1978, p. 213.

Bruce, T.A., and Douglas, J.E.: Dynamic cardiac performance. *In* Frohlich, E.D. (ed.): Pathophysiology, 2nd Ed. Philadelphia, J.B. Lippincott, 1976. p. 5.

Crowell, J.W., and Smith, E.E.: Oxygen deficit and irreversible hemorrhagic shock. *Am. J. Physiol., 206*:313, 1964.

Dodge, H.T., and Kennedy, J.W.: Cardiac output, cardiac performance, hypertrophy, dilatation, valvular disease, ischemic heart disease, and pericardial disease. *In* Sodeman, W.A., Jr., and Sodeman, T.M. (eds.): Pathologic Physiology: Mechanisms of Disease, 6th Ed. Philadelphia, W.B. Saunders, 1979, p. 271.

Donald, D.E., and Shepherd, J.T.: Response to exercise in dogs with cardiac denervation. *Am. J. Physiol., 205*:393, 1963.

Guyton, A.C.: Determination of cardiac output by equating venous return curves with cardiac response curves. *Physiol. Rev., 35*:123, 1955.

Guyton, A.C.: Essential cardiovascular regulation—the control linkages between bodily needs and circulatory function. *In* Dickinson, C.J., and Marks, J. (eds.): Developments in Cardiovascular Medicine. Lancaster, England, MTP Press, 1978, p. 265.

Guyton, A.C., *et al.*: Circulation: Overall regulation. *Annu. Rev. Physiol., 34*:13, 1972.

Guyton, A.C., *et al.*: Cardiac Output and Its Regulation. Philadelphia, W.B. Saunders, 1973.

Jamieson, G.A., and Greenwalt, T.J. (eds.): Blood Substitutes and Plasma Expanders. New York, A.R. Liss, 1978.

Kovach, A.G.B., and Sandor, P.: Cerebral blood flow and brain function during hypotension. *Annu. Rev. Physiol., 38*:571, 1976.

IX

THE BODY FLUIDS
AND THE KIDNEYS

30

Capillary Membrane Dynamics, Body Fluids, and the Lymphatic System

---

Overview

The *fluid in the body* is distributed approximately as follows: (1) *total body water, 40 L;* (2) total fluid in the cells, called the *intracellular fluid, 25 L;* (3) total fluid in the spaces between the cells, called the *interstitial fluid, 12 L;* (4) fluid in the *blood plasma, 3 L;* and (5) *extracellular fluid volume,* which is the sum of interstitial fluid volume and plasma volume, *15 L.*

The *extracellular fluid* diffuses readily in both directions through the pores in the capillary membranes, back and forth between the plasma of the blood and the fluid in the interstitial spaces. In this way, the extracellular fluid is continually mixed and transported by the blood throughout the body, thus also transporting nutrients to the cells and excreta away from the cells.

Despite the rapid diffusion of extracellular fluid through the capillary walls, the *plasma volume and interstitial fluid volume remain almost exactly constant.* The reason for this is that the forces that cause movement of fluid in each direction through the *capillary membrane pores* are almost exactly balanced. The forces that tend to cause fluid to leak out of the capillaries are (1) *capillary pressure,* equal to about 17 mm Hg, which pushes fluid outward through the capillary membranes; (2) *colloid osmotic pressure caused by proteins in the interstitial fluid,* about 5 mm Hg, which pulls fluid by osmosis outward through the capillary membranes; and (3) *negative fluid pressure in the interstitial fluid,* about −6 mm Hg, which also pulls fluid outward through the capillary membrane. Adding these together yields 28 mm Hg. Opposing these outward forces is an inward force caused by the *colloid osmotic pressure of the plasma proteins* inside the capillaries, equal to about 28 mm Hg, which pulls fluid into the capillaries by the mechanism of osmosis. Thus, the inward and outward forces are equal, a phenomenon called the *law of the capillaries.*

Besides the blood circulatory system, the body has still another fluid flow system, the *lymphatic system.* This system begins in a vast array of very minute *lymphatic capillaries* that lie between the blood capillaries in the tissues. Fluid filters from the tissue spaces into these lymphatic capillaries and is then called *lymph.* The lymph flows into progressively larger lymphatics, most of it finally entering the *thoracic duct*, a vessel about 5 mm in diameter that begins in the abdomen, then passes upward through the chest, and empties into the *left subclavian vein* in the neck, thus returning the lymph into the blood.

The lymphatic vessels all have large numbers of *lymphatic valves* that are oriented so that fluid can flow only in the direction away from the tissues. Every movement of the body, whether this be caused by muscle movements, passive movements, or even arterial pulsations, compresses some of the lymphatics and makes the fluid move past one valve after another until it finally enters the venous system. This is called the *lymphatic pump*.

The most important function of the lymphatic system is to *return plasma protein from the interstitial fluid back to the blood circulation*. A small amount of plasma protein leaks continuously through the capillary pores into the interstitial fluid. If this were not returned to the circulating blood, the plasma colloid osmotic pressure would fall too low to keep fluid in the circulation. This return of the proteins normally requires only a minute rate of lymph flow, equal to only *2 to 3 L of lymph each day*.

Occasionally, abnormalities develop in the capillary fluid exchange mechanism that cause *edema*, which means excessive leakage of fluid out of the plasma into the interstitial fluid with consequent swelling of the tissues. The causes of this are (1) *high capillary pressure*, usually resulting from heart failure or blockage of the veins flowing out of the tissues; (2) *low concentration of plasma protein*, usually caused by loss of proteins through the kidneys or failure to form enough protein because of poor nutrition; (3) *increased permeability of the capillary pores*, usually caused by toxic factors affecting the capillaries, which allows excessive amounts of protein to leak out of the circulation into the interstitial fluid; and (4) *blockage of the lymphatic system*, which prevents return of the protein from the interstitium, thus allowing the plasma protein concentration to fall too low and the interstitial protein concentration to rise too high, both of which cause excessive transudation of fluid into the tissues.

From discussions in several of the previous chapters, especially Chapters 1 and 5, we have talked many times about the fluids of the body, including the blood, the extracellular fluid, the intracellular fluid, the cerebrospinal fluid, the fluids of the eye, and others. In this and the next few chapters we will discuss mainly the extracellular fluid, which is the all-important internal environment of the body that was mentioned in Chapter 1. First, we will consider the physiological mechanisms that determine the *distribution of extracellular fluid* between the blood and interstitium and the role of the lymphatic system in this function. In the following few chapters, we will be concerned mainly with the *composition* of the extracellular fluid and particularly with the many important functions of the kidney in controlling this composition. To introduce these subjects, let us first define some of the fluids in the different compartments of the body.

BODY WATER, INTRACELLULAR FLUID, AND EXTRACELLULAR FLUID

The total water in the average 70-kg man is about 40 L, or 57 percent of his total body weight. Of this, about 25 L are intracellular fluid in the *intracellular compartment* and about 15 L are extracellular fluid in the *extracellular compartment*. These relationships are illustrated in Figure 30–1.

Blood

Blood is composed of two portions, the *cells* and the fluid between the cells, which is called *plasma*. The plasma portion of the blood is typical extracellular fluid except that it has a higher concentration of protein than do the extracellular fluids elsewhere in the body; this protein is

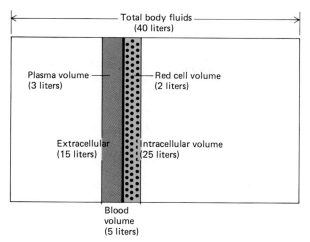

FIGURE 30-1 Diagrammatic representation of the body fluids, showing the extracellular fluid volume, blood volume, and total body fluids.

FIGURE 30-2 The dilution principle for measuring the volume of a fluid compartment.

very important for keeping the plasma inside the circulatory system, as will be discussed later in this chapter. The *cells* are of two types, *red blood cells* and *white blood cells*. The number of white blood cells is only $\frac{1}{500}$ the number of red blood cells so that for many purposes one can consider blood to be mainly a mixture of plasma and red blood cells. The normal total volume of plasma is 3 L and of red cells 2 L.

Interstitial Fluid

The interstitial fluid is the extracellular fluid that lies outside the capillaries between the tissue cells. The volume of the interstitial fluid is equal to the total extracellular fluid volume minus the plasma volume, or 15 − 3 = 12 L of interstitial fluid volume in the normal, average adult man. Interstitial fluid is usually considered to include also such special fluids as those in the cerebrospinal fluid system, the chambers of the eyes, the intrapleural space, the peritoneal cavity, the pericardial cavity, the joint spaces, and the lymph.

MEASUREMENT OF FLUID VOLUMES OF THE BODY

The Dilution Principle. Frequently it is important for the physiologist to measure the fluid volumes in different compartments of the body. To do this, he utilizes a basic principle called the *dilution principle*, as follows:

Figure 30–2 illustrates a fluid chamber in which a small quantity of dye or other foreign substance is injected. After the substance has mixed thoroughly with the fluid in the chamber, its concentration has become very dilute and

equal in all areas of the chamber (Fig. 30–2 B). Then a sample of the fluid is removed and the quantity of the substance in each milliliter is analyzed by chemical, photoelectric, or any other means. The volume of the chamber can then be calculated very simply by dividing the quantity in each milliliter of sample fluid into the total quantity initially injected. That is,

$$\text{Volume in ml} = \frac{\text{Quantity of test substance injected}}{\text{Quantity per ml of sample fluid}}$$

It should be noted that all one needs to know is (1) the *total quantity of the test substance* injected into the chamber and (2) the *quantity of the substance in each milliliter of the fluid after dispersion.*

Measurement of Total Body Water. To measure total body water one can inject *radioactive water*, water containing tritium, a radioactive isotope of hydrogen, into a person and then 30 minutes to an hour later remove a sample of blood and measure the quantity of radioactive water per milliliter in the blood or plasma. Since the tritiated water disperses throughout the body, into the cells and everywhere else, in exactly the same way as normal water, one can use this measurement to calculate the total body water. Let us assume that 50 ml of tritiated water has been injected into the person and that its quantity in each milliliter of plasma an hour later, after dispersing throughout the body, is found to be 0.001 ml. Dividing this quantity into the total amount injected (using the above formula), we find that the total body water of this person is 50 L, which is a value slightly higher than the value of 40 L in the normal adult man.

Measurement of Extracellular Fluid Volume. To measure the extracellular fluid volume one uses a different test substance, one that will diffuse everywhere in the extracellular fluid compartment but will not go through the cell membranes into the intracellular fluid compartment. Substances of this type include *radioactive sodium, thiocyanate ions*, and *inulin*, all of which can be used for determining the extracellular fluid volume. The procedure is almost identical to that used for measuring total body water. The test substance is injected intravenously, and a sample of plasma is removed about 30 minutes later, after appropriate mixing throughout the entire extracellular compartment has occurred. The same formula as above is used to calculate the extracellular fluid volume.

Measurement of Plasma Volume. The plasma volume is measured by injecting some substance that will stay in the plasma compartment of the circulating blood. Such a substance frequently used is a dye called *T-1824*, which, upon intravenous injection, combines almost immediately with the plasma proteins. Since plasma proteins do not leak readily out of the plasma compartment, neither will the dye. After appropriate mixing of the dye in the blood has occurred, within about 10 minutes, a sample of blood is removed, the plasma is separated from the red cells in a centrifuge and the concentration of dye in the plasma is measured. Then, again using the dilution formula, one determines the plasma volume.

Calculation of Blood Volume. If one knows the percentage of the blood that is blood cells, he can calculate the total blood volume from the plasma volume. To determine the percentage of the blood that is plasma, blood is centrifuged at a high speed for 15 to 30 minutes, which separates the plasma from the cells so that the percentages of these two can be measured directly. The percentage of blood cells is called the *hematocrit* of the blood. From the hematocrit and the measured plasma volume

one calculates total blood volume using the following formula:

$$\text{Blood volume} = \frac{100}{100 - \text{Hematocrit}} \times \text{Plasma volume}$$

CAPILLARY MEMBRANE DYNAMICS—EXCHANGE OF FLUID BETWEEN THE PLASMA AND THE INTERSTITIAL FLUID

The whole purpose of the blood circulation is to transport substances to and from the tissues. Therefore, it is important for both water and dissolved substances to pass interchangeably between the plasma and the interstitial fluids and to bathe the tissue cells. This is achieved mainly by diffusion of water and dissolved molecules in both directions through the capillary membrane. Usually this diffusion process is so effective that any nutrient that enters the blood will be distributed evenly among all the interstitial fluids within 10 to 30 minutes.

Fortunately, the rates of fluid diffusion in the two directions through the capillary membranes are almost exactly equal, so that the volumes of plasma and interstitial fluid remain essentially constant. Yet, under special circumstances the rate of diffusion in one direction may become greater than in the other direction, in which case the plasma and interstitial volumes change accordingly, often becoming very abnormal. Therefore, it is important that we consider the dynamics of this net exchange of fluid between the plasma and interstitial fluid and the mechanisms by which normal volumes are ordinarily maintained.

THE CAPILLARY SYSTEM AND CAPILLARY PRESSURE

Figure 30–3 illustrates a typical capillary bed, showing blood entering the capillaries from the

FIGURE 30–3 A typical capillary system. (Drawn from a figure by Zweifach: *Factors Regulating Blood Pressure.* Josiah Macy, Jr., Foundation, 1950.)

arteriole, passing through the metarteriole, then into the capillaries, and finally to the venule. The **arteriole** has a muscular coat that allows it to contract or relax in response to stimuli that come mainly from the sympathetic nerves. The **metarteriole** has sparse muscle fibers, and the openings to the **true capillaries** are usually guarded by small **muscular precapillary sphincters**, which can open and close the entryway into the capillaries. The muscles of the metarterioles and the precapillary sphincters are controlled mainly in response to local conditions in the tissues. For instance, lack of oxygen allows these muscles to relax, which increases the flow of blood through the capillary bed and in turn increases the amount of oxygen available to the tissues.

The Capillary Wall and Its "Pores."
Figure 30–4 illustrates the ultramicroscopic structure of the capillary wall. Note that the wall is composed of a unicellular layer of endothelial cells surrounded on the outside by a thin collagen and proteoglycan layer called the **basement membrane**. The total thickness of the capillary wall, including both the endothelial cell and the basement membrane together, is about 0.5 microns (μm). The diameter of the capillary is 5 to 9 μm, barely large enough for red blood cells and other blood cells to squeeze through.

Note especially in the figure two minute passageways connecting the interior and the exterior of the capillary. One of these is the *intercellular cleft*, which is a thin slit that lies between adjacent endothelial cells. Most water-soluble ions and molecules pass between the inside and the outside of the capillary through these "slit-pores."

Also present in the endothelial cells are many *pinocytic vesicles*. These originate on one surface of the cell, caused by the cell membrane entrapping small volumes of extracellular fluid. Then the vesicles diffuse through the cytoplasm to the opposite surface of the cell, where they discharge their contents; in this way these vesicles often carry large molecules and even solid particles through the capillary membrane. Occasionally, pinocytic pores coalesce to form a *pinocytic channel* all the way through the membrane, as illustrated to the right in Figure 30–4. However, it is doubtful whether large amounts of substances pass through the capillary membrane via such "pores" as these.

Finally, though the basement membrane appears rather solid in Figure 30–4, the spaces between the fibers within this basement membrane are large enough that any substance that can go through the intercellular clefts can also go through the basement membrane.

Pressures in the Capillary System.
Though the pressure at the arterial end of a capillary varies tremendously depending upon the state of contraction or relaxation of the arteriole, metarteriole, and precapillary sphincters, the normal mean pressure at the arterial end of the capillary is about 30 mm Hg. Likewise, the average pressure at the venous end of the capillary is about 10 mm Hg, and the average mean pressure in the capillary is about 17 mm Hg. Pressures have been measured at different points in capillaries using minute pipets protruding into the capillary lumens, connected to special micro pressure manometers.

INTERSTITIAL FLUID PRESSURE

Interstitial fluid pressure is the pressure of fluid in the spaces between the cells. The quantitative level of interstitial pressure has been difficult to measure because the interstitial spaces generally have a width of less than 1 μm (micrometer), and introduction of any needle or pipet to measure the pressure can cause an abnormal reading. However, during the past few years we have studied this problem in our own laboratory in a completely different way, by implanting into tissues perforated capsules such as the one illustrated in Figure 30–5. The tissue grows inward through the holes, and new tissue then lines the inner wall of the capsule. A blood vascular system also develops in the new tissue. Yet, a large fluid space remains in the middle of the capsule, and special experiments have shown that the fluid in this space communicates directly through the capsule perforations with the surrounding interstitial fluid. A needle connected to

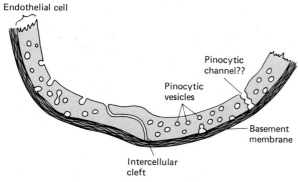

FIGURE 30–4 Structure of the capillary wall. Note especially the *intercellular cleft* at the junction between adjacent endothelial cells; it is believed that most water-soluble substances diffuse through the capillary membrane along this cleft.

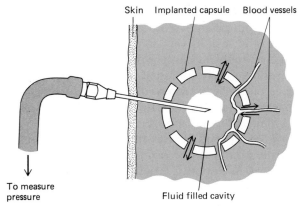

FIGURE 30–5 The perforated capsule method for measuring tissue fluid pressure.

a manometer system can be inserted through one of the perforations into the cavity and the pressure measured. Interstitial fluid pressures measured in this manner have been about −6 mm Hg—that is, 6 mm Hg less than atmospheric pressure—which demonstrates that the tissue spaces actually have a partial vacuum in them.

Pressure Difference Across the Capillary Membrane. If the average mean pressure in the capillary is 17 mm Hg and the average pressure outside the capillary, the interstitial fluid pressure, is −6 mm Hg, the total pressure difference between the two sides of the membrane would be 17 − (−6) or a total of 23 mm Hg. That is, the pressure inside the capillary is 23 mm Hg greater than the pressure outside the capillary. This pressure difference between the two sides of the membrane makes fluid tend to move out of the capillary into the tissue spaces. Fortunately, however, *colloid osmotic pressure*, another force operating at the capillary membrane, opposes this tendency for fluid to move out of the capillaries. Therefore, let us discuss this.

COLLOID OSMOTIC PRESSURE AT THE CAPILLARY MEMBRANE

The Principle of Osmotic Pressure. The principles of osmosis and osmotic pressure were discussed in Chapter 5. To review briefly, if two solutions are placed on either side of a semipermeable membrane so that water molecules can go through the membrane pores but solute molecules cannot, water will move by the process of osmosis from the more dilute solution into the more concentrated solution. However, if pressure is applied to the more concentrated solution, this movement of water by osmosis can be slowed or halted. The amount of pressure

that must be applied to stop completely the process of osmosis is called the *osmotic pressure*. For a much deeper understanding of why these effects take place, the student is now referred back to Chapter 5.

Colloid Osmotic Pressure Caused by Plasma Proteins. The amount of osmotic pressure that develops at a membrane is determined to a great extent by the size of the pores in the membrane, because osmosis occurs only when dissolved particles cannot go through the pores. In the case of the cell membrane, such particles as sodium ions, chloride ions, glucose molecules, urea molecules, and others, all cause osmotic pressure. However, at the capillary membrane the pore diameter is roughly 10 times that of the cell membrane—8 nanometers (nm) (or 80 Ångströms) in contrast to .7 nm (7 Å)—so that sodium ions, glucose molecules, and essentially all other constituents of extracellular fluid pass directly through the capillary pores without causing osmotic pressure. However, the proteins in the plasma and in the interstitial fluid are an exception, because their molecular sizes are large enough that almost all of them fail to penetrate even the large pores of the capillary membrane. Therefore, the proteins are the only solutes in the plasma and interstitial fluid that cause osmotic effects at the capillary membrane. Because solutions of plasma protein appeared to early chemists to be colloidal suspensions rather than true solutions, this protein osmotic pressure at the capillary membrane is called *colloid osmotic pressure*. However, some physiologists also call this pressure *oncotic pressure*.

Plasma has a normal protein concentration of 7.3 g percent, while interstitial fluid has a concentration of about 2.0 g percent, making a difference between the two sides of the capillary membrane of about 5.3 g percent. Because the larger concentration is inside the capillary, osmosis of fluid tends to occur always from the interstitial fluid into the capillaries.

Colloid Osmotic Pressure of Plasma and Interstitial Fluids. If pure plasma were placed on one side of a capillary membrane and pure water on the other side, the colloid osmotic pressure that would develop would be about 28 mm Hg. If pure interstitial fluid were placed on one side of the membrane and pure water on the opposite side, the colloid osmotic pressure that would develop would be about 5 mm Hg. Therefore, we say that the colloid osmotic pressure of plasma is 28 mm Hg and of interstitial fluid 5 mm Hg.

The *colloid osmotic pressure difference* between the two sides of the membrane is equal to the difference between the colloid osmotic pressures of the two fluids on the two sides of the membrane. Therefore, the colloid osmotic pressure difference at the capillary membrane is 28 − 5, or 23 mm Hg.

EQUILIBRATION OF PRESSURES AT THE CAPILLARY MEMBRANE— THE LAW OF THE CAPILLARIES

One will note from the foregoing discussion that under normal conditions the fluid pressure difference across the capillary membrane (23 mm Hg) is equal to the colloid osmotic pressure difference across the membrane (23 mm Hg). However, the fluid pressure tends to move fluid out of the capillary, whereas the colloid osmotic pressure tends to move fluid into the capillary. This balance between the two forces explains how it is possible for the circulation to keep its blood volume constant even though the capillary pressure is considerably higher than the interstitial fluid pressure. Were it not for the colloid osmotic pressures, fluid would be lost continually from the circulation until eventually the blood volume would be insufficient to maintain cardiac output.

The normal state of equilibrium between the pressures tending to make fluid leave the capillaries and those tending to return fluid to the capillaries is called the *law of the capillaries,* and it is illustrated mathematically in Figure 30–6. The mean fluid pressure in the capillary is shown to be 17 mm Hg and the interstitial fluid pressure −6 mm Hg, making a total fluid pressure difference between the inside and the outside of 23 mm Hg, tending to move fluid out of the capillary. On the other hand, the colloid osmotic pressure in the capillary is 28 mm Hg and in the interstitial fluid 5 mm Hg, making a net difference also of 23 mm Hg, but this time tending to move fluid into the capillary. Thus, the two pressures are mathematically in balance so that the fluid volume of neither the plasma nor the interstitial spaces will be changing.

Effect of Nonequilibrium at the Capillary Membrane.

Occasionally the pressures at the capillary membrane lose their state of equilibrium because one or more of them changes to a new value. When this occurs, fluid transudes through the membrane very rapidly until a new state of equilibrium develops.

Let us consider, for example, an increase in capillary pressure from the normal value of 17 up to 25 mm Hg. This would increase the net pressure across the capillary membrane from zero up to a value of 8 mm Hg in the outward direction, which would cause rapid transudation of fluid into the tissue spaces. The loss of fluid (but not of plasma proteins) out of the plasma would cause the *capillary pressure to fall* because of decreasing blood volume and would cause the *plasma colloid osmotic pressure to rise* because of increasing concentration of the plasma proteins. In addition, the *interstitial fluid pressure would rise* because of the increasing volume of interstitial fluid, and the interstitial fluid *colloid osmotic pressure would fall* because of dilution of the interstitial fluid proteins. Thus, after a few minutes the capillary pressure would be 22 mm Hg, the plasma colloid osmotic pressure 29 mm Hg, the interstitial fluid colloid osmotic pressure 4 mm Hg, and the interstitial fluid pressure −3 mm Hg. If the student will add these pressures, while also working out the directions of their action at the capillary membrane, he will see that a new state of equilibrium has been established.

In a similar manner, any other change in any one of the pressure values on the two sides of the membrane will cause rapid movement of fluid through the capillary membrane until equilibrium is reestablished, usually within a few minutes. However, this fluid movement can sometimes be so great that the blood volume or the interstitial fluid volume becomes either abnormally large or abnormally small.

Flow of Fluid Through the Tissue Spaces

In addition to very rapid diffusion of water and dissolved substances through the capillary membrane, there is a small amount of actual *flow* of fluid through the capillary membrane and tissue spaces. The distinction between diffusion and flow is the following: Diffusion means movement of each molecule along its own pathway as a result of its kinetic motion irrespective of all other molecules, whereas flow means that large quantities of molecules all move together in the same direction.

FIGURE 30–6 Mean capillary dynamics.

FIGURE 30–7   Flow of fluid through the tissue spaces.

Flow of fluid through the tissue spaces is caused by fluid movement through the arterial and venous ends of the capillaries, as follows: If we observe Figure 30–7, we will see that the fluid pressure in the capillary at the arterial end is 30 mm Hg, while at the venous end it is 10 mm Hg. Now, let us study the pressures that cause fluid movement through the capillary membrane at both ends of the capillary. A sum of all the pressures acting at the arterial end of the capillary gives a net pressure of 13 mm Hg in favor of movement of fluid *out of* the capillary, which is called the *filtration pressure* at the arterial ends of the capillaries.

At the venous ends of the capillaries the greatly decreased fluid pressure in the capillary causes a net pressure in the opposite direction. This time the sum of the pressures is 7 mm Hg, tending to move fluid *inward* rather than outward. This net pressure difference is called the *absorption pressure*. Thus, it is clear that a small amount of fluid flows continually from the arterial ends of the capillaries through the tissue spaces toward the venous ends of the capillaries.

Note that the filtration and absorption pressures are not equal, but the difference is mainly counterbalanced by the fact that the venous ends of the capillaries are about 50 percent larger and are also much more permeable. Therefore, almost equal amounts of fluid pass out of the arterial end of the capillary and return by the venous end. A small portion of the fluid, however, between $\frac{1}{10}$ and $\frac{1}{100}$ of the total, returns to the circulation by way of the lymphatics, as we shall discuss in the following pages, rather than by absorption into the venous ends of the capillaries.

THE LYMPHATIC SYSTEM

The lymphatic system, illustrated in Figure 30–8, is an accessory system of vessels that drains fluid directly from the tissue spaces between the cells and empties the fluid into the venous system. The volume of fluid that flows each minute from the entire lymphatic system averages only $\frac{1}{2000}$ the volume of blood that flows in the arteries and veins. Even so, the lymphatic system is absolutely essential to life for the following reasons.

First, small amounts of fluid leak continuously from the blood capillaries. If this fluid is not removed from the spaces between the cells, called the *interstitial spaces*, it will collect forever and the person would eventually die of *edema.*

Second, some of the constituents of the fluid that leaks from the blood into the interstitial spaces cannot return to the blood circulation in any other way than through the lymphatics. This is especially true of the proteins in the blood, and it is also true of any blood cells that pass through the vascular walls into the tissue spaces. Therefore, the lymphatic system is even more essential for returning these to the circulating blood than for returning the fluid itself, most of which could possibly be returned in other ways.

Third, at times it is important to remove toxic products from the tissues or to remove cellular debris. This is especially true when infection occurs in tissues, because bacteria release many lethal toxins and also cause tissue cells to break down. Without a mechanism for removing these products, once an infection should occur, the tissue could never recover.

Cervical nodes
Sentinel node
Subclavian vein
R. lymph duct
Thoracic duct
Axillary nodes
Cisterna chyli
Abdominal nodes
Inguinal nodes
Peripheral lymphatics

FIGURE 30–8 The lymphatic system.

Fourth, the lymphatic system provides not only for drainage of fluid from the tissues but also for filtration of the fluid before it returns to the circulating blood. This is achieved by passing the fluid through one or more **lymph nodes** each of which has an internal structure that filters debris from the fluid and also special internal cells that digest and destroy both toxins and tissue debris.

The Major Lymph Channels of the Body

In contrast to the blood circulatory system that carries blood both to and from the tissues, the lymph vessels constitute only a one-way system, beginning in the tissue spaces and eventually emptying into the venous system.

In Figure 30–8, note the *many very small lymphatic vessels leading directly from the peripheral tissues*. These coalesce to become progressively larger channels almost all of which lead to the lower neck where the lymph finally empties into the left and right subclavian veins.

A very small amount of lymph empties into some of the veins of the abdominal region as well. Some of the more important lymphatic structures are the following:

1. *The Cisterna Chyli and Thoracic Duct.* The **thoracic duct,** illustrated in Figure 30–9, is the principal lymphatic trunk through which more than three quarters of the lymphatic drainage from the entire body passes before entering the venous system. This duct originates in the abdomen from the upper end of a lobulated saclike vesicle called the **cisterna chyli.** The cisterna chyli, located on the posterior wall of the abdominal cavity below the liver, is a collecting point for literally hundreds of smaller lymphatics leading from the liver, the intestines, the pelvis, and even from the legs. Thus, most of the lymph from the lower half of the body drains into the cisterna chyli and thence into the thoracic duct. After leaving the cisterna chyli, the thoracic duct ascends through the thorax lying anterior to the vertebral column but behind the esopha-

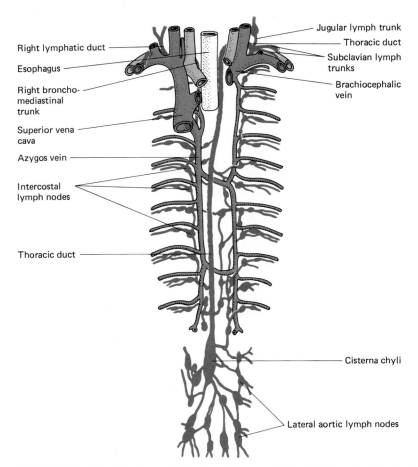

Right lymphatic duct
Esophagus
Right broncho-
mediastinal
trunk
Superior vena
cava
Azygos vein
Intercostal
lymph nodes
Thoracic duct

Jugular lymph trunk
Thoracic duct
Subclavian lymph
trunks
Brachiocephalic
vein

Cisterna chyli

Lateral aortic lymph nodes

FIGURE 30–9   The thoracic duct, cisterna chyli, right lymphatic duct, and the points where the lymphatic system empties into the junctures of the internal jugular and subclavian veins.

gus. On entering the neck, it arches to the left, backward, and downward to *empty into the junction of the internal jugular and subclavian veins.* In its course through the thorax, the thoracic duct collects lymph from the posterior chest wall, and, in the neck, it receives lymph vessels from the left side of the neck and head and from the left arm.

2. *The Right Lymphatic Duct and the Right Bronchomediastinal Trunk.* There is no large lymphatic trunk on the right side equivalent to the thoracic duct. However, a much smaller duct, the **right lymphatic duct,** does empty into the junction of the *right* internal jugular and subclavian veins. This duct receives lymph drainage from the right side of the head, the right neck, and the right arm. Another moderate-sized lymphatic trunk, the **right bronchomediastinal trunk** extends inferiorly from the right lymphatic duct into the right side of the mediastinum and collects a major share of the lymph from the mediastinal structures and lungs.

THE LYMPH FILTRATION SYSTEM— THE LYMPH NODES

Note in Figures 30–8 and 30–9 the large number of lymph nodes associated with the lymphatic drainage system. None of the peripheral lymphatic vessels drain into the thoracic duct without the lymph first passing through a lymph node, and in many instances the lymph passes through a series of nodes. It is in these lymph nodes that the fluid is filtered.

Figure 30–10 illustrates the structure of a typical lymph node. Lymph enters the node through its outer capsule via 6 to 10 small **afferent lymph vessels,** and it leaves the node through the **hilus** of the node via one or more somewhat larger **efferent lymph vessels.**

Everywhere inside the lymph node is a very fine meshwork of *collagen fibrillae* and *endothelial cells* (cells similar to the endothelial cells lining the blood vessels); this meshwork is called the **reticulum.** As the lymph percolates through the reticulum, the reticular meshwork filters

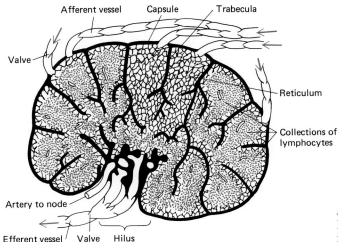

Afferent vessel   Capsule   Trabecula

Valve

Reticulum

Collections of
lymphocytes

Artery to node

Efferent vessel Valve Hilus

FIGURE 30–10 The internal structure of the lymph node. Lymphocytes have been omitted from part of the node to show the reticulum.

particulate matter from the lymph. Also entrapped in the meshwork are vast numbers of **macrophages** and **lymphocytes.** The macrophages phagocytize the foreign particles in the lymph, and lymphocytes become activated in response to bacteria or toxins and provide an *immune system* for destroying the invading agents. These protective and immune functions of the lymph nodes will be discussed in Chapters 33 and 34.

Ordinarily the lymph node is so effective as a filter that almost no bacteria pass with the lymph into the circulating blood. When many bacteria are filtered, the node often swells to a diameter three to four times normal because of proliferating lymphocytes and other cells derived from the lymphocytes, all helping to destroy the invading bacteria or their toxins. The swollen lymph node is usually inflamed and painful.

Finally, note the many small *valves* in both the afferent and the efferent lymph vessels. These prevent backflow of the lymph, thus making the lymph move always toward the venous system.

Role of Lymph Nodes in Preventing the Spread of Cancer.

A tumor is said to be *malignant* and is called a *cancer* if cells break away from the tumor and spread throughout the body. However, when cells do break away from the cancer and attempt to spread, they must first pass through lymph nodes. Almost invariably these cells become trapped in the nodes, and many if not most of them are destroyed in the nodes. But some of the trapped cells begin to proliferate in the node itself, and a new cancer is formed. From this new source still more cells empty into the lymphatic tree, and some eventu-

ally find their way into the blood and thence throughout the body.

However, for a long time after the onset of a cancer, the lymph nodes provide a protective barrier. Therefore, if the cancer is removed early enough and all of the cancerous nodes are found and eliminated at the same time, one can expect cure. For this reason, all cancer surgeons learn very precisely the anatomy of the lymphatic vessels and their principal drainage routes so that they will know where to look for the cancerous nodes.

Illustrative examples of peripheral lymphatics and their routes of drainage through lymph nodes are shown in Figures 30–11 and 30–12. Note in Figure 30–11 that almost all of the *lymphatics of the leg* drain through the *inguinal nodes* located in the groin. Therefore, for almost any cancerous condition of the leg, one immediately examines the nodes in the groin to determine whether or not any have become enlarged. And at the time of surgery these nodes are often removed whether they are enlarged or not because of the likelihood that a small undetectable cancerous growth might have already started.

Figure 30–12 illustrates an even more common site for spread of cancer to lymph nodes. This shows the lymphatic drainage from the breasts, most of which passes through the **axillary nodes** located in the armpit. From here the lymph courses upward through *nodes in the neck*, the **cervical nodes.** Therefore, when a breast is removed for breast cancer, most surgeons at the same time dissect very carefully both the axilla and the lower neck to remove as many lymph nodes as possible, thus giving the patient additional chance of survival. However, it is never possible to remove all the lymph nodes

FIGURE 30–11 Lymphatic drainage from the leg through the inguinal lymph nodes.

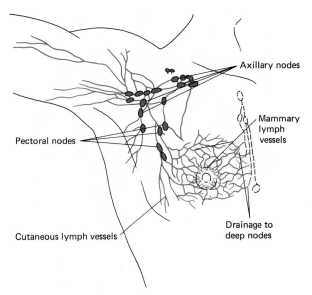

FIGURE 30–12 Lymphatic drainage from the breast through the axillary nodes, neck nodes, and deep mediastinal nodes.

in the drainage pathways from the breast, especially the **thoracic cavity nodes** where lymphatic channels penetrate through the chest wall and into the mediastinum. Therefore, after the operation, the patient is usually irradiated with x-rays or gamma rays. Cancer cells are usually especially sensitive to such rays, and many times both the nodes and any contained cancer cells can be destroyed even though they have not been removed surgically.

Aside from the more peripheral nodes illustrated in Figures 30–11 and 30–12, large numbers of **pelvic** and **abdominal cavity nodes** (illustrated in Figures 30–8 and 30–9) are often the site of cancer growth that has spread from one of the pelvic or abdominal organs.

Lymphatic Capillaries and Lymphatic Valves. Fluid from the tissues enters the lymphatic system through **lymphatic capillaries,** two of which are shown in Figure 30–13. And the relationship of a lymphatic capillary to both the tissue cells and a blood capillary is illustrated in Figure 30–14, showing movement of fluid in all directions between the blood capillary and the tissues, with a small amount of the fluid eventually finding its way into the lymphatic where it is then called *lymph.* From the lymphatic capillaries, the lymph flows into **collecting lymphatics**

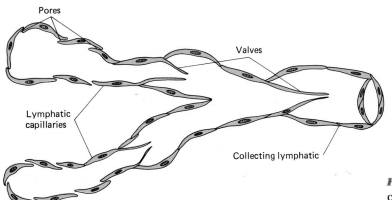

FIGURE 30–13 Structure of lymphatic capillaries and a collecting lymphatic.

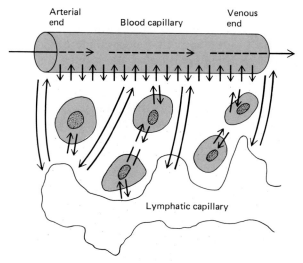

FIGURE 30–14 Relation of the lymphatic capillary to the tissue cells and to a blood capillary.

that then coalesce progressively to form the larger lymph vessels.

Lymphatic capillaries are very similar to blood capillaries. In fact, their walls are constructed of endothelial cells, the same as the blood capillaries. However, they are different in two ways: First, the lymphatic capillaries are closed at one end. Second, the lymphatic capillaries are far more porous than the blood capillaries so that both fluid and small particles can pass with ease from the tissue spaces into the capillaries. The *pores* through which these pass are illustrated in Figure 30–13. They are actually large openings between the endothelial cells. These cells overlap each other very loosely and can be pushed apart with ease when fluid or particles are pushing their way into the capillaries, as will be explained in more detail.

The Lymphatic Valves. Also shown in Figure 30–13 are three **lymphatic valves.** These occur everywhere in the lymphatic system, beginning at the point where the lymphatic capillaries empty into the collecting lymphatics and occurring periodically at all levels in the larger lymphatics, even including the thoracic duct. These valves obviously prevent backflow of lymph toward the tissues.

Physiological Functions of the Lymphatics

Return of Proteins to the Circulation. The single most important function of the lymphatics is to return proteins to the circulation when they leak out of the blood capillaries. Some of the pores in the capillaries are so large that small amounts of protein leak continuously,

amounting each day to approximately one half of the total protein in the circulation. If these proteins were not returned to the circulation, the person's plasma colloid osmotic pressure would fall so low and he would lose so much blood volume into the interstitial spaces that he would die within 12 to 24 hours. Furthermore, no other means is available by which proteins can return to the circulation except by way of the lymphatics. Later in the chapter, when we discuss edema, we will see how serious it is for proteins not to be returned to the circulation even from a single part of the body.

Flow of Lymph Along the Lymphatics. Two principal factors determine the rate of lymph flow along the lymphatics: (1) the interstitial fluid pressure and (2) the degree of "pumping" by the lymphatic vessels, a mechanism called the *lymphatic pump.*

Whenever the interstitial fluid pressure rises above normal, which occurs when the interstitial fluid volume becomes too great, fluid flows easily from the interstitial spaces into the lymphatic capillaries because the "pores" in the lymphatic capillaries are almost wide open—quite different from the blood capillaries. Therefore, the greater the tissue pressure, the greater also is the quantity of lymph formed each minute.

The Lymphatic Pump. The *lymphatic pump*, the second factor that affects lymph flow, is a mechanism of the lymph vessels for pumping lymph along the vessels. To explain this mechanism, it is first necessary to point out that all lymph vessels contain *lymphatic valves* of the type illustrated in Figure 30–13. These valves are almost identical to those in the veins, and they are all oriented centrally so that lymph will flow only toward the point where the lymph vessels empty into the circulation and never backward toward the tissues.

A major cause of lymphatic pumping is periodic contraction of the lymph vessels themselves—about once every 6 to 10 seconds. When a lymph vessel becomes stretched with excess lymph, it automatically contracts. The contraction pushes the lymph past the next lymphatic valve. The lymph then distends the new segment of the lymph vessel, causing it also to contract. This effect reoccurs at each successive segment of the vessel until the lymph is pumped all the way back into the circulation.

In addition to intrinsic contraction of the lymph vessels themselves, lymph pumping can also be caused by motion of the tissues surrounding the lymph vessels. For instance, skeletal muscle contraction adjacent to a lymph vessel can squeeze the vessel and move lymph forward. Likewise, arterial pulsations or movement of tissues because of active or passive movement of the limbs can squeeze lymphatics and cause fluid to move forward along the lymph channels. Under some conditions, these mechanisms are possibly as important as the intrinsic contraction of the lymph vessels themselves, especially during exercise.

Role of the Lymphatic Capillary in Pumping Lymph. Expansion and compression of the lymphatic capillary also plays a major role in lymph flow. The lymphatic capillary has a special structure, illustrated in Figure 30–15, that is especially important to the pumping mechanism. Note that the endothelial cells are anchored to the surrounding tissues by *anchoring ligaments*. When a tissue fills with excess fluid, the entire tissue swells, and the swelling tissue pulls on the anchoring ligaments to open the lymphatic vessel; thus, fluid is also pulled into the vessel. Note also in the figure that the endothelial cells lining the lymphatic capillary overlap each other. Because of this arrangement, fluid flows into the lymphatic very easily because the edge of each endothelial cell simply flaps inward, allowing a direct opening into the vessel. Then when the lymphatic capillary contracts, the overlapping edge of the endothelial cell flaps backward to close the opening. Thus, the overlapping edges of the endothelial cells are actually *valves* that allow fluid to flow into the lymphatic capillary but not to flow backward. Consequently, contraction of the lymphatic capillary or compression of it by the surrounding tissues squeezes the lymph forward along the lymph vessel.

The combination of all these lymph pumping mechanisms gives the lymphatic system its primary characteristic, namely, a tendency to return at all times any excess fluid that collects in the tissues back to the circulation. Indeed, under normal conditions, the pump actually creates a partial vacuum in the tissues, which is the basic cause of the negative pressure of −6 mm Hg in the interstitial fluids.

Rate of Lymph Flow, and Return of Protein to the Circulation. Lymph flow varies tremendously from time to time, but in the average person, the total lymph flow in all vessels is approximately 100 ml per hour, or 1 to 2 ml per minute. One can readily see that this is a very low rate of flow, but it nevertheless is still sufficient to remove the excess fluid and especially the excess protein that tends to accumulate in the tissue spaces. The average lymph contains 3 to 4 percent protein (2 percent from peripheral tissues and as high as 5 to 6 percent from the liver, where a major share of the lymph is formed).

THE NORMAL "DRY" STATE OF THE TISSUES—DEVELOPMENT OF EDEMA WHEN THIS FAILS

Importance of Negative Pressure in the Interstitial Spaces. Up to the present point very little has been said about the importance of the negative pressure in the interstitial fluid spaces of normal tissues. Its main importance is that it serves to pull the cells and other tissue elements together and, consequently, to keep the amount of fluid in the tissues to a minimum. That is, the lymphatic system keeps pulling fluid out of the tissue spaces until these spaces become as small as possible, and it is in this state that the tissues normally operate. Therefore, we can say that, from a relative point of view, the interstitial spaces are normally "dry," even though they still have 12 L of fluid in them. This statement will become more meaningful as we discuss edema in the succeeding paragraphs. This state of minimal fluid in the interstitial spaces is particularly beneficial to the diffusion of nutrients from the capillaries to the cells, for one of the basic principles of diffusion is that the rate of transport of substances by this mechanism becomes progressively greater the shorter the distance that the substances must diffuse.

When the body fails to maintain the "dry" state in the interstitial spaces, but instead collects large quantities of *extra* fluid in these spaces, the condition is called *edema*. The transport of nutrients to the tissues then becomes impaired, sometimes seriously enough to cause

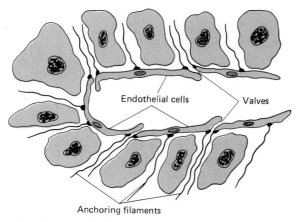

FIGURE 30–15 Special structure of the lymphatic capillaries that permits passage of substances of high molecular weight into the lymph.

gangrene of the tissues. For instance, prolonged swelling of the feet frequently causes gangrenous ulcers of the skin, this resulting at least partially from too great a distance for the nutrients to diffuse from the capillaries to the tissue cells.

The Basic Causes of Edema

Positive Interstitial Fluid Pressure as the Cause of Edema. Figure 30–16 illustrates the compactness of the cells in normal tissues and the spreading apart of the cells when the interstitial fluid pressure rises and causes edema. Recent experiments have shown that as long as the pressure remains negative, the cells remain sucked together in a compact state, but just *as soon as the interstitial fluid pressure rises above the zero level and becomes positive, the cells spread apart.* At pressures of +3 mm Hg the interstitial fluid volume is often increased to as much as 3 to 4 times the normal amount, and at +8 mm Hg the interstitial fluid volume, at least in some tissues, often rises to as much as 20 times normal, causing very great increases in sizes of the interstitial spaces, with resultant increase in distances for nutrients to diffuse from the capillaries to the cells.

Since it is positive interstitial fluid pressure that causes edema, one can readily understand that any change in the dynamics of fluid transfer at the capillary membrane that will increase the interstitial fluid pressure from its normal negative value of −6 mm Hg to a positive value can cause edema. Four basic abnormalities of capillary membrane dynamics often lead to edema. These are (1) elevated fluid pressure in the capillaries, (2) decreased colloid osmotic pressure in the capillaries, (3) increased interstitial fluid colloid osmotic pressure, and (4) increased permeability of the capillaries.

Edema Caused by Elevated Fluid Pressure in the Capillaries. Though the normal mean capillary pressure is about 17 mm Hg, this can rise to values as high as 40 to 50 mm Hg in some abnormal states of the circulation. For instance, when the *heart fails,* the blood volume increases because poor blood flow to the kidneys causes kidney retention of fluid. Also, blood dams up in the venous system and causes very high back pressure in the capillaries. The rise in the mean pressure in the capillaries in turn causes increased interstitial fluid pressure and resultant edema.

Edema Caused by Low Plasma Colloid Osmotic Pressure. Another cause of edema is a decrease in plasma colloid osmotic pressure, which results from diminished plasma protein concentration. This occurs very frequently in persons with some types of *kidney disease* who lose tremendous quantities of plasma proteins into the urine or in persons with *severe burns* who exude large amounts of proteinaceous fluid through their denuded skin. Also, lack of adequate protein in the diet, such as occurs in *famine areas* of the world, can cause failure of formation of adequate plasma proteins. When the plasma protein concentration falls below about 2.5 percent, the normal negative pressure in the interstitial spaces becomes lost and, instead, the pressure rises to a positive value. Here again, very severe edema can result.

Edema Caused by Increased Interstitial Fluid Colloid Osmotic Pressure. Increased interstitial colloid osmotic pressure can also cause high interstitial fluid pressure and edema. The most severe cause of this is *blockage of the lymphatics,* which prevents the normal return of proteins to the circulation. The proteins that leak through the capillary walls gradually accumulate in the tissue spaces until the interstitial colloid osmotic pressure approaches the plasma colloid osmotic pressure. As a result, the capillaries lose their normal osmotic advantage of holding fluid in the circulation so that fluid now accumulates abundantly in the tissues. This condition can cause edema of the greatest proportions.

Lymphatic blockage commonly occurs in the South Sea Island disease called *filariasis,* in which *filariae* (a type of nematode worm) become entrapped in the lymph nodes and cause so much growth of fibrous tissue that lymph flow through the nodes becomes totally or almost totally blocked. As a result, certain areas of the body, such as a leg or an arm, swell so greatly that the swelling is called "elephantiasis." A single leg with this condition can weigh as much as

Tissue pressure = −6 mm Hg

Tissue pressure = 3 mm Hg

Tissue pressure = 8 mm Hg

FIGURE 30–16 Effect of various tissue pressures on the amount of edema developing in the interstitial spaces.

the entire remainder of the body, all because of the extra fluid in the tissue spaces.

Edema Caused by Increased Capillary Permeability. Some abnormal conditions cause tremendous increase in permeability of the capillaries; that is, the capillary pores become greatly enlarged. When this occurs, not only does fluid leak rapidly from the capillaries into the tissue spaces, but also proteins are lost from the plasma while at the same time accumulating in great excess in the interstitial spaces. Therefore, increased capillary permeability can cause severe edema for at least three different reasons: (1) excess leakage of fluid through the enlarged pores into the interstitial spaces, (2) low colloid osmotic pressure in the plasma owing to loss of protein, and (3) increased colloid osmotic pressure in the interstitial fluid because of protein accumulation.

Increased capillary permeability frequently occurs in toxic states. For instance, in anaphylaxis, a condition described in Chapter 29 that results from reaction of antibodies with toxic proteins, capillary permeability sometimes increases so greatly that localized edema occurs at many places in the body within minutes. Another example is the action of some poisonous war gases to increase the permeability of the pulmonary capillaries so greatly that severe pulmonary edema occurs and causes the person to die because his lungs fill with fluid.

SPECIAL FLUID SYSTEMS OF THE BODY

The Potential Fluid Spaces

There are a few spaces in the body that normally contain only a few milliliters of fluid but may contain tremendous quantities of fluid under abnormal conditions. These spaces, called *potential spaces*, include especially the *pleural space*, illustrated in Figure 30–17, which is the space between the lungs and the chest wall; the *pericardial space*, which is the space between the heart and the pericardial sac in which it resides; the *peritoneal space*, which is the space between the gut and the abdominal wall; and the *joint spaces*. Ordinarily, all of these spaces are totally collapsed except for the presence of a small amount of highly viscid lubricating fluid that enables the tissue surfaces that surround the spaces to slip over each other, allowing free movement of the lungs in the chest cavity, free movement of the heart in the pericardial cavity, free movement of the joints, and so forth.

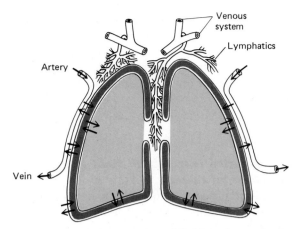

FIGURE 30–17 Dynamics of fluid exchange in the intrapleural spaces.

The Viscid Fluid in the Potential Spaces. If one will think for a few moments about the function of the potential spaces, these all exist where two tissue surfaces slide back and forth over each other—for instance, sliding of the lungs in the chest cavity during the process of breathing, sliding of the surfaces of the heart in the pericardial cavity during the heart beat, sliding of the joint surfaces over each other, and so forth. To increase the ease with which this sliding effect occurs, the potential spaces contain minute quantities of a very slick, highly viscid lubricating fluid. The essential ingredient of this fluid is *hyaluronic acid* conjugated with a small amount of protein to form a mucopolysaccharide that gives the fluid a consistency almost the same as that of the slick mucus found in saliva.

The concentration of hyaluronic acid in the fluids of the potential spaces is usually 1 to 2 percent. It is interesting that this substance is also present in essentially all interstitial spaces of the body, not merely in the potential spaces. One of its roles in the interstitial spaces is perhaps the same as that in the potential spaces, namely, to allow the different segments of the tissues to slide over each other with ease, such as sliding of the skin over the loose subcutaneous tissues.

Dynamics of Fluid Exchange Through the Surfaces of the Potential Spaces. The dynamics of fluid exchange between the potential spaces and the capillaries in the tissues that surround the spaces are almost identical to the dynamics of fluid exchange at the usual capillary membrane. The reason for this is that the linings of these spaces are mainly permeable to both proteins and fluids so that fluid flows with ease between the potential spaces and the spaces of the adjacent tissues where the capillaries lie.

Figure 30–17 illustrates the intrapleural space, showing a capillary in the tissue adjacent to the cavity. This figure shows diffusion of fluid back and forth between the cavity and the capillary and also diffusion of fluid through the visceral pleura into the tissue of the lungs. Obviously, with such free diffusion, the pleural space is actually nothing more than an enlarged tissue space. There is negative pressure in the intrapleural space just as there is negative pressure elsewhere in the tissues. Therefore, normally, the intrapleural space is kept "dry," except for a few milliliters of viscid fluid in the whole chest cavity, in exactly the same way that the other tissue spaces of the body are kept "dry."

"Edema" of the Potential Spaces. Under abnormal conditions, a person can develop "edema" of the potential spaces in exactly the same way that he develops edema in his tissue spaces. Thus, if the capillary pressure rises too high, if the plasma colloid osmotic pressure falls too low, if the colloid osmotic pressure in the pleural cavity rises too high, or if the capillaries become too permeable, fluids will transude into the space in exactly the same way that they transude into the tissue spaces in edema. "Edema" of a potential space is called an *effusion*, though physiologically it is exactly the same as edema.

Lymphatic Drainage of the Potential Spaces. All potential spaces have lymphatic drainage systems similar to that of the tissue spaces. Figure 30–17 illustrates the lymphatic drainage from the pleural cavity. This drainage normally removes the proteins that leak into the intrapleural space and, therefore, keeps the colloid osmotic pressure in the space at a low level.

One of the most serious causes of effusion in a potential space is *infection*, for this usually causes the lymphatics to become plugged with large masses of white blood cells and tissue debris caused by the infection. Also, infection increases the porosity of the surrounding capillaries, making them leak large amounts of protein. The proteins cannot leave the space even though excessive quantities of new proteins are leaking into the space. As a result, the colloid osmotic pressure increases tremendously, and, as one would expect, fluid transudes into the space in massive amounts. Therefore, an infected joint swells tremendously, an infected pleural cavity develops large amounts of pleural effusion, and an infected abdominal cavity develops large amounts of abdominal effusion, which in this case is called *ascites*.

QUESTIONS

1. Explain the dilution principle for measuring body fluid volumes.
2. Give the average pressures in the capillary and in the interstitial fluid surrounding the capillary. Give also the colloid osmotic pressures inside and outside the capillary.
3. If the capillary pressure is 25 mm Hg, the interstitial fluid pressure −2 mm Hg, the plasma colloid osmotic pressure 35 mm Hg, and the interstitial fluid colloid osmotic pressure 2 mm Hg, which way will fluid be moving as a result of the pressure balance at the capillary membrane?
4. Explain why there is net movement of fluid out of the arterial ends of the capillaries and net absorption of fluid at the venous ends.
5. Why is the most important function of the lymphatic system to return protein to the circulation?
6. Explain the mechanism of lymphatic pumping.
7. Why does edema not occur in normal tissues?
8. What causes edema to develop? What are some of its common causes?
9. Explain what is meant by a "potential fluid space." Discuss the dynamics of the fluids in these spaces.

REFERENCES

Bill, A.: Blood circulation and fluid dynamics in the eye. *Physiol. Rev.*, 55:383, 1975.

Fishman, A.P., and Renkin, E.M. (eds.): Pulmonary Edema. Baltimore, Williams & Wilkins, 1979.

Guyton, A.C., and Lindsey, A.W.: Effect of elevated left atrial pressure and decreased plasma protein concentation on the development of pulmonary edema. *Circ. Res.*, 7:649, 1959.

Guyton, A.C., et al.: Interstitial fluid pressure. *Physiol. Rev.*, 51:527, 1971.

Guyton, A.C., et al.: Circulatory Physiology II. Dynamics and Control of the Body Fluids. Philadelphia, W.B. Saunders, 1975.

Landis, E.M., and Pappenheimer, J.R.: Exchange of substances through the capillary walls. *In* Hamilton, W.F. (ed.): Handbook of Physiology. Sec. 2. Vol. 2. Baltimore, Williams & Wilkins, 1963, p. 961.

Nicoll, P.A., and Taylor, A.E.: Lymph formation and flow. *Annu. Rev. Physiol.*, 39:73, 1977.

Rothschild, M.A., et al.: Albumin synthesis. *In* Javitt, N.B. (ed.): International Review of Physiology: Liver and Biliary Tract Physiology I. Vol. 21. Baltimore, University Park Press, 1980, p. 249.

Shulman, K. (ed.): Intracranial Pressure IV. New York, Springer-Verlag, 1980.

Staub, N.C. (ed.): Lung Water and Solute Exchange. New York, Marcel Dekker, 1978.

Yoffey, J.M., and Courtice, F.C. (eds.): Lymphatics, Lymph and Lymphomyeloid Complex. New York, Academic Press, 1970.

31

The Kidneys and Excretion of Urine

Overview

The *principal functions of the kidneys* are (1) *to remove waste products* from the body and (2) *to control the concentrations of most of the ionic substances* in the extracellular fluid, including such ions as *sodium, potassium,* and *hydrogen ions.*

The functional unit of the kidneys is the *nephron.* There are two million nephrons in the two kidneys of the human being. The nephron is divided into two functionally distinct parts: (1) the *renal corpuscle* and (2) the *renal tubule.* The renal corpuscle, in turn, is composed of (a) the *glomerulus,* which is a network of capillaries, and (b) *Bowman's capsule,* which surrounds the glomerulus. The blood pressure inside the glomerulus is about 60 mm Hg, a very high pressure that causes large quantities of fluid, called the *glomerular filtrate,* to filter outward through the capillary walls into Bowman's capsule. In this glomerular filtrate are most of the waste products that need to be eliminated from the body fluids.

The glomerular filtrate leaves Bowman's capsule through the renal tubule, which is divided into four separate sequential segments: (1) the *proximal tubule,* (2) the *loop of Henle,* (3) the *distal tubule,* and (4) the *collecting duct.* As the filtrate passes through this tubular system, those portions of the filtrate needed by the body—mainly the nutrients such as glucose and amino acids, the water, and most of the ions—are reabsorbed from the tubules back into the *peritubular capillaries,* which surround the tubules. On the other hand, the waste products either are not reabsorbed at all or are only partially reabsorbed, so that most of them pass on through the tubular system into the urine.

Thus, the theory of kidney function is to filter large quantities of fluid from the plasma and then to reabsorb those constituents that are needed, yet fail to reabsorb those that are not needed. About *180 L of glomerular filtrate are formed each day, but less than 1 percent of this, about 1.5 L per day, passes into the urine.* Even so, this small volume still contains most of the waste products in highly concentrated form. Some of the more important waste products include *urea, uric acid, creatinine, phosphates, sulfates,* and *excess acids.*

Two of the more common kidney abnormalities that lead to depressed excretion of urinary products include

1. *Destruction or loss of whole nephrons.* This frequently results from *pyelonephritis,* which means infection in the kidneys. When more than three quarters of the nephrons have been lost, the person then begins to have difficulty eliminating waste products, and he accumu-

lates excess urea, uric acid, hydrogen ions, creatinine, and other substances in the body fluids, leading to a severely toxic state called *uremia.*

2. *Glomerulonephritis.* This means *inflammation of the glomeruli,* which causes blockage of many of the glomerular capillaries, making it impossible to filter adequate quantities of glomerular filtrate. Sometimes the patient develops uremia, as occurs following loss of large numbers of nephrons. In other instances, the failure to form enough filtrate causes accumulation of water and salt in the blood and increased blood volume, with consequent increase in the arterial blood pressure to hypertensive levels. When the blood pressure rises high enough, this causes enough glomerular filtrate then to be formed to provide normal urinary output, but from then on the person has *hypertension.*

Excretion of urine by the kidneys serves three major functions:

1. to *eliminate most of the waste products of metabolism* that accumulate in the blood and body fluids;
2. to *eliminate substances that are absorbed in excessive amounts* into the body from the food—such as water, salt, and toxins; and
3. to *control the concentrations of most of the ionic substances in the extracellular fluid,* including sodium, potassium, chloride, bicarbonate, and hydrogen ions. In turn, the control of hydrogen ions also controls the degree of acidity of the body fluids.

The Major Structures of the Urinary System

The principal structures of the urinary system are shown in Figure 31–1. These are

1. the **right** and **left kidneys,**
2. the **right** and **left ureters,**
3. the **urinary bladder,** and
4. the **urethra.**

This figure illustrates the urinary system as seen in an anterior view of the abdomen after all other intra-abdominal organs have been removed. Even the posterior lining of the abdominal cavity, the posterior parietal peritoneum, has been removed except in a few areas around the edges. Therefore, it is clear that the entire urinary system lies *retroperitoneally,* which means behind the peritoneal cavity where most of the gastrointestinal organs are found. When these other abdominal organs are in place, the right kidney lies posterior and immediately inferior to the right lobe of the liver, and the left kidney lies posterior to the stomach and spleen.

The Vascular Supply to the Kidneys.
Note in Figure 31–1 that both the aorta and the inferior vena cava are also in the retroperitoneal space, passing between the two kidneys. As was explained in Chapter 25, the large **right** and **left renal arteries** leave the sides of the aorta a short distance below the origin of the superior mesenteric artery, then pass laterally for only a few centimeters to reach the kidneys. The renal artery enters the kidney at its **hilum,** which is the indented area of the kidney on its medial border.

The **renal vein** leaves the hilum of the kidney anterior to the renal artery and then runs the short distance to the side of the inferior vena cava. The left renal vein passes anterior to the aorta slightly inferior to the origin of the superior mesenteric artery.

The Internal Structure of the Kidney

The internal anatomy of the kidney is shown in Figure 31–2, which shows a frontal section of the kidney as seen from the anterior.

The functional unit of the kidney is the **nephron,** several of which are illustrated in the figure. However, the diameters of the nephrons (but not their lengths) in this figure are drawn extremely large in comparison with the true nephrons, for there are more than 1 million nephrons in each kidney. It is in the nephrons that fluid and other substances leave the blood and are processed to form the urine. But, first, let us describe the gross structures of the kidney.

The Internal Vasculature of the Kidney. In the hilum of the kidney, the renal artery divides into multiple **interlobar arteries.** These then pass between the lobes of the kidney toward the kidney surface. About two thirds of the distance through the kidney they suddenly

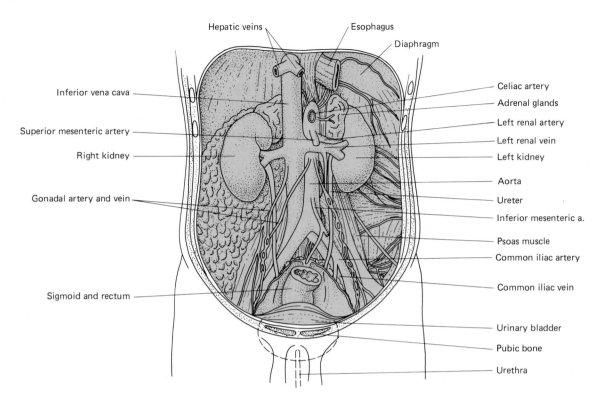

FIGURE 31–1   General organization of the urinary system.

curve to form large arterial arches called **arcuate arteries,** as illustrated in Figure 31–2; these then run parallel to the surface of the kidney. Most of the ends of the arcuate arteries anastomose with adjacent arcuate arteries, which provides some degree of safety when one or more of the interlobar arteries become occluded. From the arcuate arteries multiple small **interlobular arteries** again extend outward toward the surface of the kidney; these branch profusely to supply blood to the nephrons.

As also illustrated in Figure 31–2, the internal veins of the kidney form a pattern that is almost identical to that of the arteries, with **inter-**

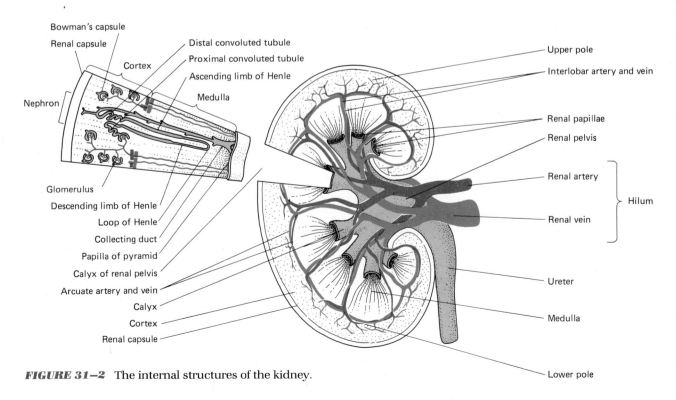

FIGURE 31–2   The internal structures of the kidney.

TABLE 31–1
Structures of the Kidney

Structure	Description or function
General	
Capsule	Strong fibrous covering of kidney
Hilum	Medial entry area to kidney of renal artery, renal vein, lymphatics, nerves, and ureteral system
Cortex	Outer portion of renal tissue containing the renal corpuscles, the proximal tubules, the distal tubules, and outer portions of the loops of Henle and collecting ducts
Medulla	Inner portion of renal tissue containing inner portions of the loops of Henle and the collecting ducts
Papillae	Innermost portions of medulla in form of inverted cones protruding into the minor calyces through which the collecting ducts empty
Urinary Outflow Tract	
Minor calyces	Collects urine from collecting ducts in papillae
Major calyces	Collects urine from 2 or more minor calyces
Renal pelvis	Collects urine from 4 to 8 major calyces
Ureter	Conducts urine from renal pelvis to urinary bladder
Vasculature	
Renal artery	
Interlobar arteries	
Arcuate arteries	Preglomerular vessels, control blood flow
Interlobular arteries and branches	
Afferent arterioles	
Glomerular capillaries	Filtration membrane for forming glomerular filtrate
Efferent arterioles	Control pressure in glomerular capillaries and help to control blood flow
Peritubular capillaries	Absorb fluid from tubules
Venules	
Interlobular veins	
Arcuate veins	Postcapillary vessels
Interlobar veins	
Renal vein	
Nephron	
Renal corpuscle	
Glomerulus	Tuft of glomerular capillaries, covered by epithelial cells, through which glomerular filtrate filters
Bowman's capsule	Collecting chamber for glomerular filtrate
Renal tubules	
Proximal tubule	Located in renal cortex, most of it convoluted; conducts fluid from Bowman's capsule
Loop of Henle	Long tubular loop that conducts tubular fluid from proximal tubule into medulla and then back to cortex
Distal tubule	Located in cortex, most of it convoluted; conducts fluid from loop of Henle
Collecting duct	(Anatomically, not considered to be part of the nephron; but, physiologically, functions with multiple nephrons); conducts fluid from distal tubules through medulla to renal minor calyces.
Juxtaglomerular apparatus	Structure at vascular pole of renal corpuscle where distal tubule passes in angle between afferent and efferent arterioles; helps to control glomerular blood flow and pressure

Note: All tubules, including collecting ducts, reabsorb fluid and electrolytes but fail to reabsorb much of the waste products lobular veins, arcuate veins, and interlobar veins running similar courses to those of the corresponding arteries, all these eventually coalescing in the hilum to form the renal vein.

The Urine Collection System of the Kidney—The Pelvis and the Calyces. After urine is formed by the nephrons, it is collected in the renal pelvis from which it then passes into the ureter and urinary bladder. The renal pelvis arises from the hilum of the kidney posterior to the renal artery and vein. Deep inside the hilum, the renal pelvis branches into four to eight large funnel-shaped chambers called major renal calyces, and each major calyx drains two or more smaller minor calyces. A conical extension of the functional portion of the kidney, called a renal papilla, projects into each minor calyx. It is through the papillae that the *collecting ducts* from the nephrons empty their urine into the calyces. Then the urine passes from the calyces into the pelvis and thence into the ureter and bladder. Several collecting ducts leading from the nephrons into the calyces are illustrated in Figure 31–2.

The Renal Capsule, Renal Cortex, and Renal Medulla. Around the outside of the kidney is a strong fibrous capsule called simply the renal capsule, and inside this is the functional renal tissue, called the renal parenchyma. In turn, the renal parenchyma is divided into the renal cortex and the renal medulla.

The renal cortex is that portion of the renal parenchyma that lies mainly peripheral to the arcuate arteries and veins. It is in this portion of the kidney that fluid is filtered from the blood in the *glomeruli* of the nephrons and that much of the processing of this fluid occurs for the formation of urine.

The medulla is the portion of the renal parenchyma that lies mainly central to the arcuate vessels. It is through this area that the *collecting ducts* pass, carrying urine from the nephrons to the calyces. In addition, a portion of each nephron dips into the medulla, a portion called the *loop of Henle.* This loop plays an essential role in controlling the concentration of urine, which will become clear as we discuss the nephron itself.

THE NEPHRON

A typical nephron is illustrated in Figure 31–3. This is divided into two parts:

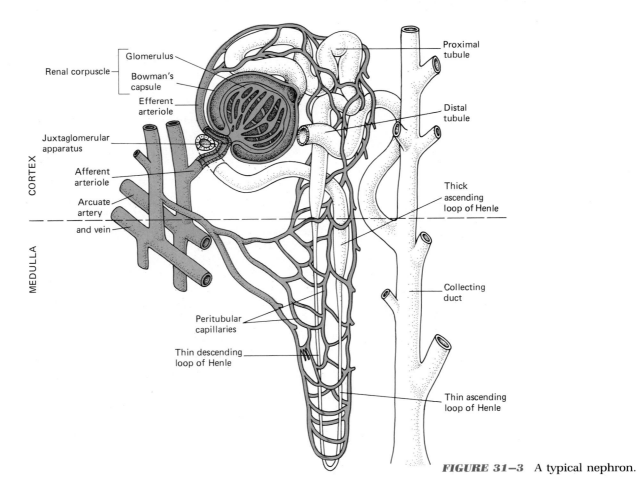

FIGURE 31–3 A typical nephron.

1. the **renal corpuscle** and
2. the **renal tubules.**

The basic principles of nephron function are: First, fluid is filtered from the blood in the renal corpuscle. Then this filtered fluid passes through the renal tubules, where those substances that are needed by the body are reabsorbed back into the blood from the tubular system, whereas those substances not needed by the body—especially the waste products—are not reabsorbed but instead pass into the calyces of the kidney as urine.

THE RENAL CORPUSCLE

The filtration function of the renal corpuscle is dependent on two structures:

1. the **glomerulus,** and
2. **Bowman's capsule.**

The *glomerulus* is a branching and anastomosing network of about 30 parallel capillaries. Blood flows into the glomerulus through the **afferent arteriole** at the **pole of the renal corpuscle,** and it leaves the pole through the **efferent arteriole.** The blood pressure inside the glomerular capillaries is about 60 percent as high as the arterial pressure itself, or about 60 mm Hg. Therefore, extremely large quantities of fluid filter through the thin walls of these capillaries. This fluid is called the *glomerular filtrate.*

Bowman's capsule surrounds the outside of the entire glomerulus, and it is into this capsule that the glomerular filtrate filters. Then, from the side of Bowman's capsule opposite to the pole of the renal corpuscle, the filtrate flows into the renal tubular system.

THE RENAL TUBULAR SYSTEM

This tubular system is divided into four sequential anatomical parts:

1. The **proximal tubule** lies entirely in the renal cortex and is very long but highly *convoluted* (which means folded into a closely packed area). About two thirds of the fluid of the glomerular filtrate is reabsorbed by the proximal tubule, but most of the waste products in this filtrate are not reabsorbed; instead they pass down the tubular system to be emptied eventually into the urine.
2. The **loop of Henle,** after leaving the proximal tubule, loops downward from the renal cortex to different depths in the renal medulla and then back again to the cortex. The loop of Henle itself is divided into two limbs: the **descending limb,** which is the portion that passes downward into the medulla, and the **ascending limb,** the portion that passes back to the cortex. The loop of Henle is also divided into a **thin segment,** the wall of which is constructed of very thin, flat epithelial cells, and a **thick segment** constructed of thick columnar cells. The thin segment includes most of the descending limb and usually the deeper part of the ascending limb; the thick segment is the upper portion of the ascending limb of the loop of Henle before it passes back into the renal cortex. The loop of Henle plays a special role in determining whether the urine will be diluted or concentrated. However, this is such a complex mechanism that it will have to await full explanation until Chapter 32.
3. The **distal tubule** receives fluid from the thick ascending limb of the loop of Henle and then runs a convoluted course in the renal cortex similar to that of the proximal tubule.
4. The **collecting duct** receives fluid from many distal tubules. It begins in the renal cortex and extends straight downward through the medulla on its way to the tip of one of the renal papillae where it empties into one of the minor renal calyces. In the distal tubule and collecting duct, most of the final 15 to 20 percent of the glomerular filtrate is absorbed, leaving about 1 percent that becomes the urine. As is also true for the proximal tubules, most of the waste products are not reabsorbed in the other portions of the tubular system as well; instead, they pass on into the urine in a concentrated form since most of the water in the filtrate is reabsorbed. Also, in the distal tubule and collecting duct selective reabsorption of such substances as sodium, potassium, phosphate, and other elements occurs for the purpose of controlling their concentrations in the extracellular fluid (as will be explained in Chapter 32).

THE PERITUBULAR CAPILLARIES, AND ABSORPTION OF FLUID FROM THE TUBULAR SYSTEM

Surrounding all the tubules is a capillary network called the **peritubular capillaries.** These capillaries originate from the efferent arteriole, then envelop the entire tubular system, and finally empty into the renal venous system.

In contrast to the high blood pressure in the glomerulus, where fluid filters out of the capillaries, the pressure in the peritubular capillaries is low, only about 12 mm Hg. This low pressure allows osmotic reabsorption of fluid from the tubules because of the colloid osmotic pressure of the plasma proteins.

The Juxtaglomerular Apparatus. Now, note in Figure 31–3 a small structure at the pole of the renal corpuscle called the **juxtaglomerular apparatus.** At this point, the distal tubule passes in the angle between the afferent and the efferent arterioles, abutting against both of these arterioles. This provides a structure where feedback control signals can pass from the distal tubule to the afferent and efferent arterioles to control the degrees of constriction of these arterioles. This feedback mechanism controls the rate of filtration of fluid into the tubular system from the glomerulus so that the fluid can be processed at the appropriate rate required for proper formation of the urine. This control mechanism is very important to kidney function and will be described in detail later in the chapter.

FUNCTION OF THE NEPHRON

In discussing the function of the nephron it is desirable to use the simplified diagram shown in Figure 31–4. This shows the "functional nephron," with an **afferent arteriole** supplying blood to the glomerulus and the blood then flowing from the glomerulus through an **efferent arteriole** into the **peritubular capillaries** and finally into the vein. Also shown are the **glomerular membrane, Bowman's capsule,** the **tubules,** and the **kidney pelvis.**

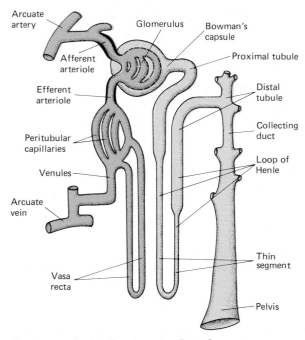

FIGURE 31–4 The functional nephron.

Basic Theory of Nephron Function

The basic function of the nephron is to clean, or "clear," the blood plasma of unwanted substances as it passes through the kidney, while retaining in the blood those substances that are still needed by the body. For instance, the end-products of metabolism such as urea and creatinine especially are cleared from the blood. And sodium ions, chloride ions, and other ions are also cleared when they accumulate in excess in the plasma.

The nephron clears the plasma of unwanted substances in two different ways: (1) It *filters* a large amount of plasma, usually about 125 ml per minute, through the glomerular membranes of the nephrons. Then as this filtered fluid flows through the tubules, the unwanted substances fail to be reabsorbed and pass on into the urine, and the wanted substances are selectively reabsorbed back into the plasma. (2) A few substances are cleared by the process of *secretion.* That is, the tubular walls actively remove substances from the blood and secrete them into the tubules.

Thus, the urine that is eventually formed is composed of both *filtered* substances and *secreted* substances.

Glomerular Filtration

The Glomerular Membrane. The capillaries in the glomerulus together with their membranous coverings are collectively called the **glomerular membrane.** Figure 31–5 illustrates the outside surfaces of several of these capillaries, showing that they are covered by epithelial cells that envelop them with a multitude of **foot processes.** These processes interdigitate with each other, leaving slits between the processes through which fluid can filter from the capillaries into Bowman's capsule.

Figure 31–6 shows three layers of the glomerular membrane from the inside to the outside: (1) the **endothelial cell layer** of the capillary itself, (2) a **basement membrane** of collagenous and proteoglycan fibers, and (3) the **epithelial cell layer** on the outside consisting of the foot processes described above. Note the many holes, called **fenestrae,** in the capillary endothelial cell layer and also the multiple **slit pores** between the adjacent foot processes of the epithelial cell layer. The basement membrane between these two layers is also highly porous because it is simply a meshwork of fibers. Thus, it is through the fenestrae, the slit pores, and the spaces within the basement membrane

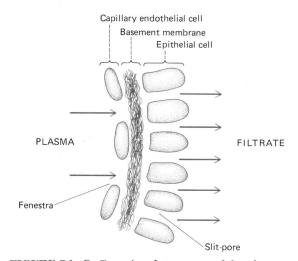

FIGURE 31-5 Scanning electron micrograph of a glomerulus. Note the finger-like projections, called "foot processes," that interdigitate with each other, leaving slits between them. Filtration occurs through these slits. (From Brenner and Rector: *The Kidney.* Philadelphia, W. B. Saunders, 1976.)

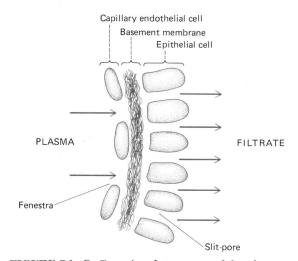

FIGURE 31-6 Functional structure of the glomerular membrane.

that fluid filters out of the glomerular capillaries into Bowman's capsule. This fluid is the glomerular filtrate, and it passes from Bowman's capsule into the tubular system.

The glomerular membrane is several hundred times as permeable to water and small molecular solutes as the usual capillary membrane elsewhere in the body, but otherwise the same principles of fluid dynamics apply as to other capillary membranes. Like other capillary membranes, the glomerular membrane is almost completely impermeable to plasma protein, and it is also impermeable to the blood cells.

However, the pressure in the glomerulus is very high, about 60 mm Hg, in contrast to the low pressures, between 15 and 20 mm Hg, in capillaries elsewhere in the body. Because of this high pressure, fluid leaks continually out of all portions of the glomerular membrane into Bowman's capsule. We shall see that most of the fluid that leaks out of the glomerular membrane is later reabsorbed from the renal tubules into the peritubular capillaries. That which is not reabsorbed becomes urine.

Fluid Dynamics at the Glomerular Membrane, and the Filtration Pressure. The fluid pressures in the normal nephron are illustrated in Figure 31-7. This figure shows that the *glomerular pressure* is normally 60 mm Hg, whereas the colloid pressure in the glomerulus is normally 32 mm Hg. The pressure in Bowman's capsule is about 18 mm Hg, and the col-

FIGURE 31-7 Normal pressures at different points in the nephron and the normal filtration pressure.

loid osmotic pressure in this capsule is essentially zero. Therefore, the pressure tending to force fluid out the glomerulus is 60 mm Hg, whereas the total pressure tending to move fluid in the opposite direction into the glomerulus from Bowman's capsule is 32 + 18, or 50 mm Hg. The difference between the outward pressure of 60 and the inward pressure of 50, or 10 mm Hg, is the net pressure pushing fluid into Bowman's capsule; this is called the *filtration pressure.*

The Glomerular Filtration Rate. The rate at which fluid filters from the blood into Bowman's capsule, called the *glomerular filtration rate,* is directly proportional to the filtration pressure. Therefore, any factor that changes any one of the pressures on the two sides of the glomerular membrane will also change the glomerular filtration rate. Thus, an increase in glomerular pressure increases the rate of glomerular filtrate formation; an increase in either the glomerular colloid osmotic pressure or the pressure in Bowman's capsule decreases the rate of filtrate formation.

The normal rate of formation of glomerular filtrate in both kidneys of the human being is 125 ml per minute. This is approximately 180 L each day, or 4.5 times the amount of fluid in the entire body, which illustrates the magnitude of the renal mechanism for purifying the body fluids. Fortunately, almost 179 L of this 180 L is reabsorbed by the tubules so that only slightly more than 1 L of this total filtrate is lost into the urine.

Effect of Afferent Arteriolar Constriction on Filtration. The major effect of constricting the *afferent arteriole* is a drastic decrease in the pressure in the glomerulus. And this in turn causes an even more drastic decrease in glomerular filtration rate. For instance, a 10-mm decrease in glomerular pressure can almost stop glomerular filtration.

The afferent arterioles are controlled partly by sympathetic nerves and partly by an automatic control mechanism intrinsic in the nephron itself, called *autoregulation,* which will be discussed in detail later in the chapter. Sympathetic stimulation constricts the arterioles and lowers the glomerular pressure, thereby decreasing glomerular filtration. On the other hand, diminished sympathetic stimulation allows afferent arteriolar dilatation and, consequently, increased glomerular filtration.

Characteristics of Glomerular Filtrate. The filtrate entering Bowman's capsule, the *glomerular filtrate,* is an ultrafiltrate of plasma. The glomerular membrane is porous enough so that water and essentially all of the dissolved constituents of plasma except proteins can filter through. Therefore, glomerular filtrate is almost identical to plasma except that only a very minute quantity of protein (0.03 percent) is present in the filtrate, and the protein concentration in plasma is approximately 7 percent, or more than 200 times as great.

Tubular Reabsorption

After the glomerular filtrate enters Bowman's capsule, it passes into the tubular system where each day all but slightly greater than 1 L of the 180 L of glomerular filtrate is reabsorbed into the blood. The remaining 1 L passes into the renal pelvis as urine.

Figure 31–8 illustrates a microscopic cross-section of a tubular region of the kidney, showing the close proximity of the tubules to the peritubular capillaries. The tubular fluid is reabsorbed first into the interstitial spaces and then from these spaces into the capillaries. Some of the substances are reabsorbed through the tubular epithelium by the process of *active reabsorption;* other substances are reabsorbed by the process of *diffusion and osmosis.*

Active Reabsorption. The term *active reabsorption* was discussed in Chapter 5. It means transport of substances through the tubular epithelial cells by special chemical transport mechanisms that have the capability of transporting the substances against concentration differences between the tubular and interstitial fluids. From the interstitial spaces, the substances then pass directly into the peritubular capillaries.

Some of the substances reabsorbed by active transport are *glucose, amino acids, proteins, uric acid,* and *most of the ions—sodium, potassium, magnesium, calcium, chloride,* and *bicarbonate.*

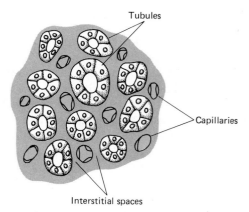

FIGURE 31–8 Cross-section of the tubules and adjacent capillaries in the kidney.

Active Reabsorption of Nutrients From the Tubular Fluid. Obviously, it is important to preserve the nutrients in the body fluids and not allow these to be wasted in the urine. To achieve this, *glucose, amino acids,* and *proteins* are all almost entirely reabsorbed even before the tubular fluid has passed all the way through the proximal tubules, the first portion of the tubular system. The active reabsorption processes for glucose, amino acids, and proteins are so powerful that ordinarily almost none of these substances are lost in the urine.

Active Reabsorption of Ions—Especially Sodium Chloride (Salt). Reabsorption of the ions from the tubular fluid is somewhat different from reabsorption of the nutrients. The body needs to conserve a certain proportion of the ions but also to eliminate the excesses. Fortunately, special control mechanisms, several of which will be discussed in Chapter 32, determine the amount of each ion that is to be reabsorbed. When the quantity already in the blood is too great, the ion is mainly excreted; but when the quantity in the blood is too low, much larger proportions of the ion will be reabsorbed.

The substance that is actively reabsorbed from the tubules to the greatest extent of all is sodium chloride, the total quantity reabsorbed each day being approximately 1200 g, which is about three fourths of all the substances actively reabsorbed in the entire tubular system. The reabsorption of sodium chloride is regulated partially by the hormone *aldosterone*, which is secreted by the adrenal cortex. This regulatory mechanism is discussed in Chapter 32.

Figure 31–9 illustrates the mechanism of active sodium reabsorption. The serrated border of the tubular epithelial cell, called the *brush border*, is extremely permeable to sodium and allows sodium to *diffuse* rapidly from the lumen of the tubule to the inside of the cell. At the base and sides of the cell, on the other hand, the membrane has entirely different properties. Here the membrane is almost completely impermeable to diffusion of sodium, but it does *actively transport* sodium in the outward direction from the cell into the peritubular fluid. This active transport probably occurs in the manner explained in Chapter 5 as follows: It is believed that the sodium ion combines with a *carrier* that is dissolved in the cell membrane. The combined sodium-plus-carrier then diffuses to the opposite side of the membrane, where the sodium is released into the peritubular fluid. Enzymes in or adjacent to the inner surface of the cell membrane cause the necessary reactions to take place, and the energy system in the cyto-

FIGURE 31–9 Mechanism for active transport of sodium from the tubular lumen into the peritubular capillary, illustrating active transport at the base and sides of the epithelial cell and diffusion through the luminal border of the cell.

plasm of the cell supplies the energy required to make the reactions occur.

Active transport of other ions occurs in the same manner. However, each ion has its own specific carrier and its own set of enzymes for catalyzing the reactions.

Absorption Against a Concentration Difference. One of the most important features of active reabsorption is that it can cause absorption of a substance even when its concentration is less in the tubule than in the peritubular fluid. To do this, however, the tubular epithelial cells must expend much energy. Therefore, these cells require tremendous amounts of nutrition, and their metabolic systems are so geared that they can transform the potential energy of their nutrients into the energy required to transport substances against the concentration difference. These cells sometimes expend as much as 90 percent of their energy for active transport alone.

Reabsorption by Diffusion. At this point we need to recall the basic principles of diffusion because this plays a major role in the reabsorption of water and a few other substances from the tubules. Diffusion means the random movement of molecules in a fluid, and it is caused by kinetic movement of all fluid molecules. In other words, each water molecule or each dissolved molecule in the water is constantly bouncing among all the others, wending its way from place to place, going first in one direction and then another. If a large enough pore is present in a membrane, the molecule can pass through the membrane—that is, the membrane is permeable to the molecule. The tubular epithelium is permeable to certain types of molecules, including water molecules. Therefore, water can diffuse from the tubules into the interstitial spaces of the kidney.

Reabsorption of Water from the Tubules. The principal method for reabsorption of water from the tubules is by *osmotic diffusion*, which may be explained as follows:

First, let us recall what is meant by *osmosis*. Osmosis means net diffusion of water through a membrane caused by a greater concentration of nondiffusible substances on one side of the membrane than on the other side. The basic principles of osmosis were discussed in Chapter 5. However, we need to see now how the principles of osmosis apply to the transport of water from the tubular lumen.

When ions, glucose, and other substances are actively transported from the tubules into the interstitial spaces of the kidneys, the concentrations of these substances become decreased in the tubular fluid and increased in the interstitial spaces. Consequently, a very large total concentration difference of all these combined solutes develops across the epithelial membrane. The low concentration of solutes in the tubular fluid means that the water concentration in the tubule is relatively high. The higher concentration of the solutes in the peritubular capillaries means that the water concentration is somewhat less. Obviously, therefore, water will now diffuse from its high concentration area in the tubules toward its low concentration area in the peritubular capillaries. This is the phenomenon of *osmosis*.

Thus, when the dissolved substances are actively transported across the epithelial membrane, this automatically causes osmotic transport of water across the membrane as well. That is, the water "follows" the solutes.

Failure of Reabsorption of Unwanted Substances—Urea, Creatinine, Uric Acid, Phosphates, Sulfates, and Nitrates

Some of the substances in the glomerular filtrate are undesirable in the body fluids; these substances, in general, are reabsorbed either not at all or very poorly by the tubules. For instance, *urea*, an end-product of protein metabolism, has no functional value to the body and must be removed constantly if protein metabolism is to continue. This substance is not actively reabsorbed, and the pores of the tubules are so small that urea diffuses through the tubular membrane much less easily than water. Therefore,

whereas water is being osmotically reabsorbed, only 50 percent of the urea is reabsorbed. The other half of the urea remains behind and passes on into the urine.

Thus, the primary function of the kidney is this separation process in the tubules, the tubules reabsorbing those substances that are needed by the body such as amino acids, electrolytes, and water, while, at the same time, allowing urea, a substance that is not needed by the body, to pass on into the urine.

Other substances that have a fate similar to that of urea include *creatinine, phosphates, sulfates, nitrates, uric acid*, and *phenols*, all substances that are end-products of metabolism and would damage the body if they remained in the body fluids in excessive amounts.

Active Tubular Secretion

A few substances are actively secreted from the blood into the tubules by the tubular epithelium. Active secretion occurs by the same mechanism as active reabsorption but in the reverse direction.

Substances that are actively secreted include *potassium ions, hydrogen ions, ammonia*, and many toxic substances that often enter the body. Also, a number of drugs such as penicillin, Diodrast, Hippuran, and phenolsulfonphthalein are removed from the blood primarily by active secretion rather than by glomerular filtration. However, in normal function of the kidneys, tubular secretion is important only to help in regulation of potassium and hydrogen ion concentrations in the body fluids, which will be discussed in the following chapter.

Recapitulation of Nephron Function—Concentration of Substances in the Urine

Now that both glomerular filtration and tubular reabsorption have been discussed, it will be valuable to review in still a different way the total function of the nephron, as follows:

The total blood flow into all the nephrons of both kidneys is approximately 1200 ml per minute. Approximately 650 ml of this are plasma and the remaining 550 ml are red blood cells. About one fifth of the plasma filters through the glomerular membranes of all the nephrons into Bowman's capsules, forming an average of 125 ml of glomerular filtrate per minute. The glomerular filtrate is actually plasma minus the proteins. As the glomerular filtrate passes downward through the tubules, approximately 65 percent of the water and ions are reabsorbed in the proximal tubules, and essentially all of the glucose, proteins, and amino acids are reabsorbed. As the remaining 35 percent of the glomerular filtrate passes through the loop of Henle, distal tubules, and collecting tubules, variable amounts of the remaining water and ions are absorbed, depending on the need of the body for these substances, as discussed in the following chapter. The pH of the tubular fluid may rise or fall depending on the relative amounts of acidic and basic ions reabsorbed by the tubular walls. Also, the osmotic pressure of the tubular fluid may rise or fall depending on whether large quantities of ions or great amounts of water are reabsorbed. Thus, the pH of the finally formed urine may vary anywhere from 4.5 to 8.2, whereas the total osmotic pressure may be as little as $\frac{1}{4}$ that of plasma or as great as four times that of plasma.

Rate of Urine Flow. The final quantity of urine formed is normally about 1 ml per minute or $\frac{1}{125}$ the amount of glomerular filtrate filtered each minute. This 1 ml of urine contains about $\frac{1}{2}$ the urea that is in the original glomerular filtrate, all of the creatinine, and large proportions of the uric acid, phosphate, potassium, sulfates, nitrates, and phenols. Thus, even though almost all the water and salt in the tubular fluid is reabsorbed, a very large proportion of the waste products in the original glomerular filtrate is never reabsorbed, but instead passes into the urine in a highly concentrated form.

Regulation of Rate of Fluid Processing by the Tubules—the Phenomenon of Glomerular Filtration Autoregulation

Function of the Juxtaglomerular Apparatus. The reabsorption of water, salts, and other substances from the tubules depends greatly upon the rate at which glomerular filtrate flows into the tubular system. If the rate is very fast, none of the constituents is reabsorbed adequately before the fluid finally empties at the other end of the tubular system into the urine. On the other hand, when very little glomerular filtrate is formed each minute, essentially all of the filtrate is reabsorbed, including the water in the filtrate, the urea, and all the other solutes. Therefore, for optimum effectiveness in reabsorbing water and salts while not reabsorbing too much urea and other end-products of metabolism, the glomerular filtration rate in each nephron must be very exactly controlled. This is called *glomerular filtration autoregulation.* And

the need for very precise autoregulation of glomerular filtration is so important that there are two separate mechanisms for controlling the rate of glomerular filtration. These are (1) an afferent arteriolar vasoconstrictor feedback mechanism to decrease filtration when it is too great and (2) an efferent arteriolar vasoconstrictor feedback mechanism to increase filtration when it is too low. However, before we can explain these mechanisms, it is first necessary to discuss more fully a special structure of the nephron, the **juxtaglomerular apparatus.**

Referring back to Figure 31–3, note that the distal tubule, where it comes up from the loop of Henle, passes between the afferent and efferent arterioles where they enter the glomerulus. This area of the nephron, called the *juxtaglomerular apparatus,* is illustrated in much greater detail in Figure 31–10, which shows the distal tubule lying in intimate contact with both the afferent and efferent arterioles. Where this contact is made, the epithelial cells of the distal tubule are dense and increased in number. Therefore, this portion of the distal tubule is called the *macula densa.* Some of the smooth muscle cells of the adjacent afferent arteriole, and to a lesser extent of the efferent arteriole, contain granules; these granulated cells are called *juxtaglomerular cells.* The granules contain a precursor form of the hormone *renin,* which plays an important role in efferent arteriolar constriction during the glomerular filtration autoregulation process, as we shall see below. This renin also plays a role in the control of arterial pressure, which was discussed in Chapter 28.

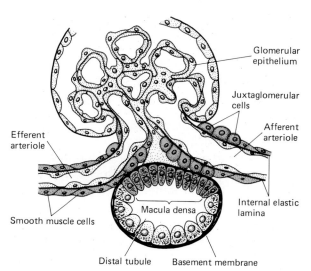

FIGURE 31–10 Structure of the juxtaglomerular apparatus, illustrating its possible feedback role in the control of nephron function. (Modified from Ham: *Histology.* Philadelphia, J.B. Lippincott, 1974.)

The Afferent Arteriolar Vasoconstrictor Mechanism. When the rate of formation of glomerular filtrate is too great, the filtered fluid flows through the tubular system too rapidly for it to be processed adequately in the proximal tubule and the loop of Henle. As a result, the concentration of chloride ions in the fluid flowing from the loop of Henle into the distal tubule becomes too high. These chloride ions then, in some way that is not yet understood, cause the afferent arteriole to constrict, thus reducing the glomerular filtration rate back toward normal and also decreasing the chloride ion concentration at the macula densa back toward normal. Thus, this afferent vasoconstrictor mechanism automatically helps to limit the rate of formation of glomerular filtrate to that amount of fluid that can be properly processed by the tubular system.

The Efferent Arteriolar Vasoconstrictor Mechanism. Constriction of the efferent arterioles causes exactly the opposite effects as afferent arteriolar constriction causes on glomerular filtration rate. That is, efferent constriction prevents easy run-off of blood from the glomerulus and therefore *increases* the glomerular pressure and glomerular filtration rate. When glomerular filtration into the tubular system is too low, an automatic mechanism causes constriction of the efferent arterioles, thereby raising the glomerular filtration rate back toward normal. This mechanism is the following: Slow flow of fluid from the loop of Henle up to the macula densa of the distal tubule, for reasons that are not yet understood, causes the juxtaglomerular cells in the afferent and efferent arterioles to secrete renin. This renin then acts on *renin substrate* (a globulin type of protein) in the surrounding interstitial fluids of the kidney to form *angiotensin*. The angiotensin in turn causes efferent arteriolar constriction, which increases the glomerular filtration rate, thereby returning the fluid flow through the tubular system back toward the normal mean value.

Thus, two different mechanisms work together to maintain a rate of glomerular filtration that is almost exactly right—not too much, not too little—for the filtrate to be processed properly by the tubular system.

Effect of Arterial Pressure on Rate of Urine Formation

An increase in arterial pressure steps up the rate of urine formation in two different ways. First, despite the autoregulation mechanism, increased arterial pressure still causes a slight rise in glomerular pressure because the autoregulation mechanism is not perfect. This rise in pressure, in turn, *increases* the glomerular filtration rate. Second, an elevation in arterial pressure also increases very slightly the peritubular capillary pressure, which tends to *decrease* the rate of reabsorption of fluid from the tubules. Therefore, because of increased flow of glomerular filtrate into the tubules and yet a tendency for decreased reabsorption of this fluid, the effect on urine output is multiplied, so that an increase in arterial pressure has marked effect on urinary output.

An increase in arterial pressure from the normal value of 100 mm Hg up to 120 mm Hg approximately doubles urinary output, and an increase to 200 mm Hg boosts urinary output six- to eightfold. Conversely, a decrease in arterial pressure to 60 mm Hg almost completely stops urinary output.

This effect of arterial pressure on urinary output is exceedingly important for the control of the arterial pressure itself, as was pointed out in Chapter 28. That is, when the arterial pressure rises too high, this automatically increases the rate of urine flow, a process that continues until the person is dehydrated enough so that his arterial pressure returns to its original value. Conversely, a decrease in arterial pressure below normal causes retention in the body of ingested fluids and electrolytes until the arterial pressure increases enough to make the kidneys once again excrete an amount of fluid equal to the daily intake. Recent experiments have demonstrated that this is the most important of all the long-term pressure control mechanisms of the body.

THE CONCEPT OF CLEARANCE

The function of the kidney is actually to *clean* or to *clear* the extracellular fluids of substances. Every time a small portion of plasma filters through the glomerular membrane, passes down the tubules, and then is reabsorbed into the blood, the plasma is "cleared" of the unreabsorbed substances. For instance, out of the 125 ml of glomerular filtrate formed each minute, approximately 60 ml of that reabsorbed leaves its urea behind. In other words, 60 ml of plasma are cleared of urea each minute by the kidneys. In the same way, 125 ml of plasma are cleared of creatinine each minute; 12 ml, of uric acid; 12 ml, of potassium; 25 ml, of sulfate; 25 ml, of phosphate; and so forth.

Calculation of Renal Clearance. The method by which one determines how much plasma is cleared of a particular substance each minute is to take simultaneous samples of blood and urine while also measuring the volume of urine excreted each minute. From the samples the quantity of the substance in each milliliter of blood is analyzed chemically, and the quantity of the substance appearing in the urine each minute is also determined. By dividing the quantity of the substance in each milliliter of plasma into the quantity passing into the urine each minute, one can calculate the milliliters of plasma cleared per minute. That is,

Plasma clearance =

$$\frac{\text{Milligrams secreted in urine per minute}}{\text{Milligrams in each milliliter of plasma}}$$

As an example, if the concentration of urea in the plasma is 0.2 mg in each milliliter and the quantity of urea entering the urine per minute is 12 mg, then the amount of plasma that loses its urea during that minute is 60 ml. To express this another way, the *plasma clearance* of urea is 60 ml per minute.

Plasma Clearance As a Test of Kidney Function. Since the primary function of the kidneys is to clear the plasma of undesired substances, one of the best means for testing overall kidney function is to measure the clearance of different substances. For instance, the clearance of urea has often been used as a test of renal function. As calculated above, normal plasma clearance of urea is approximately 60 ml per minute; the clearance is less than this if the kidneys are damaged, and the amount that it is depressed below normal gives one a measurement of the degree of kidney damage.

Measurement of Glomerular Filtration Rate by Measuring Plasma Clearance of Inulin. The rate of clearance of the substance *inulin* by the kidneys is exactly equal to the glomerular filtration rate for the following reasons: Inulin is a polysaccharide with a molecular weight of about 5200 that filters through the glomerular membrane as easily as water, so that the concentration of inulin in the glomerular filtrate is exactly equal to that in the plasma. However, inulin is not reabsorbed or secreted even in the minutest degree by the tubules. Therefore, all of the inulin of the glomerular filtrate appears in the urine. In other words, all of the originally formed glomerular filtrate is cleared of inulin, which means that the rate of inulin clearance is equal to the rate of glomerular filtrate formation. As an example, a small quantity of inulin is in-jected into the bloodstream of a person, and after mixing with the plasma its concentration is found to be 0.001 mg in each milliliter of plasma. The amount of inulin appearing in the urine is 0.125 mg per minute. On dividing 0.125 by 0.001 we find that the plasma clearance of inulin is 125 ml per minute. This, therefore, is also the amount of glomerular filtrate formed each minute.

Estimation of Renal Blood Flow by Clearance Methods. The amount of blood flow through the two kidneys can be estimated from the clearance of either Diodrast or para-aminohippuric acid. These two substances, when injected into the blood in small quantities, are about 90 percent cleared by active tubular secretion. Therefore, if the plasma clearance of Diodrast is found to be 600 ml per minute, one can estimate that 600 ml is equal to 90 percent of the plasma that flowed through the kidneys during that minute, or a plasma flow of $600 \times \frac{100}{90} =$ 667 ml per minute. And if the blood hematocrit is 40, then the renal blood flow is $667 \times \dfrac{100}{100 - 40} =$ 1111 ml per minute.

ABNORMAL KIDNEY FUNCTION

Almost any type of kidney damage decreases the ability of the kidney to cleanse the blood. Therefore, kidney abnormalities usually cause an excess of unwanted metabolic waste products in the body fluids as well as poor regulation of the electrolyte and water composition of the fluids.

Kidney Shutdown. Several types of kidney damage can cause the kidneys to stop functioning suddenly and completely. Three of the most common are (1) poisoning of the nephrons by mercury, uranium, gold, or other heavy metals; (2) plugging of the kidney tubules with hemoglobin following a transfusion reaction; and (3) destruction of the kidney tubules because of prolonged circulatory shock. In addition to these specific causes of shutdown, almost any other disease of the kidneys can cause either gradual or rapid shutdown.

Following kidney shutdown the concentrations of urea, uric acid, and creatinine, all of which are metabolic waste products, may reach ten or more times normal levels. Also, the body fluids may become extremely acidic because of failure of the kidneys to excrete sufficient quantities of acid; or the person may develop hyperkalemia (excess potassium in the body fluids) because of failure to excrete potassium, and if the person continues to drink water and eat salt, he will become very edematous because of failure to rid himself of the ingested fluid and ions. The person passes into coma within a few days, mainly because of acidosis. If the shutdown is complete, he will die in 8 to 14 days if not treated with an artificial kidney.

Kidney Abnormalities That Cause Loss of Nephrons.   Many kidney diseases destroy large numbers of whole nephrons at a time. For instance, infection of the kidney can destroy large areas of the kidney; trauma can destroy part of or an entire kidney; occasionally a person is born with congenitally abnormal kidneys in which many of the nephrons are already destroyed; or nephron destruction can result from poisons, toxic diseases, or arteriosclerotic blockage of renal blood vessels.

As many as two thirds of the nephrons in the two kidneys can usually be destroyed before the composition of the person's blood becomes greatly abnormal. The reason for this large margin of safety is that the undamaged nephrons then begin to function much more rapidly than usual. The amount of blood flowing into each remaining nephron becomes greatly increased, and the glomerular filtration rate per nephron often rises to two times normal. This increased nephron activity compensates to a great extent for the lost nephrons, allowing the metabolic waste products to be removed in sufficient quantity to maintain normal body fluid composition. However, these persons usually are treading a thin line of safety because bouts of excess metabolism caused by exercise, fever, or even ingestion of too much food may present the kidneys with far more waste products than they can handle.

Obviously, as the degree of kidney destruction progresses still further the derangements of the extracellular fluid become progressively more severe until finally the condition approaches that of kidney shutdown. The person then develops extreme edema, acidosis, and eventually coma and death.

Acute and Chronic Glomerulonephritis.   A very common kidney disease is acute glomerulonephritis. This is an inflammatory disease caused by toxins of certain streptococcal bacteria. Almost all episodes of acute glomerulonephritis occur approximately 2 weeks after a severe streptococcal sore throat or some other streptococcal infection. The glomeruli become acutely inflamed, swollen, and engorged with blood. Blood flow through the glomeruli almost ceases, and the glomerular membranes of those glomeruli that still do have a blood flow become extremely porous, allowing both red blood cells and protein to leak in large quantities into the tubules. A person with acute glomerulonephritis has decreased kidney function sometimes to the extent of complete kidney shutdown. If any urine is still formed, it usually contains large quantities of red blood cells and proteins.

In many instances of acute glomerulonephritis, the inflammation of the glomeruli regresses within 2 to 3 weeks, but even so the disease usually permanently destroys or damages many of the nephrons. Repeated small bouts of acute glomerulonephritis may damage more and more nephrons, causing *chronic glomerulonephritis.* This can run a course of a few to many years, leading eventually to edema, coma, and death.

Use of the Artificial Kidney

When the kidneys are damaged so severely that they can no longer maintain normal composition of the extracellular fluid, it is sometimes necessary to remove waste products from the extracellular fluid by use of an artificial kidney, such as the one shown in Figure 31–11. The artificial kidney is nothing more than a semiporous cellophane membrane arranged so that blood flows on one side and a *dialyzing solution* on the other side of the membrane surface. The membrane is porous to all substances in the blood except the plasma proteins and red blood cells. Therefore, almost all the blood substances can diffuse into the dialyzing solution, and the substances in the solution can also diffuse into the blood.

The dialyzing fluid contains none of the waste products of metabolism. Consequently, the waste products of metabolism diffuse from the blood into the fluid. On the other hand, the dialyzing fluid does contain ions such as sodium, chloride, and so forth in concentrations almost as great as those found in normal plasma. Therefore, these ions diffuse in both directions but with slightly greater amounts leaving the blood than entering, thus allowing removal of the excess ions.

Unfortunately, the artificial kidney requires that the person's blood be rendered incoagulable during its use by the administration of the drug *heparin* that blocks the coagulation process, and a small amount of blood is lost each time the artificial kidney is applied. Therefore, the artificial kidney is usually used for only a few hours out of every several days. Even so, some patients with no kidney function at all have been kept alive by the artificial kidney for 10 years or more.

Stretcher
scale

Radial artery

Superficial
vein

Bubble
trap

Kill dialyzer

Fresh dialyzing
solution

Constant
temperature
bath

Used dialyzing
solution

Polypropylene
plates

Dialyzing
solution

Cellophane
sheets

Dialyzing unit
(cross-section)

FIGURE 31–11 Schematic
diagram of the artificial
kidney.

QUESTIONS

1. Describe the gross structures of the kidney, including its vascular system.
2. Give the structure of a nephron.
3. Explain the mechanism for formation of glomerular filtrate and control of the glomerular filtration rate.
4. Explain the role of active reabsorption from the tubules.
5. Explain the role of absorption by diffusion from the tubules.
6. What substances are reabsorbed from the tubules by active reabsorption, and what substances by diffusion?
7. What substances are actively secreted by the tubules?
8. Explain the afferent arteriolar and the efferent arteriolar mechanisms for autoregulation of glomerular filtration rate.
9. If urea is present in the plasma in a concentration of 0.5 mg per milliliter and the quantity of urea excreted into the urine per minute is 25 mg, how much plasma is cleared of urea each minute?
10. Describe the structure and function of the artificial kidney.

REFERENCES

Andersson, B.: Regulation of body fluids. *Annu. Rev. Physiol.,* 39:185, 1977.

Aukland, K.: Renal blood flow. *Int. Rev. Physiol.,* 11:23, 1976.

Barger, A.C., and Herd, J.A.: Renal vascular anatomy and distribution of blood flow. *In* Orloff, F., and Berliner, R.W. (eds.): Handbook of Physiology. Sec. 8. Baltimore, Williams & Wilkins, 1973, p. 249.

Brenner, B.M., *et al.*: Determinations of glomerular filtration rate. *Annu. Rev. Physiol.* 38:9, 1976.

Brenner, B.M., *et al.*: Transport of molecules across renal glomerular capillaries. *Physiol. Rev.,* 56:502, 1976.

Giebisch, G., and Stanton, B.: Potassium transport in the nephron. *Annu. Rev. Physiol.,* 41:241, 1979.

Gottschalk, C.W., and Lassiter, W.E.: Micropuncture methodology. *In* Orloff, F., and Berliner, R.W. (eds.): Handbook of Physiology. Sec. 8. Baltimore, Williams & Wilkins, 1973, p. 129.

Katz, A.I., and Lindheimer, M.D.: Actions of hormones on the kidney. *Annu. Rev. Physiol.,* 39:97, 1977.

Knox, F.G., and Diaz-Buxo, J.A.: The hormonal control of sodium excretion. *Int. Rev. Physiol.,* 16:173, 1977.

Moses, A.M., and Share, L. (eds.): Neurohypophysis. New York, S. Karger, 1977.

Pitts, R.: Physiology of the Kidney and Body Fluids, 3rd Ed. Chicago, Year Book Medical Publishers, 1974.

Renkin, E.M., and Robinson, R.R.: Glomerular filtration. *N. Engl. J. Med.,* 290:785, 1974.

Stephenson, J.L.: Countercurrent transport in the kidney. *Annu. Rev. Biophys. Bioeng.,* 7:315, 1978.

Vander, A.J.: Renal Physiology. New York, McGraw-Hill, 1980.

Wright, F.S., and Briggs, J.P.: Feedback control of glomerular blood flow, pressure and filtration rate. *Physiol. Rev.,* 59:958, 1979.

32

Regulation of Body Fluid Constituents and Volumes; and the Urinary Bladder and Micturition

Overview

The kidneys play an essential role in regulating the concentrations of most of the constituents of the *extracellular fluid*. Some of the specific mechanisms for regulation are as follows:

1. *Regulation of sodium ion concentration.* Sodium ion concentration is controlled mainly by controlling the amount of water in the body. When the sodium concentration becomes too high, this causes two effects that increase the body water: (a) *secretion of antidiuretic hormone* by the *posterior pituitary gland,* this hormone then acting on the kidneys to cause water retention, and (b) *stimulation of thirst,* which makes the person drink large amounts of excess water. The increased water dilutes the sodium in the extracellular fluid and returns its concentration back toward normal. However, this also increases the blood volume, which in turn increases the arterial pressure, causing the kidneys to excrete both the excess water and along with it much of the excess sodium.

2. *Regulation of potassium ion concentration.* When the potassium ion concentration becomes too great, two negative feedback control mechanisms function to return this concentration back to normal. The first is a *direct effect of the excess potassium on the epithelial cells of the renal tubules* to cause increased transport of potassium out of the peritubular capillaries and into the tubular lumen, from whence the potassium ions are lost into the urine. The second feedback mechanism is the following: The *high potassium ion concentration stimulates the adrenal cortex to secrete greatly increased quantities of aldosterone.* The aldosterone then also stimulates the tubular epithelial cells to transport potassium ions into the tubular lumen and thence into the urine. Thus, the extracellular fluid potassium ion concentration is normally controlled within very narrow limits.

3. *Regulation of acid-base balance.* The regulation of acid-base balance really means regulation of the *hydrogen ion concentration.* When the hydrogen ion concentration is higher than normal, the person is said

to have *acidosis*. When less than normal, he is said to have *alkalosis*. Three separate mechanisms function together to control the hydrogen ion concentration: (a) All of the body fluids contain *acid-base buffers*. These are chemical substances that will bind with large numbers of hydrogen ions when these are in excess in the fluids or release large numbers of hydrogen ions when their concentration is too low. The most important of these buffer systems is the *bicarbonate buffer*, consisting of *carbonic acid* and *bicarbonate ion*. (b) *When the hydrogen ion concentration rises too high this stimulates respiration*, and the increased breathing in turn eliminates large quantities of carbon dioxide from the blood. Since carbon dioxide reacts with water to form carbonic acid, loss of the carbon dioxide also decreases the carbonic acid and its hydrogen ions back toward normal. (c) If the first two mechanisms do not return the hydrogen ion concentration to normal, then the kidneys can still correct this by increasing or decreasing the rate of excretion of hydrogen ions into the urine. *All the renal tubules have the capability of secreting hydrogen ions,* and the rate of secretion increases in direct proportion to the hydrogen ion concentration in the blood, thus eliminating any excess hydrogen ions and returning their extracellular fluid concentration back to normal.

4. *Regulation of blood volume.* When the blood volume becomes too great, the arterial pressure begins to rise after a few hours. This in turn increases the rate of glomerular filtration and also the rate of volume loss in the urine, thus decreasing the extracellular fluid volume. Part of this loss of extracellular fluid represents loss of plasma volume, thus returning the blood volume also back toward normal.

The Urinary Transport System and Micturition. Urine formed by the nephrons is emptied through the *collecting ducts* into the *pelvis of the kidney*. From here, strong peristaltic contractions in the *ureter* move the urine into the *urinary bladder* even when the pressure in the bladder is very high.

Emptying of the bladder through the *urethra* is called *micturition*. When the bladder becomes overfilled, nerve stretch receptors in the wall of the bladder are excited, and sensory signals are transmitted into the caudal end of the spinal cord. From here, reflex signals return back to the bladder wall, causing it to contract and force the urine into the opening of the urethra. This is called the *micturition reflex*. If it occurs at an opportune time for the person to urinate, he voluntarily relaxes his *external urethral sphincter* and allows urination to take place.

Now that the function of the kidneys has been explained, it is possible to discuss the mechanisms by which most of the body fluid constituents are regulated. The kidneys play a special role in this regulation by controlling (1) ion concentrations of the extracellular fluid, (2) osmotic pressure of all the body fluids, (3) acidity of the fluid, and (4) volumes of both extracellular fluid and blood. In the regulation of acidity, the respiratory system also plays a major role. All of these interrelationships are discussed in this chapter.

REGULATION OF THE ION CONCENTRATIONS AND BODY FLUID OSMOLALITY

Regulation of Sodium Ion Concentration and Body Fluid Osmolality in the Extracellular Fluid

Over 90 percent of the positively charged ions (which are called "cations") in the extracellular fluid are sodium. Furthermore, the total concentration of the positively charged ions automatically controls the concentration of the negatively charged ions (the "anions") as well, because for each positive ion there must also be a negative ion. Therefore, for all practical purposes, when the sodium ion concentration is regulated, the concentration of more than 90 percent of all the ions in the body fluids is regulated at the same time. Also, it is these ions that cause almost all of the osmotic pressure of the fluids. Therefore, it follows that sodium ion concentration and osmolality of the fluids usually go hand in hand. That is, whenever the concentration of sodium ions increases, there is an almost exact corresponding increase in extracellular fluid osmolality. Conversely, whenever the sodium ion concentration becomes greatly decreased, there is an almost exact corresponding decrease in extracellular fluid osmolality.

Now, finally, to place the whole picture in perspective, it should be recalled from the discussion in Chapters 5 and 30 that a change in osmolality of the extracellular fluids causes a simultaneous and equal change in osmolality of the intracellular fluids.

For all of these reasons, one can see that the mechanisms that control the sodium ion concentration of the extracellular fluids also control the osmotic pressure of both the extracellular and the intracellular fluids.

Dilution of the Body Fluids by Water— the Mechanism for Controlling Sodium Ion Concentration and Body Fluid Osmolality. The *concentration* of sodium ions in the extracellular fluid as well as the fluid osmolality is controlled mainly by increasing or decreasing the quantity of water in the body fluids. That is, whenever the fluids become too concentrated, water is automatically accumulated, thus diluting the fluids and thereby decreasing both the sodium ion concentration and the osmolality. Conversely, when the concentration of the sodium becomes too little, the water accumulated in the body is decreased, and this automatically increases both the sodium ion concentration and the body fluid osmolality back toward normal.

There are two separate mechanisms for regulating the degree of water dilution of the body fluids. These are (1) the antidiuretic hormone mechanism for controlling the rate of excretion of water through the kidneys and (2) the thirst mechanism for controlling the rate of intake of water. These two mechanisms work hand in hand, and they are regulated by an integrated neurologic mechanism located in the hypothalamus of the brain. Let us first discuss the antidiuretic mechanism for controlling the rate of excretion of water by the kidneys.

CONTROL OF WATER EXCRETION BY THE KIDNEYS

Function of Antidiuretic Hormone to Control Water Reabsorption by the Distal Tubules and the Collecting Ducts. Before we can discuss the overall mechanism for controlling water excretion, it is first necessary to understand the function of antidiuretic hormone in controlling water reabsorption by the tubules. In the previous chapter it was pointed out that when ions and other solutes are absorbed by the tubular epithelium, osmotic forces are created that tend to cause water absorption along with the ions and other solutes. However, for water reabsorption to occur, it is also essential for the tubular membrane to be permeable to water. In the proximal tubule and in the early portions of the loop of Henle, the tubular membrane is very permeable to water. Therefore, in these areas of the tubular system, water is reabsorbed in direct proportion to the reabsorption of solutes from the tubules. However, the distal tubule and collecting duct are sometimes permeable to water and sometimes not; this depends upon the concentration of *antidiuretic hormone* in the body fluids.

Antidiuretic hormone, which is secreted by the posterior pituitary gland, controls the permeability of the distal tubules and collecting ducts to water. In the absence of antidiuretic hormone, these tubules are almost entirely impermeable to water. Therefore, water fails to be reabsorbed and instead passes on through these tubules and is expelled into the urine. In the presence of antidiuretic hormone, however, the distal tubules and collecting ducts become highly permeable to water so that water is then easily reabsorbed back into the blood from the tubules; consequently, the amount of water that passes into the urine becomes greatly diminished. The rate of loss of water from the body fluids is thus determined by the presence or absence of antidiuretic hormone.

Mechanism for Excreting a Dilute Urine. Now that we have explained the effect of antidiuretic hormone on the distal tubules and collecting ducts, it becomes very easy also to explain how the kidney can excrete a dilute urine and thereby remove excess water from the body. This is achieved in the following way: First, let us refer to Figure 32–1, which shows the glo-

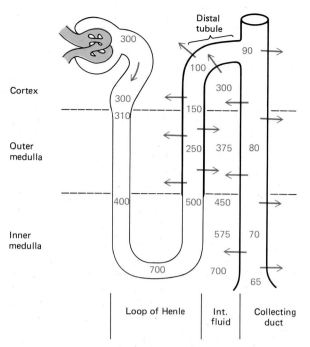

Cortex

Outer
medulla

Inner
medulla

Distal
tubule

Loop of Henle | Int. fluid | Collecting duct

FIGURE 32–1 The renal mechanism for forming a dilute urine. The darkened walls of the distal portions of the tubular system indicate that these portions of the tubules are relatively impermeable to the reabsorption of water in the absence of antidiuretic hormone. The solid arrows indicate active processes for absorption of most of the solutes besides water and the urinary waste products. (Numerical values are in milliosmoles per liter.)

merulus and the tubular system of the nephron. Note that the tubular fluid passes downward into a *descending limb* of the loop of Henle and then back upward through an *ascending limb*. This ascending limb of the loop of Henle is very impermeable to water. On the other hand, it has very powerful active transport mechanisms for absorbing ions from the tubular fluid, especially chloride and sodium ions, which are the most abundant ions. The greater proportion of the ions is thus absorbed back into the blood but the water remains in the tubular fluid. Thus, the fluid leaving the ascending limb of the loop of Henle and entering the distal tubule has already become diluted. This is illustrated by the numbers shown inside the tubules in Figure 32–1, beginning with an osmolality of 300 milliosmoles per liter (mOsm/L) at the origin of the tubular system but falling to 100 mOsm/L in the distal tubules. When the posterior pituitary gland is not secreting antidiuretic hormone, this dilute fluid, with its great excess of water, passes on through the distal tubules and collecting ducts directly into the urine. In fact, still more ions are absorbed in the distal tubule and collecting duct

so that the fluid becomes even more dilute, falling to 65 mOsm/L, before it issues forth as urine.

Thus, in the absence of antidiuretic hormone, the kidney excretes a very dilute urine, thereby removing large quantities of excess water from the body fluids.

Renal Mechanism for Excreting a Concentrated Urine—the "Countercurrent" Mechanism for Conserving Water in the Body. The kidney can also, at appropriate times, excrete a concentrated urine, thereby retaining water in the body while excreting large quantities of ions and other solutes into the urine. However, this mechanism is somewhat more difficult to understand than is the mechanism by which the kidney excretes a dilute urine. Let us study Figure 32–2 to help explain this mechanism.

Note in the figure that the upper part of the diagram represents the cortex of the kidney where the glomeruli, the proximal tubules, and the distal tubules are located. The lower part of the diagram represents the medulla into which many of the loops of Henle protrude and through which the collecting duct must also pass on its way to the pelvis of the kidney. The numbers on the diagram represent the total osmolar concentration of solutes at different points in the kidney in milliosmoles per liter. These concentrations are very high in the medulla, and they become progressively greater toward the lower medulla, rising to as high as 1200 mOsm/L, where the lower tip of the loop of Henle lies and where the collecting duct empties into the renal pelvis. The cause of this very high concentration of solutes in the medulla is twofold. First, large quantities of solutes are absorbed into the medulla from the ascending limbs of the loops of Henle and also from the collecting ducts. Second, a so-called "countercurrent" mechanism operating in the medulla makes it difficult for the blood flowing through the medulla to remove these solutes, thus causing them to accumulate in the medulla until their concentrations rise very high. The countercurrent mechanism may be explained as follows:

Note to the left in Figure 32–2 the loop called *vasa recta*. This represents capillary blood vessels that pass down from the cortex deep into the medulla and then return again to the cortex. When the blood first enters the vasa recta, it has a solute concentration of about 300 mOsm/L. However, because the vasa recta are very permeable, the osmolar concentration of the blood in the vasa recta quickly approaches that in the medulla, becoming higher and higher as the

| Vasa recta | Int. fluid | Loop of Henle | Int. fluid | Collecting duct |

FIGURE 32–2 The counter-current mechanism for concentrating the urine. (Numerical values are in milliosmoles per liter.)

blood passes farther down the vasa recta into the medulla and reaching a concentration of 1200 mOsm at the tip of the medulla. Then as the blood passes back up the ascending limb of the vasa recta, rapid diffusion of solutes out of and water into the vasa recta causes the concentration to fall once again nearly to 300 mOsm/L. Thus, the rapid diffusion of solutes and water back and forth between the medullary interstitium and the blood makes it difficult for the blood to carry many of the solutes away from the medulla. This is called the countercurrent mechanism for maintaining a high concentration of solutes in the medulla. It is represented by the arrows in the vasa recta.

Now let us understand the importance of these concentrated fluids in the medulla for excreting a concentrated urine.

Role of Antidiuretic Hormone in Excreting a Concentrated Urine. It was noted above that in the absence of antidiuretic hormone the kidney will excrete a very dilute urine. We shall now see that in the presence of large quantities of antidiuretic hormone the kidney will excrete a very concentrated urine. This results in the following way:

It will be recalled that antidiuretic hormone makes the collecting ducts very permeable to water. Therefore, in the presence of antidiuretic hormone, the high concentration of solutes in both the medullary interstitium and the vasa

recta will cause rapid osmosis of water from collecting ducts into the interstitium and the blood. Consequently, the solutes in the ducts become progressively more concentrated until they approach the concentration in the medullary interstitium and vasa recta. A concentrated urine is thus excreted, with loss of large quantities of solutes but very little loss of water. This obviously increases the water in the body fluids and decreases the solutes, thereby diluting the body fluids. For reasons discussed earlier, this decreases the sodium ion concentration at the same time.

The Osmo-Sodium Receptors of the Hypothalamus and the Antidiuretic Hormone Feedback Control System. Located in the anterior part of the hypothalamus, as illustrated in Figure 32–3, are the *supraoptic nuclei*, which contain nerve cells sensitive to the concentration especially of sodium ions in the extracellular fluid, but to a slight extent to that of other solutes as well. These cells are called *osmo-sodium receptors* because they transmit large numbers of impulses down the pituitary stalk to the posterior pituitary gland when the concentration of sodium or other osmotically active substances becomes too great. On reaching the posterior pituitary gland, the nerve impulses cause the release of antidiuretic hormone (ADH) into the circulating blood. The blood in turn carries the antidiuretic hormone to the kid-

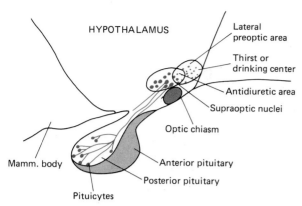

HYPOTHALAMUS
Lateral preoptic area
Thirst or drinking center
Antidiuretic area
Supraoptic nuclei
Optic chiasm
Mamm. body
Anterior pituitary
Posterior pituitary
Pituicytes

FIGURE 32–3 The supraoptico-pituitary antidiuretic system and its relationship to the thirst center in the hypothalamus.

neys. (The hormonal mechanism of the posterior pituitary gland will be presented in more depth in Chapter 45.)

Figure 32–4 illustrates the effect of antidiuretic hormone once it reaches the kidneys. In the presence of antidiuretic hormone, as already explained, the kidney excretes a very concentrated urine, thus conserving water in the body. In the absence of antidiuretic hormone, the kidney excretes a very dilute urine, thus losing large quantities of water from the body fluids.

The antidiuretic hormone system, therefore, represents a very important feedback control system for controlling both the sodium ion concentration and the osmotic concentration of the body fluids. When the sodium and osmotic concentrations become too great, increased secretion of antidiuretic hormone causes water retention in the body and thereby dilutes these concentrations. When the concentration of so-

dium or other osmotic elements becomes too low, decreased secretion of antidiuretic hormone causes water to be lost from the fluid, and their concentrations rise back toward normal level.

THE THIRST MECHANISM FOR CONTROL OF SODIUM AND OSMOTIC CONCENTRATIONS

The thirst mechanism controls the intake of water at the same time that the antidiuretic mechanism controls the output of water. Returning again to Figure 32–3, one can see located lateral to, slightly above, and forward of the antidiuretic area in the supraoptic nuclei a center called the *thirst or drinking center*. At the same time that the osmo-sodium receptors of the supraoptic nuclei are stimulated, so also are neurons stimulated in the thirst or drinking center. Therefore, when antidiuretic hormone causes the kidneys to conserve water in the body, the drinking center simultaneously causes the person to drink large amounts of water. For both of these reasons the quantity of body water increases.

Conversely, when the osmo-sodium receptors are not stimulated, so also are the neurons of the drinking center unstimulated, and the person has no desire to drink. Therefore, there is essentially no intake of water even though at the same time the kidneys excrete large amounts of water, causing rapid depletion of water in the body.

The thirst and drinking mechanisms thus operate synergistically with the kidney mechanism for control of body water, and thereby also for control of extracellular sodium ion and osmotic concentrations.

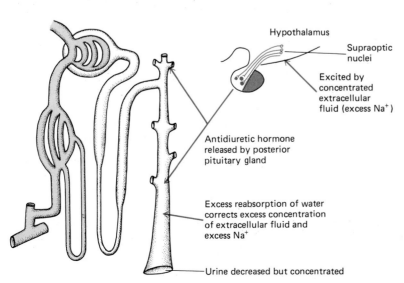

Hypothalamus
Supraoptic nuclei
Excited by concentrated extracellular fluid (excess Na⁺)
Antidiuretic hormone released by posterior pituitary gland
Excess reabsorption of water corrects excess concentration of extracellular fluid and excess Na⁺
Urine decreased but concentrated

FIGURE 32–4 Control of extracellular fluid osmolality and sodium ion concentration by the osmosodium receptor–antidiuretic hormone feedback control system.

Regulation of Potassium Ion Concentration

From earlier discussions in this text, it will be recalled that the function of both nerve and muscle is highly dependent upon a well-regulated potassium ion concentration in the extracellular fluid. For instance, when the potassium concentration becomes too high, the membrane potential of both nerve and muscle decreases, thereby diminishing the voltage of the action potential and reducing the effectiveness of nerve impulse transmission as well as the strength of muscle contraction. Indeed, a high potassium ion concentration can sometimes make the heart so weak that the person dies of heart failure.

Potassium ion concentration in the extracellular fluids is regulated in a manner similar to that of sodium ion concentration, except that the hormone aldosterone, rather than antidiuretic hormone, plays the major role.

Effect of Aldosterone on the Transport of Sodium and Potassium in the Distal Tubules and Collecting Ducts.
Aldosterone directly affects the epithelium of both the distal tubule and the collecting duct to promote, simultaneously, *active absorption of sodium* from the tubules and *active secretion of potassium* into the tubules. These effects result from the fact that aldosterone increases the tubular epithelial cell enzymes and other components of the carrier mechanisms for sodium absorption and potassium secretion.

Since aldosterone causes increased tubular absorption of sodium, one would think that this hormone would be especially important for controlling the sodium ion concentration in the extracellular fluid. However, this turns out not to be true, the reason being that the antidiuretic hormone–thirst mechanism is about 10 times as potent for control of sodium ion concentration as is the aldosterone mechanism and therefore normally overrides this mechanism. The quantity of aldosterone in the body thus can increase and decrease rather markedly with only small changes in the sodium ion concentration of the extracellular fluid.

On the other hand, the effect of aldosterone on potassium secretion does play an extremely important role in the control of potassium ion concentration.

The Aldosterone Feedback Mechanism for Control of Extracellular Fluid Potassium Ion Concentration.
Figure 32–5 illustrates the aldosterone mechanism for control of potassium ion concentration. An increase in potassium ion

FIGURE 32–5 Postulated mechanisms for control of potassium concentration in the extracellular fluid.

concentration has a very potent effect on the adrenal cortex to boost its rate of secretion of aldosterone. The aldosterone then circulates in the blood until it reaches the kidneys. Here the hormone directly affects the distal tubules and collecting ducts to increase potassium secretion. Consequently, large quantities of potassium are lost into the urine, and the potassium ion concentration in the extracellular fluid decreases back toward normal.

Conversely, when the potassium concentration becomes too low, aldosterone secretion by the adrenal cortex becomes greatly decreased. Now the rate of potassium secretion into the urine is diminished almost to zero. And since the daily food intake contains a large amount of potassium, potassium accumulates in the extracellular fluid, thereby returning the potassium ion concentration once again toward normal.

Thus, the aldosterone mechanism serves as a typical feedback mechanism for control of the potassium ion concentration.

In addition to this feedback control of potassium ion concentration, an increase in potassium ions in extracellular fluids also has a direct effect on the tubular epithelial cells, independent of the aldosterone effect, of causing increased potassium ion secretion. This, too, helps in the feedback control of potassium ion concentration in the extracellular fluid.

Regulation of Other Ions in the Extracellular Fluids

Regulation of Chloride and Bicarbonate Ions.
The regulation of chloride and bicarbonate ion concentrations is mainly secondary to that of sodium concentration. The reason for this is the following: When sodium is reabsorbed

from the kidney tubules, the positively charged sodium ions are transferred out of the tubular fluid into the interstitial fluids of the kidneys, creating a state of electronegativity in the tubules and electropositivity in the interstitial fluids. The negativity in the tubules repels negative ions from the tubules, whereas the positivity of the interstitial fluids attracts the negative ions. Therefore, when sodium ions are reabsorbed, chloride and bicarbonate ions are usually also reabsorbed. Sometimes more chloride than bicarbonate ions are reabsorbed; at other times more bicarbonate ions are reabsorbed. Which one it will be is determined by the acid-base balance of the extracellular fluid, which is discussed later in the chapter.

Regulation of Calcium, Magnesium, and Phosphate Concentrations. The details of the mechanisms by which calcium, magnesium, and phosphate concentrations are regulated by the kidneys are less clear than for sodium, potassium, chloride, and bicarbonate. Yet, it is known that too high a concentration of any one of these substances in the extracellular fluid causes the tubules to reject it and to pass it on into the urine. On the other hand, a low concentration causes the opposite effect, that is, rapid reabsorption of the substance until its concentration in the extracellular fluids returns to normal.

REGULATION OF ACID-BASE BALANCE

Regulation of acid-base balance actually means regulation of hydrogen ion (H^+) concentration in the body fluids. When the hydrogen ion concentration is great, the fluids are *acidic;* when hydrogen concentration is low, the fluids are *basic* (or *alkaline*). However, before we discuss acid-base balance, it is necessary to present some of the terminology that is often used.

First, an *acid* is a substance that has large numbers of free hydrogen ions (H^+) when dissolved in water. On the other hand, a *base* is a substance that has large numbers of hydroxyl ions (OH^-).

Second, a base is frequently called an *alkali.* Thus, when a person has excess base in his blood, he is said to have *alkalosis.* Conversely, when he has excess acid in his blood, he is said to have *acidosis.*

Third, acids and bases tend to neutralize each other. The reason for this is that the hydrogen ion combines with the hydroxyl ion to form water, as expressed by the following equation:

$$H^+ + OH^- \rightarrow H_2O$$

When there is a great excess of hydrogen ions in a solution, they wll neutralize essentially all of the hydroxyl ions so that almost none of these will be present. Conversely, when there is an excess of hydroxyl ions, they will neutralize essentially all of the hydrogen ions so that the hydrogen ion concentration becomes extremely low. To state this another way, hydrogen ions and hydroxyl ions are mainly mutually exclusive in the same fluid. When there are approximately equal amounts of acidic and basic substances in the body fluids, the hydrogen ions of the acids and the hydroxyl ions of the bases almost completely neutralize each other—that is, they combine with each other to form water—and the fluids are then said to be *neutral.* The normal body fluids are very nearly neutral, though slightly on the basic side.

Fourth, the normal concentration of hydrogen ions in the body fluids is only one part in approximately 9 billion, or to express this in chemical terms, it is 4×10^{-8} equivalents per liter. However, this is such a difficult way to express the hydrogen ion concentration that it is usually expressed in terms of pH, which is the logarithm of the reciprocal of the hydrogen ion concentration, according to the following formula:

$$pH = \log_{10}\left(\frac{1}{H^+ \text{ Conc.}}\right)$$

The normal pH of extracellular fluid and blood, therefore,

$$\left(\text{the logarithm of } \frac{1}{4 \times 10^{-8}}\right)$$

is 7.4. A very strong acid will have a pH of less than 1.0, and a very strong base will have a pH of about 14.0.

The acid-base regulatory systems of the body are all geared toward maintaining a normal pH of approximately 7.4 in the extracellular fluid. Even in disease conditions it only rarely becomes more acidic than 7.0 or more basic than 7.8.

Effects of Acidosis and Alkalosis on Bodily Functions

The hydrogen ion concentration, despite its very low level, is one of the most important controlling factors in most metabolic reactions of the cells. Therefore, an increase or a decrease in H^+ concentration above or below normal causes serious derangements in the overall function of the body.

Acidosis generally depresses mental activity, and in severe states can lead to coma and death. This occurs in many instances of acidosis resulting from severe diarrhea or from diabetes mellitus, as will be discussed in more detail later in the chapter. Usually the afflicted person will pass into coma when the pH of the extracellular fluid falls below approximately 6.9, which represents a hydrogen ion concentration almost four times the normal level.

On the other hand, alkalosis in which the hydrogen ion concentration has been decreased to less than one-half normal frequently leads to very severe overexcitability of the nervous system, often resulting in excessive initiation of signals in many areas of the brain and peripheral nerves. These signals can produce tetanic contraction of the muscles, or they can actually kill a person by causing epileptic convulsions or other derangements of nervous activity.

Regulation of Acid-Base Balance by Chemical Buffers

All of the body fluids contain *acid-base buffers*. These are chemicals that can combine readily with any acid or base in such a way that they keep the acid or base from changing the pH of the fluids greatly.

The Bicarbonate Buffer. One of the chemical systems that perform this function is the *bicarbonate buffer*, which is present in all body fluids. The bicarbonate buffer is a mixture of carbonic acid (H_2CO_3) and bicarbonate ion (HCO_3^-). When a strong acid is added to this mixture, its hydrogen ion (H^+) combines immediately with the bicarbonate ion to form carbonic acid. Carbonic acid is an extremely weak acid in which the hydrogen atom is tightly bound in the molecule so that relatively few hydrogen ions remain free in the solution. That is, "ionization" of carbonic acid as represented by the equation

$$H_2CO_3 \rightleftharpoons H^+ + HCO_3^-$$

is very slight; the reaction proceeds mainly in the left-hand direction. Therefore, the buffer system changes the strong acid (many hydrogen ions) into a weak one (few hydrogen ions) and keeps the fluids from becoming strongly acid. On the other hand, when a strong base—a substance that contains large numbers of hydroxyl ions (OH^-) but very few hydrogen ions—is added to this mixture, the OH^- ions immediately combine with the carbonic acid to form water and bicarbonate ions, according to the following equation:

$$OH^- + H_2CO_3 \longrightarrow H_2O + HCO_3^-$$

Loss of the weak acid (carbonic acid) hardly affects the hydrogen ion concentration in the body fluids. Thus, it can be seen that this mixture of carbonic acid and bicarbonate ion protects the body fluids from becoming either too acidic or too basic.

Other Buffers. Other important buffers are *phosphate* and *protein buffers*. These are especially important for maintaining normal hydrogen ion concentration in the intracellular fluids, because their concentrations inside the cells are several times as great as the concentration of the bicarbonate buffer.

In essence, the buffers of the body fluids are the first line of defense against changes in hydrogen ion concentration, for any acid or base added to the fluids immediately reacts with these buffers to prevent marked changes in the acid-base balance.

Respiratory Regulation of Hydrogen Ion Concentration

Carbon dioxide is continually formed by all cells of the body as one of the end-products of metabolism. This carbon dioxide combines with water to form carbonic acid in accord with the following reaction:

$$CO_2 + H_2O \rightleftharpoons H_2CO_3$$

In other words, even the normal metabolic processes are always pouring carbonic acid along with its hydrogen ions into the body fluids.

On the other hand, one of the functions of respiration is to expel carbon dioxide through the lungs into the atmosphere. Normally, respiration removes carbon dioxide at the same rate that it is being formed. If, however, pulmonary ventilation decreases below normal, carbon dioxide will not be expelled normally but instead will pile up in the body fluids, causing the concentration of carbonic acid to increase. As a result, the H^+ concentration rises. If the rate of pulmonary ventilation rises above normal, however, the opposite effect occurs; carbon dioxide is blown off through the lungs at a more rapid rate than it is being formed, thereby decreasing the carbon dioxide and carbonic acid concentrations. Complete lack of breathing for a minute will increase the H^+ about twofold; this corresponds to a decrease in the pH of the extracellular fluid from the normal level of 7.4 down to about 7.1. Conversely, very active overbreathing can decrease the H^+ to about one-half normal (an increase of the pH to 7.7) in about 1 minute.

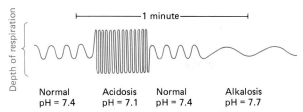

FIGURE 32-6 Effect of acidosis and alkalosis on respiration.

Thus the hydrogen ion concentration of the body can be changed greatly by over- or under-ventilation of the lungs.

Control of Respiration by the Hydrogen Ion Concentration of the Body Fluid. In the preceding paragraph the effect of respiration on the H^+ was discussed. In this paragraph we will discuss the effect of H^+ on respiration. A high concentration of H^+ stimulates the respiratory center in the medulla of the brain, greatly enhancing the rate of ventilation. Conversely, a low H^+ depresses the rate of ventilation. These effects afford an automatic mechanism for maintaining a fairly constant H^+ in the body fluids. That is, an increase in H^+ boosts the rate of ventilation. This in turn removes carbonic acid from the fluids, which decreases the H^+ back toward normal.

This respiratory mechanism for regulating hydrogen ion concentration reacts within a few seconds to a few minutes when the extracellular fluids become either too acidic or too basic. Figure 32-6 illustrates the effect on respiration of adding first a small amount of acid to the blood, and then later a small amount of alkali (base). The height of the curves represents the depth of respiration, and the rate of repetition of the curves represents the frequency of respiration. Note that acidosis greatly increases both the depth and rate of respiration, and alkalosis depresses respiratory function greatly so that the depth of respiration becomes very slight and the rate very slow. This respiratory mechanism is effective for regulating acid-base balance to the extent that it usually can return the H^+ of the body fluids about two thirds of the way to normal within a minute after an acid or alkali has been administered.

Renal Regulation of Acid-Base Balance

A number of other acids besides carbonic acid are also continually being formed by the metabolic processes of the cells, and these can be eliminated from the body only by the kidneys. They include phosphoric, sulfuric, uric, and keto acids, all of which, on entering the extracellular fluids, can cause acidosis. The kidneys normally rid the body of these excess acids as rapidly as they are formed, preventing an excessive buildup of hydrogen ions. These acids are generally called *metabolic acids*, in contradistinction to carbonic acid, which is called a *respiratory acid*.

In rare instances too many basic compounds enter the body fluids rather than too many acidic compounds. In general, this occurs when basic compounds are injected intravenously or when the person ingests a large quantity of alkaline foods or drugs.

Mechanisms by Which the Kidneys Regulate Acid-Base Balance. The kidneys regulate acid-base balance by (1) excreting hydrogen ions into the urine when the extracellular fluids become too acidic and (2) excreting basic substances, particularly sodium bicarbonate, into the urine when the extracellular fluids become too alkaline.

Secretion of Hydrogen Ions and Reabsorption of Sodium. The tubular epithelium throughout the renal tubular system continually secretes hydrogen ions into the tubular fluid. These ions react with sodium salts in the tubular fluid to form weak acids and to free the sodium ion that is bound in the salt. The sodium in turn is absorbed through the tubular wall back into the extracellular fluid. Thus, there is a net exchange of hydrogen for sodium ions, the hydrogen ions passing into the urine and the sodium ions being reabsorbed. Obviously, loss of the hydrogen ions from the extracellular fluid and their replacement by sodium ions decreases the hydrogen ion concentration. This tends to make the fluid more basic, thus helping to overcome the continual formation of acid by the metabolic processes of the body.

Control of Hydrogen Ion Secretion. The rate of hydrogen ion secretion into the tubules is approximately proportional to the degree of acidosis. Therefore, the more acidic the fluids, the greater is the rate of secretion of hydrogen ions and, therefore, the more rapidly the kidney eliminates the excess acid.

Ammonia Secretion and Its Combination with Hydrogen Ions in the Tubules. Sometimes so much metabolic acid is formed in the body and so many hydrogen ions are secreted into the tubules that the tubular fluid becomes very acidic (pH as low as 4.5) despite the presence of chemical buffers in this fluid. When this occurs, further secretion of hydrogen ions ceases. However, to allow continued secretion of the excess quantities of hydrogen ions, the tubular acidosis causes the tubular epithelial cells to begin secreting large quantities of ammonia. The ammonia in turn combines with the hydrogen ions to form ammonium ions, which prevents the tubular fluid from becoming too acidic because the hydrogen ions are removed from the fluid and the ammonium ions form neutral salts in their place.

Excretion of Large Amounts of Bicarbonate Ion in Alkalosis. In alkalosis the alkaline substances in the body fluids combine with carbonic acid to form bicarbonate salts, as discussed earlier. Therefore, the extracellular bicarbonate ion concentration becomes greatly increased. These extra bicarbonate ions pass into the tubules in large quantities as part of the glomerular filtrate, and a large share of them pass on into the urine in the form of sodium bicarbonate. Since sodium bicarbonate is the alkaline half of the bicarbonate buffer system, its excretion actually represents loss of alkaline substances from the body.

To recapitulate, when the body fluids become too alkaline, the alkaline substances react with carbonic acid to form mainly sodium bicarbonate. The sodium bicarbonate in turn is lost into the urine, which represents loss of alkaline substances from the body. Thus, this mechanism helps to correct alkalosis.

Summary of Renal Regulation of Acid-Base Balance. In summary, the renal mechanisms for regulating acid-base balance remove hydrogen ions from the extracellular fluids when the hydrogen ion concentration becomes too great, and they remove sodium and bicarbonate ions when the hydrogen ion concentration becomes too low. This principle is illustrated in Figure 32–7, which shows at point A a pH of about 7.5 in the extracellular fluid. Since this is on the alkaline side, the pH of the urine

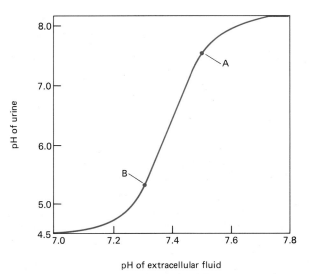

FIGURE 32–7   Effect of extracellular fluid pH on urine pH.

becomes alkaline (pH 7.7) because of excretion of alkaline substances from the body fluids. On the other hand, at point B the extracellular pH has fallen to 7.3, and the pH of the urine is very acidic (pH 5.6) because of excretion of large quantities of acidic substances from the body fluids. In both of these instances the excretion of alkaline or acidic substances returns the pH toward normal.

Abnormalities of Acid-Base Balance

Many disorders of the respiratory system, of the kidneys, or of the metabolic systems for forming acids and bases can cause serious derangement of the acid-base balance. Some of the effects of these on extracellular pH are shown in Figure 32–8. For instance, the normal pH of the blood is 7.4, but intense *overventilation* (overbreathing) can cause the pH to rise sometimes to as high as

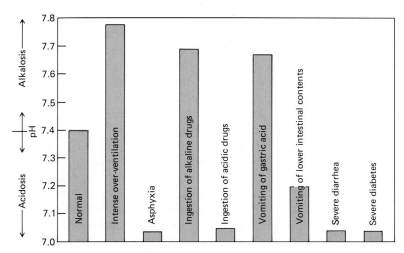

FIGURE 32–8   pH of the extracellular fluid in various acid-base disorders.

7.8 because of excess expiration of carbon dioxide through the lungs and therefore loss of carbonic acid from the blood. On the other hand, *asphyxia*, which means extreme decrease in ventilation, causes a buildup of carbon dioxide and carbonic acid in the body fluids, thereby promoting acidosis in which the pH of the blood sometimes falls to as low as 7.0.

A common cause of alkalosis is *ingestion of alkaline drugs* used for treatment of gastritis or stomach ulcers. The drugs sometimes are absorbed into the body fluids in quantities greater than the kidneys can remove, resulting in alkalosis.

Loss of large quantities of fluids from the intestinal tract at times causes acidosis but at other times alkalosis. For instance, vomiting large quantities of hydrochloric acid from the stomach gradually depletes the acid reserves of the extracellular fluid and causes alkalosis. On the other hand, vomiting of fluids from the lower intestine or loss of fluids as a result of diarrhea usually causes acidosis. This is because secretions from the lower intestine contain large quantities of sodium bicarbonate. When this basic salt is lost from the body it is mainly replaced by carbonic acid formed by the reaction of carbon dioxide with water. Therefore, the net effect is loss of sodium ions and gain of hydrogen ions, causing acidosis.

Finally, Figure 32–8 shows extreme acidosis that sometimes results from *severe diabetes mellitus*. In diabetes many of the fats normally used by the body for energy are not completely metabolized but instead are broken into substances called *keto acids* that build up in the body fluids, producing extreme acidosis.

REGULATION OF BLOOD VOLUME

In earlier chapters in which the circulatory system was discussed, the importance of blood volume as one of the major determinants of both cardiac output and arterial pressure regulation was emphasized. Indeed, only a small long-term increase in blood volume can lead to a very large increase in arterial blood pressure.

The normal blood volume of the adult person is almost exactly 5000 ml, and it rarely rises or falls more than a few hundred milliliters from this value. Two principal mechanisms for maintaining this constancy are (1) the capillary fluid shift mechanism and (2) the kidney mechanism.

The Capillary Fluid Shift Mechanism. When the blood volume becomes too great, the

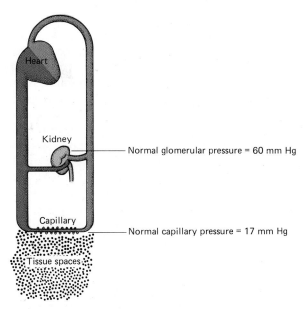

FIGURE 32–9 The capillary and kidney systems for regulating blood volume.

pressures in all the vessels throughout the body, including that in the capillaries, increase. The normal pressure in the capillaries is about 17 mm Hg, as illustrated in Figure 32–9. When the capillary pressure rises above this value, fluid automatically leaks into the tissue spaces, as we explained in Chapter 30. Therefore, the blood volume decreases back toward normal. When the capillary pressure has returned to normal, further loss of fluid into the tissue spaces ceases. Conversely, when the blood volume falls too low, the capillary pressure falls, and fluid is then absorbed from the interstitial spaces, increasing the blood volume again back toward normal. However, this mechanism for blood volume control has its limitations because the tissue spaces cannot expand indefinitely, nor can they supply an inexhaustible amount of fluid when the blood requires it.

The Kidney Mechanism. As was explained in the previous chapter, when the glomerular pressure in the kidneys rises, the amount of glomerular filtrate and the amount of urine formed by the kidneys increase greatly. Also, increased pressure in the peritubular capillaries surrounding the tubules *decreases* fluid reabsorption from the tubules, which further *increases* urine flow. Figure 32–9 shows that the normal glomerular pressure is about 60 mm Hg. An increased blood volume raises this to a higher level by two different mechanisms: First, the increased volume elevates the arterial pressure, which causes greater flow of blood through the afferent arterioles into the kidney, thus raising all intrarenal pressures. Second, the increased vascular volume stretches the atria of the heart, and the walls of these are supplied with stretch receptors called *volume receptors*. Stretching these initiates a nervous reflex that causes the renal afferent arterioles to dilate, thus increasing the blood flow into the kidney and correspondingly increasing the amount of urine formed. The volume receptors also inhibit antidiuretic hormone secretion by the posterior pituitary gland, and this too increases the urine output.

One will immediately recognize all of these effects to be feedback mechanisms by which the blood volume can be regulated; that is, an increase in blood volume initiates an increase in urinary output, which automatically decreases the blood volume back to normal. Conversely, a decrease in blood volume reduces urinary output, and the ingested fluids and salts accumulate until the blood volume rises back to normal.

THE URINARY TRANSPORT SYSTEM, AND MICTURITION

The term *micturition* means emptying the urinary bladder of urine; the control of this process is much more complex than one might expect. Before attempting to explain this, let us first discuss the anatomy of the urinary transport system.

The Urinary Transport System— The Ureter, Urinary Bladder, and Urethra

Figure 31–1 in the previous chapter illustrated the locations of the kidneys, the ureters, and the urinary bladder in the abdomino-pelvic cavity, and Figure 32–10 illustrates in more detail the emptying of the ureters into the urinary bladder, the structure of the bladder itself, and the urethra through which the urine empties to the exterior.

The **ureter** is the tube through which urine flows from each kidney to the bladder. It arises from the pelvis of the kidney behind the renal artery and vein, then it passes downward along the posterior abdominal wall anterior to the lateral border of the psoas muscle. As it enters the pelvis, it crosses anterior to the common iliac artery and vein and winds around the posterolateral wall of the pelvis to enter the bladder from its posterolateral border.

The **urinary bladder** lies posterior to the symphysis pubis in the anterior part of the pelvis. It, like the other urinary organs, is located

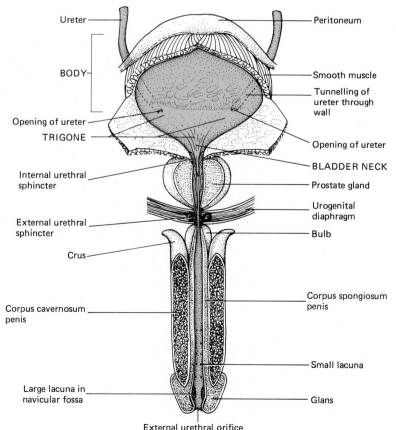

FIGURE 32–10 The urinary bladder and its relationship to the ureters and urethra.

extraperitoneally. The rectum passes posterior to the bladder and between the two ureters.

The urethra leaves the bladder from its lowermost point and passes to the exterior posterior to the symphysis pubis. In the male, it passes through the prostate gland and penis. In the female, it runs a very short course and opens between the labia minora about 1 cm anterior to the vaginal opening.

Anatomy and Function of the Ureter. The ureter is about 5 mm in diameter, and its wall is composed mainly of *smooth muscle.* When urine flows into it from the pelvis of the kidney, it becomes stretched, and a local stretch reaction occurs, causing the muscle to constrict like a ring around the lumen of the ureter. Then, during the next few seconds this ring of constriction moves downward along the ureter toward the bladder. This process is called *ureteral peristalsis*, and it forces the urine into the bladder. Ordinarily, the urine is transported the entire distance from the pelvis to the bladder in a few seconds.

Occasionally, severe infection or congenital abnormalities destroy the ability of the ureteral wall to contract. As a result, urine begins to collect in the kidney pelvis, causing it to swell, which in turn promotes infection that may extend into the kidney. Also, the stagnation of urine may lead to precipitation of various calcium compounds—calcium phosphate, calcium carbonate, calcium oxalate, and so forth—the crystalline precipitates of which can grow eventually into large *calculi*, or *renal stones*, that partially or totally fill the pelvis. These stones often cause extreme pain and further obstruction to urine flow, particularly when one of them passes into the ureter.

The Urinary Bladder. The urinary bladder is a globular-shaped chamber, the wall of which is composed largely of *smooth muscle*, that can stretch from a few milliliters volume up to 400 to 800 ml. When the bladder is almost empty, the smooth muscle layer of the bladder wall is very thick because of the contracted state of the bladder. But as the bladder fills with urine, stretching of the bladder wall thins it to only a few millimeters.

The bladder is divided into three major parts:

1. The **body of the urinary bladder.** This represents most of the bladder; it is this part that expands as the bladder fills. The smooth muscle of the body of the bladder, called the **detrusor muscle,** contracts to expel the urine when one urinates.

2. The **bladder neck.** This is the funnel-shaped lower end of the bladder where it opens into the urethra.

3. The **trigone.** This is the triangular area on the posteroinferior wall of the bladder that brackets the two openings of the ureter as well as the neck of the bladder. This portion of the bladder contracts much less than the body of the bladder, and it serves to prevent excessive stretch on the ureters or urethra.

At the points where the ureters enter the bladder, they tunnel diagonally through the muscle in the bladder wall for a centimeter or so before finally reaching the interior of the bladder at the upper poles of the trigone. This tunneling has a functional significance because excessive pressure inside the bladder compresses the walls of the distal ends of the ureters, which prevents reflux of urine backward up the ureters toward the kidneys despite the bladder pressure. This is especially important in minimizing backward spread of infection from the bladder to the kidneys.

The Urethral Sphincters. It is important to control emptying of the bladder so that urination will occur at a time that is convenient. For this purpose, the bladder is equipped with two separate sphincters that prevent outflow of urine except upon command.

1. The **internal urethral sphincter.** This is located at the point where the bladder neck fuses with the upper end of the urethra. It is actually a thickened portion of the *smooth muscle* wall of the bladder neck itself. This internal sphincter remains contracted all of the time except when commanded by appropriate nerve signals to relax, as we shall discuss later.

2. The **external urethral sphincter.** This is a *skeletal muscle sphincter*. It is the part of the urogenital diaphragm that bridges the pubic arch just inferior to the muscle floor of the pelvis, which was described in Chapter 8. The external urethral sphincter is also called the *voluntary sphincter* because it can be controlled by the will of the person; that is, it can be contracted when it is not convenient to urinate or relaxed when urination is desired, as we shall also discuss later.

In the male, the prostate gland lies between the neck of the bladder and the urogenital diaphragm, and the urethra passes through the center of this gland. This portion of the urethra is called the prostatic urethra. It plays a special role in the male functions of emission and ejaculation.

The Membranous Linings of the Urinary System and Lower Urinary Tract Infection.
From the calyces of the kidneys to the tip of the urethra, the urinary system is lined with an epithelium that varies from a few cell layers thick to many layers. This lining prevents the leakage of urine into the tissues, and it also resists passage of bacteria from the urine into the walls of the urinary tract. But, once bacteria do succeed in infecting the epithelium, the resulting inflammation in the urethra usually causes severe burning upon urination, and infection in the bladder can cause pain and especially an unending and uncontrollable desire to urinate.

Transport of Urine to the Bladder.
Urine, formed by each kidney, first passes into the *pelvis* of the kidney, shown in Figure 31–1, and then through the *ureter* to the *urinary bladder*, shown in Figure 32–10. Urine is forced along the ureter by *peristalsis*, an intermittent wavelike constriction beginning at the kidney pelvis and spreading downward along the ureter toward the bladder. The constriction squeezes the urine ahead of it. Ordinarily, the urine is transported the entire distance from the pelvis to the bladder in less than 30 seconds.

Storage of Urine in the Bladder.
The urinary bladder is a storage reservoir designed to prevent constant dribbling of urine. Until the bladder has filled to a volume of 200 to 400 ml, the intrabladder pressure does not increase greatly. This results from the ability of the smooth-muscle bladder wall to stretch tremendously without building up any significant tension in the muscle. However, once the bladder fills beyond 200 to 400 ml, the pressure does begin to rise, and it sometimes reaches as much as 40 mm Hg when the bladder fills to 600 to 700 ml.

Emptying of the Bladder—the Micturition Reflex.
In the human being micturition is caused by a combination of involuntary and voluntary nervous activity which may be explained as follows: When the volume in the bladder is greater than 200 to 400 ml, special nerve endings in the bladder wall called "stretch receptors" become excited. These transmit nerve impulses through the *visceral afferent nerve pathway* into the spinal cord (see Figure 32–11), initiating a subconscious reflex called the *micturition reflex*. The nervous centers for the subconscious reflex are in the lower tip of the cord. The reflex signal is then transmitted from here by the parasympathetic nerves (the *pelvic nerves*) to both the bladder wall and the internal urethral sphincter. The bladder wall contracts to build up pressure in the bladder, and this in turn creates a conscious

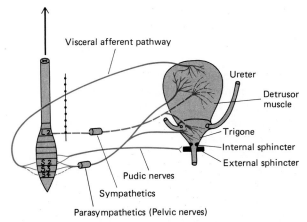

FIGURE 32–11 The nervous pathways for control of micturition.

desire to urinate. At the same time, the micturition reflex also relaxes the internal sphincter. Then, the only impediment to urination is the still contracted external urethral sphincter. If the time and place are propitious for urination, the conscious portion of the brain will relax this external sphincter by inhibiting the normal impulses to the sphincter through the *pudendal nerve*, and urination will take place.

If a person wishes to urinate before a micturition reflex has occurred, he can usually initiate the reflex by contracting the abdominal wall, which pushes the abdominal contents down against the bladder and momentarily excites some of the stretch receptors in the bladder wall, thus initiating the reflex.

Many times it is not convenient to urinate when a micturition reflex takes place. In this case, the reflex usually subsides within a minute or so, and the person loses his desire to urinate. The reflex then remains inhibited for another few minutes to as much as an hour before it returns again. If the reflex is again prevented from causing urination, it will once again become dormant for another period of time. However, as the bladder becomes more and more filled, the micturition reflex finally becomes so powerful that it is essential to urinate.

Babies, who have not developed voluntary control over the external urethral sphincter, automatically urinate every time the bladder fills. Also, adults whose spinal cords have been severed from the brain cannot keep the external sphincter contracted, so that they, too, urinate automatically when the bladder fills. However, these persons can often initiate this reflex by scratching the genital region, in this way controlling the time that the bladder will empty rather than waiting for the automatic and unannounced emptying.

QUESTIONS

1. How are the sodium ion concentration and osmolality of the extracellular fluid both regulated at the same time?
2. Explain the role of antidiuretic hormone in the control of water excretion by the kidneys.
3. Explain the countercurrent mechanism for excreting a concentrated urine.
4. What is the role of the hypothalamus in the control of urinary output? In the control of fluid intake?
5. Why is aldosterone especially important for control of potassium ion concentration but less important for control of sodium ion concentration?
6. The concentration of hydrogen ions in a solution is 8×10^{-8} equivalents per liter. Calculate the pH of the solution.
7. Explain the relative roles of the body fluid buffers, the respiratory system, and the kidneys in regulating acid-base balance.
8. How does the carbon dioxide–bicarbonate buffer system function in respiratory control of acid-base balance and in renal control of acid-base balance?
9. What is the role of ammonia in renal control of acid-base balance?
10. Explain how the capillary fluid shift mechanism and the kidney mechanism operate together to regulate blood volume.
11. Give the anatomy of the urinary transport system.
12. Explain the nervous control of micturition.

REFERENCES

Brenner, B.M., and Stein, J.H. (eds.): Acid-Base and Potassium Homeostasis. New York, Churchill Livingstone, 1978.

Brenner, B.M., and Stein, J.H. (eds.): Hormonal Function and the Kidney. New York, Churchill Livingstone, 1979.

Earley, L.E., and Gottschalk, C.W. (eds.): Strauss and Welt's Diseases of the Kidney, 3rd Ed. Boston, Little, Brown, 1979.

Ehrlich, E.N.: Adrenocortical regulation of salt and water metabolism: Physiology, pathophysiology, and clinical syndromes. *In* DeGroot, L.J., *et al.* (eds.): Endocrinology, Vol. 3. New York, Grune & Stratton, 1979, p. 1883.

Fitzsimons, J.T.: Thirst. *Physiol. Rev.*, 52:468, 1972.

Goldberg, M.: The renal physiology of diuretics. *In* Orloff, F., and Berliner, R.W. (eds.): Handbook of Physiology. Sec. 8. Baltimore, Williams & Wilkins, 1973, p. 1003.

Goldberger, E.: A Primer of Water, Electrolyte, and Acid-Base Syndromes. Philadelphia, Lea & Febiger, 1980.

Guyton, A.C., *et al.*: Theory for renal autoregulation by feedback at the juxtaglomerular apparatus. *Circ. Res.*, 14:187, 1964.

Jones, N.L.: Blood Gases and Acid-Base Physiology. New York, B.C. Decker, 1980.

Kim, Y., and Michael, A.F.: Idiopathic membranoproliferative glomerulonephritis. *Annu. Rev. Med.*, 31:273, 1980.

Kincaid-Smith, P., *et al.* (eds.): Progress in Glomerulonephritis. New York, John Wiley & Sons, 1979.

Kirschenbaum, M.A.: Renal Disease. Boston, Houghton Mifflin, 1978.

Knox, F.G. (ed.): Textbook of Renal Pathophysiology. Hagerstown, Md., Harper & Row, 1978.

Smith, K.: Fluids and Electrolytes: A Conceptual Approach. New York, Churchill Livingstone, 1980.

Tannen, R.L.: Control of acid excretion by the kidney. *Annu. Rev. Med.*, 31:35, 1980.

Valtin, H.: Renal Dysfunction: Mechanisms Involved in Fluid and Solute Imbalance. Boston, Little, Brown, 1979.

Weitzman, R., and Kleeman, C.R.: Water metabolism and the neurohypophysial hormones. *In* Bondy, P.K., and Rosenberg, L.E. (eds.): Metabolic Control of Disease, 8th Ed. Philadelphia, W.B. Saunders, 1980, p. 1241.

Zerbe, R., *et al.*: Vasopressin function in the syndrome of inappropriate antidiuresis. *Annu. Rev. Med.*, 31:315, 1980.

X

BLOOD CELLS, IMMUNITY, AND BLOOD COAGULATION

33

The Blood Cells, Hemoglobin, and Resistance to Infection

Overview

The function of the *red blood cells* is mainly to transport oxygen from the lungs to the tissues, and the function of the *white blood cells* is to destroy invading organisms and other agents that are damaging to the body. Though the bone marrow forms large numbers of both red blood cells and white blood cells, most of the white blood cells leave the blood stream and enter the tissues. Therefore, only about one out of every 500 of the circulating blood cells is a white blood cell.

The structure of the red blood cell is that of a very loose cell membrane bag only partly filled with cytoplasm but containing a very high concentration, about 34 percent, of *hemoglobin. Oxygen combines very loosely with the hemoglobin* as the blood passes through the lungs, and it is released from the hemoglobin to the tissues as the blood then passes through the peripheral capillaries. One of the main constituents of hemoglobin is *iron*; in fact, about two thirds of all the iron in the body is in the hemoglobin of the red cells. Because iron is present in the food in only very small amounts, the body has developed a special system for handling iron. Iron is transported in the blood in combination with a protein called *transferrin*. When more iron is absorbed from the intestines than can be used immediately for formation of hemoglobin, the excess is stored in the liver in the form of *ferritin*. When there is too much iron, both in the blood and in the storage areas of the liver, the transferrin will no longer accept iron from the intestinal mucosa, so that this becomes an automatic mechanism to prevent further absorption of iron. Other substances besides iron that are important for the formation of red blood cells include *vitamin B₁₂, folic acid,* and *other vitamins.*

Many people develop *anemia*, which means lack of sufficient numbers of red blood cells. The different causes are (1) *deficiency of iron*; (2) *deficiency of vitamin B₁₂* or other vitamins; (3) *loss of blood*; (4) *destruction of the bone marrow*, thus preventing the formation of red blood cells; and (5) *formation of abnormal cells* that cannot stand the trauma of passing through small capillaries so that the cells rupture easily and disappear from the blood.

The different types of white blood cells include three cells called *granulocytes*: (1) **neutrophils**, (2) **eosinophils**, and (3) **basophils**; and two *nongranulated cells*: the (4) **monocytes** and (5) **lymphocytes**. The lymphocytes are part of the immunity system that is discussed in Chapter 34. The other white blood cells are all *phagocytic cells*, which play an essential role in protecting the body against invasion by bacteria and other foreign agents that can damage the body.

When an infection occurs anywhere in the body, the granulocytes and monocytes migrate by ameboid motion out of the blood capillaries into the infected tissue area. This is caused by *chemotactic substances* released by the damaged tissues that attract the white blood cells. At first, the *neutrophils* are especially abundant because they can migrate into the infected tissue much more rapidly than the other cells. Therefore, they provide the first line of defense against the infectious agent. However, after 12 hours or more, large numbers of *monocytes* also appear on the scene. On entering the tissues, the monocytes swell to extremely large size and are then called **macrophages**. These can phagocytize several times as many bacteria as can the neutrophils and therefore provide an especially potent second line of defense against the infection.

Macrophages are present in many tissues of the body even normally, especially where foreign agents tend to invade the body, such as (1) in the walls of the *alveoli* of the lungs, (2) in the *lymph nodes*, to remove foreign agents from the lymph before the lymph flows into the blood, (3) in the *liver sinusoids*, to remove bacteria and other foreign substances from the portal blood arriving from the intestines, (4) in the *spleen* and *bone marrow*, to remove foreign agents that have succeeded in entering the general blood circulation, and (5) beneath the skin, to remove agents that invade through the skin. This system of macrophages is frequently called the **reticuloendothelial system.**

Figure 33–1 illustrates the different types of cells in the blood, showing that most of them are **red blood cells.** The red appearance of these cells is caused by the presence of large quantities of hemoglobin, which has a bright red color. The other cells in the figure are called **white blood cells,** or *leukocytes,* because when they are not stained with special dyes they have a whitish or clear appearance. The ratio of white blood cells to red blood cells is highly overrepresented in Figure 33–1 simply for the purpose of being able to illustrate all the different types of these cells. Actually, only about 1 out of every 500 cells is a white blood cell. White blood cells have several different functions, but the most important of these is to protect the body against invasion by disease organisms, as we shall discuss later in this chapter.

Note also in the figure the many small darkly staining structures called **platelets** floating among the larger cells. These are sometimes classified as white blood cells, but they are not really cells at all. Instead they are very small fragments of a special type of bone marrow cell called the **megakaryocyte.** The platelets are essential for the clotting of blood, which will be explained in Chapter 35.

THE RED BLOOD CELLS

The major function of red blood cells is to transport hemoglobin, which in turn carries oxygen from the lungs to the tissues.

In Figure 33–1 the red blood cells are shown to be round in shape. However, they are actually small disks that are all lying flat in the figure. These disks are depressed in the middle of each surface so that they are normally described as *biconcave disks*, having a mean diameter of approximately 8 μm (micrometers) and a thickness at the thickest point of 2 μm and in the center of 1 μm or less. If the student will study the red blood cells in Figure 33–1 carefully, he will see that their centers are much lighter in color than their peripheries because of their thinness in the center, thus giving the appearance of small doughnuts.

The shapes of red blood cells can change remarkably as the cells pass through capillaries. Actually, the red blood cell is a "bag" that can be deformed into almost any shape. Furthermore, because the normal cell has a great excess of cell membrane for the quantity of material inside, deformation does not stretch the membrane,

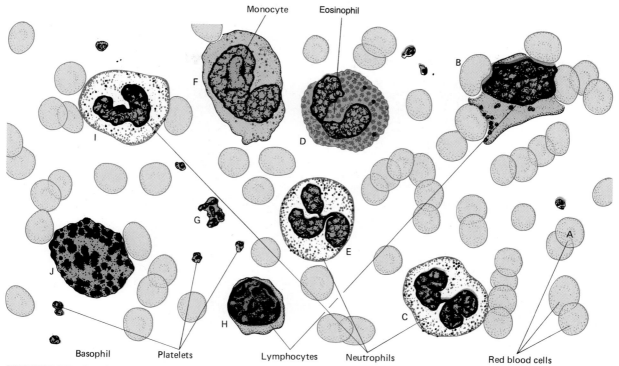

FIGURE 33-1 The blood cells.

and consequently does not rupture the cell as would be the case with many other cells.

The average number of red blood cells per cubic millimeter of blood is about 5 million; women have a few percent below this value, and men a few percent above it. The altitude at which the person lives and to a slight extent the degree of exercise the person pursues affect the number of red blood cells; these effects are discussed later.

Quantity of Hemoglobin in the Cells and Transport of Oxygen. Normal red blood cells contain approximately 34 percent hemoglobin. However, when hemoglobin formation is deficient in the bone marrow, the percentage of hemoglobin in the cells may fall to as low as 20 percent or less.

In normal blood, approximately 40 percent of the blood volume is cells and 60 percent is plasma. The percentage that is cells is called the *hematocrit*, so that the normal hematocrit is 40 percent, though this can fall to as low as 10 percent or rise to as high as 80 percent in disease conditions.

When the hematocrit is normal and the quantity of hemoglobin in each respective cell is also normal, whole blood contains an average of 15 g of hemoglobin per 100 ml. As will be discussed in more detail in connection with respiration and the transport of oxygen in Chapter 37, each gram of hemoglobin is capable of combining with approximately 1.33 ml of oxygen. There-

fore, 20 ml of oxygen can normally be carried in combination with hemoglobin in each 100 ml of blood.

GENESIS OF THE RED BLOOD CELL

Except for the lymphocytes, which are formed mainly in the lymph nodes, as will be discussed in the following chapters, all the other blood cells, both red and white, are produced in the bone marrow, which is illustrated in Figure 33-2. All the cells derive from a single cell type known as the hemocytoblast. To the left in Figure 33-3 is shown the genesis of the red blood cells, beginning with the hemocytoblast and eventual formation of **erythrocytes** (the red blood cells). The hemocytoblast passes through several stages of development, becoming first a *basophil erythroblast*, then a *polychromatophil erythroblast*, a *normoblast*, a *reticulocyte*, and finally an *erythrocyte*. During the earlier stages the cells divide many times, and they change color, as illustrated in the figure, because of the progressive formation of more and more hemoglobin. In the normoblastic stage, the nucleus degenerates, becomes extremely small, and is extruded, thus forming the reticulocyte. Then the cell usually leaves the bone marrow in the reticulocyte stage. The reticulocyte still contains small strands of basophilic reticulum mixed in with the hemoglobin in the cytoplasm. This reticulum is chiefly

FIGURE 33—2   Histological section of a typical area in the red bone marrow.

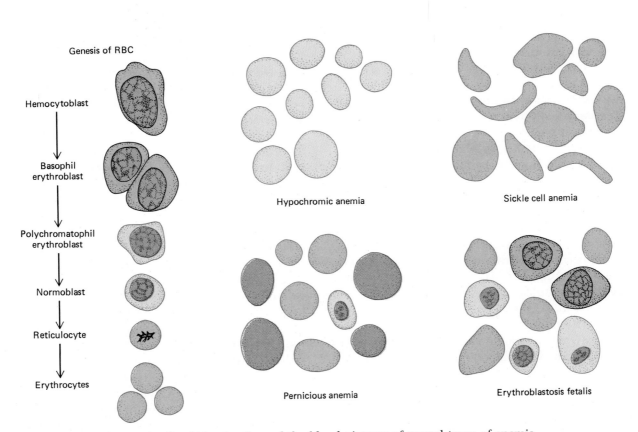

FIGURE 33—3   Genesis of red blood cells, and the blood pictures of several types of anemia.

the remains of the endoplasmic reticulum, and it continues to produce small amounts of hemoglobin in the early red blood cell even after it begins to circulate in the blood. However, this reticulum is usually dissolved within 2 days after release of the reticulocyte from the bone marrow, and the cell becomes the mature red blood cell, the erythrocyte. The erythrocyte then circulates in the blood for a period of approximately 120 days before it is destroyed. The proportion of the early red blood cells, the reticulocytes, in the circulating blood is usually slightly less than 1 percent.

The bone marrow in all bones of the body participates in the genesis of both red and white blood cells during fetal life when it is necessary to form large numbers of new blood cells rapidly. However, as a person ages through childhood, blood cell production in the shafts of the long bones ceases, and, in the adult, only the bone marrows in the flat and globular bones, such as the ilia, the vertebrae, the skull, and the sternum, produces blood cells. Partially because of this progressive decrease in bone marrow areas in which blood cells are formed, the numbers of cells in the circulating blood also decrease in old age.

Regulation of Red Blood Cell Production

The total number of red blood cells in the circulatory system is regulated within very narrow limits, so that an adequate number of cells is always available to provide sufficient tissue oxygenation and, yet, so that the cells are not overly concentrated to the extent that they impede blood flow. Thus, when a person becomes extremely *anemic* as a result of hemorrhage or any other condition, the bone marrow immediately begins to produce large quantities of red blood cells. Also, at very *high altitudes* ,where the quantity of oxygen in the air is greatly decreased, insufficient oxygen is transported to the tissues, and red cells are then also produced so rapidly that their number in the blood increases considerably. Finally, the degree of physical activity of a person determines to a slight extent the rate at which red blood cells will be produced. The athlete will often have a red blood cell count as high as 5.5 million per cubic millimeter (mm^3) whereas the asthenic person may have a count as low as 4.5 million per mm^3.

Erythropoietin, Its Response to Hypoxia, and Its Function in Regulating Red Blood Cell Production. Despite the very marked effect of hypoxia on red blood cell production, hypoxia does not have a direct effect on the bone marrow. Instead, the hypoxia causes the kidneys to secrete a hormone called *erythropoietin*, probably formed by either the juxtaglomerular apparatus or the glomerular membrane. This hormone is a glycoprotein having a molecular weight of about 34,000. Once released into the blood it circulates for 1 to 2 days before being destroyed, but during this time it acts on the bone marrow to stimulate red blood cell formation.

Only minute quantities of erythropoietin can be formed in the whole body in the complete absence of the kidneys. Therefore, persons whose kidneys have been destroyed and who are being kept alive by use of the artificial kidney usually have very severe anemia, usually with red blood cell counts of less than one-half normal.

Though erythropoietin begins to be formed almost immediately upon placing an animal or person in an atmosphere of low oxygen, very few new red blood cells appear in the circulation for the first 2 to 4 days; and it is only after 5 or more days that the maximum rate of new red cell production is reached. Thereafter, cells continue to be produced as long as the person remains in the low oxygen state or until he has produced enough red blood cells to carry adequate amounts of oxygen to his tissues despite the low oxygen.

The action of the erythropoietin is mainly to cause the formation of large numbers of hemocytoblasts in the bone marrow and to cause proliferation of the hemocytoblasts themselves to form erythroblasts. The other stages of erythrogenesis are also accelerated, but this probably results mainly from the initial stimulatory effect of the erythropoietin at the early levels. In the complete absence of erythropoietin, very few red blood cells are formed by the bone marrow. At the other extreme, when extreme quantities of erythropoietin are formed, the rate of red blood cell production can rise to as high as 8 to 10 times normal.

Vitamins Needed for Formation of Red Blood Cells

The Maturation Factor—Vitamin B$_{12}$ (Cyanocobalamin). Vitamin B$_{12}$ is an essential nutrient for all cells of the body, and growth of tissues in general is greatly depressed when this vitamin is lacking. This results from the fact that vitamin B$_{12}$ is required for conversion of ribose nucleotides into deoxyribose nucleotides, one of the essential steps in DNA formation. Therefore, lack of this vitamin causes failure of nuclear maturation and division and greatly inhibits the rate of red blood cell production.

Maturation Failure Caused by Poor Absorption of Vitamin B₁₂—Pernicious Anemia. The most common cause of maturation failure is not a lack of vitamin B_{12} in the diet but instead failure to absorb vitamin B_{12} from the gastrointestinal tract. This often occurs in the disease called *pernicious anemia*, in which the basic abnormality is an *atrophic gastric mucosa* that fails to secrete normal gastric secretions. These secretions contain a substance called *intrinsic factor*, which combines with vitamin B_{12} of the food and makes the B_{12} available for absorption by the gut. It does this by protecting the B_{12} from digestion by the gastrointestinal enzymes until it can be absorbed in the lower small intestine.

Once vitamin B_{12} has been absorbed from the gastrointestinal tract, it is stored in large quantities in the liver and then released slowly as needed to the bone marrow and other tissues of the body. The total amount of vitamin B_{12} required each day to maintain normal red cell maturation is less than 1 microgram (μg), and the store in the normal liver is about 1000 times this amount.

Relationship of Folic Acid (Pteroylglutamic Acid) to Red Cell Formation. Occasionally maturation failure anemia results from folic acid deficiency instead of vitamin B_{12} deficiency. Folic acid is part of the vitamin B complex as we shall discuss in Chapter 32, and it, like B_{12}, is required for formation of DNA but in a different way. It promotes the methylation of deoxyuridylate to form deoxythymidylate, one of the nucleotides required for DNA synthesis.

Formation of Hemoglobin

Synthesis of hemoglobin begins in the erythroblasts and continues through the reticulocyte stage. Even when young red blood cells, the reticulocytes, leave the bone marrow and pass into the blood stream, they continue to form hemoglobin for another day or so.

Figure 33–4 gives the basic chemical steps in the formation of hemoglobin. From tracer studies with isotopes it is known that hemoglobin is synthesized from *α-ketoglutaric acid*, one of the breakdown products mainly of carbohydrates and fats, and *glycine*, an amino acid derived from proteins. These two together form a *pyrrole* compound. In turn, four pyrrole compounds combine to form *protoporphyrin III*, which then combines with iron to form the *heme* molecule. Finally, four heme molecules

I.   2 α-ketoglutaric acid + glycine ⟶

II. 4 pyrrole ⟶ protoporphyrin III   (pyrrole)

III. protoporphyrin III + Fe ⟶ heme

IV. 4 heme + globin ⟶ hemoglobin

FIGURE 33–4   Formation of hemoglobin.

combine with one molecule of *globin*, a globulin, to form hemoglobin, the formula for which is shown in Figure 33–5. Hemoglobin has a molecular weight of 68,000.

Combination of Hemoglobin with Oxygen. The most important characteristic of the hemoglobin molecule is its ability to combine loosely and reversibly with oxygen. This ability will be discussed in detail in Chapter 37 in relation to respiration, for the primary function of hemoglobin in the body depends upon its ability to combine with oxygen in the lungs and then to release this oxygen readily in the tissue capillaries where the gaseous tension of oxygen is much lower than in the lungs.

Oxygen *does not* combine at the two positive valences of the ferrous iron in the hemoglobin molecule. Instead, it binds loosely at two of the six "coordination" valences of the iron atom. This is an extremely loose bond, so that the combination is easily reversible, thus allowing rapid pickup of oxygen in the lungs and equally rapid release of oxygen in the tissues.

FIGURE 33–5   The hemoglobin molecule.

Iron Metabolism

Because iron is absolutely necessary for the formation of hemoglobin, myoglobin, and other substances essential to body function such as the cytochromes, cytochrome oxidase, peroxidase, and catalase, it is important to understand the means by which iron is utilized in the body.

The total quantity of iron in the body averages about 4 g, approximately 65 percent of which is present in the form of hemoglobin. About 4 percent is in the form of *myoglobin*, 1 percent in the form of the various heme compounds (*cytochromes* and *cytochrome oxidase*) that control intracellar oxidation, 0.1 percent in the form of *transferrin* in the blood plasma, and 15 to 30 percent stored mainly in the form of *ferritin* in the liver.

Transport and Storage of Iron. A schema for transport, storage, and metabolism of iron in the body is illustrated in Figure 33–6, which may be explained as follows: When iron is absorbed from the small intestine, it immediately combines in the blood plasma with a beta globulin, *transferrin*. The iron is very loosely combined in this compound, and consequently can be released to any of the tissue cells at any point in the body. Excess iron in the blood is deposited in all cells of the body *but especially in the liver cells*, where about 60 percent of the excess is stored. There it combines with a protein of very large molecular weight, *apoferritin*, to form *ferritin*. This iron stored in the form of ferritin is called *storage iron*.

When the quantity of iron in the plasma falls very low, iron is removed from ferritin quite easily. The iron is then transported again by transferrin to those tissues of the body where it is needed.

When red blood cells have lived their life span and are destroyed, the hemoglobin released from the cells is ingested by the reticuloendothelial cells (the white blood cells and certain tissue cells—these will be discussed later in this chapter). These cells in turn digest the hemoglobin and liberate free iron that can then either be restored in the ferritin pool or be reused for formation of new hemoglobin.

Daily Loss of Iron. About 0.6 mg of iron is excreted each day by the male, mainly into the feces. Additional quantities of iron are lost whenever bleeding occurs. Thus, in the female, the menstrual loss of blood brings the average iron loss to a value of approximately 1.3 mg per day.

Obviously, the average quantity of iron derived from the diet each day must at least equal that lost from the body.

Absorption of Iron from the Gastrointestinal Tract. Iron is absorbed almost entirely in the upper part of the small intestine, primarily in the duodenum. It is absorbed by an active process, though the precise mechanism of this active absorption is unknown.

Regulation of the Total Body Iron— Role of Intestinal Absorption. As is true of almost all essential substances in the body, it is exceedingly important to maintain an appropriate amount of iron in the body—not too little, not too much. This is achieved in the following way. When the apoferritin in the liver and other organs where iron is stored becomes saturated with iron, transferrin then cannot release iron from the plasma into the storage depots. Consequently, the transferrin also becomes saturated with iron, and the active transport processes of the intestinal mucosal cells then cannot transport iron into the blood. Therefore, the iron handling system simply turns off the intestinal absorption of iron so that most of the iron in the food fails to be absorbed and instead is excreted in the feces.

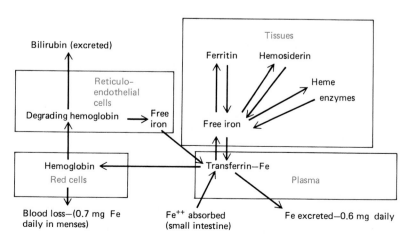

FIGURE 33–6 Iron transport and metabolism.

On the other hand, whenever the iron stores become depleted, this mechanism reverses itself completely, and the intestinal absorptive process transports iron actively from the intestinal lumen into the blood. Thus, the system is normally geared to keep the iron storage depots almost full of iron but not overly filled.

Destruction of Red Blood Cells

When red blood cells are delivered from the bone marrow into the circulatory system they normally circulate an average of 120 days before being destroyed. Even though red cells do not have a nucleus, they do still have cytoplasmic enzymes for metabolizing glucose and other substances and for utilizing oxygen, but many of these metabolic systems become progressively less active with time. As the cells become older they also become progressively more fragile, presumably because their life processes simply wear out.

Once the red cell membrane becomes very fragile, it may rupture during passage through some tight spot of the circulation. Many of the cells break up in the spleen where they squeeze through the red pulp of the spleen. When the spleen is removed, the number of abnormal cells and old cells circulating in the blood increases considerably.

Destruction of Hemoglobin. The hemoglobin released from the cells when they burst is phagocytized and digested almost immediately by reticuloendothelial cells, releasing iron back into the blood to be carried by transferrin either to the bone marrow for production of new red blood cells or to the liver and other tissues for storage in the form of ferritin. The heme portion of the hemoglobin molecule is converted by the reticuloendothelial cell, through a series of stages, into the bile pigment *bilirubin*, which is released into the blood and later secreted by the liver into the bile; this will be discussed in relation to liver function in Chapter 41.

Anemia

Anemia means a deficiency of red blood cells, with hematocrits sometimes as low as 10 percent, which can be caused either by too much blood loss or by too slow production of red blood cells. Some of the common types of anemia were illustrated in Figure 33–3. These are often caused by

1. *Blood loss.*
2. *Bone marrow destruction.* Common causes of this are drug poisoning or gamma ray irradia-

tion—for instance, exposure to radiation from a nuclear bomb blast.
3. *Failure of red blood cells to mature* because of lack of vitamin B_{12} or folic acid, as was previously explained and as occurs in pernicious anemia.
4. *Hemolysis of red cells*—that is, rupture of the cells—resulting from many possible causes, such as (a) drug poisoning, (b) hereditary diseases that make the red cell membranes friable (for example, sickle cell anemia), and (c) erythroblastosis fetalis, a disease of the newborn in which antibodies from the mother destroy red cells in the baby.

As long as an anemic person's level of muscular activity is low, he often can live without fatal hypoxia of the tissues, even though his concentration of red blood cells may be reduced to one-fourth normal. However, when he begins to exercise, his heart will not be capable of pumping enough blood to the tissues to supply the needed oxygen. Consequently, during exercise, which greatly increases the demands for oxygen, extreme tissue hypoxia results, and acute heart failure often ensues.

WHITE BLOOD CELLS (LEUKOCYTES), AND RESISTANCE OF THE BODY TO INFECTION

The body is constantly exposed to bacteria, these occurring especially in the mouth, the respiratory passageways, the colon, the mucous membranes of the eyes, and even the urinary tract. Many of these bacteria are capable of causing disease if they invade the deeper tissues. In addition, a person is intermittently exposed to highly virulent bacteria and viruses from outside the body that can cause specific diseases such as pneumonia, streptococcal infections, and typhoid fever.

On the other hand, a group of tissues including the **white blood cells** and the **reticulo-endothelial system** constantly combats any infectious agent that tries to invade the body. These tissues function in two different ways to prevent disease: (1) by actually destroying invading agents by the process of phagocytosis and (2) by forming *antibodies* against the invading agent, the antibodies in turn destroying the invader; this process is called *immunity*. The present discussion is concerned with phagocytic destruction of the invading agents; the following chapter is concerned with immunity.

The white blood cells, also called **leukocytes,** are the *mobile units* of the body's protective system. They are formed partially in the bone marrow and partially in the lymph nodes, but after formation they are transported in the blood to the different parts of the body where they are to be used.

The Types of White Blood Cells. Five different types of white blood cells are normally found in the blood. Each of these is illustrated in Figure 33–7. They are **polymorphonuclear neutrophils, polymorphonuclear eosinophils, polymorphonuclear basophils, monocytes,** and **lymphocytes.** In addition, there are large numbers of **platelets,** which are fragments of a sixth type of white blood cell, the **megakaryocyte.** The three types of polymorphonuclear cells have a granular appearance, as illustrated in the figure, for which reason they are called *granulocytes,* or in clinical terminology they are often called simply "polys."

The granulocytes and the monocytes protect the body against invading organisms by ingesting them—that is, by the process of *phagocytosis.* Also, a function of another type of white cell, the lymphocyte, is to attach to specific invading organisms and to destroy them; this is part of the immunity system and will be dis-

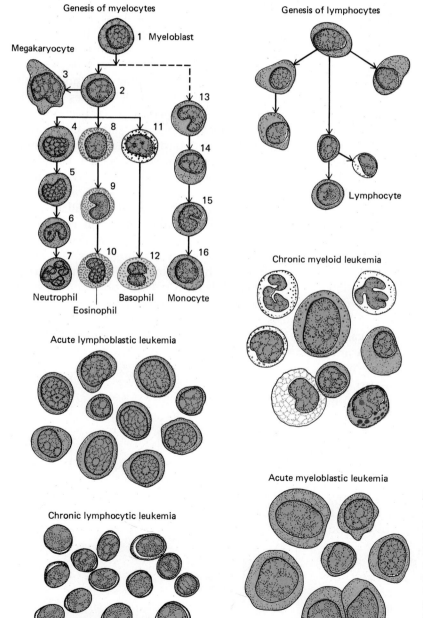

FIGURE 33–7 Genesis of white blood cells, and the blood pictures in several types of leukemia. Cell number 1 is a *myeloblast,* and number 3 is a *megakaryocyte.* Cells 4, 5, 6, and 7 illustrate the formation of *neutrophils;* 8, 9, and 10, the formation of *eosinophils;* 11 and 12, the formation of *basophils;* and 13, 14, 15, and 16, the formation of *monocytes.* (Redrawn in part from Piney: *A Clinical Atlas of Blood Diseases.* Blakiston.)

cussed in the following chapter. Finally, the function of platelets is to activate the blood clotting mechanism. All these functions are protective mechanisms of one type or another.

Concentrations of the Different White Blood Cells in the Blood.   The adult human being has approximately 7000 white blood cells per cubic millimeter of blood. The normal percentages of the different types of white blood cells are approximately the following:

Polymorphonuclear neutrophils	62.0%
Polymorphonuclear eosinophils	2.3%
Polymorphonuclear basophils	0.4%
Monocytes	5.3%
Lymphocytes	30.0%

The number of platelets in each cubic millimeter of blood is normally about 300,000.

Genesis of the White Blood Cells

The upper left portion of Figure 33–7 illustrates the stages in the development of the white blood cells. The polymorphonuclear cells and monocytes are normally formed only in the bone marrow. On the other hand, lymphocytes are produced in the various lymphogenous organs, including the lymph glands, the spleen, the thymus, the tonsils, and various lymphoid tissues in the gut and elsewhere.

Some of the white blood cells formed in the bone marrow, especially the granulocytes, remain stored within the marrow until they are needed in the circulatory system. Then when the need arises, various factors that are discussed later cause them to be released.

As also illustrated in Figure 33–7, megakaryocytes are also formed in the bone marrow and are part of the myelogenous group of bone marrow cells. These megakaryocytes break up into very small fragments in the bone marrow—the small fragments, the *platelets*, passing then into the blood.

Properties of White Blood Cells

Phagocytosis.   The most important function of the *neutrophils* and *monocytes* and, to a lesser extent, of some of the other white blood cells is phagocytosis.

Obviously, the phagocytes must be selective in the material that is phagocytized, or otherwise some of the structures of the body itself would be ingested. Whether or not phagocytosis will occur depends on three selective procedures. First, if the surface of a particle is rough, the likelihood of phagocytosis is increased. Sec-

ond, most natural substances of the body have electronegative surface charges that repel the phagocytes, which also carry electronegative surface charges. On the other hand, dead tissues and foreign particles are frequently electropositive and are therefore subject to phagocytosis. Third, the body has a means for promoting phagocytosis of specific foreign materials by first combining them with protein "antibodies" called *opsonins*. After the opsonin has combined with the particle, it allows the phagocyte to adhere to the surface of the particle, and this promotes phagocytosis. The special features of opsonization and its relationship to immunity are discussed in the following chapter.

Enzymatic Digestion of the Phagocytized Particles.   Once a foreign particle has been phagocytized, lysosomes in the cell cytoplasm immediately come in contact with the phagocytic vesicle, and their membranes fuse with those of the vesicle, thereby dumping the digestive enzymes of the lysosomes into the vesicle. Thus, the phagocytic vesicle now becomes a *digestive vesicle*, and digestion of the phagocytized particle begins immediately. This was discussed in more detail in Chapter 3.

Diapedesis.   White blood cells can move out of the blood into the tissue spaces. They do this by squeezing through the capillary pores and even through holes in some endothelial cells of the blood vessels. This process is called *diapedesis*. That is, even though a pore is much smaller than the size of the cell, a small portion of the cell slides through the pore at a time, the portion sliding through being momentarily constricted to the size of the pore, as illustrated in Figure 33–8.

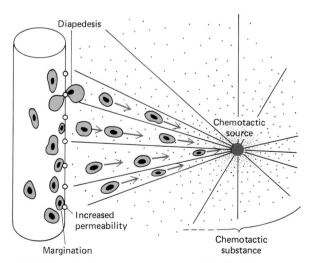

FIGURE 33–8   Movement of neutrophils by the process of *chemotaxis* toward an area of tissue damage.

Ameboid Motion. Once the cells have entered the tissue spaces, the polymorphonuclear leukocytes especially, and the large lymphocytes and monocytes to a lesser degree, move through the tissues by ameboid motion, which was described in Chapter 3. Some of the cells can move through the tissues at rates as great as 40 μ per minute—that is, they can move at least three times their own length each minute.

Chemotaxis. Different chemical substances in the tissues cause the leukocytes to move either toward or away from the source of the chemical. This phenomenon, known as chemotaxis, is also illustrated in Figure 33–8. Degenerative products of inflamed tissues, especially tissue polysaccharides and also one of the reaction products of a complex of substances called "complement" (discussed in the following chapter), can *cause neutrophils and monocytes to move toward the area of inflammation.* In addition, a number of bacterial toxins can also cause chemotaxis of leukocytes.

THE TISSUE MACROPHAGE SYSTEM (THE RETICULOENDOTHELIAL SYSTEM)

In the above paragraphs we have described the monocytes as mobile cells that are capable of filtering out of the blood into the tissues and there performing their usual phagocytic functions. In the tissues these cells are called **macrophages** because they grow to be very large and also to have extreme phagocytic capabilities. Under normal circumstances most of these macrophages become attached to the tissues and remain attached for months or even years. They are then called **tissue macrophages**, and they continue their phagocytic properties to protect the tissues from invaders. The combination of the monocytes, the mobile macrophages, and the fixed tissue macrophages is commonly called the **reticuloendothelial system**. However, this name is actually a misnomer that came about because it was formerly believed that a major share of the blood vessel endothelial cells could perform phagocytic functions similar to those performed by this macrophage system.

The tissue macrophages in various tissues differ in appearance because of environmental differences, and they are known by different names: **Kupffer cells** in the liver; **reticulum cells** in lymph nodes, spleen, and bone marrow; **alveolar macrophages** in the alveoli of the lungs; **tissue histiocytes**, or **fixed macrophages**, in the subcutaneous tissues; and **microglia** in the brain.

Tissue Macrophages in the Skin and Subcutaneous Tissues (Histiocytes). Though the skin is normally impregnable to infectious agents, this no longer holds true when the skin is broken. When infection does begin in the subcutaneous tissues and local inflammation ensues, the tissue macrophages can divide *in situ* and form many more macrophages. Then they perform the usual functions of attacking and devouring the infectious agents, as described earlier.

Macrophages of the Lymph Nodes (Reticulum Cells). Essentially no particulate matter that enters the tissues can be absorbed directly through the capillary membranes into the blood. Instead, if the particles are not destroyed locally in the tissues, they enter the lymph and flow through the lymphatic vessels to the lymph nodes located intermittently along the course of the lymphatics, as described in Chapter 30. The foreign particles are trapped there in a meshwork of sinuses lined by tissue macrophages called *reticulum cells.*

Figure 33–9 illustrates the general organization of the lymph node, showing lymph entering by way of the *afferent lymphatics*, flowing through the *medullary sinuses*, and finally passing out of the *hilus* into the *efferent lymphatics*. Large numbers of reticulum cells line the sinuses, and if any particles enter the sinuses, these cells phagocytize them and prevent general dissemination throughout the body.

Alveolar Macrophages. Another route by which invading organisms frequently enter the body is through the respiratory system. Fortunately, large numbers of tissue macrophages are present as integral components of the alveo-

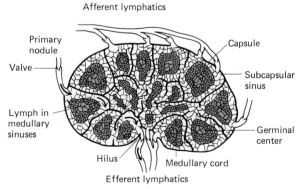

FIGURE 33–9 Functional diagram of a lymph node. (Redrawn from Ham: *Histology.* Philadelphia, J.B. Lippincott, 1974.)

lar walls. These can phagocytize particles that become entrapped in the alveoli. If the particle is not digestible the macrophages often form a "giant cell" capsule around the particle until such time, if ever, that it can be slowly dissoluted. Such capsules are frequently formed around tubercle bacilli, silica dust particles, and even carbon particles.

Tissue Macrophages (Kupffer Cells) in the Liver Sinuses.   Still another favorite route by which bacteria invade the body is through the gastrointestinal tract. Large numbers of bacteria constantly pass through the gastrointestinal mucosa into the portal blood. However, before this blood enters the general circulation, it must pass through the sinuses of the liver; these sinuses are lined with tissue macrophages called *Kupffer cells*, illustrated in Figure 33–10. These cells form such an effective particulate filtration system that almost none of the bacteria from the gastrointestinal tract succeeds in passing from the portal blood into the general systemic circulation. Indeed, motion pictures of phagocytosis by Kupffer cells have demonstrated phagocytosis of single bacteria in less than $\frac{1}{100}$ second.

Macrophages (Reticulum Cells) of the Spleen and Bone Marrow.   If an invading organism does succeed in entering the general circulation, there still remain other lines of defense by the tissue macrophage system, especially by *reticulum cells* of the spleen and bone marrow. In both of these tissues, macrophages have become entrapped by the reticular meshwork of the two organs, and when foreign particles come in con-

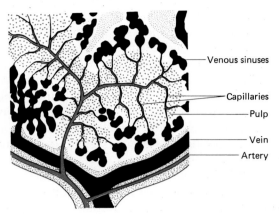

FIGURE 33–11   Functional structures of the spleen. (Modified from Bloom and Fawcett: Textbook of Histology, 10th. Ed. Philadelphia, W.B. Saunders, 1975.)

tact with the reticulum cells in this meshwork the particles are phagocytized.

The spleen is similar to the lymph nodes, except that blood, instead of lymph, flows through the substance of the spleen. Figure 33–11 illustrates the spleen's general structure, showing a small peripheral segment. Note that small arteries penetrate from the splenic capsule into the *splenic pulp* and terminate in small capillaries. The capillaries are highly porous, allowing large numbers of whole blood cells to pass out of them into the *red pulp*. These cells then gradually *squeeze* through the tissue substance of this pulp and eventually return to the circulation through the endothelial walls of the *venous sinuses*. The red pulp is loaded with macrophages (reticulum cells), and in addition the venous sinuses are also lined with macrophages. This peculiar passage of blood through the cords of the red pulp provides an exceptional means for phagocytosis of unwanted debris in the blood, especially old and abnormal red blood cells. The spleen is also an important organ for phagocytic removal of abnormal platelets, blood parasites, and any bacteria that might succeed in entering the general circulating blood.

In a similar way, reticulum cells of the bone marrow also help to remove unwanted debris and pathologic agents from blood.

Inflammation, and Function of Neutrophils and Macrophages

Inflammation is a complex of sequential changes in the tissue in response to injury. When tissue injury occurs, whether it is caused by bacteria, trauma, chemicals, heat, or any other phenomenon, large quantities of *histamine, bradykinin, serotonin,* and other substances are liberated by the damaged tissue into the surrounding fluids.

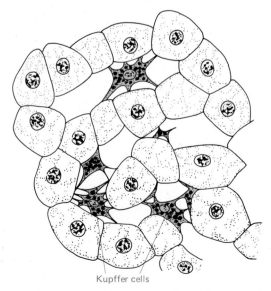

Kupffer cells

FIGURE 33–10   The Kupffer cells of the liver sinusoids. These are typical reticuloendothelial cells. (Redrawn from Smith and Copenhaver: *Bailey's Textbook of Histology.* Baltimore, Williams and Wilkins.)

These, especially the histamine, increase the local blood flow and also increase the permeability of the venous capillaries and venules, allowing large quantities of fluid and protein to leak into the tissues. Local extracellular edema results, and the extracellular fluid and lymphatic fluid both clot because of the coagulating effect of tissue exudates on the clotting factor *fibrinogen* in the leaking fluid. Thus, *brawny edema* develops in the spaces surrounding the injured cells.

The "Walling Off" Effect of Inflammation. It is clear that the above-described effects of the inflammation process "wall off" the area of injury from the remaining tissues. The tissue spaces and the lymphatics in the inflamed area are blocked by the fibrinogen clots so that fluid barely flows through the spaces. Therefore, walling off the area of injury delays the spread of bacteria or toxic products.

THE MACROPHAGE AND NEUTROPHIL RESPONSE TO INFLAMMATION

Soon after inflammation begins, the inflamed area becomes invaded by both neutrophils and macrophages, and these set about performing their scavenger functions to rid the tissue of infectious or toxic agents. However, the macrophage and neutrophil responses occur in several different stages.

The Tissue Macrophages as the First Line of Defense. The macrophages that are already present in the tissues, whether they are the histiocytes in the subcutaneous tissues, the alveolar macrophages in the lungs, the microglia in the brain, or so forth, immediately begin their phagocytic actions. Therefore, they are the first line of defense.

Neutrophilia and Neutrophil Invasion of the Inflamed Area—the Second Line of Defense. The term *neutrophilia* means an increase in the number of neutrophils in the blood. The term "leukocytosis" is also often used to mean the same as neutrophilia, though this term actually means excess number of all white cells, whatever their types.

Within a few hours after the onset of acute inflammation, the number of neutrophils in the blood sometimes increases as much as four- to fivefold—to as high as 15,000 to 25,000 per cubic millimeter. This results from a combination of chemical substances that are released from the inflamed tissues, collectively called *leukocytosis-inducing factor.* This factor diffuses from the inflamed tissue into the blood and it is carried to the bone marrow. There it is believed to dilate the venous sinusoids of the marrow, thus causing release of many leukocytes, especially neutrophils, that are stored in these venous sinusoids. In this way large numbers of neutrophils are almost immediately transferred from the bone marrow storage pool into the circulating blood.

Movement of Neutrophils to the Area of Inflammation. Products from the inflamed tissues also cause neutrophils to move from the circulation into the inflamed area. They do this in three ways, which were illustrated in Figure 33–8.

First, they damage the capillary walls and thereby cause neutrophils to stick, which is the process called *margination.*

Second, they greatly increase the permeability of the capillaries and small venules, and this allows the neutrophils to pass by *diapedesis* into the tissue spaces.

Third, the phenomenon of *chemotaxis* causes the neutrophils to migrate toward the injured tissues, as was described earlier.

Thus, within several hours after tissue damage begins, the area becomes well supplied with neutrophils. Since neutrophils are already mature cells, they are ready to begin their scavenger functions immediately for removal of foreign matter from the inflamed tissues.

Macrophage Proliferation and the Monocyte Response—the Third Line of Defense. Still a third line of defense is a slow but long-continuing increase in the number of macrophages as well. This results partly from reproduction of the already present tissue macrophages and also from migration of large numbers of monocytes into the inflamed area. Though the monocytes are still immature cells and are not capable of phagocytosis when they first enter the tissues, over a period of 8 to 12 hours they swell markedly, form greatly increased quantities of cytoplasmic lysosomes, exhibit increased ameboid motion, and move chemotactically toward the damaged tissues.

Next, the rate of production of monocytes by the bone marrow also increases. This (as well as increased production of neutrophils) results from stimulation by yet undefined stimulating factors. In long-term chronic infection there is progressively increasing production of monocytes, which increases the ratio of macrophages to neutrophils in the tissues. Therefore, the long-term chronic defense against infection is mainly a macrophage response rather than a neutrophil response.

The macrophages can phagocytize far more bacteria than can the neutrophils, and they can also ingest large quantities of necrotic tissue.

FORMATION OF PUS

When the neutrophils and macrophages engulf large numbers of bacteria and necrotic tissue, essentially all of the neutrophils and many if not most of the macrophages themselves eventually die. After several days, a cavity is often excavated in the inflamed tissues, containing varying portions of necrotic tissue, dead neutrophils, and dead macrophages. Such a mixture is commonly known as pus.

Ordinarily, pus formation continues until all infection is suppressed. Sometimes the pus cavity eats its way to the surface of the body and in this way empties itself. But if this does not happen, the dead cells and necrotic tissue in the pus gradually autolyze over a period of days, and the end-products of autolysis are usually absorbed into the surrounding tissues until most of the evidence of tissue damage is gone.

The Eosinophils

The eosinophils normally comprise 1 to 3 percent of all the blood leukocytes. They are weak phagocytes, and they exhibit chemotaxis. They also have a special propensity to collect at sites of antigen-antibody reactions in the tissues and a special capability to phagocytize and digest the combined antigen-antibody complex after the immune process has performed its function. Also, the total number of eosinophils increases greatly in the circulating blood during allergic reactions, following foreign protein injections, and during infections with parasites. It is possible that eosinophils help to remove foreign proteins, whatever their source.

The Basophils

The basophils in the circulating blood are very similar, though maybe not identical, to the large *mast* cells located immediately outside many of the capillaries in the body. These cells liberate *heparin* into the blood, a substance that can prevent blood coagulation. It is probable that the basophils in the circulating blood perform similar functions within the blood stream, or it is even possible that the blood simply transports basophils to tissues where they then become mast cells and perform the function of heparin liberation.

The mast cells and basophils also release histamine as well as smaller quantities of brady-kinin and serotonin. Indeed, it is mainly the mast cells in inflamed tissues that release these substances during inflammation.

AGRANULOCYTOSIS

A clinical condition known as agranulocytosis occasionally occurs, in which the bone marrow stops producing white blood cells, leaving the body unprotected against bacteria and other agents that might invade the tissues. The cause is usually *drug poisoning* or *irradiation following a nuclear bomb blast*. Within two days after the bone marrow stops producing white blood cells, ulcers appear in the mouth and colon, or the person develops some form of severe respiratory infection. Bacteria then rapidly invade the surrounding tissues and the blood. Without treatment, death usually ensues 3 to 6 days after agranulocytosis begins.

The Leukemias

Ordinarily, leukemias are divided into two general types: the *lymphogenous leukemias* and the *myelogenous leukemias*. The lymphogenous leukemias are caused by cancerous production of lymphoid cells, beginning first in a lymph node, and the myelogenous leukemias begin by cancerous production of young myelogenous cells (early forms of neutrophils, monocytes, or others) in the bone marrow.

EFFECTS OF LEUKEMIA ON THE BODY

The leukemic cells of the bone marrow may reproduce so greatly that they invade the surrounding bone, causing pain and eventually a tendency for the bone to fracture. Almost all leukemias spread to the spleen, the lymph nodes, the liver, and other especially vascular regions, regardless of whether the origin of the leukemia is in the bone marrow or in the lymph nodes.

Very common effects in leukemia are the development of infections, severe anemia, and a bleeding tendency caused by thrombocytopenia (lack of platelets). These effects result mainly from displacement of the normal bone marrow by the leukemic cells.

Finally, the leukemic tissues reproduce new cells so rapidly that tremendous demands are made on the body for foodstuffs, especially for amino acids and vitamins. Thus, while the leukemic tissues grow, the other tissues are debilitated. Obviously, after metabolic starvation has continued long enough, that alone is sufficient to cause death.

QUESTIONS

1. What is the significance and function of the "baglike" structure of the red blood cell?
2. Discuss the genesis of red blood cells.
3. Give the mechanism for regulating the total number of red blood cells in the circulatory system.
4. What are the roles of vitamin B_{12} and folic acid in the formation of red blood cells?
5. Discuss iron metabolism and the role of iron in the formation of hemoglobin.
6. What causes the ultimate destruction of red blood cells?
7. List four different causes of anemia.
8. Contrast the genesis of white blood cells to that of red blood cells.
9. Which white blood cells have the greatest capability for phagocytizing bacteria?
10. Discuss the role of white blood cells in the inflammatory process.
11. Describe the function of macrophages in the different tissues of the body.
12. What is the cause of the leukemias?

REFERENCES

Allison, A.C., *et al.*: Inflammation. New York, Springer-Verlag, 1978.

Erslev, A.J., and Gabuzda, T.G.: Pathophysiology of Blood. Philadelphia, W.B. Saunders, 1979.

Escobar, M.R., and Friedman, H. (eds.): Macrophages and Lymphocytes; Nature, Functions and Interaction. New York, Plenum Press, 1979.

Friedman, H., *et al.* (eds.): The Reticuloendothelial System. New York, Plenum Press, 1979.

Gowans, J.L. (in honour of): Blood Cells and Vessel Walls: Functional Interactions. Princeton, N.J., Excerpta Medica, 1980.

Houck, J.C. (ed.): Chemical Messengers of the Inflammatory Process. New York, Elsevier/North-Holland, 1980.

Kass, L: Bone Marrow Interpretation. Philadelphia, J.B. Lippincott, 1979.

Kelemen, E., *et al.*: Atlas of Human Hemopoietic Development. New York, Springer-Verlag, 1979.

Kokubun, Y., and Kobayashi, N. (eds.): Phagocytosis, Its Physiology and Pathology. Baltimore, University Park Press, 1979.

Konigsberg, W.: Protein structure and molecular dysfunction: Hemoglobin. *In* Bondy, P.K., and Rosenberg, L.E. (eds.): Metabolic Control of Disease, 8th Ed. Philadelphia, W.B. Saunders, 1980, p. 27.

Lichtman, M.A. (ed.): Hematology and Oncology. New York, Grune & Stratton, 1980.

Peschle, C.: Erythropoiesis. *Annu. Rev. Med., 31*:303, 1980.

Platt, W.R.: Color Atlas and Textbook of Hematology. Philadelphia, J.B. Lippincott, 1978.

Quastel, M.R. (ed.): Cell Biology and Immunology of Leukocyte Function. New York, Academic Press, 1979.

Spivak, J.L. (ed.): Fundamentals of Clinical Hematology. Hagerstown, Md., Harper & Row, 1980.

34

Immunity and Allergy

Overview

The term *immunity* means the ability of the body to protect itself against *specific foreign agents* such as specific bacteria, viruses, toxins, or foreign tissue cells. There are two different types of immunity systems both of which are based on the function of lymphocytes: (1) the "B"-lymphocyte system and (2) the "T"-lymphocyte system.

The "B"-Lymphocyte System. This system forms antibodies that destroy the invading agent. The mechanism of this is the following: On first exposure to a specific foreign agent, some of its chemical constituents, especially its *proteins* or *polysaccharides* (these are called *antigens*), react with *"B"-lymphocytes* in the lymph nodes. These in turn form *plasma cells*, which multiply profusely and generate large numbers of protein molecules called *antibodies*. The antibodies in turn have a specific capability of reacting with exactly the same type of antigen as that which caused their formation, but they usually will not react with other tissue substances. Therefore, the antibodies will destroy only the foreign invader. They cause destruction especially by (1) *agglutination* (which means clumping together), (2) *lysis* (which means rupturing the membrane of the agent), (3) *neutralization* (which means blocking its toxic effects), or (4) *opsonization* (which means making the agent susceptible to phagocytosis by neutrophils and macrophages).

The "T"-Lymphocyte System. In this system an entirely different set of lymphocytes in the lymph nodes, called *T-cells*, reacts with antigens in the same way that the B-cell system reacts. However, instead of forming antibodies, this system forms *sensitized T-cells*. These have reactive sites on their cell membranes that are similar to the reactive sites of antibodies. Therefore, the sensitized T-cells can attach to invading agents in much the same way that antibodies can. However, there are several different types of T-cells, the most important of which are the following: (1) *cytotoxic T-cells*, which attach directly to the invading agent and destroy it; (2) *helper T-cells*, which interact with the B-cell immune system in the lymph nodes to cause B-cell production of antibodies against the antigens that sensitize the T-cells; and (3) *suppressor T-cells*, which help to control the entire immune process, mainly preventing runaway immune reactions.

Normally, even though all tissues of the human body contain proteins, our immunity system is adapted *not to* form antibodies or sensitized lymphocytes that can attack the body's own tissues. This is called *immune tolerance.* However, under some conditions, more often in old age than at a young age, the immune system becomes abnormal, and some antibodies or sensitized lymphocytes are then formed that do attack the person's own tissues. This leads to a host of different *autoim-*

566

mune diseases, some of which are *rheumatic fever, rheumatoid arthritis, thyroid disease,* the paralytic disease *myasthenia gravis,* and an autoimmune disease called *lupus erythematosus* that attacks most of the tissues of the body and is lethal if untreated.

Allergy is another condition that results from an abnormality of the immunity system. Some people have a *hereditary type* of allergy. Such persons form large quantities of very large antibodies, called *IgE antibodies,* that have several times as many reactive sites as normal antibodies. These abnormal antibodies attach themselves very firmly to many of the body's tissue cells, especially to the basophils. Then, when an immune reaction occurs between these antibodies and the antigen (in this case called an *allergen*) that had caused them to develop, the basophils rupture and release several toxic substances that have potent effects on the body. Among these are *histamine,* which can cause *hay fever* and *urticaria* (hives), and *slow-reacting substance of anaphylaxis,* which can cause *asthma.*

IMMUNITY

In the previous chapter we discussed the roles of white blood cells and of the tissue macrophage system (the reticuloendothelial system) to protect the body against organisms that tend to damage tissues and organs. However, the body has yet another system for protection, a system that operates not only against organisms but against many toxins as well. This is called the system of *acquired immunity.* It results from the formation of *antibodies* and *sensitized lymphocytes* that attack and destroy the invading organisms or toxins.

The immune system does not resist the initial invasion by organisms. Nor does it resist the effects of toxins upon first exposure. However, within a few days to a few weeks after initial exposure, the immune system then develops its extremely powerful resistance to the invader. Furthermore, this resistance is generally specific for the particular invader and for no other one. For instance, immune protection against the paralytic toxin of botulinum bacteria or the tetanizing toxin of tetanus bacteria can be produced against doses as high as 100,000 times those that would be lethal without immunity. This is the reason that the process known as *vaccination* is so extremely important in protecting human beings against disease and toxins, as will be explained in the course of this chapter.

Unfortunately, the immune process does not always work exactly the way that it should. Sometimes elements of the immune system attack the person's own tissues rather than a specific invader. Under these conditions, serious damaging effects can occur as a result of *autoim-* munity or *allergy,* both of which also will be explained later in the chapter.

Two Basic Types of Immunity— Humoral Immunity and Cellular Immunity

Two basic, but closely allied, types of immunity occur in the body. In one of these the body develops circulating *antibodies,* which are globulin molecules that are capable of attacking the invading agent. This type of immunity is called *humoral immunity.* The second type of immunity is achieved through the formation of large numbers of highly specialized lymphocytes that are specifically sensitized against the foreign agent. These *sensitized lymphocytes* have the special capability of attaching to the foreign agent and destroying it. This type of immunity is called *cellular immunity* or, sometimes, *lymphocytic immunity.*

We shall see shortly that both the antibodies and the sensitized lymphocytes are formed in the lymphoid tissue of the body. First, let us discuss the initiation of the immune process by *antigens.*

Antigens

Each toxin or each type of organism contains one or more specific chemical compounds in its make-up that are different from all other compounds. In general, these are proteins, large polysaccharides, or large lipoprotein complexes, and it is one or more of these compounds that cause the immunity. These substances are called *antigens.*

Essentially all toxins secreted by bacteria are proteins, large polysaccharides, or mucopolysaccharides, and they are highly antigenic. Also, the bodies of bacteria or viruses usually contain several antigenic chemical compounds. Likewise, animal tissues such as a transplanted heart from another human being also contain numerous antigens that can elicit the immune process and cause subsequent destruction.

For a substance to be antigenic it usually must have a high molecular weight, 8000 or greater. Furthermore, the process of antigenicity depends upon regularly recurring patterns of specific types of atoms on the surface of the large molecule, which perhaps explains why proteins and many polysaccharides are antigenic, for they both have this characteristic.

Role of Lymphoid Tissue in Immunity

Immunity is absolutely dependent on function of the body's lymphoid tissue. In persons who have a genetic lack of lymphoid tissue, no acquired immunity whatsoever can develop. And almost immediately after birth such a person dies of a sudden and severe infection unless treated by heroic measures.

The lymphoid tissue is located most extensively in the lymph nodes, but it is also found in special lymphoid tissue such as in the spleen, in submucosal areas of the gastrointestinal tract, and, to a slight extent, in the bone marrow. The lymphoid tissue is distributed very advantageously in the body to intercept the invading organisms or toxins before they can spread too widely.

Two Types of Lymphocytes that Promote, Respectively, Cellular Immunity and Humoral Immunity— "T" Lymphocytes and "B" Lymphocytes

Though most of the lymphocytes in normal lymphoid tissue look alike when studied under the microscope, these cells are distinctly divided into two separate populations. One of the populations is responsible for forming the sensitized lymphocytes that provide cellular immunity and the other for forming the antibodies that provide humoral immunity.

Both of these types of lymphocytes are derived originally in the embryo from *lymphocytic stem cells*. The descendants of the stem cells eventually migrate to the lymphoid tissue. Before doing so, however, those lymphocytes that are eventually destined to form sensitized lymphocytes first migrate to and are preprocessed in the *thymus* gland, for which reason they are called "T" lymphocytes. These are responsible for cellular immunity.

The other population of lymphocytes—those that are destined to form antibodies—is processed in some unknown area of the body, but probably mainly in the fetal liver. This population of cells was first discovered in birds in which the preprocessing occurs in the *bursa of Fabricius*, a structure not found in mammals. For this reason this population of lymphocytes is called the "B" lymphocytes, and they are responsible for humoral immunity.

Figure 34–1 illustrates the two separate lymphocyte systems for the formation, respec-

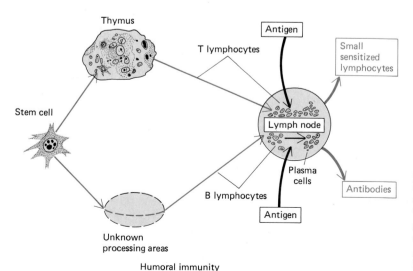

FIGURE 34–1 Formation of antibodies and sensitized lymphocytes by a lymph node in response to antigens. This figure also shows the origin of *thymic* ("T") and *bursal* ("B") lymphocytes, which are responsible for the cellular and humoral immune processes of the lymph nodes.

tively, of the sensitized lymphocytes and the antibodies.

Role of the Thymus Gland for Preprocessing the "T" Lymphocytes.

Most of the preprocessing of the "T" lymphocytes of the thymus gland occurs shortly before birth and for a few months after birth of the baby. Therefore, beyond this period of time, removal of the thymus gland usually will not seriously impair the "T" lymphocytic immunity system, the system necessary for cellular immunity. However, removal of the thymus several months before birth can completely prevent the development of all cellular immunity. Since it is the cellular type of immunity that is mainly responsible for rejection of transplanted organs such as transplanted hearts and kidneys, one can transplant organs with little likelihood of rejection if the thymus is removed from an animal a reasonable period of time before birth.

Role of the Bursa of Fabricius for Preprocessing "B" Lymphocytes in Birds.

It is also during the latter part of fetal life that the bursa of Fabricius preprocesses the "B" lymphocytes and prepares them to manufacture antibodies. Here again, this process continues for a while after birth. In mammals, recent experiments indicate that it is probably lymphoid tissue mainly in the fetal liver that performs this same function.

Spread of Processed Lymphocytes to the Lymphoid Tissue.

After formation of processed lymphocytes in both the thymus and the bursa, these first circulate freely in the blood and gradually filter into the tissues. Then they enter the lymph and are carried to the lymphoid tissue. The lymphoid tissue is constructed of a fine meshwork of interwoven collagen fibers. This filters the lymphocytes from the lymph, thereby entrapping them in the lymphoid tissue. Thus, the lymphocytes do not originate primordially in the lymphoid tissue, but instead are transported to this tissue by way of the preprocessing areas of the thymus and probably the fetal liver. But, once in the nodes, they continue to form new lymphocytes throughout life.

Mechanisms for Determining Specificity of Sensitized Lymphocytes and Antibodies— Lymphocyte Clones

Earlier in this chapter it was pointed out that the lymphocytes of the lymphoid tissue can form sensitized lymphocytes and antibodies that can react with high specificity against particular types of invading agents. This effect is believed to occur in the following way:

It is thought that literally hundreds or thousands of different types of precursor lymphocytes preexist in the lymph nodes. Each one of these is theoretically capable of forming a specific type of sensitized lymphocyte or a specific antibody. However, this sensitized lymphocyte or antibody will not be formed in significant quantity until the lymph node is exposed to the appropriate antigen. When it is exposed, the corresponding precursor lymphocyte for that particular antigen begins to proliferate madly, forming tremendous numbers of progeny; and these in turn lead to the formation of large quantities of antibodies if the precursor lymphocyte is a "B" lymphocyte, or to the formation of numerous sensitized lymphocytes if the precursor lymphocyte is a "T" lymphocyte.

The large mass of new lymphocytes, all of the same type, that are formed in response to a single specific antigen is called a *clone* of lymphocytes. Thus, the lymphoid tissue is capable of forming literally hundreds or thousands of different types of *clones*, each of which is capable of forming a specific antibody or a specific type of sensitized lymphocyte.

Specific Attributes of Humoral Immunity—the Antibodies

Formation of Antibodies by Plasma Cells.

Prior to exposure to a specific antigen, the "B" precursor lymphocytes remain dormant in the lymphoid tissue. However, upon entry of a foreign antigen, the lymphocytes specific for that antigen immediately enlarge and become a new type of cell called a *plasmablast*. This further differentiates to form in about four days a total population of about 500 *plasma cells* for each original plasmablast. The mature plasma cell then produces gamma globulin antibodies at an extremely rapid rate—about 2000 molecules per second for each cell. The antibodies are secreted into the lymph and are carried to the circulating blood. This process continues for several more days until death of the plasma cells.

THE NATURE OF THE ANTIBODIES

The antibodies are all proteins of the gamma globulin type and are called *immunoglobulins*. They have molecular weights between approximately 150,000 and 900,000.

All of the immunoglobulins are composed of combinations of *light* and *heavy polypeptide*

FIGURE 34–2 Structure of the typical antibody, showing it to be composed of two heavy polypeptide chains and two light polypeptide chains. The antigen binds at two different sites on the variable portions of the chains.

chains, and most are a combination of two light and two heavy chains, as illustrated in Figure 34–2.

This figure also shows a designated end of each of the light and each of the heavy chains, called the "variable portion." The remainder of each chain is called the "constant portion."

Specificity of Antibodies. Each antibody that is specific for a particular antigen has a different organization of amino acid residues in the variable portions of both the light and heavy chains. These have a specific steric shape for each different antigen so that when that specific antigen comes in contact with it, the prosthetic radicals of the antigen fit as a mirror image with those of the antibody, thus allowing a rapid and tight chemical bond between the antibody and the antigen.

Note, especially, in Figure 34–2 that there are two identical sides of the antibody; this allows the antibody to attach to two separate antigen molecules. Thus, most antibodies are *bivalent*. A small proportion of the antibodies are constructed of more than two light and heavy chains and have more than two reactive sites. However, there are never fewer than two reactive sites.

MECHANISMS OF ACTION OF ANTIBODIES

Antibodies can act in three different ways to protect the body against invading agents: (1) by direct attack on the invader, (2) by activation of the complement system that then destroys the invader, or (3) by activation of the anaphylactic system that changes the local environment around the invading antigen and in this way reduces its virulence.

Direct Action of Antibodies on Invading Agents. Because of the bivalent nature of the antibodies and the multiple antigen sites on most invading agents, the antibodies can inactivate the invading agent in one or more of several ways, as follows:

1. *Agglutination*, in which multiple antigenic agents are bound together into a clump by the antibodies.
2. *Precipitation*, in which the complex of antigen and antibody becomes insoluble and precipitates.
3. *Neutralization*, in which the antibodies cover the toxic sites of the antigenic agent.
4. *Lysis*, in which some very potent antibodies are capable of directly attacking membranes of cellular agents and thereby causing rupture of the cells.

However, the direct actions of antibodies attacking the antigenic invaders probably, under normal conditions, are not strong enough to play a major role in protecting the body against the invader. Most of the protection probably comes through the *amplifying* effects of the complement and anaphylactic effector systems described below.

The Complement System for Antibody Action. Complement is a system of nine different enzyme precursors (designated C-1 through C-9) plus several other associated substances that are found normally in the plasma and other body fluids, but the enzymes are normally inactive. However, when an antibody combines with an antigen, a reactive site on the "constant" portion of the antibody becomes uncovered, or activated, and this in turn sets into motion a "cascade" of sequential reactions in the complement system, illustrated in Figure 34–3. The activated enzymes then attack the agent to which the antibody is attached in several different ways, as well as initiate local tissue reactions that also provide protection against damage by the invader. Among the more important effects that occur are the following:

1. *Lysis.* The proteolytic enzymes of the complement system digest portions of the cell membrane, thus causing rupture of cellular agents such as bacteria or other types of invading cells.
2. *Opsonization and phagocytosis.* The complement enzymes attack the surfaces of bacteria and other antigens to which the antibody is attached, making these highly susceptible to phagocytosis by neutrophils and tissue macrophages. This process is called *opsonization*. It often enhances the number of bacteria that can be destroyed many hundredfold.
3. *Chemotaxis.* One or more of the complement

FIGURE 34–3 Cascade of reactions during activation of the classic pathway of complement. (From Alexander and Good: Fundamentals of Clinical Immunology. Philadelphia, W.B. Saunders, 1977.)

products cause chemotaxis of neutrophils and macrophages, thus greatly enhancing the number of these phagocytes in the local region of the antigenic agent.

4. *Agglutination.* The complement enzymes also change the surfaces of some of the antigenic agents so that they adhere to each other, thus causing agglutination.

5. *Neutralization of viruses.* The complement enzymes frequently attack the molecular structures of viruses and thereby render them nonvirulent.

6. *Inflammatory effects.* The complement products elicit a local inflammatory reaction, leading to hyperemia, coagulation of proteins in the tissues, and other aspects of the inflammation process, thus preventing movement of the invading agent through the tissues.

Activation of the Anaphylactic System by Antibodies. Some of the antibodies attach to the membranes of cells in the tissues and blood. Among the most important cells are the *mast cells* in tissues surrounding the blood vessels and the *basophils* circulating in the blood. When an antigen then reacts with one of the antibody molecules attached to the cell, there is immediate swelling and rupture of the cell, with the release of a large number of factors that affect the local environment. Such factors include especially *histamine* but also other substances that cause local inflammatory reactions. These effects in turn are believed to help immobilize the antigenic invader.

Special Attributes of the "T" Lymphocyte System—Cellular Immunity and Sensitized Lymphocytes

Release of Sensitized Lymphocytes from Lymphoid Tissue. Upon exposure to the proper antigens, sensitized lymphocytes are released from lymphoid tissue in ways that parallel antibody release. The only real difference is that instead of releasing antibodies, whole sensitized lymphocytes are formed and released into the lymph. These then pass into the circulation, where they remain a few minutes to a few hours at most, filtering out of the circulation into all the tissues of the body.

Mechanism of Sensitization of "T" Lymphocytes. It is believed that "T" lymphocytes become sensitized against specific antigens by forming on their surfaces a type of "antibody." This antibody is composed of a *variable unit* similar to the variable portion of the humoral antibody, but it has no constant portion. Instead, multiple variable units are attached directly to the cell membrane of the "T" lymphocyte.

Persistence of Cellular Immunity. An important difference between cellular immunity and humoral immunity is its persistence. Humoral antibodies rarely persist more than a few months, or at most, a few years. On the other hand, sensitized lymphocytes probably have an indefinite life span and seem to persist until they eventually come in contact with their specific antigen. There is reason to believe that such sensitized lymphocytes might persist as long as ten years in some instances.

Types of Organisms Resisted by Sensitized Lymphocytes. Although the humoral antibody mechanism for immunity is especially efficacious against the more acute bacterial diseases, the cellular immunity system is activated much more potently by the more slowly developing bacterial diseases such as tuberculosis, brucellosis, and so forth. Also, this system is active against cancer cells, cells of transplanted organs, and fungus organisms, all of which are far larger than bacteria.

MULTIPLE TYPES OF T-CELLS AND THEIR FUNCTIONS

Though it was first believed that there was only one type of sensitized T-cell, several different

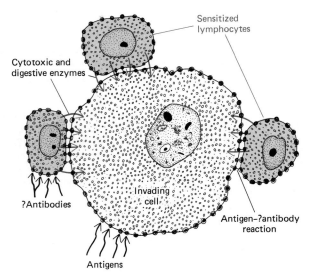

FIGURE 34–4 Destruction of an invading cell by sensitized lymphocytes.

types are now known. Among these are

1. Cytotoxic T-cells.
2. Delayed hypersensitivity T-cells.
3. Helper T-cells.
4. Suppressor T-cells.

Direct Destruction of the Invader by Cytotoxic T-Cells. Figure 34–4 illustrates sensitized cytotoxic T-cells that are bound with antigens in the membrane of an invading cell such as a cancer cell, a heart transplant cell, or a parasitic cell of another type. The immediate effect of this attachment is swelling of the sensitized lymphocyte and release of cytotoxic substances from the lymphocyte to attack the invading cell. The cytotoxic substances are probably mainly lysosomal enzymes manufactured in the T-cell.

Aside from the direct cellular destructive effects of the substances released by the cytotoxic T-cells, some of these substances also promote chemotaxis of tremendous numbers of macrophages into the area, and these in turn cause even more tissue destruction.

When a person's cytotoxic T-cells have been destroyed with special drugs, transplantation of organs is much more successful than when they are present. But, also, a person is several times more likely to develop cancer when these cells are missing, because many beginning cancers initiate cellular immunity and the cytotoxic T-cells that are formed then destroy the cancer.

Cytotoxic Effects of Delayed Hypersensitivity T-Cells. Some types of allergy are caused by the so-called hypersensitivity T-cells. These have cytotoxic effects similar to but less

severe than those of the cytotoxic cells. They are formed in response to certain allergic substances, as will be discussed later in the chapter, and cause such allergies as drug allergies and the allergic reaction to poison ivy.

Stimulation of Humoral Antibodies by Helper T-Cells. The helper T-cells play a specific role in increasing the formation of humoral antibodies. Some types of antigens can activate the T-cell system but not the B-cell system. After they do activate the T-cell system and helper T-cells are formed, however, this type of T-cell then reacts with the "B" lymphocytes of the lymph nodes in some yet unexplained way to cause them to form antibodies against the specific antigen that had activated the T-cell system. Thus, these cells increase the scope of the humoral antibody system in protecting the body against foreign agents.

Role of the Suppressor T-Cells. The specific role of the suppressor T-cells is not well understood, but they do suppress some aspects of the T-cell immune response. They are believed to play an important regulatory role in preventing runaway immune reactions that might cause serious harm to the body.

Thus, the T-cell system functions in a number of different ways to provide immunity and to control the immune response. Its special attributes are: (1) sensitized T-cells can last for many years, perhaps as long as 10 to 20 years in some instances, and (2) the T-cell system reacts against certain types of antigens that do not activate the B-cell system.

Tolerance of the Acquired Immunity System to One's Own Tissues— Role of the Thymus and the Bursa

Obviously, if a person should become immune to his own tissues, the process of acquired immunity would destroy his own body. Fortunately, the immune mechanism normally "recognizes" the person's own tissues as being completely distinctive from those of invaders, and his immunity system forms neither antibodies nor sensitized lymphocytes against his own antigens. This phenomenon is known as *tolerance* to the body's own tissues.

Mechanism of Tolerance. It is possible that tolerance to one's own tissues is determined genetically—that is, via absence of genes to form sensitized lymphocytes and antibodies against the person's own tissues. However, there is much more reason to believe that tolerance develops during the processing of the lymphocytes in the thymus and in the "B" lymphocyte processing area. The reason for this belief is that injecting a strong antigen into a fetus at the time that the lymphocytes are being processed in these two areas will prevent the development of precursor lymphocytes in the lymphoid tissue that are specific for the injected antigen.

Therefore, it believed that during the processing of lymphocytes in the thymus and in the "B" lymphocyte processing area, all those precursor lymphocytes that are specific for the body's own tissues are self-destroyed because of their exposure to the body's own antigens.

Failure of the Tolerance Mechanism— Autoimmune Diseases.
Unfortunately, persons frequently lose some of their immune tolerance to their own tissues. This usually results from destruction of some of the body's tissues, which releases antigens that then circulate in the body fluids in significant quantity. For instance, the proteins of the cornea do not seem to circulate in the fluids of the fetus; this is also true of the thyroglobulin molecule of the thyroid; therefore, tolerance to these never develops. But when damage occurs to either of these two tissues in later life, these protein molecules can then elicit immunity, and the immunity in turn can attack the cornea or the thyroid gland to cause corneal opacity or destructive thyroiditis.

Other diseases that result from autoimmunity include: *rheumatic fever*, in which the body becomes immunized against tissues in the heart and joints following exposure to a specific type of streptococcal toxin; *acute glomerulonephritis*, in which the person becomes immunized against the glomeruli, resulting from exposure to another specific type of streptococcal toxin; *myasthenia gravis*, in which immunity develops against muscles and thereby causes paralysis; and *lupus erythematosus*, in which the person becomes immunized against many different body tissues at the same time, a disease that causes extensive damage and often rapid death.

Vaccination

The process of vaccination has been used for many years to cause acquired immunity against specific diseases. A person can be vaccinated by injecting dead organisms that are no longer capable of causing disease though still have their chemical antigens. This type of vaccination is used to protect against typhoid fever, whooping cough, diphtheria, and many other types of bacterial diseases. Also, immunity can be

achieved against toxins that have been treated with chemicals so that their toxic nature has been destroyed even though their antigens for causing immunity are still intact. This procedure is used in vaccinating against tetanus, botulism, and other similar toxic diseases. And, finally, a person can be vaccinated by infecting him with live organisms that have been "attenuated." That is, these organisms either have been grown in special culture mediums or have been passed through a series of animals until they have mutated enough that they will not cause disease but will still carry the specific antigens. This procedure is used to protect against poliomyelitis, yellow fever, measles, smallpox, and many other viral diseases.

Passive Immunity

Thus far, all the acquired immunity that we have discussed has been *active immunity*. That is, the person's body develops either antibodies or sensitized lymphocytes in response to invasion of the body by a foreign antigen. However, it is possible also to achieve temporary immunity in a person without injecting any antigen whatsoever. This is done by injecting antibodies, sensitized lymphocytes, or both from someone else or from some other animal that has been actively immunized against the antigen. The immune protection that is thus achieved is called *passive immunity*. The antibodies will last for 2 to 3 weeks, and during that time the person is protected against the invading disease. Sensitized lymphocytes will last for a few weeks if transfused from another person, and for a few hours to a few days if transfused from an animal.

ALLERGY

One of the important side effects of immunity is the development, under some conditions, of allergy. There are at least three different types of allergy, two of which can occur in any person, and a third that occurs only in persons who have a specific allergic tendency.

Allergies that Occur in Normal People

Delayed-Reaction Allergy.
This type of allergy frequently causes skin eruptions in response to certain drugs or chemicals, particularly some cosmetics and household chemicals, to which one's skin is often exposed. Another

example of such an allergy is the skin eruption caused by exposure to poison ivy.

Delayed-reaction allergy is caused by sensitized lymphocytes of the hypersensitivity T-cell type (discussed earlier) and not by antibodies. In the case of poison ivy, the toxin of poison ivy in itself does not cause much harm to the tissues. However, upon repeated exposure it does cause the formation of the sensitized T-cells. Then the T-cells diffuse into the skin in sufficient numbers to combine with the poison ivy toxin and elicit a cellular immunity type of reaction. Remembering that cellular immunity can cause release of many toxic substances from the T-cells, some of which cause extensive invasion of the tissues by macrophages and their subsequent effects, one can well understand that the eventual result of some delayed-reaction allergies can be extensive inflammation and sometimes serious tissue damage.

Allergies Caused by Reaction Between Antibodies and Antigens. When a person becomes strongly immunized against an antigen and has developed a very high titer of antibodies, subsequent sudden exposure of that person to a high concentration of the same antigen can cause a serious tissue reaction. The antigen-antibody complex that is formed precipitates, and some of it deposits as granules in the walls of the small blood vessels. These granules also activate the complement system, setting off extensive release of proteolytic enzymes. The result of these two effects is severe inflammation and destruction of the small blood vessels.

One manifestation of this type of reaction is *serum sickness*. Serum injected into a person can cause subsequent formation of antibodies. When these begin to appear, they react with the protein of the injected serum and elicit a widespread antigen-antibody reaction throughout the body. Fortunately, this reaction occurs slowly over a period of days as the antibodies are formed, and usually it is not lethal. However, it can be lethal on occasion, and on other occasions it can cause widespread inflammation and edema throughout the body with development of a circulatory shocklike syndrome.

Allergies in the "Allergic" Person

Some persons have an "allergic" tendency. This phenomenon is genetically passed on from parent to child. It is characterized by the presence of large quantities of *IgE antibodies* in the circulating blood; these are very large protein molecules having many reactive sites because of multiple light and heavy chains in the antibody molecules. The IgE antibodies are called *reagins* or *sensitizing antibodies* to distinguish them from the more common IgG antibodies. When an *allergen* (defined as an antigen that reacts specifically with a specific type of IgE reagin antibody) enters the body, an allergen-reagin reaction takes place, initiating a subsequent allergic response.

The IgE antibodies (the reagins) have a special propensity to attach to cells throughout the body, especially to mast cells and basophils; therefore, the allergen-reagin reaction damages the cells. The result is *anaphylactoid types of immune reactions*. These result primarily from the rupture of the mast cells and basophils with consequent release of *histamine, slow-reacting substance of anaphylaxis, eosinophil chemotactic substance, lysosomal enzymes*, and other less important substances.

Among the different types of allergic reaction of this type are:

Anaphylaxis. When a specific allergen is injected directly into the circulation it can react in widespread areas of the body with the basophils of the blood and the mast cells located immediately outside the small blood vessels. Therefore, the anaphylactic type of reaction occurs everywhere. The histamine released into the circulation causes widespread peripheral vasodilatation as well as increased permeability of the capillaries and marked loss of plasma from the circulation. Often, persons experiencing this reaction die of circulatory shock within a few minutes unless treated with norepinephrine to oppose the effects of the histamine.

Urticaria. Urticaria results from allergens entering specific skin areas and causing localized anaphylactoid reactions. *Histamine* released locally causes (1) vasodilatation, which induces an immediate *red flare* and (2) increased permeability of the capillaries, which leads to swelling of the skin in another few minutes. The swellings are commonly called "hives." Administration of antihistamine drugs to a person prior to exposure will prevent the hives.

Hay Fever. In hay fever, the allergen-reagin reaction occurs in the nose. *Histamine* released in response to this causes local vascular dilatation with resultant increased capillary pressure, and it also causes increased capillary permeability. Both of these effects cause rapid fluid leakage into the tissues of the nose, and the nasal linings become swollen and secretory. Here again, use of antihistaminic drugs can prevent this swelling reaction.

Asthma. In asthma, the allergen-reagin reaction occurs in the bronchioles of the lungs.

Here, the most injurious product released from the mast cells seems to be *slow-reacting substance of anaphylaxis,* which causes spasm of the bronchiolar smooth muscle. Consequently, the person has difficulty breathing. Unfortunately, administration of antihistaminics has little effect on the course of asthma, because histamine does not appear to be the major factor eliciting the asthmatic reaction.

QUESTIONS

1. What is meant by immunity?
2. How do antigens activate the immunity mechanism?
3. What are the two different types of lymphocytes that play important roles in the immune process?
4. What is the function of lymphocyte clones in determining the specificity of sensitized lymphocytes and antibodies?
5. What are antibodies, and how are they formed?
6. How do antibodies react with antigens?
7. What are some of the functions of the complement system?
8. What are the different types of T-cells?
9. Describe the mechanism by which cytotoxic T-cells destroy invading organisms.
10. What is meant by immune tolerance?
11. How is vaccination used to protect a person against disease?
12. Distinguish between the types of allergies that can occur in normal people and those that occur in the "allergic" person.

REFERENCES

Amos, D.B., *et al.* (eds.): Immune Mechanisms and Disease. New York, Academic Press, 1979.

Bellanti, J.A.: Immunology II. Philadelphia, W.B. Saunders, 1978.

Benacerraf, B., and Unanue, E.R.: Textbook of Immunology. Baltimore, Williams & Wilkins, 1979.

Cohen, A.S. (ed.): Rheumatology and Immunology. New York, Grune & Stratton, 1979.

Fudenburg, H.H., and Smith, C.L. (eds.): The Lymphocyte in Health and Disease. New York, Grune & Stratton, 1979.

Ham, A.W., *et al.*: Blood Cell Formation and the Cellular Basis of Immune Responses. Philadelphia, J.B. Lippincott, 1979.

Johnson, F. (ed.): Allergy, Including IgE in Diagnosis and Treatment. Chicago, Year Book Medical Publishers, 1979.

Kaplan, J.G. (ed.): The Molecular Basis of Immune Cell Function. New York, Elsevier/North-Holland, 1979.

Nahmias, A.J., and O'Reilly, R. (eds.): Immunology of Human Infection. New York, Plenum Press, 1979.

Pernis, B., and Vogel, H.J. (eds.): Cells of Immunoglobulin Synthesis. New York, Academic Press, 1979.

Schiff, G.M.: Active immunization for adults. *Annu. Rev. Med., 31:*441, 1980.

Schwartz, L.M.: Compendium of Immunology. New York, Van Nostrand Reinhold, 1979.

Sercarz, E.E., and Cunningham, A.J. (eds.): Strategies of Immune Regulation. New York, Academic Press, 1979.

Stuart, F.P., and Fitch, F.W. (eds.): Immunological Tolerance and Enhancement. Baltimore, University Park Press, 1979.

35

Blood Coagulation, Transfusion, and Transplantation of Organs

---

Overview

Blood clotting does not occur in the normal circulation. Yet, when a blood vessel breaks and bleeding occurs, a blood clot normally forms within a few minutes at the site of the break and usually stops the bleeding. Blood clotting results from a series of chemical reactions involving mainly a series of specific plasma protein enzymes called *clotting factors*. The different stages in the clotting mechanism are the following:

1. The first stage is the *formation of prothrombin activator*. This results from one of two separate chemical processes called, respectively, (a) the extrinsic mechanism and (b) the intrinsic mechanism. The *extrinsic mechanism* begins with rupture of the blood vessel and exposure of the blood to the torn tissues surrounding the vessel. Two factors from the torn tissues, *tissue factor* and *tissue phospholipids*, then initiate a reaction in the blood plasma causing factors *V, VII,* and *X* all to react together under the catalyzing influence of calcium ions to form prothrombin activator. The *intrinsic mechanism* is initiated by trauma to the blood itself, leading to a series of reactions involving *factors V, VIII, IX, X, XI,* and *XII,* as well as calcium ions, and again the final product is prothrombin activator.
2. *Once prothrombin activator is formed, it changes prothrombin into thrombin.*
3. The *thrombin then acts as an enzyme to convert fibrinogen into fibrin threads* that enmesh red blood cells and plasma to form the clot itself.

A small amount of thrombin is formed all of the time in the circulating blood. However, still another substance, *heparin,* has a potent effect to prevent the clotting effects of this thrombin. Therefore, ordinarily no clotting occurs until greater than threshold values of thrombin are suddenly formed because of broken vessels or severely traumatized blood.

Transfusion of blood from one person to another is fraught with two types of dangers: first, *clotting of the blood* during the transfusion process and, second, *immune reactions* against the transfused blood *caused by antibodies* in the recipient's blood. The blood clotting problem is solved by adding *citrate ions* to the transfusion blood as it is withdrawn from the donor. This reacts with the calcium ions in the donor blood to bind them in a nonionic form, and without calcium ions clotting will not occur for many weeks if the blood is kept refrigerated.

Most *transfusion reactions* involving the immune mechanism are caused by two separate systems of *antigens*, called *agglutinogens*, that are attached to the red blood cell membranes: (1) the A-B-O system and (2) the Rh system. In the *A-B-O system* there are four different red blood cell types: *type O, type A, type B, and type AB.* If the donor and recipient bloods are mismatched, *agglutination* of the red blood cells is likely to occur because of antibodies, called *agglutinins,* in the recipient blood, and this is followed by *hemolysis* of the agglutinated cells a few hours later. In the *Rh system,* persons who have Rh factor on their red blood cells are said to be *Rh positive,* and the person without Rh factor is *Rh negative.* When Rh positive blood is transfused into an Rh negative person, a transfusion reaction similar to that caused by the A-B-O system frequently occurs, especially if the recipient has been exposed previously to Rh positive blood.

When a baby with Rh positive blood develops in the uterus of a mother with Rh negative blood, *anti-Rh antibodies* formed by the mother frequently are transferred through the placenta into the baby and cause severe destruction of the baby's red blood cells, which is the condition called *erythroblastosis fetalis.*

Transplantation of organs and tissues obeys essentially the same principles as transfusion of blood. However, an additional antigen system called the *HLA antigens* is especially important in causing immune reactions against the transplanted organ, usually leading to death of the organ within a week or so unless specific drugs are used to suppress the immune system.

HEMOSTASIS

The term hemostasis means prevention of blood loss. Whenever a vessel is severed or ruptured, hemostasis is achieved by a succession of different mechanisms including (1) vascular spasm, (2) formation of a platelet plug, (3) blood coagulation, and (4) growth of fibrous tissues into the blood clot to close the hole in the vessel permanently.

VASCULAR SPASM

Immediately after a blood vessel is cut or ruptured, the trauma to the vessel wall itself causes the vessel to contract; this instantaneously reduces the flow of blood from the vessel rupture. The more the vessel is traumatized, the greater is the degree of spasm; this means that a sharply cut blood vessel usually bleeds much more than does a vessel ruptured by crushing. This local vascular spasm lasts for as long as 20 minutes to an hour, during which time ensuing processes of platelet plugging and blood coagulation can take place.

FORMATION OF A PLATELET PLUG

The second event in hemostasis is an attempt by the platelets to plug the rent in the vessel. To understand this it is important that we first understand the nature of platelets themselves.

Platelets are minute round or oval discs about 2 microns (micrometers, or μ) in diameter. They are fragments of *megakaryocytes,* which are extremely large white blood cells formed in the bone marrow. The megakaryocytes disintegrate into platelets while they are still in the bone marrow and release the platelets into the blood. The normal concentration of platelets in the blood is between 200,000 and 400,000 per cubic millimeter (mm^3).

Mechanism of the Platelet Plug. Platelet repair of vascular openings is based on several important functions of the platelet itself: When platelets come in contact with a *wettable* surface, such as the collagen fibers in the vascular wall, they immediately swell and become sticky so that they stick to the collagen fibers and to each other. They also release large quantities of ADP, and the ADP, in turn, acts on nearby platelets to activate them as well, and the stickiness of these additional platelets causes them also to adhere to the originally activated platelets. Thus, very large numbers of platelets accumulate to form a *platelet plug*. If the rent in a vessel is small, the platelet plug by itself can stop blood loss completely, but if there is a large hole, a blood clot in addition to the platelet plug is required to stop the bleeding.

The platelet plugging mechanism is extremely important to close the minute ruptures in very small blood vessels that occur hundreds of times daily. Therefore, a person who has very few platelets develops literally hundreds of small hemorrhagic areas under his skin and throughout his internal tissues, but this does not occur in the normal person.

CLOTTING IN THE RUPTURED VESSEL

The third mechanism for hemostasis is formation of the blood clot. The clot begins to develop in 15 to 20 seconds if the trauma of the vascular wall has been severe, and in 1 or 2 minutes if the trauma has been minor. Activator substances both from the traumatized vascular wall and from platelets and blood proteins adhering to the collagen of the traumatized vascular wall initiate the clotting process. The physical events of this process are illustrated in Figure 35–1, and the chemical events will be discussed in detail later.

Within 3 to 6 minutes after rupture of a vessel, the entire cut or broken end of the vessel is

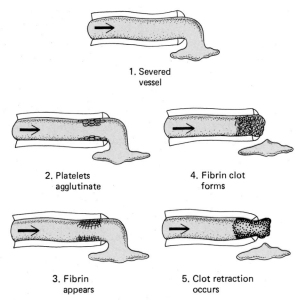

1. Severed
vessel

2. Platelets
agglutinate

4. Fibrin clot
forms

3. Fibrin
appears

5. Clot retraction
occurs

FIGURE 35–1 Stages of clot development and clot retraction following injury to a blood vessel. (Redrawn from Seegers and Sharp: Hemostatic Agents. Springfield, Ill., Charles C Thomas.)

filled with clot. After 20 minutes to an hour, the clot retracts; this closes the vessel still further.

FIBROUS ORGANIZATION OF THE BLOOD CLOT

Once a blood clot has formed, it usually becomes invaded by fibroblasts, which subsequently form fibrous tissue all through the clot. This begins within a few hours after the clot is formed, and fibrous closure of the hole in the vessel is complete within approximately 7 to 10 days.

Mechanism of Blood Coagulation

Almost all research workers in the field of blood coagulation agree that clotting takes place in three essential steps:

First, a substance called *prothrombin activator* is formed in response to rupture of the vessel or damage to the blood itself.

Second, the prothrombin activator catalyzes the conversion of prothrombin into *thrombin*.

Third, the thrombin acts as an enzyme to convert fibrinogen into *fibrin threads* that enmesh red blood cells and plasma to form the clot itself.

Let us first discuss the latter two of these steps—that is, the steps for forming the clot itself, beginning with the conversion of prothrombin to thrombin; then we will come back to the initiating stages in the clotting process that lead to the formation of prothrombin activator.

CONVERSION OF PROTHROMBIN TO THROMBIN

After prothrombin activator has been formed as a result of rupture of the blood vessel or as a result of damage to the blood itself, this activator causes conversion of prothrombin to thrombin, which in turn causes polymerization of fibrinogen molecules into fibrin threads within the next 10 to 15 seconds.

Prothrombin is a plasma protein having a molecular weight of 68,700. It is an unstable protein that can split easily into smaller compounds, one of which is *thrombin*, having a molecular weight of 33,700—almost exactly half that of prothrombin.

Prothrombin is formed continually by the liver, and if the liver fails to produce prothrombin, within 24 hours its concentration in the plasma falls too low to provide normal blood

FIGURE 35–2 Schema for conversion of prothrombin to thrombin, and polymerization of fibrinogen to form fibrin threads.

coagulation. Vitamin K is required by the liver for this formation of prothrombin; therefore, either lack of vitamin K or the presence of liver disease that prevents normal prothrombin formation can often decrease the prothrombin level so low that a bleeding tendency results.

Effect of Prothrombin Activator to Form Thrombin from Prothrombin. Figure 35–2 illustrates the conversion of prothrombin to thrombin under the influence of prothrombin activator and calcium ions. The rate of formation of thrombin from prothrombin is almost directly proportional to the quantity of prothrombin activator available, which in turn is approximately proportional to the degree of trauma to the vessel wall or blood. And the rapidity of the clotting process is proportional to the quantity of thrombin formed.

CONVERSION OF FIBRINOGEN TO FIBRIN—FORMATION OF THE CLOT

Fibrinogen. Fibrinogen is a high molecular weight (340,000) protein, most if not all of which is formed in the liver. Liver disease occasionally decreases the concentration of circulating fibrinogen, as it does the concentration of prothrombin, which was pointed out previously.

Action of Thrombin on Fibrinogen to Form Fibrin. Thrombin is a protein *enzyme* with proteolytic capabilities. It acts on fibrinogen to split away two small peptides. This forms an activated molecule called *fibrin monomer*, which has the automatic capability of polymerizing with other fibrin monomer molecules. Therefore, many fibrin monomer molecules combine within seconds into *long fibrin threads* that form the *reticulum* of the clot. Then another plasma globulin factor, *fibrin stabilizing factor*, acts as an enzyme to cause stronger bonding between the fibrin monomer molecules and also bonding between adjacent polymer chains. This adds tremendously to the strength of the threads.

The Blood Clot. The clot is composed of the meshwork of fibrin threads running in all directions with entrapped blood cells, platelets, and plasma. The fibrin threads adhere to damaged surfaces of blood vessels; therefore, the blood clot prevents blood loss.

Initiation of Coagulation: Formation of Prothrombin Activator

Now that we have discussed the clotting process itself, we can return to the more complex mechanisms that initiate the clotting process. Clotting can be initiated by (1) trauma to the tissues, (2) trauma to the blood, or (3) contact of the blood with special substances such as collagen outside the blood vessel endothelium. Each instance leads to the formation of *prothrombin activator*, which then causes the actual clotting.

There are two basic ways in which prothrombin activator is formed: (1) by the *extrinsic pathway* that begins with trauma to the blood vessel and the tissue outside the blood vessel or (2) by the *intrinsic pathway* that begins with trauma to the blood itself.

THE EXTRINSIC MECHANISM FOR INITIATING CLOTTING

The extrinsic mechanism for initiating the formation of prothrombin activator begins with blood coming in contact with torn vascular walls or other traumatized tissues and occurs according to the basic steps illustrated in Figure 35–3.

1. *Release of Tissue Factor and Tissue Phospholipids.* The traumatized tissue releases two factors that set the clotting process into motion. These are (a) *tissue factor*, which is a proteolytic enzyme or a mixture of enzymes, and (b) *tissue phospholipids*, which are mainly phospholipids from the tissue cell membranes.

2. *Conversion of Protein Clotting Factors in Plasma to Form Prothrombin Activator.* The tissue factor and tissue phospholipids released in the first step now react with several protein *clotting factors* that are present in the plasma, called *factor V*, *factor VII*, and *factor X*. The product of this reaction is *prothrombin activator*.

As has already been discussed, the prothrombin activator then causes conversion of prothrombin to thrombin, and a complete blood

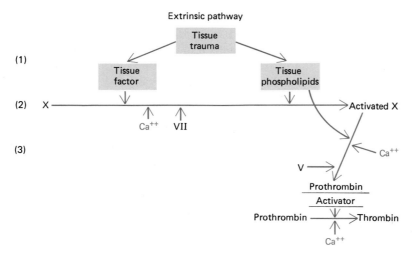

FIGURE 35–3 The extrinsic pathway for initiating blood clotting.

clot can develop within the next 10 to 15 seconds. Thus, any time a blood vessel is ruptured, the trauma to the wall of the vessel itself can initiate the clotting process and thereby stop the bleeding.

THE INTRINSIC MECHANISM FOR INITIATING CLOTTING

The second mechanism for initiating the formation of prothrombin activator, and therefore for initiating clotting, begins with trauma to the blood itself and continues through the series of cascading reactions illustrated in Figure 35–4.

Here again, a number of protein clotting factors in the plasma are involved: *factor V, factor VIII, factor IX, factor X, factor XI,* and *factor XII.* Damage of almost any type to the blood or contact of the blood with collagen activates one or more of these factors, and this in turn activates the next, then the next, until eventually the final product is again *prothrombin activator.* And, as is true for the extrinsic pathway of blood clotting, the prothrombin activator then causes

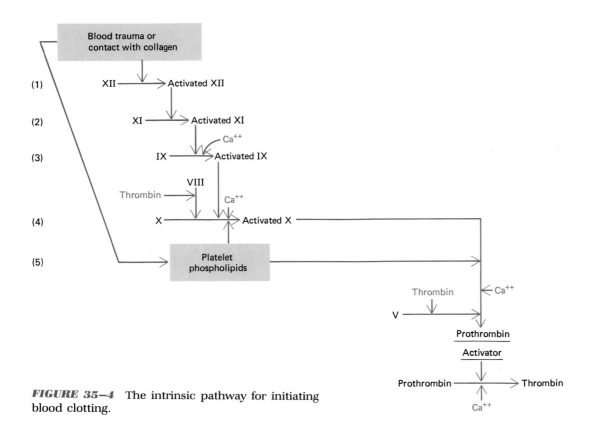

FIGURE 35–4 The intrinsic pathway for initiating blood clotting.

the actual clotting process. Note especially in Figure 35–4 the important role that platelets play in the intrinsic mechanism to supply phospholipids for the reaction, a function that is not required of the platelets for the extrinsic mechanism.

ROLE OF CALCIUM IONS IN THE INTRINSIC AND EXTRINSIC PATHWAYS

Except for the first two steps in the intrinsic pathway, calcium ions are required for promoting all of the reactions in both clotting pathways. Therefore, in the absence of calcium ions, blood clotting will not occur. Fortunately, in the living body the calcium ion concentration rarely falls low enough to affect significantly the kinetics of blood clotting.

SUMMARY OF BLOOD CLOTTING INITIATION

From the above schemas of the extrinsic and intrinsic systems for initiating blood clotting, one can see that clotting is initiated after rupture of blood vessels by both of the pathways. The tissue factor and tissue phospholipids initiate the extrinsic pathway, whereas contact of factor XII and the platelets with collagen in the vascular wall initiates the intrinsic pathway. In contrast, when blood is removed from the body and maintained in a test tube, it is the intrinsic pathway alone that must elicit the clotting.

An especially important difference between the extrinsic and intrinsic pathways is that the extrinsic pathway is explosive in nature; once initiated, its speed of occurrence is limited only by the amount of tissue factor and tissue phospholipids released from the traumatized tissues, and by the quantities of factors X, VII, and V in the blood. With severe tissue trauma, clotting can occur in as little as 15 seconds. On the other hand, the intrinsic pathway is much slower to proceed, usually requiring 1 to 3 minutes to cause clotting.

Prevention of Blood Clotting in the Normal Vascular System

Obviously, it is important that blood not clot in the normal circulation. Fortunately, several important anticlotting mechanisms prevent this from occurring. The three most important are the following:

Endothelial Surface Factors. Probably the most important factor for preventing clotting in the normal vascular system is the nature of the endothelial lining of the blood vessels themselves. The endothelial cells are extremely smooth, which prevents contact activation of the clotting system. But perhaps even more important, a thin molecular layer of negatively charged protein is adsorbed to the inner surface of the endothelium, and this repels the negatively charged protein clotting factors as well as the platelets. This, too, prevents the onset of clotting. However, when the endothelial wall becomes damaged, both of these effects are lost, and clotting usually begins immediately.

Antithrombin-Heparin Cofactor. One of the most important anticoagulants in the blood is a protein called *antithrombin-heparin cofactor.* When only a small amount of thrombin is formed, even before this can cause a clot to develop, it combines with antithrombin-heparin cofactor, which first blocks the effect of the thrombin on the fibrinogen and subsequently inactivates the bound thrombin during the next 12 to 20 minutes. Therefore, only when the rate of thrombin formation exceeds a critical level will blood clotting occur.

Heparin. Small amounts of heparin, a powerful anticoagulant, are normally present in the blood. Heparin is a conjugated polysaccharide found in the cytoplasm of many types of cells, including even the cytoplasm of unicellular animals. Especially large quantities of heparin are formed by the basophilic *mast cells* located in the pericapillary connective tissue throughout the body, and the heparin then diffuses into the circulatory system. The *basophil cells* of the blood, which seem to be functionally almost identical to the mast cells, also release some heparin into the plasma.

Mast cells are extremely abundant in the tissue surrounding the capillaries of the lungs and to a lesser extent the capillaries of the liver. It is easy to understand why large quantities of heparin might be needed in these areas, for the capillaries of the lungs and liver receive many embolic clots formed in the slowly flowing venous blood; sufficient formation of heparin can prevent further growth of the clots.

Mechanism of Heparin Action. Heparin prevents blood coagulation almost entirely by combining with antithrombin-heparin cofactor, which makes this factor combine with thrombin 1000 times as rapidly as normally. Therefore, in the presence of an excess of heparin, the removal of thrombin from the circulating blood is almost instantaneous. This complex of heparin and antithrombin-heparin cofactor also reacts in a similar way with several other activated coagulation factors of both the intrinsic and extrinsic pathways, thus inactivating their proteolytic (and blood clotting) functions as well.

Conditions that Cause Excessive Bleeding in Human Beings

Excessive bleeding can result from deficiency of any one of the many different clotting factors. However, deficiency of a few of these is especially likely to cause clinically serious bleeding tendencies, as follows:

Hemophilia. Hemophilia results from genetic lack of *factor VIII, factor IX,* or *factor XI* in the blood. (About 85 percent of the patients lack factor VIII.) Many persons with hemophilia die in early life because of severe bleeding. However, if a person with hemophilia is given a transfusion of normal plasma, this will provide the missing clotting factor, and clotting will occur normally for several days thereafter. This is a method for treating the serious episodes of bleeding in these patients. Also, treatment with purified clotting factor (factor VIII, for instance) is usually used prophylactically to prevent the bleeding episodes.

Thrombocytopenia. Thrombocytopenia means lack of adequate numbers of platelets in the circulating blood. This disease is most often caused by antibodies against the platelets that attack and destroy the platelets. Thus, it is one of the "autoimmune diseases." It occasionally also results from poisoning by toxins or drugs.

The patient with thrombocytopenia usually has large numbers of *minute hemorrhages* both in the skin and in the deep tissues because the platelet plugging method for stopping small bleeding points in the vasculature becomes deficient. The hemorrhages in the skin cause *purplish blotches* all over the surface of the body, usually 0.5 to 1 cm in diameter. The bleeding tendency can be stopped for a period of a few days by transfusing fresh whole blood into the patient or by transfusing separated platelets, though both of these are very difficult procedures.

Bleeding in Persons with Liver Disease or Lack of Vitamin K. In liver disease, or when vitamin K is not available to the liver, several of the clotting factors fail to be produced, especially *prothrombin* and *factors VII, IX,* and *X.* Therefore, any serious liver disease is likely to lead also to serious bleeding tendency.

Prevention of Clotting of Blood Removed from the Body

When blood is removed for transfusion, it is necessary to prevent its clotting. This can be done for a few hours by adding a large amount of *heparin,* though this is not practical for most transfusion purposes. Instead, the usual procedure is to remove the calcium ions from the blood. As discussed earlier, calcium ions are necessary to promote most of the steps in the clotting process. The usual method for removing calcium ions is to add either *citrate ions* or *oxalate ions.* The citrate ions combine with the calcium ions to form calcium citrate, a soluble but un-ionized substance, and this obviously prevents clotting thereafter. The oxalate combines with the calcium to form a precipitated calcium oxalate, and this too prevents clotting.

When blood is to be used for transfusion, citrate ion instead of oxalate ion must be used because the recipient's body can metabolize citrate and remove it from the transfused blood, whereas oxalate is very toxic. After the citrate has been metabolized, the blood will again clot normally.

BLOOD COAGULATION TESTS

Bleeding Time. When a sharp knife is used to pierce the tip of the finger or lobe of the ear, bleeding ordinarily lasts 3 to 6 minutes.

Clotting Time. A method widely used for determining clotting time is to collect blood in a chemically clean glass test tube and then to tip the tube back and forth approximately every 30 seconds until the blood has clotted. By this method, the normal clotting time ranges between 5 and 8 minutes.

Prothrombin Time. A test to determine the quantity of prothrombin in the blood is performed by first rendering the blood incoagulable with oxalate, which precipitates the ionic calcium. Then large amounts of calcium, tissue extract, factor V, and factor VII are all mixed suddenly with the sample of blood. The only additional factor needed for the formation of thrombin that is not added to the blood is prothrombin. Therefore, the length of time required for the blood to clot depends on the amount of prothrombin already in the blood. The longer the time required, the less the amount of prothrombin available. The normal prothrombin time is approximately 12 seconds—that is, this is the time required for the clot to appear after mixing has occurred. If the prothrombin time is as long as 20 to 30 seconds, the person can be expected to be a bleeder.

The quantities of factor V, factor VII, and several of the other factors that enter into blood coagulation can be estimated by similar procedures.

TRANSFUSION

Often a person loses so much blood that to save his life he must be given a transfusion of new blood immediately. At other times transfusions are given to treat anemia or some other blood deficiency such as hemophilia or thrombocytopenia. Unfortunately, the bloods of different people are not all exactly alike, and failure to transfuse the appropriate type of blood is likely to cause death of the recipient. For this reason it is very important to understand the physiology of blood grouping and blood matching prior to giving transfusions.

The principal reason one person's blood may not be suitable for another person is that the recipient may be immune to some of the proteins in the blood cells of the donor. Antibodies in the recipient's plasma can cause *agglutination* (clumping) and *hemolysis* (rupture) of the injected cells, thus plugging some of the vessels and releasing large quantities of hemoglobin out of the cells into the circulation.

O-A-B Blood Groups

The A and B Antigens—the Agglutinogens. Two different but related antigens—type A and type B—occur in the red cell membranes of different persons. Because of the way these antigens are inherited, a person may have neither of them in the cells, or may have one or both simultaneously.

As will be discussed below, some bloods also contain strong antibodies called *agglutinins* that react specifically with either the type A or type B antigens in the cell membranes, causing agglutination and hemolysis. Because the type A and type B antigens in the cells make the cells susceptible to agglutination, these antigens are called *agglutinogens*.

The Four Major O-A-B Blood Groups. In transfusing blood from one person to another, the bloods of donors and recipients are normally classified into four major O-A-B groups, as illustrated in Table 35–1, depending on the presence or absence of the two agglutinogens.

When neither A nor B agglutinogen is present, the blood group is *group O*.

When only type A agglutinogen is present, the blood is *group A*.

When only type B agglutinogen is present, the blood is *group B*.

When both A and B agglutinogens are present, the blood is *group AB*.

The Agglutinins. When type A agglutinogen *is not present* in a person's red blood cells,

TABLE 35–1

The Blood Groups and Their Constituent Agglutinogens and Agglutinins

Blood groups	Agglutinogens	Agglutinins
O	None	Anti-A and Anti-B
A	A	Anti-B
B	B	Anti-A
AB	A and B	None

antibodies known as "anti-A" agglutinins develop in the plasma. Also, when type B agglutinogen *is not present* in the red blood cells, antibodies known as "anti-B" agglutinins develop in the plasma.

Thus, referring once again to Table 35–1, it will be observed that group O blood, though containing no agglutinogens, does contain both *anti-A* and *anti-B agglutinins*.

Group A blood contains type A agglutinogens and *anti-B agglutinins*.

Group B blood contains type B agglutinogens and *anti-A agglutinins*.

Group AB blood contains both A and B agglutinogens but no agglutinins at all.

Blood Typing. Prior to giving a transfusion, it is necessary to determine the blood group of the recipient and the group of the donor blood so that the bloods will be appropriately matched.

The usual method of blood typing is the slide technique illustrated in Figure 35–5. In using this technique a drop or more of blood is removed from the person to be typed. This is then diluted approximately 50 times with saline to provide a suspension of red blood cells. Two separate drops of this suspension are placed on a microscope slide, and a drop of anti-A aggluti-

FIGURE 35–5 Agglutination of some cells and lack of agglutination of other cells in the process of blood typing.

TABLE 35-2

Blood Typing—Showing Agglutination of Cells of the Different Blood Groups with Anti-A and Anti-B Agglutinins

Red blood cells	Sera	
	Anti-A	Anti-B
O	−	−
A	+	−
B	−	+
AB	+	+

nin serum is mixed with one of the drops of cell suspension while a drop of anti-B agglutinin serum is mixed with the second drop. After allowing several minutes for the agglutination process to take place, the slide is observed under the microscope to determine whether or not the cells have clumped. If they have clumped, one knows that an immune reaction has resulted between the serum and the cells.

Table 35-2 illustrates the reactions that occur with each of the four different types of blood, as follows:

Group O red blood cells have no agglutinogens and, therefore, do not clump with either the anti-A or the anti-B serum.

Group A blood has A agglutinogens and therefore does clump with anti-A agglutinins.

Group B blood has B agglutinogens and clumps with anti-B serum.

Group AB blood has both A and B agglutinogens and clumps with both types of serum.

The Rh Blood Types

In addition to the O-A-B blood group system, several other systems are sometimes important in the transfusion of blood, the most important of which is the Rh system. The one major difference between the O-A-B system and the Rh system is the following: In the O-A-B system, the agglutinins responsible for causing transfusion reactions develop spontaneously, whereas in the Rh system anti-Rh antibodies almost never occur spontaneously. Instead, persons must first be massively exposed to the Rh factor, usually by transfusion of blood into them, before they will develop enough agglutinins to cause significant transfusion reaction.

The Rh Antigens—"Rh Positive" and "Rh Negative" Persons. There are six common types of Rh antigens, each of which is called an Rh factor, but only three of these—known as C, D, and E Rh antigens—are usually antigenic enough to cause significant development of anti-Rh antibodies that are capable of causing transfusion reactions. Therefore, anyone who has any one of these three antigens, or any combination of them, is said to be *Rh positive*. A person who has no C, D, or E antigens is said to be *Rh negative*.

The Rh Immune Response. Formation of Anti-Rh Agglutinins. When red blood cells containing Rh factor are injected into an Rh negative person, anti-Rh agglutinins develop slowly, the maximum concentration of agglutinins occurring approximately 2 to 4 months later. This immune response occurs to a much greater extent in some people than in others. On multiple exposure to the Rh factor, the Rh negative person eventually becomes strongly "sensitized" to the Rh factor, and thereafter he will have a serious transfusion reaction if transfused with additional Rh positive blood.

Erythroblastosis Fetalis. Erythroblastosis fetalis is a disease of the fetus and newborn infant characterized by progressive destruction of the baby's blood.

This disease occurs in babies when the mother is Rh negative and the father Rh positive. If the baby inherits the Rh positive characteristic from the father, the mother is likely to become immunized against the baby's Rh positive blood cells. This is especially true when the mother has had several Rh positive babies one after another. Once the mother has become immunized against the Rh factor, *anti-Rh agglutinins* formed by the mother's immune system then diffuse through the placenta from the mother into the baby and begin to destroy the baby's cells. In many instances this kills the baby prior to birth, and the baby is stillborn. In other instances the baby is born extremely anemic because of destruction of the red blood cells. Also, the baby is severely *jaundiced*, which means that its skin has a very yellow color because the macrophage system converts hemoglobin released from the destroyed red blood cells into *bilirubin*, which is a yellow pigment that colors the skin.

The treatment for the newborn jaundiced and anemic child is to replace its blood with Rh negative blood that does not have anti-Rh agglutinins. Otherwise, the baby is very likely to die.

Technique for Transfusion

The usual technique for transfusion is to remove blood from a donor into a sterile bottle containing a citrate solution. The citrate de-ionizes the calcium so that the blood remains unclotted. This blood can then be kept for several weeks at 4°C until it is needed. Also, fresh blood can be transfused by using heparin as an anticoagulant.

Transfusion Reactions Resulting from Mismatched Blood Groups

Hemolysis of Red Cells Following Transfusion Reactions. When there is an antibody-antigen reaction involving one of the blood groups, the usual effect is for the red blood cells to *agglutinate.* Clumps of agglutinated cells then become lodged in the small capillaries throughout the body as shown in Figure 35–6. Over a period of hours these cells gradually break up and release their cellular contents, mainly hemoglobin, into the blood. Therefore, the net result of the transfusion reaction is almost always to release large amounts of hemoglobin into the circulating blood. Sometimes the antigen-antibody reaction is so strong that it actually causes direct hemolysis of the red cells without first causing agglutination; this direct hemolysis occurs within minutes rather than over a period of hours.

Acute Kidney Shutdown Following Transfusion Reactions. The large amount of circulating hemoglobin following a transfusion reaction can often cause acute kidney shutdown, and this in turn sometimes leads to death within a week to 12 days as a result of *uremia.*

The kidney shutdown is caused mainly by blockage of the kidney tubules with hemoglobin, as shown in the lower part of Figure 35–6. When the hemoglobin filters through the glomerulus into the renal tubules, it is in the dissolved form. However, large quantities of fluid are then reabsorbed back into the blood in the tubules, and this concentrates the hemoglobin so greatly that it precipitates. Consequently, the tubules become totally plugged with hemoglobin, causing acute renal failure.

TRANSPLANTATION OF TISSUES AND ORGANS

In this modern age of surgery, many attempts are being made to transplant tissues and organs from one person to another, or, occasionally, from lower animals to the human being. Most of the different antigenic proteins of red blood cells that cause transfusion reactions plus still many more are present in the other cells of the body. Consequently, any foreign cells transplanted into a recipient can cause immune responses and immune reactions. In other words, most recipients are just as able to resist invasion by cells from another person as to resist invasion by bacteria.

In the case of transplants from one person to another, *except when special drug therapy is employed,* immune reactions almost always occur, causing death of all the cells in the transplant within 3 to 10 weeks after transplantation. This process is called "rejection" of the transplant.

Transplants from one identical twin to another are an exception in which transplants are almost always successful without drug therapy. The reason for this is that the antigenic proteins of both twins are determined by identical genes derived originally from the single fertilized ovum.

Some of the different tissues and organs that have been transplanted from one person to another are skin, kidney, heart, liver, glandular tissue, bone marrow, and lung. When special procedures have been employed to prevent the antibody reactions that cause rejection, kidney transplants have been successful for longer than 10 years, and liver and heart transplants for 1 to 8 years.

Procedures to Overcome Rejection of Transplanted Tissue

Tissue Typing. In the same way that red blood cells can be typed to prevent reactions between recipient and donor, so also is it possible to "type" tissues to help prevent graft rejection, though thus far this procedure has met with far less success than has been achieved in red blood cell typing. The most important antigens that cause graft rejection are a group of antigens called the HLA antigens. These are a group of 50 or more different antigens in the tissue cell membranes. The best success has been

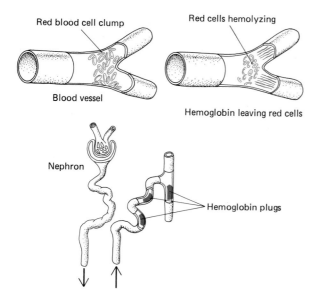

FIGURE 35–6 Mechanism of transfusion reactions caused by mismatched blood.

tissue-type matches between members of the same family. Of course, the match in identical twins is exact.

Use of Anti-Lymphocyte Serum. It was pointed out in the last chapter that rejected transplanted tissues are usually destroyed by lymphocytic T-cells that have become sensitized against the transplant. These cells invade the transplant and gradually destroy the cells. The cells of the transplant begin to swell, their membranes become very permeable, and finally they rupture. Simultaneously, massive numbers of macrophages move in and help in the destruction. Within a few days to a few weeks after this process begins, the tissue is completely destroyed.

Therefore, one of the most effective procedures to prevent rejection of transplanted tissues has been to infuse into the recipient *antilymphocyte serum*. This serum is made in horses by injecting human lymphocytes into them; the antibodies that develop in these animals will then attack and destroy human lymphocytes. When serum from one of these animals is injected into the transplanted recipient, the number of circulating small lymphocytes can be decreased to as little as 5 to 10 percent of the normal number, and there is corresponding decrease in the likelihood of the rejection reaction.

Suppression of Antibody and Sensitized Lymphocyte Formation. Occasionally, a human being has naturally suppressed antibody formation resulting from a hereditary depression of his immune system. Transplants of tissues into such individuals can sometimes be successful without drug therapy, or at least their destruction is delayed. Also, irradiative destruction of most of the lymphoid tissue by either x-rays or gamma rays renders a person much more susceptible than usual to a transplant. Treatment with certain drugs, such as azathioprine (Imuran), cyclosporin A, and glucocorticoid hormones, all of which suppress the immune response, also increases the likelihood of success with transplants. Unfortunately, all of these procedures also leave the person unprotected by the immune system against disease.

To summarize, transplantation of living tissues in human beings up to the present is still partially experimental, though some degree of success is now being recorded, especially for kidney transplants. But when someone succeeds in blocking the rejection response of the recipient to a donor organ without at the same time destroying the recipient's specific immunity for disease, this story will change overnight.

QUESTIONS

1. What role does vascular spasm play in hemostasis?
2. What role do platelet plugs play in hemostasis?
3. Give the functions of thrombin and fibrinogen in the clotting process.
4. Describe the extrinsic mechanism for initiating clotting. What is the role of tissue damage in this mechanism?
5. What is the difference between the intrinsic mechanism and the extrinsic mechanism for initiating clotting?
6. How is clotting prevented in the normal vascular system?
7. How do the following conditions cause bleeding tendencies: hemophilia, thrombocytopenia, and liver disease?
8. What are the agglutinogens?
9. What are the agglutinins?
10. How does the Rh factor sometimes cause transfusion reactions and at other times cause erythroblastosis fetalis?
11. Discuss the problems in the transplantation of organs and tissues.

REFERENCES

Ballantyne, D.L., and Converse, J.M.: Experimental Skin Grafts and Transplantation Immunity: A Recapitulation. New York, Springer-Verlag, 1979.

Chatterjee, S.N. (ed.): Renal Transplantation: A Multidisciplinary Approach. New York, Raven Press, 1980.

Cunningham, B.A.: The structure and function of histocompatibility antigens. *Sci. Am.*, 234(4):96, 1977.

Engelbert, H.: Heparin. New York, S. Karger, 1978.

Gowans, J.L. (in honour of): Blood Cells and Vessel Walls: Functional Interactions. Princeton, N.J., Excerpta Medica, 1980.

Hirsh, J., *et al.*: Concepts in Hemostasis and Thrombosis. New York, Churchill Livingstone, 1979.

Hubbell, R.C. (ed.): Advances in Blood Transfusion. Arlington, Va., American Blood Commission, 1979.

Kline, D.L., and Reddy, K.N.N. (eds.): Fibrinolysis. Boca Raton, Fla., CRC Press, 1980.

Lewis, J.H., *et al.*: Bleeding Disorders. Garden City, N.Y., Medical Examination Publishing Co., 1979.

McDuffie, N.M. (ed.): Heparin: Structure, Cellular Functions, and Clinical Application. New York, Academic Press, 1979.

Mohn, J.F., *et al.* (eds.): Human Blood Groups. New York, S. Karger, 1977.

Murano, G., and Bick, R.L. (eds.): Basic Concepts of Hemostasis and Thrombosis: Clinical Laboratory Evaluation of Thrombohemorrhagic Phenomena. Boca Raton, Fla., CRC Press, 1980.

Thomson, J.M. (ed.): Blood Coagulation and Hemostasis: A Practical Guide. London, Churchill Livingstone, 1979.

Touraine, J.L., *et al.* (eds.): Transplantation and Clinical Immunology. New York, Elsevier/North-Holland, 1980.

Wall, R.T., and Harker, L.A.: The endothelium and thrombosis. *Annu. Rev. Med.*, 31:361, 1980.

36

Anatomy of the Respiratory System, and the Larynx

Overview

The anatomical structures of the respiratory system are the (1) **nose,** (2) **pharynx,** (3) **larynx,** (4) **trachea,** (5) **bronchi,** (6) **bronchioles,** (7) **respiratory lobule,** and (8) **pulmonary vascular system.**

When air is inspired through the nose, it is immediately humidified, and most of the particulate matter in the air is removed, which is called the "air-conditioning" function of the nasal cavities. *Humidification* is caused by the extensive moist surface of the **nasal septum** and the **conchae** that protrude from the lateral walls into the nasal cavities. *Removal of the particles* results mainly from turbulence of the air caused by the conchae, the turbulence causing precipitation of the particles against the **nasal mucous membrane.**

The *trachea* and *bronchi* are all partially encircled by **cartilage plates** that prevent compression of these respiratory passageways by mechanical factors or excess pressure in the thoracic cavity.

The *bronchioles* are minute branching respiratory tubes constructed mainly of smooth muscle. Though the bronchioles normally remain dilated, various abnormal states, such as the allergic reaction that occurs in asthma, can cause bronchiolar spasm and make it difficult for a person to breathe.

The *respiratory lobule* receives air from the most peripheral muscular bronchiole called the **terminal bronchiole.** It is comprised of **respiratory bronchioles** (which are mainly nonmuscular), **alveolar ducts, atria, alveolar sacs,** and **alveoli.** All of these structures have extremely thin walls through which oxygen and carbon dioxide can diffuse readily into or out of the blood in the pulmonary capillaries.

An extensive array of **pulmonary capillaries** surrounds all walls of the respiratory lobule, thus allowing ready transport by the blood of oxygen away from the respiratory lobule and carbon dioxide to this lobule.

Ordinarily, the blood pressure in the pulmonary capillaries is very low so that the osmotic forces of the plasma proteins cause continual osmosis of fluid out of the lung tissues and thereby keep the respiratory passages of the lungs "dry." However, under abnormal conditions, the pulmonary capillary pressure can rise to very high levels, or the colloid osmotic pressure of the blood plasma can fall too low, so that fluid then transudes out of the capillaries and causes *pulmonary edema;* this sometimes occurs so rapidly that it is lethal within less than an hour.

589

The principal muscle of normal respiration is the **diaphragm,** but all other muscles that elevate or depress the anterior thoracic wall or that compress the abdomen, thus pushing the diaphragm upward, can also contribute to the pulmonary ventilatory process, especially during deep respiration. Contraction of the diaphragm elongates the lungs and thereby causes inspiration. Elevating the anterior thoracic wall also causes inspiration by raising the ribs from an inferiorly slanting position to a horizontal position that increases the anteroposterior diameter of the chest.

The *larynx*, which is responsible for *phonation*, is constructed of four major cartilages plus several smaller cartilages. Inside the larynx are two **vocal cords,** each of which is actually a fold of tissue called the **vocal fold** extending medially from each lateral wall of the larynx. The cartilages of the larynx serve as hinged levers operated by the laryngeal muscles to control the vocal cords. They stretch or relax the cords and also bring them closer together or farther apart. In addition, slips of muscle within the vocal cords can make the cords either thin or thick. Thus, in several different ways, both the frequency and the quality of phonation can be changed—from low pitch to high pitch, from whisper to guttural sound, and so forth.

The principal function of the respiratory system is to oxygenate the pulmonary blood and to remove carbon dioxide. However, auxiliary functions provide us with our voice and even control of the acidity of our body fluids by increasing or decreasing the level of carbonic acid in the circulating blood, as was discussed in Chapter 32.

Figure 36–1 illustrates the general plan of the respiratory system and the pulmonary vasculature. From the discussions of Chapters 23 and 24 it is already clear that the sole function of the right heart is to pump blood through the lungs. We shall say more about the blood flow system of the lungs later in this chapter, but, first, let us discuss the anatomy of the system for aerating the lungs.

The Nasal Cavities, and Their "Air-Conditioning" Function. In the nose are two nasal cavities, one of which is illustrated in Figure 36–2; these have extensive surfaces to which the

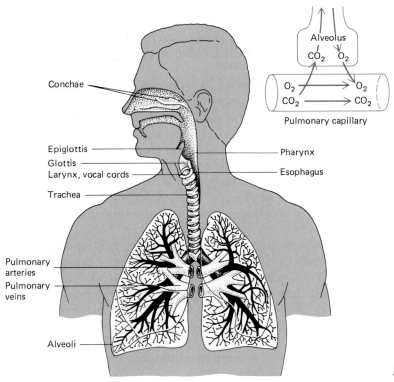

FIGURE 36–1   The respiratory system.

TABLE 36–1
The Respiratory Structures

Structure	Function
Nose	
Conchae and septum	1. Warm air
	2. Moisten air
	3. Precipitate particles
Nasal mucous glands	Moisten surfaces of nasal cavity
Ciliated epithelium	Moves mucus and precipitated particles toward pharynx
Olfactory epithelium	Sensory area for smell
Nasopharynx	
Uvula and soft palate	Reflex closure of posterior nasopharynx when food enters pharynx
Pharynx	Passageway for food to esophagus and air to larynx and trachea
Larynx	
Epiglottis	Helps to block food from trachea
Vocal cords (or vocal folds)	1. Phonation
	2. Reflex blockage of food from trachea
Muscles of larynx	
1. Cricothyroid (2)	Stretch vocal folds
2. Posterior cricoarytenoid (2)	Abduct vocal folds (open cords)
3. Lateral cricoarytenoid (2)	Adduct vocal folds (close cords)
4. Transverse arytenoid	Adduct vocal folds
5. Thyroarytenoid (2)	Loosens vocal folds and shapes them
Trachea	
Tracheal cartilages	C-shaped, rigid partial rings that prevent collapse of trachea
Lungs	
Right	
Upper, middle, and lower lobes	Gas exchange with blood
Left	
Upper and lower lobes	
Bronchi	Branching air passageways of lung with cartilage plates in walls
Bronchioles	Muscular walled branching air passageways smaller than 1 to 2 mm in diameter
Respiratory lobule	Area for exchange of gas with blood
Successive segments	
1. Respiratory bronchioles	
2. Alveolar ducts	
3. Atria	
4. Alveolar sacs	
5. Alveoli	
Respiratory membrane	Membrane between air in respiratory lobule and blood in pulmonary capillaries, about 70 m^2
Pulmonary capillaries	Arching network enveloping all outside surfaces of respiratory lobule
Muscles of respiration	
Inspiratory	
Diaphragm	
External intercostals	Elongates thoracic cavity
Sternocleidomastoids and other anterior neck muscles	Lift rib cage, which increases anteroposterior diameter of thoracic cavity
Scaleni	
Anterior serrati	
Expiratory	
All abdominal muscles	Compress abdominal contents against diaphragm to decrease length of thoracic cavity
Rectus abdominis muscles	Depress rib cage, decreasing anteroposterior diameter of thoracic cavity
Internal intercostals	

Frontal sinus
Olfactory epithelium
Superior concha
Middle concha
External nares
Inferior concha
Middle meatus
Hard palate
Tongue
Hyoid bone
Vocal cord
Thyroid cartilage
Cricoid cartilage
Larynx
Thyroid gland

Olfactory bulb
Sphenoid sinus
Soft palate
Auditory tube

Nasopharynx
Uvula
Palatine tonsil
Pharynx
Epiglottis
Glottis
Esophagus
Trachea

FIGURE 36–2 The upper respiratory passageways.

inspired air is exposed before it enters the pharynx, trachea, and lungs. All these surfaces are covered by a membrane called the **nasal mucosa,** which itself is lined with *ciliated, pseudostratified columnar epithelium,* with each surface cell having as many as a hundred *cilia* that beat forcefully and continually toward the pharynx.

Between the two nasal cavities is a partition called the **nasal septum.** This contributes considerably to the surface area of the cavities, but even more important are the three winglike projections shown in Figure 36–2 protruding medially from the lateral wall of each nasal cavity, the **superior, middle,** and **inferior conchae.** The total area of all these nasal surfaces in both nasal cavities is approximately 160 cm². As the air passes through the nasal cavities and is exposed to these extensive surfaces it is "conditioned" before it enters the lungs in the following ways:

First, it is warmed to within 1 degree of the body temperature before it enters the pharynx. This results from the warmth of rapidly flowing blood in an extensive vasculature of the nasal mucosa.

Second, the air is humidified to within a percent or so of full saturation. This prevents drying of the pharynx, trachea, and lungs. To provide the needed moisture for humidifying the air, the nasal mucosa is abundantly supplied with small **mucous glands** that secrete a *mucoid fluid* inside the nose. For this reason, this mucosa is frequently also called the *nasal mucous membrane.*

Third, the nose removes all large particles from the inhaled air and even most of the minute particles as well. To subserve this function, the nose has a special air filtration system described in the following section.

The Air Filtration Function of the Nose— Function of the Conchae. The hairs at the entrance to the nostrils (the entrances to the nasal cavities) obviously help to remove large particles from the inhaled air. But small particles are removed by an entirely different process, by *turbulent precipitation.* That is, as the air passes through the nasal passageways, it contacts many obstructing surfaces, especially the conchae and the posterior pharyngeal wall. Each time it contacts one of these obstructions it must change its direction of movement. However, the particles suspended in the air cannot, because of their momentum, change direction of travel as rapidly as can the air. Therefore, they continue forward and strike the surfaces of the obstructions, thus becoming embedded in the mucus on the mucosa. Then, this mucus and the entrapped particles are moved toward the pharynx at a rate of about 1 cm per minute by the *cilia* that beat continually on the surfaces of the epithelial cells. Once the mucus and particles reach the pharynx they are swallowed.

This nasal turbulence mechanism for removing particles from the inspired air is so effective that almost no particles larger than 2 to 5 μ in diameter enter into the lungs through the nose. This size is smaller than the size of red blood cells. Of the still smaller particles, many do eventually enter the lungs and can cause serious lung disease. For instance, serious lung disease frequently occurs in coal miners because of coal dust or silica dust that lodges in the air spaces of the lungs.

The Olfactory Epithelium. Note in the upper part of the nasal cavity illustrated in Figure 36–2 the area labeled **olfactory epithelium.** This is the area where the sensory receptors for *smell* are located. This epithelium extends inferiorly about 1 cm over each superior concha and also about 1 cm inferiorly on each side of the septum. The nerve signals from the olfactory epithelium are transmitted by way of short **olfactory nerves** that pass through multiple small foramina in the *cribriform plate of the ethmoid bone* and then enter the **olfactory bulb** on the superior surface of this plate. From here, the signals travel posteriorly through the olfactory tract into the olfactory region of the brain.

The Pharynx; Separation of Air from Food. The pharynx, illustrated in Figure 36–2, is commonly called the *throat.* An upward projection of the pharynx, the **nasopharynx,** extends superiorly to the base of the skull and posterior to the nasal cavities. The inferior border of the pharynx is the *larynx* anteriorly and the opening of the *esophagus* posteriorly, also illustrated in Figure 36–2. The larynx is the entryway for air into the trachea, the tube that conducts air to the lungs, and the esophagus is the entryway for food into the digestive tract. The food passing from the mouth to the esophagus and the air from the nose to the lungs both pass through the pharynx. At two separate points the food and air must be separated as follows.

First, food must be prevented from refluxing from the oral part of the pharynx into the nasopharynx. This is achieved by the soft palate. The anterior two thirds of the roof of the mouth is supported by the bony processes of the maxilla and palatine bone and is called the **hard palate.** The posterior part of the palate has no bone and is the **soft palate;** instead, it has a muscular core known as the **palatine muscles,** covered on both the nasal and the oral surfaces by mucous membrane. The posterior tip of this soft palate extends posteriorly as a small tongue-like extension called the **uvula.** The muscles of the soft palate are so arranged that when they contract they pull the palate upward against the posterior wall of the nasopharynx, thus closing the nasopharynx from the oral part of the pharynx. Nerve reflexes from the mouth and pharyn-

geal wall activate this closure whenever food is propelled backward into the pharynx from the mouth.

Second, food must be prevented from entering the trachea. This is achieved by the function of two structures of the larynx:

1. The **epiglottis,** illustrated in Figure 36–2, is an upward extension from the larynx. It normally lies against the base of the tongue, thus allowing unobstructed passage of the air to and from the trachea. However, when one swallows, the tongue rolls backward, which swings the epiglottis posteriorly over the opening of the larynx. The upper surface of the epiglottis then serves as a "chute" over which some of the food glides directly into the esophagus while the remainder passes along the sides of the epiglottis. The swallowing process also causes the hyoid bone and the entire larynx to be drawn superiorly and anteriorly. This pulls the larynx away from the esophagus, and the tug on the esophagus opens the *upper esophageal constrictor* so that the food passes downward unimpeded.

2. The **vocal cords** are muscular folds on the inside lateral walls of the larynx. These provide a second line of defense against passage of food and fluid into the lower respiratory system. Though the vocal cords are normally used to provide the voice, which will be discussed later in the chapter, they are also capable of closing together tightly from the two sides of the larynx to cause airtight closure of the opening. Even the slightest amount of fluid or food entering the vocal cord area causes powerful reflex spasm of the muscles controlling the vocal cords, thus completely blocking entry to the trachea.

The Thoracic Cage, the Pleural Cavities, and Expansion of the Lungs

Figure 36–3 illustrates the **thoracic cage** (also called the *chest cage*) and its relationship to the lungs. The thoracic cage is the bony enclosure composed of ribs, rib cartilages, sternum, and thoracic vertebrae. Inside the cage are two **pleural cavities,** the right and left cavities, in which the right and left lungs are suspended. For ease of description, these two cavities together are frequently referred to as the **thoracic** or **chest cavity.** The lower border of each pleural cavity is the **diaphragm,** and the compartment in the middle of the thorax between the two pleural cavities is the **mediastinum.** It will be recalled

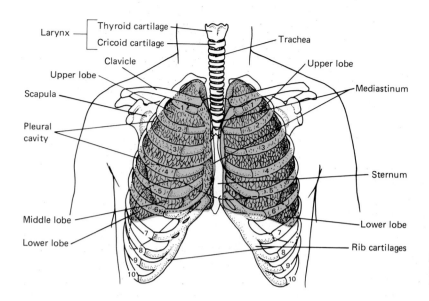

FIGURE 36–3 The chest cage and its relationship to the lungs and trachea.

from Chapter 23 that the heart lies in the lower part of the mediastinum. The inner surface of each pleural cavity is lined by a thin epithelial membrane called the **parietal pleura.** The surface of each lung is covered with a similar membrane called the **visceral pleura** except where the lung is attached to the mediastinum. The thin space between the two pleurae is the **intrapleural space.** Both the visceral and the parietal pleurae secrete small amounts of slick *pleural fluid* having a consistency similar to that of thin mucus. This fluid provides lubrication that allows the lungs to slide freely inside the pleural cavities.

Since the outer surfaces of the lungs are not connected with the walls of the thoracic cage by any direct physical attachments, what is it that keeps the lungs expanded? The answer to this is that whenever either fluid or air enters the intrapleural space, it is absorbed into the blood by the pleurae, thus causing suction inside the space that pulls the lungs against the walls of

the cavities. During respiration the lungs slide freely in the pleural cavities so that lung expansion occurs evenly in all areas at the same time.

THE INSPIRATORY AND EXPIRATORY MUSCLES

There are two basic ways in which the volume of the thoracic cavity can be expanded: (1) by downward movement of the diaphragm to lengthen the cavity, and (2) by upward movement of the entire anterior thoracic wall, which increases the anteroposterior diameter of the thoracic cavity—that is, the ribs then extend horizontally rather than slanting inferiorly, which extends the anterior chest wall anteriorly as well.

Inspiration, and the Muscles of Inspiration. The configuration of the thoracic cavity during inspiration is illustrated to the right in Figure 36–4. The length of the cavity has been increased by downward contraction of the dia-

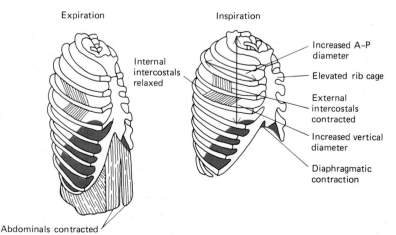

FIGURE 36–4 Expansion and contraction of the thoracic cage during expiration and inspiration, illustrating especially diaphragmatic contraction, elevation of the rib cage, and function of the intercostals.

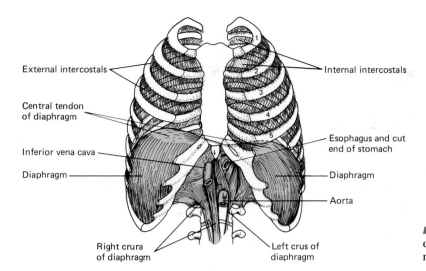

FIGURE 36–5 Anatomy of the diaphragm and of the intercostal muscles.

phragm, and the anteroposterior diameter has been increased by contraction of the muscles that elevate the rib cage.

The **diaphragm,** illustrated in more detail in Figure 36–5, is the most important of all the respiratory muscles. In fact, during normal quiet breathing, respiration is caused almost entirely by the diaphragm and only slightly by the other respiratory muscles. Contraction of the diaphragm enlarges the thoracic cavity, causing a decrease in the intrathoracic pressure that draws air inward. Then, during normal expiration the diaphragm simply relaxes, and elastic recoil of both the lungs and the walls of the thoracic cavity forces the air outward without the action of any muscles. About half of this elastic recoil is caused by elasticity of the lungs themselves. The other half is caused by the elasticity of the fascial and other tissues in the chest wall surrounding the lungs.

The diaphragm is a dome-shaped muscle attached around its edges to the lower, internal margins of the entire rib cage, to the inside of the sternum, and very strongly to the anterior surfaces of the lumbar vertebral column by way of the **right** and **left crura of the diaphragm** as illustrated in Figure 36–5. The central dome of the diaphragm is not muscle but is instead a fibrous sheet of fascia called the **central tendon of the diaphragm.** It is mainly on this central tendon that the two lungs rest; it is the muscular borders of the diaphragm that contract.

The diaphragm is innervated by the **phrenic nerves.** These arise in the neck from the cervical spinal nerve plexus, with the phrenic nerve fibers originating mainly from cervical spinal nerves C3 and C4, with a few fibers from C5 as well. The two phrenic nerves pass downward along the two lateral sides of the

mediastinum and then branch to all areas of the diaphragm. The fact that the phrenic nerves arise from the spinal cord in the neck region makes it possible for a person to continue breathing after the spinal cord has been totally transected at levels below the fifth cervical vertebra. Most quadriplegic patients (patients who are paralyzed in all four limbs) are persons who have broken their necks between the fifth and seventh cervical vertebrae and have transected their cords. This paralyzes more than 95 percent of the body, including all or almost all of the muscles in the upper extremities as well as all trunk muscles and muscles of the lower extremities. Yet these persons can continue to breathe and live for many years thereafter.

Other Muscles of Inspiration. Still other muscles besides the diaphragm can contribute to inspiration. All of these function by elevating the rib cage, which increases the anteroposterior diameter of the thorax. Though this is not important during normal quiet breathing, during heavy exercise it can account for the intake of as much as 40 percent of the inspired air. These muscles are as follows:

1. The **external intercostal muscles** pass between the ribs, lying on the outer surfaces of the internal intercostal muscles; they angle anteriorly from the lower border of each upper rib and attach to the upper border of the rib below. When they contract they pull the upper rib anteriorly and the lower rib superiorly; this moves the entire anterior chest wall anteriorly and superiorly as illustrated to the right in Figure 36–4.

2. All the **muscles of the neck that elevate the rib cage.** These include (a) the **sternocleidomastoid muscles** that elevate the sternum,

(b) the **anterior serrati** that elevate the upper ribs, and (c) the **scaleni** that elevate the first two ribs, thus also elevating the entire rib cage because of the attachments of the ribs to the sternum. Since some of the neck muscles are supplied by nerve fibers from the first few cervical spinal nerves as well as cranial nerve XI, these muscles, too, help to provide respiration in quadriplegic patients.

Expiration, and the Muscles of Expiration. Expiration is illustrated to the left in Figure 36–4. The chest has been flattened by pulling the rib cage inferiorly, and the internal length of the chest cavity has been decreased by upward excursion of the diaphragm. Both these effects are caused mainly by contraction of the abdominal muscles as follows:

1. The **rectus abdominis** muscles pull downward on the anterior margin of the rib cage as was explained in Chapter 8. This decreases the anteroposterior diameter of the chest.
2. Contraction of the muscles that encircle the abdomen, including the **external oblique,** the **internal oblique,** and the **transverse abdominal muscles,** causes the abdominal contents to be pushed upward against the inferior surface of the diaphragm, thus also pushing the diaphragm against the lungs to decrease the length of the thoracic cavity.

Another group of muscles that plays a smaller role in expiration is the **internal intercostal muscles.** These angle posteriorly from the lower border of each rib to the upper border of the rib below. Therefore, contraction of the internal intercostals pulls the upper ribs inferiorly and posteriorly, thus having the same effect as contracting the rectus abdominis muscles in decreasing the anteroposterior diameter of the chest cavity.

The Lung Lobes, Trachea, Bronchi, and Bronchioles

The Lung Lobes. Referring again to Figure 36–3 and also to Figure 36–6, one can see that each of the two lungs is divided into lobes, the **left lung** into two lobes, the *upper* and *lower lobes*, and the **right lung** into three lobes, the *upper, middle,* and *lower lobes.* The interlobar surfaces are covered with visceral pleura where the lobes contact one another and are lubricated with the same thin mucoid pleural fluid that lubricates the outer surfaces of the lungs; therefore, the lobes slide against one another in the same way that the entire lungs slide within the thoracic cavity.

The fact that the lungs are divided into lobes has considerable clinical importance, for lung infections, especially pneumonia, frequently are confined to only one lobe because the barriers between the lobes help to prevent extension of the infection.

The Trachea and Bronchi. The **trachea** is the main tube leading from the larynx toward the lungs. At about 3 cm above the heart it bifurcates (divides) to form two main bronchi, the **right bronchus** and **left bronchus,** each of

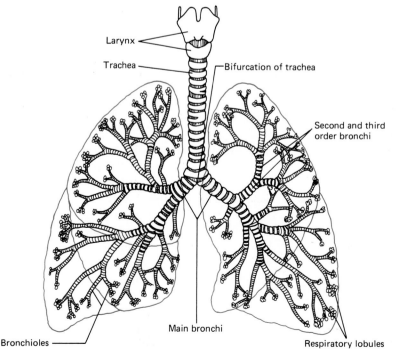

Larynx

Trachea

Bifurcation of trachea

Second and third order bronchi

Main bronchi

Bronchioles

Respiratory lobules

FIGURE 36–6 The lung lobes and the bronchial tree.

which enters the central, medial border of the respective lung. This area of entry into the lung is called the **hilum** of the lung. It is also through the hilum that the pulmonary blood vessels, lymphatic vessels, and nerves enter or leave the lung.

On entering the lung, each main bronchus branches many times to form secondary, tertiary, and higher-order bronchi, these branches giving the appearance of limbs of a tree spreading throughout all parts of the lungs as illustrated in Figure 36–6.

The Cartilages of the Trachea and Bronchi. It must be remembered that the process of respiration is achieved by drawing air into the lungs, and everyone who has tried to draw a milk shake through a wet straw knows that the straw simply collapses and no beneficial results are achieved. Therefore, it is essential that the air passageways not collapse during this suction. This is insured by the presence of cartilage in the walls of the trachea and bronchi. Figure 36–7 illustrates a cross-section of the trachea, showing a C-shaped partial **cartilaginous ring** that extends five sixths of the distance around the trachea. Anteriorly, the cartilage is continuous; posteriorly, the gap between the two ends of the cartilage is bridged by a web of smooth muscle covered by connective tissue. The cartilaginous rings in the trachea prevent tracheal collapse even during the highest degrees of suction by the lungs.

In the larger bronchi, partial cartilaginous rings also extend much of the distance around the bronchi, as can be seen in Figure 36–6. But as the bronchi become progressively smaller, the cartilage becomes many small **cartilaginous plates** that extend only short distances around the bronchi. In the aggregate, though, these still make even the smallest bronchi semirigid so that

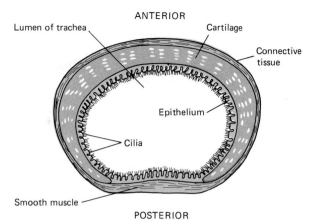

ANTERIOR

Lumen of trachea — Cartilage

Connective tissue

Epithelium

Cilia

Smooth muscle

POSTERIOR

FIGURE 36–7 Cross-sectional view of the trachea.

they too will not collapse under high degrees of suction. Even so, the bronchi are pliable enough that they can bend and stretch to accommodate the enlargement of the lung during inspiration.

The Bronchioles. When the bronchi become smaller than 1 to 2 mm in diameter, they lose their cartilage, and the subsequent branches are then the **bronchioles.** These are illustrated in Figure 36–8. The walls of the bronchioles are composed of smooth muscle and connective tissue. Though the smooth muscle of the bronchioles normally remains relaxed so that the bronchioles stay open, particles of other irritating substances entering these passageways can cause bronchiolar spasm. This happens in patients who have asthma, for these persons are allergic to pollens or other substances in the air called *allergens*. When an allergic reaction occurs in the bronchioles, it causes local spasm, making it extremely difficult for the asthmatic person to breathe. Thus, he or she literally chokes from internal airway obstruction.

The bronchioles, like the bronchi, branch several times. The last bronchiolar branch that

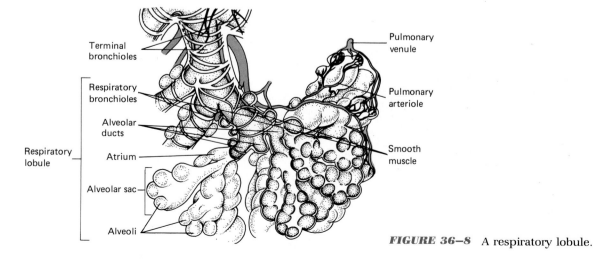

Terminal bronchioles

Respiratory bronchioles

Alveolar ducts

Respiratory lobule

Atrium

Alveolar sac

Alveoli

Pulmonary venule

Pulmonary arteriole

Smooth muscle

FIGURE 36–8 A respiratory lobule.

still has a complete muscular coat is called a **terminal bronchiole.** But this, too, branches into several *respiratory bronchioles* that have little smooth muscle and are·the entryways into the *respiratory lobules*, the final air spaces of the lungs, as will be explained subsequently.

The Cilia and Mucous Glands Lining the Bronchial Tree.   The inner surfaces of the trachea and bronchial tree all the way peripherally to the terminal bronchioles are lined with **ciliated respiratory epithelial cells** similar to the surface cells in the nasal cavities. However, the cilia in the bronchial tree beat in the opposite direction to those in the nose, that is, upward toward the pharynx. Also, the surface membrane of the bronchial tree is well supplied with small mucous glands similar to those of the nasal mucosa. The secreted mucus is slowly transported by the beating cilia along the surface of the bronchial tree toward the pharynx, and on entering the pharynx it is swallowed. In this way most dust particles, bacteria, or other debris that enter the lungs are removed from the lung passageways.

THE RESPIRATORY LOBULE

The respiratory exchange unit of the lung is the **respiratory lobule,** illustrated in Figure 36–8. The respiratory lobule is the portion of the lung that branches from each terminal bronchiole. It consists of

1. several **respiratory bronchioles** that branch from the terminal bronchiole;
2. multiple **alveolar ducts** branching from each respiratory bronchiole; each duct then branches several times to form

3. **atria,** which then branch again to form several
4. **alveolar sacs,** and each of these has a dozen or more outcropping pouches called
5. **alveoli.**

The Respiratory Membrane.   The walls of all parts of the respiratory lobule are collectively called the **respiratory membrane.** This membrane is extremely thin, so thin that oxygen and carbon dioxide can pass freely between the air inside the lobule and the blood in the capillaries surrounding the lobule. Indeed, the thickness of this membrane, from the air space to the blood, in most areas is less than $1\,\mu$, which is only one eighth the diameter of a red blood cell. Also, the total surface area of the entire respiratory membrane of both lungs is about $70\,\text{m}^2$, about the surface area of one side of a tennis court. Therefore, one can readily understand why oxygen and carbon dioxide can pass with ease between the air of the lungs and the blood of the pulmonary capillaries.

The Pulmonary Capillaries.   Figure 36–9 illustrates the envelopment of the respiratory membrane by **pulmonary capillaries,** and Figure 36–10 shows an actual photograph of the capillary network in the wall of an alveolus. The coloring of the capillaries in Figure 36–9 shows that as the blood flows from the pulmonary artery (which is colored blue) toward the pulmonary vein (which is colored red) the blood becomes oxygenated, changing from the dusky bluish-red color of deoxygenated blood to the bright red color of oxygenated blood.

Note especially in Figure 36–10 the extensive branching and anastomosing of the pulmonary capillaries. Indeed, this branching is so ex-

Pulmonary arteriole

Alveoli

Capillaries

Pulmonary venule

From pulmonary artery

To pulmonary vein

FIGURE 36–9   Envelopment of the alveoli by pulmonary capillaries.

FIGURE 36–10 Surface view of capillaries in an alveolar wall. (From Maloney and Castle: *Resp. Physiol.*, 7: 150, 1969. Reproduced by permission of ASP Biological and Medical Press, North-Holland Division).

tensive that it is frequently said that a "sheet of blood" literally flows through the alveolar walls.

THE PULMONARY CIRCULATION

The physiologic anatomy of the pulmonary circulation, illustrated in Figure 36–11, is very simple. The right ventricle pumps blood into the **pulmonary artery.** From there, the blood flows into the **right** and **left pulmonary arteries,** then into progressively smaller branches, and finally through the **pulmonary arterioles** before entering the pulmonary capillaries. From here, the blood returns to the left atrium through the **pulmonary venules** and **pulmonary veins.**

Because blood flows around a continuous circuit composed of the systemic and pulmonary circulatory systems, the same amount of

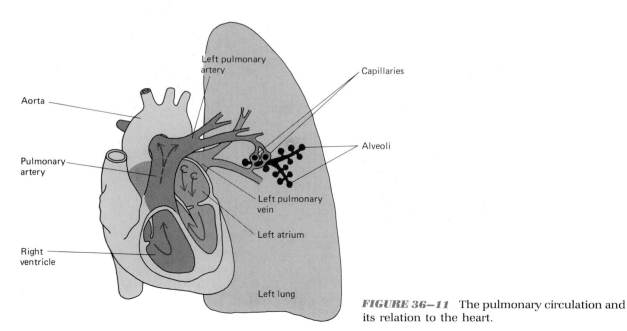

FIGURE 36–11 The pulmonary circulation and its relation to the heart.

blood must flow through the lungs as through the systemic circulation. The vessels of the lungs are very expansible, so that whenever the rate of blood flow into the lungs increases, the pulmonary vessels automatically stretch to allow the more rapid flow demanded in the pulmonary circulation. This allows facile transport of blood through the lungs under widely varying conditions.

Regulation of Blood Flow Through Separate Areas of the Lungs—Response to Low Oxygen. Since all portions of the lungs perform essentially the same function, there is little need for extensive regulation of the pulmonary blood vessels. However, one feature of blood flow regulation in the lungs is distinctly different from the regulation in the remainder of the body, as follows: It is quite important for blood to flow only through those portions of the lungs that are adequately aerated and not through portions that are not aerated. Research studies have shown that when the concentration of oxygen in any area of the lungs falls too low, and remains low for longer than a minute, the local vessels constrict. Therefore, when a bronchus becomes blocked and a local area of the lungs can no longer receive enough oxygen from the air, the vessels of this lung area constrict, thereby forcing the blood to flow instead through other areas of the lungs that are still aerated.

Pulmonary Vascular Pressures

The resistance to blood flow in the pulmonary circulation is so little that the *mean pulmonary arterial pressure* averages only *13 mm Hg*, and the *systolic pressure* (the highest pressure during contraction of the ventricle) averages only *22 mm Hg*, whereas the *diastolic pressure* (the lowest pressure between contractions) is about *8 mm Hg*. Thus, the mean pulmonary arterial pressure is only one seventh the mean systemic arterial pressure, which is about 100 mm Hg, as was discussed in Chapter 28.

The *pulmonary venous pressure* is about *2 mm Hg*, which is also the pressure in the left atrium. The pulmonary capillary pressure has never been measured directly, but it must be greater than the pulmonary venous pressure and less than the pulmonary arterial pressure— that is, somewhere between 2 and 13 mm Hg. Indirect measurements of *pulmonary capillary pressure* indicate that it is probably about *7 mm Hg*.

The mean pressure gradient from the pulmonary artery to the pulmonary vein is 13 mm Hg minus 2 mm Hg, or 11 mm Hg. This compares with a pressure gradient in the systemic circulation of approximately 100 mm Hg. To express this differently, the same amount of blood flows through the lungs as through the systemic circulation with about one ninth the force propelling it. This means also that the total resistance of the pulmonary circulation is normally only one ninth the total resistance of the systemic circulation.

Pressure in the Pulmonary Artery During Exercise. When one exercises, the rate of blood flow through the systemic circulation sometimes increases to as high as five to seven times normal, which means that the rate of flow through the lungs must also increase this much. Still, the pulmonary arterial pressure increases only about 50 percent because the vessels of the lungs passively enlarge to accommodate whatever amount of blood needs to flow. This obviously keeps the right heart from becoming overloaded during exercise.

Pulmonary Capillary Dynamics. Only about one tenth of the blood of the lungs is actually in the capillaries at any one time, and the blood flows through them so rapidly that it remains in the capillaries at most only 1 to 2 seconds. During exercise, when blood flow is even more rapid, the blood may remain in the pulmonary capillaries only $\frac{1}{4}$ to $\frac{1}{2}$ second. Yet, the respiratory membrane is so extremely permeable to oxygen and carbon dioxide that even during this very short period of time the blood can become almost completely oxygenated and can eliminate the necessary amount of carbon dioxide.

The Pulmonary Capillary Mechanism for Maintaining Dry Lungs. Another extremely important feature of pulmonary capillary dynamics is a mechanism for keeping the air spaces of the lungs dry. The plasma of the blood has a colloid osmotic pressure of about 28 mm Hg, which causes a continual tendency for fluid to be absorbed by osmosis into the capillaries. At the same time, the mean pulmonary capillary pressure, which tends to force fluid outward through the pores of the capillaries, is only 7 mm Hg. Therefore, the force causing fluid to be absorbed into the capillaries is about 21 mm Hg greater than that causing fluid to filter out of the capillaries. Thus, this excess plasma colloid osmotic pressure continually promotes fluid absorption from the tissues and alveoli of the lungs, and if a small amount of fluid does enter the alveoli, it is usually absorbed within a few minutes.

Abnormalities of the Pulmonary Circulation

Pulmonary congestion means too much blood and fluid in the lungs. It results from too high a pressure in the pulmonary circulation, which in turn is caused most frequently by failure of the left heart to pump blood adequately from the lungs into the systemic circulation. Once the pulmonary capillary pressure rises above the plasma colloid osmotic pressure, about 28 mm Hg, fluid leaks out of the capillaries extremely rapidly, and the alveoli no longer remain dry. Then, severe *pulmonary edema* (excess fluid in the lungs) develops, sometimes developing so rapidly that it causes death within 30 minutes to 2 hours.

Atelectasis means collapse of a lung or part of a lung. This usually occurs when something blocks one or more of the bronchi, cutting off air flow. When this happens, the air in the blocked alveoli is absorbed into the blood within a few hours, causing this area of the lung to collapse. Simultaneously, the lack of oxygen in the collapsed alveoli causes vascular constriction, a mechanism that was discussed earlier in this chapter; in addition, the collapse of the lung tissue causes compression and kinking of the vessels. As a result of these combined effects, the blood flow through the collapsed lung decreases to about one fifth its previous value, which causes most of the blood to flow through the uncollapsed areas of the lungs where adequate oxygen is available.

Surgical removal of large portions of the lungs causes excessive blood flow through the remaining lung tissue. Ordinarily blood can flow through each lung up to four times as rapidly as normal before the pulmonary arterial pressure begins to rise significantly. Therefore, a person can lose one whole lung, which increases the blood flow through the opposite lung about twofold, without causing any serious disability. Yet, if this person attempts too heavy exercise, he rapidly reaches an upper limit to the rate of pulmonary blood flow before pulmonary hypertension and failure of the right heart develop. This same effect also occurs in emphysema, a disease of the lungs in which large portions of the lung are destroyed, most often the result of smoking.

Effect of Congenital Heart Disease on the Pulmonary Circulation

Congenital heart disease means an abnormality of the heart or of closely allied blood vessels that is present at birth. Many types of congenital heart disease affect the pulmonary circulation; two of these, *patent ductus arteriosus* and *tetralogy of Fallot*, are illustrated in Figures 36–12 and 36–13.

Patent Ductus Arteriosus. During normal fetal life most of the blood pumped by the right ventricle bypasses the lungs, flowing from the pulmonary artery through the **ductus arteriosus** (Fig. 36–12) into the aorta. Immediately after birth, expansion of the lungs dilates the pulmonary blood vessels so that blood can then flow through them with ease. As a result, the pulmonary arterial pressure falls, causing the blood flow to reverse in the ductus, now flowing backward from the aorta into the pulmonary artery. This backward flow causes oxygenated blood to be carried from the aorta into the ductus arteriosus, which promotes immediate contraction of the ductus wall because its muscle is highly sensitive to oxygen. This closes the ductus within minutes; then, over a period of weeks,

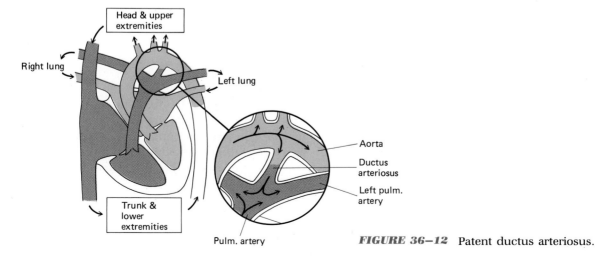

FIGURE 36–12 Patent ductus arteriosus.

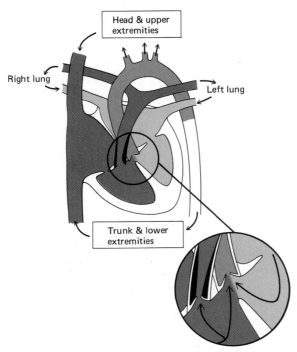

FIGURE 36–13 Tetralogy of Fallot.

fibrous tissue normally grows into the ductus to occlude it permanently.

However, in about 1 out of every 2000 babies, the ductus fails to close, which is the condition called **patent ductus arteriosus,** and this vessel connecting the aorta and the pulmonary artery remains open throughout life, as illustrated in Figure 36–12. As the child grows older and his heart becomes stronger, the pressure in the aorta becomes much greater than the pressure in the pulmonary artery, which forces tremendous quantities of blood from the aorta into the pulmonary artery. Then, the blood flows a second time through the lungs and through the left heart, and maybe through the lungs again, often passing around and around this circuit two or more times before finally entering the systemic circulation. In other words, greatly excessive amounts of blood are pumped through the lungs. Obviously, this places a great strain on the heart, but it does not cause extensive damage to the lungs in early life. However, in later life the very high pressure in the pulmonary vessels often promotes fibrosis of the vessels and congestion of the lungs. Finally, the person may have such difficulty aerating his blood that this will cause early death, though a few patients live almost a normal life span.

Tetralogy of Fallot—the "Blue Baby." Figure 36–13 shows the congenital abnormality called tetralogy of Fallot. This disease presents four abnormalities of the heart:

1. a *hole in the septum* (the partition) between the right and left ventricles;
2. a greatly *constricted pulmonary artery;*
3. *displacement of the aorta to the right,* with the aorta then overlying the hole in the left ventricles; and
4. a *hypertrophied right ventricular muscle.*

Very little blood passes from the right ventricle through the constricted pulmonary artery. Instead, most of it is forced through the hole in the septum and then into the aorta. Thus, most of the blood pumped by the right ventricle bypasses the lungs and is not aerated. Because nonaerated blood has a dusky bluish color, the baby's whole body assumes this hue. Therefore, tetralogy of Fallot is one of the congenital abnormalities of the heart that cause the so-called "blue baby."

Surgical Treatment. In the past few years many advances have been made in the surgical repair of cardiac abnormalities. For instance, any patient with a patent ductus arteriosus can be treated very simply by tying a tight ligature around the ductus. Treatment of tetralogy of Fallot is more complicated but is usually accomplished reasonably satisfactorily by open heart surgery in which extensive repair of the heart defects is performed inside the heart itself.

THE LARYNX AND VOCALIZATION

Speech obviously involves the respiratory system and the larynx, but it also involves (1) specific speech control centers in the cerebral cortex, which were discussed in Chapter 18; (2) respiratory control centers of the brain stem, discussed in Chapter 38; and (3) structures of the mouth and nasal cavities that provide for articulation and resonance. Basically, speech requires two separate mechanical functions:

1. *phonation,* which is achieved by the larynx, and
2. *articulation,* which is achieved by the structures of the mouth and nose.

The Larynx. Figure 36–14A illustrates an anterior view of the larynx and Figure 36–14B a posterior view. The larynx has a skeletal structure all its own, but composed of cartilage and ligaments rather than bone and ligaments. There are four major cartilages in the larynx and several others of less importance. The four major ones are

1. the **thyroid cartilage,**
2. the **cricoid cartilage,** and
3 and 4. two **arytenoid cartilages.**

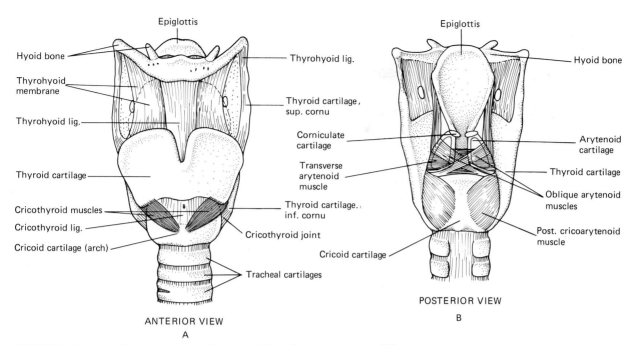

FIGURE 36-14 The larynx: Anterior view (A) and posterior view (B).

The *cricoid cartilage* forms the base of the larynx. Inferiorly, it attaches to the upper end of the trachea, and it completely encircles the larynx. As illustrated in Figure 36–14B, its posterior part is about two times as broad as its anterior part.

The *thyroid cartilage* is a large double-winged structure that spreads over the anterior surface of the larynx. It does not encircle the posterior part of the larynx at all. From its two lateral margins it extends superiorly to form two *superior horns* and inferiorly to form two *inferior horns.* The inferior horns attach by synovial joints to the two sides of the cricoid cartilage, forming two lateral **cricothyroid joints.** These allow the thyroid cartilage to swing with a limited degree of motion anteriorly or posteriorly. The **cricothyroid muscles** that connect the cricoid and thyroid cartilages anteriorly, when contracted, swing the thyroid cartilage anteriorly, while several smaller posterior muscles swing it posteriorly. The superior portions of the thyroid cartilages are suspended from the hyoid bone by the thyrohyoid membrane. In turn, contraction of the suprahyoid muscles from the mandible to the hyoid bone elevates both the hyoid bone and the entire laryngeal structure superiorly and anteriorly; this opens the esophagus and at the same time closes the epiglottis over the superior opening of the larynx, as was discussed earlier.

The two *arytenoid cartilages* are illustrated on the posterior side of the larynx in Figure 36–14B. They are much smaller cartilages than the thyroid and cricoid cartilages, but they are exceedingly important in controlling the voice. Each arytenoid cartilage is triangular in shape, with the base of the triangle resting on the superior surface of the posterior part of the cricoid cartilage. The superior angle of the arytenoid is called the **vocal process;** it is to this that the posterior end of the vocal cord attaches. The medial angle of the base of the arytenoid forms a joint with the cricoid cartilage, allowing the vocal processes of the arytenoids to swing either (1) medially or laterally, thus closing or opening the cords, or (2) anteriorly or posteriorly, thus loosening or tightening the cord. In addition, the arytenoid can be rotated about a vertical axis through the cricoarytenoid joint, allowing the medial border to move anteriorly or posteriorly, which can also loosen or tighten the cord. The lateral angle of the base of the arytenoid is called the muscular process; it is to this that several muscles attach and control the movements of the arytenoids.

The Vocal Cords and Phonation. The vibrating elements of the larynx are the **vocal cords.** These are not actual "cords" as their name would imply but instead are "folds" on the lateral walls of the interior larynx, as illustrated earlier in Figure 36–2, and they are also called **vocal folds.** The medial border of each vocal fold encloses a strong band of elastic tissue called the **vocal ligament.** This attaches posteriorly to the vocal process of the arytenoid cartilage and anteriorly to the posterior surface of the

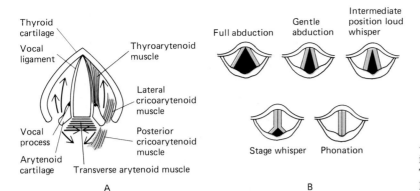

FIGURE 36–15 Laryngeal function in phonation. (Modified from Greene: The Voice and Its Disorders. 3rd Ed. Philadelphia, J.B. Lippincott, 1972.)

thyroid cartilage. Lateral to the vocal ligament, also in the vocal fold, is the *thyroarytenoid muscle,* a thin muscle that helps to control the shape and tension of the vocal fold, as will be discussed later. And covering both the muscle and the vocal ligament is a thin epithelium.

Figure 36–15 illustrates the functional structure of the larynx. The most important of the intrinsic muscles of the larynx that position and control the degree of stretch of the vocal folds are also shown; these are as follows:

1. The **posterior cricoarytenoid muscles** pull inferiorly on the lateral angles of the arytenoids; this causes the vocal processes to move laterally and thereby cause separation (abduction) of the vocal folds. Thus, the air passage is opened widely to allow free passage of air.
2. The **lateral cricoarytenoid muscles** act oppositely to the posterior cricoarytenoid muscles, pulling laterally on the lateral angles of the arytenoids, which causes the vocal processes to move medially, thereby adducting the folds (bringing the folds together).
3. The **transverse arytenoid muscle** pulls the arytenoid cartilages together and, therefore, approximates the two vocal folds so that they vibrate in a stream of expired air.
4. The **thyroarytenoid muscles** lie in the vocal folds lateral to the vocal ligaments as discussed before. These are made up of many small slips of muscle controlled separately by different nerve fibers. Contraction of the lateral portions of the muscles pulls the vocal processes of the arytenoid cartilages anteriorly, which loosens the vocal ligaments and allows a lower frequency of phonation. The medial portions of the thyroarytenoid muscles can change the shape of the vocal folds— to cause thick or thin, sharp or blunt edges, and so forth—during different types of pho-

nation. Thick but loose vocal folds will cause a low frequency sound. Other shapes of the folds are responsible for the whisper and various types of special intonations as illustrated in Figure 36–15B.

In addition to the intrinsic laryngeal muscles that attach to the arytenoid cartilages, the cricothyroid muscles on the anterior surface of the larynx can swing the entire thyroid cartilage anteriorly, which provides still another way to tense the vocal folds and change the frequency of phonation.

Articulation and Resonance. The three major organs for articulation are the lips, the tongue, and the soft palate. These need not be discussed in detail because all of us are familiar with their movements during speech and other vocalizations.

The resonators include the mouth, the nasal cavity and its associated paranasal air sinuses, the pharynx, and the thoracic cavity. Here again, we are all very familiar with the resonance qualities of these different structures. For instance, the function of the nasal resonators is illustrated by the change in quality of the voice when a person has a severe cold.

QUESTIONS

1. Describe the anatomy of the nasal cavities as it relates to the air conditioning function of the nose.
2. What are the anatomical structures responsible for separating air from food in the pharynx?
3. Describe the thoracic cage, the pleural cavities, and the relationship of the lungs to these.
4. List the muscles of inspiration and expiration, and tell how each functions to cause inspiration or expiration.
5. Give the anatomical structure of the diaphragm, emphasizing its relationships to the pleural cavities, to the heart, and to the liver.
6. Describe the trachea and the bronchial tree.

7. Where in the respiratory system are ciliated epithelial cells found, and what is their function?
8. List and describe the separate divisions of the respiratory lobule.
9. What are the characteristics of the respiratory membrane and its relationship to the pulmonary capillaries?
10. Describe the cartilages of the larynx and the functions of the laryngeal muscles that control phonation.

REFERENCES

Anderson, J.E.: The thorax, *In* Grant's Atlas of Anatomy. Baltimore, Williams & Wilkins, 1978, 1-1 to 1-91.

Copenhaver, W.M., Kelly, D.E., and Wood, R.L.: The respiratory system. *In* Bailey's Textbook of Histology, 17th Ed. Baltimore, Williams & Wilkins, 1978, pp. 552–576.

Figge, F.H.J.: The respiratory system. *In* Sobotta/Figge Atlas of Human Anatomy. Vol. II. Baltimore, Urban & Schwarzenberg, 1977, pp. 145–192.

Goss, C.M.: The respiratory system. *In* Gray's Anatomy of the Human Body, 29th Ed. Philadelphia, Lea & Febiger, 1973, pp. 1112–1154.

Ham, A.W., and Cormack, D.H.: The respiratory system. *In* Histology, 8th Ed. Philadelphia, Lippincott, 1979, pp. 725–752.

Hammersen, F.: The respiratory system. *In* Sobotta/Hammersen Histology. Baltimore, Urban & Schwarzenberg, 1980, pp. 139–146.

Langman, J., and Woerdeman, M.W.: Thorax. *In* Atlas of Human Anatomy, Philadelphia, W.B. Saunders, 1978, pp. 1–68.

Leeson, T.S., and Leeson, C.R.: The respiratory system. *In* A Brief Atlas of Histology. Philadelphia, W. B. Saunders, 1979, pp. 175–192.

Leeson, T.S., and Leeson, C.R.: The respiratory system. *In* Histology, 4th Ed. Philadelphia, W. B. Saunders, 1981, pp. 401–427.

Netter, F.H.: Respiratory system. *In* The CIBA Collection of Medical Illustrations. Vol. 7, Summit, N.J., CIBA Medical Education Division, 1979.

Pernkopf, E.: The lungs. *In* Atlas of Topographical and Applied Human Anatomy. Vol. II. Philadelphia, W.B. Saunders, 1980, pp. 92–136.

Pernkopf, E.: Larynx. *In* Atlas of Topographical and Applied Human Anatomy. Vol. I. Philadelphia, W.B. Saunders, 1980, pp. 313–339.

Weiss, L., and Greep, R.O.: The respiratory system. *In* Histology, 4th Ed. New York, McGraw-Hill, 1977, pp. 765–830.

Wheater, P.R., Burkitt, H.G., and Daniels, V.G.: Respiratory system. *In* Functional Histology. New York, Churchill Livingstone, 1979, pp. 161–170.

Williams, P.L., and Warwick, R.: The respiratory system. *In* Gray's Anatomy, 36th British Edition. Philadelphia, W. B. Saunders, 1980, pp. 1228–1266.

(See also General Anatomical References, Chapter 1.)

37

Mechanics of Respiration; Pulmonary Blood Flow; and Transport of Oxygen and Carbon Dioxide

Overview

There are no physical attachments between the lungs and the chest wall. Instead, the lungs are held tightly against this wall by a slight vacuum in the *intrapleural space,* the very thin space between the lungs and the wall. When the chest cavity enlarges, this vacuum causes the lungs to expand at the same time. Expansion of the lungs in turn creates a slight negative pressure in the alveoli, which sucks air into the lungs, causing *inspiration.* During *expiration,* the *intra-alveolar pressure* becomes slightly positive and pushes the air outward.

The volume of air inspired with each breath is the *tidal air;* this is normally about $\frac{1}{2}$ liter (L). The *respiratory rate* averages about 12 per minute. During very deep breathing, the maximum tidal air that can be achieved, called the *vital capacity,* is about 4.5 L in the normal person or as great as 6.5 L in the athlete.

The concentrations of the different gases in the alveoli are expressed in terms of the pressure exerted by each gas, called the *partial pressure of the gas.* The approximate partial pressure of important respiratory gases in the alveoli when a person is at sea level are the following: *oxygen,* 104 mm Hg; *carbon dioxide,* 40 mm Hg; *water vapor,* 47 mm Hg; and *nitrogen,* 569 mm Hg.

The pressure of oxygen in the blood entering the pulmonary capillaries is low, only about 40 mm Hg. Therefore, oxygen diffuses into the pulmonary blood, increasing its pressure to equal the 104 mm Hg partial pressure of the oxygen in the alveolar air. On the other hand, the pressure of carbon dioxide in the blood entering the pulmonary capillaries is high, about 45 mm Hg, so that carbon dioxide diffuses out of this blood into the alveoli until the carbon dioxide pressure falls to equal the 40 mm Hg partial pressure of the carbon dioxide in the alveoli. Thus, the pulmonary blood absorbs oxygen and releases carbon dioxide.

When the systemic arterial blood enters the peripheral tissue capillaries, oxygen diffuses into the tissue cells because these cells are continually using oxygen, keeping the cellular oxygen pressure at only a few millimeters of mercury. Conversely, these cells are continually forming carbon dioxide so that the cellular carbon dioxide pressure is considerably greater than that of the tissue capillary blood; therefore, carbon dioxide diffuses out of the cells into the blood and is transported to the lungs.

About *97 percent of the oxygen transported in the blood* from the lungs to the peripheral tissues is carried in *chemical combination with hemoglobin,* with only 3 percent transported in the dissolved state in the fluid of the blood. However, oxygen binds only loosely with the hemoglobin so that it can be displaced easily from the hemoglobin in the peripheral capillaries and released to the cells.

About 7 percent of the *carbon dioxide* transported by the blood is carried in the dissolved state. The remainder is *carried mainly by combining with water inside the red blood cells to form bicarbonate ions,* a process catalyzed by a red blood cell enzyme called *carbonic anhydrase.* A smaller portion of the carbon dioxide combines with the hemoglobin molecule inside the red blood cell and is transported in this form. Thus, red cells are important for transport of both oxygen to the tissues and carbon dioxide from the tissues.

The first stage in the respiratory process is a mechanical one, to move air into and out of the lungs, which is caused by the respiratory muscles discussed in the last chapter. However, it must be recalled again that the lungs are not attached in any way to the chest walls. Instead, they simply float in the chest cavity, and as the chest cavity enlarges so also do the lungs enlarge. The reason for this is that enlargement of the chest creates a suction between the parietal and visceral pleurae, thus holding the lungs tightly against the inside of the chest wall even though there are no actual physical attachments. This mechanism obviously serves the very useful purpose of allowing the lungs to slide freely in the chest cavity and thus to expand uniformly throughout, as was also discussed in the previous chapter.

Therefore, movement of air into and out of the lungs is caused by slight suction during inspiration and slight excess pressure during expiration. Let us discuss briefly some of the important details of the pulmonary pressure changes that cause air movement.

Pulmonary Pressures

Alveolar Pressure. One will recall from the basic laws of physics that when a volume of gas is increased, its pressure falls. Thus, during inspiration the enlargement of the chest cage decreases the pressure in the alveoli to about −3 mm Hg—that is, to 3 mm Hg less than atmos-

pheric pressure—and it is this negative pressure that pulls air through the respiratory passageways into the alveoli.

During expiration exactly the opposite effects occur; compression of the chest cage around the lungs increases the alveolar pressure to approximately +3 mm Hg, which obviously pushes air out of the alveoli to the atmosphere.

If a person breathes with maximal effort but with his nose and mouth closed so that air cannot flow into or out of his lungs, the alveolar pressure can be decreased to as low as −80 mm Hg or increased to as high as +100 mm Hg. Thus, the maximum strength of the muscles of respiration is far greater than that needed for normal quiet respiration. This provides a tremendous reserve respiratory ability that can be called upon in times of need for maximal respiratory activity, such as during heavy exercise.

Intrapleural Pressure. The space between the visceral pleura of the lungs and the parietal pleura of the chest cavity is called the *intrapleural space,* and the pressure in this space is the *intrapleural pressure.* This pressure is always a few millimeters of mercury less than that in the alveoli. Thus, Figure 37–1 illustrates that during inspiration the intra-alveolar pressure is about −3 mm Hg and the intrapleural pressure is about −8 mm Hg. During normal expiration, the intra-alveolar pressure rises to +3 mm Hg; the intrapleural pressure rises to −2 mm Hg. It will be noted that the intrapleural pressure remains about 5 mm Hg less than the

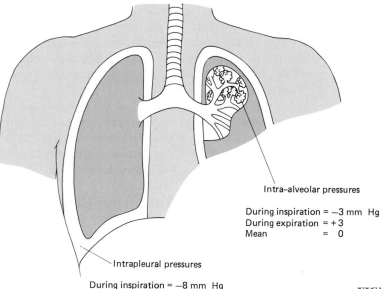

Intra-alveolar pressures

During inspiration = -3 mm Hg
During expiration = $+3$
Mean = 0

Intrapleural pressures

During inspiration = -8 mm Hg
During expiration = -2
Mean = -5

FIGURE 37-1   Alveolar and intrapleural pressures during normal breathing.

intra-alveolar pressure. The reason for this is that the lungs are always pulling away from the chest wall because of two effects: First, the surface tension of fluid lining the inside surfaces of the alveoli makes the alveoli try to collapse in the same manner that soap bubbles will collapse if an opening allows air to escape. Second, elastic fibers spread in all directions through the tissues of the lungs, and these also tend to contract the lungs. Both of these effects pull the lungs away from the outer walls of the pleural cavity, creating an average negative pressure in the intrapleural space of about -5 mm Hg.

Surfactant in the Alveoli.   If the alveoli were lined with water instead of normal alveolar fluid, the surface tension would be so great that it would cause the alveoli to remain collapsed almost all the time. However, a substance called a *surface active agent* or simply *surfactant* is secreted into the alveoli by epithelial cells of the alveolar walls. This substance acts like a detergent, greatly decreasing the surface tension of the fluid lining the alveoli and thus preventing lung collapse. However, there are times when surfactant is not secreted in sufficient quantities. For instance, many premature newborn babies secrete so little surfactant that they cannot expand their lungs after birth, and death results from asphyxia. This condition is called the *respiratory distress syndrome*.

Collapse of the Lung Caused by Pneumothorax.   When an opening is made in the chest wall the elastic forces in the lungs cause them to collapse immediately, sucking air through the opening into the chest cavity. This is called a "pneumothorax." Then, when the person tries to breathe, instead of the lungs expanding and contracting, air flows in and out of the hole in the chest. Thus, a wound of the chest can kill a person by suffocation; yet the condition can be treated very easily by sucking air out of the pleural cavity and plugging the hole.

Artificial Respiration

When the respiratory muscles fail, a person can be kept alive only by some means of artificial respiration. The simplest method is to force air into and out of his mouth or nose. Several devices have been made for this purpose, one of which is called the *resuscitator*. To use this apparatus a mask is placed over the mouth and nose, and intermittent blasts of air fill the lungs. Between blasts, air is either pulled back out of the lungs into the atmosphere or allowed to flow out because of elastic recoil of the chest wall and lungs.

For prolonged artificial respiration the person is usually placed inside an artificial tank respirator, in the manner shown in Figure 37-2. This is a large tank provided with some means for repetitively increasing and decreasing the pressure and for controlling the extent of the rise and fall in pressure. In the respirator of Figure 37-2 the air pressure is alternated by inward and outward movement of a large leather diaphragm at the foot end of the tank, and the positive and negative pressures are controlled by relief valves shown on top of the tank. Because respiration normally is accomplished mainly by

FIGURE 37–2 The artificial respirator.

the inspiratory muscles rather than by the expiratory muscles, the pressure inside the tank is usually adjusted to give more of a vacuum cycle than a pressure cycle. A vacuum of about -10 mm Hg causes approximately normal inspiration by pulling outward on the chest cage and abdominal wall, thereby sucking air into the lungs. Then, the respirator changes from the vacuum cycle to the pressure cycle and applies 2 to 3 mm Hg of positive pressure. This causes expiration by pushing against the abdomen and chest.

Spirometry and the Volume of Respiratory Air

Figure 37–3 illustrates a *spirometer*, an apparatus that can be used to record the flow of air into and out of the lungs. It consists of a drum inverted in a tank of water, with a tube extending from the air space in the drum to the mouth of the person to be tested. The drum is suspended from pulleys and counterbalanced by a weight. As the person breathes in and out, the drum moves up and down, and the counterweight balancing the drum also rides up and down, recording on a moving paper the changing volume of the chamber. Figure 37–4 illustrates a typical spirometer recording for successive breath cy-

cles, showing different depths of inspiration and expiration.

Tidal Volume, Respiratory Rate, and Minute Respiratory Volume. The air that passes into and out of the lungs with each respiration is called the *tidal air*, and the volume of this air in each breath is called the *tidal volume*. The normal tidal volume is about 500 ml, and the normal rate of respiration for an adult is usually about 12 times per minute. Therefore, a total of about 6 L of air normally passes into and out of the respiratory passageways each minute. This amount is called the *minute respiratory volume*.

The Inspiratory Capacity. In Figure 37–4, after three normal breaths, the person breathes inward as deeply as possible. The amount of air that he can pull into his lungs beyond that already in his lungs at the beginning of the breath is called the *inspiratory capacity*. This quantity is approximately 3000 ml in the normal person.

Expiratory Reserve Volume. After several more normal respirations, the person expires as much as possible (see Figure 37–4). The amount of air that he is capable of expiring beyond that which he normally expires is called the *expiratory reserve volume*. This is usually about 1100 ml.

Residual Volume and the Functional Residual Capacity. In addition to the expiratory reserve volume, there is air in the lungs that cannot be expired even by the most forceful exhalation. The volume of this air is about 1200 ml, and it is called the *residual volume*. The sum of the expiratory reserve volume plus the residual volume is called the *functional residual capacity;* this is the amount of air remaining in the respiratory system at the end of a normal expiration. It is this air that allows oxygen and carbon dioxide transfer into and out of the blood to continue even between breaths.

Vital Capacity. Finally, to the far right in Figure 37–4 is shown the spirogram of a maxi-

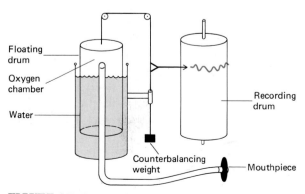

Floating drum

Oxygen chamber

Water

Recording drum

Counterbalancing weight

Mouthpiece

FIGURE 37–3 The spirometer.

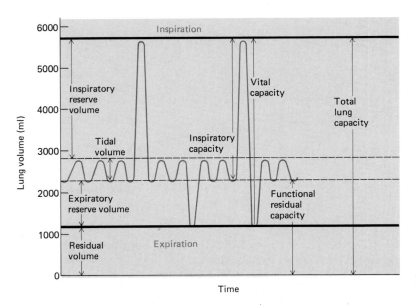

FIGURE 37–4 A spirogram, showing the divisions of the respiratory air.

mal inspiratory effort followed immediately by a maximal expiratory effort. The total change in pulmonary volume between these two extremes is called the *vital capacity*. The vital capacity of the normal person is approximately 4500 ml. A well-trained male athlete may have a vital capacity as great as 6500 ml; a small female frequently has a vital capacity no greater than 3000 ml.

The vital capacity is a measure of the person's overall ability to inspire and expire air, and it is determined mainly by two factors: (1) the strength of the respiratory muscles and (2) the elastic resistance of the chest cage and lungs to expansion and contraction. Disease processes that either weaken the muscles—such as poliomyelitis—or decrease the expansibility of the lungs—such as tuberculosis—can decrease the vital capacity. For this reason, vital capacity measurements are an invaluable tool in assessing the functional ability of the mechanical breathing system.

DEAD SPACE

Much of the air pulled into the respiratory passages with each breath never reaches the alveoli because it merely fills the passageways such as the nose, the pharynx, the trachea, and the bronchi. Then this air is expired without ever entering the alveoli. This air is useless from the point of view of oxygenating the blood. Consequently the respiratory passageways are called *dead space*. The total volume of this space is normally about 150 ml, which means that during inspiration of a normal tidal volume of 500 ml, only 350 ml of new air actually enters the alveoli.

Alveolar Ventilatory Rate. The most important measure of the effectiveness of a per-

son's respiration is his *alveolar ventilatory rate*, which is the total quantity of new air that enters his alveoli each minute. With only 350 ml of air reaching the alveoli with each breath and with a normal respiratory rate of 12 times per minute, the alveolar ventilatory rate is approximately 4200 ml per minute. During maximal respiratory effort this can be increased to higher than 120 L per minute, and at the opposite extreme a person can remain alive, at least for a few hours, with an alveolar ventilatory rate as low as 1200 ml per minute.

Exchange of Alveolar Air with Atmospheric Air

At the end of each expiration approximately 2300 ml of air still remains in the lungs. This is called the *functional residual capacity* as was explained previously. With each breath, about 350 ml of new air is brought into this air, and with each expiration this same amount of air is removed. It is obvious, then, that each breath does not exchange all the old alveolar air for new air, but instead *exchanges only about one seventh of the air*. This effect is illustrated in Figure 37–5, which shows an alveolus containing a quantity of some foreign gas at the beginning of the series of breaths. After the first breath a small portion of this gas has been removed, after the second breath a little more, after the third a little more, and so forth. Note that even after the 16th breath a small portion of the foreign gas is still in the alveolus.

At a normal alveolar ventilation of 4200 ml per minute, approximately one-half the alveolar gases are replaced by new air every 23 seconds. This slow turnover of air in the alveoli keeps the

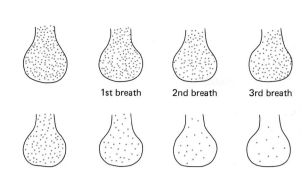

1st breath   2nd breath   3rd breath

4th breath   8th breath   12th breath   16th breath

FIGURE 37–5 Expiration of a foreign gas from the alveoli with successive breaths.

alveolar gaseous concentrations from rising and falling greatly with each individual breath.

ALVEOLAR AIR

Alveolar air is a mixture of inspired air, water vapor from the respiratory passageways, and carbon dioxide excreted from the blood. Also, because oxygen is continually being absorbed into the blood, the oxygen concentration of alveolar air is considerably less than that of the atmosphere. However, before it will be possible to understand how the concentrations of oxygen and carbon dioxide are controlled in the alveolar air we will need to review briefly the principle of partial pressures.

Partial Pressures

The partial pressure of a gas is the amount of pressure exerted by each type of gas in a mixture of gases. For instance, let us assume that we have a mixture of 50 percent oxygen (in terms of numbers of molecules) and 50 percent nitrogen and that the total pressure of the mixture is 100 mm Hg. The portion of this pressure that is caused by oxygen is 50 mm Hg, and the portion caused by nitrogen is also 50 mm Hg. Therefore, it is said that the partial pressure of each of these two gases is 50 mm Hg. But now let us assume that the composition is changed to 75 percent nitrogen and 25 percent oxygen, and at the same time the total pressure is changed to 1000 mm Hg. Now the partial pressure exerted by the nitrogen will be 750 mm Hg, and that of the oxygen will be 250 mm Hg.

Thus, the partial pressure of a gas is a measure of the total force that each gas alone exerts against the walls surrounding it. Consequently, the penetrating power of a gas through a membrane, such as the respiratory membrane between the alveoli and the blood, is also di-

rectly proportional to its partial pressure. That is, the greater the partial pressure of a gas in the alveolus, the greater is its tendency to pass through this membrane into the blood.

The partial pressures of the important respiratory gases are represented by the following symbols:

Nitrogen	P_{N_2}
Oxygen	P_{O_2}
Carbon dioxide	P_{CO_2}
Water vapor	P_{H_2O}

Composition of Alveolar Air. Table 37–1 shows the partial pressures and percent concentrations of nitrogen, oxygen, carbon dioxide, and water vapor in atmospheric air and alveolar air at sea level. This table shows that normal atmospheric air is composed almost entirely of only two gases, about four-fifths nitrogen and one-fifth oxygen, with almost negligible quantities of carbon dioxide and water vapor. In contrast, alveolar air contains considerable quantities of both carbon dioxide and water vapor, whereas the oxygen concentration is considerably less than that of atmospheric air. These differences can be explained as follows:

Humidification of Air as It Enters the Respiratory Passages. When air is inspired, it is humidified immediately by moisture from the linings of the respiratory passages. At normal body temperature the partial pressure of water vapor in the lungs is 47 mm Hg, and as long as the body temperature remains constant, this partial pressure also remains constant.

The mixing of water vapor with the incoming atmospheric air dilutes the air so that the pressure of the other gases becomes slightly decreased. This explains why the nitrogen partial pressure in the alveoli is slightly less than its partial pressure in atmospheric air.

Oxygen and Carbon Dioxide Partial Pressures in the Alveoli. Alveolar air continually loses oxygen to the blood, and this oxygen is replaced by carbon dioxide diffusing out of the blood into the alveoli. This explains why the oxygen pressure in alveolar air is much less than in

TABLE 37–1
Partial Pressures (mm Hg) and Percent Concentrations of Respiratory Gases in the Atmosphere and in the Alveoli

Gas	Atmospheric air	Alveolar air
N_2	597.0 (78.62%)	569.0 (74.9%)
O_2	159.0 (20.84%)	104.0 (13.6%)
CO_2	0.15 (0.04%)	40.0 (5.3%)
H_2O	3.85 (0.5%)	47.0 (6.2%)

atmospheric air, and it also explains why the carbon dioxide pressure is considerably greater in the alveolar air even though it is almost non-existent in atmospheric air.

The normal partial pressure of oxygen in the alveoli is approximately 104 mm Hg in comparison with 159 mm Hg in the atmosphere. That of carbon dioxide is 40 mm Hg in comparison with almost zero in the atmosphere. However, these values change greatly from time to time depending on the rate of alveolar ventilation and the rate of oxygen and carbon dioxide transfer into and out of the blood. For instance, a high alveolar ventilatory rate provides greater amounts of oxygen to the alveoli and increases the alveolar oxygen pressure. Also, a high alveolar ventilatory rate removes carbon dioxide from the alveoli more rapidly than usual, thereby decreasing the carbon dioxide pressure.

TRANSPORT OF GASES THROUGH THE RESPIRATORY MEMBRANE

The Respiratory Membrane

The *respiratory membrane*, which is also called the *pulmonary membrane*, is composed of all the pulmonary surfaces that are thin enough to allow gases to diffuse into the pulmonary blood.

These include, as illustrated in Figure 37–6, the membranes of the *respiratory bronchioles*, the *alveolar ducts*, the *atria*, the *alveolar sacs*, and the outpouchings of the sacs, the *alveoli*.

The total area of the respiratory membrane is approximately 70 m^2, which is about equal to the floor area of a moderate-sized classroom. Less than 100 ml of blood is in the capillaries at any one time. If one will think for a moment of this small amount of blood spread out evenly over the entire floor of a classroom, he can quite readily understand that very large quantities of gas can enter and leave this blood in only a fraction of a second.

Figure 37–7 illustrates diagrammatically an electron micrographic cross-section of the respiratory membrane, showing also a red blood cell in a capillary in close proximity to the membrane. This membrane is composed of several layers, including (1) a thin lining of fluid on the surface of the alveolus containing the substance called "surfactant" that was discussed earlier in the chapter; (2) a lining of epithelial cells that is the surface of the alveolus; (3) a small interstitial space containing small amounts of connective tissue and basement membrane; and (4) a layer of capillary endothelial cells that forms the wall of the capillary. However, all layers of this membrane are extremely thin, so much so that most

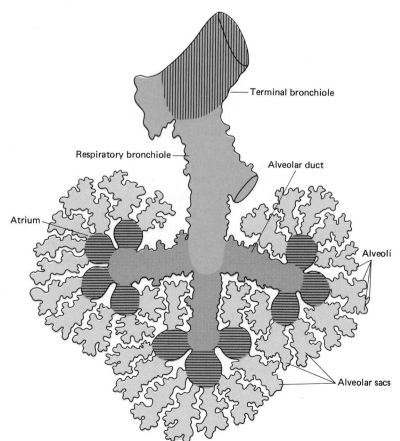

FIGURE 37–6 The respiratory unit, illustrating the anatomy of the respiratory surfaces. (From Miller: The Lung. Springfield, Ill., Charles C Thomas.)

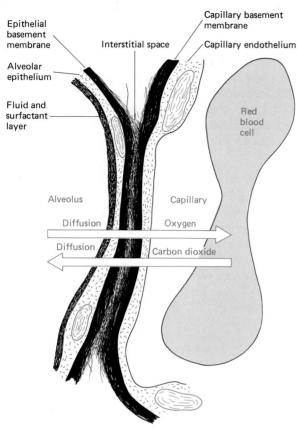

FIGURE 37-7 Ultrastructure of the respiratory membrane.

Epithelial basement membrane
Alveolar epithelium
Fluid and surfactant layer
Interstitial space
Capillary basement membrane
Capillary endothelium
Red blood cell
Alveolus
Capillary
Diffusion — Oxygen
Diffusion — Carbon dioxide

Solution and Diffusion of Gases in Water

To understand the transport of gases through the pulmonary membrane, one must first be familiar with the physical principles of gaseous solution and diffusion in water. Figure 37–8 shows a chamber containing water at the bottom and normal air at the top. Molecules of nitrogen and oxygen are continually bouncing against the surface of the water, and some of the molecules enter the water to become dissolved. The dissolved molecules then bounce among the water molecules in all directions, some of them eventually reaching the surface again to bounce back into the gaseous space. After the air has remained in contact with the water for a long time, the number of molecules passing out of the solution equals exactly the number of molecules passing into the solution. When this

areas of the respiratory membrane are less than 1μ thick, which is only a fraction of the thickness of a red blood cell itself. The thinness of the membrane obviously allows rapid diffusion of oxygen and carbon dioxide between the alveolar air and the blood.

$N_2 = 74.2\%$	—	$P_{N_2} = 564$ mm Hg
$O_2 = 19.7\%$	—	$P_{O_2} = 149$ mm Hg
$H_2O = 6.2\%$	—	$P_{H_2O} = 47$ mm Hg
Total = 100%	—	Total = 760 mm Hg
Dissolved N2	—	$P_{N_2} = 564$ mm Hg
Dissolved O2	—	$P_{O_2} = 149$ mm Hg
Water		

FIGURE 37-8 A sample of air exposed to water, showing equilibration between gases in the gaseous phase and in the dissolved state.

state has been reached, the gases in the gaseous phase are said to be *in equilibrium* with the dissolved gases.

In the equilibrium state, the partial pressure tending to force nitrogen molecules into the water is 564 mm Hg, and the pressure of nitrogen molecules pushing outward from the water is also 564 mm Hg. If the nitrogen pressure becomes greater in the gaseous phase than in the dissolved phase, more nitrogen molecules will move into the solution than out. If the pressure becomes less in the gas than in the solution, more nitrogen molecules will leave the solution than will enter it. In other words, *the pressure of a gas is the force with which it attempts to move from its present surroundings.* Whenever the pressure is greater at one point than at another, whether this be in a solution or in a gaseous mixture, more molecules will move toward the low pressure area than toward the high pressure area.

Diffusion of Gases Through Tissues and Fluids. Figure 37–9 shows in a chamber a solution containing dissolved gas. At end A of the chamber is a high concentration of dissolved gas molecules, while at end B is a low concentration. The molecules are bouncing at random in all directions because of their kinetic activity. Obviously, because of the higher concentration at A than at B, it is easier for large numbers of mole-

Dissolved gas molecules

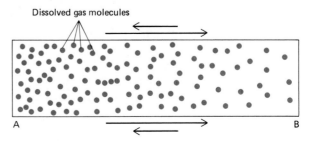

A   B

FIGURE 37-9 Diffusion of gas molecules through a liquid.

cules to move from A to B than in the opposite direction, which is indicated by the lengths of the arrows. Eventually the numbers of molecules at both ends of the chamber will become approximately equal. Movement of molecules in this manner from areas of high concentration toward areas of low concentration is called *diffusion.*

The rate of diffusion of gas molecules in the body fluids and tissues is determined by the pressure differences between the different points. Far more gas molecules are at point A in Figure 37–9 than at point B. Consequently, the pressure of the gas at point A is also far greater than at point B. The difference between the two pressures is called simply the *pressure difference,* and the net rate of diffusion between the two points is directly proportional to the pressure difference.

Diffusion of Gases Through the Respiratory Membrane. With the above background information, we can now discuss all the factors that determine the rate of gas diffusion through the pulmonary membrane. These are the following:

1. The greater the *pressure difference* between one side of the membrane and the other, the greater will be the rate of gaseous diffusion. If the pressure of a gas in the alveolus is 100 mm Hg while that in the blood is 99 mm Hg, the pressure difference will be only 1 mm, and the rate of gaseous diffusion will be very slight. If the pressure in the blood suddenly falls to 0 mm Hg, the pressure difference rises to 100 mm, and the gaseous diffusion increases to 100 times its former rate.

2. The greater the *area of the respiratory membrane,* the greater will be the quantity of gas that can diffuse in a given period of time. In some pulmonary diseases, such as *emphysema,* large portions of the lungs are destroyed and the total area of the respiratory membrane is greatly decreased. Often the decrease is enough to cause continual respiratory embarrassment.

3. The *thinner the membrane,* the greater will be the rate of gaseous diffusion. The normal respiratory membrane is sufficiently thin so that venous blood entering the pulmonary capillary can be brought almost to complete gaseous equilibrium with the alveolar air within

approximately one fifth of a second. Occasionally, though, the thickness of the membrane and of the fluid on its surface increases manyfold, especially when the lungs become edematous because of pulmonary congestion, pneumonia, or some other lung disease. The person then may die simply because gases cannot diffuse through the thick membrane rapidly enough.

4. The greater the *solubility of the gas in the respiratory membrane,* the greater also will be the rate of gas diffusion. The reason for this is that when greater quantities of gas dissolve in a given area of the membrane, larger numbers of molecules can then traverse the membrane at the same time. Considering the solubility of oxygen to be 1, the relative solubilities of the three important respiratory gases are the following:

$$Oxygen = 1$$
$$Carbon\ dioxide = 20$$
$$Nitrogen = 0.5$$

Therefore, carbon dioxide diffuses through the respiratory membrane approximately 20 times as easily as oxygen, whereas oxygen diffuses approximately 2 times as easily as nitrogen.

5. A gas diffuses through the membrane approximately in *inverse proportion to the square root of its molecular weight.* The above three usual respiratory gases all have square roots of their molecular weights that are almost equivalent. However, for gases with much smaller molecular weights, such as helium, this can be a significant factor in determining the rate of diffusion.

Putting all of these factors together, one derives the formula shown at the bottom of the page for rate of diffusion through the respiratory membrane.

TRANSPORT OF OXYGEN TO THE TISSUES
Diffusion of Oxygen from the Alveoli into the Blood and from the Blood into the Tissues

The overall transport of oxygen from the alveolus to the tissue cell requires three different events: (1) diffusion of oxygen from the alveolus into the

$$Rate\ of\ Diffusion \propto \frac{(Pressure\ A - Pressure\ B) \times Surface\ Area \times Solubility}{Membrane\ Thickness \times \sqrt{Molecular\ Weight}}$$

TABLE 37–2
Gaseous Pressures in the Alveoli and the Blood

Gas	Alveolar air	Venous blood	Arterial blood
O_2	104	40	100
CO_2	40	45	40
N_2	569	569	569

pulmonary blood, (2) transport of the blood through the arteries to the tissue capillaries, and (3) diffusion of oxygen from the capillaries to the tissue cell.

The data of Table 37–2 explain why oxygen diffuses from the alveoli into the pulmonary blood. The oxygen pressure (Po_2) of "venous" blood entering the lungs is only 40 mm Hg in comparison with the alveolar air Po_2 of 104 mm Hg. The pressure difference, therefore, is 64 mm Hg, which causes extremely rapid diffusion of oxygen into the blood. During the very short time that the blood remains in the capillary, only about 1 second, it attains a Po_2 of approximately 100 mm Hg, almost as much as that of the alveolar air itself.

The aerated blood then flows from the lungs to the tissue capillaries, where oxygen diffuses through the capillary membrane and tissue spaces into the tissue cells. The reason for this direction of diffusion is that the oxygen pressure in the cells is very low, which may be explained as follows: When oxygen enters a cell it reacts very readily with sugars, fat, and proteins to form carbon dioxide and water, as illustrated in Figure 37–10. As a result, much of the oxygen is immediately removed, thus keeping the intracellular Po_2 usually at about 20 mm Hg.

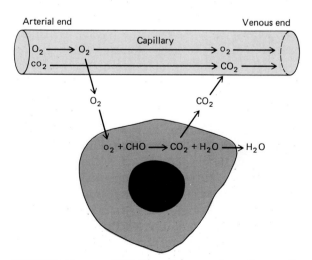

FIGURE 37–10 Diffusion of oxygen from the capillary to the tissue cell, formation of carbon dioxide inside the cell, and diffusion of carbon dioxide back to the tissue capillary.

The high pressure, 100 mm Hg, of the oxygen in the arterial blood as it enters the capillaries, causes it to diffuse into the interstitial fluid and thence through the cellular membrane to the interior of the cells.

In summary, oxygen movement, both into the blood of the lungs and from the blood to the tissues, is caused by diffusion. The oxygen pressure difference from the lungs to the pulmonary blood is sufficient to keep oxygen always diffusing into the blood, and the pressure difference from the capillary blood to the tissue cells is also always sufficient to keep oxygen diffusing from the blood to the cells.

Transport of Oxygen in the Blood by Hemoglobin

When oxygen diffuses from the lungs into the blood, a small proportion of it becomes dissolved in the fluids of the plasma and red cells, but approximately 60 times as much combines immediately with the hemoglobin of the red blood cells in the manner explained in Chapter 33 and is carried in this combination to the tissue capillaries. Indeed, without the hemoglobin the amount of oxygen that could be carried to the tissues would be only a fraction of that required to maintain life.

When the blood passes through the tissue capillaries, oxygen splits away from the hemoglobin and diffuses into the tissue cells. Thus, hemoglobin acts as a *carrier of oxygen*, increasing the amount of oxygen that can be transported from the lungs to the tissues to some 60 times more than that which could be transported only in the dissolved state.

Combination of Oxygen with Hemoglobin: the Oxygen-Hemoglobin Dissociation Curve. Figure 37–11 illustrates the so-called *oxygen-hemoglobin dissociation curve*. This curve shows the percentage of the hemoglobin that is combined chemically with oxygen at each oxygen pressure level. Aerated blood leaving the lungs usually has an oxygen pressure of about 100 mm Hg. Referring to the curve, it is seen that at this pressure *approximately 97 percent of the hemoglobin is combined with oxygen.* As the blood passes through the tissue capillaries, the oxygen pressure falls normally to about 40 mm Hg. At this pressure *only about 70 percent* of the hemoglobin is combined with oxygen. Thus, about 27 percent of the hemoglobin normally loses its oxygen to the tissue cells. Then, on returning to the lungs it combines with new oxygen and transports this once again to the tissue cells.

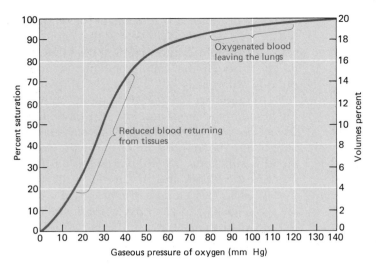

FIGURE 37–11   The oxygen-hemoglobin dissociation curve.

Note in Figure 37–11 another scale on the right-hand side of the graph, in addition to the percent saturation scale. This second scale gives "volumes percent," which means the number of milliliters of oxygen actually bound with each 100 ml of normal blood at the different percent levels of saturation. When the blood is oxygenated to the normal arterial level of 97 percent saturation, about 19 ml of oxygen will be bound with the hemoglobin. Then, when the blood loses oxygen to the tissues and the hemoglobin saturation falls to 70 percent, the amount of oxygen that remains bound with the blood entering the veins is still 14 ml per 100 ml of blood. Therefore, each 100 ml of blood passing through the tissues normally delivers about 5 ml of oxygen to the cells. During heavy exercise, this delivery can increase to as much as 15 to 18 ml for each 100 ml of blood passing through the tissues, as we shall discuss more fully in the following section.

Utilization Coefficient and the Hemoglobin Reserve Capacity.   The proportion of the hemoglobin that loses its oxygen to the tissues during each passage through the capillaries is called the *utilization coefficient.* In the normal person the utilization coefficient is 27 percent, or expressed another way, approximately one fourth of the hemoglobin is used to transport oxygen to the tissues under normal conditions.

When the tissues are in extreme need of oxygen, the oxygen pressure in the tissues falls to extremely low values, allowing oxygen to diffuse from the capillary blood much more rapidly than usual. As a result, the saturation of the hemoglobin in the tissue capillaries can fall to as low as 10 to 20 percent instead of the normal level of 70 percent, and the utilization coefficient sometimes rises to as high as 80 to 90 percent.

Therefore, without even increasing the rate of blood flow, the amount of oxygen transported to the tissues in times of serious need can be increased more than threefold.

If one also remembers that the cardiac output can increase as much as five- to sevenfold in times of stress, then it is clear that the amount of oxygen that can be transported to the tissues can be raised to as much as 15 to 20 times normal, part of this being caused by an increase in the utilization coefficient and even more by the increase in cardiac output.

Hemoglobin as a Tissue Oxygen Buffer.   For cellular function to continue at a normal pace, the concentrations of all substances in the extracellular fluid must remain relatively constant at all times. One of the functions of hemoglobin is to keep the oxygen pressure in the tissues almost always between the limits of 20 and 45 mm Hg. This may be explained as follows: As blood flows through the capillaries, 27 percent of the oxygen is usually removed from the hemoglobin, making the hemoglobin saturation fall to about 70 percent. Referring again to Figure 37–11, it can be seen that to remove this much oxygen the oxygen pressure in the tissues can never rise above 45 mm Hg. On the other hand, whenever the oxygen pressure falls to 20 mm Hg, more than three fourths of the oxygen is then released from the hemoglobin, and this rapid release usually keeps the tissue oxygen pressure from falling any lower except in times of extreme oxygen need. Thus, hemoglobin automatically releases oxygen to the tissues so that the tissue P_{O_2} remains almost always between the limits of 20 and 45 mm Hg. If it were not for this buffering function of hemoglobin, a person could not survive when the oxygen pressure in the atmosphere becomes either too high or too low, as often occurs either when breathing oxygen or, at the other extreme, ascending only a few thousand feet up a mountain.

Carbon Monoxide as a Hemoglobin Poison. Carbon monoxide combines with hemoglobin in a manner almost identical to that of oxygen, except that the combination of carbon monoxide with hemoglobin is approximately 210 times as tenacious as the combination of oxygen with hemoglobin. Also, because the two substances combine with hemoglobin at the same point on the molecule, they cannot both be bound to the hemoglobin at the same time. Therefore, carbon monoxide mixed with air in a concentration of only 0.1 percent, 200 times less than that of oxygen in normal air, will cause half of the hemoglobin to combine with carbon monoxide and therefore leave only half of the hemoglobin to combine with oxygen. When the concentration of carbon monoxide rises above approximately 0.2 percent, which is still 100 times less than the normal concentration of oxygen, the quantity of hemoglobin available to transport oxygen becomes so slight that death ensues.

When death is about to occur from carbon monoxide poisoning, this can frequently be prevented by administering pure oxygen. The pure oxygen, on entering the alveoli, has a partial pressure of approximately 600 mm Hg, six times the normal alveolar oxygen pressure. The force with which oxygen can combine with hemoglobin is therefore increased sixfold, which forces the carbon monoxide off the hemoglobin molecule six times as rapidly as would occur without treatment.

TRANSPORT OF CARBON DIOXIDE

Referring once again to Figure 37–10, it can be seen that, when oxygen is used by the cells for metabolism, carbon dioxide is formed. The pressure of carbon dioxide (P_{CO_2}) in the cells becomes high, about 50 mm Hg, and a pressure difference develops between the cells and the blood in the capillary. This causes carbon dioxide to diffuse out of the cell, into the interstitial fluid, and from there into the capillary blood. It is transported in the blood to the lungs. Thus, the P_{CO_2} of the blood entering the tissue capillaries is about 40 mm Hg, but this rises to 45 mm Hg in these capillaries as carbon dioxide enters from the cells. Therefore, the P_{CO_2} of the blood

entering the lungs is also approximately 45 mm Hg, whereas the P_{CO_2} in the air of the alveolus is only 40 mm Hg. A pressure difference of 5 mm Hg thus exists between the blood and the alveolus, causing carbon dioxide to diffuse out of the blood. Furthermore, because of the extreme solubility of carbon dioxide in the respiratory membrane, the P_{CO_2} of the pulmonary blood falls to almost complete equilibrium with the alveolar P_{CO_2}, that is, to 40 mm Hg, before it leaves the pulmonary capillary.

Chemical Combinations of Carbon Dioxide with the Blood

Figure 37–12 illustrates the different ways that carbon dioxide is transported in the blood. These are: (1) as *dissolved carbon dioxide*, (2) in the form of *bicarbonate ion* (HCO_3^-), and (3) *combined with hemoglobin* ($Hgb \cdot CO_2$).

Only about 7 percent of the carbon dioxide is transported in the blood in the dissolved state. About 93 percent of it diffuses from the plasma into the red cell where it undergoes two chemical reactions: First, the carbon dioxide reacts with water to form carbonic acid. Red blood cells contain an enzyme called *carbonic anhydrase*, which speeds this reaction approximately 250-fold. Therefore, about 70 percent of the carbon dioxide combines with water to form carbonic acid inside the red blood cells, whereas only small quantities react in the plasma. The carbonic acid in the cells immediately reacts with the acid-base buffers of the cells and becomes mainly *bicarbonate ion* (HCO_3^-).

Second, about 23 percent of the total carbon dioxide entering the red cell combines di-

FIGURE 37–12 Transport of carbon dioxide in the blood.

rectly with hemoglobin to form a compound called *carbaminohemoglobin*. This reaction does not occur at the same point on the hemoglobin molecule as the reaction between oxygen and hemoglobin. Therefore, hemoglobin can combine with both carbon dioxide and oxygen at the same time. In this way hemoglobin acts as a carrier not only of oxygen but also of carbon dioxide. However, the reaction occurs relatively slowly so that this method of carbon dioxide transport is much less important than transport in the form of bicarbonate ion.

To recapitulate, carbon dioxide is transported in the blood in three principal forms. First, only a very small quantity is transported as dissolved carbon dioxide. Second, by far the largest proportion is transported in the form of bicarbonate ion. Third, a few percent is transported in the form of carbaminohemoglobin.

When the blood enters the pulmonary capillaries, all the chemical combinations of carbon dioxide with blood are reversed, and the carbon dioxide is released into the alveoli.

QUESTIONS

1. How does alveolar pressure change during the respiratory cycle?
2. What is the function of surfactant in the alveoli?
3. Explain what is meant by tidal volume, respiratory rate, minute respiratory volume, functional residual capacity, and vital capacity. Give average values for each.
4. Explain why dead space occurs in the respiratory airways and tell its significance.
5. Explain what is meant by "partial pressure." What are the approximate partial pressures of the important gases in the alveoli?
6. Explain the factors that determine the rate of transport of a gas through the respiratory membrane.
7. Discuss the overall transport of oxygen from the atmosphere to the tissue cells.
8. Explain the role of hemoglobin in oxygen transport.
9. Explain the transport of carbon dioxide from the tissues to the blood and thence through the lungs to the atmosphere.

REFERENCES

Bartels, H., and Baumann, R.: Respiratory function of hemoglobin. *Int. Rev. Physiol.*, *14*:107, 1977.

Bauer, C., *et al.* (eds.): Biophysics and Physiology of Carbon Dioxide. New York, Springer-Verlag, 1980.

Bradley, G.W.: Control of the breathing pattern. *Int. Rev. Physiol.*, *14*:185, 1977.

Ellis, P.D., and Billings, D.M.: Cardiopulmonary Resuscitation: Procedures for Basic and Advanced Life Support. St. Louis, C.V. Mosby, 1979.

Fishman, A.P.: Assessment of Pulmonary Function. New York, McGraw-Hill, 1980.

Fishman, A.P.: Hypoxia in the pulmonary circulation. *Circ. Res.*, *38*:221, 1976.

Fishman, A.P., and Pietra, G.: Hemodynamic pulmonary edema. *In* Fishman, A.P., and Renkin, E.M. (eds.): Pulmonary Edema. Baltimore, Waverly Press, 1979, p. 79.

Guyton, A.C., *et al.*: Forces governing water movement in the lung. *In* Pulmonary Edema. Washington, D.C., American Physiological Society, 1979, p. 65.

Jones, N.L.: Blood Gases and Acid-Base Physiology. New York, B.C. Decker, 1980.

Macklem, P.T.: Respiratory mechanics. *Annu. Rev. Physiol.*, *40*:157, 1978.

Murray, J.F.: The Normal Lung. Philadelphia, W.B. Saunders, 1976.

Parker, J.C., *et al.*: Pulmonary transcapillary exchange and pulmonary edema. *In* Guyton, A.C., and Young, D.B. (eds.): International Review of Physiology: Cardiovascular Physiology III, Vol. 18. Baltimore, University Park Press, 1979, p. 261.

Randall, D.J.: The Evolution of Air Breathing in Vertebrates. New York, Cambridge University Press, 1980.

Singh, R.P.: Anatomy of Hearing and Speech. New York, Oxford University Press, 1980.

Thurlbeck, W.M.: Structure of the lungs. *Int. Rev. Physiol.*, *14*:1, 1977.

Wagner, P.D.: Diffusion and chemical reaction in pulmonary gas exchange. *Physiol. Rev.*, *57*:257, 1977.

Wyman, R.J.: Neural generation of the breathing rhythm. *Annu. Rev. Physiol.*, *39*:417, 1977.

38

Regulation of Respiration; and the Physiology of Respiratory Abnormalities

Overview

The basic rhythm of respiration is provided by the *respiratory center*, located in the reticular substance of the medulla and pons of the brain stem. This center in turn is comprised of three major groups of neurons called (1) the *inspiratory area*, (2) the *expiratory area*, and (3) the *pneumotaxic area*.

In normal quiet breathing, the inspiratory area is activated about once every 5 seconds and causes *inspiration that lasts for about 2 seconds*. Thus, normal respiration is caused almost entirely by contraction of the inspiratory muscles with little contribution from the expiratory muscles. However, in heavy breathing, the expiratory center becomes active during periods between inspiratory activity, and the expiratory muscles then contribute about as much as the inspiratory muscles to the respiratory process. The pneumotaxic center controls the depth of breathing as well as the time interval between breaths.

During normal respiration a person breathes a total of about 6 L of air per minute, called the *minute respiratory volume.* During very heavy exercise, this can increase to as high as 150 L per minute.

The rate and depth of respiration are controlled by four different factors: (1) the *pressure of carbon dioxide* (Pco_2) in the blood, (2) the *concentration of hydrogen ions* (pH) in the blood, (3) the *pressure of oxygen* (Po_2) in the blood, and (4) *nervous signals from the muscle-controlling areas of the brain.* For controlling respiration, the pressure of carbon dioxide and the concentration of hydrogen ions in the blood are considerably more important than the pressure of oxygen, which is contrary to what might be expected. However, this is fortunate because the concentrations of both carbon dioxide and hydrogen ions in the tissue cells are determined almost entirely by the ability of the lungs to expire carbon dioxide (which also decreases the blood carbonic acid and reduces the hydrogen ion concentration at the same time). On the other hand, the *hemoglobin* in the blood functions as a very powerful *"oxygen buffer"* that helps to regulate the tissue oxygen concentration, so that exact control of respiration is not required for maintaining normal

619

amounts of oxygen in the tissue cells. During *heavy exercise,* when one requires greatly increased respiration, nervous signals from the muscle-controlling areas of the brain have a direct stimulatory effect on the respiratory center to increase respiration, and this helps to keep the concentrations of carbon dioxide, hydrogen ions, and oxygen almost exactly normal in the blood during exercise.

Some of the more important respiratory abnormalities are the following:

1. *Hypoxia,* which means low oxygen content and can be caused by (a) low partial pressure of oxygen in the air, (b) abnormalities of the lungs that diminish oxygen diffusion into the pulmonary blood, (c) decreased amounts of hemoglobin in the blood to carry oxygen to the tissues, (d) inability of the heart to pump adequate quantities of blood to the tissues, or (e) inability of the tissues to utilize the oxygen even if it should get there.
2. *Dyspnea,* which means "air hunger." This most often results from excess carbon dioxide in the blood, which stimulates the dyspneic sensation. However, some persons develop "psychic dyspnea" because of a neurotic state.
3. *Pneumonia,* which means infection in the lungs. This causes the alveoli to fill with infectious exudates that block the absorption of air from the alveoli into the pulmonary blood.
4. *Pulmonary edema,* which is most often caused by high pulmonary capillary pressure resulting from failure of the left heart. Fluid transudes out of the capillaries into the lung tissues and alveoli and blocks the transfer of oxygen and carbon dioxide through the respiratory membrane.
5. *Emphysema,* which occurs in most instances as a result of smoking. In this condition, as much as four fifths of the alveolar walls may be destroyed so that the person actually has as little as one-fifth normal functioning lung tissue.
6. *Asthma,* which results from spasm of the terminal bronchioles of the lungs, usually resulting from allergic stimulation of the bronchiolar smooth muscle.

THE RESPIRATORY CENTER AND THE BASIC RHYTHM OF RESPIRATION

The nervous system adjusts the rate of alveolar ventilation almost exactly to the demands of the body, so that the blood oxygen pressure (Po_2) and carbon dioxide pressure (Pco_2) are hardly altered even during strenuous exercise or other types of respiratory stress.

The respiratory control system is illustrated in Figure 38–1. It is composed of three separate groups of neurons located bilaterally in the medulla and pons of the brain stem:

1. The *inspiratory area.*
2. The *expiratory area.*
3. The *pneumotaxic area.*

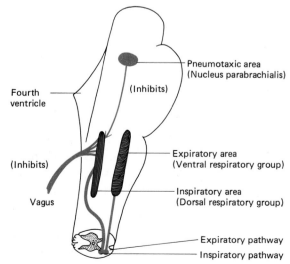

FIGURE 38–1 Organization of the respiratory center.

Both the inspiratory and expiratory areas are located in the reticular substance of the medulla, the inspiratory area is located dorsolaterally in each side of the medulla, and the expiratory area is located ventrolaterally. The pneumotaxic area, on the other hand, is located in the reticular substance of the upper one third of the pons.

THE INSPIRATORY AREA AND ITS RHYTHMICAL OSCILLATION

The basic rhythm of respiration is generated in the inspiratory area. Every few seconds this area becomes excited and transmits nerve signals to the inspiratory muscles, especially to the diaphragm. At the onset of each cycle of excitation, the signals start off very weak, but they increase progressively during the next 1 or 2 seconds, causing the muscles of inspiration to contract more and more forcefully to draw air into the lungs. Then, these inspiratory signals cease abruptly, and all of the inspiratory muscles suddenly relax. In normal respiration the elastic recoil of the lungs and chest cage causes the lungs to deflate during the next 2 to 3 seconds back to their normal size, thus promoting normal expiration. Thus, the respiratory cycle is completed, lasting about 2 seconds for inspiration and 3 seconds for expiration. Then the inspiratory center becomes excited again and the cycle is repeated, this occurring throughout the life of the person.

FUNCTION OF THE PNEUMOTAXIC CENTER TO CONTROL DEPTH AND RATE OF RESPIRATION

Stimulation of the pneumotaxic center increases the rate of respiration but simultaneously decreases the depth of respiration by an almost equal amount. Therefore, the total volume of air breathed per minute changes very little. One might ask: What could be the purpose of changing to rapid shallow breathing? Remember, however, that in some lower animals such as the dog the method for keeping the animal cool in hot climates is to "pant," which means to breathe with rapid, shallow breaths that evaporate maximum amounts of moisture from the upper respiratory surfaces, thus cooling the whole animal. The pneumotaxic center is closely related to the so-called *panting center*, which is also located in the upper portion of the brain stem and causes similar rapid, shallow breathing.

THE EXPIRATORY AREA

The expiratory neurons are almost entirely dormant during normal quiet respiration, because quiet respiration is achieved by contraction only of inspiratory muscles, as discussed before. On the other hand, when the respiratory drive becomes much greater than normal, especially during heavy exercise, signals then spill over into the expiratory area and cause powerful excitation of the expiratory muscles during the expiratory phase of the respiratory cycle. Thus, during very heavy breathing, air is not only pulled into the lungs by the inspiratory muscles, but it is pushed outward as well by the expiratory muscles.

LIMITATION OF LUNG FILLING BY LUNG STRETCH RECEPTORS— THE HERING-BREUER REFLEX

Located in the walls of the bronchi and bronchioles throughout the lungs are nervous stretch receptors that are excited when the lungs become overinflated. These receptors send signals through the vagus nerves to the inspiratory center, which instantaneously limits further inspiration. This is called the *Hering-Breuer reflex*. This reflex has the same effect as the pneumotaxic center in increasing the rate of respiration because it decreases the depth of respiration but at the same time increases the rate to make up the difference.

The purpose of the Hering-Breuer reflex is not to control respiration. Instead, it is a protective mechanism to prevent excess lung inflation, which in turn prevents damage to the lungs.

Failure of the Respiratory Center

Occasionally the oscillating mechanism of the respiratory center fails. One of the more frequent causes of this is cerebral concussion or some other intracerebral abnormality that causes excess pressure on the brain medulla. The pressure collapses the blood vessels supplying the respiratory center, which blocks all activity of the medulla and therefore stops the respiratory rhythm. Another common cause of failure is acute poliomyelitis, which sometimes destroys neuronal cells in the reticular substance of the hindbrain, thereby depressing the respiratory center. Finally, one of the most frequent of all causes of respiratory failure is attempted suicide with sleeping drugs. These anesthetize the respiratory neurons and in this way stop the rhythm of respiration.

Failure of the respiratory rhythm is one of the most difficult of all abnormalities of the body to treat, except by giving artificial respiration. In general, very few drugs can be used to excite the respiratory center, and those that are available—caffeine, picrotoxin, and a few others—are so weak in their effects on the respiratory center that they are of little value. Fortunately, respiratory depression caused by poliomyelitis, pressure on the brain, or sleeping drugs is often reversible provided artificial respiration is maintained long enough.

REGULATION OF ALVEOLAR VENTILATORY RATE

When one needs great amounts of respiratory air, both the inspiratory and expiratory centers become strongly excited, and great volumes of air are exchanged. Not only does the depth of respiration increase but the rate increases as well—the depth sometimes increasing from the normal of 0.5 L per breath to more than 3 L, and the rate from the normal of 12 per minute to as rapid as 50 per minute. Therefore, the total volume of air breathed each minute, the *minute respiratory volume*, can rise from 6 L to more than 150 L, a 25-fold increase, and the amount of new air that actually reaches the alveoli, the *alveolar ventilatory rate*, can increase almost as much.

Many different factors contribute to the control of respiration, but by far the most important of these are the following four:

1. The *pressure of carbon dioxide* (Pco_2) in the blood.
2. The *concentration of hydrogen ions* (pH) in the blood.
3. The *pressure of oxygen* (Po_2) in the blood.
4. Nervous *signals from the muscle controlling areas* of the brain.

Effect of Carbon Dioxide and Hydrogen Ions on the Respiratory Center to Increase Alveolar Ventilation

Among the most powerful stimuli known to affect the respiratory center are excess concentrations of carbon dioxide and hydrogen ions in the blood. Both stimulate respiration by exciting a so-called respiratory *chemosensitive area*, illustrated in Figure 38–2, that is located bilaterally

FIGURE 38–2 Stimulation of the inspiratory area by the *chemosensitive area* located bilaterally in the medulla, lying only a few microns beneath the ventral medullary surface. Note also that hydrogen ions stimulate the chemosensitive area and mainly carbon dioxide in the fluid gives rise to the hydrogen ions.

and ventrally in the substance of the medulla. This area in turn sends excitatory signals to both the inspiratory and expiratory centers.

Response of the Chemosensitive Neurons to Hydrogen Ions. The primary stimulus for exciting the neurons of the chemosensitive area is the concentration of hydrogen ions *inside* the cell bodies of these neurons. Therefore, whenever the hydrogen ion concentration in the blood supplying the respiratory center increases above normal, this also causes the hydrogen ion concentration inside the chemosensitive neurons to increase and therefore to excite respiration.

However, hydrogen ions in the blood are not as powerful a stimulus of respiration as one might expect, for the following reason: The cell membranes of the neurons are only slightly permeable to hydrogen ions. Therefore, the concentration of hydrogen ions inside these cells does not change nearly as much as the hydrogen ion concentration in the blood. Even so, the response that does occur is enough that the blood hydrogen ion concentration is one of the more important controllers of respiration.

Effect of Carbon Dioxide in Stimulating the Chemosensitive Area. Carbon dioxide, unlike hydrogen ions, can diffuse very rapidly into the neuronal cells because carbon dioxide is highly soluble in cell membranes. Therefore, . whenever the carbon dioxide concentration in the blood increases, so also does the carbon dioxide concentration in the neurons of the chemosensitive area increase at the same time. But, how does this carbon dioxide excite the neurons? The answer is: The carbon dioxide reacts with the water inside the cell to form carbonic acid and this in turn dissociates into hydrogen and bicarbonate ions. Recalling that it is hydrogen ions that are the primary stimulus to the neuronal cells, one can readily see that the net effect of the excess carbon dioxide is to stimulate these neurons very powerfully.

Therefore, even though hydrogen ions are the primary stimulus of the chemosensitive neurons, carbon dioxide, strangely enough, has an effect about twice as powerful as hydrogen ions in stimulating the respiratory center.

Stimulation of the Chemosensitive Area by Both Blood Carbon Dioxide and Cerebrospinal Fluid Carbon Dioxide. The chemosensitive area is located very near to the surface of the medulla. For this reason, excess carbon dioxide can enter the neurons of this area from two sources, either directly from the blood capillaries in the substance of the medulla or by diffusion of carbon dioxide from the cerebrospinal fluid through the surface of the medulla and into the chemosensitive area. This double mechanism for transporting carbon dioxide to the chemosensitive area is important for a special reason: Transport of enough carbon dioxide from the blood capillaries into the respiratory center to cause maximal respiratory stimulation requires as much as a minute's time; this amount of delay in increasing respiration could be very serious during stressful conditions such as heavy exercise. Fortunately, though, the blood flow to the arachnoidal blood vessels that intermingle with the cerebrospinal fluid is great enough to increase the cerebrospinal fluid carbon dioxide to a very high level within a few seconds. It is this carbon dioxide that stimulates respiration immediately and very powerfully long before the carbon dioxide arriving directly from the blood capillaries can do so.

Control of Carbon Dioxide Concentration in the Body Fluids by the Carbon Dioxide—Respiratory Feedback Mechanism. It is extremely important that the respiratory system is excited by the concentration of carbon dioxide in the blood because this provides a method by which the carbon dioxide concentration can in turn be controlled in all fluids of the body. This can be explained as follows:

The concentration of carbon dioxide in the blood is controlled by the rate of alveolar ventilation. That is, increased ventilation causes the lungs to blow off greater amounts of carbon dioxide from the blood. Therefore, when excess carbon dioxide excites the respiratory center, the resulting increase in ventilation causes the excess carbon dioxide concentration to return to near normal. Conversely, when carbon dioxide concentration falls too low, the resulting decrease in ventilation allows the carbon dioxide concentration to rise again back toward normal.

The body has no other significant means for controlling carbon dioxide concentration in the blood and body fluids, which makes this mechanism all-important. Should carbon dioxide concentration rise too high, it would stop essentially all chemical reactions in the body, because carbon dioxide is one of the end-products of almost all metabolic reactions. On the other hand, if the carbon dioxide concentration should fall too low, other dire consequences would occur, such as development of alkalosis caused by loss of carbonic acid from the body fluids. The alkalosis then would cause increased irritability of the nervous system, sometimes resulting in tetany or even epileptic convulsions.

Importance of the Hydrogen Ion Feedback Mechanism for Regulating Respiration. Regulation of respiration by the blood hydrogen ion concentration helps to maintain normal acid-base balance in the body fluids, which was discussed in more detail in Chapter 32. When carbon dioxide is blown off through the lungs, loss of this carbon dioxide from the body fluids shifts the chemical equilibria of the acid-base buffers in such a way that much of the carbonic acid of the body fluids dissociates into water and carbon dioxide. This removes much of the carbonic acid portion of the body's acid and thereby decreases the hydrogen ion concentration. Thus, stimulation of the respiratory center by increased hydrogen ion concentration initiates a feedback mechanism that automatically decreases the hydrogen ion concentration back toward normal. Conversely, reduced hydrogen ion concentration decreases alveolar ventilation, which allows carbonic acid to accumulate in the fluids, thus raising the hydrogen ion concentration once again back toward normal.

Therefore, this hydrogen ion control of respiration is one of the mechanisms for control of the acid-base balance of the body, as was presented in detail in Chapter 32.

Regulation of Alveolar Ventilation by Oxygen Deficiency

On first thought most students intuitively believe that respiration is regulated mainly by the need of the body for oxygen. However, the rate of ventilation normally has little effect on the amount

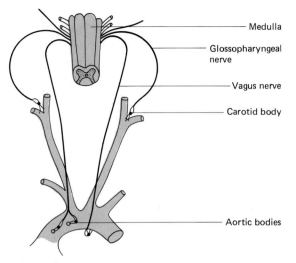

FIGURE 38–3 The chemoreceptor system for stimulating respiration.

of oxygen delivered to the tissues. The reason for this is the following: At ordinary ventilatory rates, or even at a rate as low as one-half normal, the hemoglobin in the blood becomes almost completely saturated with oxygen as the blood passes through the lungs. Increasing the ventilatory rate to an infinite level will not further saturate the hemoglobin because all of the hemoglobin that is available to combine with oxygen has already combined. Therefore, extreme increases or moderate decreases in alveolar ventilation make very little difference in the amount of oxygen carried from the lungs by the hemoglobin. There is thus no need for very sensitive regulation of respiration to maintain constant oxygen concentration in the blood.

On rare occasions, however, the oxygen concentration in the alveoli does fall too low to supply adequate quantities of oxygen to the hemoglobin. This occurs especially when one ascends to high altitudes where the oxygen con-

centration in the atmosphere is very low. Or it occurs when a person contracts pneumonia or some disease that reduces the oxygen in the alveoli. Under these conditions the respiratory system needs to be stimulated by *oxygen deficiency*. A mechanism for this is the *chemoreceptor system*, shown in Figure 38–3. Minute *aortic* and *carotid bodies*, each only a few millimeters in diameter, lie adjacent to the aorta and carotid arteries in the chest and neck, each of these having an abundant arterial blood supply and containing neuronal receptor cells called *chemoreceptors*, which are sensitive to the lack of oxygen in the blood. When stimulated, these receptors send signals along the vagus and glossopharyngeal nerves to the medulla, where they stimulate the respiratory center to increase alveolar ventilation.

COMPARATIVE RESPIRATORY STIMULATORY EFFECTS OF HYDROGEN IONS, CARBON DIOXIDE, AND OXYGEN LACK

Figure 38–4 shows that maximal oxygen lack can increase alveolar ventilation to only 1.6 times normal in comparison with an increase of 10 times caused by excess carbon dioxide and 5 times caused by excess hydrogen ions. This illustrates that under normal conditions the oxygen lack system for control of respiration is a very weak one.

However, Figure 38–4 is to some extent misleading, because in those special conditions in which oxygen lack occurs at the same time that there is even slight excess of carbon dioxide and hydrogen ions in the body fluids, the oxygen lack mechanism does then become a powerful stimulator of respiration. Furthermore, under special conditions, such as after a person has been at high altitude for several days, the carbon

FIGURE 38–4 Effect on alveolar ventilation of maximal increase in hydrogen ions, maximal oxygen lack, maximal excess of CO_2, and maximal exercise.

dioxide mechanism loses its potency for control of respiration, and the oxygen lack mechanism becomes more potent, increasing alveolar ventilation then as much as five- to sevenfold.

Effect of Exercise on Alveolar Ventilation

Alveolar ventilation increases almost directly in proportion to the amount of work performed by the body during exercise, reaching as high as 120 L per minute in the most strenuous exercise. This value is more than 20 times that during normal quiet respiration, as illustrated to the right in Figure 38–4.

Physiologists still have difficulty explaining the extreme increase in pulmonary ventilation that occurs during exercise. Indeed, despite the extremely rapid production of carbon dioxide during exercise and simultaneous rapid usage of oxygen, alveolar ventilation increases so greatly that this prevents the blood concentrations of these gases from changing significantly from normal. Therefore, it is almost certain that the increase in respiration during exercise is not caused by either or both of these two chemical factors.

If it is not chemical factors that increase ventilation during exercise, it must be some stimulus entering the respiratory center by way of nerve pathways. Two such nerve pathways illustrated in Figure 38–5 have been discovered: (1) At the same time that the cerebral cortex transmits signals to the exercising muscles it also transmits parallel signals into the respiratory center to increase the rate and depth of respiration. (2) Movement of the limbs and other parts of the body during exercise sends sensory signals up the spinal cord to excite the respiratory center. Therefore, it is believed that these two signals, one from the cerebral cortex and the other from the moving parts of the body, are the factors that increase respiration during exercise. If these two factors fail to increase respiration adequately, only then do carbon dioxide and hydrogen ions begin to collect in the body fluids and oxygen begins to diminish; these then excite the respiratory center as a second line of defense to increase ventilation.

Other Factors that Affect Alveolar Ventilation

Effect of Arterial Pressure. Arterial pressure is another factor that helps to regulate alveolar ventilation. This operates through the *baroreceptor system*, which was described in

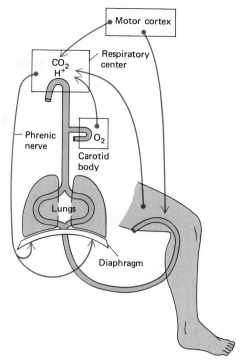

FIGURE 38–5 Mechanisms by which muscle activity stimulates respiration.

Chapter 28. When the arterial pressure is high, signals from this system depress the respiratory center, and ventilation is reduced correspondingly. Conversely, when the pressure is low and blood flow to the tissues is poor, ventilation is increased; this partially compensates for the poor blood flow by allowing improved oxygenation and better removal of carbon dioxide in the lungs.

Effect of Psychic Stimulation. Impulses initiated by psychic stimulation of the cerebral cortex can also affect respiration. For instance, anxiety states frequently can lead to intense hyperventilation, sometimes so much so that the person makes his body fluids alkalotic by blowing off too much carbon dioxide, thus precipitating alkalotic tetanic contraction of muscles throughout the body.

Effect of Sensory Impulses. Sensory impulses from all parts of the body can affect respiration. The effect of entering a cold shower is well known; this causes an intense inspiratory gasp followed by a period of prolonged inspiration and then rapid, forceful breathing. Even a pinprick can cause sudden changes in the rate and depth of respiration. However, these factors affect alveolar ventilation only temporarily, for the chemical factors mentioned previously soon become dominant over these aberrant effects.

Effect of Speech. The speech centers of the brain also control respiration at times. When one talks, it is as important to control the flow of air between the vocal cords as to control the vocal cords themselves. Therefore, whenever nerve signals are transmitted from the brain to the vocal cords, collateral signals are sent simultaneously into the respiratory system.

PHYSIOLOGY OF RESPIRATORY ABNORMALITIES

Hypoxia

One of the most important effects of most respiratory diseases is hypoxia, which means diminished availability of oxygen to the cells of the body. The principal types of hypoxia are *hypoxic hypoxia, stagnant hypoxia, anemic hypoxia,* and *histotoxic hypoxia,* the mechanisms of which are shown in Figure 38–6.

Hypoxic Hypoxia. Hypoxic hypoxia means failure of oxygen to reach the blood of the lungs. The obvious causes of hypoxic hypoxia are (1) *too little oxygen in the atmosphere,* (2) *obstruction of the respiratory passages,* (3) *thickening of the pulmonary membrane,* and (4) *decreased area of the pulmonary membrane.*

Stagnant Hypoxia. Stagnant hypoxia means failure to transport adequate oxygen to the tissues because of too little blood flow. The most common cause of stagnant hypoxia is low cardiac output caused by *heart failure* or *circulatory shock.* Immediately after a heart attack the blood flow to the tissues may be so slight that the person may die of the stagnant hypoxia itself. At other times the transport of oxygen will be barely enough to keep him alive, but his tissues nevertheless will suffer severely from oxygen deficiency.

Anemic Hypoxia. Anemic hypoxia means too little hemoglobin in the blood to transport oxygen to the tissues. This may be caused by any one of three different abnormalities: First, the person may have *anemia,* which means too few red blood cells. Second, he may have a sufficient number of red blood cells but *too little hemoglobin in the cells.* Third, he may have plenty of hemoglobin, but much of it may have been *poisoned by carbon monoxide* or some other poison so that it cannot transport oxygen; for instance, breathing air with a content of only 0.2 percent carbon monoxide can decrease the hemoglobin available to transport oxygen to as low as one-third normal, which is a lethal level.

Histotoxic Hypoxia. This means failure of the tissues to utilize oxygen even though adequate quantities are transported to them. The classic cause of serious histotoxic hypoxia is *cyanide poisoning;* this blocks the enzymes responsible for use of oxygen in the cell. Mild forms of histotoxic hypoxia frequently occur with *vitamin deficiencies,* for lack of certain vitamins often results in diminished quantities of oxidative enzymes in the cells.

Oxygen Therapy

In most types of hypoxia, breathing purified oxygen can be tremendously beneficial, the benefit occurring in three separate ways, which may be described as follows:

First, the normal partial pressure of oxygen in the alveoli is about 104 mm Hg, but when a person breathes pure oxygen, this can rise to as high as 600 mm Hg. Such an elevation can increase as much as sixfold the pressure that causes oxygen to diffuse through the pulmonary membrane.

Second, even when there is insufficient hemoglobin in the blood to carry enough oxygen to the tissues, oxygen therapy can still be beneficial. The reason for this is that when the partial pressure of oxygen in the alveoli becomes 600 mm Hg, 2 ml of oxygen dissolves in the fluid of every 100 ml of blood and is transported in this form to the tissues. This effect is illustrated in Figure 38–7. At a normal alveolar oxygen partial pressure of about 100 mm Hg, almost no oxygen is dissolved in the fluids. Yet as the alveolar oxygen partial pressure rises to 600 mm Hg, one sees by the shaded area of the figure that reasonable quantities of oxygen begin to dissolve in the fluids. This oxygen is the difference between life and death in some hypoxic persons.

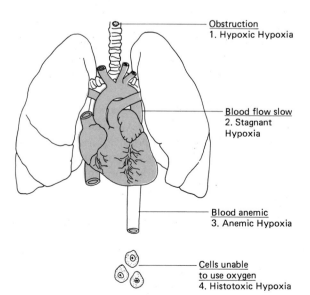

Obstruction
1. Hypoxic Hypoxia

Blood flow slow
2. Stagnant Hypoxia

Blood anemic
3. Anemic Hypoxia

Cells unable to use oxygen
4. Histotoxic Hypoxia

FIGURE 38–6 The types of hypoxia.

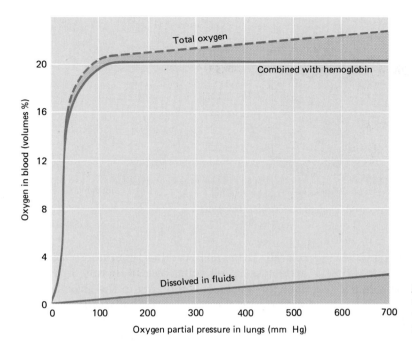

FIGURE 38-7 Effect of elevated alveolar partial pressure of oxygen on the total oxygen transported by each 100 ml of blood.

The third means by which oxygen therapy can benefit some types of hypoxia is by decreasing the volume of gases that must flow in and out of the lungs through the trachea. If the respiratory passageways are partially obstructed, the blood can still be adequately oxygenated with only one-fifth normal alveolar ventilation if the person breathes pure oxygen rather than air because the concentration of oxygen in air is only 20 percent. Yet the decreased alveolar ventilation will still cause a buildup of carbon dioxide in the body fluids, sometimes enough to produce serious acidosis even though the tissues become adequately oxygenated.

Dyspnea

Dyspnea means *air hunger*, that is, a psychic feeling that more ventilation is needed than is being attained. Most instances of dyspnea occur when some respiratory abnormality causes too much carbon dioxide to collect in the body fluids. The respiratory center then becomes overly excited, and signals are transmitted to the conscious portions of the brain, apprising the psyche of the need for increased ventilation. This causes a feeling of air hunger or dyspnea.

However, many persons develop *psychic dyspnea* because of neuroses. In these instances the neurosis causes the person to become unduly conscious of his respiration, making him feel that he is not receiving adequate quantities of air. One of the neuroses that most frequently causes this type of dyspnea is cardiac neurosis, for many who fear cardiac disease have been acquainted with cardiac patients who have had episodes of dyspnea because of pulmonary edema. Therefore, the neurotic person who believes that he has heart disease often develops psychic dyspnea.

Pneumonia

Pneumonia is caused by infection of the lung with bacteria such as the pneumococcus bacterium or with a virus. The infection causes the walls of the alveoli to become inflamed and edematous and the spaces in the alveoli to become filled with fluid and blood cells. This is illustrated in the center of Figure 38-8. Pneumonia causes hypoxic hypoxia for two reasons: first, because fluid and blood cells fill many alveoli so that they are not aerated at all, and, second, because the membranes of those alveoli that are still aerated are frequently so thickened with edema that oxygen cannot diffuse with ease.

Pulmonary Edema

Pulmonary edema means collection of fluid in the interstitial spaces of the lungs and in the alveoli; this affects respiration in the same way pneumonia affects it. Generalized pulmonary edema is usually caused by failure of the left heart to pump blood from the pulmonary circulation into the systemic circulation. This can result from mitral or aortic valvular disease or from failure of the left ventricular muscle. In any event, blood is dammed up in the pulmonary circulation, and the pulmonary capillary pres-

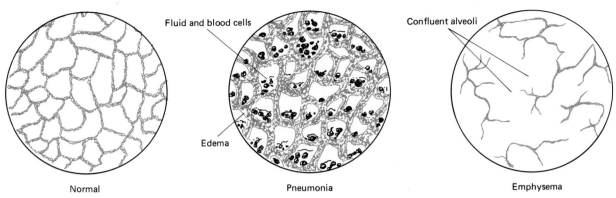

FIGURE 38–8 Histologic appearance of the normal lung, of the lung with pneumonia, and of the emphysematous lung.

sure rises. When this pressure becomes greater than the colloid osmotic pressure of the blood, 28 mm Hg, fluid transudes very rapidly from the plasma into the alveoli and interstitial spaces of the lungs, causing hypoxic hypoxia. Sometimes acute failure of the left heart causes pulmonary edema to occur so rapidly that the person dies of hypoxia in only 20 to 40 minutes.

Emphysema

Emphysema is a disease caused most frequently by chronic bronchial and alveolar infection resulting from *smoking*. In this condition, large portions of the alveolar walls are destroyed; this is illustrated to the right in Figure 38–8 and even more dramatically in Figure 38–9, which contrasts the emphysematous lung on the right with the normal lung on the left. Therefore, the total surface area of the pulmonary membrane becomes greatly diminished, which also diminishes the aeration of the blood. Hypoxic hypoxia results, and the quantity of carbon dioxide in the body fluids also becomes increased.

The emphysematous person also usually develops pulmonary hypertension, for every time the wall of an alveolus is destroyed, the blood vessels in the wall are also destroyed. This increases the pulmonary resistance, which in turn elevates the pulmonary arterial pressure and eventually overloads the right heart.

Atelectasis

Atelectasis means collapse of a portion of the lung or of an entire lung. A common cause of atelectasis is a chest wound that allows air to leak into the pleural cavity. The surface tension of the fluid in the alveoli and the elastic fibers in the interstitial spaces of the lungs cause the lungs to collapse to a small size, the alveoli los-

ing all their air. Another common cause of atelectasis is plugging of a bronchus, in which case the air beyond the plug in the alveoli is absorbed into the blood, thus collapsing the alveoli.

In atelectasis, not only do the alveoli collapse, but the blood vessels also collapse simultaneously. Consequently, blood flow through the collapsed lung decreases greatly, allowing the major portion of the pulmonary blood to flow through those areas of the lungs that are still aerated. Because of this shift of blood flow, atelectasis of a whole lung often does not greatly diminish the aeration of the blood.

Asthma

Asthma is usually caused by an allergic reaction to pollens in the air. The reaction causes spasm of the bronchioles, thus impeding air flow into and out of the lung. Furthermore, the outward flow of air through the bronchioles is obstructed more than the inflow, which often allows the asthmatic person to inspire with ease but to expire only with difficulty. As a result, the lungs become progressively more distended, and with repeated asthmatic attacks year in and year out, the prolonged distention of the thoracic cage causes the chest to become barrel-shaped. Asthma rarely becomes severe enough to cause serious hypoxia, but it does cause severe dyspnea. Asthma can usually be treated by administering drugs that relax the bronchiolar musculature.

The Cough and Sneeze Reflexes

A means for keeping the respiratory passages clean is to force air very rapidly outward by either coughing or sneezing. The cough reflex is initiated by any irritant touching the surface of

FIGURE 38—9 Contrast of the emphysematous lung (above) with the normal lung (below), showing extensive alveolar destruction. (Reproduced with permission of Patricia Delaney and the Department of Anatomy, The Medical College of Wisconsin.)

the glottis, the trachea, or a bronchus. Sensory signals are transmitted to the medulla of the brain, and, in turn, appropriate motor signals are transmitted back to the respiratory system and larynx to cause the cough. The expiratory muscles first contract very strongly, building up high pressure in the lungs, while at the same time the vocal cords remain clamped tightly closed. Then suddenly the vocal cords open, allowing the pressurized air in the lungs to flow out with a blast. Sometimes the air flows as rapidly as 70 miles per hour. In this way unwanted foreign matter such as particles, mucus, or other substances is expelled from the respiratory passageways.

The sneeze reflex is very similar to the cough reflex except that it is initiated by irritants in the nose. Impulses pass from the nose to the medulla and then back to the respiratory system. A sudden forceful expiration blows air outward while the soft palate varies its position to allow rapid flow of air successively through both the nose and the mouth. In this way the sneeze is capable of clearing the nasal passageways in the same manner that the cough reflex clears many of the lower passageways.

QUESTIONS

1. What causes the normal rhythmic nerve signals that promote inspiration and expiration of air?
2. Explain the role of and the mechanisms by which carbon dioxide controls alveolar ventilation.
3. Explain the relationship between hydrogen ion concentration and the regulation of alveolar ventilation.
4. Why is oxygen not the normal regulator of alveolar ventilation, and under what conditions does it play an important role?
5. What is there about exercise that causes an exceptionally great increase in alveolar ventilation?
6. What are the four major types of hypoxia?
7. Explain why oxygen therapy is of value in each of the types of hypoxia. How much so?
8. Explain the cough and sneeze mechanisms.

REFERENCES

Avery, M.E., *et al.:* The lung of the newborn infant. *Sci. Am., 228*:74, 1973.

Cohen, M.I.: Neurogenesis of respiratory rhythm in the mammal. *Physiol. Rev., 59*:1105, 1979.

Dosman, J.A., and Cotton, D.J. (eds.): Occupational Pulmonary Disease: Focus on Grain Dust and Health. New York, Academic Press, 1979.

Fishman, A.P., and Pietra, G.G.: Primary pulmonary hypertension. *Annu. Rev. Med., 31*:421, 1980.

Guenter, C.A., *et al.:* Clinical Aspects of Respiratory Physiology. Philadelphia, J.B. Lippincott, 1978.

Guyton, A.C., *et al.:* Basic oscillating mechanism of Cheyne-Stokes breathing. *Am. J. Physiol., 187*:395, 1956.

Hall, W.J., and Douglas, R.G., Jr.: Pulmonary function during and after common respiratory infections. *Annu. Rev. Med., 31*:233, 1980.

Hechtman, H.B. (ed.): Acute Respiratory Failure: Etiology and Treatment. West Palm Beach, Fla., CRC Press, 1979.

Irsigler, G.B., and Severinghaus, J.W.: Clinical problems of ventilatory control. *Annu. Rev. Med., 31*:109, 1980.

Moser, K.M. (ed.): Pulmonary Vascular Diseases. New York, Marcel Dekker, 1979.

Putnam, S.J. (ed.): Advances in Pulmonary Medicine. Philadelphia, W.B. Saunders, 1978.

Tisi, G.M.: Pulmonary Physiology in Clinical Medicine. Baltimore, Williams & Wilkins, 1980.

Von Euler, C., and Lagercrantz, H. (eds.): Central Nervous Control Mechanisms in Breathing. New York, Pergamon Press, 1980.

Wolfe, W.G., and Sabiston, D.C.: Pulmonary Embolism. Philadelphia, W.B. Saunders, 1980.

39

Aviation, Space, and Deep Sea Diving Physiology

Overview

The greatest problem in *aviation and other high altitude physiology* is the progressively diminishing partial pressure of oxygen in the air as one ascends to higher altitudes. Because of this, at an altitude of 23,000 ft, only half of the hemoglobin in the arterial blood is combined with oxygen; the normal person is likely to pass into *coma* within a few minutes of breathing air at this altitude. However, if the person is *acclimatized,* which causes several changes in the respiratory system, including a marked increase in number of red blood cells, he can sometimes breathe at the altitude of Mt. Everest, about 29,000 ft, for as long as half an hour without developing coma.

When a person *breathes pure oxygen,* he can ascend to an altitude of about 47,000 ft. The reason for this difference is that normal air contains almost four-fifths nitrogen and only one-fifth oxygen. When the nitrogen is replaced with oxygen, despite the very low total pressure (the *barometric pressure*) at the higher altitudes, the fact that all of the gas inspired is oxygen rather than a mixture of oxygen and nitrogen allows adequate blood oxygenation.

In *space physiology,* the respiratory problem is solved by providing appropriate pressures of oxygen and nitrogen in the cabin of the spaceship. Therefore, the major physiological problem is not respiratory but instead the *acceleratory* and *deceleratory forces* that occur during take-off and landing. These require appropriate positioning of the person in the spaceship so that these forces are applied mainly horizontally to the body rather than along the vertical axis. When such forces are applied vertically, the blood is displaced into one end of the body or the other, causing either loss of all of the blood to the bottom of the body so that none returns to the heart, and the circulation collapses, or forcing so much blood into the head that blood vessels rupture in the brain and eyes.

During space travel, the person also experiences *weightlessness,* which means that he literally floats inside the spaceship. Despite much apprehension at the outset of the space program about the physiological effects of this on the body, weightlessness has not proved to be a serious problem, except that the person's muscles and bones become deconditioned because of lack of muscle effort to overcome gravity.

In *deep sea diving,* serious physiological problems result from the *very high pressures of both oxygen and nitrogen* in the air that is breathed. For instance, at a depth 300 ft below the sea, the partial pressure of each of these gases is increased 10-fold.

The *very high pressures of oxygen* will cause *oxygen poisoning* in all persons who are exposed long enough. The most debilitating effect of this occurs in the brain, leading eventually to *convulsions* and *coma*.

The *very high nitrogen pressures* lead to excessive absorption of nitrogen in the body fluids. At a sea level depth of 200 ft, the amount of nitrogen absorbed within 15 minutes to 2 hours is enough to cause drowsiness, and even total anesthesia can occur at still deeper levels. Also, when the person ascends back to the surface, the nitrogen that has become dissolved in the tissue fluids during the high pressure period escapes from the dissolved state and becomes small *bubbles of gaseous nitrogen* in the tissues because it is no longer compressed by the pressure of the sea around the body. These bubbles can literally rupture the tissues, causing immediate pain. This condition is commonly called the *bends*. Permanent damage is likely to occur in the nervous system, leaving the person *paralyzed* or *mentally impaired*.

One of the most important problems in aviation, space, and deep sea diving physiology is the altered barometric pressure to which the aviator, astronaut, or diver is exposed. Therefore, many of the physical principles applicable to respiration are also applicable to aviation, space flights, or dives deep beneath the sea.

Problems also arise in aviation, space, and diving physiology because of mechanical stresses on the body or because of extremes of climate. For instance, extreme acceleration during takeoff of a spaceship can cause the astronaut to weigh as much as 1500 lb, a force that the body can withstand only when in the proper position. Also, in the upper atmosphere the air temperature is sometimes 20° to 50°C below zero, and, deep beneath the sea, pressures on the outside of the body can sometimes become so intense that the entire chest wall actually collapses.

EFFECTS OF LOW BAROMETRIC PRESSURE

Oxygen Deficiency at High Altitudes

At high altitudes the barometric pressure is low, as shown in Table 39–1, and the partial pressure of oxygen is reduced correspondingly. This decreases the amount of oxygen absorbed into the blood and leads to hypoxia. Table 39–1 also shows the effect of low oxygen partial pressure on the saturation of arterial hemoglobin in the blood. For instance, when breathing air at an altitude of approximately 23,000 ft only half of the arterial hemoglobin is saturated with oxygen. Obviously, this decreases the effectiveness of oxygen transport to the tissues at least 50 percent, so that a person at this altitude will likely be debilitated because of tissue hypoxia.

It will be also noted from Table 39–1 that the alveolar oxygen partial pressure when

TABLE 39–1

Effects of Low Atmospheric Pressures on Alveolar Oxygen Concentrations and on Arterial Oxygen Saturation

			Breathing air		Breathing pure oxygen	
Altitude (ft)	Barometric pressure (mm Hg)	P_{O_2} in air (mm Hg)	P_{O_2} in alveoli (mm Hg)	Arterial oxygen saturation (%)	P_{O_2} in alveoli (mm Hg)	Arterial oxygen saturation (%)
0	760	159	104	97	673	100
10,000	523	110	67	90	436	100
20,000	349	73	40	70	262	100
30,000	226	47	21	20	139	99
40,000	141	29	8	5	58	87
50,000	87	18	1	1	16	15

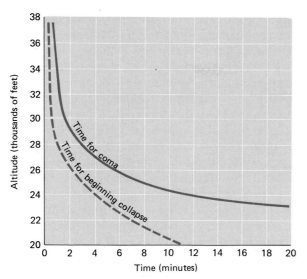

FIGURE 39–1 Time required at different altitudes for oxygen deficiency to cause collapse or coma. (From Armstrong: Principles and Practice of Aviation Medicine. Baltimore, Williams & Wilkins, 1900.)

breathing air at an altitude of 50,000 ft is only 1 mm Hg and that the arterial oxygen saturation is only 1 percent. Therefore, at this altitude the small quantities of oxygen stored in the tissues actually diffuse backward into the blood and then from the blood into the lungs. Obviously, a person would have no more than a minute or two to live under these conditions.

Effects of Oxygen Deficiency. Figure 39–1 shows graphically the time required at various altitudes for oxygen deficiency to cause either collapse or coma in a normal *unacclimatized* person. Above 30,000 ft an unacclimatized person lapses into coma in about 1 minute. At 20,000 ft the person usually does not go into coma, but after 10 or more minutes he exhibits signs of collapse such as weakness, mental haziness, and other deficiencies. He usually does not pass into coma until he ascends to altitudes of 22,000 to 24,000 ft, which is called his *ceiling*.

The first symptom of oxygen deficiency occurs in vision. Even at altitudes as low as 6000 to 10,000 ft, the sensitivity of the eyes to light begins to decrease—not enough to be noticed usually, but nevertheless enough to be measured by instruments. When the aviator rises to altitudes of 12,000 to 15,000 ft he is likely to experience alterations in psychic behavior. Some aviators tend to go to sleep; others develop euphoric exaltation. In almost all instances the acuity of reasoning becomes greatly depressed. In short, the aviator at this altitude is likely to develop a sleepy slap-happiness, and as he rises to 18,000 to 24,000 ft these symptoms often become so severe that control of the aircraft is endangered. For this reason safety regulations require that an aviator breathe oxygen when he ascends to dangerous altitudes whether he feels need for the oxygen or not. On occasion, he rises too high before applying his oxygen, and by that time his mental acuity might have become so depressed that he no longer knows his need. This often results in a vicious cycle of depressed mentality, for he rises still higher without knowing what he is doing; his mentality becomes further depressed, he rises higher, and so forth until coma develops.

Effect of Water Vapor and Carbon Dioxide on Alveolar Oxygen Partial Pressure. Were it not for water vapor and carbon dioxide in the alveoli, the aviator could go to considerably higher heights than he actually can before developing oxygen deficiency. Regardless of the altitude, the alveolar pressure of water vapor does not change at all, and the pressure of carbon dioxide does not change greatly. Therefore, unless the barometric pressure is considerably greater than the combined pressures of these two gases, little space will be left in the alveoli for other gases. Figure 39–2 shows this effect. At sea level the combined pressure of all the gases in the alveoli is 760 mm Hg. Carbon dioxide is continually excreted from the blood into the alveolar

FIGURE 39–2 Effects of low barometric pressures at different altitudes on alveolar gas concentrations.

air, keeping the P_{CO_2} at a level of 40 mm Hg, and water vapor continually evaporates from the surfaces of the alveoli, creating a water vapor pressure of 47 mm Hg. This pressure remains constant as long as the body temperature is normal, regardless of changes in barometric pressure. Thus, the combined pressure of water vapor and carbon dioxide at sea level is 40 plus 47 or 87 mm Hg. Subtracting this from 760 mm Hg, the remaining pressure available in the alveolus for both nitrogen and oxygen is 673 mm Hg. This is more than adequate, so that the oxygen partial pressure is high enough to keep the arterial blood saturated with oxygen.

Now, let us see what effect the water vapor pressure and carbon dioxide partial pressure have on alveolar function at an altitude of 50,000 ft. Here, the total barometric pressure is only 87 mm Hg. The water vapor pressure remains 47 mm Hg because the body temperature does not change. The P_{CO_2} has fallen to about 24 mm Hg because of increased respiration. Therefore, the combined pressure of water vapor and carbon dioxide is now 71 mm Hg, which when subtracted from 87 mm Hg leaves a total partial pressure for both nitrogen and oxygen of only 16 mm Hg. Furthermore, the person is so hypoxic that his blood absorbs oxygen out of the alveoli almost as rapidly as it enters. Therefore, the P_{O_2} falls to about 1 mm Hg, which is so slight that the person will become unconscious in only a few seconds and will die within another minute or so. Were it not for the carbon dioxide and water vapor, it would be possible at 50,000 ft to supply much more oxygen to the blood.

Respiratory Compensation for Oxygen Deficiency—the Process of Acclimatization. The chemoreceptor mechanism described in the previous chapter increases alveolar ventilation when a person develops oxygen deficiency at high altitudes. This mechanism unfortunately is not very powerful, for it normally can increase alveolar ventilation only about 60 percent. Yet even this amount allows the aviator to ascend several thousand feet higher than he could otherwise.

However, when a person is exposed to high altitudes for several days at a time, the oxygen deficiency, for reasons not well understood, stimulates alveolar ventilation more and more until it increases to as much as five to seven times normal. This slow process of adjusting to high altitudes is called *acclimatization.*

Another factor that favors acclimatization to high altitudes is an increase in the number of red blood cells, for oxygen deficiency causes rapid production of these cells by the bone marrow, as explained in Chapter 33. Unfortunately, the formation of many new cells is a slow process, requiring several weeks or even months to help acclimate a person to high altitudes. Persons who live at high altitudes all the time often develop red blood cell counts as high as 7 to 8 million per cubic millimeter, which is 50 percent or more greater than normal.

Acclimatization to high altitudes is not of special importance in aviation because the aviator rarely remains aloft long enough for acclimatization to occur. It is far more important to the mountain climber, who must become slowly acclimatized if he is to succeed in ascending to the top of the highest mountains. This explains why ascension is normally accomplished in slow stages over a period of weeks, thereby allowing the body to become progressively acclimatized. By using this procedure of acclimatization, conquerors of Mt. Everest have been able to remove their oxygen masks atop this highest mountain of the world at an altitude over 29,000 ft, though this would cause coma in a normal person in a minute or two.

Oxygen Breathing at High Altitudes

A person can ascend to far higher altitudes when he breathes pure oxygen than when he breathes air, because oxygen then occupies the space in the alveoli normally occupied by nitrogen in addition to the usual oxygen. This allows the alveolar partial pressure of oxygen to remain quite high even when the barometric pressure falls to a low value. Referring again to Figure 39–2, it will be noted that at 20,000 ft the combined alveolar pressure of oxygen and nitrogen when breathing air is 278 mm Hg, and the alveolar P_{O_2} is only 40 mm Hg, low enough to cause serious hypoxia. If the nitrogen were replaced with oxygen, the oxygen pressure would be 278 mm Hg, and the arterial hemoglobin would be 100 percent saturated with oxygen, though when breathing air the saturation is only 67 percent.

Figure 39–3 presents graphically the percentage saturation of hemoglobin in the arterial blood at different altitudes when breathing pure oxygen as opposed to breathing air. It is evident from this figure that when breathing pure oxygen the arterial blood remains 100 percent saturated at all altitudes below approximately 33,000 ft. Above that point, however, the barometric pressure becomes so low that even with nitrogen eliminated from the alveoli, the pressure of oxygen is still not sufficient to keep the blood

FIGURE 39–3 Relative effects of breathing air or pure oxygen on the saturation of arterial hemoglobin with oxygen at different altitudes.

saturated. The lowest level of arterial oxygen saturation at which a nonacclimatized person can remain alive for more than a few hours is about 50 percent. Therefore, the ceiling for a person breathing pure oxygen is approximately 47,000 ft, in comparison with a ceiling of 23,000 ft for one breathing air.

Pressurized Cabins

Obviously, all the problems of low barometric pressure at high altitudes can be overcome by pressurizing the airplane. Usually the air inside the cabin of passenger planes is pressurized to maintain approximately the same pressure as that at 5000 ft.

Explosive Decompression. One of the major problems of pressurizing high altitude equipment is the possibility that the chamber might explode. Experiments have shown, fortunately, that sudden decompression usually does not cause serious damage to the body because of the decompression itself. The danger lies in exposure to the low partial pressures of oxygen in the rarefied atmosphere. Referring to Figure 39–1, one sees that at altitudes above 30,000 ft the person has only a minute or more of consciousness, during which time he must institute appropriate lifesaving measures, such as putting on an oxygen mask or parachuting to safety. If he parachutes without the aid of a special automatic parachute opening device, it is important to open the chute as soon as possible, because he is likely to lapse into coma and fall to earth without benefit of the parachute. On the other hand, if he opens the chute he is likely to lapse into coma and die because of the altitude. Fortunately, modern high-flying jets are equipped

with special devices that eject the aviator out of the plane and automatically delay the opening of the chute until an appropriate altitude is reached.

ACCELERATION EFFECTS OF AVIATION

Airplanes move so rapidly and change their direction of motion so frequently that the body is often subjected to severe physical stress caused by the sudden changes in motion. When the *velocity* of motion changes, the effect is called *linear acceleration.* When the *direction* of motion is changed, the effect is called *centrifugal acceleration.* In general, the forces induced by linear acceleration during the normal flight of an airplane are not sufficient to cause major physiologic effects. But when an airplane turns, dives, or loops, the centrifugal forces caused by centrifugal acceleration are frequently sufficient to promote serious derangements of bodily function.

Centrifugal Acceleration

Positive G. Figure 39–4 illustrates an airplane going into a dive and then pulling out. While the airplane is flying level, the downward force of the aviator against his seat is exactly equal to his weight. However, as he begins to come out of the dive, he is pushed against the seat with much greater force than his weight because of centrifugal force. At the lowest point of the dive the force in the example of Figure 39–4 is six times that which would be caused by

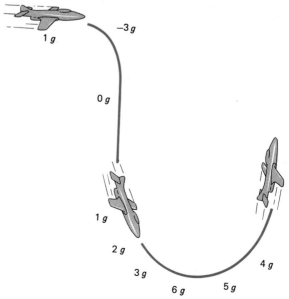

FIGURE 39–4 Positive and negative g during a dive and pullout of an airplane.

normal gravitational pull. This effect is called *positive centrifugal acceleration,* and the person is said to be under the influence of +6 g force, or, in other words, a force six times that of gravity.

Negative G. At the beginning of the dive in Figure 39–4, the airplane changes from level flight to a downward direction, which throws the pilot upward against his seat belt. He then is not exerting any force at all against his seat but instead is being held down by the seat belt with a force equal to three times his weight. This effect is called *negative centrifugal acceleration,* and the amount of force exerted is said to be −3 g.

Effect of Centrifugal Acceleration on the Circulatory System. The most important effects of centrifugal acceleration on the body occur in the circulation. Positive g causes the blood in the vascular system to be centrifuged toward the lower part of the body. The veins of the abdomen and legs distend greatly, storing far more blood than usual, so that little or none of it returns uphill to the heart. The cardiac output falls to very low values or even to zero. If the positive g force is too great, the arterial pressure falls and the person lapses into coma.

Negative g causes opposite effects; so much blood is centrifuged into the upper part of the body that the cardiac output rises and the arterial pressure rises. Occasionally the person develops such high pressure in his head vessels that brain edema results, and, rarely, vessels rupture in both the brain and the eyes, causing serious mental impairment or visual damage.

Figure 39–5 depicts graphically the time required for blackout (coma) to ensue at various degrees of positive acceleration, when a person is in the sitting position. A person can usually withstand +4 g almost indefinitely; he might develop a certain amount of dizziness, but, in general, his circulatory control system can maintain sufficient cardiac output to prevent blackout. However, at acceleration values above +4 g, blackout usually occurs quite rapidly, at 5 g in about 8 seconds; at 12 g in about 3 seconds; and at 20 g in about $2\frac{1}{2}$ seconds. Obviously, the amount of time that an aviator can remain conscious at high rates of acceleration limits the rapidity with which he can come out of a dive, and it also limits the sharpness of turns that he can make. Most airplanes are designed to withstand 20 or more positive g. Therefore, the maneuverability of an airplane is dependent more on what the aviator himself can stand than on the strength of the airplane.

Antiblackout Measures. Application of pressure to the outside of the legs and lower

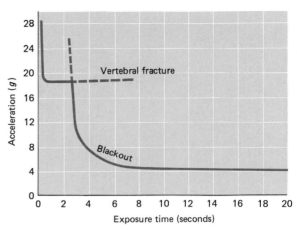

FIGURE 39–5 Effect of exposure time to different degrees of positive acceleration in producing blackout or vertebral fracture. (From Armstrong: Principles and Practice of Aviation Medicine. Baltimore, Williams & Wilkins, 1900.)

abdomen can prevent pooling of blood during positive acceleration.

Another means for partially accomplishing the same effect is for the aviator himself to tighten his abdominal muscles as much as possible during the pullout from a dive. This enables him to withstand a little more acceleration than he otherwise could and allows him several seconds extra exposure time before blacking out.

Different types of "anti-g" suits for accomplishing this have also been devised. One of these utilizes air bags applied to the abdomen and legs. When the airplane pulls out of the dive, the bags inflate so that positive pressure is applied to the legs and abdomen. In this way the aviator is not uncomfortably subjected to pressure all the time but only when it is needed.

Because of the speed of the modern jet airplane, the tremendous acceleratory forces that can develop make it impossible for an aviator to turn sharply without blacking out. One of the solutions to this problem will probably be for the aviator to lie prone while flying the plane, for the body can withstand +15 g in the horizontal position in comparison with only +4 g in the sitting position.

Effect of Positive Acceleration on the Skeleton. Also illustrated in Figure 39–5 is the effect of positive acceleration in causing vertebral fracture. This graph shows that with the aviator in the sitting position, positive acceleration greater than 20 g is likely to rupture a vertebra because of the intense force of the upper body pressing downward. Only a split second of the intense force can cause the fracture, for the stress that a bone can withstand does not depend on how long the stress is applied, but on how much stress is applied. Similar rupture of

FIGURE 39–6 Effect of distance of fall on the velocity attained. (From Armstrong: Principles and Practice of Aviation Medicine. Baltimore, Williams & Wilkins, 1900.)

other tissues can also occur, such as tearing of the supporting structures of the heart with consequent internal hemorrhage and death.

Deceleration Forces During Parachute Descent. When a person jumps from an airplane, gravity causes him to begin falling toward earth. Figure 39–6 illustrates the velocity of fall after different distances of fall (if the person has not opened his parachute). By the time he has fallen 1500 ft his velocity will have reached approximately 175 ft per second. At this speed, the air resistance exactly opposes the tendency for gravity to increase his velocity even more. Therefore, this velocity of fall, 175 ft per second, is the maximum that will be attained, and it is called the *terminal velocity*. (At very high altitudes the terminal velocity is considerably greater than 175 ft per second because of the rarefied atmosphere, but it will slow down to 175 ft per second when the person falls into the heavier atmosphere.)

One of the deceleratory forces caused by parachute descent is the *opening shock* that occurs when the parachute opens. The force of the shock is roughly proportional to the square of the velocity of fall at the time of opening. Parachutes are designed so that even after the person has reached the terminal velocity of fall the shock will not be sufficient to harm the body.

On landing, the parachutist approaches earth at a velocity of about 15 mi per hour. This is equivalent to jumping without a parachute from a wall approximately 8 ft high. Even this velocity can be dangerous if the parachutist is not ready to cope with the landing jolt. If he lands stiff-legged, he will almost certainly suffer a fracture of a leg, the pelvis, or a vertebra. On the other hand, if he does not tense his muscles when landing, he will pancake on the ground and likely suffer other injuries. Therefore, it is important that he land with his legs flexed but tense.

SPACE PHYSIOLOGY

The physiologic principles of traveling in spaceships are much the same as those applicable to aviation physiology, except that the importance of some of the factors is intensified. The special problems in space physiology include (1) weightlessness, (2) intense linear acceleration forces, (3) supply of oxygen and other nutrients, and (4) special environmental hazards, especially radiation.

Weightlessness. Prior to the advent of the spaceship, persons had experienced the phenomenon of weightlessness for not more than a few seconds at a time, this state having been created in the cabins of airplanes undergoing a trajectory path, upward and over a hump, calculated for exact course and velocity to provide zero g forces for a period of 10 to 25 seconds. Yet it was known that once a spaceship was in orbit or was traveling from planet to planet, the spaceship itself would be subjected to exactly the same gravitational forces as the astronaut inside the ship. Therefore, the person would experience the phenomenon of prolonged weightlessness; that is, there would be no force pulling him in any direction toward any part of the spaceship.

Prior to the first space flight, many physiologists were concerned about how a human being would react to weightlessness of long duration. Fortunately, now that man has experienced this phenomenon, it has proved to have little significant effect on the physiology of the body, except that the astronaut frequently has difficulty readjusting to the "weightful" state on returning to earth.

Weightlessness does necessitate the use of a few operational devices for flying the spaceship and for other activities within the spaceship. First, since the astronaut "floats" inside the spaceship, he must be either strapped in place or have appropriate hand holds. Second, his food must be in special containers that have closed tops, because the food will literally float off any plate, and fluids will float out of any glass. Often, foods are prepared in tubes of the toothpaste type, and the food is simply squeezed into the mouth. Likewise, water is sucked from a tube. Excreta must also be forced into a container rather than left to float freely in the spaceship. Otherwise, one can imagine what the atmosphere inside the spaceship would be like after a few days of flying. Except for these simple problems, the state of weightlessness is actually very peaceful and pleasant, which is what one might expect from his own experience of floating in water.

Return to Earth After a Period of Weightlessness in Space. When an astronaut has remained in the weightlessness state for more than 1 to 3 weeks and then returns to earth, he frequently faints when he first stands. The reason for this is that his circulatory system has become adjusted to the weightlessness state, and it must now become readjusted to the "weightful" state, which requires 2 to 5 days for full readjustment. Two factors seem to be responsible for these adaptive changes in the circulation: First, the blood volume decreases slightly during the space sojourn so that when the person first stands after return to earth, displacement of blood into the lower legs by gravity leads to reduced cardiac output and a tendency to faint. Second, the blood vessels of the circulation, especially those of the lower body, seem to become more relaxed than normal during a long period of weightlessness. This, too, allows excessive quantities of blood to "pool" in the lower body when the astronaut first stands on returning to earth. It is interesting to note that these same two effects occur in patients who have laid in bed for long periods of time; they, too, often faint after a long period of recumbency.

Linear Acceleration of a Spaceship. At takeoff, a high velocity is reached in only a few minutes while putting the spaceship into orbit. This is achieved using rocket propulsion from one or more boosters. Figure 39–7 shows the intensity of forward acceleration at different times during the 5 minutes required to put a spaceship into orbit, illustrating that linear acceleration increases instantaneously to 2 g when the first booster begins to fire. As that booster becomes lighter and lighter because of decreasing fuel, the acceleration increases to as much as 9 to 10 g. Then, after the first booster has dropped, the second booster goes through a similar pattern of acceleratory changes. Also, the final stage, which carries the space capsule, accelerates for another minute or so before its rocket propulsion completely ceases. Upon cessation of all propulsion at the end of 5 minutes, the astronaut enters the state of weightlessness, which is designated in the figure by zero g acceleration.

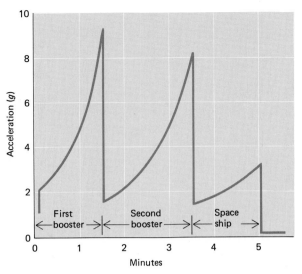

FIGURE 39–7 Acceleratory forces during takeoff of a spaceship.

In the sitting position the astronaut cannot withstand more than about 4 g, but in the reclining position, the position assumed in a reclining easy chair, the astronaut can stand 12 to 15 g, which is greater than any of the forces illustrated in Figure 39–7. Therefore, this is the position that the astronaut assumes during takeoff.

Linear Deceleration When Returning to Earth. One of the greatest problems in all of space physiology is the effect of slowing the spaceship as it reenters the atmosphere. The tremendous amount of kinetic energy stored in the momentum of the moving spaceship must be dissipated as the spaceship decelerates. This energy becomes heat and obviously could increase the temperature inside the spaceship to very high levels. Therefore, special precautions are taken to provide a heat dissipation shield at the front of the spaceship that absorbs most of this heat rather than allowing it to be transmitted to the cabin of the ship and also to provide adequate heat insulation of the cabin itself.

A spaceship traveling at a speed of Mach 100, a speed 100 times the velocity of sound and about 3 times that required for orbiting the earth, but which might well be used in interplanetary space flights, requires a distance of about 10,000 mi for safe deceleration. Any more rapid deceleration than this would create more g forces than the astronaut could stand.

Replenishment of Oxygen and Other Nutrients. It is a simple matter to provide all the oxygen and nutrients needed by the astronaut if he is to remain in space for not more than a few days to a few weeks. However, interplanetary travel will require that an astronaut remain in space for many months or even years, for which reason it becomes physically impossible to carry sufficient oxygen and other nutrients within the confines allowable in the spaceship. Therefore, many different procedures have been attempted to create a complete *life cycle* within the spaceship that will continually replenish the spaceship's oxygen and also supply adequate nutrients to the astronaut. For instance, algae are capable of living on human excreta, utilizing both carbon dioxide and fecal excreta to form oxygen, carbohydrates, proteins, and fats. Theoretically, therefore, algae can be used as food, and the oxygen can be rebreathed. Unfortunately, though, the amount of algae necessary to provide this complete life cycle for astronauts is far greater than space limitations in the spaceship will allow. Furthermore, the foods developed from algae have proved to be neither palatable nor totally life sustaining. Yet future research might provide an adequate means for finally completing this life cycle.

Other methods for completing the life cycle are based on chemical or electrochemical procedures for separating oxygen from carbon dioxide and for resynthesizing certain types of foodstuffs. Here again, the success has not been completely rewarding, so that today's space flights are still limited by the amount of oxygen and nutrients that can be carried from earth at the time of takeoff.

Radiation Hazards in Space Physiology. Early in the space program, scientists began to realize that enough radiation of different types—gamma rays, x-rays, electrons, cosmic rays, and so forth—exists at certain points in space that these rays alone could be lethal. Figure 39–8 illustrates two major belts of such radiation, called *Van Allen radiation belts*, that extend entirely around the earth and that are especially prominent in the equatorial plane of the earth. One of these belts begins at an altitude of about 300 mi and extends to about 3000 mi. An outer belt then begins at about 6000 mi and extends to 20,000 mi.

The intensity of radiation in either one of the two Van Allen belts is great enough that a person in a spaceship orbiting the earth within one of these belts for any prolonged period of time could receive enough radiation to cause death. Since significant radiation does not begin until an altitude of about 300 mi is reached, one can understand why space travel around the earth is normally confined to altitudes below 300 mi.

Other Problems of Space Flights. Other hazards at high altitudes include some of the same hazards of aviation but to much severer degree, such as (1) exposure of the astronaut to ultraviolet radiation, which can cause severe sunburn or can blind him; (2) exposure to extreme heat when in the direct pathway of sunrays and to extreme cold when on the opposite side of a spaceship or the opposite side of the moon from the sunrays; and (3) exposure to zero barometric pressure in case of decompression of the spaceship. All of these problems require special engineering such as pressurized suits, protective filters for the eyes, and very intricate heat control systems for the space capsule.

DEEP SEA DIVING AND HIGH PRESSURE PHYSIOLOGY

The human body is occasionally exposed to extremely high barometric pressures, such as those which occur when diving deeply beneath the sea or when working in a compression chamber—for instance, in a tunnel beneath a river filled with compressed air to prevent cave-in. The very high barometric pressure increases the amount of gases that pass from the alveoli

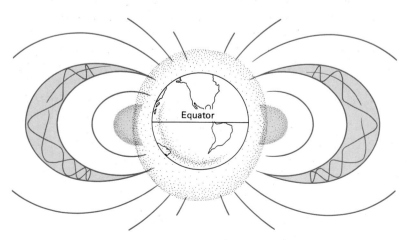

FIGURE 39–8 The hazardous Van Allen radiation belts around earth. (From Newell, H.E.: *Science, 131*:385, 1960.)

into the blood and become dissolved in all the body fluids. This extra dissolved gas, unfortunately, often leads to serious physiologic disturbances.

Pressures at Different Depths of the Sea. A person on the surface of the sea is exposed to the pressure of the air above the earth. This pressure is approximately 760 mm Hg; or it is also stated to be 1 atmosphere (atm) pressure. The weight of 33 ft of sea water is equal to the weight of the atmosphere, so that at a depth of 33 ft beneath the sea the pressure is double that at the surface, or 2 atm of pressure. Thus, the pressure increases as one descends beneath the sea approximately in accordance with the following table:

Feet	Atmosphere(s) pressure
0	1
33	2
67	3
100	4
200	7
300	10

Effect of High Gas Pressures on the Body

Figure 39–9 illustrates the pressures of each of the alveolar gases when a diver is breathing (1) air at sea level, (2) compressed air at 33 ft below sea level, and (3) compressed air at 100 ft below sea level. Note that the pressure of water vapor and carbon dioxide remain the same regardless of how far below the surface the person descends. The reason for this is that water vapor pressure is directly dependent on the temperature of the body, as explained earlier in the chapter, and the pressure of carbon dioxide is dependent on the rate of carbon dioxide release from the blood into the lungs, which continues to be the same regardless of depth. On the other hand, the pressures of nitrogen and oxygen increase almost directly in proportion to the depth below sea level to which the person descends.

Oxygen Poisoning. Ordinarily a person breathing compressed air can descend to about 200 ft below the sea without danger from excess oxygen absorption. At levels deeper than this, his tissues begin to be damaged very rapidly by the oxygen if he still breathes compressed air. The reason for this is the following: Hemoglobin normally releases oxygen to the tissues at pressures ranging between 20 and 45 mm Hg. In this way the hemoglobin automatically "buffers" the oxygen in the tissues, as was explained in Chapter 37, maintaining a relatively constant oxygen pressure that also allows a relatively constant rate of cellular oxidation. When there are extreme oxygen pressures in the lungs, however, large quantities of oxygen dissolve in the fluids of the blood, and the oxygen transported in this manner, in contrast to the oxygen transported by the hemoglobin, can be released to the tissues at pressures that are much greater than 45 mm Hg, indeed sometimes as high as several thousand mm Hg, supplying far more than normal amounts of oxygen to the cells. This causes deranged cellular metabolism, often damaging the cells themselves or leading to abnormal cellular function.

The most debilitating effect of oxygen poisoning occurs in the brain, usually manifested by nervous twitchings, convulsions, and coma. Obviously, the person often loses control of himself, which might make him descend too deeply in the sea or fail to operate his breathing apparatus correctly, which will lead to his death.

The dangers of oxygen poisoning can be eliminated by supplying progressively less percentage of oxygen in the diver's breathing mixture as he goes to greater depths.

Nitrogen Poisoning. High pressures of nitrogen, also, can seriously affect a diver's mental functions. Even though this gas does not enter into any metabolic reactions in the body, when it dissolves in the body fluids in high concentrations, it exerts an anesthetic effect on the central nervous system. Ordinarily, a person can

Pressure on outside of body

760 mm Hg	1520 mm Hg	3040 mm Hg
$P_{N_2} = 572$ mm Hg $P_{O_2} = 101$ " $P_{CO_2} = 40$ " $P_{H_2O} = 47$ "	$P_{N_2} = 1173$ mm Hg $P_{O_2} = 260$ " $P_{CO_2} = 40$ " $P_{H_2O} = 47$ "	$P_{N_2} = 2375$ mm Hg $P_{O_2} = 578$ " $P_{CO_2} = 40$ " $P_{H_2O} = 47$ "
Sea level	33 ft below sea level	100 ft below sea level

FIGURE 39–9 Effect of different sea depths on the gas pressures in the alveoli.

descend to approximately 200 ft below sea level before this anesthetic effect becomes serious, which is about the same maximum safe depth as that for the prevention of oxygen poisoning. However, at depths slightly deeper than this the diver will actually fall asleep because of the increasing amounts of nitrogen dissolved in his fluids. Also, in the early stages of this "nitrogen narcosis," the person frequently develops a sense of extreme exhilaration associated with seriously depressed mental acuity, a condition called "raptures of the depths." This state is comparable to drunkenness from alcohol, and it can cause the diver to descend much deeper than he should or to perform other dangerous acts that will lead to his death.

To avoid the dangers of nitrogen poisoning, the nitrogen is frequently replaced by helium in the breathing mixture. Helium does not cause an anesthetic effect, and it has an additional advantage of diffusing out of the body fluids more rapidly than nitrogen will when the diver ascends.

The Problem of Carbon Dioxide Washout from the Diving Mask. If high concentrations of carbon dioxide are allowed to collect in a diver's mask, he will soon develop respiratory acidosis and perhaps even coma. Therefore, carbon dioxide must be washed out of the mask rapidly enough so that its concentration will never rise to any significant level. To do this, the same *volume* of gas must flow through the mask at all times. When the diver descends to 200 ft below the sea, his air must be compressed seven times to withstand the pressure. This also reduces the volume seven times and requires that the compressor pump seven times as much mass of air for carbon dioxide washout as is required at sea level. It can be seen, therefore, that the rate at which air is supplied to the diver must be increased directly in proportion to the depth at which he is working. Indeed, it is this factor of carbon dioxide washout that determines how much air must be pumped to the diver rather than the amount of oxygen that he needs.

SCUBA Diving. The term "SCUBA" means "self-contained underwater breathing apparatus." (A typical SCUBA system is illustrated in Figure 39–10.) That is, in contrast to the diving helmet or the diving bell, there are no air hoses to an attendant boat on the surface of the sea. The major factor that limits the time a person can remain under water using a SCUBA system is the problem of carbon dioxide washout from the alveoli. The deeper the diver goes, the less becomes the *volume* of each unit mass of gas liberated from his tank. Therefore, the greater

FIGURE 39–10 A SCUBA apparatus.

must be the *mass* of gas that flows out of his tank to maintain carbon dioxide washout from his lungs. A tank of compressed air thus will last only one-seventh as long at 200 ft as at a few feet below the surface of the sea. For this reason, a SCUBA diver can descend to a 200-foot depth for only 15 to 20 minutes at a time. Other than for this limitation, the physical principles of SCUBA diving are essentially the same as those of diving using a diving bell or diving helmet.

Decompression Sickness

Solution of Nitrogen in the Body Fluids. In addition to the anesthetic effect of nitrogen, this gas can also cause very serious damage because of gas bubble formation in the body fluids when the diver returns toward the surface. The amount of nitrogen normally dissolved in the entire body when a person breathes air at sea level pressure is about 1 L. But if he remains 100 ft below the sea for several hours and breathes compressed air all this time, that is, at 4 atm pressure, the amount of dissolved nitrogen increases to four times as much, and at 7 atm pressure to seven times as much.

Bubble Formation During Decompression. After a diver has remained at a depth of 200 ft for several hours, he will have dissolved about 7 "sea-level" L of nitrogen in his body fluids. This amount of dissolved nitrogen exerts a pressure of 3918 mm Hg. When the diver is still at the 200-foot depth, the pressure of the water is about 5300 mm Hg, which is far more than enough pressure to keep the nitrogen gas compressed in the body fluids. However, if the diver then suddenly rises to sea level, where the air pressure on the outside of the body is only 760 mm Hg, a situation has now developed in which the pressure of the 7 L of dissolved nitrogen in the body fluids is about four times as great as the pressure on the outside of the body. The nitrogen attempts to escape by any route possible, pushing against the inside of the blood vessels, skin, and cells. As a result, small bubbles of nitrogen develop everywhere in the body, and these obviously can cause very severe structural damage to almost any part of the body.

Effect of Bubble Formation on the Body. The effects of bubble formation on the body are known by many different names: *decompression sickness*, the *bends*, *diver's paralysis*, *caisson disease*, and others. The most distressing effects occur when bubbles develop in the central nervous system. They sometimes occur inside neuronal cells, but mostly in the interstitial tissues of the brain and spinal cord, causing mechanical rupture of nerve fiber pathways. The damage can lead to serious mental disorders and very frequently to permanent paralysis because of ruptured fiber pathways to the muscles.

Decompression sickness usually causes severe pain, which may result from bubble damage in the central nervous system, in peripheral nerves, or in pain-sensitive tissues such as the joints and bones or from distention of the gastrointestinal gases, causing severe bloating of the gut.

In short, many different symptoms are produced by bubble formation and gas expansion in the body, though the most distressing and permanent of these result from damage to the nervous system.

Prevention of Decompression Sickness. The bubbles of decompression sickness can be prevented by allowing the diver to come to the surface very slowly or by decompressing him in a decompression chamber. In either event, the pressure around his body must be decreased so gradually that the excess gases dissolved in his fluids can leave through his lungs before a damaging quantity of bubbles develops.

QUESTIONS

1. How does low barometric pressure affect oxygen transport to the tissues?
2. What are the respiratory compensations for oxygen deficiency?
3. Approximately how much negative g and positive g can a person stand during flight maneuvers?
4. How much of a hazard is the state of weightlessness in space to the function of the circulation?
5. Explain the problems of linear acceleration as they relate to spaceship travel.
6. What causes the pressure of gases in the lungs to increase deep beneath the sea?
7. What are the toxic effects of high pressures of oxygen? Of high pressures of nitrogen?
8. Explain the role of nitrogen in the development of decompression sickness.

REFERENCES

Bennett, P.B., and Elliott, D.H.: The Physiology and Medicine of Diving and Compressed Air Work, 2nd Ed. Baltimore, Williams & Wilkins, 1975.
Bullard, R.W.: Physiological problems of space travel. *Annu. Rev. Physiol., 34*:205, 1972.
Fisher, A.B., et al.: Oxygen toxicity of the lung: Biochemical aspects. *In* Fishman, A.P., and Renkin, E.M. (eds.): Pulmonary Edema. Baltimore, Williams & Wilkins, 1979, p. 207.
Frisancho, A.R.: Functional adaptation to high altitude hypoxia. *Science, 187*:313, 1975.
Gamarra, J.A.: Decompression Sickness. Hagerstown, Md., Harper & Row, 1974.
Grover, R.F., et al.: High-altitude pulmonary edema. *In* Fishman, A.P., and Renkin, E.M. (eds.): Pulmonary Edema. Baltimore, Williams & Wilkins, 1979, p. 229.
Hempelman, H.V., and Lockwood, A.P.M.: The Physiology of Diving in Man and Other Animals. London, Edward Arnold, 1978.
Lahiri, S.: Physiological responses and adaptations to high altitude. *Int. Rev. Physiol. 15*:217, 1977.
Pace, N.: Respiration at high altitude. *Fed. Proc., 33*:2126, 1974.
Reeves, J.T., et al.: Physiological effects of high altitude on the pulmonary circulation. *In* Robertshaw, D. (ed.): International Review of Physiology: Environmental Physiology III. Vol. 20. Baltimore, University Park Press, 1979, p. 289.
Shilling, C.W., and Beckett, M.W. (eds.): Underwater Physiology IV. Bethesda, Md., Federation of American Societies for Experimental Biology, 1978.
Sloan, A.W.: Man in Extreme Environments. Springfield, Ill., Charles C Thomas, 1979.

XII

THE DIGESTIVE AND METABOLIC SYSTEMS

40

Anatomy of the Digestive System

Overview

The digestive system begins with the **mouth** and extends through the **pharynx, esophagus, stomach, small intestine** (comprised of the **duodenum, jejunum,** and **ileum**), and **large intestine** (comprised of the **cecum, colon, rectum,** and **anal canal**).

The *pharynx* and *esophagus* serve specifically the swallowing function.

One of the most important roles of the *stomach* is to store large quantities of food after a meal and to release this food a little at a time to the remainder of the digestive tract. The outlet of the stomach is the **pylorus,** and the rate of release of the stomach contents is controlled by the degree of contraction of the **pyloric sphincter muscle** that encircles the pylorus.

The *small intestine* is the locus of most digestion of the food, and it is also here that almost all of the products of digestion are absorbed. To enhance absorption, multiple circular folds of mucosa, the **plicae circulares,** protrude into the intestinal lumen, increasing the total absorptive area by about threefold. In addition, the entire small intestine mucosa is covered with **intestinal villi,** several hundred to each square centimeter of surface area; it is through the epithelium of these that most absorption occurs.

The *large intestine* is about two times as large in diameter as the small intestine; it serves two functions: first, to absorb as much fluid and electrolytes as possible from the intestinal waste, thus conserving these in the body, and, second, to store fecal material until a time appropriate for its elimination.

Along the entire course of the digestive tract are multiple **glands** that secrete (1) *digestive juices* for digesting the food and (2) *mucus* for lubricating the passage of the food along the digestive tract. Some of these are complex glands that lie outside the digestive tract and transport their secretions into the digestive tract through **ducts;** these include the **salivary glands,** the **pancreas,** and the **liver.** In addition, in the walls of all parts of the digestive tract, from the esophagus through the large intestine, are much less complex **mucosal glands.**

The wall of the digestive tube is constructed of multiple layers, the more important of which are (1) the **mucosa,** which lines the inner surface of the entire tube and contains many *mucosal glands;* (2) the **submucosa,** which is a layer of loose connective tissue containing many blood vessels and the deeper portions of the mucosal glands; and (3) two layers of smooth muscle, a **circular muscle layer** and a **longitudinal**

muscle layer, which cause gastrointestinal contractions. Between the two layers of muscle is an especially important nervous plexus, the **myenteric plexus,** and in the submucosa is a less extensive plexus, the **submucosal plexus.** These control the muscular movements of the gastrointestinal tract as well as secretion by most of the mucosal glands.

The portions of the gastrointestinal tract that hang freely in the abdominal cavity are attached to the posterior wall by a strong weblike membrane called the **mesentery.** The organs of the gastrointestinal tract, as well as the mesentery, are all lined on their surfaces by **visceral peritoneum;** the outer wall of the abdomino-pelvic cavity is lined with **parietal peritoneum.**

The obvious function of the **digestive system,** also called the **alimentary tract** or **gastrointestinal tract,** is to begin processing the food for the body. The food must eventually be converted into chemical substances that the cells can utilize for their metabolic functions. This process begins with chewing the food and then propelling it along the digestive tract at an appropriate pace so that it can be digested, which means splitting it chemically into smaller molecular substances. The products of digestion are then absorbed from the digestive tract into the blood and transported wherever they are needed in the body.

The purpose of this chapter will be to describe the principal organs of the digestive tract. In the following three chapters we shall trace the successive modifications of the different foods and the nutrients derived from these foods as they are processed and used by the body.

THE ORGANS OF THE DIGESTIVE SYSTEM

Figure 40–1 illustrates the principal organs of the digestive system:

1. The **mouth** and **pharynx** are the entryway to the digestive system.
2. The **salivary glands** secrete *saliva* into the oral cavity.
3. The **esophagus** functions as a conduit for rapid transport of food from the pharynx into the stomach.
4. The **stomach** stores food for several hours and also is where much early digestion of the food occurs.
5. The **small intestine** is comprised successively of the **duodenum,** the **jejunum,** and the **ileum;** it is in these areas that final digestion of the food occurs and where almost all of the digestive products are absorbed.

6. The **liver** and **gallbladder.** The liver is both a very large secretory gland and also an organ for modifying the chemical structures of many of the nutrients before they are carried by the blood to other parts of the body. The liver secretes *bile*, which is stored temporarily in the **gallbladder.** Then, during the digestion process, the bile is emptied into the duodenum to help promote digestion, especially digestion of fats.
7. The **pancreas.** This is a large gland that secretes the most important of all the digestive juices for digesting the three major foods of the diet: the *proteins*, the *carbohydrates*, and the *fats.*
8. The **large intestine** includes (a) the **cecum,** (b) the **colon,** (c) the **rectum,** and (d) the **anal canal.** The colon is mainly a large storage and absorptive chamber. Those portions of the food that have not been digested and absorbed are stored in the colon for 8 to 24 hours before defecation. During this holding period, the last remnants of usable material from the ingested food and drink are absorbed into the blood—especially almost all of the remaining water, plus the sodium, potassium, chloride, and other ions, all of which must be conserved by the body if one is to survive periods of starvation or dehydration.

THE MOUTH, PHARYNX, AND ASSOCIATED STRUCTURES

Figure 40–2 illustrates the principal structures of the *mouth* and *pharynx* that relate to digestive functions. Anatomically, the mouth extends from the inside of the lips to a posterolateral constriction called the **fauces** that separates the mouth from the pharynx. The pharynx is a vertical chamber posterior to the fauces, about 12 cm long, that extends superiorly to the base of the

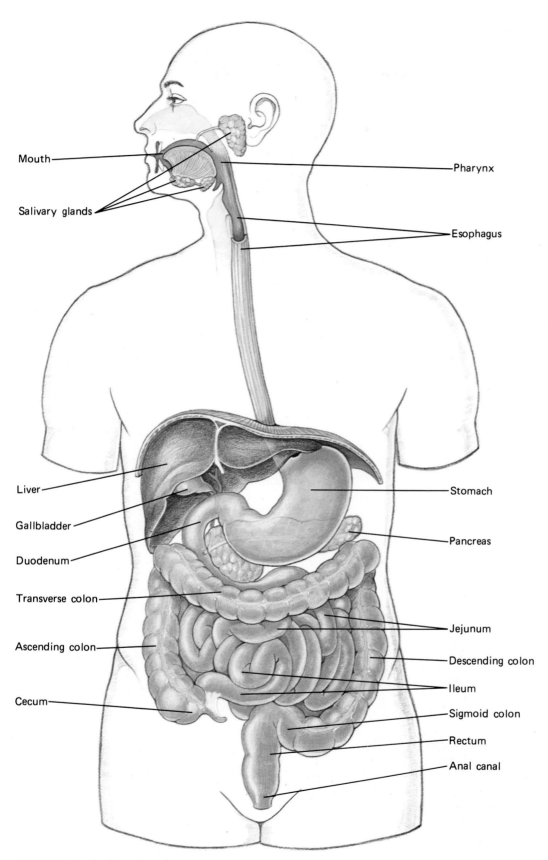

Mouth

Salivary glands

Pharynx

Esophagus

Liver

Gallbladder

Duodenum

Transverse colon

Ascending colon

Cecum

Stomach

Pancreas

Jejunum

Descending colon

Ileum

Sigmoid colon

Rectum

Anal canal

FIGURE 40–1 The digestive system.

TABLE 40–1
The Digestive Tract

Structure	Description or Function
Mouth	
Hard palate	Anterior two thirds of roof of mouth
Soft palate and uvula	Posterior one third of roof of mouth; close nasopharynx from oral pharynx during swallowing
Tongue	
Papillae	Located on anterior two thirds of tongue; locus of most taste buds, the sensory receptor organs of taste
Circumvallate	
Foliate	
Fungiform	
Filiform	
Pharynx	
Nasopharynx	Located above soft palate and posterior to nasal cavities
Oral pharynx	Located posterior to mouth
Laryngeal part of pharynx	Located posterior to larynx and root of tongue
Fauces	Narrowed region between mouth and pharynx; narrows further during swallowing or gagging
Palatine tonsil	Lymphoid body located on each side of fauces; helps to provide immunity against pharyngeal infectious agents
Esophagus	Conducts food and drink from pharynx to stomach in 5 to 8 seconds during swallowing
Lower esophageal sphincter	Thickened circular musculature at lower end of esophagus; normally prevents reflux of stomach contents but opens during swallowing
Stomach	
Body	Upper three fourths of stomach; stores swallowed food; secretes highly acidic gastric juice containing enzymes for protein digestion
Fundus	Domelike part of body of stomach lying superior to and to the left of the esophageal opening into the stomach
Pyloric region	Lower one fourth of stomach distal to the angular notch; mixes food and digestive juices to form chyme; provides strong peristalsis to empty stomach
Pylorus and pyloric sphincter	Opening from stomach into duodenum and thickened circular muscle sphincter around opening that controls stomach emptying
Small intestine	
Duodenum	First part, 20 to 30 cm long, in shape of a "C"; bile from liver and pancreatic juice enters duodenum about 8 cm distal to pylorus
Jejunum and Ileum	Second and third parts, 3 and 4 m long, respectively; locus of most digestion of food and absorption of nutrients
Special structures	
Plicae circulares	Circular folds of mucous membrane, more numerous and larger in proximal small intestine; increase absorptive area about threefold
Intestinal villi	One-millimeter projections, several hundred per cm² of surface area; absorb most of intestinal fluids and nutrients into their blood and lymphatic capillaries
Ileocecal valve	Opening from ileum into large intestine; prevents reflux of cecal contents into ileum
Large intestine	Absorbs most remaining fluid and electrolytes from chyme; stores remaining fecal contents until propitious time for defecation
Cecum	Bulbous pouch at and inferior to ileocecal valve
Ascending colon	Ascends along right abdominal cavity wall
Transverse colon	Crosses under liver and stomach
Descending colon	Descends along left abdominal cavity wall
Sigmoid colon	Tortuous connecting link from descending colon to rectum
Rectum	Pelvic end of large intestine from which feces are defecated
Anal canal	Opening through which feces are defecated
Internal anal sphincter	Thickened circular smooth muscle around anal canal; provides involuntary control of defecation
External anal sphincter	Skeletal muscle constrictor around anal canal; provides voluntary control of defecation

TABLE 40–1 **(Continued)**

Structure	Description or Function
Intrinsic nervous system	Network of nerve cell bodies and nerve fibers that controls intestinal movements and secretion
Myenteric plexus	Lies between circular and longitudinal muscle layers
Submucosal plexus	Located in submucosal layer
Glands	
Mucosal and submucosal	Secrete mucus in all areas of digestive tract and digestive enzymes in stomach and small intestine
Salivary	Secrete saliva into mouth
Parotid	Located anteroinferior to ear
Submandibular	Located inferior to lateral tongue
Sublingual	Located inferior to anterior tongue
Pancreas	Secretes pancreatic juice through duodenal papilla into duodenum
Liver	Secretes bile through duodenal papilla into duodenum

skull posterior to the nasal cavities and inferiorly to the entrance of the esophagus. Anatomically it is divided into three parts:

1. The **nasal pharynx,** also called the **nasopharynx,** is posterior to the nasal cavities and above the soft palate.
2. The **oral pharynx** is posterior to the mouth.
3. The **laryngeal part** is posterior to the larynx.

The Palate and the Uvula. The palate is divided into the **hard palate** and the **soft palate.** The *hard palate* constitutes most of the roof of the mouth. This part of the palate is hard because it is backed up by the palatine process of the maxilla and the horizontal plate of the palatine bone. The *soft palate* is a muscular structure covered by mucous membrane, as described in Chapter 36. It projects backward from the hard palate into the pharynx to separate the nasopharynx and the oral pharynx. Its function is to close the opening between the oral pharynx and the nasopharynx during swallowing, as was also described in Chapter 36.

The **uvula** is a tongue-like projection from the posterior part of the soft palate. It helps to provide final closure of the opening between the mouth and nasopharynx.

The Fauces. The fauces is the narrowed opening between the mouth and the oral pharynx. Its important structures are as follows:

1. The **palatoglossal arches,** one on each side of the oral cavity, are muscular and mucosal folds curving inferiorly from the lateral borders of soft palate to the two sides of the base of the tongue.
2. The **palatopharyngeal arches,** also muscular and mucosal folds similar to the palatoglossal arches, are located 1 to 2 cm

posteriorly, and curve laterally and inferiorly from the posterolateral margins of the soft palate near the uvula to blend into the sides of the oral pharynx.

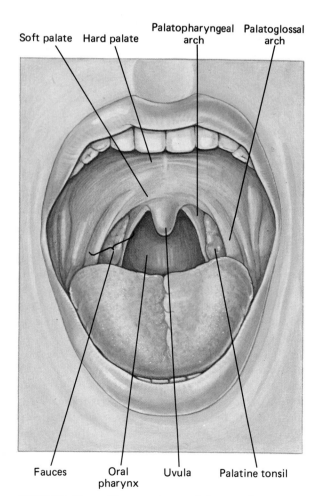

FIGURE 40–2 The structures of the mouth and oral pharynx.

3. One **palatine tonsil** lies on each side of the fauces in a depression called the **tonsillar fossa** between the palatoglossal arch and the palatopharyngeal arch.

When the muscles in the palatoglossal and palatopharyngeal arches contract, especially the palatoglossal arches, the sides of the fauces close nearly together, thus narrowing the opening between the mouth and the oral pharynx. This is what happens when a person gags. This narrowing is often a protective function to prevent unwanted objects from passing from the mouth into the pharynx. For instance, when a person tries to swallow some rough object such as a sharp bone, nervous reflexes originating from the circle of mucous membrane around the fauces will elicit the gag reflex, and almost invariably this will prevent the object from passing farther. Also, even during normal swallowing the palatoglossal arches close part way, leaving a narrowed opening through which the food is swallowed. If the food has not been chewed adequately, this small opening will impede its passage.

The *palatine tonsils* are part of the *lymphoid system* that was described in Chapters 30 and 34. They help to filter bacteria and toxins from fluids that are absorbed from the mouth. And they also help to develop immunity against these foreign agents. Frequently, though, the tonsils themselves become the subject of infection by bacteria, which is the clinical condition called *tonsillitis.* The tonsils then enlarge to several times normal size, from a normal diameter of about 1.5 cm to 3.5 cm, often large enough to severely interfere with the swallowing process. Also, the tonsils become very painful, and the infectious process can lead to high fever.

The Teeth and Mastication. The **teeth** are obviously of major importance to break the food into small particles, which is the process called *mastication.* The teeth and their relation to the bone structures of the maxilla and mandible were described in Chapter 6.

The Pharyngeal Constrictor Muscles. A distinctive functional element of the pharynx is the pharyngeal constrictor muscles—the superior, middle, and inferior constrictors—that lie immediately under the pharyngeal mucosa on the lateral and posterior walls of the pharynx, as was described in Chapter 8. These muscles contract during the first stage of swallowing to force the food from the pharynx into the esophagus, as will be explained further in the following chapter.

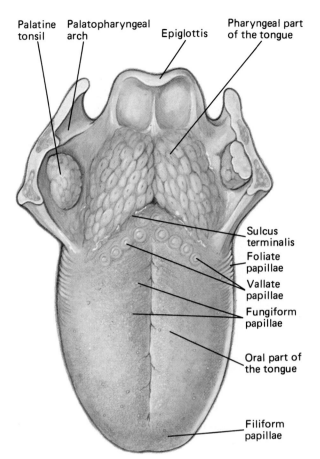

FIGURE 40-3 The tongue.

THE TONGUE

The **tongue,** illustrated in Figure 40-3, is a muscular organ that is important to many functions, including (a) positioning food on the teeth, (b) initiation of swallowing, (c) speech, and (d) taste. The muscular structure and nervous control of the tongue have been described in earlier chapters. Some of its surface features are described here:

1. The **sulcus terminalis** is a V-shaped groove separating the anterior two thirds of the tongue, called the **anterior tongue,** from the posterior one third, the **posterior tongue.**
2. There are many **papillae,** which are minute projections, on the superior and lateral surfaces of the anterior tongue. The tongue has four different types of papillae:
   a. The **circumvallate papillae** are large, circular, flat-topped structures several millimeters in diameter with grooves around them. These are arranged in a "V" formation along the anterior border of the sulcus terminalis.

b. The **foliate papillae** are elongated projections located along the sides of the tongue.

c. The **fungiform papillae** are 1 mm-sized raised structures sprinkled over the anterior tongue.

d. The **filiform papillae** are the very fine, sharp projections on all areas of the anterior tongue except where the foliate and fungiform papillae are located. It is these filiform papillae that give the tongue its fuzzlike appearance.

Most of the **taste buds,** which are the sensory organs of taste, are located on the papillae. The taste buds of the circumvallate papillae are especially important, because they mainly detect bitter substances, which in nature are usually poisonous. When these are detected they causing gagging and retching, and the food is usually expelled from the mouth rather than being swallowed.

THE SALIVARY GLANDS

The *salivary glands*, illustrated in Figure 40–4, secrete saliva into the mouth, more than a liter each day. Three salivary glands are present on each side of the mouth:

1. The **parotid gland** is located over the posterior aspect of the ramus of the mandible anterior and inferior to the ear. The parotid gland empties its secretion through the **parotid duct** that opens through a small raised **parotid duct papilla** located on the inside of the cheek lateral to the second upper molar tooth. One can see this papilla by pulling the angle of the mouth upward to expose the inner surface of the cheek. It is the parotid gland that swells greatly when one has the mumps; it then hurts terribly when sour foods are put into the mouth because nervous reflexes from the taste buds stimulate it to secrete saliva into its swollen and blocked ducts.

2. The **submandibular gland** lies medial to the body of the mandible and inferior to the posterolateral floor of the mouth. It empties its saliva through the **submandibular duct** that runs anteriorly from the gland to the underside of the tongue, emptying into the angle between the tongue and floor of the mouth about 5 mm lateral to the midline.

3. The **sublingual gland** lies along the inner margin of the mandible inferior to the floor of the anterior mouth. It empties through several small ducts that enter the floor of the mouth inferior to the tongue and posterior to the opening of the submandibular duct.

Cellular Structure of the Salivary Glands. Figure 40–5 illustrates the *secretory cells* of a typical salivary gland (in this instance, the submandibular gland). As was discussed in Chapter 2, the salivary glands are *compound*

Mucous membrane (cut edge)

Opening of submandibular duct

Sublingual gland

Submandibular duct

Submandibular gland

Parotid gland

Accessory part of parotid gland

Parotid duct

Body of mandible

FIGURE 40–4 The salivary glands and their ducts.

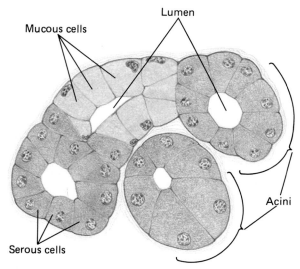

FIGURE 40–5 The cellular structure of the submandibular salivary gland, illustrating both serous and mucous secretory cells as they appear in several acini of the gland.

acinous glands in which the secretory cells are organized into literally thousands of separate small spheroidal secretory units called **acini.** The secretory cells secrete into the central lumens of the acini, and these lumens coalesce to form progressively larger ducts, finally leading to the salivary duct that empties into the mouth.

Note especially in Figure 40–5 that there are two different types of secretory cells in the submandibular gland:

1. **Serous cells** secrete a thin fluid containing the important digestive enzyme *ptyalin* for beginning the digestive process.
2. **Mucous cells** secrete *mucus*, a very slick and viscous fluid that lubricates the food as it is swallowed.

The parotid gland, on the other hand, contains only serous cells, and the sublingual gland contains a preponderance of mucous cells.

THE ESOPHAGUS

Figure 40–6 illustrates a cross-section of the esophagus. It is important to study this cross-section carefully because the entire gastrointestinal tract has essentially the same basic structure, though organized somewhat differently in each section of the tract. The layers of the esophageal wall from the inside toward the outside are as follows:

1. The **mucosa,** or **mucous membrane,** consists of the following three layers:
   a. The **epithelium** is similar to that of the mouth; that is, it is of the stratified squa-

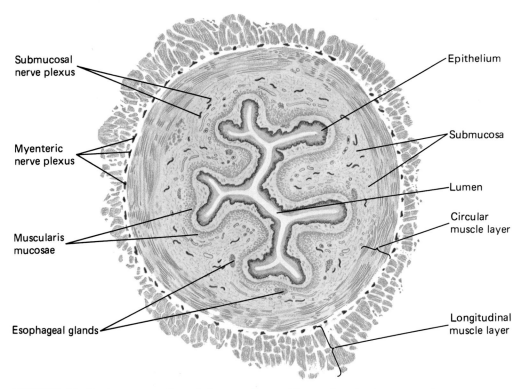

FIGURE 40–6 A cross-section of the esophagus, illustrating the important layers of its wall. The same layers, but with varying details, are found in all other parts of the gastrointestinal tract as well.

mous type, which allows the esophagus to resist damage by hot, corrosive, or abrasive foods. However, at all lower levels of the digestive tube—from the stomach to the anal canal—the epithelium is of the simple columnar type.

  b. The **lamina propria** is a layer of loose connective tissue.
  c. The **muscularis mucosae** is a layer of smooth muscle. When this muscle contracts, it causes folds in the mucosa, as illustrated in the figure.

2. The **submucosa** is a layer of loose connective tissue that lies under the mucosa and contains many blood vessels.
3. A **circular muscle layer** is composed of muscle fibers that extend in the circular direction around the esophagus.
4. A **longitudinal muscle layer** is composed of muscle fibers that extend lengthwise along the esophagus.

In the lower half of the esophagus and in the entire lower portion of the digestive tract, the muscle layers are all composed of smooth muscle. However, in the upper one third to one half of the esophagus, most of the muscle is skeletal muscle that is actually an extension of the pharyngeal constrictor muscles downward from the pharynx into the upper part of the esophagus.

The Esophageal Glands. Note in the mucosa and submucosa a number of small **esophageal glands.** These glands secrete mucus into the esophagus for the purpose of lubrication.

The Intrinsic Nervous System of the Esophagus (and of All Lower Portions of the Digestive Tube). The entire digestive tube, from the origin of the esophagus to the anus, has its own **intrinsic nervous system.** This is comprised of an extensive network of *nerve cell bodies* and *nerve fibers* located in the wall of the tube. It contains both *sensory fibers* and *motor fibers*, and it is capable of sustaining local reflexes that originate in *sensory receptors* of the gut wall itself and cause either contraction of the smooth muscle of the wall or secretion by the mucosal glands. When the nerves from the brain and spinal cord to the digestive tract have been cut, this intrinsic nervous system still functions almost normally to control the local muscular and secretory functions of the digestive tract.

The intrinsic nervous system is divided into two separate but interconnected nerve plexuses:

1. The **myenteric nerve plexus.** This plexus lies between the longitudinal and circular muscle layers. It is a network of many nerve cell bodies and nerve fibers that extends the entire length of the digestive tract beginning at the level of the esophagus. It receives nerve connections from both the parasympathetic nervous system and the sympathetic system. In general, stimulation by the parasympathetic nervous system increases the activity of the myenteric plexus while the sympathetic system inhibits it. This plexus is especially concerned with controlling the circular and longitudinal muscle layers of the gut wall. For instance, it controls the forward moving ring-like constrictions that cause forward movement of food, which is the process called *peristalsis.*

2. The **submucosal nerve plexus.** This is a less extensive plexus that lies within the submucosa and is particularly concerned with sensory nerve signals arising from the mucosal epithelium and the resultant reflexes that cause secretion by the gastrointestinal glands.

The Lower Esophageal Sphincter. Beginning 2 to 3 cm above the stomach and extending upward for another 5 cm, the circular muscular layer of the esophagus is considerably thickened, as illustrated in Figure 40–9. This thickened muscle is called the **lower esophageal sphincter.** Normally this sphincter remains contracted and prevents reflux of food from the stomach into the esophagus. However, during swallowing it must be opened. This is one of the control functions of the myenteric plexus; the same myenteric neuronal mechanism that causes the peristaltic movements that force the food down the esophagus also causes relaxation of the lower esophageal sphincter ahead of the peristaltic wave, thus allowing the food to pass unimpeded into the stomach.

THE ABDOMINO-PELVIC CAVITY

The **abdomino-pelvic cavity** extends from beneath the diaphragm to the recesses of the pelvis. The portion of this cavity above the brim of the bony pelvis is called the **abdominal cavity,** and the portion within the pelvis is called the **pelvic cavity.** However, in common parlance, the whole cavity is often referred to simply as the abdominal cavity.

Figure 40–7A illustrates an anterior view of the abdominal cavity as it looks when it is first opened widely, showing the liver lying under the diaphragm, the stomach inferolateral to the liver, and the intestines posterior to a large fibrous-

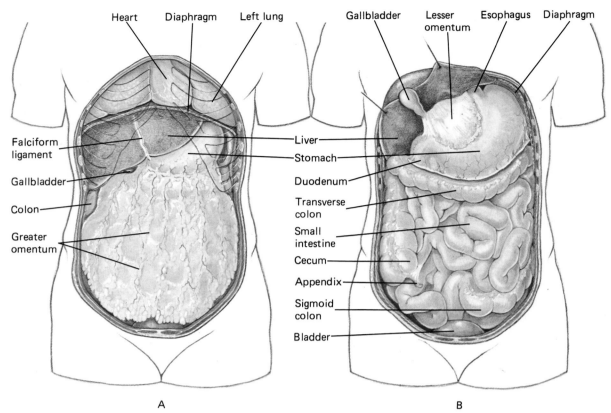

Heart   Diaphragm   Left lung   Gallbladder   Lesser   Esophagus   Diaphragm
                                                        omentum

Falciform
ligament   Liver
                                                                  Stomach
Gallbladder   Duodenum

Colon   Transverse
                                                                  colon
Greater   Small
omentum   intestine

                                                                  Cecum

                                                                  Appendix

                                                                  Sigmoid
                                                                  colon

                                                                  Bladder

A   B

FIGURE 40-7 A, The abdominal cavity as it appears when opened anteriorly. B, The abdominal cavity as it appears after removal of the greater omentum and with the liver and diaphragm lifted upward.

fatty sheet called the *greater omentum* hanging from the greater curvature of the stomach. Figure 40–7B illustrates the same view but now with the liver reflected upward and the greater omentum removed. Note especially how compactly the abdomen is filled with loose coils of intestines.

The Peritoneum.   All surfaces within the abdomino-pelvic cavity are lined by a thin membrane called the **peritoneum,** which is illustrated in Figure 40–8 in a longitudinal sectional view of the abdomino-pelvic cavity. The peritoneum is divided into two parts:

1. The **visceral peritoneum** lines the outer surfaces of the organs in the abdomino-pelvic cavity.
2. The **parietal peritoneum** lines the peripheral surfaces (called the "parietal" surfaces) of the abdomino-pelvic cavity.

Some of the organs of the abdominal cavity lie deep to the parietal peritoneum, between it and the wall of the abdomen. These organs are said to be *extraperitoneal.* They include such structures as the two kidneys, the urinary bladder, the aorta, the vena cava, the pancreas, and

portions of the intestines, including part of the duodenum and the ascending and descending colons.

The thin space between the parietal and visceral peritoneums is called the **peritoneal space.** Normally, the organs are so compacted together that this space is minimal, filled with only about 100 ml of viscid fluid that contains a mucoprotein material including *hyaluronic acid* that serves as a lubricant. This allows the abdominal viscera to slide freely over one another, which is especially important when peristaltic waves travel along the digestive tube.

The Mesentery and the Omentum.   The stomach and much of the intestines hang freely in the peritoneal cavity, supported from the posterior wall of the abdomino-pelvic cavity by means of a fibrous membrane called the **mesentery,** which is also illustrated in Figure 40–8. The mesentery is covered on both surfaces by visceral peritoneum. Within the mesentery are the blood vessels, lymphatic vessels, and nerves that supply the stomach and intestines.

The portion of the mesentery that suspends the stomach is called the omentum; this is divided into two parts:

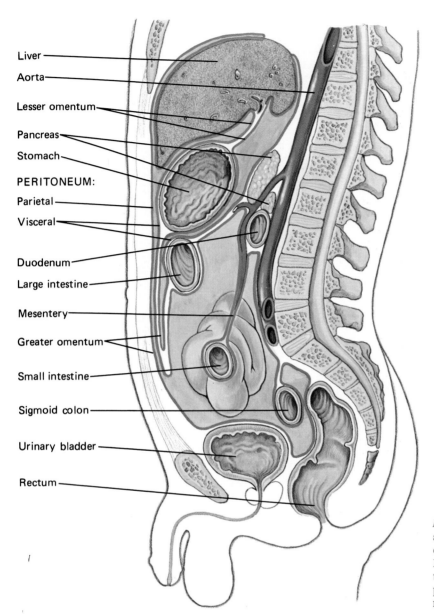

Liver
Aorta
Lesser omentum
Pancreas
Stomach
PERITONEUM:
Parietal
Visceral
Duodenum
Large intestine
Mesentery
Greater omentum
Small intestine
Sigmoid colon
Urinary bladder
Rectum

FIGURE 40–8 A longitudinal section of the abdomino-pelvic cavity as seen from the side. Note that much of the digestive tube is suspended from the posterior abdominal wall by mesentery.

1. The **lesser omentum** attaches the lesser curvature of the stomach to the hilum of the liver.
2. The **greater omentum** is an extensive sheet suspended from the greater curvature of the stomach and draped inferiorly over the transverse colon and small intestinal coils, as was depicted in Figure 40–7A. It is also adherent to the transverse colon, thus providing support for the inferior border of the stomach. Large amounts of fat can accumulate in the greater omentum, and this fat can be reabsorbed within days when it is needed to provide energy during periods of starvation. The greater omentum also sends projections that fill in many of the recesses between the intestines. When infection occurs in the abdominal cavity, the omentum often becomes ad-

herent to the tissues around the infected area and prevents spread of the infection to other parts of the abdomen.

THE STOMACH

The **stomach** serves two principal functions: (1) to store food until it can be properly processed by the remainder of the gastrointestinal tract and (2) to aid in the digestive process. The structural features of the stomach that allow it to perform these functions are illustrated in Figure 40–9. Both anatomically and physiologically the stomach is divided into two parts, separated by a slight constriction called the **angular notch.** These parts are the following:

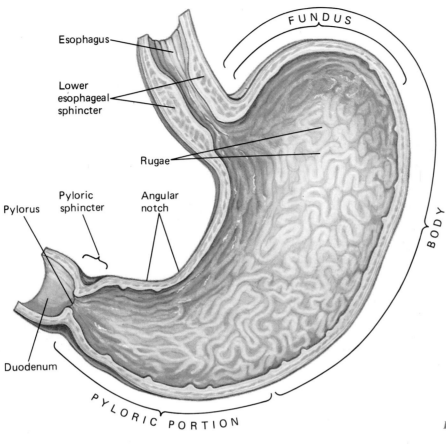

Esophagus

Lower
esophageal
sphincter

FUNDUS

Rugae

BODY

Pyloric
sphincter

Pylorus

Angular
notch

Duodenum

PYLORIC PORTION

FIGURE 40–9   The stomach.

1. The **body of the stomach.** The body is approximately the first three quarters of the stomach, extending from the inferior surface of the diaphragm to the angular notch. The superiormost portion of the body is the **fundus,** which lies to the left of and superior to the opening of the esophagus into the stomach, which itself is called the **cardiac orifice.** The body is the major storage area of the stomach, and its mucosal glands secrete most of the stomach digestive juices. When greatly stretched, the stomach can store as much as a liter of food and fluid, but half a liter usually gives a person a sense of fullness.

2. The **pyloric portion of the stomach** extends from the angular notch to the point where the stomach empties into the duodenum. The dilated chamber in this portion of the stomach is called the **pyloric antrum.** In this region, the food becomes thoroughly mixed with the stomach secretions. To achieve this, the muscular wall of the pyloric portion of the stomach is much thicker than that of the body, and it engages in powerful intermittent contractions that knead the food and stomach juices until they mix. The resulting mixture is called *chyme.*

The Pylorus and the Pyloric Sphincter. The opening between the stomach and the duodenum is the **pylorus,** and the infolded mucosal and muscular ridge around the pylorus are called the **pyloric valve** because this acts as a valve to control the movement of the chyme from the stomach into the small intestine. The circular muscular layer of the pyloric valve is extremely thick and powerful and is the **pyloric sphincter.** Nervous and hormonal signals arising mainly from the duodenum, but also partly from the stomach, control the degree of contraction of the pyloric sphincter and thereby control the rate at which the chyme is emptied from the stomach into the duodenum. This control function of the pyloric valve is so important to the function of the digestive tract that it will be discussed in detail in Chapter 41.

THE STOMACH WALL AND THE GASTRIC MUCOSA

Figure 40–10 shows the several layers of the stomach wall. These are all basically the same as those discussed earlier for the esophagus except for the mucosa. The mucosa is especially

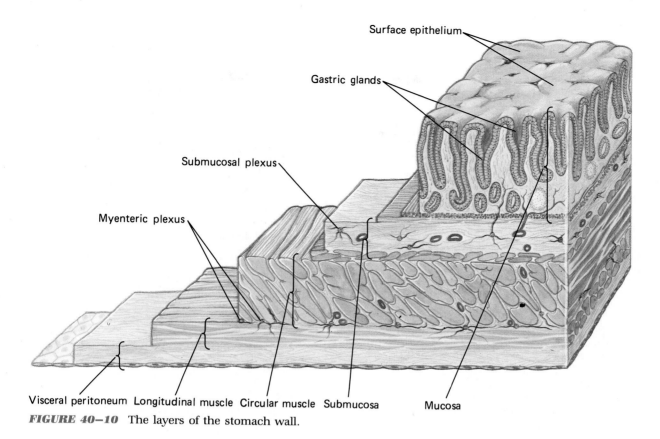

Surface epithelium

Gastric glands

Submucosal plexus

Myenteric plexus

Visceral peritoneum Longitudinal muscle Circular muscle Submucosa Mucosa

FIGURE 40—10 The layers of the stomach wall.

adapted for the functions of the stomach. Its special features are listed here:

1. The mucosa is not a flat sheet; instead, as illustrated in Figure 40–9, it crops up into multiple longitudinal folds called **rugae** that project into the lumen of the stomach. When the stomach is empty, the rugae are many and large, but they become progressively flattened as the stomach becomes more filled.

2. The epithelial surface of the mucosa is highly resistant to destruction. Almost all of the epithelial cells that line the surface secrete *mucus* and are called **mucous cells.** The mucus is especially viscous and highly resistant to both the digestive juices and the acid secreted by the stomach itself.

3. The mucosa is very thick, several millimeters in thickness. This provides a high degree of protection for the deeper layers of the stomach wall from excoriation by rough or caustic foods as well as by the digestive juices.

4. Protruding deeply into the gastric mucosa over its entire surface are a multitude of **gastric glands.** Each day these secrete about 2 L of highly acidic gastric juice that contains important enzymes for digesting protein.

Another special feature of the stomach wall is that both the *myenteric nerve plexus* lying between the circular and longitudinal muscle layers and the *submucosal nerve plexus* in the submucosal layer are very prominent. These play exceptionally important roles in controlling both the movements of the stomach and the secretion by the gastric glands, as will become evident in our discussion of these functions in the following chapter.

THE SMALL INTESTINE

The small intestine, illustrated in Figures 40–1 and 40–7B, is a continuous tube 3 to 6 cm in diameter and about 7 m long after death when it has lost all of its muscular tone (but only 3 to 5 m during life). It extends from the pylorus of the stomach all the way to the large intestine. The wall of the small intestine, illustrated in Figure 40–11, has the same general layers as those elsewhere in the gastrointestinal tract, but several special features are listed here:

1. The **plicae circulares.** These are multiple circular folds of mucosa that protrude from the intestinal wall into the intestinal lumen.

Mucosa

Submucosa

Muscle:
Circular
Longitudinal

Visceral peritoneum

Mesentery

Circular folds

Myenteric nerve plexus

Mesentery

Submucosal nerve plexus   Epithelial lining   Submucosal gland   Mucosal gland   Muscularis mucosae

A

B

FIGURE 40–11 The principal anatomical structures of the small intestine.

They increase the surface area of the mucosa by approximately threefold, which obviously increases the area both for secretion of intestinal juices and for absorption of the digestion products.

2. The **intestinal villi.** These protrude from the surface of the mucosa as shown in Figure 40–12. Each of these villi is approximately 1 mm in length and $\frac{1}{5}$ mm in diameter, and there are several hundred of them to each square centimeter of surface area. Inside the villi are blood capillaries and lymphatics into which nutrients are absorbed from the intestinal lumen. The epithelium on the surface of the villi is especially adapted to absorb the nutrients.

3. The **intestinal glands.** These are short pitlike structures located between the villi that penetrate into the mucosa. They secrete large quantities of fluid, about 2 L a day, that mix

with the chyme and then are reabsorbed by the villi, thus providing a flow of fluid that carries the nutrients into the blood and lymphatics. They also secrete mucus, which helps to lubricate the movement of the chyme through the intestines.

THE THREE DIVISIONS OF THE SMALL INTESTINE

The small intestine is divided into three separate divisions:

1. the **duodenum,** which is the first 20 to 30 cm of the small intestine;
2. the **jejunum,** the next 3 m;
3. the **ileum,** the remaining 4 m.

The Duodenum. The duodenum and some of its related structures are illustrated in Figure 40–13. Unlike the remainder of the small

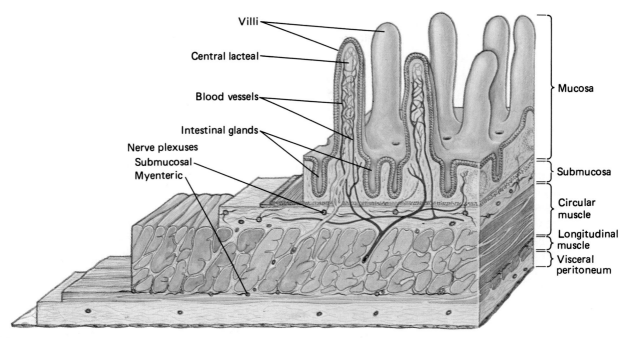

FIGURE 40–12 Structure of the small intestinal wall, illustrating especially the villi and the intestinal glands.

intestine, the duodenum, except for its first part, does not hang freely in the abdominal cavity by a mesentery; instead, it is attached directly to the posterior abdominal wall and is covered only on its anterior surface by peritoneum. Thus, most of it is an extraperitoneal structure. One can also

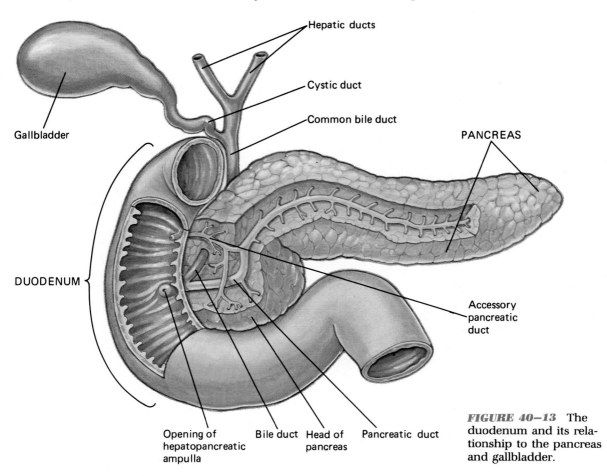

FIGURE 40–13 The duodenum and its relationship to the pancreas and gallbladder.

observe in Figure 40–13 that the duodenum is C-shaped, with the head of the pancreas lying within the lesser curvature of the "C." Two especially important features of the duodenum are the following:

1. The secretions of both the pancreas and the liver are delivered into the duodenum. The **common bile duct** from the *liver* and *gallbladder* and the **pancreatic duct** from the *pancreas* usually join each other in the head of the pancreas to form a combined **bile-pancreatic duct** that empties into the duodenum through a small hill-like projection, the **duodenal papilla,** on the posteromedial wall of the duodenum about 8 cm below the pylorus. A much smaller amount of pancreatic secretion usually enters the duodenum through an **accessory pancreatic duct,** also illustrated in Figure 40–13.

2. In the wall of the first 8 to 12 cm of the duodenum are many glands of a special type not found in the other divisions of the small intestine. These are the **duodenal glands** (also called **Brunner's glands**), illustrated in Figure 40–14, that penetrate especially deep into the submucosa and have many tubular branches. They secrete large amounts of slightly alkaline mucus that protects the duodenal wall from damage by the acid juice from the stomach before the acid can be neutralized by the pancreatic juice.

The Jejunum and Ileum. Both the **jejunum** and the **ileum** are suspended in the abdominal cavity from an extensive mesentery.

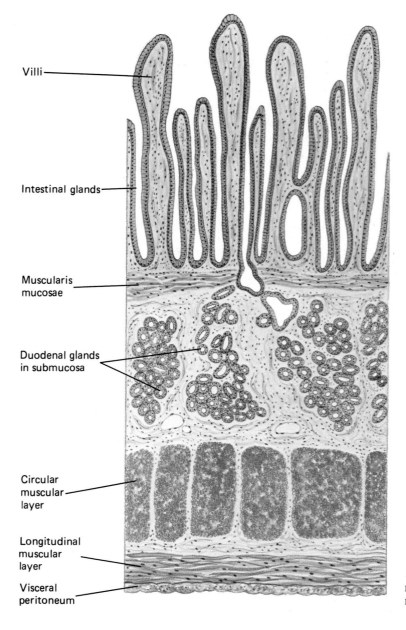

Villi

Intestinal glands

Muscularis mucosae

Duodenal glands in submucosa

Circular muscular layer

Longitudinal muscular layer

Visceral peritoneum

FIGURE 40–14 The wall of the duodenum, illustrating especially the duodenal glands.

These parts of the small intestine form multiple loops, and the loops move freely among one another as different segments of the intestine become successively filled with chyme.

There is no distinct demarcation boundary between the jejunum and the ileum. However, the jejunum has far more plicae circulares and villi than the ileum. This is in keeping with the respective functions of the jejunum and ileum, namely, absorption of most of the nutrients from the chyme in the jejunum and upper ileum and absorption of most of the remaining water and electrolytes in the distal portions of the ileum, which does not require such an extensive surface area.

THE PANCREAS

The **pancreas,** illustrated in Figure 40–13, is the most important of the digestive glands. It lies in the lesser curvature of the duodenum and posterior to the inferior border of the stomach. Note that it has a distinct **head** adjacent to the duodenum and a long **tail** posterior to the stomach. Its glandular structure is very similar to that of the salivary glands except that it is about as large as all the salivary glands together. The pancreas contains literally millions of *acini* similar to those illustrated in Figure 40–5 for the submandibular gland. All of the cells of the pancreatic acini are of the serous type, secreting entirely digestive enzymes and no mucus. The pancreas, like the salivary glands, also has many small ducts that coalesce to form the large pancreatic duct, and often also the smaller accessory pancreatic duct. Both of these empty into the upper portion of the duodenum as described earlier.

The pancreas is different from the salivary glands in one very important aspect: it not only secretes a digestive juice into the digestive tract, which is called its *exocrine secretion*, but also produces an additional secretion, called its *endocrine secretion*, which is taken up directly into the blood capillaries of the pancreas.

The Exocrine Secretion. An exocrine secretion is one that is secreted through a duct or other opening either to the exterior of the body, as is the case for the sweat glands in the skin, or into some internal body cavity, as is the case for the glands that secrete digestive juices into the gastrointestinal tract. The exocrine secretion of the pancreas is the *pancreatic juice* secreted through the pancreatic ducts into the duodenum. This is the most important of all the digestive juices. Its total quantity is more than a liter a day, and its enzymes are responsible for digesting about 50 percent of all carbohydrates,

50 percent of all proteins, and 90 percent of the fats, as we shall learn in Chapter 42. Therefore, when a person's pancreas is destroyed, his capability of obtaining nutrition from food is seriously diminished.

In addition, the pancreatic secretion is very alkaline, and this allows it to neutralize the acid from the stomach.

The Endocrine Secretion. An endocrine secretion of a gland is secreted directly into the blood capillaries of the gland and then rapidly transported in the blood throughout the entire body. The principal constituent of the endocrine secretion of the pancreas is *insulin*, but it also contains several other hormones (see Chapter 46).

The pancreatic endocrine secretions are formed by the **islets of Langerhans,** which are small segregated collections of special cells that are entirely different from the pancreatic acinar cells. Each islet is less than 1 mm in diameter, but thousands of islets are dispersed among the acini. They have no means for secreting into the pancreatic duct system; instead, their secretions are taken up directly into the blood vessels.

THE LIVER

The **liver** is the largest internal organ of the body, which is evident from its size in Figures 40–1 and 40–7. Note also its relationship to the other intra-abdominal structures and its location directly inferior to the diaphragm.

The structure of the liver is illustrated in more detail in Figure 40–15. At the top are shown the **coronary, right triangular,** and **left triangular ligaments** that suspend the liver from the diaphragm. The liver is also suspended from the midline of the anterior abdominal wall by the **falciform ligament.** That portion of the liver to the left of the falciform ligament, which represents about one quarter of the entire liver, is the **left lobe of the liver.** Immediately to the right of the falciform ligament are two small lobes, each about 7 cm wide, the **caudate lobe** located on the dorsal surface of the liver and the **quadrate lobe** located on the visceral surface. These together constitute about another quarter of the liver. Then to the right side of these two lobes is the largest of all the lobes, the **right lobe,** which represents approximately half of the entire liver. The **gallbladder** is located on the visceral surface of the liver in the groove between the quadrate lobe and right lobe.

Note also in Figure 40–15 the relationship of the left lobe of the liver to the stomach, which

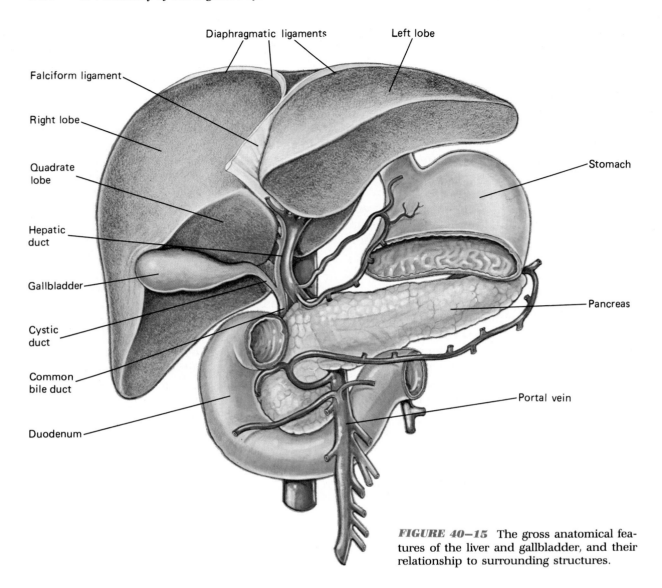

Falciform ligament

Right lobe

Quadrate lobe

Hepatic duct

Gallbladder

Cystic duct

Common bile duct

Duodenum

Diaphragmatic ligaments   Left lobe

Stomach

Pancreas

Portal vein

FIGURE 40–15 The gross anatomical features of the liver and gallbladder, and their relationship to surrounding structures.

lies posterior and inferior to this lobe. Finally, note the entire area on the visceral surface of the liver where the blood vessels and nerves enter and the bile ducts and lymphatics exit. This area is called the **hilum of the liver.**

The Multiple Functions of the Liver. One of the functions of the liver is to secrete bile into the duodenum; the bile in turn helps in the digestion of fats. This is only one function of the liver, however, and indeed one that a person can live without. The other functions of the liver are so essential to life that a person can survive for no longer than 24 to 48 hours after his liver has been totally destroyed. These other functions will be discussed at many places throughout this book, but, in brief, some of them are

1. to *store many of the nutrients* required for life, including carbohydrates, fats, proteins, vitamins, and some important minerals such as iron;

2. to *detoxify many substances* that are poisonous to the body, including many drugs and especially the large amounts of very toxic ammonia that is formed during protein metabolism—this ammonia is converted by the liver into urea which has almost no toxicity;

3. to *synthesize the plasma proteins* that are essential for keeping the fluid of the blood from leaking out of the circulation into the tissues;

4. to *synthesize most of the body fat* before it is transported to and stored in the fat tissues;

5. to *synthesize many of the amino acids* that are required for synthesis of the proteins in the body; and

6. for many other metabolic functions.

The Internal Structure of the Liver. To perform its many functions, the liver is comprised of many billions of large **hepatic cells** (*liver cells*) that are literally bathed in flowing blood. These cells are arranged in small functional units called **hepatic lobules,** of which there are close to a million in the liver. Figure 40–16 illustrates the internal structure of a single lobule, the important features of which are as follows:

1. Great numbers of liver cells are arranged in single or double layers called **hepatic plates.**
2. **Portal venules** are the entry points for portal blood from the gastrointestinal tract, carrying absorbed nutrients, many of which will be stored in the hepatic cells as the blood flows through the liver.
3. Large numbers of **venous sinusoids** branch from the portal venules and carry the portal blood between the hepatic plates.
4. A **central vein** drains the blood from the venous sinusoids. From the central veins, the blood flows into the hepatic venous system that leads directly into the inferior vena cava.
5. **Hepatic arterioles** supply about one quarter of the blood flow into the lobule, mainly supplying the connective tissue structures with the nutrients required for maintaining their integrity.
6. The **bile ductule system** begins with many

small **bile canaliculi** that lie between the adjacent liver cells. At the periphery of the lobule, the canaliculi empty into a **bile ductule** that in turn joins many other ductules eventually to form the hepatic duct that leaves the hilum of the liver.
7. The **lymphatic system of the liver** begins in the **spaces of Disse** that lie between the surfaces of the hepatic plates and the endothelial walls of the venous sinusoids. The spaces of Disse empty into small lymphatic vessels that lie alongside the portal venules.

Thus, the liver is literally a large mass of hepatic cells. These form the bile that is secreted into the gut, but much more important is their relationship to the blood flowing through the sinusoids, for the hepatic cells continually remove substances from the blood and process them metabolically. Then the cells return the processed substances to the blood as they are needed by the body.

Bile Secretion and the Gallbladder. The liver secretes between 600 and 1200 ml of bile each day. This secretion leaves the liver by way of the **hepatic duct.** This duct joins the **cys-**

FIGURE 40–16 Structure of a liver lobule. (Modified from Elias. *In* Williams and Warwick. Gray's Anatomy. 36th British Edition. Philadelphia, W.B. Saunders, 1980.)

Bile canaliculi · Bile ductule · Portal venule · Hepatic arteriole · Space of Disse · Lymphatic vessel · Hepatic cells · Hepatic plates · Venous sinusoid · Central vein

tic duct from the **gallbladder** to form the **common bile duct** about 6 cm after leaving the liver. As shown in Figure 40–15, the common bile duct then leads to the duodenum. Normally, almost all of the bile secretion in the hepatic duct flows backward through the cystic duct into the gallbladder where it is stored temporarily; also it becomes concentrated, because the gallbladder can absorb as much as 90 percent of the water and salts from the bile. After a meal, particularly a meal containing a large amount of fat, the gallbladder contracts and empties the bile back down the cystic duct but this time through the common bile duct into the duodenum.

Now, observing Figure 40–13 again, recall the relationship of the common bile duct to the pancreatic duct. These two ducts usually join each other about 1 cm before emptying into the duodenum and then empty together through the combined bile-pancreatic duct passing through the duodenal papilla. This duct has a small smooth muscle constrictor around it called the **sphincter of Oddi** that remains constricted most of the time. However, after one eats a meal, this sphincter relaxes at the same time that the gallbladder contracts, thus allowing the bile as well as the pancreatic juice to flow freely into the duodenum.

THE LARGE INTESTINE

A liter or more of chyme empties from the small intestine into the large intestine each day. Almost all of the nutrients of the food have been removed from the chyme by this time, but it still contains fluid and salts that need to be reabsorbed before the feces are finally expelled. Otherwise, under very dry conditions or in many geographical areas where there are not enough salts in the food, the person would perish. In addition, for both human beings and other mammals, it is important to eliminate the feces only periodically rather than continually, so that these can be disposed of in a manner that will not attract undue attention or cause continual spread of disease.

Therefore, the function of the large intestine is to reabsorb all possible amounts of water and salts and also to store the fecal material until an appropriate time for expulsion.

The anatomy of the large intestine is illustrated in Figure 40–17. It is comprised of six important successive divisions:

1. The **cecum** is the first part of the large intestine into which the ileum empties. Most of the cecum is the portion of the large intestine that extends inferiorly from the entry point of the ileum as a bulbous enlargement.
2. The **ascending colon** passes upward from the cecum to the level of the liver.
3. The **transverse colon** crosses from the right side of the abdominal cavity, inferior to the liver, all the way over to the left side where it passes inferior to the stomach. In its passage from right to left, it lies anterior to the duodenum and to the first part of the jejunum.
4. The **descending colon** extends down the left side of the abdominal cavity to the outer rim of the pelvis.
5. The **sigmoid colon** extends from the outer rim of the pelvis to the posterior wall of the pelvis anterior to the sacrum.
6. The **rectum** lies along the posterior wall of the pelvis.
7. The **anal canal** empties at the anus.

All portions of the large intestine are extraperitoneal except the transverse colon and the sigmoid colon. These two parts hang freely in the abdominal cavity and are supported by mesenteries in the same manner as the stomach and most of the small intestine.

Structure of the Wall of the Large Intestine. Except for the rectum, the wall structure of the large intestine is essentially the same throughout. Its layers are basically the same as those in the other parts of the alimentary tract except for three special features:

First, the diameter of the large intestine is about two times that of the small intestine.

Second, the mucosa of the large intestine differs from that of the small intestine in two principal ways: (a) It has no villi and therefore is not adapted for absorbing extreme amounts of fluid and nutrients. (b) The intestinal glands of the large intestine, though having an appearance almost identical to that of the intestinal glands of the small intestine, secrete no digestive enzymes. Instead, they secrete large amounts of *mucus* and small amounts of *sodium bicarbonate*. The role of the mucus is to protect the large intestinal wall from excoriation by the fecal matter, and the role of the bicarbonate is to neutralize the acid generated by bacteria in the feces.

Third, the longitudinal muscle of the large intestine is collected into three narrow linear strips, each about 1 cm in width, called **teniae coli.** These three strips extend the entire length of the colon, from the cecum through the sigmoid colon. When the teniae contract, the portions of the large intestine that do not contain this longitudinal muscle bulge outward. Then

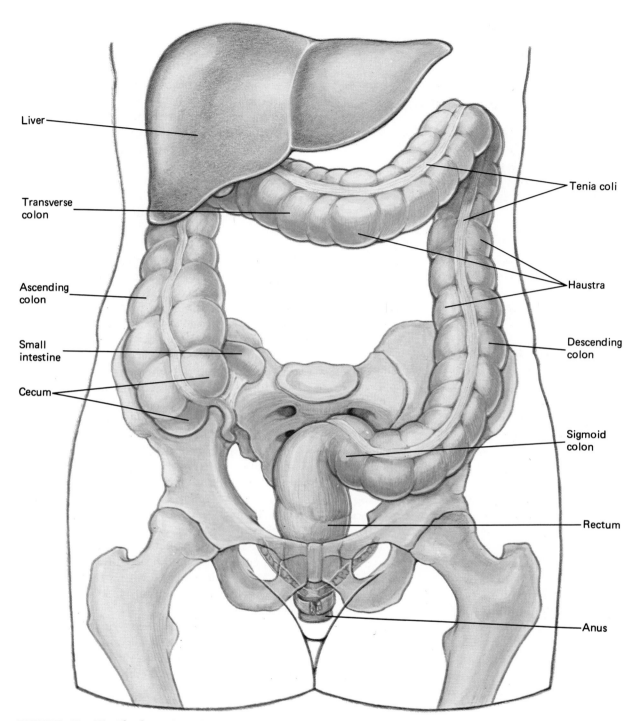

FIGURE 40–17   The large intestine.

the circular muscle also contracts in concentric rings 8 to 12 cm apart. This makes the bulges in the colon have pouchlike appearances. These pouches are called **haustra,** many of which are seen in Figure 40–17.

The Absorptive Function of the Large Intestine.   The liter or more of chyme that empties into the cecum from the small intestine each day is still in a very fluid state. Therefore,

most of the fecal matter in the cecum and ascending colon is also still mainly fluid. However, these two divisions of the large intestine absorb most of the fluid and salts before the fecal matter enters the transverse colon. Therefore, in the transverse colon, the fecal matter is mainly a mush, and in the descending colon and sigmoid it becomes progressively more dehydrated until it reaches the normal consistency of feces. In

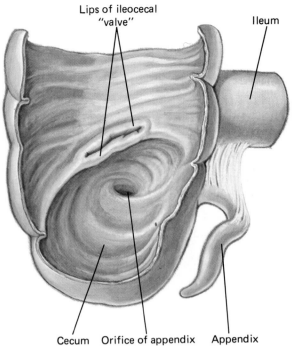

Lips of ileocecal "valve"

Ileum

Cecum   Orifice of appendix   Appendix

FIGURE 40–18 The ileocecal valve and the appendix.

general, the first half of the colon is called the *absorbing colon*, and the second half the *storage colon*.

The Ileocecal Valve and Emptying of Chyme from the Ileum into the Cecum. Figure 40–18 illustrates the juncture between the ileum and the cecum, showing that the ileum actually protrudes inward through the wall of the cecum to form an opening called the **ileocecal valve.** The lips of the ileocecal valve project about 1 cm into the cecum. These normally prevent backflow of cecal contents into the ileum, because when the cecum contracts and the intracecal pressure builds up, the cecal contents press against the outer surfaces of the lips and close these together rather than passing backward into the ileum. This valve function of the ileocecal valve is mainly important because the contents of the large intestine are loaded with bacteria, especially colon bacilli. This is in contrast to the ileum and the other portions of the digestive tract all the way back to the stomach that are much freer of bacteria.

The wall of the ileum for several centimeters proximal to the ileocecal valve is thicker than usual. This portion of the ileum normally remains partially contracted, which prevents too rapid emptying of the small intestinal contents into the cecum, thus allowing sufficient time for absorption of the nutrients.

The Appendix and Appendicitis. Note the **appendix** in Figure 40–18. This is a small wormlike extension from the inferior end of the cecum. In some lower mammals, the appendix is very large like the bulbous lower end of the cecum itself and plays an important role in digestion and absorption from the gastrointestinal tract. However, in the human being it is a "vestigial" remnant, which means an inherited structure from an earlier level of animal that now has no known function or value. Nevertheless, it does have a small lumen several millimeters in diameter that connects with the cecum, and infection frequently spreads into the appendix from the cecum to cause *appendicitis.* In the appendectomy operation, the appendix is tied off where it exits from the cecum and removed beyond the tie.

The Rectum. The rectum, illustrated in Figure 40–19, has a structure somewhat different from that of the remainder of the large intestine. Most importantly, its muscular coat is thicker, and the longitudinal muscle is no longer collected into the narrow teniae coli strips.

The rectum also has several large horizontal folds of mucosa extending from its inner wall into its lumen, usually three in number. These folds are called **rectal valves.** Once feces have passed beyond these folds into the distal rectum, the resulting distension of the rectal walls elicits a nerve reflex called the **defecation reflex.** This is initiated by stretch receptors in the rectal wall, and signals are sent to the spinal cord, then back again to the rectum to cause strong contraction of the rectal wall to expel the feces. Thus, the rectum will normally make an automatic effort to empty itself soon after it has become filled.

The Anal Sphincters. Surrounding the anal canal are two separate muscular sphincters, shown in Figure 40–19: (1) the **internal anal sphincter** and (2) the **external anal sphincter.**

The *internal anal sphincter* is comprised of about 2 cm of thickened circular smooth muscle of the anal canal. It normally remains constricted until a defecation reflex occurs. When this reflex causes rectal contraction, it simultaneously relaxes the internal anal sphincter so that the feces can be expelled.

The *external anal sphincter,* on the other hand, is composed of a circular ring of skeletal muscle around the canal immediately distal to the internal anal sphincter. Under normal conditions, nerve signals from the spinal cord maintain continual constriction of the external sphincter, thus helping the internal anal sphinc-

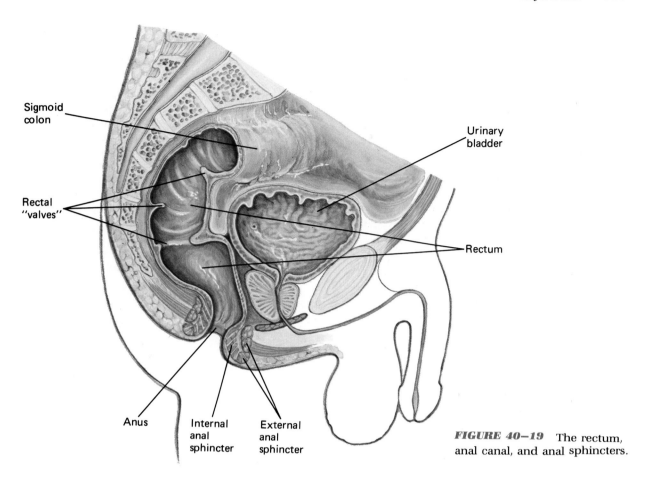

Sigmoid colon

Rectal "valves"

Anus

Internal anal sphincter

External anal sphincter

Urinary bladder

Rectum

FIGURE 40-19 The rectum, anal canal, and anal sphincters.

ter to guard against inadvertent fecal leakage. However, when the defecation reflex occurs, the person can voluntarily control this external anal sphincter (which he cannot do for the internal sphincter), and he has the choice whether to allow defecation to occur or not. If the time and place are propitious, he will relax this sphincter voluntarily and also help the defecation process along by contracting the abdominal muscles, but otherwise he will constrict the sphincter even more powerfully and prevent defecation until the defecation reflex goes away.

QUESTIONS

1. Describe the important structures of the mouth and oral pharynx, and discuss the function of the fauces to limit the passage of large objects into the pharynx.
2. Describe the salivary glands and their ducts, as well as the nature of the cells in their acini.
3. Describe the layers of the esophagus, and discuss the function of each.
4. Locate the major organs of the digestive tract in the abdomino-pelvic cavity.
5. Explain the manner in which the stomach, the transverse colon, the sigmoid colon, and most of the small intestine are suspended in the abdominal cavity.
6. Give the functional parts of the stomach, and what are the differences between the stomach mucosa and the mucosa of the esophagus?
7. What are the two parts of the intrinsic nervous system of the gut, and how are they arranged in the gut wall?
8. Describe the mucosal surface of the small intestine, including the intestinal villi.
9. Explain the relationship of the duodenum to the pancreas and gallbladder, and trace the hepatic, cystic, and pancreatic ducts on their way to the duodenum.
10. Give the anatomical structure of a liver lobule.
11. List the sections of the large intestine, and give the differences between the large intestine and the small intestine.
12. Describe the ileocecal valve and its anatomical relationship to the cecum and appendix.
13. Explain the functional features of the rectum, anal canal, and anal sphincters.

REFERENCES

Anderson, J.E.: The abdomen. *In* Grant's Atlas of Human Anatomy. Baltimore, Williams & Wilkins, 1978, pp. 2-1 to 2-125.

Bloom, W., and Fawcett, D.W.: The digestive system. *In* A Textbook of Histology, 10th Ed. Philadelphia, W. B. Saunders, 1975, pp. 598-742.

Copenhaver, W.M., Kelly, D.E., and Wood, R.L.: The digestive system. *In* Bailey's Textbook of Histology, 17th Ed. Baltimore, Williams & Wilkins, 1978, pp. 455–551.

Figge, F.H.J.: The digestive system. *In* Sobotta/Figge Atlas of Human Anatomy. Vol. II. Baltimore, Urban & Schwarzenberg, 1977, pp. 61–143.

Goss, C.M.: The digestive system. *In* Gray's Anatomy of the Human Body, 29th Ed. Philadelphia, Lea & Febiger, 1973, pp. 1155–1264.

Ham, A.W., and Cormack, D.H.: The digestive system. *In* Histology, 8th Ed. Philadelphia, Lippincott, 1979, pp. 645–724.

Hammersen, F.: The digestive system. *In* Sobotta/Hammersen Histology. Baltimore, Urban & Schwarzenberg, 1980, pp. 114–138.

Langman, J., and Woerdeman, M.W.: Abdomen. *In* Atlas of Human Anatomy. Philadelphia, W.B. Saunders, 1978, pp. 69–143.

Leeson, T.S., and Leeson, C.R.: The digestive tract. *In* Histology, 4th Ed. Philadelphia, W. B. Saunders, 1981, pp. 328–400.

Leeson, T.S., and Leeson, C.R.: The digestive tract. *In* A Brief Atlas of Histology. Philadelphia, W. B. Saunders, 1979, pp. 145–174.

Netter, F.H.: Digestive system. *In* The CIBA Collection of Medical Illustrations. Vol. 3. Summit, N.J., CIBA Medical Education Division, 1964.

Pernkopf, E.: The abdomen. *In* Atlas of Topographical and Applied Human Anatomy. Vol. II. Philadelphia, W.B. Saunders, 1980, pp. 173–333.

Weiss, L., and Greep, R.O.: The digestive tract. *In* Histology, 4th Ed. New York, McGraw-Hill, 1977, pp. 643–700.

Williams, P.L., and Warwick, R.: The alimentary system. *In* Gray's Anatomy, 36th British Edition. Philadelphia, W. B. Saunders, 1980, pp. 1267–1384.

(See also General Anatomical References, Chapter 1.)

41

Gastrointestinal Movements and Secretion, and Their Regulation

Overview

The gastrointestinal (GI) movements can be divided into two different functional types: (1) *peristalsis* and (2) *mixing movements*. Peristalsis propels the GI contents through the GI tract; it consists of circular constrictions around the gut that move along the gut wall, thus squeezing the contents inside the gut in the forward direction. Peristalsis is controlled mainly by intrinsic nervous action in the *myenteric plexus* in the gut wall, which is the controlling nervous system of the GI tract itself.

The *mixing movements* are different in different parts of the GI tract. In the stomach, they are mainly local peristaltic movements that move only a short distance along the stomach wall but dig deeply into the stomach contents and therefore mix these with the stomach secretions. In the small intestine, mixing is achieved mainly by multiple rings of muscle contraction occurring one approximately every 6 to 10 cm along the intestine and recurring several times a minute, thus chopping the intestinal contents repetitively.

To initiate swallowing, the tongue thrusts upward against the palate and then rolls backward to force food toward the pharynx. As the food passes through the fauces, sensory receptors in this area transmit signals into the brain stem to elicit the *swallowing reflex*. This in turn (1) closes the nasopharynx from the pharynx; (2) closes the opening into the trachea; (3) pulls the larynx forward, which opens the upper end of the esophagus; and (4) initiates constriction of the pharynx, beginning in the upper pharynx and spreading to the lower pharynx, thus forcing the food into the esophagus. Once in the esophagus, the food stretches the esophageal wall, which stimulates the myenteric plexus to institute peristaltic waves that force the food the remaining distance to the stomach within 5 to 8 seconds.

Emptying of food from the stomach is controlled mostly in response to the food already in the duodenum. Distention of the duodenum causes a myenteric reflex back to the stomach called the *enterogastric reflex* that *inhibits stomach contractions* and tightens the *pyloric sphincter,* thus preventing further emptying of food into the duodenum until the duodenal distention disappears. Excess acid in the duodenum also elicits the enterogastric reflex and prevents further stomach emptying until the acid has been neutralized by pancreatic and intestinal secretions. This is especially important because the gastric secretions are highly acidic and will digest the wall of the duodenum if they are not neutralized.

Another highly specialized movement of the GI tract is the process of *defecation*. When feces distend the rectum, this excites stretch receptors in its wall, initiating a *defecation reflex* that passes first through *sensory nerves* into the lower end of the spinal cord, then back from the cord through *parasympathetic nerves* to the rectum, sigmoid, and descending colon to cause strong peristaltic waves. The person then determines whether or not he will allow defecation to occur; he does this by voluntarily constricting the *external anal sphincter* to prevent defecation or relaxing the sphincter to allow it.

Secretion by the GI glands occurs in response to food passing through the digestive tract. The important secretions are

1. *Salivary secretion,* which contains large amounts of *mucus* and also the enzyme *ptyalin,* which begins the digestion of the starches.
2. *Gastric secretion,* which contains large amounts of *hydrochloric acid* and the enzyme *pepsin,* both of which are especially important for beginning the digestion of proteins.
3. *Pancreatic secretion,* which contains large amounts of *trypsin* for further digestion of proteins, *amylase* for digesting carbohydrates, and *pancreatic lipase* for digesting fats. In addition, pancreatic secretion contains a high concentration of *sodium bicarbonate,* which neutralizes the hydrochloric acid entering the duodenum from the stomach.
4. *Liver secretion,* which contains large amounts of *bile salts* that mix with fats in the diet to help in their digestion and absorption.
5. *Small intestinal secretion,* which contains mucus and large quantities of water and electrolytes. In addition, the epithelial cells covering the villi are filled with *peptidases* for final digestion of proteins; *sucrase, lactase,* and *maltase* for final digestion of the carbohydrates; and small amounts of *intestinal lipase* for further digestion of fats.

GASTROINTESTINAL MOVEMENTS

There are two important purposes of the GI movements: (1) to keep the food moving along the gut at a pace at which it can be digested and absorbed and (2) to keep the food continually mixed with the GI secretions so that all parts of the food will be digested and the digestive end-products constantly exposed to the gut wall for maximum absorption. To achieve these goals, two basic types of movement occur in the GI tract: *propulsive movements* and *mixing movements.* The propulsive movements keep food moving along the gut, and the mixing movements mix the food with the gastrointestinal secretions. The movements in different parts of the GI tract exhibit differences that need to be described separately for certain portions of the gut. However, let us first consider the general characteristics of the two types of movements.

Propulsive Movements of the Gastrointestinal Tract—Peristalsis

Food is moved along the GI tract by *peristalsis,* which is caused by slow advancement of a circular constriction, as illustrated in Figure 41–1. This has very much the same effect as encircling one's fingers tightly around a thin tube full of paste and then pulling the fingers along the tube. Any material in front of the fingers will be squeezed forward.

Mechanism of Peristalsis. Peristalsis is caused by nerve impulses that move along the myenteric nerve plexus. Stimulation of any single point of plexus causes a progressive signal to travel around the gut and also lengthwise in both directions. The impulses traveling around the gut constrict it, and those traveling lengthwise cause the constriction to move along the

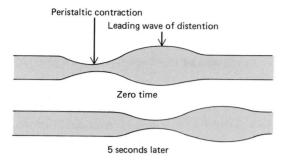

Peristaltic contraction

Leading wave of distention

Zero time

5 seconds later

FIGURE 41—1 Peristalsis.

gut. The usual rate of movement of this constriction is a few centimeters per second.

The usual stimulus that initiates peristalsis is *distention* of the gut, for this excites the local nerve plexus, causing a circular constriction to begin and to advance in a lengthwise direction as well.

Law of the Gut. Even though peristalsis can move in both directions along the gut, it most frequently moves toward the anus. The probable reason for this is that the myenteric nerve plexus itself is "polarized" in this direction. That is, when a portion of the gut becomes distended, it causes constriction of the gut on the headward side of the distention and relaxation on the anal side. The upstream constriction pushes the food forward and the relaxation downstream allows easy forward movement; the movement, therefore, is normally analward rather than headward. This is called the *law of the gut.*

Mixing Movements in the Gastrointestinal Tract

The mixing movements consist of two basic types: (1) *peristaltic movements*, which fail to move the food forward but nevertheless do succeed in mixing the intestinal contents adjacent to the wall of the gut, and (2) *segmental movements*, which are isolated constrictions that occur at many points along the gut at the same time. The segmental movements occur rapidly, several times each minute, and each time "chop" the food into new segments. The precise characteristics of the mixing movements are quite different in the different parts of the gut, for which reason they will be described specifically for each part of the gastrointestinal tract.

Let us now begin at the upper end of the gastrointestinal tract and describe both the propulsive and mixing movements as the food passes analward.

Swallowing

After the food has been properly masticated, the first act of swallowing is to thrust the anterior part of the tongue upward against the palate. This squeezes the food toward the back of the mouth in the form of a bolus. Contact of this bolus with the posterior surfaces of the mouth and the surfaces of the throat elicits nervous signals that pass to the **medulla oblongata** in the hindbrain. Here a sequence of neurogenic events called the **swallowing reflex** occurs, sending nerve signals over the **vagus nerves** to cause the following events in swallowing:

1. The sides of the throat between the mouth and the pharynx are brought medially to form a slit. This slit is narrow enough that only the food that has been properly masticated can pass.
2. The soft palate and the uvula are pulled upward by extrinsic muscles of the soft palate to close the opening from the nose into the pharynx.
3. The entire larynx is pulled upward toward the lower jaw. This pulls the larynx away from the esophagus, thus also pulling the anterior border of the esophagus forward, a maneuver that opens the upper end of the esophagus and at the same time causes the epiglottis to move backwards over the glottis, thus protecting the airway from entry of food and also forming a chute for the food to pass into the esophagus.
4. The muscles of the upper pharynx then contract, constricting the pharyngeal walls around the food. The constriction then spreads downward to the lower pharynx, squeezing the food progressively toward and into the esophagus.
5. The larynx is then dropped back to its original position, which again closes the upper end of the esophagus; at the same time, the epiglottis opens to allow air to enter the trachea once more while the soft palate moves downward to open the air passage through the nose.

Obviously, during all these events, respiration ceases, but the entire sequence takes place in less than 2 seconds, so one hardly knows that respiration has been interrupted.

Nervous Pathways For Control of Swallowing. Figure 41—2 illustrates the nervous pathways involved in the swallowing mechanism, showing that the swallowing receptors transmit impulses from the posterior mouth and

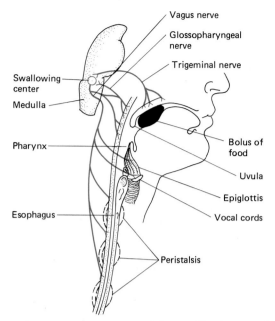

FIGURE 41–2 Anatomy of the swallowing mechanism.

throat mainly through the *trigeminal nerve* into the reticular substance of the *medulla oblongata* where the swallowing center is located. Once this center has been activated, the sequence of muscular reactions just listed above occurs automatically and usually cannot be stopped. Nerve signals go mainly by way of the *vagus nerves* to the pharyngeal and laryngeal muscles, and contraction of these moves the bolus of food into the upper esophagus. Then impulses from the vagi activate the proximal portion of the esophagus to push the food on toward the stomach.

Esophageal Stage of Swallowing. The musculature of the pharynx and of the upper one third of the esophagus is different from that in the rest of the gastrointestinal tract, for this muscle is skeletal muscle, controlled directly by nerves from the brain; in contrast, the remainder of the muscle in the esophagus and GI tract is smooth muscle and is only indirectly controlled by the central nervous system through the effects of the autonomic nervous system on the myenteric nerve. Therefore, all contractions of the pharynx and upper one third of the esophagus are initiated directly by vagal nerve impulses, and without the vagal nerves the act of swallowing becomes paralyzed.

However, once food has reached the middle third of the esophagus, the distention elicits a typical peristaltic wave called the *secondary peristaltic wave* of the esophagus that pushes the food the rest of the way into the stomach.

The entire time required for food to pass from the pharynx to the stomach is 5 to 8 seconds.

Function of the Lower Esophageal Constrictor. Several centimeters above the point where the esophagus empties into the stomach, the wall of the esophagus is thickened, and the muscle coat is considerably stronger than elsewhere in the esophagus. This portion of the esophagus is called the *lower esophageal constrictor,* and it remains mildly constricted under normal conditions to help prevent reflux of gastric contents into the esophagus.

When swallowed food passes down the esophagus, a "leading wave" of relaxation is transmitted through the myenteric nerve plexus of the esophageal wall to the lower esophageal constrictor and causes it to relax. This allows the food to pass on into the stomach. On the other hand, when food in the stomach attempts to move backward up the esophagus, the constrictor normally does not relax but, instead, prevents any backward flux.

However, the lower esophageal constrictor operates differently in two abnormal conditions. First, in vomiting, reverse peristalsis in the stomach will then open the constrictor to allow food to be vomited. Second, in rare persons, the myenteric nerve plexus of the esophagus is poorly formed or absent. Because of this, the constrictor will not open normally during swallowing, so that food tends to collect above the constrictor. This condition, called *achalasia,* causes tremendous enlargement of the esophagus, and the collected food in the esophagus often becomes infected with bacteria. The infection often then spreads to the esophageal wall, causing ulcers to develop. The esophagus sometimes holds as much as a full liter of swallowed food for many hours. It also causes severe pain in the chest or in the deep throat.

Motor Functions of the Stomach

The motor functions of the stomach are threefold: (1) storage of large quantities of food immediately after a meal, (2) mixing of this food with gastric secretions, and (3) emptying of the food from the stomach into the small intestine. The basic functional parts of the stomach are illustrated in Figure 41–3. Physiologically the stomach can be divided into two major parts; the *body* and the *antrum.*

Storage Function of the Stomach. Food, on emptying from the esophagus into the stomach, first enters the body, which is an ex-

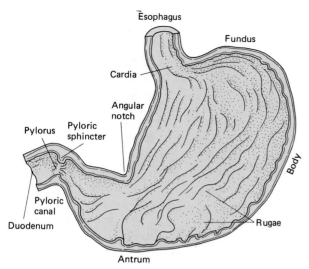

Esophagus
Fundus
Cardia
Angular notch
Pylorus
Pyloric sphincter
Pyloric canal
Duodenum
Antrum
Rugae
Body

FIGURE 41–3 Physiologic anatomy of the stomach.

tremely elastic bag that can store very large quantities of food. Furthermore, the tone of the body is normally slight, so that even as much as half a liter of food usually does not increase the pressure in the stomach greatly. Thus, the body is mainly a receptive organ for holding food until it can be utilized by the remainder of the GI tract.

Mixing in the Stomach and Formation of Chyme. The gastric glands, which cover most of the mucosa of the body, secrete large quantities of digestive juices that come into contact with the stored food. Weak, rippling peristaltic waves, called *tonus waves* or *mixing waves*, pass along the stomach wall approximately once every 20 seconds. These begin in any part of the body and spread for variable distances toward the antrum, and they become more intense when food is present in the stomach than when it is not present.

The mixing waves mix the gastric secretions with the outermost layer of food, gradually eating away the stored food, and these waves also gradually move the mixture toward the antral part of the stomach. On entering the antrum, the waves become stronger, and the food and gastric secretions become mixed to an even greater degree of fluidity. As the food becomes thoroughly mixed with the gastric secretions, the mixture takes on a milky-white sludge appearance and is then called *chyme.*

Propulsion of Chyme Through the Pylorus and Emptying of the Stomach. The opening from the stomach into the duodenum is known as the *pylorus.* At this point, the muscular coat is greatly hypertrophied, forming a very

strong muscular sphincter called the *pyloric sphincter.* The tonus or mixing waves are rarely strong enough to push chyme through the pylorus into the duodenum. However, occasionally very powerful peristaltic contractions occur, beginning either in the body or antrum, and these generate as much as 50 mm Hg pressure in the prepyloric portion of the antrum. This is usually enough pressure to push open the pyloric sphincter and propel the chyme on into the duodenum.

Regulation of Stomach Emptying. Emptying of the stomach is controlled mainly by the intensity of the strong peristaltic waves. Among the different factors that determine whether or not the peristaltic waves will succeed in pushing the chyme through the pylorus are the following:

Degree of Fluidity of the Chyme. Obviously, the better the food has become mixed with gastric secretions the more easily it can flow through the narrow passageway of the pylorus. Therefore, food ordinarily will not pass out of the stomach until it has been thoroughly mixed.

Quantity of Chyme Already Present in the Small Intestine. When a large amount of chyme has already emptied into the small intestine, particularly when a large portion of this is still present in the duodenum, a reflex called an *enterogastric reflex* spreads backward through the myentric nerve plexus from the duodenum to the stomach to inhibit peristalsis and to a slight extent also to increase the degree of tonic contraction of the pyloric sphincter. In this way, the duodenum keeps itself from becoming overfilled.

Presence of Acids and Irritants in the Small Intestine. The gastric secretions, as will be discussed later in the chapter, are highly acidic, but the acid in the chyme entering the duodenum is ordinarily neutralized by alkaline pancreatic secretions that also empty into the duodenum. Until the acid becomes neutralized by pancreatic juice, irritation of the wall of the duodenum elicits an enterogastric reflex similar to that elicited by distention; this too inhibits the peristaltic waves in the stomach and tightens the pyloric sphincter, thus stopping gastric emptying. In this way, the duodenum protects itself from too much acidity. Likewise, any other irritant also causes an enterogastric reflex that will do the same thing.

Presence of Fats in the Small Intestine. When fats enter the small intestine from the stomach, they extract from the mucosa of the duodenum and jejunum several hormones, including cholecystokinin, small amounts of secretin, and probably others. These are absorbed into the blood and carried to the stomach. Here they inhibit stomach peristalsis and slow stomach emptying. This mechanism allows adequate time for fat digestion to occur in the small intestine. Proteins and carbohydrates, on the other hand, have much less inhibitory effect on stomach emptying. Fortunately, both of these are digested much more easily in the intestine than is fat.

In summary, gastric emptying is determined mainly by fluidity of the contents in the stomach and the state of the duodenum. If the duodenum is already filled, has irritant substances in it, or has fat in it, emptying will proceed very slowly, but if the duodenum is empty and the contents of the stomach are very fluid, emptying will proceed rapidly.

Movements of the Small Intestine

Propulsive Movements. It is in the small intestine that the most typical peristalsis occurs, for distention of any portion of the small intestine with chyme initiates a peristaltic wave. Peristalsis is far more intense when the parasympathetic nerves from the central nervous system stimulate the myenteric plexus, and sympathetic stimulation can inhibit greatly or totally block peristalsis.

Mixing Contractions—Segmentation. The presence of chyme in the small intestine initiates a type of contraction called *segmentation*, which is illustrated in Figure 41–4. When the small intestine becomes distended, many constrictions occur either regularly or irregularly along the distended area. As shown in the figure, the intestine becomes "chopped" into small sausage-like vesicles. The constrictions then relax,

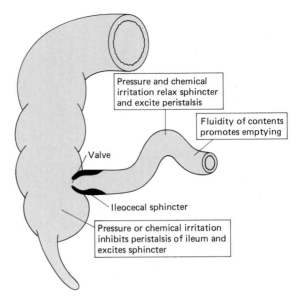

FIGURE 41–5   Emptying at the ileocecal valve.

but others occur at different points a few seconds later. Thus, repetitive "chopping" of the chyme keeps it mixed continually while it is in the small intestine.

Both the segmentation and peristaltic movements of the small intestine are controlled by the myenteric nerve plexus.

Emptying of Intestinal Contents at the Ileocecal Valve. Figure 41–5 shows the *ileocecal valve*, where the small intestine empties into the large intestine. Note that the small intestine actually protrudes forward into the colon. This projection of the valve prevents the contents of the colon from regurgitating into the small intestine; instead the lips of the valve simply close on themselves when pressure builds up in the colon.

Emptying of the small intestine at the ileocecal valve occurs in very much the same way that the stomach empties; that is, peristaltic waves in the small intestine build up pressure behind the valve and push chyme forward into the colon, but if the colon has become too full, myenteric nerve signals can inhibit peristalsis and thereby slow or stop the emptying.

Movements of the Colon

The functions of the colon, illustrated in Figure 40–17, are (1) absorption of water and electrolytes from the chyme and (2) storage of fecal matter until it can be expelled. The first half of the colon is concerned mainly with absorption and the distal half with storage. Except when the

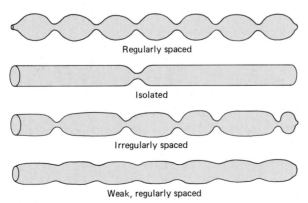

FIGURE 41–4   Segmentation movements of the small intestine.

Regularly spaced

Isolated

Irregularly spaced

Weak, regularly spaced

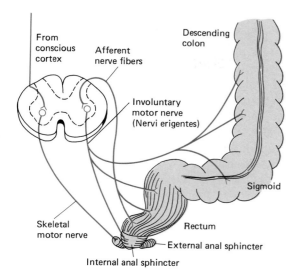

FIGURE 41-6 Anatomy involved in the defecation reflex.

bowels are to be emptied, the movements of the colon are usually very sluggish.

Mixing Movements. The mixing movements of the colon are similar to the segmentation movements of the small intestine, but they occur much more slowly. Concentric contractions of the colon divide the colon into large pockets called *haustrations*, which are illustrated in Figures 41-5 and 41-6. The circular constrictions last for about 30 seconds, and after another few minutes new constrictions occur again in nearby but not the same areas. Thus, the fecal material is slowly "dug" into and rolled over in much the same manner that one spades the earth.

Ordinarily, a liter or more of chyme empties into the colon each day, and of this, most of the water and electrolytes are reabsorbed before defecation takes place, leaving an average volume of feces of 100 to 200 ml each day.

Propulsive Movements. The typical peristaltic movements that occur in the small intestine do not occur in the colon. In fact, the colon has no peristaltic movements at all 95 to 99 percent of the time. Yet when the colon becomes overfilled, strong peristaltic movements, called *mass movements*, occur every 2 to 4 minutes and usually continue to occur for only about 15 minutes. But in this time these movements propel the fecal material long distances, sometimes all the way from the ascending colon to the descending colon. Then the mass movements cease, but reappear again many hours later when some part of the colon becomes overfilled again.

Defecation. When the mass movements of the colon have succeeded in moving fecal

material into the rectum, a special reflex called the *defecation reflex* occurs. This causes emptying of the rectum and lower parts of the colon. Figure 41-6 shows that filling the rectum excites nerve endings that send signals into the lower part of the spinal cord. These cause reflex signals to be transmitted back through the sacral parasympathetic nerves to the descending colon, sigmoid, rectum, and *internal anal sphincter,* causing contraction of the gut but relaxation of the sphincter. This reflex thus causes emptying of the bowels if the *external anal sphincter* is also relaxed. However, the external anal sphincter is a skeletal muscle that guards the outer opening of the anus, controlled by voluntary skeletal nerves, and can be relaxed or tightened according to the will of the person. If the time is not propitious for emptying the bowels, tightening of the external anal sphincter prevents defecation despite the defecation reflex. On the other hand, relaxation of the sphincter allows defecation to take place.

If a person prevents defecation when the defecation reflex occurs, the reflex usually dies out after a few minutes but returns a few hours later. Also, a person can frequently initiate a defecation reflex at will by tightening his abdominal muscles, which compresses the rectal wall and elicits the typical reflex. Unfortunately, this elicited reflex is usually much weaker than the natural reflex, so that defecation is less efficacious under these conditions than when the reflex has been elicited naturally.

Special Types of Gastrointestinal Movements

Antiperistalsis and Vomiting. Occasionally, some intensely irritating substance enters the GI tract. An immediate effect of this is an increase in the rate of local secretion of mucus, which helps to protect the interior of the gut. Simultaneously, the local gut wall contracts intensely. For reasons not too well understood, these intense local contractions elicit *antiperistalsis,* meaning peristalsis backward toward the mouth, rather than forward peristalsis. Food usually cannot move backward from the colon into the small intestine because of the ileocecal valve, but it can move all the way from the tip of the small intestine back to the stomach.

On reaching the stomach, the irritative material is then rapidly expelled by the vomiting process, as follows: The intense irritation of the gut initiates signals transmitted through the visceral sensory nerves into the brain. These cause a sensation of *nausea*, and, if the signals are strong enough, they will also cause an automatic reflex integrated in the reticular substance of the medulla oblongata called the *vomiting reflex.* This reflex first closes the airway into the trachea, then causes relaxation of the lower esophageal sphincter but at the same time very tight contraction of both the diaphragm and the abdominal muscles. The squeezing action of these muscles on the stomach pushes food out of the stomach upward through the esophagus and mouth. This is the vomiting process.

Gastrocolic and Duodenocolic Reflexes. Almost everyone is familiar with the natural desire to defecate following either a heavy meal or the first meal of the day. The cause of this is the *gastrocolic* and *duodenocolic reflexes*—mainly the latter. These reflexes are elicited by increased filling of the stomach and duodenum, which in turn transmit signals downward along the myenteric nerve plexus all the way to the colon. This causes increased excitability of the entire colon, initiating both mass movements and defecation reflexes.

The Peritoneal Reflex. Irritation of the peritoneum in any way, whether caused by cutting the peritoneum during an abdominal operation, by infection of the peritoneum, or even by a severe blow to the abdomen that causes trauma of the peritoneum, will elicit a *peritoneal reflex* that strongly excites the sympathetic nerves to the gut. These nerves in turn *inhibit* GI activity and thereby stop or slow movement of chyme along the intestinal tract. Obviously, lack of movement aids in the repair of the peritoneal damage, but it also causes severe constipation and sometimes even *intestinal obstruction* so that no food movement can occur.

Mucosal Reflexes. Irritation inside the gut or distention of the gut ordinarily excites rather than inhibits the intrinsic nervous system of the gut. Reflexes occur mainly locally, causing increased local secretion and increased local motor activity. The secretion dilutes the irritating factor, and if the irritation is not strong enough to initiate antiperistalsis and vomiting, then the motor activity instead moves the irritant on through the gut. When the irritation is great it causes *diarrhea.* And if the degree of irritation increases still more, it can cause diarrhea from the lower part of the intestinal tract while causing vomiting from the upper part. Obviously, these reflexes are protective in nature and prevent irritating substances or infectious processes inside the intestinal tract from causing severe permanent damage.

Gastrointestinal Secretions

Glands are present throughout the GI tract to secrete chemicals that mix with the food and digest it. These secretions are of two types: first, mucus which protects the wall of the GI tract, and, second, enzymes and allied substances that break the large chemical compounds of the food into simple compounds.

Mucus. Mucus is secreted in every portion of the GI tract. It contains a large amount of mucoprotein that is resistant to almost all digestive juices. Mucus also lubricates the passage of food along the mucosa, and it forms a thin film everywhere to prevent the food from excoriating the mucosa. It is *amphoteric,* which means it is capable of neutralizing either acids or bases. All these properties of mucus make it an excellent substance to protect the mucosa from physical damage and to prevent the digestion of the wall of the gut by the digestive juices.

Salivary Secretion

Saliva is secreted by the *parotid, submaxillary, sublingual,* and smaller glands in the mouth. Saliva is about half mucus and half a solution of the enzyme *ptyalin.* The function of the mucus is to provide lubrication for swallowing. Without mucus one can hardly swallow. If one mixes food with water to take the place of the mucus, approximately 10 times as much water as mucus is necessary to provide the same degree of lubrication.

The function of the ptyalin in the saliva is to begin the digestion of starches and other carbohydrates in the food. Ordinarily the food is not exposed to saliva in the mouth long enough for more than 5 to 10 percent of the starches to become digested. However, the mixed saliva and food is usually stored in the body of the stomach for 30 minutes to several hours before it is mixed with stomach secretions. During this time the saliva may digest more than 50 percent of the starches.

Regulation of Salivary Secretion. The *superior and inferior salivatory nuclei* located in the brain stem, shown in Figure 41–7, control secretion by the salivary glands. These nuclei in turn are controlled mainly by taste impulses and tactile sensory impulses from the mouth. Foods that have a pleasant taste ordinarily cause the secretion of large quantities of saliva, whereas some unpleasant foods may decrease salivary secretion so greatly that swallowing is made very difficult. Also, the tactile sensation of smooth-

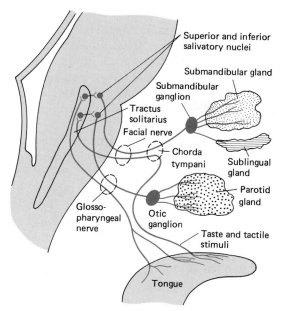

FIGURE 41–7 Nervous regulation of salivary secretion.

textured foods inside the mouth increases salivation, and the sensation of roughness decreases salivation. This effect presumably allows those foods that will not abrade the mucosa to be swallowed with ease and causes the rejection of abrasive foods.

The Phases of Salivary Secretion. In addition to the salivation that occurs while food is actually in the mouth, salivation frequently occurs even before food enters the mouth—that is, when a person is thinking about or smelling pleasant food—and it continues to occur even after the food has been swallowed. Therefore, salivary secretion can be divided into three phases: the *psychic phase*, the *gustatory phase*, and the *gastrointestinal phase*. The psychic phase presumably makes the mouth ready for food and aids in the secretion of saliva as the food is presented to the mouth. The gustatory phase supplies the saliva that mixes with the food while one is chewing, and the GI phase continues the secretion of saliva even after the food has passed for storage into the stomach. Secretion during the GI phase is especially likely to be abundant when one has swallowed irritant foods because nerve signals from the stomach then excite the salivatory nuclei. The saliva, on being swallowed, helps to neutralize the irritant substance, thereby relieving the irritation of the stomach.

Esophageal Secretions

The esophagus secretes only *mucus.* Normally, food passes through the esophagus all the way from the mouth to the stomach in about 7 seconds. This food has not been subjected to the mixing movements of the GI tract and, therefore, is in its most abrasive state. Fortunately the esophagus is supplied with a great abundance of mucous glands that secrete mucus to protect its mucosa from excoriation.

Gastric Secretions

The entire surface of the stomach mucosa is covered with gastric glands. Figure 41–8 illustrates one of these glands from the body of the stomach. Note that it is tubular in structure and extends from the surface of the mucosa all the way to the depths of the submucosa. This gland has three different types of secretory cells. The cells nearest the mouth of the gland secrete mucus and are called *mucous neck cells.* Deeper in the gland are very large numbers of *peptic cells* that secrete the digestive enzyme pepsinogen, which will be discussed later. Interspersed among the peptic cells are fewer but larger *oxyntic cells*, which secrete hydrochloric acid; these, too, will be discussed later.

Mucus. The primary function of the gastric secretions is to begin the digestion of proteins. Unfortunately, though, the wall of the stomach is itself constructed mainly of smooth muscle which itself is mainly protein. Therefore, the surface of the stomach must be exceptionally well protected at all times against its own digestion. This function is performed mainly by mucus that is secreted in great abundance in all parts of the stomach. The entire surface of the stomach is covered by a layer of very small *mucous cells*, which themselves are composed al-

FIGURE 41–8 A gastric gland from the body of the stomach.

most entirely of mucus; this mucus prevents gastric secretions from ever touching the deeper layers of the stomach wall. In addition, the gastric glands that secrete the stomach digestive enzymes also secrete a major amount of mucus at the same time. And, in the antral region of the stomach where the powerful peristaltic movements occur and where excoriation of the stomach wall is particularly likely to occur, mucus is secreted in especially high concentration by the antral glands. In the absence of mucus secretion, holes are eaten into the wall of the stomach in only a few hours. These holes are called *stomach ulcers.*

Digestive Substances.   The major digestive substances secreted by the gastric glands are *hydrochloric acid* and *pepsinogen*. The hydrochloric acid activates the pepsinogen to form *pepsin*, which is an enzyme that begins the digestion of proteins.

Less abundant enzymes secreted by the stomach are *gastric lipase* for beginning the digestion of fats, and *rennin* for aiding in the digestion of *casein*, one of the proteins in milk. These enzymes are secreted in such minor quantities that they are of almost no importance.

The total quantity of stomach secretion each day is about 2000 ml.

Regulation of Gastric Secretion.   Neurogenic Mechanisms.   Stomach secretion is regulated by both neurogenic and hormonal mechanisms, as shown in Figure 41–9. Some of the neurogenic mechanisms are quite similar to those regulating salivary secretion. For instance, food in the stomach can cause local nervous reflexes (called submucosal reflexes) that occur entirely in the wall of the stomach itself to cause

local secretion. Also, signals from the stomach mucosa to the medulla of the brain can cause reflexes back to the stomach through the vagus nerves to cause secretion. In addition, secretory signals from the medulla to the stomach can be excited by impulses originating in various other areas of the brain, particularly in the cerebral cortex.

The "Gastrin" Mechanism.   Gastric secretion is also regulated by a hormone called *gastrin*. When meats and certain other foods reach the antral portion of the stomach, they cause the hormone *gastrin*, a large polypeptide, to be extracted from the antral mucosa and to be absorbed into the blood stream. This hormone then passes by way of the blood to the fundic glands of the stomach and causes them to secrete a strongly *acidic* gastric juice. The acid, in turn, greatly aids in the digestion of the meats that first initiated the gastrin mechanism. In this way the stomach helps to tailor-make the secretion to fit the particular type of food that is eaten.

The Phases of Gastric Secretion.   Large amounts of stomach juices are often secreted when one simply thinks of pleasant food or particularly when food is smelled. This is called the *cephalic phase* of stomach secretion. It prepares the stomach for food that is to be eaten. The second phase of gastric secretion is the *gastric phase*, which is the secretion that occurs while the food is in the stomach itself. This is caused mainly by reflexes initiated by food in the stomach and by the gastrin mechanism. Finally, even after the food has left the stomach, gastric secretions continue for several hours. This is called the *intestinal phase* of gastric secretion. It is caused by myenteric nerve signals from the in-

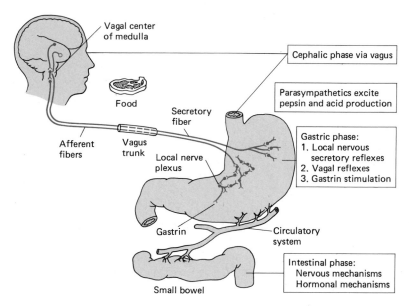

Vagal center
of medulla

Cephalic phase via vagus

Food

Parasympathetics excite
pepsin and acid production

Secretory
fiber

Gastric phase:
1. Local nervous
   secretory reflexes
2. Vagal reflexes
3. Gastrin stimulation

Afferent   Vagus
fibers   trunk

Local nerve
plexus

Gastrin

Circulatory
system

Intestinal phase:
   Nervous mechanisms
   Hormonal mechanisms

Small bowel

FIGURE 41–9   The phases of gastric secretion and their regulation.

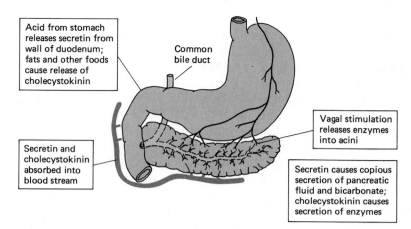

Acid from stomach releases secretin from wall of duodenum; fats and other foods cause release of cholecystokinin

Common bile duct

Vagal stimulation releases enzymes into acini

Secretin and cholecystokinin absorbed into blood stream

Secretin causes copious secretion of pancreatic fluid and bicarbonate; cholecystokinin causes secretion of enzymes

FIGURE 41–10 Regulation of pancreatic secretion.

testine to the stomach and also by hormones that pass to the stomach in the blood after being extracted from the intestinal mucosa by the food. Ordinarily, the amount of secretion during the intestinal phase is only about 10 percent of the total secretion.

Pancreatic Secretions

The pancreas, shown in Figure 41–10, is a large gland located immediately beneath the stomach. It empties about 1200 ml of secretions each day into the upper portion of the small intestine, a few centimeters beyond the pylorus. These secretions contain large quantities of *amylase* for digesting carbohydrates, *trypsin* and *chymotrypsin* for digesting proteins, *pancreatic lipase* for digesting fats, and other less important enzymes. It is obvious from this list that the pancreatic secretions are as important for digesting the food as any others of the entire GI tract.

In addition to the digestive enzymes, pancreatic secretions contain large amounts of *sodium bicarbonate*, which react with the hydrochloric acid emptied into the duodenum in the chyme from the stomach to form sodium chloride and carbonic acid. The carbonic acid then is absorbed into the blood, becomes water and carbon dioxide, and the carbon dioxide is expired through the lungs. The net result is an increase in the quantity of sodium chloride, a neutral salt, in the intestine. Thus, pancreatic secretions neutralize the acidity of the chyme coming from the stomach. This is one of the most important functions of pancreatic secretion.

***Regulation of Pancreatic Secretion.
The "Secretin" Mechanism and Neutralization of Chyme.*** When chyme enters the upper small intestine it causes a hormone called *secretin* to be released from the intestinal mucosa; the quantity of secretin released is especially abundant when the chyme is highly acidic. The secretin in turn is absorbed into the blood and then carried to the glandular cells of the pancreas. There it causes the cells to secrete large quantities of fluids containing extra large amounts of sodium bicarbonate. The bicarbonate then reacts with the acid of the chyme to neutralize it. Thus, the secretin mechanism is an automatic process to prevent excess acid in the upper small intestine.

When satisfactory neutralization does not occur, the acidic chyme, containing also large quantities of the protein-digesting enzyme pepsin, is likely to eat into the wall of the duodenum and cause *duodenal ulcers*. In fact, ulcer of the duodenum is about four times as common as ulcer of the stomach, because this area is not as well protected by mucous glands as is the stomach.

The "Cholecystokinin" Mechanism to Cause Enzyme Secretion. At the same time that secretin is extracted from the intestinal mucosa, another hormone, *cholecystokinin*, also is extracted mainly in response to fats, but to a lesser extent in response to proteins and carbohydrates. Cholecystokinin, like secretin, passes by way of the blood to the pancreas, but, unlike secretin, it causes the secretory cells to secrete large quantities of digestive enzymes instead of sodium bicarbonate. These enzymes, on entering the duodenum, begin digesting the foods.

Vagal Regulation of the Pancreas. Stimulation of the vagus nerve also causes the secretory cells of the pancreas to secrete highly concentrated enzymes. The quantity of fluid secreted, however, is usually so small that the enzymes remain in the ducts of the pancreas and later are floated into the intestinal tract by the copious secretion of fluid that follows secretin stimulation.

Vagal stimulation of pancreatic secretion seems to be a by-product of the vagal reflexes to the stomach. That is, some of the reflex impulses initiated by food in the stomach return to the pancreas rather than to the stomach. This allows preliminary formation of pancreatic enzymes even before the food enters the intestine. However, vagal stimulation of pancreatic secretion is much less important than the hormonal stimulation of secretion by secretin and cholecystokinin.

Liver Secretion

The liver, shown in Figure 41–11, secretes a solution called *bile* that contains a large quantity of *bile salts*, a moderate quantity of *cholesterol*, a small quantity of the green pigment *bilirubin* (which is a waste product of red blood cell destruction), and a number of other less important substances. The only substance in bile that is of importance to the digestive functions of the GI tract is the bile salts. The remaining contents are actually waste products being *excreted* from the body by this route. The amount of bile secreted each day averages about 800 ml.

The bile salts are not enzymes for digesting foods, but they act as a powerful *detergent* (a substance that lowers the surface tension at the surface between water and fats). This helps the mixing movements of the intestine break the large fat globules of the food into small globules, thus allowing the lipases of the intestinal tract, which are water soluble, to attack larger surface areas of the fat and to digest it. Without this action of bile, less than half of the fats in the food are digested. Bile salts also help in the absorption of the fat digestion products, as will be discussed more fully in the following chapter.

Regulation of Bile Secretion. Liver secretion, unlike secretion by other GI glands, continues steadily throughout the day and does not increase and decrease significantly in response to food in the intestine. The only hormone known to affect it is secretin, which can increase the output of bile as much as 80 percent, but the increase in secretion is caused mainly by increased water and sodium bicarbonate but not of the other contents of bile.

Storage of Bile in the Gallbladder. Even though bile secretion is a continuous process, the flow of bile into the GI tract is not continuous. As illustrated in Figure 41–11, a circular muscle around the outlet where the common bile duct empties into the duodenum, called the *sphincter of Oddi*, normally blocks the flow of bile into the gut. Instead, the bile flows into the *gallbladder*, which is attached to the side of the common bile duct. Much of the fluid and electrolytes of the bile are then reabsorbed into the blood by the gallbladder mucosa. This concentrates by as much as 12-fold the bile salts, cholesterol, and bilirubin, all of which cannot be reabsorbed, and allows the gallbladder, even though it has a maximum volume of only 50 ml, to accommodate the active components (bile salts) of most of an entire day's liver secretion of bile.

Emptying of the Gallbladder—the "Cholecystokinin" Mechanism. When food enters the small intestine, two mechanisms simultaneously cause the gallbladder to empty its contents into the small intestine. First, the *cholecystokinin* extracted from the wall of the duodenum by fats and other foods of the chyme passes through the blood of the gallbladder and causes the muscular wall to contract and also causes at least some degree of relaxation of the sphincter

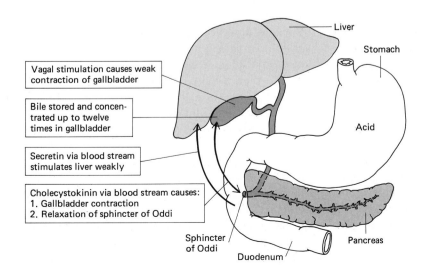

Vagal stimulation causes weak contraction of gallbladder

Bile stored and concentrated up to twelve times in gallbladder

Secretin via blood stream stimulates liver weakly

Cholecystokinin via blood stream causes:
1. Gallbladder contraction
2. Relaxation of sphincter of Oddi

Liver

Stomach

Acid

Sphincter of Oddi

Duodenum

Pancreas

FIGURE 41–11 Bile secretion, bile storage in the gallbladder, and the cholecystokinin mechanism for promoting gallbladder emptying.

of Oddi. This is the same cholecystokinin that causes the pancreas to secrete large quantities of enzymes. Second, the presence of food in the duodenum causes duodenal peristalsis, and the peristaltic waves send periodic inhibitory nerve signals to the sphincter of Oddi through the nervous system of the gut to open it. This combination of gallbladder contraction and opening of the sphincter of Oddi allows the stored bile to empty into the intestine, and the bile salts immediately begin their emulsifying action on the fats.

Gallstones. Gallstones are caused mainly by the fatty waste product *cholesterol*, which is excreted in the bile. Cholesterol is relatively insoluble, but it is normally held in solution in the bile by physical attraction to the bile salts. Often, though, too much water or too much of the bile salts is absorbed from the bile in the gallbladder. Then the cholesterol will no longer remain in solution. Crystals of cholesterol begin to precipitate, and these grow, thus forming gallstones. Sometimes the gallstones fill the entire bladder.

A means for preventing the formation of gallstones is to eat a diet low in fat, for cholesterol is formed by the liver in great abundance in response to a high fat diet. Once gallstones have formed, the usual treatment is removal of the stones, or, preferably, removal of the gallbladder itself along with the stones. Absence of the gallbladder does not greatly affect the digestion of fats, because bile continues to be excreted into the intestine, though now it enters the gut almost all the time rather than periodically.

Secretion in the Small Intestine

The small intestinal mucosa secretes the enzymes *sucrase*, *maltase*, and *lactase* for splitting disaccharides into monosaccharides, the final digestion products of carbohydrates. Also secreted are large quantities of *peptidases* for performing the final steps in protein digestion, and small quantities of *lipases* for splitting fats.

However, the secretion of enzymes in the small intestine does not occur in the usual manner. Instead, the digestive enzymes are formed in the epithelial cells lining the intestinal wall, and much of the digestive process occurs either inside these cells or on their surfaces. Also, some of the cells slough off into the intestinal lumen, then break up and release small amounts of the enzymes that act directly on the food of the chyme.

Secretion of Water and Electrolytes by the Small Intestinal Glands. Shallow tubular glands called *crypts of Lieberkühn* occur

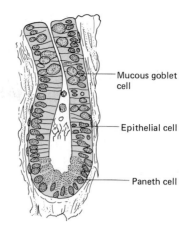

FIGURE 41–12 A crypt of Lieberkühn, found in all parts of the small intestine between the villi, which secretes almost pure extracellular fluid.

throughout the small intestine in the spaces between the villi. One such gland is illustrated in Figure 41–12. These glands have a variety of different types of cells. Though the specific functions of most of these cells are not clear, it is known that the glands secrete almost pure extracellular fluid, that is, a watery solution of the usual extracellular electrolytes; about 2 L are secreted each day. But why should these glands secrete this watery secretion without any enzymes? The answer is that this fluid is required for absorption of many of the end-products of digestion. This works in the following way: The extracellular fluid, on entering the lumen of the gut, dissolves many of the digestive end-products and acts as a vehicle to transport these to the villi and thence through the villus membrane into the blood and lymph. Thus, a continuous circulation of fluid occurs from the intestinal wall into the lumen of the intestine and then from there back into the blood or lymph, carrying with it the digestive end-products.

Mucus Secretion in the Small Intestine. The small intestine also secretes along its entire surface large quantities of mucus, which provide the same protective function in this part of the GI tract as in the stomach, the esophagus, and elsewhere. In the first few centimeters of the duodenum, especially abundant amounts of mucus are secreted by large mucous glands, called *Brunner's glands*, lying deep in the mucosa. The function of this secretion is to protect this portion of the intestinal tract from the powerful digestive action of pepsin and hydrochloric acid in the chyme newly arrived from the stomach. Once the chyme has been neutralized by pancreatic juice, however, it no longer has such a strong tendency to digest the wall of the intestine, which explains why Brunner's glands are

needed only in this uppermost region of the intestinal tract.

Quantity of Secretion. The total amount of secretion of the small intestine is about 2000 ml per day, which compares with about 1200 ml of saliva, 2000 ml of gastric juice, 1200 ml of pancreatic juice, and 800 ml of bile.

Regulation of Secretion in the Small Intestine. Most secretion in the small intestine is regulated mainly by local nervous reflexes. That is, food distending the intestinal tract or irritating the intestinal mucosa initiates reflexes in the intrinsic nerve plexus of the gut to stimulate secretion by the intestinal mucosa.

However, there probably is also a hormonal mechanism for regulating intestinal secretion. Though not as important quantitatively as the nervous reflex mechanism, it is believed that food in the small intestine extracts a mixture of hormones from the mucosa, and these hormones then stimulate increased intestinal secretion.

Secretions of the Large Intestine

The large intestine, like the esophagus, performs no digestive functions. Therefore, its only significant secretion is mucus. The entire mucosa is coated with mucous cells that provide lubrication for the passage of feces from the ileocecal valve to the anus and that also protect the large intestine from digestion by the enzymes emptied from the small intestine. The portion of the large intestine near the ileocecal valve is protected against the digestive hormones better than the distal portion. Consequently, during severe diarrhea the rapid flow of digestive enzymes from the small intestine into the distal colon is very likely to cause extreme irritation. Prolonged and severe diarrhea is sometimes associated with a condition called *ulcerative colitis*, which occasionally leads to holes (ulcers) in the colon that cause death. The mucus secreted in the large intestine normally protects against this.

CELLULAR MECHANISMS OF SECRETION

Thus far, we have discussed the GI glands and their secretions without stating the cellular mechanisms involved in glandular secretion. Let us now explain some of these mechanisms.

Secretion of Organic Substances. Each glandular cell has its own peculiar mechanism for secretion. However, the basic mechanism for secreting organic substances such as the protein

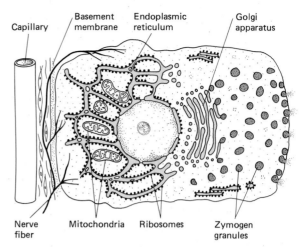

FIGURE 41–13 Typical function of glandular cell in formation and secretion of enzymes or other secretory substances.

digestive enzymes is illustrated in Figure 41–13. The protein enzymes are synthesized by the ribosomes attached to the outer surfaces of the endoplasmic reticulum, and as they are formed they are delivered to the inner lumen of this reticulum. During the next 20 to 30 minutes, they migrate through the reticulum to the Golgi apparatus in the manner described in Chapter 3. The Golgi apparatus in turn packages the enzymes in the form of small granules called *zymogen granules* and extrudes these into the cytoplasm toward the secretory pole of the cell. Thus, large numbers of these granules accumulate underneath the cell surface and remain there until an appropriate secretory signal causes them to be emptied from the cell.

The signal that causes secretion can be either hormonal or nervous. In either case, the signal is believed to cause ions to be actively transported into the base of the glandular cell. These ions in turn cause water to be absorbed into the cell by osmosis. Thus, a flow of water and electrolytes occurs into the cell and streams toward the cell apex. In some instances this actually ruptures the apical membrane and washes the zymogen granules outward with the flowing fluid. But, in other instances, the secretory signal probably causes some type of attractive force that makes the zymogen granules migrate to the cell membrane and then to extrude their enzymes. In some cells, it is believed that movement of calcium ions into the cell causes this to occur.

Secretion of Hydrochloric Acid by the Oxyntic Cells of the Gastric Glands. Some glands secrete highly specialized substances, for example, the hydrochloric acid secreted by the

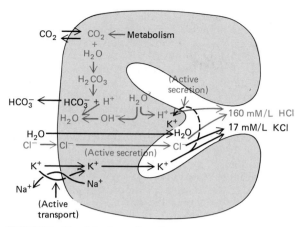

FIGURE 41–14 Postulated mechanism for the secretion of hydrochloric acid.

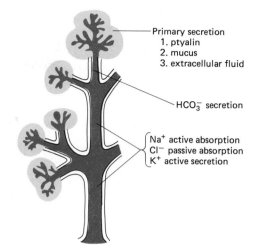

FIGURE 41–15 Formation and secretion of saliva by a salivary gland.

oxyntic cells of the gastric glands. The oxyntic cells have a special anatomical structure, which is illustrated diagrammatically in Figure 41–14. From the secretory pole of the cell, a system of small tubules, called *canaliculi*, penetrates deeply into the cell. The membranes of these canaliculi actively secrete both hydrogen ions and chloride ions; these then combine with each other in the canaliculi to form hydrochloric acid. In this way, a very strong solution of hydrochloric acid, having a pH of less than 1.0, is secreted by the oxyntic cells into the lumen of the gastric gland and from there into the stomach itself.

The chemical reactions illustrated in Figure 41–14 show that the eventual secretion of hydrochloric acid by the oxyntic cells is the product of many successive steps within the secretory cell. The most important of these steps is the provision of energy, which is released from ATP, for the active secretion of both hydrogen and chloride ions. In fact, as much as 80 percent of the total energy consumed by glandular cells goes for this express purpose of active transport during the secretory process.

Secretion of Saliva by the Salivary Gland. Figure 41–15 illustrates salivary secretion representing a still more complex method of secretion. In the salivary glands a so-called *primary secretion* is formed in the acini of the gland. Among the contents of the primary secretion are *ptyalin* (a digestive enzyme for starches), *mucus*, and *extracellular fluid*. Then, as the primary secretion passes through the ductules leaving the acini, *bicarbonate ion* is secreted into this primary secretion. Finally, as this secretion passes through the larger ducts of the salivary gland, both sodium and chloride ions are reabsorbed and potassium ions are secreted into the duct. Therefore, one sees that the primary secre-

tion is modified considerably as it passes through the ductules and larger ducts of the salivary glands before being discharged into the mouth in the form of saliva.

The other glands of the GI tract synthesize and secrete their products in similar ways, though of course the details of the processes are different for each type of gland.

QUESTIONS

1. Explain the function of both the propulsive movements and the mixing movements of the GI tract.
2. Explain nervous control of the swallowing process.
3. What are the motor functions of the stomach, including especially the control of stomach emptying?
4. How do the movements of the small intestine and of the colon differ from each other and also from movements in other portions of the GI tract?
5. How does vomiting occur?
6. How are the different phases of salivary secretion controlled?
7. What are the different secretions of the stomach? How are they regulated?
8. What are the components of pancreatic secretion? How is each of these controlled?
9. Give the characteristics of bile. Explain how its release into the intestinal tract is regulated.
10. How do the secretions of the small intestine differ from those of the large intestine? How are both controlled?
11. Explain the cellular mechanisms of secretion.

REFERENCES

Atanassova, E., and Papasova, M.: Gastrointestinal motility. *Int. Rev. Physiol.,* 12:35, 1977.
Binder, H.J. (ed.): Mechanisms of Intestinal Secretion. New York. A.R. Liss, 1979.

Brooks, F.P. (ed.): Gastrointestinal Pathophysiology. New York, Oxford University Press, 1978.

Cohen, S., *et al.*: Gastrointestinal motility. *In* Crane, R.K. (ed.): International Review of Physiology: Gastrointestinal Physiology III. Vol. 19. Baltimore, University Park Press, 1979, p. 107.

Davenport, H.W.: A Digest of Digestion, 2nd Ed. Chicago, Year Book Medical Publishers, 1978.

Gall, E.A., and Mostofi, F.K. (eds.): The Liver. Huntington, N.Y., R.E. Krieger, 1980.

Glass, G.B. (ed.): Gastrointestinal Hormones. New York, Raven Press, 1980.

Grossman, M.I.: Neural and hormonal regulation of gastrointestinal function: An overview. *Annu. Rev. Physiol., 41*:27, 1979.

Hendrix, T.R., and Paulk, H.T.: Intestinal secretion. *Int. Rev. Physiol., 12*:257, 1977.

Jenkins, G.N.: The Physiology and Biochemistry of the Mouth. Philadelphia, J.B. Lippincott, 1978.

Jones, R.S., and Myers, W.C.: Regulation of hepatic biliary secretion. *Annu. Rev. Physiol., 41*:67, 1979.

Mason, D.K.: Salivary Glands in Health and Disease. Philadelphia, W.B. Saunders, 1975.

Paumgartner, G., *et al.* (eds.): Biological Effects of Bile Acids. Baltimore, University Park Press, 1979.

Phillips, S.F., and Devroede, G.J.: Functions of the large intestine. *In* Crane, R.K. (ed.): International Review of Physiology: Gastrointestinal Physiology III. Vol. 19. Baltimore, University Park Press, 1979, p. 263.

Rappaport, A.M.: Hepatic blood flow: Morphologic aspects and physiologic regulation. *In* Javitt, N.B. (ed.): International Review of Physiology: Liver and Biliary Tract Physiology I. Vol. 21. Baltimore, University Park Press, 1980, p. 1.

Rehfeld, J.F.: Gastrointestinal hormones. *In* Crane, R.K. (ed.): International Review of Physiology: Gastrointestinal Physiology III. Vol. 19. Baltimore, University Park Press, 1979, p. 291.

Soll, A., and Walsh, J.H.: Regulation of gastric acid secretion. *Annu. Rev. Physiol., 41*:35, 1979.

Van Der Reis, L. (ed.): The Esophagus. New York, S. Karger, 1978.

42

Digestion and Assimilation of Carbohydrates, Fats, and Proteins

Overview

The most important *carbohydrates* in the diet are *starches, glycogen, sucrose* (cane sugar), and *lactose* (the sugar in milk). The starches and glycogen are both large polymers of *glucose*. They are first digested by ptyalin in the salivary secretions and *amylase* in pancreatic secretion into the disaccharide *maltose*. The maltose, in turn, is split by *maltase* in the epithelial cells of the intestinal villi to form glucose. Similarly, *sucrose is digested by sucrase* to form glucose and fructose, and *lactose is digested by lactase* to form glucose and galactose. The glucose, fructose, and galactose are all then *absorbed through the epithelial membrane of the intestinal villi into the portal blood.* As this portal blood passes through the liver, the fructose and galactose are mainly converted into glucose. Thus, essentially all carbohydrates are delivered to the tissue cells in the form of glucose. Uptake of glucose by most cells is controlled mainly by the hormone *insulin* which, therefore, is the major controller of carbohydrate metabolism in the body. Once in the cell, the glucose is used principally to provide metabolic energy.

Neutral fat, which is the great bulk of fatty substances in the diet, is composed of *glycerol* bound with *three fatty acids.* The fatty acids are split away from the glycerol in the small intestine mainly by the digestive action of *pancreatic lipase.* However, the *bile salts* in the bile are also necessary for full fat digestion. These exert a *detergent action* on the fat globules that helps to break the fat into very small particles that can then be digested. Also, as fatty acids are removed from the fat during the digestive process, they attach to the bile salts and are *"ferried"* in this form to the villi where they are absorbed. On passing through the villi, the fatty acids recombine with glycerol to form still new small fatty globules coated by a layer of protein and called *chylomicrons.* These enter the lymph in the villi and are carried with this lymph upward *through the thoracic duct* to be emptied in the blood and finally deposited in the *fat cells* throughout the body. When this stored fat is to be used for energy, it is split in the fat cells again into fatty acids and glycerol, and the fatty acids are transported in the blood to be used by tissue cells everywhere, mainly for energy. Much of the fat is also used by the liver to form other substances, especially *cholesterol* and *phospholipids,* that are needed in abundance throughout the body.

All *proteins* are composed of long sequences of *amino acids.* In the stomach, under the influence of the digestive enzyme *pepsin,* the pro-

685

teins are split into proteoses, peptones, and very large polypeptides. In the small intestine these are further digested under the influence of *trypsin, chymotrypsin,* and *carboxypolypeptidase*—all secreted in pancreatic juice—into small polypeptides. These are finally split by *peptidases* from the epithelial cells of the villi into amino acids. The amino acids are then *absorbed into the blood in the intestinal villi.* Many of the amino acids are temporarily stored in the liver until they are needed elsewhere, but eventually they are transported to all cells of the body to be converted into tissue proteins or used for energy.

Also absorbed by the gastrointestinal (GI) tract are large quantities of ions, especially *sodium ions, potassium ions, calcium ions, iron ions, chloride ions, phosphate ions, bicarbonate ions,* and *magnesium ions.* Most of these are *actively absorbed by the intestinal mucosa,* which means that they are carried by active transport through the intestinal epithelium in the manner that was explained in Chapter 5.

The term *digestion* means the splitting of large chemical compounds in the foods into simpler substances that can be used by the body. The term *assimilation* includes several functions: (1) absorption of the digestive end-products into the body fluids, (2) transport of these to the cells where they will be used, and (3) chemical change of some of them into other substances that are specially needed for various purposes. The function of the digestive and assimilative processes is to provide nutrients for the chemical reactions of metabolism.

DIGESTION, ABSORPTION, AND DISTRIBUTION OF CARBOHYDRATES

Carbohydrates are composed of carbon, hydrogen, and oxygen. The basic unit of a carbohydrate is a *monosaccharide,* the most common of which in food is *glucose,* and which has the following chemical formula:

Two other important monosaccharides in food are *fructose* and *galactose,* the formulas for which are the same as that of glucose except that some of the "H" and "OH" radicals are transposed.

In food, glucose and the other monosaccharides are usually *polymerized* (combined) to form larger chemical compounds such as *starches, glycogens, pectins,* and *dextrins.* By far the most common carbohydrate of the diet is starch, which is a polymer of glucose. The glucose molecules in starches are joined together in the manner illustrated in Figure 42–1. The successive molecules of glucose combine with each other by a *condensation* process, which means that one glucose molecule loses a hydrogen ion and the next loses a hydroxyl ion. The hydrogen and hydroxyl ions combine to form water, and the two glucose molecules connect together at the points where the ions were removed.

In addition to starches, another common source of carbohydrates is the *disaccharides,* which are combinations of only two molecules of monosaccharides. The common disaccharides in the diet are maltose, isomaltose, sucrose, and lactose. *Maltose* and *isomaltose* are each a combination of two glucose molecules, and they are derived mainly by splitting starches into their disaccharide components, all of which are either maltose or isomaltose. These differ from each other only in the manner in which the glucose molecules are joined. *Sucrose* is a combination of one molecule of glucose and one molecule of fructose. It is the same as cane sugar, or common table sugar. *Lactose* is a combination of one molecule of glucose and one

FIGURE 42–1 This is the manner in which glucose molecules are polymerized to form the starch molecule, often involving thousands of glucose molecules.

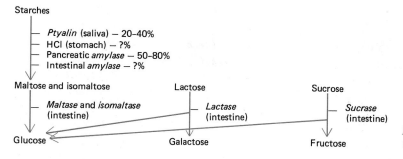

Starches
— *Ptyalin* (saliva) — 20-40%
— HCl (stomach) — ?%
— Pancreatic *amylase* — 50-80%
— Intestinal *amylase* — ?%

Maltose and isomaltose   Lactose   Sucrose

— *Maltase* and *isomaltase* (intestine)   — *Lactase* (intestine)   — *Sucrase* (intestine)

Glucose   Galactose   Fructose

FIGURE 42–2 Digestion of carbohydrates.

molecule of galactose. It is the sugar present in milk.

Basic Mechanism of Carbohydrate Digestion— The Process of Hydrolysis

Carbohydrate digestion breaks the starch or other carbohydrate polymers into their component monosaccharides. To do this, one molecule of water must be added to the compound at each point where two successive monosaccharides are joined. This is a process of *hydrolysis,* which is opposite to the condensation process by which the successive monosaccharides are bound to each other. The secretions of the digestive tract contain enzymes that catalyze this hydrolysis process.

Scheme of Digestion of Carbohydrates.

The schema in Figure 42–2 shows the digestion of the most common carbohydrates in the diet: the starches, lactose, and sucrose. Starches and other large carbohydrates are digested principally by *ptyalin* in the saliva and *amylase* in the pancreatic juice, but perhaps to a slight extent also by *hydrochloric acid* in the stomach and *intestinal amylase* in the small intestine. The re-

sulting products of these reactions are the disaccharides maltose and isomaltose.

The intestinal epithelial cells contain the enzymes *maltase, isomaltase, lactase,* and *sucrase,* which split maltose, isomaltose, lactose, and sucrose into their respective monosaccharides at the same time that these are absorbed through the intestinal wall into the blood. The final products of carbohydrate digestion are *glucose, galactose,* and *fructose,* as shown by the schema. Because all the monosaccharides derived from maltose are glucose and half of those derived from the other two carbohydrates are glucose, it is evident that this substance is by far the most abundant end-product of carbohydrate digestion. On the average, about 80 percent of the monosaccharides formed by digestion is glucose, 10 percent galactose, and 10 percent fructose.

Absorption of Monosaccharides

The Absorptive Epithelium of the Intestine.

Figure 42–3A illustrates a longitudinal section through a segment of the intestine. It shows the mucosa protruding into the lumen of the intestine in the form of large folds called

Villi

Valvulae conniventes

A

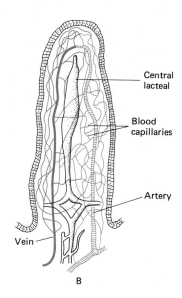

Central lacteal

Blood capillaries

Artery

Vein

B

FIGURE 42–3 (A) Distribution of villi and mucosal folds on the inner surface of the small intestine. (B) Structure of a villus, showing the blood vessel system and the central lacteal.

FIGURE 42–4 Brush border of the gastrointestinal epithelial cell, showing also pinocytic vesicles, mitochondria, and endoplasmic reticulum lying immediately beneath the brush border. (Courtesy of Dr. Wm. Lockwood.)

valvulae conniventes (or *folds of Kerckring*). Also, on the surface of the mucosa everywhere are literally millions of small *villi* only a millimeter or so in length. The structure of a villus is shown in Figure 42–3B. On the surface of the villus is an epithelial lining and within its substance is a small artery, multiple blood capillaries, a small vein, and a large central lymphatic capillary called the central *lacteal*. It is into the blood capillaries and this central lacteal that substances are absorbed.

The epithelial cells on the lumenal surface of each villus have a "brush border" (Fig. 42–4), which comprises thousands of minute *microvilli* only 0.1 μ (micrometer) in diameter. These microvilli provide a tremendous surface area through which substances can be absorbed into the interior of the cell.

Because of (1) the mucosal folds, (2) the villi, and (3) the microvilli on the surface of the epithelial cells, the total absorptive area of the intestine is about 600 times as great as it would be without these structures, giving a total absorptive area of about 550 m^2 for the entire small intestine, which is equal to the floor area of about two houses.

Route of Absorption of Monosaccharides. The monosaccharides are absorbed through the intestinal epithelium into the blood of the villus capillaries. The blood then empties into the portal venous system and finally flows through the liver into the general circulation. As the monosaccharides pass through the liver, they are partially processed prior to reaching the peripheral cells for metabolism, as we shall discuss later.

Mechanism for Absorption of Monosaccharides. Monosaccharides are absorbed from the GI tract by an active absorption mechanism

called *sodium cotransport*. The mechanism of this process was discussed in Chapter 5 and is also essentially the same as that for absorption from the tubules of the kidney, which was discussed in Chapter 31. Briefly, the monosaccharides combine with a carrier substance in the epithelial cells and sodium binds with the same carrier. Then, because sodium is moved through the cell by its own active transport mechanism, this indirectly pulls the monosaccharide along with it. For this process to occur, energy must be expended by the epithelial cells, which is the reason why the process is called "active" absorption.

Active absorption of the monosaccharides is very important because it allows absorption to occur even when monosaccharides are present in the intestine in extremely small concentrations, concentrations even smaller than those in the blood itself.

Fate of the Monosaccharides in the Body

Glucose in the Blood and Extracellular Fluid—Conversion of Fructose and Galactose to Glucose. When a person eats the usual diet that is high in carbohydrates, approximately 80 percent of the monosaccharides absorbed from the gut is glucose, and essentially all of the remainder is fructose and galactose. However, almost immediately, both these other monosaccharides are also converted into glucose. The fructose is mainly converted as it is absorbed through the intestinal epithelial cells because of a metabolic interconversion that takes place in these cells. The galactose is absorbed very rapidly by the liver cells, converted into glucose, and then returned to the blood. Thus, for practical purposes, all carbohydrates finally reach the individual tissue cells in the form of glucose, which then supplies a major share of the cellular energy.

The concentration of glucose in the blood and extracellular fluid is approximately 90 mg in each 100 ml, whereas the concentrations of fructose and galactose are usually very slight because of their rapid conversion to glucose.

Transport of Glucose Through the Cell Membrane—Effect of Insulin. Before glucose can be used by the cells it must be transported through the cell membrane. Unfortunately, the pores of the cell membrane are too small to allow glucose to enter by the process of simple diffusion. Here again the glucose must be transported by a chemical process, called *facilitated diffusion*, the general principles of which were

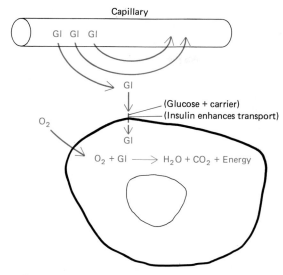

FIGURE 42–5 Transport of glucose from the capillary into the extracellular fluid and then from the extracellular fluid into the cell to be utilized for energy.

discussed in Chapter 5 and are shown again in Figure 42–5. Glucose first combines with a carrier, a protein, in the cell membrane. Then it is transported to the inside of the cell where it breaks away from the carrier.

In some way that has not yet been completely explained, the hormone *insulin* greatly enhances this facilitated transport of glucose through the cell membrane. Some possible ways in which insulin might do this are (1) by catalyzing the reaction between glucose and the carrier, (2) by removing glucose from the carrier on the inside of the cell, or (3) by operating as one of the components of the carrier mechanism itself. Regardless of which of these might be correct, the rate at which glucose can be transported through the cell membrane is determined mainly by the amount of insulin available. When the pancreas fails to secrete insulin, as much as five times less glucose enters than the cell needs. When insulin is secreted in excessive abundance, glucose enters so rapidly that glucose metabolism becomes much greater than normal. It is obvious, therefore, that the rate of carbohydrate metabolism is regulated in accordance with the rate of insulin secretion by the pancreas.

Regulation of Blood Glucose Concentration. Buffer Effect of the Liver. After a meal, large quantities of monosaccharides are absorbed into the blood, and the glucose in the portal blood coming from the intestines rises from the normal concentration of 90 mg per 100 ml to as high as double this. However, this portal blood flows through the liver before it reaches the general circulation, and the liver removes about two thirds of the excess glucose. In this way the liver usually keeps the concentration of glucose in the general blood circulation from rising above 120 to 140 mg per 100 ml, despite very rapid absorption from the intestines.

The mechanism by which the liver removes the glucose from the portal blood is the following: Glucose is first absorbed through the cellular membranes into the liver cells. Then it is converted to *glycogen*, a polymer of glucose, and stored until a later time. When the blood glucose level falls to lower values several hours after a meal, the glycogen is split back into glucose, which is transferred out of the liver into the blood.

In essence, then, the liver is a "buffer" organ for blood glucose regulation, for it keeps the blood glucose level from rising too high and from falling too low.

Insulin Production by the Pancreas as a Means For Controlling Blood Glucose Concentration. After a person eats a large meal, the rise in blood glucose concentration stimulates the pancreas to produce large quantities of insulin. The insulin in turn promotes rapid transport of glucose into the cells, thus decreasing the blood glucose level back toward normal. Therefore, in addition to the liver buffer mechanism, this pancreatic production of increased quantities of insulin also aids in preventing excessive rises in blood glucose concentration.

Effect of Epinephrine, Sympathetic Stimulation, and Glucagon in Preventing Low Blood Glucose Concentration. A low blood glucose level stimulates the sympathetic centers of the brain, causing secretion of norepinephrine and epinephrine by the adrenal glands and excitation of all the sympathetic nerves throughout the body. Also, the low glucose concentration directly stimulates the pancreas to secrete the hormone *glucagon*. This glucagon, the norepinephrine, epinephrine, and sympathetic stimulation all cause liver glycogen to split into glucose, which is then emptied into the blood. This returns the blood glucose concentration back toward normal, acting as a protective mechanism against low blood glucose levels.

Gluconeogenesis. Another effect that occurs when the blood glucose level falls too low is formation of glucose from proteins, this occurring by a series of chemical reactions in the liver cells. This phenomenon is called *gluconeogenesis*. The importance of gluconeogenesis is that it provides glucose to the blood even during periods of starvation. Glucose, unfortunately, is not stored to a major extent in the body, for only 300 g at most are stored in the form of glycogen in both the liver and all the remainder of the body, mainly the muscles. Ordinarily this amount is not sufficient by itself to maintain the blood glucose concentration at normal values for more than 24 hours. However, as the blood glucose level falls below normal, gluconeogenesis begins in the liver cells and continues until an adequate supply of glucose is available again.

Later in this chapter it will be noted that most of the cells of the body can utilize fats for energy when glucose is not available. However, the neurons of the brain are unable to utilize fats, and without an adequate supply of glucose these cells begin to die. This is the major reason why it is very important that the blood glucose concentration be kept at a normal level even during long periods of starvation.

ENERGY FROM GLUCOSE

The major function of glucose in the body is to provide energy, though some glucose molecules are used as building stones for synthesis of other needed compounds. Energy is derived from glucose by two means: by splitting the molecules of glucose into smaller compounds and by oxidizing these to form water, which liberates an extremely large amount of energy. These mechanisms of energy release are discussed in the following chapter.

DIGESTION, ABSORPTION, AND DISTRIBUTION OF NEUTRAL FATS

Neutral fats, like carbohydrates, are composed of carbon, hydrogen, and oxygen, though the relative abundance of oxygen in fats is considerably less than in carbohydrates. A representative molecule of neutral fat and its digestive end-products are illustrated in Figure 42–6. It is evident from the formulas in the figure that a fat molecule contains two major components: first, a glycerol nucleus and, second, three fatty acid radicals. Each fatty acid radical is combined with the glycerol in a condensation process, that

is, removal of a hydroxyl radical from the glycerol and a hydrogen ion from the fatty acid, with formation of a water molecule and bonding of the fatty acid and the glycerol at the points of removal. This mechanism was noted previously as the means by which monosaccharides also combine with each other to form complicated carbohydrates.

The differences among various fats lie in the composition of the fatty acids in the molecule. Most fats in the human body have fatty acids with 16 or 18 carbon atoms in their chains. One of the most common of the fatty acids is *stearic acid*, which has 18 carbon atoms and is illustrated in Figure 42–6. The fats containing the longer chain fatty acids are more solid than those containing the shorter fatty acids. Other than this, the chemical and physical properties of most fats do not vary greatly from one to the other.

Some of the fatty acids in the body and in the diet are *unsaturated*, which means that at various points in the carbon chain the atoms are bonded together by double bonds rather than single bonds, and that there is a corresponding lack of two hydrogen atoms. The unsaturated fats are needed to form a few special structures of the cells, but otherwise even these perform the same principal function as the saturated fats, which is to provide energy for the metabolic processes.

Digestion of Fats

The digestion of fats, like that of carbohydrates, is a *hydrolysis* process. This is catalyzed by the enzyme called *lipase*. Almost all of the lipase is secreted by the pancreas, though small amounts are secreted by the stomach and small intestine. The diagram in Figure 42–7 shows the complete schema of fat digestion, which occurs almost entirely in the small intestine. The end-products of fat digestion are *fatty acids, glycerol,* and *glycerides*. Glycerides are composed of a glycerol nucleus with one or two of the fatty acid chains still attached. Though the end result of complete

$$CH_3-(CH_2)_{16}-\overset{\overset{\displaystyle O}{\|}}{C}-O-CH_2$$

$$CH_3-(CH_2)_{16}-\overset{\overset{\displaystyle O}{\|}}{C}-O-CH \xrightarrow{\text{lipase}} 3CH_3-(CH_2)_{16}-\overset{\overset{\displaystyle O}{\|}}{C}-O-H \quad + \quad \begin{matrix} HO-CH_2 \\ | \\ HO-CH \\ | \\ HO-CH_2 \end{matrix}$$

$$CH_3-(CH_2)_{16}-\overset{\overset{\displaystyle O}{\|}}{C}-O-CH_2$$

Tristearin   Stearic acid   Glycerol

FIGURE 42–6 A neutral fat, tristearin, and its digestion by *lipase* into the fatty acid stearic acid and glycerol.

FIGURE 42–7 Digestion of fats.

fat digestion is the splitting of the fats entirely into fatty acids and glycerol, the process goes to completion for only about 40 percent of the fat molecules, leaving many glycerides, mainly the monoglyceride (glycerol still attached to one fatty acid), still among the digestive products.

Because monoglycerides pass through the intestinal membrane with almost the same ease as glycerol and fatty acids, the digestive process is quite adequate for absorption to occur.

Role of Bile Salts in Fat Digestion. The bile salts secreted by the liver are essential for complete digestion of fat in the intestine even though they perform no digestive enzyme function. In the absence of bile salts, as much as 50 percent of the fat passes undigested all the way through the gastrointestinal tract and is expelled in the feces.

The bile salts play two principal roles in fat digestion. The first of these is its effect of *acting as a detergent;* that is, it greatly decreases the surface tension of the fatty globules in the food. This allows the mixing movements of the intestines to break the fat globules into very finely emulsified particles, thereby providing greatly increased surface area on which the water-soluble digestive enzymes, the lipases, can act.

The second way in which bile salts increase fat digestion is *to "ferry" the end-products of digestion*, the fatty acids and the glycerides, away from the fat globules as the digestive process proceeds. The bile salt molecules themselves aggregate to form colloidal particles called *micelles.* These have a fatty core, but they can remain in colloidal solution in the fluids of the intestines because the surfaces of the micelles are ionized, which is a property that promotes water solubility. The fatty acids and the glycerides become absorbed in the fatty portions of these micelles as they are split away from the fat globules, and in this form they are ferried from the fat globules to the intestinal epithelium, where absorption occurs.

Absorption of the End-Products of Fat Digestion

Once the micelles have ferried the products of fat digestion to the intestinal epithelium, the fatty acids and glycerides are released from the micelles and, like the monosaccharides, are also absorbed by the villi of the intestinal mucosa. But, unlike the monosaccharides, they are *absorbed into the central lacteal*, a lymph vessel in the center of the villus, as shown in Figure 42–3B, instead of into the blood. The mechanism by which this fat absorption occurs is the following:

The fatty acids and glyceride molecules are very soluble in the brush border of the epithelial cells lining the surfaces of the villi. Therefore, they diffuse readily from the intestinal lumen into the interior of these cells. Then, the endoplasmic reticulum inside the cell resynthesizes new molecules of neutral fat and expels the newly formed fat into the interstitial fluid of the villi in the form of small fat globules called *chylomicrons;* these are immediately picked up by the central lacteal.

Transport of Fat Through the Lymphatics. Lymph is "milked" from the central lacteals into the abdominal lymphatics by rhythmic contraction of the villi. This contraction is stimulated by a hormone, *villikinin*, which is released from the intestinal mucosa when fats are in the chyme. After leaving the central lacteals, the chylomicrons are transported upward through the *thoracic duct*, the major lymphatic channel of the body, to empty into the blood circulation at the juncture of the internal jugular and subclavian veins.

Chylomicrons. The fat globules absorbed into the central lacteal, the *chylomicrons*, are about $1\ \mu$ in diameter. Their surfaces are covered with a layer of protein that is formed in the epithelial cells of the villi as the fat is absorbed. Because the protein is hydrophilic, this keeps the chylomicrons suspended in the lymph and prevents them from sticking to each other or to the walls of the lymphatics or blood vessels. It is in this form that fats are transported through the lymphatics and finally into the blood.

After a fatty meal the level of chylomicrons in the circulating blood reaches a maximum in approximately 2 to 3 hours, sometimes becoming as much as 1 to 2 percent of the blood, but within another 2 to 3 hours almost all of them will have been deposited in the fat tissue of the body or in the liver.

Fat Tissue

Fat tissue is a special type of connective tissue that has been modified to allow storage of neutral fat. It is found beneath the skin, between the muscles, between the various organs, and in almost all spaces not filled by other portions of the body. The cytoplasm of fat cells sometimes contains as much as 95 percent neutral fat. These cells store fat until it is needed to provide energy elsewhere in the body.

Fat tissue provides a *buffer* function for fat in the circulating fluids. After a fatty meal the high concentration of fat in the blood is very soon lowered by deposition of the extra fat in the fat tissue. Then, when the body needs fat for energy or other purposes, it can be mobilized from the fat tissues and returned to the circulating blood, as will be explained below.

Because of this buffer function of the fat tissues, the fat in the fat cells is in a constant state of flux. Ordinarily, half of it is removed every 8 days, and new fat is deposited in its place.

Transport of Fats in the Body Fluids

Free Fatty Acids. Most fat is transported in the blood from one part of the body to another in the form of *free fatty acids.* These usually exist in the blood as a loose combination of free fatty acids with albumin, one of the plasma proteins. Every fat cell contains large quantities of the fat digestive enzyme *lipase.* However, this remains in an inactive form except when there is need for release of fat from the fat tissue. Several hormones, especially *cortisol* from the adrenal cortex and *epinephrine* from the adrenal medulla, can activate the lipase. This then digests the neutral fat in the fat cell into glycerol and fatty acids. The fatty acids diffuse out of the cell and immediately combine with albumin in the blood and are transported in this form to other tissues of the body where they are released from the albumin. Some combine with glycerol in fat tissue to form new neutral fat. Others enter other tissue cells where they are split into smaller molecules and used to supply energy, as is explained below.

Concentration of Free Fatty Acids. Even though almost all fat transport in the body is in the form of free fatty acids, these are normally present in the blood in a concentration of only about 10 mg per 100 ml of blood, or one-tenth the concentration of glucose. However, the fatty acids are transported to their destination within a few seconds to a few minutes, remaining in the blood only a short time, and can therefore account for tremendous amounts of available energy to the cells. When fats are being used in great quantities by the cells, the blood concentration of fatty acids increases as much as fourfold or more.

Lipoproteins. Lipoproteins are minute fatty particles covered by a layer of adsorbed protein. These are suspended in a colloid form in the plasma and to a lesser extent in other extracellular fluids. The *chylomicrons* are a type of lipoprotein because they are composed of lipid substances (neutral fat, phospholipids, and cholesterol) and a layer of adsorbed protein. However, in addition to the chylomicrons, large numbers of much smaller lipoprotein particles are also present in the blood. These are formed almost entirely in the liver, and their function is to transport neutral fat, phospholipids, and cholesterol from the liver to the different cells of the body.

Synthesis of Fat from Glucose and Proteins

Much of the fat in the body is not derived directly from the diet but instead is synthesized in the body. The fat cells themselves are capable of synthesizing small amounts of fat, but most fat is synthesized in the liver and then transported to the fat cells. Both glucose and amino acids derived from proteins can be converted into fat, but by far the most important source is glucose.

When an excess of glucose is in the diet and sufficient insulin is secreted by the pancreas to cause all the glucose to enter the cells, almost all the extra glucose not used immediately for energy passes mainly into the liver cells but to some extent also into the fat cells to be converted into fat. The fat formed in the liver is then transported in the lipoproteins to the fat cells. Thus, the fat tissues provide a means for storing energy derived from carbohydrates and proteins as well as from fats. This conversion of other foods to fats explains why eating any type of food, whether it be fat, carbohydrate, or protein, can increase the amount of fat tissue.

Functions of the Liver in the Utilization of Fat

The liver is undoubtedly the most important organ of the body for controlling fat utilization. The liver, in addition to converting much of the excess glucose into fat, converts fat into substances that can be used elsewhere in the body for special purposes. For example, some of the fats must be desaturated to provide the unsatu-

FIGURE 42–8 Alternate oxidation of the fatty acid molecule to form multiple molecules of acetic acid.

rated fats required by all cells of the body for their metabolic processes; some must be converted into the fatty substances *cholesterol* and *phospholipids* needed for cellular structures; and others are broken into smaller molecules that can be used easily by the cells for energy. The liver performs all these functions.

When the body is depending mainly on fats instead of glucose for energy, the quantity of fat in the liver gradually increases. This is enhanced by adrenocortical hormones, for they cause the fat tissue cells to mobilize their fat.

Energy from Fats

Fatty acids can be utilized for energy by almost all cells of the body with the exception of the neuronal cells of the brain. However, about 40 percent of the fatty acids used for energy are first split in the liver into *acetoacetic acid* and then transported to the other tissue cells to be used for energy, as explained in the following chapter.

The first stage in the utilization of fats for energy is to split the neutral fat into glycerol and fatty acids by the lipases in the fat cells, and then to transport these products to the other cells. The glycerol, being very similar to some of the breakdown products of glucose, can then be used for energy in very much the same manner as glucose. However, by far the major amount of energy in the fat molecule is in the fatty acid chains, and before these can be used for energy they must be split into still smaller chemical compounds. Ordinarily, this is accomplished by a chemical process called *alternate oxidation* of the carbon chain, which is illustrated in Figure 42–8.

The net result is the formation of many molecules of *acetic acid*. These are then oxidized in the cells in a manner almost identical to the oxidation of glucose, giving tremendous amounts of energy to the cells, as will be explained in the following chapter.

In the liver, most of the acetic acid molecules formed from the breakdown of fatty acids condense two at a time to form *acetoacetic acid*, as shown in Figure 42–9.

The acetoacetic acid is called a *keto acid*, and it in turn can change into several other closely related forms of keto acids, or even into acetone.

The keto acids are highly diffusible through cellular membranes. Therefore, as shown in Figure 42–10, the keto acids formed in the liver diffuse immediately into the blood and are transported to all cells throughout the body. There the keto acids are oxidized in the same manner as glucose to provide energy for cellular functions, as will be explained further in the following chapter.

Fat-Sparing Effect of Carbohydrates.
As long as sufficient glucose is available to supply the energy needs of the cells, this glucose is metabolized for energy in preference to fatty acids and keto acids, and if more than enough glucose is available, the excess is converted into fat. Thus, when glucose is available, the metabolism of fat stops; for this reason glucose is said to be a *fat sparer.*

Conversely, whenever the available glucose is very slight, the body automatically shifts its metabolic system to derive energy from fat instead of carbohydrates. This conversion to fat utilization is caused by two major hormonal changes: First, the decrease in blood glucose concentration causes the pancreas to decrease its rate of insulin secretion, and this in turn causes the fat cells to release greatly increased quantities of fatty acids into the circulating blood. Second, lack of sufficient carbohydrates indirectly causes the adrenal glands to secrete increased quantities of cortisol, one of the adrenocortical hormones. This has a direct effect on fat cells of activating the cellular lipase and causing still additional release of fatty acids into the circulating blood, therefore increasing the utilization of free fatty acids for energy.

FIGURE 42–9 Condensation of two molecules of acetic acid to form acetoacetic acid.

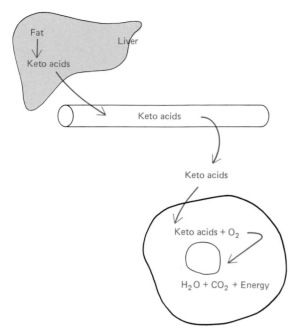

FIGURE 42–10 Formation of keto acids in the liver, their transport to the cells, and their utilization for energy.

Phospholipids and Cholesterol

Two additional substances, *phospholipids* and *cholesterol*, have physical properties similar to those of neutral fats. These serve especially as structural components of cellular and intracellular membranes. The phospholipids are composed of glycerol, fatty acids, and a phosphate side chain. Cholesterol is composed mainly of a sterol nucleus that is synthesized from acetic

A phospholipid

Cholesterol

FIGURE 42–11 Two substances similar to and derived from fats, a phospholipid and cholesterol.

acid, the chief end-product of fatty acid degradation. The chemical structures of these substances are shown in Figure 42–11. Both phospholipids and cholesterol are fat-soluble but only very slightly water-soluble. They are synthesized in all cells of the body, though to a much greater extent in the liver cells than in other cells. Both are transported in lipoproteins from the liver to other parts of the body.

Both phospholipids and cholesterol are major constituents of cell membranes and membranes of intracellular structures such as the nuclear membrane, the membranes of the endoplasmic reticulum, the membranes of mitochondria, of lysosomes, and so forth.

The precise functions of the phospholipids and cholesterol in the membranes are not entirely understood, but they do play roles in determining membrane permeability and its transport properties.

DIGESTION, ABSORPTION, AND DISTRIBUTION OF PROTEINS

Proteins are large molecules usually made up of hundreds to thousands of *amino acids* joined together. Amino acids, in turn, are small organic compounds that have an amino radical, —NH_2, and an acidic radical, —COOH, both on the same molecule. Twenty important amino acids are known to be present in the body proteins; the formulas of these are shown in Figure 42–12.

Some of the amino acids can be synthesized in the body from other amino acids, but 10 of them cannot. These 10 are called *essential amino acids*, for they must be provided in the diet in order for the human body to form the proteins necessary for life.

Amino acids combine with each other to form proteins by means of *peptide linkages*, an example of which is illustrated in Figure 42–13. It will be noted that the product of the two combined acids, which is called a *peptide*, still has an amino radical and an acid radical, which can provide reactive points for combinations with still additional amino acids. The nature of the protein is determined by the types of amino acids in the protein and also by the pattern in which they are joined.

Digestion of Proteins to Form Amino Acids

Referring again to the equation in Figure 42–13, it is evident that a peptide linkage is another example of *condensation*, which is the same means

Amino Acids

Glycine

$$H - \overset{\overset{\displaystyle H}{|}}{\underset{\underset{\displaystyle NH_2}{|}}{C}} - COOH$$

Proline

Alanine

Serine

Cysteine

Aspartic acid

Glutamic acid

Asparagine

Glutamine

Tyrosine

Essential Amino Acids

Threonine

Lysine

Methionine

Arginine

Valine

Phenylalanine

Leucine

Tryptophan

Isoleucine

Histidine

FIGURE 42–12 The amino acids, showing the 10 essential amino acids, which cannot be synthesized in the body either at all or in sufficient quantity.

FIGURE 42–13 Combination of two amino acids by *peptide linkage.*

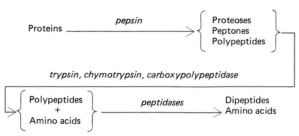

FIGURE 42–14 Digestion of proteins.

by which the component parts of fats and carbohydrates are combined. Therefore, the digestion of protein, like that of carbohydrates and fats, is accomplished by a process of *hydrolysis*, which is the opposite of condensation.

The schema in Figure 42–14 shows the sequence for protein digestion, which begins with *pepsin* action in the stomach. Pepsin is secreted in the form of *pepsinogen*, a substance that has no digestive properties, but once it comes in contact with the hydrochloric acid also secreted by the same gastric glands, it is soon activated to form pepsin. The hydrochloric acid also provides an appropriate reactive medium for pepsin, for it can split proteins only in acid surroundings.

Protein is digested in the stomach into *proteoses*, *peptones*, and *polypeptides*, all of which are smaller combinations of amino acids than proteins—the proteoses are nearly as large as proteins, the peptones are intermediate in size, and the polypeptides are combinations of only a few amino acids. After entering the small intestine, these substances are further split by *trypsin, chymotrypsin,* and *carboxypolypeptidase* of the pancreatic juice into small polypeptides and some amino acids. Then the small polypeptides are finally split by *peptidases* of the pancreatic and intestinal juices into amino acids with a few remaining dipeptides. Thus, the final products of protein digestion are the basic components of proteins, the *amino acids*, along with a few dipeptides (two amino acids bound together).

Absorption of Amino Acids

Amino acids (and a few dipeptides) are absorbed from the gastrointestinal tract in almost exactly the same manner as monosaccharides, that is, by active transport into the blood of the intestinal villi, utilizing the sodium cotransport mechanism. The transport is carrier-mediated in the manner discussed in Chapter 5, and energy expenditure is required for it to occur; these are also characteristics of monosaccharide transport.

After absorption through the intestinal mucosa, the amino acids pass into the capillaries of the villi and thence into the portal blood, flowing through the liver before entering the general circulation.

Amino Acids in the Blood

All of the different amino acids circulate in the blood and extracellular fluid in small quantities. However, their total concentration is only about 30 mg in each 100 ml of fluid; or, to express this another way, the total concentration of all the 20 different amino acids together is only about one-third that of glucose. The reason for this small concentration is that the amino acids, on coming in contact with cells, are absorbed very rapidly.

Buffer Action of the Liver and Tissue Cells for Regulating Blood Amino Acid Concentration. The liver acts as a buffer for amino acids in the same manner that it acts as a buffer for glucose. When the blood concentration of amino acids rises high, a large proportion of them is absorbed into the liver cells, where they can be temporarily stored, probably combined with each other to form small protein molecules. When the amino acid concentration in the blood falls below normal, the stored amino acids pass back out of the liver cells into the blood, to be used as needed elsewhere in the body.

Most other cells of the body also have this ability to store amino acids to at least some extent and to release these into the blood when the blood content of amino acids falls. As a result, amino acids are in a state of continual flux from one part of the body to another. If the amount of amino acids in the cells of one tissue falls too low, then amino acids will enter these cells from the blood, and they in turn will be replaced by amino acids released from other cells. This continual flux of the amino acids among the various cells is illustrated by the diagram of Figure 42–15.

Effect of Cortisol on Amino Acid Flux. Recent research has shown that cortisol, one of the hormones secreted by the adrenal cortex, aids in the movement of amino acids from one area of the body to another. Though the precise nature of this mobilization function of cortisol is yet unknown, it is believed that it increases the rate of transport of amino acids through the cellular membranes and thereby promotes their rapid transfer from tissue to tissue. When an area of the body is damaged and is in need of amino acids for repair of its cells, the rate of cortisol secretion becomes greatly increased, and

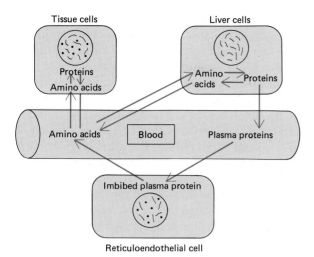

FIGURE 42–15 Reversible equilibrium between the tissue proteins, plasma proteins, and plasma amino acids.

the resultant mobilization of amino acids helps to supply the needed materials for repair.

The Tissue Proteins and Their Synthesis

The proteins of the cells perform two major functions. First, they provide most of the structural elements of the cells, and, second, they are the enzymes that control the different cellular chemical reactions. Therefore, the types of proteins in each cell determine the cell's functions. Each cell is capable of synthesizing its own proteins, and this synthesis is controlled by the genes of the cell's nucleus in the manner described in Chapter 4. Basically, this process is the following:

Regulation of Protein Synthesis by the Genes. The nucleus of each cell of the human body contains 46 chromosomes arranged in 23 pairs. Each of these chromosomes contain thousands of *deoxyribose nucleic acid* molecules, many of which are genes. The function of each gene is determined by its intrinsic chemical structure and also by its position in the chromosome thread.

Each gene of the nucleus controls the formation of a corresponding type of *ribose nucleic acid* that is transported to the cytoplasm of the cell. This has a slightly different chemical composition from the deoxyribose nucleic acid of the nuclear gene, and it in turn acts as a "template" to control the formation of a protein by the ribosomes in the cytoplasm. Since it is the proteins that perform the structural and enzymatic functions in the cell, the nuclear genes

regulate, in a roundabout way, the entire function of the cell.

Formation of Plasma Proteins. The proteins in the plasma are of three different types: *albumin*, which provides the colloid osmotic pressure in the plasma; *globulins*, which provide the antibodies; and *fibrinogen*, which is used in the process of blood clotting. Almost all these are formed in the liver and then released into the blood, though a small portion of the globulins in particular are formed by the lymphoid tissue as will be explained in the discussion of immunity in Chapter 45.

Whenever the concentration of proteins in the plasma falls too low to maintain normal colloid osmotic pressure, the production of plasma proteins by the liver increases markedly. Though the means by which this control function works is unknown, it obviously is of great value in maintaining normal circulatory dynamics, for, if the colloid osmotic pressure of the blood should ever vary too far above or below normal, the transfer of fluids through the capillary membranes into and out of the interstitial spaces would become abnormal.

Conversion of Proteins into Amino Acids. Most of the body's cells synthesize more proteins than are absolutely necessary to maintain life of the cells. Therefore, if amino acids are needed elsewhere in the body, some of the cellular proteins can be reconverted into amino acids and then transported in this form. The reconversion process is catalyzed by enzymes, called *cathepsins*, that are in all cells, stored normally in the lysosomes.

The quantity of proteins in a cell is determined by a balance between their rate of synthesis and their rate of degradation. Even the plasma proteins circulating in the blood are subject to reconversion into amino acids, for they can be imbibed by reticuloendothelial cells and other cells and then split into amino acids by the intracellular enzymes.

The constant balance between amino acids and proteins in cells, and between amino acids and the plasma proteins, is shown in Figure 42–15. By this constant interchange of amino acids, the proteins in all parts of the body are maintained in reasonable equilibrium with each other. If one tissue suffers loss of proteins or if the blood suffers loss of plasma proteins, many of the proteins in the remainder of the body will soon be converted into amino acids, which are transported to the appropriate point to form new protein. For example, in widespread cancer that is using extreme quantities of amino acids for formation of new cancer cells, the amino acids are derived continually from the tissue proteins, leading to serious debility. Also, when large quantities of blood are lost, the plasma proteins are replenished to normal within approximately 7 days by the transfer of amino acids from the tissue proteins to the liver, where new plasma proteins are formed.

Use of Amino Acids to Synthesize Needed Chemical Substances

Most metabolic reactions in the cells require special chemical substances to keep them operating, and most of these chemicals are synthesized from amino acids. For instance, the muscles require large quantities of *adenine* and *creatine* to cause muscular contraction, the blood cells require large amounts of *heme* to form hemoglobin, and the kidneys require large quantities of *glutamine* to be used in forming ammonia. All of these substances are synthesized from amino acids.

Also, many of the hormones secreted by the endocrine glands are synthesized from amino acids. These include *norepinephrine, epinephrine,* and *thyroxine,* which are synthesized from *tyrosine; histamine,* which is synthesized from histidine; and several *pituitary hormones, parathyroid hormone,* and *insulin,* which are all small proteins.

Derivation of Energy from Amino Acids

In addition to the use of amino acids for synthesizing new proteins or other chemical substances, some of them are also used for energy, as shown in Figure 42–16. The first step in using proteins for energy is to remove the amino radical. This process is called *deamination,* and it *occurs in the liver.* In the process, the removed amino radical is converted into *ammonia,* which

in turn combines with carbon dioxide to form *urea,* all of these reactions also occurring in the liver. The urea is excreted by the kidneys into the urine. Thus, once again the liver is extremely important for one of the metabolic processes.

Referring back to the formulas of the amino acids, it will be evident that removal of the amino radical from certain of these acids still leaves very complicated chemical compounds. A few of these cannot then be utilized by the body because of their nature, and, therefore, are mainly excreted by the liver into the gastrointestinal tract in the bile and lost in the feces. But most of the deaminated amino acids have sufficiently simple chemical structures so that they can enter into the same cellular reactions as glucose and keto acids. These are often directly oxidized to form water and carbon dioxide, liberating *energy* in the process; or, if energy is not needed at the moment, they can be *converted into fat or carbohydrate* and later utilized for energy in the form of keto acids or glucose.

Conversion of Proteins to Fats and Carbohydrates. Ordinarily the body's cells must synthesize about 45 g of new proteins each day to replace the proteins being destroyed by the natural processes of wear and tear. If extra quantities of amino acids above the amount needed for this purpose are eaten, these normally are deaminated and converted to fats or carbohydrates, or are used for energy.

Protein-Sparing Effect of Carbohydrates and Fats During Starvation. When an insufficient quantity of food is eaten, the major portion of the energy needed for the chemical processes of the cells is derived from carbohydrates and fats as long as these are available, and the proteins are spared. This is called the *protein-sparing* effect of these substances. However, when the stores of carbohydrates and fats are finally depleted, amino acids then begin to be mobilized and deaminated to be used for energy. One can live for another few days on this energy derived from the proteins, but this final process rapidly depletes the cells of their functional elements and soon leads to death.

ABSORPTION OF IONS AND WATER

Absorption of Ions. Ions are absorbed from the gastrointestinal tract in almost exactly the same manner as from the tubules in the kidneys as described in Chapter 26. Sodium, for instance, is *actively absorbed;* that is, it combines with a carrier in the epithelial cells and is trans-

FIGURE 42–16 Complete schema for utilization of proteins in the body.

ported through the intestinal membrane in this form to be released on the opposite side into the blood.

Though less definitive experiments are available for absorption of other electrolytes from the GI tract, it is known that potassium, calcium, magnesium, chloride, phosphates, and iron are all also actively absorbed in a similar manner.

Absorption of Water. Water absorption from the GI tract is controlled almost entirely by osmotic forces that operate as follows: When the monosaccharides, amino acids, and ions are absorbed from the small intestine by active absorption, the osmotic pressure of the intestinal fluids becomes very slight because of loss of the solutes. On the other hand, the osmotic pressure of the interstitial fluid on the opposite side of the epithelial membrane becomes increased. As a result, an osmotic pressure gradient develops across the intestinal membrane, which causes water to be absorbed by osmosis from the intestinal lumen into the extracellular fluids. It is in this way that 8 or more liters of gastrointestinal fluid are normally absorbed from the GI tract each day.

QUESTIONS

1. Explain the role that hydrolysis plays in the digestion of carbohydrates, fats, and proteins.
2. Give the mechanisms for absorption of monosaccharides and proteins.
3. How does the absorption of lipids differ from the absorption of monosaccharides?
4. Explain why glucose is the most important of all the monosaccharides in bodily metabolism.
5. How is blood glucose concentration controlled? What is the role of insulin in this control?
6. Explain the role of bile salts in fat digestion and absorption.
7. What is the relative importance of free fatty acids, chylomicrons, and lipoproteins for transport of fats in the body fluids?
8. What functions do phospholipids and cholesterol subserve in the body?
9. Explain how amino acids can be converted into glucose or fatty acids or can be used for energy.
10. Explain the absorption of ions and of water.

REFERENCES

Bender, D.A.: Amino Acid Metabolism. New York, John Wiley & Sons, 1978.

Butler, T.M., and Davies, R.E.: High-energy phosphates in smooth muscle, *In* Bohr, D.F., *et al.* (eds.): Handbook of Physiology. Sec. 2, Vol. 2. Baltimore, Williams & Wilkins, 1980, p. 237.

Coleman, J.E.: Metabolic interrelationships between carbohydrates, lipids and proteins. *In* Bondy, P.K., and Rosenberg, L.E. (eds.): Metabolic Control and Disease, 8th Ed. Philadelphia, W.B. Saunders, 1980, p. 161.

Davenport, H.W.: A Digest of Digestion, 2nd Ed. Chicago, Year Book Medical Publishers, 1978.

Esmann, V. (ed.): Regulatory Mechanisms of Carbohydrate Metabolism. New York, Pergamon Press, 1978.

Friedmann, H.C. (ed.): Enzymes. Stroudsburg, Pa., Dowden, Hutchinson & Ross, 1980.

Frizzell, R.A., and Schultz, S.G.: Models of electrolyte absorption and secretion by gastrointestinal epithelia. *In* Crane, R.K. (ed.): International Review of Physiology: Gastrointestinal Physiology III. Vol. 19. Baltimore, University Park Press, 1979, p. 205.

Goldfarb, S.: Regulation of hepatic cholesterogenesis. *In* Javitt, N.B. (ed.): International Review of Physiology: Liver and Biliary Tract Physiology I. Vol. 21. Baltimore, University Park Press, 1980, p. 317.

Gross E., and Meienhofer, J. (eds.): The Peptides. New York, Academic Press, 1979.

Halsted, C.H.: Intestinal absorption and malabsorption of folates. *Annu. Rev. Med.*, 31:79, 1980.

Havel, R.J., et al.: Lipoproteins and lipid transport. *In* Bondy, P.K., and Rosenberg, L.E. (eds.): Metabolic Control and Disease, 8th Ed. Philadelphia, W.B. Saunders, 1980, p. 393.

Levy, R.I. (ed.): Nutrition, Lipids, and Coronary Heart Disease. New York, Raven Press, 1979.

Lund-Andersen, H.: Transport of glucose from blood to brain. *Physiol. Rev.*, 59:305, 1979.

Matthews, D.M.: Intestinal absorption of peptides. *Physiol. Rev.*, 55:537, 1975.

McCarty, R.E.: How cells make ATP. *Sci. Am.*, 238(3):104, 1978.

Miller, G.J.: High density lipoproteins and atherosclerosis. *Annu. Rev. Med.*, 31:97, 1980.

Rankow, R.M., and Polayes, I.M.: Diseases of the Salivary Glands. Philadelphia, W.B. Saunders, 1976.

Robinson, A.M., and Williamson, D.H.: Physiological roles of ketone bodies as substrates and signals in mammalian tissues. *Physiol. Rev.*, 60:143, 1980.

Rosenberg, L.E., and Scriver, C.R.: Disorders of amino acid metabolism. *In* Bondy, P.K., and Rosenberg, L.E. (eds.): Metabolic Control and Disease, 8th Ed. Philadelphia, W.B. Saunders, 1980, p. 583.

Ross, R., and Kariya, B.: Morphogenesis of vascular smooth muscle in atherosclerosis and cell structure. *In* Bohr, D.R., *et al.* (eds.): Handbook of Physiology. Sec. 2, Vol. 2. Baltimore, Williams & Wilkins, 1980, p. 69.

Rothschild, M.A.: Albumin synthesis. *In* Javitt, N.B. (ed.): International Review of Physiology: Liver and Biliary Tract Physiology I. Vol. 21. Baltimore, University Park Press, 1980, p. 249.

Silk, D.B.A., and Dawson, A.M.: Intestinal absorption of carbohydrate and protein in man. *In* Crane, R.K. (ed.): International Review of Physiology: Gastrointestinal Physiology III. Vol. 19. Baltimore, University Park Press, 1979.

Watson, D.W., and Sodeman, W.A., Jr.: The small intestine. *In* Sodeman, W.A., Jr., and Sodeman, T.M. (eds.): Pathologic Physiology: Mechanisms of Disease, 6th Ed. Philadelphia, W.B. Saunders, 1979, p. 824.

43

Energetics of Foods and Nutrition

Overview

The intracellular substance used to energize almost all cellular functions is *adenosine triphosphate* (ATP). Two of the phosphate radicals on the adenosine triphosphate molecule are combined with the remainder of the molecule by so-called *high energy bonds,* each of which contains about 8000 calories of energy per mole of adenosine triphosphate. This amount of energy is sufficient to energize almost any chemical reaction necessary for cell function. The chemical reaction for release of the energy from the ATP is

$$\text{Adenosine triphosphate} \longrightarrow \text{Adenosine diphosphate}$$
$$+ \; PO_4 — + \; 8000 \; \text{calories}$$

In turn, the energy used to synthesize adenosine triphosphate in the cells is derived from the *glucose, fatty acids,* and *amino acids* in the food. The cells first split the glucose, fatty acids, and most of the amino acids into *acetic acid.* This then enters the mitochondria where it in turn is split into *carbon dioxide* and *hydrogen atoms.* The hydrogen atoms then combine with oxygen to form water, but release at the same time tremendous amounts of energy that cause phosphate ions to bind with adenosine diphosphate, thus creating new ATP.

About *40 percent* of the energy derived from the food in the average American diet is from *carbohydrate, 45 percent* from *fat,* and *15 percent* from *protein.* The amount of energy in the different foods is expressed in terms of *Calories* (with a capital "C"), which actually means *kilocalories.* One gram of carbohydrates supplies 4.1 Calories of energy; 1 g of fat, 9.3 Calories; and 1 g of protein, 4.1 Calories. Therefore, on a per gram basis, fat supplies more than two times as many Calories as does either carbohydrate or protein. The normal energy requirements of an average person for quiet daily living is about 1800 Calories, whereas a person performing very heavy work may require as many as 6000 to 8000 Calories per day.

Some of the special nutrients required by the body include

1. *Vitamin A,* which is especially important to maintain the health of the different epithelial structures of the body.
2. *Thiamine,* which is necessary to form the enzyme *decarboxylase,* which removes carbon dioxide from various foodstuffs during cellular metabolic processes; lack of this vitamin will cause the disease *beriberi.*
3. *Niacin* and *riboflavin,* which help to promote the oxidative metabo-

lism processes required for formation of adenosine triphosphate; in the absence of these vitamins, the disease *pellagra* occurs.

4. *Vitamin B₁₂* and *folic acid,* which are required by the bone marrow for formation of red blood cells; lack of these will cause the disease called *pernicious anemia.*

5. *Vitamin C,* which is necessary for the formation of healthy connective tissue throughout the body, and in the absence of which the disease *scurvy* occurs.

6. *Vitamin D,* which is necessary for absorption of calcium from the intestinal tract, and in the absence of which the disease *rickets* occurs.

7. *Vitamin K,* which is required for the formation of several of the blood clotting factors, and in the absence of which a person can bleed severely.

The ultimate function of all the digestive and metabolic processes of the body is to provide nutrients for the body, and by far the major portion of these supplies the energy for performing the various bodily functions. Energy is required to lift an arm, to move a leg, or to do any activity employing muscular contraction. It is needed for the secretion of digestive juices, the development of membrane potentials in nerves and other cells, the synthesis of new chemical compounds, and active absorption of substances from the gastrointestinal tract or kidney tubules. In short, almost all functions performed by the body require energy that in turn must be supplied by the ingested food. The final steps for release of energy from the foods—that is, the end stages of metabolism—are described in this chapter.

ADENOSINE TRIPHOSPHATE: THE COMMON PATHWAY OF ALMOST ALL ENERGY

The cells do not use the actual foods for their immediate supply of energy. Instead, they use almost entirely a chemical compound called *adenosine triphosphate* (ATP) for this energy. The foods, in turn, are then used to synthesize more ATP. The importance of this compound to the

function of the cell was pointed out in Chapter 3. The present chapter explains the role of ATP in the overall utilization of energy by the body.

The formula for ATP is shown in Figure 43–1. Extremely large amounts of energy are stored in this molecule at the *bonds* where the last two phosphate radicals join with the remainder of the molecule. These bonds (~) are called *high energy phosphate bonds.* Every time a cell needs energy, a phosphate radical is broken away from ATP at the high energy bond, and this liberates the needed energy. Each mole of ATP releases 8000 calories of energy for each of the high energy bonds that is broken.

In short, there is a storehouse of ATP in each cell that provides the necessary energy for muscular contraction, for development of membrane potentials, for active absorption, and so forth, but this adenosine triphosphate must be replenished continually.

Formation of Adenosine Triphosphate

Use of Energy from Carbohydrates to Form Adenosine Triphosphate. In the preceding chapter it was noted that carbohydrates are digested to form glucose, or are changed into glucose after absorption. Then the glucose is

FIGURE 43–1 Adenosine triphosphate (ATP).

used by the cells for energy. Part of the energy is released from glucose by a process called *glycolysis* that does not require oxygen, but by far the major amount of energy is released when the glucose is **oxidized,** which is called *oxidative metabolism.*

Glycolysis. In glycolysis, the glucose molecule, which has six carbon atoms, is split by a series of cellular enzymes into two smaller molecules having only three carbon atoms. Then the three-carbon molecules are further modified to form another three-carbon molecule, *pyruvic acid,* which has the following formula:

$$CH_3 - \overset{\displaystyle \overset{O}{\|}}{C} - COOH$$

During glycolysis a small amount of energy is released from the glucose molecule, and this energy is used to form adenosine triphosphate. By splitting glucose to form two pyruvic acid molecules, energy is liberated without the expenditure of any oxygen. This is called energy liberation by *anaerobic metabolism,* and it is illustrated by the first stage of the reaction in Figure 43–2.

Oxidative Release of Energy from Carbohydrates. After the glucose has been split into pyruvic acid molecules, these are then metabolized with oxygen to form carbon dioxide and water. This reaction is shown by the second stage in Figure 43–2. The oxidative metabolism of pyruvic acid provides about 18 times as much energy as the glycolytic breakdown of glucose to form pyruvic acid. Therefore, by far the major amount of energy liberated from carbohydrates for the performance of cellular function is derived from *oxidative metabolism.*

The chemical reactions by which pyruvic acid is oxidized to supply energy have been worked out in great detail, the general principles of which are shown in Figure 43–3. The reactions of the first stage, called the *citric acid cycle* or *Krebs cycle,* split the pyruvic acid molecule into carbon dioxide and hydrogen; the carbon dioxide is removed by enzymes called *decarbox-*

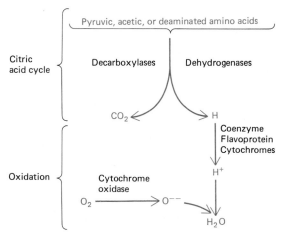

FIGURE 43–3 Splitting of pyruvic acid, acetic acid, or deaminated amino acids into carbon dioxide and hydrogen in the tricarboxylic acid cycle, and oxidation of the released hydrogen atoms by the cellular oxidative enzymes.

ylases, and hydrogen atoms are removed by *dehydrogenases.* In the second stage, called *oxidation,* the hydrogen reacts with oxygen to form water.

When hydrogen atoms are split away from pyruvic acid by the dehydrogenases, they immediately combine with a substance called *coenzyme.* Then, under the influence of other enzymes, hydrogen atoms are passed to *flavoprotein molecules* and finally to *cytochrome molecules.* During this process, the hydrogen atoms are released into the surrounding fluid as *hydrogen ions.* Simultaneously, dissolved oxygen that has been carried to the tissues by hemoglobin is changed into *oxygen ions* by *cytochrome oxidase.* The presence of ionic hydrogen and ionic oxygen in the same solution provides two highly reactive substances that immediately form water molecules. Thus, the hydrogen atoms removed from the pyruvic acid become oxidized with oxygen to form water.

Formation of Adenosine Triphosphate During Glucose Metabolism. One might ask why it is necessary for the hydrogen and oxygen to go through the complicated stages of the above reactions, for it is well known that hydrogen and oxygen can combine with each other very rapidly simply by being burned together in a fire. The answer to this question is that the indirect procedure is required to channel the released energy in the proper direction to form new ATP.

The quantity of ATP formed in the different stages of glucose metabolism is the following: For each molecule of glucose metabolized, 2 molecules of ATP are formed during glycolysis, 2

FIGURE 43–2 Derivation of energy from glucose by glycolysis and by oxidation.

are formed in the citric acid cycle, and 34 are formed during oxidation of hydrogen, making a total of 38 molecules of ATP for each molecule of glucose metabolized.

The total amount of energy in each mole of glucose is 686,000 calories. Of this amount, 266,000 become stored in the form of ATP. The remainder is lost as *heat* caused by the chemical reactions. Thus, the overall *efficiency* of energy transfer from glucose to ATP is 39 percent, the remaining 61 percent of the energy becoming heat, which represents wasted energy.

Use of Energy from Fats and Proteins to Form Adenosine Triphosphate. It was pointed out in the preceding chapter that fatty acids are split into acetic acid, and that amino acids derived from proteins are deaminated to form deaminated amino acids. The same decarboxylases and dehydrogenases that remove carbon dioxide and hydrogen from pyruvic acid do the same for the acetic acid and most of the deaminated amino acids; the hydrogen atoms are then oxidized as explained above for the carbohydrates. Large amounts of energy are released, especially during oxidation of the hydrogen atoms, to synthesize ATP.

The quantity of energy derived in this manner from fats and proteins represents about 80 percent of the total energy derived by the cells in contrast to only 20 percent from oxidation of glucose. The reason for this is that half or more of the carbohydrate eaten by a person is first stored in the body as fat and then later used for energy in the form of fatty acids.

Regulation of Rate of Food Metabolism by the Presence of Adenosine Diphosphate (ADP). The reactions that result in the formation of ATP cannot occur unless *adenosine diphosphate* is available from which the ATP can be formed. Therefore, the rate of breakdown of foodstuffs is controlled to a great extent by the presence or absence of ADP.

Every time ATP is used by the cells for energy it loses one phosphate radical and becomes ADP. The newly formed ADP immediately initiates reactions with the foodstuffs to cause release of new energy that is used to convert the ADP back into ATP. These interrelationships are shown in Figure 43–4. Then, when all the ADP has been resynthesized into ATP, the metabolism of the food ceases. It is in this way that the amount of food used for energy is automatically geared to the needs of the body.

Interaction of Adenosine Triphosphate with Creatine Phosphate

Another substance that contains high energy phosphate bonds, *creatine phosphate*, is present in the cells in quantities several times as great as those of ATP. When adequate amounts of ATP are available, much of it is used to form creatine phosphate in accordance with the reactions shown in Figure 43–4, and, as rapidly as the ATP is used for this purpose, still more ATP is formed. This results in the buildup of large quantities of creatine phosphate. Then when the cell demands energy rapidly and in large amounts, energy is released from ATP directly to the functional elements of the cells. The energy stored in the creatine phosphate is used immediately to reform new ATP. This reaction occurs in a fraction of a second, much more rapidly than the oxidative reactions, and it provides a very rapid source of extra energy that can be used to keep the cell functioning at an extremely high rate of

FIGURE 43–4 Overall schema of energy transfer from foods to the adenylic acid system and then to the functional elements of the cells. (Modified from Soskin and Levine: Carbohydrate Metabolism. University of Chicago Press.)

metabolism for a short period of time, even though the oxidative methods for reconstituting ATP are much slower to respond.

An Overall Schema for the Energy Mechanisms of the Cell. Figure 43–4 also illustrates an overall schema for the chemical reactions that provide energy for cellular functions. This schema shows that the breakdown of ATP to ADP releases the needed energy for muscular contraction, glandular secretion, neuronal activity, intermediary metabolism, and other energy functions of the cell. Within a few seconds, creatine phosphate also breaks down, providing energy for resynthesis of much of the ATP. During the ensuing minutes, the ATP and creatine phosphate are both resynthesized by energy from the foodstuffs, part of it by glycolysis and other anaerobic procedures, but 90 percent or more by the oxidation of pyruvic acid, acetic acid, deaminated amino acids, and a few other substances such as alcohol, glycerol, and lactic acid.

The Oxygen Debt. For a few seconds at a time a person can perform very strenuous feats of exercise requiring energy release many times that which can be sustained over a long period of time. This immediate burst of energy is provided to a great extent by the ATP and creatine phosphate stored in the cells. During the next few minutes, while the ATP and creatine phosphate are being resynthesized by the oxidative processes of metabolism, an extra quantity of oxygen above the usual amount must be utilized. Also, some of the stores of oxygen in the hemoglobin of the blood, in the myoglobin of muscle, and in the dissolved state in the body fluids will have been used during the rapid burst of energy release, and this, too, must be replenished after the exercise is over. The extra oxygen that must be used to restore completely normal conditions after exercise is over is called the *oxygen debt*.

In essence, our ability to develop an oxygen debt explains why an athlete continues to breathe very hard for a minute or so after running a race, but even more important, it explains why we can perform feats involving tremendous amounts of activity for a few seconds, even though this activity cannot be sustained for long.

NUTRITION

The term *nutrition* means the supplying of foods that are required to keep one alive and healthy. These foods include carbohydrates and fats, which supply most of the body's energy, and proteins, vitamins, and minerals, which are required for synthesis of special structures and special chemical compounds needed by the body.

Foods Supplying Energy

People in various parts of the world have widely differing diets, and some, in comparison with others, even obtain different proportions of their energy from different types of foodstuffs. Approximately 40 percent of the energy in the food of the average American diet is in the form of carbohydrate, 45 percent fat, and 15 percent protein. In other less prosperous parts of the world, the abundance of fat and protein in the diets is often less than half these values, while the energy derived from carbohydrate sometimes rises to as much as 80 percent of the total.

The Calorie As a Measure of Energy. The energy of a food is measured in terms of the amount of heat liberated by complete breakdown of the food into its metabolic end-products, and this is expressed in *Calories*, a unit for measuring heat. One Calorie (spelled with a capital "C") is the amount of heat required to raise the temperature of 1 kg of water 1 degree Celsius.

The Calorie is a very good measure of food energy, for most of the energy released in the body eventually becomes heat anyway. For example, the chemical reactions for extracting energy from the foods are so inefficient that 61 percent of the energy becomes heat as ATP is formed, and by the time the ATP is used to perform cellular functions, 75 to 90 percent of its energy also becomes heat. The remainder is converted into muscular action or other functional activities of the body. Then as these functions are performed, almost all of this remaining energy finally becomes heat too. As an example, a large amount of energy is used to pump blood around the circulatory system. As the blood flows through the blood vessels, the energy that has been imparted to the blood is converted into heat because of friction between the blood and the walls of the vessels. In this way, all the energy expended by the heart eventually becomes heat. Likewise, almost all the energy expended by the skeletal muscles of the body eventually becomes heat, because most of it is used to overcome the friction of the joints and the viscosity of the tissues, and these two effects in turn convert the energy into heat.

Energy Content of the Different Types of Foods. The amount of energy released to the body by oxidation of 1 g of each of the three major types of foods is the following:

	Calories
Carbohydrate	4.1
Fat	9.3
Protein	4.1

From these figures it is evident that 1 g of fat supplies more than twice as much energy as 1 g of either carbohydrate or protein. For this reason, fats in the diet are very deceptive. Often one thinks that he is deriving very little of his energy from fat and yet may be obtaining as much from fat as from carbohydrate. A second reason why fat is deceptive is that it often occurs in a pure form in foods, whereas carbohydrates and proteins are generally diluted several times over with water. When one eats potatoes with butter, the fat of the butter generally contains almost as much energy as the entire potato, for two reasons: First, the butter supplies two and one-quarter times as much energy per gram as the starch of the potato, and, second, the starch comprises only about one-sixth the bulk of the potato, because most of the potato is water.

Daily Energy Requirements

An average man of 70 kg who lies in bed all day long and does nothing else except eat and exist usually requires about 1650 Calories of energy each day. If he simply sits in a chair, another 200 or more Calories are required. Therefore, about 1800 Calories per day is the normal basal amount of energy required simply for living. In addition, any type of exercise requires still more energy, which is shown by the values in Table 43–1 for different types of activity. From this table it is evident that walking upstairs requires approximately 17 times as much energy as lying in bed asleep. However, this tremendous rate of energy utilization can be continued only for short periods of time. Over long periods, a well-conditioned worker can average as much as 6000 to 8000 Calories of energy expenditure each 24 hours, or, in other words, as much as four times the basal rate.

Protein Requirements of the Body

However low the concentration of amino acids might fall to in the blood, the liver still deaminates some of them all the time, and these then are used mainly for energy. Because of this continual loss of amino acids, the proteins of the body diminish constantly unless they are replenished by proteins in the diet. Normally, even when more than enough carbohydrates and fats are present in the diet to provide their protein-

TABLE 43–1

*Energy Expenditure Per Hour During Various Types of Activity for a 70-Kg Man**

Form of Activity	Calories per hour
Sleeping	65
Awake lying still	77
Sitting at rest	100
Standing relaxed	105
Dressing and undressing	118
Tailoring	135
Typewriting rapidly	140
"Light exercise"	170
Walking slowly (2.6 miles per hour)	200
Carpentry, metal working, industrial painting	240
"Active exercise"	290
"Severe exercise"	450
Sawing wood	480
Swimming	500
Running (5.3 miles per hour)	570
"Very severe exercise"	600
Walking very fast (5.3 miles per hour)	650
Walking up stairs	1100

*Extracted from data compiled by Professor M.S. Rose.

sparing effect, a minimum of 45 g of protein still is required each day to replace this continual loss.

Partial Proteins. Certain types of proteins do not contain all the essential amino acids in the proportion in which each is needed in the body. When one eats such a protein his requirement will be considerably above 45 g per day. For instance, the proteins from vegetables or grains have ratios of the various amino acids different from those found in the human body. On the other hand, the proteins of animal origin, in general, have almost exactly the same amino acid compositions as those of the human being. Therefore, the person who requires 45 g of animal protein in his diet each day might require as much as 65 g of vegetable protein, the exact amount depending on the type of vegetable eaten. The proteins that cannot supply the right proportions of the different amino acids are called *partial proteins*, because they supply only part of the needed variety of amino acids.

An especially serious example of nutritional deficiency that occurs because of too much partial protein and too little complete protein in the diet is the disease condition called *kwashiorkor*, occurring mostly in young African children. The major diet of these children is a cornmeal mush. The corn contains a moderate amount of protein, but this protein is very deficient in one of the essential amino acids, tryptophan. Therefore, only small amounts of animal proteins can be synthesized from the amino acids derived from this diet because of the missing tryptophan; instead, by far the greater majority of the amino acids are simply degraded and used for energy. As a result, the child becomes very protein-deficient, which causes greatly stunted growth and, in many instances, death. Those children who survive are frequently mentally deficient for the remainder of their lives because their brains have not developed properly.

Special Need for Highly Unsaturated Fat in the Diet

A small amount of highly unsaturated fat is essential for nutrition of animals. The body cannot desaturate fat sufficiently by its own metabolic processes to supply this. The types of unsaturated fatty acids usually required in the diet are *arachidonic, linoleic,* and *linolenic* acids. These are believed to be needed by the cells to form specific cellular chemicals, among which are the prostaglandins that are used by cells to control internal cell function. Without unsaturated fatty acids, animals develop skin sores, mental changes, and other evidence of general cellular debility. Whether or not these same effects would occur in the human being is not known, because such a small amount of these substances is required in the diet that no human being has been proved to be suffering from a deficiency of them.

Vitamin Requirements of the Body

The vitamins are chemical compounds needed in only minute quantities by the body to perform special functions. The daily requirements of each of the vitamins is given in Table 43–2, and the amounts of the more common vitamins in the different foods (except for vitamin D, which is present in quantity only in liver and irradiated milk) are given in Table 43–3. In general, eating a balanced diet will provide an adequate quantity of all the different vitamins. Occasionally, though, an abnormality makes it impossible to utilize one of the vitamins, in which case a vitamin deficiency disease can occur even in the presence of a normally satisfactory diet.

The precise chemical functions of many of the vitamins in the body are only partially known, but from the physiologic effects caused by lack of them in the diet, their functions can at least be speculated upon as follows.

TABLE 43–2
Daily Requirements of the Vitamins

A	5000 IU*
Thiamine	1.5 mg
Riboflavin	1.8 mg
Niacin	20 mg
Ascorbic acid	45 mg
D	400 IU
E	15 IU
K	none
Folic acid	0.4 mg
B_{12}	3 μg
Pyridoxine	2 mg
Pantothenic acid	unknown

*IU, international units.

Vitamin A.   Vitamin A is used by the eyes to synthesize the light-sensitive retinal pigments, rhodopsin and the color-sensitive pigments, used by the rods and cones for vision. This was discussed in detail in Chapter 21. Also, lack of vitamin A causes lack of growth of all tissues of the body and also causes the epithelial structures, such as the skin, the intestinal mucosa, and the germinal epithelium of the ovaries and testes, to become highly *keratinized,* or horny. These effects lead to scaliness of the skin, failure of growth of young animals, failure of reproduction, and even hardening of the cornea, occasionally causing corneal opacity and blindness.

Thiamine—Beriberi.   Thiamine forms a compound in the cells called *thiamine pyrophosphate,* which is part of a *decarboxylase* that removes carbon dioxide from pyruvic acid and other substances. Without thiamine the oxidative processes for metabolizing especially carbohydrates become deficient, and this can cause almost any type of abnormality in the body. It especially affects the functions of the nervous system, the heart, and the gastrointestinal system.

Lack of thiamine causes pathologic changes in neurons and in the myelin sheaths of nerve fibers, often resulting in actual destruction of the cells or irritation or degeneration of peripheral nerves. Frequently the fibers of peripheral nerves become so irritable that a condition called *polyneuritis* results, which manifests itself by excruciating pain along the course of the nerves.

In the heart, thiamine deficiency decreases the strength of the muscle. The heart becomes greatly dilated and pumps with little force, resulting in *congestive heart failure.*

In the GI tract, thiamine deficiency causes weakness of the intestinal muscle, poor secre-

TABLE 43–3
*Vitamin Content of Various Foods**

Vitamin A units/100 g	Thiamine mg/100 g	Riboflavin mg/100 g	Niacin mg/100 g	Vitamin C mg/100 g
Apricots 5000	Barley 0.450	Dried beans 0.30	Asparagus 1.2	Asparagus 45
Broccoli 4000	Dried beans 0.540	Almonds 0.50	Barley 4.7	Brussel sprouts 95
Butter 3500	Buckwheat 0.450	Beef 0.20	Soybeans 4.0	Cabbage 65
Carrots 6000	Dried cowpeas 0.900	Cheese 0.55	Dried beans 3.0	Butter 70
Collards 8000	Egg yolk 0.320	Chicken 0.30	Beef 5.4	Cauliflower 82
Chard 9000	Whole wheat 0.585	Collards 0.25	Chicken 8.5	Lemon juice 53
Endive 15,000	Pork meat 1.0	Eggs 0.40	Collards 2.3	Orange juice 45
Kale 20,000	Lamb meat 0.330	Lamb 0.24	Corn 1.3	Fish 20
Liver 70,000	Beef meat 0.120	Liver 2.60	Lamb 8.0	Potatoes 10
Mustard	Liver 0.400	Milk 0.18	Liver 17.0	Lean meats 5
greens 10,000	Millet 0.700	Peanuts 0.45	Mackerel 2.1	Liver 25
Sweet potatoes 3000	Brown rice 0.370	Pork 0.24	Pork 7.0	
Pumpkin 7000	Wheat germ 2.0	Grains 0.10	Salmon 6.2	
Spinach 25,000			Wheat 4.3	

	Pantothenic acid mg/100 g	Pyridoxine mg/100 g	Folic acid mg/100 g
Meats	0.8	0.1	0.15
Eggs	1.4	0.02	0.90
Milk	0.3	0.13	0.005
Cereals	0.4	0.05	0.10
Fruits	0.2	0.05	0.05
Vegetables	0.5	0.15	0.08

*Modified from W.H. Eddy and G. Dalldorf.

tion of digestive juices, and poor maintenance of the intestinal mucosa. Severe indigestion, constipation, lack of appetite, and other symptoms very often develop.

A thiamine-deficient person who has neuritis, enlargement of the heart, and gastrointestinal symptoms all at the same time is said to have *beriberi.*

Niacin and Riboflavin—Pellagra. Niacin and riboflavin are utilized in the body to form respectively *dehydrogenase coenzyme* and *flavoprotein.* Earlier in the chapter it was pointed out that these two substances perform necessary functions in the oxidation of hydrogen atoms after they are removed from pyruvic acid and other substrates. Without niacin and riboflavin, therefore, the oxidation of foods becomes deficient, and the cells fail to receive adequate quantities of energy.

Niacin deficiency leads especially to discoloration of the skin, which becomes darkened on exposure to sunlight. Along with this, severe muscular weakness, diarrhea, and one or more psychoses often occur. This is the clinical picture of the condition called *pellagra.*

Deficiency of riboflavin is not as likely to lead to such severe difficulties as deficiency of

niacin, because other types of flavoproteins that can perform almost the same functions as the protein derived from riboflavin are present in the body. Nevertheless, riboflavin deficiency, when added to deficiencies of niacin or thiamine, can greatly intensify the symptoms observed in pellagra or beriberi. A common result of riboflavin deficiency alone is cracking at the angles of the mouth, called *cheilosis.*

Vitamin B_{12} and Folic Acid—Pernicious Anemia. These two substances are needed by the bone marrow to form red blood cells and are also needed in all the other tissues of the body for adequate growth. When they are lacking, the red cells released into the blood are few in number, and those that are released are usually larger than normal, poorly formed, and very fragile. Therefore, the person develops very severe anemia, called *pernicious anemia.*

On studying the bone marrow, one finds that the new red cells being formed have abnormal structures, for which reason it is clear that vitamin B_{12} and folic acid are necessary for the formation of the structural elements of the cells. However, lack of these vitamins does not affect the formation of hemoglobin.

Red cells are affected more than other cells

of the body by lack of these two substances probably because these cells are produced much more rapidly than most other cells. This is supported by the fact that cancer cells, which are also rapidly reproduced, fail to reproduce in the absence of these vitamins. In fact, purposeful deficiency of these two vitamins can actually be used as a means for slowing the growth of most cancers.

Less Important Vitamin B Compounds. For approximately the first 20 years after beriberi was discovered to be a nutritional disease, it was believed that the vitamin extract used to treat this disease contained only one single vitamin, and this was called *vitamin B.* Later, many different vitamins were discovered in the extract, and these have come to be known as the *vitamin B complex.* The vitamins included in this complex are thiamine (vitamin B_1), riboflavin (vitamin B_2), niacin, vitamin B_{12}, folic acid, *pyridoxine, pantothenic acid, biotin, inositol, choline,* and *para-aminobenzoic acid.* The functions of some of the less well-known members of the vitamin B complex are not well known, but some of the functions of pyridoxine and pantothenic acid, two of the more important of them, are the following:

Pyridoxine. Pyridoxine is needed for synthesis of some of the amino acids and perhaps other compounds. Dietary lack of this vitamin can cause *dermatitis, decreased rate of growth, anemia,* different types of *mental symptoms,* and gastrointestinal disturbances. However, pyridoxine deficiency is rare in the human being and almost never occurs without simultaneous deficiency of other vitamins of the B complex.

Pantothenic Acid. Pantothenic acid is used in the body to form a special chemical called *coenzyme A,* which catalyzes the acetylation of many substances in the cells. For example, acetylation steps are required for the oxidation of almost all foodstuffs, including pyruvic acid from glucose, acetic acid from fats, and deaminated amino acids from protein. Therefore, little energy metabolism can occur in the body in the absence of pantothenic acid.

Pantothenic acid deficiency has never been known to develop in the human being, because this substance is widespread in almost all foods of the diet. In animals, deficiency can be created artificially, and the results are retarded growth, failure of reproduction, graying of the hair, dermatitis, fatty liver, and many other effects. All these effects testify to the importance of pantothenic acid to the metabolism of the body.

Ascorbic Acid (Vitamin C)—Scurvy. The major function of ascorbic acid is to maintain normal intercellular substances throughout the body. These include the connective tissue fibers that hold the cells together, the intercellular cement substance between the cells, the matrix of bone, the dentin of the teeth, and other substances excreted by the cells into the intercellular spaces.

Deficiency of ascorbic acid in the diet causes failure of wounds to heal because new fibers and new cement substance are not deposited. Also, it causes bone growth to cease and blood vessels to become so fragile that they bleed on the slightest provocation. In short, lack of vitamin C leads to loss of integrity of many of the tissues of the body, and the resulting picture, characterized especially by bleeding of the gums, splotchy hemorrhages beneath the skin, and many internal abnormalities, is the disease called *scurvy.*

Vitamin D—Rickets. Vitamin D promotes calcium absorption from the gastrointestinal tract. Without adequate quantities of vitamin D the bones become decalcified, a process called *rickets,* and in very severe vitamin D deficiency the level of ionic calcium in the blood may fall so low that muscular tetany develops. All these effects of vitamin D deficiency will be discussed in relation to calcium metabolism and parathyroid hormone in Chapter 47.

Vitamin E. In lower animals, lack of vitamin E can cause the male germinal epithelium in the testes to degenerate, causing sterility; in the female it can cause reabsorption of a fetus even after conception. For these reasons vitamin E is sometimes called the *antisterility* vitamin. Severe lack of the vitamin in animals can also cause muscle degeneration and paralysis.

Vitamin E is believed to function by preventing oxidation of unsaturated fatty acids in the cells. In the absence of vitamin E, the lack of adequate quantities of these unsaturated fatty acids causes abnormal structure and function of such cellular organelles as the mitochondria, the lysosomes, and even the cell membrane.

Vitamin K. Vitamin K is required for the formation of prothrombin, and several other blood clotting factors, by the liver. Therefore, deficiency of vitamin K depresses blood coagulation so that there is excessive bleeding. This subject was discussed in greater detail in Chapter 35. Vitamin K usually is not in the diet in large quantities. Yet the normal person does not experience vitamin K deficiency because a large amount of this substance is synthesized in the colon by bacteria and is then absorbed. If the bacteria of the colon are destroyed by administration of antibiotic drugs, vitamin K deficiency usually develops with the next few days.

TABLE 43–4
Content in Grams of a 70-Kg Adult Man

TABLE 43–4
Content in Grams of a 70-Kg Adult Man

Water	41,400	Mg	21
Fat	12,600	Cl	85
Protein	12,600	P	670
Carbohydrate	300	S	112
Na	63	Fe	3
K	150	I	0.014
Ca	1,160		

Mineral Requirements of the Body

Table 43–4 gives the amounts of different minerals and some other substances in the entire body, and Table 43–5 gives the daily requirements of minerals. Some of these minerals have already been discussed in relation to other phases of the physiology of man. For instance, sodium, chloride, and calcium are major constituents of the extracellular fluid, and potassium, phosphate, and magnesium are major constituents of the intracellular fluid. These minerals are responsible for development of electrical potentials at the cell membrane and for the maintenance of proper osmotic equilibria between the extracellular and intracellular fluids. In addition, calcium and phosphate are major constituents of bone, and phosphate forms a great number of different chemicals used for a myriad of functions inside all cells, some of which were discussed earlier in this chapter. The remaining minerals that need special comment are iron, iodine, cobalt, copper, zinc, and fluorine.

Iron. About two thirds of the iron of the body is in the hemoglobin in the blood; most of the remainder is stored in the liver in the form of *ferritin*. The iron in ferritin can be mobilized when needed and carried in the blood to the bone marrow, where it is used to form hemoglobin.

Iron is also present in some of the enzymes of the cells—especially in the cytochromes. Therefore, a second function of iron is to aid in the oxidation of food in the tissue cells.

Iodine. Iodine is used by the thyroid gland in the synthesis of thyroxine, a hormone that increases the rate of metabolism of the body. The functions of thyroxine and its relation to iodine metabolism will be discussed in Chapter 45.

Copper and Cobalt. Copper and cobalt both affect the formation of red blood cells. Copper, in some way yet unknown, helps to catalyze the formation of hemoglobin. Very rarely does copper deficiency exist, however, so that this is almost never a cause of hemoglobin deficiency. Cobalt is an essential element in vitamin B_{12}, and in this way is essential for maturation of red blood cells. When cobalt in other forms besides vitamin B_{12} is in the diet in excess, the bone marrow, for reasons not yet understood, produces too many red blood cells, causing polycythemia.

Zinc. Zinc forms part of the structure of the enzyme *carbonic anhydrase*, which is present in many parts of the body, especially in the red blood cells and in the epithelium of the kidney tubules. This substance catalyzes the reaction of carbon dioxide and water to form carbonic acid and also catalyzes the same reaction in the reverse direction. It causes carbon dioxide to combine with water about 210 times as rapidly as it would otherwise, which allows the blood to transport carbon dioxide far more easily than could be possible in the absence of this enzyme.

In the kidney tubules, carbonic anhydrase catalyzes the reactions that cause the secretion of hydrogen ions into the tubular fluid, thereby helping to regulate the acid-base balance of the body fluids.

In addition to being utilized in carbonic anhydrase, zinc is also a component of many other enzymes, including especially many of the peptidases that are required for digestion of proteins in the intestines.

Fluorine. Fluorine in the diet protects against *carious* teeth. It does not make the teeth any stronger but is believed to inactivate bacterial secretions that can cause tooth decay. Only a small trace of this substance in the drinking water of the growing child usually provides a major degree of protection against tooth decay throughout life, which is the reason why city water supplies are now usually fluorinated.

TABLE 43–5
Daily Requirements of Minerals

Na	1.0 g	I	250.0 μg
K	2.5 g	Mg	unknown
Cl	2.5 g	Co	trace
Ca	1.0 g	Cu	trace
PO₄	1.5 g	Zn	trace
Fe	12.0 g	F	trace

REGULATION OF FOOD INTAKE

Food intake is regulated by the sensations of *hunger* and *appetite*. Hunger means craving for food, and the term appetite is often used in the same sense as hunger except that it usually means a desire for specific food instead of for food in general.

The term *satiety* means the opposite of hunger, a feeling of complete fulfillment in the quest for food. Satiety results from a filling meal.

Neural Centers for Regulation of Food Intake. Stimulation of the *lateral hypothalamus* causes an animal to eat voraciously, whereas stimulation of the *medial nuclei of the hypothalamus* causes complete satiety even in the presence of highly appetizing food. Therefore, we can call the lateral hypothalamus the *hunger center* or the *feeding center*, and we can call the medial hypothalamus a *satiety center.*

In addition to the hypothalamic centers, however, several conscious centers of the cerebral cortex and portions of the amygdaloid nuclei also enter into regulation of food intake—not so much into the regulation of total quantity of intake, but into the specific choice of food, for it is in these areas that memories of previously eaten pleasant or unpleasant foods are stored, and these memories adjust the appetite for different foods accordingly.

Long-Term and Short-Term Regulation of Food Intake. There are two entirely different types of food intake regulation, called respectively "long-term regulation" and "short-term" regulation.

Long-term regulation means regulation of food intake in relation to the amount of nutritive stores in the body. For instance, a person who has been underfed for many weeks has intensified hunger until his normal nutritive stores have been replenished. Conversely, an animal that has been force-fed until it is greatly overweight has almost no hunger, a condition that can last for weeks until its normal body weight has returned.

The precise mechanism by which the nutritive stores affect hunger is not known, though it is very clear that a decreased level of glucose in the body fluids, which usually goes along with decreased stores of other nutrients in the body, increases a person's hunger. Decreased quantities of amino acids and fatty acids in the body fluids seem also to cause the same effect, in this way controlling the degree of hunger.

Short-term regulation means regulation of dietary intake in relation to the amount of food that can be processed by the gastrointestinal system in a given period of time. For instance, if a person overeats, he can so overload his GI tract that he will become sick from this alone. Therefore, during the process of eating, two principal mechanisms prevent such overeating. These are (1) "metering" of food as it passes through the mouth and (2) reflexes caused by distention of the upper GI tract. Metering of food means that

sensory receptors in the mouth and pharynx detect the amount of chewing, salivation, swallowing, and tasting and thereby quantitate the amount of food that passes through the mouth. In some way that is not understood, this information passes to the hypothalamic feeding center to inhibit hunger for up to 30 minutes or an hour, but no longer. Likewise, as food fills the stomach and other regions of the upper gastrointestinal tract, visceral sensory impulses, caused mainly by distention of the gut, are transmitted to the feeding center and inhibit it. In this way, overfilling of the GI tract is prevented until the food that is already there has had time to be digested.

Obesity

Much obesity is caused simply by overeating as a result of poor eating habits. For instance, many people eat three meals a day simply because of habit rather than because of hunger at the time of the meal.

Other instances of obesity result from *inherited* imbalance between the hunger and satiety centers of the hypothalamus. For instance, when an exceedingly obese person forces himself to diet until he reduces to many pounds below his obese weight, he develops voracious hunger and, if left to his own means, will regain weight almost exactly to his original obese level. Once he reaches this level, his hunger becomes essentially the same as that of a normal person. Thereafter, he eats merely enough to maintain his weight rather than to gain still more. This is analogous to setting the thermostat in a house to a high level. In other words, the "hungerstat" is set at a higher level in these persons than in others, causing excessive eating until the person becomes very obese.

Starvation

During starvation, essentially all the stored carbohydrates in the body, which amount to about 300 g of glycogen in the liver and muscles, become used up within the first 12 to 24 hours. Thereafter, the person exists on his stored fats, and finally on stored proteins as well. For the first 2 to 4 weeks, almost all of the energy used by the body is derived from the stored fats. But eventually even these are almost depleted, so that finally the proteins must also be used. Most tissues can give up as much as one half of their proteins before cellular death begins. Therefore, for at least another few days to a week, the body can derive its energy from proteins. But, finally, death ensues because proteins are the necessary chemical elements for performance of cellular functions. This usually occurs 4 to 7 weeks after starvation begins.

QUESTIONS

1. Give reasons why adenosine triphosphate is so important to cellular metabolism.
2. Explain how carbohydrates and fats are utilized in the formation of adenosine triphosphate.
3. Explain the role of creatine phosphate in cellular metabolism.
4. What is the energy equivalent (in Calories per gram) of carbohydrate, fat, and protein?
5. What are the approximate daily energy requirements (in Calories) at rest and during work?
6. What are the daily requirements for proteins and fats?
7. What are the effects of deficiency of each of the different vitamins?
8. Review the different mineral requirements of the body, and give the function of each of the minerals.
9. Explain the mechanisms of and the differences between long-term and short-term regulation of food intake.

REFERENCES

Alfin-Slater, R.B., and Kritchevsky, D. (eds.): Nutrition and the Adult: Macronutrients. New York, Plenum Press, 1979.

Bray, G.A., and York, D.A.: Hypothalamic and genetic obesity in experimental animals: An autonomic and endocrine hypothesis. *Physiol. Rev.,* 59:719, 1979.

DeLuca, H.F. (ed.): The Fat-Soluble Vitamins. New York, Plenum Press, 1978.

Felig, P.: Starvation. *In* DeGroot, L.J., *et al.* (eds.): Endocrinology. Vol. 3. New York, Grune & Stratton, 1979, p. 1927.

Festing, M.F.W. (ed.): Animal Models of Obesity. New York, Oxford University Press, 1979.

Frieden, E.: The chemical elements of life. *Sci. Am.,* 227:52, 1972.

Hodges, R.E.: Nutrition in Medical Practice. Philadelphia, W.B. Saunders, 1979.

Hunt, S.M., *et al.*: Nurition: Principles and Clinical Practice. New York, John Wiley & Sons, 1980.

Keesey, R.E.: Neurophysiologic control of body fatness. *In* Lauer, R.M., and Shekelle, R.B. (eds.): Childhood Prevention of Atherosclerosis and Hypertension. New York, Raven Press, 1980.

Kharasch, N. (ed.): Trace Metals in Health and Disease. New York, Raven Press, 1979.

Salans, L.B.: Obesity and the adipose cell. *In* Bondy, P.K., and Rosenberg, L.E. (eds.): Metabolic Control and Disease, 8th Ed. Philadelphia, W.B. Saunders, 1980, p. 495.

Shoden, R.J., and Griffin, W.S.: Fundamentals of Clinical Nutrition. New York, McGraw-Hill, 1980.

Suitor, C.W., and Hunter, M.F.: Nutrition: Principles and Application in Health Promotion. Philadelphia, J.B. Lippincott, 1980.

Thompson, C.I.: Controls of Eating. New York, Spectrum Publications, 1979.

XIII

BODY TEMPERATURE

44

Body Heat, and Temperature Regulation

Overview

The *metabolic rate* of the body is the rate at which energy is released from metabolism of nutrients in the entire body. Under very quiet conditions, this rate is 60 to 70 Calories per hour, which is called the *basal metabolic rate.* However, it can rise to as much as 20 times this much during heavy exercise, which enhances the metabolic rate more than any other stimulus. Other factors that can increase the metabolic rate to a much less extent include (1) *sympathetic stimulation,* about 2 times; (2) the effect of *thyroid hormone* on cell metabolism, also about 2 times; and (3) *fever,* about 2 times for each 8°C rise in body temperature.

Except under special conditions, all of the energy released from the nutrients in the body eventually becomes *heat.* Even most of the energy that causes muscular activity becomes heat because of friction of the joints, vicious movement of the muscle tissue itself, and so forth. It is this heat that keeps the body warm.

The *body temperature* is very exactly controlled by balancing the *rate of heat loss* against *rate of heat production.* Heat is lost in three ways: (1) by *radiation,* (2) by *conduction* to air and to solid objects, and (3) by *evaporation* of water from the lungs and skin, especially when a person sweats.

Nervous centers in the *hypothalamus* called the *hypothalamic thermostat* control the body temperature by controlling both heat loss and heat production. When the body temperature rises to above the *normal temperature of 37° C,* the rate of heat loss becomes greater than the rate of heat production. Conversely, at temperatures below this level, the hypothalamic thermostat reduces heat loss and promotes heat production.

Heat loss is controlled by (1) controlling the *rate of blood flow* in the skin, which in turn controls the rate of heat transfer from the central body core to the body surface, and (2) controlling the *rate of sweating,* which in turn controls the rate of evaporation from the skin. The rate of heat production is controlled (1) by *sympathetic stimulation* throughout the body, which increases cellular metabolism, (2) by *increasing the muscle tone and causing shivering,* both of which greatly increase the rate of heat production by the muscles, and, over long periods of time, (3) by *control of the rate of thyroid hormone secretion,* which increases the rate of metabolism in all cells of the body.

Fever means a body temperature that is elevated beyond the normal range. It usually occurs when abnormal *proteins* or *polysaccharides* are released into the blood during disease processes. For example, as little as $\frac{1}{1000}$ g of polysaccharides derived from certain bacteria can cause extreme fever. The polysaccharide or abnormal protein causes *resetting of the hypothalamic thermostat,* making it control the temperature at a high level rather than normal level.

Experiments have shown that fever helps the body resist the devastating effects of many infectious diseases.

HEAT PRODUCTION IN THE BODY

Metabolic Rate. In the preceding chapter it was pointed out that essentially all of the energy released from foods eventually becomes heat. Therefore, the rate of heat production by the body is a measure of the rate at which energy is released from foods. This is called the *metabolic rate.*

The metabolic rate is measured in Calories, which is the same term used to express the amount of energy in foods. When a normal individual is in a very quiet state, the metabolic rate may be as low as 60 to 70 Calories per hour. On the other hand, it may be as high as 1000 to 2000 Calories per hour for a few minutes at a time, or as high as 200 to 300 Calories per hour for several hours at a time.

Factors that Affect the Metabolic Rate

Any factor that increases the rate of energy release from foods also increases the metabolic rate. Some of the more important of these are the following:

Exercise. Exercise is perhaps the most powerful stimulus for increasing the metabolic rate. When muscles contract, a tremendous quantity of adenosine triphosphate (ATP) is degraded to adenosine diphosphate (ADP) in the muscle cells, and this then enhances the rate of oxidation of the foodstuffs. During very strenuous exercise lasting only a moment or two the metabolic rate can actually increase to as much as 40 times that during rest.

Effect of Sympathetic Stimulation and Norepinephrine on Metabolic Rate. When the sympathetic nervous system is stimulated, norepinephrine is released directly into the tissues by the sympathetic nerve endings. Also, large quantities of this hormone and of epinephrine are released into the blood by the adrenal medullae. These two hormones then exert a direct effect on all cells to raise their metabolic rates. They enhance the breakdown of glycogen

into glucose and also increase the rates of some of the enzymatic reactions that promote oxidation of foods.

Strong sympathetic stimulation can increase the metabolic rate as much as 100 percent, but the metabolic rate remains elevated for only a few minutes after the sympathetic stimulation ceases. By controlling the activity of the sympathetics, the central nervous system has a means for regulating the rates of activity of all the cells of the body, increasing their activity when this is required and decreasing their activity when the need no longer exists.

Effect of Thyroid Hormone on Metabolic Rate. Thyroid hormone has an effect on all cells of the body similar to that of norepinephrine and epinephrine, except that the thyroid hormone requires several days to act fully but then continues to act for as long as 4 to 8 weeks after its release from the thyroid gland, rather than for only a few minutes. Thyroid hormone secreted in very large amounts can, like norepinephrine and epinephrine, increase the metabolic rate as much as 100 percent. On the other hand, complete lack of secretion by the thyroid gland causes the metabolic rate to fall to as low as 50 percent below normal. In other words, the overall span from total lack of thyroid hormone to great excess can increase the metabolic rate as much as fourfold.

The precise mechanism by which thyroid hormone causes its metabolic effect on cells is not known, but it is known to increase the quantities of most of the cellular enzymes, which might explain its metabolic effect. The actions of thyroid hormone will be discussed at further length in the following chapter.

Most of the other endocrine hormones besides norepinephrine, epinephrine, and thyroxine have only minor effects on the overall metabolic rate, though insulin from the pancreas, growth hormone from the anterior pituitary gland, testosterone from the testes, and adrenocortical hormones all can increase the general rate of metabolism as much as 5 to 15 percent.

Effect of Body Temperature on Metabolic Rate. The higher the temperature of a chemically reactive medium, the more rapid are the chemical reactions. This effect is also observed in the chemical reactions of the cells in the body. Each degree Celsius increase in temperature raises the metabolic rate approximately 10 percent. Therefore, in a person with very high fever the metabolic rate may be as much as twice normal because of the fever itself.

Specific Dynamic Action of Foods on Metabolic Rate. After a meal, the metabolic rate usually rises and remains elevated for the ensuing 2 to 10 hours. In general, a meal containing large quantities of fats and carbohydrates will increase the metabolic rate only 4 to 5 percent. In contrast, a meal containing large quantities of proteins can increase the metabolic rate as much as 30 percent. This effect is called the *specific dynamic action of foods.*

The specific dynamic action of foods is believed to be caused at least partially by the greater metabolism required for digestion, absorption, and assimilation of the foods. But in addition to this effect, proteins probably increase the metabolic rate an extra amount because of direct stimulatory effects of some of the amino acids and other breakdown products of proteins.

Basal Metabolic Rate

Because of the many different factors that can affect the metabolic rate, it is extremely difficult to compare metabolic rates from one person to another. To have any valid comparison at all, a person's rate of energy utilization must be measured when he is in a so-called *basal* state. This means that (1) he is not exercising and has not been exercising for at least 30 minutes to an hour; (2) he is at complete mental rest so that his sympathetic nervous system is not overactive; (3) the temperature of the air is completely comfortable so that his sympathetic nervous system is not unduly stimulated by this factor; (4) he has not eaten any food within the last 12 hours that could cause a specific dynamic action on his metabolic rate; and (5) his body temperature is normal, to avoid the effect of fever on metabolism.

Under these *basal conditions* most of the factors that affect the metabolic rate are controlled, but this basal state does not remove the effect of the constantly secreted thyroid hormone. Therefore, the *basal metabolic rate* in essence is determined by two major factors: first, the inherent rates of chemical reactions of the cells, and, second, the amount of thyroid hormone acting on the cells. Because the inherent rate of cellular activity is relatively constant from one person to another, an abnormal basal metabolic rate is usually caused by an abnormal secretory rate of thyroid hormone. For this reason basal metabolic rates are often measured to assess the degree of thyroid activity.

METHODS FOR MEASURING BASAL METABOLIC RATE

Direct Method for Measuring Basal Metabolic Rate. Since the basal metabolic rate is actually a measure of the amount of heat produced by the body, it can be determined by measuring the heat given off from the body in a known period of time. To do this, the person is placed in a large chamber called a *human calorimeter.* This chamber is cooled by water flowing through a radiator system. The heat given off from the body is picked up by the cooling system and then measured by appropriate physical apparatus. This is called the *direct* method for measuring the basal metabolic rate because it measures the heat output directly. Obviously, it is a very cumbersome method, but it has been a very important tool in experimental studies.

Indirect Method for Measuring Basal Metabolic Rate. An indirect method for measuring the basal metabolic rate is based on the amount of oxygen burned by the body in a given period of time. From this the rate of energy release can be calculated.

The amount of energy released when 1 L of oxygen burns with carbohydrates is 5.05 Calories. When 1 L of oxygen burns with fats the amount of energy released is 4.70 Calories. When 1 L of oxygen burns with proteins the amount is 4.60 Calories. It is obvious from these figures that every time 1 L of oxygen is burned, almost the same amount of energy is released regardless of which one of the three different foods is being used for energy. Therefore, it is reasonable to use an approximate average of these values, 4.825 Calories, as the amount of energy released in the body every time 1 L of oxygen is burned. When calculating the basal metabolic rate in this way, the value obtained is never more than 4 percent in error even when a great excess of one type of food or the other is being used for metabolism, and 4 percent is far less than the error of measurement anyway. Therefore, to determine the amount of energy being generated in the body, one needs only to determine the amount of oxygen being used. This is done using a respirometer.

FIGURE 44—1 A respirometer for measuring the rate of oxygen utilization.

The Respirometer. Figure 44–1 shows a respirometer. The subject places a mouthpiece in his mouth and breathes in and out of a large inverted, counterbalanced can that rides up and down in a water bath. The respirometer contains pure oxygen, which is breathed back and forth into the lungs. The oxygen is gradually absorbed into the blood, and in its place carbon dioxide is expired. Then soda lime in the respirometer reacts chemically with the carbon dioxide to remove it from the respiratory gases. The net result, therefore, is a continual loss of oxygen from the respirometer; this loss causes the inverted can to sink deeper and deeper into the water, and the recording pen also falls lower and lower, giving a record of the rate of oxygen utilization.

Method for Expressing the Basal Metabolic Rate. The basal metabolic rate is generally expressed in terms of Calories per square meter (m^2) of body surface area per hour. The reason for expressing it in terms of body surface area is to allow comparisons between persons of different sizes. Experimental studies have shown that the basal metabolic rate varies from one normal person to another approximately in proportion to the body surface area and not in proportion to the weight. Thus, someone who weighs 200 lb does not have a basal metabolic rate twice that of someone weighing 100 lb, but only about 30 percent greater. His surface area also is about 30 percent greater.

A representative calculation of basal metabolic rate might be the following: A person is found to use 1.8 L of oxygen in 6 minutes. Therefore, in 1 hour he uses 18 L of oxygen, and his total basal metabolic rate for this period would be 18 times 4.825 or 86.85 Calories per hour. To express this in Calories per m^2 per hour one uses the chart in Figure 44–2A, which shows the square meters of surface area in relation to

weight and height. If the person weighs 70 kg and is 180 cm tall, his surface area is 1.9 m^2. Therefore, his basal metabolic rate is 86.85 Calories per hour divided by 1.9, or 45.7 Calories per square meter per hour.

The basal metabolic rate is often then expressed in percentage of normal. To do this, one refers to the chart in Figure 44–2B, which shows the normal basal metabolic rate for males and females at different ages. If this person is a boy aged 18, his normal basal metabolic rate is 40 Calories per square meter per hour. However, his actual basal metabolic rate is 5.7 Calories greater than the normal value, or 14 percent greater than normal. His basal metabolic rate, therefore, is stated to be +14. If his basal metabolic rate had been 25 percent less than normal, it would have been expressed as −25, rather than plus.

When the thyroid gland is secreting an extreme quantity of thyroxine, the basal metabolic rate can at times go as high as +100. On the other hand, when the thyroid gland is secreting almost no thyroxine, the basal metabolic rate falls to as low as −40 to −50.

Major Loci of Heat Production in the Body

Though all the tissues of the body produce heat, those that have rapid chemical reactions produce the major amounts. In the resting state, the liver, heart, brain, and most of the endocrine glands produce large amounts of heat. This causes their temperatures to be a degree or so higher than that of most of the other tissues.

During rest the amount of heat produced by each skeletal muscle is not very great, but, because half of the entire mass of the body is composed of muscles, heat production of all the

FIGURE 44–2 (A) Diagram for determining the body surface area when the weight and height are known. (From DuBois: Basal Metabolism in Health and Disease. Philadelphia, Lea & Febiger.) (B) Normal basal metabolic rates for males and females at different ages.

skeletal muscles together nevertheless accounts for 20 to 30 percent of all the body's heat production even at rest. Therefore, a slight increase or decrease in the degree of muscular tone can have considerable effect on the amount of heat produced. During severe exercise the amount of heat produced by the muscles can rise for about a minute to as great as 40 times that produced by all the remaining tissues together. For this reason, changes in the degree of muscular activity constitute one of the most important means by which the body regulates its temperature. This is discussed in detail later in this chapter.

HEAT LOSS FROM THE BODY

The heat produced in the body must be removed continually or otherwise the body temperature would continue rising indefinitely. The body loses heat to its surroundings in three ways: radiation, conduction, and evaporation.

Heat Loss by Radiation

About 60 percent of the heat loss from a nude person sitting in a room at 70°F (21°C) is by radiation, as shown in Figure 44–3. Heat loss in this manner is based on the principle that objects facing each other are always radiating heat toward one another. The human being radiates heat toward the walls, and the walls radiate heat

FIGURE 44–3 Methods by which heat is lost from the nude body. The percentage values are those for heat loss by each method when the person is in a room at 70°F.

toward him. However, since his body temperature is usually greater than that of the walls, he normally radiates more heat to the walls than he receives from them.

The principle of radiation is used in radiant heating systems in commercial buildings and in many homes. The floors, ceilings, and walls of the rooms are heated to temperatures between 70 and 85°F (21 and 29.4°C). These are temperatures approximately comfortable for the human being, because they permit him to radiate heat at a rate that allows his body temperature to remain constant. A person is often comfortable in a radiantly heated room even though the air temperature may be in the 50's. This fact illustrates very forcefully how important radiation can be as a means of heat exchange.

Heat Loss by Conduction

Approximately 18 percent of the heat lost by the nude person in Figure 44–3 leaves his body by conduction, 15 percent by conduction to the air and 3 percent to the floor and stool. These values are only average, for the colder the air and adjacent solid objects, the greater is the conduction of heat into them.

Effect of Convection Air Currents on Loss of Heat by Conduction. Even though the temperature of the air should remain constant at 70°F (21°C), the rate at which heat will be conducted into the air depends on how rapidly the air is moving. If it is flowing past the body, every time the air adjacent to the body becomes warmed it is carried away and replaced by cooler air, and the more rapidly the air moves, the greater is the amount of heat conducted from the body. For this reason it is sometimes said that large quantities of heat are lost from the body by *convection*. This is not actually true; the heat is still being lost to the air by conduction, though the convection currents carry the heated air away.

Heat Loss by Evaporation

A small amount of water continually diffuses through the skin and evaporates, and evaporation of each gram of water removes approximately $\frac{1}{2}$ Calorie of heat from the body. Therefore, even normally, about 22 percent of the heat formed in the body is removed by evaporation. Evaporation of only 150 ml of water each hour would remove all the heat produced in the body under basal conditions. This shows how important evaporation can be as a cooling mechanism.

The Sweat Mechanism. In addition to the continual diffusion of water through the skin, the sweat glands produce large quantities of sweat when the body becomes overheated. The sweat obviously increases the amount of heat that can be lost by evaporation. Under extreme conditions about 1.5 L of sweat can be secreted each hour, which, if it all evaporates, will remove as much as 800 Calories of heat from the body, or about 12 times the basal level of heat production.

Necessity for Evaporation in Tropical Climates. Evaporation as a means of heat loss is extremely important in tropical climates, for when the temperature of the air and of the surroundings rises above the temperature of the body, heat cannot be lost by either radiation or conduction. Instead, the body gains heat by these means. However, evaporation of sweat can still keep the body temperature below that of the surroundings. Therefore, one's only means for maintaining normal body temperature when the environmental temperature is above body temperature is by evaporation. Those few unfortunate persons who are born without sweat glands must forever live in temperate or cold climates or otherwise periodically douse themselves with water.

Effect of Air Convection on Evaporation. Air currents have much the same effect on the removal of heat by evaporation as by conduction, for water evaporating from the skin quickly saturates the air immediately adjacent to the skin. If this air does not move away rapidly, the evaporation process will cease. However, if new air continually replaces the old, the air next to the skin never becomes saturated with moisture, and evaporation can continue unabated. This explains why in hot climates a fan is an important means for keeping cool; it also explains why one is usually much cooler outdoors under the shade of a tree, where the air is not moving, even though the air temperature may be the same.

Effect of Clothing on Heat Loss

Clothing is a barrier to the transfer of heat from the body to the surroundings. Heat radiated or conducted from the skin is absorbed by the inner surface of the clothing, and before it can be radiated or conducted to the surroundings it must be conducted through the mesh of the cloth. Therefore, the inside of the clothing becomes warm in comparison with the outside; the inside warmth decreases the rate of heat loss from the body to the clothing and even radiates much of the heat back to the body.

The insulation properties of most clothing result mainly from air entrapped within the clothing mesh and not from the actual material composing the clothing. A furred animal utilizes this principle in the winter time, for his hair grows long and stands on end, entrapping large amounts of air that become warm and act as a body insulator.

When clothing becomes wet, the spaces are filled with water, and this conducts heat many hundred times as rapidly as does air. As a result, the clothing is no longer an adequate heat insulator; instead, heat is conducted almost as if the clothing did not exist. Therefore, wet clothing has almost no value for keeping the body warm in cold climates, and one of the most important lessons to be learned for arctic survival is always to keep the clothing dry whatever the circumstances.

Effect of Clothing in the Tropics. In the tropics, clothing must be designed for two special purposes: (1) to protect one from exposure to the sun's heat and (2) to allow maximum evaporation of sweat. Almost any clothing that blocks light rays also blocks the heat rays of the sun, but black clothing absorbs light rays and converts these into heat, whereas white clothing reflects the rays. This explains why white clothing is far cooler than black clothing. Also, different types of clothing have vastly different effects on evaporation. Those types that rapidly absorb water, such as the cottons and the linens, usually do not depress the evaporative cooling of the body, because sweat is absorbed into the clothing and evaporated from its surface in the same manner as from the body. This is not true of wool and even less so of plastic materials, for they are not absorptive enough to allow satisfactory evaporative cooling.

THE BODY TEMPERATURE

The body temperature is determined by the balance between heat production and heat loss, as shown by Figure 44–4. If these two are exactly equal, the body temperature neither rises nor falls. When heat production is greater than heat loss, the body temperature rises; conversely, when heat loss is greater than heat production, the body temperaure falls. Appropriate regulatory systems are always at work in the body to keep heat production and heat loss approximately equal, thereby maintaining a normal body temperature.

1. Basal metabolism
2. Muscular activity
—shivering—
3. Thyroxine effect on cells
4. Epinephrine effect on cells
5. Temperature effect on cells

1. Radiation
2. Evaporation
—convection—
3. Conduction
—convection—

Heat production   Heat loss

FIGURE 44–4 Balance between heat production and heat loss, illustrating that the body temperature remains constant as long as these two are equal.

Normal Values of Body Temperature. The normal body temperature varies no more than a degree or so from one person to another. However, when a person is exposed to extremely cold or hot weather, his overall body temperature may vary as much as 1°F (0.6°C). Also, when he experiences extreme emotions that cause excessive stimulation of the sympathetic nervous system, the amount of heat production may become great enough to raise the body temperature a degree or so. Finally, extremely hard exercise can sometimes increase the body temperature as much as 5 to 6°F (2 to 3°C), though usually within 10 to 20 minutes after the exercise is over the temperature will have fallen back to that of the basal state.

Figure 44–5 depicts the normal temperature under different conditions as measured both in the mouth and rectally. The average normal oral temperature is considered to be between 98.0 and 98.6°F (36.6 and 37°C); the average

Oral   °F   °C   Rectal

104 — 40

102 — 39   Hard exercise

                       Emotion or moderate exercise
Hard work, emotion   A few normal adults
A few normal adults   38   Many active children
Many active children   100

Usual range   37   Usual range
of normal   98   of normal

                      36
Early morning   Early morning
Cold weather, etc.   96   Cold weather, etc.

FIGURE 44–5 Ranges of body temperature under different normal conditions. (From DuBois: Fever and the Regulation of Body Temperature. Springfield, Ill., Charles C Thomas.)

normal rectal temperature is generally about 1°F (0.6°C) higher. The reason for the higher rectal temperature is that the mouth is continually cooled through the facial surfaces and by evaporation in the mouth and nose.

REGULATION OF BODY TEMPERATURE

Hypothalamic Regulation of Temperature

The Preoptic Temperature-Sensitive Center. Located in the anteriormost portion of the hypothalamus, in the *preoptic area*, is a group of neurons that respond directly to temperature. When the temperature of the blood increases, the rates of discharge of these cells likewise increase. When the temperature decreases, the rates of discharge decrease.

From this preoptic temperature-sensitive area, signals radiate to various other portions of the hypothalamus to control either heat production or heat loss. In general, the hypothalamus can be divided into two major heat control divisions: an anterior *heat-losing center* which, when stimulated, reduces the body heat and a posterior *heat-promoting center* which, when stimulated, increases the body heat. These areas are shown in Figure 44–6. The anterior center is composed mainly of hypothalamic parasympathetic nervous centers, whereas the posterior center mainly operates through the sympathetic nervous system.

Function of the Hypothalamic Heat-Promoting Center to Increase Body Temperature. Whenever blood colder than normal passes into the preoptic region of the hypothalamus, the preoptic heat-sensitive cells are inhib-

ited, and this in turn strongly activates the *hypothalamic heat-promoting center* in the posterior hypothalamus. Activation of this center automatically initiates several different mechanisms to increase the body heat. These are the following:

Vasoconstriction of the Skin to Increase Body Temperature. Stimulation of the heat-promoting center excites the sympathetics to cause the blood vessels of the skin to constrict powerfully. This lessens the flow of warm blood from the internal structures to the skin, decreasing the transfer of heat to the body surface from those internal organs producing most of the heat. Very little heat is transported directly from the internal structures to the body surface by means other than through the flowing blood because the fat beneath the skin is a very adequate heat insulator. Therefore, when vasoconstriction occurs, the temperature of the skin falls to approach the temperature of the surroundings; this causes heat loss to be greatly diminished, allowing the internal body to retain its heat and the temperature to rise.

Sympathetic Stimulation of Metabolism to Increase Body Temperature. Stimulation of the sympathetic nerves releases norepinephrine throughout the body tissues and also causes both epinephrine and norepinephrine to be secreted into the blood by the adrenal medullae. These hormones in turn raise the rates of metabolism of all cells as much as 100 percent for a few minutes at a time or as much as 15 to 20 percent chronically, thereby increasing heat production. This, too, tends to elevate the body temperature.

Shivering to Increase Body Temperature. Stimulation of the heat-promoting center increases the degree of wakefulness, and also causes transmission of strong signals into the bulboreticular formation and red nucleus of the hindbrain. Signals passing through these regions increase the tone of all the muscles, which in turn increases the amount of heat produced by the muscles. In addition, facilitation of the spinal cord neuronal circuitry for the stretch reflex causes this reflex to oscillate; that is, a slight movement stretches a muscle, which elicits the stretch reflex and causes the muscle to contract. The contraction stretches the antagonistic muscle, which then develops a stretch reflex itself. Its contraction then stretches the first muscle, and the cycle becomes repetitive so that continual shaking develops. All this muscular activity can increase the rate of heat production by 300 to 400 percent. Consequently, when the body is exposed to extreme cold, shivering is a very pow-

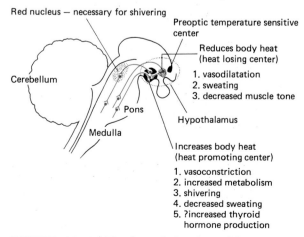

FIGURE 44–6 The hypothalamic and brain stem mechanisms for regulating body temperature.

erful force to maintain normal body temperature.

Piloerection to Increase Body Temperature. Piloerection means that hairs stand on end. This occurs when the sympathetic centers are stimulated for the sympathetic nerves excite the small arrector pili muscles located at the bases of the hairs, which causes the hairs to "stand on end," a process called **piloerection.** In the human being this mechanism does not protect against heat loss because of the scarcity of hairs, but in lower animals piloerection entraps large quantities of air in the zone adjacent to the body and provides increased insulation against cold.

Increased Thyroid Hormone Production to Increase Body Temperature. If the body is exposed to cold for several weeks, as at the beginning of winter, the thyroid gland enlarges and begins to produce greater quantities of thyroid hormone. This is caused by formation of a neurosecretory hormone in the preoptic region of the hypothalamus (see Chapter 45) that passes in the blood to the anterior pituitary gland to augment its production of thyrotropic hormone. This hormone in turn excites the thyroid gland. Under extreme conditions of cold, increase in thyroid hormone production over a period of weeks can step up the rate of heat production as much as 20 to 30 percent, thus allowing one to withstand the prolonged cold.

Function of the Hypothalamic Heat-Losing Center to Decrease the Body Temperature. When the temperature of the preoptic temperature-sensitive region rises too high, the neurons of this area become overly excited, which also stimulates the *anterior hypothalamic heat-losing* center. Because of reciprocal innervation between this center and the posterior hypothalamic heat-promoting center, the latter becomes inhibited. As a result, all the mechanisms of the heat-promoting center that tend to increase the body temperature become inoperative. For example, instead of vasoconstriction, the blood vessels to the skin dilate, allowing the skin to become very warm so that heat can be lost rapidly. Also, the increased metabolism caused by sympathetic stimulation ceases, and muscular tone greatly decreases; the production of thyroid hormone diminishes gradually. The reversal of all of these effects allows the rate of heat loss to increase and the rate of heat production to decrease, thus making the body temperature fall.

However, in addition to reversing the heat-promoting effects, stimulation of the heat-losing center causes two effects of its own to decrease the temperature. These are sweating and panting.

Sweating to Decrease Body Temperature. If the reversal of the heat-promoting effects is not sufficient to bring the body temperature back to normal, the anterior hypothalamus initiates the sweating process by sending sweat-promoting signals to all the sweat glands of the body through the sympathetic nerves. As much as 1.5 L of sweat can be poured onto the surface of the skin in an hour, and under favorable conditions a large proportion of this will evaporate. By this means, a refrigeration mechanism is initiated to reduce the body temperature when it tends to rise too high.

Panting. In lower animals, though not in the human being, excessive stimulation of the hypothalamus by heat initiates a neurogenic mechanism in the pons of the hindbrain to cause panting. The animal breathes very rapidly but very shallowly, so that a tremendous quantity of air passes into and out of the respiratory passageways. Evaporation of water from the tongue, mouth, nose, and other upper respiratory passageways is greatly increased, allowing a major amount of heat loss. Many lower animals—dogs, for instance—do not have well developed sweat mechanisms, and panting is often their only means of regulating body temperature in hot climates.

Because of the shallowness of breathing during panting, most of the air entering the alveoli is dead space air, so that the alveolar ventilation of the animal remains essentially normal. This prevents overventilation despite the great turnover of air in the upper respiratory passageways.

Summary of Hypothalamic Functions in Body Temperature Regulation. The hypothalamus, when exposed to a temperature above normal, decreases heat production and increases heat loss. Conversely, when exposed to a temperature below normal, it increases heat production and decreases heat loss. Figure 44–7 illustrates these effects, showing an experiment in which the internal body temperature was progressively changed while heat production and evaporative heat loss were measured. Note that when body temperature was below normal—that is, below 37.0°C—heat production increased markedly; when above normal, heat loss by evaporation increased markedly.

Thus, the hypothalamus acts as a thermostat, maintaining the internal body temperature usually within one-half degree of the normal average value. The efficiency of the hypothalamus as a thermostat is indicated by the fact that

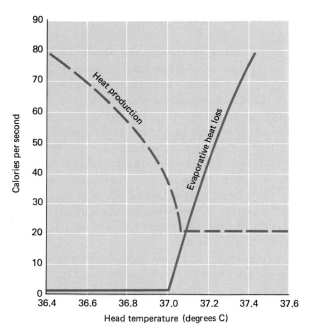

FIGURE 44–7 Effect of hypothalamic temperature on (A) evaporative heat loss from the body and (B) heat production caused primarily by muscular activity and shivering. This figure demonstrates the extremely critical temperature level at which increased heat loss begins and increased heat production stops. (Drawn from data in Benzinger, Kitzinger, and Pratt: Temperature, Part III. Ed. by Hardy, p. 637. Reinhold Publishing Corp.)

the nude body can be exposed for several hours to dry air temperatures as low as 50°F (10°C) or as high as 150°F (65°C) without changing the internal body temperature more than 1 to 2°F (0.6 to 1°C).

Effect of Skin Thermoreceptors on Body Temperature Control. Though it is the hypothalamus that is the basic temperature controller of the body, signals transmitted to the hypothalamus from thermoreceptors in all areas of the skin can modify the "thermostatic setting" of the hypothalamus. That is, when the air temperature falls very low, which excites the cold receptors of the skin, the "setting" of the hypothalamic thermostat is automatically increased to a temperature level several tenths of a degree above normal body temperature. Conversely, exposure of the skin to hot air reduces the setting to slightly below normal.

These effects play an important role in helping the body adapt to extremes of air temperature, for the following reason: If the body becomes exposed to an extreme of temperature and no readjustment is made in the control system, the internal body temperature will have to become abnormal before the hypothalamus will react. Fortunately, the thermal signals from the skin initiate the hypothalamic reaction *before*

the internal body temperature becomes abnormal, thus helping to maintain a very even internal temperature.

FEVER

Fever means a body temperature that is elevated beyond the normal range. It occurs in many disease states. Therefore, measurement of body temperature is of extreme importance in assessing the severity of a person's affliction. The most frequent cause of fever is severe bacterial or viral infection such as occurs in pneumonia, typhoid fever, tuberculosis, diphtheria, measles, yellow fever, mumps, poliomyelitis, and so forth. A less frequent cause of fever is the destruction of body tissues by some means other than infection. For instance, after a severe heart attack a person usually develops fever, and a patient exposed to destruction of his tissues by x-ray or nuclear radiation may have fever for the ensuing few days.

Resetting of the Hypothalamic Thermostat by Proteins or Polysaccharides as a Cause of Fever

Fever is usually caused by abnormal proteins or polysaccharides released into the body fluids during disease processes. These substances have either direct or indirect effects on the hypothalamic thermostat to raise its "setting" to a higher temperature level. For example, less than 0.001 g of a polysaccharide derived from the bodies of typhoid bacilli can reset the level of the thermostat from normal up to as high as 106°F (41°C). When this happens, all the heat-promoting mechanisms for increasing the body temperature become active and remain active until the body temperature reaches 106°F. Then the mechanisms for heat loss and heat production becomes equalized again, and the temperature continues to be regulated at this elevated level as long as the abnormal polysaccharide is present. In other words, the body temperature is still regulated during fever, but it is regulated at a temperature level considerably above normal because of resetting of the thermostat.

Chills and Sweating in Feverish Conditions. Figure 44–8 shows the effect of disease on the course of the body temperature. At the beginning of this chart, the oral temperature is 98.6°F, but after a few hours the disease suddenly sets the thermostat to a level of 103°F. During the

FIGURE 44–8 Fever caused by a disease process; development of chills when the temperature is rising and development of sweating when the temperature is falling.

ensuing hours, all the heat-promoting mechanisms for increasing body temperature operate at full force. These include skin vasoconstriction, increased metabolism caused by norepinephrine and epinephrine secretion, and shivering. Even when the body temperature reaches 101°F, it still has not reached the setting of the thermostat, and the heat-promoting phenomena continue to take place. Therefore, even though the temperature is high, the skin remains cold, and shivering occurs. This is called a *chill;* when someone is having a chill it is quite certain that his body temperature is actively rising.

After a number of hours the body temperature reaches the setting of the thermostat, and the chills disappear, but the temperature remains regulated at the high thermostatic setting until some factor breaks the disease process.

In Figure 44–8 the disease process is corrected after another few hours, and the thermostat is suddenly set back to its normal level of 98.6°F. However, at first the body temperature still has not fallen. Therefore, the heat-losing mechanisms, including especially skin vasodilatation and sweating, become powerfully activated. The person's skin suddenly becomes warm, and he begins to sweat, which is known as the *crisis.* When this happens one knows that the patient's temperature is beginning to fall. In the days before the "miracle" drugs (the antibiotics), many, many persons with bacterial diseases died, and the duty of the doctor was mainly to keep the sick person comfortable until he either died or a crisis appeared. If the crisis did appear, the doctor left the sickroom and announced to the family that all was now going to be well. The family in turn revered the doctor for his miraculous achievement. Unfortunately for the reputa-

tion of the present-day doctor, he too frequently cures the patient long before a crisis can develop.

Possible Value of Fever in Disease. One wonders what the value of fever in disease might be. At present no clear-cut answer is available, but suggested values are (1) many bacterial and viral agents do not survive as well at high temperatures as at normal body temperatures. Therefore, elevation of the temperature might well be a means for combating these infections. For example, high temperature has been shown to be especially lethal to gonococcal and syphilitic bacteria. (2) Because high rates of chemical reactions occur in the cells at high temperatures, it is possible that these increased rates allow the cells to repair damage much more rapidly than they could otherwise.

In lower animals experiments have shown that artificial prevention of fever in some otherwise mild diseases can cause death. Also, the administration of aspirin to a child with a viral disease such as chicken pox or influenza to reduce fever can unfortunately predispose the child to very severe brain damage from the virus.

Effect of Temperature Itself on the Body

When the body temperature rises to between 108° and 110°F (42 and 43°C), it becomes very difficult, and often impossible, for the temperature regulatory mechanisms to return the body temperature to normal again. One reason for this is that at these high temperatures the rates of cellular metabolism become greatly increased because of the temperature itself so that often the regulatory systems cannot overcome this very rapid rate of heat production.

When the body temperature rises above 108°F (42°C), the metabolic rates of the cells become so great that the cells begin to "burn themselves out"; when the body temperature rises to 112 to 114°F (44 to 45°C), death almost always ensues simply because of the heat itself. The most damaging effect of a very high body temperature occurs in the neuronal cells of the brain, sometimes causing permanent destruction of these cells even though the person recovers.

When the body temperature falls far below normal, below about 92°F (33°C), heat regulation again becomes impaired, but for opposite reasons. The low body temperature causes such slow rates of chemical reactions in the cells that no amount of regulation can bring the rate of heat production up high enough to return the body temperature to normal. The cold temperature decreases the rate of cellular metabolism, and this allows the temperature to fall still lower, which decreases the rate of metabolism more, creating a vicious cycle until the body temperature falls so low that the person dies. Usually death occurs when the body temperature reaches about 75°F (24°C).

Body temperatures down to about 85°F (29°C) do not cause significant damage to the body, though the bodily functions become so greatly slowed that the person remains in a state of suspended animation until he is rewarmed.

QUESTIONS

1. Explain how each of the following factors affects metabolic rate: exercise, sympathetic stimulation, thyroid hormone, body temperature, and foods.
2. What is meant by basal metabolic rate? How is it determined?
3. Explain how heat is lost by radiation, conduction, and evaporation.
4. What is the role of the hypothalamus in body temperature regulation?
5. Explain how the body increases heat production when the temperature falls too low.
6. How does the temperature-controlling mechanism reduce heat production and increase heat loss when the body temperature becomes too great?
7. Explain how resetting of the hypothalamic thermostat causes fever.
8. What are the effects on the body of either too high a body temperature or too low a body temperature?

REFERENCES

Bennett, A.F.: Activity metabolism of the lower vertebrates. *Annu. Rev. Physiol., 40:*447, 1978.

Benzinger, T.H.: Heat regulation: Homeostasis of central temperature in man. *Physiol. Rev., 49:*671, 1969.

Cena, K., and Clark, J.A.: Transfer of heat through animal coats and clothing. *In* Robertshaw, D. (ed.): Internal Review of Physiology: Environmental Physiology III. Vol. 20. Baltimore, University Park Press, 1979, p. 1.

Hardy, R.N.: Temperature and Animal Life. Baltimore, University Park Press, 1979.

Heller, H.C., *et al.:* The thermostat of vertebrate animals. *Sci. Am., 239*(2):102, 1978.

Hensel, H.: Neural processes in thermoregulation. *Physiol. Rev., 53:*948, 1973.

Hensel, H.: Thermoreceptors. *Annu. Rev. Physiol., 36:*233, 1974.

Herman, R.H., *et al.* (eds.): Metabolic Control in Mammals. New York, Plenum Press, 1979.

Homsher, E., and Kean, C.J.: Skeletal muscle energetics and metabolism. *Annu. Rev. Physiol., 40:*93, 1978.

Klachko, D.M., *et al.* (ed.): Hormones and Energy Metabolism. New York, Plenum Press, 1978.

Kluger, M.J.: Temperature regulation, fever, and disease. *In* Robertshaw, D. (ed.): International Review of Physiology: Environmental Physiology III. Vol. 20. Baltimore, University Park Press, 1979, p. 209.

Mitchell, D.: Physical basis of thermoregulation. *Int. Rev. Physiol., 15:*1, 1977.

Underwood, L.S., and Tieszen, L.L. (eds.): Comparative Mechanisms of Cold Adaptation. New York, Academic Press, 1979.

Wyndham, C.H.: The physiology of exercise under heat stress. *Annu. Rev. Physiol., 35:*193, 1973.

XIV

GENERAL ENDOCRINOLOGY

45

Introduction to Endocrinology: the Endocrine Glands, the Pituitary Hormones, and Thyroxine

Overview

The *endocrine glands* secrete *hormones* into the circulating blood, and these hormones in turn act on *target cells* elsewhere in the body. There are six very important endocrine glands and several other less important ones. The important ones are (1) the *pituitary gland,* which secretes eight important hormones; (2) the *thyroid gland,* three important hormones; (3) the *parathyroid glands,* one important hormone; (4) the *adrenal glands,* four important hormones; (5) the *islets of Langerhans* of the pancreas, two important hormones; and (6) the *ovaries* in the female, two important hormones, and the *testicles* in the male, one important hormone.

Hormones affect cell function in either of two general ways: (1) by *activating the cyclic AMP mechanism* or (2) by *activating genes.* In the cyclic AMP mechanism, the activating hormone combines with a specific *receptor substance* on the surface of the cell membrane. This activates the enzyme *adenyl cyclase* in the membrane, which in turn converts some of the ATP inside the cell into *cyclic adenosine monophosphate (cyclic AMP).* This substance has an activating effect on many intracellular chemical reactions, causing the cell to increase its specific functional activities. In the genetic mechanism of hormonal control, the activating hormone reacts with a *receptor substance in the cell cytoplasm,* and the combination of the hormone and the receptor then migrates into the nucleus where it *activates one or more specific genes;* these then promote specific functional effects within the cell.

The *pituitary gland* is divided into two separate parts that are considered to be separate glands: (1) the *anterior pituitary gland* and (2) the *posterior pituitary gland.*

The *anterior pituitary gland* secretes six major hormones: (1) *growth hormone,* (2) *thyroid-stimulating hormone,* (3) *adrenocorticotropic hormone,* (4) *prolactin,* (5) *follicle-stimulating hormone,* and (6) *luteinizing hormone.* Most of these function to control other glands and will be discussed in later chapters in relation to these control functions. Secretion of all of the anterior pituitary hormones is controlled by

729

the *hypothalamus.* Specific nuclei in the hypothalamus secrete *releasing* or *inhibiting factors,* one for control of each of the anterior pituitary hormones. These factors are secreted into the blood in the hypothalamic capillaries. This blood is then conducted to the anterior pituitary gland through *hypothalamic-hypophyseal portal veins.*

Growth hormone has a generalized effect on all cells of the body. It increases the transport of amino acids into all cells, increasing both the numbers of cells and the sizes of cells, thus promoting generalized growth of all body tissues. It is excess secretion of growth hormone by the anterior pituitary gland that causes an occasional person to become a giant.

The *posterior pituitary gland* secretes two important hormones: (1) *antidiuretic hormone,* which decreases the amount of water excreted by the kidneys into the urine, and (2) *oxytoxin,* which contracts the uterus, helping to expel the baby at the time of birth.

The most important hormone secreted by the *thyroid gland* is *thyroxine.* Thyroxine *increases the rates of almost all chemical reactions in all cells* of the body, but the precise mechanism by which it does so is not known. However, either all or almost all of the *enzymes* in the cells are increased in quantity. In addition, the numbers of *mitochondria* in all cells are increased. Whether these increases in enzymes and mitochondria are the cause of the increased cellular metabolism or the result of this increase is still disputed.

In total absence of thyroid secretion of thyroxine, which is a condition called *hypothyroidism,* the body's overall metabolic rate decreases to about one-half normal. When great excesses of thyroxine are secreted, a condition called *hyperthyroidism,* the metabolic rate often increases to double normal.

Thyroxine secretion is controlled by the hypothalamus and anterior pituitary gland. The hypothalamus secretes *thyrotropin-releasing factor,* which is then carried in the hypothalamic-hypophyseal portal veins to the anterior pituitary gland, where it stimulates the production of *thyroid-stimulating hormone.* This hormone is carried by way of the blood to the thyroid gland to increase thyroxine secretion.

The Nature of Hormones and their Function

A hormone is a chemical substance elaborated by one part of the body that controls or helps to control some function elsewhere in the body. In general, hormones are divided into two types: first, the *local hormones,* which affect cells in the vicinity of the organ secreting the hormone, including such hormones as acetylcholine, histamine, and the gastrointestinal hormones, all of which have been discussed at different points in this text; and, second, the *general hormones,* which are emptied into the blood by specific **endocrine glands** and then flow throughout the entire circulation to affect cells and organs in far distant parts of the body. Some general hormones affect all cells almost equally; others affect specific cells far more than others. For example, growth hormone secreted by the pituitary gland and thyroxine secreted by the thyroid gland affect all cells of the body. On the other hand, the pituitary gland also produces gonadotropic hormones, which affect the sex organs much more than other tissues, even though these hormones are secreted into the general circulation.

Figure 45–1 illustrates the more important of the endocrine glands and shows their locations in the body. These are

1. The **pituitary gland,** located in a bony cavity, the *sella turcica,* immediately beneath the hypothalamus of the brain.
2. The **thyroid gland,** lying in the anterior part of the neck immediately below the larynx and overlying the trachea.
3. Four very minute **parathyroid glands,** lo-

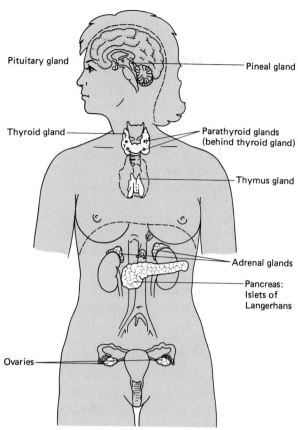

Pituitary gland

Thyroid gland

Ovaries

Pineal gland

Parathyroid glands (behind thyroid gland)

Thymus gland

Adrenal glands

Pancreas: Islets of Langerhans

FIGURE 45–1 The more important of the endocrine glands and their locations in the body.

cated behind the four lobes of the thyroid gland.

4. Two **adrenal glands,** one draped over the superior pole of each kidney.
5. The **pancreas,** located behind and beneath the stomach.
6. Two **ovaries,** present only in the female and located in the pelvic cavity on either side of the uterus.
7. Two **testicles,** present only in the male and located in the scrotum.
8. The **pineal gland,** a small structure of about 0.5 g mass that protrudes from the posterior surface of the diencephalic part of the brain.
9. The **thymus gland,** which is part of the lymphoid system of the body and was discussed in Chapter 34 in relation to immunity.

Regulatory Functions of Hormones. Some examples of the regulatory functions of the general hormones are the following: (1) The **anterior pituitary gland** secretes six different hormones that regulate respectively the rate of growth of all tissues of the body, the rate of secretion of thyroid hormone by the thyroid gland, the rate of secretion of adrenocortical hormones

by the adrenal gland, the rate of formation of milk by the mammary glands, and the rates of secretion of several different sex hormones. (2) Thyroxine secreted by the **thyroid gland** controls the rate of metabolism of all cells. (3) The adrenal medullary hormones epinephrine and norepinephrine, which have been discussed in connection with the sympathetic nervous system, cause the same reactions throughout the body as stimulation of the general sympathetic nervous system. (4) Hormones secreted by the **adrenal cortex** regulate reabsorption of sodium by the kidneys and also regulate some aspects of carbohydrate, fat, and protein metabolism. (5) Insulin secreted by the **pancreas** regulates the utilization of glucose throughout the body. (6) Hormones secreted by the **testes** control the sexual and reproductive functions of the male. (7) Hormones secreted by the **ovaries** and the **placenta** control the sexual and reproductive functions of the female. (8) Hormones secreted by the **parathyroid glands** regulate the concentration of calcium in the body fluids.

In the present chapter and the following few chapters, the functions of the general hormones are considered in detail. The study of these and their functions is called *endocrinology*, and the glands that secrete the general hormones are called *endocrine glands.* The word *endocrine* means secretion to the interior of the body as opposed to *exocrine*, which means secretion to the exterior, as is true of the intestinal glands, the sweat glands, and others.

CHEMICAL NATURE OF THE HORMONES

All of the general hormones are of two different chemical types:

1. *Small proteins* or *derivatives of proteins.*
2. *Steroid compounds.*

All of the hormones except the adrenocortical and sex hormones belong in the category of small proteins or derivatives of proteins such as polypeptides, polypeptide amines, or chemical compounds derived from one or more amino acids.

The adrenocortical and sex hormones are all steroids. A typical steroid is cholesterol, the chemical formula for which was presented in Chapter 42. The steroid hormones have a chemical structure similar to that of cholesterol and in most instances can be synthesized by the adrenal cortex and sex glands from cholesterol itself.

MECHANISMS OF HORMONAL ACTION

The mechanism by which the different hormones control activity levels of target tissues differs from one hormone to another. However, two important general mechanisms are (1) activation of the *cyclic AMP system* of cells, which in turn elicits specific cellular functions, and (2) *activation of the genes* of the cell, which then cause formation of intracellular proteins that in turn initiate specific cellular functions. These two mechanisms may be described as follows:

Cyclic AMP, an Intracellular Hormone Mediator

Most of the protein and peptide hormones exert their effects on cells by first causing cyclic AMP (*cyclic 3',5'-adenosine monophosphate*) to be formed in the cell. Once formed, the cyclic AMP initiates the desired activities inside the cell. Thus, cyclic AMP is said to be an *intracellular hormonal mediator*. It is also frequently called the "second messenger" for hormone mediation—the "first messenger" being the original stimulating hormone.

Figure 45–2 illustrates the function of the cyclic AMP system in more detail. The first event in this mechanism of hormonal control is the action of the *stimulating hormone* on a specific *receptor* in the cell membrane. In fact, whether or not the hormone will affect a particular cell is determined by the presence or absence of this specific receptor for that hormone.

Second, once the hormone binds with the receptor, an enzyme in the cell membrane called *adenyl cyclase* is activated.

Third, the adenyl cyclase then converts some of the adenosine triphosphate (ATP) inside the cell cytoplasm into cyclic AMP. The cyclic AMP persists in the cell for a few seconds to many minutes before it is reconverted into ATP. However, as long as the stimulating hormone remains active at the receptor, still more cyclic AMP will continue to be formed.

Fourth, once generated inside the cell, the cyclic AMP in turn performs any number of different physiologic functions, including:

1. Activating enzymes
2. Altering cell permeability
3. Altering the degree of smooth muscle contraction
4. Activating protein synthesis
5. Causing secretion by the cell

The specific effect that occurs in each individual cell is determined by the characteristics of the cell. Thus, if the cell is a glandular cell, it will form its specific secretion; or if the cell is a smooth muscle cell, it will contract or relax depending upon whether the cyclic AMP is excitatory or inhibitory in that particular cell.

The cyclic AMP mechanism has been shown to act as an intracellular hormonal mediator system for at least some of the functions of each of the following hormones:

1. *Adrenocorticotropic hormone*
2. *Thyroid-stimulating hormone*
3. The *gonadotropic hormones*, which stimulate the sex glands
4. *Antidiuretic hormone*
5. *Parathyroid hormone*, which controls extracellular fluid calcium
6. *Glucagon*, which helps to control blood glucose concentration
7. *Epinephrine*, which controls contraction of many smooth muscles
8. The *hypothalamic releasing factors*, which control secretion of most of the anterior pituitary hormones

Hormonal Control by Activation of Genes

A second important way in which hormones act is to activate one or more genes in the nucleus; these in turn cause synthesis of proteins in the target cells. This is the mechanism of hormonal

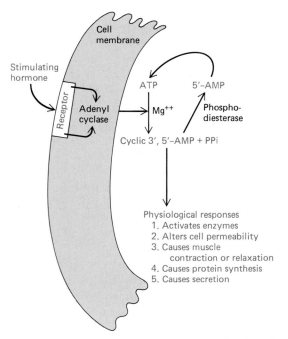

FIGURE 45–2   The cyclic AMP mechanism by which many hormones exert their control of cell function.

control utilized by the steroid hormones secreted by the adrenal cortex, the ovaries, and the testicles.

The general mechanism of steroid hormone function is the following: First, the hormone enters the cytoplasm of the cell where it combines with a specific receptor protein. Second, after several additional steps that occur in the cytoplasm and nucleus, one or more specific genes are activated in the nucleus. Third, the genes cause the formation of specific messenger RNA molecules. Fourth, the RNA diffuses to the cytoplasm and causes the formation of specific proteins. Fifth, the proteins then increase specific activities of the cells. For instance, in the renal tubules, protein enzymes are formed in response to aldosterone stimulation to cause sodium absorption and potassium secretion.

THE ANTERIOR PITUITARY HORMONES

The *pituitary gland*, shown in Figure 45–3 and known also as the *hypophysis*, is about the size of the tip of the little finger, and it lies in a small bony cavity, the pituitary fossa, beneath the base of the brain. It is divided into two completely separate parts, the *posterior pituitary gland*, known also as the *neurohypophysis*, which is connected by a stalk with the hypothalamus of the brain, and the *anterior pituitary gland*, called

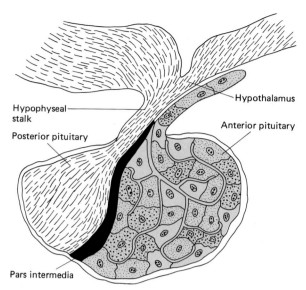

FIGURE 45–3 The pituitary gland.

also the *adenohypophysis*, which lies anterior to the posterior pituitary gland and has functions not related to those of the posterior pituitary gland.

Figure 45–3 shows the gross structure of the anterior pituitary gland, and Figure 45–4 illustrates its cellular structure, showing it to be composed of multiple types of cells each of which secretes one or more different hormones. All of these hormones are either small proteins or large polypeptides. Though 30 or more anterior pituitary hormones have been postulated by

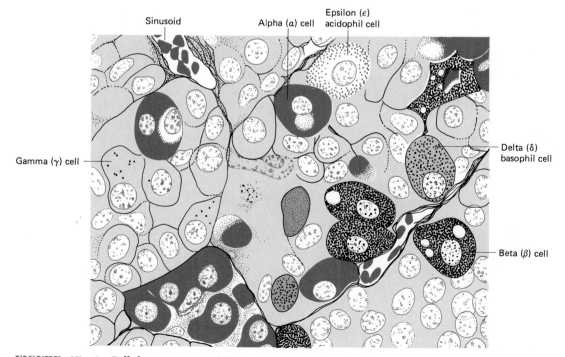

FIGURE 45–4 Cellular structure of the anterior pituitary gland.

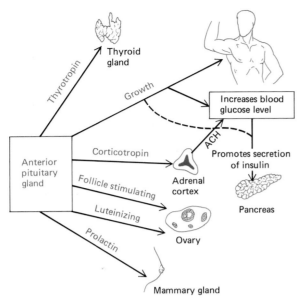

FIGURE 45–5 Metabolic functions of the anterior pituitary hormones.

many different research workers, only 6 have thus far proved to have major significance. These were listed earlier in the chapter but bear repeating here.

1. Growth hormone
2. Thyroid-stimulating hormone (thyrotropin)
3. Adrenocorticotropic hormone (adrenocorticotropin)
4. Prolactin
5. Follicle-stimulating hormone
6. Luteinizing hormone

The last two are called *gonadotropic hormones* because they regulate the functions of the sex glands.

The general functions of these six hormones are illustrated in Figure 45–5.

Growth Hormone

Growth hormone is a small protein containing 191 amino acids in a single chain and having a molecular weight of 22,005. It is secreted by the anterior pituitary gland throughout life even though most growth in the body stops at adolescence.

The function of growth hormone during the growing phase of an animal's life is to promote development and enlargement of all bodily tissues. It increases the sizes of all cells and also their number. Consequently, each tissue and each organ becomes larger under the influence of growth hormone. The bones enlarge and lengthen, the skin thickens, and the soft tissues,

including the heart, liver, tongue, and all the other internal organs, increase in size. In other words, growth hormone is exactly what its name implies; it causes the person to grow.

Function of Growth Hormone After Adolescence. After adolescence, growth hormone secretion diminishes slightly but never stops. The hormone continues to function as before, that is, to promote protein synthesis and formation of other cellular elements. However, by this age most of the body's bones have grown as much as they can. Therefore, growth of body stature ceases. A few bones, though, especially those of the lower jaw and nose, do continue to grow, increasing the prominence of these with age even in the normal person.

Basic Metabolic Effects of Growth Hormone. Growth hormone is known to have the following basic effects on the metabolic processes of the body:

1. Increased rate of protein synthesis in all cells of the body
2. Decreased rate of carbohydrate utilization by all or most cells
3. Increased mobilization of fats and use of fats for energy

The most conspicuous influence of growth hormone is to promote growth, which occurs mainly because the hormone causes widespread protein synthesis. Yet, it also has effects on both carbohydrates and fat metabolism, conserving carbohydrates while at the same time using up the fat stores.

Stimulation of Growth of Cartilage and Bone—Role of "Somatomedin." The growth of cartilage and bone is not a direct effect of growth hormone on these structures. Instead, growth hormone causes the liver, and perhaps the kidney to a lesser extent, to form a substance called *somatomedin*. This substance then acts directly on the cartilage and bone to make them grow. Somatomedin is necessary for deposition of collagen and ground substance, both of which are essential for growth of the cartilage and bone.

The reason that growth stops at adolescence is that the *epiphyseal cartilages*, discussed in Chapter 10, that separate the ends, or "epiphyses," of the long bones from the shafts of these bones exhaust their potential for growth at this age. Consequently, the epiphyses then unite directly with the shafts of the bones at the point where the growth cartilages had previously existed. Increased linear growth of these bones therefore cannot occur beyond this time.

Stimulation of Protein Deposition by Growth Hormone. Aside from promoting the growth of cartilage and bone, growth hormone also causes growth of most other tissues as well because of its effect of increasing the formation of protein. The precise mechanism or mechanisms by which it stimulates protein formation are not clear, but it is known that growth hormone has the following effects, all of which could be important in increasing the total body protein:

1. Increased transport of amino acids through cell membranes, thereby providing an appropriate intracellular supply of the building blocks from which proteins are formed.
2. Increased formation of RNA, which in turn causes greater formation of protein molecules.
3. Activation of the ribosomes to increase the rate at which they synthesize protein.
4. Decreased breakdown of proteins once they have been formed, thus allowing proteins to accumulate in greater quantities in the cells.

Effect of Growth Hormone on Fat and Carbohydrate Metabolism. Growth hormone causes release of fatty acids from adipose tissue and thus increases the overall utilization of fat by the energy-consuming processes of the body. At the same time, it decreases the utilization of carbohydrate in most cells, so that glycogen begins to accumulate in the cells and uptake of glucose from the blood decreases. Therefore, growth hormone enhances the circulating fats in the blood because of their increased release from the fat storage areas, and it also augments the circulating glucose in the blood because of diminished utilization of the glucose.

Unfortunately, the mechanisms by which growth hormone causes these effects on both fats and carbohydrates are still unknown. However, they are important to the overall metabolism of the body mainly because they tend to decrease the deposition of fat stores at the same time that they enhance the protein and carbohydrate stores.

CONTROL OF GROWTH HORMONE SECRETION

The secretion of growth hormone by the anterior pituitary gland is controlled by the hypothalamus, which secretes a *growth hormone–releasing factor.* This factor is carried directly to the anterior pituitary gland from the hypothalamus through small veins called the *hypothalamic-hypophyseal portal system,* which will be discussed fully later in the chapter. This growth hormone–releasing factor then acts on the anterior pituitary gland to cause growth hormone secretion. However, various factors can affect the hypothalamus to cause increased or decreased secretion of growth hormone.

The rate of growth hormone secretion varies markedly from day to day depending on the metabolic needs of the body, which is quite contrary to the earlier belief that its secretion was constant in children, as a continuous stimulus to growth, but absent in adults. One situation that causes marked increase in growth hormone secretion is generalized nutritional deficiency. It is presumed that this in some way results from depressed formation of proteins, which in turn has an effect on the hypothalamus to trigger growth hormone secretion. The increased growth hormone then helps to promote deposition of new protein stores, thereby relieving cellular protein deficiency. In this way, the growth hormone mechanism could act as a controller of the protein stores of some or all the cells in the body.

Abnormalities of Growth Hormone Secretion. A decrease in growth hormone secretion causes *dwarfism;* an increase causes overgrowth of the person, which can result in *giantism* or in the condition called *acromegaly.*

Dwarfs. A person whose anterior pituitary gland fails to secrete growth hormone fails to grow. This causes the so-called "pituitary" type of dwarf (though most dwarfs have small stature because of heredity rather than because of anterior pituitary failure). The pituitary type of dwarf remains childlike in all physical respects. His organs as well as bones fail to grow, and his degree of sexual development remains that of a young child even after he has reached the age of an adult. He may never grow to more than twice the height of a newborn baby.

In the past it has been impossible to treat pituitary dwarfism adequately because of scarcity of human growth hormone. But, now, genetic engineering has made it possible for bacteria to synthesize an abundance of this hormone.

Giants. Secretion of excess growth hormone, if it occurs before adolescence, can cause a giant. After adolescence the growing parts of most bones have "fused" and are no longer capable of growing longer regardless of the amount of growth hormone available. Giantism almost always results from a tumor of the acid-staining cells, the *acidophils,* of the pituitary gland, this tumor secreting tremendous quantities of growth hormone.

Acromegaly. If a growth hormone–secreting tumor develops after adolescence, the excessive secretion of growth hormone cannot cause increased height; nevertheless, it can still cause enlargement of the soft tissues. It can also cause thickening of the bones, and a few of the bones, mainly the so-called "membranous bones," can even continue to grow. As a result, disproportionate growth occurs in some parts of the body, resulting in the condition called *acromegaly*. In acromegaly the lower jaw grows excessively long, causing the chin sometimes to protrude a half-inch or more beyond the remainder of the face. The nose and lips also enlarge, and the internal organs such as the tongue, the liver, and the various glands enlarge. The bones of the feet and hands, unlike most bones of the body, continue to grow, so that the hands and feet become very large. In short, an acromegalic person is the same as a giant, except that failure of his already "fused" long bones to grow causes his height to be normal while other features of his body exhibit excessive, disproportionate growth.

Thyroid-Stimulating Hormone

Thyroid-stimulating hormone, also known as *thyrotropin*, is another one of the hormones secreted by the anterior pituitary gland. It controls the amount of hormone secreted by the thyroid gland by increasing the number of thyroid cells, the size of these cells, and also their rate of thyroxine production. When the anterior pituitary fails to secrete thyroid-stimulating hormone, the thyroid gland becomes so incapacitated that it secretes almost no hormone. In other words, the thyroid gland is almost completely controlled by thyroid-stimulating hormone. Further relationships of the thyroid gland to thyroid-stimulating hormone are discussed later in the chapter in connection with thyroxine.

Adrenocorticotropic Hormone

A third hormone secreted by the anterior pituitary, *adrenocorticotropic hormone*, also known as *ACTH* or *corticotropin*, controls secretion of adrenocortical hormones by the adrenal cortices in much the same manner that thyroid-stimulating hormone controls secretion by the thyroid gland. Adrenocorticotropic hormone increases both the number of cells in the adrenal cortex and their degree of activity, resulting in increased output of adrenocortical hormones. The relationship of adrenocorticotropic hormone to the different adrenocortical hormones is discussed in the following chapter.

Prolactin

Prolactin is a hormone secreted by the anterior pituitary gland during pregnancy and during the entire period of milk production after birth of the baby. This hormone stimulates both breast growth and secretory functions of the breasts. These effects will be discussed in relation to reproduction in Chapter 50.

The Gonadotropic Hormones

The functions of the two gonadotropic hormones are summarized briefly in the following paragraphs; their relationship to sexual function will be presented in detail in Chapter 49.

Follicle-Stimulating Hormone. In the female, follicle-stimulating hormone initiates growth of *follicles* (fluid chambers) in the ovaries. In each of these a single ovum develops in preparation for fertilization. This hormone also helps to cause the ovaries to secrete *estrogens*, one of the female sex hormones. In the male, follicle-stimulating hormone stimulates growth of the germinal epithelium in the testes, thus promoting the development of sperm that can then fertilize the female ovum.

Luteinizing Hormone. In the female, luteinizing hormone joins with follicle-stimulating hormone to cause estrogen secretion. It also causes the follicle to rupture, allowing the ovum to pass into the abdominal cavity and then through a fallopian tube, within which time fertilization may take place. It also helps to cause the ovary to secrete *progesterone*. In the male, luteinizing hormone causes the testes to secrete the male sex hormone *testosterone*.

Regulation of Anterior Pituitary Secretion—The Hypothalamic-Hypophyseal Portal System

The anterior pituitary gland is a highly vascular organ that receives blood supply from two sources: (1) the usual arteriolar source and (2) the so-called *hypothalamic-hypophyseal portal system*, which is illustrated in Figure 45–6. The hypothalamus, which was discussed in more detail in Chapter 28, controls many of the automatic functions of the body. After the blood passes through the capillaries of the hypothalamus, particularly through the lower part called

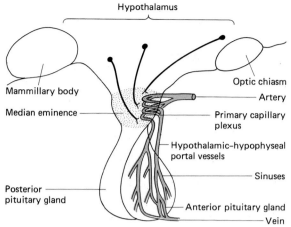

FIGURE 45–6 The hypothalamic-hypophyseal portal system.

the *median eminence*, it leaves by way of small *hypothalamic-hypophyseal portal veins* that course down the anterior surface of the pituitary stalk to the anterior pituitary gland. At this point, the veins dip into the gland where the blood flows through large numbers of *venous sinuses* that bathe the anterior pituitary cells.

The hypothalamus secretes a series of different *neurosecretory substances* called *hypothalamic releasing* and *inhibitory factors*. These factors are secreted into the blood of the hypothalamic-hypophyseal portal system and are then transported to the venous sinuses in the anterior pituitary gland, where they regulate secretion of the various anterior pituitary hormones. The five factors most important for control of anterior pituitary secretion are the following:

1. *Thyrotropin-releasing factor* (TRF), which causes release of thyroid-stimulating hormone.
2. *Corticotropin-releasing factor* (CRF), which causes release of adrenocorticotropic hormone.
3. *Growth hormone–releasing factor* (GRF), which causes release of growth hormone.
4. *Luteinizing hormone–releasing factor* (LRF), which causes release of both luteinizing hormone and follicle-stimulating hormone.
5. *Prolactin inhibitory factor* (PIF), which causes inhibition of prolactin secretion.

Note that four of these factors cause release (secretion) of the respective anterior pituitary hormones. In the absence of these four releasing factors, these hormones are secreted in only minute quantities by the anterior pituitary gland.

On the other hand, there is one important inhibitory factor secreted by the hypothalamus, prolactin inhibitory factor. In the absence of this factor the anterior pituitary gland secretes an excess of prolactin.

Because of the powerful effects of these releasing and inhibitory factors on anterior pituitary secretion, most control of the anterior pituitary gland is carried out through stimulation or inhibition of the many neuronal control centers in the hypothalamus that in turn regulate the output of the hypothalamic releasing and inhibitory factors. For instance, it was pointed out earlier in this chapter that protein deficiency in some way excites the hypothalamus to induce increased secretion of growth hormone by the anterior pituitary gland. Later in this chapter we will see that thyroid hormone secretion is also controlled almost entirely by stimulation of hypothalamic centers that in turn control thyroid-stimulating hormone secretion by the anterior pituitary gland. In succeeding chapters we shall see that control of many of the other endocrine hormones is effected through a similar hypothalamic-pituitary system.

THE POSTERIOR PITUITARY HORMONES

Though the posterior pituitary gland is located immediately behind the anterior pituitary gland, the two glands do not have any proved direct relationship. Also, in a sense, the posterior pituitary gland is not a gland at all, because it only stores hormones rather than secreting them. It stores the two hormones *antidiuretic hormone* (also called *vasopressin*) and *oxytocin*. Antidiuretic hormone has a very important function of helping to control water excretion by the kidney, which was discussed in detail in Chapter 32. Oxytocin stimulates contraction of muscle in the uterus and breasts and therefore plays important roles in birth of the baby and in expelling milk from the mother's breast. These functions of oxytocin will be discussed in Chapter 50.

Both antidiuretic hormone and oxytocin are *polypeptides*, each composed of eight amino acids. They are secreted by neuronal cells in the anterior hypothalamus and then are conducted through nerve axons to the posterior pituitary gland, where both hormones are stored. Figure 45–7 illustrates the anatomy of this mechanism. Antidiuretic hormone is formed in the neurons of the *supraoptic nucleus*. Then, over a period of hours and days, this hormone is transported

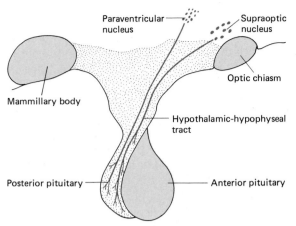

FIGURE 45—7 Hypothalamic control of the posterior pituitary.

slowly down the axoplasmic center of the nerve fibers by way of the hypothalamic-hypophyseal nerve fiber tract and is stored at the tips of these nerve fibers in the posterior pituitary gland. Oxytocin, on the other hand, is formed in the neurons of the *paraventricular nucleus* and transported in a similar manner to the posterior pituitary gland.

Once these two hormones have reached the posterior pituitary gland, they remain there until nerve impulses are transmitted from either the supraoptic nucleus to cause release of antidiuretic hormone or from the paraventricular nucleus to cause release of oxytocin.

Though both of these hormones are discussed in greater detail elsewhere in this text, let us briefly review their functions here as well.

ANTIDIURETIC HORMONE AND CONTROL OF WATER REABSORPTION IN THE TUBULES OF THE KIDNEY

Antidiuretic hormone prevents the body fluids from becoming too concentrated in the following way. When the concentration of sodium and other osmotically active substances in the extracellular fluids rises, the increased osmotic pressure of these fluids causes special neurons in the supraoptic nucleus called *osmoreceptors* to send impulses to the posterior pituitary gland. The impulses cause some of the antidiuretic hormone stored in the gland to be released. This in turn passes by way of the blood to the collecting ducts of the kidney and causes increased quantities of water to be reabsorbed while allowing large quantities of sodium and other tubular solutes to pass on into the urine.

The mechanism by which antidiuretic hormone increases water reabsorption from the collecting ducts is to increase the pore size in the epithelial cells enough for water molecules to diffuse through, but not large enough for most other substances in the tubular fluid to pass through. Thus, water is returned to the body fluids, whereas the sodium and other solutes are lost into the urine. In this way, overconcentration of the extracellular fluids is corrected.

Pressor Function of Antidiuretic Hormone. Antidiuretic hormone is also frequently called *vasopressin* because injection of large quantities of purified hormone causes the arterial pressure to rise. Also, even normal amounts of circulating antidiuretic hormone in the blood help to keep the arterial pressure from falling. Therefore, antidiuretic hormone plays at least some part in daily control of arterial pressure. When the circulatory system is subjected to severe stress, such as during traumatic surgical operations or following hemorrhage, antidiuretic hormone is then secreted in massive quantities and is very important in helping to maintain the arterial pressure in a normal range.

OXYTOCIN

An "oxytocic" agent is a substance that causes the uterus to contract, and this is one of the primary effects of *oxytocin*. Oxytocin is secreted in moderate quantities during the latter part of pregnancy and in especially large quantities at the time that the baby is born. Its effect to contract the uterus aids in the expulsion of the baby. It is well known that a mother whose oxytocin-secreting mechanism has been destroyed has considerable difficulty in delivering her baby. This will be discussed further in relation to childbirth in Chapter 50.

Effect of Oxytocin on Milk Ejection. Oxytocin also has an important function in helping the mother to provide milk for the newborn infant as follows: Milk formed by the *glandular cells* of the breasts is secreted continually into the *alveoli* of the breasts but remains there until the baby nurses. For approximately the first minute after nursing begins the baby receives essentially no milk, but the suckling stimulus excites the mother's nipple and transmits nerve signals upward through the spinal cord and finally to the hypothalamus where these signals cause secretion of oxytocin. This hormone in turn flows by way of the blood to the breast where it causes many small *myoepithelial cells* surrounding the alveoli to contract, thereby squeezing the milk from the alveoli into the ampullae and into the ducts so that the baby can remove it by suckling, as will be explained more fully in the discussion of lactation in Chapter 50.

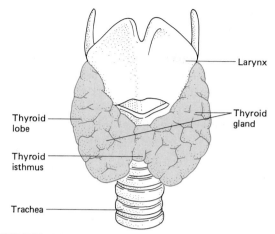

FIGURE 45-8 Gross structure of the thyroid gland and its relationship to the larynx and trachea.

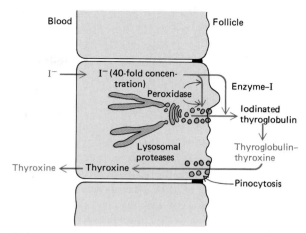

FIGURE 45-10 Mechanisms of iodine transport, thyroxine formation, and thyroxine release into the blood. (Triiodothyronine formation and release parallels that of thyroxine.)

THYROXINE

Formation of Thyroxine by the Thyroid Gland. Figure 45-8 illustrates the gross structure of the thyroid gland, and Figure 45-9 illustrates its microscopic appearance, showing it to be composed of large follicles. The cells lining the follicles secrete thyroxine to the interior of the follicles, where it is temporarily stored.

The mechanism of thyroid secretion is illustrated in Figure 45-10, which shows that iodine is first transported into the follicular cell in the form of ionic iodine (I^-). This transport results from the action of a so-called *iodide pump* in the cell membrane that concentrates the iodine in the cell about 40-fold. The follicular cell then converts the iodide ion to *elemental iodine*. At the same time, the cell also secretes a protein called *thyroglobulin* into the follicle. Under the influence of an appropriate enzyme called

iodinase, elemental iodine reacts with the thyroglobulin to convert much of the amino acid tyrosine in the molecule of the thyroglobulin into *thyroxine*. The chemical reaction for this process is shown in Figure 45-11.

Another hormone very similar to thyroxine but with one less iodine in its structure, *triiodothyronine*, is formed in smaller amounts at the same time that thyroxine is formed. It has almost exactly the same effects in the body as thyroxine except that it acts several times as rapidly as thyroxine. Because the greater proportion of the thyroid hormone is thyroxine, we shall discuss its functions in the subsequent paragraphs, but it should be remembered that smaller amounts of triiodothyronine function in an almost parallel manner to the thyroxine.

Storage and Release of Thyroxine. Thyroxine formed in the above manner usually remains stored in the follicle for several weeks as part of the thyroglobulin molecule before being released back through the follicular cells into the

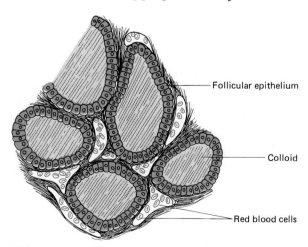

FIGURE 45-9 Microscopic appearance of the thyroid gland, showing the secretion of thyroglobulin into the follicles.

$$I_2 + HO - \langle \rangle - CH_2 - CHNH_2 - COOH \xrightarrow{\text{iodinase}}$$

Tyrosine

$$HO - \langle \rangle - O - \langle \rangle - CH_2 - CHNH_2 - COOH$$

Thyroxine

and

$$HO - \langle \rangle - O - \langle \rangle - CH_2 - CHNH_2 - COOH$$

3,5,3'-Triiodothyronine

FIGURE 45-11 Chemistry of thyroxine and triiodothyronine formation.

circulating blood. At the time of release, thyroxine is broken away from the thyroglobulin by the action of *proteases* released into the follicle by lysosomes in the thyroid cells. Then the thyroxine diffuses backward through the thyroid cells and enters the blood.

In the blood the thyroxine immediately combines with a plasma protein from which it is released over a period of several days to the tissue cells. Thus, this overall process assures a steady but very slow flow of thyroxine into the tissues. Because of this slowness, some of the effects of thyroxine are still apparent as long as 6 to 8 weeks after it is first formed in the thyroid gland.

Mechanism of Action of Thyroxine

Effect of Thyroxine on Cellular Metabolism and on the Cellular Enzymes. Thyroxine steps up the rate of metabolism of all cells. Though the precise means by which this is accomplished are not known, studies have shown that tissues exposed to thyroxine develop greatly enhanced quantities of most of their enzymes. At least 13 different cellular enzymes have been shown to be greatly increased under the influence of thyroxine, and because enzymes are the regulators of chemical reactions in the cells it is quite easy to understand how this could raise the metabolic rates of the cells. Because thyroxine affects all but a very few tissues of the body, it regulates the overall rate of activity of all functions of the body.

We still do not know the intracellular mechanism by which thyroxine augments the quantities of enzymes. However, since so many different enzymes are involved, and since the rate of enzyme formation is controlled primarily by genetic mechanisms of the cell, it is presumed that thyroxine activates specific genes to increase these enzymes. It also increases the number and sizes of the mitochondria, but whether this is a primary or a secondary effect is unknown.

EFFECTS OF THYROXINE ON SPECIFIC FUNCTIONS OF THE BODY

Effect on Total Body Metabolism. Total lack of thyroxine production by the thyroid gland decreases the metabolic rate to about one-half normal. On the other hand, secretion of very large quantities of thyroxine can increase the rate of metabolism to as much as two times normal. Therefore overall, the thyroid gland can change the rate of metabolism as much as four-fold. Measurements of the basal metabolic rate can be used to estimate the degree of activity of the thyroid gland.

Thyroxine causes the body to burn its available carbohydrates very rapidly, and then to make additional deep inroads on the stores of fats. Therefore, a person with excess production of thyroxine usually loses weight, sometimes very rapidly. On the other hand, a person with less than normal production of thyroxine often develops obesity.

Effect on the Cardiovascular System. Thyroxine affects the cardiovascular system in two ways. First, because the metabolic rate rises, all the tissues of the body require increased quantities of nutrients. This leads to *vasodilatation* in all the tissues and causes the heart to pump greater quantities of blood than usual. Second, thyroxine has a *direct effect on the heart*, increasing its rate of metabolism as well as its rate and forcefulness of contraction. These effects also help to boost the cardiac output.

Effect on the Nervous System. Thyroxine greatly increases the activity of the nervous system. The reflexes become very excitable with excess thyroxine, but very sluggish with diminished thyroxine. Thyroxine increases a person's degree of wakefulness, whereas lack of thyroxine sometimes makes him sleep as many as 12 to 15 hours per day.

A special effect of thyroxine on the nervous system is to cause a continuous *tremor* of the muscles. The tremor is usually very fine but rapid, having a frequency of 10 to 16 times per second, which is considerably faster than the tremor caused by basal ganglia or cerebellar disease.

Effect on the Gastrointestinal Tract. Thyroxine increases the motility of the gastrointestinal tract and promotes copious flow of digestive juices. If these activities are enhanced sufficiently, diarrhea may develop. On the other hand, lack of thyroxine causes the opposite effects—sluggish motility and greatly diminished secretion—resulting in constipation. Excess production of thyroxine also causes a voracious appetite because of the rapid rate of metabolism. The person eats a large amount of food, digests it rapidly, absorbs large quantities of nutrients, but metabolizes these as rapidly as they become available.

Regulation of Thyroxine Production

Earlier in the chapter it was pointed out that thyroxine production is regulated almost entirely by thyroid-stimulating hormone from the

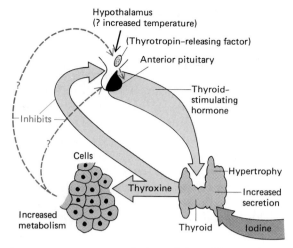

FIGURE 45–12 Regulation of thyroid secretion.

anterior pituitary gland. In turn, as illustrated in Figure 45–12, the rate of secretion of thyroid-stimulating hormone is regulated by *thyrotropin-releasing factor* secreted by the hypothalamus. Therefore, to describe the regulation of thyroxine production, we need only to discuss the factors that regulate secretion of thyrotropin-releasing factor.

The primary stimuli controlling the rate of secretion of thyrotropin-releasing factor are the level of circulating thyroxine in the blood and the level of metabolism in the body. If either the thyroxine level or metabolism falls to a low value, the rate of secretion of thyrotropin-releasing factor increases automatically, in turn increasing the secretion of thyrotropin and consequently of thyroxine. The thyroxine then raises the metabolic rate of the body back toward normal.

On the other hand, if the body's rate of metabolism rises above normal, the hypothalamus decreases its secretion of thyrotropin-releasing factor, and an opposite sequence of events reduces the secretion of thyroxine, thereby reducing both the thyroxine and the metabolic rate back toward normal. Thus, as demonstrated in Figure 45–12, the thyroid-stimulating hormone–thyroxine mechanism ordinarily acts as a feedback system to regulate body metabolism at the normal mean value.

Effect of Cold Weather on Thyroxine Production. When a person or an animal is exposed to very severe cold, the hypothalamus secretes greater quantities of thyrotropin-releasing factor. Over a period of 3 to 4 weeks, the thyroid gland gradually enlarges, and the rate of thyroxine secretion increases. The basal metabolic rate can be boosted by as much as 20 to 30 percent by this mechanism, which helps to warm the body and partially compensate for the cooling effect of the weather.

Abnormalities of Thyroid Secretion

Hyperthyroidism. Failure of the anterior pituitary–thyroid regulatory system to function properly frequently leads to greatly increased production of thyroxine, sometimes to as much as 15 times normal. Most frequently the increased production is caused by a substance called *long-acting thyroid stimulator* (LATS) that is believed to be an antibody formed by the immune system and that specifically stimulates the thyroid cells. The excess production of thyroxine resulting from this stimulation then causes the rate of secretion of thyroid-stimulating hormone by the hypothalamus to be reduced rather than elevated.

In a few hyperthyroid persons, overproduction of thyroxine is caused by a small *thyroid adenoma*, a tumor of the thyroid gland that secretes thyroxine independently of control by the anterior pituitary gland or other stimuli. In either event the excess production of thyroxine causes *hyperthyroidism.*

In hyperthyroidism the basal metabolic rate rises very high—sometimes to as much as twice normal—the person loses weight, he develops diarrhea, he becomes highly nervous and tremulous, his heart rate is greatly increased, and the heart often beats so hard that he feels it palpitating in his chest. The state of hyperthyroidism is often so severe and so prolonged that it actually "burns out" the tissues, leading to degenerative processes in many parts of the body. One of the most common areas of degeneration is the muscle of the heart itself.

The usual methods for treating hyperthyroidism are (1) administration of a drug that suppresses thyroid function or destroys the thyroid gland or (2) surgical removal of a major portion of the gland. Administration of the drug *propylthiouracil* blocks the chemical reaction of tyrosine and iodine to form thyroxine, and hyperthyroidism can often be controlled with this drug.

Exophthalmos (Protrusion of the Eyeballs). Most patients with severe hyperthyroidism develop edematous and overgrown tissues in the eye sockets behind the eyes, thus causing the eyeballs to protrude, a condition called *exophthalmos* (Fig. 45–13). It is believed that thyroxine has little to do with the protrusion of the eyeballs but instead that this is caused by the same autoimmune mechanism that overstimulates the thyroid gland in hyperthyroidism. Because exophthalmos results at least partially from increased quantity of tissue behind the eyes, elimination of the hyperthyroidism will not

FIGURE 45–13 A hyperthyroid person with exophthalmos.

eliminate all the exophthalmos, which remains throughout life thereafter.

Hypothyroidism. Diminished production of thyroxine is called *hypothyroidism*. A person can live for many years with complete lack of thyroxine production, but the rate of metabolism in all his tissues is decreased to only slightly more than one-half normal. He is extremely lethargic, sleeping sometimes as much as 12 to 15 hours a day. He usually is constipated, his mental reactions are sluggish, and he often becomes fat. In addition to increased deposition of fat throughout his body, in very severe hypothyroidism, illustrated in Figure 45–14, a gelatinous

FIGURE 45–14 Patient with myxedema. (Courtesy of Dr. Herbert Langford.)

mixture of mucoprotein and extracellular fluid is deposited in the spaces between the cells, giving an edematous appearance. When this occurs the condition is called *myxedema*.

Goiter. In hyperthyroidism the overactive thyroid gland usually enlarges two- to threefold and is then called a *goiter*. In hypothyroidism the gland frequently enlarges also, and here again the enlarged gland is also called a goiter. Therefore, the term goiter is not synonymous with either hyper- or hypothyroidism but instead means simply enlargement of the thyroid gland.

Hypothyroidism is usually caused by some abnormality of the thyroid gland that makes it impossible for the gland, even when stimulated by thyroid-stimulating hormone, to secrete enough thyroxine. Yet the poorly secreting gland enlarges more and more in a futile attempt to produce adequate quantities of thyroxine, and large amounts of colloid substance containing almost no thyroxine are secreted into the follicles. For this reason this type of enlarged thyroid gland is called a *colloid goiter*.

Sometimes a colloid goiter becomes 15 times as large as the normal thyroid gland, weighing 500 or more grams and occupying a space in the neck as large as 0.05 L. Obviously, a gland this large can obstruct breathing and swallowing.

Endemic Goiter. Persons residing in areas of the world where the food contains very little iodine cannot produce an adequate quantity of thyroxine. As a result, their circulating level of thyroxine as well as their metabolic rate fall below normal, and this enhances the output of thyroid-stimulating hormone by the anterior pituitary gland, which in turn stimulates the thyroid gland in an attempt to produce increased quantities of thyroxine. Unfortunately, even this stimulus cannot enhance the output of thyroxine when iodine is lacking. But the anterior pituitary gland continues to produce large amounts of thyroid-stimulating hormone, so that the thyroid continues to enlarge, becoming progressively filled with colloid that contains almost no thyroxine. The enlarged gland is called an *endemic goiter* because everyone in the iodine-deficient geographic region develops an enlarged gland. Endemic goiter was formerly widely prevalent in those regions of the world, such as the Great Lakes region of the United States and the Swiss Alps, where iodine is not present in the soil. More recently, however, a small amount of iodine has been added to most commercial table salts, so that now an inadequate intake of iodine is very rare.

QUESTIONS

1. Explain the role of cyclic AMP as an intracellular hormone mediator.
2. Explain how hormones sometimes control cellular function by activating genes.
3. How does growth hormone stimulate growth of cartilage and bone?
4. How does growth hormone promote protein deposition?
5. What are the causes of dwarfs, giants, and acromegaly?
6. List the six most important anterior pituitary hormones.
7. What are the hypothalamic releasing and inhibitory factors?
8. What functions do antidiuretic hormone and oxytocin subserve?
9. Give the cellular and chemical mechanisms for the formation of thyroxine.
10. What are the effects of thyroxine on cellular metabolism? What is the mechanism of these effects?
11. What are the causes and effects of hyperthyroidism and hypothyroidism?

REFERENCES

Austin, C.R., and Short, R.V. (eds.): Mechanisms of Hormone Action. New York, Cambridge University Press, 1979.

Baxter, J.D., and MacLeod, K.M.: Molecular basis for hormone action. *In* Bondy, P.K., and Rosenberg, L.E. (eds.): Metabolic Control and Disease, 8th Ed. Philadelphia, W.B. Saunders, 1980, p. 104.

Besser, G.M. (ed.): The Hypothalamus and Pituitary. *Clin. Endocrinol. Metab.*, 6(1), 1977.

Chiodini, P.G., and Liuzzi, A.: The Regulation of Growth Hormone Secretion. St. Albans, Vt., Eden Medical Research, 1979.

DeGroot, L.J. (ed.): Endocrinology. New York, Grune & Stratton, 1979.

DeGroot, L.J.: Thyroid hormone action. *In* DeGroot, L.J., *et al.* (eds.): Endocrinology. Vol. 1. New York, Grune & Stratton, 1979, p. 357.

Dillon, R.S.: Handbook of Endocrinology: Diagnosis and Management of Endocrine and Metabolic Disorders, 2nd Ed. Philadelphia, Lea & Febiger, 1980.

Dumont, J.E., and Vassart, G.: Thyroid gland metabolism and the action of TSH. *In* DeGroot, L.J., *et al.* (eds.): Endocrinology. Vol. 1. New York, Grune & Stratton, 1979, p. 311.

Ekins, R., *et al.* (eds.): Free Thyroid Hormones. New York, Excerpta Medica, 1979.

Ezrin, C., *et al.* (eds.): Pituitary Diseases. Boca Raton, Fla., CRC Press, 1980.

Goss, R.J.: The Physiology of Growth. New York, Academic Press, 1977.

Li, C.H. (ed.): Thyroid Hormones. New York, Academic Press, 1978.

McClung, M.R., and Greer, M.A.: Treatment of hyperthyroidism. *Annu. Rev. Med.*, *31*:385, 1980.

Ontjes, D.A., *et al.*: The anterior pituitary gland. *In* Bondy, P.K., and Rosenberg, L.E. (eds.): Metabolic Control and Disease, 8th Ed. Philadelphia, W.B. Saunders, 1980, p. 1165.

Robbins, J., *et al.*: The thyroid and iodine metabolism. *In* Bondy, P.K., and Rosenberg, L.E. (eds.): Metabolic Control and Disease, 8th Ed. Philadelphia, W.B. Saunders, 1980, p. 1325.

Ryan, W.G., *et al.* (eds.): Endocrine Disorders: A Pathophysiologic Approach, 2nd Ed. Chicago, Year Book Medical Publishers, 1980.

Schulster, D., and Levitski, A. (eds.): Cellular Receptors for Hormones and Neurotransmitters. New York, John Wiley & Sons, 1980.

Stanbury, J.B. (ed.): Endemic Goiter and Endemic Cretinism. New York, John Wiley & Sons, 1980.

Tepperman, J.: Metabolic and Endocrine Physiology: An Introductory Text. Chicago, Year Book Medical Publishers, 1980.

46

Adrenocortical Hormones, Insulin, and Glucagon

Overview

The **adrenal cortex** secretes two very important hormones, (1) *aldosterone* and (2) *cortisol,* both of which are essential to life.

Aldosterone is called *mineralocorticoid* because it specifically alters the concentrations of ions (the minerals) in the body fluids. The most important effect of aldosterone is to increase the rate of *reabsorption of sodium ions* from the renal tubules while at the same time increasing the rate of *secretion of potassium ions* from the blood into these same tubules. The net result is *retention of sodium in the body* and *loss of potassium.* During *Addison's disease,* in which the adrenal glands secrete no aldosterone, the extracellular potassium ion concentration sometimes increases to more than double normal, which has a direct effect on the heart to decrease its pumping effectiveness. At the same time, too little sodium is retained in the body to maintain adequate extracellular fluid volume, eventually causing circulatory shock. These two factors together lead to death within a few days.

When aldosterone secretion is excessive, usually caused by an *aldosterone secreting tumor,* sodium and water are retained in the body, and the *extracellular fluid volume becomes greater than normal.* This in turn *increases the cardiac output* and leads to *hypertension.*

Cortisol is called a *glucocorticoid* because it affects the *metabolism of glucose* in the body. However, it has equally potent effects on *fat and protein metabolism* as well, so that it is in reality a general metabolic hormone. Two basic metabolic effects of cortisol are (1) to *mobilize both fats and proteins from tissues* and (2) to utilize these to supply much of the energy required for body metabolism. At the same time, cortisol also *decreases the rate of utilization of carbohydrates* for energy.

Another important effect of cortisol is to *stabilize the membranes of lysosomes,* that is, to diminish the ease with which lysosomes rupture inside cells and release their intracellular digestive enzymes. This effect of cortisol greatly *reduces the amount of tissue inflammation* that occurs in many diseases, especially in the *autoimmune diseases* such as *rheumatic fever, rheumatoid arthritis,* and some *acute kidney diseases.*

The **islets of Langerhans in the pancreas** secrete two important hormones, (1) *insulin* and (2) *glucagon,* both of which have profound effects on glucose metabolism.

The most important primary effect of *insulin* is to *increase the transport of glucose through the cell membrane* into most cells of the body. This in turn *increases the rate of metabolism of glucose* by the cells, including increase in the use of glucose for energy, increase in the storage of glycogen in both the muscles and liver, and increase in the conversion of glucose into fats by both the liver and the fat cells. In the absence of insulin, which is the clinical condition called *diabetes,* very little glucose can be used for any of these purposes. Instead, the cells utilize mainly fats and proteins to supply their energy needs. One consequence is greatly increased circulating fatty substances in the blood, especially *fatty acids, cholesterol,* and a breakdown product of fatty acids called *acetoacetic acid.* The excess acid in the blood can lead to *diabetic coma,* and the excess cholesterol can lead to early development of *atherosclerosis* and severe *heart attacks.*

The primary function of *glucagon* is to *increase the blood glucose concentration.* It achieves this in two ways: (1) Glucagon has a direct and very rapid action to *split the glycogen in the liver cells into glucose molecules,* which are then released into the blood. (2) Glucagon also *promotes the conversion of amino acids* by the liver cells *into glucose,* a process called *glucogenesis,* thus providing still more glucose for release into the circulatory blood.

THE ADRENAL GLANDS AND THE ADRENOCORTICAL HORMONES

An **adrenal gland** drapes like a cap over the upper pole of each kidney. The gross appearances of these two glands and their locations on top of the kidneys are illustrated in Figure 45–1 of the previous chapter. Each gland is about 5 m across and 1 m thick, and, as illustrated in Figure 46–1, it has two distinct parts:

1. The **adrenal medulla.** This is a thin central core of the gland consisting almost entirely of modified neuronal cells derived from the sympathetic nervous system. These cells secrete *epinephrine* and *norepinephrine* in response to sympathetic stimuli and therefore play a very important role in the functions of the sympathetic system, as was discussed fully in Chapter 28.

2. The **adrenal cortex.** The adrenal cortex surrounds the medulla on all sides, which is also shown in Figure 46–1. The glandular cells of this cortex are illustrated in Figure 46–2; these are large cells containing considerable amounts of fatty substances with high proportions of cholesterol. The hormones secreted by these cells are all steroids having chemical structures very similar to this cholesterol, and in fact the cells can synthesize the adrenocortical hormones from cholesterol. Though as many as 30 different steroid hormones have been isolated from the adrenal cortex, they can all be divided into three different categories on the basis of their functions:

   a. *Mineralocorticoids,* which control the kidney excretion of sodium and potassium. By far the most important of the mineralocorticoids is *aldosterone.*

   b. *Glucocorticoids,* which help to control the metabolism of proteins, fats, and glucose. By far the most important of these hormones is *cortisol.*

   c. *Androgens,* which cause masculinizing effects in the body similar to the effects caused by the male testicular hormone testosterone.

The adrenal cortex is also divided into three distinct layers, which are illustrated in both Figures 46–1 and 46–2. These are

a. The **zona glomerulosa,** which secretes aldosterone almost exclusively.

Zona glomerulosa secretes aldosterone

Zona fasciculata

Zona reticularis

Cortisol and androgens

Medulla

Cortex

Magnified section

FIGURE 46–1 Secretion of adrenocortical hormones by the different zones of the adrenal cortex.

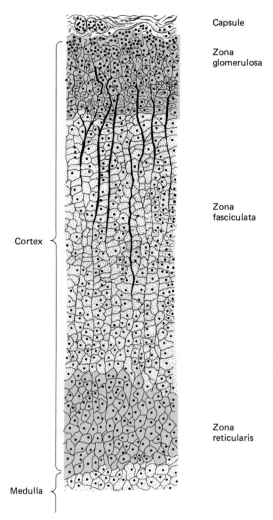

Capsule

Zona glomerulosa

Cortex

Zona fasciculata

Zona reticularis

Medulla

FIGURE 46–2 Cellular structure in the different zones of the adrenal cortex.

b. The **zona fasciculata,** which makes up by far the greater portion of the adrenal cortex and secretes most of the cortisol. Androgens are probably also secreted by the zona fasciculata.

c. The **zona reticularis,** which is the deepest layer of the adrenal cortex, lying adjacent to the adrenal medulla. This layer probably also contributes to the secretion of both cortisol and the androgens.

FUNCTION OF THE MINERALOCORTICOIDS—ALDOSTERONE

The function of the mineralocorticoids is to regulate the quantities and concentrations of both sodium and potassium ions in the extracellular fluid. Their effect is to *increase* the sodium ions and *decrease* the potassium ions.

The adrenal cortex secretes at least three different hormones that can be classified as min-eralocorticoids: *aldosterone, corticosterone,* and minute quantities of *deoxycorticosterone.* However, aldosterone accounts for at least 95 percent of the total mineralocorticoid activity.

The function of aldosterone in relation to electrolyte absorption from the kidney tubules has already been presented in Chapter 32; its overall function can be summarized briefly as follows:

Effect on Sodium. Aldosterone has a direct effect on the epithelial cells of the distal tubules and collecting ducts to increase the rate of sodium reabsorption. When large quantities of aldosterone are secreted, essentially all of the sodium entering the glomerular filtrate is reabsorbed back into the blood, and almost no sodium passes into the urine. On the other hand, when minute amounts of aldosterone are secreted, much less sodium is reabsorbed, and as many as 20 to 30 g of sodium may appear in the urine each day. Therefore, it is evident that aldosterone is essential to prevent rapid depletion of sodium in the body. Thus, aldosterone secretion represents a method for controlling the total quantity of sodium in the extracellular fluids.

Basic Mechanism by Which Aldosterone Increases Sodium Reabsorption. When the rate of aldosterone secretion is suddenly increased, this does not cause an immediate increase in sodium reabsorption. Instead, the aldosterone first binds with a *receptor protein* in the renal epithelial cell cytoplasm. During the ensuing few minutes this combination of aldosterone and receptor migrates to the nucleus where it activates the genetic mechanism to produce enzymes or carrier proteins or both for transport of sodium ions through the tubular epithelium. After about 45 minutes to an hour enough of these enzymes or carrier substances appear in the epithelial cells to begin increasing the rate of sodium transport, and maximum activity occurs in about 3 hours. Then, when aldosterone secretion ceases, the activity of the sodium transport mechanism decreases back to a low level within another 2 to 4 hours.

Effect on Potassium. At the same time that aldosterone increases sodium reabsorption from the tubules, it also causes simultaneous increase in potassium secretion into the tubules and therefore marked excretion of potassium into the urine. Thus, aldosterone has exactly the opposite effect on the quantity of potassium in the extracellular fluid to its effect on sodium.

The increase in potassium excretion results mainly from the fact that the sodium and potassium transport mechanisms in the tubular epithelial cells are a partial exchange process, with potassium from the extracellular fluids being exchanged for much of the sodium from the tubules. In other words, the transport mechanism of the tubular epithelial cells is the same sodium-potassium "pump" that we discussed many times in the text for transport of sodium in one direction through a cell membrane and simultaneous transport of potassium in the opposite direction. It is this sodium-potassium pump of the renal tubules that is stimulated by aldosterone, thus increasing the quantity of sodium ions in the extracellular fluid while at the same time increasing potassium excretion into the urine and consequently decreasing the potassium in the extracellular fluid.

Effect on Chloride Ion. Even though aldosterone causes an exchange of sodium and potassium through the tubular epithelial cells, usually far more sodium is reabsorbed from the tubules than potassium is excreted. Also, since the absorbed sodium ions carry positive charges out of the tubules and into the extracellular fluids, this creates a strong electronegative charge in the lumen of the tubules. This electronegativity in turn has an especially powerful effect on chloride transport through the tubular membrane. The chloride ions, being themselves electronegative, are repelled from the negative tubular fluids and into the surrounding extracellular fluid. Thus, a secondary effect of mineralocorticoid action is to decrease the amount of chloride lost in the urine and to increase the quantity of chloride ions in the extracellular fluid. The net result is greater sodium chloride (salt) in the extracellular fluid and reduced potassium.

Effect on Water Reabsorption from the Tubules and on Extracellular Fluid Volume. The great increase in sodium and chloride reabsorption from the tubules results in an increase in the reabsorption of water as well. The reason for this is simply an osmotic effect that can be explained as follows: Transport of the sodium and chloride ions from the tubules into the extracellular fluid diminishes the concentrations of these in the tubular fluid and thus also reduces the tubular osmotic pressure. Therefore, an osmotic gradient develops across the tubular membrane to cause greatly increased absorption of water by osmosis into the peritubular fluids.

The increase in both sodium chloride and water in the extracellular fluids often raises the extracellular fluid volume as much as 10 to 15 percent.

Effect on Blood Volume and on Circulatory Dynamics. The excess extracellular fluid enters both the interstitial spaces and the blood, increasing both the interstitial fluid volume and the blood volume. The greater blood volume in turn forces a larger than normal quantity of blood toward the heart, which boosts the *cardiac output* a small amount; and this in turn often causes *elevated arterial pressure* as well. Therefore, excess aldosterone is sometimes a cause of high blood pressure.

Regulation of Aldosterone Secretion

The regulation of aldosterone secretion has already been discussed in Chapter 32 in connection with the effects of aldosterone on renal function. However, let us review this regulation briefly.

Effect of Potassium Concentration on Aldosterone Secretion. The most potent long-term stimulator of aldosterone secretion is a rise in potassium ion concentration in the plasma. The increased secretion of aldosterone in turn causes the kidneys to excrete large amounts of potassium from the body, thereby reducing the potassium concentration in the extracellular fluid back toward normal. Research studies have shown that this is a very powerful mechanism by which the body regulates the long-term level of potassium concentration in the extracellular fluid.

Effect of Angiotensin on Aldosterone Secretion. Next to potassium ions, the factor that causes the greatest acute increase in aldosterone secretion is angiotensin. It will be recalled from Chapter 38 that when the blood flow through the kidneys decreases below normal, large quantities of the substance *renin* are secreted by the kidneys into the blood, and the renin in turn acts as an enzyme to cleave the vasoconstrictor hormone angiotensin from one of the plasma proteins. The angiotensin then causes powerful constriction of the arterioles throughout the body, thus raising the arterial pressure.

At the same time that angiotensin causes arteriolar constriction, it also acts directly on the adrenal cortex to increase the rate of secretion of aldosterone. The stimulation of aldosterone secretion is considerable during the first 8 to 24 hours of increased angiotensin but becomes much weaker over the ensuing days and weeks. This is in contrast to the effect of potassium, which continues to exert its effect on aldosterone secretion indefinitely.

Effect of Decreased Body Sodium and Decreased Extracellular Fluid Volume on Aldosterone Secretion. Though an acute decrease in body sodium has rather little effect on aldosterone secretion, when the body sodium remains diminished as long as a day or more, the rate of aldosterone secretion increases markedly. Likewise, decreased extracellular fluid volume over a long period of time will also increase aldosterone secretion. Yet, the mechanism by which both of these factors increase aldosterone secretion is still greatly disputed. One of the ways is through the effect of decreased sodium or decreased extracellular fluid volume to reduce renal blood flow and stimulate the formation of angiotensin. This in turn stimulates aldosterone secretion. Second, some research studies have demonstrated that decreased body sodium probably causes secretion of some yet uncharacterized hormone from the anterior pituitary gland that stimulates the adrenal cortex to secrete aldosterone.

Effect of Adrenocorticotropic Hormone on Mineralocorticoid Secretion. In the following paragraphs we shall see that one of the anterior pituitary hormones, *adrenocorticotropic hormone* (also called *corticotropin* or *ACTH*), is by far the most important factor for controlling the secretion of glucocorticoid hormones by the adrenal cortex. This hormone also causes a slight to moderate increase in the secretion of mineralocorticoids, though these mineralocorticoids are likely to be other adrenal steroids besides aldosterone that have the same effect. Because of this action of adrenocorticotropic hormone, patients who have excess secretion of this hormone by the pituitary will usually develop excess retention of sodium, and, if the condition persists for long periods of time, hypertension will occur. This is the disease called *Cushing's disease* also having other effects that will be discussed more fully later in this chapter.

Glucocorticoids—Cortisol

The functions of the glucocorticoids are less well understood than those of the mineralocorticoids. However, many of the body's metabolic systems become greatly deranged when glucocorticoids are absent, and the person becomes unable to resist almost any trauma or disease condition that tends to destroy tissues. Therefore, the most important function of glucocorticoids is to enhance resistance to physical "stress," though the means by which this is effected are yet very vague.

Several different adrenocortical hormones exhibit glucocorticoid activity, but by far the most abundant one is *cortisol* (also known as *hydrocortisone*). Other much less important glucocorticoids are *corticosterone* and *cortisone*.

Effect of Cortisol on Glucose Metabolism. The earliest discovered effect of glucocorticoids was their ability to increase the concentration of glucose in the blood. This is caused by two different functions of cortisol. First, cortisol depresses utilization of glucose by the tissue cells. Consequently, the glucose accumulates in the extracellular fluid instead of being utilized fully by the tissue cells.

Second, cortisol causes the liver cells to convert proteins and the glycerol portion of fats into glucose, a process called *gluconeogenesis*. It is believed that this results partly because cortisol increases the liver enzymes that promote gluconeogenesis. However, gluconeogenesis is increased also because cortisol has a specific effect to cause amino acids to be mobilized from the protein stores of the body and fat also to be mobilized from the fat stores. The excess supply in the blood of amino acids and glycerol (derived from the fat) provides material that the liver can convert into glucose.

Gluconeogenesis is very important during starvation because it supplies a continual source of glucose in the blood even when the person is not ingesting glucose. This glucose is necessary to provide nutrition for the neurons in the brain because these cells normally can use only glucose for energy.

Effect of Cortisol on Protein Metabolism. One of the most important effects of cortisol is that of *decreasing* the quantity of protein in most tissues of the body, with the exception of the liver. It does this both by suppressing the rate of protein formation in the nonliver cells and also by causing breakdown of proteins already present in the cells into amino acids and then release of these into the blood.

The blood concentration of amino acids increases considerably under the influence of cortisol, both because of the decreased use of amino acids by the cells to form proteins and because of release of amino acids from the cells into the blood. This effect to increase the blood amino acids is especially important in times of bodily stress because it makes amino acids available for use wherever in the body they are needed.

A final effect of cortisol on protein metabolism is to *increase* the rate of protein formation in the liver cells. Since it is the liver cells that produce most of the plasma proteins, cortisol also increases the quantity of plasma proteins.

Effect of Cortisol on Fat Metabolism. Cortisol mobilizes fat from the fat depots in much the same manner that it mobilizes amino acids from the cells. The net result is a decrease in the amount of fat in the storage areas and an increase in use of fat for energy and other purposes. It is believed, for example, that the mobilization of fat from the storage areas during periods of starvation is caused mainly by increased production of cortisol.

During periods of rapid fat mobilization, the liver splits much of the fat into *keto acids.* If these are used immediately by the cells for energy, they cause no significant physiologic effect, but if they are not used immediately, their concentration in the extracellular fluid can become great enough to cause acidosis. This is occasionally one of the untoward effects of excessive cortisol secretion.

Effect of Cortisol on Lysosome Stabilization. When cells are damaged either because of trauma or because of disease, the lysosomes in the cytoplasm begin to break open, releasing their digestive enzymes into the interior of the cell. These enzymes in turn autolyze internal structures of the cells, which diminishes cellular function and sometimes actually kills the cells. Therefore, it is often important to prevent this process. Cortisol has a special ability to stabilize the membranes of lysosomes and thereby to prevent their dissolution. Fortunately, in disease states, large quantities of cortisol are almost invariably secreted into the blood, and it is believed that one of the most important functions of cortisol is to help a person resist the devastating effects of some diseases by preventing rupture of the lysosomes inside the cells.

REGULATION OF CORTISOL SECRETION— THE ROLE OF "STRESS"

Figure 46–3 summarizes the regulation of adrenal secretion of cortisol and other glucocorticoids. The mechanism can be explained as follows: The primary stimulus that initiates glucocorticoid secretion is called "stress," which includes almost any type of damage to the body. For instance, a painful contusion of some part of the body, a broken bone, severe damage to large tissue areas by some disease condition, or any other destruction of parts of the body usually

FIGURE 46–3 Mechanism for regulation of glucocorticoid secretion.

sets off a sequence of events that leads to cortisol secretion. The stress probably causes these reactions by initiating nerve impulses that are transmitted from the periphery into the hypothalamus. The hypothalamus then secretes the substance *corticotropin-releasing factor* (CRF), which passes by way of the hypothalamic-hypophyseal portal system into the anterior pituitary gland. Here, this factor causes the cells of the gland to secrete adrenocorticotropic hormone, which flows in the blood to the adrenal cortex where it elicits *cortisol* secretion. The cortisol then mobilizes protein and fat from all over the body and also causes gluconeogenesis. The increased availability of amino acids, fats, and glucose in the blood helps in the repair of the damage, thus attenuating the initial stimulus that had set off the sequence of events leading to cortisol secretion. In addition, the cortisol prevents breakdown of the lysosomes, thus also preventing further destruction of the tissues.

Androgens

Androgens are hormones that cause the development of male sex characteristics, as we shall see in Chapter 48. Though the male testes are the primary source of these hormones, the adrenal cortex also secretes minute quantities, quantities so small that in the normal person they have no significant effect. However, an *androgen-producing tumor* of the adrenocortical cells develops occasionally, which secretes very large quantities of male sex hormones that cause serious masculinizing effects even in the female body. This is explained below.

Abnormalities of Adrenocortical Secretion

Hyposecretion of Adrenocortical Hormones—Addison's Disease. The adrenal cortices are occasionally destroyed by disease, or sometimes they simply atrophy. Sometimes, excessive overstimulation of the adrenal glands by stress also causes them to become, first, greatly enlarged, then hemorrhagic, and finally replaced almost completely by fibrous tissue. In each instance hyposecretion or, rarely, complete lack of secretion of adrenocortical hormones occurs, leading to a serious condition called *Addison's disease.*

Complete failure of the adrenal cortices usually causes death within 3 to 5 days unless the person is appropriately treated. This early death is caused by the lack of aldosterone, because the kidney's sodium reabsorptive and potassium secretory mechanisms are highly dependent on this hormone. Without adequate sodium reabsorption, the extracellular fluid volume decreases so greatly within only a few days that death ensues. This effect can be offset to some extent by having the person eat extra quantities of salt, and it can be overcome entirely by having him ingest by mouth each day extremely minute amounts of one of the mineralocorticoids, for instance, 0.2 mg of fludrocortisone, a synthetic mineralocorticoid.

Even if the life of the person with Addison's disease is saved by administering a mineralocorticoid, he still remains unable to resist stress, and even a slight respiratory infection may prove lethal. He also has little energy. Usually, therefore, to provide completely adequate function of the body, treatment with cortisol (about 30 mg daily) or some other glucocorticoid is necessary in addition to the mineralocorticoid.

Hypersecretion of Adrenocortical Hormones. Hypersecretion by the adrenal cortex can result either from a tumor in one part of an adrenal gland, or because of increased production of adrenocorticotropic hormone by the anterior pituitary, this in turn stimulating the adrenal cortices.

The effects of hyperadrenalism depend upon which types of cells of the adrenal gland are secreting excessive quantities of hormone. In general, three different patterns of hyperadrenalism often occur.

Primary Aldosteronism. If the increased secretion occurs in the outer layer of the adrenal cortex, in the *zona glomerulosa*, the effects are those of excessive aldosterone secretion. Therefore, the concentration of potassium in the blood decreases, the body sodium and extracellular fluid volume increase, the blood volume increases, cardiac output increases, and the arterial pressure rises, sometimes severely so. This condition is called *primary aldosteronism.*

Cushing's Disease. If there is generalized hypertrophy of the adrenal cortex, it is usually cortisol that is secreted in the greatest quantity. This causes the clinical condition called *Cushing's disease*, which has the following characteristics: First, there is excessive mobilization of proteins and fats from their storage areas. The protein mobilization causes *weakness of the muscles* and sometimes weakness of other protein structures such as the fibers that hold the tissues together beneath the skin. This allows laxity of the skin and predisposes to tears in the subcutaneous tissues, which can be observed as long, linear, *purplish striae.* Second, the excess mobilization of proteins and fats also causes *increased gluconeogenesis* and raises the blood glucose concentration, sometimes enough to cause a very high blood sugar level—a condition known as *adrenal diabetes.* Last, cortisol has a moderate amount of mineralocorticoid activity in addition to its glucocorticoid activity so that some of the same effects that are seen in primary aldosteronism, especially the development of hypertension, also occur.

Adrenogenital Syndrome. Occasionally some of the cells of the adrenal gland form an *androgen-secreting tumor.* When this occurs, the person, even a child or female, develops masculine characteristics such as growth of hair on the face, deepening of the voice, sometimes baldness, changes in portions of the female sexual organs to resemble the sexual organs of the male, atrophy of the female breasts, and considerably enhanced muscular development. This condition is called the adrenogenital syndrome.

INSULIN

Secretion of Insulin by the Pancreas. The pancreas is a long gland that lies immediately beneath the stomach, a picture of which was shown in Figure 40–15 in Chapter 40. Also, a microscopic picture of a small section of the pancreas is shown in Figure 46–4. It is composed of two different types of tissue. One type is the *acini*, which secrete digestive juices into the intestines; this was discussed in detail in Chapter 41. The other type is the *islets of Langerhans,*

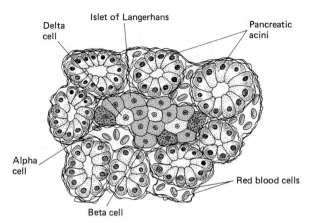

FIGURE 46–4 Physiologic anatomy of the pancreas.

which secrete hormones directly into the blood; this secretion into the blood is an *endocrine* function of the pancreas, whereas the secretion of digestive juices is an *exocrine* function.

The pancreas has many thousands of *islets of Langerhans*, each of which is less than 1 mm in size. The islets are composed of several different types of cells, the most important of which are the **alpha** and **beta cells.** The beta cells secrete the hormone *insulin*, and the alpha cells secrete the hormone *glucagon*, which is discussed later in the chapter. Both of these hormones are small proteins, insulin having a molecular weight of 5808 and glucagon 3485.

Function of Insulin in the Body

Insulin has profound effects on metabolism of all the major foodstuffs—carbohydrates, fats, and protein. Without insulin, an animal or human being cannot grow, partly because it cannot utilize more than a small percentage of the carbohydrate that is eaten but also because the cells cannot synthesize protein. Also, in the absence of insulin the cells utilize extreme amounts of fat, and this results in such debilitating conditions as loss of weight, acidosis, and even coma, so that the person with complete lack of insulin usually has only a short time to live. Therefore, let us see how insulin affects the metabolism of each of the major foodstuffs and why its effects are so very important.

EFFECT OF INSULIN ON GLUCOSE METABOLISM

Effect on Glucose Transport into Cells.
The most important effect of insulin is to promote glucose transport into almost all cells of the body, especially into muscle cells, fat cells, and liver cells.

The mechanism by which insulin promotes glucose transport through the membranes of most cells is illustrated in Figure 46–5. This shows that glucose combines with a carrier substance in the cell membrane and then diffuses to the inside of the membrane where it is released to the interior of the cell. The carrier is then used again and again to transport additional quantities of glucose. This transport process is one of *facilitated diffusion*, which was discussed in Chapter 5; this means simply that combining glucose with the carrier makes it easy for the glucose to diffuse through the membrane, but the transport mechanism can never increase the glucose concentration inside the cell to a level greater than its concentration on the outside.

The effect of insulin on glucose transport is to *activate the facilitated diffusion mechanism*. Within seconds to minutes after insulin combines with the cell membrane, the rate of glucose diffusion into the cell is often increased as much as 15- to 20-fold, suggesting a rapid direct action of insulin either on the cell membrane itself or on the glucose carrier system.

Transport of glucose into liver cells depends on a still different mechanism. The liver cell membrane is so permeable that glucose can diffuse easily through this membrane even in the absence of facilitated transport. However, it can diffuse in both directions—both into and out of the cell. But in the presence of insulin several enzymes in the liver cells are strongly activated, which causes "trapping" of glucose inside the cells. These enzymes are *glucokinase*, which causes the glucose to combine with the phosphate ion, and *glycogen synthetase*, which then

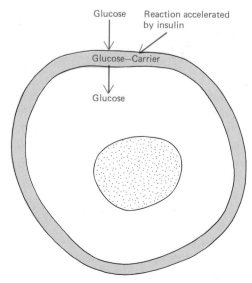

FIGURE 46–5 Basic mechanism of insulin action.

causes large numbers of glucose molecules to combine with each other to form *glycogen*, the large molecular weight polymer of glucose. In the absence of insulin, this trapping mechanism ceases and another enzyme, *phosphorylase*, becomes activated and depolymerizes the glycogen back into glucose, allowing the glucose to leave the cell.

Effect of Insulin to Increase the Use of Glucose for Energy. As we shall learn later in this chapter, when the blood glucose concentration rises above normal the pancreas secretes excessive amounts of insulin. This insulin, in turn, causes rapid transport of the glucose into the cells and makes it available for cellular function. Therefore, one of the obvious effects of insulin is to cause rapid utilization of glucose by most cells of the body for energy.

Effect of Insulin to Promote Glycogen Storage in the Liver and Muscle. After a meal, when there is both excess glucose and excess insulin, glucose is often transported into both the liver cells and the muscle cells at rates much greater than it can be utilized for energy. Therefore, much of it is then stored in the form of glycogen, the glycogen concentration in liver cells sometimes rising to as high as 5 to 6 percent and in muscle cells to over 1 percent.

During periods between meals, the muscle cells reconvert the glycogen back into glucose and use this for energy. Also, as was pointed out above, in the liver cells the enzyme phosphorylase becomes activated and depolymerizes the glycogen back into glucose, releasing the glucose back into the circulating blood to be used wherever in the body it is needed. Thus, the liver is a major temporary storehouse of glucose.

Conversion of Glucose into Storage Fat. After the liver and muscle stores of glycogen have been filled, all the remaining glucose that still cannot be used immediately is stored as fat in the fat tissues. About nine tenths of this fat is synthesized in the liver from the great excess of glucose that is transported into these cells under the influence of insulin. Then, this fat is released into the blood in the form of lipoproteins, which were discussed in Chapter 42, and transported to the *fat cells* (the *adipose cells*) in the fat tissue. In addition, another one tenth of the storage fat is synthesized in the fat cells themselves. Insulin promotes glucose transport into these cells in the same way that it acts elsewhere in the body.

In summary, the effect of insulin on glucose metabolism is to increase its utilization for energy or to cause the glucose to be stored either in the form of glycogen or in the form of fat.

Effect of Insulin on Blood Glucose. In the presence of large amounts of insulin, the rapid transport of glucose into the cells throughout the body decreases the blood glucose concentration. Conversely, lack of insulin causes glucose to dam up in the blood instead of entering the cells. Complete lack of insulin usually produces a rise in blood glucose concentration from a normal value of 90 mg per 100 ml up to about 350 mg per 100 ml. On the other hand, a great excess of insulin can decrease the blood glucose to as low as 25 mg per 100 ml, or to about one-fourth the normal level.

EFFECT OF INSULIN ON FAT METABOLISM

The effect of insulin on fat metabolism is almost exactly opposite to its effect on glucose metabolism. That is, insulin greatly *inhibits* almost all aspects of fat metabolism except the synthesis and storage of fat from glucose, as described above. There are two ways in which insulin inhibits fat metabolism:

First, when glucose is present in excess in cells, the cells have a preference for using glucose over the use of fatty acids for metabolism because of the nature of the enzyme systems in the cells. Therefore, the rate of degradation of the fatty acids becomes greatly decreased.

Second, before the fat that has been stored in fat cells can be utilized for energy, this must be released from the cells. The mechanism for release is for an enzyme in the fat cells themselves called *hormone-sensitive lipase* to cause fatty acids to split away from the stored fat and then to diffuse into the blood. However, insulin strongly inhibits hormone-sensitive lipase so that once fat has been stored in the fat cells, as long as insulin is present in sufficient quantity the fat will not be released.

Thus, in the presence of insulin fat becomes relatively unavailable for metabolism whereas the availability of glucose becomes greatly enhanced.

Effect of Insulin Lack on Fat Metabolism. In the absence of insulin, essentially all phases of fat metabolism are greatly accelerated. First, the hormone-sensitive lipase of the fat cells becomes strongly activated, and large quantities of fatty acids are released into the blood. Second, these fatty acids then become readily available to cells everywhere in the body, and many of them are used almost immediately for energy, especially by the muscle cells. Third, large proportions of the fatty acids are transported into the liver and are converted into *triglycerides, phospholipids*, and *cholesterol*. Most of these three

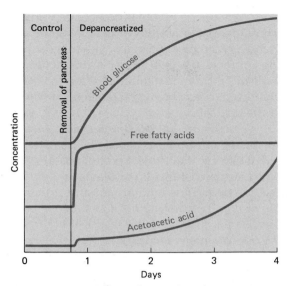

FIGURE 46–6 Effect of removing the pancreas on the concentrations of blood glucose, plasma free fatty acids, and acetoacetic acid.

substances are then released into the blood in lipoproteins, thus greatly increasing the concentration of blood lipids. Fourth, the very rapid metabolism of fatty acids in the liver causes the formation of tremendous quantities of *aceto-acetic acid*, which is subsequently released into the blood. Much of this is utilized by the tissue cells for energy, but some of it remains in the blood and can cause severe generalized acidosis, often leading to *acidotic coma* and death, which are the most dire results of severe diabetes.

Figure 46–6 illustrates the rapid increase in fatty acids in the blood when there is sudden loss of insulin secretion caused by removal of the pancreas. Note also the progressive increase in acetoacetic acid in the blood, which results from prolonged excess fat metabolism, and the slow build-up of blood glucose resulting from failure of the cells to utilize carbohydrates.

EFFECT OF INSULIN ON PROTEIN METABOLISM AND ON GROWTH

Insulin is almost as potent as growth hormone in causing protein deposition in cells. It accomplishes this by both direct and indirect effects on protein metabolism.

The direct effects of insulin on protein metabolism are threefold:

1. Insulin increases the rate of transport of most of the amino acids through the cell membranes, thereby increasing the quantities of amino acids available in the cells for synthesizing proteins.
2. Insulin increases the formation of RNA in cells.

3. Insulin increases the formation of proteins by the ribosomes.

Thus, insulin has a strong direct effect of promoting protein formation in all or most cells of the body.

The indirect effect of insulin on protein metabolism results from its promotion of glucose utilization by the cells. When glucose is available to be utilized for energy, a *protein-sparing* effect is provided, because carbohydrates are then used in preference to proteins. On the other hand, in the absence of insulin this protein-sparing effect is lost, so that large quantities of protein instead of carbohydrates are utilized along with fats for energy.

Effect of Insulin on Growth. Because insulin promotes the formation of proteins, as well as makes available large amounts of energy from carbohydrates, this hormone has a powerful effect on growth. Indeed, its absence in an animal causes as much stunting of growth as does the absence of growth hormone from the anterior pituitary gland.

Regulation of Insulin Secretion

When the blood glucose level rises, the pancreas begins secreting insulin within a few minutes. This is caused by a direct effect of glucose on the beta cells in the pancreatic islets, causing them to produce greatly increased quantities of insulin. The insulin in turn causes the excess glucose to be transported into the cells where it can be used for energy, stored as glycogen, or converted into fat.

Thus, the insulin mechanism provides a feedback means for controlling the concentration of glucose in the blood and extracellular fluids. That is, too high a glucose level initiates insulin secretion, which then causes increased utilization of glucose and return of the blood glucose level back to or toward normal. Conversely, when the blood glucose level falls too low, the rate of insulin secretion decreases, and glucose is now conserved in the body fluids until its level returns to normal.

We shall note later in the chapter that the hormone glucagon also helps to control the blood glucose concentration.

Diabetes Mellitus

Diabetes mellitus is the disease that results from failure of the pancreas to secrete insulin. It is caused by *degeneration or inactivation of the beta cells* of the islets of Langerhans, but the basic causes of these effects usually are not known. In some diabetic persons, particularly those who develop diabetes very early in life, the disease is caused by *inheritance* from a parent (or some preceding ancestor) of beta cells that are highly prone to degenerate. In other persons, *antibodies* develop against the beta cells and cause their destruction, which is an example of an autoimmune disease process. In still others, antibodies sometimes develop against the insulin itself and destroy the insulin before it can act elsewhere in the body; the quantity of insulin secreted may be entirely normal, but the insulin never reaches its destination.

Pathophysiologic Effects of Diabetes.
The primary abnormality in diabetes is failure to utilize adequate quantities of glucose for energy but instead to utilize excessive quantities of fats. This causes the blood glucose level to rise, often to as high as 3 times normal and rarely as high as 10 times normal. Large quantities of glucose are lost into the urine because the kidney tubules cannot reabsorb all that enters them in the glomerular filtrate each minute. The excess tubular glucose also creates a tremendous amount of osmotic pressure in the tubules, and this diminishes the reabsorption of water. As a result, the diabetic person loses large quantities of water as well as glucose into the urine. In extreme cases the excess output of urine causes extracellular dehydration, which can in itself be harmful.

Failure of the diabetic person to utilize glucose for energy deprives him of a major portion of the energy in his food. He loses weight and becomes weakened because of excess burning of his fat and protein stores. As a result of the nutrient deficiency in diabetes, the diabetic person usually becomes very hungry, so that he often eats voraciously even though the carbohydrate portion of his food contributes little to his nutrition.

Ketosis and Diabetic Coma.
The extremely rapid metabolism of fats in diabetes sometimes increases the quantity of keto acids in extracellular fluids to as high as 10 milliequivalents per liter, which is 10 times normal. On occasion this becomes sufficient to make the pH of the body fluids fall from its normal value of 7.4 to as low as 7.0, or rarely as low as 6.9. This degree of acidosis is incompatible with life for more than a few hours. The person breathes extremely rapidly and deeply to blow off carbon dioxide, which helps to offset the metabolic acidosis, but despite this the acidosis often becomes severe enough to cause coma. Unless the person is treated he usually dies in less than 24 hours. Treatment requires immediate administration of large quantities of insulin. Sometimes glucose is administered along with the insulin to help promote the shift from fat to glucose metabolism. Intravenous administration of alkaline solutions can also be of great benefit to neutralize the acidosis.

Treatment of Diabetes with Insulin.
The diabetic person can usually be treated quite adequately by daily injections of insulin. Two principal types of insulin available for this purpose are *crystalline zinc insulin* and *protamine zinc insulin*. The duration of action of crystalline zinc insulin after injection is 4 to 6 hours, and that of protamine zinc insulin is 24 to 30 hours. Usually a person who has severe diabetes must take an injection of crystalline zinc insulin at mealtimes when his blood glucose concentration is likely to rise very high temporarily, and he must take an injection of protamine zinc insulin each morning to provide a steady rate of glucose inflow into his cells throughout the day.

Atherosclerosis in Diabetes.
Prolonged diabetes usually leads to early development of atherosclerosis, and this subsequently causes heart disease, kidney damage, cerebral vascular accidents, or other circulatory disorders. The reason for the development of atherosclerosis is that, even with the best possible treatment of diabetes, glucose metabolism can never be maintained at a sufficiently high level to prevent excess fat metabolism, and cholesterol deposition in the walls of the blood vessels is always an unfortunate accompaniment of rapid fat metabolism. Because of this, the person who develops diabetes in childhood usually has a shortened life, regardless of how well he is treated.

Hyperinsulinism

Hyperinsulinism occasionally develops because of either overtreatment of a diabetic person with insulin or too much secretion of insulin by a pancreatic islet tumor. In either case low blood glucose concentration ensues. This in turn causes overexcitability of the brain at first and then coma. The brain neurons require a constant supply of glucose because they cannot utilize significant amounts of fats or proteins for energy. Furthermore, the rate of glucose uptake by the neurons, unlike that of other cells, is dependent mainly on the blood glucose concentration rather than on the amount of insulin available. Whenever excess insulin is available, the blood glucose becomes very low so that the neurons no longer receive the amounts of glucose needed to maintain their metabolism. This causes them to become first excessively excitable and later depressed. In the excitement stage convulsions may occur, but in the depressed stage the person develops coma not unlike the coma that occurs in untreated diabetes. Indeed, it is sometimes a problem to diagnose the cause of coma in a diabetic. It may result from too little insulin secretion, in which case it is diabetic coma, or from too much treatment with insulin, in which case the abnormality is hyperinsulinism.

GLUCAGON

The alpha cells of the islets of Langerhans secrete a hormone called *glucagon.* Many of the functions of glucagon are opposite to those of insulin, although others complement the functions of insulin.

Glucagon raises the blood glucose level; insulin reduces it. On the other hand, both insulin and glucagon increase the availability of glucose to the cells for their utilization. Glucagon does this by mobilizing glucose from the liver; insulin does so by increasing the transport of glucose into the cells. For instance, during heavy exercise both of these hormones work together to promote utilization of glucose by the muscles.

Basic Mechanisms of Glucagon Function. Glucagon elevates the blood glucose concentration in two ways: First, it increases the breakdown of liver glycogen into glucose, making this available for transport into the blood. Glucagon achieves this effect by activating the enzyme *adenyl cyclase* in the liver cell membranes, which in turn increases the quantity of cyclic AMP in the liver cells. The cyclic AMP then activates the enzyme phosphorylase, which causes *glycogenolysis* (breakdown of the liver glycogen to glucose).

Second, glucagon increases *gluconeogenesis* (conversion of proteins to glucose) by the liver. It does this mainly by activating the liver cell enzymatic system that is responsible for this process.

The blood glucose concentration can rise as much as 20 percent within a few minutes after injection of glucagon.

Control of Glucagon Secretion, and Function of Glucagon in the Body. Glucagon secretion is controlled in almost exactly the opposite manner to the control of insulin. That is, when the blood glucose concentration falls below normal, the pancreas begins to secrete

increased quantities of glucagon into the blood. Indeed, when the blood glucose concentration falls to as low as 60 mg per 100 ml of blood (about 30 percent below normal), the pancreas literally pours glucagon into the blood. This effect of low blood glucose concentration on glucagon secretion results from direct stimulation of the *alpha* cells in the islets of Langerhans. The glucagon in turn causes almost immediate release of glucose from the liver, thereby rapidly increasing blood glucose concentration back up toward the normal level of 90 to 100 mg per 100 ml.

Thus, the glucagon mechanism, like the insulin mechanism, helps to regulate the blood glucose concentration, but with one difference: The glucagon mechanism acts to keep the blood glucose concentrations from falling too low whereas insulin keeps it from rising too high. The glucagon mechanism is especially activated during severe exercise and during starvation, both of which tend to decrease blood glucose.

An especially important function of glucagon is to keep the glucose concentration high enough for normal function of the brain neurons and therefore to prevent hypoglycemic convulsions or hypoglycemic coma, which were discussed above.

QUESTIONS

1. Review the effects of aldosterone on electrolyte and water metabolism.
2. Discuss the important factors that regulate aldosterone secretion.
3. What are the metabolic effects of cortisol on carbohydrates, fats, and proteins?
4. What is the effect of cortisol on lysosomes?
5. What role does "stress" play in the regulation of cortisol secretion?
6. Give the effects of insulin on glucose and fat metabolism.
7. Explain the effects of insulin on protein metabolism and on growth.
8. What role does insulin play in the control of blood glucose concentration?
9. What are the effects of the disease diabetes on cellular metabolism? Why do acidosis and diabetic coma frequently occur?
10. How do the functions of glucagon differ from those of insulin?

REFERENCES

Bondy, P.K.: The adrenal cortex. *In* Bondy, P.K., and Rosenberg, L.E. (eds.): Metabolic Control and Disease, 8th Ed. Philadelphia, W.B. Saunders, 1980, p. 1427.

Fajans, S.S.: Diabetes mellitus: Description, etiology and pathogenesis, natural history and testing procedures. *In* DeGroot, L.J., *et al.* (eds.): Endocrinology. Vol. 2. New York, Grune & Stratton, 1979, p. 1007.

Fitzgerald, P.J., and Morrison, A.B. (eds.): The Pancreas. Baltimore, Williams & Wilkins, 1980.

Genazzani, E., *et al.* (eds.): Pharmacological Modulation of Steroid Action. New York, Raven Press, 1980.

Guyton, J.R., *et al.*: A model of glucose-insulin homeostasis in man that incorporates the heterogeneous fast pool theory of pancreatic insulin release. *Diabetes*, *27*:1027, 1978.

Hall, J.E., *et al.*: Control of arterial pressure and renal function during glucocorticoid excess in dogs. *Hypertension*, *2*:139, 1980.

Hedeskov, C.J.: Mechanism of glucose-induced insulin secretion. *Physiol. Rev.*, *60*:442, 1980.

James, V.H. (ed.): The Adrenal Gland. New York, Raven Press, 1979.

Klachko, D.M., *et al.* (eds.): The Endocrine Pancreas and Juvenile Diabetes. New York, Plenum Press, 1979.

Lund-Anderson, H.: Transport of glucose from blood to brain. *Physiol. Rev.*, *59*:305, 1979.

Matschinsky, F.M., *et al.*: Metabolism of pancreatic islets and regulation of insulin and glucagon secretion. *In* DeGroot, L.J., *et al.* (eds.): Endocrinology. Vol. 2. New York, Grune & Stratton, 1979, p. 935.

Nelson, D.H.: The Adrenal Cortex: Physiological Function and Disease. Philadelphia. W.B. Saunders, 1979.

Notkins, A.L.: The causes of diabetes. *Sci. Am.*, *241*(5):62, 1979.

Podolsky, S. (ed.): Clinical Diabetes: Modern Management. New York, Appleton-Century-Crofts, 1980.

Raisz, L.G., *et al.*: Hormonal regulation of mineral metabolism. *Int. Rev. Physiol.*, *16*:199, 1977.

Tan, S.Y., and Mulrow, P.J.: Aldosterone in hypertension and edema. *In* Bondy, P.K., and Rosenberg, L.E. (eds.): Metabolic Control and Disease, 8th Ed. Philadelphia, W.B. Saunders, 1980, p. 1501.

Unger, R.H., and Orci, L: Glucagon: secretion, transport, metabolism, physiologic regulation of secretion, and derangement in diabetes. *In* DeGroot, L.J., *et al.* (eds.): Endocrinology. Vol. 2. New York, Grune & Stratton, 1979, p. 959.

Young, D.B., and Guyton, A.C.: Steady state aldosterone dose-response relationships. *Circ. Res.*, *40*(2):138, 1977.

47

Calcium Metabolism, Bone, Parathyroid Hormone, and Physiology of Teeth

Overview

Calcium ions are essential for the function of all cells of the body, partly because they have a potent effect to decrease the *permeability of cell membranes,* but also because they *activate many of the intracellular enzymes.* One of the most serious effects that occurs in the absence of adequate numbers of calcium ions in the blood and tissue fluid is *tetany,* which means spurious generation of action potentials in the peripheral nerves that causes tetanic contractions of the muscles. This results from extreme leakiness of the nerve membrane to sodium ions, allowing them to leak to the inside of the fiber and thus generate repetitive action potentials.

In bone, calcium is combined with *phosphate* and other ions to form *hydroxyapatite,* which is the principal *bone salt.* In fact, at least 99 percent of all the calcium in the body is in the bones. Hydroxyapatite is about as hard as marble, and it provides the *compressional strength* of bone. Bone also contains vast numbers of very strong *collagen fibers* that provide the *tensional strength* of bone.

Vitamin D is essential for adequate *absorption of calcium* from the gastrointestinal tract. *Calcium is actively transported* through the epithelial membrane of the intestine, and this transport is enhanced manyfold by vitamin D. In the absence of vitamin D, newly forming bone fails to deposit normal amounts of hydroxyapatite crystals so that the bone is soft and bends into contorted forms, a condition called *rickets.*

The *calcium ion concentration* in the extracellular fluid is controlled by two separate hormones: (1) *parathyroid hormone,* secreted by four small **parathyroid glands** that lie behind the thyroid gland, and (2) *calcitonin,* secreted by the thyroid gland. By far the more important of these is parathyroid hormone. Parathyroid hormone has three major effects that increase the blood calcium ion concentration. First, it activates large numbers of cells called **osteoclasts** in the cavities of bone. These in turn literally eat their way into the bone and release calcium ions into the blood. Second, parathyroid hormone stimulates the *reabsorption of calcium by the renal tubules,* thus decreasing the loss of calcium ions in the urine. Third, this hormone also *increases the rate of calcium absorption* from the intestines, mainly because parathyroid hormone is essential to activate the absorptive function of vitamin D.

Whenever the blood calcium ion concentration falls below normal, this has a direct effect on the parathyroid gland cells to increase their rate of secretion of parathyroid hormone. Normally, within a few hours the increased amounts of parathyroid hormone will bring the calcium ion concentration back up to normal. Loss of all four parathyroid glands, which sometimes occurs inadvertently when the thyroid gland is removed for treating hyperthyroidism, will lead to lethal *hypoparathyroid tetany* within a few days.

Calcitonin has almost exactly the opposite effects on calcium ion concentration to those of parathyroid hormone. When the calcium ion concentration becomes too high, this has a direct effect on the thyroid gland to increase the secretion of calcitonin. In turn, the calcitonin causes increased deposition of calcium in the bones, thus decreasing the calcium ion concentration back to normal. However, calcitonin is a much weaker hormone than parathyroid hormone and one that the body can lack without serious consequences.

The *teeth* are highly specialized bone structures, comprised like other bone tissue of (1) strong *collagen fibers* and other similar fibers to provide tensional strength and (2) *hydroxyapatite crystals* to provide compressional strength. The main body of a tooth is composed of *dentine*, which has approximately the same composition and strength as most other compact bone of the body. The outer surfaces of the teeth are covered by a layer of *enamel* that is exceedingly hard and resistant to mechanical abrasion as well as to acids and enzymes created by bacterial action in the mouth. The reason for this extreme resistance is twofold: (1) enamel contains *very large and very dense hydroxyapatite crystals* that are very resistant to chemical destruction, and (2) it also contains a fine meshwork of *extremely strong special protein fibers* that also are almost completely resistant to all types of destruction.

CALCIUM METABOLISM

The adult human body contains about 1200 g of calcium, at least 99 percent of which is deposited in the bones, but a very small and extremely important portion of the calcium is dissolved in the blood plasma and interstitial fluid. Approximately one-half of that in the plasma is ionized, and the remaining half is bound with the plasma proteins. It is the ionized calcium that diffuses into the interstitial fluid and enters into chemical reactions.

The normal concentration of calcium in the plasma is 2.4 millimoles (mM) per liter and the concentration of calcium ions is 1.2 mM per liter.

Functions of Calcium Ions. **Effect on Cell Membranes and the Nervous System.** One of the principal functions of calcium ions is their effect on the cell membrane. In unicellular animals, calcium decreases the permeability and increases the strength of the membrane, and without calcium the membrane becomes very friable (easily ruptured).

In the human being this basic effect of calcium on the membrane is not so obvious, but some of the secondary effects of its action on the cellular membrane are very important. For instance, a *decrease* in calcium ion concentration to one-half the normal level causes the membranes of nerve fibers to become very leaky to sodium ions and, therefore, to become partially depolarized and to transmit repetitive and uncontrolled impulses. This causes spasm of the skeletal muscles, a condition called *hypocalcemic tetany.*

On the other hand, a great increase in the concentration of calcium ions depresses neuronal activity, especially in the central nervous system. This presumably occurs because the membranes will not depolarize with normal ease.

Effect of Calcium Ions on the Heart. A second effect of decreased calcium ion concentration is weakness of the heart muscle. Decreased calcium causes the duration of cardiac systole to decrease, and the heart dilates excessively during diastole. An excess of calcium promotes overcontraction of the heart, causing the muscle to contract much too forcefully during systole and not to relax satisfactorily during diastole.

These effects of high and low calcium concentrations on the heart can be explained by the basic mechanism of muscle contraction, which was discussed in Chapter 11. When the cardiac impulse passes over the cardiac muscle, a small amount of calcium ions is released into the sarcoplasm of the muscle fibers, part of the ions coming from the sarcoplasmic reticulum and part from the extracellular fluid through the walls of the T tubules. It is these calcium ions that cause the contractile process. When only small amounts of calcium are available in the extracellular fluid, the intensity of contraction is reduced, whereas excess calcium in the extracellular fluids causes overcontraction of the heart.

Effect of Calcium Ions on Blood Coagulation. Another important function of calcium ions is to promote blood coagulation. It will be recalled from the discussion of blood coagulation in Chapter 35 that calcium enters into most of the chemical reactions of the clotting process. Fortunately, though, the calcium ion concentration only rarely falls low enough or rises high enough to cause serious abnormalities of clotting.

Reaction of Calcium Ions with Phosphate Ions to Form Bone Salts

One of the most important functions of calcium ions is to combine with phosphate ions to form bone salts. Calcium (Ca^{++}) and phosphate (HPO_4^{--}) ions react together to form *calcium phosphate* ($CaHPO_4$), a relatively insoluble compound. The mathematical product of the concentrations of calcium and phosphate ions in a solution can never be greater than a critical value called the *solubility product*, or they

will precipitate in the form of calcium phosphate crystals. Therefore, the greater the concentration of calcium in the solution, the less can be the concentration of phosphate; or the greater the concentration of phosphate, the less can be the concentration of calcium.

Thus, calcium and phosphate are inextricably related to each other in the deposition and reabsorption of bone, because the salts of bone are mainly calcium phosphate compounds. Every time calcium is deposited, phosphate is deposited also; and every time bone is reabsorbed, both calcium and phosphate are absorbed into the body fluids at the same time.

Figure 47–1 illustrates the absorption, utilization, and excretion of calcium and phosphate. Note that both calcium and phosphate are absorbed from the intestines and then are excreted mainly by way of the kidneys. Vitamin D and parathyroid hormone play especially important roles in controlling the rate of calcium absorption from the intestines, the rate of calcium excretion by the kidneys, and the rate of absorption of calcium from bones. We will discuss all these effects in subsequent sections of the chapter.

In general, absorption and excretion of phosphate follow essentially the same pathways as those for calcium, as we shall see in succeeding paragraphs.

Function of Vitamin D in Calcium and Phosphate Absorption

In the absence of vitamin D, almost no calcium is absorbed from the intestinal tract, but in the presence of this vitamin calcium absorption can be very great. When the rate of calcium absorption is increased, usually the rate of phosphate absorption is increased as well because move-

FIGURE 47–1 Absorption, utilization, and excretion of calcium and phosphate.

ment of calcium out of the intestines releases large amounts of phosphates from insoluble or un-ionized calcium phosphate compounds of the food. Since this released phosphate is highly absorbable by the intestinal mucosa, it simply follows along when the calcium is absorbed.

Source of Vitamin D. There are several different vitamin D compounds, but the most important of these is the substance *cholecalciferol*. This is usually formed in the skin of animals by ultraviolet irradiation of 7-dehydrocholesterol, a fatty substance in the skin. Consequently, a person sufficiently exposed to sunlight needs no vitamin D in his diet. If he does not receive sufficient sunlight, then generally vitamin D must be provided in his food. Usually only foods of animal origin contain vitamin D. A cow exposed to sunlight forms vitamin D continually in its skin, and some of it is secreted into the milk. Also, most animals store vitamin D in great quantities in the liver; therefore, liver is usually an excellent source of this vitamin. An artificial source of vitamin D is irradiated milk, for milk contains steroids which can be converted into vitamin D by irradiation with ultraviolet light.

Basic Mechanism of Function of Vitamin D. Vitamin D itself has almost no direct effect on increasing calcium absorption from the intestines. Instead, it must first be converted into an active product, *1,25-dihydroxycholecalciferol*. It is this compound that directly influences the intestinal epithelium to promote calcium absorption. The conversion of vitamin D to 1,25-dihydroxycholecalciferol occurs in the following steps, which are also illustrated in Figure 47–2.

1. The liver first converts the vitamin D (*cholecalciferol*) into *25-hydroxycholecalciferol*. As the quantity of this substance builds up in the liver, it inhibits further conversion of vitamin D, thus providing a feedback regulation of the amount of 25-hydroxycholecalciferol available at any one time.
2. The kidney converts the 25-hydroxycholecalciferol into 1,25-dihydroxycholecalciferol, which is the final active product. However, for this conversion to take place, *parathyroid hormone* is required. This is one major mechanism, perhaps the most important, by which parathyroid hormone exerts its effects on calcium metabolism, as we shall see in greater detail later in the chapter.

Mechanism by Which 1,25-Dihydroxycholecalciferol Increases Calcium Absorption. Figure 47–2 illustrates three different mechanisms by which 1,25-dihydroxycholecal-

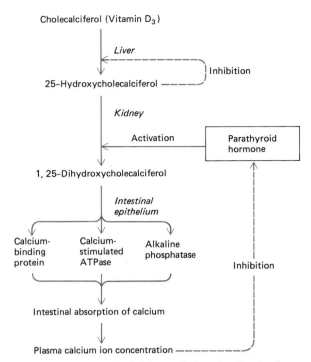

FIGURE 47–2 Activation of vitamin D_3 to form *1,25-dihydroxycholecalciferol* and the role of vitamin D in controlling the plasma calcium concentration.

ciferol enhances the absorption of calcium from the intestines. Probably the most important of these is the effect of this "hormone" to increase the quantity of a special protein in the intestinal epithelium, called *calcium-binding protein*; this in turn combines with calcium and causes it to be transported through the epithelium. Though the precise mechanism of calcium transport is yet unknown, it is abundantly clear that the role played by 1,25-dihydroxycholecalciferol is an essential part of the transport mechanism.

It should also be noted in Figure 47–2 that once the plasma calcium ion concentration becomes too high, the rate of formation of parathyroid hormone is inhibited, which then decreases the rate of activation of vitamin D, thereby shutting off the absorption of calcium from the intestines. This is one of the important mechanisms for control of the plasma calcium ion concentration, a subject that will be discussed in much greater detail after we have considered the relationship of bone to calcium metabolism.

BONE AND ITS FORMATION

Figure 47–3 illustrates the structure of a long bone, showing an articular surface at each end where it is jointed with the other bones, and a hollow shaft that is especially designed for resisting mechanical stresses. To the right in the

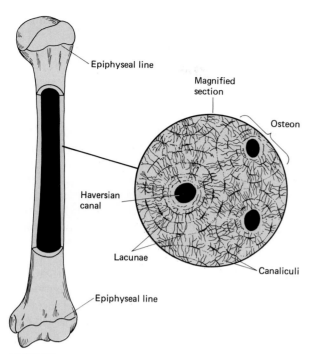

FIGURE 47–3 Structure of bone.

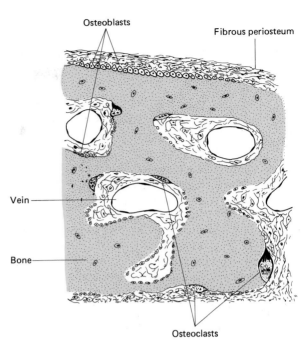

FIGURE 47–4 Osteoblastic deposition of bone and osteoclastic reabsorption of bone.

figure is shown a greatly magnified microscopic cross-section of bone; the colored portion of the figure represents deposits of bone salts and the black areas represent spaces that contain blood vessels and tissue fluids. Bone, like other tissue, is continually supplied with an adequate flow of nutrients in the blood.

Chemical Composition of Bone. Bone is composed of two major constituents, (1) a strong protein *matrix* having a consistency almost like that of leather and (2) *bone salts* deposited in this matrix to make it hard and nonbendable. By far the major portion of the salts of bone have the following approximate chemical composition:

$$[Ca_3(PO_4)_2]_3 \cdot Ca(OH)_2$$

This is the chemical formula for *hydroxyapatite*, a hard marble-like compound. Small amounts of calcium carbonate ($CaCO_3$) are also present but probably are not of primary importance.

The protein matrix prevents the bone from breaking when tension is applied, and the salts prevent the bone from crushing when pressure is applied. The matrix, therefore, is analogous to steel in reinforced concrete structures, and the salts are analogous to the concrete itself.

Deposition of Bone

Figure 47–4 illustrates both bone deposition and bone absorption. Note on the upper surface of the bone and also in some of the cavities a type of cell called *osteoblasts*. These are the cells that deposit bone. Bone deposition occurs in two stages, as follows:

First, the osteoblasts secrete the matrix. This contains a protein substance that polymerizes to become very strong *collagenous fibers*, which then constitute by far the major portion of the new matrix.

Second, after the protein matrix has formed, calcium salts precipitate in the matrix, making it the hard structure that we know to be bone. Deposition of the salts requires (1) combination of calcium and phosphate to form calcium phosphate, $CaHPO_4$, and (2) slow conversion of this compound into hydroxyapatite over a period of several weeks or months. However, the concentrations of calcium and phosphate in the extracellular fluids are normally not sufficient to cause automatic precipitation of calcium phosphate crystals. Therefore, it is believed that the newly formed collagenous fibers of the bone matrix have a special affinity for calcium phosphate, causing calcium phosphate crystals to deposit even though the mathematical product of calcium ion and phosphate ion concentrations in the surrounding fluids is not as great as the solubility product.

Regulation of Bone Deposition. Deposition of bone is regulated partially by the amount of strain applied to bone. That is, the greater the weight applied and the greater the bending of the bone the more active become the osteoblasts, the reason for which is yet unknown. Nevertheless, a bone subjected to continuous and excessive loads usually grows thick and strong, whereas bones not used at all, such as the bones of a leg in a plaster cast, waste away.

Another important cause of bone production is a break in a bone. Injured osteoblasts at the site of the break become extremely active and proliferate in all directions, secreting large quantities of protein matrix to cause deposition of new bone. As a result, the break is normally repaired within a few weeks.

Reabsorption of Bone

Figure 47–4 also shows, in addition to osteoblasts, several giant cells, called *osteoclasts*, each of which contains many nuclei. These cells are present in almost all the cavities of the bone and have the ability to cause bone reabsorption. They do this by secreting enzymes and probably acidic substances that digest the protein matrix and dissolve the bone salts so that they will be absorbed into the surrounding fluid. Thus, as a result of osteoclastic activity, both calcium and phosphate are released into the extracellular fluid, while the bone is literally eaten away.

Balance Between Osteoclastic Reabsorption and Osteoblastic Deposition of Bone. Osteoclastic reabsorption of bone occurs all the time in multiple small cavities throughout the bone, as shown in Figure 47–4, but this is offset by continued osteoblastic deposition of new bone. Indeed, new bone is usually being deposited on the opposite side of the same cavity. The strength of the bone depends on the relative rates of the two processes. If the rate of osteoblastic activity is greater than that of osteoclastic activity, the bone will be increasing in strength. This occurs in athletes and in others who subject their bones to excessive strain. On the other hand, osteoclastic activity is usually greater than osteoblastic activity when the bones are out of use, thereby causing the bones to become weakened.

One might wonder why old bone is continually reabsorbed and new bone deposited in its place. However, in those persons in whom this does not occur, as is true of a few disease conditions, the bone becomes very brittle and then breaks easily. Therefore, it is believed that this continual turnover of bone keeps the tensile strength of the matrix strong. That is, new and strong collagenous fibers take the place of old and weakened fibers. It is especially evident that

the bones become more and more brittle in older people, in whom this process of bone reabsorption and redeposition occurs progressively more slowly with advancing age.

An interesting effect of continual absorption and redeposition of bone is the tendency for crooked bones to become straight over a period of years. Compression of the inner curvature of a bent bone seems to promote rapid deposition of new bone, whereas stretching of the outer curvature seems to promote reabsorption. As a result, the inside of the curvature becomes filled with new bone while the outside is absorbed. In a child a broken bone may initially heal with many degrees of angulation and yet will become essentially straight within a few years.

The "Osteocytic Membrane" System and a Rapid Phase of Calcium and Phosphate Absorption. Though the osteoclastic mechanism can cause massive absorption of calcium and phosphate from the bones, it takes several days to increase osteoclastic activity greatly. Yet, at times increased amounts of calcium are required in the blood within minutes or hours to prevent the lethal effects of hypocalcemia. Fortunately, there is still another system that is capable of causing rapid removal of calcium and phosphate from the bones. This is the *osteocytic membrane system*. It consists of a vast array of bone cells, the osteocytes, that interconnect with each other throughout the substance of all bone. Referring again to the microscopic cross-section of bone illustrated in Figure 47–3, note the many small cavities called *lacunae*. It is in these that the osteocytes reside, and they project multiple tentacles into the many tubules leading outward from the lacunae. The tentacles of the separate osteocytes join with those of adjacent osteocytes, thus forming a continuous membranous system throughout the entire bone structure. This is the *osteocytic membrane*.

In ways not entirely understood, activation of the osteocytic membrane by parathyroid hormone, as will be discussed subsequently, can cause very rapid removal of both calcium and phosphate from the adjacent bone even though the matrix of the bone is not reabsorbed. It is believed that this results from a calcium pumping mechanism, with the osteocytes literally pumping the calcium out of the "fluid" of the bones and into the blood capillaries in the haversian canals. The low calcium concentration in the bone fluid then causes the more soluble salts of bone, such as newly deposited calcium phosphate, to be dissolved and this, too, to be pumped into the capillaries.

Parathyroid Hormone and Its Regulation of Calcium Metabolism

PARATHYROID HORMONE AND ITS REGULATION OF CALCIUM METABOLISM

Secretion by the Parathyroid Glands. *Parathyroid hormone,* also called *parathormone,* is a small protein, molecular weight approximately 9500, secreted by the **parathyroid glands.** It causes release of calcium salts from bones as well as increased calcium absorption from the intestine and kidney tubules. Its primary function is to regulate the concentration of ionic calcium in the extracellular fluids.

Normally there are four separate very minute parathyroid glands, two lying behind each lobe of the thyroid gland, as shown in Figure 47–5. Each of these glands weighs only 0.04 g. A histologic cross-section of a parathyroid gland is shown in Figure 47–6, illustrating two types of cells in the gland, **chief cells** and **oxyphil cells.** The parathyroid glands of some animals contain only chief cells, for which reason it is assumed

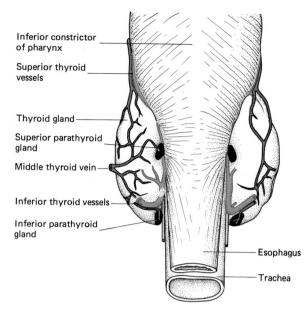

FIGURE 47–5 Location of the four parathyroid glands on the posterior surfaces of the two lobes of the thyroid gland.

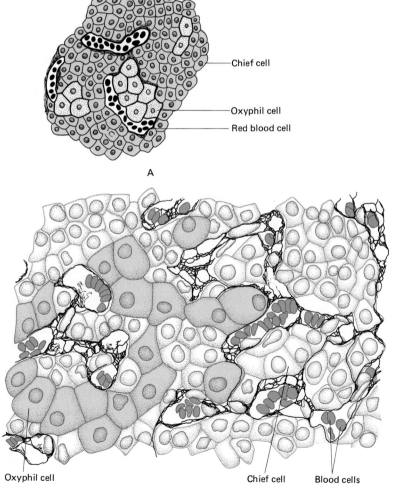

FIGURE 47–6 Histologic structure of a parathyroid gland.

that the chief cells are the ones that secrete parathyroid hormone. The oxyphil cells may be the same as the chief cells but in a different stage of development.

Effect of Parathyroid Hormone on Increasing the Extracellular Fluid Concentration of Calcium Ions

Parathyroid hormone increases the extracellular fluid calcium ion concentration by two different basic mechanisms, one of which is a very rapidly acting mechanism; the other is much more important for long-term regulation.

The short-term mechanism is based on the ability of parathyroid hormone to cause calcium reabsorption from the bone whereas the long-term mechanism is based on the ability of parathyroid hormone to increase absorption of calcium both from the intestine and from the kidney tubules.

Increase in Calcium Ion Concentration When Parathyroid Hormone Causes Bone Absorption. Parathyroid hormone causes calcium absorption from bones by two entirely separate mechanisms, one of which begins to act within minutes and the other of which takes several days to become fully effective. The first of these stimulates the osteocytic membrane, as discussed earlier in the chapter, to transport calcium ions actively out of the bone fluid, delivering these calcium ions into the blood capillaries. Within 15 to 30 minutes one can already see a beginning increase in blood calcium ion concentration; within 2 to 3 hours this initial effect of parathyroid hormone to elevate the calcium ion concentration is fully developed. The top curve in Figure 47–7 shows the progressive rise in calcium ion concentration for the first few hours after infusion of parathyroid hormone is started.

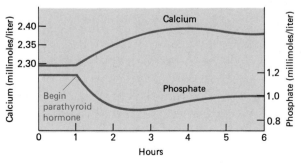

FIGURE 47–7 Approximate changes in calcium and phosphate concentrations during the first five hours of parathyroid hormone infusion at a moderate rate.

The second mechanism by which parathyroid hormone increases the extracellular fluid calcium ion concentration is to activate the osteoclastic cells in the bones. These cells literally eat their way through the bones and release both calcium and phosphate ions into the blood. Activation of the osteoclastic system occurs in two stages: (1) immediate activation of the osteoclasts that are already formed and (2) formation of new osteoclasts from so-called *osteoprogenitor cells.* Usually several days of excess parathyroid hormone secretion causes the osteoclastic system to become well developed, but it can continue to grow for literally months under the influence of very strong parathyroid hormone stimulation. As a result, the rate of calcium absorption from the bones can become so great that the bones can be almost completely decalcified within a few months.

Bones contain about 1000 times as much calcium as that found in all of the extracellular fluid of the body. Therefore, even a very slight amount of bone absorption causes a very large percentage increase in the calcium ion concentration of the extracellular fluid. Thus, if ever the calcium ion concentration of the blood falls too low, greater secretion of parathyroid hormone can usually correct this within a few hours by increasing the rate of calcium absorption from bone.

Increase in Calcium Ion Concentration Caused by Parathyroid Hormone Stimulation of Intestinal and Kidney Tubular Absorption of Calcium Ions. Earlier in the chapter it was pointed out that parathyroid hormone is one of the factors that activates vitamin D. The active form of vitamin D then promotes rapid absorption of calcium from the intestinal tract. A similar effect occurs in the kidney tubules as well. Therefore, increased quantities of parathyroid hormone enhance the input of calcium ions to the extracellular fluid from the intestines while at the same time decreasing excretion of calcium by the kidneys. However, the usual daily turnover of calcium by the intestines and the kidneys is only about 1 g. Therefore, this is a very slow mechanism. Nevertheless, it is the mechanism that determines the long-term balance between total body intake and output of calcium. For this reason, it is also the mechanism that in the long run determines the total amount of calcium in the body, both in the bones and in the extracellular fluid.

Effect of Parathyroid Hormone on Extracellular Fluid Phosphate Concentration. Parathyroid hormone has a special effect on phosphate metabolism that it does not have on calcium metabolism: It causes *increased* kidney excretion of phosphate ions. Therefore, at the same time that parathyroid hormone increases the concentration of calcium ions in the body fluids, it tends to *decrease* the phosphate concentration a small amount. This effect is illustrated by the lower curve of Figure 47–7, which shows that the phosphate ion concentration decreases by about 20 percent in the extracellular fluids within 1 to 2 hours after beginning an infusion of parathyroid hormone at a moderate rate.

This decrease in phosphate ion concentration reduces the likelihood that calcium will precipitate with phosphate at unwanted places in the body, an effect that does occur in the blood, muscle, tendons, and elsewhere when both the calcium and phosphate concentrations rise too high at the same time.

Regulation of Parathyroid Secretion, and Feedback Control of Calcium Ion Concentration

Almost any factor that decreases the calcium ion concentration of the extracellular fluid will cause the parathyroid glands to secrete increased quantities of parathyroid hormone. If the decrease in calcium ion concentration continues for a long period of time, the parathyroid glands will even enlarge, sometimes as much as tenfold, thereby multiplying the formation of parathyroid hormone many times.

Thus, the parathyroid hormone system plays an essential role in both hour-by-hour and year-by-year control of calcium ion concentration in the extracellular fluid. That is, when the ionic level of calcium falls too low, parathyroid hormone is secreted and calcium ions are absorbed from the bone, the intestines, and the kidneys to increase calcium ion concentration level in the extracellular fluid. In this way, a relatively constant concentration of calcium ions is maintained at all times.

The Value of Parathyroid Hormone to the Body. If the calcium ion concentration falls more than 50 percent below normal, the person develops immediate tetany and dies because of spasm of his respiratory muscles. On the other hand, if the calcium concentration becomes too great, he is likely to have severe mental or cardiac disturbances. Therefore, it is necessary that the calcium ion concentration remain almost exactly constant all the time. Of course, the bones have almost a thousand times the calcium

in all the extracellular fluid, so that the slight amount of calcium mobilized from the bones from day to day usually is not missed. The main value of parathyroid hormone, then, is to regulate calcium ion concentration in the extracellular fluid, so that the other functions of this ion besides that of bone deposition can continue normally all the time. The bone, on the other hand, can wait several months if necessary until adequate calcium is available to replenish its lost stores.

Abnormalities of Parathyroid Hormone Secretion

Hypersecretion of Parathyroid Hormone. Occasionally a parathyroid tumor develops in one of the glands, and extreme secretion of parathyroid hormone causes tremendous overgrowth of the osteoclasts. Sometimes these cells grow so large and so numerous that they cause large honeycomb cavities in the bone, and even combine to form large masses that resemble tumors. The result is often bones so weakened that even walking on a leg can cause it to break. Indeed, most hyperparathyroid persons first become aware of their disease through a broken bone.

Hypersecretion also increases the calcium ion concentration in the body fluids, but if the hypersecretion is not too great, the level of calcium ions will not rise enough to cause untoward effects. Rarely, though, the level becomes great enough to cause precipitation of calcium phosphate in tissues other than the bone, such as in the lungs, muscles, and heart, sometimes leading to death within a few days.

Hyposecretion of Parathyroid Hormone. The most common cause of deficient parathyroid secretion is surgical removal of all the parathyroid glands. This usually occurs inadvertently when a surgeon is removing the thyroid gland, because of the close proximity of the parathyroid glands to this other gland.

Loss of parathyroid secretion allows the calcium ion concentration to fall so low that tetany occurs within about 3 days, and unless the person is treated he dies almost immediately. However, administration of parathyroid hormone or large quantities of certain vitamin D compounds can mobilize sufficient calcium from the bones to restore normal function within a few hours.

A rare person has hereditary hyposecretion of parathyroid hormone. Usually the degree of hyposecretion is not sufficient to cause tetany, but it may be sufficient to cause chronically depressed osteoclastic activity in the bones. This results in brittle bones, probably because without osteoclastic activity the constant absorption and re-formation of new bone ceases, and the protein matrix becomes old and brittle and is not replaced as often as needed to maintain adequate bone strength.

Rickets

Rickets is a disease caused either by prolonged dietary calcium deficiency or by lack of vitamin D. Lack of dietary calcium for a short time will never cause rickets, for when the calcium ion concentration in the extracellular fluid falls below normal, large quantities of parathyroid hormone are secreted and calcium is automatically absorbed from the bones, thereby reestablishing an adequate calcium ion concentration. However, if insufficient calcium is absorbed from the gut for many months, then all or most of the calcium in the bones will have been absorbed, and little more is available. The calcium concentration of the extracellular fluids then falls to very low values. At this point the person has fully developed rickets, the two major effects of which are (1) depletion of calcium salts from the bones and, therefore, severe weakening of the bones, often culminating in fractures or deformed bones, and (2) tetany caused by the diminished extracellular fluid calcium ion concentration.

The most frequent cause of rickets is a deficiency of vitamin D rather than lack of calcium in the diet. The usual child living in temperate climates receives far too little sunlight during the winter months to provide the amount of vitamin D needed for absorption of calcium from the gut. Fortunately, the stores of vitamin D in the child's liver are usually sufficient to provide adequate calcium absorption for the early months of winter. Therefore, rickets usually develops in the early spring, before the child has been exposed to the sun again.

CALCITONIN—A CALCIUM-DEPRESSING HORMONE

A recently discovered hormone called *calcitonin* causes exactly the opposite effect on blood calcium ion concentration to that caused by parathyroid hormone, that is, decrease in calcium ion concentration rather than increase. Calcitonin is secreted by the thyroid gland, not by the follicular cells that secrete thyroxine but by "parafollicular cells" that lie in the angles between the follicles. It is a large polypeptide with a molecular weight of about 3000.

Calcitonin decreases calcium ion concentration in three distinct ways:

1. Within minutes it greatly suppresses the activity of the osteoclasts and probably also the calcium pump of the osteocytic membrane and therefore reduces the rate of release of calcium from the bone into the blood.
2. It increases osteoblastic activity within about 1 hour, and this causes greater deposition of calcium in the bones and therefore removes calcium from the extracellular fluid.
3. It causes a prolonged reduction in the rate of formation of new osteoclasts.

The action of calcitonin is different from that of parathyroid hormone in a very important respect: It begins to act several times as rapidly as parathyroid hormone and therefore can help to correct acute hypercalcemia very rapidly.

Secretion of calcitonin is greatly enhanced when blood calcium concentration rises above normal. The calcitonin in turn causes some of the blood calcium to deposit in the bones, thereby returning the calcium ion concentration toward normal. Obviously, therefore, the calcitonin mechanism is a rapidly acting feedback mechanism, even more rapid than that of the parathyroid mechanism, which helps to stabilize calcium ion concentration in the extracellular fluid. However, it has far less effect quantitatively than does the parathyroid mechanism, and its long-term effect is even less significant, so that for practical purposes one can consider the long-term regulation of calcium ion concentration to be almost entirely a parathyroid function.

PHYSIOLOGY OF TEETH

The teeth cut, grind, and mix food. To perform these functions, the jaws have extremely powerful muscles capable of providing an occlusive force of as much as 50 to 100 lb between the front teeth and as much as 150 to 200 lb between the jaw teeth. Also, the upper and lower teeth are provided with projections and facets that interdigitate so that each set of teeth fits with the other. This fitting is called *occlusion*, and it allows even small particles of food to be caught and ground between the tooth surfaces.

Function of the Different Parts of Teeth

Figure 47–8 illustrates a lengthwise section of a tooth, showing its major functional parts, *enamel, dentine, cementum,* and *pulp.* The tooth can also be divided into the *crown,* which is the portion that protrudes above the gum into the mouth, and the *root,* which is the portion in the bony socket of the jaw. The collar between the crown and the root, where the tooth is surrounded by the gum, is called the *neck.*

Dentine. The main body of the tooth is composed of dentine, which has a very strong body structure. Dentine is made up principally of calcium salts of phosphate and carbonate (hydroxyapatite crystals) embedded in a strong meshwork of *collagen fibers.* In other words, the principal constituents of dentine are very much the same as those of bone. The major difference is its structure, for dentine does not contain any osteoblasts, osteoclasts, or spaces for blood vessels or nerves. Instead, it is deposited and nourished by a layer of cells called *odontoblasts,* which line the inner surface of the pulp cavity.

The calcium salts in dentine make it extremely resistant to compressional forces; the collagen fibers make it tough and resistant to tensional forces that might result when the teeth are struck by solid objects.

Enamel. The outer surface of the tooth is covered by a layer of enamel that is formed prior to eruption of the tooth by special epithelial cells called *ameloblasts.* Once the tooth has erupted, no more enamel is formed. Enamel is composed of large and very dense crystals of hydroxyapatite embedded in a fine meshwork of extremely strong fibers composed of a special protein similar to the keratin of hair. The crystalline structure of the calcium salts makes the enamel extremely hard, much harder than dentine. Also, the special protein meshwork makes enamel very resistant to acids, enzymes, and other corrosive agents, because this protein of enamel is one of the most insoluble and resistant proteins known.

Cementum. Cementum is a bony substance secreted by the *periodontal membrane* that lines the tooth socket. It forms a thin layer between the tooth and the inner surface of the socket. Many collagenous fibers pass directly from the bone of the jaw, through the periodontal membrane, and into the cementum. It is these collagenous fibers and the cementum that hold the tooth in place. When the teeth are exposed to excessive strain, the layer of cementum increases in thickness and strength. It also increases in thickness and strength with age, causing the teeth to become progressively more firmly seated in the jaws as one reaches adulthood and beyond.

Pulp. The inside of each tooth is filled with pulp, which is composed of connective tissue and an abundant supply of nerves, blood vessels, and lymphatics. The cells lining the surface of the pulp cavity are the *odontoblasts,* which, during the formative years of the tooth, lay down the dentine but at the same time encroach more and more on the pulp cavity, making it smaller. In later life the dentine stops growing and the pulp cavity remains essentially constant in size. However, the odontoblasts are still viable and send projections into small *dentinal tubules* that penetrate all the way through the dentine; these are of importance for providing nutrition.

Dentition

Human beings and most other mammals develop two sets of teeth during a lifetime. The first teeth are called the *deciduous teeth* or *milk teeth,* and they number 20 in the human being. These erupt between the seventh month and second year of life and last until the sixth to the thirteenth year. After each deciduous tooth is lost, a permanent tooth replaces it. Also, an additional 8 to 12 molars appear posteriorly in the jaw, making the total number of permanent teeth 28 to 32, depending on whether the person

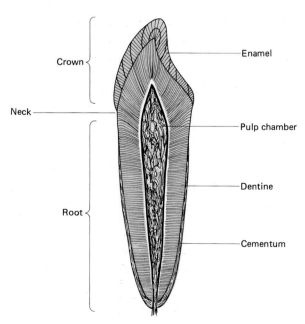

FIGURE 47–8 Functional parts of a tooth.

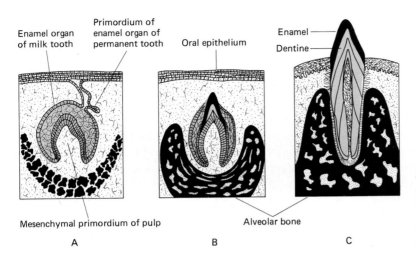

FIGURE 47–9 (A) Primordial tooth organ. (B) The developing tooth. (C) The erupting tooth. (Modified from Bloom and Fawcett: A Textbook of Histology, 10th Ed., Philadelphia, W.B. Saunders, 1975.)

finally grows his 4 *wisdom teeth*, which do not appear in everyone.

Formation of Teeth. Figure 47–9 shows the formation and eruption of teeth. Figure 47–9A shows protrusion of the oral epithelium into the *dental lamina*, this to be followed by the development of a tooth-producing organ. The upper epithelial cells form ameloblasts, which secrete the enamel on the outside of the tooth. The lower epithelial cells grow upward to form a pulp cavity and also to form the odontoblasts that secrete dentine. Thus, enamel is formed on the outside and dentine on the inside, developing an early tooth as illustrated in Figure 47–9B.

Eruption of Teeth. During early childhood, the teeth begin to protrude upward from the jawbone through the oral epithelium into the mouth, as in Figure 47–9C. The cause of eruption is unknown, though several theories have been offered. The most likely theory is that the bone underneath the tooth grows continually and in so doing shoves the tooth forward.

Development of Permanent Teeth. During embryonic life a tooth-forming organ also develops in the dental lamina for each permanent tooth that will be needed after the deciduous teeth are gone. These tooth-producing organs slowly form the permanent teeth throughout the first 6 to 20 years of life. When each permanent tooth becomes fully formed, it, like the deciduous tooth, pushes upward through the bone of the jaw. In so doing, it erodes the root of the deciduous tooth and eventually causes it to loosen and fall out. Soon thereafter the permanent tooth erupts to take the place of the original one.

Effect of Metabolic Factors on Tooth Development. The rate of development and the speed of eruption of teeth can be accelerated by both thyroid and growth hormones. Also, the deposition of salts in the early forming teeth is affected considerably by various factors of metabolism, such as the availability of calcium and phosphate in the diet, the amount of vitamin D present, and the rate of parathyroid hormone secretion. When all these factors are normal, the dentine and enamel will be healthy. When these factors are deficient, the calcification of the teeth also may be defective, so that the teeth will be abnormal throughout life.

Mineral Exchange in Teeth

The salts of teeth, like those of bone, are composed basically of *hydroxyapatite* with adsorbed carbonates and various cations bound together in a hard crystalline substance. Also, new salts are constantly being deposited while old salts are being reabsorbed from the teeth, as also occurs in bone. However, deposition and reabsorption occurs mainly in the dentine and cementum; very little occurs in the enamel. Most of that which does occur in the enamel occurs by exchange of minerals with the saliva instead of with the fluids of the pulp cavity. The cementum has characteristics almost identical to those of usual bone, including the presence of osteoblasts and osteoclasts, so that its mineral exchange and remodelling are the same as in bone. However, the mechanism by which minerals are deposited and reabsorbed from the dentine is not clear. It is probable that the small processes of the odontoblasts that protrude into the tubules of the dentine are capable of absorbing salts and then of providing new salts to take the place of the old.

In summary, rapid mineral exchange occurs in the dentine and cementum of teeth, though the mechanism of this exchange in dentine is unclear. On the other hand, enamel exhibits extremely slow mineral exchange so that it maintains most of its original mineral complement throughout life.

Dental Abnormalities

The two most common dental abnormalities are *caries* and *malocclusion*. Caries means erosions of the teeth, and malocclusion means failure of the projections of the upper and lower teeth to interdigitate properly.

Caries and the Role of Fluorine. It is generally agreed by all research investigators of dental caries that caries results from the action of bacteria on the teeth, the most common of which is *Streptococcus mutans*. The first event in the development of caries is the deposit of *plaque*, a film of precipitated products of saliva and food, on the teeth. Large numbers of bacteria inhabit this plaque and are readily available to cause caries. However, these bacteria depend to a great extent on carbohydrates for their food. When carbohydrates are available, the bacteria are strongly activated and multiply. In addition, they form acids, particularly lactic acid, and proteolytic enzymes. The acids are the major culprit in the causation of caries, because the calcium salts of teeth are slowly dissolved in a highly acid medium. Once the salts have become absorbed, the remaining organic matrix is rapidly digested by the proteolytic enzymes.

Enamel is far more resistant to demineralization by acids than is dentine, primarily because the crystals of enamel are very dense and also are about 200 times as voluminous as the dentine crystals. Therefore, the enamel of the tooth is the primary barrier to the development of caries. Once the carious process has penetrated through the enamel to the dentine, it then proceeds many times as rapidly because of the high degree of solubility of the dentine salts.

Some teeth are more resistant to caries than others. Studies show that teeth formed in children who drink water containing small amounts of *fluorine* develop enamel that is more resistant to caries than the enamel in children who drink water not containing fluorine. Fluorine does not make the enamel harder than usual, but instead it displaces hydroxyl ions in the hydroxyapatite crystals which, in turn, makes the enamel several times less soluble. It is also believed that the fluorine might be toxic to some of the bacteria as well. Regardless of the precise means by which fluorine protects the teeth, it is known that small amounts of fluorine deposited in enamel make teeth about 3 times as resistant to caries as are teeth without fluorine.

Malocclusion. Malocclusion is usually caused by a hereditary abnormality that causes the teeth of one jaw to grow to an abnormal position. In malocclusion, the teeth cannot perform their normal grinding or cutting action adequately. Occasionally malocclusion also results in abnormal displacement of the lower jaw in relation to the upper jaw, causing such undesirable effects as pain in the mandibular joint or deterioration of the teeth.

The orthodontist can often correct malocclusion by applying prolonged gentle pressure against the teeth with appropriate braces. The gentle pressure causes absorption of alveolar jaw bone on the compressed side of the tooth and deposition of new bone on the tensional side of the tooth. In this way the tooth gradually moves to a new position as directed by the applied pressure.

QUESTIONS

1. What are the functions of calcium in cellular metabolism and in the formation of bone?
2. How do phosphate ions interact with calcium ions?
3. Explain the role of vitamin D in calcium absorption from the intestines, and give the schema for its function.
4. Review the composition of bone and its deposition.
5. What is the difference between osteoclastic and osteoblastic functions of bone cells?
6. In what ways does parathyroid hormone increase blood calcium concentration?
7. Discuss the relative roles of parathyroid hormone and calcitonin in the control of blood calcium concentration.
8. Explain the effects of hyper- and hyposecretion of parathyroid hormone.
9. What are the parts of teeth? How do they resemble bone?
10. How does mineral exchange occur in teeth?
11. Explain the development of caries and the role of fluorine in protecting against caries.

REFERENCES

Avioli, L.V., and Raisz, L.G.: Bone metabolism and disease. *In* Bondy, P.K., and Rosenberg, L.E. (eds.): Metabolic Control and Disease, 8th Ed. Philadelphia, W.B. Saunders, 1980, p. 1709.

Barzel, U.S. (ed.): Osteoporosis II. New York, Grune & Stratton, 1979.

Bringhurst, F.R., and Potts, J.T., Jr.: Calcium and phosphate distribution, turnover, and metabolic actions. *In* DeGroot, L.J., *et al.* (eds.): Endocrinology. Vol. 2. New York, Grune & Stratton, 1979, p. 551.

Copp, D.H.: Calcitonin: Comparative endocrinology. *In* DeGroot, L.J., *et al.* (eds.): Endocrinology, Vol. 2. New York, Grune & Stratton, 1979, p. 637.

Dennis, V.W., *et al.*: Renal handling of phosphate and calcium. *Annu. Rev. Physiol.*, 41:257, 1979.

Fraser, D.R.: Regulation of the metabolism of vitamin D. *Physiol. Rev., 60:*551, 1980.

Ham, A.W.: Histophysiology of Cartilage, Bone, and Joints. Philadelphia, J.B. Lippincott, 1979.

Jaros, G.C., *et al.*: Model of short-term regulation of calcium ion concentration. *Simulation, 32:*193, 1979.

Lawson, D.E.M.: Vitamin D. New York, Academic Press, 1978.

Mayer, G.P.: Parathyroid hormone secretion. *In* De-Groot, L.J., *et al.* (eds.): Endocrinology. Vol. 2. New York, Grune & Stratton, 1979, p. 607.

Myers, H.M.: Fluorides and Dental Fluorosis. New York, S. Karger, 1978.

Nellans, H.N., and Kimberg, D.V.: Intestinal calcium transport: Absorption, secretion, and vitamin D.

In Crane, R.K. (ed.): International Review of Physiology: Gastrointestinal Physiology III. Vol. 19. Baltimore, University Park Press, 1979, p. 227.

Newman, H.N.: Dental Plaque: The Ecology of the Flora on Human Teeth. Springfield, Ill., Charles C Thomas, 1980.

Norman, A.W.: Vitamin D: The Calcium Homeostatic Steroid Hormone. New York, Academic Press, 1979.

Parsons, J.A.: Physiology of parathyroid hormone. *In* DeGroot, L.J., *et al.* (eds.): Endocrinology. Vol. 2. New York, Grune & Stratton, 1979, p. 621.

Talmage, R.V., and Cooper, C.W.: Physiology and mode of action of calcitonin. *In* DeGroot, L.J., *et al.* (eds.): Endocrinology. Vol. 2. New York, Grune & Stratton, 1979, p. 647.

TABLE 48–1
The Male Organs of Reproduction

Structure	Function or description
Testis (one on each side)	
Seminiferous tubules	Formation of sperm
Spermatogonia	Original cells from which the sperm are formed
Sertoli cells	Supporting cells in seminiferous tubules; provide nutrients for and control of sperm formation
Interstitial cells of Leydig	Cells lying outside seminiferous tubules; secrete testosterone
Epididymis: head, body, and tail (one on each side)	Coiled tube, about 6 m long, that conducts sperm from seminiferous tubules and in which the sperm mature
Vas deferens (one on each side)	Stores sperm and transports sperm
Ampulla of vas deferens	Enlarged portion of end of vas; stores additional sperm
Seminal vesicle (one on each side)	Secretes seminal fluid, a mucoid fluid that contains nutrients for sperm
Ejaculatory duct (one on each side)	Transports sperm from vas deferens and seminal vesicle to urethra
Prostate gland	Secretes prostatic fluid, a milky, alkaline fluid
Penis	
Erectile tissue	Causes erection of penis
Corpora cavernosa	Two lateral bodies of erectile tissue that provide most of the erection
Corpus spongiosum	Medial, dorsal erectile tissue around urethra
Glans	Bulbous tip of penis, highly sensitive, contains erectile tissue of corpus spongiosum
Prepuce	Covers glans
Urethra	Conducts semen or urine to exterior
Urethral glands	Secrete mucus into urethra
Bulbourethral gland (also called Cowper's gland; one on each side)	Secretes mucus into urethra, especially during intercourse

passage about 1.5 cm in diameter called the **inguinal canal.** After the spermatic cord enters the abdomen, the blood vessels and nerves separate from the vas deferens and connect with appropriate pelvic and abdominal vascular and nerve trunks. Then the vas abruptly turns inferiorly into the pelvis, running alongside the posterolateral border of the urinary bladder to the prostate gland.

Now, let us explain the origin of this peculiar anatomy of the spermatic cord and its course to the testis. The testis is formed in the embryo on the posterior wall of the abdominal cavity. However, during growth of the fetus, the testis migrates out of the abdominal cavity, coursing through the anterior abdominal wall to form the inguinal canal, finally passing into the scrotum. This explains why the testis is connected to abdominal blood and nerve supplies by way of the spermatic cord. It also explains several different abnormalities that frequently occur:

1. *Undescended testis.* Occasionally in a male either one or both of the testes fails to descend from the abdomen into the scrotum during fetal life. In this case the testis remains in the abdomen, and it usually also fails to form normal sperm, if it forms any, because the higher temperature in the abdomen is

destructive to the germinal epithelium of the testis. Therefore, the individual is usually sterile. In some instances surgical transposition of the testis into the scrotum will prevent the sterility, but usually only if the surgery is performed prior to puberty.

2. *Inguinal hernia.* When the testis descends from the abdomen into the scrotum, it must penetrate through the abdominal wall musculature and fascia above the inguinal ligament. This forms the tunnel in the abdominal wall that becomes the inguinal canal. However, in a few percent of male children this canal is much larger than normal, which allows some of the abdominal contents in addition to the cord to protrude through the canal. When this happens, the bulging abdominal structures appear underneath the skin in the inguinal region and sometimes even penetrate all the way down into the scrotum. This is called an *inguinal hernia.* On occasion, a loop of bowel herniates through the canal and causes bowel obstruction, which may lead to a serious emergency requiring immediate surgery.

The Prostate Gland—Its Relation to the Other Sex Organs. Figure 48–2 shows the *prostate gland* encasing the initial portion of the urethra as the urethra leaves the neck of the uri-

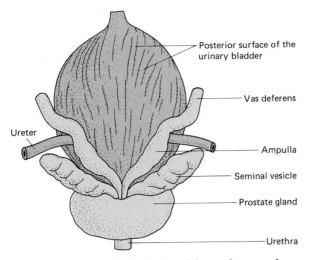

FIGURE 48–2 Relationships of the male reproductive structures to the urinary bladder and urethra (shown from the posterior side).

nary bladder. This portion of the urethra is called the **prostatic urethra.**

Note also in this figure the relationship of the prostate gland to the two ampullae of the vasa deferentia and the two seminal vesicles. The terminal end of the ampulla joins the duct of the seminal vesicle to form the ejaculatory duct that then passes through the substance of the prostate gland to empty into the prostatic urethra. Figure 48–3 illustrates the orifices of the two ejaculatory ducts where they empty into the urethra.

The *prostatic fluid* secreted by the prostate gland is highly alkaline and has a thin consistency and milky appearance. It is secreted directly into the prostatic urethra through large numbers of small prostatic gland ductules rather than through a single large duct. The

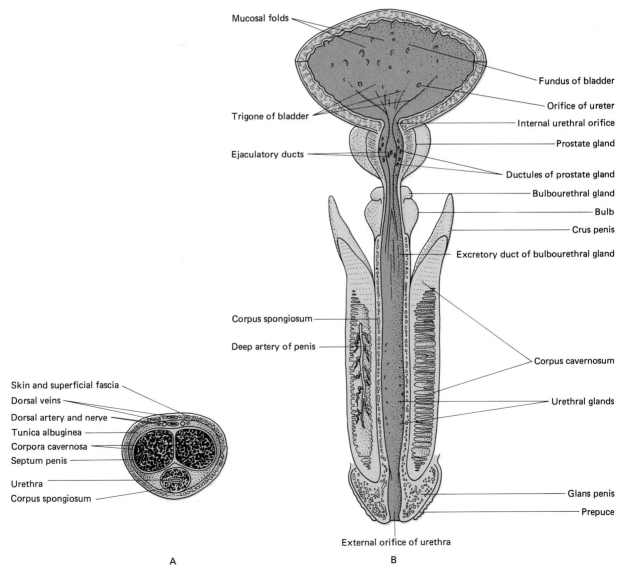

A   B

FIGURE 48–3 (A) Cross-section of the penis. (B) Internal structures of the penis and prostate gland.

openings of these ductules into the urethra also are illustrated in Figure 48–3.

THE PENIS

The basic structure of the **penis** is illustrated in cross-section in Figure 48–3A and in a lengthwise section in Figure 48–3B. Its principal parts are the following:

1. The **erectile tissue of the penis** consists of the two **corpora cavernosa** and the **corpus spongiosum.** These three bodies are comprised of literally thousands of cavernous spaces that contain blood. It is the pressure of the blood filling these spaces that swells the penis and results in erection. The proximal portions of all of these erectile bodies extend posteriorly into the urogenital triangle from the base of the penis. The extension from each corpus cavernosus is called a **crus penis,** and the extension from the corpus spongiosum, which is enlarged to form a bulb, is called the **bulb of the penis.** Three muscles of the urogenital triangle overlie these extensions, as illustrated in Figure 48–4. There are

   a. two **ischiocavernosus muscles,** each of which is attached to the medial margin of the ischiopubic ramus and envelops a crus penis, and

   b. the **bulbospongiosus muscle.** This muscle wraps around the bulb of the penis. When the erectile bodies are erected, contraction of these three muscles further increases the turgor of the penis.

2. The **glans penis** is the bulbous enlargement at the tip of the penis. This is comprised mainly of the enlarged end of the corpus spongiosum. Therefore, it shares in the erectile process.

3. The **prepuce** (foreskin) is the skin that covers the glans penis. It is the prepuce that is removed when a person is circumcised.

4. The **urethra** passes through the whole length of the penis in the substance of the corpus spongiosum. Multiple small **urethral glands** lie immediately underneath the urethral mucosa and secrete mucus through short ducts into the urethra.

5. The **tunica albuginea** is a strong fibrous encasement of the erectile tissue. It is this tunic that limits the degree of swelling by the erectile tissue during erection, thus causing the penis to become turgid.

6. The **dorsal nerves** transmit mainly the sexual sensations from the glans penis.

Associated with the penis, but not part of it, are the **bulbourethral glands,** one of which lies on each side of the urethra immediately inferior to the prostate gland. The duct of each of these glands passes about 2.5 cm through the substance of the corpus spongiosum before entering the lateral margin of the urethra. These glands secrete mucus into the urethra, especially during sexual stimulation.

INTERNAL STRUCTURES OF THE TESTIS AND EPIDIDYMIS

Figure 48–5 illustrates the internal structures of the testis and epididymis, as well as the connections of the epididymis with both the testis and the vas deferens.

The testis is comprised of about 250 *lobules*, each of which contains 2 to 3 seminiferous tubules, or a total of about 900 seminiferous tubules in the entire testis. It is in the seminiferous tubules that the sperm are formed. Each of these tubules is about 75 cm long and is compacted in the form of coils in the lobule.

The seminiferous tubules empty the newly formed sperm into a branching network of 20 to 30 secondary tubules, called the **rete testis,** and these in turn empty into the epididymis through 12 to 15 short **efferent ductules.**

The **epididymis** is a single long tube about 600 cm long coiled up into a compact mass ap-

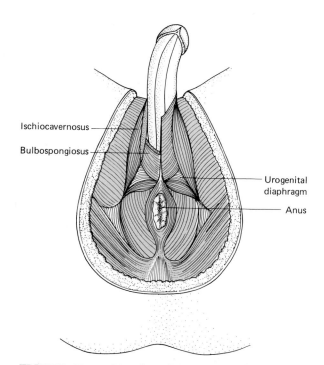

Ischiocavernosus

Bulbospongiosus

Urogenital diaphragm

Anus

FIGURE 48–4 Muscles of the male perineum and their relationship to the erectile tissue of the penis.

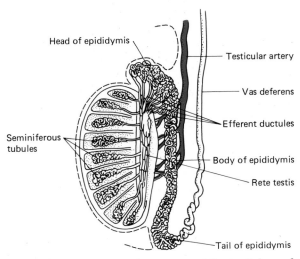

FIGURE 48–5 Internal structure of the testicle, and relationship of the testicle to the epididymis.

plied to the surface of the testis. It is comprised of three parts:

1. the **head,** which lies at the upper pole of the testis and into which the efferent ductules empty;
2. the **body,** which lies along the posterior surface of the testis; and
3. the **tail,** which is attached by fibrous tissue to the lower pole of the testes. The tail of the epididymis in turn is continuous with the vas deferens.

The Seminiferous Tubule and Sperm Formation (Spermatogenesis)

Figure 48–6 illustrates the general cross-sectional organization of a single seminiferous tubule, and Figure 48–7 shows the specific stages

in the formation of the sperm, which is called *spermatogenesis.* The outside of the seminiferous tubule is surrounded by connective tissue, but immediately inside this are large numbers of cells called **spermatogonia.** These are also known as male germinal cells because it is from these that all sperm eventually derive. The spermatogonia divide continually to form still more spermatogonia as well as to form the cells that eventually become spermatozoa, which are the sperm.

The first stage in spermatogenesis is formation of **primary spermatocytes** from the spermatogonia; each of these then divides to form two **secondary spermatocytes,** each of which in turn divides to form two **spermatids.** The spermatid then changes into a sperm by, first, losing some of its cytoplasm, second, reorganizing the chromatin material of its nucleus to form a compact head, and, third, collecting the remaining cytoplasm and cell membrane at one end of the cell to form a tail.

Located among the germinal cells of the seminiferous tubule are many large cells called **Sertoli cells.** The bases of these lie against the outer wall of the tubule, and they extend all or most of the way to the lumen of the tubule. Projecting extensions of the Sertoli cells envelop the spermatocytes and spermatids, and even the final spermatozoa remain attached to the Sertoli cells until they are fully formed. For obvious reasons, therefore, the Sertoli cells are frequently called "nurse cells." Some of the roles that they play in the development of the sperm are to

1. provide a special local environment for division and metabolism of the cells,
2. probably provide special nutrients and perhaps local hormones that are needed for

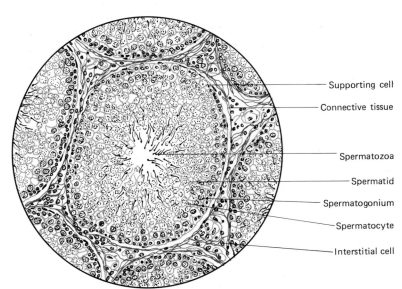

FIGURE 48–6 Cross-sectional view of a single seminiferous tubule.

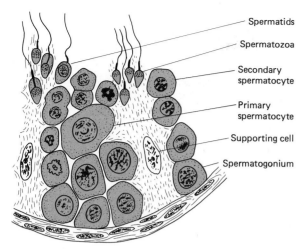

FIGURE 48–7 Development of spermatozoa by the germinal epithelium. (Modified from Arey: Developmental Anatomy, 7th ed. Philadelphia, W. B. Saunders, 1974.)

sperm development, and

3. remove most of the cytoplasm from the spermatids to cause compaction of the head of the sperm and formation of its tail.

Division of Chromosomes During Sperm Formation.

When the primary spermatocytes divide, the cell division is not of the usual type, for each of the 23 pairs of chromosomes are split apart, allowing 23 unpaired chromosomes to pass into half of the secondary spermatocytes and the other 23 to pass into the other half. In other words, only half of the genes enter each secondary spermatocyte, so that only half of the genes from the parent cell are also in each respective sperm. This type of cell reproduction is called *meiosis;* reproduction of other cells throughout the body is termed *mitosis*, which was described in Chapter 4. Meiosis involves two successive cell divisions. In the first division, from primary to secondary spermatocytes, the chromosome number is halved, and this is called the *reduction division.* In the second division, from secondary spermatocytes to spermatids, the half (or haploid) number of chromosomes is retained, and this is called the *equational division.*

A similar division of chromosomes occurs in the ovum. Thus, each parent makes an equal contribution to the hereditary characteristics of the child.

Sex Determination by the Sperm.

Whether a baby will be male or female is determined by the sperm for the following reasons: One of the 23 pairs of chromosomes in all human cells is called the *sex pair* because these two chromosomes determine whether the person will be male or female. One type of sex chromosome is known as a *Y chromosome*, or *male*

chromosome. Another is known as an *X chromosome*, or *female chromosome*. A female has two X chromosomes. A male has one X chromosome and one Y chromosome. When the spermatocytes undergo meiosis the X-Y pair of sex chromosomes splits; the Y chromosome passes into one secondary spermatocyte and the X chromosome passes into the other, and consequently half of the derived sperm will each contain a single Y chromosome and the other half a single X chromosome. The unfertilized ovum always contains only a single X chromosome because the female cells have no Y chromosomes. Therefore, if a sperm containing an X chromosome fertilizes an ovum, the fertilized ovum then has a pair of X chromosomes. This combination leads to formation of a female. On the other hand, combination of a sperm containing a Y chromosome with an ovum creates an X-Y pattern in the fertilized ovum and causes the development of a male child.

Characteristics of Sperm.

Figure 48–8 shows the structure of the sperm, which itself is a single cell constructed principally of a **head** and a **tail.** The head is composed of the nucleus of the cell with only a very thin cytoplasmic and cell membrane layer around its surface. However, on the outside of the anterior two thirds of the head is a thick cap called the **acrosome.** This contains a number of enzymes similar to those found in lysosomes of the typical cell, in-

FIGURE 48–8 Structure of a sperm.

cluding *hyaluronidase* that can digest proteogly-can filaments of tissues and a *proteolytic enzyme* similar to trypsin that can digest proteins. These enzymes probably play important roles in allowing the sperm to fertilize the ovum, as we shall discuss in Chapter 50.

The tail of the sperm has three major components: (1) a central skeleton constructed of multiple microtubules, collectively called the **axoneme**—the structure of this is similar to that of cilia as described in Chapter 3; (2) a thin cell membrane covering the axoneme; and (3) a collection of mitochondria surrounding the axoneme in the proximal portion of the tail (called the body of the tail).

To and fro movement of the tail (flagellar movement) provides motility for the sperm. This movement is believed to result from a rhythmic longitudinal sliding motion between the tubules that make up the axoneme, and energy for this process is believed to be supplied in the form of ATP synthesized by the mitochondria in the body of the tail. Normal sperm move in a straight line at a velocity of about 1 to 4 mm per minute. This allows them to move through the female genital tract in quest of the ovum.

Maturation and Storage of the Sperm.
After formation in the seminiferous tubules, the sperm begin a long passage through the 6-m-long *epididymis*, usually requiring several days for this passage. While the sperm are still in the seminiferous tubules or even while still in the head of the epididymis, they are unable to fertilize an ovum, but after they have passed through most of the epididymis and have reached its distal end, they then become capable of fertilization; this is the process called *sperm maturation*.

Some of the sperm remain stored in the epididymis until the time of emission and ejaculation, but others pass on into the vas deferens and even into the ampulla of the vas deferens where they are stored. In this storage state the sperm can remain fertile for as long as a month. However, so long as they remain in storage they also remain inactive because the fluids in the storage tubes are slightly acidic, which inhibits sperm activity.

The Male Sexual Act

The basic elements of the male sexual act include *erection* followed at the height of sexual stimulation by *emission* and *ejaculation*. This entire act can be performed in some males whose spinal cords have been cut in the low thoracic region, which illustrates that the basic mechanism for the act can be controlled by the lower spinal cord alone. However, in normal performance of the sexual act, psychogenic factors involving the highest levels of the cerebrum appear to be equally as important as the spinal cord mechanisms.

Initiation of Erection—Psychogenic and Local Stimulatory Factors.
Erection is caused by *parasympathetic nerve signals* carried to the penis from the sacral part of the spinal cord (S2 to S4) by way of the *pelvic nerve plexus*. These signals dilate the arteries that supply the erectile tissue, which allows a tremendous amount of blood to enter this tissue under high pressure, inflating it like a balloon. This causes the penis to become greatly enlarged, to become turgid, and to extend forward, which is the act of erection.

The parasympathetic neural stimulation that causes the erection can itself be initiated in either one of two ways or both: (1) by psychic stimulation—in fact, even erotic thoughts or dreams can lead to full erection, and (2) by stimulation of the external genital areas or even the internal genital organs—by far the most potent stimulation results from stimulation of the *glans penis*. The afferent nerve signals from the glans are carried in the dorsal nerves of the penis to the pudendal nerve and then to the sacral cord. Other sexually excitatory signals are carried through visceral afferents in the parasympathetic nerves to the sacral cord and visceral afferents in the sympathetic nerves to the thoracolumbar cord.

Importance of Lubrication During the Sexual Act.
To and fro movement of the penis in the vagina during intercourse creates a massaging effect that is the necessary condition for maximal sexual stimulus. However, this is effective only in case there is adequate lubrication, because an abrasive effect inhibits male sexual excitation and blocks completion of the act. Most of the lubrication is provided by the female, as we shall discuss in the following chapter. However, even in the male, at the same time that parasympathetic stimulation causes erection, it also excites mucus secretion by the bulbourethral glands located at the proximal end of the urethra and also by the small urethral glands along the course of the urethra. Some of this mucus is expelled during intercourse and helps to lubricate the sexual movements.

Emission and Ejaculation.
Emission and ejaculation are the culmination of the male sexual act. When sexual excitement becomes extremely intense, the sexual reflex centers of the spinal cord send an entirely new set of nerve signals to the genital organs, but this time

through the sympathetic nerves from the upper lumbar portion of the cord rather than through the parasympathetic nerves. These signals cause contraction of the epididymis, the vas deferens, and the ampulla of the vas deferens to cause expulsion of the sperm into the prostatic urethra. Then, contractions of the walls of the seminal vesicles and of smooth muscle in the prostate gland expel seminal fluid and prostatic fluid also into the prostatic urethra, forcing the sperm forward. All of these fluids together constitute the *semen*. The process to this point is called *emission*.

The filling of the internal urethra then elicits additional afferent signals that are transmitted mainly to the sacral region of the cord. In turn, nerve signals from the sacral cord stimulate further rhythmic contraction of the internal genital organs as well as of the ischiocavernosus and bulbocavernosus muscles that encase the bases of the penile erectile tissue; all these effects together force the semen through the urethra to the exterior. This is the process of *ejaculation*. At the same time, rhythmic contraction of the pelvic muscles and even of the muscles of the trunk cause thrusting movements of the pelvis and penis that help to propel the semen into the deepest recesses of the vagina and perhaps even into the cervix of the uterus.

This entire period of emission and ejaculation is called the *orgasm*. At its termination, sexual excitement disappears almost entirely within a few minutes, and erection ceases.

The total quantity of semen ejaculated is usually about 3.5 ml, with each milliliter containing about 120 million sperm, or a total of about 400 million sperm. The alkaline fluid from the prostate gland neutralizes the acidic fluid from the epididymis and vas deferens, which releases the sperm from their inactive state and stimulates their motility.

Male Sterility. Approximately 1 male out of every 20 to 25 is sterile. Frequent causes of this are (1) congenitally deficient testes that are incapable of producing sperm or that produce abnormal sperm with two heads, two tails, or low mobility and (2) previous infection in the male genital ducts. Also, occasionally the seminiferous tubules of the testes may have been partially or totally destroyed by mumps infection, typhus infection, x-ray irradiation, or nuclear radiation.

Male sterility can also occur when the number of sperm in the ejaculate falls too low, even though all the sperm are normal. Often a male is sterile when this number falls below 10 to 20 million per milliliter of ejaculate or the vol-

ume of semen falls below 2 ml. It is difficult to understand why sterility should exist in this instance, since only one sperm is required to fertilize the ovum. However, it is believed that a large number of sperm are necessary to provide enzymes or other substances that help the single fertilizing sperm to reach the ovum. The acrosome of the sperm head releases *hyaluronidase* and several proteinases that are believed to help disperse the granulosal cells that cover the surface of the ovum when it is expelled by the ovary. This action is believed to allow the sperm to reach the ovum.

HORMONAL REGULATION OF MALE SEXUAL FUNCTIONS

Role of the Anterior Pituitary Gonadotropic Hormones

Puberty and Luteinizing Hormone-Releasing Factor. The testes of the child remain dormant until they are stimulated at the age of 10 to 14 by gonadotropic hormones from the pituitary gland. At that age, the hypothalamus begins to secrete *luteinizing hormone-releasing factor*, which in turn causes the anterior pituitary gland to secrete both *follicle-stimulating hormone* and *luteinizing hormone*. These cause an upsurge of testicular growth and function, causing the male sex life to begin. This stage of development is called *puberty*.

Follicle-Stimulating Hormone. Follicle-stimulating hormone causes proliferation of the spermatogonia, thus initiating the process of sperm formation. However, this alone is not sufficient to carry spermatogenesis to completion, which requires the action of testosterone as well, as we shall note subsequently.

Luteinizing Hormone. Luteinizing hormone causes the testes to secrete testosterone (and very small amounts of several other steroid hormones that have similar actions). The testosterone is produced in special glandular cells called **interstitial cells of Leydig** that are located in the triangular interstitial areas between the adjacent seminiferous tubules, as illustrated in Figures 48–6 and 48–9. Testosterone secretion increases rapidly at puberty and reaches its peak in early adulthood but declines to about 20 percent of its peak at the age of 80.

Functions of Testosterone

Effect on Spermatogenesis. Testosterone causes the testes to enlarge. Also, it must be present along with follicle-stimulating hor-

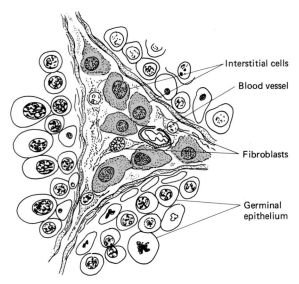

FIGURE 48–9   The interstitial cells of the testis. (Modified from Bloom and Fawcett: A Textbook of Histology, 8th ed. Philadelphia, W. B. Saunders, 1974.)

mone before spermatogenesis will proceed through the spermatocyte and spermatid stages to form the spermatozoa. Unfortunately, though, the precise action of testosterone in spermatogenesis is not yet clear.

Effect on Male Sex Characteristics. After a male embryo begins developing inside its mother's uterus, its testes begin to secrete testosterone when it is only a few weeks old. This testosterone then causes the embryo to develop male sexual organs and male secondary characteristics. That is, it causes the formation of a penis, a scrotum, a prostate, the seminal vesicles, the vas deferens, and other male sexual organs. In addition, testosterone causes the testes to descend through the inguinal canal from the abdominal cavity into the scrotum; if production of testosterone by the fetus is insufficient, the testes will fail to descend but instead remain in the abdominal cavity in the same manner that the ovaries remain in the abdominal cavity in the female.

Testosterone secretion by the fetal testes is caused by a hormone called *chorionic gonadotropin* that is formed in the placenta during pregnancy, as will be discussed in more detail in Chapter 50. Immediately after birth of the child, loss of connection with the placenta removes this stimulatory effect, so that the testes then stop secreting testosterone. Consequently, the male sexual characteristics cease developing from birth until puberty. At puberty the reinstitution of testosterone secretion because of hypothalamic and pituitary stimulation causes the male sex organs to begin growing again. The testes, scrotum, and penis then enlarge about 10-fold.

Effect on Secondary Sex Characteristics. In addition to the effects on the genital organs, testosterone exerts other general effects throughout the body to give the adult male his distinctive characteristics. It promotes growth of hair on his face and also along the midline of his abdomen, on his pubis, and on his chest. On the other hand, it causes baldness in those male individuals who also have a hereditary predisposition to baldness. It increases the growth of the larynx so that the male, after puberty, develops a deeper pitch to his voice. It causes an increase in the deposition of protein in his muscles, bones, skin, and other parts of his body, so that the male adolescent becomes generally larger and more muscular than the female. Also, testosterone sometimes promotes abnormal secretion by the sebaceous glands of the skin, this leading to acne on the face of the postpubertal male.

QUESTIONS

1. Name the primary and accessory male sexual organs, and give the basic functions of each.
2. Describe the spermatic cord, and explain the causes of an inguinal hernia.
3. Describe the erectile tissue of the male penis, and explain the mechanics of erection.
4. Trace the course of movement of sperm from the seminiferous tubules to the urethra, and explain the function of the epididymis.
5. Describe the formation of sperm and their functional characteristics.
6. Explain the neural control of erection, emission, and ejaculation.
7. Give the mechanisms of emission and ejaculation.
8. What initiates puberty?
9. What are the roles of follicle-stimulating hormone and luteinizing hormone in the male?
10. How does testosterone affect the male sexual organs?
11. How does testosterone affect the secondary sexual characteristics of the male?

REFERENCES

See the references at the end of Chapter 49.

49

The Female Reproductive System, and the Female Hormones

Overview

The principal structures of the female reproductive system are (1) the **vagina**, (2) the **uterus**, (3) two **uterine tubes**, and (4) two **ovaries.** Approximately once each month during the reproductive period of the female, from about the age of 14 to 45, one of the two ovaries releases an ovum that is then transported through one of the uterine tubes to the uterus. If the ovum is fertilized by a sperm on its way, it will grow to become a baby.

Surrounding the opening of the vagina is the same type of **erectile tissue** that is found in the penis of the male. During sexual intercourse, this tissue, as well as most of the other tissues of the external genitalia, becomes engorged with blood and tightens around the male penis. At the anterior margin of the vagina is the **clitoris**, which erects like the male penis and has a **glans** the same as that of the penis but much smaller. Stimulation of the nerve endings in this glans, and to a less extent in other areas around the vagina, elicits powerful sexual sensations during intercourse. These send nerve signals to the lower end of the spinal cord and thence back to the sexual organs to cause swelling of the genitalia, secretion of mucoid fluid into the vagina for the purpose of lubrication, and rhythmic contractions of the vagina and uterus at the time of the *orgasm.*

Monthly cyclic changes occur in the ovaries, in the rates of secretion of the female hormones, and in the inner lining of the uterus; these changes are collectively called the *female sexual cycle.* At the time of birth, several hundred thousand immature **ova** called **primary oocytes** are already deposited in the ovaries immediately underneath the ovarian surfaces. At the onset of each sexual cycle, **follicular cells** surrounding several of these oocytes begin to proliferate, and **vesicular follicles** filled with fluid develop around the ova. One of the follicles grows much larger than the others and ruptures on about the fourteenth day of the cycle, releasing the ovum into the pelvic cavity. This is called *ovulation.* The ovum is swept rapidly into one of the **uterine tubes** because large numbers of **cilia** on the **fimbria** of the uterine tubes beat continually toward the **ostium** of the uterine tube. If the female has recently had intercourse, a sperm traveling up the uterine tube may fertilize the ovum at approximately the time that it enters this tube.

During approximately the first 2 weeks of each sexual cycle, the follicular cells of the growing follicles secrete large quantities of *estro-*

gen. Then, after ovulation, the follicular cells of the ruptured follicle enlarge several fold and become the **corpus luteum,** which continues to secrete estrogen but now also secretes very large quantities of *progesterone* as well. If a fertilized ovum does not implant in the uterus, the corpus luteum involutes after 10 to 12 days, and the estrogen and progesterone secretion rates decrease to very low levels.

Growth of the follicles is caused mainly by *follicle-stimulating hormone,* which is secreted by the anterior pituitary gland, and ovulation is caused mainly by *luteinizing hormone,* also secreted by the anterior pituitary gland. The rhythm of the female sexual cycle is caused by a complex series of negative feedback interplays between the two ovarian hormones, estrogen and progesterone, and the two anterior pituitary hormones, follicle-stimulating hormone and luteinizing hormone.

During the first half of each sexual cycle, the lining of the uterus, the **endometrium,** thickens several fold because of marked proliferation of its cells and growth of the endometrial glands. These effects are caused by the estrogen secreted during the early part of the cycle. During the latter half of the cycle, under the influence of both estrogen and progesterone, the endometrium thickens another twofold because of the swelling of the endometrial tissues; at the same time the endometrial glands become highly secretory. The endometrium at this time is prepared for implantation of a fertilized ovum. However, if a fertilized ovum does not implant, involution of the corpus luteum and loss of its secretion of estrogen and progesterone causes the blood vessels in the endometrium to become spastic, and the blood supply to the endometrium almost ceases. As a result, the endometrium degenerates and sloughs off inside the uterine cavity, which is the process of *menstruation.* The sloughed products along with some blood, which together are called the *menstrum,* are expelled by peristaltic contractions of the uterus that can sometimes cause *cramps.* Then the endometrial cycle begins again under the influence of a new cycle of hormone secretion by new follicles in the ovaries.

THE FEMALE ORGANS OF REPRODUCTION

Figure 49–1 illustrates the overall structure of the female reproductive system. Its principal organs are the following:

1. The **vagina** is the tubular opening leading from the external genitalia to the uterus.
2. The **uterus** is the muscular structure in which the fetus develops and grows.
3. One of the two **uterine tubes** leads from each superior corner of the uterus, then curves laterally and posteriorly around the wall of the pelvis. It is through one of the uterine tubes that an ovum passes from an ovary during each monthly sexual cycle into the uterus.
4. One of the two **ovaries** is suspended underneath the distal opening of each uterine tube. During the reproductive period of the female, from about the age of 14 to 45, one of the two ovaries releases an ovum at approximately monthly intervals; if fertilized, it may develop to become a baby.

Note also in Figure 49–1 the relationship of the vagina and uterus to the urinary bladder and rectum. The external end of the vagina lies immediately inferior to the urethra, and the posterior part of the vagina lies inferior to the urinary bladder. Also, along its entire course, the vagina lies anterior to the rectum.

Observe also the manner in which the distal end of the uterus protrudes into the posterior vagina. This protruding portion of the uterus is called the **cervix.**

THE EXTERNAL GENITALIA

The female external genitalia, illustrated in Figure 49–2, are together called the **vulva.** The separate structures of the vulva are as follows:

1. The **mons pubis** is the protruding pad anterior to the vagina where the pubic hair grows. The protruding nature of the mons pubis is caused by fat deposits in the tissue overlying the symphysis pubis. The same protruding fat pad is not present in the male.

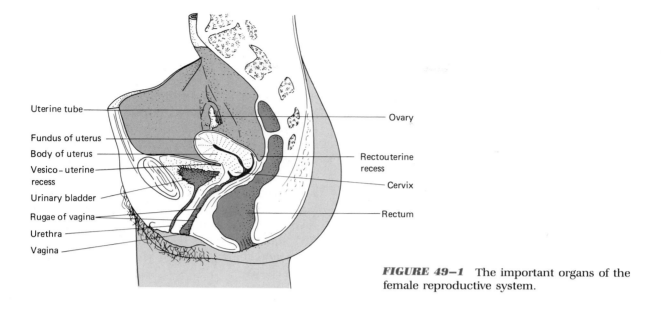

Uterine tube

Fundus of uterus

Body of uterus

Vesico-uterine recess

Urinary bladder

Rugae of vagina

Urethra

Vagina

Ovary

Rectouterine recess

Cervix

Rectum

FIGURE 49-1   The important organs of the female reproductive system.

TABLE 49-1
The Female Organs of Reproduction

Structure	Function or description
Ovary (one on each side)	
Follicles	Provide for development of ova and secretion of estrogen
Oocytes	Cells from which mature ova develop
Follicular cells (granulosa and thecal cells)	Cells that secrete estrogens
Corpus luteum	Structure that develops from ovulated follicle; secretes large quantities of progesterone and estrogen
Uterine tube (also called fallopian tube; one on each side)	Transports ovum to uterus in about 3 days; lumen epithelium partially lined with ciliated cells; secretes nutrient fluids for developing morula
Fimbria of uterine tube	Finger-like extensions of lateral end of uterine tube that usually wrap around ovary; ciliated epithelium moves ovulated ovum into uterine tube
Uterus	
Body and fundus	Principal parts of uterus in which fetus develops
Cervix and cervical canal	Provides entryway from body into vagina; cervical glands secrete mucus
Endometrium	Internal lining of body and fundus; contains many endometrial glands that secrete nutrients for early fetal tissues; mostly sloughs off during menstruation
Myometrium	Smooth muscle wall of uterus; increases 30-fold during pregnancy
Vagina	Conduit from uterus to exterior; its epithelium transudes mucoid fluid during intercourse
External genitalia	
Mons pubis	Protuberance over pubis covered by hair
Labia majora	Large lateral lips
Labia minora	Thin protruding lips medial to labia majora
Vestibule of vagina	Recessed cleft between labia minora
Clitoris and glans of clitoris	Small homologue of male penis located at anterior margin of vestibule
Erectile tissue	
Bulb of the vestibule, ventral body of clitoris, and glans of clitoris	Continuous erectile body; tightens opening of vagina and helps to cause erection of clitoris
Crura and lateral bodies of the clitoris	Two lateral erectile bodies; provide most of erection of clitoris
Bulbourethral gland (also called Bartholin's gland; one on each side)	Secretes mucus into vestibule, especially during intercourse

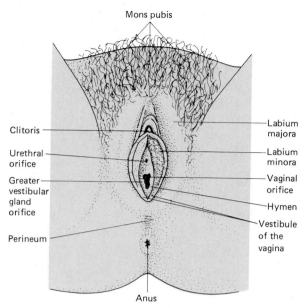

Mons pubis

Clitoris

Urethral orifice

Greater vestibular gland orifice

Perineum

Labium majora

Labium minora

Vaginal orifice

Hymen

Vestibule of the vagina

Anus

FIGURE 49–2   The female external genitalia.

2. The two **labia majora** are the two broad ridges, or "lips," that extend posteriorly along the two lateral sides of the vagina from the mons pubis.
3. The two **labia minora** are two folds lying inside the labia majora but lateral to both the vaginal and the urethral openings.
4. The **vestibule of the vagina** is the depressed cleft between the two labia minora and into which both the vagina and the urethra open.
5. The **vaginal orifice** is located in the posterior recess of the vestibule. Surrounding this orifice is the **hymen,** which, in the virgin, is usually a ringlike membrane that partially covers the vaginal orifice. Soon after sexual activity has begun usually only remnants of the hymen remain.
6. The **clitoris** is the homologue of the male penis but much smaller than in the male. At its tip is a small bulbous protrusion protruding beyond the point where the anterior ends of the labia minora come together; this is the **glans of the clitoris** that corresponds to the glans of the male penis but is much smaller. The glans is a richly innervated structure that elicits powerful sexual sensations when massaged or when stimulated during sexual intercourse.

Another anatomical landmark associated with the external genitalia, but not itself a genital organ, is the **urethral orifice,** which is the external opening of the urethra located midway between the clitoris and the vaginal orifice.

THE VAGINA

The **vagina,** illustrated in Figure 49–1, is the canal about 8 cm long and 3 cm in diameter that leads from the vaginal orifice to the uterus. Its inner surface is lined with a *mucous membrane,* and its deeper wall has both a *circular* and a *longitudinal layer* of *smooth muscle* similar to the layers of the intestinal wall. Note in the figure the circular folds in the mucous membrane. These are called the **rugae** of the vagina. They are especially important during intercourse to cause sexual stimulation of the male glans penis. Also, the mucous membrane is lined with *stratified squamous epithelium,* which allows it to withstand the abrasion of intercourse and also to resist infection by such organisms as gonorrheal bacteria.

The exterior opening of the vagina is somewhat more constricted than the remainder and is called the **introitus.** At the opposite end of the vagina, comprising the superior one half to two thirds of its extent, the diameter is 50 to 100 percent larger than that of the introitus. This area is called the **vaginal vault,** and at the dome of this the inferiormost portion of the uterus, the **uterine cervix,** protrudes about 1 cm into the vagina. The anterior, posterior, and lateral recesses between the cervix and the walls of the vagina are known respectively as the **anterior** and **posterior vaginal fornices** and the **left** and **right lateral fornices.**

The vagina is bordered anteriorly by the urethra and urinary bladder and posteriorly by the rectum, as also illustrated in Figure 49–1. These relationships are especially important when one is performing surgical operations in this area.

The Erectile Tissue and Other Structures Around the Vagina. The female has erectile tissue similar to that of the male, but in a different formation as illustrated in Figure 49–3. This is divided into three parts, the same as in the male:

1. the **bulb of the vestibule,** which itself is divided into two halves that straddle the two sides of the vagina lying deep to the two labia majora, and
2 and 3. two **crura of the clitoris,** one of which lies alongside the inferior medial border of each ischiopubic arch.

The anterior ends of the two crura come together anterior to the vagina to form the two **corpora cavernosa** that make up the main body of the clitoris; the anterior ends of the two halves of the bulb of the vagina come together to

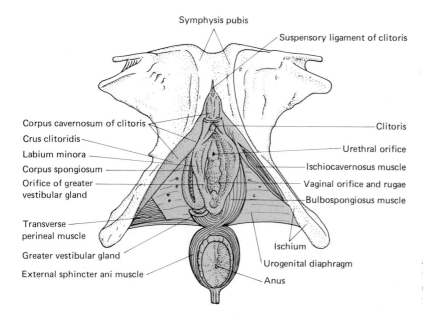

Symphysis pubis

Suspensory ligament of clitoris

Corpus cavernosum of clitoris

Crus clitoridis

Labium minora

Corpus spongiosum

Orifice of greater vestibular gland

Transverse perineal muscle

Greater vestibular gland

External sphincter ani muscle

Clitoris

Urethral orifice

Ischiocavernosus muscle

Vaginal orifice and rugae

Bulbospongiosus muscle

Ischium

Urogenital diaphragm

Anus

FIGURE 49–3 The deep structures around the vaginal opening, emphasizing especially the erectile tissue and their associated muscles.

form the **ventral body** and **glans of the clitoris.** During sexual intercourse, all of this erectile tissue becomes swollen and hard the same as in the male. This results in erection of the clitoris and also causes swelling of the bulb around the vaginal orifice to provide a firm but pliable opening that serves to massage the glans of the penis during intercourse.

Three muscles lie superficial to the erectile tissue, as also occurs at the base of the penis in the male:

1. The **bulbospongiosus muscle** lies over the bulb of the vestibule deep to the labia majora, thus also surrounding the vaginal opening. Contraction of this muscle further tightens the vaginal opening during intercourse.
2. One of two **bulbocavernosus muscles** overlies each crus of the clitoris. Contraction of these two muscles squeezes the blood forward in the two corpora cavernosa and causes additional erection of the body of the clitoris.

Lubrication of the sexual act is aided by *mucus* secreted by the two **greater vestibular glands,** illustrated in Figure 49–3, that lie underneath the bulbospongiosus muscle, deep and posterior to the end of each labium majus. The duct from each gland courses forward alongside the vaginal vestibule for 1 to 2 cm and then empties into the vestibule posterolateral to the vaginal orifice immediately outside the hymen, at a point most propitious for lubricating the sexual act.

The Urogenital Diaphragm. Note also in Figure 49–3 the triangular structure composed of muscle and fascia that lies between the two ischiopubic arches and deep to the erectile tissue and their associated muscles. This is called the **urogenital diaphragm;** it is through this that the urethra and the vagina pass. The muscles in the inner core of the urogenital diaphragm are the **sphincter urethrae** and the **deep transverse perineal muscles.** During birth of a baby, the urogenital diaphragm is often severely torn. If it is not repaired, lack of appropriate support of the posterior vaginal wall will in later life allow the bladder and the superior wall of the vagina to protrude outward through the vaginal opening, a condition called *cystocele.* This can make it difficult for the woman to urinate, or at other times it can cause continual dribbling of urine through the urethra, a phenomenon called *incontinence.*

THE UTERUS

The **uterus** is illustrated in Figures 49–1, 49–4, and 49–5. It is divided into three separate parts:

1. the **body,** which comprises the major mass of the uterus;
2. the **fundus,** the domed top of the uterus; and
3. the **cervix,** the portion of the uterus that protrudes into the posterior vagina.

The uterus contains an inner space, the **uterine cavity,** shown in Figure 49–5; this opens inferiorly into the vagina through the **cervical canal.** Note in the figure that the cervical canal has numerous folds. Also, in the upper two thirds of the canal, large numbers of deep **glandular follicles** secrete a very thick mucus that plugs the canal during early pregnancy. In the

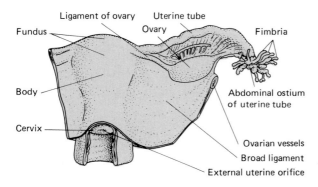

FIGURE 49—4 Posterior view of the uterus and its relationship to the ovary and uterine tube.

occasional female this mucus has physical or chemical characteristics that also impede sperm penetration following intercourse or destroy the sperm, which can lead to partial or total sterility.

The Uterine Wall. The wall of the uterus is divided into three layers:

1. The **endometrium** is an inner connective tissue and epithelial lining of the uterine cavity. It is several millimeters thick and contains many thousands of coiled **endometrial glands;** these secrete nutrient material for the early developing embryo, as we shall discuss further in the following chapter. It is the inner portion of the endometrium that is sloughed off and expelled in the menstrum during menstruation, which will be discussed later in the chapter.
2. The **myometrium** is the smooth muscle component of the uterine wall and constitutes most of the mass of the uterus. During pregnancy, the myometrial mass increases about 30-fold. Contraction of this muscle is mainly responsible for expulsion of the fetus at the time of delivery. It is also this muscle that causes the cramping muscle contractions that expel the menstrum at the time of menstruation.
3. The **peritoneum** is the outer covering of the uterus; this is continuous with the peritoneum of the entire abdomino-pelvic cavity.

The Suspensory Ligaments of the Uterus. The uterus is suspended in the pelvic cavity by four important ligaments on each side:

1. The **broad ligament** illustrated in Figures 49–4 and 49–5 is a broad fold of peritoneum and fibrous tissue attached between the entire lateral side of the uterus and the lateral wall of the pelvic cavity. Within this ligament the uterine artery passes along the lateral border of the uterus after its origin near the cervix from the internal iliac artery. Normally, this artery is only a few millimeters in diameter, but during pregnancy its diameter increases to as much as 7 to 8 ml, which allows each uterine artery to carry more than half a liter of blood each minute to the placenta.
2. The **round ligament** is a condensation of fibrous tissue within the superior border of the broad ligament. Medially it attaches to the uterus just below and anterior to the uterine tube, and laterally it attaches to the anterolateral pelvic wall near the inguinal ligament.
3. The **cardinal ligament** is a condensation of fascia around the uterine vessels in the base of the broad ligament, attaching laterally from the cervical region of the uterus to the anterolateral wall of the pelvic diaphragm.
4. The **uterosacral ligament** attaches posteriorly from the cervical region of the uterus to the deep fascia over the sacrum.

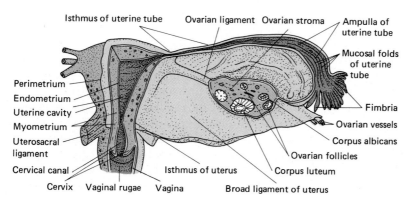

FIGURE 49—5 The internal structures of the uterus, and ovum, and a uterine tube.

During pregnancy, the fundus of the uterus enlarges tremendously and eventually extends most of the way up the abdominal cavity; yet the ligaments hold only the lower end of the uterus in place. Therefore, the upper part of the uterus then literally floats among the other pelvic and abdominal contents.

THE UTERINE TUBES

Figure 49–4 illustrates one of the **uterine tubes,** which is a tubular extension from the uppermost corner of the uterine fundus at the point called the **horn of the uterus.** Along its entire extent, the uterine tube is attached to the superior border of the broad ligament by a thin peritoneal fold called the **mesosalpinx.** At the lateral end of the tube, its lumen opens directly into the pelvic cavity; this opening is the **abdominal ostium** of the uterine tube. Around the ostium are numerous finger-like projections called **fimbria.** Though Figures 49–4 and 49–5 show the ostium and the fimbria to extend laterally from the end of the uterine tube, in life these normally curve downward toward the ovary, and the fimbria wrap partially around the ovary. We shall see later that this allows maximum opportunity for an ovum to enter the ostium of the uterine tube after it is released from the ovary.

Internal Structure of the Uterine Tube. Figure 49–4 illustrates the inside of the uterine tube. The lumen of the tube is considerably enlarged near the fimbriated end; this portion of the tube is called the **ampulla.** But, at the uterine end of the uterine tube the canal becomes narrow; this area is the **isthmus.** When an infection occurs in the uterine tube, especially infection caused by gonorrhea, the isthmus frequently becomes permanently occluded; if this occurs on both sides, it causes permanent sterility. In addition, a small number of females are born with congenital uterine tube occlusion; this often occurs also at the isthmus.

Note the many **mucosal folds** inside the uterine tube. These are supplied with **glands** that secrete nutrients for development of the fertilized ovum. Many of the epithelial cells lining the uterine tube, as well as those lining the fimbriae, are ciliated. The *cilia* beat continually toward the uterus and thereby move the ovum to the uterine cavity.

The Ovaries and Ovulation

Figure 49–4 shows that each **ovary** lies on the posterior surface of the broad ligament about 2 cm inferior to the uterine tube. The ovary is loosely attached to this ligament by a peritoneal fold called the **mesovarium.** It is also suspended by two stronger fibrous ligaments:

1. The **suspensory ligament of the ovary** attaches from the lateral pole of the ovary to the posterolateral wall of the pelvic cavity. The ovarian vessels pass along this ligament from their vascular connections midway up the abdomen.
2. The **ligament of the ovary** attaches from the medial pole of the ovary to the lateral side of the uterus immediately inferior to the uterine horn.

The Internal Structure of the Ovary. A longitudinal section of the ovary is shown in Figure 49–5. Its principal parts are as follows:

1. The **ovarian stroma.** This is the connective tissue that constitutes the main body of the ovary. It is within this stroma that the ova and the female hormone secreting cells develop.
2. The **ovarian follicles.** In each ovary are thousands of follicles, each of which contains an *immature ovum* surrounded by *epithelioid follicular cells.* The follicles undergo progressive stages of development leading to release of an ovum approximately once each month from the surface of the ovary, which is the process of *ovulation.* Figure 49–5 illustrates multiple follicles at different stages of development, which will be described more fully later.
3. The **corpus luteum.** After a mature follicle releases its ovum from the ovary, the follicular cells change into a hormone-secreting organ called the corpus luteum, one of which is illustrated in Figure 49–5. After 2 weeks of secretory life (if pregnancy does not begin, as will be explained later) the corpus luteum degenerates and is invaded by fibrous tissue, forming a *corpus albicans;* then this too is resorbed during the next few weeks.

Growth of the Follicle, and Ovulation. During early formation of the female embryo, germ cells migrate from the yolk sac into the early developing ovaries and come to rest just underneath the ovarian surfaces. These cells proliferate to form about 300,000 immature ova in the two ovaries (called **primary oocytes** at this stage). From the time of birth of the female until the time of puberty, the status of these oocytes remains mainly unchanged, but then the anterior pituitary gland begins to secrete *gonadotropic hormones* that promote growth of the follicles and release of approximately one ovum each month throughout the child-bearing years.

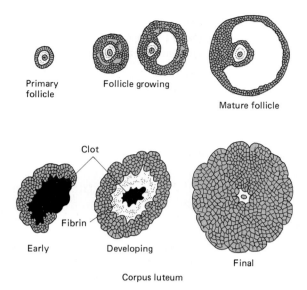

Primary
follicle

Follicle growing

Mature follicle

Clot

Fibrin

Early

Developing

Final

Corpus luteum

FIGURE 49—6 Growth of the follicle, and formation of the corpus luteum. (Modified from Arey: Developmental Anatomy, 7th ed. Philadelphia, W.B. Saunders Co.)

The follicles undergo the following stages of development.

First, at the beginning, each primary oocyte in the ovary is surrounded by a layer of epithelioid follicular cells called **granulosa cells;** these along with the enclosed oocyte are the **primary follicle,** one of which is illustrated in the upper left corner of Figure 49–6.

Second, under the stimulus of the gonadotropic hormones from the pituitary gland, a few of the primary follicles enlarge during each monthly sexual cycle of the female. This enlargement is characterized by marked proliferation of the granulosa cells to form multiple layers, as illustrated by the "growing follicles" in the upper part of Figure 49–6. In addition, many adjacent stromal cells also take on epithelioid characteristics and are then called **thecal cells.** These join the mass of granulosa cells to become part of the follicular cell mass. The follicular cells then secrete fluid within the mass of cells, creating a cavity called an **antrum,** also shown in Figure 49–6. At this stage the follicle is a **vesicular follicle.** The antrum continues to enlarge in these follicles until the follicles actually balloon outward from the surface of the ovary. Several such vesicular follicles are shown in the ovary to the right in Figure 49–5.

Third, one of the vesicular follicles balloons from the ovarian surface much more than the others; this is the **mature follicle.** Inside this, a single ovum is encased in a protruding mound of granulosa cells.

Fourth, within a few hours after the follicle becomes mature, its surface breaks open and releases the vesicular fluid into the pelvic cavity. Within another few minutes, the walls of the ruptured follicle contract and extrude the mound of granulosa cells along with its enclosed ovum also into the pelvic cavity. This is the process of *ovulation.*

Fifth, immediately after ovulation occurs, all the vesicular follicles besides the one that has ovulated suddenly stop growing and begin to resorb without ever rupturing. Presumably the release of the follicular fluid from the ruptured follicle into the pelvic cavity and rapid absorption of its hormones by the peritoneum in some way cause involution of these other follicles. Regardless of the exact cause, only one ovum is usually expelled into the pelvic cavity during each monthly sexual cycle, though rarely a second or more are expelled before the remaining follicles start to involute. This is one of the causes of multiple births, accounting for the so-called *fraternal twins* (nonidentical twins), which are about three fourths of all twins.

Maturation of the Oocytes, and Reduction of the Chromosomes. Figure 49–7 shows the development of the mature ovum from the primitive germ cells. After multiple division of the germ cells during embryonic life, some of them eventually become **primary oocytes,** which are the immature ova of the primary ovarian follicles. The primary oocyte does not divide further until the follicle approaches maturity. Then, shortly before ovulation, the nucleus of the primary oocyte does divide but without re-

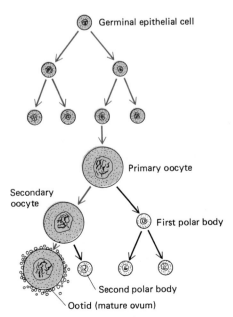

Germinal epithelial cell

Primary oocyte

Secondary
oocyte

First polar body

Second polar body

Ootid (mature ovum)

FIGURE 49—7 Formation of the mature ovum from the germinal epithelium. (Modified from Arey: Developmental Anatomy, 7th ed. Philadelphia, W.B. Saunders Co.)

producing the chromosomes. Instead, each pair of chromosomes splits apart so that 23 unpaired chromosomes remain in the oocyte, which is now called a **secondary oocyte;** the other 23 chromosomes are expelled in the so-called *first polar body.*

It is while the oocyte is in this secondary oocyte stage that ovulation occurs. Then if fertilization follows, immediately after the sperm enters the ovum the female nucleus of the fertilized ovum divides again, and a *second polar body* is expelled, but this time the chromosomes also divide so that 23 unpaired chromosomes still remain in the female nucleus. Once the ovum is fertilized, the 23 chromosomes of the male nucleus from the sperm combine with the 23 chromosomes of the female nucleus to form the normal cell complement of 46 chromosomes. Thus, half of the chromosomes, and consequently half of the genes as well, of the fertilized ovum are supplied by the mother and half by the father.

Transport of the Ovum in the Uterine Tube. Normally, the abdominal ostia of the uterine tubes lie in close proximity to the ovaries, and the cilia lining the epithelial surfaces of the fimbria and the lumens of the uterine tubes are always beating toward the uterus, causing fluids from the region of the ovary to move constantly out of the pelvic cavity into the tubes. The ovum, on being expelled into the pelvic cavity from the ruptured follicle, is thus transported into one of the uterine tubes along with this current of fluid. The lining of the uterine tube, however, is pocked along its entire extent with large cavities that obstruct the movement of the ovum, so that usually 3 to 4 days are required for its passage to the uterus.

The ovum must be fertilized within 8 to 24 hours after it is released from the ovary or it dies. Therefore, fertilization usually takes place in the upper portion of the uterine tube, and the fertilized ovum begins to divide while it is still in the tube. The uterine tube epithelium secretes substances that provide nutrition for the developing mass of cells, called at this stage the **morula.** However, some of the nutrition is also supplied by the large amount of cytoplasm already in the ovum itself.

Female Sterility. About 1 female out of 20 is sterile. As is true in the male, many of these instances of sterility are caused by previous infection. The infection sometimes blocks the uterine tubes, or at other times involves the ovaries in a mass of scar tissue. The most common infection causing female sterility, as is true also in the male, is gonorrhea. Sterility is often also caused by an inborn failure of the ovaries to de-

velop and expel ova into the abdominal cavity. Frequently this results from insufficient stimulation by the anterior pituitary gonadotropic hormones. At other times it is caused by an excessively thick capsule on the surface of the ovary, or occasionally by congenitally abnormal ovaries.

The Female Sexual Act

During intercourse the female helps the male reach a sufficient degree of sexual stimulation for emission and ejaculation to occur. She does this by providing appropriate conditions for maximum masculine stimulation. This is achieved mainly by swelling of most of the female genital organs and secretion of adequate lubricant.

Swelling of the Female Genital Organs—Role of Psychic and Local Stimulatory Factors. During the early stages of the female sexual act, most of the female genital organs become greatly swollen, which results from generalized vasodilatation and resultant congestion with blood. This includes swelling of the labia majora, the labia minora, the wall of the vaginal canal, and even the uterus. In addition, the erectile tissue of the bulb of the vestibule becomes tightly engorged with blood, which helps to provide a tight though distensible opening around the vaginal canal. At the same time engorgement of the erectile tissue in the clitoris causes erection of the clitoris in the same manner that this occurs in the penis, which brings the glans of the clitoris to the anterior border of the vestibule of the vagina, where it becomes most exposed to stimulation.

The neural excitation that leads to these vasocongestive phenomena can be elicited by either psychic or genital stimulation or both, the same as in the male. For the female to participate adequately in the sexual experience it is especially important for her to have the appropriate psychic desire, but, superimposed onto this, afferent nerve impulses from all regions of the genitals play equally as great a role in increasing and maintaining sexual excitation during the period of intercourse. By far the most sensitive structure for eliciting the sexual excitation is the glans of the clitoris, partly because it has an exceedingly high concentration of sensory receptors and partly because its erected position exposes it most propitiously to the massaging movements of the penis.

The *afferent nervous signals* from the genitals that cause sexual excitation are carried mainly in the *pudendal nerve* to the *sacral por-*

tion of the cord, but some signals are probably also carried in both the parasympathetic and the sympathetic visceral afferents to the cord. On the other hand, the *efferent signals* that cause the vasocongestion of the genitals are transmitted almost entirely through the *pelvic parasympathetic nerves* that originate from the sacral cord at levels S2 through S4.

Lubrication of the Vagina. The same parasympathetic nervous signals that cause the vasocongestion also cause lubrication of the vagina. This results mainly from transudation of a mucoid fluid through the epithelium of the vaginal mucosa. The formation of this fluid has not been completely explained because there are not obvious large mucous glands in the vaginal epithelium. An additional amount of mucus is secreted into the vestibule by the two greater vestibular glands, but this mucus, as well as that secreted by the male as discussed in the previous chapter, is probably of much less importance in lubricating the sexual act than the lubrication provided in the vagina itself.

The Female Orgasm. At the peak of sexual excitement, the female has an orgasm that is the counterpart of emission and ejaculation in the male. The components of the orgasmic response include: (a) rhythmic contractions of the wall of the vagina near its opening, (b) rhythmic contraction of the uterus, (c) contraction of the pelvic and other muscles, similar to the contractions that occur in the male though usually less intense, and (d) an intense psychic state that tends to override other sensory sensations.

The *efferent neuronal signals* that cause these orgasmic events are believed to be transmitted through the *sympathetic nerves* originating from approximately L1 and L2 spinal cord segments, the same as for eliciting orgasm in the male. Thus, the early stages of the sexual act in both the male and the female are elicited principally by the parasympathetic nervous system, whereas the culmination of the act, the orgasm, requires function of the sympathetic nervous system.

HORMONAL REGULATION OF FEMALE SEXUAL FUNCTIONS

Relation of the Anterior Pituitary Gonadotropic Hormones to the Monthly Sexual Cycle

The anterior pituitary gland of the female child, like that of the male child, secretes essentially no gonadotropic homones until the age of 10 to 14 years. However, at that time the anterior pitui-

FIGURE 49–8 Plasma concentrations of the gonadotropic hormones and ovarian hormones during the normal female sexual cycle. (Estradiol is the principal estrogen.)

tary gland begins to secrete two **gonadotropic hormones.** At first it secretes mainly **follicle-stimulating hormone,** which initiates the beginning of sexual life in the growing female child, but later it secretes **luteinizing hormone** as well, which helps to control the monthly female cycle. The rates of secretion of these hormones as well as of estrogen and progesterone, the two ovarian hormones, during the different phases of the female cycle are shown in Figure 49–8.

Follicle-Stimulating Hormone and Stimulation of Follicle Growth. It is follicle-stimulating hormone that causes the few primary follicles of the ovary to begin growing each month, promoting very rapid proliferation of the follicular cells surrounding the primary oocyte as was illustrated in Figure 49–6. These cells then begin to secrete estrogens, one of the two major female sex hormones. Thus, the two functions of follicle-stimulating hormone are (1) to cause proliferation of the ovarian follicular cells and (2) to cause secretory activity by these cells. Consequently, the follicular antria develop and the follicles enlarge greatly. As soon as the follicles grow to about one half their maximum size, the anterior pituitary gland then begins to secrete greatly increased quantities of luteinizing hormone along with the follicle-stimulating hormone.

Luteinizing Hormone, Stimulation of Ovulation, and Formation of the Corpus Luteum. Luteinizing hormone increases still more the rate of secretion by the follicular cells and soon makes one follicle grow so large that it ovulates, expelling its ovum into the pelvic cavity, as explained earlier. Immediately after ovulation the luteinizing hormone also causes the individual follicular cells to grow in size and to develop a fatty, yellow appearance. These cells are then known as **lutein cells.** The entire mass of lutein cells develops within 1 to 2 days into the corpus luteum, which means a "yellow body." This change of follicular cells into lutein cells and growth of the corpus luteum is illustrated in the lower part of Figure 49–6. The corpus luteum continues to secrete *estrogen*, as had the follicular cells earlier in the cycle, but it secretes large quantities of *progesterone* as well.

THE OVARIAN HORMONES— ESTROGEN AND PROGESTERONE

The two ovarian hormones, **estrogen** and **progesterone,** are responsible for sexual development of the female and also for the female monthly sexual changes. These hormones, like the adrenocortical hormones and the male hormone testosterone, are both steroid compounds, and they are formed mainly from the fatty substance cholesterol. Estrogen is actually several different hormones called estradiol, estriol, and estrone, the most important of which is estradiol. However, they have almost identical functions and almost but not exactly identical chemical structures. For this reason they are considered together as if they were a single hormone.

Functions of Estrogen. Estrogen causes the cells in several parts of the body to proliferate—that is, to increase in number. For example, it causes the smooth muscle cells of the uterus to proliferate, making the female uterus, after puberty, become about two to three times as large as that of a child. Also, estrogen causes enlargement of the vagina, development of the labia surrounding the vagina, growth of hair on the pubis, broadening of the hips, conversion of the pelvic outlet into an ovoid shape rather than the funnel shape of the male, growth of the breasts, proliferation of the glandular elements of the breasts, and, finally, deposition of fatty tissues in characteristic areas such as the thighs and hips. In summary, essentially all of the characteristics that distinguish the female from the male are caused by estrogen, and the basic reason for the development of these characteristics is the ability of estrogen to promote proliferation of the respective cellular elements in certain regions of the body.

Estrogen also increases the growth rate of all bones immediately after puberty, but this also causes the growing portions of the bones to "burn out" within a few years, so that growth then stops. As a result, the female grows very rapidly for the first few years after puberty and then ceases growing entirely. On the other hand, the male child continues to grow even beyond this time and grows to a taller height than that of the female—not because of more rapid growth but because of more prolonged growth.

Estrogen also has very important effects on the internal lining of the uterus, the endometrium, which is discussed below in relation to the menstrual cycle.

Functions of Progesterone. Progesterone has little to do with the development of the female sexual characteristics; instead, it is concerned principally with preparing the uterus for acceptance of a fertilized ovum and preparing the breasts for secretion of milk. Specifically, progesterone causes the glandular cells of both the uterine endometrium and the breasts to enlarge and become highly secretory. This effect on the endometrium is discussed subsequently in relation to the menstrual cycle. Finally, progesterone inhibits the contractions of the uterus and prevents the uterus from expelling a fertilized ovum that is trying to implant or a fetus that is developing.

Regulation of the Female Sexual Cycle

The Reproductive Life of the Female— Puberty and Menopause. Puberty is the beginning of reproductive life. This occurs in most girls between the ages of 11 and 16. It is at this time that a girl begins to have monthly cycles of sex hormone secretion, culminating at the end of each sexual cycle in menstruation.

The onset of puberty is caused by the beginning secretion in the hypothalamus of *luteinizing hormone releasing factor*, which was discussed in Chapter 45. This factor is transported in the hypothalamic-hypophyseal portal system to the anterior pituitary gland, where it stimulates the secretion of both *follicle-stimulating hormone* and *luteinizing hormone*. Why the hypothalamus fails to secrete this releasing factor prior to the time of puberty is unknown, but perhaps it is because of immaturity of some of the hypothalamic neuronal cells that do not become functional until puberty.

The end of the reproductive life of the female, that is, the time when she stops having monthly cycles, occurs in the average woman at about age 45. This stage of her life is called the *menopause*. Its cause is the following: After about 30 years of developing follicles and secreting ovarian hormones, the ovaries eventually stop their cycles because almost all the primary follicles have either grown into mature follicles and ruptured or have degenerated. This causes the monthly cycles to cease because the ovaries no longer have enough follicular cells to secrete significant amounts of estrogen and progesterone, even though the anterior pituitary gland continues to secrete large quantities of follicle-stimulating hormone for the remainder of the woman's life.

Control of the Monthly Female Sexual Cycle—Oscillation Between the Anterior Pituitary and Ovarian Hormones.

The monthly female sexual cycle is caused by alternating secretion of follicle-stimulating and luteinizing hormones by the anterior pituitary gland and estrogens and progesterone by the ovaries, as illustrated in Figure 49–8. The cycle of events that causes this alternation is the following:

1. At the beginning of the monthly cycle, that is, when menstruation begins, the anterior pituitary gland begins to secrete increasing quantities of follicle-stimulating hormone along with smaller quantities of luteinizing hormone. These two together cause several follicles to grow in the ovaries, as well as causing considerable secretion of estrogen.

2. The estrogen then is believed to have two sequential effects on the anterior pituitary gland secretion. First, it has a negative feedback effect to inhibit secretion of both follicle-stimulating hormone and luteinizing hormone, their secretory rates falling to a low ebb at about the tenth day of the cycle. Then, suddenly, the anterior pituitary gland begins to secrete very large quantities of both follicle-stimulating hormone and luteinizing hormone, but especially the latter. This phase of secretion is called the *luteinizing hormone surge,* and it is believed to result from a second action of the estrogen, namely, a positive feedback effect. It is this surge that causes very rapid final development of one of the ovarian follicles and causes it to ovulate within about 2 days.

3. The process of ovulation, occurring at approximately the fourteenth day of the normal 28-day cycle, then leads to the development of the corpus luteum, as described earlier in the chapter. The corpus luteum secretes large quantities of progesterone and also continues to secrete considerable amounts of estrogen as well.

4. The estrogen and progesterone secreted by the corpus luteum inhibit the anterior pituitary gland once again and diminish greatly its rate of secretion of both follicle-stimulating hormone and luteinizing hormone. Without these hormones to stimulate it, the corpus luteum involutes so that the secretion of both progesterone and estrogen now falls to a very low ebb.

5. At this time, the anterior pituitary gland, being no longer inhibited by estrogen and progesterone, begins to secrete large quantities of follicle-stimulating hormone once again, thus initiating the cycle of the next month. This process continues throughout the entire reproductive life of the female.

The Endometrial Cycle and Menstruation

It is evident from Figure 49–8 that during the first half of the monthly cycle the only ovarian hormone secreted in large quantity is estrogen. Then during the latter half both estrogen and progesterone are secreted.

Estrogen causes the endometrium, that is, the lining of the inside of the uterus, to grow in thickness. Both the epithelial cells on the surface and the deeper cells of the endometrium proliferate approximately threefold. Also, the glands of the endometrium increase greatly in depth and tortuosity. These changes constitute the *proliferative phase of endometrial development* and are illustrated in the first portion of Figure 49–9. This phase lasts for about 11 days after menstruation is over.

At about the middle of the monthly cycle the corpus luteum begins secreting progesterone, which then causes further thickening of the endometrium while also initiating several special effects: (1) The endometrial glands begin secreting a nutrient fluid that can be used by a fertilized ovum before it implants; (2) large quantities of fatty substances and glycogen deposit in the endometrial cells; and (3) blood flow to the endometrium increases. This complex of changes is called the *secretory phase of the endometrial cycle.* Thus, the major function of progesterone is to make available an adequate supply of nutrients for an embryo to develop should it implant in the endometrium.

Menstruation.

If an ovum becomes fertilized and begins to divide, a special hormone called *chorionic gonadotropin* (to be discussed more fully in the following chapter) is then released by the developing mass of embryonic cells, and this in turn stimulates the corpus luteum to continue to produce estrogen and progesterone, and pregnancy begins. However, if toward the end of the monthly cycle no such fertilization and growth of embryonic tissue has occurred, no such hormone appears either. Without the stimulatory effect of this hormone, the corpus luteum involutes, and the production of both estrogen and progesterone almost ceases. The sudden lack of these two hormones causes the blood vessels of the endometrium to become spastic so that blood flow to the surface layers of the endometrium ceases. As a result, much of the endometrial tissue dies and sloughs into the uterine cavity. Also, blood oozes from the denuded endometrial wall, causing a blood loss of about 50 ml during the next few days. The sloughed endometrial tissue plus the blood and much serous exudate from the denuded uterine surface, which is called the *menstrum*, is gradually expelled by intermittent contractions of the uterine muscle for about 3 to 5 days. This process is called *menstruation*. The uterine contractions can cause cramps if they occur with too much force.

The deepest pits of the endometrial glands remain intact during menstruation, despite the sloughing of all the surface layers of the endometrium. During the first few days after menstruation, new epithelium grows outward from the pits of the glands to cover the inner surface of the uterus. Then, under the influence of renewed estrogen production by the ovaries, the endometrial cycle begins again. These sequential changes during the entire cycle are shown in Figure 49–9.

The Period of Fertility During the Sexual Cycle

An ovum can be fertilized by a sperm for a period of 8 to 24 hours after ovulation. Also, sperm can live in the female genital tract usually for 24 to 48 hours. Therefore, for successful fertilization, sexual exposure must occur either shortly before ovulation so that sperm are already available when ovulation occurs, or within a few hours after ovulation.

Ovulation occurs almost exactly 14 days before menstruation begins. Therefore, in any woman who has a normal 28-day sexual cycle, ovulation will usually occur on the fourteenth day after the beginning of menstruation. However, many women, instead of having normal 28-day cycles, have cycles as short as 21 days or as long as 40 days. If these women are regular, they can still calculate that ovulation will normally occur about 14 days before menstruation begins. However, in those women who are irregular, it becomes impossible to predict the time of ovulation. To be on the safe side, one usually allows 4 to 5 days on either side of the calculated day of ovulation when using the rhythm method of contraception.

HORMONAL SUPPRESSION OF FERTILITY— THE LUTEINIZING HORMONE SURGE AND "THE PILL"

It has long been known that either estrogen or progesterone given in sufficient quantity during the early part of the monthly female cycle can prevent ovulation. It will be recalled from earlier discussions in this chapter that ovulation occurs approximately 1½ days after a massive surge of luteinizing hormone secretion by the anterior pituitary gland, which itself occurs on the twelfth to thirteenth day of the normal cycle. It is believed that either estrogen or progesterone, or both given together, suppress ovulation by inhibiting the anterior pituitary secretion of luteinizing hormone and thereby preventing the luteinizing hormone surge. This obviously prevents conception as well.

The problem in devising methods for hormonal suppression of ovulation has been to develop appropriate combinations of estrogen and progesterone that will suppress ovulation but

| Proliferative phase (11 days) | Secretory phase (12 days) | Menstrual phase (5 days) |

FIGURE 49–9 Endometrial changes during the monthly sexual cycle, showing especially the process of menstruation.

that will not cause unwanted effects of these two hormones. For instance, too much of either of the hormones can cause abnormal menstrual bleeding patterns. Fortunately, several synthetic estrogens and progestins have been developed that have maximum capability to suppress the luteinizing hormone surge but will not affect the menstrual cycle severely. Appropriate combinations of these synthetic hormones are marketed as "the pill."

Yet, most hormone regimens for contraception still have at least some undesirable side effects in the body, particularly causing salt retention by the kidneys, which can, over a period of years, lead to high blood pressure in many women. Therefore, successful control of conception in ways other than by use of the pill is much to be desired.

QUESTIONS

1. List and describe the separate structures of the female external genitalia.
2. Describe the uterus and its relationships to the vagina and the uterine tubes.
3. Trace the development of the ovum, its maturation, and the process of ovulation.
4. Give a description of the uterine tube, and explain how the ovum is transported to the uterus.
5. In what ways does the female sexual act resemble the male sexual act, and in what ways is it dissimilar?
6. What are the effects of estrogen on the body?
7. What are the effects of progesterone on the body?
8. What causes the onset of puberty, and what causes the menopause?
9. How do the gonadotropic hormones from the anterior pituitary gland interact with the ovarian hormones to cause the rhythmic monthly sexual cycle?
10. Describe the endometrial cycle and menstruation.
11. Explain the mechanism of hormonal suppression of fertility.

REFERENCES

Benirschke, K.: The endometrium. *In* Yen, S.S.C., and Jaffe, R.B. (eds.): Reproductive Endocrinology. Philadelphia, W.B. Saunders, 1978, p. 241.

Catt, K.J., and Pierce, J.G.: Gonadotropic hormones of the adenohypophysis (FSH, LH and prolactin). *In* Yen, S.S.C., and Jaffe, R.B. (eds.): Reproductive Endocrinology. Philadelphia, W.B. Saunders, 1978, p. 34.

Channing, C.P., *et al.*: Ovarian follicular and luteal physiology. *In* Greep, R.O. (ed.): International Review of Physiology: Reproductive Physiology III. Vol. 22. Baltimore, University Park Press, 1980, p. 117.

Crighton, D.B., *et al.*: Control of Ovulation. Boston, Butterworth, 1978.

Diamond, M.C., and Korenbrot, C.C. (eds.): Hormonal Contraceptives and Human Welfare. New York, Academic Press, 1978.

Ewing, L.L., *et al.*: Regulation of testicular function: A spatial and temporal view. *In* Greep, R.O. (ed.): International Review of Physiology: Reproductive Physiology III. Vol. 22. Baltimore, University Park Press, 1980, p. 41.

Griffin, J.E., and Wilson, J.D.: The testis. *In* Bondy, P.K., and Rosenberg, L.E. (eds.): Metabolic Control and Disease, 8th Ed. Philadelphia, W.B. Saunders, 1980, p. 1535.

Hafez, E.S.E. (ed.): Human Reproduction: Conception and Contraception. Hagerstown, Md., Harper & Row, 1979.

Hall, P.F.: Testicular hormones: Synthesis and control. *In* DeGroot, L.J., *et al.* (eds.): Endocrinology, Vol. 3. New York, Grune & Stratton, 1979, p. 1511.

Huff, R.W., and Pauertstein, C.J.: Physiology and Pathophysiology of Human Reproduction. New York, John Wiley & Sons, 1978.

Kase, N.G., and Speroff, L.: The ovary. *In* Bondy, P.K., and Rosenberg, L.E. (eds.): Metabolic Control and Disease, 8th Ed. Philadelphia, W.B. Saunders, 1980, p. 1579.

Midgley, A.R., and Sadler, W.A. (eds.): Ovarian Follicular Development and Function. New York, Raven Press, 1979.

Mishell, D.R., Jr., and Davajan, V. (eds.): Reproductive Endocrinology, Infertility, and Contraception. Philadelphia, F.A. Davis Co., 1979.

Peters, H., and McNatty, K.P.: The Ovary: A Correlation of Structure and Function in Mammals. Berkeley, University of California Press, 1980.

Phillips, D.M.: Spermiogenesis. New York, Academic Press, 1974.

Reiter, R.J. (ed.): The Pineal and Reproduction. New York, S. Karger, 1978.

Richards, J.S.: Maturation of ovarian follicles: Actions and interactions of pituitary and ovarian hormones on follicular cell differentiation. *Physiol. Rev.,* 60:51, 1980.

Savoy-Moore, R.T., and Schwartz, N.B.: Differential control of FSH and LH secretion. *In* Greep, R.O. (ed.): International Review of Physiology: Reproductive Physiology III. Vol. 22. Baltimore, University Park Press, 1980, p. 203.

Steinberger, A., and Steinberger, E. (eds.): Testicular Development, Structure and Function. New York, Raven Press, 1980.

Styne, D.M., and Grumbach, M.M.: Puberty in the male and female: Its physiology and disorders. *In* Yen, S.S.C., and Jaffe, R.B. (eds.): Reproductive Endocrinology, Philadelphia, W.B. Saunders, 1978, p. 189.

Thomas, J.A., and Singahl, R.L. (eds.): Sex Hormone Receptors in Endocrine Organs. Baltimore, Urban & Schwarzenberg, 1980.

50

Reproduction and Fetal Physiology

Overview

Fertilization of the ovum usually occurs in the first portion of the fallopian tube. A *single sperm* penetrates through the membrane of the ovum, carrying with it its 23 unpaired chromosomes. These immediately pair up with 23 unpaired chromosomes of the ovary, giving a *full complement of 46 chromosomes* arranged in 23 pairs. This sets into motion the *cell division* process that leads to the formation of a baby.

Approximately 7 days after fertilization, the dividing mass of cells, now called a **blastocyst,** *implants itself into the endometrium* of the uterus. The outer cells of the blastocyst, the **trophoblasts,** form the **fetal membranes** and the **placenta;** the inner cells form the **embryo,** which will develop into the fetus.

During the first few weeks after implantation of the ovum, *nutrition* is supplied to the blastocyst by *trophoblastic digestion and phagocytosis* of the endometrium. However, by approximately the twelfth week of pregnancy, the *placenta* has developed to the extent that it thereafter supplies essentially all of the required nutrients. The placenta is composed of a *maternal part* and a *fetal part.* The maternal part consists of multiple large chambers called **placental sinuses** through which the mother's blood flows continually. The fetal part consists mainly of a great mass of **placental villi,** which protrudes into the placental sinuses and through which the fetus's blood passes. *Nutrients diffuse* from the mother's blood through the **placental villus membrane** into the fetal blood and are carried through the **umbilical cord** to the fetus. In turn, the *excreta* from the fetus, such as carbon dioxide, urea, and other substances, diffuse from the fetal blood into the mother's blood and then are excreted to the exterior by the mother's excretory organs.

The *placenta secretes* extremely large quantities of *estrogen* and *progesterone,* about 30 times as much estrogen as is secreted by the corpus luteum and about 10 times as much progesterone. These hormones are very important in promoting the *development of the fetus.* During the first few weeks of pregnancy, another hormone, *chorionic gonadotropin,* also secreted by the placenta, stimulates the corpus luteum and makes it continue to secrete estrogen and progesterone during the early weeks of pregnancy. These hormones from the corpus luteum are essential for continuance of pregnancy for the first 8 to 12 weeks, but after that time the placenta secretes enough estrogen and progesterone to keep the pregnancy going.

At the end of approximately 9 months of growth and development, the fully formed baby is expelled from the uterus, a process called *partu-*

rition. Though the exact cause of parturition is not known, it undoubtedly results from such factors as (1) *mechanical stimulation* of the uterus by the enlarging baby and (2) changes in the rates of secretion of the *placental hormones,* especially estrogen and progesterone.

The *basic structures of all the fetal organs* are developed during the *first 3 months of pregnancy.* However, these organs are not fully mature until the end of the full 9-month gestation period. When a baby is born before this time, it is said to be *premature.* Most *regulatory functions* of the premature baby are *still underdeveloped* so that the baby has difficulty controlling its body temperature, its nutritional status, its body fluids, and so forth. Also, many premature babies have not yet begun to secrete *surfactant,* a substance secreted in mature lungs that has a detergent-like effect and is essential to allow expansion of the lungs. As a result, many of these babies die of the condition called *respiratory distress syndrome.*

During pregnancy, mainly under the influence of estrogen and progesterone secreted by the placenta and *prolactin* secreted by the anterior pituitary gland, the *breasts* enlarge severalfold, and their glandular structures become fully developed. However, both the estrogen and the progesterone prevent milk formation until after birth of the baby. Loss of the placenta at the time of birth removes these two hormones, and then under the influence of continued secretion of prolactin by the anterior pituitary gland, the breasts produce copious quantities of milk.

FERTILIZATION OF THE OVUM AND EARLY GROWTH

Entry of the Sperm into the Ovum. After the ovum is expelled from the ovary, it remains viable for 8 to 24 hours. During this time it usually moves approximately one quarter of the distance down the uterine tube toward the uterus. Obviously, then, fertilization must occur either in the abdominal cavity before the ovum enters the tube or in the upper portion of one of the tubes.

Sperm travel at a velocity of only 1 to 4 mm per minute, and the total length of a uterine tube is approximately 15 cm, which would require approximately 40 minutes for a sperm to reach the ovum. However, in lower animals this often occurs within a few minutes, suggesting that contractions of the uterus and uterine tubes induced by the female climax during intercourse might play an important role in the fertilization process.

The ovum, when released from the ovary, carries on its surface a large number of granulosa cells that collectively are called the **corona radiata,** shown in Figure 50–1. It is believed that the corona radiata is at least partially dispersed by enzymes secreted by the sperm, especially *hyaluronidase* and several *proteinases.* These enzymes supposedly digest the linkages that hold the cells together and allow the cells to break away from the ovum so that the sperm can attack and enter the ovum. In addition, the alkaline secretions of the uterine tube also play an important role in removing the corona radiata.

After a single sperm has entered the ovum, the membrane of the ovum suddenly takes on new properties and inactivates any additional sperm as they attempt to enter. This effect prevents more than one set of male chromosomes from combining with the female chromosomes of the ovum.

Shortly after a sperm has entered the ovum, the head of the sperm begins to swell,

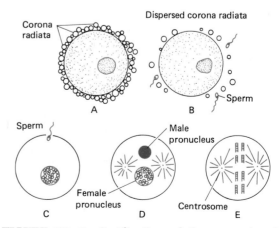

FIGURE 50–1 Fertilization of the ovum by the sperm and beginning cleavage. (Modified from Arey: *Developmental Anatomy,* 7th Ed. Philadelphia, W.B. Saunders, 1974.)

forming a **male pronucleus** as shown in Figure 50–1D; the original nucleus of the ovum is also still present and is called the **female pronucleus.** It will be recalled from the previous two chapters that each of these pronuclei has only 23 *unpaired* chromosomes. However, the two pronuclei soon combine so that the fertilized ovum now has 46 *paired* chromosomes in a single nucleus, as do all other cells of the human body. A few hours later the chromosomes reproduce and split apart, initiating division of the ovum into two daughter cells.

Cell Multiplication and Division

Early Division of the Ovum. The first cleavage of the fertilized cell occurs approximately 30 hours after the sperm has entered the ovum, and the next few generations occur every 10 to 15 hours. At the time the ovum reaches the uterus, the total number of cells in the cellular mass is usually about 16 to 32. By this time the cells have actually begun to *differentiate*, which means that some of them have developed different characteristics from the others.

Differentiation. Figure 50–2 illustrates various stages of cleavage of the fertilized mammalian ovum, the darkened cells representing those that will eventually form the embryo and the lighter cells those that will eventually form the fetal membranes. Note the differences in size and other characteristics of the cells as cell division proceeds during the first few days after fertilization. This is the process of cell differentiation. Obviously, still many more changes in characteristics of the cells will take place before the final human being is formed. Unfortunately, we do not know all the causes of differentiation, but let us discuss the theories of how the cells change their characteristics to form all the different organs and tissues of the body.

The earliest and simplest theory for explaining differentiation was that the genetic composition of the nucleus undergoes changes during successive generations of cells in such a way that one daughter cell develops one set of characteristics whereas another daughter cell develops entirely different characteristics. This theory probably explains a few stages of differentiation, because highly differentiated cells, when grown in tissue culture, cannot revert all the way back to the original primordial state. Yet most cells can revert at least two or three steps backward in differentiation, which indicates that other factors besides simple changes in genetic potency probably play the dominant roles in differentiation.

The cytoplasm, in addition to the nucleus, is known also to participate in differentiation, because cells without nuclei can divide and even differentiate to some extent for a few stages before the cells die. However, the long-term function of the cytoplasm in differentiation is probably controlled by the nucleus.

Embryologic experiments show that certain cells in an embryo control the differentiation of adjacent cells. For instance, a central axis develops in the embryo, and it in turn, as a result of complex *inductions* in the surrounding tissues, causes formation of essentially all the organs of the body.

Another instance of induction occurs when the developing eye vesicles deep in the head come in contact with the outer layer of the head, the ectoderm, and cause it to thicken into a lens plate that folds inward to form the lens of the eye. It is possible that most of the embryo develops as a result of complex inductions, one part of the body affecting another part, and this part affecting still other parts.

Thus our understanding of cell differentiation is still hazy. We know many different control mechanisms by which differentiation *could* occur. Yet the overall basic controlling factors in cell differentiation are yet to be discovered; when learned, they will make a tremendous difference in our understanding of bodily development.

Implantation of the Ovum

Figure 50–3 shows the *blastocyst* stage of a developing human ovum $1\frac{1}{2}$ weeks after fertilization. It illustrates the method by which the mass of cells attaches itself to the inner wall of the uterus. At this stage the cells on the outside of the blastocyst, called *trophoblastic cells*, secrete large quantities of proteolytic enzymes that digest the endometrium, and then the trophoblastic cells phagocytize the digested products. Thus, they literally eat their way into the wall of the endometrium. The trophoblastic cells also

A　　B　　C　　　　D

FIGURE 50–2 Differentiation of cells in the early embryo during gestation: (A) two blastomeres, (B) six blastomeres, (C) a hemisected early blastocyst, (D) a later blastocyst stage showing the early fetus at the top. (Modified from Arey: Developmental Anatomy, 7th Ed. Philadelphia, W.B. Saunders, 1974.)

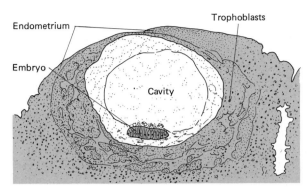

FIGURE 50-3 Implantation of a 1½-week old human ovum in the endometrium. The colored portion is the developing embryo. (Courtesy of Dr. Arthur Hertig.)

grow and divide very rapidly, and soon they and adjacent cells begin forming the placenta and fetal membranes while the embryo develops on the inside of these membranes, on one wall of the cavity of the blastocyst.

NUTRITION OF THE FETUS IN THE UTERUS

The Trophoblastic Phase of Nutrition

During the first few weeks after implantation of the ovum, the placenta and its blood supply are not sufficiently formed to supply the fetus with nutrients. During this phase, nutrition is provided only by trophoblastic digestion and phagocytosis of the endometrium. It will be recalled that prior to implantation of the ovum, the cells of the endometrium store large quantities of proteins, lipid materials, and glycogen. Also, small quantities of iron and vitamins are stored awaiting phagocytosis by the developing ovum. Even so, the embryo can obtain nutrition in this manner for only the first few weeks of its development. By that time the placenta will have developed to a stage such that it can begin supplying nutrition. Figure 50-4 shows that the trophoblastic period of phagocytic nutrition lasts until approximately the twelfth week of pregnancy.

Anatomy of the Pregnant Uterus, and Placental Nutrition of the Fetus

Figure 50-5 illustrates the anatomy of the **uterus** and the **fetus** shortly before birth of the baby; Figure 50-6 gives a more diagrammatic view. Note in Figure 50-5 that both the body

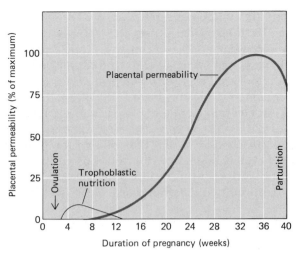

FIGURE 50-4 Periods of fetal nutrition: first, trophoblastic phagocytosis of the endometrium lasting for the first few weeks of pregnancy and, second, diffusion through the placenta during the remainder of pregnancy.

and the fundus of the uterus have enlarged immensely, with an increase in muscular mass of about thirty-fold. Note also that the baby and the uterus occupy about one half of the entire abdominal cavity, shoving most of the intestines upward and also pressing downward against the urinary bladder and rectum. The head-downward position of the baby in the uterus as shown in both figures occurs in about 19 out of 20 pregnancies; the buttocks-downward position, called the *breech position*, occurs in about 1 out of every 20 pregnancies and invariably causes a more difficult delivery.

Figure 50-5 illustrates the **placenta** at the very top of the uterus, but the place of attachment of the placenta can be anywhere on the inner wall of the uterus. Thus, Figure 50-6 shows the placenta on the side wall. Note also in Figure 50-6 the **umbilical cord** through which the umbilical arteries and vein carry blood to and fro between the placenta and the baby. Shown too are the **fetal membranes** (also called the amnion). The fetal membranes are a large membranous bag that surrounds the entire fetus and attaches to the entire inner surface of the uterus. Inside this bag is a cavity called the **amniotic cavity**, which contains about 1½ L of fluid in addition to the fetus. The fetus is literally surrounded by this *amniotic fluid*.

Anatomy of the Placenta. Figure 50-7 illustrates the gross organization of the placenta and, at the bottom, a cross-section of a **placental villus**. The mother's arterial blood flows into large placental chambers called **placental sinuses**. The fetal portion of the placenta is com-

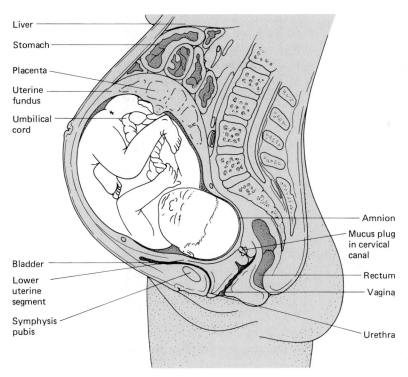

Liver

Stomach

Placenta

Uterine
fundus

Umbilical
cord

Amnion

Mucus plug
in cervical
canal

Bladder

Lower
uterine
segment

Symphysis
pubis

Rectum

Vagina

Urethra

FIGURE 50—5 Pregnancy shortly
before birth of the baby.

posed of many small cauliflower-like growths
that project into the placental sinuses. Each of
these in turn is covered with tremendous numbers of small villi containing fetal blood capillaries. The baby's blood flows into the capillaries of
the villi where it receives nutrients from the
mother's blood, and then it passes back through
the umbilical vein to the fetus.

The cross-section of a villus in the lower
part of Figure 50–7 shows the close proximity of
the mother's blood in the placental sinuses to
the baby's blood in the fetal capillaries. The
mother's blood flows so rapidly through the sinuses that the sinus blood maintains very high

concentrations of the nutrients needed by the
fetus.

***Diffusion of Nutrients Through the
Villi.*** Nutrients pass through the placental villi
into the fetal blood almost entirely by the process of diffusion. The gaseous pressure of oxygen
in the blood of the maternal sinuses is normally
about 50 mm Hg, whereas the gaseous pressure
of oxygen in the fetal capillaries is only
30 mm Hg. Because of this pressure difference,
oxygen simply diffuses through the membrane
of the villus from the mother's blood to the
baby's blood. Likewise, glucose, amino acids,
fats, many of the vitamins, and most of the minerals are present in greater concentration in the
mother's blood than in the baby's blood because
the baby utilizes these substances almost as rapidly as they enter his blood. As a consequence,
all of them diffuse into the baby's blood and
thereby supply the fetus with nutrition.

Figure 50–4 shows the progressive increase in placental permeability as pregnancy
proceeds. The total amount of nutrients that can
be transported through the placenta reaches a
maximum 32 to 36 weeks after pregnancy begins,
about 6 weeks before the baby is born. At this
point the placental tissues begin to grow old and
degenerate. As a result, even though the fetus
still requires progressively increasing quantities
of nutrients, the ability of these nutrients to
enter the baby becomes lessened. Fortunately,
birth of the baby occurs soon, and the baby assumes independent existence.

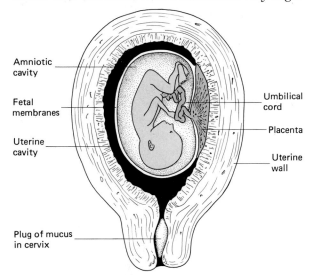

Amniotic
cavity

Fetal
membranes

Uterine
cavity

Umbilical
cord

Placenta

Uterine
wall

Plug of mucus
in cervix

FIGURE 50—6 The fetus, the fetal membranes, and
the placenta in the pregnant uterus.

PLACENTA

VILLUS

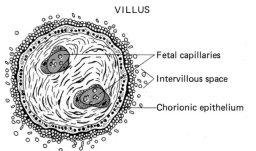

FIGURE 50—7 Gross anatomic structure of the placenta and microscopic cross-section of a placental villus. (Modified from Goss: Gray's Anatomy of the Human Body. Philadelphia, Lea & Febiger, 1973; and from Arey: Developmental Anatomy, 7th Ed. Philadelphia, W.B. Saunders, 1974.)

Excretion Through the Placenta by Diffusion. In addition to the diffusion of nutrients into the baby, excretory products diffuse through the placenta from the fetal blood to the maternal blood. For instance, the baby's metabolism continually forms large quantities of carbon dioxide, urea, uric acid, creatinine, phosphates, sulfates, and other normal excretory products. The concentration of each of these substances rises in the baby's blood until it is greater than in the mother's blood. Then, because of this reverse concentration gradient, these substances diffuse backward through the villi, and the mother excretes them through her kidneys, lungs, and gastrointestinal tract.

Active Transport Through the Villi. The epithelium that covers the surface of the placental villi is formed from the trophoblastic cells of the early fetus. These cells continue their phagocytic activity throughout the life of the placenta, so that, in addition to the large amount of nutrients that pass by diffusion through the pla-

cental membrane, small amounts of nutrients are actively transported through the villi all through the period of gestation. In the first few weeks of placental development this aids greatly in increasing the amount of amino acids, fatty substances, and some minerals that can be supplied to the baby, but after 12 to 20 weeks the amount of nutrients supplied in this manner becomes insignificant.

THE HORMONES OF PREGNANCY

In the previous chapter, the vital roles played by the different sex hormones in the prepregnancy phase of sex life became clear. However, hormones play an equally important, if not more important, role during pregnancy as before. Most of these hormones are secreted by the placenta itself. Two of these are *estrogen* and *progesterone*, the same female sex hormones secreted by the ovaries during the normal female monthly cycle. However, two others that are also important and even necessary for pregnancy to proceed are *chorionic gonadotropin* and *human chorionic somatomammotropin*. These hormones affect both the mother and the fetus. In the mother, they help to control changes in the uterus and breasts that are necessary to carry the fetus to birth and to provide milk production. Also, they help to regulate the development of the fetus itself, especially the sexual organs.

Secretion of Estrogen and Progesterone During Pregnancy

From the last chapter we recall that moderate quantities of both estrogen and progesterone are secreted by the corpus luteum during the latter half of the normal female monthly sexual cycle. During the first few weeks of pregnancy, estrogen and progesterone continue to be secreted by the corpus luteum. That is, instead of degenerating at the end of the month as it normally does, the corpus luteum grows even larger and produces two to three times as much estrogen and progesterone as during the normal cycle for an additional 15 to 20 weeks. This growth of and secretion by the corpus luteum is caused by the hormone *chorionic gonadotropin*, a hormone formed by the early developing fetal tissues that we will discuss in more detail later in this chapter.

Even the amounts of estrogen and progesterone secreted by the enlarged corpus luteum are themselves small in comparison with the amounts of these two hormones that are subse-

FIGURE 50-8 Rates of secretion of estrogens, progesterone, and chorionic gonadotropin at different stages of pregnancy.

quently secreted by the placenta itself. Placental secretion of these two hormones begins within a few weeks after the onset of pregnancy and increases especially rapidly after about the sixteenth week of pregnancy, reaching a maximum shortly before birth of the baby. Estrogen secretion rises to about 30 times and progesterone secretion about 10 times its secretion during the normal monthly sexual cycle. Figure 50-8 illustrates the progressively increasing secretion of estrogen and progesterone during pregnancy, showing very small rates of secretion during the normal monthly cycle (shown as the first 4 weeks on the chart) and extreme rates of secretion of both estrogen and progesterone by the end of pregnancy.

Functions of Estrogen During Pregnancy. In the mother, estrogen causes (1) rapid proliferation of the uterine musculature; (2) greatly increased growth of the vascular system supplying the uterus; (3) enlargement of the external sex organs and of the vaginal opening, providing an appropriately enlarged passageway for birth of the baby; and (4) probably also some relaxation of the pelvic ligaments that allows the pelvic opening to stretch as the baby is born.

In addition to the effects on the reproductive organs, estrogen also causes the breasts to grow rapidly. The ducts, especially, enlarge, and the glandular cells proliferate. Finally, estrogen causes a pound or more of extra fat to deposit in the breasts.

Another effect of estrogen not yet completely understood is the effect on the fetus itself. It is believed that estrogen causes much of the rapid proliferation of fetal cells and also aids in the differentiation of some of these cells into special organs. In particular estrogen is believed to control development of some of the female sex characteristics.

Functions of Progesterone During Pregnancy. The first function of progesterone during pregnancy is to make increased quantities of nutrients available in the early endometrium for use by the developing ovum. It does this by causing the endometrial cells to store glycogen, fat, and amino acids. In addition, progesterone has a strong inhibitory effect on the uterine musculature, causing it to remain relaxed throughout pregnancy. It is believed that this effect allows pregnancy to continue until the fetus is large enough to be born and live an independent existence.

Progesterone complements the effects of estrogen on the breasts. It causes the glandular elements to enlarge further and to develop a secretory epithelium, and it promotes deposition of nutrients in the glandular cells, so that when milk production is required the appropriate materials will be available.

Secretion and Functions of Chorionic Gonadotropin During Pregnancy

If the corpus luteum degenerates or is removed from the ovary at any time during the first 2 to 3 months of pregnancy, the loss of estrogen and progesterone secretion by this corpus luteum causes the fetus to stop developing and to be expelled within a few days. For this reason it is important that the corpus luteum remain active at least during the first third of the pregnancy. Beyond that time removal of the corpus luteum usually does not affect pregnancy because the placenta by then is secreting many times as much estrogen and progesterone as the corpus luteum.

It will be recalled from the previous chapter that the corpus luteum normally degenerates and is absorbed at the end of each female monthly cycle. To keep the corpus luteum intact when the ovum implants, a special hormone, a small glycoprotein called *chorionic gonadotropin*, is secreted by the developing fetal tissues—by the trophoblasts. This hormone has almost exactly the same properties as luteinizing hormone. It not only keeps the corpus luteum from involuting, but also actually stimulates it so that it enlarges severalfold during the first 2 to 4 months of pregnancy.

Chorionic gonadotropin begins to be formed from the day that the trophoblasts implant in the uterine endometrium. Its concentration is highest approximately at the eighth week of pregnancy, as shown in Figure 50-8. Thus, the concentration is greatest at the time of pregnancy when it is essential to prevent involution of the corpus luteum. In the middle and latter parts of pregnancy the secretion of chorionic gonadotropin falls to very low values. Its only known function at that time of pregnancy is to stimulate production of testosterone by the testes of the male fetus, which was discussed in the previous chapter and which plays an important role in the development of the male fetus.

Secretion and Functions of Human Chorionic Somatomammotropin

Recently, a hormone called *human chorionic somatomammotropin* has been discovered. This is a small protein that begins to be secreted about the fifth week of pregnancy and increases progressively throughout the remainder of pregnancy.

Research studies on chorionic somatomammotropin have shown that, when given in very large quantities, it can increase the development of the breasts, for which reason it was first named *placental lactogen*. However, this function in the human being is now believed to be exceedingly weak, which accounts for the change in the name of the hormone.

A second effect of the hormone is to promote growth of the fetus, similar to the growth-promoting effect of growth hormone produced by the anterior pituitary gland. However, this effect also is weak.

Finally, recent research has suggested that this hormone has its most important actions on glucose and fat metabolism in the mother rather than having its most important effects in the fetus. The hormone decreases the utilization of glucose by the mother and therefore makes more glucose available for the fetus. At the same time, it causes increased mobilization of fatty acids from the fat tissues of the mother so that she can use this fat to provide her own energy in place of glucose. Since glucose is the major substrate used by the fetus for energy, the importance of these hormonal effects is obvious.

Thus, it now appears that human chorionic somatomammotropin is primarily important in helping to provide nutrition for the fetus.

FETAL PHYSIOLOGY

In general, the physiology of the growing fetus during the last 3 months of pregnancy is not far different from that of the normal child. All the organs of the body assume their final anatomic form, with minor exceptions, by 4 to 5 months

after the beginning of pregnancy, and most of them can function almost normally by 6 months after the beginning of pregnancy. For example, long before birth the kidneys of the fetus become at least partially active, the gastrointestinal tract imbibes and absorbs fluid from the amniotic cavity, breathing is attempted even though it cannot be accomplished because the fetus is immersed in amniotic fluid, the heart pumps blood quite normally from approximately the third month on, and the metabolic systems operate very much the same as they do after birth.

Growth of the Fetus

Figure 50–9 depicts the increase in length and weight of the fetus during the 40 weeks of pregnancy. The length of the fetus increases almost directly in proportion to its age, whereas the weight increases in proportion to the third power of the age. Thus, the weight is almost infinitesimal until the twelfth to sixteenth week, but at that time it begins to increase extremely rapidly. Two months before birth the fetus usually weighs about one-half its birth weight, and one month prior to birth it weighs approximately three-fourths its birth weight. In other words, the major gain in weight occurs during the last 2 to 3 months. This is an especially important factor when one is considering the appropriate nutrition for the mother during pregnancy, because essentially all of the nutrients required by the baby are needed during the last 3 months. However, for several months prior to that time, growth of the uterus, placenta, and fetal membranes requires additional nutrients.

Special Nutrients Required by the Fetus. The fetus requires especially large quantities of iron, calcium, phosphorus, amino acids,

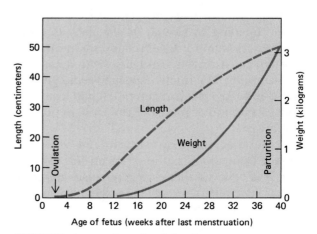

FIGURE 50–9 Growth of the fetus.

and vitamins. Iron begins to be used to form red blood cells within the first weeks of development of the fetus. Much of this early iron enters the fetus by active absorption from the endometrium by the trophoblasts. Throughout the remainder of pregnancy, still larger quantities of iron diffuse through the placental membrane to be used by the liver, spleen, and bone marrow for production of the fetus' blood.

Calcium is needed to ossify the bones. During the first two-thirds of pregnancy the fetal bones contain mainly organic matrix and almost no calcium salts. During the last 3 months of fetal development, ossification occurs very rapidly, approximately one-half occurring in the final month. It is at this time, therefore, that the mother needs an especially abundant amount of milk or other foods containing calcium and phosphate.

The large quantities of amino acids and vitamins required by the fetus provide the necessary building materials for growth of the fetal tissues. During the last 3 months of pregnancy, a mother can become depleted of protein and vitamins if these are not in the diet in adequate amounts. Also, brain development in the fetus is highly dependent on these nutrients.

The Fetal Nervous System

The major anatomic parts of the nervous system are formed during the first few months of fetal growth, but complete function of this system is not reached even by the time the child is born. A premature baby always exhibits signs of poor nervous system function. As an example, one of the most difficult problems in treating a premature baby is to keep its body temperature normal, because the hypothalamic temperature control centers of a 6- or 7-month fetus usually have not developed sufficiently to regulate body temperature. Therefore, the infant must be placed in an incubator for several weeks until its nervous system has developed to a greater degree.

By the time of birth most, if not all, of the peripheral nerve fibers are completely developed, but in the central nervous system the deposition of myelin around many of the large nerve fibers is far from complete. Though myelin is not always necessary for nerve fibers to function, it has been inferred from this lack of complete myelination that certain portions of the central nervous system probably are far from optimum function at the time of birth. On the other hand, all the neurons that will ever be formed by the child are believed to be present by birth. This means that every time a neuron is destroyed thereafter, no new neuron will take its place.

Changes in the Circulation at Birth

Figure 50–10 illustrates the circulatory system in the fetus, showing very important special vessels that are not present after birth. For instance, blood returning to the fetus from the placenta through the *umbilical vein* enters the fetal circulatory system through a special vessel called the *ductus venosus*, which passes directly through the liver. Also, blood from the portal veins of the fetus flows through the ductus venosus, bypassing the liver entirely and emptying directly into the general venous system.

On reaching the heart, fetal blood bypasses the nonaerated lungs by two routes. First, it can flow from the right atrium through an opening called the *foramen ovale*, illustrated in the figure, directly into the left atrium, thus bypassing both the right ventricle and the lungs. Second, the remainder of the blood enters the right ventricle and is pumped into the pulmonary artery. However, most of this blood, instead of going through the lungs and left heart, passes through a vessel called the *ductus arteriosus* directly into the aorta.

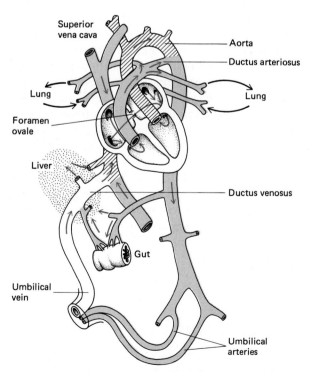

FIGURE 50–10 The fetal circulation, showing especially the ductus arteriosus, ductus venosus, and foramen ovale. (Modified from Arey: Developmental Anatomy, 7th Ed. Philadelphia, W.B. Saunders, 1974.)

Thus, by utilizing the ductus venosus, the foramen ovale, and the ductus arteriosus, blood flow through the liver and lungs occurs only to a minor extent in the fetus. This is a means for conserving the energy of the fetal heart, because the functions normally required of the fetal liver are performed instead by the mother's liver, and the fetal lungs cannot aerate the blood until after birth.

After birth all three of these openings soon close. The ductus venosus and the ductus arteriosus become gradually occluded during the first few days of life. Presumably the ductus venosus becomes occluded because it no longer carries the tremendous blood flow from the umbilical vein, and the lesser blood flow from the portal system is not sufficient to keep it open. The ductus arteriosus closes because the functioning lungs increase the oxygen in the blood, which in turn has a direct effect on the wall of the ductus to cause increased contraction of the ductus muscle.

The foramen ovale is covered by a thin valve on the left atrial side. As long as the pressure in the right atrium is greater than that in the left atrium, which is the case before birth of the fetus, blood flows from right to left, bypassing the right ventricle and lungs. However, when the lungs expand after birth, the blood vessels of the lungs also expand, and the right heart can then pump blood so easily through the lungs that the right atrial pressure falls to about 2 mm Hg less than the left atrial pressure. This backward pressure differential thereafter keeps the valve of the foramen ovale closed. In about two-thirds of all persons the valve gradually adheres to the opening so that it becomes permanently closed; in about one third it does not, but, regardless of which is true, blood ceases to flow through it after birth.

Thus, because of a series of changes in the dynamics of the circulation, blood begins to flow through both the liver and lungs immediately after birth, even though in the fetus the heart had conserved its energy by bypassing these two organs.

PHYSIOLOGY OF THE MOTHER DURING PREGNANCY

The mother's physiology is changed during pregnancy in several ways. First, accessory changes occur in her reproductive organs and breasts to provide for development of the fetus and to provide nutrition for the newborn child. Second, all her metabolic functions are in-creased to supply sufficient nutrition to the growing fetus. Third, tremendous production of certain hormones by the placenta during pregnancy causes many side effects not directly associated with reproduction.

Changes in Weight. The pregnant mother gains an average of about 22 lb (approximately 10 kg) during pregnancy. In general, this gain is accounted for in the following manner: fetus, 7 lb; uterus, 2 lb; placenta and membranes, 2 lb; breasts, 1.5 lb; and the remainder, about 9.5 lb fat and increased quantities of extracellular fluid and blood. The amount of increase in fat and blood fluids varies tremendously from one mother to another, depending upon her eating habits, especially salt and fat intake, and upon the amounts of hormones secreted during pregnancy.

Changes in Metabolism. The mother's metabolic rate in general rises approximately in proportion to her increase in weight, plus perhaps an additional 5 to 10 percent. Much of this increase is occasioned simply by the greater amount of energy required for the mother to carry the growing load. However, rapid growth of the fetus also demands heightened activity of most of the mother's functions, such as rapid intermediary metabolism in her liver, rapid pumping of blood by her heart, increased respiration, and increased digestion and assimilation of food.

Changes in the Body Fluids and Circulation. The female sex hormones and extra adrenocortical hormones produced during pregnancy cause the mother usually to gain about 5 to 7 lb of fluid, or, in other words, about 3 L. About 0.5 L of this is in the plasma, and another 0.5 L is red blood cells, making a total gain in blood volume of about 1 L. About one third of the extra blood is needed to fill the sinuses of the placenta, but the other two thirds of a liter collects in the circulation, causing blood to flow toward the heart with greater ease than usual. As a result, the mother's cardiac output becomes roughly 30 percent more than normal, with about half of the increased cardiac output flowing through the placenta.

During birth of the baby, the mother loses an average of 200 to 300 ml of blood as the placenta separates from the uterus. This ordinarily causes no physiologic inconvenience because of the extra blood that had been stored during pregnancy. After birth, loss of the estrogen and other steroid hormones produced by the placenta causes the kidneys to excrete most of the remainder of the excess fluid and salt in the next few days.

BIRTH OF THE BABY (PARTURITION)

Duration of Pregnancy and Onset of Parturition. The duration of pregnancy, from the time of the last menstrual period until birth of the baby, is normally 40 weeks, though occasionally surviving babies are born as early as 28 weeks or as late as 46 weeks. Approximately 90 percent of all babies are born within 10 days before or after the 40-week interval.

The reason for the relatively constant duration of pregnancy has never been completely understood. Presumably, growth of the baby and of the placenta to a certain size and state of maturity initiates birth. The probable factors that cause the onset of parturition are the following:

When the baby becomes large, pressure of its body inside the uterus stretches the uterine musculature, which in turn initiates uterine contractions. In addition, movements of the baby in the uterus—such as the feet and hands striking the uterine wall—also initiate contractions. Obviously, the larger the baby becomes, the more likely are these initiated contractions to become strong enough to cause birth.

Also important are several hormonal factors. The concentration of *progesterone* secreted by the placenta begins to decrease a few weeks prior to birth, and, since progesterone normally inhibits the uterus, this change perhaps allows an increase in uterine contractions. On the other hand, the concentration of *estrogen* increases up until birth, and this increases the activity of the uterus, in contrast to the inhibition caused by progesterone. These two hormonal factors probably account for some of the progressively increasing contractility of the uterus shortly before birth.

A third factor possibly helping to initiate parturition is an increase in the secretion of *oxytocin* by the hypothalamic–posterior pituitary system shortly before the end of pregnancy. This hormone causes extreme contractility of the uterine musculature, and absence of its secretion in animals usually makes parturition difficult.

Mechanism of Parturition. The uterine musculature, like almost all smooth muscle, undergoes rhythmic contractions much of the time. However, because of the influence of progesterone, these contractions are so weak during the early months of pregnancy that they can hardly be noted. During the last 3 months of pregnancy, they increase steadily. Then, a few hours before birth, the contractions suddenly become extremely intense and are then called *labor* contractions. These very strong contractions wedge the head against the cervix, which slowly over a period of hours stretches the cervical ring and vaginal canal, and expels the baby. This period, from onset of the strong contractions until the baby is born, is called the period of *labor*, and the actual expulsion of the child is called *parturition*.

Approximately 19 times out of 20 the portion of the baby that pushes against the cervix is the head, and this acts as a wedge to open the cervical and vaginal canals. In most of the remaining cases the buttocks are the presenting portion of the baby, though occasionally a leg, a shoulder, or even the side of the baby may be against the cervical canal. In all of these instances the cervix cannot be wedged open nearly as effectively as by the head, and birth of the baby is considerably more difficult. When the head comes first, the remainder of the body slips through the vaginal canal within a few seconds after the head is born; when the head is not the presenting part, the body portion of the baby may be born relatively easily, but the head, which is the largest part of the child, then has difficulty in passing through the canal. One of the problems of bottom-first birth is often a period of interrupted umbilical blood supply to the baby because the cord can become compressed against the wall of the delivery canal for a minute or more by the large head.

Theoretical Cause of the Sudden Onset of Labor. The reason the rhythmic uterine contractions suddenly become intense enough to cause labor is yet undetermined, but the following explanation has been offered as a possibility.

Irritation of the cervix is known to cause a muscular reaction over the entire uterus, making it become very excitable and causing its contractions to increase. It is presumed that when the natural contractions of the uterus reach a certain level of intensity, which occurs at the end of approximately 9 months of pregnancy, they push the baby's head against the cervix, which stretches it and thereby makes the entire uterus contract more forcefully. This contraction in turn pushes the baby's head into the cervix still more and initiates still further intensity of uterine contraction. Thus, a vicious cycle develops, as illustrated in Figure 50–11, and the uterine contractions become stronger and stronger until finally the baby is expelled.

The increased uterine excitability caused by stretching or irritating the cervix is believed to result from two mechanisms. First, cervical stimulation is believed to transmit impulses upward

1. Baby's head stretches cervix . . .
2. Cervical stretch excites fundic contraction
3. Fundic contraction pushes baby down and stretches cervix some more . . .
4. Cycle repeats over and over again . . .

FIGURE 50–11   A postulated mechanism for initiation of labor.

through the uterine musculature itself, thus resulting in uterine contraction. Second, in lower animals, and therefore presumably in women, cervical irritation has been shown to transmit nerve impulses to the hypothalamus, causing greater secretion of *oxytocin*, which in turn increases the contractility of the uterus, as was explained in Chapter 45.

PRODUCTION OF MILK (LACTATION)

The Breasts and the Mammary Glands

Figure 50–12 illustrates a **breast.** The glandular portion of the breast is called the **mammary gland.** Prior to pregnancy most of the mass of the breast is composed of fat tissue and connective tissue. Interspersed among these is the im-

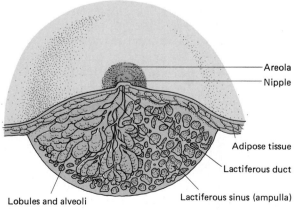

Areola
Nipple
Adipose tissue
Lactiferous duct
Lactiferous sinus (ampulla)
Lobules and alveoli

FIGURE 50–12   The breast and its mammary gland.

mature structure of the mammary gland. During pregnancy, under the influence of tremendous amounts of estrogen and progesterone secreted by the placenta and also large amounts of prolactin secreted by the anterior pituitary gland, the mammary gland enlarges immensely and becomes the major portion of the breast.

Glandular Structure of the Mammary Gland.   The individual portions of the mammary gland, illustrated in the lower half of Figure 50–12, are

1. The **lobules** and **alveoli.** Each breast contains hundreds of lobules, and each of these is divided into large numbers of small sacs called *alveoli* lined with glandular cells known as the **secretory epithelium.** It is this epithelium that secretes the milk.
2. The **lactiferous ducts.** The ducts leading from the lobules coalesce into progressively larger ducts, forming eventually about 15 large *lactiferous ducts* that empty through the nipple.
3. The **lactiferous sinuses** (also called *ampullae*), which are bulbous enlargements of the lactiferous ducts immediately before they empty through the nipple. When the baby suckles the mother's breast, special muscle-like cells called **myoepithelial cells** surrounding the alveoli contract and force the milk from the alveoli and lobules into the sinuses. The baby in turn sucks the milk from the sinuses. Without the contraction of the myoepithelial cells there would be no milk in the sinuses for the baby. This complex process is controlled by the special hormone *oxytocin* as will be explained later in this chapter.

Hormonal Control of Lactation— Estrogen, Progesterone, and Prolactin

The secretion of the *estrogen* and *progesterone* after puberty in a girl begins to prepare the breasts for lactation. The breasts enlarge and the glandular elements begin to develop. However, this early glandular development is slight compared with that achieved during pregnancy. The tremendous quantities of estrogen and *human chorionic somatomammotropin* secreted by the placenta during pregnancy and *prolactin* secreted by the anterior pituitary gland cause rapid development of the glandular structure of the breasts, and the large quantities of progesterone change the glandular cells into actual secreting cells. By the time the baby is born, the breasts will have reached a degree of development capable of producing milk.

Initiation of Lactation— Function of Prolactin

Though estrogen and progesterone are essential for the physical development of the breasts during pregnancy, both these hormones also have a specific effect to inhibit the actual secretion of milk. On the other hand, the hormone *prolactin* has exactly the opposite effect, promotion of the secretion of milk. This hormone is secreted by the mother's pituitary gland, and its concentration in her blood rises steadily from the fifth week of pregnancy until birth of the baby, at which time it has risen to very high levels, usually about 10 times the normal nonpregnant level. This is illustrated in Figure 50–13. In addition, the placenta secretes large quantities of human chorionic somatomammotropin, which also has mild lactogenic properties, thus supporting the prolactin from the mother's pituitary. Even so, only a few milliliters of fluid are secreted each day until after the baby is born. This fluid is called *colostrum;* it contains essentially the same concentration of proteins and lactose as milk but almost no fat, and its maximum rate of production is about 1/100 the subsequent rate of milk production.

This absence of lactation during pregnancy is caused by the overriding suppressive effects of progesterone and estrogen, which are secreted in tremendous quantities as long as the placenta is still in the uterus and which completely subdue the lactogenic effects of both prolactin and human chorionic somatomammotropin. However, immediately after the baby is born, the sudden loss of both estrogen and progesterone secretion by the placenta now allows the lactogenic effect of the prolactin from the mother's pituitary gland to assume its natural role, and within 2 or 3 days the breasts begin to secrete copious quantities of milk instead of colostrum.

Following birth of the baby, the *basal level* of prolactin secretion returns during the next few weeks to the nonpregnant level, as shown in Figure 50–13. However, each time the mother nurses her baby, nervous signals from the nipples to the hypothalamus cause approximately a 10-fold surge in prolactin secretion lasting about 1 hour, which is also shown in the figure. The prolactin in turn acts on the breasts to provide the milk for the next nursing period. If this prolactin surge is absent, if it is blocked as a result of hypothalamic or pituitary damage, or if nursing does not continue, the breasts lose their ability to produce milk within a few days. On the other hand, milk production can continue for several years if the child continues to suckle, but the rate of milk formation normally decreases considerably within 7 to 9 months.

Hypothalamic Control of Prolactin Secretion. Though secretion of most of the anterior pituitary hormones is enhanced by neurosecretory *releasing* factors transmitted from the hypothalamus to the anterior pituitary gland through the hypothalamic–hypophysial portal system, the secretion of prolactin is controlled by an exactly opposite effect. That is, the hypothalamus synthesizes a *prolactin-inhibitory factor* (PIF). Under normal conditions, large amounts of PIF are continuously transmitted to the anterior pituitary gland so that the normal rate of prolactin secretion is slight. However, during pregnancy and during lactation the formation of PIF itself is suppressed, thereby allowing the anterior pituitary gland to secrete prolactin and thus control milk production.

Milk Ejection—Role of Oxytocin. When a baby sucks on the nipples, it usually obtains no milk from the breasts for approximately the first 45 seconds to a minute. Then suddenly milk appears in the ducts of both breasts even though suckling occurs on only one breast, indicating that some general phenomenon has occurred to cause milk to flow toward the nipples.

Experiments have shown that suckling causes sensory nerve signals to pass first into the spinal cord, then upward through the brain stem, and finally into the hypothalamus to stim-

FIGURE 50–13 Changes in rates of secretion of estrogens, progesterone, and prolactin for 8 weeks prior to parturition and for 36 weeks thereafter. Note especially the decrease of prolactin secretion back to basal levels within a few weeks, but also the intermittent periods of marked prolactin secretion (for about 1 hour at a time) during and after periods of nursing.

ulate the production of the hormone *oxytocin* by the posterior pituitary gland. This hormone then circulates through the blood to the breasts, where it causes *myoepithelial cells* surrounding the alveoli to contract and expel the milk collected in the alveoli into the lactiferous sinuses and ducts leading to the nipples. This process is called *milk ejection.*

This milk ejection mechanism can be adversely affected tremendously by psychic factors. For example, a mother's fear that she might not be able to nurse her baby can actually keep her from doing so. Also, disturbances caused by other children in the family or by overly concerned relatives may lead to difficulty in milk ejection and cause the breasts to fail to empty. Failure to empty in turn causes the anterior pituitary gland to cease its production of prolactin, and the breasts stop secreting milk.

Composition of Milk

Milk contains the usual substances needed for energy and growth by the baby. These include two types of protein, *casein* and *lactalbumin;* an easily digested sugar, *lactose,* composed of one molecule of glucose and one of galactose; and large quantities of fatty substances such as *butter fats, cholesterol,* and *phospholipids.*

In addition, milk contains small quantities of vitamins and large quantities of calcium phosphate. Yet, it has a conspicuous lack of iron. This lack usually is not detrimental to the baby in early life, because a sufficient quantity of iron is stored in the fetal liver prior to birth to continue the formation of hemoglobin for about 2 months. Beyond this time, however, iron must be in the baby's diet, or it will develop progressive anemia.

Table 50–1 shows the relative compositions of human and cow's milk. When cow's milk is substituted for human milk, the baby's metabolic systems must adjust to a large increase in

TABLE 50–1
Composition of Milk

	Human	*Cow*
Water	88.5	87.0
Fat	3.3	3.5
Sugar	6.8	4.8
Casein	0.9	2.7
Lactalbumin and other protein	0.4	0.7
Ions (including calcium phosphate)	0.2	0.7

ions—to the phosphates in particular—which can occasionally lead temporarily to serious ion disturbances in the baby. Also, it is desirable to fortify cow's milk with extra quantities of easily digested sugar such as pure glucose (dextrose).

Effect of Lactation on the Mother

Production of milk by the mother in some ways is as great a drain on her metabolic systems as pregnancy itself. Particularly does she lose large amounts of stored proteins and fats during lactation. Also, if large quantities of calcium phosphate are not in her diet, her parathyroid glands become greatly overactive. This causes reabsorption of her bones, releasing the necessary calcium and phosphate for milk formation. Fortunately, though, the amount of calcium phosphate stored in the bones of the average mother is tremendous in comparison with the amount that the baby will need during the first few months of life. Therefore, only when the mother already has some degree of decalcification will this loss cause her any difficulty.

Frequently mothers develop extensive dental caries during pregnancy, which has often been ascribed to the loss of calcium phosphate from the teeth. Experiments, however, have shown that no significant amount of calcium phosphate leaves the teeth, but that the caries are probably caused by enhanced bacterial growth in the mouth during pregnancy.

QUESTIONS

1. Describe the fertilization of the ovum and early growth of the fetus.
2. Explain the process of implantation and early nutrition of the fetus.
3. Describe the organization and function of the placenta for nutrition of the fetus.
4. What are the functions of estrogen during pregnancy?
5. What are the functions of progesterone during pregnancy?
6. What are the functions of chorionic gonadotropin during pregnancy?
7. What are the functions of human chorionic somatomammotropin during pregnancy?
8. What are some of the important nutritional needs of the growing fetus?
9. Describe the changes in the circulation of the fetus at birth.
10. Describe the following changes in the mother during pregnancy: changes in weight, changes in metabolism, and changes in body fluids and circulation.
11. What is the mechanism of parturition? What factors cause it to begin?

12. Explain the functions of estrogen, progesterone, human chorionic somatomammotropin, prolactin, and oxytocin in milk production.

REFERENCES

Battaglia, F.C., and Meschia, G.: Principal substrates of fetal metabolism. *Physiol. Rev.,* 58:499, 1978.

Buster, J.E., and Marshall, J.R.: Conception, gamete and ovum transport, implantation, fetal-placental hormones, hormonal preparation of parturition and parturition control. *In* DeGroot, L.J., *et al.* (eds.): Endocrinology. Vol. 3. New York, Grune & Stratton, 1979, p. 1595.

Challis, J.R.G.: Endocrinology of late pregnancy and parturition. *In* Greep, R.O. (ed.): International Review of Physiology: Reproductive Physiology III. Vol. 22. Baltimore, University Park Press, 1980, p. 277.

Chamberlain, G., and Wilkinson, A. (eds.): Placenta Transfer. Baltimore, University Park Press, 1979.

Cowie, A.T., *et al.*: Hormonal Control of Lactation. New York, Springer-Verlag, 1980.

Epel, D.: The program of fertilization. *Sci. Am.,* 237(5): 128, 1977.

Fenichel, G.M.: Neonatal Neurology. New York, Churchill Livingstone, 1980.

Grant, N.F., and Worley, R.: Hypertension in Pregnancy: Concept and Management. New York, Appleton-Century-Crofts, 1980.

Grundmann, E., and Kirsten, W.H. (eds.): Perinatal Pathology. New York, Springer-Verlag, 1979.

Haymond, M.W., and Pagliara, A.S.: Endocrine and metabolic aspects of fuel homeostasis in the fetus and neonate. *In* DeGroot, L.J., *et al.* (eds.): Endocrinology. Vol. 3. New York, Grune & Stratton, 1979, p. 1779.

Hogarth, P.J.: Biology of Reproduction. New York, John Wiley & Sons, 1978.

Li, C.H. (ed.): The Chemistry of Prolactin. New York, Academic Press, 1980.

Nathan, D.G., and Oski, F.A. (eds.): Hematology of Infancy and Childhood, 2nd Ed. Philadelphia, W.B. Saunders, 1980.

Rudolph, A.M.: Fetal and neonatal pulmonary circulation. *Annu. Rev. Physiol.,* 41:383, 1979.

Sinclair, J.C. (ed.): Temperature Regulation and Energy Metabolism in the Newborn. New York, Grune & Stratton, 1978.

Smith, M.S.: Role of prolactin in mammalian reproduction. *In* Greep, R.O. (ed.): International Review of Physiology: Reproductive Physiology III. Vol. 22. Baltimore, University Park Press, 1980, p. 249.

Thorburn, G.D., and Challis, J.R.G.: Endocrine control of parturition. *Physiol. Rev.,* 59:863, 1979.

Vorherr, H. (ed.): Human Lactation. New York, Grune & Stratton, 1979.

XVI

SPORTS PHYSIOLOGY

Muscle blood flow can increase as much as *25-fold* during the most strenuous exercise. To provide this increased flow through the many contracting muscles, the cardiac output can increase as much as sixfold in a well-trained athlete such as a marathoner. During the training process of the *marathon runner,* the *size of the heart chambers* and the *heart mass both increase about 40 percent.* In maximal exercise, the heart pumps blood at about 90 percent of its full pumping capacity. Therefore, the pumping capacity of the heart is much more of a limiting factor in delivering adequate oxygen to the muscles during endurance athletics than is the respiratory system.

Tremendous amounts of heat are generated inside the body during exercise. Therefore, when exercising in hot and humid conditions or without ventilated clothing, a person is likely to develop *heat stroke.* Also, he can lose as much as 5 to 10 lb of body fluid in 1 hour because of sweating, leading to muscle cramps, weakness, or even circulatory collapse. Heat stroke can be lethal if not treated instantly. Among the best replacement fluids during exercise are *fruit juices.*

It is fitting to end this text with a chapter on sports physiology because there are no normal stresses to which the body is exposed that even nearly approach the extreme stresses of heavy exercise. In fact, if some of the extremes of exercise were continued for even slightly prolonged periods of time, they might easily be lethal. Therefore, in the main, sports physiology is a discussion of the ultimate limits to which most of the bodily mechanisms can be stressed. To give one simple example: In a person who has extremely high fever, approaching the level of lethality, the body metabolism increases to about 100 percent above normal. By comparison, the metabolism of the body during a marathon race increases to 2000 percent above normal.

THE FEMALE AND THE MALE ATHLETE

Most of the quantitative data that will be given in this chapter will be for the young male adult, not because it is desirable to know only these values but because it is only in this class of athletes that relatively complete measurements have been made. However, for those measurements which have been made in the female, almost identically the same basic physiological principles apply to women equally as to men except for quantitative differences caused by differences in body size, body composition, and the presence or absence of the male sex hormone testosterone. In general, most quantitative values—such as muscle strength, pulmonary ventilation, and cardiac output, all of which are related mainly to the muscle mass—will vary between two thirds and three quarters of the values recorded in men. However, this does not translate into this much differential in athletic performance because the body size is correspondingly smaller. A good indication of the relative performance capabilities of the female versus the male athlete comes from the relative times required for running the marathon race. In a recent comparison, the top female performer had a running time about 12 percent less than that of the top male performer. On the other hand, for some endurance events, women have proved to have capabilities superior to those of men. As an example, the record for the two-way swim across the English channel is presently held by a woman, not a man. Part of the reason for this has been reputed to be that the female has extra fat in her subcutaneous tissues to insulate her from the cold of the channel waters, but this certainly is not all of the reason, and it may be more the wishful thinking of the male ego rather than fact.

The hormonal differences between woman and man certainly account for a large part if not most of the differences in athletic performance. *Testosterone* secreted by the male testicles has a powerful anabolic effect, which means that it causes greatly increased deposition of protein everywhere in the body, especially in the muscles. In fact, even the male who participates in very little sports activity but who nevertheless is well-endowed with testosterone will have muscles that grow to sizes 40 percent or more greater than those of his female counterpart and with a corresponding increase in strength. Thus, the male who begins to train for sports activity already has a running start on the female.

The female sex hormone *estrogen* probably also accounts for some of the difference between female and male performance, though not nearly so much as the effect of testosterone. Estrogen is known to increase the deposition of fat in the female, especially in certain tissues such as the breasts, the hips, and the subcutaneous tissue. At least partly for this reason, the average nonathletic female has about 26 percent body fat composition in contrast to the nonathletic male, who has about 15 percent. In marathon runners who have trained themselves to the least amount of excess fat, the male runner has about 4 percent body fat composition and the female, 6 percent. Thus, either in the untrained or the trained state, the female usually averages about 50 percent more body fat than the male. This obviously is a detriment to the highest levels of athletic performance in those events in which performance is dependent upon speed or bodily strength, but on the other hand it could be an aid in grueling endurance athletic events that require the fat for energy.

Estrogen plays another more insidious role in athletics, for it is the estrogen secreted by the female ovaries after puberty that makes the female stature smaller than that of the male. Immediately after puberty, the surge in estrogen secretion causes a rapid spurt of growth that usually makes the postpubertal female grow more rapidly than her male counterpart. On the other hand, this growth is short-lived because the epiphyseal cartilages of the long bones, which is where the growth occurs, rapidly run their course and actually disappear, allowing the epiphyses to unite with the shafts of the long bones—therefore, no further growth. As a result, the female frequently reaches her full height at an age of perhaps 15 to 17 years, whereas the male may continue to grow until the age of 19 to 21. This difference obviously looms large in most athletic events because the very design of athletic competition often gives the edge to those of greater body size.

Finally, one cannot neglect the effect of the sex hormones on temperament. There is no doubt that testosterone promotes aggressiveness and that estrogen is associated with a more mild temperament. Certainly a large part of competitive sports is the aggressive spirit that drives a person to his maximum effort, often at the expense of judicious restraint.

THE MUSCLES IN EXERCISE

Strength, Power, and Endurance of Muscles

The final common denominator in athletic events is what the muscles can do for you— what strength they can give when it is needed, what power they can achieve in the performance

of work, and how long they can continue in their activity.

The strength of a muscle is determined mainly by its size, with *a maximum contractile force between 2.5 kg and 3.5 kg per cm^2* of muscle cross-sectional area. Thus, the male who is well laced with testosterone and therefore has correspondingly enlarged muscles will be much stronger than those persons without the testosterone advantage. Also, the athlete who has hypertrophied his muscles through an exercise training program likewise will have increased muscle strength because of increased muscle size.

To give an example of muscle strength, a world-class weight lifter might have a quadriceps muscle with a cross-sectional area as great as 150 cm^2. This would translate into a maximum contractile strength of 525 kg (or 1155 lb), with all this force applied to the patellar tendon. Therefore, one can readily understand how it is possible for this tendon to be ruptured or actually to be avulsed from its insertion into the tibia below the knee. Also, when such forces occur in tendons that span a joint, similar forces are also applied to the surfaces of the joints, or sometimes to ligaments spanning the joints, thus accounting for such happenings as displaced cartilages, compression fractures about the joint, or torn ligaments.

Yet, to make matters still worse, the *holding strength* of muscles is about 40 percent greater than the contractile strength. That is, if a muscle is already contracted and a force then attempts to stretch out the muscle, this requires about 40 percent more force than can be achieved by a shortening contraction. Therefore, the force of 525 kg calculated previously for the patellar tendon becomes 735 kg (1617 lb). This obviously further compounds the problems of the tendons, joints, and ligaments. It can also lead to internal tearing in the muscle itself. In fact, stretching out of a maximally contracted muscle is one of the best ways to insure the highest degree of muscle soreness.

The *power* of muscle contraction is different from muscle strength, for power is a measure of the amount of work that the muscle can perform in a given period of time. This is determined not only by the strength of muscle contraction but also by its *velocity of contraction* and the number of times that it contracts each minute.

Muscle power is generally measured in *kilogram-meters (kg-m)/minute.* That is, a muscle that can lift a kilogram weight to a height of 1 m or that can move some object laterally against a force of 1 kg for a distance of a meter in 1 minute is said to have a power of 1 kg-m/minute. The maximum power that all of the muscles in the body of a highly trained athlete with all of the muscles working together can achieve is approximately the following:

First 10 to 15 seconds	7000 kg-m/minute
Next 1 minute	4000 kg-m/minute
Next half hour	1700 kg-m/minute

Thus, it is clear that a person has the capability of an extreme power surge for a short period of time, such as during a 100-m dash that can be completed entirely within the first 10 seconds, whereas for long-term endurance events the power output of the muscles is only one fourth as great as during the initial power surge. Yet, this does not mean that one's athletic performance is four times as great during the initial power surge as it is for the next half hour, because the efficiency for translation of muscle power output into athletic performance is often much less during rapid activity than during less rapid but sustained activity. Thus, the velocity of the hundred meter dash is only $1\frac{3}{4}$ times as great as the velocity of the 30-minute race despite the 4-fold difference in short-term versus long-term muscle power capability.

The final characteristic of muscle performance is *endurance*. This, to a great extent, depends on the nutritive support for the muscle—more than anything else on the amount of glycogen that has been stored in the muscle prior to the period of exercise. A person on a high carbohydrate diet stores far more glycogen in his muscles than a person on either a mixed diet or a high fat diet. Therefore, endurance is greatly enhanced by a high carbohydrate diet. When athletes run at speeds typical for the marathon race, their endurance as measured by the time that they can sustain the race until complete exhaustion is approximately the following:

High carbohydrate diet	240 minutes
Mixed diet	120 minutes
High fat diet	85 minutes

The corresponding amounts of glycogen stored in the muscle are approximately the following:

High carbohydrate diet	33 g per kg of muscle.
Mixed diet	17.5 g per kg of muscle.
High fat diet	6 g per kg of muscle.

The Muscle Metabolic Systems in Exercise

The same basic metabolic systems are present in muscle as in all other parts of the body; these were discussed in detail in Chapters 31 and 32. However, special quantitative measures of the activities of three metabolic systems are exceedingly important in understanding the limits of physical activity.

THE PHOSPHAGEN SYSTEM

Adenosine Triphosphate. The basic source of energy for muscle contraction is adenosine triphosphate (ATP), which has the following basic formula.

$$\text{Adenosine}—PO_4 \sim PO_3 \sim PO_3$$

The bonds attaching the last two phosphate radicals to the molecule, designated by the symbol ~, are so-called *high energy phosphate bonds.* Each of these bonds stores about 8000 calories of energy per mole of ATP. Therefore, when one phosphate radical is removed from the molecule, 8000 calories of energy that can be used to energize the muscle contractile process are released. Then, when the second phosphate radical is removed, still another 8000 calories become available. Removal of the first phosphate converts the ATP into *adenosine diphosphate* (ADP), and removal of the second converts this ADP into *adenosine monophosphate* (AMP).

Unfortunately, the amount of ATP present in the muscles, even in the well-trained athlete, is sufficient to sustain maximal muscle power for only 5 or 6 seconds, maybe enough for a 50-m dash. Therefore, except for a few seconds at a time, it is essential that new adenosine triphosphate be formed continuously, even during the performance of athletic events. Figure 51–1 illustrates the overall metabolic system, showing the breakdown of ATP first to ADP and then to AMP, with the release of energy to the muscles for contraction. To the left-hand side of the figure are illustrated the three different metabolic mechanisms that are responsible for reconstituting a continuous supply of adenosine triphosphate in the muscle fibers. These are the following:

Release of Energy from Phosphocreatine. Phosphocreatine is another chemical compound that has a high energy phosphate bond, with the following formula:

$$\text{Creatine} \sim PO_3$$

This can decompose to *creatine* and *phosphate ion,* as illustrated to the left in Figure 51–1, and in doing so releases large amounts of energy. In fact, the high energy phosphate bond of phosphocreatine has slightly more energy than the bond of ATP. Therefore, the phosphocreatine can

I. Phosphocreatine ⟶ Creatine + PO₃

II. Glycogen ⟶ Lactic acid

III. Glucose / Fatty acids / Amino acids + O₂ ⟶ CO₂ + H₂O + Urea

ATP / ADP / AMP ⟶ Energy for muscle contraction

FIGURE 51–1 The three important metabolic systems that supply energy for muscle contraction.

easily provide enough energy to reconstitute the high energy bonds of the ATP. Furthermore, most muscle cells have two to three times as much phosphocreatine as ATP.

A special characteristic of energy transfer from phosphocreatine to ATP is that it occurs within a small fraction of a second. Therefore, in effect, all the energy stored in the muscle phosphocreatine is instantaneously available for muscle contraction, just as is the energy stored in the ATP.

The cell phosphocreatine plus its ATP are called the *phosphagen energy system*. These together can provide maximal muscle power for a period of 10 to 15 seconds, barely enough for the 100-m run. Thus, the energy from the phosphagen system is used for maximal short bursts of muscle power.

THE GLYCOGEN–LACTIC ACID SYSTEM

The stored glycogen in muscle can be split into glucose and the glucose then utilized for energy. The initial stage of this process, called *glycolysis*, occurs entirely without use of oxygen and therefore is said to be *anaerobic metabolism* (see Chapter 42). During glycolysis, each glucose molecule is split into two *pyruvic acid molecules*, and energy is released to form several ATP molecules. Ordinarily the pyruvic acid then enters the mitochondria of the muscle cells and reacts with oxygen to form still many more ATP molecules. However, when there is insufficient oxygen for this second stage (the oxidative stage) of glucose metabolism to occur, most of the pyruvic acid is converted into *lactic acid*, which then diffuses out of the muscle cells into the interstitial fluid and blood. Therefore, in effect, much of the muscle glycogen becomes lactic acid, but in doing so considerable amounts of adenosine triphosphate are formed entirely without the consumption of oxygen.

Another characteristic of the glycogen–lactic acid system is that it can form ATP molecules about two and one-half times as rapidly as can

the oxidative mechanism of the mitochondria. Therefore, when large amounts of adenosine triphosphate are required for moderate periods of muscle contraction, this anaerobic glycolysis mechanism can be used as a rapid source of energy. It is not as rapid as the phosphagen system, but about half as rapid.

Under optimal conditions the glycogen–lactic acid system can provide 30 to 40 seconds of maximal muscle activity in addition to the 10 to 15 seconds provided by the phosphagen system.

THE AEROBIC SYSTEM

The aerobic system means the oxidation of foodstuffs in the mitochondria to provide energy. That is, as illustrated to the left in Figure 51–1, glucose, fatty acids, and amino acids from the foods—after some intermediate processing—combine with oxygen to release tremendous amounts of energy that are used to convert AMP and ADP into ATP, as was discussed in Chapter 42.

In comparing this aerobic mechanism of energy supply with the glycogen–lactic acid system and the phosphagen system, the relative maximum rates of power generation in terms of ATP utilization are the following:

Aerobic system	1 *M* of ATP per minute.
Glycogen–lactic acid system	2.5 *M* of ATP per minute.
Phosphagen system	4 *M* of ATP per minute.

On the other hand, when comparing the systems for endurance, the relative values are the following:

Phosphagen system	10 to 15 seconds.
Glycogen–lactic acid system	30 to 40 seconds.
Aerobic system	unlimited time (as long as nutrients last).

Thus, one can readily see that the phosphagen system is the one utilized by the muscle for power surges, and the aerobic system is required for prolonged athletic activity. In between is the glycogen–lactic acid system, which is especially important for giving extra power during such intermediate races as the 200- to 800-m runs.

WHAT TYPES OF SPORTS UTILIZE WHICH ENERGY SYSTEMS?

By considering the vigor of a sports activity and its duration, one can estimate very closely which of the energy systems are used for each activity. The following are various approximations:

Almost entirely phosphagen system:

100-m dash
jumping
weight lifting
diving
football dashes

Phosphagen and glycogen–lactic acid systems:

200-m dash
basketball
baseball home run
ice hockey

Mainly glycogen–lactic acid system:

400-m dash
100-m swim
tennis
soccer

Glycogen–lactic acid and aerobic systems:

800-m dash
200-m swim
1500-m skating
boxing
2000-m rowing
1500-m run
1-mi run
400-m swim

Aerobic system:

10,000-m skating
cross-country running
marathon run (26.2 mi, 42.2 km)
jogging

RECOVERY OF THE MUSCLE METABOLIC SYSTEMS AFTER EXERCISE

Performance in athletic events is often determined by how rapidly the athlete can recover strength between surges of activity, and in general this means how rapidly the energy systems can recover. Each of these systems has its own characteristic rate of recovery, as follows:

The Phosphagen System. The total amount of energy in the phosphagen system in all the musculature of the well-trained male athlete's body is equivalent to about $0.6\,M$ of ATP (about $0.3\,M$ for the female), and this can be almost completely depleted in an average of 10 to 15 seconds of maximal muscle activity. However, the glycogen–lactic acid system can replenish this phosphagen system as rapidly as $2.5\,M$ of ATP per minute, and the aerobic system can replenish it as rapidly as $1\,M$ per minute. Therefore, in theory, it would be possible for these other energy systems to replenish fully the phosphagen system within 15 to 30 seconds after its full depletion, which would mean that a person could run a second 100-m dash in less than 1 minute after the first. However, in practice, this does not work quite that way because the other systems function at full force to replenish the phosphagen only when the phosphagen system is almost totally depleted. Instead, the phosphagen normally is replenished with a *half-time* of about 20 to 30 seconds. This means that for those events using the phosphagen system alone, such as high jumping, one could reasonably expect to have full replenishment of this system within about 3 to 5 minutes.

The Glycogen–Lactic Acid System. The limitation in the use of this system for energy is mainly the amount of lactic acid that the person can tolerate in his muscles and body fluids. Lactic acid causes extreme *fatigue*, which serves as a self limitation to further use of this system for energy. The amount of time required for replenishing this system, therefore, is determined by how rapidly the person can eliminate lactic acid from his body. Under most conditions, this is achieved with a half-time of about 20 to 30 minutes; therefore, as long as an hour after an athletic event has utilized the glycogen–lactic acid system to its fullest extent, this metabolic system still will not have achieved full recovery.

The Aerobic System—Short-Term Recovery and the "Oxygen Debt." Recovery of the aerobic system has a short-term phase and a long-term phase, one lasting about an hour and the other several days. The short-term phase of recovery is a function of the so-called *oxygen debt*, illustrated in Figure 51–2. The oxygen debt is defined as the extra amount of oxygen that must be taken into the body after an athletic event to restore all the metabolic systems back to their full normal state. Oxygen debt can be accu-

FIGURE 51–2 Rate of oxygen uptake by the lungs during maximal exercise for 4 minutes and then for almost 1 hour after the exercise is over. This figure demonstrates the principle of *oxygen debt.*

mulated in two different ways: First, part of this debt results from usage of oxygen that is already stored in different parts of the body. For instance, even normally about 0.3 L of oxygen is stored in the muscles themselves combined with myoglobin, an oxygen-binding chemical substance similar to hemoglobin that is present in muscle fibers. In addition, almost 1 L of oxygen is normally combined with all the hemoglobin in the blood, and another 0.5 L is in the air of the lungs as well as about 0.25 L dissolved in all the body fluids. Most of this oxygen can be used by the muscles during exercise and therefore must be replenished after the exercise is over.

Second, oxygen debt can be accumulated by depletion of both the phosphagen and the glycogen–lactic acid systems. As much as 2 L of oxygen are required to replenish a fully depleted phosphagen system, and as much as 8 L to replenish a depleted glycogen–lactic acid system.

In toto, a person can develop an oxygen debt up to 10 to 12 L, and this oxygen is repaid to the body for as long as an hour or more after periods of exhaustive exercise. Figure 51–2 illustrates that the repayment occurs in two different stages. First, that portion of the debt not related to the accumulation of lactic acid, called *alactacid oxygen debt,* which means the oxygen needed to replenish the oxygen stores of the body as well as to replenish the phosphagen system, is usually fully repaid within 2 to 3 minutes. On the other hand, removal of the lactic acid from all the body fluids requires an hour or more, so that the *lactic acid oxygen debt,* which is usually by far the greater proportion of the total debt, continues to be repaid very slowly for at least an hour. Therefore, for those sports that

deplete the glycogen–lactic acid metabolic system, one should allow for full recovery at least an hour and preferably 2 hours between events.

The mechanism of recovery of the glycogen–lactic acid system is simply to remove the lactic acid from the blood and other body fluids. This is achieved in two ways: First, some of the lactic acid is converted back into pyruvic acid and then metabolized directly by all the body tissues. Second, much of the lactic acid is converted into glucose by the liver, and the glucose in turn is used mainly to replenish the glycogen stores of the muscles.

The Aerobic System—Long-Term Recovery; Importance of Muscle Glycogen Stores. Earlier in the chapter we discussed the importance of stored muscle glycogen for muscle endurance. This is true because glycogen is the food substrate of choice not only for the glycogen–lactic acid system but also for the aerobic oxidative energy system as well. The muscle endurance that can be achieved may be as long as 4 hours of exhaustive exercise in the athlete, who has a high concentration of muscle glycogen, or as little as 1.5 hours in the athlete with minimal muscle glycogen.

Recovery from exhaustive muscle glycogen depletion is not a simple matter, requiring hours to days rather than the seconds or minutes required for the phosphagen and glycogen–lactic acid metabolic systems. Figure 51–3 illustrates this recovery process under three different conditions: first, in persons on a high carbohydrate diet; second, in persons on a high fat/high protein diet; and, third, in persons with no food. Note that on the high carbohydrate diet, full recovery occurred in approximately 2 days, whereas persons on the high fat/high protein diet or on no food at all showed extremely little recovery even after periods as long as 5 days. The message of this study is that it is important for an athlete not to participate in exhaustive exercise during the last 24 to 48 hours prior to a grueling athletic event.

NUTRIENTS USED DURING MUSCLE ACTIVITY

Though we have emphasized the importance of a high carbohydrate diet and large stores of muscle glycogen for maximal athletic performance, this does not mean that only carbohydrates are used for muscle energy—it means simply that carbohydrates are used by preference. Actually, the muscles often use large amounts of fat for energy in the form of *fatty*

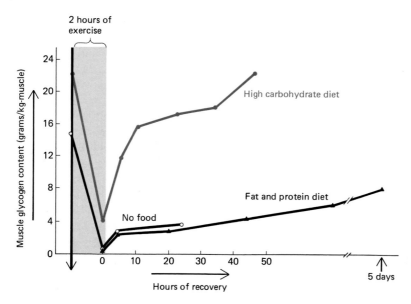

FIGURE 51-3 Effect of diet on the rate of muscle glycogen replenishment following prolonged exercise. (Reprinted from Fox: *Sports Physiology.* Philadelphia, Saunders College Publishing, 1979.)

acids and *acetoacetic acid* (see Chapter 42) and also use to a much less extent proteins in the form of *amino acids.* In fact, even under the best conditions, in those endurance athletic events that last longer than 4 to 5 hours, the glycogen stores of the muscle become depleted and are then of little further use for energizing muscle contraction. Instead, the muscle now depends upon glucose that can be absorbed from the blood, which is limited, or upon energy from other sources, mainly from fats.

Figure 51-4 illustrates the approximate relative usage of carbohydrates and fat for energy during prolonged exhaustive exercise under three different dietary conditions: high carbohydrate diet, mixed diet, and high fat diet. Note that most of the energy is derived from carbohy-

drate during the first few seconds or minutes of the exercise, but at the time of exhaustion, as much as 50 to 80 percent of the energy is being derived from fats rather than carbohydrates.

Not all of the energy from carbohydrates comes from the stored muscle glycogen. In many persons almost as much glycogen is stored in the liver as in the muscles, and this can be released into the blood in the form of glucose, then taken up by the muscles as an energy source. In addition, glucose solutions given to an athlete to drink during the course of an athletic event (in optimal concentrations of 2 to 2.5 percent) can provide as much as 30 to 40 percent of the energy required during the event.

In essence, then, if muscle glycogen and blood glucose are available, these are the energy nutrients of choice for intense muscle activity. Yet, even so, for a real endurance event one can expect fat to supply more than 50 percent of the required energy after about the first 3 to 4 hours.

EFFECT OF ATHLETIC TRAINING ON MUSCLES AND MUSCLE PERFORMANCE

Importance of Resistance Training.
One of the cardinal principles of muscle development during athletic training is the following: Muscles that function under no load, even if they are exercised for hours upon end, increase little in strength. At the other extreme, muscles that contract at or near their maximal force of contraction will develop strength very rapidly even if the contractions are performed only a few times each day. Utilizing this principle, experiments on muscle building have shown that 6 maximal or nearly maximal muscle contractions performed in three separate sets 3 days out of

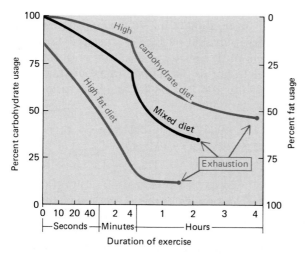

FIGURE 51-4 Effect of duration of exercise as well as type of diet on relative percentages of carbohydrate or fat used for energy by muscles. (Based partly on data in Fox: *Sports Physiology.* Philadelphia, Saunders College Publishing, 1979.)

FIGURE 51–5 Approximate effect of optimal resistive exercise training on increase in muscle strength over a training period of 10 weeks.

each week gives approximately optimal increase in muscle strength and without producing chronic muscle fatigue. The upper curve in Figure 51–5 illustrates the approximate percentage increase in strength that can be achieved in the previously untrained person by this optimal resistive training program, showing that the muscle strength increases about 30 percent during the first 6 to 8 weeks but reaches a plateau after that time. Along with this increase in strength is approximately an equal percentage increase in muscle mass, which is called *muscle hypertrophy.*

Muscle Hypertrophy. The basic size of a person's muscles is determined mainly by heredity plus the level of testosterone secretion, which, in the male, causes considerably larger muscles than in the female. However, with training, the muscles can be hypertrophied perhaps an additional 30 to 60 percent. Most of this hypertrophy results from increased diameter of the muscle fibers, but this is not entirely true because greatly enlarged muscle fibers can split down the middle along their entire length to form entirely new fibers, thus increasing the numbers of fibers as well.

The changes that occur inside the hypertrophied muscle fibers themselves include (1) increased numbers of myofibrils, proportionate to the degree of hypertrophy; (2) increased numbers and sizes of mitochondria; (3) as much as 25 to 40 percent increase in the components of the phosphagen metabolic system, including both ATP and phosphocreatine, (4) as much as 100 percent increase in stored glycogen, and (5) as much as 75 to 100 percent increase in stored triglyceride (fat). In addition, the enzymes required for the oxidative metabolic system are increased, increasing the maximum oxidation rate and efficiency of the oxidative metabolic system as much as 45 percent.

FAST TWITCH AND SLOW TWITCH MUSCLE FIBERS

In the human being, all muscles have varying percentages of *fast twitch* and *slow twitch muscle fibers.* For instance, the gastrocnemius muscle has a higher preponderance of fast twitch fibers, which gives it the capability of very forceful and rapid contraction of the type used in jumping. On the other hand, the soleus muscle has a higher preponderance of slow twitch muscle fibers and therefore is said to be the muscle that is used to a greater extent for prolonged lower leg muscle activity.

The basic differences between the fast twitch and the slow twitch fibers are the following:

1. Fast twitch fibers are about two times as large in diameter.
2. The enzymes that promote rapid release of energy from the phosphagen and glycogen–lactic acid energy systems are two to three times as active in fast twitch fibers as in slow twitch fibers, thus making the maximal power that can be achieved by fast twitch fibers as great as two times that of slow twitch fibers.
3. Slow twitch fibers are mainly organized for endurance, especially for generation of aerobic energy. They have far more mitochondria than the fast twitch fibers. In addition, they contain considerably more myoglobin, a hemoglobin-like protein that combines with oxygen within the muscle fiber; and even more important, myoglobin increases the rate of diffusion of oxygen throughout the fiber by shuttling oxygen from one molecule of myoglobin to the next. In addition, the enzymes of the aerobic metabolic system are considerably more active in slow twitch fibers than in fast twitch fibers.
4. The number of capillaries per mass of fibers is greater in the vicinity of slow twitch fibers than in the vicinity of fast twitch fibers.

In summary, fast twitch fibers can deliver extreme amounts of power for short periods of time. On the other hand, slow twitch fibers provide endurance, delivering prolonged strength of contraction over much longer periods of time.

Hereditary Differences Among Athletes for Fast Twitch Versus Slow Twitch Muscle Fibers. Some persons have considerably more fast twitch than slow twitch fibers, and others have more slow twitch fibers; this obviously could determine to some extent the athletic capabilities of different individuals. Unfortunately, athletic training has not been shown to change the relative proportions of fast twitch and slow twitch fibers, however much an athlete might wish to develop one type of athletic prowess over another. Instead, this is an aspect of genetic inheritance that helps to determine which area of athletics is most suited to each person; some people are born to be marathoners; others are born to be sprinters and jumpers. For example, the following are recorded percentages of fast twitch versus slow twitch fiber in the quadriceps muscles of different types of athletes:

	Fast Twitch	Slow Twitch
Marathoners	18	82
Swimmers	26	74
Average Male	55	45
Weight Lifters	55	45
Sprinters	63	37
Jumpers	63	37

RESPIRATION IN EXERCISE

Though one's respiratory ability is of relatively little concern for the performance of sprint types of athletics, it is critical for maximal performance in endurance athletics. Let us see how important it is:

Oxygen Consumption and Pulmonary Ventilation in Exercise. Normal oxygen consumption for a young adult male at rest is about 250 ml per minute. However, under maximal conditions this can be increased to approximately the following average levels:

Untrained average male	3600 ml per minute
Athletically trained average male	4000 ml per minute
Male marathon runners	5100 ml per minute

Figure 51–6 illustrates the relationship between oxygen consumption at different degrees of exercise and *total pulmonary ventilation.* It is clear from this figure, as would be expected, that there is a linear relationship. In round numbers, both oxygen consumption and total pulmonary ventilation increase about 20-fold between the resting state and maximum intensity of exercise.

The Limits of Pulmonary Ventilation. How severely do we stress our respiratory system during exercise? This can be answered by the following comparison for the normal male:

Pulmonary ventilation at maximal exercise	100 to 110 L per minute
Maximal breathing capacity	150 to 170 L per minute

FIGURE 51–6 Effect of exercise on oxygen consumption and ventilatory rate. (From Gray: Pulmonary Ventilation and Its Physiological Regulation. Springfield, Ill., Charles C Thomas.)

Thus, the maximal breathing capacity is about 50 percent greater than the actual pulmonary ventilation during maximal exercise. This obviously provides an element of safety for the athlete, giving him extra ventilation that can be called on in such conditions as (1) exercise at high altitudes, (2) exercise under very hot conditions, and (3) abnormalities in the respiratory system.

The important point is that the respiratory system is not normally the most limiting factor in the delivery of oxygen to the muscles during maximal muscle aerobic metabolism. We shall see shortly that the ability of the heart to pump blood to the muscles is a much greater limiting factor.

Effect of Training on $\dot{V}_{O_2}$ Max. The abbreviation for the rate of oxygen usage under maximal aerobic metabolism is $\dot{V}_{O_2}$ Max. Figure 51–7 illustrates the progressive effect of athletic

FIGURE 51–7 Increase in $\dot{V}_{O_2}$ Max over a period of 7 to 13 weeks of athletic training. (Reprinted from Fox: *Sports Physiology.* Philadelphia, Saunders College Publishing, 1979.)

training on $\dot{V}_{O_2}$ Max recorded in a group of subjects beginning at the level of no training and then pursuing the training program for 7 to 13 weeks. In this study, it is surprising that the $\dot{V}_{O_2}$ Max increased only about 10 percent. Furthermore, the frequency of training, whether two times or five times per week, made little difference on the increase in $\dot{V}_{O_2}$ Max. Yet, as was pointed out earlier, the $\dot{V}_{O_2}$ Max of marathoners is about 45 percent greater than that of the untrained person. Part of this greater $\dot{V}_{O_2}$ Max of the marathoner is genetically determined; that is, it is those persons who have greater chest sizes and stronger respiratory muscles who select themselves to become marathoners. However, it is also very likely that the very prolonged training of the marathoner does increase the $\dot{V}_{O_2}$ Max by values considerably greater than the 10 percent that has been recorded in short-term experiments such as that in Figure 51–7.

The O₂ Diffusing Capacity of Athletes.
The *O₂ diffusing capacity* is a measure of the rate at which oxygen can diffuse from the alveoli into the blood. This is expressed in terms of *milliliters of oxygen that will diffuse for each millimeter of mercury difference between alveolar partial pressure of oxygen and pulmonary blood oxygen pressure.* That is, if the partial pressure of oxygen in the alveoli is 91 mm Hg while the pressure in the blood is 90 mm Hg, the amount of oxygen that diffuses through the respiratory membrane each minute is the diffusing capacity. The following are measured values for different diffusing capacities:

Nonathlete at rest	23 ml per minute
Nonathlete during maximum exercise	48 ml per minute
Speed skaters during maximum exercise	64 ml per minute
Swimmers during maximum exercise	71 ml per minute
Oarsmen during maximum exercise	80 ml per minute

The most startling fact about these results is the almost threefold increase in diffusing capacity between the resting state and the state of maximum exercise. This results from the fact that blood flow through many of the pulmonary capillaries is very sluggish or even dormant in the resting state, whereas in exercise increased blood flow through the lungs causes all of the pulmonary capillaries to be perfused at their maximum level, thus providing far greater surface area through which oxygen can diffuse into the pulmonary capillary blood.

It is also clear from the above values that those athletes who require greater amounts of oxygen per minute have higher diffusing capacities. Is this because persons with naturally greater diffusing capacities choose these types of

sports, or is it because something about the training procedures increases the diffusing capacity? The answer to this is not known, but one must believe that training does play some role in this, particularly the endurance types of training.

The Blood Gases During Exercise.
Because of the great usage of oxygen by the muscles in exercise, one would expect the oxygen pressure of the arterial blood to decrease markedly and the carbon dioxide pressure of the venous blood to increase far above normal. However, this is not the case. Both of these remain nearly normal, illustrating the extreme ability of the respiratory system to provide very adequate aeration of the blood even in heavy exercise. This illustrates another very important point. The blood gases do not have to become abnormal for respiration to be stimulated in exercise. Instead, respiration is stimulated mainly by neurogenic mechanisms. Part of this stimulation results from direct stimulation of the respiratory center by the same nervous signals that are transmitted from the brain to the muscles to cause the exercise. Part is believed to result from sensory signals transmitted into the respiratory center from the contracting muscles and moving joints. All this nervous stimulation of respiration is normally sufficient to provide almost exactly the proper increase in pulmonary ventilation to keep the blood respiratory gases—the oxygen and the carbon dioxide—almost normal.

Effect of Smoking on Pulmonary Ventilation in Exercise.
It is widely stated that smoking can decrease an athlete's "wind." This is a very true statement for many reasons. First, one effect of nicotine is to cause constriction of the terminal bronchioles of the lungs, which increases the resistance of air flow into and out of the lungs. Second, the irritating effects of smoke cause increased fluid secretion in the bronchial tree, as well as some swelling of the epithelial linings. Third, nicotine paralyzes the cilia on the surfaces of the respiratory epithelial cells that normally beat continuously to remove excess fluids and foreign particles. As a result, much debris accumulates in the respiratory passageways and adds further to the difficulty of breathing. Putting all these factors together, even the light smoker will feel respiratory strain during maximal exercise, and his level of performance obviously may be reduced.

Much more severe are the effects of chronic smoking, because there is hardly any chronic smoker who does not eventually develop some degree of emphysema. In this disease, the following occur: (1) chronic bronchitis,

(2) obstruction of many of the terminal bronchioles, and (3) destruction of many alveolar walls. In severe emphysema, as much as four fifths of the respiratory membrane can be destroyed; then even the slightest exercise can cause respiratory distress. In fact, many such patients cannot even perform the athletic feat of walking across the floor of a single room without gasping for breath. Such is the indictment of smoking.

THE CARDIOVASCULAR SYSTEM IN EXERCISE

Muscle Blood Flow. The final common denominator of cardiovascular function in exercise is to deliver oxygen and other nutrients to the muscles. For this purpose, the muscle blood flow increases drastically during exercise. Figure 51–8 illustrates a recording of muscle blood flow in the leg calf of a person for a period of 6 minutes during strong intermittent contraction. Note the great increase in flow—about 13-fold—but note also that the flow decreased during each muscle contraction. There are two points that can be made from this study: (1) The actual contractile process itself temporarily decreases muscle blood flow because the contracting muscle compresses the intramuscular blood vessels; therefore, strong tonic contractions can cause rapid muscle fatigue because of lack of delivery of enough oxygen and nutrients during the continuous contraction. (2) The blood flow to muscles during exercise can increase markedly. The following comparison illustrates the maximum increase in blood flow that can occur in the well-trained athlete:

Resting blood flow	3.6 ml per 100 g of muscle per minute
Blood flow during maximal exercise	90 ml per 100 g of muscle per minute

FIGURE 51–8 Effects of muscle exercise on blood flow in the calf of a leg during strong rhythmic contraction. The blood flow was much less during contraction than between contractions. (From Barcroft and Dornhorst: *J. Physiol., 109*:402, 1949.)

Thus, muscle blood flow can increase a maximum of about 25-fold during the most strenuous exercise. About half of this increase in flow results from intramuscular vasodilation caused by the direct effects of increased muscle metabolism, as was explained in Chapters 26 and 27. The other half results from multiple factors, the most important of which is probably the moderate increase in arterial blood pressure that occurs in exercise, usually about a 30-percent increase. The increase in pressure not only forces more blood through the blood vessels, but it also stretches the walls of the arterioles and further reduces the vascular resistance. Therefore, a 30-percent increase in blood pressure can often more than double the blood flow—in addition to the great increase in flow already caused by the metabolic vasodilation.

Work Output, Oxygen Consumption, and Cardiac Output During Exercise. Figure 51–9 illustrates the interrelationships between work output, oxygen consumption, and cardiac output during exercise. It is not surprising that all of these are directly related to each other, as

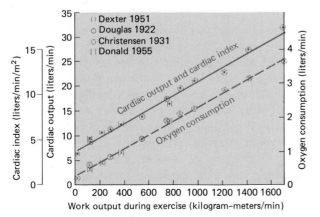

FIGURE 51–9 Relationship between cardiac output and work output (solid curve) and between oxygen consumption and work output (dashed curve) during different levels of exercise. (From Guyton, Jones, and Coleman: Circulatory Physiology: Cardiac Output and Its Regulation. Philadelphia, W.B. Saunders, 1973.)

shown by the linear functions, because the muscle work output increases oxygen consumption, and oxygen consumption in turn dilates the muscle blood vessels, thus increasing venous return and cardiac output. Typical cardiac outputs at several levels of exercise are the following:

Average young adult male at rest	5.5 L per minute
Maximum output during exercise in young untrained male	23 L per minute
Maximum output during exercise in male marathoner	30 L per minute

Thus, the normal untrained person can increase his cardiac output a little over fourfold, and the well-trained athlete can increase his output about sixfold. Individual marathoners have been clocked at cardiac outputs as great as 35 to 40 L per minute.

Effect of Training on Heart Hypertrophy and on Cardiac Output. From the above data, it is clear that marathoners can achieve maximum cardiac outputs about 40 percent greater than those achieved by the untrained person. This results mainly from the fact that the heart chambers of marathoners enlarge about 40 percent, and, along with enlargement of the chambers, the heart mass enlarges 40 percent or more as well. Therefore, it is not only the skeletal muscles but also the heart that hypertrophy during athletic training. However, heart enlargement and increased pumping capacity occur only in the endurance types, not in the sprint types, of athletic training.

Even though the heart of the marathoner is considerably larger than that of the normal person, his resting cardiac output is almost exactly the same as the normal. However, this normal cardiac output is achieved by a large stroke volume at a reduced heart rate. Comparisons between the untrained person and the marathoner are the following:

		Stroke volume	Heart rate
Resting:	untrained	75 ml	75 beats per minute
	trained	105 ml	50 beats per minute
Maximum:	untrained	110 ml	195 beats per minute
	trained	162 ml	185 beats per minute

Thus, the heart pumping effectiveness of each heart beat is 40 to 50 percent greater in the highly trained athlete than in the untrained person, but there is a corresponding decrease in heart rate at rest.

Role of Stroke Volume and Heart Rate in Increasing the Cardiac Output. Figure 51–10 illustrates the approximate changes in stroke volume and heart rate as the cardiac output increases from its resting level of about 5.5 L per minute to 30 L per minute in the marathon runner. The *stroke volume* increases from 105 ml to 162 ml, an increase of about 50 percent, while the heart rate increases from 50 to 185 beats per minute, an increase of 270 percent. Therefore, the heart rate increase accounts for by far a greater proportion of the increase in cardiac output than does the increase in stroke volume during strenuous exercise. The stroke volume reaches its maximum by the time the cardiac output has increased only half way to its maximum. Any further increase in cardiac output must occur by increasing the heart rate.

Relationship of Cardiovascular Performance to $\dot{V}_{O_2}$ Max. During maximal exercise, both the heart rate and the stroke volume are increased to about 95 percent of their maximal levels. Since the cardiac output is equal to stroke volume *times* heart rate, one finds that the cardiac output is about 90 percent of the maximum that the person can achieve. This is in

FIGURE 51–10 Approximate stroke volume output and heart rate at different levels of cardiac output in a marathon athlete.

contrast to about 65 percent of maximum for pulmonary ventilation. Therefore, one can readily see that the cardiovascular system is normally much more limiting on $\dot{V}_{O_2}$ Max than is the respiratory system. For this reason, it is frequently stated that the performance that can be achieved by the marathoner is mainly dependent on his heart, for this is the most limiting link in the delivery of adequate oxygen to the exercising muscles. Therefore, the 40 percent advantage in maximum cardiac output that the marathoner has over the average untrained male is probably the most important physiological benefit of the marathoner's training program.

Effect of Heart Disease and Old Age on Athletic Performance. Because of the critical limitation that the cardiovascular system places on maximal performance in endurance athletics, one can readily understand that any type of heart disease that reduces the maximum cardiac output will cause an almost corresponding decrease in achievable muscle power. Therefore, a person with congestive heart failure frequently has difficulty achieving even the muscle power required to climb out of bed, much less to walk across the floor.

The maximum cardiac outputs of older persons also decrease considerably—as much as a 50-percent decrease between the teens and the age of eighty. Again, one finds that the maximum achievable muscle power is greatly reduced.

BODY HEAT IN EXERCISE

Almost all the energy released by the internal metabolism of nutrients is eventually converted into body heat. This even applies to the energy that causes muscle contraction, for the following reasons: First, the maximum efficiency for conversion of nutrient energy into muscle work, even under the best of conditions, is only 20 to 25 percent; the remainder of the nutrient energy is converted into heat during the course of the intracellular chemical reactions. Second, almost all of the energy that does go into creating muscle work still becomes body heat because all but a small portion of this energy is used for (1) overcoming viscous resistance to the movement of the muscles and joints, (2) overcoming the friction of the blood flowing through the blood vessels, and (3) other similar effects—all of which convert the muscle contractile energy into heat.

Now, recognizing that the oxygen consumption by the body can increase as much as 20- to 25-fold in the well-trained athlete and that the amount of heat liberated in the body is directly proportional to the oxygen consumption (as discussed in Chapter 33), one quickly realizes that tremendous amounts of heat are injected into the internal body tissues during endurance athletic events.

Next, coupling this vast rate of heat flow into the body with a very hot and humid day so that the sweating mechanism cannot eliminate the heat, the athlete can easily develop an intolerable and even lethal situation called *heat stroke.*

Heat Stroke. During endurance athletics even under normal environmental conditions the body temperature often rises from its normal level of 98.6°F to 102 to 103°F (37°C to 40°C). But, with very hot and humid conditions or great excesses of athletic clothing, the body temperature can then easily rise as high as 106 to 108°F. At this level the elevated temperature itself becomes destructive to tissue cells, especially destructive to brain cells. When this happens, multiple symptoms begin to appear, including:

1. Extreme weakness
2. Exhaustion
3. Headache
4. Dizziness
5. Nausea
6. Profuse sweating
7. Confusion
8. Staggering gait
9. Collapse
10. Unconsciousness

This whole complex is called "heat stroke," and failure to treat immediately can lead to death. In fact, even though the person has stopped the exercise, the temperature does not easily decrease by itself. One of the reasons for this is that at these high temperatures the temperature-regulating mechanism itself often fails. A second reason is that the high temperature approximately doubles the rates of all intracellular chemical reactions, thus liberating still more heat.

The treatment of heat stroke is to reduce the body temperature as rapidly as possible. The most practical way to do this is to remove all clothing, maintain a spray of water on all surfaces of the body or continually sponge the body, and blow air over the body with a strong fan. Experiments have shown that this can reduce the temperature either as rapidly or almost as rapidly as any other procedure, though some physicians prefer total immersion of the body in ice water containing a mush of crushed ice if this could possibly be available.

BODY FLUIDS AND SALT IN EXERCISE

As much as 5- to 10-lb weight loss has been recorded in athletes in a period of 1 hour during endurance athletic events under hot and humid conditions. Essentially all of this weight loss results from loss of sweat. Loss of enough sweat to decrease body weight only 3 percent can significantly diminish a person's performance, and 5 to 10 percent rapid decrease in weight in this way can often be very serious, leading to muscle cramps, nausea, and other effects. Therefore, it is essential to replace fluid as it is lost.

Replacement of Salt and Potassium.

Sweat contains a large amount of salt, for which reason it has long been stated that all athletes should take salt tablets when performing exercise on hot and humid days. Unfortunately, overuse of salt tablets has led to more harm than good. Furthermore, if an athlete will become acclimatized to the heat by progressive increase in athletic exposure over a period of 1 to 2 weeks rather than performing maximal athletic feats on the first day, the sweat glands will also become acclimatized so that the amount of salt lost in the sweat is only a small fraction of that prior to acclimatization. This sweat gland acclimatization results mainly from increased aldosterone secretion by the adrenal cortex. The aldosterone in turn has a direct effect on the sweat glands to increase the reabsorption of sodium chloride from the sweat before it issues forth onto the surface of the skin. Once the athlete is acclimatized, only rarely do salt supplements need to be considered during athletic events.

On the other hand, recent experience by armed forces suddenly exposed to heavy exercise in the desert has demonstrated still another electrolyte problem—the problem of potassium loss. This results partly from the fact that the increased secretion of aldosterone during heat acclimatization increases the loss of potassium in the urine as well as some increase in potassium in the sweat. As a consequence of these new findings, some of the newer supplemental fluids for athletics are beginning to contain properly proportioned amounts of potassium, usually in the form of fruit juices.

DRUGS AND ATHLETES

Without belaboring this issue, let us list some of the effects of drugs in athletics:

First, *caffeine* can increase athletic performance. In one experiment on a marathon runner, his running time for the marathon was reduced by 7 percent by judicious use of caffeine in amounts similar to those found in a cup or so of coffee.

Second, use of *male sex hormones* to increase muscle strength probably can increase athletic performance under some conditions, though actual experiments have been inconclusive in proving this. Unfortunately, some of the synthetic testosterone preparations can cause liver damage, and, in the male, any type of male sex hormone preparation can lead to decreased testicular function including both decreased formation of sperm and decreased secretion of the person's own natural testosterone. In the female, even more dire effects can occur because she is not normally adapted to the male sex hormone.

Other drugs, such as *amphetamines* and *cocaine*, have been reputed to increase one's athletic performance. However, it is equally true that overuse of these drugs can lead to deterioration of performance. Furthermore, actual experiments have failed to prove the value of such drugs. Some athletes have been known to die during athletic performance because of interaction between such drugs and the norepinephrine and epinephrine released by the sympathetic nervous system during exercise. One of the causes of death under these conditions is overexcitability of the heart, leading to ventricular fibrillation, which is lethal within seconds.

QUESTIONS

1. Discuss the differences between the female and the male athlete.
2. What is the relationship between muscle cross-sectional area and muscle strength?
3. How does muscle power differ from muscle strength?
4. Characterize the three important metabolic systems that supply energy in exercise.
5. Explain the mechanisms for recovery of each of the above three metabolic systems after they are depleted.
6. How do the different nutrients contribute to muscle energy during endurance athletics?
7. Explain the principles of muscle building and the changes in the muscle fibers during muscle hypertrophy.
8. What are the differences between fast twitch and slow twitch fibers?
9. Approximately how much can the untrained person and the trained person increase their rates of oxygen consumption above the resting level during maximum exercise?
10. What are the effects of maximum exercise on blood oxygen and carbon dioxide? Why do these effects occur?

11. How much can muscle blood flow increase during maximum exercise? What are the causes of this increase?

12. Explain the relationship between work output, oxygen consumption, and cardiac output during exercise.

13. What are the relationships of both respiratory and cardiovascular performance to $\dot{V}_{O_2}$ Max?

14. Discuss the perils of excess body heat during exercise.

15. Discuss the problems of fluid and electrolyte loss during exercise. Discuss their replacement.

REFERENCES

Appenzeller, O., and Atkinson, R. (eds.): Health Aspects of Endurance Training. New York, S. Karger, 1978.

Apple, D.F., Jr., and Cantwell, J.D.: Medicine for Sport. Chicago, Year Book Medical Publishers, 1979.

Baer, H.P., and Drummond, G.I. (eds.): Physiological and Regulatory Functions of Adenosine and Adenine Nucleotides. New York, Raven Press, 1979.

Basmajian, J.V.: Muscles Alive; Their Functions Revealed by Electromyography. Baltimore, Williams & Wilkins, 1978.

Bevegard, B.S., and Shepherd, J.T.: Regulation of the circulation during exercise in man. *Physiol. Rev.*, 47:178, 1967.

Clarke, D.H.: Exercise Physiology. Englewood Cliffs, N.J., Prentice-Hall, 1975.

Esmann, V. (ed.): Regulatory Mechanisms of Carbohydrate Metabolism. New York, Pergamon Press, 1978.

Fox, E.L.: Sports Physiology. Philadelphia, W.B. Saunders, 1979.

Guyton, A. C., *et al.*: Circulatory Physiology: Cardiac Output and Its Regulation, 2nd Ed. Philadelphia, W.B. Saunders, 1973.

Holloszy, J.O., and Booth, F.W.: Biochemical adaptations to endurance exercise in muscle. *Annu. Rev. Physiol.*, 38:273, 1976.

Homsher, E., and Kean, C.J.: Skeletal muscle energetics and metabolism. *Annu. Rev. Physiol.*, 40:93, 1978.

Klachko, D.M., *et al.* (eds.): Hormones and Energy Metabolism. New York, Plenum Press, 1978.

Northrip, J.W., *et al.*: Introduction to Biomechanic Analysis of Sport. Dubuque, Iowa, W.C. Brown, 1979.

Parmley, W.W., and Talbot, W.: Heart as a pump. *In* Berne, R.M., *et al.* (eds.): Handbook of Physiology. Sec. 2, Vol. 1. Baltimore, Williams & Wilkins, 1979, p. 429.

Rasch, P.J., and Burke, R.K.: Kinesiology and Applied Anatomy: The Science of Human Movement. Philadelphia, Lea & Febiger, 1978.

Strauss, R.H. (ed.): Sports Medicine and Physiology. Philadelphia, W.B. Saunders, 1979.

Winter, D.A.: Biomechanics of Human Movement. New York, John Wiley & Sons, 1979.

Wyndham, C.H.: The physiology of exercise under heat stress. *Annu. Rev. Physiol.*, 35:193, 1973.

Index*

*Boldface page numbers indicate pages on which the index term is defined; italic page numbers indicate an illustration.